Medical Biomethods Handbook

Medical Biomethods Handbook

Edited by

John M. Walker

and

Ralph Rapley

University of Hertfordshire, Hatfield, UK

HUMANA PRESS TOTOWA, NEW JERSEY

999 Riverview Drive, Suite 208
Totowa, New Jersey 07512

www.humanapress.com

This publication is printed on acid-free paper. ∞
ANSI Z39.48-1984 (American Standards Institute)

Permanence of Paper for Printed Library Materials.

Cover illustration: Figures 2A–D, Chapter 29, "*In Situ* Hybridization," by A. D. Watters.

Cover design by Patricia F. Cleary.

For additional copies, pricing for bulk purchases, and/or information about other Humana titles, contact Humana at the above address or at any of the following numbers: Tel.: 973-256-1699; Fax: 973-256-8341; E-mail: humana@humanapr.com; or visit our Website: www.humanapress.com

Printed in the United States of America. 10 9 8 7 6 5 4 3 2 1

E-ISBN 1-59259-870-6
Library of Congress Cataloging in Publication Data
Medical biomethods handbook / edited by John M. Walker and Ralph Rapley.
p. ; cm.
Includes bibliographical references and index.
ISBN 1-58829-288-6 (hardcover : alk. paper) — ISBN 1-58829-334-3 (pbk. : alk. paper)
1. Molecular biology—Handbooks, manuals, etc. 2. Molecular diagnosis—Handbooks, manuals, etc.
[DNLM: 1. Genetic Techniques. 2. Chemistry, Analytical—methods. 3. Clinical Laboratory Techniques. 4. Immunologic Techniques. QZ 52 M489 2005] I. Walker, John M., 1948- II. Rapley, Ralph. III. Molecular biomethods handbook.
QH506.M425 2005
616'.0028—dc22 2004010386

Preface

There have been numerous advances made in many fields throughout the biosciences in recent years, with perhaps the most dramatic being those in our ability to investigate and define cellular processes at the molecular level. These insights have been largely the result of the development and application of powerful new techniques in molecular biology, in particular nucleic acid and protein methodologies.

The purpose of the *Medical Biomethods Handbook* is to introduce the reader to a wide-ranging selection of those molecular biology techniques that are most frequently used by research workers in the field of medical and clinical research. Clearly, within the constraints of a single volume, we have had to be selective. However, all of the techniques described are core methods and in regular research use. We have aimed to describe both the theory behind, and the application of, the techniques described. A companion volume, the *Molecular Biomethods Handbook*, published in 1998, provides similar details on a range of other basic molecular biology techniques. For those who require detailed laboratory protocols, these can be found in the references cited in each chapter and in the laboratory protocol series *Methods in Molecular Biology*™ and *Methods in Molecular Medicine*™ published by Humana Press.

The *Medical Biomethods Handbook* should prove useful to undergraduate students (especially project students), postgraduate researchers, and all research scientists and technicians who wish to understand and use new techniques, but do no yet have the necessary background to set up specific techniques. In addition, it will be useful for all those wishing to update their knowledge of particular techniques. All chapters have been written by well-established research scientists who run their own research programs and who use the methods on a regular basis. In sum, then, our hope is that this book will prove a useful source of information on all the major molecular biomedical techniques in use today, as well as a valuable text for those already engaged in or just entering the field of molecular biology.

John M. Walker
Ralph Rapley

Contents

Contributors

MARILENA AQUINO DE MURO • *CABI Bioscience UK Centre, Egham, Surrey, UK*
GUY T.N. BESLEY • *Willink Biochemical Genetics Unit, Royal Manchester Children's Hospital, Manchester, UK*
IOANNIS BOSSIS • *Developmental Endocrinology Branch, National Institute of Child Health and Human Development, National Institutes of Health, Bethesda, MD*
SYLVAIN BRISSE • *Unit Biodiversity of Emerging Bacterial Pathogens, INSERM U389, Institut Pasteur, Paris, France*
JULIO E. CELIS • *Department of Proteomics in Cancer, Institute of Cancer Biology, The Danish Cancer Society, Copenhagen, Denmark*
YANN-JANG CHEN • *Center of General Education, National Yang-Ming University, Taiwan*
ZHONG CHEN • *University of Utah Medical Center, Salt Lake City, UT*
RAHUL CHODHARI • *Department of Paediatrics and Child Health, Royal Free and University College Medical School, University College London, London, UK*
FRANK CHRISTIANSEN • *Department of Clinical Immunology and Biochemical Genetics, Royal Perth Hospital; School of Surgery and Pathology, University of Western Australia, Perth, Australia*
EDDIE CHUNG • *Department of Paediatrics and Child Health, Royal Free and University College Medical School, University College London, London, UK*
JOHAN T. DEN DUNNEN • *Center for Human and Clinical Genetics, Leiden University Medical Center, Leiden, Nederland*
JÖRG DÖTSCH • *Department of Pediatrics, University of Erlangen, Erlangen, Germany*
FAYE A. EGGERDING • *Huntington Medical Research Institutes, Pasadena, CA*
PHILLIP G. FEBBO • *Duke Institute for Genome Sciences and Policy, Duke University, Durham, NC*
PATRICIA A. FETSCH • *Laboratory of Pathology, National Cancer Institute, National Institutes of Health, Bethesda, MD*
MARA GIORDANO • *Department of Medical Science, Universita del Piemonte Orientale Amedeo Avogadro, Novara, Italy*
H. GOOSSENS • *Department of Medical Microbiology, Universitaire Instelling Antwerpen, Wilrijk, Belgium; Leiden University Medical Center, Leiden, The Netherlands*
SIMON G. GREGORY • *Center for Human Genetics, Duke University Medical Center, Durham, NC*
PAVEL GROMOV • *Department of Proteomics in Cancer, Institute of Cancer Biology, The Danish Cancer Society, Copenhagen, Denmark*
IRINA GROMOVA • *Department of Proteomics in Cancer, Institute of Cancer Biology, The Danish Cancer Society, Copenhagen, Denmark*
SAMIR HANASH • *Fred Hutchinson Cancer Research Center, Seattle, WA*
BRONWEN HARVEY • *GE Healthcare Bio-Sciences, Amersham, UK*
MARION HILL • *Departments of Clinical Chemistry and Haematology, Queen's Medical Centre, Nottingham, UK*

ANDREAS HOCHHAUS • *III. Medizinische Klinik, Fakultät für Klinische Medizin Mannheim der Universität Heidelberg, Mannheim , Germany*

RACHEL E. IBBOTSON • *Molecular Biology Department, Royal Bournemouth Hospital, Bournemouth, UK*

M. IEVEN • *Department of Medical Microbiology, Universitaire Instelling Antwerpen, Wilrijk, Belgium*

LOUISE IZATT • *Department of Clinical Genetics, Guy's and St. Thomas' Hospitals, London, UK*

GARETH J. S. JENKINS • *Swansea Clinical School, University of Wales, Swansea, Wales, UK*

MARGARET A. JENKINS • *Division of Laboratory Medicine, Austin Health, Heidelberg, Victoria, Australia*

WILLIAM J. JORDAN • *Division of Oral Biology, University of Medicine and Dentistry of New Jersey, Newark, NJ*

RICHARD KITCHING • *Department of Laboratory Medicine and Pathobiology, Sunnybrook and Women's College Health Sciences Centre, University of Toronto, Toronto, Ontario, Canada*

CHI-HUNG LIN • *Institute of Microbiology and Immunology, National Yang-Ming* University, Taiwan

BENNY K. C. LO • *St. Edmund's College, University of Cambridge, Cambridge, UK*

KATHERINE LOENS • *Department of Medical Microbiology, Universitaire Instelling Antwerpen, Wilrijk, Belgium*

GARETH J. MAGEE • *Centre for Cutaneous Research, Barts and the London, Queen Mary's School of Medicine and Dentistry, University of London, London, UK; Biomedical Research Centre, Ninewells Hospital and Medical School, Dundee, Scotland, UK*

CYRIL MAMOTTE • *Department of Clinical Immunology and Biochemical Genetics, Royal Perth Hospital, Perth, Australia*

HERBERT OBERACHER • *Institute of Legal Medicine, Innsbruck Medical University, Innsbruck, Austria*

CARLA OSIOWY • *National Microbiology Laboratory, Canadian Science Centre for Human and Animal Health, Winnipeg, Manitoba, Canada*

LYLE J. PALMER • *Western Australian Institute for Medical Research, Centre for Medical Research, School of Population Health, University of Western Australia, Perth, Australia*

ANTON E. PARKER • *Molecular Biology Department, Royal Bournemouth Hospital, Bournemouth, UK*

WALTHER PARSON • *Institute of Legal Medicine, Innsbruck Medical University, Innsbruck, Austria*

DORIEN J. M. PETERS • *Center for Human and Clinical Genetics, Leiden University Medical Center, Leiden, Netherlands*

RALPH RAPLEY • *Department of Biosciences, University of Hertfordshire, Hatfield, UK*

WOLFGANG RASCHER • *Department of Pediatrics, University of Erlangen, Erlangen, Germany*

SUJIVA RATNAIKE • *Division of Laboratory Medicine, Austin Health, Heidelberg, Victoria, Australia*

BERND H. A. REHM • *Institute of Molecular BioSciences, Massey University, Palmerston North, New Zealand*

FRANK REINECKE • *Institut für Molekulaire Mikrobiologie und Biotechnologie, Westfälische Wilhelms-Universität Münster, Münster, Germany*

MARK P. RICHARDS • *Agricultural Research Service, Animal and Natural Resources Institute, US Department of Agriculture, Beltsville, MD*

J. H. ROELFSEMA • *Center for Human and Clinical Genetics, Leiden University Medical Center, Leiden, Netherlands*

ELIZABETH RUGG • *Department of Dermatology, University of California, Irvine, CA*

BEATRIZ SANCHEZ-VEGA • *University of Texas MD Anderson Cancer Center, Houston, TX*

AVERY A. SANDBERG • *Department of DNA Diagnostics, St. Joseph's Hospital and Medical Center, Phoenix, AZ*

ELLEN SCHOOF • *Department of Pediatrics, University of Erlangen, Erlangen, Germany*

ARUN SETH • *Department of Laboratory Medicine and Pathobiology, Sunnybrook and Women's College Health Sciences Centre, University of Toronto, Toronto, Ontario, Canada*

PATRICIA SEVERINO • *Albert Einstein Institute of Research and Education, Albert Einstein Hospital, Sao Paulo, Brazil*

STELLA B. SOMIARI • *Windber Research Institute, Winbder, PA*

PIRKKO SOUNDY • *GE Healthcare Bio-Sciences, Amersham, UK*

MARYALICE STETLER-STEVENSON • *Laboratory of Pathology, National Cancer Institute, National Institutes of Health, Bethesda, MD*

JOHN F. STONE • *Department of DNA Diagnostics, St. Joseph's Hospital and Medical Center, Phoenix, AZ*

CONSTANTINE A. STRATAKIS • *Developmental Endocrinology Branch, National Institute of Child Health and Human Development, National Institutes of Health, Bethesda, MD*

DAVID SUGDEN • *Centre for Reproduction, Endocrinology, and Diabetes, School of Biomedical Sciences, King's College London, London, UK*

ANU SUOMALAINEN • *Programme of Neurosciences and Department of Neurology, Helsinki University, Helsinki, Finland*

ANN-CHRISTINE SYVÄNEN • *Department of Medical Sciences, Uppsala University, Uppsala, Sweden*

BIMAL D. M. THEOPHILUS • *Department of Haematology, Birmingham Children's Hospital NHS Trust, Birmingham, UK*

D. URSI • *Department of Medical Microbiology, Universitaire Instelling Antwerpen, Wilrijk, Belgium*

ANTONIOS VOUTETAKIS • *National Institute for Dental and Craniofacial Research, National Institutes of Health, Bethesda, MD*

IGOR VORECHOVSKY • *Division of Human Genetics, University of Southampton School of Medicine, Southampton, UK*

JOHN M. WALKER • *Department of Biosciences, University of Hertfordshire, Hatfield, UK*

PENG-HUI WANG • *Department of Obstetrics and Gynecology, Taipei Veterans General Hospital; and Institute of Clinical Medicine, National Yang-Ming University, Taiwan*

AMANDA WATTERS • *Department of Pathology, Western Infirmary, Glasgow, Scotland, UK*
JAMES L. WEAVER • *Division of Applied Pharmacology Research, OTR, CDER, Food and Drug Administration, Silver Spring, MD*
NEIL V. WHITTOCK • *Institute of Biomedical and Clinical Science, Peninsula Medical School, Exeter, UK*

Value-Added eBook/PDA on CD-ROM

This book is accompanied by a value-added CD-ROM that contains an eBook version of the volume you have just purchased. This eBook can be viewed on your computer, and you can synchronize it to your PDA for viewing on your handheld device. The eBook enables you to view this volume on only one computer and PDA. Once the eBook is installed on your computer, you cannot download, install, or e-mail it to another computer; it resides solely with the computer to which it is installed. The license provided is for only one computer. The eBook can only be read using Adobe® Reader® 6.0 software, which is available free from Adobe Systems Incorporated at www.Adobe.com. You may also view the eBook on your PDA using the Adobe® PDA Reader® software that is also available free from Adobe.com.

You must follow a simple procedure when you install the eBook/PDA that will require you to connect to the Humana Press website in order to receive your license. Please read and follow the instructions below:

1. Download and install Adobe® Reader® 6.0 software

   You can obtain a free copy of the Adobe® Reader® 6.0 software at www.adobe.com

   *Note: If you already have the Adobe® Reader® 6.0 software installed, you do not need to reinstall it.

2. Launch Adobe® Reader® 6.0 software

3. Install eBook: Insert your eBook CD into your CD-ROM drive

   PC: Click on the "Start" button, then click on "Run"

   At the prompt, type "d:\ebookinstall.pdf" and click "OK"

   *Note: If your CD-ROM drive letter is something other than d: change the above command accordingly.

   MAC: Double click on the "eBook CD" that you will see mounted on your desktop. Double click "ebookinstall.pdf"

4. Adobe® Reader® 6.0 software will open and you will receive the message

   "This document is protected by Adobe DRM" Click "OK"

   *Note: If you have not already activated the Adobe® Reader® 6.0 software, you will be prompted to do so. Simply follow the directions to activate and continue installation.

Your web browser will open and you will be taken to the Humana Press eBook registration page. Follow the instructions on that page to complete installation. You will need the serial number located on the sticker sealing the envelope containing the CD-ROM.

If you require assistance during the installation, or you would like more information regarding your eBook and PDA installation, please refer to the eBookManual.pdf located on your cd. If you need further assistance, contact Humana Press eBook Support by e-mail at ebooksupport@humanapr.com or by phone at 973-256-1699.

1

Basic Techniques in Molecular Biology

Ralph Rapley

1. Introduction

There have been many developments over the past three decades that have led to the efficient manipulation and analysis of nucleic acid and proteins. Many of these have resulted from the isolation and characterization of numerous DNA-manipulating enzymes, such as DNA polymerase, DNA ligase, and reverse transcriptase. However, perhaps the most important was the isolation and application of a number of enzymes that enabled the reproducible digestion of DNA. These enzymes, termed *restriction endonucleases* or *restriction enzymes*, provided a turning point for not only the analysis of DNA but also the development of recombinant DNA technology and provided a means for the detection and identification of disease and disease markers.

1.1. Enzymes Used in Molecular Biology

Restriction endonucleases recognize specific DNA target sequences, mainly 4–6 bp in length, and cut them, reproducibly, in a defined manner. The nucleotide sequences recognized are of an inverted repeat nature (typically termed *palindromic*), reading the same in both directions on each strand *(1)*. When cut or cleaved, they produce a flush or blunt-ended, or a staggered, cohesive-ended fragment depending on the particular enzyme as indicated in **Fig. 1**. An important property of cohesive ends is that DNA from different sources (e.g., different organisms) digested by the same restriction endonuclease will be complementary (termed "sticky") and so will join or anneal to each other. The annealed strands are held together only by hydrogen-bonding between complementary bases on opposite strands. Covalent joining of nucleotide ends on each of the two strands may be brought about by the introduction of the enzyme DNA ligase. This process, termed *ligation*, is widely exploited in molecular biology to enable the construction of recombinant DNA molecules (i.e., the joining of DNA fragments from different sources). Approximately 500 restriction enzymes have been characterized to date that recognise over 100 different target sequences. A number of these, termed *isoschizomers*, recognize different target sequences but produce the same staggered ends or overhangs. In addition to restriction enzymes a number of other enzymes have proved to be of value in the manipulation of DNA and are indicated at appropriate points within this chapter.

2. Extraction and Separation of Nucleic Acids

2.1. DNA Extraction Techniques

The use of DNA for medical analysis or for research-driven manipulation usually requires that it be isolated and purified to a certain degree. DNA is usually recovered from cells by methods that include cell rupture but that prevent the DNA from fragmenting by mechanical shearing. This is generally undertaken in the presence of EDTA, which chelates the magnesium ions needed

From: *Medical Biomethods Handbook*
Edited by: J. M. Walker and R. Rapley © Humana Press, Inc., Totowa, NJ

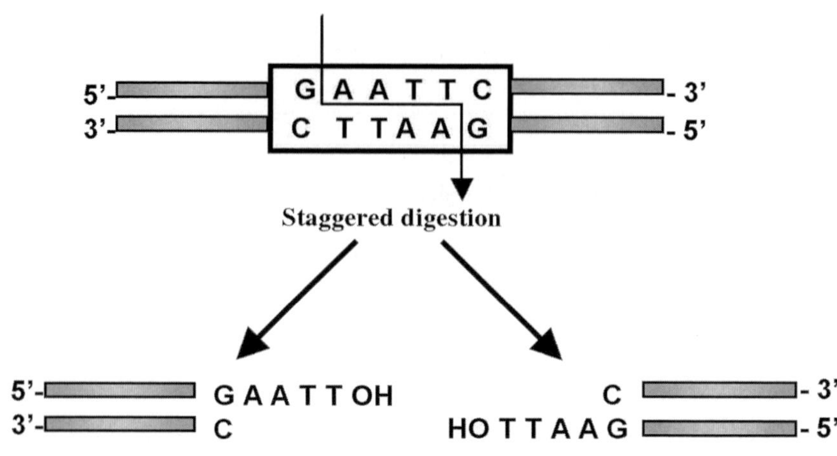

Fig. 1. The cleavage of a DNA strand with a target site for the restriction enzyme *Eco*R1 indicating the ends of the DNA formed following digestion.

as cofactors for enzymes that degrade DNA, termed *DNase*. Ideally, cell walls, if present, should be digested enzymatically (e.g., lysozyme in the bacteria or bacterial). In addition the cell membrane should be solubilized using detergent. Indeed, if physical disruption is necessary, it should be kept to a minimum and should involve cutting or squashing of cells, rather than the use of shear forces. Cell disruption and most of the subsequent steps should be performed at 4°C, using glassware and solutions that have been autoclaved to destroy DNase activity.

After release of nucleic acids from the cells, RNA can be removed by treatment with ribonuclease (RNase) that has been heat treated to inactivate any DNase contaminants; RNase is relatively stable to heat as a result of its disulfide bonds, which ensure rapid renaturation of the molecule on cooling. The other major contaminant, protein, is removed by shaking the solution gently with water-saturated phenol, or with a phenol/chloroform mixture, either of which will denature proteins but not nucleic acids. Centrifugation of the emulsion formed by this mixing produces a lower, organic phase, separated from the upper, aqueous phase by an interface of denatured protein. The aqueous solution is recovered and deproteinized repeatedly, until no more material is seen at the interface. Finally, the deproteinized DNA preparation is mixed with 2 vol of absolute ethanol, and the DNA is allowed to precipitate out of solution in a freezer. After centrifugation, the DNA pellet is redissolved in a buffer containing EDTA to inactivate any DNases present. This solution can be stored at 4°C for at least a month. DNA solutions can be stored frozen, although repeated freezing and thawing tends to damage long DNA molecules by shearing. A flow diagram summarizing the extraction of DNA is given in **Fig. 2**. The above-described procedure is suitable for total cellular DNA. If the DNA from a specific organelle or viral particle is needed, it is best to isolate the organelle or virus before extracting its DNA, because the recovery of a particular type of DNA from a mixture is usually rather difficult. Where a high degree of purity is required, DNA may be subjected to density gradient ultracentrifugation through cesium chloride, which is particularly useful for the preparation plasmid DNA. It is possible to check the integrity of the DNA by agarose gel electrophoresis and determine the concentration of the DNA by using the fact that 1 absorbance unit equates to 50 μg/mL of DNA:

$50A_{260}$ = Concentration of DNA sample (μg/mL)

The identification of contaminants may also be undertaken by scanning ultraviolet (UV)-spectrophotometry from 200 nm to 300 nm. A ratio of 260 nm : 280 nm of approx 1.8 indicates that the sample is free of protein contamination, which absorbs strongly at 280 nm.

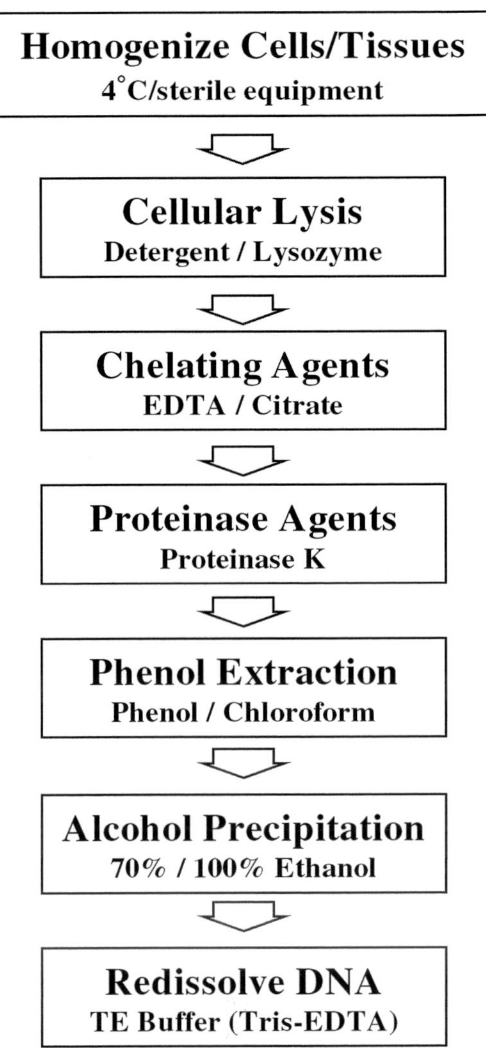

Fig. 2. General steps involved in extracting DNA from cells or tissues.

2.2. RNA Extraction Techniques

The methods used for RNA isolation are very similar to those described above for DNA; however, RNA molecules are relatively short and, therefore, less easily damaged by shearing, so cell disruption can be rather more vigorous *(2)*. RNA is, however, very vulnerable to digestion by Rnases, which are present endogenously in various concentrations in certain cell types and exogenously on fingers. Gloves should therefore be worn, and a strong detergent should be included in the isolation medium to immediately denature any RNases. Subsequent deproteinization should be particularly rigorous, because RNA is often tightly associated with proteins. DNase treatment can be used to remove DNA, and RNA can be precipitated by ethanol. One reagent in particular that is commonly used in RNA extraction is guanadinium thiocyanate, which is both a strong inhibitor of RNase and a protein denaturant. It is possible to check the integrity of an RNA extract by analyzing it by agarose gel electrophoresis. The most abundant RNA species, the rRNA molecules, are 23S and 16S for prokaryotes and 18S and 28S for eukaryotes. These appear as discrete bands on the agarose gel and indicate that the other RNA components are likely to be intact. This is usually carried out under denaturing conditions to prevent secondary structure formation in the RNA. The concentration of the RNA may be estimated by using UV spectrophotometry in a similar manner to that used for DNA. However in the case of RNA at 260 nm, 1 absorbance unit equates to 40 μg/mL of RNA. Contaminants may also be identified in the same way by scanning UV spectrophotometry; however, in the case of RNA, a 260 nm : 280 nm ratio of approx 2 would be expected for a sample containing little or no contaminating protein *(3)*.

In many cases, it is desirable to isolate eukaryotic mRNA, which constitutes only 2–5% of cellular RNA from a mixture of total RNA molecules. This may be carried out by affinity chromatography on oligo(dT)-cellulose columns. At high salt concentrations, the mRNA containing poly(A) tails binds to the complementary oligo(dT) molecules of the affinity column, and so mRNA will be retained; all other RNA molecules can be washed through the column by further high-salt solution. Finally, the bound mRNA can be eluted using a low concentration of salt *(4)*. Nucleic acid species may also be subfractionated by more physical means such as electrophoretic or chromatographic separations based on differences in nucleic acid fragment sizes or physicochemical characteristics.

2.3. Electrophoresis of Nucleic Acids

In order to analyze nucleic acids by size, the process of electrophoresis in an agarose or polyacrylamide support gel is usually undertaken. Electrophoresis may be used analytically or preparatively and can be qualitative or quantitative. Large fragments of DNA such as chromosomes may also be separated by a modification of electrophoresis termed *pulsed field gel electrophoresis* (PFGE), which uses alternating directions of DNA migration *(5)*. The easiest and most widely applicable method is electrophoresis in horizontal agarose gels as indicated in **Fig. 3**. In order to visualize the DNA, staining has to be undertaken, usually with a dye such as ethidium bromide. This dye binds to DNA by insertion between stacked base-pairs, termed *intercalation*, and exhibits a strong orange/red fluorescence when illuminated with UV light. Alternative stains such as SYBRGreen or Gelstar, which have similar sensitivities, are also available and are less hazardous to use.

In general, electrophoresis is used to check the purity and intactness of a DNA preparation or to assess the extent of an enzymatic reaction during, for example, the steps involved in the cloning of DNA. For such checks "mini-gels" are particularly convenient, because they need little preparation, use small samples, and provide results quickly. Agarose gels can be used to separate molecules larger than about 100 basepairs (bp). For higher resolution or for the effective separation of shorter DNA molecules, polyacrylamide gels are the preferred method. In recent years, a number of acrylic gels have been developed which may be used as an alternative to agarose and polyacrylamide.

When electrophoresis is used preparatively, the fragment of gel containing the desired DNA molecule is physically removed with a scalpel. The DNA is then recovered from the gel fragment in various ways. This may include crushing with a glass rod in a small volume of buffer, using agarase to digest the agrose leaving the DNA, or by the process of electroelution. In this method, the piece of gel is sealed in a length of dialysis tubing containing buffer and is then placed between two electrodes in a tank containing more buffer. Passage of an electrical current between the electrodes causes DNA to migrate out of the gel piece, but it remains trapped within the dialysis tubing and can, therefore, be recovered easily.

3. Nucleic Acid Blotting and Gene Probe Hybridization

3.1. Nucleic Acid Blotting

Electrophoresis of DNA restriction fragments allows separation based on size to be conducted, however, it provides no indication as to the presence of a specific, desired fragment among the complex sample. This can be achieved by transferring the DNA from the intact gel onto a piece of nitrocellulose or Nylon membrane placed in contact with it. This provides a

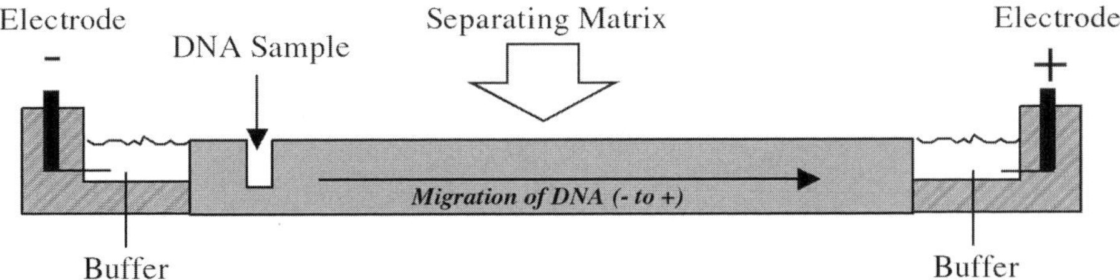

Fig. 3. Schematic illustration of a typical horizontal gel electrophoresis setup for the separation of nucleic acids.

more permanent record of the sample because DNA begins to diffuse out of a gel that is left for a few hours. First the gel is soaked in alkali to render the DNA single stranded. It is then transferred to the membrane so that the DNA becomes bound to it in exactly the same pattern as that originally on the gel *(6)*. This transfer, named a Southern blot (see Chapter 4) after its inventor Ed Southern, is usually performed by drawing large volumes of buffer by capillary action through both gel and membrane, thus transferring DNA from the gel to the membrane. Alternative methods are also available for this operation such as electrotransfer or vacuum-assisted transfer. Both are claimed to give a more even transfer and are much more rapid, although they do require more expensive equipment than the capillary transfer system. Transfer of the DNA from the gel to the membrane allows the membrane to be treated with a labeled DNA gene probe. This single-stranded DNA probe will hybridize under the right conditions to complementary single-stranded DNA fragments immobilized onto the membrane.

3.2. Hybridization and Stringency

The conditions of hybridization are critical for this process to take place effectively. This is usually referred to as the stringency of the hybridization and it is particular for each individual gene probe and for each sample of DNA. Two of the most important components are the temperature and the salt concentration. Higher temperatures and low salt concentrations, termed *high stringency*, provide a favorable environment for perfectly matched probe and template sequences, whereas reduced temperatures and high salt concentrations, termed *low stringency*, allow the stabilization of mismatches in the duplex. In addition, inclusion of denaturants such as formamide allow the hybridization temperatures to be reduced without affecting the stringency.

A series of posthybridization washing steps with a salt solution such as SSC, containing sodium citrate and sodium chloride, is then carried out to remove any unbound probe and control the binding of the duplex. The membrane is developed using either autoradiography if the probe is radiolabeled or by a number of nonradioactive methods. The precise location of the probe and its target may be then visualized. The steps involved in Southern blotting are indicated in **Fig. 4**. It is also possible to analyze DNA from different species or organisms by blotting the DNA and then using a gene probe representing a protein or enzyme from one of the organisms. In this way, it is possible to search for related genes in different species. This technique is generally termed Zoo blotting.

A similar process of nucleic acid blotting can be used to transfer RNA separated by gel electrophoresis onto membranes similar to that used in Southern blotting. This process, termed *Northern blotting*, allows the identification of specific mRNA sequences of a defined length by hybridization to a labeled gene probe *(7)*. It is possible with this technique to not only detect specific mRNA molecules, but it may also be used to quantify the relative amounts of the specific mRNA present in a tissue or sample. It is usual to separate the mRNA transcripts by gel electrophoresis under denaturing conditions because this improves resolution and allows a more accurate estimation of the sizes of the transcripts. The format of the blotting may be altered from transfer from a gel to direct application to slots on a specific blotting apparatus containing

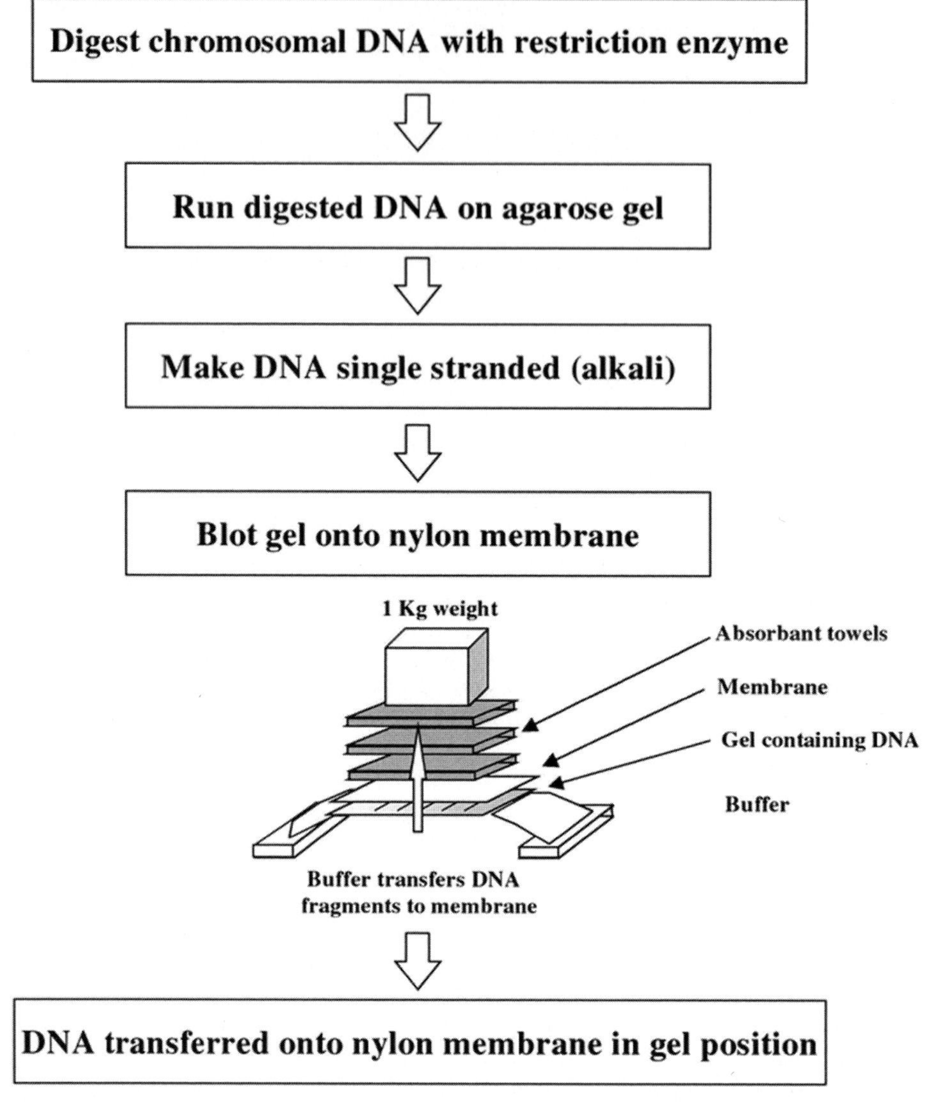

Fig. 4. The procedure involved in a typical Southern blot indicating the construction of a simple capillary transfer setup.

the Nylon membrane. This is termed *slot* or *dot blotting* and provides a convenient means of measuring the abundance of specific mRNA transcripts without the need for gel electrophoresis, it does not, however, provide information regarding the size of the fragments.

Hybridization techniques are essential to many molecular biology experiments; however, the format of the hybridization may be altered to improve speed sensitivity and throughput. One interesting alternative is termed *surface plasmon resonance* (SPR). This is an optical system based on difference between incident and reflected light in the presence of absence of hybridization. Its main advantage is that the kinetics of hybridization can be undertaken in real time and without a DNA label. A further exciting method for hybridization is also in use which uses arrays of single-stranded DNA molecules tethered to small hybridization chips. Hybridization to a DNA sample is detected by computer, allowing DNA mutations to be quickly and easily identified.

3.3. Production of Gene Probes

The availability of a gene probe is essential in many molecular biology techniques; yet, in many cases, it is one of the most difficult steps (see Chapter 2). The information needed to produce a gene probe may come from many sources, but with the development and sophistication of genetic databases, this is usually one of the first stages *(8)*. There are a number of genetic databases such as those at Genbank and EMBL and it is possible to search these over the Internet and identify particular sequences relating to a specific gene or protein. In some cases, it is possible to use related proteins from the same gene family to gain information on the most useful DNA sequence. Similar proteins or DNA sequences but from different species may also provide a starting point with which to produce a so-called heterologous gene probe. Although, in some cases, probes are already produced and cloned, it is possible, armed with a DNA sequence from a DNA database, to chemically synthesize a single-stranded oligonucleotide probe. This is usually undertaken by computer-controlled gene synthesizers, which link dNTPs together based on a desired sequence. It is essential to carry out certain checks before probe production to determine that the probe is unique, is not able to self-anneal, or is self-complementary, all of which may compromise its use.

Where little DNA information is available to prepare a gene probe, it is possible in some cases to use the knowledge gained from analysis of the corresponding protein. Thus, it is possible to isolate and purify proteins and sequence part of the N-terminal end of the protein. From our knowledge of the genetic code, it is possible to predict the various DNA sequences that could code for the protein and then synthesize appropriate oligonucleotide sequences chemically. Because of the degeneracy of the genetic code, most amino acids are coded for by more than one codon, therefore, there will be more than one possible nucleotide sequence that could code for a given polypeptide. The longer the polypeptide, the larger is the number of possible oligonucleotides that must be synthesized. Fortunately, there is no need to synthesize a sequence longer than about 20 bases, as this should hybridize efficiently with any complementary sequences and should be specific for one gene. Ideally, a section of the protein should be chosen that contains as many tryptophan and methionine residues as possible, because these have unique codons and there will therefore be fewer possible base sequences that could code for that part of the protein. The synthetic oligonucleotides can then be used as probes in a number of molecular biology methods.

3.4. DNA Gene Probe Labeling

An essential feature of a gene probe is that it can be visualized by some means. In this way, a gene probe that hybridizes to a complementary sequence may be detected and identify that desired sequence from a complex mixture. There are two main ways of labeling gene probes, traditionally this has been carried out using radioactive labels, but gaining in popularity are nonradioactive labels. Perhaps the most used radioactive label is phosphorous-32 (^{32}P), although for certain techniques sulfur-35 (^{35}S) and tritium (^{3}H) are used. These may be detected by the process of autoradiography where the labeled probe molecule, bound to sample DNA, located, for example, on a Nylon membrane, is placed in contact with an X-ray-sensitive film. Following exposure, the film is developed and fixed just as a black-and-white negative and reveals the precise location of the labeled probe and, therefore, the DNA to which it has hybridized.

3.5. Nonradioactive DNA Labeling

Nonradioactive labels are increasingly being used to label DNA gene probes. Until recently, radioactive labels were more sensitive than their nonradioactive counterparts. However, recent developments have led to similar sensitivities, which, when combined with their improved safety, have led to their greater acceptance.

The labeling systems are either termed direct or indirect. Direct labeling allows an enzyme reporter such as alkaline phosphatase to be coupled directly to the DNA. Although this may alter the characteristics of the DNA gene probe, it offers the advantage of rapid analysis because

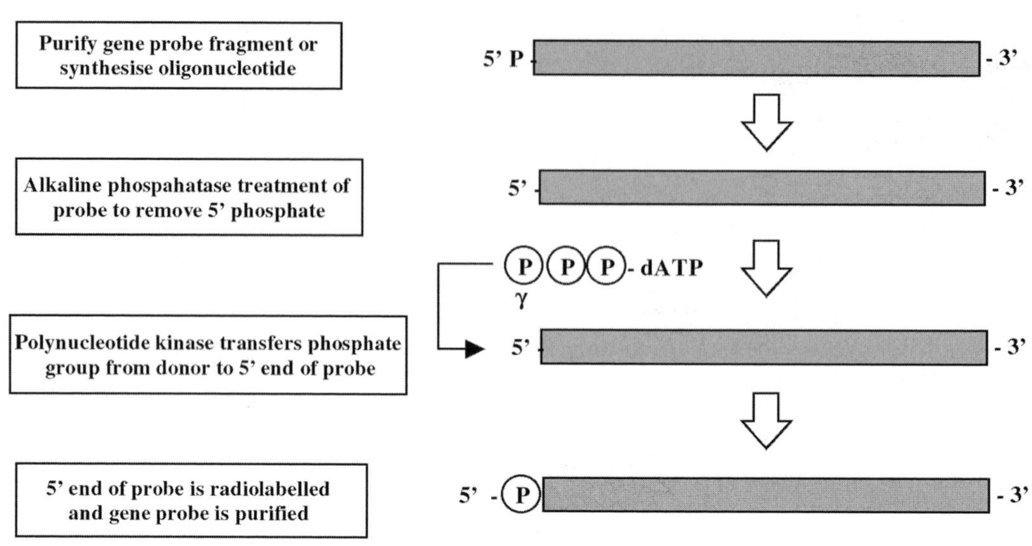

Fig. 5. End labeling of a gene probe at the 5' end with alkaline phosphatase and polynucleotide kinase.

no intermediate steps are needed. However indirect labeling is, at present, more popular. This relies on the incorporation of a nucleotide that has a label attached. At present, three of the main labels in use are biotin, flourescein, and digoxigenin. These molecules are covalently linked to nucleotides using a carbon spacer arm of 7, 14, or 21 atoms. Specific binding proteins may then be used as a bridge between the nucleotide and a reporter protein such as an enzyme. For example, biotin incorporated into a DNA fragment is recognized with a very high affinity by the protein streptavidin. This may either be coupled or conjugated to a reporter enzyme molecule such as alkaline phosphatase or horseradish peroxidase (HRP). This is usually used to convert a colorless substrate into a colored insoluble compound and also offers a means of signal amplification. Alternatively, labels such as digoxigenin incorporated into DNA sequences may be detected by monoclonal antibodies, again conjugated to reporter molecules, including alkaline phosphatase. Thus, rather than the detection system relying on autoradiography, which is necessary for radiolabels, a series of reactions resulting in either a color or a light or a chemiluminescence reaction takes place. This has important practical implications because autoradiography may take 1–3 d, whereas color and chemiluminescent reactions take minutes. In addition, no radiolabeling and detection minimize the potential health and safety hazards encountered when using radiolabels.

3.6. End Labeling of DNA

The simplest form of labeling DNA is by 5'- or 3'-end labeling. 5'-End labeling involves a phosphate transfer or exchange reaction, where the 5' phosphate of the DNA to be used as the probe is removed and in its place a labeled phosphate, usually ^{32}P, is added. This is usually carried out by using two enzymes, the first, alkaline phosphatase, is used to remove the existing phosphate group from the DNA. Following removal of the released phosphate from the DNA, a second enzyme polynucleotide kinase is added that catalyzes the transfer of a phosphate group (^{32}P labeled) to the 5' end of the DNA (*see* **Fig. 5**). The newly labeled probe is then purified, usually by chromatography through a Sephadex column and may be used directly.

Using the other end of the DNA molecule, the 3' end, is slightly less complex. Here, a new dNTP, which is labeled (e.g., ^{32}PαdATP or biotin-labeled dNTP), is added to the 3' end of the DNA by the enzyme terminal transferase as indicated in **Fig. 6**. Although this is a simpler

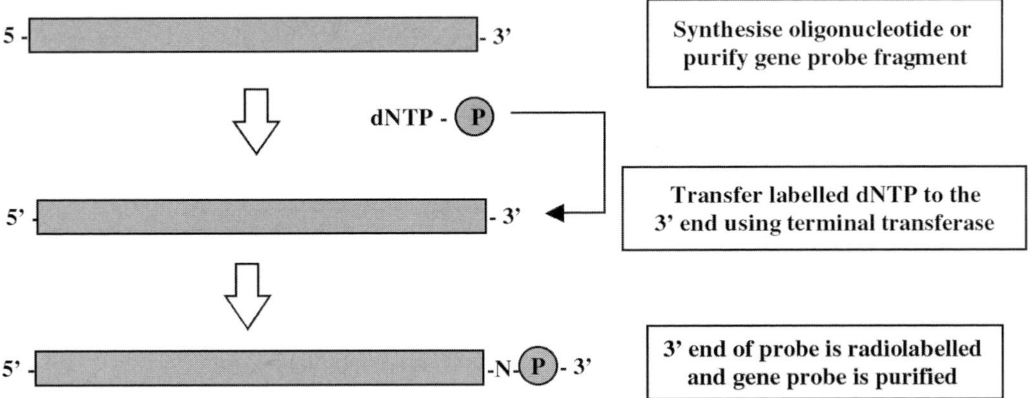

Fig. 6. End labeling of a gene probe at the 3' end using terminal transferase. Note that the addition of a labeled dNTP at the 3' end alters the sequence of the gene probe.

reaction, a potential problem exists because a new nucleotide is added to the existing sequence and so the complete sequence of the DNA is altered, which may affect its hybridization to its target sequence. End labeling methods also suffer from the fact that only one label is added to the DNA, so these methods are of a lower activity in comparison to methods that incorporate labels along the length of the DNA.

3.7. Random Primer Labeling of DNA

The DNA to be labeled is first denatured and then placed under renaturing conditions in the presence of a mixture of many different random sequences of hexamers or hexanucleotides. These hexamers will, by chance, bind to the DNA sample wherever they encounter a complementary sequence and, thus, the DNA will rapidly acquire an approximately random sprinkling of hexanucleotides annealed to it. Each of the hexamers can act as a primer for the synthesis of a fresh strand of DNA catalyzed by DNA polymerase because it has an exposed 3' hydroxyl group, as seen in **Fig. 7**. The Klenow fragment of DNA polymerase is used for random primer labeling because it lacks a 5'–3' exonuclease activity. This is prepared by cleavage of DNA polymerase with subtilisin, giving a large enzyme fragment that has no 5' to 3' exonuclease activity, but which still acts as a 5' to 3' polymerase. Thus, when the Klenow enzyme is mixed with the annealed DNA sample in the presence of dNTPs, including at least one that is labeled, many short stretches of labeled DNA will be generated. In a similar way to random primer labelling, polymerase chain reaction (PCR) may also be used to incorporate radioactive or nonradioactive labels.

3.8 Nick Translation Labeling of DNA

A traditional method of labeling DNA is by the process of nick translation. Low concentrations of DNase I are used to make occasional single-strand nicks in the double-stranded DNA that is to be used as the gene probe. DNA polymerase then fills in the nicks, using an appropriate deoxyribonucleoside triphosphate (dNTP), at the same time making a new nick to the 3' side of the previous one. In this way, the nick is translated along the DNA. If labeled dNTPs are added to the reaction mixture, they will be used to fill in the nicks, as indicated in **Fig. 8**. In this way, the DNA can be labeled to a very high specific activity.

4. Medical Applications of Basic Molecular Techniques: Restriction Mapping of DNA Fragments

Restriction mapping involves the size analysis of restriction fragments produced by several restriction enzymes individually and in combination. Comparison of the lengths of fragments

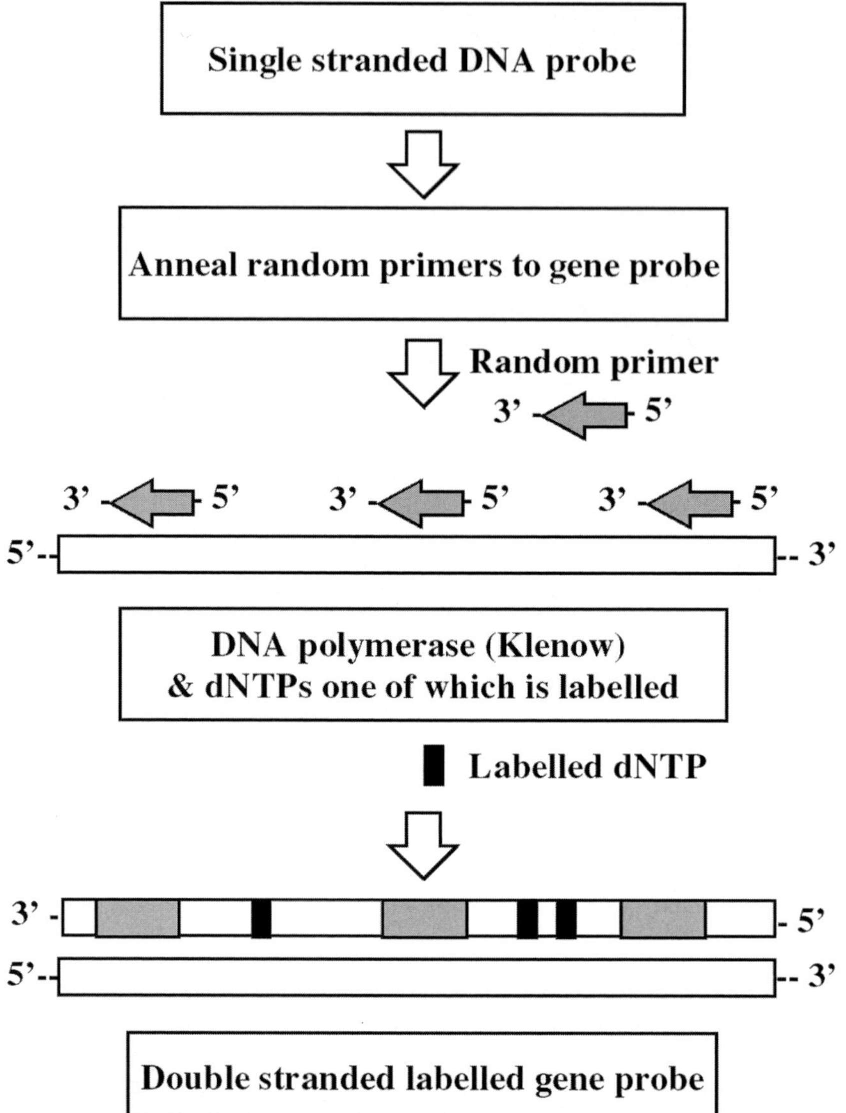

Fig. 7. Random primer gene probe labeling. Random primers are incorporated and used as a start point for Klenow DNA polymerase to synthesize a complementary strand of DNA while incorporating a labeled dNTP at complementary sites.

obtained allows their relative positions within the DNA fragment to be deduced. Any mutation that creates, destroys, or moves the recognition sequence for a restriction enzyme leads to a restriction fragment length polymorphism (RFLP) *(9)*. An RFLP can be detected by examining the profile of restriction fragments generated during digestion. Conventionally, this required the purification of the original starting DNA sample before digestion with single or multiple restriction enzymes. The resultant fragments were then size-separated by gel electrophoresis and visualized by staining with ethidium bromide. Routine RFLP analysis of genomic DNA samples generally also involves hybridization with labeled gene probes to detect a specific gene fragment. The first useful RFLP was described for the detection of sickle cell anemia. In this case, a difference in the pattern of digestion with the restriction endonuclease *Hha*I could

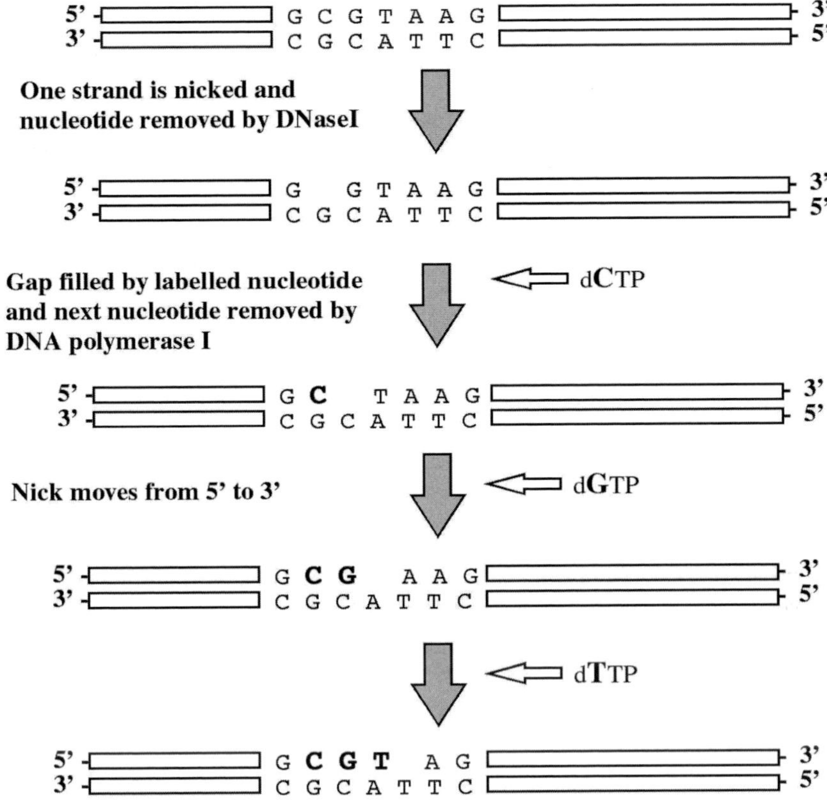

Fig. 8. Nick translation. The removal of nucleotides and their subsequent replacement with labeled nucleotides by DNA polymerase I makes the gene probe more labeled as nick translation proceeds.

be identified between DNA samples from normal individuals and patients with the disease. This polymorphism was later shown to be the result of a single base substitution in the gene for β-globin that changed a codon, GAG specific for the amino acid glutamine to GUG, which encoded valine. A further example is the use of the restriction enzyme *Mnl*I to detect factor V Leiden. This arises from a point mutation in exon 13 of the factor V gene and, in particular, constitutes the most frequent genetic risk factor for venous thrombosis. Although this is a good example of how a basic molecular biology technique could be used for the identification of a genetic mutation, the occurrence is infrequent and thus of limited use.

RFLPs may arise by a number of different means that alter the relative position of restriction endonuclease recognition sequences. In general, most polymorphisms are randomly distributed throughout a genome; however, there are certain regions where many polymophisms exist. These are termed *hypervariable regions* and have been found in regions flanking structural genes from several sources. These were first identified as differences in the numbers of short repeated sequences, termed *minisatellites*. These occurred within the genomes from different individuals, as evidenced by RFLP analysis using specific gene probes. Several types of minisatellite sequence such as variable numbers of tandem repeats (VNTR) have now been described.

An additional source of polymorphic diversity present in the human genome is termed *single-nucleotide polymorphism* (SNP) (pronounced snip). SNPs are substitutions of one base at a precise location within the genome. Those that occur in coding regions are termed cSNPs.

Estimates indicate that an SNP occurs once in every 300 bases and there are thought to be approx 10 million in the human genome. Interest in SNPs lies in the fact that these differences can account for the differences in disease suseptability, drug metabolism, and response to environmental factors between individuals. There are now a number of initiatives to identify SNPs and produce a genome SNP map. A number of maps have been partially completed and a number of bioinformatics resources have been developed, such as the SNP consortium ***(10,11)***.

There are numerous disorders that arise as a result of point mutations or deletions/insertions, and it is possible given an appropriate gene probe complementary to the genetic lesion to identify and detect it successfully following DNA blotting. However, the development and refinement of the PCR has led to many Southern-blot-based methods now being superseded by amplification techniques and will no doubt be ultimately overtaken by the continued development of microarray technology ***(12)***.

References

1. Smith, H. O. and Wilcox, K. W. (1970) A restriction enzyme from Haemophilus influenzae: purification and general characterization. *J. Mol. Biol.* **51,** 379.
2. Jones, P., Qiu J., and Rickwood, D. (1994) *RNA Isolation and Analysis*, Bios Scientific, Oxford.
3. Heptinstall, J. (2000) Spectrophotometric analysis of nucleic acids, in *The Nucleic Acid Protocols Handbook* (Rapley, R., ed.), Humana, Totowa, NJ.
4. Bryant, S. and Manning, D. L. (2000) Isolation of mRNA by affinity chromatography, in *The Nucleic Acid Protocols Handbook* (Rapley, R., ed.), Humana, Totowa, NJ.
5. Maule, J. (2000) Pulsed-field gel electrophoresis, in *The Nucleic Acid Protocols Handbook* (Rapley, R., ed.), Humana, Totowa, NJ.
6. Southern, E. M. (1975) Detection of specific sequences among DNA fragments separated by gel electrophoresis. *J. Mol. Biol.* **98,** 503.
7. Alwine, J. C., Kemp, D. J., and Stark, G. R. (1977) Method for the detection of specific RNAs in in agarose gels by transfer to diazobenzyloxymethyl paper and hybridization with DNA probes. *Proc. Natl. Acad. Sci. USA* **74,** 5350.
8. Aquino de Muro, M. and Rapley, R. (2002) *Gene Probes: Principles and Protocols*, Humana, Totowa, NJ.
9. Enayat, M. S. (2000) Restriction fragment length polymorphism, in *The Nucleic Acid Protocols Handbook* (Rapley, R., ed.), Humana, Totowa, NJ.
10. Wheeler, D. L., Church, D. M., Federhen, S., et al. (2003) Database resources of the National Center for Biotechnology. *Nucleic Acids Res.* **31,** 28–33.
11. Syvanen, A. C. (2001) Accessing genetic variation: genotyping single nucleotide polymorphisms. *Nat. Rev. Genet.* **2,** 930–942.
12. Wikman, F. P., Lu, M. L., Thykjaer, T., et al. (2000) Evaluation of the performance of a p53 sequencing microarray chip using 140 previously sequenced bladder tumor samples. *Clin. Chem.* **46,** 1555–1561.

2

Probe Design, Production, and Applications

Marilena Aquino de Muro

1. Introduction

A probe is a nucleic acid molecule (single-stranded DNA or RNA) with a strong affinity with a specific target (DNA or RNA sequence). Probe and target base sequences must be complementary to each other, but depending on conditions, they do not necessarily have to be exactly complementary. The hybrid (probe–target combination) can be revealed when appropriate labeling and detection systems are used. Gene probes are used in various blotting and *in situ* techniques for the detection of nucleic acid sequences. In medicine, they can help in the identification of microorganisms and the diagnosis of infectious, inherited, and other diseases.

2. Probe Design

The probe design depends on whether a gene probe or an oligonucleotide probe is desired.

2.1. Gene Probes

Gene probes are generally longer than 500 bases and comprise all or most of a target gene. They can be generated in two ways. Cloned probes are normally used when a specific clone is available or when the DNA sequence is unknown and must be cloned first in order to be mapped and sequenced. It is usual to cut the gene with restriction enzymes and excise it from an agarose gel, although if the vector has no homology, this might not be necessary.

Polymerase chain reaction (PCR) is a powerful procedure for making gene probes because it is possible to amplify and label, at the same time, long stretches of DNA using chromosomal or plasmid DNA as template and labeled nucleotides included in the extension step (*see* **Subheadings 2.2.** and **3.2.3.**). Having the whole sequence of a gene, which can easily be obtained from databases (GenBank, EMBL, DDBJ), primers can be designed to amplify the whole gene or gene fragments (see Chapter 28). A considerable amount of time can be saved when the gene of interest is PCR amplified, for there is no need for restriction enzyme digestion, electrophoresis, and elution of DNA fragments from vectors. However, if the PCR amplification gives nonspecific bands, it is recommended to gel purify the specific band that will be used as a probe.

Gene probes generally provide greater specificity than oligonucleotides because of their longer sequence and because more detectable groups per probe molecule can be incorporated into them than into oligonucleotide probes *(1)*.

2.2. Oligonucleotide Probes

Oligonucleotide probes are generally targeted to specific sequences within genes. The most common oligonucleotide probes contain 18–30 bases, but current synthesizers allow efficient synthesis of probes containing at least 100 bases. An oligonucleotide probe can match perfectly

From: *Medical Biomethods Handbook*
Edited by: J. M. Walker and R. Rapley © Humana Press, Inc., Totowa, NJ

its target sequence and is sufficiently long to allow the use of hybridization conditions that will prevent the hybridization to other closely related sequences, making it possible to identify and detect DNA with slight differences in sequence within a highly conserved gene, for example.

The selection of oligonucleotide probe sequences can be done manually from a known gene sequence using the following guidelines ***(1)***:

- The probe length should be between 18 and 50 bases. Longer probes will result in longer hybridization times and low synthesis yields, shorter probes will lack specificity.
- The base composition should be 40–60% G-C. Nonspecific hybridization may increase for G-C ratios outside of this range.
- Be certain that no complementary regions within the probe are present. These may result in the formation of "hairpin" structures that will inhibit hybridisation to target.
- Avoid sequences containing long stretches (more than four) of a single base.
- Once a sequence meeting the above criteria has been identified, computerized sequence analysis is highly recommended. The probe sequence should be compared with the sequence region or genome from which it was derived, as well as to the reverse complement of the region. If homologies to nontarget regions greater than 70% or eight or more bases in a row are found, that probe sequence should not be used.

However, to determine the optimal hybridization conditions, the synthesized probe should be hybridized to specific and nonspecific target nucleic acids over a range of hybridization conditions.

These same guidelines are applicable to design forward and reverse primers for amplification of a particular gene of interest to make a gene probe. It is important to bear in mind that, in this case, it is essential that the 3' end of both forward and reverse primers have no homology with other stretches of the template DNA other than the region you want to amplify. There are numerous software packages available (LaserDNA™, GeneJockey II™, etc.) that can be used to design a primer for a particular sequence or even just to check if the pair of primers designed manually will perform as expected.

3. Labeling and Detection

3.1. Types of Label

3.1.1. Radioactive Labels

Nucleic acid probes can be labeled using radioactive isotopes (e.g., ^{32}P, ^{35}S, ^{125}I, ^{3}H). Detection is by autoradiography or Geiger–Muller counters. Radiolabeled probes used to be the most common type but are less popular today because of safety considerations as well as cost and disposal of radioactive waste products. However, radiolabeled probes are the most sensitive, as they provide the highest degree of resolution currently available in hybridization assays ***(1,2)***. High sensitivity means that low concentrations of a probe–target hybrid can be detected; for example, ^{32}P-labeled probes can detect single-copy genes in only 0.5 μg of DNA and Keller and Manak ***(1)*** list a few reasons:

- ^{32}P has the highest specific activity.
- ^{32}P emits β-particles of high energy.
- ^{32}P-Labeled nucleotides do not inhibit the activity of DNA-modifying enzymes, because the structure is essentially identical to that of the nonradioactive counterpart.

Although ^{32}P-labeled probes can detect minute quantities of immobilized target DNA (<1 pg), their disadvantages is the inability to be used for high-resolution imaging and their relatively short half-life (14.3 d); ^{32}P-labeled probes should be used within a week after preparation.

The lower energy of ^{35}S plus its longer half-life (87.4 d) make this radioisotope more useful than ^{32}P for the preparation of more stable, less specific probes. These ^{35}S-labeled probes, although less sensitive, provide higher resolution in autoradiography and are especially suitable for *in situ* hybridization procedures. Another advantage of ^{35}S over ^{32}P is that the ^{35}S-labeled nucleotides present little external hazard to the user. The low-energy β-particles barely penetrate the upper dead layer of skin and are easily contained by laboratory tubes and vials.

Similarly, ^{3}H-labeled probes have traditionally been used for *in situ* hybridization because the low-energy β-particle emissions result in maximum resolution with low background. It has the longest half-life (12.3 yr).

The use of ^{125}I and ^{131}I has declined since the 1970's with the availability of ^{125}I-labeled nucleoside triphosphates of high specific activity. ^{125}I has lower energies of emission and a longer half-life (60 d) than ^{131}I, and are frequently used for *in situ* hybridization.

3.1.2. Nonradioactive Labels

Compared to radioactive labels, the use of nonradioactive labels have several advantages:

- Safety.
- Higher stability of probe.
- Efficiency of the labeling reaction.
- Detection *in situ.*
- Less time taken to detect the signal.

Concern over laboratory safety and the economic and environmental aspects of radioactive waste disposal have been key factors in their development and use. Some examples are as follows:

- Biotin: This label can be detected using avidin or streptavidin which have high affinities for biotin. Because the reporter enzyme is not conjugated directly to the probe but is linked to it through a bridge (e.g., streptavidin–biotin), this type of nonradioactive detection is known as an indirect system. Usually, biotinylated probes work very well, but because biotin (vitamin H) is a ubiquitous constituent of mammalian tissues and because biotinylated probes tend to stick to certain types of Nylon membrane, high levels of background can occur during hybridizations. These difficulties can be avoided by using nucleotide derivatives, including digoxigenen-11-UTP, -11-dUTP, and -11-ddUTP, and biotin-11-dUTP or biotin-14-dATP. After hybridization, these are detected by an antibody or avidin, respectively, followed by a color or chemiluminescent reaction catalysed by alkaline phosphatase or peroxidase linked to the antibody or avidin ***(1,2)***.
- Enzymes. The enzyme is attached to the probe and its presence usually detected by reaction with a substrate that changes color. Used in this way, the enzyme is sometimes referred to as a "reporter group," Examples of enzymes used include alkaline phosphatase and horseradish peroxidase (HRP). In the presence of peroxide and peroxidase, chloronaphtol, a chromogenic substrate for HRP, forms a purple insoluble product. HRP also catalyzes the oxidation of luminol, a chemiluminogenic substrate for HRP ***(2,3)***.
- Chemiluminescence. In this method, chemiluminescent chemicals attached to the probe are detected by their light emission using a luminometer. Chemiluminescent probes (including the above enzyme labels) can be easily stripped from membranes, allowing the membranes to be reprobed many times without significant loss of resolution.
- Fluorescence chemicals attached to probe fluoresce under ultraviolet (UV) light. This type of label is especially useful for the direct examination of microbiological or cytological specimens under the microscope—a technique known as fluorescent *in situ* hybridization (FISH). Hugenholts et al. have some useful considerations on probe design for FISH ***(4)***.
- Antibodies. An antigenic group is coupled to the probe and its presence detected using specific antibodies. Also, monoclonal antibodies have been developed that will recognize DNA–RNA hybrids. The antibodies themselves have to be labeled, using an enzyme, for example.
- DIG system. It is the most comprehensive, convenient, and effective system for labeling and detection of DNA, RNA, and oligonucleotides. Digoxigenin (DIG), like biotin, can be chemically coupled to linkers, and nucleotides and DIG-substituted nucleotides can be incorporated into nucleic acid probes by any of the standard enzymatic methods. These probes generally yield significantly lower backgrounds than those labeled with biotin. An anti-digoxigenin an-

Table 1
Types of Label

Radioactive labels
^{32}P
^{35}S
^{125}I
^{131}I
^{3}H
Nonradioactive labels
Biotin
Chemiluminescent enzyme labels (acridinium ester, alkaline phosphatase, β-D-galactosidase, horseradish peroxidase [HRP], isoluminol, xanthine oxidase)
Fluorescence chemicals (fluorochromes)
Antibodies
Digoxigenin system

tibody–alkaline phosphatase conjugate is allowed to bind to the hybridized DIG-labeled probe. The signal is then detected with colorimetric or chemiluminescent alkaline phosphatase substrates. If a colorimetric substrate is used, the signal develops directly on the membrane. The signal is detected on an X-ray film (as with ^{32}P- or ^{35}S-labeled probes) when a chemiluminescent substrate is used. Roche Biochemicals has a series of kits for DIG labeling and detection, as well as comprehensive detailed guides ***(5,6)*** with protocols for single-copy gene detection of human genome on Southern blots, detection of unique mRNA species on Northern blots, colony and plaque screening, slot/dot blots, and *in situ* hybridization.

The one area in which nonradioactive probes have a clear advantage is *in situ* hybridization. When the probe is detected by fluorescence or color reaction, the signal is at the exact location of the annealed probe, whereas radioactive probes can only be visualized as silver grains in a photographic emulsion some distance away from the actual annealed probe *(7)*.

3.2. Labeling Methods

The majority of radioactive labeling procedures rely upon enzymatic incorporation of a nucleotide labeled into the DNA, RNA, or oligonucleotide.

Table 1 summarizes the various types of label *(2)*.

3.2.1. Nick Translation

Nick translation is one method of labeling DNA, which uses the enzymes pancreatic Dnase I and *Escherichia coli* DNA polymerase I. The nick translation reaction results from the process by which *E. coli* DNA polymerase I adds nucleotides to the 3'-OH created by the nicking activity of Dnase I, while the 5' to 3' exonuclease activity simultaneously removes nucleotides from the 5' side of the nick. If labeled precursor nucleotides are present in the reaction, the preexisting nucleotides are replaced with labeled nucleotides. For radioactive labeling of DNA, the precursor nucleotide is an [α-^{32}P]dNTP. For nonradioactive labeling procedures, a digoxigenin or a biotin moiety attached to a dNTP analog is used ***(2)***.

3.2.2. Random-Primed Labeling (or Primer Extension)

Gene probes, cloned or PCR-amplified, and oligonucleotide probes can be random-primed labeled with radioactive isotopes and nonradioactive labels (e.g., DIG). Random-primed labeling of DNA fragments (double- or single-stranded DNA) was developed by Feinberg and Volgestein ***(8,9)*** as an alternative to nick translation to produce uniformly labeled probes.

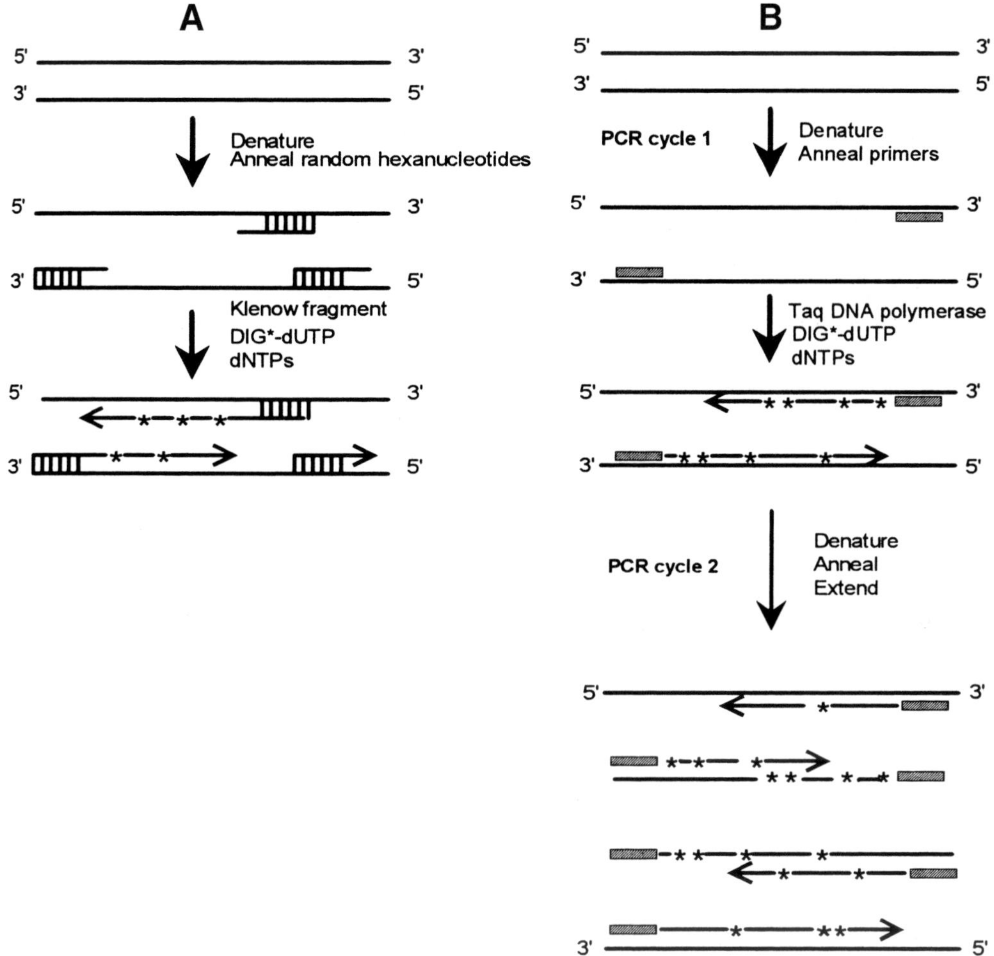

Fig. 1. Steps involved in the following: **(A)** random-primed DIG labeling: Double-stranded DNA is denatured and annealed with random oligonucleotide primers (6-mers); the oligonucleotides serve as primers for the 5' to 3' Klenow fragment of *E. coli* DNA polymerase I, which synthesizes labeled probes in the presence of DIG–dUTP. **(B)** PCR–DIG labelling: DIG–dUTP is incorporated during PCR cycles into the DNA strands amplified from the DNA target. The asterisk represents the digoxigenin molecule incorporated along the DNA strands.

Double-stranded DNA is denatured and annealed with random oligonucleotide primers (6-mers). The oligonucleotides serve as primers for the 5' to 3' polymerase (the Klenow fragment of *E. coli* DNA polymerase I), which synthesizes labeled probes in the presence of a labeled nucleotide precursor. **Figure 1A** shows the steps involved in random-primed DIG labeling as an example.

3.2.3. DIG–PCR Labeling

A very robust method for labeling a gene probe with DIG uses PCR. The probe is PCR-amplified using the appropriate set of primers and thermocycling parameters, however, the dNTP mixture has less dTTP because the labeled DIG–dUTP will also be added to the reaction. (Similarly, when this method is used with [α-^{32}P]dCTP, the dNTP mixture will not have dCTP.) The advantage of PCR–DIG labeling, over random-primed DIG labeling, is the incorporation of a higher number of DIG moieties along the amplified DNA strands during the PCR cycles. It is worth noting that the random incorporation of large molecules of DIG–dUTP along the DNA strands during the PCR cycles makes the amplified fragment run slower on an agarose gel. A control PCR reaction, without DIG–dUTP, should also be prepared at the same time to verify whether the size of the amplified fragment with incorporated DIG (labeled probe) corresponds to the desired gene fragment. **Figure 1B** shows the steps involved in PCR–DIG labeling, and refs. ***(10–12)*** describe successful examples of use of PCR–DIG labeling.

3.2.4. Photobiotin Labeling

Photobiotin labeling is a chemical reaction, not an enzymatic one. Biotin and DIG can be linked to a nitrophenyl azido group that is converted by irradiation with UV or strong visible light to a highly reactive nitrene that can form stable covalent linkages to DNA and RNA ***(2)***. The materials for photobiotin labeling are more stable than the enzymes needed in nick translation or oligonucleotide labeling and are less expensive, and it is a method of choice when large quantities of probe but not very high sensitivities ***(3,13)***.

3.2.5. End Labeling

End labeling of probes for hybridization is mainly used to label oligonucleotide probes (for a review, see **ref. *14***).

Roche Biochemicals ***(6)*** has developed three methods for labeling oligonucleotides with digoxigenin:

- The 3'-end labeling of an oligonucleotide 14–100 nucleotides in length with 1 residue of DIG-11-ddUTP per molecule
- The 3' tailing reaction, where terminal transferase adds a mixture of unlabeled nucleotides and DIG-11-dUTP, producing a tail containing multiple digoxigenin residues
- The 5' end labeling in a two-step synthesis with first an aminolinker residue on the 5' end of the oligonucleotide, and then after purification, a digoxigenin-*N*-hydroxy-succinimide ester is covalently linked to the free 5'-amino residue.

Oligonucleotides can also be labeled with radioisotopes by transferring the γ-^{32}P from [γ-^{32}P]ATP to the 5' end using the enzyme bacteriophage T4 polynucleotide kinase. If the reaction is carried out efficiently, the specific activity of such probes can be as high as the specific activity of [γ-^{32}P]ATP itself ***(2)***.

Promega has a detailed guide ***(15)*** with protocols on radioactive and nonradioactive labeling of DNA. The choice of probe labeling method will depend on the following:

- Target format: Southern, Northern, slot/dot, or colony blot (*see* **Subheading 4.**)
- Type of probe: gene or oligonucleotide probe
- Sensitivity required for detection: single-copy gene or detection of PCR-amplified DNA fragments

For example, 3'- and 5'-end labeling of oligonucleotides give good results on slot and colony hybridization in contrast with poor sensitivity when using Southern blotting.

4. Target Format

4.1. Solid Support

A convenient format for the hybridization of DNA to gene probes or oligonucleotide probes is immobilization of the target nucleic acid (DNA or RNA) onto a solid support while the probe is free in solution. The solid support can be a nitrocellulose or Nylon membrane, Latex or magnetic beads, or microtiter plates. Nitrocellulose membranes are very commonly used and produce low background signals; however, they can only be used when colorimetric detection will be performed and no probe stripping and reprobing is planned. For these purposes, positively charged Nylon membranes are recommended, and they also ensure an optimal signal-to-noise ratio when the DIG system is used. Although nitrocellulose membranes are able to bind large quantities of DNA, they become brittle and gradually release DNA during the hybridization step. Activated cellulose membranes, on the other hand, are more difficult to prepare, but they can be reused many times because the DNA is irreversibly bound *(2)*.

After size fractionation of nucleic acids by electrophoresis, they are transferred to a filter membrane, which is then probed. The presence of target is confirmed by the detection of a probe on the filter membrane, for example, radiolabeled probe can be detected by autoradiography and the location of the target sequence in the bands in the original gel determined.

The different immobilization techniques include the following: Southern blots, when whole or digested chromosomal DNA is electrophoresed in an agarose gel, denatured, and blotted onto a membrane; Northern blots, when the same procedure is used for RNA; slot blots, when whole RNA or denatured DNA is loaded under vacuum into slots onto membranes (similar procedure for dot blots); colony blots, when colonies are treated with lysozyme on plates and further treatment with protease, and denaturation and neutralization solutions are applied and the procedure adjusted according the microorganism's peculiarities. The great advantage of colony blotting over slot blotting is that strains with a specific sequence can be rapidly detected from plates and the DNA preparation procedure can then be done only for the strains of interest. In a similar way, the slot blotting procedure has the advantage of quickly highlighting which DNA sample has the gene sequence of interest when a gene probe is hybridized to whole-DNA samples. The Southern blotting procedure, which involves the digest of DNA with restriction enzymes and gel electrophoresis, takes a longer preparation time than slot blotting but can provide information on the size and position of the gene as well as grouping the samples based on the similar patterns when different restriction enzymes are used to digest the samples. Schleicher and Schuell has a detailed manual on solid supports and DNA transfer *(16)*.

4.2. In Solution

Both the probe and the target are in solution. Because both are free to move, the chances of reaction are maximized and, therefore, this format generally gives faster results than others.

4.3. In Situ

In this format, the probe solution is added to fixed tissues, sections, or smears, which are then usually examined under the microscope (see Chapter 29). The probe label (e.g., a fluorescent marker) produces a visible change in the specimen if the target sequence is present and hybridization has occurred. However, the sensitivity might be low if the amount of target nucleic acid present in the specimen is low. This can be used for the gene mapping of chromosomes and for the detection of microorganisms in specimens.

5. Hybridization Conditions

Many methods are available to hybridize probes in solution to DNA or RNA immobilized on nitrocellulose membranes (see Chapter 4). These methods can differ in the solvent and temperature used, the volume of solvent and the length of time of hybridisation, the method of agitation, when required, and the concentration of the labeled probe and its specificity, the stringency of the washes after the hybridization. However, basically, target nucleic acid immobilized on membranes by the Southern blot, Northern blot, slot or dot blot, or colony blot procedures are hybridized in the same way. The membranes are first prehybridized with hybridization buffer minus the probe. Nonspecific DNA binding sites on the membrane are saturated with carrier DNA and synthetic polymers. The prehybridization buffer is replaced with the hybridization buffer containing the probe and incubated to allow hybridization of the labeled probe to the target nucleic acid. The optimum hybridization temperature is experimentally determined, starting with temperatures 5°C below the melting temperature (T_m). The T_m is defined as the temperature corresponding to the midpoint in transition from helix to random coil and depends on length, nucleotide composition and ionic strength for long stretches of nucleic acids. G-C pairs are more stable than A-T pairs because G and C form three H bonds as

Table 2
Examples of the Applications of Nucleic Acid Probes in Medical Research

Application	Ref.
Detection of tumor suppressor genes in human bladder tumors	**21**
Identification of *Leishmania* parasites	**22**
Detection of malignant plasma cells of patients with multiple myeloma	**23**
Diagnosis of human papillomavirus	**24**
Visual gene diagnosis of HBV and HCV	**25**
Detection and identification of pathogenic *Vibrio parahaemolyticus*	**26**
Detection of *Vibrio cholerae*	**27–29**
Molecular analysis of tetracycline resistance in *Salmonella enterica*	**30**
Identification of fimbrial adhesins in necrotoxigenic *E. coli*	**31**
Epidemiological analysis of *Campylobacter jejuni* infections	**32**
Molecular analysis of NSP4 gene from human rotavirus strains	**33**
Physical mapping of human parasite *Trypanosoma cruzi*	**34**
Detection and identification of African trypanosomes	**35**
Changes related to neurological diseases (Alzheimer's, Huntington's)	**36**
Detection and identification of pathogenic *Candida* spp.	**37,38**
Identification of *Mycobacterium* spp.	**39**
Detection of rifampin resistance in *Mycobacterium tuberculosis*	**40**
Identification of *Staphylococcus aureus* directly from blood cultures	**41**
Detection of rabies virus genome in brain tissues from mice	**42**

opposed to two between A and T. Therefore, double-stranded DNA rich in G and C has a higher T_m (more energy required to separate the strands) than A-T rich DNA. For oligonucleotide probes bound to immobilized DNA, the dissociation temperature, T_d, is concentration dependent. Stahl and Amman *(17)* discussed in detail the empirical formulas used to estimate T_m and T_d.

Following the hybridization, the unhybridized probe is removed by a series of washes. The stringency of the washes must be adjusted for the specific probe used. Low-stringency washing conditions (higher salt and lower temperature) increases sensitivity; however, these conditions can give nonspecific hybridization signals and high background. High-stringency washing conditions (lower salt and higher temperature, closer to the hybridization temperature) can reduce background and only the specific signal will remain. The hybridization signal and background can also be affected by probe length, purity, concentration, sequence, and target contamination *(1)*.

In aqueous solution, RNA–RNA hybrids are more stable than RNA–DNA hybrids, which are, in turn, more stable than DNA–DNA ones. This results in a difference in T_m of approx 10°C between RNA-RNA and DNA-DNA hybrids. Consequently, more stringent conditions should be used with RNA probes *(8)*.

In general, the hybridization rate increases with probe concentration. Also, within narrow limits, sensitivity increases with increasing probe concentration. The concentration limit is not determined by any inherent physical property of nucleic acid probes, but by the type of label and nonspecific binding properties of the immobilization medium involved.

6. Applications in Medical Research

At least three basic applications of nucleic acid probes in medical research can be mentioned: (1) detection of pathogenic microorganisms, (2) detection of changes to nucleic acid sequences, and (3) detection of tandem repeat sequences. **Table 2** presents only a few examples of recently published literature on applications of nucleic acid probes in medical research.

6.1. Detection of Pathogenic Microorganisms

The application of nucleic acid probes has particularly been evident in microbial ecology, where probes can be used to detect unculturable microorganisms and pathogens in the environment or simply provide rapid identification of species and group levels. Through the development of DNA–DNA and RNA–DNA hybridization procedures and recombinant DNA methodology, the isolation of species-specific gene sequences is readily achieved ***(18,19)***. Oligonucleotide hybridization probes complementing either small ribosomal subunits, large ribosomal subunits, or internal transcribed spacer regions have now been developed for a wide variety of microorganisms ***(20)***, such as *Actinomyces*, *Bacteriodes*, *Borrelia*, *Clostridium*, *Campylobacter*, *Candida*, *Haemophilus*, *Helicobacter*, *Lactococcus*, *Mycoplasma*, *Neisseria*, *Proteus*, *Rickettsia*, *Vibrio*, *Streptococcus*, *Plasmodium*, *Pneumocystis*, *Trichomonas*, *Desulfovibrio*, *Streptomyces*, including some uncultivated species such as marine proteobacteria and thermophilic cyanobacterium, and *Chlamydia* species, *Rickettsia* species, *Trypanosoma* species, *Treponema pallidum*, *Pneumocystis carinii*, and *Mycobacterium* species to mention only a few examples with medical relevance. Detection of a nucleic acid sequence unique to a particular microorganism would demonstrate its presence in a specimen and, perhaps, confirm an infectious disease.

6.2. Detection of Changes to Nucleic Acid Sequences

A change to the DNA sequence is a mutation, which could involve deletion, insertion, or substitution. Changes in certain gene sequences can cause inherited diseases such as cystic fibrosis, muscular dystrophies, phenylketonuria, apolipoprotein variants, and sickle cell anemia, and they can be diagnosed by probe detection. Nucleic acid probes have successfully been used to detect those mutations. With some inherited diseases, more than one type of mutation can cause the disease, in which case, a probe might have to be used under low stringency (to allow hybridization to a range of sequences) or several probes might be used to ensure hybridization to all target sequences.

6.3. Detection of Tandem Repeat Sequences

Tandem repeat sequences are usually 30–50 bp in length. Their size and distribution are distinctive for an individual. They can be detected using nucleic acid probes and PCR. They are the basis of so-called "DNA fingerprinting," which can be used in forensic science to confirm the identity of a suspect from specimens (any body fluid, skin, and hair) left at the scene of a crime. This technique can also be used for paternity tests, sibling confirmation, and tissue typing.

References

1. Keller, G. H. and Manak, M. M. (1989) *DNA Probes*, Stockton, New York.
2. Sambrook, J. and Russell, D. W. (2001) *Molecular Cloning: A Laboratory Manual*, 3rd ed., Cold Spring Harbor Laboratory Press, Cold Spring Harbor, NY.
3. Karcher, S. J. (1995) *Molecular Biology: A Project Approach*, Academic, San Diego, CA.
4. Hugenholtz, P., Tyson, G. W., and Blackall, L. L. (2002) Design and evaluation of 16S rRNA-targeted oligonucleotide probes for fluorescence *in situ* hybridization, in *Gene Probes: Principles and Protocols* (Aquino de Muro, M. and Rapley, R., eds.), Humana, Totowa, NJ, pp. 29–42.
5. Boehringer Mannheim GmbH (1995) *The DIG System User's Guide for Filter Hybridisation*, Boehringer Mannheim, Mannheim, Germany.
6. Boehringer Mannheim GmbH (1996) *Nonradioactive* In Situ *Hybridisation Manual: Application Manual*, 2nd ed. Boehringher Mannheim GmbH, Mannheim, Germany.
7. Alphey, L. and Parry, H. D. (1995) Making nucleic acid probes, in *DNA cloning 1: Core Techniques* (Glover, D. M. and Hames, B. D., eds.), IRL, Oxford, pp. 121–141.
8. Feinberg, A. P. and Vogelstein, B. (1983) A technique for radiolabeling DNA restriction endonuclease fragments to high specific activity. *Analy. Biochem.* **132**, 6–13.
9. Feinberg, A. P. and Vogelstein, B. (1984) Addendum. *Analy. Biochem.* **137**, 266–267.
10. Aquino de Muro, M. and Priest, F. G. (1994) A colony hybridization procedure for the identification of mosquitocidal strains of *Bacillus sphaericus* on isolation plates. *J. Invertebr. Pathol.* **63**, 310–313.

11. Aquino de Muro, M. and Priest, F. G. (2000) Construction of chromosomal integrants of *Bacillus sphaericus* 2362 by conjugation with *Escherichia coli*. *Res. Microbiol.* **151**, 547–555.
12. Garratt, L. C., McCabe, M. S., Power, J. B., and Davey, M. R. (2002) Detection of single-copy genes in DNA from transgenic plants, in *Gene Probes: Principles and Protocols* (Aquino de Muro, M. and Rapley, R., eds.), Humana, Totowa, NJ, pp. 211–222.
13. Hilario, E. (2002) Photobiotin labeling, in *Gene Probes: Principles and Protocols* (Aquino de Muro, M. and Rapley, R., eds.), Humana, Totowa, NJ, pp. 19–22.
14. Hilario, E. (2002) End labeling procedures, in *Gene Probes: Principles and Protocols* (Aquino de Muro, M. and Rapley, R., eds.), Humana, Totowa, NJ, pp. 13–18.
15. Promega Corp. (1996) *Protocols and Applications Guide*, 3rd ed., Promega Corp., Madison, WI.
16. Schleicher & Schuell, Inc. (1995) *Blotting, Hybridization and Detection: An S&S Laboratory Manual*, 6th ed., Schleicher & Schuell, Inc., Keene, NH.
17. Stahl, D. A. and Amman, R. (1991) Development and application of nucleic acid probes, in *Nucleic Acid Techniques in Bacterial Systematics* (Stackebrandt, E. and Goodfellow, M., eds.), Wiley, Chichester, pp. 205–244.
18. Brooker, J. D. Lockington, R. A. Attwood, G. T., and Miller, S. (1990) The use of gene and antibody probes in identification and enumeration of rumen bacterial species, in *Gene Probes for Bacteria* (Macario, A. J. L. and Conway de Macario, E., eds.), Academic, San Diego, CA, pp. 390–416.
19. Stahl, D. A. and Kane, M. D. (1992) Methods in microbial identification, tracking and monitoring of function. *Curr. Opin. Biotechnol.* **3**, 244–252.
20. Ward, D. M., Bateson, M. M., Weller, R., and Ruff-Roberts, A. L. (1992) Ribosomal RNA analysis of micro-organisms as they occur in nature. *Adv. Microb. Ecol.* **12**, 219–286.
21. Orlow, I. and Cordon-Cardo, C. (2002) Evaluation of alterations in the tumor suppressor genes INK4A and INK4B in human bladder tumors, in *Gene Probes: Principles and Protocols* (Aquino de Muro, M. and Rapley, R., eds.), Humana, Totowa, NJ, pp. 43–59.
22. Mendoza-Leon, A., Luis, L., and Martinez, C. (2002) The β-tubulin gene region as a molecular marker to distinguish *Leishmania* parasites, in *Gene Probes: Principles and Protocols* (Aquino de Muro, M. and Rapley, R., eds.), Humana, Totowa, NJ, pp. 61–83.
23. Brown, R. D. and Joy Ho, P. (2002) Detection of malignant plasma cells in the bone marrow and peripherical blood of patients with multiple myeloma, in *Gene Probes: Principles and Protocols* (Aquino de Muro, M. and Rapley, R., eds.), Humana, Totowa, NJ, pp. 85–91.
24. Nuovo, G J. (2002) Diagnosis of human papillomavirus using *in situ* hybridization and *in situ* polymerase chain reaction, in *Gene Probes: Principles and Protocols* (Aquino de Muro, M. and Rapley, R., eds.), Humana, Totowa, NJ, pp. 113–136.
25. Wang, Y., Pang, D., Zhang, Z., Zheng, H., Cao, J., and Shen J. (2003) Visual gene diagnosis of HBV and HCV based on nanoparticle probe amplification and silver staining enhancement. *J. Med. Virol.* **70(2)**, 205–211.
26. Cook, D. W., Bowers, J. C., and DePaola, A. (2002) Density of total and pathogenic (tdh+) *Vibrio parahaemolyticus* in Atlantic and Gulf Coast molluscan shellfish at harvest. *J. Food Protect.* **65(12)**, 1873–1880.
27. Dalsgaard, A., Serichantalergs, O., Forslund, A., et al. (2001) Clinical and environmental isolates of *Vibrio cholerae* serogroup O141 carry the CTX phage and the genes enconding the toxin-coregulated pili. *J. Clin. Microbiol.* **39(11)**, 4086–4092.
28. Kondo, S., Kongmuang, U., Kalnauwakul, S., Matsumoto, C., Chen, C. H., and Nishibuchi, M. (2001) Molecular epidemiologic analysis of *Vibrio cholerae* O1 isolated during the 1997-8 cholera epidemic in southern Thailand. *Epidemiol. Infect.* **127(1)**, 7–16.
29. Nair, G. B., Bag, P. K., Shimada, T., et al. (1995) Evaluation of DNA probes for specific detection of *Vibrio cholerae* O139 Bengal. *J. Clin. Microbiol.* **33(8)**, 2186–2187.
30. Frech, G. and Schwarz, S. (2000) Molecular analysis of tetracycline resistance in *Salmonella enterica* subsp. *enterica* serovars Typhimurium, Enteritidis, Dublin Choleraesuis, Hadar and Saintpaul: construction and application of specific gene probes. *J. Appl. Microbiol.* **89(4)**, 633–641.
31. Mainil, J. G., Gerardin, J., and Jacquemin, E. (2000) Identification of the F17 fimbrial subunit- and adhesin-enconding (f17A and f17G) gene variants in necrotoxigenic *Escherichia coli* from cattle, pigs and humans. *Vet. Microbiol.* **73(4)**, 327–335.
32. Fujimoto, S., Umene, K., Saito, M., Horikawa, K., and Blaser, M. J. (2000) Restriction fragment length polymorphism analysis using random chromosomal gene probes for epidemiological analysis of *Campylobacter jejuni* infections. *J. Clin. Microbiol.* **38(4)**, 1664–1667.

33. Kirkwood, C. D., Gentsch, J. R., and Glass, R. I. (1999) Sequence analysis of the NSP4 gene from human rotavirus strains isolated in the United States. *Virus Genes* **19(2)**, 113–122.
34. Santos, M. R. M., Lorenzi, H., Porcile, P., et al. (1999) Physical mapping of a 670-kb region of chromosomes XVI and XVII from the human protozoan parasite *Trypanosoma cruzi* encompassing the genes for two immunodominant antigens. *Genome Res.* **9(12)**, 1268–1276.
35. Radwanska, M., Magez, S., Perry-O'Keefe, H., et al. (2002) Direct detection and identification of African trypanosomes by fluorescence in situ hybridization with peptide nucleic acid probes. *J. Clin. Microbiol.* **40(11)**, 4295–4297.
36. Higgins, G. A. and Mah, V. H. (1989) *In situ* hybridisation approaches to human neurological disease, in *Gene Probes* (Conn, P. M., ed.), Academic, San Diego, CA, pp. 183–196.
37. Rigby, S., Procop, G. W., Haase, G., et al. (2002) Fluorescence *in situ* hybridization with peptide nucleic acid probes for rapid identification of *Candida albicans* directly from blood culture bottles. *J. Clin. Microbiol.* **40(6)**, 2182–2186.
38. Oliveira, K., Haase, G., Kurtzman, C., Hyldig-Nielsen, J. J., and Stender, H. (2001) Differentiation of *Candida albicans* and *Candida dubliniensis* by fluorescent in situ hybridization with peptide nucleic acid probes. *J. Clin. Microbiol.* **39(11)**, 4138–4141.
39. Cloud, J. L., Neal, H., Rosenberry, R., et al. (2002) Identification of *Mycobacterium* spp. by using a commercial 16S ribosomal DNA sequencing kit and additional sequencing libraries. *J. Clin. Microbiol.* **40(2)**, 400–406.
40. El Hajj, H. H., Marras, S. A. E., Tyagi, S., Kramer, F. R., and Alland, D. (2001) Detection of rifampin resistance in *Mycobacterium tuberculosis* in a single tube with molecular beacons. *J. Clin. Microbiol.* **39(11)**, 4131–4137.
41. Oliveira, K., Procop, G. W., Wilson, D., Coull, J., and Stender, H. (2002) Rapid identification of *Staphylococcus aureus* directly from blood cultures by fluorescence in situ hybridization with peptide nucleic acid probes. *J. Clin. Microbiol.* **40(1)**, 247–251.
42. Reddy, C. C., Jayakumar, R., Kumanan, K., and Nainar, A. M. (2002) Detection of rabies virus genome in brain tissues by using *in situ* hybridization. *Indian J. Anim. Sci.* **72(1)**, 3–5.

3

Restriction Enzymes

Tools in Clinical Research

Gareth J. S. Jenkins

1. Restriction Enzymes

Restriction enzymes (or restriction endonucleases) are bacterial enzymes capable of cleaving double-stranded DNA. Even though the enzymes are bacterial in origin, because of the universal nature of DNA they can digest DNA from any species, including humans. Importantly, restriction enzymes (REs) carry out this cleavage at specific sites in DNA governed by the sequence context (so-called recognition sequences). Hence, REs are known to be extremely sequence-specific; subtle alterations in the recognition sequence render the sites indigestible. This fact is the basis of their usefulness in clinical research and diagnostics. **Table 1** is a list of a few of the common REs showing their sequence specificities. **Figure 1** shows the interaction of a RE with DNA. The RE interacts with DNA via multiple hydrogen bonds (typically 10–15) plus numerous van der Waals interactions. Only when the RE–DNA complex is tightly bound does the catalytic domain cause DNA cleavage *(2)*.

Restriction enzymes were first discovered in the 1950s and their subsequent isolation in the 1970s paved the way for modern recombinant DNA technologies *(3)*. In fact, without REs, gene cloning technologies (e.g., the construction of genetically modified organisms) would not have become as ubiquitous as they currently are. Therefore, REs have a central place in current DNA manipulation methodologies, but as will be seen here, they are also being used to answer questions about the integrity of DNA in clinical specimens. The clinical use of REs relies on their extreme sequence specificity.

Restriction enzymes can be grouped into three main groups, known as type I, type II, and type III. This chapter focuses on the role of the type II REs, which are the most commonly used REs in DNA manipulation. Type II REs cleave DNA within the same sequences that are recognized by the enzyme, hence an internal digestion. The other two types of RE are more complex (e.g., type I REs cleave DNA outside of this recognition sequence). There are currently well over 1000 type II REs characterized, with over 200 commercially available. A searchable database exists online (rebase.neb.com) containing the specific details of a large number of available RE's and the companies that sell them *(4)*.

2. Function of REs

Restriction enzymes were first identified in bacterial strains that were shown to be resistant to certain bacteriophages (virus equivalents in bacteria). This phenomenon was termed "host-controlled restriction" and was later shown to be caused by the presence of specific REs within these bacteria, which destroyed the bacteriophage DNA before it could insert itself into the

From: *Medical Biomethods Handbook*
Edited by: J. M. Walker and R. Rapley © Humana Press, Inc., Totowa, NJ

Table 1
List of Six Commonly Used Restriction Enzymes

Name	Host bacteria	Recognition sequence	Digestion temperature
*Hae*III	*Haemophilis parainfluenzae*	GGCC	37°C
*Pvu*II	*Proteus vulgaris*	CAGCTG	37°C
*Eco*RI	*Escherichia coli*	GGATCC	37°C
*Hha*I	*Haemophilus haemolyticus*	GCGC	37°C
*Msp*I	*Moraxella species*	CCGG	37°C
*Taq*I	*Thermus aquaticus*	TCGA	65°C

Note: Included are details of the bacteria from which they were isolated and some of their reaction characteristics, including their recognition sequences.

bacterial genome. Obviously, the bacteria's own DNA would normally be susceptible to similar digestion, but for the presence of DNA-modifying enzymes, which modify the bacterial genome and protect it. Hence, particular species of bacteria produce REs and modifying enzymes (actually DNA methylases) that recognize the same DNA sequences. The bacterial DNA is methylated by the methylase enzyme, protecting it from RE-mediated digestion while incoming unmethylated bacteriophage DNA is destroyed.

3. Sequence Specificity of REs

Restriction enzymes usually recognize four to six basepair sequences, often in the form of palindromes (read same sequence in the 5' to 3' direction on both strands). Each species of bacteria produces a different RE recognizing a different DNA sequence. In fact, REs are named after the bacterial species from which they were isolated (*see* **Table 1**). Where bacteria produce more than one RE, they are known as *Pvu*I, *Pvu*II, and so forth. REs produced by different bacteria but with the same recognition sequence are known as isoschizomers.

If one assumes that the human genome contains a random sequence of the A, C, G, and T bases and that all four bases are present at the same frequency, it can be estimated that REs cut human DNA every 256–4096 bp (4-bp cutters, 6-bp cutters, respectively). Hence, in the 3 billion bases of the human genome, there would be, on average, 10 million sites for a 4-base cutting RE and on average 700,000 sites for a 6-base cutting RE. Given the availability of over 200 different REs, it is easy to see how frequent RE sites are in human DNA; in fact, RE sites are estimated to cover 50% of the genome *(5)*.

4. Role of REs in Clinical Research and Diagnostics

It is the sequence specificity of REs that is exploited in clinical research. Alteration of a single base within a RE site removes the ability of the RE to digest that particular stretch of DNA. Hence, DNA sequence changes can be inferred by the loss of corresponding RE sites. Given that up to 50% of the genome is covered by 1 or another of the 200 available REs, it is relatively straightforward to monitor the integrity of these RE sites for the presence of clinically related DNA alterations (mutation, deletion). The alteration of known RE sites can be readily examined in large numbers of clinical specimens simultaneously, allowing the study of DNA alterations linked to a particular clinical condition. For example, if mutation of a single base leads to loss of a particular RE site and this is linked to the occurrence of a disease as a result of the altered protein produced by this gene, then RE analysis is able to rapidly screen large numbers of samples for this particular sequence change. In the following subsections are listed several molecular approaches involving REs currently used for the analysis of clinical specimens.

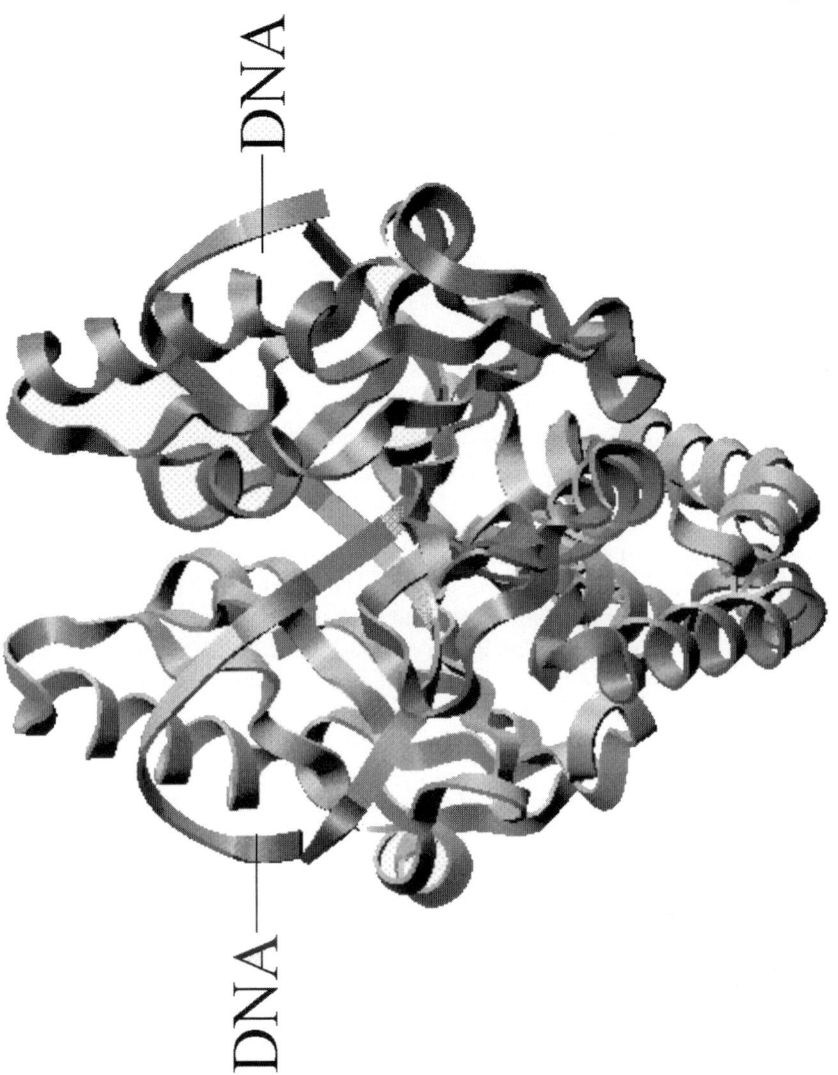

Fig. 1. Restriction enzyme (*Pvu*II) interaction with DNA. (Taken from the nucleic acid database [http://ndbserver.rutgers.edu/] deposited by the authors of **ref. *1*.**)

4.1. Restriction Fragment Length Polymorphism

Restriction fragment length polymorphism (RFLP) analysis is widely used to examine the sequences of large numbers of samples simultaneously. RFLP involves the digestion of individual DNA samples with a particular RE, followed by separation of the daughter products by electrophoresis. This allows the study of the integrity of the recognition sites of one particular RE at a time and, thereby, the integrity of the DNA sequence contained within the RE site. Hence, alterations in the distribution of the RE sites for each RE in turn can be assessed in large numbers of samples. This process has been particularly useful in studying population differences in individuals of many species, not just humans.

In the past, RFLP analysis was carried out as follows: Genomic DNA was initially digested with each RE in turn; the DNA was then electrophoresed and probed for specific sequences by Southern blotting. However, more recently, with the advent of polymerase chain reaction

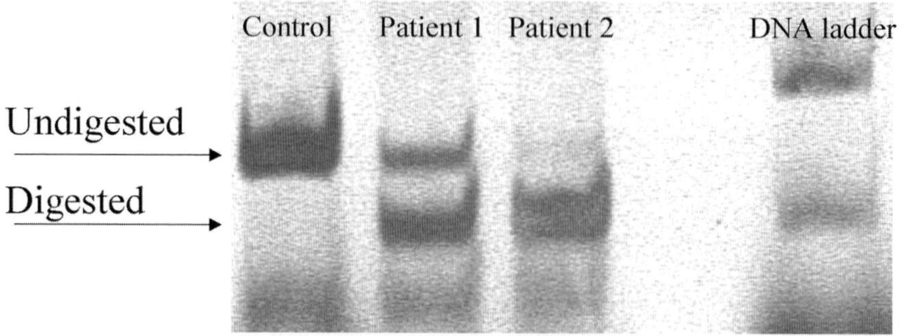

Fig. 2. RFLP analysis. PCR-amplified DNA is digested with a RE. Left lane shows undigested PCR product. Right lane shows DNA size ladder. Patient 1 shows a heterozygous individual (one sequence digests, one sequence does not digest). Patient 2 shows individual homozygous for the RE site sequence (both sequences digest).

(PCR), a simpler approach has become available. Currently, the gene of interest (or gene region) is initially PCR-amplified from the genomic DNA; this PCR product is then digested with the RE and the RE fragments are separated and visualized by electrophoresis. **Figure 2** shows the result of this process. In **Fig. 2**, the digested PCR products from two different individuals are shown. It should be borne in mind that every human somatic cell contains two copies of every gene. Hence, analysis of the integrity of a particular gene sequence is actually carried out in duplicate each time. In **Fig. 2** the right-hand digested PCR product is from an individual who is homozygous for the RE site sequence (both DNA sequences digest, hence both RE sites intact), whereas the left-hand digested PCR product is from an individual who is heterozygous for the sequence of this particular RE site (one copy digests, one copy does not). RFLP patterns can be easily produced for many RE sites (up to 50% of genome) within a gene of interest and these patterns of digestion can be compared among large numbers of clinical samples for differences. By this method, mutations at particular RE sites that lead to loss of RE sites and are important in clinical conditions have been identified.

Such RFLP analysis can actually detect two separate molecular events in the clinical samples, namely, mutations and deletions. Point mutations within the particular RE site under study will be detected by the loss of digestion at one site leading to loss of two daughter strands post electrophoresis. As was seen above, this process is monitored in both copies of each gene simultaneously (heterozygous vs homozygous changes). Conversely, the development of a new RE site within a particular PCR product by mutation can lead to an extra pair of daughter fragments on the gel. These new RE sites are identified by the presence of smaller than expected digestion fragments. Furthermore, deletions of DNA tracts will be detected by the loss of specific RE sites. These deletions can also sometimes be seen by the size of the amplified PCR product changing. RE analysis can also be used to study specific deletions in heterozygous individuals. This is particularly useful in cancer research, where tumor suppressor genes are often deleted during tumor evolution. Where heterozygosity (one cut band, one uncut band) is present in the normal tissue but is lost in the tumor (either cut band or uncut band alone), this indicates that one copy of the particular sequence has been deleted. Therefore, choosing RE sites in tumor suppressor genes for which the individual is heterozygous is the key for this type of analysis.

In both cases (mutation and deletion), RFLP analysis is used as an initial screen for alterations in large numbers of samples. The alterations are always confirmed by DNA sequencing.

The use of DNA sequencing provides important information on the type of mutation (G to A, A to G, etc.) present in the DNA and the position of such mutations (base and codon preference). In the case of deletion events, sequencing can identify the size of the deleted region and

Table 2
Some of the Polymorphisms Present in Cancer-Related Genes that Have Been Studied by RE Analysis

Gene	Cancer type	Region	RE	Ref.
DCC gene	Colon	Codon 201	SalI	6
p53 gene	All	Intron 7	ApaI	7
CYP1A1	All	3'UTR	MspI	8
H-ras	Bladder	Intron 4	BstEII	9
CYP17	Breast	5' Promoter	MspI	10

Note: UTR = untranslated region.

the points at which the deletion occurred. In terms of mutations, much can be gained from studying the actual mutation types in clinical tissues as a result of the fact that chemical mutagens that lead to the mutation being induced produce a characteristic pattern of mutations (types and position). Therefore, causative mutagens can be retrospectively identified from the pattern of mutations seen in clinical tissues. This can provide important information on mutagen exposures, potentially leading to a reduction in such exposures in the future.

The advantage of RFLP analysis lies in its simplicity, high-throughput nature, and low cost. Sequencing the gene of interest from a large number of individuals would be the ideal way of looking for sequence changes that might be clinically important. However, this would be an extremely time consuming and expensive process. Hence, we use alternative screening methods like RFLP to reduce the amount of sequencing necessary.

There are two research areas in which RFLP analysis can be particularly useful in clinical research. These are the study of DNA polymorphisms and the study of tumor mutations.

4.1.1. DNA polymorphisms

Polymorphisms are natural sequence variations that can lead to interindividual phenotypic differences. These polymorphisms (p/ms) occur on average once in every 1000 bp. Most p/ms do not produce profound phenotypic differences, as selection has acted during evolution to rid our gene pool of these disadvantageous changes. However, many p/ms do modulate an individual's risk of developing certain diseases, hence the current focus on mapping large numbers of p/ms in large numbers of individuals in order to study their effect on disease etiology. For example, cytochrome P450 enzymes (P450s) metabolize exogenous chemicals that enter the body in an effort to detoxify them, occasionally, this metabolism increases the toxicity or carcinogenicity of the parent compound. Hence, overactivity or underactivity of particular P450s can be linked to cancer risk if the metabolic product or the parent compound is carcinogenic, respectively. Therefore, p/ms that affect P450 activity can modulate cancer risks in individuals (*see* **Table 2**). RFLP analysis is a very suitable methodology for scanning a large number of p/ms in cancer-modulating genes in large numbers of individuals. Examples of gene p/ms important in increasing cancer risks are shown in **Table 2**.

4.1.2. Tumour Mutations

Tumors result from genetic damage accumulated in normal tissue during an individual's life-span. This genetic damage often occurs in genes involved in the control of cell division, such that mutant clones are produced that divide uncontrollably. The cell division genes targeted are known as oncogenes (promote division) and tumor suppressor genes (suppress division). The genetic damage inflicted in these genes includes point mutations and deletions and often occurs early in tumor development such that mature tumor tissue contains cells in which the genetic damage (e.g., mutation) is ubiquitous. This means that the mutations are readily detectable in tumor tissue by methods such as RFLP and DNA sequencing. Indeed, REs have been widely used to study the activation of oncogenes such as the *K-ras* oncogene, which is activated through mutation at codon 12 ***(11,12)***. Furthermore, the *p53* gene, commonly mutated in tumor development, has been shown to acquire mutations most frequently at codon 248 (www.iarc.fr/p53), which contains a *Msp*I site (CCGG), hence allowing RE-based analysis to provide information on tumor-specific p53 mutations. Rarer mutations (present in <10% of the cells) such as those present in precancerous tissue would not be detectable by RFLP or sequencing. This is a consequence of the limited sensitivity of the detection step in RFLP analysis (i.e., the identification of a band on a gel). For these rare mutations, more sensitive mutation detection methods are needed.

4.2. Restriction Site Mutation

Restriction site mutation (RSM) employs REs to detect mutations, particularly those involved in tumor formation. RSM has been developed to detect mutations when they occur very infrequently, when RFLP is not suitable (for a review, *see* ***13***). RSM is capable of detecting point mutations when they are present in a 10,000-fold excess of nonmutated DNA ***(14,15)***. Therefore, RSM is highly suited to detecting cancer-causing mutations early in tumor evolution (i.e. in pre-malignant tissue). RSM is able to detect such rare mutations because of a reversal of the digestion and PCR steps compared to RFLP (*see* **Fig. 3**). With RSM, the DNA is digested first in order to destroy nonmutated sequences, mutations arising in target RE sites render those particular RE sites resistant to digestion. Subsequent PCR amplification of digested DNA ensures that only undigested (i.e., mutated) sequences are amplified. This arrangement is the basis of the sensitivity of RSM.

We developed RSM in our laboratory over 10 years ago ***(16)*** and have since been continually validating and optimizing the methodology ***(13,15)***. We have used RSM to detect the action of mutagenic chemicals through the analysis of induced mutations in tissues and cells exposed to putative mutagens. In addition, we have recently studied a range of early premalignant tissues for the presence of initiating mutations (e.g., of the *p53* tumor suppressor gene). **Table 3** is a summary of some of the recent research carried out by ourselves and others, using RSM to detect DNA mutations in chemically exposed cells or in premalignant tissue. **Figure 4** illustrates the potential of RSM to detect p53 mutations in premalignant clinical samples. This RSM approach has also been used by other groups looking at the presence of early cancer-causing mutations in clinical tissue ***(20)***. Our aims in these clinical studies have been threefold. First, how early are p53 mutations in these tumor types? Second, by comparison to in vitro studies on putative mutagens, can we identify causative mutagens based on the mutation patterns? For example, as shown in **Table 3**, we have recently identified a similarity between the premalignant mutations present in esophageal and gastric tissue and those mutations induced experimentally by reactive oxygen species (ROS), suggesting a role for ROS in gastrointestinal (GI) tract carcinogenesis ***(14,17,19)***. Finally, as p53 is a key tumor suppressor in humans whose loss coincides with tumor development, can early *p53* mutations predict cancer progression in individual cases? We have found early *p53* mutations in a subset of premalignant clinical tissue samples by use of RSM (esophageal, gastric, colon, and bladder). We are now following up these patients closely in order to determine if the presence of an early *p53* mutation predicts which patients progress to cancer fastest. If this proves to be the case, then RSM analysis of early mutations in genes such as *p53* may be useful in assessing cancer risk on an individual basis.

4.3. DNA Methylation Analysis

DNA methylation is a frequent modification in the DNA of genomes ***(21)***. It is widely accepted that methylation (which occurs at 5' CG 3' sequences in mammals) is responsible for the silencing of unwanted genes in specific cell types. The loss of this methylation pattern will, therefore, switch on genes that should be silent, whereas *de novo* methylation will switch off genes that should be switched on. The involvement of DNA methylation abnormalities in cancer development is now well established, often seen as the silencing of tumor suppressor genes

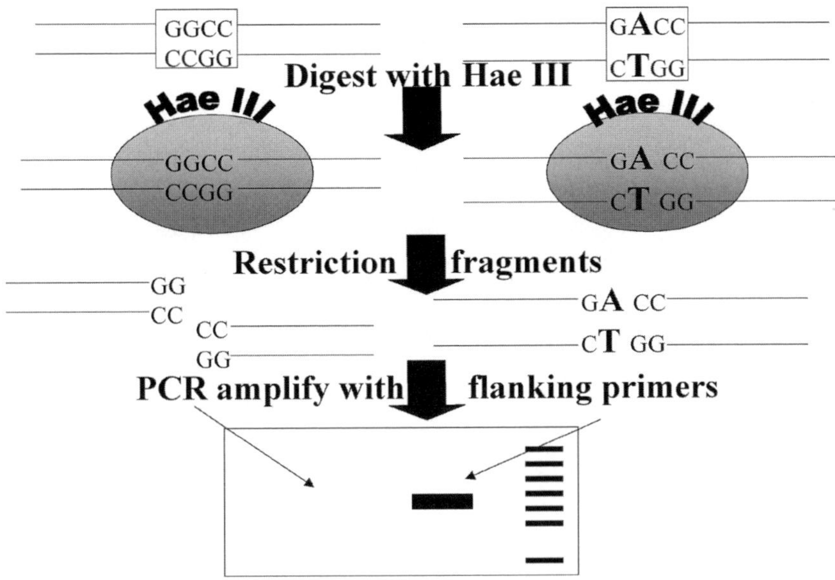

Fig. 3. The principle of RSM. DNA is initially digested with the RE whose site is under mutation analysis (*Hae*III in this case). This digestion cleaves the unmutated DNA copies (as seen on the left), but leaves the mutated copies intact (as seen on the right). Subsequent PCR amplifies the mutated DNA, producing a band after electrophoresis, digested DNA will not PCR, hence no band on the gel. A second digestion step is often included to remove any amplified un-mutated DNA after PCR.

Table 3
A Summary of Published RSM Data Showing the Detection of Mutations in Both Clinical Samples and In Vitro Mutagen Experiments

Subject analyzed	Gene	Codon	Mutation types identified	Ref.
Fibroblasts treated with oxidizing agent 4-NQO	*p53*	248	GC to AT	17
Fibroblasts treated with reactive oxygen species generator	*p53*	248	GC to TA, GC to AT	18
Fibroblasts treated with reactive oxygen species generator	*p53*	248	GC to AT	14
Premalignant esophageal tissue	*p53*	248	GC to AT	15
Premalignant gastric tissue	*p53*	248	GC to AT	19

Note: These data show the similarity between the mutation induced in vitro by oxidative agents and the mutations identified in upper gastrointestinal (GI) tract tissues, suggesting a role of oxygen free radicals in upper GI tract cancer.

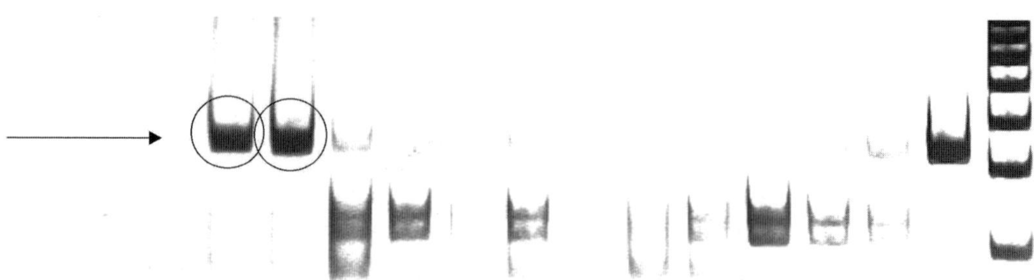

Fig. 4. RSM analysis of premalignant esophageal tissue from 12 individuals for p53 mutations at codon 248 (*Msp*I restriction site). On the right of the gel is a DNA size ladder and next to this is a positive control for the PCR amplification, showing the expected band size. Lanes 1 and 2 show an undigested PCR product of the correct size as a result the presence of a p53 mutation at codon 248 in the *Msp*I restriction site. These two samples are actually from the same individual.

by *de novo* methylation ***(22)***. Therefore, methods to study DNA methylation are being developed ***(23)***. REs are particularly useful in studying DNA methylation ***(24)***. This is the result of the fortuitous existence of pairs of REs (isoschizomers) that are differentially sensitive to DNA methylation. For example, *Hpa*II and *Msp*I both recognize the CCGG sequence, but *Hpa*II will not digest this sequence if the internal C is methylated. *Msp*I will digest both methylated and unmethylated sequences. This feature of pairs of REs can be exploited to monitor the methylation status of tumor suppressor genes in clinical samples. Because CG sites are methylated in mammals, RE sites containing CG sequences such as *Hpa*II are particularly useful. **Figure 5** is the outline of how RE analysis can measure DNA methylation.

5. Concluding Remarks

In conclusion, REs are remarkable enzymes that have been exploited in molecular biology for over 30 yr. In that time, REs have played a paramount role in genetic manipulation (cloning, etc.) as well as in clinical research. The basis of the usefulness of REs is their sequence specificity. The fact that each RE stringently digests only its own recognition sequence and fails to recognize this sequence even if only 1-bp changes has led to their widespread use with clinical samples. In fact, the sequence specificity of REs is such that they can also recognize chemical modifications of normal DNA bases (so-called DNA adducts). Hence, REs can be employed to detect modified DNA after exposure to DNA-damaging chemicals resulting from the failure of REs to digest these modified sequences ***(25,26)***.

As has been seen here, the use of REs in clinical research is widespread. REs are regularly used to analyze the DNA of clinical samples for the presence of mutations, deletions, and methylation abnormalities. RFLP analysis of human pathogens (bacteria, viruses, and other micro-organisms) has also been important in the study of human disease, through the identification, at the DNA level, of virulent strains and strains resistant to drug therapy.

Importantly, hundreds of REs are now available with different sequence combinations, thus allowing their application to many different DNA sequences. Furthermore, in an effort to increase the application of REs in mutational analysis, PCR-based methods have been developed that artificially create RE sites at specific sequences. These methods employ mismatched primers during the PCR step, hence producing PCR products containing unique RE sites for mutation analysis ***(11,12)***. It is hoped that future research involving REs will benefit from new RE's being discovered, with new recognition sequences. Importantly, advances in protein engineering can also, in the near future, allow the design of new REs with tailor made recognition sequences not currently available.

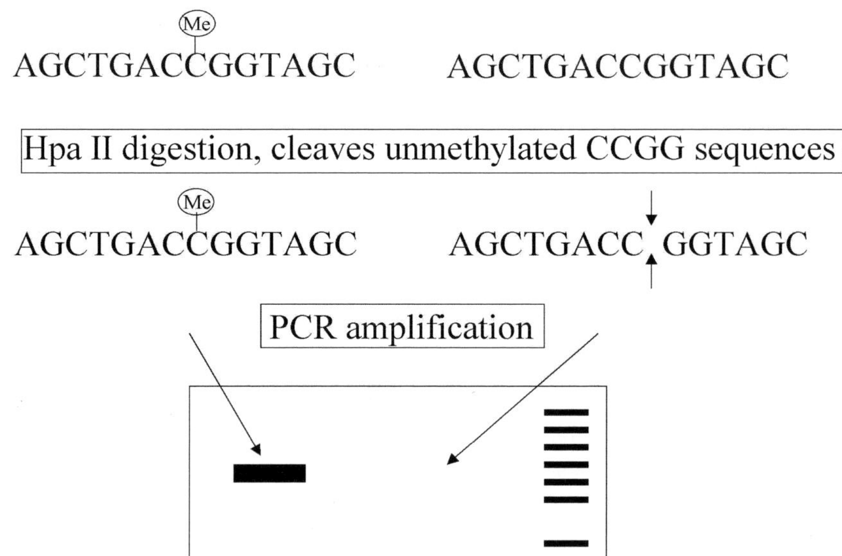

Fig. 5. Mammalian DNA is methylated at the 5 carbon position of the cytosine base in a CG sequence context, as shown. To detect this methylation, DNA is initially digested with a methylation sensitive RE (*Hpa*II in this case). This RE cuts unmethylated DNA, but fails to cut methylated DNA (marked Me). Subsequent PCR amplification allows the methylated (undigested) DNA to be amplified. Hence, methylation is ultimately detected by the presence of a PCR band on a gel.

References

1. Horton, J. R., Nastri, H. G., Riggs, P. D., Cheng, X. (1998) Asp34 of PvuII endonuclease is directly involved in DNA minor groove recognition and indirectly involved in catalysis *J. Mol. Biol.* **284**, 1491–1504.
2. Pingoud, A. and Jeltsch, A. (2001) Structure and function of type II restriction endonucleases. *Nucleic Acids Res.* **29**, 3705–3727.
3. Brown, T. A. (1990) *Gene Cloning*, 2nd ed., Chapman & Hall, London.
4. Roberts, R. J. and Macelis, D. (2001) REBASE: restriction enzymes and methylases. *Nucleic Acids Res.* **29**, 268–269.
5. Cotton, G. H. (1989) Detection of single base changes in nucleic acids. *Biochem. J.* **263**, 1–10.
6. Minami, R., Aoyama, N., Honsako, Y., Kasuga, M., Fujimura, T., and Maeda, S. (1997) Codon 201 (arg/gly) polymorphism of DCC (deleted in colorectal cancer) gene in flat and polypoid type colorectal tumours. *Dig. Dis. Sci.* **42**, 2446–2452.
7. Prosser, J. and Condie, A. (1991) Biallelic Apa I polymorphism of the human p53 gene (TP53). *Nuc. Acids. Res.* **19**, 4799.
8. Garte, S. (1998) The role of ethnicity in cancer susceptibility gene polymorphisms: example of CYP1A1. *Carcinogenesis* **19**, 1329–1332.
9. Cohen, J. B. and Levinson, A. D. (1988) A point mutation in the last intron responsible for increased expression and transforming activity of the c-Ha-ras oncogene. *Nature* **334**, 119–124.
10. Ye, Z. and Parry, J. M. (2002) The CYP17 MspA1 polymorphism and breast cancer risk: a meta-analysis. *Mutagenesis* **17**, 119–126.
11. Levi, S., Urbano-Ispizua, Gill, R., et al. (1991) Multiple K-ras codon 12 mutations in cholangiocarcinomas demonstrating a sensitive polymerase chain reaction technique *Cancer Res.* **51**, 3497–3502.
12. Ward, R., Hawkins, N., O'Grady, R., et al. (1998) Restriction endonuclease mediated selective polymerase chain reaction. *Am. J. Pathol.* **153**, 373–379.

13. Jenkins, G. J. S., Suzen, H. S., Sueiro, R. A., and Parry, J. M. (1999) The restriction site mutation (RSM) assay. A review of the methodology development and the current status of the technique. *Mutagenesis* **14**, 439–448.
14. Jenkins, G. J. S., Morgan, C., Parry, E. M., Baxter, J. N., and Parry J. M. (2001) The detection of mutations induced *in vitro* in the human p53 gene by a model reactive oxygen species (ROS) employing the restriction site mutation (RSM) assay. *Mutat. Res.* **498**, 135–144.
15. Jenkins G. J. S., Doak, S. H., Griffiths, A. P., Baxter, J. N., and Parry, J. M. (2003) Early p53 mutations in non-dysplastic Barrett's tissues detected by the restriction site mutation (RSM) methodology. *Br. J. Cancer* **88**, 1271–1276 .
16. Parry, J.M., Shamsheer, M., and Skibinski, D. (1990) Restriction site mutation analysis, a proposed methodology for the detection and study of DNA base changes following mutagen exposure. *Mutagenesis*, **5**, 209–212.
17. Jenkins G. J. S., Chalestori, M. H., Song H., and Parry, J. M. (1998) Mutation analysis using the restriction site mutation (RSM) assay. *Mutat. Res.* **405**, 209–220
18. Perwez Hussain, S. P., Aguilar, F. Amstad, P., and Cerutti, P. (1994) Oxyradical induced mutagenesis of hotspot codons 248 and 249 of the human p53 gene. *Oncogene* **9**, 2277–2281.
19. Morgan, C., Jenkins, G. J. S., Ashton, T., et al. (2003) The detection of p53 mutations in pre-cancerous gastric tissue using the RSM assay. *Br. J. Cancer* **89**, 1314–1319.
20. Perwez Hussain, S., Amstad, P., Raja, K., et al. (2000) Increased p53 mutation load in noncancerous colon tissue from ulcerative colitis: a cancer prone chronic inflammatory disease, *Cancer Res.* **60**, 333–3337.
21. Ehrlich, M. and Wang, R. H. Y. (1981) 5-Methylcytosine in eukaryotic DNA. *Science* **212**, 1350–1357.
22. Toyota, M. and Issa, J. P. J. (2000) The role of DNA hypermethylation in human neoplasia. *Electrophoresis* **21**, 329–333.
23. Fraga, M. E. and Esteller, M. (2002) DNA methylation: a profile of methods and applications *Biotechniques* **33**, 632.
24. Oakeley, E. J. (1999) DNA methylation analysis: a review of current methodologies. *Pharmacol. Ther.* **84**, 389–400.
25. Puvvada, M. S., Hartley, J. A., Jenkins, T. C., and Thurston, D. E. (1993) A quantitative assay to measure the relative DNA-binding affinity of pyrrolo[2,1-c][1,4]benzodiazepine (pbd) antitumor antibiotics based on the inhibition of restriction-endonuclease BamHI. *Nucleic Acids Res.* **21**, 3671–3675.
26. Denissenko, M. F., Venkatachalam, S., Ma, Y. H., and Wani, A. A. (1996) Site-specific induction and repair of benzo[a]pyrene diol epoxide DNA damage in human H-ras protooncogene as revealed by restriction cleavage inhibition *Mutation Res.* **36**, 27–42.

4

Southern Blotting as a Diagnostic Method

Bronwen Harvey and Pirkko Soundy

1. Introduction

Nucleic acid hybridization is a process in which complementary single strands of nucleic acids combine to achieve a stable double-stranded nucleic acid molecule. This action has been utilized to establish a molecular or genetic relatedness between organisms and to characterize their genomes. Furthermore, this technique is one of several diagnostic tools useful for detecting a wide variety of conditions. Since the determination of the basic principles of duplex formation and stability in the 1950s, many variations of the hybridization techniques have been developed. Southern first transferred DNA fragments from agarose, after electrophoretic separation, onto nitrocellulose *(1)*. The technique is known as Southern blotting. Alwine et al. *(2)* described shortly afterward a similar Northern blotting technique in which separated RNA strands are transferred from an agarose gel to a suitable solid support. A logical extension of these blotting techniques has been the dot or slot blot in which the sample is applied directly to the solid support without prior size separation *(3)*. Over the years, these techniques have been further developed and modified extensively by many researchers across the world. The application of these methods is as varied as the procedures used (e.g., to determine the changes in the nutritional state of an environment, to establish taxa genetically, to distinguish pathogenic from nonpathogenic viruses, to analyze gene structure).

This chapter restricts itself to the application of Southern blotting to provide information relating to genetic diseases. The DNA sample, which can include blood, tissue biopsies, buccal scraps, amniotic fluid, cultured cells, and so forth, is generally digested using a restriction endonuclease and then subjected to electrophoresis in a horizontal agarose gel. After sufficient time has elapse to achieve adequate separation of the required fragments, the gel is soaked in an alkali solution to achieve denaturation of the double-stranded nucleic acid, then neutralized and prepared for transfer. Transfer to a nitrocellulose, polyvinylidene (PVDF), or Nylon membrane is achieved by a process of blotting in which buffer is drawn through the gel and the membrane. The fragments carried with the buffer are retained on the surface of the membrane. The retention is made more permanent through a fixation process. The blot can then be used in a hybridization with labeled probes to identify the fragments of interest.

There are many variations on this basic theme *(4,5)*. Those procedures requiring the identification of fragments in excess of 10,000 bases advocate the use of a depurination step to improve the efficiency of large-fragment transfer. Methodologies using positively charged Nylon membranes often omit the neutralization step and advocate the use of alkaline transfer buffer, which can also serve as a fixative. There are many ways to achieve the transfer of DNA from the gel to the solid support. Southern's original method *(1)* describes the use of a capillary

From: *Medical Biomethods Handbook*
Edited by: J. M. Walker and R. Rapley © Humana Press, Inc., Totowa, NJ

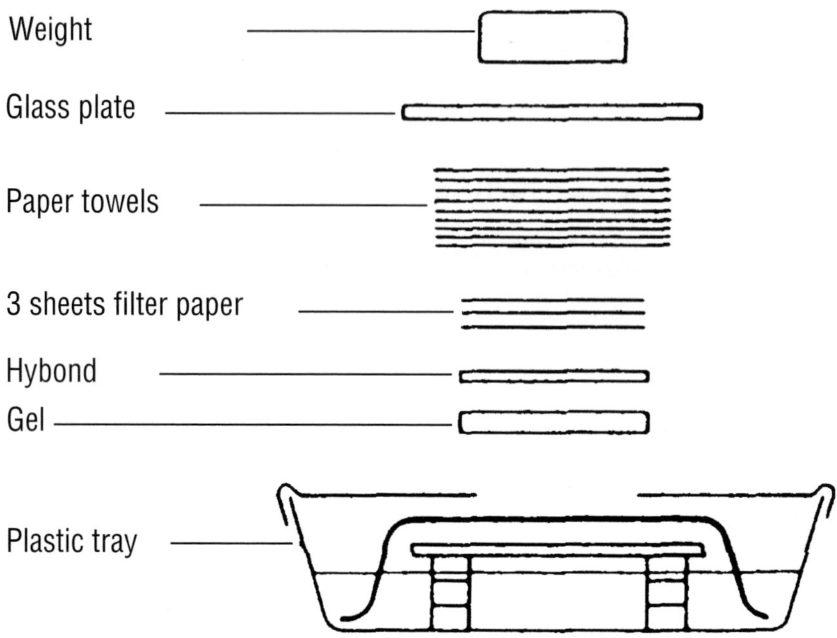

Fig. 1. Diagrammatic representation of capillary transfer apparatus; suitable for the transfer of DNA or RNA fragments separated by gel electrophoresis to a suitable Nylon or nitrocellulose membrane for example. Hybond is available from GE Healthcare Bio-Sciences.

transfer procedure, and this remains the most widely used technique by far because of its low cost and convenience, transfer is often an overnight step (*see* **Fig. 1**). If speed is a requirement, the transfer process can be shortened by using specialized vacuum blotting apparatus or electroblotting devices. Such techniques allow transfer in 30–60 min compared to several hours for capillary transfer.

The membrane of choice is determined by the sensitivity required and the detection method to be used. The quantity of sample has a significant effect on both of these. The use of nitrocellulose usually results in low backgrounds and is recommended when the level of target is high. Nitrocellulose membrane is available in supported and unsupported forms, depending on the manufacturing method employed; however, the handling characteristics of the latter can be poor. Unsupported membranes are produced when the active substrate is cast as a pure sheet. Because of their fragile nature, unsupported membranes should be handled with care. Supported membranes are those for which the active substrate is cast onto an inert "web" or support.

Nitrocellulose membrane can bind 80–125 μg nucleic acid/cm^2, which is significantly less than the binding capacity of 400–600 μg/cm^2 for a Nylon membrane. Its ability to bind small molecules (<400 nts) is also poor, and transfer buffers must contain high salt concentrations to ensure efficient nucleic acid binding. Nylon membranes are available in uncharged and positively charged supported forms. Charged Nylon has a higher binding capacity and is particularly useful when working with low-molecular-weight nucleic acid. Binding to a Nylon membrane is independent of the ionic strength of the transfer buffer. However, backgrounds can be elevated. Where repeated use of a membrane in hybridization assays is needed, the use of Nylon membranes is strongly advised. Nylon is also recommended for use with medium- or low-abundance targets and is, in general, the membrane of choice when working with nucleic acids. PVDF membranes behave similarly to uncharged Nylon, but because of its hydrophobic nature, use in nucleic acid blotting is limited.

Table 1
Factors Affecting Hybridization Rate and Stringency

Factor	How hybridization is affected
Temperature	High temperature increases hybridization stringency; temperatures below T_m are recommended (for RNA long probe, 10–15°C below T_m; for DNA long probe, approx 25°C below T_m)
Ionic strength	Optimal hybridization in the presence of 1.5 *M* Na^+; lowering ionic strength increases stringency.
Destabilizing agents (or T_m modifiers)	For example, formamide and urea; used to lower the effective hybridization temperature
Mismatched basepairs	Mismatches lower hybridization rate
Duplex length	Hybridization rate increases with increased probe length
Viscosity	Increases rate of filter hybridization

Fixation bonds the target nucleic acid to the membrane. Suboptimal fixation will lower sensitivity by reducing target concentration and is particularly harmful if the blot is to be used more than once. The principal fixation methods of heat and ultraviolet (UV) light can be used with all types of membrane. Heat fixation is very reproducible but requires a vacuum oven for nitrocellulose. UV crosslinking, performed using an UV crosslinker (constant energy setting), is faster than heat. Alkali provides a third alternative method when charged Nylon membranes are used

Original methods describe the use of a DNA probe radioactively labelled by random priming or nick translation with [^{32}P]dCTP to detect specific nucleic acid fragments immobilized on nitrocellulose membrane. Since then, many different methods for labeling nucleic acid probes ranging from short oligonucleotides to longer DNA or RNA fragments have been developed ***(6,7)***. Nonradioactive labeling kits and reagents are also available, finding favor in a number of niche areas ***(8)***. The role of the hybridization buffer is to provide conditions that promote hybridization between the labeled probe molecules and its complementary sequence immobilized on the membrane, and to simultaneously limit hybridization between sequences that are not perfectly matched ***(6)***. **Table 1** lists factors affecting the hybridization rate and stringency ***(4–6)***. Many different formulations of hybridization buffers have been developed, containing inorganic salts and blocking agents such as Denhardt's solution (mixture of bovine serum albumin [BSA], Ficoll 400, and polyvinyl pyrrolidone), denatured DNA from salmon sperm (or other species), and heparin ***(4–6)***. A short prehybridization step in hybridization buffer is usually carried out to reduce nonspecific background hybridization before adding the labeled probe. This is especially important in genomic Southern blots that contain all genomic sequences on the membrane. Hybridization with the probe is usually carried out for several hours to allow hybridization between low-abundance sequences. Although various rapid-hybridization buffers containing volume excluders are available to speed up this step. After hybridization, the blot is washed to remove unhybridized probe. The stringency of washing is usually controlled by stepwise reductions in the ionic strength of wash buffer and/or by temperature ***(4–6)***. The replacement of the old plastic bag technology with specialized temperature-controlled rotisserie devices for performing hybridization and washes has resulted in more consistent results and safer handing of radioactivity. After washing, the blot is subjected to autoradiography with X-ray film to visualize the bound probe ***(9)***.

2. Southern Blotting in the Diagnosis of Human Disease

Southern blotting has been applied to the diagnosis of many human diseases at the molecular level. These genetic diseases are caused by point mutations, gene rearrangements, or the amplification of genes or specific sequences within the genome. These methods have in common that restriction-digested genomic DNA is size-separated in agarose gel electrophoresis, transferred onto the membrane, and hybridized with gene-specific probes.

Restriction fragment length polymorphism (RFLP) analysis was one of the early methods to diagnose point mutations implicated in genetic diseases (see Chapter 3). This method was based on the observation that if a point mutation changes a restriction fragment recognition sequence, it is possible to detect this change by Southern blotting analysis in which the affected restriction enzyme is used to cut genomic DNA prior to analysis. The change in the size of detected fragments with a gene-specific probe signals the presence of mutation in the analyzed gene. For example, this method has been applied to the diagnosis of hemophilia A, which is the most common inherited bleeding disorder, affecting approx 1 in 5000 males worldwide. Hemophilia A is an X-linked, recessively inherited bleeding disorder that results from a deficiency of procoagulant factor VIII (FVIII). Affected males suffer from joint and muscle bleeds and easy bruising, the severity of which is closely correlated with the level of activity of coagulation factor VIII (FVIII:C) in their blood. Gitschier et al. demonstrated using Southern blotting that it was possible to diagnose the disease in 42% of affected families ***(10)***. Having identified the *Bcl*I polymorphism, X-linked inheritance was demonstrated in three generations of a Utah family. DNA from a family member was restricted with *Bcl*I, electrophoresed on a 0.8% agarose gel, and blotted on to Nylon membrane and probe with a complementary sequence within the factor VIII gene labeled with radioactivity. Twelve bands were observed by autoradiography. Eleven hybridizing bands remained constant, and one varied in position at either approx 0.9 or 1.1 kb in size.

The presence of gene point mutations has also been diagnosed with Southern blotting using allele-specific probes. These are usually short oligonucleotides in which the mismatched sequence is situated in the middle. By carefully controlling the hybridization and wash conditions (temperature and ionic strength of wash buffers), it is possible to distinguish between the binding of oligonucleotides differing by only one nucleotide. This method has been applied, for example, to distinguish between the normal human β-globin gene and the sickle cell β-globin gene ***(11,12)***. Sickle cell anemia is caused by a single base change, resulting in the replacement of glutamic acid residue at position 6 of the protein (hemoglobin) by a valine residue.

Gene rearrangements can be diagnosed with Southern blotting if a probe hybridizing to the affected areas is used. Rearrangement is detected by observing change in the size and pattern of hybridized genomic restriction fragments. This type of analysis has been applied to the diagnosis of acute promyelocytic leukemia (APL), a subtype of the cancer acute myeloid leukemia. The disease is characterized by abnormal, heavily granulated promyelocytes, a form of white blood cells. APL results in the accumulation of these atypical promyelocytes in the bone marrow and peripheral blood, and they replace normal blood cells. At the molecular level, the disease involves translocation between the retinoic acid receptor-α (*RAR*-α) on chromosome 17 and the promyelocytic leukemia locus (*PML*) on chromosome 15. This results in the transcription of novel fusion messenger RNAs. By separating restriction-digested genomic DNA in pulse field gel electrophoresis followed by hybridization with probes derived from *PML* and *RAR*-α genes, it was possible to detect translocation events that correlated with disease progression ***(13)***.

Gene amplifications are implicated in many diseases. Charcot–Marie–Tooth (CMT) syndrome is a common autosomal-dominant neuromuscular disorder. The disease is characterized by a slowly progressive degeneration of the muscles in the foot, lower leg, hand, and forearm and a mild loss of sensation in the limbs, fingers, and toes. The first sign of CMT is generally a high arched foot or gait disturbances. Other symptoms of the disorder may include foot bone abnormalities such as high arches and hammer toes, problems with hand function and balance, occasional lower leg and forearm muscle cramping, loss of some normal reflexes, occasional partial sight and/or hearing loss, and, in some patients, scoliosis (curvature of the spine). Genetically, the disorder is usually characterized by duplication of a region on chromosome 17 through unequal crossover. As a result, affected patients carry three copies of this region. One diagnostic approach to CMT, type 1A exploits Southern blot hybridization and the relative intensity for three polymorphic *Msp*I RFLP bands within the duplicated area to judge whether patients have two or three copies of this region using a region-specific probe. In order to normalize the observed intensity of the signal resulting from the CMT gene probe, another probe derived from unconnected sequences is used ***(14)***.

The most significant changes in the use of Southern blotting in diagnosis have been seen since the introduction of primer-specific polymerase chain reaction (PCR) and automated non-radioactive sequencing techniques. Mutation and gene deletions once detected via Southern blot analysis are now routinely analyzed with these rapid and inexpensive methods, which are often fully automated. Cystic fibrosis, Duchenne muscular dystrophy, sickle cell anemia and thalassaemia, to name a few, are now diagnosed by polymerase chain reaction (PCR). PCR methods can be completed in as little as 4 h, whereas Southern blotting can take up to 5 d to achieve the same result. More important, the amount of DNA required for analysis is significantly less with PCR amplification methods. The Southern blotting diagnosis method generally requires 5–10 µg of genomic DNA. The introduction of a PCR-based method is generally only achieved once the gene defect has been characterized at the molecular level. Hence, research into disorders where the defect is unknown or further information is required still utilizes Southern blot analysis as an important research tool. In the routine laboratory, the use of Southern blotting is restricted to those diseases that require additional information the Southern blot can provide. One such disease is Fragile X; however prescreening using PCR analysis is common.

2.1. Fragile X Syndrome

Fragile X syndrome is a common genetic disease. This inherited form of mental retardation affects 1 in 4000 males and 1 in 8000 females ***(15)***. Males with fragile X often exhibit characteristic physical features and accompanying autistic and attention-deficit behaviors. Individual IQs are in the range 35–70 ***(16–18)***. Approximately 30% of females with full mutations are mentally retarded, and their level of retardation is, on average, less severe than that seen in males.

In 1943, Martin and Bell ***(19)*** were able to link the cognitive disorder to an unidentified mode of X-linked inheritance. In 1967, Lubs ***(20)*** discovered excessive genetic material extending beyond the low arm of the X chromosome in affected males. Diagnosis was originally based on cytogenetic analysis of metaphase spreads, but less than 60% of the affected cells in affected individuals showed a positive result. With this variability in the test, the carrier status of individuals could not be determined. Interpretation of the result is further complicated by the presence of other fragile sites in the same region of the X chromosome.

The fragile X gene (*FMR1*) is located in chromosomal band Xq27.3 and encodes an RNA-binding protein, which was initially characterized in 1991 ***(21–23)*** and contains a tandemly repeated trinucleotide sequence (CGG) end at its 5' end. The disease is caused by the absence of a functional *FMR1* gene product ***(24)***. A small number of individuals classified as fragile X cytogentically have expansion at the nearby *FRAXE* locus, which also contains an unstable CGG repeat ***(25–28)***. The normal distribution of the repeat in the unaffected population varies from 6 to 50. Affected individuals are classified into one of two major groups; premutations of approx 50–200 repeats and full mutations with more than 200 repeats. Some alleles with approx 45–55 copies of the repeat are unstable and expand from generation to generation; others are stably inherited ***(29–31)***.The larger the size of the female premutation, the greater the risk of expansion during meiosis ***(32)***. Individual male or females carrying a premutation are unaffected ***(20,29,30)***. Males pass on the mutation relatively unchanged to all their daughters, all of whom are unaffected.

In some cases of premutation, an accurate estimate of the size of the expansion is required, most especially in family studies. This increase in resolution is achieved by Southern blot assay, using *Pst*I digestion of genomic DNA and detection with a probe close to the repeat array. Characteristic of the full mutation is methylation of the promotor region of the gene (CpG island) ***(33)***; this correlates directly to gene inactivation. This is an important event in the disease pathogenesis, its effect on clinical severity is, however, unpredictable, especially in females.

Today, two main approaches are used in diagnosis of fragile X syndrome: PCR ***(36–39)*** and Southern blot analysis ***(40)***. The use of flanking primers allows the amplification of the region of DNA containing the repeats. The size of the PCR product is therefore indicative of the number of repeats present. However, the efficiency of the PCR reactions inversely relates to the number of repeats; hence, large mutations are more difficult to amplify and can fail to yield a PCR product. False negatives by PCR can also be an issue caused by the presence of normal and full mutations in some male patients ***(41)***. No information as to the extent of methylation can be determine by a PCR-based assay. Southern blotting allows both the size of the repeat segment and methylation status to be assayed simultaneously. Methylation-sensitive restriction enzymes such as *Eag*I or *Nru*I can be used to distinguish between methylated and unmethylated alleles. When combined with *Eco*RI, these enzymes give fragment sizes of 2.6 kb from normal unmethylated *FMR-1* genes. Methylated normal genes are not cut by these enzymes and yield 5.1-kb *Eco*RI fragments. Methylated and unmethylated expansions are indicated by the presence of bands or smears above the 5.1-kb and 2.6-kb fragments, respectively.

It is common practice in diagnosis laboratories to use PCR for prescreening and only to proceed to Southern blotting for those samples that fail to amplify (males) or show a single normal allele (females). If the etiology of mental impairment is unknown, DNA analysis for fragile X syndrome should be performed as part of a comprehensive genetic evaluation, which includes routine cytogenetic analysis. Cytogenetic studies are critical because constitutional chromosome abnormalities have been identified as frequent or more frequently than fragile X mutations in mentally retarded individuals referred for fragile X testing. For individuals who are at risk because of an established family history of fragile X syndrome, DNA testing alone is sufficient. If the diagnosis of the affected relative was based on previous cytogenetic testing for fragile X syndrome, then it is advised that at least one affected relative should be included in the DNA testing profile. Prenatal testing ***(42)*** of a fetus is indicated following a positive carrier test in the mother. When the mother is a known carrier, DNA testing can be offered to determine whether the foetus inherited the normal or mutant *FMR1* gene. Results must be interpreted with caution because the methylation status of the *FMR1* gene is often not yet established in chorionic villi at the time of sampling. Follow-up amniocentesis might be necessary to resolve an ambiguous result. In a very small number of patients, deletions or point mutations ***(43–48)*** rather than trinucleotide expansion are responsible for the syndrome. In these cases, linkage ***(31,49,50)*** or rare mutation studies are more useful.

References

1. Southern, E.M. (1975) Detection of specific sequences among DNA fragments separated by gel electrophoresis. *J. Mol. Biol.* **98**, 503–517.
2. Alwine, J. C., Kemp, D. J., and Stark, G. R. (1977) Method for detection of specific RNAs in agarose gels by transfer to diazobenzyloxymethyl-paper and hybridization with DNA probes. *Proc. Natl. Acad. Sci. USA* **74**, 5350–5354.
3. Kafatos, F. C., Jones, C. W., and Efstratiadis, A. (1979) Determination of nucleic acid sequence homologies and relative concentrations by a dot hybridization procedure. *Nucleic Acid Res.* **7(6)**, 1541–1552.
4. Rapley R (ed.) (2000) *The Nucleic Acid Protocols Handbook*, Humana, Totowa, NJ.
5. Sambook, J., Russel, D. W., and Sambrook, J. (2000) *Molecular Cloning: A Laboratory Manual*, Cold Spring Harbor Laboratory Press, Cold Spring Harbor, NY.
6. Hames, B. D. and Higgins, S. J. (1985) *Nucleic Acid Hybridisation: A Practical Approach*, IRL, Oxford, UK.
7. Cunningham, M. W., Harris, D. W., and Mundy, C. (1990) In vitro labelling: nucleic acids, in *Radioisotopes in Biology* (Salter, R.J., ed.), IRL, Oxford, UK pp. 137–189.
8. Isaac, P. G. (1994) *Protocols for Nucleic Acid Anaylsis by Nonradioactive Probes*, Humana, Totowa, NJ.

9. Laskey, R.A. (1990) Radioisotope detection using X ray film, in *Radioisotopes in Biology* (Salter, R.J., ed.), IRL, Oxford, UK.
10. Gitschier, J., Drayna, D., Tuddenham, E. G., White, R. L., and Lawn, R. M. (1985) Genetic mapping and diagnosis of haemophilia A achieved through a BclI polymorphism in the factor VIII gene. *Nature* **314**, 738–740.
11. Conner, B. J., Reyes, A. A., Morin, C., Itakura, K., Teplitz, R. L., and Wallace, R. B. (1983) Detection of sickle cell beta S-globin allele by hybridization with synthetic oligonucleotides. *Proc. Natl. Acad. Sci. USA* **80**, 278–82.
12. Feldenzer, J., Mears, J. G., Burns, A. L., Natta, C., and Bank, A. (1979) Heterogeneity of DNA fragments associated with the sickle-globin gene. *J. Clin. Invest.* **64**, 751–755.
13. Xiao, Y. H., Miller, W. H., Jr., Warrell, R. P., Dmitrovsky, E., and Zelenetz, A. D. (1983) Pulsed-field gel electrophoresis analysis of retinoic acid receptor alpha and promyelocytic leukemia rearrangements. Detection of the t(15;17) translocation in the diagnosis of acute promyelocytic leukemia. *Am J Pathol.* **143**, 1301–1311.
14. Chen, K. L, Wang, Y. L., Dodson, L. A., et al. (1996) Normalized Southern hybridization to enhance testing for Charcot–Marie–Tooth disease, Type 1A. *Mol. Diagn.* **1**, 65–71.
15. Murray, J., Cuckle, H., Taylor, G., and Hewison, J. (1997). Screening for fragile X syndrome. *Health Technol. Assess.* **1**, 1–71.
16. deVries, B. B. A., Halley, D. J. J., Oostra, B. A., and Niermeijer, M. F. (1998) The Fragile X syndrome. *J. Med. Genet.* **35(7)**, 579–589.
17. Hagerman, R. J. and Silverman, A. C. (1991) *Fragile X syndrome: Diagnosis, Treatment, and Research*, Johns Hopkins University Press, Baltimore, MD.
18. Warren, S. T. and Nelson, D. L. (1994) Advances in molecular analysis of fragile X syndrome. *JAMA* **271**, 536–542.
19. Martin, J. P. and Bell, J. (1943) A pedigree of mental defect showing sex-linkage. *J. Neurol. Psychiatry* **6**, 154–157.
20. Lubs, H. (1969) A marker X chromosome. *Am. Hum. Genet.* **21**, 231.
21. Heitz, D., Rousseau, F., Devys, D., et al. (1991) Isolation of sequences that span the fragile-X and identification of a fragile-X related CpG island. *Science* **251**, 1236–1239.
22. Kremer, E. J., Pritchard, M., Lynch, M., et al. (1991) Mapping of DNA instability at the Fragile-X to a trinucleotide repeat sequence P(CCG)N. *Science* **252**, 1711–1714.
23. Verkerk, A., Pierelt, M., Sutcliff, J. S., et al. (1991) Identification of a gene (FMR-1) containing a CGG repeat coincident with a breakpoint cluster region exhibiting length variation in fragile-X syndrome. *Cell* **65**, 905–914.
24. Hagerman, R. J., Hull, C. E., Safanda, J. F., et al. (1994) High functioning fragile-X males—demonstration of an unmethylated fully expanded Fmr-1 mutation associated with protein expression. *Am. J. Med. Genet.* **51**, 298–308.
25. Gu, Y. H., Shen, Y., Gibbs, R. A., and Nelson, D. J. (1996) Identification of FMR2, a novel gene associated with the FRAXE CCG repeat and CpG island. *Nature Genet.* **13**, 109–113
26. Gecz, J., Gedeon, A. K., Sutherland, G. R., and Mulley, J. C. (1996) Identification of the gene FMR2, associated with FRAXE mental-retardation. *Nature Genet.* **13**, 105–108.
27. Knight, S. J. L., Flannery, A., Hirst, M. C., et al. (1993) Trinucleotide repeat amplification and hypermethylation of a CpG island in FRAXE mental-retardation. *Cell* **74**, 127–134.
28. Knight, S. J. L., Voelckel, M. A., Hirst, M. C., et al. (1994) Triplet repeat expansion at the FRAXE locus and X-linked mild mental handicap. *Am. J. Hum. Genet.* **55**, 81-86.
29. Abramowicz, M. J., Parma, J., and Cochaux, P. (1996) Slight instability of a FMR-1 allele over 3 generations in a family from the general-population. *Am. J. Med. Genet.* **64**, 268–269.
30. Mornet, E., Chateau, C., Hirst, M. C., et al. (1996) Analysis of germline variation at the FMR1 CFF repeat shows variation in the normal-premutated borderline range. *Hum. Mol. Genet.* **5**, 821–825.
31. Zhong, N., Ju, W., Pietrofesa, J., Wang, D., Dobkin, C., and Brown, W. T. (1996) Fragile-X "gray zone" alleles—AGG patterns, expansion risks, and associated haplotypes. *Am. J. Med. Genet.* **64**, 261–265.
32. Fu, Y. H., Kuhl, D. P., Pizzuti, A., et al. (1991) Variation of the CGG repeat at the fragile-X site results in genetic instability—resolution of the Sherman paradox. *Cell* **67**, 1047–1058.
33. Fisch, G. S., Snow, K., Thibodeaa, S. N., et al. (1995) The fragile-X premutation in carriers and its effect on mutation size in offspring. *Am. J. Hum. Genet.* **56**, 1147–1155.
34. Heitz, D., Rousseau, F., Devys, D., et al. (1991) Isolation of sequences that span the fragile-X and identification of a fragile-X related CpG island. *Science* **251**, 1236–1239.

35. Haddad, L. A., Mingroninetto, R. C., Viannamorgante, A. M., and Pena, S. D. J. (1996) A PCR-based test suitable for screening for fragile-X syndrome among mentally-retarded males. *Hum. Genet.* **97**, 808–812.
36. Brown, W. T., Nolin, S., Houck, G., Jr., et al. (1996) Prenatal-diagnosis and carrier screening for fragile-X by PCR. *Am. J. Med. Genet.* **64**, 191–195.
37. Goonewardena, P. and Zhang, J. A. (1995) Single tube nonradioactive PCR assay for the detection of the full spectrum or FMR1 CGG repeats seen in the normal, carrier and fragile-X syndrome individuals. *Am. J. Hum. Genet.* **57**, 1914–1914.
38. Wang, Q., Green, E. P., Mathew, C. G., and Bobrow, M. (1994) Nonradioactive PCR based screening for fragile-X syndrome. *J. Med. Genet.* **31**, 173–173.
39. Brown, W. T., Houck G. E., Jr., Jeziorowska, A., et al. (1993) Rapid fragile-X carrier screening and prenatal-diagnosis using a nonradioactive PCR test. JAMA **270**, 1569–1575.
40. Gold, B., Radu, D., Balanko, A., and Chiang, C.S. (2000) Diagnosis of fragile X syndrome by Southern blot hybridization using a chemiluminescent probe: a laboratory protocol. *Mol. Diagn.* **5**, 169–178.
41. Bullock, S., et al. (1995) Molecular DNA testing of a family manifesting fragile-X syndrome in both the fragile X—a full mutation and mosaic forms. *J. Med. Genet.* **32**, 153.
42. Yamauchi, M., Nagata, S., Seki, N., et al. (1993) Prenatal diagnosis of fragile X syndrome by direct detection of the dynamic mutation due to an unstable DNA sequence. *Clin. Genet.* **44**, 169–172.
43. Mannermaa, A., Pulkkinen, L., Kajanoja, E., Ryynanen, M., and Saarikoski, S. (1996) Deletion in the FMR1 Gene in a Fragile-X Male. *Am. J. Med. Genet.* **64**, 293–295.
44. Deboulle, K., et al. (1993) A point mutation in the FMR-1 gene associated with fragile-X mental-retardation. *Nature Genet.* **3**, 31–35.
45. Hart, P. S., Olson, S. M., Crandall, K., and Tarleton, J. (1995) Fragile-X syndrome resulting from a 400 basepair deletion within the FMR1 gene. *Am. J. Hum. Genet.* **57**, 1395–1395.
46. Hirst, M., Grewal, P., Flannery, A., et al. (1995) Two new cases of FMR1 deletion associated with mental impairment. *Am. J. Hum. Genet.* **56**, 67–74.
47. Dreyfuss, J. C. (1992) Clinical fragile-X syndrome due to total or partial gene deletion. *Med. Sci.* **8**, 878–878.
48. Hammond, L. S., Macia, M. M., Tarleton, J. C., and Shashidhar Pai, G. (1997) Fragile X syndrome and deletions in FMR1: new case and review of the literature. *Am. J. Med. Genet.* **72**, 430–434.
49. Hirst, M. C., Grewal, P. K., and Davies, K. E. (1994) Precursor Arrays For Triplet Repeat Expansion At the Fragile-X Locus. *Hum. Mol. Genet.* **3**, 1553–1560.
50. Eichler, E. E., Macpherson, J. N., Murray, A., Jacobs, P. A., Chakravarti, A., and Nelson, D. L. (1996) Haplotype and interspersion analysis of the FMR1 CGG repeat identifies 2 different mutational pathways for the origin of the fragile-X syndrome. *Hum. Mol. Genet.* **5**, 319–330.

5

Western Blotting as a Diagnostic Method

Pirkko Soundy and Bronwen Harvey

1. Introduction

Protein blotting, the transfer of proteins from a separating gel onto a thin uniform support matrix, first appeared in 1979. Continuing the geographic theme following Southern's publication of his method for the identification of specific DNA fragments *(1)* in 1975 and the introduction of Northern blotting *(2)* not long after, the technique became known as Western blotting. Today, the original article by Towbin et al. *(3)* is cited many thousands of times a year. The technique itself has been modified and extended over the years *(4)*. Once on a solid support, procedures that would otherwise proved difficult or impossible in the gel can be undertaken. A blot allows for rapid staining and destaining protocols of the separated proteins. Low concentrations of sample are more easily detected because they are not spread throughout the thickness of the gel but are "concentrated" on the surface; also, membranes are easier to handle and manipulate.

2. Protein Blotting Techniques

In a classic Western blotting *(3)*, protein samples are separated in an electrophoretic gel *(5,6)* and then electroblotted onto a support matrix. Once retained, proteins can be probed with almost anything that can selectively bind peptide motifs in order to identify the presence or absence of particular entities. Commonly, detection is achieved via an enzyme suitable for use with a variety of colorimetric, fluorescent, or chemiluminescent substrates. Not unexpectedly, there is substantial commercial interest in the technique.

2.1. Transfer Techniques

Electroblotting is by far the most widely used Western blotting technique. Capillary blotting and diffusion blotting provide alternative techniques; however, these are relatively slow procedures. The basic equipment for the separation and transfer of proteins using polyacrylamide gel electrophoresis is commercially available from several companies. There are two basic types of transfer systems: (1) semidry and (2) tank buffer apparatus. By far the most widely used are the latter tank buffer systems (for reviews, *see* **refs.** *7–9*). The transfer of protein from the gel is achieved by applying an electric potential across the gel and the membrane, which are in contact. The membrane, nitrocellulose, polyvinylidene difluoride (PVDF), or Nylon, provides a more stable matrix than the gel, making subsequent manipulations much easier. In tank systems, an arrangement of sponge pad, filter paper, gel, membrane, filter paper, and sponge pad soaked in transfer buffer are held together, in that order, by a ridge plastic cassette. This cassette is place into a vertical tank filled with transfer buffer between platinum electrodes (*see* **Fig. 1**). The arrangement of these electrodes ensures a uniform voltage gradient over the whole

From: *Medical Biomethods Handbook*
Edited by: J. M. Walker and R. Rapley © Humana Press, Inc., Totowa, NJ

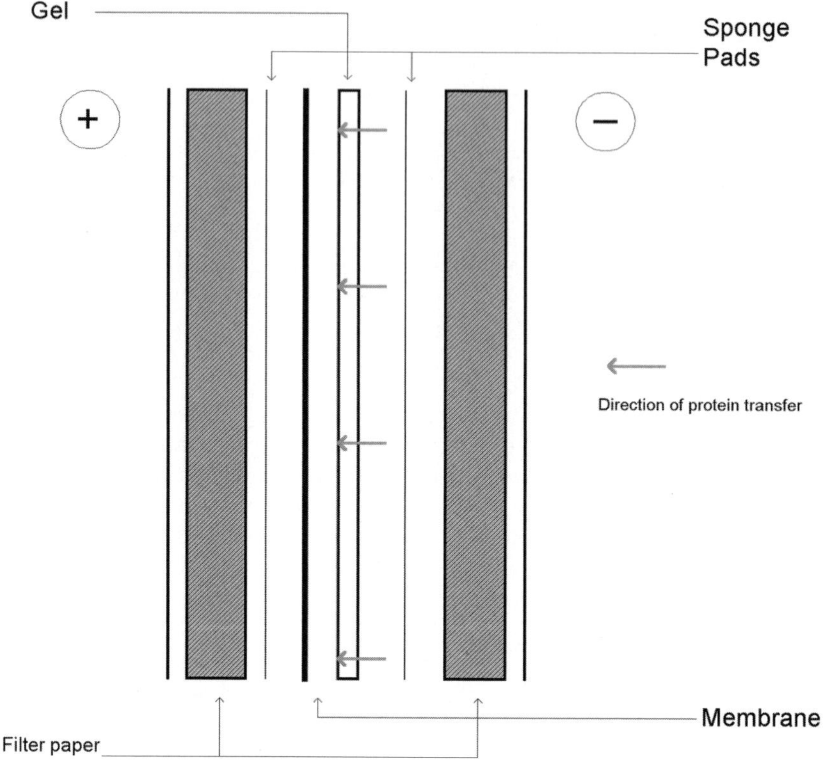

Fig. 1. Diagrammatic representation of electroblotting.

gel. The cassette must be placed in such a way to ensure that the membrane is between the gel and the anode. With semidry blotting apparatus, a similar arrangement of buffer-soaked filter paper, gel, membrane, and filter paper is placed between two large graphite electrodes. However, in this case, no additional liquid is present; hence the term semidry. Again, the membrane layer should be between the gel and the anode. In both cases, transfer times are short, varying from 30 min to a few hours. Modifications to the composition of the transfer buffer ***(3,10,11)*** and the use of cooling and circulation devices in the case of tank systems improve the effectiveness and reproducibility of the process. The transfer conditions required depend on the nature of the experiment and are dependent on factors such as the type of gel matrix, the nature of the proteins to be transferred, the type of membrane used, and so forth. A review of these criteria influencing this key step is provided in **Table 1**.

A critical factor in achieving successful protein blotting is the quality of the sample. Proteolysis leads to the appearance of smears, producing erroneous molecular-weight determination and/or false negatives. Proteolysis is the fragmentation of intact protein and can commonly be caused by proteolytic enzymes (proteasease) and/or physical forces such as shearing to which the protein is exposed during the extraction process. Key factors to consider in sample preparation include temperature, avoidance of freeze/thaw cycles (which subject the protein to excessive physical forces), the use of nonionic detergents (which can disrupt the protein conformation), use of protease inhibitors (which prevent proteolysis), and the use of subcellular fractionation of the sample. The latter assists in a more effective separation of the protein mixture by reducing its complexity.

Traditionally, proteins were transferred onto nitrocellulose sheets. In 1986, Millipore Corporation introduced PVDF ***(12)***, which is now widely used. This membrane offers the advantages of a higher mechanical strength than unsupported nitrocellulose, a high protein-binding

Table 1
Factors Influencing Protein Transfer

Parameter	Comments
Gel composition	The larger the pore size (low concentrations of monomer), the faster the transfer ***(10)***. The larger the protein the slower the transfer. Gradient gels are more effective at resolving protein mixtures, containing a wide range of molecular weights.
Membrane selection	The properties of the membrane must be considered when determining transfer conditions.
Ionic strength	Dilute buffers allow the use of high transfer voltages, minimizing heating. Reuse of transfer buffers is not advised.
Buffer type	Different buffers lead to different transfer efficiency. Tris buffers are more efficient that phosphate or acetate buffers. Proteins near their isoelectric point will transfer poorly.
pH	Effects buffer conductivity; alternation can lead to initial current drain and decreased resistance.
Methanol	Methanol is essential to improve binding of proteins to nitrocellulose. Its presence can reduce protein mobility because of fixation of the protein within the gel.
SDS	Sodium dodecyl sulfate (SDS) improves transfer by increasing protein mobility; however, resulting denaturation can reduce recognition via antibodies. SDS can lead to reduced binding to the nitrocellulose membrane. Its use depends on the nature of the proteins requiring detection.
Physical parameters	Poor contact between gel and membrane reduces the efficacy of transfer. Uniform conductivity and temperature within the transfer tank improve transfer.
Temperature	Temperature can alter buffer resistance and, subsequently, the power delivered, as well as the state of denaturation of the protein being transferred.
Transfer time	Heating should be minimized during long transfer times.
Power	Use of high currents achieves rapid elution of the proteins out of the gel.

capacity (125–160 µg/cm^2), and chemical stability. Nylon membranes have found favor with some users; there are examples indicating that groups of proteins only bind to Nylon ***(13)*** or more strongly to Nylon than nitrocellulose ***(14)***. The high background associated with this type of membrane in this application is probably responsible for its intermittent use. Nitrocellulose with its good binding capacity and low background is probably the most popular membrane and is available as a pure ester in both supported and unsupported forms. Unsupported membranes are produced when the active substrate is cast as a pure sheet. Because of their fragile nature, unsupported membranes should be handled with care. In supported membranes, the active substrate is cast on either side of an inert "web" or support, usually a polyester. Membranes with this type of internal support have a greater tensile strength, improving the handling characteristics of the membrane. Mixed membranes containing cellulose acetate and other cellulose derivatives as well as nitrocellulose are also available, but these types of membrane have a lower binding capacity than pure nitrocellulose. Most methods utilize a 0.45-µm pore size membrane. However not all proteins/peptides with a molecular weight less than 20 kDa are retained by nitrocellulose of this pore size ***(15,16)***. Membranes with 0.2 or even 0.1 μm are available for this application. PVDF is a Teflon-type polymer composed of repeating (–CH_2–CF_2–)*n*; it offers high mechanical strength and a high protein-binding capacity. Most PVDF membranes on the market are hydrophobic and unsupported. The membrane needs to be prewet in methanol prior to use. Proteins interact noncovalently through dipolar and hydrophobic interactions. The high binding capacity is the result of its open, porous polymeric structure and large surface area. The membrane is stable in many solvents, making it a more suitable membrane than nitrocellulose in some applications. For example, this characteristic has led to the use of PVDF in the harsh conditions employed in protein and peptide sequencing ***(17–19)***. Separated proteins are blotted onto PVDF, their positions are identified using a total protein stain such as Coomassie blue or Ponceau red, and the portion of the blot carrying the protein of interest is excised and further analyzed to determine its sequence. The protein is subjected to N-terminal sequencing in an appropriate instrument or, alternatively, subjected to internal sequencing. In this latter case, the protein immobilized on the membrane is first digested with specific proteases (such as trypsin, thermolysin, endoproteinase Lys-C, endoproteinase Glu-C, endoproteinase Asp-N clostripain). Hydrophilic peptides are released from the membrane into the digestion mixture and is then be separated by reverse-phase high-performance liquid chromatography (HPLC). Each peptide can then be individually sequenced. In both of these examples, sequencing is achieved using Edman chemistry, which removes amino acid residues from the N-terminus of the protein or peptide, one at a time in sequence. Each cycle consists of the following steps:

1. Coupling with phenyl isothiocynate (PITC) under mild alkaline conditions, forming a phenylthiocarbamyl peptide.
2. Cleavage of the residue as a anilinothiazolinone (ATZ), an amino acid derivative.
3. Conversion of the cleaved ATZ derivative to a stable phenylthiohydantoin (PTH) derivative.

This stabilized PTH amino acid derivative is then identified by reverse-phase HPLC. With direct sequencing, the membrane on which the target protein is retained is subjected directly to Edman chemistry within the instrument. The methodology is limited to the identification of approx 30 residues. In addition, gaps in the sequence often result, particularly when the level of target protein is low. Often, the proteins are also blocked at the N-termini in the course of the process, leading to a reduction in the target available for sequencing. The frequency of blockage increase with time, suggesting that other processes are at work ***(20,21)***.

2.2. Detection of Proteins/Peptides

In Western blotting following transfer, detection of target proteins is general achieved using a specific antibody. The attachment of specific antibodies to their targets can be readily visualized by using methods based on direct or indirect immunoassays. These methods involve the use of various labels conjugated to an antibody (*see* **Fig. 2**). Some of these labels can directly produce a detectable signal (e.g., fluorescent molecules, such as fluorescein, or an radioisotope such as iodine-125), but more commonly, today, indirect immunoassay systems based on enzymes such as horseradish peroxidase (HRP) or alkaline phosphatase (AP) are used. In this case, the detectable signal is generated by the action of the enzyme on a particular substrate; hence, the systems are described as indirect. Chromogenic, chemiluminescent, or fluorescent enzyme substrates are available. Some of the more commonly used substrates with some useful websites are given in **Table 2**. Chromogenic substrates result in the deposition of a colored product at the site of the conjugated enzyme and, hence, the protein of interest. Chemiluminescent and fluorescent substrates both produce light that can be detected as appropriate via X-ray film, charge-coupled device (CCD) cameras, or scanners. At present, the most widely used signal generation system in Western blotting is chemiluminescence. This has several advantages, including the ability to achieve high levels of sensitivity and fast processing times. Chemiluminescence is the generation of electromagnetic radiation, as light through the release

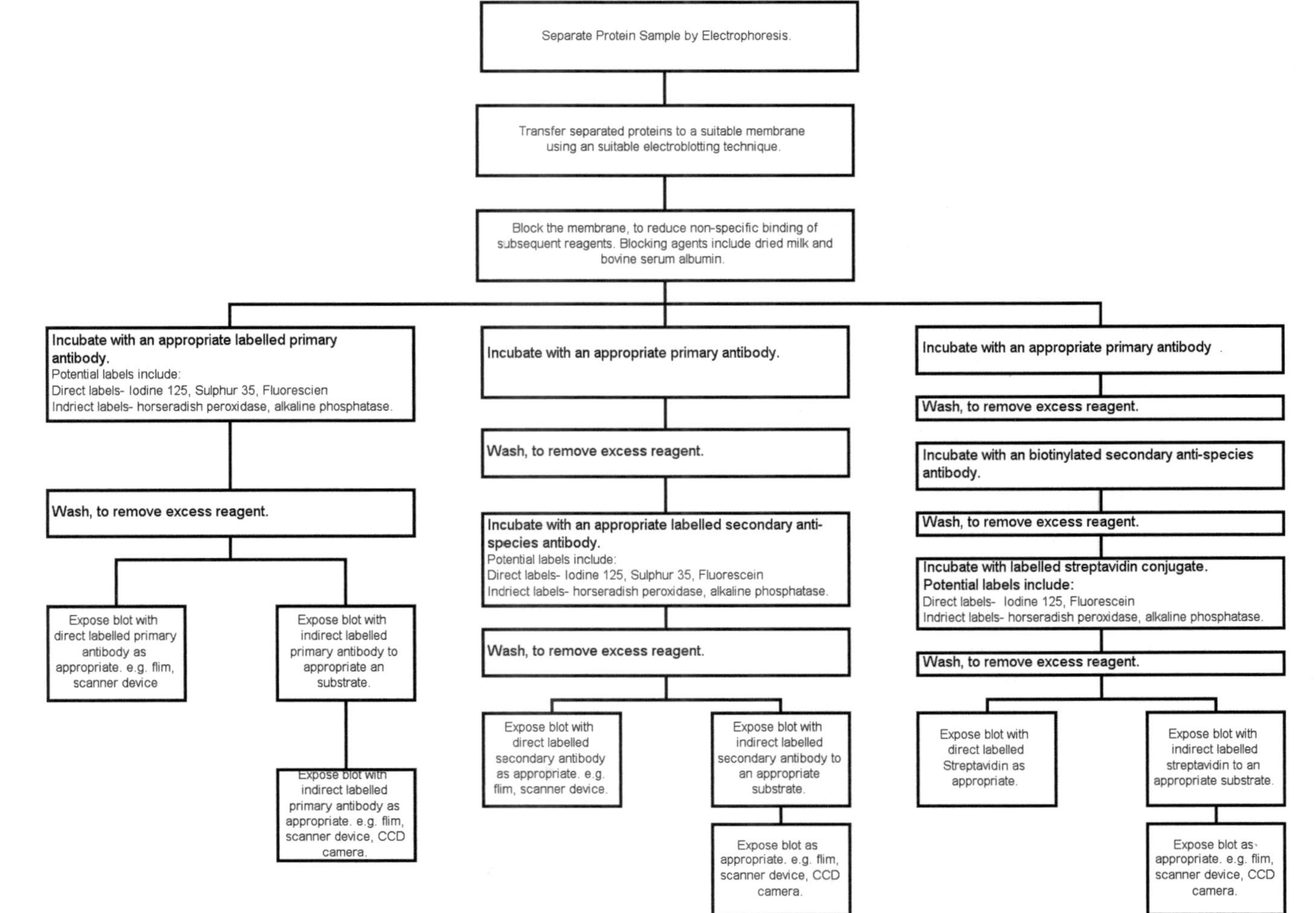

Fig. 2. Examples of Western blotting detection strategies.

Table 2
Commonly Available Enzyme Substrates for Use in Western Blotting Applications

Substrate	Enzyme	
	Peroxidase	Alkaline phosphatase
Colormetric (color produced follows each substrate listed)	3,3',5,5'-Tetramethylbenzidine (blue) 4 Chloro-1-naphthol (blue/black) Diaminobenzidine ± cobalt chloride (produces a red/brown color in the absence of cobalt chloride, black when present)	5-Bromo-4-chloro-3-indolylphosphate/nitroblue-tetrazolium (black/purple) Naphthol phosphate/Fast Red (red)
Fluorescent	AmplexGold™ (www.molecularprobes.com) FluorBlot™ (www.piercenet.com)	2'-[2-Benzothiazoyl]-6'-hydroxybenzothiazole phosphate (www.jblsci.com) 2-Hydroxy-3-naphtoic acis-2'-phenylaniide phosphate (www.roche-applied-science.com) DDAO phosphate (www.molecularprobes.com)
Chemiluminescent	Luminol/peroxide (www.amershambiosciences.com) Acridans (e.g., 2',3',6'-trifluorophenyl 10 methylacridan-9-carboxylate) (www.lumigen.com)	Dioxetanes (e.g., disodium 3-[4 methoxyspiro {1,2 dioxetane-3,2'-tricyclo [3.3.1.1] decan}-4-yl phenyl] phosphate [www.tropix.com])

Note: Full details of the substrate mode of action and recommended applications and condition can be found at the website cited. In general, additional information about all enzyme substrate for Western blotting can be found at these sites.

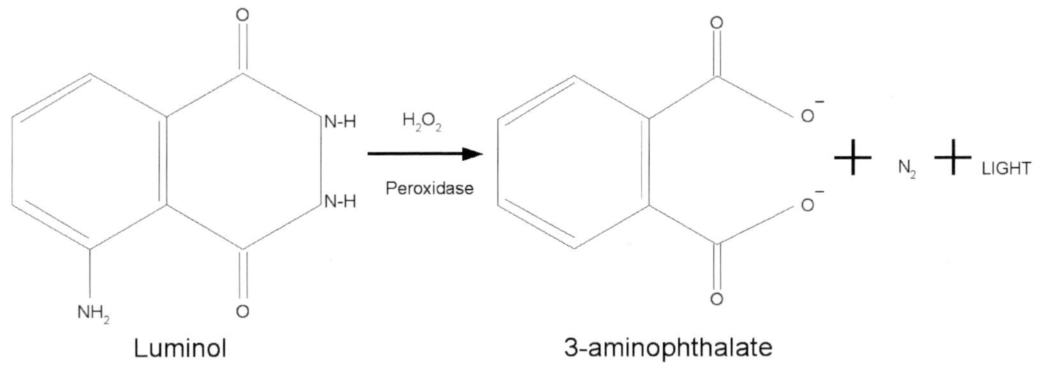

Fig. 3. Oxidation of luminol.

of energy from a chemical reaction. Usually, these reactions involve the cleavage of a relatively weak O–O bond, which liberates a large amount of energy. HRP systems generally are based on the oxidation of the cyclic diacylhydrazide luminol in the presence of hydrogen peroxide (*see* **Fig. 3**) *(22)*. Immediately following oxidation the luminol is an excited state, which decays back to the ground state via a light-emitting pathway. The presence of phenolic enhancers cause the light emission to be increased 1000-fold *(23)*. Often, the process is referred to as enhanced chemiluminescence. AP systems are almost exclusively based on the use of 1,2-dioxetane substrates *(24)* (*see* **Fig. 4**).

Key to any methodology is efficient blocking of the membrane. This will minimize background or nonspecific binding of the antibodies and reagents used for detection and, therefore, maximize the specific signal-to-noise ratio. A number of protein and detergent solutions are routinely used for this purpose. These include nonfat dried milk, Tween-20®, bovine serum albumin (BSA), whole serum, and gelatine. Care must be taken to ensure that the blocking agent is compatible with the detection reagents used. The most common in use today are nonfat dried milk and BSA *(3,25)* **Figure 5** outlines a few of the available detection methodologies in Western blotting; many variations on these themes exist in the reagents and kits that are available from suppliers. In a single-layer system, the primary (or recognition) antibody is conjugated either to an enzyme or a direct label. A double-layer system involves the use of an anti-immunogobulin capable of recognizing the primary antibody. This anti-immunogobulin is conjugated to an enzyme or a direct label. Another method utilizes a biotin-tagged anti-immunogobulin to introduce a third layer. The biotin molecule binds a streptavidin or avidin molecule which is conjugated to a suitable label, usually an enzyme. Increasing the number of layers within the detection procedure increases the sensitivity of the system. However, the label itself and the type of substrate employed also contribute the sensitivity of the system.

3. Western Blot Assays in the Diagnosis of Viral Infections

Western blotting has mostly found use in detecting immunogenic responses elicited by infectious agents, such as bacteria, viruses, and parasites, where the presence of these agents is difficult to detect with methods that aim to isolate and culture the infectious agent from patient samples. Other diagnostic uses for Western blot assays include detection of the presence of abnormal cellular proteins such as the prion protein. In general, Western blot assays do not provide truly quantitative information. In these applications, the technique is usually used as a secondary method to confirm initial results obtained with enzyme immunoassays (EIAs) or enzyme-linked immunosorbent assays (ELISAs) or provides additional information not obtainable with these procedures. Often, many different antigens can be assayed together, thus providing a wider picture of the antigenic response being studied.

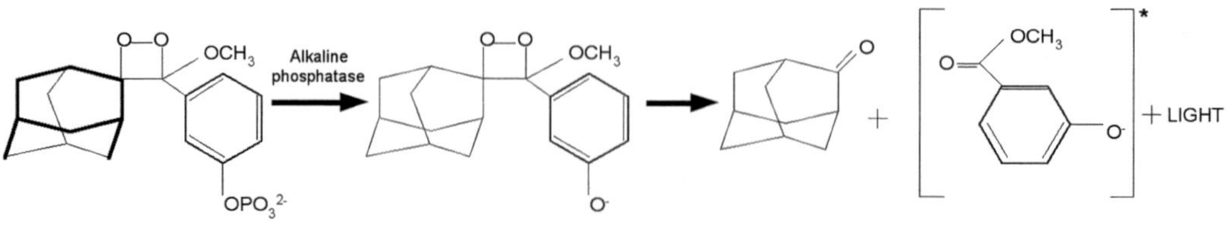

Fig. 4. Example of 1,2 dioxetane hydrolysis by alkaline phosphate.

In addition to many in-house-developed diagnostic Western blot systems, several companies are manufacturing commercial kits for use in serological diagnostics of different infectious diseases. Confirmation of human immunodeficiency virus (HIV) infections is an important application area in viral diagnosis. Detection of immunological responses in Lyme disease is a major application for bacterial diagnosis. Cysticercosis represents a parasitic disease where the Western blot technique demonstrates the presence of the parasite *Taelia cysticercosis*, which is difficult to prove by other means. Western blotting methods have proved useful for showing the presence of disease-related human proteins in autoimmune diseases. How Western blot assays are currently applied to these clinical areas will now be discussed.

3.1. Western Blot Assays in the Diagnosis of Viral Infection

3.1.1. HIV-I Diagnostics

HIV-1 and HIV-2 are the two causative agents of autoimmune deficiency disease (AIDS). Whereas HIV-2 infections are mostly restricted to Africa, HIV-1 has spread worldwide. Following infection by either virus, the patient develops antibodies against viral proteins. Detection of these antibodies with EIAs is used in the routine detection of HIV infection in patients and screening of blood donors. Western blotting is recommended for use as a secondary confirmatory test to further analyze samples that have repeatedly produced positive EIA results (*26*, and references therein). Anti-HIV antibodies can be found in most body fluids, including serum, saliva, and urea. All of these can be used as samples for testing for the disease. These Western blot kits contain specific HIV proteins fractionated according to their molecular weight by electrophoresis on a polyacrylamide gel in the presence of sodium dodecylsufate (SDS). These separated proteins are electrotransferred from the gel to a nitrocellulose membrane, which is then washed, blocked, and packaged. In the clinical laboratory, these nitrocellulose strips are incubated with HIV antibodies present in the patients serum (or other body fluid). These are the primary antibodies of our description in **Fig. 5**. The strips are washed to remove unbound material. Visualization of the human immunoglobulins specifically bound to the HIV proteins is achieved using a series of incubations in, for example, goat anti-human IgG conjugated with biotin, followed by streptavidin conjugated with HRP and then the HRP substrate 4-chloro-1-naphthol (*see* **Fig. 5C**). If antibodies to any of the target HIV proteins or antigens are present, bands corresponding to the positions of one or more of the HIV proteins result. The kit will contain suitable control material to ensure that the nitrocellulose strip and other reagents are performing correctly.

The structure and protein composition of HIV-1 and 2 viruses are well characterized, allowing identification of antigens acting in the immunological response. Following HIV infection, the earliest antibodies are against group-specific antigen (GAG) protein p24 and its precursor p55, but with later progression of the disease, these antibodies can decrease in quantity and be difficult to detect. Antibodies against envelope protein (ENV) precursor protein gp160 and its precursors gp120 and gp41 are usually found in most HIV-infected patients. Multimeric forms of gp41 can give rise to bands of 120 and/or 160 kDa *(27)*. Antibodies against HIV polymerase (POL) enzyme proteins p31, p51, and p66 can also be detected in patient samples *(27–31)*.

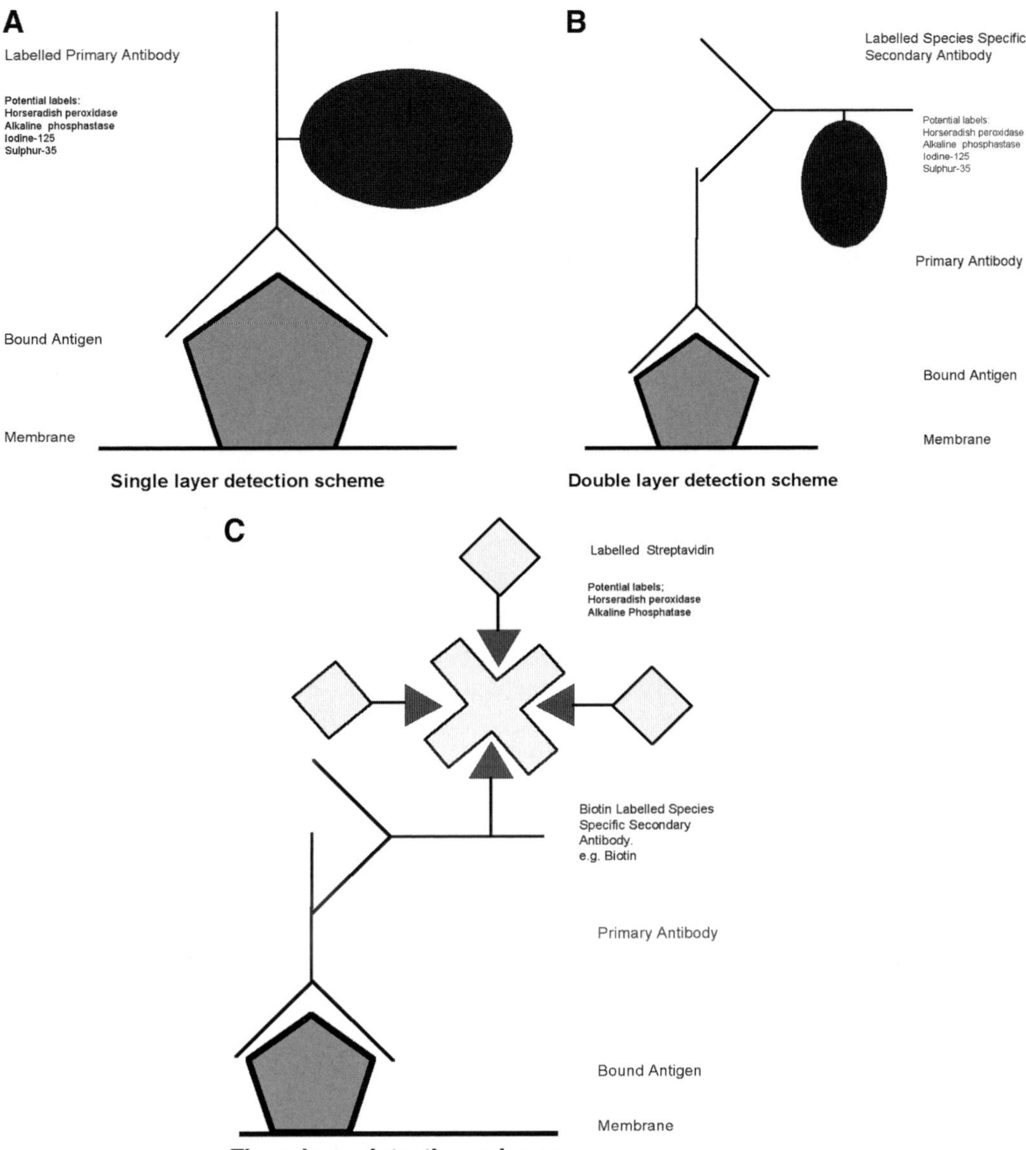

Fig. 5. Diagrammatic representation of some immunodetection regimes used in Western blotting.

In addition to the complexities in antigen patterns arising from the HIV viruses, the interpretation of blotting results is complicated by the different combinations of antibodies that may be found in patients. Antibody patterns may change during the progression of the infection ***(27–31)***. This biological variation has resulted in different guidelines being put forward for scoring Western blot results as positive for detecting HIV-1/2 virus infection. The requirement for detecting antibodies against all three major protein groups (ENV, GAG, and POL) for a positive score results in a relatively high number of indeterminate results, as only 72–79% of positive samples type contained antibodies against all three different groups of HIV proteins ***(32)***. The World Health Organisation (WHO) has recommended using the presence of at least two bands, including those corresponding to gp41, gp120, or gp160 *(33)*, as a criteria for a positive finding. A negative test result is the absence of any reactive bands on Western blot. Although detection of anti-HIV antibodies can indicate the presence of viral infection, final diagnosis should only be based on clinical examination.

Saliva and urine have been established as alternatives to serum samples for testing for HIV antibodies. These fluids offer the advantage of easy collection. Although the antibody concentration is lower than in serum, it is still suitable for detection with sensitive immunological methods such as Western blotting. Factors such as increase in urine volume following consumption of large quantities of liquids or the use of diuretics *(34)* further reduce antibody concentrations. Saliva contains secretions from salivary glands, oral microorganisms, cells, and gingival–crevicular transudate, which has been shown to contain immonoglobulin G against HIV proteins in infected individuals *(34)*.

The Federal Drug Administration (FDA) has approved three different kits for HIV testing. The Western Blot HIV-1 kits from Cambridge Biotech and Genetic Systems (Bio-Rad Laboratories) can be used with serum/plasma samples for confirmatory testing of blood donors. The Cambridge Biotech kit can also be used with urine samples and the Genetic Systems kit with dried blood spots. Organon Teknika's OraSure HIV-1Western blot kit is exclusive for use with oral fluid samples. Many other diagnostic kits exist for detecting HIV antibodies using the technique; *see* **Table 3** for details.

All of the kits for HIV diagnosis are based on the same principal design. The kit contains several identical test nitrocellulose strips to which size-separated HIV proteins have been transferred and immobilized. These strips are stable for several months if stored at cold room temperatures. The viral proteins have been obtained from virus purified from infected H-9/HTLV-IIB, T-lymphocyte cell line with ultracentrifugation (OraSure), or partially purified and inactivated by treatment with psoralen and ultraviolet (UV) light (Cambridge Biotech). In addition to HIV antigens, the blotted membrane can also contain control bands such as the staphylococcal protein (JN HIV-1/2 Kit). These controls are for detecting the presence of general IgG proteins in a sample and act as positive confirmation for correct functioning of the kit.

Before the individual membrane strips are incubated with a patient samples, they are blocked to inhibit nonspecific binding. The patient sample is diluted in sample buffer. The extent of dilution required depends on the test kit used and the sample. For example, the OraSure kit requires 150 µL of saliva diluted 1 : 7 for use; the Genelabs HIV-1 blot 1.3 kit requires 20 µL of serum diluted 1:100 for use. If anti-HIV antibodies are present in the patient sample, they bind to their cognate antigen on the test strip. After washing, the bound antibodies are detected with a secondary anti-human IgG antibody–alkaline phosphatase conjugate (OraSure), which catalyzes the conversion of nitroblue tetrazolium (NBT) into a colored deposited product in the presence of 5-bromo-4-chloro-2-indolyl-phosphate (BCIP). Alternatively, the Cambridge Biotech HIV-1 kit uses a biotin-labeled goat anti-human IgG and an avidin conjugated to HRP. Color reaction is achieved using the substrate 4-chloro-1-naphthol. All of the HIV-test kits contain negative control serum, as well as low- and high-positive control serums, which help in assessing the patient test results.

The actual protocol for scoring the bands present on membranes reacted with patient sera vary from one manufacturer to other. With the OraSure kit, the intensity of bands on the patient test strip are compared to the intensity of the gp41 band on the low-positive test strip. If bands of equal or higher intensity are detected, the band is scored as present. In the Cambridge Biotech kit, the intensity of the p24 band, seen with a weakly reactive serum, is used as reference for scoring. The overall test with both kits is considered positive only if at least two of the major bands (gp160, gp41, gp120, or p24) are present, according to the recommendations by WHO.

Antibodies against HIV-2 proteins can give positive reaction on Western blot containing HIV-1 antibodies, however, often the results are indeterminate. This is why some of the HIV-1 Western blot kits contain additional antigens derived from HIV-2. The JN HIV-1/2 Kit contains a recombinant HIV-2 membrane protein, Genelabs HIV-1 blot 2.4 contains a line of gp46

Table 3
Examples of Commercially Available HIV Western Blot Kits

Test kit	Manufacturer/supplier	Sample type	FDA approval for diagnostic purposes
Genetic Systems HIV-1 Western Blot	Bio-Rad Laboratories (www.biorad.com)	Human serum, plasma and dried blood spot	Yes
Cambridge Biotech HIV-1 Western Blot	Calypte Biomedical Corp. (www.calypte.com)	Human serum or plasma and urine	Yes
OraSure^{R} HIV-1 Western Blot Kit	Organon Teknika (www.orasure.com)	Oral fluid	Yes
Bio-Rad LAV Blot I/ Blot II	Bio-Rad Laboratories plasma	Human serum or	No
JN HIV-1/2 Kit	JN International Medical Corp. (www.jnii-usa-bharat.com)	Human serum	No
Genelabs HIV-1 blot 1.2, 1.3, 2.2, 2.4	Genelabs Diagnostics (www.genelas.com.sg)	Human serum or plasma	No
Bioblot HIV-1	Biokit, Spain (www.biokit.com)	Human serum or plasma	No

from HTLV-II, and the Bioblot HIV-1 plus kit contains a synthetic HIV-2 peptide. Native HIV-2 antigens are immobilized on BioRad products and Genelabs HIV blot 2.4.

3.1.2. Detection of Other Viral Diseases

Western blotting can be used for differentiation between related causative agents that cause different diseases. Herpes simplex virus 1 and 2, the causative agents of oral and genital herpes, respectively, are 95% similar genetically. Most antibodies raised against one species react with the other. The Western blotting assay of glycoprotein G is able to differentiate between the two viruses *(**35**)*. In the ParvoBlot Western Blot Biotrin kit, IgM molecules against parvovirus antigens VP1 and VP2 are detected. In this assay, IgG molecules present in the patient sample are absorbed with sample buffer and removed with centrifugation. Membrane-bound IgM antibodies are detected with HRP-conjugated anti-human IgM followed by diaminobenzidine substrate detection (ParvoBlot kit Biotrin).

3.2. Western Blot Assays in the Diagnosis of Bacterial Infections

Compared with viruses, bacteria are more complex organisms with a large number of proteins that can elicit immunological reactions. Similarities between closely and distantly related species can complicate detection of antibodies. Lyme disease is an example of a bacterial disease for which Western blotting has been widely used for diagnosis, but it is not without technical problems. Other bacteria that are currently detected with Western blot assays are listed in **Table 4**.

3.2.1. Diagnosis of Lyme borreliosis

The spirochete *Borrelia burgdorferi* causes Lyme disease. Several subspecies of the bacteria have been isolated in different parts of world. *Borrelia burgdorferi sensu stricto, Borrelia afzelii*, and *Borrelia garinii* are the three genospecies responsible. These bacteria are carried by ticks and transmitted following a bite. Typically, the disease has three stages: a circular skin rash reaction with virus like symptoms 7–10 d after infection, followed several weeks later with problems with nervous system and heart, and, finally, months or years later with arthritic and neurological symptoms *(**36**)*. Diagnosis can be achieved by demonstrating of the presence of *B. burgdorferi* in patient samples with bacterial culture, tissue examination, or polymerase chain reaction (PCR), but these methods are not always feasible. Serological methods for detecting antibodies against *B. burgdorferi* proteins are used for routine analysis.

These Western blot kits contain specific bacterial proteins fractionated according to their molecular weight by electrophoresis, electrotransfered onto a membrane such as nitrocellulose. In some cases, purified bacterial protein are applied directly to a membrane, as so-called dot blot. The membrane is then blocked. In the laboratory, the membrane is incubated with bacterial (primary) antibodies present in the patients serum (or other body fluid). The membranes are washed to remove unbound material. Visualization of the human immunoglobulins specifically bound to the bacterial proteins is achieved using a series of incubations. The nature of the disease progression (see below) means that these secondary antibodies can be either anti-human IgM or IgG conjugates. If antibodies to any of the target proteins or antigens are present, a band corresponding to the positions of one or more of the bacterial proteins result. The kit will contain suitable control material to ensure that the Western blot or dot blot, and other reagents are performing correctly.

Detection of antibodies in Lyme disease is hampered by biological variability in the immunological responses. The first response following exposure to *B. burgdorferi* is the appearance of type M immunoglobulins against the outer surface protein C (22–25 kDa), p39, and flagellin protein of 41 kDa. The IgM levels usually peak 3–6 wk after exposure. IgG antibodies are generally detectable after several weeks and can persist for years. Three antigens eliciting IgM antibodies, Osp A (31 kDa), Osp B (34 kDa), and p18, p28, p30, p45, p58, p66, and p93, can be detected by IgG antibodies found in patient sera *(**37**)*. Patient-to-patient variations in the timing

Table 4
Commercial Western Blot Assay Kits for Detecting Bacterial Infections

Bacteria	Test kit	Manufacturer
Borrelia burgdorferi	Lyme IgG Western Blot	Quest Biomedical
Borrelia burgdorferi	Lyme IgM Western Blot	Quest Biomedical
Borrelia burgdorferi	Borrelia burgdorferi sl IgM/IgG immunoblots	KMI Diagnostics
Borrelia burgdorferi	QualiCode™ Borrelia burgdorferi IgG and IgM Western Blot	Immunetics, Inc.
Borrelia burgdorferi	Lyme IgG Western Blot Lyme IgM Western Blot	IgeneX, Inc.
Borrelia burgdorferi	Lyme disease Western Blot Kit	JN-International Medical Corp.
Borrelia afzelii, Borrelia garinii	EU Lyme IgG MarBlot™ EU Lyme IgM MarBlot™	Trinity Biotech
Borrelia burgdorferi	Lyme IgG MarBlot™ Lyme IgM MarBlot™	Trinity Biotech
Ehrlichia. phagocytophila, or *E. ewingii*	HE IgG and IgM Western Blot	Imugen
Ehrlichia. phagocytophila, or *E. ewingii*	Human Granulocytic Ehrlichiosis (HE) Western Blot Kit	JN-International Medical Corp.
Ehrlichia. Phagocytophila, or *E. ewingii*	HE Ehrlichia IgG MarBlot™	Trinity Biotech
Helicobacter pylori	Helico Blot 2.1	Genelabs Diagnostics

and extent of the serological response to *B. burgdorferi* exposures exist. Patients with persistent infection may produce IgM antibodies longer than expected or lack IgG response ***(38)***. This poses the problem for accurate diagnosis of the disease status based on antibodies against *B. burgdorferi* antigens. A negative test result may be obtained if testing is performed shortly after infection and, conversely, positive results may indicate the presence of past infection, which could be unconnected to current disease ***(39)***. Treatment of tick borreliosis can also abrogate immune response, despite the continued presence of the bacteria in patient tissues ***(40,41)***.

Decorin-binding protein A (DpbA) is 43–63% similar among *B. burgdorferi sensu stricto, B. afzelii*, and *B. garinii*. The sequence diversity is reflected in that most patient sera only recognized DbpA from one species. Whereas most patients with neuroborreliosis and Lyme arthritis contained antibodies against DbpA, patients with an earlier stage of Lyme disease failed to do so ***(42)***. Similarly, sequence variation in OspC protein from different *Borrelia* species has been shown to result in the presence of different epitopes and variation in detection with polyclonal antisera ***(43)***. Further complications to accurate diagnosis is contributed by crossreactivity between antibodies against *B. burgdorferi* antigens and antigens derived from other spirochetes, rickettsia, ehrlichia, or bacteria such as *Helicobacter pylori* ***(44)***, and the Epstein–Barr virus ***(45)*** resulting in false positives.

It is not surprising that against this complicated background of biological variation and crossreactivity it has been difficult to determine a simple set of rules according to which a positive test is scored. The US Center for Disease Control (CDC) recommends that IgM Western blot tests for *B. burgdorferi* antibodies be scored positive if two or three bands corresponding to 23- to 25-kDa OspC or 39-kDa or 41-kDa antigens are detected on the blot. The recommendation for a positive test result of an IgG Western blot is the presence of 5 out of 10 antigen bands ***(46)***. European recommendations for IgG require the presence of two bands for *B. afzelii* subtype (PKo) and at least one band for *B. garinii* subtype (Pbi). Positive IgM result for pKo requires that p39, OspC, and p17 be detected or that a strong band is observed for p41.

For pBi, the requirement for the p17 band is omitted ***(47)***. Despite these measures, the controversy surrounding the accuracy and specificity of serological tests for Lyme borreliosis has not subsided ***(39,48–50)***. A list of currently available kits is given in **Table 4**. The immunoblot assay can, however, provide a wider analysis of the extent of immunological response and increase the sensitivity of detection and accuracy of testing ***(48)***. However, ELISA and Western blot assays are not independent tests, as both detect the same antibodies. Western blot should, therefore, be considered as a supplemental rather than confirmatory test for Lyme disease and is recommended as a secondary test to corroborate positive or indeterminate results obtained by EIA or ELISA. The interpretation of commercial Western blot assays is hampered by the finding that the templates for band patterns provided in test kits and the patterns obtained with positive control samples do not always align. Hence, assignment of detected bands to correct antigens is made difficult and subjective and can lead to a false conclusion or indeterminate results ***(49)***. These limitations mean that immunological testing should be only used in conjunction with clinical diagnosis ***(39)***.

Several groups have started to address the technical problems associated with Lyme borreliosos Western blot assays by preparing synthetic antigens for testing. This approach allows the possibility for choosing unique regions of antigens that are not conserved between the different species. Adding recombinant VlsE and PbpA antigens to whole-cell lysates of *Borrelia* proteins has been shown to improve detection ***(51)***. Gomer-Solecki and co-workers ***(52)*** have developed an immunoblotting assay in which recombinant proteins containing key parts of *Borrelia* antigens are used. This assay improved detection specificity and identification and facilitated the detection of early disease. By using p83/100, p39, OspC, p41 Flagellin, Osp7, and p58 antigens, the diagnosis of later-stage arthritic and neuroborreliosis was improved ***(53)***. By combining the use of a dot blot assay where antigens are present at fixed positions with the use of recombinant antigens improved the specificity of detection over a Western blot, but also made scoring of the test much simpler ***(49)***. These developments are likely to result in a new generation of more accurate and specific serological tests for Lyme borreliosis.

3.2.2. Diagnosis of Human Ehrlichiosis

The group of human ehrlichiosis are caused by tick-borne gram-negative *Ehrlichia* bacteria. Monocytic ehrlichiosis is attributable to *Ehrlichia chaffeensis*; granulocytic ehrlichiosis is caused by *E. phagocytophila* or *E. ewingii* ***(54–56)***. The relatedness of these causative agents makes immunodiagnostics difficult because of antibody crossreactivity with antigens derived from other species. Unver and co-workers ***(57)*** have demonstrated that in Western blot analysis, outer membrane protein of 30 kDa was only recognized by antibodies raised against *E. chaffeensis* and a protein of 44 kDa was specific to *E. phagocytophila*.

3.3. Western Blot Assays in the Diagnosis of Parasite Infections

The generation of a diagnostic test for parasite infections has the complication that the native antigens needed for preparing immunoblots may not be easily available. This is the case in cysticercosis, which is endemic in many parts of the world. It is caused by the pork tapeworm *Taenia solium*. This parasite transfers from pigs to humans in infected meat containing eggs. The larvae hatch inside the small intestine of the human host and enter the bloodstream through the intestinal wall. Once in circulation, the worms can migrate to target tissues, including the central nervous system, where they will encyst and develop into a cysticercus. Clinical symptoms include migraines and resemble epilepsy. In order to overcome the limitations of obtaining target antigen, molecular cloning strategies have been applied to obtaining recombinant *T. solium* antigens.

Taenia solium infections are diagnosed currently with Western blot assays in which lentil-lectin purified glycoproteins are used as antigens. These antigens detect antibodies against seven major *T. solium* glycoproteins ***(58)***. Saline extraction can also be used to prepare antigens from *T. solium* cysts for analysis. However, the lentil-lectin purification remains the more sensitive method ***(59)***. The primary source of *T. solium* antigens is from cysts found in naturally infected pigs, which limits the availability of these antigens for testing. Whatever the source, antigen mixtures are prepared and fractionated according to their molecular weight by electrophoresis, then electrotransferred onto a membrane and blocked. In the clinical laboratory, the membrane is incubated in the patient's serum for example, which contains primary antibodies against some or all of the target proteins. The membrane is washed to remove unbound material. Visualization of the human immunoglobulins specifically bound to the parasite proteins is achieved using a series of incubations in an anti-human immunoglobulin conjugated with an indirect label and subsequent incubation to allow for visualization. If antibodies to any of the target proteins or antigens are present, bands corresponding to the positions of one or more of the parasite proteins result. The kit will contain suitable control material to ensure that the performance of the reagents is as expected and that identification of the specific parasite protein is possible.

Greene and co-workers ***(60)*** used degenerate oligonucleotides derived from the amino acid sequence of purified 14- to 18-kDa polypeptides, corresponding to major *T. solium* antigens, to amplify cDNAs for related antigens from the *T. solium* cDNA library. One of the cDNAs, sTs14, showed promise as a specific diagnostic marker. Further research has shown that the *T. solium* diagnostic antigens of 14, 18, and 21 kDa are members of 8-kDa antigen family for which 18 unique genes have been characterized. One out of nine 8-kDa gene-coded proteins (TsRS1) was shown to be a 100% specific and sensitive as an antigen for detection of antibodies against *T. solium* in cysticercosis ***(61)***. Hubert and co-workers ***(62)*** screened a *T. solium* metacestode cDNA expression library with patient sera and identified antigen NC-3, which showed high sensitivity and specificity in serodiagnostics. **Table 5** lists other human parasite infections that are currently diagnosed with the aid of Western blot assays.

3.4. Detection of Human Proteins in Disease States

3.4.1. Detection of Circulating Autoantibodies

Autoimmune diseases are characterized by the production of antibodies against normal human proteins or other cellular molecules. Western blotting can be used with other immunological methods for detecting these antibodies. The presence of autoantibodies is included in the diagnostic criteria for systemic lupus erythematosus (anti-Smith antigen and anti-double-strand DNA antibodies), mixed connective tissue disease (anti-Ui-nuclear riboprobtein antibodies), and Sjögrens's syndrome (anti-SS-A/Ro and anti-SS-B/RoLa antibodies) ***(63)***. A limitation of the Western blot method in detecting autoantibodies is that not all antibodies can recognize the denatured and immobilized antigens presented on immunoblots. This may explain findings that ELISA methods with the same antigens can be more sensitive in detecting autoantibodies in patient sera ***(64,65)***.

Typical Western blot protocol for detecting autoantibodies involves preparation of antigens from the tissue affected. For example, solubilized cell extracts prepared from the lung adenocarcinoma cell line or lung tumors have been used to detect circulating autoantibodies in tumor patients ***(66)***. Decreased opportunity for detecting crossreacting antibodies with complex antigen preparations is achieved by further purification of the principal antigens that present main targets for autoantibodies in different diseases. Purification of SS-A and SS-B antigens with ion-exchange chromatography and differential salt extraction result in preparations that allowed further characterization of the antigens and their use in diagnostic tests ***(67)***. With the use of gene technology, diagnostic tests for autoantibodies have been further refined by creating and expressing recombinant antigens for use in tests. For example, Vitozzi and co-workers ***(65)*** combined the two major antigens of autoimmune hepatitis (cytochrome P450 2D6 and formiminotransferase cyclodeaminase) into a recombinant protein and used this expressed chimera for detecting autoantibodies directed against both proteins in one assay. Regardless of their source, the test antigens are detected via a Western blot using anti-human antibodies linked to either HRP or AP with autoantibodies from the patient's serum. Hu and co-workers ***(68)*** performed systematic optimization of the method improving all steps of the protocol. They concluded that using protein G coupled to HRP reduced false-positive reactions and increased detection of correct antibodies. Furthermore, inclusion of CHAPS (3-[3-cholamidopropyl-

Table 5
Commercial Western Blot Assays for Detecting Antibodies Against Human Parasites

Disease/causative agent	Kit	Manufacturer
Cysticercosis/*Taelia solium*	QualiCode Cysticercosis Western Blot	Immunetics
Cysticercosis/*Taelia solium*	Cysticercosis Western Blot	JN-International Medical Corp.
Hydatid disease/parasite *Echinococcus granulosus*	QualiCode Hydatid Western Blot	Immunetics
Hydatid disease/parasite *Echinococcus granulosus*	Hydatid Western Blot	JN-International Medical Corp.
Babesiosis/parasite *Babesia microti*	QualiCode Babesiosis IgG or IgM Western Blot	Immunetics
Babesiosis/parasite *Babesia microti*	Babesiosis Western Blot	JN-International Medical Corp.

dimethylammonio]-1-propanesulfonate) increased binding of protein G to antibodies and the inclusion of Tween-20 reduced background on membrane and eliminated the need to use protein-blocking agents in incubation buffers. Finally, PVDF produced higher-quality blots with lower backgrounds than a nitrocellulose membrane.

With developments in proteomics, Western blot assays have found use in elucidating the antigens that are responsible for eliciting antigenic response in various diseases. Brichory and colleagues ***(66)*** have used two-dimensional gel electrophoresis coupled with Western blotting to investigate the targets of circulating autoantibodies found in the sera of lung cancer patients. Tumor cell extracts were separated and then blotted onto the PVDF membrane and detected with the patient's sera. Two major groups of antigen spots were detected. Subsequent excision of these spots and mass spectrometry analysis revealed that these antigen spots corresponded to annexin I and II. This approach is currently being used to identify further cancer markers for use in diagnostics and aiding evaluation of disease prognosis ***(69)***.

3.4.2. Diagnosis of Prion Diseases

Western blot assays can also be used to demonstrate the presence of variant forms of normal cellular proteins. Creutzfeld–Jacob disease (CJD) is a neurodegenerative disease with aberrant metabolism of the 33–75 sialoglycoprotein, prion protein (Pr), whose exact function is not yet know. Normal cellular forms of the protein, found in neurons and other tissues, is sensitive to protease digestion. In prion diseases, a variant form of the protein, Pr^{Sc}, which is derived from the native protein by post-translational modification and conformational change, accumulates in cells, eventually resulting in the death of the affected neurons. This variant form is partially resistant to protease digestion and insoluble in detergent solutions ***(70)***. Several different Pr^{Sc} protein isoforms differing in their sensitivity to protease K digestion have been identified ***(71)***. Variant CJD also contains Pr^{Sc} protein, although the pathogenenis and clinical course of this disease is different. A Western blot assay for detecting Pr^{Sc} protein isoforms involves the concentration of the protein from brain extracts ***(71)***. Before separation, the precipitate containing Pr^{Sc} protein is digested with protease K. Protease-resistant Pr^{Sc} protein is detected with a monoclonal antibody against Pr protein, followed by a secondary antibody linked to HRP or AP. Using sensitive, enhanced chemifluorescent detection methods, Wadsworth and co-workers ***(72)*** were able to detect Pr^{Sc} protein in the tonsils, spleen, and lymph nodes, where concentrations were significantly lower than in the brain. In addition to the Prion protein, other proteins have been shown to be altered in CJD. One such protein is an isoform of 14-3-3 abundant brain protein. Cerebrospinal fluid of CJD patients contains different isoforms than normal brain or brains affected with other dementias. By using Western blot analysis, the presence of the CJD-associated isoforms can be diagnosed ***(73)***.

References

1. Southern, E. M. (1975) Detection of specific sequences among DNA fragments separated by gel electrophoresis. *J. Mol. Biol.* **98,** 503–517.
2. Alwin, J. C., Kemp, D. J., and Stark, G. R. (1977) Method for detection of specific RNAs in agarose gels by transfer to diazobenzyloxymethyl-paper and hybridization with DNA probes. *Proc. Natl. Acad. Sci. USA* **74(12),** 5350–5354.
3. Towbin, H., Straelin, T., and Gordin, J. (1979) Electrophoretic transfer of proteins from polyacrylamide gels to nitrocellulose sheets: procedure and some applications. *Proc. Natl. Acad. Sci. USA* **7(9),** 4350–4354.
4. Dunbar, B. S. (ed.) (1994) *Protein Blotting: A Practical Approach*, IRL, Oxford.
5. Laemmli, U. K. (1970) Cleavage of structural proteins during the assembly of the head of bacteriophage T4. *Nature* **227(259),** 680–685.
6. Gershoni, J. M. and Palade, G. E. (1982) Electrophoretic transfer of proteins from sodium dodecyl sulfate-polyacrylamide gels to a positively charged membrane filter. *Anal. Biochem.* **124(2),** 396–405.
7. Timmons, T. and Dunbar, B. S. (1990) *Methods Enzymol.* **128,** 679–689.
8. Bers, G. and Garfin, D. (1985) Protein and nucleic acid blotting and immunobiochemical detection. *BioTechniques* **3(4),** 276–288.
9. Gershoni, J. M. and Palade, G. E. (1983) Protein blotting: principles and applications. *Anal. Biochem.* **131(1),** 1–15.
10. Dunn, S. D. (1986) Effects of the modification of transfer buffer composition and the renaturation of proteins in gels on the recognition of proteins on Western blots by monoclonal antibodies. *Anal. Biochem.* **157(1),** 144–153.
11. Van Dam, A. P., et al. (1986) Technical problems concerning the use of immunoblots for the detection of antinuclear antibodies. *J. Immunol. Methods* **129(1),** 63–70.
12. Matsudaira, P. (1987) Sequence from picomole quantities of proteins electroblotted onto polyvinylidene difluoride membranes. *J. Biol. Chem.* **262(21),** 10,035–10,038.
13. Miribel, L. and Arnaud, P. (1988) Electrotransfer of proteins following polyacrylamide gel electrophoresis; nitrocellulose versus nylon membranes. *J. Immunol. Methods* **107,** 253–259.
14. Gershoni, J. M. and Palade, G. E. (1982) Electrophoretic transfer of proteins from sodium dodecyl sulfate-polyacrylamide gels to a positively charged membrane filter. *Anal. Biochem.* **124(2),** 396–405.
15. Lin, W and Kasamatsu H. (1983) On the electrotransfer of polypeptides from gels to nitrocellulose membranes. *Anal. Biochem.* **128(2),** 302–311.
16. Tovey, E. R., Ford, S. A., and Baldo, B. A. (1987) Protein blotting on nitrocellulose: some important aspects of the resolution and detection of antigens in complex extracts. *J. Biochem. Biophys. Methods* **14,** 1–17.
17. Hewick, R. M., Hunkapiller, M. W., Wood, L. E., and Dreyer, W. J. (1981) A gas–liquid solid phase peptide and protein sequenator. *J. Biol. Chem.* **256(15),** 7990–7997.
18. Aebersold, R. H., Leavitt, J., Saavedra, R. A., Hood, L. E., and Kent, S. B. (1987) Internal amino acid sequence analysis of proteins separated by one- or two-dimensional gel electrophoresis after in situ protease digestion on nitrocellulose. *Proc. Natl. Acad. Sci USA* **84(20),** 6970–6974.
19. Moos, M., Jr., Nguyen, N. Y., and Liu, T. Y. (1988) Reproducible high yield sequencing of proteins electrophoretically separated and transferred to an inert support. *J. Biol. Chem.* **263(13),** 6005–6008.
20. Matsudaira, P. (1987) Sequence from picomole quantities of proteins electroblotted onto polyvinylidene difluoride membranes. *J. Biol. Chem.* **262(21),** 10,035–10,038.
21. Walsh, M. J., McDougall, J., and Wittmann-Liebold, B. (1988) Extended N-terminal sequencing of proteins of archaebacterial ribosomes blotted from two-dimensional gels onto glass fiber and poly(vinylidene difluoride) membrane. *Biochemistry* **27**, 6876–6879.
22. Whitehead, T. P., Thorpe, G. H. C., Carter, T. J. N, Groncutt, C., and Kiricka L. J. (1983) Enhanced luminescence procedure for sensitive determination of peroxidase-labelled conjugates in immunoassays. *Nature* **305**, 158–159.
23. Lundin, A. and Hallander, L. O. B. (1987) *Bioluminescence and Chemiluminescence: New Perspectives*, Wiley, New York.
24. Bronstein, I., Edwards, B., and Voyta, J. C. (1989) 1,2 Dioxetanes: novel chemiluminescent enzyme substrates. Application to immunoassays. *J. Biolumines.* **4**, 99–11.
25. Johnson, D. A., Gautsch, J. W., Sportsman, R., and Elder, J. H. (1984) Improved technique utilizing non fat dry milk for analysis of proteins and nucleic acids transferred to nitrocellulose. *Gene Anal. Tech.* **1**, 3–8.

26. Center for Disease Control (1989) Interpretation and use of the Western Blot assay for serodiagnosis of human immunodeficiency virus type 1 infections. MMWR **38(S-7)**, 1–7; MMWR **51(18)**, 395–396 and references therein.
27. McDougal, J. S., Kennedy, M. S., Nicholson, J. K. A., et al. (1987) Antibody response to human immunodeficiency virus in homosexual men: relation of antibody specificity, titer, and isotype to clinical status, severity of immunodeficiency and disease progression. *J. Clin. Invest.* **80**, 316–324.
28. Lange, J. M. A., Coutinho, R. A., Krone, W. J. A., et al. (1986) Distinct IgG recognition patterns during progression of subclinical and clinical infection with lymphadenopathy associated virus/human T lymphotropic virus. *Br. Med. J.* **292**, 228–230.
29. Esteban, J. I., Shih, J. W., Tai, C. C., et al. (1985) Importance of Western blot analysis in predicting infectivity of anti-HTLV-III/LAV positive blood. *Lancet* **2**, 1083–1086.
30. Goudsmith, J., Lange, J. M. A., Paul, D. A., et al. (1987) Antigemia and antibody titers to core and envelope antigens in AIDS, AIDS-related complex and subclinical human immunodeficiency virus infection. *J. Infect. Dis.* **155**, 558–560.
31. Weber, J. N., Clapham, P. R., Weiss, R. A., et al. (1987) Human immunodeficiency virus infection in two cohorts of homosexual men: neutralizing sera and association of anti-gag antibody with prognosis. *Lancet* **1**, 119–121.
32. Martinez, P. M., Torres, A. R., de Lejarazu, R. O., Montoya, A., Martin, J. F., and Eiros, J. M. (1999) Human Immunodeficiency virus antibody testing by enzyme-linked fluorescent and Western blot assays using serum, gingival-crevicular transudate and urine samples. *J. Clin. Microbiol.* **3**, 1100–1106.
33. World Health Organization (1990) AIDS: Proposed WHO criteria for interpreting results from Western blot assays for HIV-1, HIV-2, HTLV-I/HTLV-II. *Weekly Epidemiol. Rec.* **65**, 281–283.
34. Mortimer, P. P. and Parry, J. V. (1994) Detection of antibody to HIV in saliva: a brief review. *Clin. Diagn. Virol.* **2**, 231–243.
35. Slomka, M. J., Ashley, R. L., Cowan, F. M., Cross, A., and Brown, D. W. (1995) Monoclonal antibody blocking tests for the detection of HSV-1 abd HSV-2 specific humoral responses: comparison with Western blot assay. *J. Virol. Methods* **1**, 27–35.
36. Rahn, D. W. (1998) Natural History of Lyme disease. in *Lyme Disease* (Rahn, D. W. and Evans, J., eds.), American College of Physicians, Philadelphia, PA, pp. 35–48.
37. Dressler, F., Whalen, J. A., Reinhardt, B. N., and Steere, A. C. (1993) Western blotting in the serodiagnosis of Lyme disease. *J. Infect. Dis.* **167**, 392–400.
38. Craft, J. E., Fischer, D. K., Shimamoto, G. T., and Steere, A. C. (1985) Antigens of *Borrelia burgdorferi* recognized during Lyme disease: appearance of a new immunoglobulin M response and expansion of the immunoglobulin G response late in the illness. *J. Clin. Invest.* **78**, 934–939.
39. FDA Medical Bulletin. (1999) http://www.fda.gov/medbull/summer99/Lyme.html.
40. Hauptl, T., Hahn, G., Rittig, M., et al. (1993) Persistence of *Borrelia burgdorferi* in ligamentous tissue from a patient with chronic Lyme borreliosis. *Arthritis Rheum.* **36**, 1621–1626.
41. Preac-Mursic, V., Magret, W., Busch, U., Pleterski Rigler, D., and Hagl, S. (1996) Kill kinetics of *Borrelia burgdorferi* and bacterial findings in relation to the treatment of Lyme borreliosis. *Infection* **24**, 9–18.
42. Heikkila, T., Seppala, I., Saxen, H., Panelius, J., Yrjanainen, H., and Lahdenne, P. (2002) Species-specific serodiagnosis of Lyme arthritis and neuroborreliosis due to *Borrelia burgdorferi sensu stricto*, *B. afzelii*, and *B. garinii* by using decorin binding protein A. *J. Clin. Microbiol.* **40**, 453–460.
43. Theisen, M., Frederiksen, B., Lebech, A. M., Vuust, J., and Hansen, K. (1993) Polymorphism in ospC gene of Borrelia burgdorferi and immunoreactivity of OspC protein: implications for taxonomy and for use of OspC protein as a diagnostic antigen. *J. Clin. Microbiol.* **31,** 2570–2576.
44. Magnarell, L. A., Anderson, J. F., and Johnson, R. C. (1987) Cross-reactivity in serological tests for Lyme disease and other sprirochetal infections. *J. Infect. Dis.* **156,** 183–188.
45. Goossens, H. A., van den Bogaard, A. E., and Nohlmans, M. K. (1999) Evaluation of fifteen commercially available serological tests for diagnosis of Lyme borreliosis. *Eur. J. Clin. Microbiol. Infect. Dis.* **18,** 551–560.
46. Centre for Disease Control (1995) Recommendations for test performance and interpretation from the second national conference on the serological diagnosis of Lyme disease. *MMWR* **44**, 590–591.
47. Hauser, U., Lehnert, G., and Wilske, B. (1999) Validity of interpretation criteria for standardized Western blots (immunoblots) for serodiagnosis of Lyme borreliosis based on sera collected throughout Europe. *J. Clin. Microbiol.* **37,** 2241–2247.

48. Craven, R. B., Quan, T. J., Bailey, R. E., et al. (1996) Improved serodiagnostic testing for Lyme disease: Results of a multicenter serologic evaluation. *Emerg. Infect. Dis.* **2,** 136–140.
49. Fawcett, P., Rose, C. D., Gibney, K. M., and Doughty, R. A. (1998) Comparison of immunoblot and Western blot assasy for diagnosing Lyme borreliosis. *Clin. Diag. Lab. Immunol.* **5,** 503–506.
50. Brenner, C. (2003) Explanation of the Lyme disease Western Blot. Lymenet. www.lymenet.de/labtests/brenner.htm.
51. Schulte-Spechtel, U., Lehnert, G., Liegl, G., et al. (2003) Significant improvement of the recombinant Borrelia-specific immunoglobulin G immunoblot test by addition of VlsE and DbpA homologue derived from *Borrelia garinii* for diagnosis of early neuroborreliosis. *Clin. Microbiol.* **41,** 1299–1303.
52. Gomes-Solecki, M. J., Dunn, J. J., Luft, B. J., et al. (2000) Recombinant chimeric Borrelia proteins for diagnosis of Lyme disease. *J. Clin. Microbiol.* **38,** 2530–2535.
53. Wilske, B., Habermann, C., Fingerle, V., et al. (1999) An improved recombinant IgG immunoblot for serodiagnosis of Lyme borreliosis. *Med. Microbiol. Immunol.* **188,** 139–144.
54. Rikihisa, Y. (1991) The tribe Ehrlichieae and ehrlichial diseases. *Clin. Microbiol. Rev.* **4,** 286–308.
55. Buller, R. S., Arens, M., Hmiel, S. P., et al. (1999) *Ehrlichia ewingii*, a newly recongnized agent of human ehlichiosis. *N. Engl. J. Med.* **341,** 148–155.
56. Chen, S. M, Dumler, J. S., Bakken, J. S., and Walker, D. H. (1994) Identification of a granulocytotrophich Ehrlichia species as the etiologic agent of human disease. *J. Clin. Microbiol.* **32,** 589–595.
57. Unver, A., Felek, S., Paddock, C. D., et al. (2001) Western blot analysis of sera reactive to human monocytic ehrlichiosis and human granulocytic ehrlichiosis agents. *J. Clin. Microbiol.* **39,** 3982–3986.
58. Tsang, V. C., Brand, J. A., and Boyer, A. E. (1989) An enzyme-linked immunoelectrotransfer blot assay and glycoprotein antigens for diagnosing human cysticercosis (*Taenia solium*). *J. Infect. Dis.* **159,** 50–59.
59. Rodriguez-Canul, R., Allan, J. C., Fletes, C., Sutisna, I. P., Kapti, I. N., and Craig, P. S. (1997) Comparative evaluation of purified Taenia solium glycoproteins and crude metacestode extracts by immunoblotting for the serodiagnosis of human *T. solium* cysticercosis. *Clin. Diagn. Lab Immunol.* **4,** 579–582.
60. Greene, R. M., Hancock, K., Wilkins, P. P., and Tsang, V. C. (2000) *Taenia solium*: molecular cloning and serologic evaluation of 14- and 18-kDa related, diagnostic antigens. *J. Parasitol.* **86,** 1001–1007.
61. Hancock, K., Khan, A., Williams, F. B, et al. (2003) Characterization of the 8-kilodalton antigens of Taenia solium metacestodes and evaluation of their use in an enzyme-linked immunosorbent assay for serodiagnosis. *J. Clin. Microbiol.* **41,** 2577–2586.
62. Hubert, K., Andriantsimahavandy, A., Michault, A., Frosch, M., and Muhlschlegel, F. A., et al. (1999) Serological diagnosis of human cysticercosis by use of recombinant antigens from *Taenia solium* cysticerci. *Clin. Diagn. Lab. Immunol.* **6(4),** 479–482.
63. Muhlen, C. A. and Tan, E. M. (1995) Autoantibodies in the diagnosis of systemic rheumatic diseases. *Semin. Arthritis Rheum.* **24,** 323–358.
64. Yago M., Belmonte, M. A., Olmos, M. J., Beltran, J., Teruel, C., and Segarra, M. (1999) Detecting anti-SSA and anti-SSB antibodies in routine analysis: a comparison between double immunodiffusion and immunoblotting. *Ann. Clin. Biochem.* **36,** 365–371.
65. Vitozzi, S., Lapierre, P., Djilali-Saiah, I., and Alvarez F. (2002) Autoantibody detection in type 2 autoimmune hepatitis using a chimera recombinant protein. *J. Immunol. Methods* **262,** 103–110.
66. Brichory, F. M., Misek, D. E., Yim, A-M., et al. (2001) An immune response manifested by the common occurrence of annexin I and II autoantibodies and high circulating levels of IL-6 in lung cancer. *Proc. Natl. Acad. Sci. USA* **98,** 9824–9829.
67. Lieu, T-S., Jiang, M., Steigerwald, J. C., and Tan, E. M. (1984) Identification of the SS-A/Ro intracellular antigen with autoimmune sera. *J. Immunol. Methods* **71,** 217–228.
68. Hu, G. R., Harrop, P., Warlow, R., Gacis, M. L., and Walls, R. S. (1997) High resolution autoantibody detection by optimized protein-G horseradish peroxidase immunostaining in a Western blot assay. *J Immunol Methods.* **202,** 113-121.
69. Naour, F. L, Brichory, F., Beretta, L., and Hanash, S. M. (2002) Identification of tumor-associated antigens using proteomics. *Technol. Cancer Res. Treat.* **1,** 257–262.
70. Prabhakar, S. and Bhatia, R. (2001) Diagnosis of Creutzfeld-Jacob disease. *Neurol. India* **49,** 325–328.

71. Safar, J., Wille, H., Itri, V. et al. (1998) Eight Prion strains have Pr^{Sc} molecules with different conformations. *Nature Med.* **4,** 1157–1165.
72. Wadsworth, J. D. F., Joiner, S., Hill, A. F., et al. (2001) Tissue distribution of protease resistant prion protein in variant Creutzfeld–Jacob disease using a highly sensitive immunoblotting assay. *Lancet* **358,** 171–180.
73. Wiltfang, J., Otto, M., Baxter, H. C., et al. (1999) Isoform pattern of 14-3-3 proteins in the cerebrospinal fluid of patients with Creutzfeld–Jacob disease. *J. Neurochem.* **73,** 2485–2490.

6

Principles and Medical Applications of the Polymerase Chain Reaction

Bimal D. M. Theophilus

1. Introduction

The polymerase chain reaction (PCR) is currently one of the mainstays of medical molecular biology. One of the reasons for the wide adoption of PCR is the elegant simplicity of the way in which the reaction proceeds and the relative ease of the practical manipulation steps. Indeed, combined with the relevant bioinformatics resources for the practical design and for the determination of the required experimental conditions, it provides a rapid means for DNA diagnostic identification and analysis. It has also opened up the investigation of cellular and molecular processes to those outside the field of molecular biology and also contributed in part to the successful sequencing of the human genome project. Polymerase chain reaction is an in vitro amplification method able to generate a relatively large quantity (about 10^5 copies, or approx 0.25–0.5 µg) of a specific DNA sequence from a small amount of a heterogeneous DNA target, often comprising the total cellular genome. The sensitivity of PCR is such that successful amplification can be achieved from a single cell, as in single-sperm typing and preimplantation diagnosis, or from a minority DNA population that is present among an excess of background DNA (e.g.. from viruses infecting only a few cells, or from low levels of "leaky" RNA transcription from nonexpressing tissues). The PCR process consists of incubating a reaction sample containing the DNA substrate and required reactants repeatedly between three different temperature incubations: denaturation, annealing, and extension. Many current investigators, the author included, recall their initial introduction to PCR as comprising the laborious transfer of sample tubes between three water baths heated to different temperatures representing the three incubations. Fortunately and no doubt instrumental to the wide implementation of PCR in numerous areas of fundamental research and applications is the automation of the process. This was the result of the development of programmable thermal cylcers that were developed only a few years subsequent to the invention of PCR by Mullis in 1983. Today, the technology has advanced to modern thermal cyclers that use Peltier heating and cooling elements to produce fast temperature changes or ramp rates of around 3°C/s and that maintain accurate temperature control across the entire sample block. In addition PCR undertaken in 96-well microtiter plate systems is now possible, which is useful for high-throughput analysis of clinical samples.

2. Elements of PCR

2.1. PCR Practical Procedures

Polymerase chain reaction achieves near-exponential amplification of a DNA sequence, the length of which is defined by a pair of oligonucleotide primers, complementary to 20–25 bp of sequences at the 5' and 3' ends of the target molecule, respectively. It is relatively straightfor-

From: *Medical Biomethods Handbook*
Edited by: J. M. Walker and R. Rapley © Humana Press, Inc., Totowa, NJ

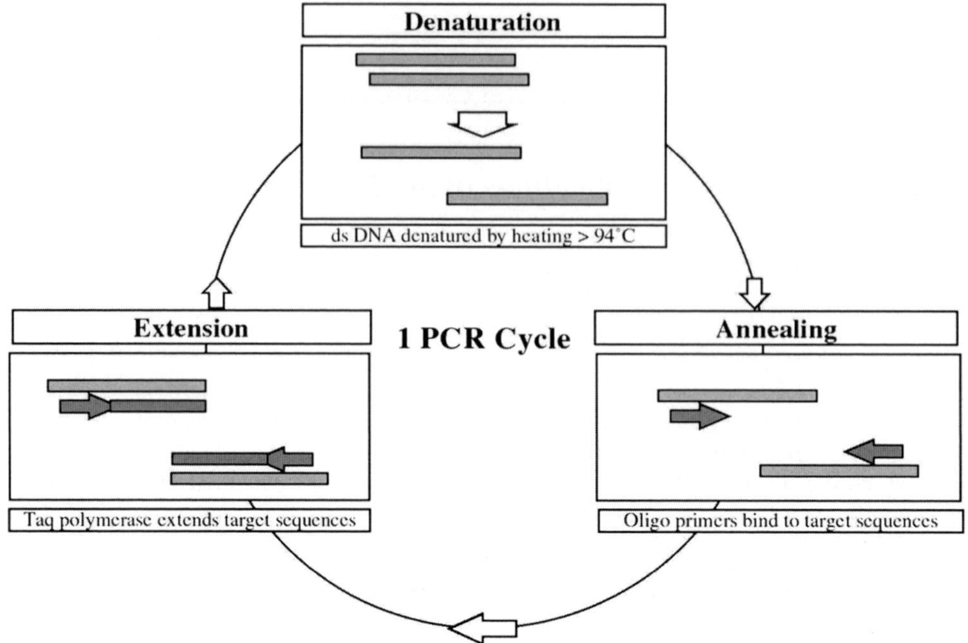

Fig. 1. Simplified scheme of one PCR cycle that involves denaturation, annealing and extension.

ward for the PCR to amplify up to approx 2 kb of sequences, but special modifications such as the use of a cocktail of enzymes are required to amplify larger sequences of up to 40 kb in a process termed "long-range PCR." Typical sources of target DNA include blood, mouthwashes or buccal scrapes, chorionic villus (for antenatal diagnosis), one to two cells from an eight-cell embryo (for preimplantation diagnosis), hair, and Guthrie spots. Extracted DNA does not need to be particularly pure and could even be partially fragmented, as in the case of archival material such as that derived from paraffin-embedded tissues. In the case of blood samples, simple boiling releases enough DNA for successful amplification. A PCR reaction mix is usually 5–50 µL in volume and is set up in 0.2-mL or 0.5-mL thin-walled reaction tubes or, as indicated, in 96-well microtiter plates. It comprises reaction buffer (optimized for magnesium chloride), deoxynucleotide triphospates (dNTPs), the oligonucleotide primers, a thermostable DNA polymerase (usually *Taq* DNA polymerase), which synthesizes DNA by incorporating dNTPs and extending the annealed primers, and, finally, the template to be amplified.

2.2. Steps Involved in a PCR Cycle

The initial step of PCR involves incubation of around 94°C for 5 min to denature the target genomic DNA into single strands. This step is not usually necessary when using cDNA derived from RNA as the template. The subsequent and main part of the process comprises the sequential incubation of the sample at three different temperatures, which together constitute one PCR cycle (*see* Fig. 1). The first step in a cycle involves incubation at 94–96°C for 10 s to 1 min to denature newly synthesized template DNA. Subsequent incubation is at a temperature determined by the melting properties of the primers, ideally around 56°C, optimized to allow them to anneal only to their specific target sequences. The third incubation is usually 72°C for 20 s to 2 min, during which *Taq* DNA polymerase extends the primers mediating the synthesis of DNA using the target sequence as the template. Normally, 30–35 such cycles are performed to comprise a PCR reaction, which typically takes about 3 h to complete. In the first cycle of PCR, primers can only anneal to the original target DNA. In this case, the primer defines the 5' end of each newly synthesized strand, but the 3' end is undefined and variable. However, in subsequent cycles, as newly synthesized DNA from one cycle becomes a template in the next cycle, the synthesized DNA will be defined at both ends by each primer. Because the synthesis of the latter species increases exponentially while DNA synthesized from the original target only increases linearly, the predominant product (termed *amplicon*) resulting from PCR will be defined by the location of the forward and reverse primers (*see* Fig. 2). Whereas amplification is exponential in the earlier cycles, it plateaus out in later cycles because of the exhaustion of reaction components. For this reason, quantitative comparisons or estimates of PCR products should be based on the exponential phase of the reaction using "real time PCR" (see Chapters 23 and 25). It is safe to leave the finished reaction in the thermal cycler at room temperature for several hours or overnight, although, typically, a final incubation "hold" temperature of 4–12°C is programmed after the final PCR cycle.

2.3. Primer Design and PCR

The specificity of PCR lies in the design of the two oligonucleotide primers. These not only have to be complementary to sequences flanking the target DNA but must not be self-complementary or bind each other to form dimers because both prevent DNA amplification (*see* Fig. 3). They also have to be matched in their GC content and have similar annealing temperatures. The increasing use of bioinformatics resources such as Oligo, Generunner, and Genefisher in the design of primers makes the design and the selection of reaction conditions much more straightforward. These resources allow the sequences to be amplified and the primer length, product size, GC content, and so forth, to be input, and, following analysis, they provide a choice of matched primer sequences. Indeed, the initial selection and design of primers without the aid of bioinformatics would now be unnecessarily time-consuming. With careful consideration of primer sequences and reaction conditions, it is possible to amplify more than one region in a single PCR. This process is termed *multiplex PCR* and is used extensively in molecular-based diagnostics, although it does require a degree of optimization.

2.4. Sensitivity and Contamination in PCR

The enormous sensitivity of the PCR is also one of its main drawbacks because the very large degree of amplification makes the reaction vulnerable to contamination. Even a trace of foreign DNA, such as that contained in dust particles, may be amplified to significant levels and may give false-positive results. Hence, cleanliness is paramount when carrying out PCR, and dedicated equipment and, in some cases, laboratory areas and even laboratories are used. It is possible that previously amplified products may also contaminate PCR. However, this may be overcome by a number of methods including ultraviolet (UV) irradiation to damage the already amplified products so that they cannot be used as templates. A further interesting solution is to incorporate uracil into PCR and then treat the products with the enzyme uracil-N-glycosylase (UNG), which degrades any PCR amplicons with incorporated uracil, rendering them useless as templates. In addition, most PCRs are now undertaken using hotstart. Here, the reaction mixture is physically separated from the template or the enzyme. When the reaction begins, mixing occurs and thus avoids any mispriming that might have arisen.

3. Medical Applications and PCR

The main areas of medical diagnosis to which PCR is applied are the detection of infectious pathogens (e.g., associated with sexually transmitted diseases or respiratory tract infections) and the identification of mutations in genes that are either responsible or constitute risk factors for human disease. In some tests, the result is simply found in the presence, size, or color of the PCR product, whereas in other situations, PCR is used to generate sufficient specific starting material on which to perform a number of post-PCR manipulations to obtain the result.

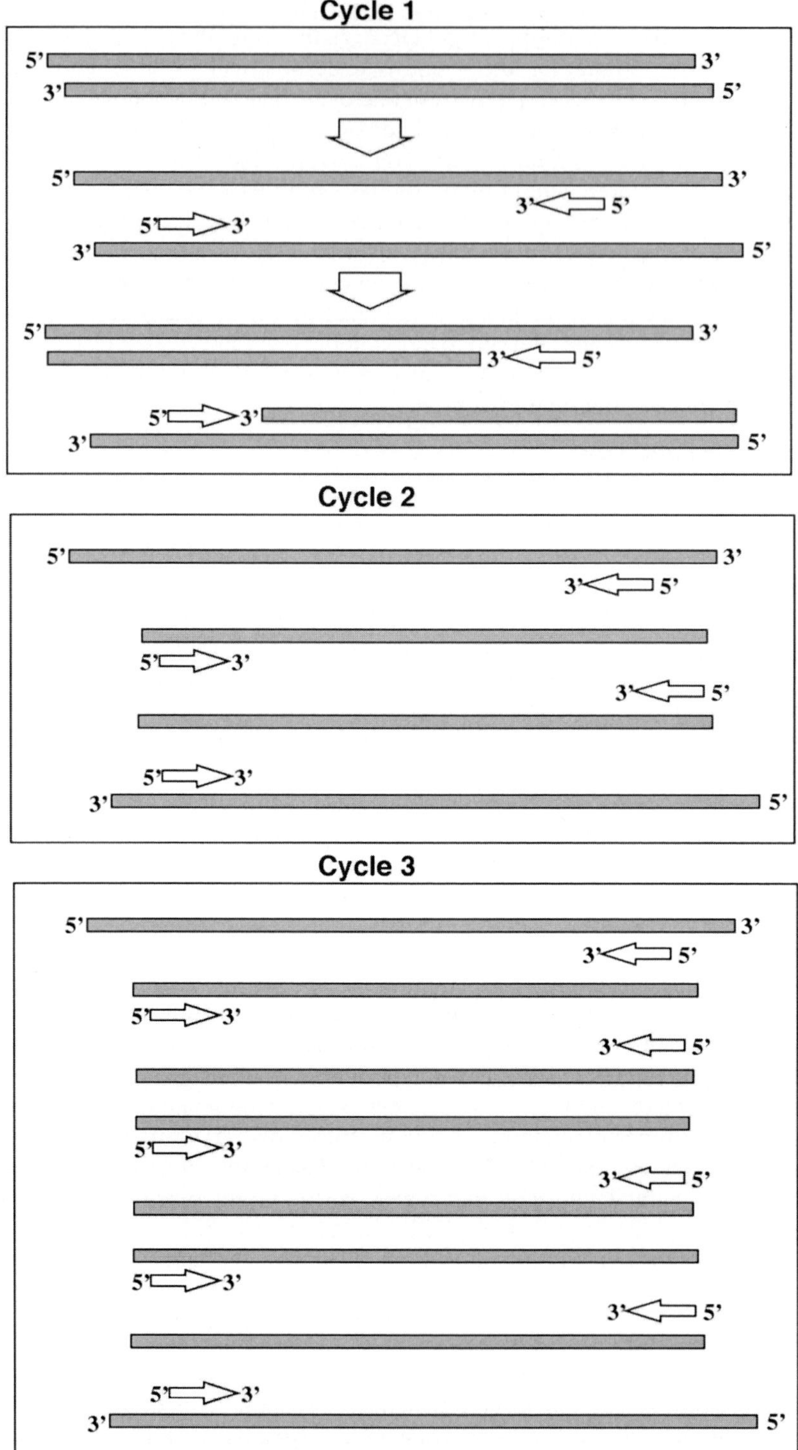

Fig. 2. Representation of further cycles in the PCR. Note the original template strands can also be copied, whereas newly synthesised PCR products are amplified exponentially.

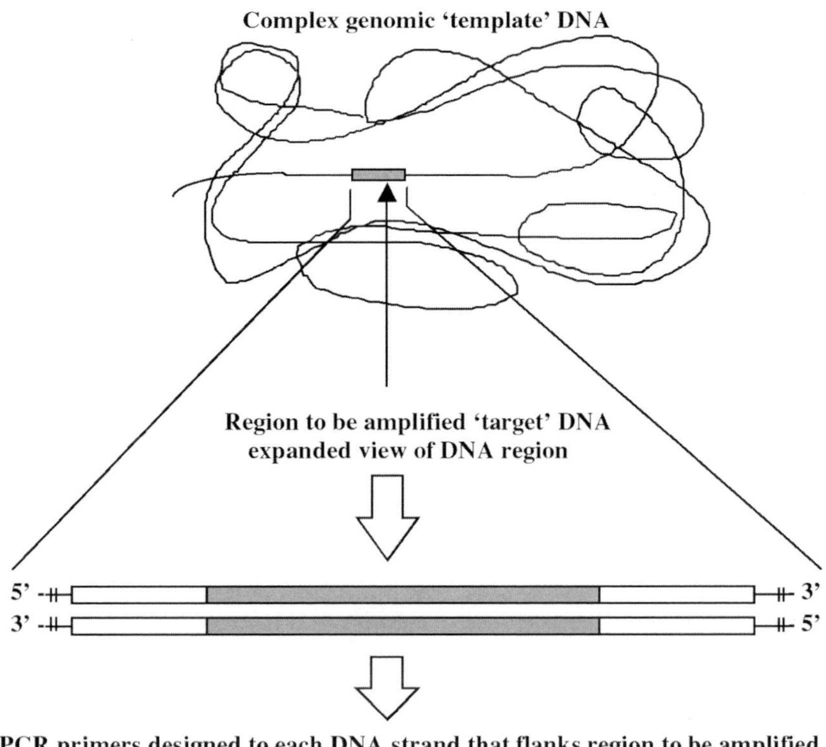

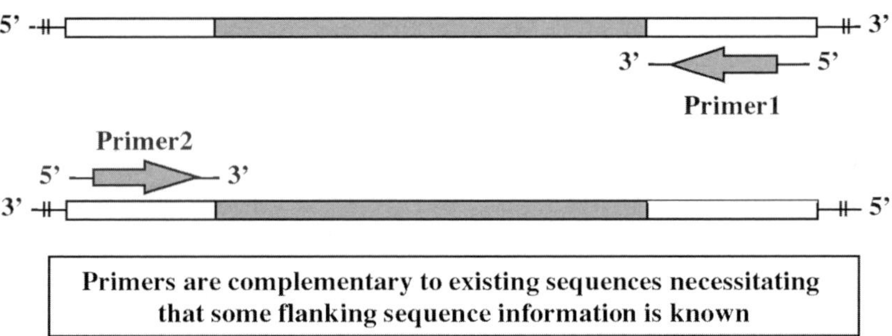

Fig. 3. The location of PCR primers. PCR primers designed to sequences adjacent to the region to be amplified, allowing a region of DNA (e.g., a gene) to be amplified from a complex starting material of genomic template DNA

3.1. Polymorphism Detection: RFLPs, VNTRs, and STRs

Before specific genes and associated mutations were fully characterized, genetic diagnosis was often carried out using restriction fragment length polymorphisms (RFLPs) (see Chapter 3). RFLPs are variations in the length of restriction fragments between alleles arising from the presence or absence of specific restriction enzyme sites. RFLPs within or closely linked to an affected gene were therefore used as markers to track inheritance of a faulty gene through a family pedigree. VNTRs (variable number of tandem repeat sequences) and STRs (short tandem repeats) are polymorphic sequences applied in a manner similar to RFLPs, but arise from variations in the number of tandem repeat sequences at a given locus between different alleles. They are performed by PCR using primers flanking the polymorphic region, incubation with a restriction enzyme (in the case of RFLPs), and analysis of the fragment sizes following gel electrophoresis. Subsequent research and the completion of the human genome project means that the underlying basis of most common single-gene disorders is now understood and so these methodologies are less widely used. However, they are still of use where the causative mutation in a particular family or disease remains elusive or where limited facilities or expertise prevents complete screening of a gene to identify a mutation.

3.2. PCR-Based Mutation Screening

One of the major clinical applications of PCR that has spawned numerous diagnostic laboratories is the screening of multiexon genes, without clearly defined founder mutations, and where, therefore, a significant number of heterogeneous mutations may be responsible for the disease phenotype in different families or populations (e.g., hemophilia, Gaucher disease, certain breast and ovarian cancers [which arise from mutations in the *BRCA1* and *BRCA2* genes], and colon cancer. Unfortunately, there is no single, quick, inexpensive, and reliable mutation screening method of universal application. The majority of techniques can only analyze 200–500 nucleotides at a time for maximum sensitivity, which corresponds to the size of most exons. Individual exons and their intron/exon boundaries (which may be the site of exon splice site mutations) are amplified using PCR primers complementary to flanking intronic sequence. Large exons are usually amplified as several overlapping segments.

3.2.1. Reverse Transcriptase PCR

A number of screening methods are able to analyze 1 kb or more at a time, in which case, PCR-amplified complementary or cDNA may be used. This is DNA synthesized using messenger RNA (mRNA) as a template mediated by the enzyme reverse transcriptase (*see* Fig. 4). This procedure, referred to as RT-PCR, enables the screening of multiple contiguous exons in a single analysis, which enables the identification of gene rearrangements and splicing defects that may remain undetected by analysis of the genomic DNA coding sequence. However, RNA has a short half-life and is easily degraded; therefore, great care must be exercised in handling it. In addition, there is evidence that mutation-containing RNA may be less stable than normal RNA, therefore, the mutant allele may be underrepresented, or absent, in the RT-PCR product from a heterozygous sample. Finally, the gene under analysis may not be expressed in tissues that are readily accessible for analysis (e.g., blood), although sufficient "ectopic" RNA may be expressed by the so-called "leaky" transcription for successful PCR and analysis. Although mutation screening and detection methods are able to identify sequence differences between a normal and a patient sample, they do not indicate whether the change is pathogenic or simply a sequence polymorphism. This distinction usually depends on the nature of the change, the location of the change with regard to functional domains of the encoded protein, conservation of the affected amino acid residue among homologous proteins, and whether the change has been observed previously (by reference to mutation databases). Many mutation-detection methods depend on differences in the melting properties of DNA during gel electrophoresis. Single-strand conformation polymorphism (SSCP) analysis ***(1)*** is based on differences in the electrophoretic mobility of single-stranded DNA conformers in a nondenaturing gel system. It is a popular choice because it is simple to perform, but it is only able to analyze PCR products up to about 200 bp and it has limited sensitivity. Conformation-sensitive gel electrophoresis (CSGE) (see Chapter 13) ***(2)*** is based on the electrophoretic mobility of DNA heteroduplexes under mildly denaturing conditions. It is only slightly more demanding to perform than SSCP, but has close to 100% detection efficiency. Denaturing gradient gel electrophoresis (DGGE) ***(3)*** identifies mismatch-containing heteroduplexes based on an abnormal denaturing profile on electrophoresis through a gradient of increasing denaturant concentration. Denaturing high-performance liquid chromatography (dHPLC) is a recently introduced method that is rapidly gaining in popularity. dHPLC distinguishes heteroduplexes from homoduplexes by ion-pair reverse-phase liquid chromatography according to differences in their melting behavior ***(4)*** (see Chapter 24). Positive ions in a buffer coat DNA in a hydrophobic layer, which interacts with a hydrophobic polystyrene matrix in a length- and sequence-specific manner. DNA is eluted from the matrix by a linear acetoni-

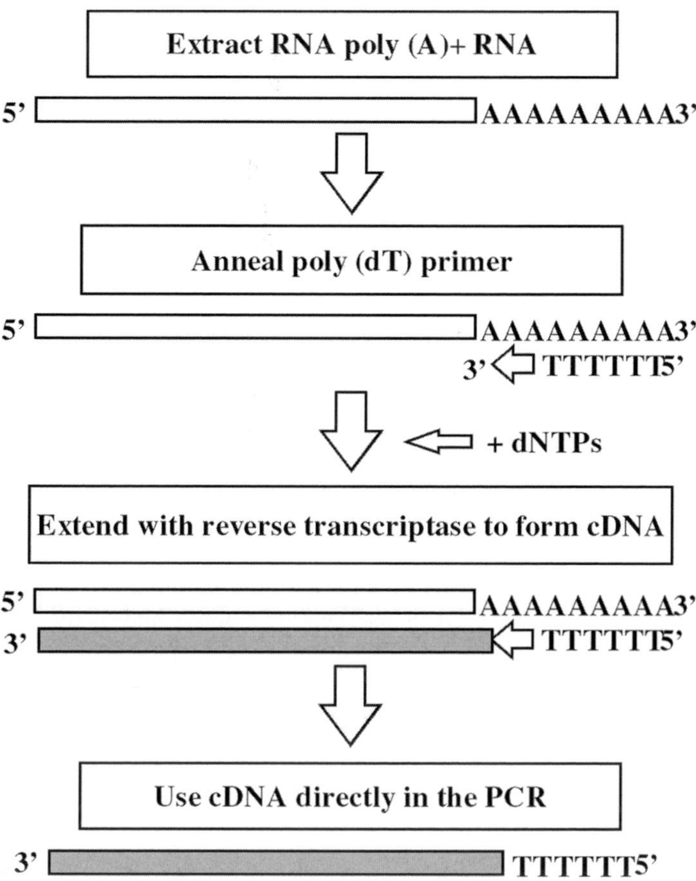

Fig. 4. Reverse-transcriptase PCR (RT-PCR). In RT-PCR, mRNA is converted to complementary DNA (cDNA) using the enzyme reverse transcriptase. The cDNA is then used directly in PCR.

trile gradient. Although preliminary work is required to optimize the conditions for each PCR product, once established it is a rapid, highly automated, and sensitive technique able to detect close to 100% of mutations.

3.2.2. PCR-Based Mismatch Detection

Some other methods are based on chemical or enzymatic cleavage of mismatches in DNA heteroduplexes ***(5,6)***. The protein truncation test (PTT) ***(7)*** (see Chapter 17) identifies frame shift, splice site, and nonsense mutations by virtue of their ability to result in premature protein truncation, and it is applied to diseases where these types of mutation predominate, such as adenomatous polyposis coli (APC). Mutation-screening methods merely indicate that a mutation is present, which must then be characterized by DNA sequencing. With the development of automated fluorescent sequencers, many laboratories now sequence the complete coding sequence of a gene in order to identify mutations without the prior application of a mutation-screening method. Although the most laborious part of this process is often visual analysis of the sequence generated, there has been significant progress in the development of sequence analysis software, which is able to highlight sequence differences between two aligned sequences. However, these programs are not perfect, and it is often necessary to check sequences manually, especially in the case of heterozygous mutations. Also, although often considered the gold standard method of mutation detection, sequencing technology itself is not 100% accurate and may miss mutations or produce artifacts that must be investigated further.

3.3. PCR-Based Mutation Detection

Some conditions arise from one or a few clearly defined mutations. In these cases, one of a number of methods can be applied that detect the presence or absence of a defined sequence change. These are generally technically simpler and more rapid tests than those used for mutation screening described in **Subheading 3.2.**

3.3.1. Restriction Enzyme Analysis

If, fortuitously, a mutation creates or destroys a restriction site, then it may be identified simply by restriction enzyme analysis as described for RFLPs. For example, PCR followed by digestion with the restriction enzyme *Mnl*I is used to detect factor V Leiden, which arises from a point mutation in exon 13 of the factor V gene and constitutes the most frequent genetic risk factor for venous thrombosis. Alternatively, for mutations that do not alter a site for a restriction enzyme, a diagnostic site may be "engineered" into a PCR product using mutagenic primers.

3.3.2. Deletion Analysis

The most common cystic fibrosis (CFTR) allele is a three-nucleotide deletion, F508del. Simply amplifying the region by PCR using flanking primers and distinguishing the normal and mutant alleles by size on gel electrophoresis can identify this allele. Similarly, 60–65% of mutations causing Duchenne muscular dystrophy (DMD) are deletions of one or more exons within the dystrophin gene, 98% of which can be identified by two multiplex PCR reactions.

3.3.3. Analysis of Gene Rearrangements

Gene rearrangements are a feature of hematological malignancies, as well as certain nonmalignant diseases, such as the intron 1 and intron 22 inversion mutations responsible for approx 50% of severe hemophilia A. In the case of hematological malignancies, the novel chimeric transcripts arising from rearrangements are identified by RT-PCR *(8)*. In hemophilia A, the intron 1 inversion mutation is detected by a standard PCR reaction, whereas the intron 22 inversion requires long-range PCR or Southern blotting for identification (see Chapter 14).

3.3.4. Allele-Specific PCR

Allele-specific oligo PCR (ASO-PCR, also known as the amplification refractory mutation system, ARMS) involves the hybridization of three primers in a single reaction (see Chapter 14). These comprise normal and mutant forward primers in which the final (3') base of the primer is homologous to either the normal or mutant sequences, respectively, together with a common reverse primer. The primer annealing temperature is optimized to ensure that primers only anneal to a perfectly matched template sequence, which is a requirement for subsequent extension by *Taq* DNA polymerase. Separate PCR reactions may be performed with the normal and mutant primers, respectively or a single reaction may be performed if the normal and mutant primers are distinguishable (e.g., with different fluorescent labels). Clearly, in order to design specific primers, the nature of the mutation to be detected must be known. An example of its use is to screen for a G to A transition at position 20210 in the 3' untranslated region of the prothrombin gene *(9)*. The A allele is associated with elevated plasma prothrombin levels and carriers have a 2.8-fold increased risk of venous thrombosis. In addition, a multiplex ARMS test has been designed and marketed in kit form to test for a panel of common cystic fibrosis mutations.

3.3.5. 5'Nuclease/TaqMan™ Assay

The 5' Nuclease or TaqMan™ assay also involves the use of three primers: two allele-specific forward primers and a common reverse primer, which in this case is labeled with a "reporter" fluorophore at its 5' end and a "quencher" fluorophore at its 3' end (see Chapter 23). As with the allele-specific PCR, only the perfectly matched primer is extended by *Taq* DNA polymerase. On encountering the labeled reverse primer, the latter will be degraded by the 5'→3' exonuclease activity of *Taq* DNA polymerase. This will result in the separation of the reporter and quencher fluorophores and a consequent increase in fluorescence. TaqMan is used for mutation detection and for screening specific subtypes of microbial pathogens. It is also used for association studies, which involve the correlation of single-nucleotide polymorphisms (SNPs) with complex disorders such as diabetes, heart disease, cancer, and mental disorders ***(10)***. SNPs are also of interest in the rapidly emerging field of pharmacogenetics, where they are analyzed for their role in determining variations between individuals in their responses to specific drugs or drug toxicity (e.g., in the cytochrome P450 genes) ***(11)***. It is anticipated that this will lead to the identification of novel drug targets and aid in the production of individually tailored "designer" drugs.

3.3.6. Quantitative and Real-Time PCR

Real-time PCR (see Chapter 23) requires the use of special DNA cyclers such as the Roche Lightcycler that couples PCR with fluorescent detection, thereby enabling the detection of PCR product as it is synthesized. In its simplist form, a DNA-binding dye such as SYBR green is included in the reaction. As amplicons accumulate, SYBR green binds the dsDNA. This binding is proportional although nonspecific and fluorescence emission is detected following excitation. Diagnostically, real-time PCR enables the determination of a viral load in infectious diseases by comparing the amount of virus-specific product to a standard curve generated from samples containing known concentrations of DNA. Real-time PCR also has many research applications such as comparisons of gene expression between different tissues or at different stages of development. Although relatively expensive in comparison to other methods for determining expression levels, it is simple, rapid, and reliable. In addition to the quantitative nature, real-time PCR systems may also be used for genotyping and for accurate determination of the amplicon melting temperature using a so-called melting curve analysis. PCR has also been extended to further developments such as sequencing by minisequencing or pyrosequencing *(12,13)*. In addition, the new technology of microarrays uses, in some cases, PCR-derived material *(14)*. It is this area that may form the future of rapid nucleic acid diagnostic; however, PCR is still and no doubt will continue to be a mainstay of genetic-based diagnostics.

References

1. Orita, M., Iwahana, H., Kanazawa, H., Hayashi, K., and Sekiya, T. (1989) Detection of polymorphisms of human DNA by gel electrophoresis as single-strand conformation polymorphisms. *Proc. Natl. Acad. Sci. USA* **86,** 2766–2770.
2. Williams, I. J. and Goodeve, A. C. (2002) Conformation-sensitive gel electrophoresis, in PCR Mutation Detection Protocols (Theophilus, B. D. M. and Rapley, R., eds.), Humana, Totowa, NJ
3. Fischer, S. G. and Lerman, L. S. (1983) DNA fragments differing by a single base-pair substitution are separated in denaturing gradient gels: correspondence with melting theory. *Proc. Natl. Acad. Sci USA* **80,** 1579–1583.
4. Huber, C. G., Oefner, P. J., and Bonn, G. K. (1995) Rapid and accurate sizing of DNA fragments by ion-pair chromatography on alkylated nonporous poly(styrenedivinylbenzene) particles. *Anal. Chem.* **67,** 578–585.
5. Cotton, R. G. H., Rodrigues, N. R., and Campbell, R. D. (1988) Reactivity of cytosine and thymine in single base-pair mismatches with hydroxylamine and osmium tetroxide and its application to the study of mutations. *Proc. Natl. Acad. Sci. USA* **85,** 4397–4401.
6. Heisler, L. and Lee, C-H. (2002) Cleavase® fragment length polymorphism analysis for genotyping and mutation detection, in PCR Mutation Detection Protocols (Theophilus, B. D. M. and Rapley, R., eds.), Humana, Totowa, NJ.
7. Roest, P. A. M., Roberts, R. G., Sugino, S., van Ommen, G-J. B., and den Dunnen, J. T. (1993) Protein truncation test (PTT) for rapid detection of translation-terminating mutations. *Hum. Mol. Genet.* **2,** 1719–1721.

8. Cotter, F. E. (1996) Molecular Diagnosis of Cancer, Humana, Totowa, NJ.
9. Poort, S. R., Bertina, R. M., and Vos, H. L. (1997) Rapid detection of the prothrombin 20210 variation by allele specific PCR. Thromb. *Haemost.* **78,** 11,157–11,163.
10. Chakravarti, A. (1999) Population genetics—making sense out of sequence. *Nature Genet.* **21,** 56–60.
11. McCarthy, J. J. and Hilfiker, R. (2000) The use of single-nucleotide polymorphism maps in pharmacogenomics. *Nature Biotechnol.* **18,** 505–508.
12. Hacia, J. G. (1999) Resequencing and mutational analysis using oligonucleotide microarrays. *Nature Genet.* **21,** 42–47.
13. Syvanen, A. C. (1999) From gels to chips: "minisequencing" primer extension for analysis of point mutations and single nucleotide polymorphisms. *Hum. Mutat.* **13,** 1–10.
14. Ronaghi, M., Uhlen, M., and Nyren, P. (1998) A sequencing method based on real-time pyrophosphate. *Science* **281,** 363–365.

7

Single-Strand Conformation Polymorphism (SSCP) Analysis

Igor Vorechovsky

1. Introduction

The identification of a large number of disease genes in recent years has led to a considerable improvement in clinical diagnostic procedures, therapeutic interventions, and prognostic projections and provided carrier or presymptomatic testing to family members of affected individuals. Because the number of gene alterations known to be linked to genetic disorders has risen dramatically over the last decade, the availability of technically simple, cost-effective, and reliable methods to detect changes in the nucleotide sequence has become increasingly important. Although we have recently seen a considerable improvement in our ability to detect DNA alterations, costly mutation analysis using nucleotide sequencing has driven a search for less expensive scanning methods.

In addition to the costs, the choice of a suitable method of mutation analysis is governed not only by the experimental sensitivity, expected mutation pattern in the target sequence, and functional consequences of these changes, but also by staff expertise, personal experience, and preference. However, the primary selection criterion for such a method is the ability of a technique to detect the presence of unknown mutations in the analyzed regions ("scanning methods") as opposed to identifying known mutations already characterized at the nucleotide level. Scanning procedures represent a cost-effective alternative to nucleotide sequencing, but usually at a price of an inferior detection rate. The former group of techniques includes procedures based on conformation polymorphism changes, denaturing gradient gel electrophoresis, constant denaturant capillary electrophoresis, mismatch repair, and RNAse cleavage methods. The latter group, exemplified by allele-specific hybridization or amplification, primer extension, oligonucleotide ligation assay, and ligase chain reaction is specific for previously identified mutations/polymorphisms or a set of nucleotide alterations.

2. Single-Strand Conformation Polymorphism Analysis

Single-strand conformation polymorphism (SSCP) analysis ***(1,2)*** is one of the simplest scanning techniques for detecting unknown mutations. Variation in a DNA sequence is detected by alterations in the conformation of denatured DNA fragments. Denatured DNA fragments are allowed to renature under conditions that prevent the formation of double-stranded DNA and allow higher-order structures to form in single-stranded fragments. These fragments are then run through nondenaturing polyacrylamide gels to detect variations in these structures that are manifested as mobility shifts ***(3)*** (*see* **Fig. 1**). The term PCR-SSCP refers to a polymerase chain reaction (PCR)-amplified product analyzed by SSCP.

The SSCP analysis is useful for detecting microlesions, such as single-base substitutions, small deletions, small insertions, or microinversions. These changes constitute the majority of

From: *Medical Biomethods Handbook*
Edited by: J. M. Walker and R. Rapley © Humana Press, Inc., Totowa, NJ

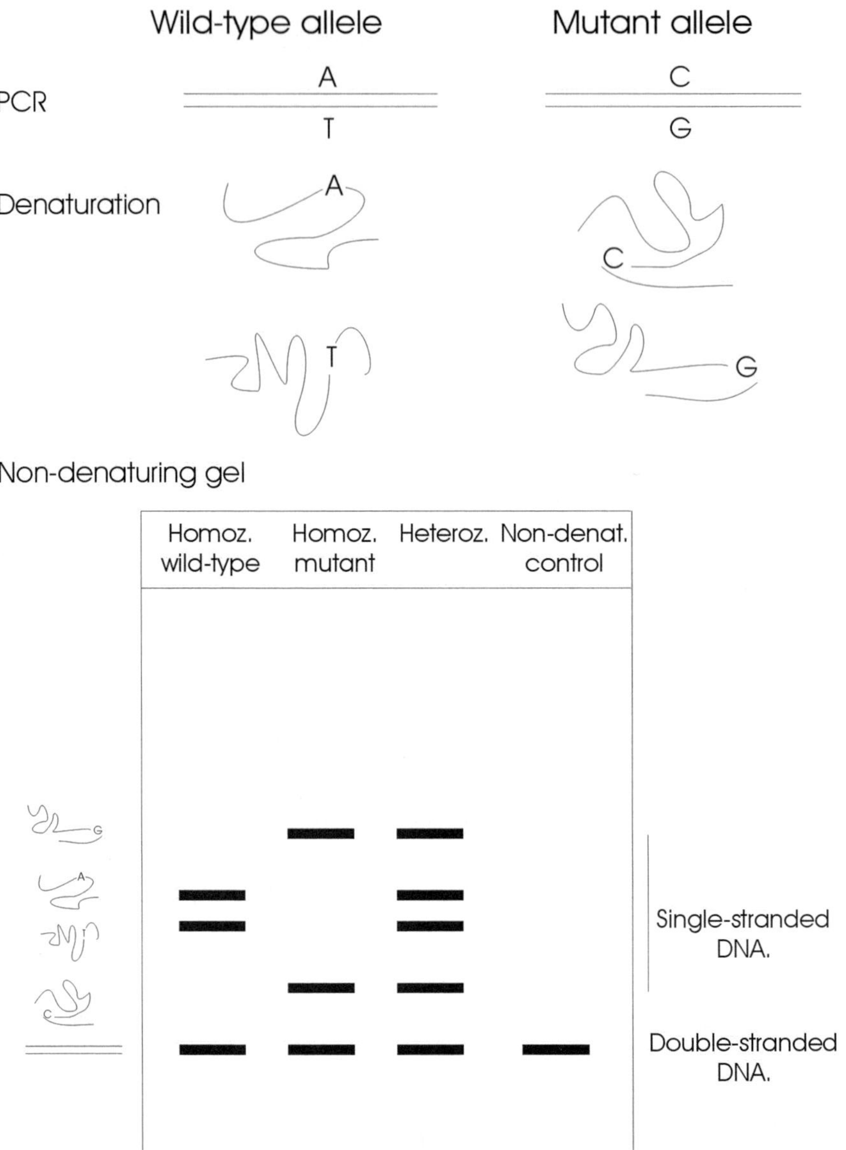

Fig. 1. Single-strand conformation polymorphism analysis.

disease-causing mutations, and this has undoubtedly contributed to the popularity of SSCP analysis among the mutation-screening methods. SSCP requires no special equipment, is simple to use, and mutant DNA fragments can be isolated for further analysis. However, the efficient mutation detection is limited to shorter fragments, and some mutations, such as large, heterozygous insertions or deletions encompassing the amplified region, are likely to go undetected.

3. Sensitivity of PCR-SSCP

Unlike denaturing gradient gel electrophoresis, there is no adequate theoretical model for predicting the three-dimensional structure of single-stranded DNA under a given set of conditions *(3)*. Is it still possible to determine how sensitive SSCP analysis is for a sequence of interest?

As a rule of thumb, the detection rate of microlesions should be more than 85% for fragments shorter than 300 bp using a single SSCP run ***(4)***. The sensitivity of PCR-SSCP depends on the following factors:

1. The mutation pattern in the target sequence.
2. Size of DNA fragments and its GC content.
3. Gel temperature.
4. Gel matrix composition.
5. Buffer composition (ionic strength and pH).
6. DNA concentration.

Disease genes containing small insertions/deletions and single-base substitutions perform generally better in SSCP analysis than those with a mutation pattern dominated by large deletions ***(5)***. The examination of β-globin, *TP53*, and rhodopsin mutations suggested that the type of mutation (transition vs transversion) did not play a major role in SSCP sensitivity ***(6)***. Although the position/type of base substitution does not seem to significantly contribute to the detection rate either ***(6,7)***, the sequence flanking mutated residues is likely to influence sensitivity ***(8)***. In tumor samples, SSCP is able to identify mutations against the background of 85–95% of the wild-type allele ***(9)***, a figure sufficient to identify clonal changes in solid tumors that contain a substantial proportion of normal cells.

Single-strand conformation polymorphism sensitivity was found to vary dramatically with the size of the DNA fragments analyzed, with the optimal size at about 150 bp and a substantial decline of sensitivity for fragments larger than 300 bp ***(4,6,7,10)***. However, the extent of such a decrease of the detection rate has been difficult to measure, because sensitivity is influenced by multiple factors. A blind quality control study of fragments larger than 450 bp reported a detection rate of 84% ***(10)***, and an even higher proportion of mutations was found by running amplicons sized between 300 and 800 bp in low-pH buffer systems ***(11)***. Because a single base change contributes less to the tertiary structure in a larger fragment than in a small one, the probability of detecting a base substitution in amplicons larger than 300 bp is generally lower than for optimally sized fragments. A restriction endonuclease digestion may help achieve a suitable fragment size if the initial amplicon is too large, but this extra step may be costly when analyzing multiple samples.

The sensitivity of SSCP detection was found to be greater in templates with high GC content, which might be the result of an intricately folded strand resulting from more hydrogen bonds in such DNA ***(12)***. The ratio cytosine/adenosine was reported to correlate with the optimal electrophoretic temperature ***(13)***, with a suggested formula for optimal gel temperature ($[80°C/(A+1)]/\{k+[C/(A+1)]\}$), in which the coefficient k may need to be experimentally derived. The use of a temperature gradient in the 15–25°C interval may favor conformational transitions in mutated samples and increase the detection rate ***(14)***. Temperature should be controlled during electrophoresis, as an excessive current may generate heat, resulting in the disappearance of mobility shifts ***(8)***. Multiple gels run at different temperatures are likely to increase sensitivity, and room and cold temperatures around 5°C have been recommended most commonly ***(4,15)***.

Optimized polyacrylamide concentrations and crosslinking should be used for SSCP gels ***(4,8)***. The most suitable matrix appears to be a crosslinked acrylamide polymer with low crosslinking values (%C, mass of the cross-linkers/mass of all monomers and crosslinkers per 100 mL volume), ranging between 1.3% and 2.6% ***(8)***. Adding glycerol to SSCP gels was found to improve the detection rate, probably by reducing the pH of the Tris-borate buffer through the reaction of glycerol and the borate ion or weakening electrostatic repulsion between the negatively charged phosphates in the DNA backbone, leading to a greater stabilization of the tertiary structure ***(11)***. The sensitivity of SSCP analysis might also be increased using low-viscosity hydroxyethylcellulose and neutral pH of the running buffer ***(14)***.

As the higher-order structure of nucleic acids depends on the entire sequence of the amplified DNA, the sensitivity of SSCP analysis is likely to vary from one fragment to another.

The finding that a shift of a mutated fragment is detected only under certain physical conditions has led to the use of multiple gels with varying running parameters to increase sensitivity ***(4,16)***. It is also convenient to combine SSCP with other methods of mutation detection, particularly heteroduplex analysis (HA) ***(17)***. HA detects mutations by virtue of the altered electrophoretic mobility of a DNA fragment containing one or more mismatched bases (heteroduplex) vs a fragment of identical sequence consisting of two complementary DNA strands (homoduplex). Not only do both techniques require similar equipment and gels, but, unlike for SSCP, the nature of the mismatched basepairs appears to be the overriding determinant for the ability to detect mutations, with the magnitude of separation G:G/C:C > A:C/T:G = A:G/T:C > A:A/T:T ***(12)***.

In practice, the most commonly altered variables are polyacrylamide and buffer concentrations, use of an alternative gel matrix such as the MDE gels, gel temperature (5–25°C), and glycerol concentration (0–10%) in the gel matrix. All of these variables are likely to interact to determine the final detection rate and should be controlled ***(12)***. Excessive amounts of loaded DNA (e.g., because of insufficient labeling of fragments), inappropriate pH of the gel, a high voltage leading to gel overheating, and PCR products with spurious bands are among the most common mistakes of inexperienced users.

Polymerase chain reaction may involve the incorporation of a label, either directly into the product or via an isotopically or nonisotopically labeled oligonucleotide primer. Alternatively, DNA fragments can be visualized after electrophoresis (e.g., by silver staining). PCR should be tested using an agarose gel before running SSCP gels to see if the product is clean. PCR products with spurious bands, particularly those larger than the amplicon, may interfere with the gel interpretation and should be avoided. A meticulous computer-assisted oligonucleotide primer design (http://www.hgmp.mrc.ac.uk/GenomeWeb/nuc-primer.html) for PCR will pay off.

Loading different amounts of PCR products may result in altered mobilities of DNA fragments and this may lead to difficulties in evaluating SSCP patterns. Differential signal intensities from individual samples are generally associated with a suboptimal detection rate and every effort should be made to normalize the well-to-well signal. If the efficiency of PCR is markedly different from sample to sample (e.g., because of the presence of enzyme inhibitors following DNA extraction from some paraffin-embedded tissues), the initial amount of the template can be further adjusted.

The absence of any shifts does not automatically translate into the absence of mutations/polymorphisms in the analyzed sample. Because SSCP does not have a very high exclusion power, negative results of PCR-SSCP mutation screening need cautious interpretation.

4. Clinical Significance of SSCP

The PCR-SSCP is a rapid, simple, and cost-effective scanning method for mutation detection. It is particularly useful for the initial screening of optimally sized PCR-amplified fragments for point mutations, small deletions, and insertions. It serves well as an inexpensive method of choice in situations in which the detection rate is not required to be absolute and in laboratories with no special equipment for identifying nucleotide alterations. SSCP has been used to detect mutations in DNA or reverse-transcribed RNA (complementary DNA) samples in a large number of human disease genes as well as in tumor-derived material. The method can be applied to both Mendelian and complex diseases and is suitable for use in prenatal diagnosis. It has been successfully used for genotyping not only human samples but also microbial and parasitic pathogens. In combination with other techniques for mutation detection discussed in this book, SSCP provides a sensitive and specific tool for testing any nucleotide sequence for mutations and polymorphisms. The wise choice of the most appropriate procedures is a crucial step for the successful identification of molecular changes underlying genetic disorders, carrier status, pathogen genotypes and susceptibility to complex traits including cancer.

Acknowledgment

This work was supported by the grant from the EC program (QLG2-1999-00876).

References

1. Orita, M., Suzuki, Y., Sekiya, Y., and Hayashi, K. (1989) Rapid and sensitive detection of point mutations and DNA polymorphisms using the polymerase chain reaction. *Genomics* **5**, 874–879.
2. Orita, M., Iwahana, H., Kanazawa, H., and Sekiya, T. (1989) Detection of polymorphism of human DNA by gel electrophoresis as single strand conformation polymorphism. *Proc. Natl. Acad. Sci. USA* **86**, 2766–2770.
3. Hayashi, K. (1996) in *Laboratory Protocols for Mutation Detection* (Landegren, U., ed.), Oxford University Press, New York, pp. 14–22.
4. Hayashi, K. and Yandell, D. W. (1993) How sensitive is PCR-SSCP? *Hum. Mutat.* **2**, 338–346.
5. Vorechovsky, I., (2002) Mutation analysis of large genomic regions in tumor DNA using single-strand conformation polymorphism. Lessons from the ATM gene. *Methods Mol. Med.* **68**, 115–124.
6. Sheffield, V. C., Beck, J. S., Kwitek, A. E., Sandstrom, D. W., and Stone, E. M. (1993) The sensitivity of single-strand conformation polymorphism analysis for the detection of single base substitutions. *Genomics* **16**, 325–332.
7. Nataraj, A. J., Olivos-Glander, I., Kusukawa, N., and Highsmith, W. E., Jr. (1999) Single-strand conformation polymorphism and heteroduplex analysis for gel-based mutation detection. *Electrophoresis* **20**, 1177–1185.
8. Glavac, D. and Dean, M. (1993) Optimization of the single-strand conformation polymorphism (SSCP) technique for detection of point mutations. *Hum. Mutat.* **2**, 404–414.
9. Wu, J. K., Ye, Z., and Darras, B. T. (1993) Sensitivity of single-strand conformation polymorphism (SSCP) analysis in detecting p53 point mutations in tumors with mixed cell populations. *Am. J. Hum. Genet.* **52**, 1273–1275.
10. Jordanova, A., Kalaydjieva, L., Savov, A., et al. (1997) SSCP analysis: a blind sensitivity trial. *Hum. Mutat.* **10**, 65–70.
11. Kukita, Y., Tahira, T., Sommer, S. S., and Hayashi, K. (1997) SSCP analysis of long DNA fragments in low pH gel. *Hum. Mutat.* **10**, 400–407.
12. Highsmith, W. E., Jr., Nataraj, A. J., Jin, Q., et al. (1999) Use of DNA toolbox for the characterization of mutation scanning methods. II: evaluation of single-strand conformation polymorphism analysis. *Electrophoresis* **20**, 1195–1203.
13. Li, W., Gao, F., Liang, J., Li, C., et al. (2003) Estimation of the optimal electrophoretic temperature of DNA single-strand conformation polymorphism by DNA base composition. *Electrophoresis* **24**, 2283–2289.
14. Gelfi, C., Vigano, A., De Palma, S., et al. (2002) Single-strand conformation polymorphism for p53 mutation by a combination of neutral pH buffer and temperature gradient in capillary electrophoresis. *Electrophoresis* **23**, 1517–1523.
15. Kaczanowski, R., Trzeciak, L., and Kucharczyk, K. (2001) Multitemperature single-strand conformation polymorphism. *Electrophoresis* **22**, 3539–3545.
16. Leren, T. P., Solberg, K., Rodningen, O. K., Ose, L., Tonstad, S., and Berg, K. (1993) Evaluation of running conditions for SSCP analysis: application of SSCP for detection of point mutations in the LDL receptor gene. *PCR Methods Appl.* **3**, 159–162.
17. Crepin, M., Escande, F., Pigny, P., et al. (2003) Efficient mutation detection in MEN1 gene using a combination of single-strand conformation polymorphism (MDGA) and heteroduplex analysis. *Electrophoresis* **24**, 26–33.

8

Denaturing Gradient Gel Electrophoresis (DGGE)

Jeroen H. Roelfsema and Dorien J. M. Peters

1. Introduction

Denaturing gradient gel electrophoresis (DGGE) is a robust method for point mutation detection that has been widely used for many years *(1)*. It is a polymerase chain reaction (PCR)-based method, the principle being the altered denaturing temperature of a PCR product with a mutation compared to the wild-type product (see Chapter 6). PCR performed on DNA of an individual with a point mutation in one of two genes will lead to a mixture of different products. PCR products from both the wild-type gene and the mutated gene will be formed. These are known as the homoduplex products. The difference in melting temperature between these two products, however, is subtle. Another type of product, heteroduplexes, consisting of a wild-type strand combined with a mutant strand of DNA, will also be formed during the last cycles of the reaction. The real strength of DGGE lies in the fact that the heteroduplex PCR products will have much lower melting temperatures compared to the homoduplex PCR products, because the heteroduplexes have a mismatch (*see* **Fig. 1**).

To visualize the different melting temperatures of these homoduplexes and heteroduplexes, the products should be run on an acrylamide gel with a gradient of denaturing agents: urea and formamide. These denaturing agents alone are not sufficient. In addition, the gel should be run at a high temperature, usually 60°C. During electrophoresis, the PCR products will run through the gel as double-stranded DNA until they reach the point where they start to denature. Once denatured, the PCR products could continue running through the gel as single-stranded DNA, but the fragments have to remain precisely where they denatured. To achieve this, a so-called GC clamp is attached, to prevent complete denaturing. This GC clamp is a string of 40–60 nucleotides composed only of guanine and cytosine and is attached to one of the PCR primers. PCR with a GC clamp results in a product with one end having a very high denaturing temperature. A PCR product running through a DGGE gel will, therefore, denature partially. The GC clamp remains double stranded. The fragment will form a Y-shaped piece of DNA that will stick firmly at its position on the gel.

2. Practical Steps

2.1. Designing the PCR Products

The melting characteristics of PCR products screened for point mutations are crucial for DGGE. It is important that the fragment, when it reaches the critical point in the gel, denatures immediately, instead of slowly denaturing at one end and progressing with this process as the product runs deeper in the gel. Such a slow-melting process will result in fuzzy bands or smears, rendering mutation detection impossible. Because the melting characteristics are vital for suc-

From: *Medical Biomethods Handbook*
Edited by: J. M. Walker and R. Rapley © Humana Press, Inc., Totowa, NJ

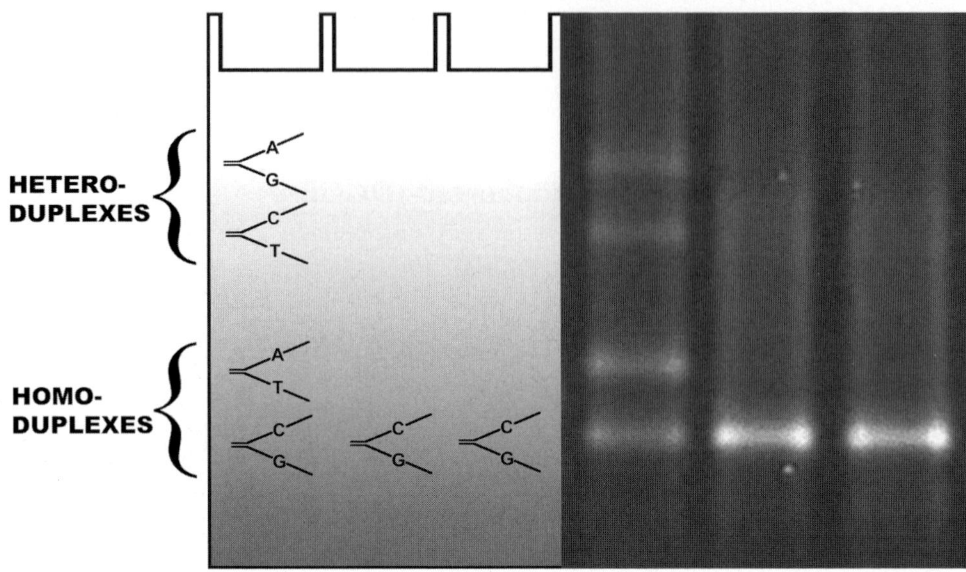

Fig. 1. Diagram of a DGGE gel and the actual result. During PCR, two homoduplex and two heteroduplex PCR products will be formed, as shown in the diagram. All four types of product will have different melting temperatures and will, therefore, melt at different positions in a gradient gel. The two heteroduplex products will melt earlier than the two homoduplex products because of their mismatch. On the gel, an example of a mutation resulting in four bands on a DGGE is visible. This is a *de novo* mutation in the gene coding for CREB-binding protein in a patient with Rubinstein–Taybi syndrome. The two adjacent lanes contain the wild-type products from the parents of the patient.

cess, primers to amplify the target should be chosen with great care. With this aim, special software that analyzes the melting curves of possible PCR products is used for primer selection. A number of programs are available for various platforms, either commercially (e.g., Winmelt from Bio-Rad Laboratories and Meltingeny from Ingeny International) or for free. There are websites where a sequence can be analyzed online as well. The experimenter will usually see a rather irregular melting curve when analyzing a target sequence. Attachment of a GC clamp of 40–60 nucleotides most often flattens this curve dramatically (*see* **Fig. 2**). The curve should be flat within a range of 1°C. Of course, the melting temperature around the GC clamp is very high. If attaching a GC clamp at one side does not flatten the curve, one should try attaching it to the other side, because for DGGE, it does not matter whether the GC clamp is attached to the forward or to the reverse primer. The selection process involves trying various combinations of forward and reverse primers to find products with a good flat curve and primers that will work well together in a PCR. DGGE products typically range from 200 to 400 bp. It is difficult to find the correct melting curve for products longer than 400 bp.

The length of the GC clamp also alters the melting behavior of the entire product. A GC clamp of 55 nucleotides (nt) is routinely used but sometimes 60 nt are necessary, especially with GC-rich sequences. Eventually, most target sequences will produce a good, flat curve in the computer analysis. However, there are troublesome sequences for which one has to use some tricks *(2)*. For instance, if the melting curve of a product goes down at the end without a GC-clamp, one can attach a second, smaller GC clamp. This is known as bipolar clamping. Strings of 8–20 nt are used in the rare cases where this is needed. Special attention is also required for GC-rich sequences for which the melting curve does not make a sharp turn where the GC clamp starts, but climbs to a higher melting temperature as it gets closer to the GC

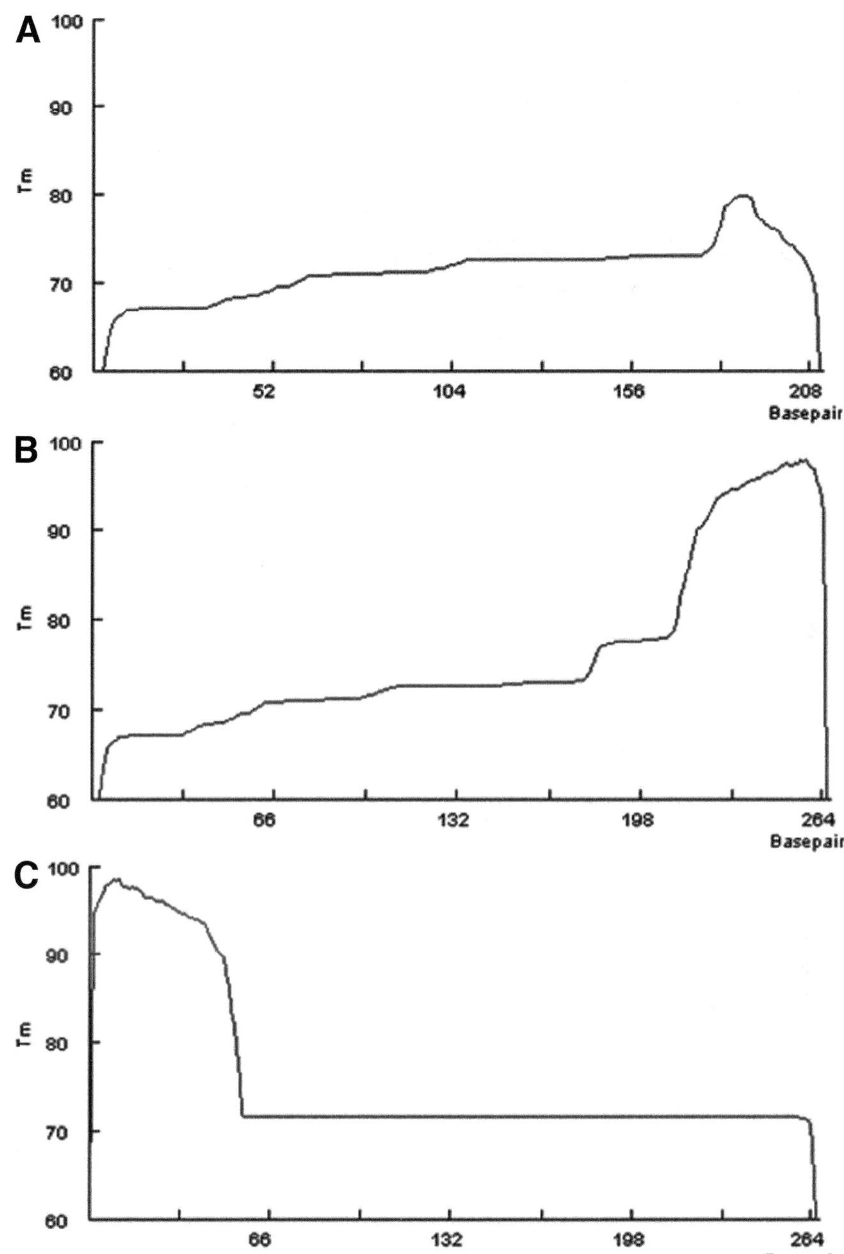

Fig. 2. Melt maps made with the program Meltingeny (Ingeny International): **(A)** The melt map of the PCR product without the attachment of a GC clamp. **(B)** The melt map with the GC clamp attached to the right side (cgcccgccgcgccccgcgcccggcccgccgcccccgcccgcgcccccggcccggg). The curve reveals a number of different melt domains. It is unlikely that this product would be successful in DGGE analysis. **(C)** The map with the GC clamp attached at the left side of the product. Now, only two melt domains remain: the high-melt domain of the sequence of the GC clamp and the lower-melt domain of the target sequence. Note that the curve is completely flat. Obviously, the GC clamp at the left side should be chosen.

clamp. Here, the addition of a few A's and T's between the GC clamp and the annealing part of the primer may cause a sharp turn in the curve.

As mentioned above, a GC clamp consists of 40–60 guanines and cytosines. Using a GC clamp with a sequence that has been proven to work in practice is recommended. Designing your own GC clamp may prove difficult, perhaps resulting in a GC clamp that forms a very stable hairpin structure during PCR. Several melting prediction programs offer a sequence of a GC clamp. Performing a PCR using primers with a GC-clamp does not require special arrangements. The protocol is the same as any other PCR and there are no changes in annealing or denaturing temperatures.

Denaturing gradient gel electrophoresis is a very suitable method for mutation screening on genomic DNA, because the majority of exons are shorter than 400 bp and can be analyzed as one fragment. Methods that can screen larger fragments thus offer little advantage for small exons. Over the years, mutation analysis in many genes has revealed a large number of mutations affecting the consensus splice sites *(3)*. Primers in the intronic sequences flanking the exons are, therefore, selected in such a way that the splice sites are screened as well as the exonic sequence. The remainder of the intron is much less likely to harbor deleterious mutations, but is more likely to contain harmless polymorphisms when compared to coding sequences. Therefore, it is wise not to include too much intronic sequence within the DGGE fragment.

2.2. Visualization of Mutations

To separate the homoduplexes and heteroduplexes, DGGE fragments are run on acrylamide gels. Gels with 9% acrylamide ensure sharp bands and are easy to handle. To pour these gradient gels, two types of stock solution are used: 9% 37.5 : 1 acrylamide/bisacrylamide in 0.5X TAE and the same stock solution with 7 *M* urea and 40% formamide. The former is called 0% denaturant and the latter is called 100% denaturant.

Gradients from 100% to 0% are rarely used because, in such a broad gradient, the denaturing points of the homoduplexes and heteroduplexes would probably be very close to each other; therefore, a range of 30% for the gradients is recommended. To select a urea/formamide gradient, the predicted melting temperature (T_m) of a product as obtained from the computer analysis is used in the empirical formula $T_m \times 3.2–182.4 = \%$ denaturant. The melting point is positioned approximately in the center of the gel by simply adding and subtracting 15% from this calculated urea/formamide concentration to obtain the desired 30% gradient. Acrylamide gels with gradients of urea and formamide are poured using a simple gradient mixer that consists of two reservoirs that are connected at the base with a short tube. Such a system is widely used for pouring all types of gradient.

The first electrophoresis run for a product is on a so-called time-travel gel on which the DGGE products are loaded at 15- to 20-min time intervals in consecutive lanes of the gel. After the electrophoresis run, the denatured products in all lanes should be at the same height in the gel, regardless of what time they were loaded. If this is not the case, then the system probably has to be redesigned (*see* **Fig. 3**). If a product does not result in sharp bands or gets stuck at a position too high or too low in the gel, the gradient used should be adjusted.

Technically, the most challenging problem with DGGE experiments is performing the electrophoresis at 60°C. There are various possible methods to achieve a temperature of 60°C, but usually the glass plates, with the acrylamide gel in between, are submerged in a tank of water that is heated to a constant temperature with the help of a thermostat. There are a number of commercially available systems with total equipment kits suitable for DGGE, in which the lower buffer chamber forms the water bath that is heated. The electrophoresis is usually performed overnight at 90 V in 0.5X TAE buffer, but shorter runs during the day with higher voltages are possible. After electrophoresis, the gels are soaked in a 0.5X TAE solution containing ethidium bromide to visualize the DNA fragments.

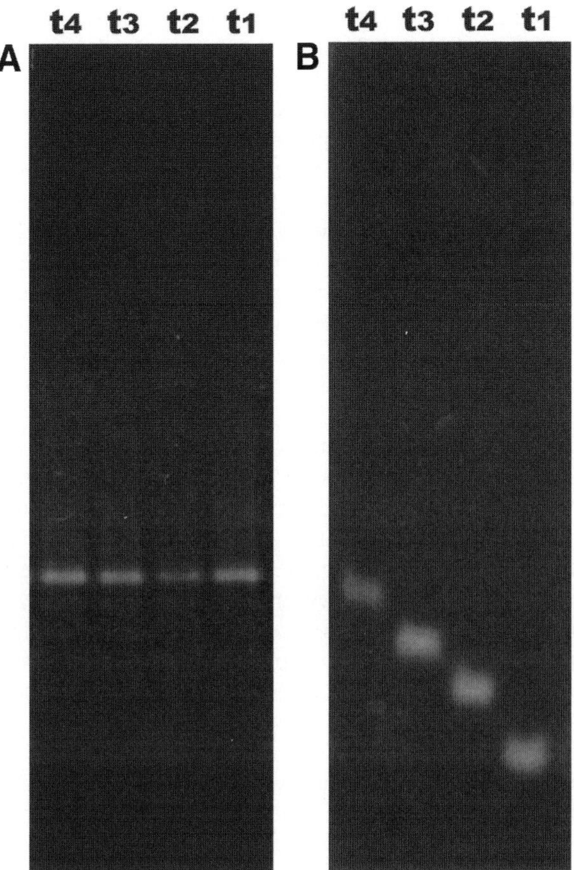

Fig. 3. Time travel gels for two different PCR products. In each gel, the same product is loaded at four different time-points. A good result is shown in **(A)**. In **(B)**, however, the bands do not stay at the same position in the gel. Note the fuzzy bands. This is a clear example of a product that denatures slowly and is, therefore, probably not suitable for DGGE analysis.

3. Applications of DGGE

Denaturing gradient gel electrophoresis is a method to identify small mutations (e.g., point mutations). The definition of a point mutation is the transition or transversion of one nucleotide into another. However, there are more types of small mutation such as deletions or insertions of one or more nucleotides that can be identified by DGGE as well. In fact, these mutations will cause a large difference in melting temperatures in both the homoduplexes and the heteroduplexes and can therefore be seen quite clearly on the gels.

As mentioned earlier, DGGE products typically range from 200 to 400 bp, making DGGE well suited for analyzing exons in genomic DNA, although DGGE can be applied to RNA screening as well. However, RNA is more vulnerable to degradation than DNA and requires conversion into complementary (cDNA). Scanning for mutations in genes involved in hereditary disorders is therefore often done on genomic DNA for both diagnostic purposes and for research. For instance, DGGE is widely applied in the analysis of the various genes involved in hereditary colorectal cancer such as APC, MSH2, MSH6, MLH1, and so forth ***(4–6)***. Presymptomatic diagnosis is particularly important with a potentially lethal disease such as colorectal cancer that can be treated. Mutation analysis has revealed that in families with colorectal cancer, the mutation is often unique. Obviously, screening a family for an unknown mutation requires a technique, such as DGGE, that is tried and proven, particularly when the stakes are very high. However, for research purposes, reliability is important as well. Investigation into types of mutation requires that the screening will reveal almost all point mutations so that an unbiased analysis of the mutation spectrum can be made.

Duchenne muscular dystrophy is caused by mutations in a huge gene, coding for dystrophin, on chromosome X. Most of these mutations are large deletions or duplications, but approx 30% of the mutations are point mutations somewhere in 1 of the more than 70 exons, or their flanking splice sites *(7)*. To increase the speed of the screening procedure, the PCR products are grouped together, ranging from three to six fragments and run in one lane. Of course, this multiplexing technique is not limited to the large genes. An example is the α-1-antitrypsin gene. The entire gene can be screened using two multiplex amplification reactions. The products of both reactions can be analyzed on the same gel, thus allowing the rapid screening of a large number of individuals ***(8)***.

The primary strength of DGGE is its ability to easily detect the heteroduplexes that will be formed if a mutation is present. During the screening for mutations in genes that are involved in dominant hereditary diseases, the heteroduplexes will be formed during PCR. However, is DGGE applicable to recessive hereditary diseases? Cystic fibrosis is one of the most frequently inherited recessive diseases known. Mutation detection performed since the identification of the CFTR gene in 1989 has revealed that a few mutations are frequent, most notably the Δ508 mutation, a deletion of three nucleotides causing an in-frame deletion of amino acid phenylanaline on position 508 ***(9)***. Such a frequent mutation is often analyzed by other methods, designed to screen for specific mutations that are known. For the remainder of the mutations that are rare and may never have been identified previously, DGGE is very well suited. The fact that cystic fibrosis is a recessive gene does not hamper screening for the simple reason that DNA obtained from the parents of an affected child will usually be screened. These parents are heterozygote carriers of a mutation. Nevertheless, screening patients can be done without many problems, because rare mutations are seldom found in both genes. However, one has to be aware that homozygosity of rare mutations will be found much more frequently in communities that are isolated by geographical or cultural conditions. Even then, careful analysis of DGGE gels will also reveal the homozygote mutations in the majority of cases. It is, however, possible to mix the DNA that is screened with DNA from unaffected individuals in order to create heteroduplexes. From a technical point of view, there is no difference between a recessive disorder and an X-linked disorder. Again, it is often the heterozygote mothers who are screened.

Point mutation detection is not limited to genetic disorders. It is also applied to tumor samples. Cancer is caused by a series of mutations in genes and these mutations vary from the loss of whole chromosome arms to point mutations. The amount of DNA from a surgically removed tumor may not be great and, depending on the tumor and the DNA isolation method, DNA may be degraded into relatively small fragments. In general, this will not be a problem for DGGE because the PCR fragments are usually small anyway. Another problem is the fact that tumor samples do not solely consist of tumor cells. Blood vessels and connective tissue may be present as well. Especially, malignant invasive tumors may lead to samples with a high percentage of unaffected cells. In our experience, and that of others, DGGE is sensitive enough to find mutations when present. Typical genes that are often screened in this type of research are *TP53* and *K-*, *N-*, and *H-RAS* ***(10,11)***. Mutation detection on the *TP53* gene is often limited to exons 5–8 that are thought to harbor the majority of mutations. However, mutation analysis of the entire gene has shown that this leads to a neglect of many mutations ***(12)***. Immunohistochemistry is often used to investigate the *p53* status in tumors. The researcher should be aware that immunohistochemistry cannot replace DGGE and that DGGE cannot replace immunohistochemistry. Both ways of looking at *p53* in tumors are complementary.

An entirely different application for DGGE is to assess the number and types of different bacteria species. The genes encoding for ribosomal RNA are used as a target because these genes are highly conserved among different species. Primers that anneal in the most conserved parts can be used to amplify the genes in completely different species. Sequence variations in the less conserved parts of the fragments can be revealed by DGGE and used to identify different species. This technique was developed first in microbiological ecology to investigate the number of species living in soil or water. The method is now also used to assess species of bacteria living in or on the human body ***(13,14)***. One application, for instance, is monitoring patients treated with antibiotics ***(15)***.

Point mutation detection is often performed using either single-strand conformation polymorphism (SSCP) analysis or DGGE ***(16)***. Both methods have been around for more than a decade and are relatively inexpensive compared to new, state-of-the-art technologies that usually require very expensive equipment. Therefore, when does one decide to use SSCP or to use DGGE? DGGE is a robust method; once optimized for a product, it will work. The results are very easy to score, because a quick glance on an ultraviolet (UV) illuminator will reveal mutations immediately. The clarity of the results is a very strong advantage of DGGE. Another strong advantage is the fact that radioactive labels are not needed. SSCP needs radioactive PCR, or silver staining, and the results are less clear. On the other hand, SSCP takes less time to optimize compared to DGGE and can be done without an investment in expensive material. It is also the number of samples that determines the method of choice. For instance, screening candidate genes with a small number of samples typically calls for SSCP. DGGE is best suited for screening a large number of samples over a longer period.

After finding an aberrant band on a gel, it is only clear that there is a sequence variation present within that specific product from that specific sample. The nature or exact location of the variation is not known. For this, sequencing the PCR product is needed. Why, if sequencing is needed at the end, is mutation screening not performed using sequencing rather than DGGE? The answer is twofold. First, DGGE results are very clear. Mutations present in DGGE products can be seen immediately, whereas sequencing requires either very sophisticated software or very tedious scrutinizing behind the computer. Second, and often the deciding factor, when large numbers of samples have to be screened, DGGE is less expensive.

Primers are selected in the intronic sequences flanking the exons to screen the entire coding region and the flanking splice sites as well. Screening exons from genomic DNA with DGGE products that encompass relatively large pieces of intronic DNA can be cumbersome with some genes because of the large amount of polymorphisms found in the intronic sequences. Screening those genes by DGGE will result in samples showing aberrant bands on a gel of which the majority turns out to be polymorphisms after sequencing. Researchers working on those genes usually resort to direct sequencing of their samples as a method for point mutation detection.

Denaturing gradient gel electrophoresis is the method of choice if one wants to screen a large number of samples for unknown mutations. Because radioactive markers are not needed, it is considered a user-friendly technique. This together with the reliability and the cost-effectiveness are the strong points for DGGE.

Acknowledgment

We thank Stefan White for critically reading the manuscript.

References

1. Sheffield, V. C., Cox, D. R., Lerman, L. S., and Meyers, R. M. (1989) Attachment of a 40-base-pair G+C-rich sequence (GC-clamp) to genomic DNA fragments by the polymerase chain reaction results in improved detection of single-base changes. *Proc. Natl. Acad. Sci. USA* **86**, 232–236.
2. Wu, Y., Hayes, V. M., Osinga, J., et al. (1998) Improvement of fragment and primer selection for mutation detection by denaturing gradient gel electrophoresis. *Nucleic Acids Res.* **26**, 5432–5440.
3. Krawczak, M., Reiss, J., and Cooper, D. N. (1992) The mutational spectrum of single base-pair substitutions in mRNA splice junctions of human genes: causes and consequences. *Hum. Genet.* **90**, 41–54.
4. van der Luijt, R. B., Khan, P. M., Vasen, H. F., et al. (1997) Molecular analysis of the APC gene in 105 Dutch kindreds with familial adenomatous polyposis: 67 germline mutations identified by DGGE, PTT, and southern analysis. *Hum. Mutat.* **9**, 7–16.

5. Wagner, A., Hendriks, Y., Meijers-Heijboer, E. J., et al. (2001) Atypical HNPCC owing to MSH6 germline mutations: analysis of a large Dutch pedigree. *J. Med. Genet.* **38**, 318–322.
6. Wijnen, J., Vasen, H., Khan, P. M., et al. (1995) Seven new mutations in hMSH2, an HNPCC gene, identified by denaturing gradient-gel electrophoresis. *Am. J. Hum. Genet.* **56**, 1060–1066.
7. den Dunnen, J. T., Grootscholten, P. M., Bakker, E., et al. (1989) Topography of the Duchenne muscular dystrophy (DMD) gene: FIGE and cDNA analysis of 194 cases reveals 115 deletions and 13 duplications. *Am. J. Hum. Genet.* **45**, 835–847.
8. Hayes, V. M. (2003) Genetic diversity of the alpha-1-antitrypsin gene in Africans identified using a novel genotyping assay. *Hum. Mutat.* **22**, 59–66.
9. Kerem, B., Rommens, J. M., Buchanan, J. A., et al. (1989) Identification of the cystic fibrosis gene: genetic analysis. *Science* **245**, 1073–1080.
10. Soussi, T. and Beroud, C. (2001) Assessing TP53 status in human tumours to evaluate clinical outcome. *Nature Rev. Cancer* **1**, 233–240.
11. Nedergaard, T., Guldberg, P., Ralfkiaer, E., and Zeuthen, J. (1997) A one-step DGGE scanning method for detection of mutations in the K-, N-, and H-ras oncogenes: mutations at codons 12, 13 and 61 are rare in B-cell non-Hodgkin's lymphoma. *Int. J. Cancer* **71**, 364–369.
12. Hartmann, A., Blaszyk, H., McGovern, R. M., et al. (1995) p53 gene mutations inside and outside of exons 5-8: the patterns differ in breast and other cancers. *Oncogene* **10**, 681–688.
13. Tannock, G. W. (2002) Analysis of the intestinal microflora using molecular methods. *Eur. J. Clin. Nutr.* **56 (Suppl. 4)**, S44–S49.
14. Favier, C. F., Vaughan, E. E., De Vos, W. M., and Akkermans, A. D. (2002) Molecular monitoring of succession of bacterial communities in human neonates. *Appl. Environ. Microbiol.* **68**, 219–226.
15. Donskey, C. J., Hujer, A. M., Das, S. M., et al. (2003) Use of denaturing gradient gel electrophoresis for analysis of the stool microbiota of hospitalized patients. *J. Microbiol. Methods* **54**, 249–256.
16. Orita, M., Suzuki, Y., Sekiya, T., and Hayashi, K. (1989) Rapid and sensitive detection of point mutations and DNA polymorphisms using the polymerase chain reaction. *Genomics* **5**, 874–879.

9

Quantitative Analysis of DNA Sequences by PCR and Solid-Phase Minisequencing

Anu Suomalainen and Ann-Christine Syvänen

1. Introduction

The PCR technique provides a highly specific and sensitive means for analyzing nucleic acids, but it does not allow their direct quantification. This limitation is because the efficiency of PCR depends on the amount of template sequence present in the sample, and the amplification is exponential only at low template concentrations *(1)*. Owing to this plateau effect of PCR, the amount of amplification product does not directly reflect the original amount of template. Moreover, subtle differences in reaction conditions, such as material from biological samples, may cause significant sample-to-sample variation in the final yield of the PCR product.

The problem of performing accurate quantitative PCR analyses has been addressed by two principal approaches. A quantitative PCR result can be obtained by "kinetic PCR," in which the amplification process is monitored at numerous times or concentration points *(2,3)* (see Chapter 25). Most conveniently, the amplification process can be monitored in real time by, for example, the homogeneous TaqMan 5' nuclease assay or molecular beacon probes using a fluorescence detection instrument *(4,5)*. The other approach, known as competitive PCR, utilizes an internal quantification standard sequence that is coamplified in the same reaction as the target sequence *(6–8)*. The efficiency of the amplification is affected by the sequence of the PCR primers as well as the size and sequence of the template. Therefore, the internal standard should be as similar to the target sequence as possible to ensure that the ratio of the two sequences remains constant throughout the amplification. An ideal PCR quantification standard differs from the target sequence only at one nucleotide position, by which the two sequences can be identified and quantified after the amplification. Determination of the relative amounts of PCR products originating from the target and standard sequences allows calculation of the initial amount of the target sequence. If two target sequences are present as a mixture in a sample, it is easy and often sufficient to measure their relative amounts. To determine the absolute amount of a target sequence, it is necessary to add a known amount of standard sequence to the sample before amplification. In this case, a measure of the amount of the analyzed sample, such as the number of cells or the total amount of DNA, RNA, or protein, is needed.

We have developed a solid-phase minisequencing method for the identification of point mutations or nucleotide variations in human genes *(9)*. This method is based on the distinct detection of two sequences that differ from each other only at a single nucleotide, making the method an ideal tool for quantitative analysis of DNA *(10)* and RNA *(11)* sequences by competitive PCR. In the solid-phase minisequencing method, a DNA fragment containing the variable nucleotide is first amplified using one biotinylated and one unbiotinylated PCR primer.

From: *Medical Biomethods Handbook*
Edited by: J. M. Walker and R. Rapley © Humana Press, Inc., Totowa, NJ

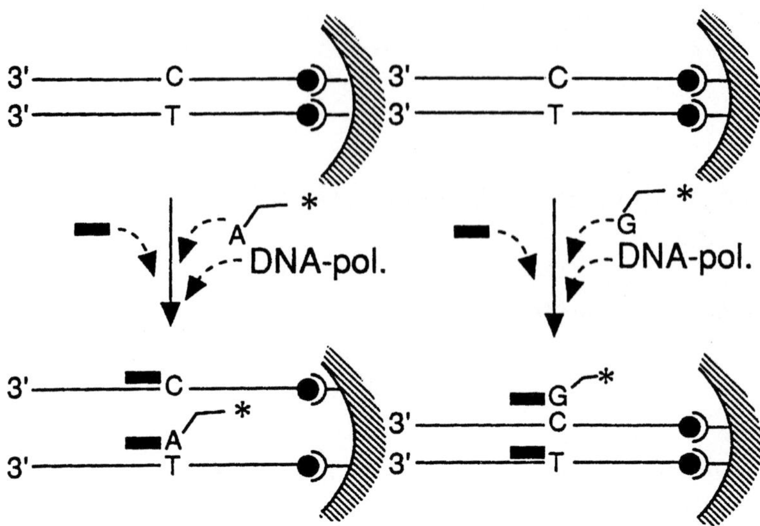

Fig. 1. Principle of the solid-phase minisequencing method. Analyses for two nucleotides performed in separate wells are shown on the left and right. **Top panel**: one PCR primer is biotinylated at its 5' end, resulting in a PCR product carrying biotin at the 5' end of one of its strands (filled circle). The product is captured in a streptavidin-coated microtiter well and denatured. **Lower panel**: A detection step primer hybridizes to the single-stranded template, 3' adjacent to the variant nucleotide. The DNA polymerase extends the primer with the [^{3}H]-labeled dNTP if it is complementary to the nucleotide present at the variable site. After washing, the sample is denatured and the eluted radioactivity expressing the amount of incorporated label is measured.

The PCR product carrying a biotin residue at the 5' end of one of its strands is captured on an avidin-coated solid support and denatured. The nucleotides at the variable site in the immobilized DNA strand are then identified by two separate primer extension reactions, in which a DNA polymerase incorporates a single-labeled deoxynucleotide triphosphate (dNTP) (*see* **Fig. 1**). Our first-generation assay format utilizes [^{3}H]dNTPs as labels and streptavidin-coated microtiter plates as a solid support ***(12)***. The results of the assay are numeric counts per minute (cpm) values expressing the amount of [^{3}H]dNTP incorporated in the minisequencing reactions. The ratio between the cpm values obtained in the minisequencing assay (*R* value) directly reflects the ratio between the two sequences in the original sample (*see* **Fig. 2**). The method allows quantitative determination of a sequence that represents less than 1% of a mixed sample; that is, the dynamic range for the quantitative analysis spans five orders of magnitude ***(10,14)***. Furthermore, because the two sequences differ from each other by a single nucleotide, they are amplified with equal efficiency during PCR, and the *R* value obtained is not affected by the amount of template present in the reaction (*see* **Fig. 3**). Consequently, the quantitative analysis can be performed irrespective of the phase of the PCR process. If two sequences are not present in the sample itself, quantitation by minisequencing can be done relative to a standard added to the sample, as described above.

2. Practical Steps

The basic materials and equipment needed for solid-phase minisequencing are available in most molecular genetics laboratories. Routine contamination-free PCR facilities are needed, as well as a PCR machine. In addition, microtiter plates with streptavidin-coated wells (e.g., BioBind assembly strip; Thermo-Labsystems, Finland) are required, and use of a multichannel

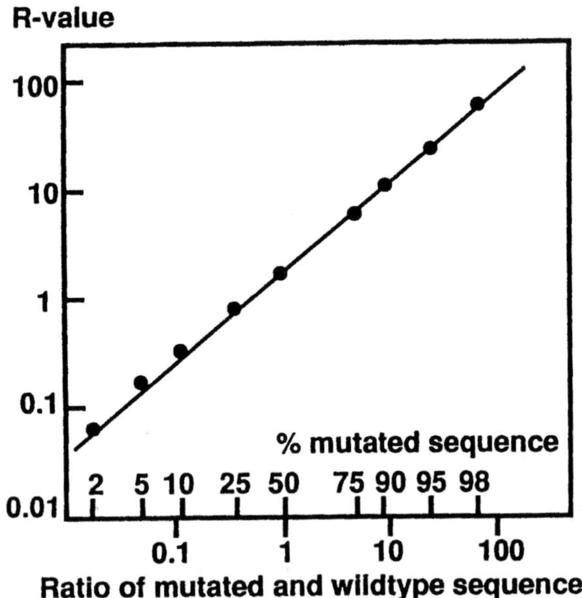

Fig. 2. Solid-phase minisequencing standard curve prepared by analyzing mixtures of two 63-mer oligonucleotides differing from each other at one nucleotide in the mitochondrial $tRNA^{Leu(UUR)}$ gene *(16)*. The C_{cpm}/T_{cpm} ratio obtained from the minisequencing reactions is plotted as a function of the original ratio between two oligonucleotides in the mixtures. (From **ref.** ***13***.)

pipet and microtiter plate washer speed up the procedure, but are optional. A shaker at 37°C and a water bath or incubator at 50°C, as well as a liquid scintillation counter are needed. Exact descriptions of the materials and methods are in given in **ref.** ***14***.

2.1. PCR Primer Design

Initially, PCR primers are designed so that one PCR primer is biotinylated at its 5' end during its synthesis, using a biotin–phosphoramidite reagent. If oligonucleotides are to be used as quantification standards (*see* **Subheading 2.2.**), the length of oligonucleotides that can be synthesized with acceptable yields sets an upper limit for the length of the PCR product at about 80–100 bp. The detection step primer for the minisequencing reaction is an oligonucleotide complementary to the biotinylated strand of the PCR product designed to hybridize with its 3' end immediately adjacent to the variant nucleotide to be detected (*see* **Fig. 1**). The detection step primer should be at least five nucleotides nested in relation to the unbiotinylated PCR primer.

2.2. Quantification Standards

To accurately quantify a sequence in a sample that contains only one sequence type, a standard should be designed to differ from the target sequence at the nucleotide to be detected in the minisequencing reaction (*see* **Fig. 1**). To construct a standard curve, a second standard identical to the target sequence is required. Oligonucleotide standards can be synthesized using a DNA synthesizer; PCR products or cloned DNA fragments also can be used. First, the molecular concentrations of the standards have to be determined. The optimal amount of the standard added to a sample depends on the abundance of the target sequence in the original sample. The ratio of the target-to-standard sequence should preferably be between 0.1 and 10. If no estimate of the target sequence is available, it might be necessary to initially titrate the optimal amount

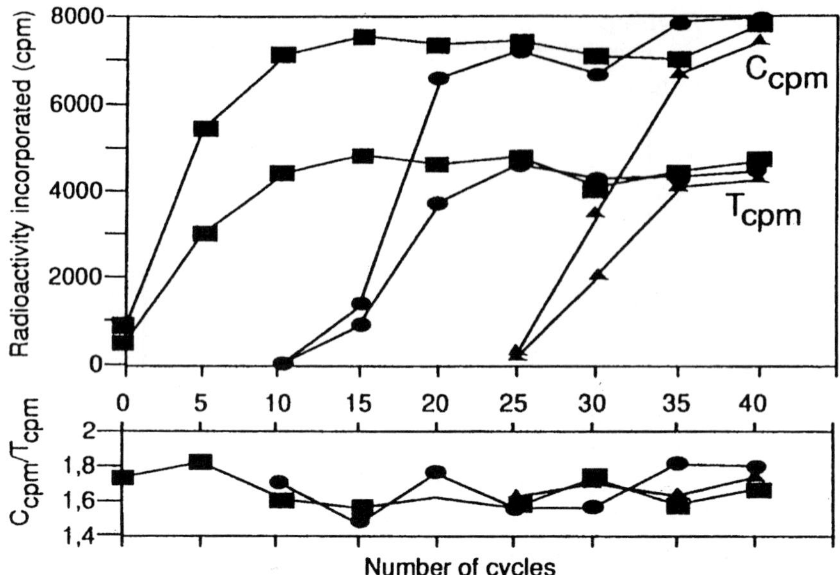

Fig. 3. Result of the solid-phase minisequencing assay obtained at different PCR cycles and amounts of template. Mixtures of equal amounts (10^3, 10^7, or 10^{11} molecules) of the same oligonucleotide as in Fig. 2 were analyzed. The upper panel shows the cpm values obtained in the minisequencing assay at different PCR cycles, and the lower panel shows the corresponding C_{cpm}/T_{cpm} ratios.

of the standard in the analysis (e.g., 10^2, 10^4, or 10^6 molecules). For accurate quantification, standards representing both sequence variants should be available, and analysis of mixtures of known amounts of the two standards should be analyzed to construct a standard curve, as demonstrated in **Figs. 1** and **2** and **Table 3**.

2.3. PCR

Polymerase chain reaction follows routine protocols except that the amount of biotin-labeled primer should be reduced, not to exceed the biotin-binding capacity of the microtiter well. If high binding capacity is required, avidin-coated polystyrene beads can also be used. The PCR should be optimized so that one-tenth of the PCR product produces a single visible band after agarose gel electrophoresis and staining with ethidium bromide.

2.4. Solid-Phase Minisequencing (see Fig. 1)

The detection of biotin-labeled PCR product starts with affinity capture: The PCR product binds from its 5' biotin to the streptavidin-coated microtiter well. Each nucleotide to be detected at the variant site needs to be analyzed in a separate well. Thus, minimally two wells are needed per PCR product. After capture, the wells are washed carefully to remove PCR-derived dNTPs. Then, the PCR product is denatured in the wells, after which only the biotin-labeled DNA strand of the PCR product remains attached to the well wall and the complementary strand is washed off. Next, the minisequencing solution is added. This consists of a thermostable DNA polymerase, such as *Taq* polymerase, in its buffer, supplemented with [^{3}H]-labeled dNTP specific to the mutation to be detected, as well as a detection step primer, which hybridizes its 3' end adjacent to the nucleotide to be detected. If the labeled dNTP matches the variant site, the polymerase extends the detection primer with this nucleotide, and the primer becomes [^{3}H] labeled. After denaturation of the primer, the amount of incorporated nucleotide is measured by a liquid scintillation counter. The ratio between the cpm values for the two nucleotides detected in sepa-

Table 1
Example of the Determination of the Allele Frequencies of a Polymorphism in the PROS1 Gene by Quantitative Analysis of Pooled DNA Samples

	$[^3H]$dNTP incorporated (cpm)[b]		R value (A_{cpm}/G_{cpm})	Allele frequency[c]	
Sample[a]	A reaction	G reaction		A allele	G allele
Pool 1390	2750	1280	2.14	0.59	0.41
Pool 860	2240	1048	2.15	0.59	0.41
Pool 920	2510	1190	2.11	0.58	0.42
Control (AA)	2100	52	40	—	—
Control (GG)	96	1930	0.050	—	—
Control (AG)	2480	1660	1.49	—	—
No DNA	64	39	—	—	—

[a]The numeral gives the number of individuals in each pool. AA, GG, and AG indicate the genotype of the individual control samples.

[b]Mean values of five (pools) or two (individual controls) parallel assays. The specific activities of $[^3H]$dATP and $[^3H]$dGTP were 58 and 32 Ci/mmol, respectively.

[c]The allele frequencies determined from 50 individual samples were 0.61 (A) and 0.39 (G) ***(12)***.

rate wells reflects the ratio between the two sequences in the original sample. Because the specific activities of $[^3H]$-labeled dNTPs vary, these have to be corrected before calculating the final ratio. Alternatively, a standard curve can be used to correct for these factors.

Our standard assay format uses $[^3H]$-labeled dNTPs because of their low radioactivity and high chemical resemblance to natural dNTPs such as polymerase substrates. However, dNTPs or dideoxy-nucleotides labeled with other isotopes or with fluorophores can also be used. Furthermore, streptavidin-coated microtiter plates made of scintillating polystyrene are available. These allow direct detection of the extended primer after the minisequencing reaction, but a scintillation counter for microtiter plates is needed ***(15)***.

3. Examples of Applications

3.1. Determination of Allele Frequencies by Quantitative Analysis of Pooled DNA Samples

We have developed a system for forensic DNA typing, in which a panel of 12 single-nucleotide polymorphisms (SNPs) is analyzed by the solid-phase minisequencing method ***(12)***. The statistical interpretation of forensic and paternity testing results requires information on allele frequencies of the analyzed markers in each particular population. To rapidly obtain this information in the Finnish population, we utilized the quantitative nature of the solid-phase minisequencing method to determine allele frequencies of polymorphic markers by analyzing pooled DNA samples derived from hundreds of individuals. The ratio between the two sequences at each polymorphic locus in the pooled DNA samples is equivalent to the allele frequencies in the population. **Table 1** shows the results from the analysis of allele frequencies of a polymorphism in the *PROS1* gene on chromosome 13. A good correlation between allele frequencies determined from the pooled samples and those determined from about 100 alleles individually were observed ***(12)***. This possibility of determining the allele frequencies of common SNPs with good accuracy in pooled samples has recently been used in many association studies as an approach to increase the throughput of genotyping SNP markers. We also have determined the carrier frequency of a rare disease allele causing recessively inherited aspartylglucosaminuria in Finland by determining the frequency of the mutant allele (**Table 2**) by quantitative analysis of large pooled DNA samples ***(10)***.

Table 2
Determination of the Copy Number of the Aspartylglucosaminidase Gene

Karyotype of sample	$C_{cpm}/(C_{cpm} + G_{cpm})$	Deduced AGA gene copy number
46,XY,-4,+der(4) t(4;12) (q31.3;pl2.2)mat	0.28–0.33[a]	1
46,XX,del(4)(q33)	0.26–0.32[a]	1
46,XX,-21,+der(21) t(4;21) (q28;p13)mat	0.63–0.70[a]	3
Controls[b]	0.50–0.54	2

[a]Range of variation of five parallel assays.

[b]DNA from individuals heterozygous for the target nucleotide in the aspartylglucosaminidase gene.

Source: **ref. *19***

3.2. Detection of Heteroplasmic Point Mutations of Mitochondrial DNA

Disease-causing point mutations of mitochondrial DNA (mtDNA) are most often heteroplasmic; that is, the tissues of patients contain both mutant and normal mtDNA. The solid-phase minisequencing method is particularly useful for detecting heteroplasmic mtDNA mutations, allowing both identification and quantification of the mutation in the same assay. Using this method, we have observed a correlation between the degree of heteroplasmy and the severity and age of disease onset in two mitochondrial disorders associated with mtDNA point mutations ***(13,16)*** (**Table 3**). Furthermore, a decline of a mutant mitochondrial DNA population was shown to occur during aging ***(17,18)***. The method is widely used in clinical DNA diagnosis laboratories for routine detection of point mutations.

3.3. Determination of Gene Copy Numbers

We have developed an alternative method to the widely used fluorescence *in situ* hybridization techniques in the determination of the copy number of human genes. The copy number of the aspartylglucosaminidase gene, located on the chromosome 4q, was determined by solid-phase minisequencing in the samples of three patients with either deletions or duplications involving the distal region of chromosome 4q. A known amount of DNA from a patient homozygous for a mutation in the marker gene was mixed with the DNA samples to be analyzed to serve as an internal standard, and the relative amount of normal sequence was determined by the solid-phase minisequencing method ***(19)***. This application demonstrates the suitability of the method for determining monosomies, trisomies, and loss of heterozygosity, provided that a DNA standard containing a suitable polymorphism is available.

3.4. Identification of Mixed Samples

The *R* values obtained by solid-phase minisequencing, when individual genomic DNA samples are analyzed for a variable nucleotide, fall into three distinct categories that unequivocally define the genotype of the sample. *R* Values falling outside of these three categories that normally differ from each other by a factor of 10 suggest the presence of contaminating DNA in the sample ***(12)***. The ability of the method to identify a mixed sample is an advantage in forensic analyses, where stain samples could contain DNA from several individuals, as well as in prenatal diagnosis, where placental biopsy samples could contain contaminating maternal DNA, and for mutation analysis in tumor samples that could contain normal DNA. Recently, we have utilized the possibility of quantitative genotyping of mixed samples to monitor the success of allogenic stem cell transplantation in leukaemia patients. In this application, the change of each patient's blood cell genotypes to that of the donor were assessed quantitatively after the transplantation ***(20)***.

Table 3
Example of the Result of a Solid-Phase Minisequencing Analysis of Mixtures of Two 63-Mer Oligonucleotides Differing from Each Other at One Nucleotide in the Mitochondrial $tRNA^{Leu(UUR)}$ gene

Ratio of wild-type sequence to mutated sequence	T_{cpm} (wt oligo)[a]	C_{cpm} (mut oligo)[b]	R value (C_{cpm}/T_{cpm})
Wild type	3110	44	0.014
50 : 1	3640	190	0.05
20 : 1	2780	420	0.15
10 : 1	2830	730	0.26
4 : 1	2520	1690	0.67
1 : 2	1650	2810	1.7
1 : 4	790	3630	4.6
1 : 10	350	3790	10.8
1 : 20	210	4760	22.7
1 : 50	120	4800	40.0
Mutant	43	4580	106.5
H_2O	41	23	—

[a]The specific activities of the [^{3}H] dNTPs: dTTP=126 Ci/mmol; dCTP=67 Ci/mmol.
[b]In this case, two [^{3}H] dCTPs were incorporated into the mutant sequence.
Source: **ref. *13***.

3.5. High-Throughput Genotyping of SNPs

Single-nucleotide polymorphisms (SNPs) are the most abundant form of genetic variation, occurring approx every 1–2 kb in the human genome ***(21)*** (see Chapter 42). New assay formats with low reagent costs are required for highly multiplexed SNP genotyping, which is becoming of increasing importance in association studies of complex disorders. Solid-phase minisequencing meets that challenge when adapted to a microchip format ***(22)***. In this format, minisequencing primers, or tags homologous to the detection primers, are attached on a glass chip, and the detection of minisequencing reaction is based on the incorporation of a fluorescent dye. The microarray-based minisequencing allows simultaneous, quantitative analysis of up to 100 SNPs in 56 samples per microscope slide, detecting 2% of the minority alleles ***(23)***.

In conclusion, solid-phase minisequencing is an accurate high-sensitivity, low-cost, high-throughput method applicable to a routine molecular genetic laboratory for single-nucleotide detection. It is widely used in DNA diagnostic laboratories for nuclear gene mutation detection as well as relative quantification of mitochondrial DNA mutations. It also is applicable to the microarray format for efficient screening and genotyping of SNPs.

References

1. Syvänen, A.-C., Bengtström, M., Tenhunen, J., and Söderlund, H. (1988) Quantification of polymerase chain reaction products by affinity-based hybrid collection. *Nucleic Acids Res.* **16**, 11,327–11,338.
2. Murphy, L. D., Herzog, C. E., Rudick, J. B., Fojo, A. T., and Bates, S. E. (1990) Use of the polymerase chain reaction in the quantitation of mdr-1 gene expression. *Biochemistry* **29**, 10,351–10,356.
3. Noonan, K. E., Beck, C., Holzmayer, T. A., et al. (1990) Quantitative analysis of MDR1 (multidrug resistance) gene expression in human tumors by polymerase chain reaction. *Proc. Natl. Acad. Sci. USA* **87**, 7160–7164.
4. Livak, K. J., Flood, S. J., Marmaro, J., Giusti, W., and Deetz, K. (1995) Oligonucleotides with fluorescent dyes at opposite ends provide a quenched probe system useful for detecting PCR product and nucleic acid hybridization. *PCR Methods Appl.* **4**, 357–362.

5. Tyagi, S. and Kramer, F. R. (1996). Molecular beacons: probes that fluoresce upon hybridization. *Nat Biotechnol.* **14**, 303–308.
6. Chelly, J., Kaplan, J. C., Maire, P., Gautron, S., and Kahn, A. (1988) Transcription of the dystrophin gene in human muscle and non-muscle tissue. *Nature* **333**, 858–860.
7. Wang, A. M., Doyle, M. V. and Mark, D. F. (1989) Quantitation of mRNA by the polymerase chain reaction. *Proc. Natl. Acad. Sci. USA* **86**, 9717–9721.
8. Gilliland, G., Perrin, S., Blanchard, K., and Bunn, H. F. (1990) Analysis of cytokine mRNA and DNA: detection and quantitation by competitive polymerase chain reaction. *Proc. Natl. Acad. Sci. USA* **87**, 2725–2729.
9. Syvänen, A.-C., Aalto-Setälä, K., Harju, L., Kontula, K., and Söderlund, H. (1990) A primer-guided nucleotide incorporation assay in the genotyping of apolipoprotein E. *Genomics* **8**, 684–692.
10. Syvänen, A.-C., Ikonen, E., Manninen, T., et al. (1992) Convenient and quantitative determination of the frequency of a mutant allele using solid-phase minisequencing: application to aspartylglucosaminuria in Finland. *Genomics* **12**, 590–595.
11. Ikonen, E., Manninen, T., Peltonen, L., and Syvänen, A.-C. (1992). Quantitative determination of rare mRNA species by PCR and solid-phase minisequencing. *PCR Methods Appl.* **1**, 234–240.
12. Syvänen, A.-C., Sajantila, A., and Lukka, M. (1993). Identification of individuals by analysis of biallelic DNA markers, using PCR and solid-phase minisequencing. *Am. J. Hum. Genet.* **52**, 46–59.
13. Suomalainen, A., Majander, A., Pihko, H., Peltonen, L., and Syvänen, A.-C. (1993) Quantification of tRNA3243(Leu) point mutation of mitochondrial DNA in MELAS patients and its effects on mitochondrial transcription. *Hum Mol Genet,* **2**, 525–534.
14. Suomalainen, A. and Syvänen, A.-C. (2000) Quantitative analysis of human DNA sequences by PCR and solid-phase minisequencing. *Mol. Biotechnol.* **15**, 123–131.
15. Ihalainen, J., Siitari, H., Laine, S., Syvänen, A.-C., and Palotie, A. (1994) Towards automatic detection of point mutations: use of scintillating microplates in solid-phase minisequencing. *BioTechniques* **16**, 938–943.
16. Suomalainen, A., Kollmann, P., Octave, J. N., Söderlund, H., and Syvänen, A.-C. (1993) Quantification of mitochondrial DNA carrying the tRNA(8344Lys) point mutation in myoclonus epilepsy and ragged-red-fiber disease. *Eur. J. Hum. Genet.* **1**, 88–95.
17. Rahman, S., Poulton, J., Marchington, D., and Suomalainen, A. (2001) Decrease of 3243 A→G mtDNA mutation from blood in MELAS syndrome: a longitudinal study. *Am. J. Hum. Genet.* **68**, 238–240.
18. Olsson, C., Johnsen, E., Nilsson, M., Wilander, E., Syvänen, A.-C., and Lagerström-Fermer, M. (2001) The level of the mitochondrial mutation A3243G decreases upon ageing in epithelial cells from individuals with diabetes and deafness. *Eur. J. Hum. Genet.* **9**, 917–921.
19. Laan, M., Grön-Virta, K., Salo, A., et al. (1995) Solid-phase minisequencing confirmed by FISH analysis in determination of gene copy number. *Hum. Genet.* **96**, 275–280.
20. Fredriksson, M., Barbany, G., Liljedahl, U., Hermanson, M., Kataja, M., and Syvänen, A.-C. (2004) Assessing hematopoietic chimerism after allogeneic stem cell transplantation by multiplexed SNP genotyping using microarrays and quantitative analysis of SNP alleles. *Leukemia* **18**, 1–12.
21. Sachidanandam, R., Weissman, D., Schmidt, S. C., et al. (2001) A map of human genome sequence variation containing 1.42 million single nucleotide polymorphisms. *Nature* **409**, 928–933.
22. Syvänen, A.-C. (1999) From gels to chips: "minisequencing" primer extension for analysis of point mutations and single nucleotide polymorphisms. *Hum. Mutat.* **13**, 1–10.
23. Lindroos, K., Sigurdsson, S., Johansson, K., Rönnblom, L., and Syvänen, A.-C. (2002) Multiplex SNP genotyping in pooled DNA samples by a four-colour microarray system. *Nucleic Acids Res.* **30**, e70.

10

Introduction to Capillary Electrophoresis of DNA

Biomedical Applications

Beatriz Sanchez-Vega

1. Principles of Capillary Electrophoresis

Capillary electrophoresis (CE) separations are carried out inside a capillary tube, which usually has a diameter of 50 μm to facilitate temperature control. The length of the capillary differs in different applications, but it is typically in the region of 20–50 cm. The capillaries most widely used are fused silica covered with an external protective coating. A small portion of this coating is removed to form a window for detection purposes. The ends of the capillary are dipped into reservoirs filled with the electrolyte. Electrodes made of an inert material such as platinum are also inserted into the electrolyte reservoirs to complete the electrical circuit. The capillary is filled with running buffer, one end is dipped into the sample, and an electric field (electrokinetic injection) or pressure is applied to introduce the sample inside the capillary. Migration through the capillary is driven by application of a high-voltage current (10–30 kV).

The molecules are detected as they pass through the window at the opposite end of the capillary. The most frequently used detector is laser-induced fluorescence (LIF), which detects fluorochromes attached to the DNA molecules; alternative detectors include ultraviolet (UV) absorbance and fluorescence. Given the short path, detection requires monitoring by sensitive equipment such as charge-coupled device (CCD) cameras. The detectors are interfaced with computers responsible not only for collecting and displaying the data but also for maintaining the timing between filter wheels and controling timing exposures, readouts of the CCD cameras, and run-time processing of the CCD images and spectral data. The molecules are detected as fluorescent peaks as they pass through the detector. An electropherogram, which is a plot of the detector response with time, is generated. Because the area of each peak is proportional to the concentration of the DNA molecule, integrated peak areas are routinely used for semiquantification due to their greater dynamic range than peak heights (*see* **Fig. 1**).

Analysis times are in the range of 3–30 min, depending on the complexity of the separation. Modern instruments are relatively sophisticated and may contain fiber-optical detection systems, high-capacity autosamplers, and temperature-control devices.

Separation of DNA fragments by CE has advantages over the classical slab gel–based separations in terms of speed and resolution, especially now that instruments that can run more than one sample at a time are available. The principle behind DNA sequencing by an instrument with many capillaries is identical to that of using a single capillary, although the design of the sheath flow cuvet and the fluorescence detection systems is considerably more complicated. At present, 8-, 16-, 48-, 96-, and 384-capillary instruments are commercially available for DNA analysis.

From: *Medical Biomethods Handbook*
Edited by: J. M. Walker and R. Rapley © Humana Press, Inc., Totowa, NJ

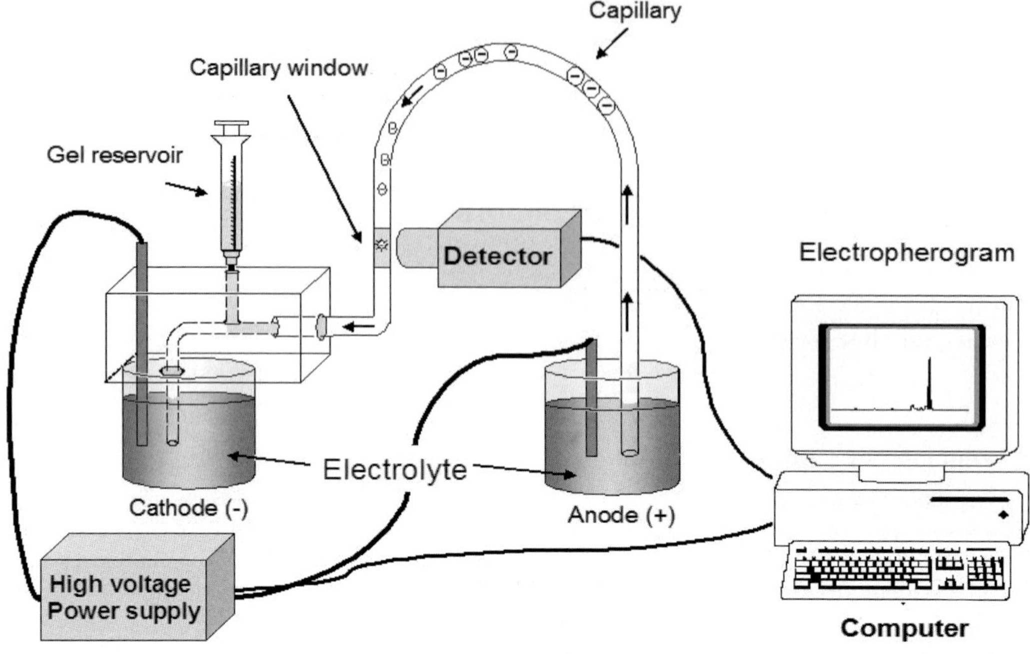

Fig. 1. Schematic diagram of a capillary electrophoresis system.

1.1. Capillary Gel Electrophoresis

There is a special situation for biopolymers such as RNA, DNA, or sodium dodecyl sulfate (SDS)–loaded proteins, which have a constant charge-to-size ratio, that is, the increase in the charge is directly related to the increase in size of the molecule. Molecules with a constant charge-to-size ratio may have very similar electrophoretic mobilities, so no electrophoretic separation occurs in free solution. In these cases, separations are performed in capillaries filled with a gel solution. In capillary gel electrophoresis (CGE), a sieving effect occurs as solutes of various sizes migrate through the gel-filled capillary toward the detector. Smaller ions are able to migrate quickly through the gel, whereas larger ions become entangled in the gel matrix, reducing their migration rate.

Initially, the gels used in CGE were polyacrylamide covalently bonded to the capillary wall. These fixed gels suffered, however, from problems of shrinkage and blockage and could have relatively short lifetimes. In addition, if sample components contaminated the gel, it could not be reused and would have to be discarded. There has been a recent tendency, therefore, to use pumpable gel solutions, which can be used to fill the capillary with a noncross-linked liquid gel matrix in which pores are created by the tangling of long linear polymers. These have the advantage of being introduced into the capillary under low pressure, extending the life of the capillary. The use of liquid gels also allows replacement of the gel between injections, reducing the contamination problems encountered with fixed gels.

1.2. Resolution

Good resolution in CE data means achieving sharp peaks and optimal separations between peaks. Peak spacing is the separation distance multiplied by the intrinsic velocity difference between two distinct molecules. This intrinsic velocity difference, in turn, is dependent on the physical properties of the molecule itself and on the separation medium (e.g., composition, concentration, ionic strength). Diffusion, thermal gradient, and initial peak width are the major contributors to the peak width. Electric field strength, capillary dimensions, and polymer concentration, in turn, mainly influence these three factors. Each of these factors is considered in detail in this chapter.

An increase in electric field strength should lead to a decrease in diffusion due to shorter run times. It should, however, also lead to an increase in peak dispersion as a result of an increase in the heat generated when the electric current is applied through the capillary (Joule heat). Thus, the electric field strength should be 150 V/cm or lower for long sequencing reads. On the other hand, the diameter of the capillary influences the efficiency in dissipating the Joule heat. The narrower the capillary, the smaller the thermal gradient and, therefore, the peak width. A reduction of capillary diameter to less than 50 μm does not improve resolution, indicating that effects other than the thermal gradient determine the peak width in narrower capillaries.

The small cross-sectional area of the capillary creates the phenomenon of electroendosmosis. Because CE is normally carried out in capillaries made from fused silica, the negatively charged silanol groups on the surface of the capillary result in a high proportion of unbound, hydrated positive ions. Although some of these cations are tightly bound to the immobilized negative charges, the application of an electric field results in movement of the remainder toward the cathode. Since they are hydrated, the consequence is a bulk flow of liquid in the same direction, termed *electroendosmotic flow* (EOF). Many of the CE techniques rely on EOF to pump the solutes toward the detector. The EOF is produced along the entire length of the capillary, generating a constant flow rate at all distances along the capillary. The consequence is that the solutes are being swept along at the same rate throughout their transport along the capillary, minimizing sample diffusion.

Separation is improved with increasing polymer concentration, but at the cost of lower mobility and, therefore, increased run times. Low polymer concentrations are useful for fast screenings at low resolution, whereas higher concentrations produce higher resolutions. Whereas in cross-linked gels the concentration can be lowered easily, in the absence of cross-linking a decrease in the concentration results in a dilute solution. To compensate for this effect, an increase in the polymer length will keep the solution well entangled, allowing for low viscosity at the same time. To cover a large range of separation sizes, it is better to use low polymer concentrations; therefore, the polymer length must be increased to ensure sufficient entanglement. This leads to a more uniform resolution in function of the DNA size and is especially important for DNA sequencing, in which long reads are required ***(1)***.

When analyzing the same DNA sample with the same parameters under native and denaturing conditions, the resolution is better in the environment that denatures the DNA molecules. If high resolution is necessary, denaturing conditions (such us the use of formamide in the sample preparation) should be used.

2. Applications of CE to Biomedicine

Small alterations in DNA sequences of genes lead to many human diseases, such as cancer, diabetes, heart disease, myocardial infarction, atherosclerosis, cystic fibrosis, and Alzheimer's disease. These alterations in DNA sequence include many types of mutations and polymorphisms, such as nucleotide substitutions, deletion or insertion of some sequence, differences in a variable number of tandem repeat (VNTR) locus, and instability of microsatellite repeat sequences. CGE is a particularly suitable method for carrying out analysis of DNA fragments for diagnostic purposes. Currently available methods for analysis of mutations and polymorphisms include some modified polymerase chain reaction (PCR) techniques, restriction fragment length polymorphism (RFLP), single-strand conformation polymorphism (SSCP), VNTR, and short tandem repeats (STRs) analysis, and hybridization techniques, as well as PCR analysis itself (*see* **Fig. 2**).

Labeled DNA fragments to be analyzed by CE are obtained by the incorporation of fluorescent-dye-labeled nucleotides or fluorescent dye-labeled primers during the PCR reaction. The range of available dyes and the reduction in costs over recent years have allowed modification

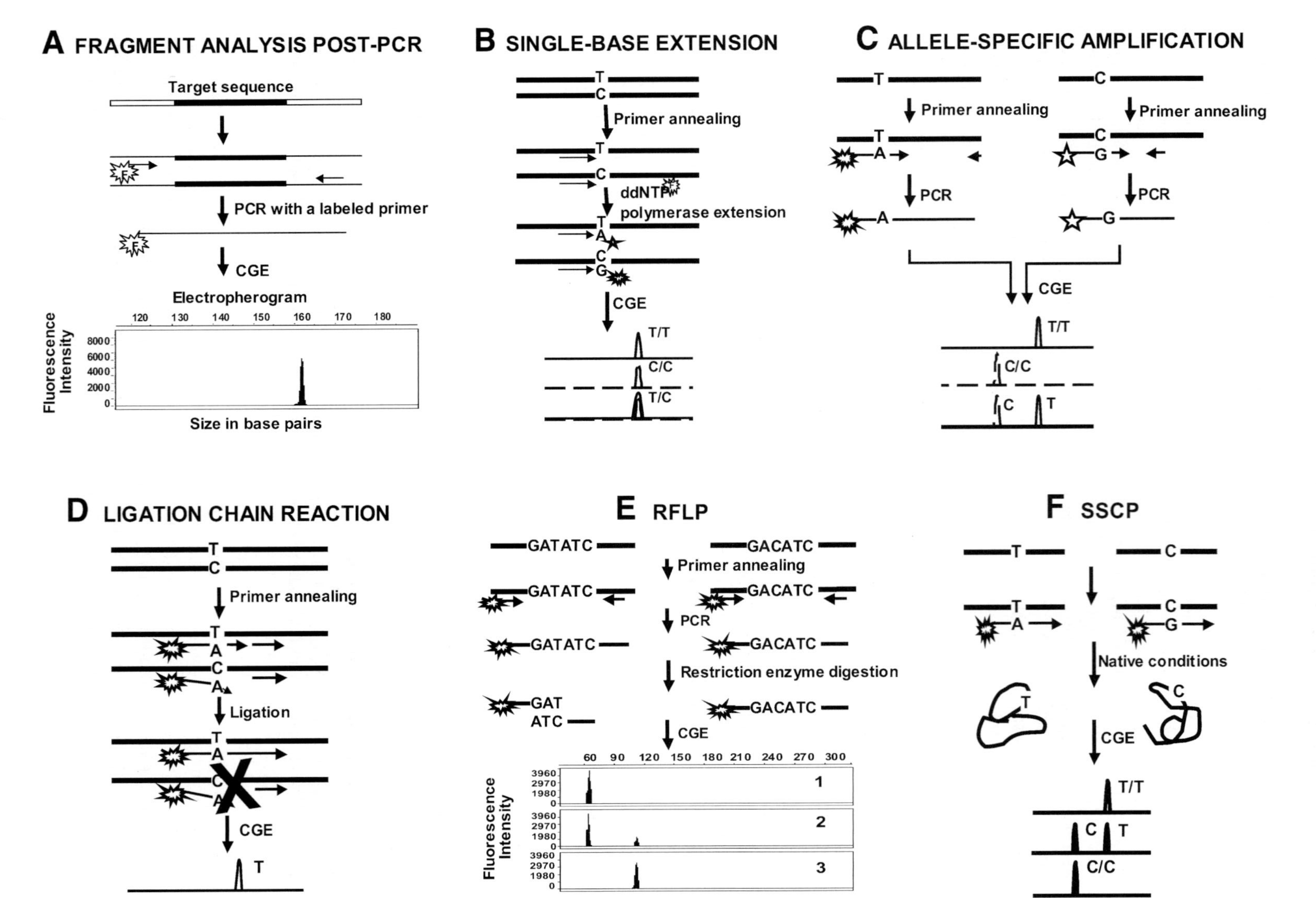

A FRAGMENT ANALYSIS POST-PCR
Target sequence
F
PCR with a labeled primer
F
CGE
Electropherogram
Fluorescence Intensity
8000
6000
4000
2000
0
120 130 140 150 160 170 180
Size in base pairs
B SINGLE-BASE EXTENSION
T
C
Primer annealing
ddNTP
polymerase extension
A
G
CGE
T/T
C/C
T/C
C ALLELE-SPECIFIC AMPLIFICATION
T
C
Primer annealing
A
G
PCR
CGE
T/T
C/C
C
T
D LIGATION CHAIN REACTION
T
C
Primer annealing
A
Ligation
CGE
T
E RFLP
GATATC
GACATC
Primer annealing
PCR
Restriction enzyme digestion
GAT
ATC
CGE
Fluorescence Intensity
3960
2970
1980
0
60 90 120 150 180 210 240 270 300
1
2
3
F SSCP
T
C
A
G
Native conditions
CGE
T/T
C
T
C/C

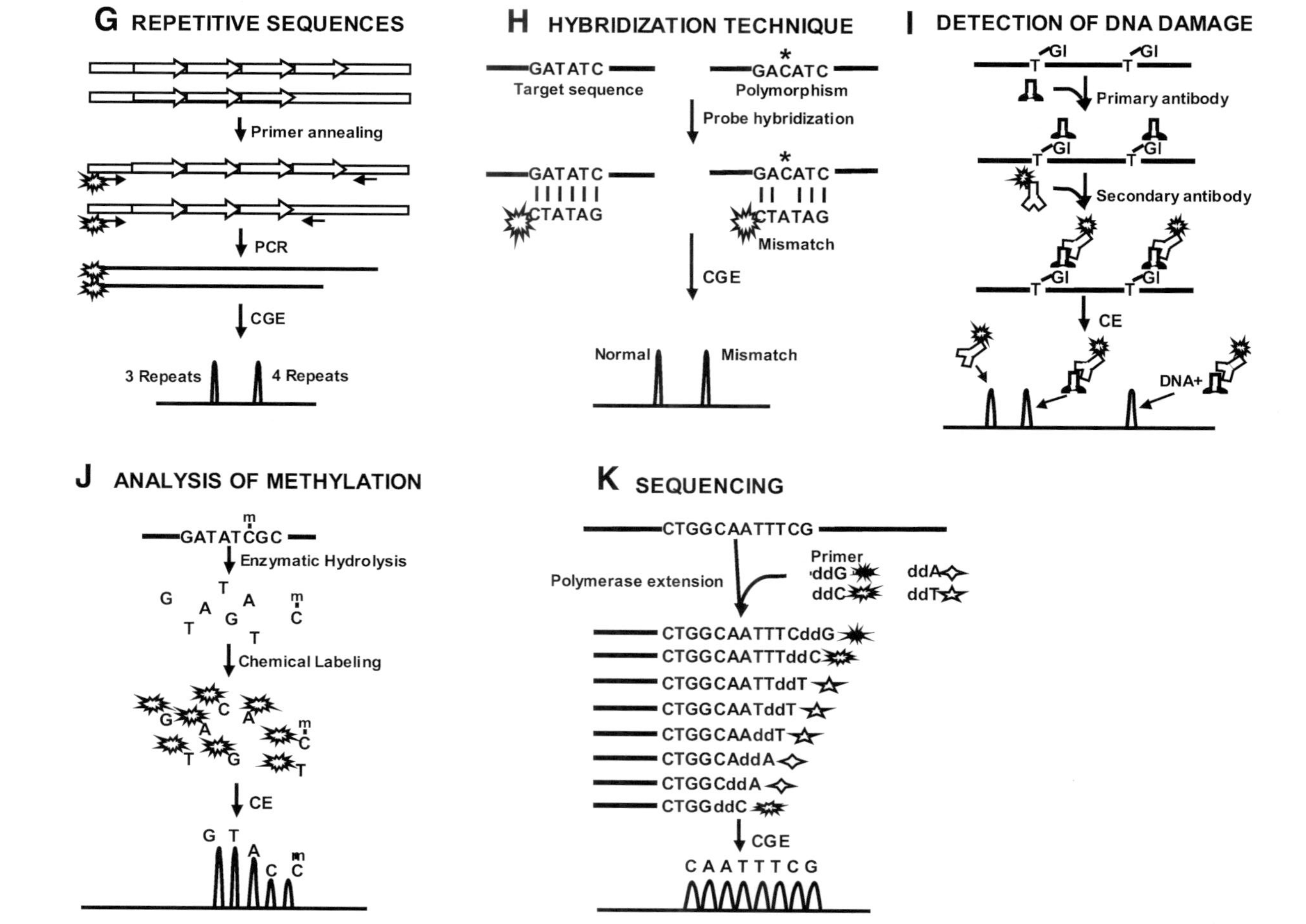

Fig. 2. Scheme of the methods applied to capillary electrophoresis: **(A)** post-PCR fragment analysis; **(B)** single-base extension; **(C)** allele-specific amplification; **(D)** ligation chain reaction; **(E)** RFLP and an example of its application to detect the D835 mutation on the *FLT3* gene in patients with acute leukemia: wild type *(1)*, heterozygote *(2)*, and homozygote *(3)* for the D835 mutation; **(F)** SSCP; **(G)** analysis of repetitive sequences; **(H)** hybridization technique; **(I)** detection of DNA damage; **(J)** analysis of methylation levels; and **(K)** sequencing.

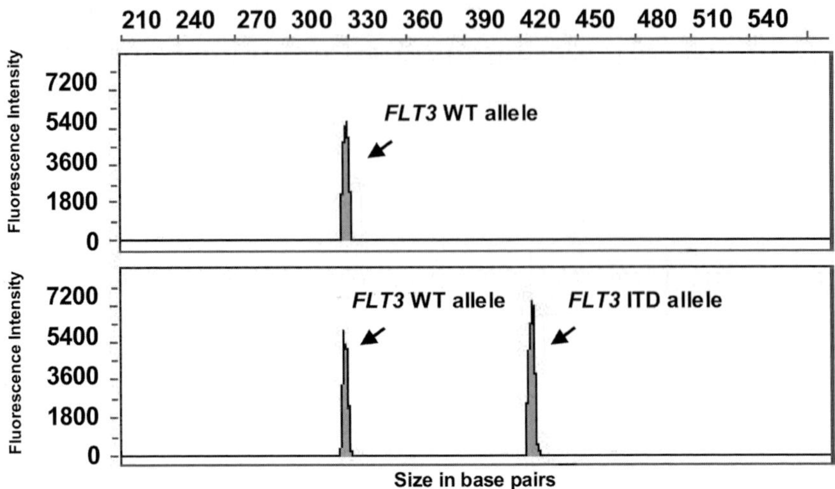

Fig. 3. Example of post-PCR detection showing the results of amplifying the ITD region of the *FLT3* wild-type (WT) allele in a healthy individual (upper panel) and in a patient with acute leukemia (lower panel).

of existing tests to take advantage of the four/five-color detection system and the degree of automation offered by commercially available CE electrophoresis systems. Four-color, high-resolution analysis is particularly useful for the rapid mapping of genetic traits, leading to gene discovery and eventual diagnostic testing. Automation is essential for rapid gene-based mutation analysis in clinical laboratories to screen a large number of DNA samples.

2.1. Post-PCR Fragment Analysis

DNA amplification by PCR has been applied to clinical diagnosis of infectious diseases, genetic disorders, and cancer. After amplification of the DNA region of interest, the sensitivity and speed of the method of analysis is crucial in clinical diagnosis. The use of CGE with LIF detection has been proved to yield better performance than conventional slab gel electrophoresis in the clinical laboratory ***(2)*** (*see* **Fig. 2A**).

2.1.1. Detection of Viral Sequences

The PCR technique is a powerful tool for detecting the presence of a virus earlier than any discernible antibody response in a person infected with a virus. Detection of specific DNA or RNA regions of a certain virus genome by multiplex PCR using CE has been applied to the detection of human immunodeficiency virus (HIV) ***(3–5)***, polio virus ***(6)***, and hepatitis C virus ***(7,8)***.

2.1.2. Detection of Intragene Deletion and Duplication Sequences

Capillary gel electrophoresis has been applied to the analysis of PCR products in the detection of small gene deletions in several genetic diseases. Examples of this application include the detection of over 98% of deletions occurring on the dystrophin gene for the diagnosis of Duchenne muscular dystrophy ***(9,10),***; an 8-bp deletion in exon 3 of the *P450c21B* gene in individuals affected by 21-hydroxylase deficiency, a recessively inherited disease ***(11)***, and the ΔF508 mutation, a 3-bp deletion in the gene *CFTR* that is the most frequently mutation found in individuals affected with cystic fibrosis ***(12)***. Another example is detection of the internal tandem duplication (ITD) in the juxtamembrane domain-coding sequence of the *FLT3* gene in acute leukemias. The mutation of this receptor is important in acute myelogenous and lymphoblastic leukemias because patients with this mutation generally have a poorer prognosis for survival than patients without it ***(13)*** (*see* **Fig. 3**).

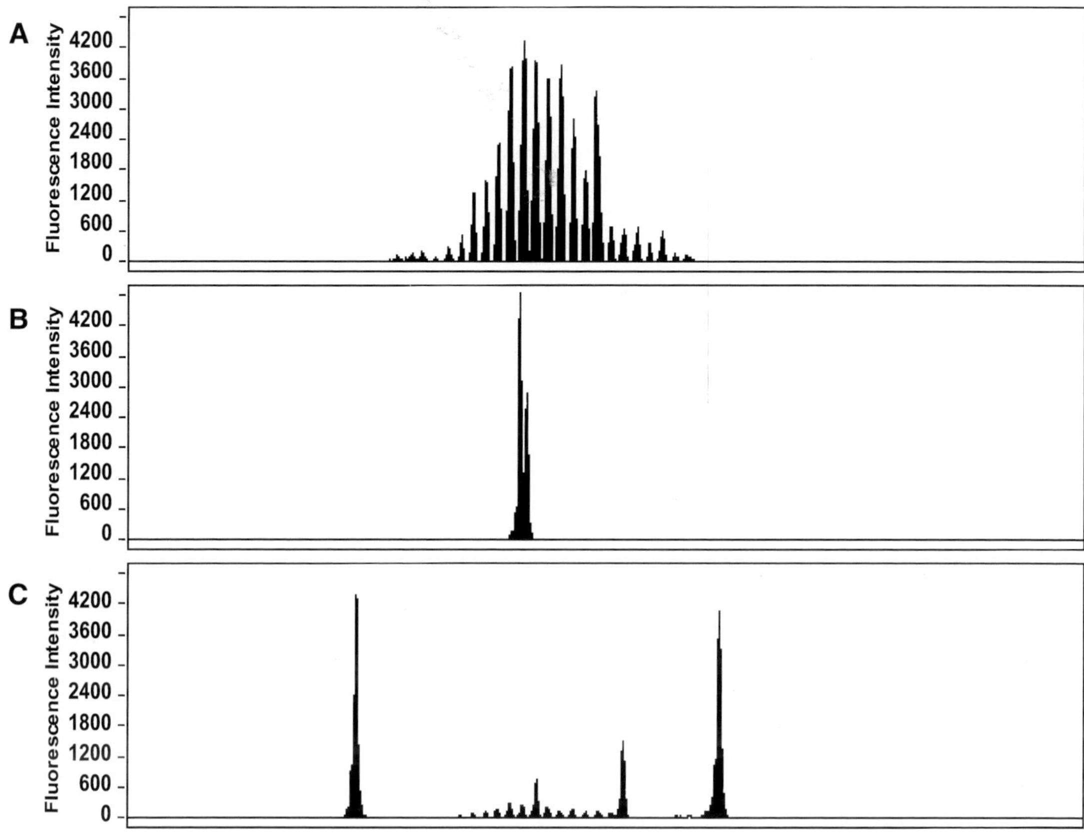

Fig. 4. *IgH* gene rearrangement results. The electropherograms show **(A)** a normal polyclonal B-cell population, **(B)** a monoclonal peak from a lymphoma patient, and **(C)** an oligoclonal population from an immunological reactive process.

2.1.3. Detection of Gene Rearrangements

Fluorescent PCR combined with CE analysis has been used to identify monoclonal rearrangements in the T-cell receptor (*TCR*) ***(14)*** and immunoglobulin heavy-chain (*IgH*) ***(15)*** genes in order to detect residual disease in lymphoproliferative disorders. Normal lymphocytes rearrange the *TCR* and *IgH* genes randomly during normal development, thus, a normal polyclonal population shows many different rearrangements. Neoplastic cells exhibit a single rearrangement of the *TCR* or *IgH* gene. These rearrangements can be detected by PCR if the amplification is conducted across the sequence where the gene rearranges. Polyclonal samples show a gaussian-like distribution of the peaks, whereas monoclonal or neoplastic cases display a single predominant peak. Four or five separate peaks are consistent with oligoclonal populations and almost always are the result of immunological reactive processes to unknown antigens (*see* **Fig. 4**).

2.1.4. Detection of Chromosomal Reorganizations

Chromosomal translocations and inversions are often detected in many types of cancer. Chromosomal translocations result in juxtaposition of segments from two different genes, giving rise to a fusion gene that encodes chimeric proteins with transforming activity. In general, both genes involved in the fusion contribute to the transforming potential of the chimeric oncoprotein *(16)*. CE has been applied to the detection of many fusion transcripts, such as the bcr/abl transcript in chronic myeloid leukemia *(17)*.

On the other hand, gene activation often results from chromosomal reorganizations that relocate a proto-oncogene close to the regulatory elements of the *IgH* or *TCR* locus. This event causes deregulation of the proto-oncogene expression, which can then lead to neoplastic transformation of the cell. One of these proto-oncogenes is the *BCL-2* gene, which is translocated close to the regulatory element of the *IgH* gene in persons with follicular lymphoma. The different PCR products can be resolved by CGE *(18)*.

2.2. Single-Base Extension or Single Nucleotide Primer Extension

Single-base extension (SBE) is a method used for the accurate detection of single-nucleotide polymorphism variants in transcripts. Standard labeled Sanger dideoxynucleotide terminators (ddNTP), which lack the 3' hydroxyl group needed for chain elongation, are used in the extension reaction; as a result, the primer is extended by a single nucleotide. The reaction mixture is subsequently separated by CGE *(19)* (**Fig. 2B**).

For example, Leber's hereditary optic neuropathy was detected by annealing a primer immediately 5' to the mutation on the template and extending the primer by one fluorescently labeled ddNTP complementary to the mutation. By using two or more differently labeled terminators, both the mutant and wild-type sequences could be detected simultaneously *(20)*. In addition, multiplexed SBE genotyping has been applied to the detection of known point mutations on the *p53* gene using three unique primers that probe for mutations of clinical importance on this gene *(21)*, and to the diagnosis of multiple endocrine neoplasia type 2 by the analysis of codons 634 and 918 on the *RET* proto-oncogene *(22)*.

A different field of application is the determination of ovine prion protein allelic variants at codons 136, 154, and 171, where the four mutations responsible for amino acid changes are typed simultaneously *(23)*.

2.3. Allele-Specific Amplification

Allele-specific amplification (ASA) allows simultaneous analysis of multiple specific alleles, known point mutations, or known polymorphisms at the same locus. Specificity is achieved in a single PCR reaction by designing one or both PCR primers so that they overlap the site of sequence difference between the amplified alleles. This results in simultaneous amplification of numerous DNA fragments differing in size, which are subsequently separated by CGE (*see* **Fig. 2C**).

This technique was applied to the detection of K329E, the most prevalent mutation in medium-chain acyl-coenzyme A dehydrogenase deficiency. The mutant allele produces a DNA fragment that differs in size from the DNA fragment generated by the normal allele *(24)*. ASA also can be used to detect multiple mutations in a locus. For example, several point mutations on the 21-hydroxylase gene of a patient with congenital adrenal hyperplasia were analyzed simultaneously by this method *(25)*.

Familial defective apolipoprotein B-100 (FDB) is a dominantly inherited disorder characterized by a decreased affinity of low-density lipoprotein (LDL) for the LDL receptor. This phenotype is a consequence of a substitution of adenine by guanine in exon 26 of the *FDB* gene in the region coding for the putative LDL receptor-binding domain of the mature protein. Mutation screening for *FDB* is performed by ASA followed by CGE *(26)*.

Another mutation that can be detected by ASA and CGE is the G1691A point mutation in the coagulation factor V gene, the most common genetic cause of thrombophilia. This mutation has been shown to cause resistance to cleavage by activated protein C and is associated with an increased thrombotic risk *(27)*.

2.4. Ligase Chain Reaction

In this method, a thermostable ligase is used to specifically bond two adjacent oligonucleotides, which hybridize to a complementary target with perfect base pairing at the junction. This approach enables single base pair differences to be detected at the ligation junction (*see* **Fig. 2D**). The design of primers with distinct labels for different base substitutions allows for simultaneous detection of several changes. This technique has been applied to the screening of the 31 most frequent mutations in the cystic fibrosis gene in a multiplex oligonucleotide ligation assay ***(28)***.

Sickle cell disease refers to a collection of autosomal recessive genetic disorders characterized by a hemoglobin variant (Hb S) which differs from hemoglobin A in having a single amino acid substitution on the β-chain. Individuals who are affected with other types of sickle cell disease are compound heterozygotes, possessing one copy of the Hb S variant and one copy of a different β-globin gene variant such as Hb C or Hb β-thalassemia. CE has been applied to the detection of both hemoglobin forms Hb S and Hb C in prenatal diagnoses of sickle cell diseases ***(29)***.

2.5. Restriction Fragment-Length Polymorphism Analysis

Restriction fragment-length polymorphism (RFLP) analysis is a powerful tool for the detection of known mutations and polymorphisms, which result in the loss or creation of a recognition site at which a particular restriction enzyme cuts. RFLPs are usually caused by mutation at a cutting site. DNA carrying the different allelic form will give different sizes of DNA fragments when digested with the appropriate restriction enzyme (*see* **Fig. 2E**) (see Chapter 3).

The E4 allele on the *APOE* gene has been proven to be a major risk factor for late-onset familial and sporadic Alzheimer's disease. DNA samples for *APOE* genotyping are prepared by PCR amplification and subsequent digestion with that *Hha*I restriction enzyme to yield constant fragments of 16, 18, and 35 base pairs and four typical polymorphic fragments of 48, 72, 83, and 91 base pairs, corresponding to alleles E1, E2, E3, and E4, respectively ***(30)***. The risk of developing Alzheimer's disease for the individual having the E4/E4 genotype is about 90%, for the individual having the E3/E3 genotype, 20%; and for the individual having the E3/E4 genotype, 47%.

Capillary gel electrophoresis has also been applied to resolve DNA fragments digested with *Bst*NI endonuclease to detect mutations in codon 12 of *K-RAS* ***(31)*** and to detect the D385 mutation in the *FLT3* gene, in which amplification of the sequence of interest is followed by *Eco*RV digestion.

A different application of the RFLP method is mitochondrial DNA typing. The polymorphic control region of mitochondrial DNA (mtDNA) is being used more and more commonly in forensic applications to differentiate among individuals in a population. Two hypervariable regions (HV1 and HV2) are often sequenced following amplification of the mtDNA by PCR. A methodology has been developed that uses restriction endonuclease digestion of the PCR-amplified mtDNA by *Rsa*I and *Mnl*I and CGE to separate and size the RFLP fragments ***(32)***.

2.6. Single-Strand Conformation Polymorphism

The SSCP method is based on the principle that the electrophoretic mobility of single-stranded DNA in a nondenaturing condition depends not only on its size but also on its sequence (see Chapter 7). Single-stranded DNA has a folded structure that is determined by intramolecular interactions. This sequence-based secondary structure affects the mobility of the DNA during electrophoresis under nondenaturing conditions. A DNA molecule containing a mutation, even a single-base substitution, will have a different secondary structure than the wild type, resulting in a different mobility shift during electrophoresis than that of the wild type (*see* **Fig. 2F**). The technological aspects of CE-SSCP as well as the potential of CE-SSCP for routine genetic analysis were reviewed by Kourkine et al. in 2002 ***(33)***.

Some examples of applications of SSCP by CGE include analysis of the *p53* tumor suppressor gene for DNA diagnosis of cancer ***(34)***, detection of mutations on the *K-RAS* oncogene ***(35)***, detection of a *Mycobacterium tuberculosis*–specific amplified DNA fragment ***(36)***, detection of mutations in the β-*globin* gene promoter ***(37)*** and the lipoprotein lipase and breast cancer 2 (*BRCA2*) genes ***(38)***, and identification of multiple mycobacterial species in a sample ***(39)***.

CE-SSCP also has been used for the separation of fully supercoiled molecules, single topoisomers, and relaxed and open circular DNA forms in research on novel anticancer molecules targeting the activity of topoisomerase I ***(40)***.

One approach that combines restriction enzyme digestion of fluorescence-labeled PCR products with SSCP by CE has been developed to screen exon 11 mutations of the breast cancer 1 (*BRCA1*) gene ***(41)***. The large size of the *BRCA1* gene and the many mutations found throughout its entire coding sequence make screening for mutations in this gene particularly challenging. Capillary RFLP-SSCP electrophoresis appeared as a technically convenient technique to test mutations in this gene, requiring amplification of fewer PCR fragments than traditional SSCP.

2.7. Analysis Of Repetitive DNA Sequences

The human genome contains large numbers of repetitive DNA sequences, some of which are arrayed as tandem repeat units. Polymorphic tandem repeats occur in two general families: VNTRs and STRs. VNTR or minisatellite DNA consists of variable numbers of a repeating unit from 15 to 100 nucleotides long. The repeated unit in STR or microsatellite DNA ranges from two to six bases.

Products of different lengths can be amplified by using primers complementary to conserved sequences flanking the tandem repeat regions; the length of each amplification product is directly proportional to the number of repeats. These amplification products can be resolved to individual alleles by CGE. Thus, a person can have at most two different alleles, one from each chromosome (*see* **Fig. 2G**).

Laser-induced fluorescence detection of fluorescence-labeled PCR products and multicolor analysis enable the rapid generation of multilocus DNA profiles.

2.7.1. VNTR

Capillary gel electrophoresis has been applied to the analysis of the human apolipoprotein B gene (*APOB*). This gene maps to the short arm of chromosome 2, and a VNTR is located immediately downstream. The *APOB* VNTR alleles generally contain 25–52 repeats of a basic 16-bp unit. Larger alleles containing more than 38 repeat units are more common in people who have myocardial infarction and show a significant association with coronary disease ***(42)***.

Capillary gel electrophoresis also has been used to determine the lung cancer risk associated with rare *HRAS1* VNTR alleles ***(43)***.

2.7.2. STR

The number of tandem repeats for a STR is highly variable from individual to individual and can be as many as 15 different alleles in a certain locus. These sequences also show high levels of heterozygosity, which make them very useful as genetic markers (see Chapter 33).

2.7.2.1. Highly Polymorphic Markers

2.7.2.1.1. Genetic Mapping. The almost random distribution of microsatellites and their high level of polymorphism greatly facilitated the construction of genetic maps and enabled subsequent positional cloning of several genes. A comprehensive genetic map of the human genome based on 5264 microsatellites ***(44)*** and a genetic map of the mouse genome based on 4006 simple sequence length polymorphisms ***(45)*** were constructed with these markers. When a trait or disease is inherited through a family and all of the affected members in the family also share a STR that is inherited together, it can be concluded that the disease and marker must be close to each other on a chromosome. This linkage is the first step in the isolation of disease-causing genes and their localization to an area of DNA. Linkage disequilibrium scanning of the human genome with many thousands of genetic markers has been used to map complex diseases such as schizophrenia ***(46)*** and autistic disorder ***(47)***.

2.7.2.1.2. Trisomies. The D21S11 locus, a highly polymorphic (90% heterozygosity) microsatellite marker, was amplified and analyzed by CGE for prenatal DNA diagnosis of Down's syndrome ***(48)***. In the analysis of this microsatellite marker, trisomic individuals are expected to fall into two major groups: those with three bands of similar intensities or those with two bands with a ratio of 2 : 1.

2.7.2.1.3. Duplication in Chromosomes. The Charcot–Marie–Tooth 1A disease is the result of a 1.5-Mb duplication in chromosome subband 17p11.2-p12. Suitable dinucleotide repeat markers within this region are amplified by using fluorescence-labeled primers and separated by CGE. The results are two peaks corresponding to two alleles with a dosage difference ***(49)***.

2.7.2.1.4. Forensic DNA Typing. In forensic work, DNA from a biological sample and from a known reference material (most often, buccal swabs or blood) are amplified by multiplex PCR with fluorescence-labeled primers and separated by CGE. As a result, unique DNA profiles are generated from both the evidence and the known reference sample and compared to determine whether the patterns are similar. If the DNA profiles match, then the suspect cannot be excluded as a source of the questioned sample. Likewise, if the DNA profiles do not match, the suspect can be excluded as the donor of the questioned sample. The analysis of 6–10 STRs provides a random match probability of approx 1 in 5 billion.

DNA typing is a useful tool in crime solving, not only for blood samples, sperm, or saliva but also for traces of DNA left on tools or pieces of clothing used in burglaries or thefts. On these kinds of samples, the sources of DNA are extremely small amounts of skin debris left after gripping tools. When a sensitive technique such as PCR coupled with CGE is used, it is possible to get a profile from these low amounts of DNA ***(50)***. CE provides efficient separation, resolution, sensitivity, and precision for a reliable genotyping of STR loci. Sizing precision of <0.16 nucleotide standard deviation is obtained with this system, thus allowing the accurate genotyping of variants that differ in length by a single nucleotide ***(51)***.

Automated fluorescent detection of a 10-loci multiplex has been applied to paternity testing ***(52)***. DNA is extracted from blood samples of the mother, child, and alleged father and subjected to STR analysis. By using 10–15 STRs, the calculated probability of paternity usually exceeds 99.99%.

2.7.2.1.5. Chimerism. Analysis of the relative amounts of donor and recipient DNA in bone marrow after bone marrow transplantation is frequently used to determine the status of the transplant. One method involves testing the patient and donor for genetic profiles before the transplant and then comparing these with the profile obtained from the patient after transplant. The analysis of chimerism in patients who have undergone allogeneic stem cell transplantation by CE systems was reviewed by Lion in 2003 ***(53)*** (*see* **Fig. 5**).

2.7.2.2. Microsatellite Instability

Microsatellite instability (MSI) has been shown to be relevant to various diseases, including cancers. Cells with mutations in one of the genes responsible for DNA mismatch repair (MMR) accumulate mutations at a very high rate. Because DNA mismatches are more likely to occur in DNA microsatellites, defective DNA MMR leads to the phenomenon of MSI, in which the progeny of the defective cells have varying lengths of a given microsatellite. The microsatellite reference panel to study MSI is BAT25, BAT26, D5S346, D2S123, and D17S250 ***(54)***.

Studies of MSI are performed routinely in patients with colorectal cancer because of the defective MMR phenotype of colorectal carcinomas. The electrophoretic profile of the amplified products is compared between tumor and normal DNA from the same patient ***(55)*** (*see* **Fig. 6**).

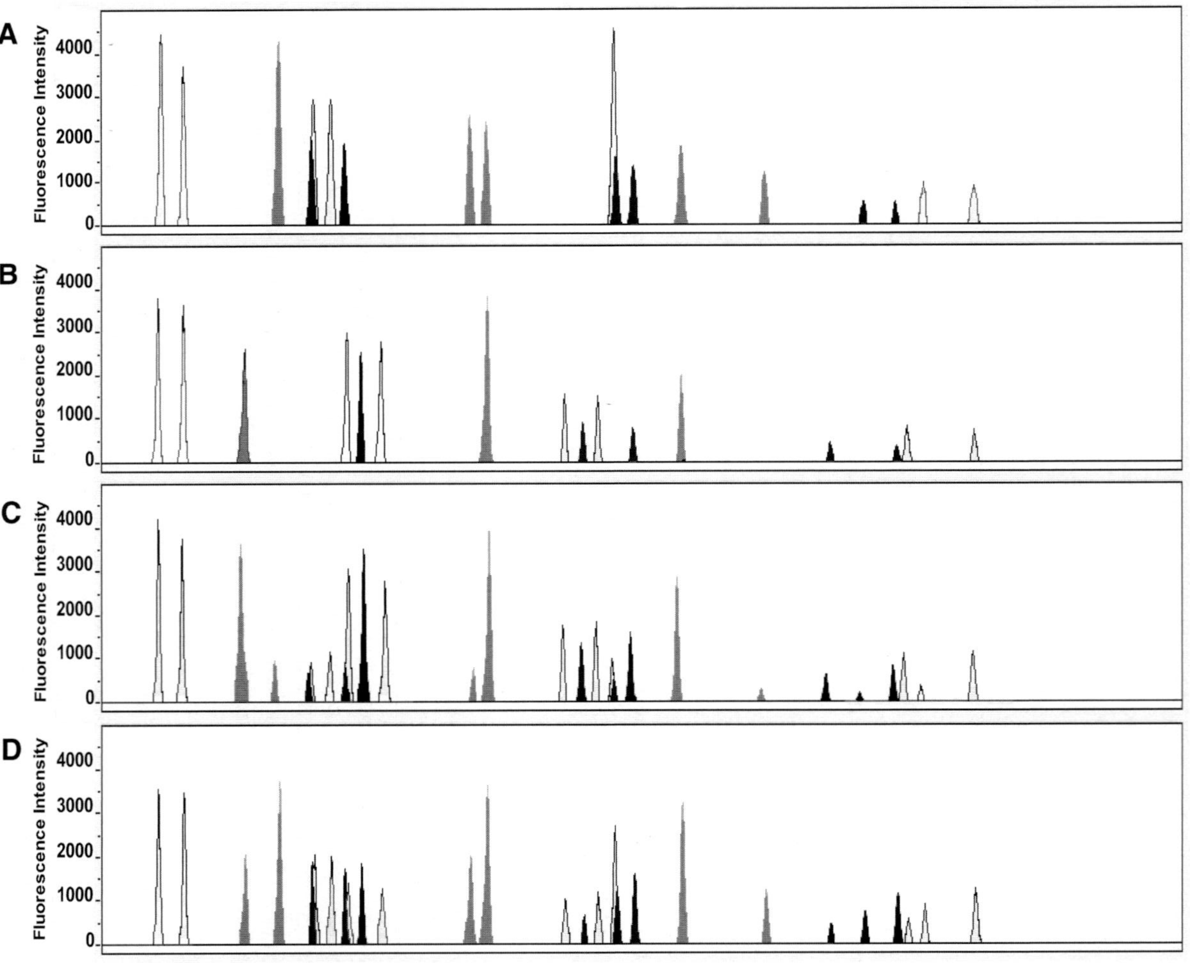

Fig. 5. Chimerism study in a patient undergoing bone marrow transplantation (BMT). The electropherograms show the genetic profiles of **(A)** the patient before BMT, **(B)** the donor, **(C)** the patient 1 mo after BMT, showing a mixed profile with patient and donor markers, **(D)** the patient 6 mo after BMT, showing an increase in the patient pre-BMT markers. This indicates a rejection of the transplant.

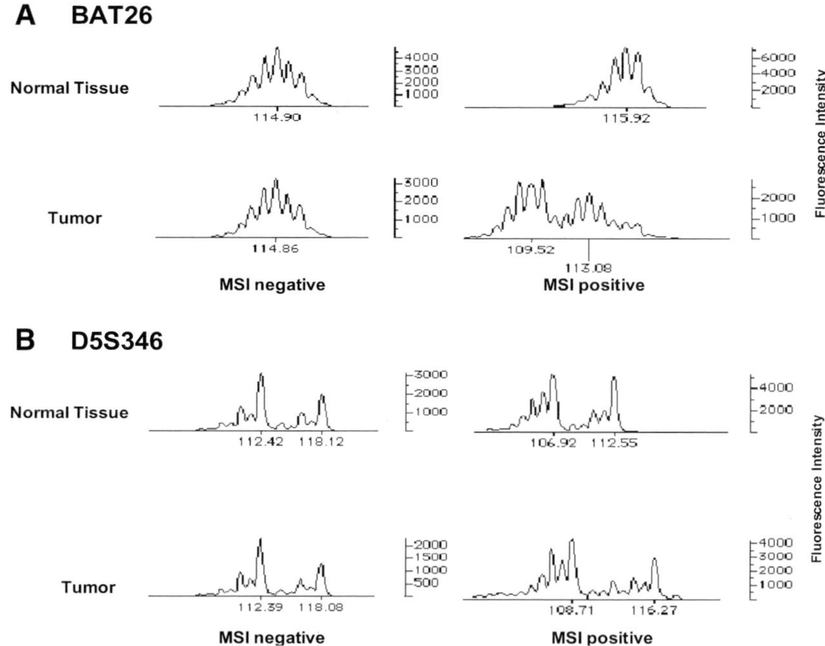

Fig. 6. Microsatellite instability results from BAT26 **(A)** and D5S346 **(B)** markers in two tumor samples. The left panels show MSI-negative tumor tissues and the right panels show MSI-positive tumor tissues when compared with normal tissue from the same patient.

2.7.2.3. Loss of Heterozygosity

Loss of heterozygosity (LOH) refers to a mutation that results in the loss of one allele at a specific locus. This phenomenon is very frequent in cancer cells. Whereas normal cells are heterozygous at many loci, tumor cells are homozygous at the same loci. LOH can be detected by CGE in bladder carcinoma ***(56)***, lung cancer ***(57)***, breast cancer ***(58)***, renal cancer ***(59)***, and other solid tumors ***(60)***.

2.7.2.4. Dynamic Mutations

Some microsatellite repeats can undergo an increase in copy number by a process of dynamic mutations, typically by expansion of an unstable trinucleotide repeat. These repeat regions are polymorphic in unaffected individuals, so the number of repeated triplets varies within a certain range. In affected individuals, the number of repeats exceeds the normal range. When, in a family, the number of repeats increases in the normal range, the phenomenon is called a permutation. In such cases, there is a strong possibility of further expansion to a full mutation during inheritance from parent to child. The high resolving power of CGE allows separation of heterozygous microsatellite bands differing by only one triplet.

Spinocerebellar ataxia type 2 results from an expansion of a stretch of polyglutamine repeats within the coding sequence of the ataxin-2 gene. The clinical phenotype arises when more than 32–35 glutamine residues are found in this region ***(61)***. Other examples are fragile X syndrome ***(62)***, myotonic muscular dystrophy ***(63)***, Kennedy's disease ***(64)***, spinocerebellar ataxia type 1 ***(65)***, and Huntington's disease ***(66)***.

2.8. Hybridization Technique

In this technique, a probe of known sequence is capable of recognizing a specific DNA sequence in a mixture of many different DNA sequences (see Chapter 4). The probe can be a synthetic oligonucleotide, a peptide nucleic acid (PNA), a DNA sequence, an RNA sequence, or DNA-based drugs such as DNA vaccines, antisense DNA, protein-binding oligonucleotides, and ribozymes. The probe is generally labeled with a fluorochrome that can be detected after the hybridization takes place.

In a DNA hybridization assay, the sample DNA is heated to separate the two DNA strands and then exposed to the probe. The degree of DNA sequence identity is detected through a hybridization reaction between the DNA and the probe. The hybridization products are then resolved by CGE (*see* **Fig. 2H**).

This method has applications in such diverse areas as analysis of pooled genomic DNA samples, detection of mutations, screening of populations for polymorphisms, and identification of species in environmental mixtures.

2.8.1. Analysis of Heteroduplex DNA

Heteroduplex DNA is formed by pairing an unknown nucleic acid strand with a reference strand of a known sequence. Single-stranded nucleic acids with complementary sequences come together and form a double-stranded hybrid molecule in which hydrogen bonds form only between complementary base pairs. The thermal stability of DNA is related to the number of hydrogen bonds between complementary base pairs; therefore, the greater the similarity in nucleotide sequences between two samples of DNA, the more hydrogen bonds will be present in the heteroduplexes produced. Heteroduplexes generated with a mismatch in the sequence between the two strands will have a different sequence-dependent electrophoretic mobility than homoduplex DNA generated with perfectly matched sequences.

Heteroduplex analysis also can be done to detect different sequences in a mixture by using different probes labeled with different colors. Multiplex analysis has been applied to the simultaneous detection of six heterozygous mutations in *BRCA1* and *BRCA2* ***(67)***, screening of the three common prothrombotic polymorphisms pl(A), factor V Leiden, and MTHFR (C677T) ***(68)***, rapid genetic screening of hemochromatosis ***(69)***, genetic diagnosis of factor V Leiden ***(70)***, clinical diagnosis of rifampin-resistant tuberculosis strains ***(71)***, and analysis of complex mutational spectra of human cells in culture ***(72)***.

Heteroduplex technology is frequently combined with other screening techniques such as SSCP or ASA. Some examples of the combination of heteroduplex analysis with SSCP are detection of mutations, polymorphisms, and variants in *BRCA1* and *BRCA2* genes ***(73)*** and detection of mutations in the *p53* gene ***(74)***. Heteroduplex analysis has been combined with ASA in order to detect specific mutations in the *BRCA1* and *BRCA2* genes ***(75)***.

2.8.2. Synthetic Oligonucleotides Probes

The use of oligonucleotides as probes for hybridization has been used to determine the specificity of PCR products. One example is the use of oligonucleotides to target specific HIV-1 sequences. When these oligonucleotides are mixed with genomic DNA amplified by PCR, the presence of the HIV-1 virus within an individual can be detected (76).

2.8.3. Peptide Nucleic Acid Probes

Peptide nucleic acids are molecules in which the bases are conjugated to a polyamide backbone. The distance between bases and the complementarity are the same as in a DNA or RNA molecule. The PNA probe is hybridized to a DNA sample amplified by PCR, denatured at low ionic strength, and resolved by CE. The neutral backbone of PNA ensures an efficient CE separation of the PNA/DNA hybrids from both double-stranded and single-stranded DNA. DNA strands fully complementary to the target PNA are retarded compared to single-nucleotide mismatched strands ***(77)***.

This method can be used to perform multiplex analyses on several mutations simultaneously when using different amplicon lengths and a set of mutation-specific PNAs. Each targeted mutation can be identified by the size of its corresponding amplicon. Its genotype is further characterized by its interaction with a specific PNA that can differentiate between wild-type and mutant alleles. This approach has been applied to the detection of R553X and R1162X single-base mutations in patients with cystic fibrosis ***(78)*** and to identification of the H63D, S65C, and C282Y mutations in the hereditary hemochromatosis gene ***(79)***.

2.8.4. Analysis of DNA-Based Drugs

Antisense oligonucleotides are synthetic oligomers that bind a target mRNA or DNA, forming duplexes and thereby blocking its translation into a peptide or triggering its degradation. Among several DNA analogs proposed as antisense DNA drugs, antisense DNA in which the backbone is modified into phosphorothioate is the most promising for therapeutic application because of its resistance to nuclease degradation.

This methodology is a useful tool for synthesis control and analysis of phosphorothioate analogs ***(80)*** and for the determination of their subcellular distribution ***(81)***, *in vitro* stability ***(82)***, tissue half-life ***(83)***, purity ***(84)***, and metabolism ***(85)***.

2.9. Temperature Gradient Capillary Electrophoresis

In temperature gradient CE (TGCE), a continuous, gradually decreasing temperature gradient is established during the electrophoresis. This approach can be applied to detection of DNA mutations based on thermodynamic stability and mobility shift during electrophoresis. By spanning a wide temperature range, it is possible to perform simultaneous heteroduplex analysis for various mutation types that have different melting temperatures ***(86)***.

Genomic DNA also can be amplified with one fluorescence-labeled primer and one GC-clamped primer and analyzed by TGCE under denaturing temperature conditions. This tactic has been applied to the detection of mutations in exon 8 of the *p53* gene from tumor samples and controls ***(87)***.

2.10. Quantification Of DNA Damage

Damage to cellular DNA is implicated in the early stages of carcinogenesis and in the cytotoxicity of many anticancer agents, including ionizing radiation. CE is a sensitive technique for measuring cellular levels of DNA damage through the detection of specific DNA lesions by specific monoclonal antibodies. The detection is done by the use of a secondary antibody labeled with a fluorochrome. This method has been applied to the determination of thymine glycol residues in DNA generated by irradiation of cells during radiation therapy. The sensitivity of detection is in the range of zeptomoles ***(88)*** (*see* **Fig. 2I**).

Another way to detect DNA damage is the evaluation of DNA laddering during apoptosis. A key step in the onset of apoptosis is cleavage of the genomic DNA between nucleosomes, resulting in polynucleosome-sized fragments of DNA that give rise to a characteristic DNA laddering pattern. A fluorescent intercalating dye is used to label the DNA molecules that will be resolved by CGE. Also, the use of a DNA standard curve allows quantification of the apoptotic DNA. The use of CGE with LIF detection permits analysis of DNA laddering with improved automation and much greater sensitivity ***(89)***.

2.11. Analysis of Methylation Status

DNA methylation refers to the addition of a methyl group to the 5-position of cytosine residues that are followed immediately by a guanine (so-called CpG dinucleotides). Small stretches of DNA known as CpG islands are rich in CpG dinucleotides. These CpG islands are frequently located within the promoter regions of human genes, and methylation within the islands has been shown to be associated with transcriptional inactivation of the corresponding gene. Changes in DNA methylation profiles are common features of development ***(90)***, and DNA methylation is also involved in X-chromosome inactivation ***(91)*** and allele-specific silencing of imprinted genes ***(92)***. Aberrant changes in methylation profiles are associated with many serious pathological consequences such as tumorigenesis ***(93)***. Methylation of CpG islands has been shown to be important in transcriptional repression of numerous genes that function to prevent tumor growth or development. This phenomenon includes both genome-wide hypomethylation and gene-specific hypermethylation.

Capillary electrophoresis has been applied to the evaluation of the relative methylation degree of genomic DNA methylation. In this approach, genomic DNA is enzymatically hydrolyzed to single nucleotides. Single nucleotides are then labeled with a fluorescent marker and separated by CE to identify cytosine and 5-methyl cytosine residues ***(94)*** (*see* **Fig. 2J**).

This method has been applied to determine DNA methylation levels and their oncogenic potential in chronic lymphocytic leukemia ***(95)***, and it could be useful for the diagnosis of genetic diseases that have specific DNA methylation pattern alterations such as Angelman and Prader–Willi syndromes.

2.12. Sequencing

DNA sequencing is the determination of the order of nucleotides in a DNA molecule. The most popular method for doing this is called the Sanger sequencing method or chain-termination method. This method uses synthetic nucleotides that lack the 3' hydroxyl group and are unable to form the 3'–5' phosphodiester bond necessary for chain elongation. These nucleotides are called dideoxynucleotides (ddNTPs).

The sequencing reaction consists in a single-stranded template DNA to which a short complementary primer is annealed and extended by a DNA polymerase. The sequencing reaction contains a low concentration of ddNTPs, each labeled with a different fluorochrome, in addition to the normal deoxynucleotides. Once a ddNTP is incorporated in the elongating chain, it blocks further chain extension; as a result, a mixture of chains of lengths determined by the template sequence is accumulated. This mixture can then be resolved by CGE. CGE resolution allows separation of chains that differ in length by only one nucleotide (*see* **Fig. 2K**).

Sequencing by using CGE has been in constant development with steady advances in speed ***(96)***, separation media ***(97)***, and data collection and large-scale instruments ***(98)***. DNA sequencing is considered the standard for identification of nucleic acids and detection of mutations, and CGE is routinely used in research and clinical situations in which panels of mutations or large numbers of samples are analyzed and when turnaround time is critical.

Direct sequencing of amplified fragment-length polymorphism bands also has been used as a polymorphism isolation strategy of population genetic parameters in genomic DNA of nongenomic model species ***(99)***.

The combination of direct cycle sequencing of PCR products with CGE also provides a simple and rapid method convenient for routine hepatitis C virus genotyping analysis ***(100)***.

2.12.1. Polymorphism Ratio Sequencing

In the polymorphism ratio sequencing (PRS) method, dideoxy-terminator extension ladders generated from a sample and a reference template are labeled with different fluorescent dyes and coinjected into a capillary for comparison of relative signal intensities. PRS allows detection and genotyping of single-nucleotide polymorphisms in the analysis of individual or multiplexed samples ***(101)***.

2.12.2. Simulseq or Simultaneous Sequencing of Multiple PCR Products

This strategy allows multiple PCR products to be sequenced in a single sequencing reaction and analyzed simultaneously in a single lane or capillary. To achieve this separation, the different primers used for the sequencing reactions have long tails of nucleotides with different lengths ***(102)***.

3. Microchips

Microchip analysis has become an attractive option in CE for clinical applications, not only because these kinds of analysis require usage of complex samples that are often limited in quantity but also because this system could allow clinicians to make quicker treatment and drug therapy decisions.

The use of CE in the microchip format allows high-speed separations of very small samples, and multiple channel systems have great potential for high-throughput analysis. The first report of DNA separation on a microchip device was published by Woolley and Mathies in 1994 ***(103)***. Since then, great progress has been made in the design of these devices, going from single-channel to multiple-channel chips, allowing analysis of several DNA samples. The design of these systems was reviewed by Gao et al. in 2001 ***(104)***.

A typical microchip device consists of one or several separation channels of 3–10 cm in length and 10–100 μm in diameter. There are also buffer, waste, and sample reservoirs. High voltages of 2–30 kV can be applied to the reservoirs by platinum electrodes because these systems disperse the heat very efficiently. The small samples used in the CE microchips require extremely sensitive detection techniques. LIF is the most commonly used detection method in CE microchips, although electrochemical, ultraviolet (UV), chemiluminiscence, and indirect fluorescence detection methods have been combined with these systems.

The applications of conventional CE can be scaled to microchip systems. Detection of mutations in hereditary hemochromatosis, STR genotyping, and rapid mitochondrial DNA sequence polymorphism analysis have all been done by these devices ***(105)***.

The high-throughput capabilities of these systems were demonstrated by the simultaneous separation of multiple STR amplicons in 96 channels in 8 min ***(106)***. Some other clinical applications carried out on microchip CE systems were reviewed by Wessagowit and South and Gawron et al., including DNA sequencing, genetic analysis, immunoassays, and protein and peptide analyses ***(107,108)***.

The applications and designs of microchip systems are constantly expanding. Integration of sample preparation and separation in a single microchip can be addressed in a few years, which will lead to a tremendous impact on high-throughput DNA analysis.

References

1. Heller, C. (2001) Principles of DNA separation with capillary electrophoresis. *Electrophoresis* **22**, 629–643.
2. Liu, M. S. and Chen, F. T. (2000) Rapid analysis of amplified double-stranded DNA by capillary electrophoresis with laser-induced fluorescence detection. *Mol. Biotechnol.* **15**, 143–146.
3. Gong, X. and Yeung, E. S. (2000) Genetic typing and HIV-1 diagnosis by using 96 capillary array electrophoresis and ultraviolet absorption detection. *J. Chromatogr. B: Biomed. Sci. Appl.* **741**, 15–21.
4. Zhang, N. and Yeung, E. S. (1998) On-line coupling of polymerase chain reaction and capillary electrophoresis for automatic DNA typing and HIV-1 diagnosis. *J. Chromatogr. B. Biomed. Sci. Appl.* **714**, 3–11.
5. Lu, W., Han, D. S., Yuan, J., and Andrieu, J. M. (1994) Multi-target PCR analysis by capillary electrophoresis and laser-induced fluorescence. *Nature* **368**, 269–271.
6. Rossomando, E. F., White, L., and Ulfelder, K. J. (1994) Capillary electrophoresis: separation and quantitation of reverse transcriptase polymerase chain reaction products from polio virus. *J. Chromatogr. B: Biomed. Sci. Appl.* **656**, 159–168.
7. Li, N., Tan, W. G., Tsang, R. Y., Tyrrell, D. L., and Dovichi, N. J. (2002) Quantitative polymerase chain reaction using capillary electrophoresis with laser-induced fluorescence detection: analysis of duck hepatitis B. *Anal. Bioanal. Chem.* **374**, 269–273.
8. Tan, W. G., Tyrrell, D. L., and Dovichi, N. J. (1999) Detection of duck hepatitis B virus DNA fragments using on-column intercalating dye labeling with capillary electrophoresis-laser-induced fluorescence. *J. Chromatogr. A* **853**, 309–319.
9. Gelfi, C., Orsi, A., Leoncini, F., et al. (1995) Amplification of 18 dystrophin gene exons in DMD/BMD patients: simultaneous resolution by capillary electrophoresis in sieving liquid polymers. *Biotechniques* **19**, 254–258, 260–263.
10. Shen, Y., Xu, Q., Han, F., et al. (1999) Application of capillary nongel sieving electrophoresis for gene analysis. *Electrophoresis* **20**, 1822–1828.
11. Guttman, A., Barta, C., Szoke, M., Sasvari-Szekely, M., and Kalasz, H. (1998) Real-time detection of allele-specific polymerase chain reaction products by automated ultra-thin-layer agarose gel electrophoresis. *J. Chromatogr. A* **828**, 481–487.

12. Gelfi, C., Righetti, P. G., Brancolini, V., Cremonesi, L., and Ferrari, M. (1994) Capillary electrophoresis in polymer networks for analysis of PCR products: detection of delta F508 mutation in cystic fibrosis. *Clin. Chem.* **40**, 1603–1605.
13. Kiyoi, H. and Naoe, T. (2002) FLT3 in human hematologic malignancies. *Leukemia Lymphoma* **43**, 1541–1547.
14. Greiner, T. C. and Rubocki, R. J. (2002) Effectiveness of capillary electrophoresis using fluorescent-labeled primers in detecting T-cell receptor gamma gene rearrangements. *J. Mol. Diagn.* **4**, 137–143.
15. Novella, E., Giaretta, I., Elice, F., et al. (2002) Fluorescent polymerase chain reaction and capillary electrophoresis for IgH rearrangement and minimal residual disease evaluation in multiple myeloma. *Haematologica* **87**, 1157–1164.
16. Knudson, A. G. (2002) Cancer genetics. *Am. J. Med. Genet.* **111**, 96–102.
17. Martinelli, G., Testoni, N., Montefusco, V., et al. (1998) Detection of bcr-abl transcript in chronic myelogenous leukemia patients by reverse-transcription-polymerase chain reaction and capillary electrophoresis. *Haematologica* **83**, 593–601.
18. Sanchez-Vega, B., Vega, F., Medeiros, L. J., Lee, M. S., and Luthra, R. (2002) Quantification of bcl-2/JH fusion sequences and a control gene by multiplex real-time PCR coupled with automated amplicon sizing by capillary electrophoresis. *J. Mol. Diagn.* **4**, 223–229.
19. Matyas, G., Giunta, C., Steinmann, B., Hossle, J. P., and Hellwig, R. (2002) Quantification of single nucleotide polymorphisms: a novel method that combines primer extension assay and capillary electrophoresis. *Hum. Mutat.* **19**, 58–68.
20. Piggee, C. A., Muth, J., Carrilho, E., and Karger, B. L. (1997) Capillary electrophoresis for the detection of known point mutations by single-nucleotide primer extension and laser-induced fluorescence detection. *J. Chromatogr. A* **781**, 367–375.
21. Vreeland, W. N., Meagher, R. J., and Barron, A. E. (2002) Multiplexed, high-throughput genotyping by single-base extension and end-labeled free-solution electrophoresis. *Anal. Chem.* **74**, 4328–4333.
22. Bugalho, M. J., Domingues, R., and Sobrinho, L. (2002) The minisequencing method: a simple strategy for genetic screening of MEN 2 families. *BMC Genet.* **3**, 8.
23. Zsolnai, A., Anton, I., Kuhn, C., and Fesus, L. (2003) Detection of single-nucleotide polymorphisms coding for three ovine prion protein variants by primer extension assay and capillary electrophoresis. *Electrophoresis* **24**, 634–638.
24. Arakawa, H., Uetanaka, K., Maeda, M., Tsuji, A., Matsubara, Y., and Narisawa, K. (1994) Analysis of polymerase chain reaction-product by capillary electrophoresis with laser-induced fluorescence detection and its application to the diagnosis of medium-chain acyl-coenzyme A dehydrogenase deficiency. *J. Chromatogr. A* **680**, 517–523.
25. Barta, C., Sasvari-Szekely, M., and Guttman, A. (1998) Simultaneous analysis of various mutations on the 21-hydroxylase gene by multi-allele specific amplification and capillary gel electrophoresis. *J. Chromatogr. A* **817**, 281–286.
26. Lehmann, R., Koch, M., Pfohl, M., Voelter, W., Haring, H. U., and Liebich, H. M. (1996) Screening and identification of familial defective apolipoprotein B-100 in clinical samples by capillary gel electrophoresis. *J. Chromatogr. A* **744**, 187–194.
27. van de Locht, L. T., Kuypers, A. W., Verbruggen, B. W., Linssen, P. C., Novakova, I. R., and Mensink, E. J. (1995) Semi-automated detection of the factor V mutation by allele specific amplification and capillary electrophoresis. *Thromb. Haemost.* **74**, 1276–1279.
28. Gomez-Llorente, M. A., Suarez, A., Gomez-Llorente, C., et al. (2001) Analysis of 31 CFTR mutations in 55 families from the South of Spain. *Early Hum. Dev.* **65 (Suppl.)**, S161–S164.
29. Day, N. S., Tadin, M., Christiano, A. M., Lanzano, P., Piomelli, S., and Brown, S. (2002) Rapid prenatal diagnosis of sickle cell diseases using oligonucleotide ligation assay coupled with laser-induced capillary fluorescence detection. *Prenat. Diagn.* **22**, 686–691.
30. Somsen, G. W., Welten, H. T., Mulder, F. P., Swart, C. W., Kema, I. P., and de Jong, G. J. (2002) Capillary electrophoresis with laser-induced fluorescence detection for fast and reliable apolipoprotein E genotyping. *J. Chromatogr. B: Anal. Technol. Biomed. Life Sci.* **775**, 17–26.
31. Mitchell, C. E., Belinsky, S. A., and Lechner, J. F. (1995) Detection and quantitation of mutant K-ras codon 12 restriction fragments by capillary electrophoresis. *Anal. Biochem.* **224**, 148–153.
32. Butler, J. M., Wilson, M. R., and Reeder, D. J. (1998) Rapid mitochondrial DNA typing using restriction enzyme digestion of polymerase chain reaction amplicons followed by capillary electrophoresis separation with laser-induced fluorescence detection. *Electrophoresis* **19**, 119–124.

33. Kourkine, I. V., Hestekin, C. N., and Barron, A. E. (2002) Technical challenges in applying capillary electrophoresis-single strand conformation polymorphism for routine genetic analysis. *Electrophoresis* **23**, 1375–1385.
34. Atha, D. H., Kasprzak, W., O'Connell, C. D., and Shapiro, B. A. (2001) Prediction of DNA single-strand conformation polymorphism: analysis by capillary electrophoresis and computerized DNA modeling. *Nucleic Acids Res.* **29**, 4643–4653.
35. Liu, M. S., Rampal, S., Hsiang, D., and Chen, F. T. (2000) Automated DNA mutation analysis by single-strand conformation polymorphism using capillary electrophoresis with laser-induced fluorescence detection. *Mol. Biotechnol.* **15**, 21–27.
36. Gillman, L. M., Gunton, J., Turenne, C. Y., Wolfe, J. and Kabani, A. M. (2001) Identification of Mycobacterium species by multiple—fluorescence PCR- single-strand conformation polymorphism analysis of the 16S rRNA gene. *J. Clin. Microbiol.* **39**, 3085–3091.
37. Glavac, D., Potocnik, U., Podpecnik, D., Zizek, T., Smerkolj, S., and Ravnik-Glavac, M. (2002) Correlation of MFOLD-predicted DNA secondary structures with separation patterns obtained by capillary electrophoresis single-strand conformation polymorphism (CE-SSCP) analysis. *Hum. Mutat.* **19**, 384–394.
38. Rozycka, M., Collins, N., Stratton, M. R., and Wooster, R. (2000) Rapid detection of DNA sequence variants by conformation-sensitive capillary electrophoresis. *Genomics* **70**, 34–40.
39. Iwamoto, T., Sonobe, T., and Hayashi, K. (2002) Novel algorithm identifies species in a polymycobacterial sample by fluorescence capillary electrophoresis-based single-strand conformation polymorphism analysis. *J. Clin. Microbiol.* **40**, 4705–4712.
40. Raucci, G., Maggi, C. A. and Parente, D. (2000) Capillary electrophoresis of supercoiled DNA molecules: parameters governing the resolution of topoisomers and their separation from open forms. *Anal. Chem.* **72**, 821–826.
41. Kringen, P., Egedal, S., Pedersen, J. C., et al. (2002) BRCA1 mutation screening using restriction endonuclase fingerprinting-single-strand conformation polymorphism in an automated capillary electrophoresis system. *Electrophoresis* **23**, 4085–4091.
42. Baba, Y., Tomisaki, R., Sumita, C., et al. (1995) Rapid typing of variable number of tandem repeat locus in the human apolipoprotein B gene for DNA diagnosis of heart disease by polymerase chain reaction and capillary electrophoresis. *Electrophoresis* **16**, 1437–1440.
43. Lindstedt, B. A., Ryberg, D., Zienolddiny, S., Khan, H., and Haugen, A. (1999) Hras1 VNTR alleles as susceptibility markers for lung cancer: relationship to microsatellite instability in tumors. *Anticancer Res.* **19**, 5523–5527.
44. Dib, C., Faure, S., Fizames, C., et al. (1996) A comprehensive genetic map of the human genome based on 5,264 microsatellites. *Nature* **380**, 152–154.
45. Dietrich, W. F., Miller, J. C., Steen, R. G., et al. (1994) A genetic map of the mouse with 4,006 simple sequence length polymorphisms. *Nature Genet.* **7**, 220–245.
46. Breen, G., Sham, P., Li, T., Shaw, D., Collier, D. A., and St Clair, D. (1999) Accuracy and sensitivity of DNA pooling with microsatellite repeats using capillary electrophoresis. *Mol. Cell. Probes* **13**, 359–365.
47. Cook, E. H., Jr., Courchesne, R. Y., Cox, N. J., et al. (1998) Linkage-disequilibrium mapping of autistic disorder, with 15q11-13 markers. *Am. J. Hum. Genet.* **62**, 1077–1083.
48. Gelfi, C., Cossu, G., Carta, P., Serra, M., and Righetti, P. G. (1995) Gene dosage in capillary electrophoresis: pre-natal diagnosis of Down's syndrome. *J. Chromatogr. A* **718**, 405–412.
49. Latour, P., Boutrand, L., Levy, N., et al. (2001) Polymorphic short tandem repeats for diagnosis of the Charcot-Marie- Tooth 1A duplication. *Clin. Chem.* **47**, 829–37.
50. Van Hoofstat, D. E., Deforce, D. L., Hubert De Pauw, I. P., and Van den Eeckhout, E. G. (1999) DNA typing of fingerprints using capillary electrophoresis: effect of dactyloscopic powders. *Electrophoresis* **20**, 2870–2876.
51. Moretti, T. R., Baumstark, A. L., Defenbaugh, D. A., Keys, K. M., Brown, A. L., and Budowle, B. (2001) Validation of STR typing by capillary electrophoresis. *J. Forensic Sci.* **46**, 661–676.
52. Laszik, A., Brinkmann, B., Sotonyi, P., and Falus, A. (2000) Automated fluorescent detection of a 10 loci multiplex for paternity testing. *Acta Biol. Hung.* **51**, 99–105.
53. Lion, T. (2003) Summary: reports on quantitative analysis of chimerism after allogeneic stem cell transplantation by PCR amplification of microsatellite markers and capillary electrophoresis with fluorescence detection. *Leukemia* **17**, 252–254.
54. Boland, C. R., Thibodeau, S. N., Hamilton, S. R., et al. (1998) A National Cancer Institute Workshop on Microsatellite Instability for cancer detection and familial predisposition: development of

international criteria for the determination of microsatellite instability in colorectal cancer. *Cancer Res.* **58**, 5248–5257.

55. Berg, K. D., Glaser, C. L., Thompson, R. E., Hamilton, S. R., Griffin, C. A., and Eshleman, J. R. (2000) Detection of microsatellite instability by fluorescence multiplex polymerase chain reaction. *J. Mol. Diagn.* **2**, 20–28.
56. Wada, T., Louhelainen, J., Hemminki, K., et al. (2000) Bladder cancer: allelic deletions at and around the retinoblastoma tumor suppressor gene in relation to stage and grade. *Clin. Cancer Res.* **6**, 610–615.
57. Yoshino, I., Fukuyama, S., Kameyama, T., Shikada, Y., Oda, S., Maehara, Y., and Sugimachi, K. (2003) Detection of loss of heterozygosity by high-resolution fluorescent system in non-small cell lung cancer: association of loss of heterozygosity with smoking and tumor progression. *Chest* **123**, 545–550.
58. Murthy, S. K., DiFrancesco, L. M., Ogilvie, R. T., and Demetrick, D. J. (2002) Loss of heterozygosity associated with uniparental disomy in breast carcinoma. *Mod. Pathol.* **15**, 1241–1250.
59. Fukunaga, K., Wada, T., Matsumoto, H., Yoshihiro, S., Matsuyama, H. and Naito, K. (2002) Renal cell carcinoma: allelic loss at chromosome 9 using the fluorescent multiplex-polymerase chain reaction technique. *Hum. Pathol.* **33**, 910–914.
60. Sell, S. M., Patel, S., Stracner, D., and Meloni, A. (2001) Allelic loss analysis by capillary electrophoresis: an accurate, automated method for detection of deletions in solid tumors. *Genet. Test* **5**, 267–268.
61. Hussey, J., Lockhart, P. J., Seltzer, W., et al. (2002) Accurate determination of ataxin-2 polyglutamine expansion in patients with intermediate-range repeats. *Genet. Test* **6**, 217–220.
62. O'Connell, C. D., Atha, D. H., Jakupciak, J. P., Amos, J. A., and Richie, K. (2002) Standardization of PCR amplification for fragile X trinucleotide repeat measurements. *Clin. Genet.* **61**, 13–20.
63. Kiba, Y. and Baba, Y. (2001) Analysis of triplet-repeat DNA by capillary electrophoresis. *Methods Mol. Biol.* **163**, 221–229.
64. Nesi, M., Righetti, P. G., Patrosso, M. C., Ferlini, A., and Chiari, M. (1994) Capillary electrophoresis of polymerase chain reaction-amplified products in polymer networks: the case of Kennedy's disease. *Electrophoresis* **15**, 644–646.
65. Dorschner, M. O., Barden, D., and Stephens, K. (2002) Diagnosis of five spinocerebellar ataxia disorders by multiplex amplification and capillary electrophoresis. *J. Mol. Diagn.* **4**, 108–113.
66. Williams, L. C., Hegde, M. R., Herrera, G., Stapleton, P. M., and Love, D. R. (1999) Comparative semi-automated analysis of (CAG) repeats in the Huntington disease gene: use of internal standards. *Mol Cell Probes* **13**, 283–289.
67. Tian, H., Brody, L. C. and Landers, J. P. (2000) Rapid detection of deletion, insertion, and substitution mutations via heteroduplex analysis using capillary- and microchip-based electrophoresis. *Genome Res.* **10**, 1403–13.
68. O'Connor, F., Fitzgerald, D. J., and Murphy, R. P. (2000) An automated heteroduplex assay for the Pi(A) polymorphism of glycoprotein IIb/IIIa, multiplexed with two prothrombotic genetic markers. *Thromb. Haemost.* **83**, 248–252.
69. Jackson, H. A., Bowen, D. J., and Worwood, M. (1997) Rapid genetic screening for haemochromatosis using heteroduplex technology. *Br. J. Haematol.* **98**, 856–859.
70. Bowen, D. J., Standen, G. R., Granville, S., Bowley, S., Wood, N. A., and Bidwell, J. (1997) Genetic diagnosis of factor V Leiden using heteroduplex technology. *Thromb. Haemost.* **77**, 119–122.
71. Thomas, G. A., Williams, D. L., and Soper, S. A. (2001) Capillary electrophoresis-based heteroduplex analysis with a universal heteroduplex generator for detection of point mutations associated with rifampin resistance in tuberculosis. *Clin. Chem.* **47**, 1195–1203.
72. Khrapko, K., Coller, H. A., Hanekamp, J. S., and Thilly, W. G. (1998) Identification of point mutations in mixtures by capillary electrophoresis hybridization. *Nucleic Acids Res.* **26**, 5738–5740.
73. Kozlowski, P. and Krzyzosiak, W. J. (2001) Combined SSCP/duplex analysis by capillary electrophoresis for more efficient mutation detection. *Nucleic Acids Res.* **29**, E71.
74. Kourkine, I. V., Hestekin, C. N., Magnusdottir, S. O., and Barron, A. E. (2002) Optimized sample preparation for tandem capillary electrophoresis single-stranded conformational polymorphism/heteroduplex analysis. *Biotechniques* **33**, 318–320, 322, 324–325.
75. Tian, H., Brody, L. C., Fan, S., Huang, Z., and Landers, J. P. (2001) Capillary and microchip electrophoresis for rapid detection of known mutations by combining allele-specific DNA amplification with heteroduplex analysis. *Clin. Chem.* **47**, 173–185.

76. Bianchi, N., Mischiati, C., Feriotto, G., et al. (1994) Capillary electrophoresis: detection of hybridization between synthetic oligonucleotides and HIV-1 genomic DNA amplified by polymerase-chain reaction. *J. Virol. Methods* **47**, 321–329.
77. Armitage, B. A. (2003) The impact of nucleic acid secondary structure on PNA hybridization. *Drug Discov. Today* **8**, 222–228.
78. Basile, A., Giuliani, A., Pirri, G., and Chiari, M. (2002) Use of peptide nucleic acid probes for detecting DNA single-base mutations by capillary electrophoresis. *Electrophoresis* **23**, 926–929.
79. Igloi, G. L. (2001) Simultaneous identification of mutations by dual-parameter multiplex hybridization in peptide nucleic acid-containing virtual arrays. *Genomics* **74**, 402–407.
80. Freudemann, T., von Brocke, A., and Bayer, E. (2001) On-line coupling of capillary gel electrophoresis with electrospray mass spectrometry for oligonucleotide analysis. *Anal. Chem.* **73**, 2587–2593.
81. McKeon, J. and Khaledi, M. G. (2001) Quantitative nuclear and cytoplasmic localization of antisense oligonucleotides by capillary electrophoresis with laser-induced fluorescence detection. *Electrophoresis* **22**, 3765–3770.
82. Gilar, M., Belenky, A., Budman, Y., Smisek, D. L., and Cohen, A. S. (1998) Study of phosphorothioate-modified oligonucleotide resistance to 3'-exonuclease using capillary electrophoresis. *J. Chromatogr. B: Biomed. Sci. Appl.* **714**, 13–20.
83. Zellweger, T., Miyake, H., Cooper, S., et al. (2001) Antitumor activity of antisense clusterin oligonucleotides is improved in vitro and in vivo by incorporation of 2'-*O*-(2-methoxy)ethyl chemistry. *J Pharmacol Exp Ther* **298**, 934-40.
84. DeDionisio, L. A. (2001) Analysis of modified oligonucleotides with capillary gel electrophoresis. *Methods Mol. Biol.* **162**, 353–370.
85. Lagu, A. L. (1999) Applications of capillary electrophoresis in biotechnology. *Electrophoresis* **20**, 3145–3155.
86. Zhu, L., Lee, H. K., Lin, B., and Yeung, E. S. (2001) Spatial temperature gradient capillary electrophoresis for DNA mutation detection. *Electrophoresis* **22**, 3683–3687.
87. Kristensen, A. T., Bjorheim, J., and Ekstrom, P. O. (2002) Detection of mutations in exon 8 of TP53 by temperature gradient 96-capillary array electrophoresis. *Biotechniques* **33**, 650–653.
88. Weinfeld, M., Xing, J. Z., Lee, J., Leadon, S. A., and Le, X. C. (2002) Immunofluorescence detection of radiation-induced DNA base damage. *Mil. Med.* **167**, 2–4.
89. Fiscus, R. R., Leung, C. P., Yuen, J. P., and Chan, H. C. (2001) Quantification of apoptotic DNA fragmentation in a transformed uterine epithelial cell line, HRE-H9, using capillary electrophoresis with laser-induced fluorescence detector (CE-LIF). *Cell Biol. Int.* **25**, 1007–1011.
90. Reik, W., Dean, W., and Walter, J. (2001) Epigenetic reprogramming in mammalian development. *Science* **293**, 1089–1093.
91. Panning, B. and Jaenisch, R. (1996) DNA hypomethylation can activate Xist expression and silence X-linked genes. *Genes Dev.* **10**, 1991–2002.
92. Li, E., Beard, C., and Jaenisch, R. (1993) Role for DNA methylation in genomic imprinting. *Nature* **366**, 362–365.
93. Jones, P. A. and Baylin, S. B. (2002) The fundamental role of epigenetic events in cancer. *Nature Rev. Genet.* **3**, 415–428.
94. Fraga, M. F., Rodriguez, R., and Canal, M. J. (2000) Rapid quantification of DNA methylation by high performance capillary electrophoresis. *Electrophoresis* **21**, 2990–2994.
95. Stach, D., Schmitz, O. J., Stilgenbauer, S., et al. (2003) Capillary electrophoretic analysis of genomic DNA methylation levels. *Nucleic Acids Res.* **31**, E2–E2.
96. Kotler, L., He, H., Miller, A. W., and Karger, B. L. (2002) DNA sequencing of close to 1000 bases in 40 minutes by capillary electrophoresis using dimethyl sulfoxide and urea as denaturants in replaceable linear polyacrylamide solutions. *Electrophoresis* **23**, 3062–3070.
97. Albarghouthi, M. N. and Barron, A. E. (2000) Polymeric matrices for DNA sequencing by capillary electrophoresis. *Electrophoresis* **21**, 4096–40111.
98. Dolnik, V. (1999) DNA sequencing by capillary electrophoresis. *J. Biochem. Biophys. Methods* **41**, 103–119 (review).

99. Nicod, J. C. and Largiader, C. R. (2003) SNPs by AFLP (SBA): a rapid SNP isolation strategy for non-model organisms. *Nucleic Acids Res.* **31**, e19.
100. Doglio, A., Laffont, C., Thyss, S., and Lefebvre, J. C. (1998) Rapid genotyping of hepatitis C virus by direct cycle sequencing of PCR-amplified cDNAs and capillary electrophoresis analysis. *Res. Virol.* **149**, 219–227.
101. Blazej, R. G., Paegel, B. M., and Mathies, R. A. (2003) Polymorphism ratio sequencing: a new approach for single nucleotide polymorphism discovery and genotyping. *Genome Res.* **13**, 287–293.
102. Murphy, K. M. and Eshleman, J. R. (2002) Simultaneous sequencing of multiple polymerase chain reaction products and combined polymerase chain reaction with cycle sequencing in single reactions. *Am. J. Pathol.* **161**, 27–33.
103. Woolley, A. T. and Mathies, R. A. (1994) Ultra-high-speed DNA fragment separations using microfabricated capillary array electrophoresis chips. *Proc. Natl. Acad. Sci. USA* **91**, 11,348–11,352.
104. Gao, Q., Shi, Y., and Liu, S. (2001) Multiple-channel microchips for high-throughput DNA analysis by capillary electrophoresis. *Fresenius J. Anal. Chem.* **371**, 137–145.
105. Medintz, I. L., Paegel, B. M., Blazej, R. G., et al. (2001) High-performance genetic analysis using microfabricated capillary array electrophoresis microplates. *Electrophoresis* **22**, 3845–3856.
106. Medintz, I. L., Berti, L., Emrich, C. A., Tom, J., Scherer, J. R., and Mathies, R. A. (2001) Genotyping energy-transfer-cassette-labeled short-tandem-repeat amplicons with capillary array electrophoresis microchannel plates. *Clin. Chem.* **47**, 1614–1621.
107. Wessagowit, V. and South, A. P. (2002) Dermatological applications of DNA array technology. *Clin. Exp. Dermatol.* **27**, 485–492.
108. Gawron, A. J., Martin, R. S., and Lunte, S. M. (2001) Microchip electrophoretic separation systems for biomedical and pharmaceutical analysis. *Eur. J. Pharm. Sci.* **14**, 1–12.

11

Mapping Techniques

Simon G. Gregory

1. Introduction

In 1920, German botanist Hans Winkler first used the term *genome*, reputedly by the fusion of GENe and chromosOME, in order to describe the complex notion of the entire set of chromosomes and all of the genes contained within an organism. A great deal of progress has since been made in the elucidation of the complex molecular interactions that underlie cellular functioning and the syntenic relationship between organisms at a nucleotide level.

The basis for these advances was the characterization of the structure of DNA by Watson and Crick in 1953 *(1)* and the realization that DNA could be decoded to provide a guide to genetic inheritance. This underpinned the concept of genetics and provided scientists with the opportunity to explore and quantify the nature and extent of the biological information passed on from one generation to the next. The characterization of biological inheritance permitted the elucidation of what it was that was being encoded and how it could determine biochemical function. Finally, extending from elucidation of the mechanisms behind inheritance of monogenic diseases, scientists are beginning to grasp how sequence is also involved in complex interactions, occasionally under the influence of environmental factors, to contribute to many (but still not all) diseases.

The speed at which the vast amount of human sequence data were generated can be attributed to the evolution of strategies and techniques developed to map and sequence organisms such as bacteria *(2)*, yeast *(3)* and the nematode worm *(4)*. The availability of such an evolutionary diverse collection of species, with the addition of the mouse *(5)* and other complex multicellular organisms, has also enabled comparisons to be made at a nucleotide level.

The first genomes that were characterized were relatively small by current standards, bacteriophage ΦX174, 5 kb *(6,7)* and bacteriophage λ, 48 kb *(8)*—but they provided the underlying techniques and strategies that are being used for the more complex organisms currently being studied. Chain termination sequencing, developed by Sanger *(8)*, is a synthetic method in which nested sets of labeled fragments are generated in vitro by a DNA polymerase reaction. Because the method is highly sensitive and robust, it has been amenable to biochemical optimization, producing long, accurate sequence reads, and also to automation, which was necessary for large-scale application of the technique. In these respects, it differed from the method of Maxam and Gilbert *(9)*, which necessitated production of all of the labeled material prior to chemical degradation to form the sequence ladders of nested fragments. As a result, the Sanger method has remained the technique by which the majority of genomic sequence from a variety of complex organisms is presently being generated (*see* **Fig. 1**). However, neither method is capable of generating single reads of greater than 2–300 nucleotides, limited in part by the sequence

From: *Medical Biomethods Handbook*
Edited by: J. M. Walker and R. Rapley © Humana Press, Inc., Totowa, NJ

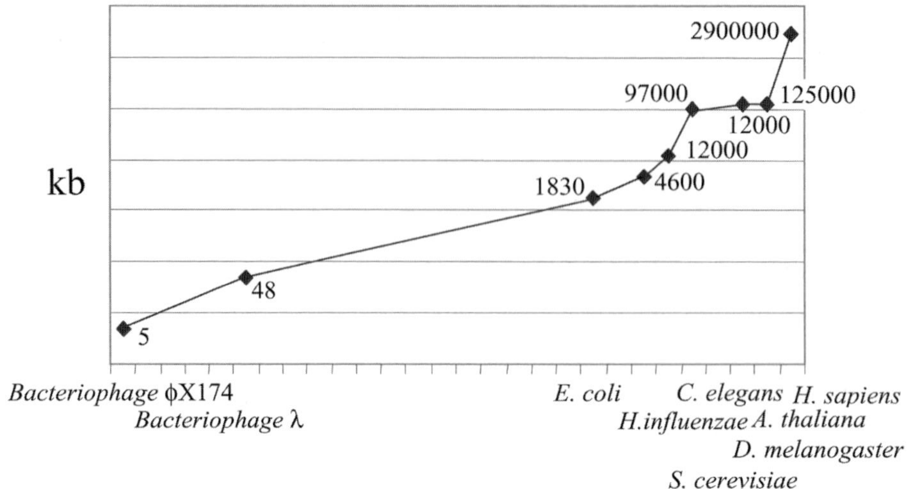

Fig. 1. A plot of the near-logarithmic increase in the complexity of genomic mapping and sequencing; from the first full genomic sequence of bacteriophage ΦK174 in 1977 to the completed human genome in April 2003.

production itself and partly by the ability to separate the sequence by gel electrophoresis at single-base resolution (even today, sequencing read lengths approaching 1 kb are rare).

The assembly of larger tracts of DNA therefore required the development of methods to reassemble a consensus sequence from multiple individual reads. Two approaches were adopted for this; first, the construction of physical maps of restriction fragments using sequence-specific restriction enzymes to order and orientate large segments of DNA from which individual units were selected for sequencing; second, the use of the information gained from each individual sequence read to order and orient each segment relative to overlapping neighbors, which required the development of advanced computer programs to make the task possible on all but the smallest scale.

A further modification of the latter was made by Anderson ***(10)***, who developed the random shotgun strategy to elucidate the mitochondrial genome, involved using a random fragmentation process by partial DNAse I digestion ***(11)***. This removed the dependence on sequence-specific restriction enzymes while still relying on sequence-based assembly of contiguous tracts of overlapping reads. The random shotgun approach, in which genomic DNA is randomly sequenced in similarly sized segments and then assembled simultaneously to provide a representation of the genomic template, provided the basis of the strategies used to assemble sequences of large inserts cloned in plasmids, lambda phage, and cosmid vectors ***(15)***, and also the later bacterial artificial chromosome (BAC) and P1-derived artificial chromosome (PAC) clones ***(17)***. The same random shotgun strategy was adopted to sequence the 1.8-Mb genome of the bacterium *Haemophilus influenzae* ***(12)***. Although the whole-genome shotgun sequencing approach has proven itself to be a successful strategy for the rapid assembly of smaller genomes, there are doubts as to whether this strategy is suitable for assembling the sequence of complex organisms.

The generation of a physical map, in which the genome is divided into bacterial clone units of 40–200 kb and assembled into contiguous stretches (contigs) of overlapping clones, is a process analogous to the sequence contig assembly process. In contrast to sequence assembly, however, the information used to compare individual clones and identify overlaps of a physical map [e.g., the *Caenorhabditis elegans* ***(4)*** and *Saccharomyces cerevisiae* ***(3)*** genome projects] use a one-dimensional fingerprint prepared by separating restriction fragments from a limit digest of each cloned DNA by electrophoresis. Overlaps between clones were detected on the basis of partially (or completely) shared fingerprint patterns. An alternative approach to identify overlapping relationships between clones was to test clones for the presence of characterized markers. Overlaps between clones could be identified on the basis that they shared a single copy sequence. The presence of the sequence was identified using a specific hybridization probe or polymerase chain reaction (PCR) assay. Given a physical map of overlapping clones, individual clones can then be selected from the map to provide maximum genomic coverage with minimal redundancy. These clones permit specific regions to be targeted for further investigation and, in particular, for the determination of the complete DNA sequence separately from other clones within the physical map. Because the source of the genomic sequence is limited to an individual clone, problems encountered with sequence assemblies are greatly reduced compared to the corresponding whole-genome assemblies.

At the time of their inception, the physical maps of the *C. elegans* ***(4)*** and *S. cerevisiae* ***(3)*** genomes were constructed to enhance the molecular genetics of the respective organisms by facilitating the cloning of known genes and to serve as an archive for genomic information. However, the data associated with the construction of the clonal physical maps—even with good alignment to the genetic map—carried only a tiny proportion of information present within the genome. Consequently, a minimum tile path of the 30-kb cosmid and 15-kb lambda clones, used to build the physical maps of the *C. elegans* and *S. cerevisiae*, respectively, were subcloned into M13 phage vectors (1.3–2 kb insert size) and sequenced on a per-clone basis. The physical maps of the two genomes, and subsequently of ***(2)***, ***(13)***, *Drosophila melanogaster* ***(14)***, and human ***(15)***, used restriction enzyme fragments in various ways to overlap clonal units for the construction of genomewide physical maps.

2. Mapping the Human Genome

The human genome is contained within 22 autosomes (numbered from 1–22, largely according to size) and 2 sex chromosomes, X and Y (female XX and male XY). Chromosomes are punctuated with centromere structures that are either located close to chromosome ends (acrocentric), toward a chromosome end (submetacentric), or centrally between ends (metacentric). The initial size estimate of the genome, 3200 Mb, was based largely on cytometric measurements ***(18)*** and has since been revised to 2900 Mb in light of the higher-resolution human draft sequence analysis ***(19)*** and is supported by observed sizes of completed chromosome sequences that suggested that the earlier figures were overestimates ***(20–22)*** (see Chapter 42). The construction of a map of the human genome was an important step toward understanding and characterizing the sequence contained within it, as it provided a means by which all of the features could be ordered and partitioned, and the task of detailed characterization and sequencing could be divided up into manageable segments.

2.1. Cytogenetic Mapping

The treatment of metaphase chromosome spreads with trypsin digestion and Giemsa staining creates differential chromosome banding patterns (see Chapter 32). The generation of light (R-bands) and dark (G-bands) bands by Giemsa staining is reliant on nucleotide content, and the staining pattern therefore reflects the base composition and correlates other properties of the different regions (*see* **Fig. 2**). However, the maximum genome-wide resolution was limited to an 850-genomewide banding pattern ***(23)***. The recognition of characteristic banding patterns of chromosomal regions provided the basis for much of the early characterization of chromosome aberrations (duplications, deletions, and translocations) that were associated with clinical phenotypes ***(24–26)***.

The ability to hybridize labeled probes containing specific sequences, and to detect their location on metaphase chromosome by autoradiographic or fluorescent detection techniques (fluorescence *in situ* hybridization [FISH]) ***(27)*** revolutionized cytogenetic mapping. Typically, FISH has utilized cloned DNA as the template for the generation of a fluorescently labeled PCR

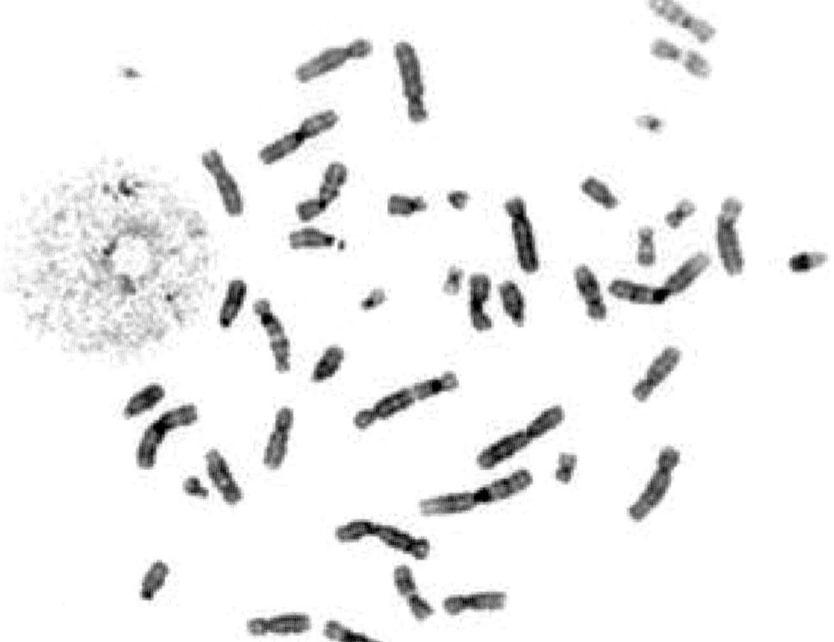

Fig. 2. Giemsa staining of human metaphase chromosomes (G-banding).

probe for hybridization to genomic DNA. The signal emitted by the fluorescent nucleotide contained within the probe is detected by using epifluorescence microscopy. Initially, the location of the probe relative to the metaphase banding pattern provided an approximate map position for the sequence represented by the probe. Pairs of markers, labeled with different fluorochromes, could be simultaneously placed relative to the cytogenetic banding and also ordered with respect to each other. The use of pairs of differentially labeled markers in combination with a third reference marker enabled FISH to be applied to chromosomal DNA in a less condensed state (in interphase nuclei). Although no banding pattern can be obtained in interphase DNA, the decondensed state of the chromatin relative to metaphase chromosomes means that increased levels of resolution could be obtained, as probes were better separated. An interprobe distance of 1–5 Mb can be resolved using metaphase FISH, 0.1–1.0 Mb by interphase FISH ***(28)***, and 5 kb by FISH using mechanical pretreatment to extend DNA into fibers ***(29)***.

2.2. Genetic Mapping

Genetic maps utilize the likelihood of recombination between adjacent markers during meiosis to calculate intermarker genetic distances, and from this to infer a physical distance. The closer two landmarks are together on a chromosome, the less likelihood there is of a recombination event occurring between, with the opposite being true for markers that are further apart. The calculation of distance and, therefore, the metric on which the genetic map is based is the length of the chromosomal segment that, on average, undergoes one exchange with a sister chromatid during meiosis, the Morgan (M). Therefore, a 1% recombination frequency is equivalent to 1 centimorgan (cM), and because the human genome covers 3000 cM and contains approx 3000 Mb, 1 cM is approximately equivalent to 1 Mb. However, recombination is known to be nonrandom, which can lead to a level of inaccuracy ***(30)*** in inferring physical distances from measurements of genetic recombination.

Table 1
A Comparison of Marker Content Within Genetic Maps

No. of markers	Ref.
100	*43*
813	*44*
2066	*45*
5840	*46*
5264	*30*
5136	*42*

The inherent limitation of primary genetic maps was the lack of availability of polymorphic markers between which genetic distances could be calculated. This was ameliorated in part by the suggested use of restriction fragment length polymorphisms (RFLPs), identified by Kan and Dozy ***(31)***, for the construction of a genomewide genetic linkage map ***(32)***. The first such map ***(33)*** was limited in its usefulness, however, because of RFLPs have a maximum heterozygosity of 50% and the low level of resolution of the 403 characterized polymorphic markers, including 393 RFLPs, covering the genome. The identification of hypervariable regions, which showed multiallelic variation ***(34)***, provided a new source of markers for genetic mapping. The variable regions contained short (11–60 bp) variable number tandem repeats (VNTRs) which showed allelic variation. However, these minisatellite markers ***(35)*** and VNTRs ***(36)*** were shown to cluster at chromosome arms and were not inherently stable ***(37)***. The identification of microsatellite markers (containing dinucleotide, trinucleotide or tetranucleotide repeats) greatly facilitated the generation of genetic maps. They were proven to be widely distributed throughout the genome, showed allelic variation ***(38,39)***, and were amenable to PCR amplification ***(40)*** by sequence-tagged-site screening ***(41)***. In a relatively short period of time, a number of genetic maps were published with increasing marker density and resolutions, culminating in the most recent deCODE genetic map that contains 5136 markers genotyped across 1257 meioses ***(42)*** (*see* **Table 1**).

2.3. Radiation Hybrid Mapping

The utilization of somatic cell hybrids to maintain human genomic fragments, such as whole chromosomes or chromosomal regions, permits the generation of another form of mapping resource to be generated—the radiation hybrid map. The modification of a technique that fragmented human chromosomes by irradiation and that were then rescued by fusion to rodent cells ***(47)*** prompted Cox et al. ***(48)*** to propose that radiation hybrid (RH) mapping could be applied to the construction of long-range maps of mammalian chromosomes.

The premise of the technique is similar to that of the genetic map; that is, the more closely related two markers are within the genome, the less likelihood there is of a radiation-induced break between them in a reference panel of cell lines and, hence, the less likely is their segregation to different chromosomal locations based on the association of the markers to different sets of fragments. As the presence of two markers within a radiation fragment gives no indication to their physical distance, a panel of radiation hybrids was required. By estimating the frequency of breakage and, thus, the distance between two markers, it is possible to determine their order. The unit of map distance is the centiray (cR) and represents a 1% probability of breakage between two markers for given a radiation dose. Unlike the level of information garnered from a genetic marker, which may or may not be informative within a varying number of meioses, the RH marker is either positive or negative for a DNA fragment, effectively digitizing PCR results. Any amplifiable single-copy sequence can therefore be placed in a RH map. The RH mapping technique has been used for the construction of high-resolution gene maps ***(49,50)*** and has also been used to supplement the construction of chromosome physical maps ***(51)***.

2.4. Physical Mapping

The generation of a physical map relies upon the construction of an ordered and orientated set of clone-based contigs. The term *contig* was coined by Staden ***(52)*** to refer to a contiguous set of overlapping segments that together represent a consensus region. These segments can be sequence, or clones, whose overlapping relationship is defined by information in common to each pair of overlapping segments. Performing a pairwise comparison of the dataset associated with each segment identifies the overlaps. Similarities that are statistically significant indicate the presence, and sometimes the extent, of overlap. Bacterial clone contigs are the most convenient route for the sequence generation of larger genomes. They present a means of coordinating physical mapping and, because of the way in which they are constructed, provide an optimal set of clones (the tile path) for sequencing.

2.4.1. YAC Maps

The main benefit of using yeast artificial chromosomes (YACs) for the construction of a physical map is that the insert size (up to 2 Mb) results in the coverage of large regions of the genome with relatively few clones. Green and Olson ***(53)*** utilized YACs to construct a physical map across the cystic fibrosis region on human chromosome 7 by overlapping YACs by sequence-tagged sites (STS) content data. Chromosome-specific ***(54,55)*** and genome wide YAC maps have also been published ***(56,57)***. Although STS content mapping is the most frequently used method to generate YAC contigs, techniques such as repeat-mediated fingerprinting, either by *Alu*-PCR ***(58)*** or by repeat content hybridization ***(43)***, have also been used. The advantages of using YACs are, however, offset by the relative difficulty of constructing YAC libraries and analyzing the cloned DNA compared to the use of bacterial cloning systems. Many YAC clones have also been found to be chimeric (i.e., to contain fragments derived from noncontiguous parts of genomic DNA being cloned) ***(59–61)***. Rather than being used as a primary sequence resource, YACs came more generally to be used to support the construction of detailed landmark maps and to underpin sequence-ready bacterial clone maps ***(62,63)***. Recently, YACs have been used to facilitate gap closure in the bacterial clone maps by linking contigs ***(64)***. The links are identified by STS content mapping. In these cases, the YACs have been sequenced directly.

2.4.2. Bacterial Clone Maps

In contrast to YACs, bacterial clone libraries are easier to make and the cloned DNA is more easily manipulated. Chimerism is low ***(65,66)*** and the supercoiled recombinant DNA can be purified readily from the host DNA. An important factor influencing the construction of bacterial clone contigs is the available genomic resources. Whereas the *C. elegans* and *S. cerevisiae* maps utilized total genomic 30-kb cosmid and 15-kb lambda libraries, in sevenfold and fivefold coverage, respectively, current bacterial clone contig construction utilizes large-insert P1-derived artificial chromosome (PAC) ***(66)*** and bacterial artificial chromosome (BAC) ***(65)*** libraries. Each BAC or PAC clone typically contains an insert of 100–300 kb and maps have been constructed from a greater than 15-fold genomic clone coverage.

Two types of strategy were used for the construction of sequence-ready bacterial maps of the human genome: the hierarchical strategy and the whole-genome fingerprinting approach. The hierarchical strategy was based on the utilization of well-characterized publicly available markers at a target density along the length of a chromosome, 15 markers/Mb. The use of RH and/or genetic mapped markers, in the form of gene fragments (expressed sequence tags [ESTs]), polymorphic microsatellite markers used for genetic mapping, and markers designed from flow-sorted small insert libraries, allowed for a region or chromosome-specific strategy for map construction and sequence generation (*see* **Fig. 3A–D**). The whole-genome fingerprinting approach of map construction relied on the *in silico* assembly of fingerprints generated from the restriction digest of large-insert bacterial clones from total-genomic PAC or BAC libraries. This process generated substantial amounts of genomic coverage, but contigs lacked

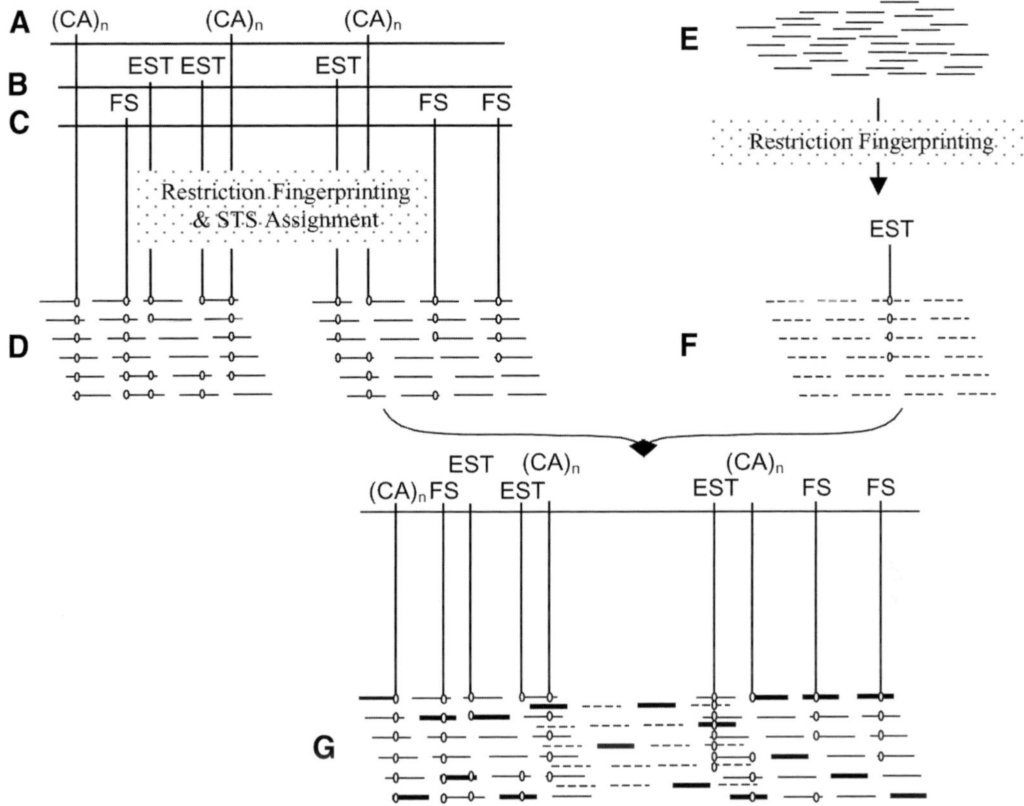

Fig. 3. A representation of the two strategies used to construct sequence-ready bacterial clone maps of the human genome. The hierarchal method utilizes polymorphic genetic markers (CA)*n* **(A)**, expresses sequence tags (ESTs) **(B)**, and markers generated by sequencing small insert libraries (FS) **(C)** that have been RH mapped and screened across genomic libraries. Restriction fingerprinting and STS assignment is performed in parallel prior to data assimilation in FPC **(D)**. The whole-genome fingerprinting approach utilizes restriction digest fingerprinting of a 15-fold redundant genomic library **(E)**, including some marker data, to establish bacterial clone coverage **(F)**. Data from both of these techniques is combined to generate contiguous map coverage **(G)**, to provide a resource for the selection of a minimum tile path clones (bold lines).

order and orientation without the supplementation of a subset of the publicly available markers used in the hierarchical strategy (*see* **Fig. 3E–F**). Ultimately, it was a combination of both techniques that were used for the construction of sequence-ready maps of the human genome. The assimilation of both sets of data generated an estimated >98% coverage of the coding (euchromatic) portion of the human genome and acted as the resource for the generation of high-quality finished sequence data (*see* **Fig. 3G**).

The relative uniformity of the physical map, coupled with the use of markers from the RH and genetic maps in its construction, permits a high-resolution comparison to be made among the three types of map. Algorithms used to construct the RH map assume random distribution of breaks along the length of the chromosome. However, experimental data ***(50)*** have shown that the high retention rate, coupled with variation of DNA fragment sizes (in comparison to the rest of the chromosome), results in an overestimation of the centiray distances between markers adjacent to chromosome centromeres. From the perspective of the genetic map, it has been proven that there are large variations in the rate of recombination along the length of a chromo-

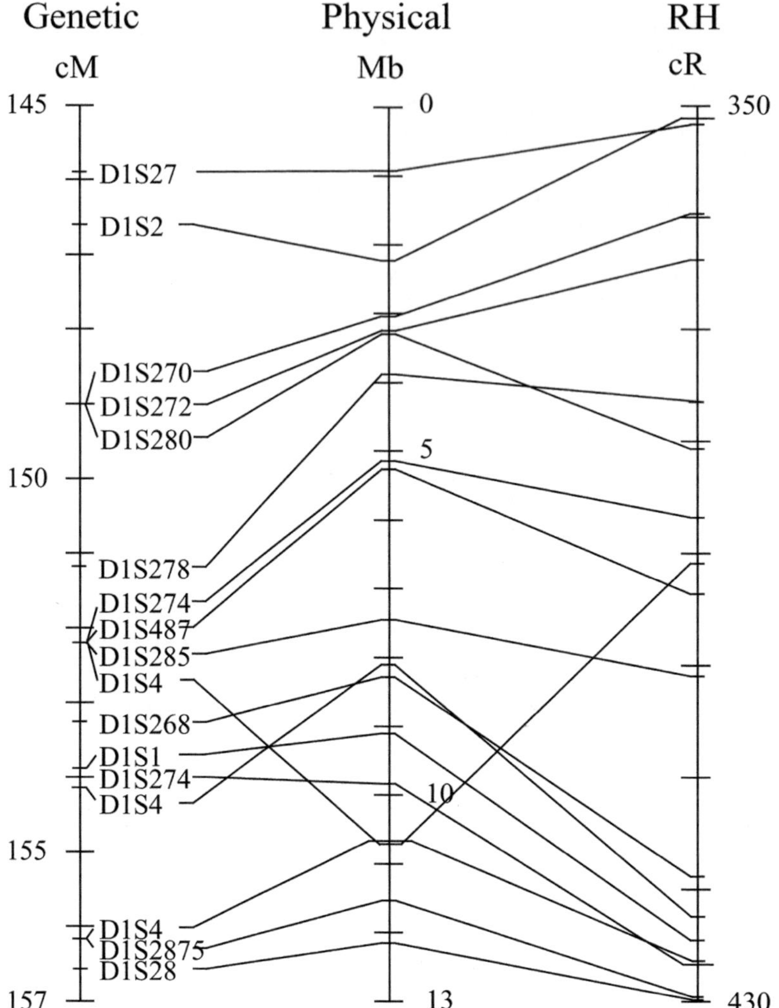

Fig. 4. A comparison of marker distribution among genetic, physical, and RH maps in a 12-Mb region of chromosome 1. Only markers that were present on all three maps have been represented.

some ***(19)***. For example, recombination rates at telomeres are greater than within chromosome arms, which are, in turn, greater than regions adjacent to the centromere. The improved resolution of markers on the physical map therefore enables correct ordering of markers with respect to their locations on the RH map and permits the separation of markers previously binned within the same genetic interval on the genetic map (*see* **Fig. 4**).

The next phase in the evolution of physical map construction was driven by the availability of ordered genomic sequence. Conservation of sequence and long-range order between organisms that are sufficiently closely related means that the genome of one species can act as the template upon which a physical map of another can be built and, in doing so, elucidate the homologous relationship between them ***(67)***. The success of the comparative physical mapping approach was demonstrated by the construction of a clone map of the mouse genome using the assembled human genome sequence as a template ***(68)***. In this study, the human genomic sequence was used to align stringently assembled BAC fingerprint contigs by matching mouse BAC end sequences (BESs) to their corresponding locations in the human genome. Ordered

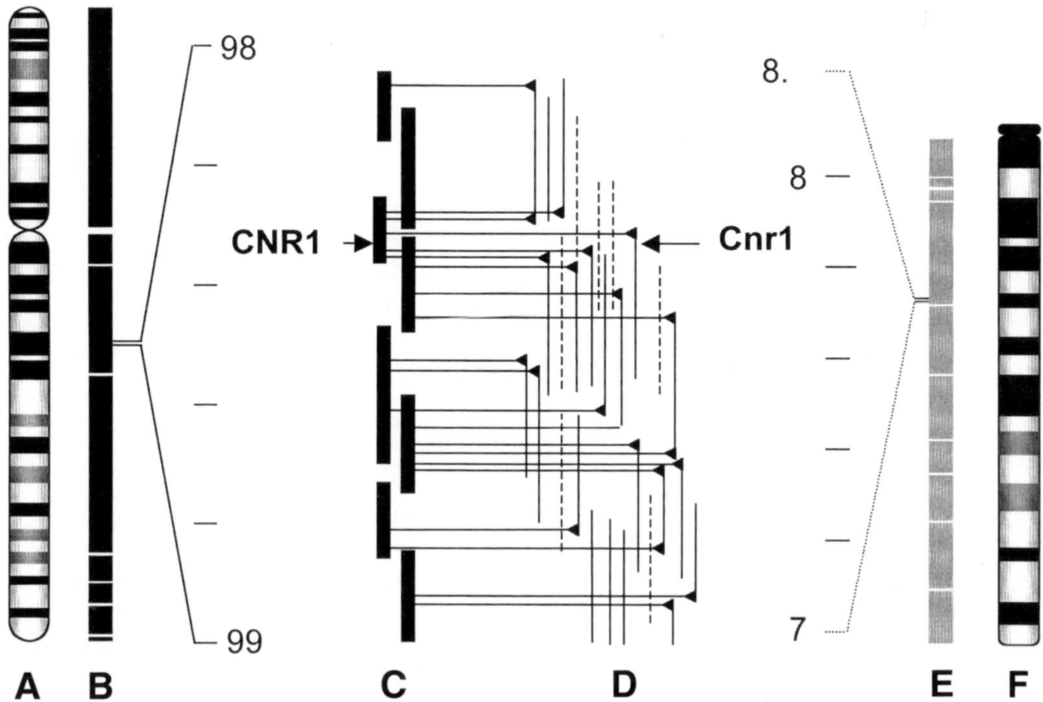

Fig. 5. Construction of the physical map of the mouse genome using the human genomic sequence. The generation of finished sequence from large-insert bacterial clones **(C)**, originating from the physical map **(B)** of human chromosome 6 **(A)**, provided the template for the alignment of end sequences of mouse BACs **(D)** that had previously been assembled into fingerprint contigs. Contig assembly using the described strategy resulted in rapid assembly of sequence-ready contig coverage **(E)** of the mouse genome, including mouse chromosome 4 **(F)**.

and orientated contigs (previously assembled by fingerprinting) were subsequently joined following further fingerprint analysis and the addition of available genetic and RH markers (*see* **Fig. 5**). The availability of BESs from a highly redundant fingerprint assembly of BAC clones and using the strategy outlined above, greatly simplified the process of contig assembly, as the majority of the 7500 contigs generated in the first fingerprinting phase were juxtaposed correctly relative to each other on the basis of homology between the two genomes. As a result, 7500 × 7500 possible joins (more than 56 million) was reduced to analysis of <10,000 putative joins. This permitted the construction of a physical map covering 98% of the 2500 Mb mouse genome, contained within 296 contigs, in approx 12 mo. The same approach could be adopted for any genome for which there is sufficient sequence homology to allow alignment of BESs (or equivalent sequence tags), plus sufficient homology between the template genome and the genome under study. The approach has important applications for genomes where the full genome sequence is anticipated and also (perhaps even more importantly) it is a cost-effective way to provide access to regions of a genome for which there are no plans to generate genomic sequence on any scale.

Although the construction of a physical map, and therefore a clone-by-clone approach, proved successful for the generation of human sequence, are physical maps required given the possible contribution of a whole-genome shotgun (WGS) approach to sequencing complex organisms? The main advantages of WGS are that the production of data is very rapid, can be highly automated, avoids cloning biases of BAC systems, and is very cost-effective. The assembly inherent from the sequence alignment also provides important mapping information that is unbiased by

Table 2
Species-Specific Genomic Fingerprint Databases

Organism	Ref.
A. thaliana	*69*
Rice	*70*
H. sapiens	*15*
M. musculus	*68*
R. rattus	http://www.bcgsc.ca/lab/mapping
C. neoformans	http://www.bcgsc.ca/lab/mapping
Bovine	http://www.bcgsc.ca/lab/mapping
G. aculeatus	http://www.bcgsc.ca/lab/mapping
Porcine	http://www.nps.ars.usda.gov/
D. rario	http://www.sanger.ac.uk/Projects/D_rerio/
Soybean	http://hbz.tamu.edu/soybean.html

additional experimental mapping systems or procedures. Although it remains true that WGS in isolation has disadvantages that prevent completion of either the map or finished sequence of a large genome, the possibility of combining the advantages of both approaches has been explored. A hybrid strategy emerged from the *Drosophila* project, and has since been adopted for the mouse genome. Sevenfold WGS coverage was generated from subcloned plasmids of varying sizes, which, when assembled with BESs, generated 96% coverage of the euchromatic portion of the mouse genome. This estimate was derived by assessing the amount of WGS coverage provided, which matched 187 Mb of finished mouse sequence. For a second, independent estimate, a genomic alignment of a curated collection of cDNAs to the WGS assembly was also used. This alignment included 96.4% of cDNA bases. Paired-end reads from large-insert plasmids and BACs provided the scaffold upon which the assembled whole-genome shotgun sequence was ordered and orientated, and it simultaneously integrated BAC clones into the sequence. A tiling path of BAC clones from the physical map is currently being used for directed finishing of the draft genomic sequence. The physical map helped to assemble the sequence scaffold, and the WGS data increased the rate of clone-based finishing ***(5)***.

If WGS sequence data can accurately place BACs via their BESs within the sequence assembly, is a restriction fingerprint database actually required? The answer is probably yes. Whereas BES localization within a WGS assembly facilitates a more optimal minimum tiling path selection, overlaps within fingerprinting contigs can link sequence assemblies [as reported by the assembly of the mouse WGS sequence ***(5)***]. Assembling plasmid and BAC end sequences in the WGS assembly generated 377 anchored 'supercontigs'. This number was reduced to 88 when 2 or more sequence supercontigs were localized within a single restriction fingerprint contig. The overlaps generated by fingerprint analysis may also be able to resolve errors in the genomic assembly, where, for example, low copy repeats may have resulted in a compression of the sequence assembly. The proven success of assembling genomewide physical maps, the cost of constructing a >15-fold genomic BAC library, and the ease with which genomewide fingerprint databases can be assembled has led to the construction of several genomic fingerprint databases (*see* **Table 2**). Whereas genomewide fingerprint maps will facilitate the large-scale characterization of many varied species, the construction of small region-specific sequence-ready maps will continue to be important for detailed interspecies sequence comparisons ***(67)***.

3. Computational Genomics

3.1. In Silico Gene Prediction

The *in silico* prediction of genes within the genomic sequence utilizes characteristic sequence motifs associated with genes (see Chapter 28). Unlike prokaryotic organisms, in which genes are located as a single tract of DNA, complex organisms, from yeast onward, contain genes that

Table 3
Genes in the Human Genome

Dataset	Gene no.	Date	Ref. or source
Hypothetical	100,000	1992	*72*
CpG islands	80,000	1993	*82*
EST clusters	60,000–70,000	1994	*83*
Unigene clusters	92,000	1996	*49*
Gene sequences	140,000	1999	IncyteGenomics[a]
Chromosome 22 sequence	43,000–61,000	1999	*20*
Chromosomes 22 and 21 sequence	44,000	2000	*21*
Tetraodon sequence	28,000–34,000	2000	*84*
ESTs in dbEST	120,000	2000	*85*
EST and mRNA	35,000	2000	*86*
Draft sequence	31,000	2001	*19*

[a]Press release available at http://incyte.com/company/news/1999/genes.shtml.

are usually segmented by introns of noncoding sequence ***(71,72)***. The spliceosome recognizes sequence motifs within the intron that leads to their excision, usually a GT dinucleotide at the 5' end of an intron (splice donor), an AG dinucleotide at the 3' end (splice acceptor), and an internal branch point ***(73)***. Localizing the site of translation initiation within genomic sequence, unlike the transcription start sites, which are predicted on the basis of the combined occurrence of CpG islands and TATA boxes flanked by regions of C-G motifs, has been facilitated by the identification of a consensus motif within the genomic sequence. The translation start site, usually an ATG, is typically found in a consensus [GCC$^{A}/_{G}$CC**ATG**G], which includes the two bases that exert the strongest effect: a G at the first base after the translation start, ATG, and a purine (preferably an A) that is located three nucleotides upstream ***(74)***. The identification of stop codons (TGA, TAA, or TAG) and polyadenylation signals ***(75)***, most commonly as AATAAA ***(76)***, has helped define the 3' ends of genes within genomic sequence.

Whereas initial computer programs were developed to identify single exons [e.g., GRAIL ***(77)*** and HEXON ***(78)***], more recently developed programs attempt to identify complete gene structures. These programs, GENSCAN ***(79)*** and FGENESH ***(80)*** predict individual exons based on codon usage and sequence signals and assemble these putative exons into candidate gene structures. The greatest problem associated with using *in silico* gene prediction programs to identify coding structures within genomic sequence is the number of over predictions that are generated. Although setting low prediction thresholds might identify all genes, the gene structures will be generated with low specificity ***(81)***. The estimated total number of genes contained within the human genome has varied considerably according to the technique that was used to evaluate it and the data that were available at the time. Estimates have been based variously on average genome and gene size, the number of observed CpG islands (and the proportion associated with genes), redundant and nonredundant EST sequences and, finally, chromosome specific totals, comparative analysis and genomewide sequence analysis (*see* **Table 3**).

3.2. Sequence Analysis

Although computer programs have been written to predict a number of features associated with coding sequences, the alignment of experimental data is critical to the validation of predicted structures. Programs such as CLUSTALW ***(87)***, which is used to align multiple nucleotide or protein sequences, DOTTER ***(88)***, which utilizes pairwise local sequence alignment strategy, and PipMaker *(89)*, which graphically represents the percentage identify between two sequences, are useful for inferring structural and functional conservation by sequence homol-

Table 4
Sequence Queries Available Using BLAST Alignment

Program	Query	Database	Comparison
blastn	DNA	DNA	DNA level
blastp	Protein	Protein	Protein level
blastx	DNA	Protein	Protein level
tblastn	Protein	DNA	Protein level
tblastx	DNA	DNA	Protein level

Source: Adapted from Brenner, S. E., Chothia, C., and Hubbard, T. J. (1998) Assessing sequence comparison methods with reliable structurally identified distant evolutionary relationships. *Proc. Natl. Acad. Sci. USA* ***95,*** 6073–6078.

ogy. Sequence homology searching can also be performed using SSAHA ***(90)***, Exonerate (Slater, unpublished), and BLAT ***(91)***, which permits homology sequences to be identified within gigabases of DNA. BLAST (Basic Local Alignment Search Tool), which measures the local similarity between two sequences ***(92,93)***, is the primary method for identifying protein and DNA sequence similarities prior to incorporation of the features into project-specific AceDB databases (http://www.acedb.org/) or genome browsers. One of the major advantages of using BLAST for sequence alignment is the flexibility with which nucleotide and amino acid sequences can be aligned (*see* **Table 4**).

Whereas BLAST alignment of human mRNA, EST, and protein sequences to predict coding structures provides a primary level of support, predicted features can also be supported by alignments with sequence from other organisms for comparison (comparative sequence analysis). The identification of sequences that are conserved between species is important because sequences that contain elements that are potentially functional are more likely to retain their sequence than nonfunctional segments, under the constraints of natural selection during evolution. The evolutionary distance between species is an important consideration. Sequence comparisons between closely related species may facilitate the identification of gene structures and regulatory elements but if the evolutionary distance between the species is relatively small, these sequences may be obscured by nonfunctional sequence conservation. Therefore, a variety of species, including more distantly related species, might be required to identify potential functional sequences using the comparative approach.

The identification of conserved sequences by comparative analysis has focused on the identification of noncoding regions ***(94–96)*** and protein coding regions ***(97–99)*** between human and mouse genomes. The alignment of sequence from multiple organisms has also been used to identify upstream regions that may affect gene expression ***(90)***. Although comparative sequence analysis may not identify all control regions associated with a gene, conserved regions may be identified that would be a candidate for further experimental investigation ***(101)***. The availability of large tracts of human genomic sequence has necessitated the development of databases (genome browsers) that provide a framework upon which the enormous amount of data associated with the human genome can be stored and displayed. The three main databases, Ensembl, developed at the Wellcome Trust Sanger Institute and the European Bioinformatics Institute (http://www.ensembl.org/Homo_sapiens/), the University of California Santa Cruz (UCSC) genome browser (http://genome.cse.ucsc.edu/), and the NCBI browser (http://www.ncbi.nlm.nih.gov/mapview/map_search.cgi?taxid=9606) each contain information pertaining to physical maps, chromosome-specific sequence assemblies, aligned mRNAs and ESTs, cross-species homologies, single-nucleotide polymorphisms (SNPs), and repeat elements. The development of generic genome browsers, as such as those hosted by Ensembl, makes possible the rapid identification of homologous sequences between comparative organisms and, in doing so, assist in identifying conserved features that might be of some functional significance.

4. Conclusions

The generation of genetic, RH and physical maps of the human genome have provided a means by which the genome's various components can be partitioned and characterized, in addition to acting as the basis for identifying segregating regions of inheritance that are associated with disease-causing genes. The construction of a sequence-ready physical map of the human genome has been central to the production of a high-quality finished genomic sequence, an essential facet of the complete identification all coding and regulatory features contained within.

In the future, substantial efforts will be made to fully characterize all of the genes in the human genome. These studies will result in a fuller understanding of the specificity and range of biochemical structures and functions that are encoded in the human genome sequence. In general, there is likely to remain a distinction between the study of functions encoded at the DNA level, which affect gene expression via transcriptional control, and the study of functions reflected at the protein level following translation, taking into account posttranslational modifications (processes which are largely genetically determined). Without a genic catalog, functional studies are necessarily limited to the investigation of a specific target—a gene, a protein, or a disease. These approaches are an essential part of fully interpreting the genome, as they provide a means by which hypotheses can be experimentally tested and which produce valid and supplementary results. However, the production of a complete gene catalog (if completion can indeed be measured or achieved) will provide the raw material for modeling whole systems. The extensive use of computational biology to suggest how such complex systems are made up of their interacting components will, in itself, enable predictions to be made of the system model. These predictions can be tested, both to determine the validity of the modeled system and to test the success of the methods used to derive the system.

A more complete knowledge of biochemical processes will yield a better understanding of complex disease and how it should be treated. At present, our knowledge is primarily based on monogenic diseases. As the problem is reduced to a single gene, hypotheses for function can be tested by biochemical assays, protein structural studies, experimental knockouts, or the study of naturally occurring mutants. The approach to complex disease centers on a similar approach (i.e., trying to identify the one or few dominant genetic factors that contribute the most significant effect to the overall phenotype). However, there is a realization that these genetic factors might not fully explain the observed phenotype and that a proportion of the remaining factors might not be identified. In t11lhese instances, a comprehensive knowledge of the systems involved will be more informative than the approach of complex disease genetics, both in how the phenotype arises and how it might be possible to intervene more effectively. This is potentially a true long-term value of the genome sequence and its interpretation in a biochemical, biological, and genetic context, for the advancement of medicine in the future.

References

1. Watson, J. D. and Crick, F. (1953) A structure for deoxyribose nucleic acid. *Nature* **171**, 171.
2. Kohara, Y., Akiyama, K., and Isono, K. (1987) The physical map of the whole *E. coli* chromosome: application of a new strategy for rapid analysis and sorting of a large genomic library. *Cell* **50**, 495–508.
3. Olson, M. V., Dutchik, J. E., Graham, M. Y., et al. (1986) Random-clone strategy for genomic restriction mapping in yeast. *Proc. Natl. Acad. Sci. USA* **83**, 7826–7830.
4. Coulson, A., Sulston, J., Brenner, S., and Karn, J. (1986) Toward a physical map of the genome of the nematode *Caenorhabditis elegans*. *Proc. Natl. Acad. Sci. USA* **83**, 7821–7825.
5. Mouse Genome Sequencing Consortium. (2002) Initial sequencing and comparative analysis of the mouse genome. *Nature* **420**, 520–562.
6. Sanger, F., Air, G. M., Barrell, B. G., et al. (1977) Nucliotide sequence of bacteriophage phi X174 DNA. *Nature* **265**, 687–695.
7. Sanger, F., Coulson, A. R., Friedmann, T., et al. (1978) The nucleotide sequence of bacteriophage phiX174. *J. Mol. Biol.* **125**, 225–246.

8. Sanger, F., Coulson, A. R., Hong, G. F., Hill, D. F., and Petersen, G. B. (1982) Nucleotide sequence of bacteriophage lambda DNA. *J. Mol. Biol.* **162**, 729–773.
9. Maxam, A. M. and Gilbert, W. A new method for sequencing DNA. (1977) *Proc. Natl. Acad. Sci. USA* **74**, 560–564.
10. Anderson, S. (1981) Shotgun DNA sequencing using cloned DNase I-generated fragments. *Nucleic Acids Res.* **9**, 3015–3027.
11. Anderson, S., Bankier, A. T., Barrell, B. G., et al. (1981) Sequence and organization of the human mitochondrial genome. *Nature* **290**, 457–465.
12. Fleischmann, R. D., Adams, M. D., White, O., et al. (1995) Whole-genome random sequencing and assembly of Haemophilus influenzae Rd. *Science* **269**, 496–512.
13. Arabidopsis Genome Initiative. (2000) Analysis of the genome sequence of the flowering plant *Arabidopsis thaliana*. *Nature* **408**, 796–815.
14. Hoskins, R. A., Nelson, C. R., Berman, B. P., et al. (2000) A BAC-based physical map of the major autosomes of *Drosophila melanogaster*. *Science* 287, 2271–2274.
15. McPherson, J. D., Marra, M., Hillier, L., et al. (2001) A physical map of the human genome. *Nature* 409, 934–941.
16. Kim, U. J., Shizuya, H., de Jong, P. J., Birren, B., and Simon, M. I. (1992) Stable propagation of cosmid sized human DNA inserts in an F factor based vector. *Nucleic Acids Res.* **20**, 1083–1085.
17. Burke, D. T., Carle, G. F., and Olson, M. V. (1987) Cloning of large segments of exogenous DNA into yeast by means of artificial chromosome vectors. *Science* **236**, 806–812.
18. Morton, N. E. (1991) Parameters of the human genome. *Proc. Natl. Acad. Sci. USA* 88, 7474-6.
19. International Human Genome Sequencing Consortium. (2001) Initial sequencing and analysis of the human genome. *Nature* **409**, 860–921.
20. Dunham, I., Shimizu, N., Roe, B. A., et al. (1999) The DNA sequence of human chromosome 22. *Nature* **402**, 489–495.
21. Hattori, M., Fujiyama, A., Taylor, T. D., et al. (2000) The DNA sequence of human chromosome 21. *Nature* **405**, 311–319.
22. Deloukas, P., Matthews, L. H., Ashurst, J., et al. (2001) The DNA sequence and comparative analysis of human chromosome 20. *Nature* **414**, 865–871.
23. Bickmore, W. A. and Sumner, A. T. (1989) Mammalian chromosome banding—an expression of genome organization. *Trends Genet.* **5**, 144–148.
24. Pinkel, D., Landegent, J., Collins, C., et al. (1988) Fluorescence in situ hybridization with human chromosome-specific libraries: detection of trisomy 21 and translocations of chromosome 4. *Proc. Natl. Acad. Sci. USA* **85**, 9138–9142.
25. Tkachuk, D. C., Westbrook, C. A., Andreeff, M., et al. (1990) Detection of bcr-abl fusion in chronic myelogeneous leukemia by in situ hybridization. *Science* **250**, 559–562.
26. Dauwerse, J. G., Kievits, T., Beverstock, G. C., et al. (1990) Rapid detection of chromosome 16 inversion in acute nonlymphocytic leukemia, subtype M4: regional localization of the breakpoint in 16p. *Cytogenet. Cell Genet.* **53**, 126–128.
27. Pinkel, D., Straume, T., and Gray, J. W. (1986) Cytogenetic analysis using quantitative, high-sensitivity, fluorescence hybridization. *Proc. Natl. Acad. Sci. USA* **83**, 2934–2988.
28. Wilke, C. M., Guo, S. W., Hall, B. K., et al. (1994) Multicolor FISH mapping of YAC clones in 3p14 and identification of a YAC spanning both FRA3B and the t(3;8) associated with hereditary renal cell carcinoma. *Genomics* **22**, 319–326.
29. Heiskanen, M., Karhu, R., Hellsten, E., Peltonen, L., Kallioniemi, O. P., and Palotie, A. (1994) High resolution mapping using fluorescence in situ hybridization to extended DNA fibers prepared from agarose-embedded cells. *Biotechniques* **17**, 928–929, 932–933.
30. Dib, C., Faure, S., Fizames, C., et al. (1996) A comprehensive genetic map of the human genome based on 5,264 microsatellites. *Nature* **380**, 152–154.
31. Kan, Y. W. and Dozy, A. M. (1978) Polymorphism of DNA sequence adjacent to human beta-globin structural gene: relationship to sickle mutation. *Proc. Natl. Acad. Sci. USA* **75**, 5631–5635.
32. Botstein, D., White, R. L., Skolnick, M., and Davis, R. W. (1980) Construction of a genetic linkage map in man using restriction fragment length polymorphisms. *Am. J. Hum. Genet.* **32**, 314–331.
33. Donis-Keller, H., Green, P., Helms, C., et al. (1987) A genetic linkage map of the human genome. *Cell* **51**, 319–337.
34. Wyman, A. R. and White, R. (1980) A highly polymorphic locus in human DNA. *Proc. Natl. Acad. Sci. USA* **77**, 6754–6758.

35. Jeffreys, A. J., Wilson, V., and Thein, S. L. (1985) Hypervariable 'minisatellite' regions in human DNA. *Nature* **314**, 67–73.
36. Nakamura, Y., Leppert, M., O'Connell, P., et al. (1987) Variable number of tandem repeat (VNTR) markers for human gene mapping. *Science* **235**, 1616–1622.
37. Royle, N. J., Clarkson, R. E., Wong, Z., and Jeffreys, A. J. (1988) Clustering of hypervariable minisatellites in the proterminal regions of human autosomes. *Genomics* **3**, 352–360.
38. Litt, M. and Luty, J. A. (1989) A hypervariable microsatellite revealed by in vitro amplification of a dinucleotide repeat within the cardiac muscle actin gene. *Am. J. Hum. Genet.* **44**, 397–401.
39. Weber, J. L. and May, P. E. (1989) Abundant class of human DNA polymorphisms which can be typed using the polymerase chain reaction. *Am. J. Hum. Genet.* **44**, 388–396.
40. Saiki, R. K., Gelfand, D. H., Stoffel, S., et al. (1988) Primer-directed enzymatic amplification of DNA with a thermostable DNA polymerase. *Science* **239**, 487–491.
41. Olson, M., Hood, L., Cantor, C., and Botstein, D. (1989) A common language for physical mapping of the human genome. *Science* **245**, 1434–1435.
42. Kong, A., Gudbjartsson, D. F., Sainz, J., et al. (2002) A high-resolution recombination map of the human genome. *Nature Genet.* **31**, 241–247.
43. Cohen, D., Chumakov, I., and Weissenbach, J. (1993) A first-generation physical map of the human genome. *Nature* **366**, 698–701.
44. Weissenbach, J., Gyapay, G., Dib, C., et al. (1992) A second-generation linkage map of the human genome. *Nature* **359**, 794–801.
45. Gyapay, G., Morissette, J., Vignal, A., et al. (1994) The 1993–1994 Genethon human genetic linkage map. *Nature Genet.* **7**, 246–339.
46. Murray, J. C., Buetow, K. H., Weber, J. L., et al. (1994) A comprehensive human linkage map with centimorgan density. Cooperative Human Linkage Center (CHLC). *Science* **265**, 2049–2054.
47. Goss, S. J. and Harris, H. (1975) New method for mapping genes in human chromosomes. *Nature* **255**, 680–684.
48. Cox, D. R., Burmeister, M., Price, E. R., Kim, S., and Myers, R. M. (1990) Radiation hybrid mapping: a somatic cell genetic method for constructing high-resolution maps of mammalian chromosomes. *Science* **250**, 245–250.
49. Schuler, G. D., Boguski, M. S., Stewart, E. A., et al. (1996) A gene map of the human genome. *Science* **274**, 540–546.
50. Deloukas, P., Schuler, G. D., Gyapay, G., et al. (1998) A physical map of 30,000 human genes. *Science* **282**, 744–746.
51. Mungall, A. J., Edwards, C. A., Ranby, S. A., et al. (1996) Physical mapping of chromosome 6: a strategy for the rapid generation of sequence-ready contigs. *DNA Seq,* **7**, 47–49.
52. Staden, R. (1980) A new computer method for the storage and manipulation of DNA gel reading data. *Nucleic Acids Res.* **8**, 3673–3694.
53. Green, E. D. and Olson, M. V. (1990) Chromosomal region of the cystic fibrosis gene in yeast artificial chromosomes: a model for human genome mapping. *Science* **250**, 94–98.
54. Chumakov, I. M., Le Gall, I., Billault, A., et al. (1992) Isolation of chromosome 21-specific yeast artificial chromosomes from a total human genome library. *Nature Genet.* **1**, 222–225.
55. Foote, S., Vollrath, D., Hilton, A., and Page, D. C. (1992) The human Y chromosome: overlapping DNA clones spanning the euchromatic region. *Science* **258**, 60–66.
56. Chumakov, I. M., Rigault, P., Le Gall, I., et al. (1995) A YAC contig map of the human genome. *Nature* **377**, 175–297.
57. Hudson, T. J., Stein, L. D., Gerety, S. S., et al. (1995) An STS-based map of the human genome. *Science* **270**, 1945–1954.
58. Coffey, A. J., Roberts, R. G., Green, E. D., et al. (1992) Construction of a 2.6-Mb contig in yeast artificial chromosomes spanning the human dystrophin gene using an STS-based approach. *Genomics* **12**, 474–484.
59. Green, E. D., Riethman, H. C., Dutchik, J. E., and Olson, M. V. (1991) Detection and characterization of chimeric yeast artificial-chromosome clones. *Genomics* **11**, 658–669.
60. Bates, G. P., Valdes, J., Hummerich, H., et al. (1992) Characterization of a yeast artificial chromosome contig spanning the Huntington's disease gene candidate region. *Nature Genet.* **1**, 180–187.
61. Slim, R., Le Paslier, D., Compain, S., et al. (1993) Construction of a yeast artificial chromosome contig spanning the pseudoautosomal region and isolation of 25 new sequence-tagged sites. *Genomics* **16**, 691–697.

62. Collins, J. E., Cole, C. G., Smink, L. J., et al. (1995) A high-density YAC contig map of human chromosome 22. *Nature* **377**, 367–379.
63. Bouffard, G. G., Idol, J. R., Braden, V. V., et al. (1997) A physical map of human chromosome 7: an integrated YAC contig map with average STS spacing of 79 kb. *Genome Res.* 7, 673–692.
64. Coulson, A., Huynh, C., Kozono, Y., and Shownkeen, R. (1995) The physical map of the *Caenorhabditis elegans* genome. *Methods Cell Biol.* **48**, 533–550.
65. Shizuya, H., Birren, B., Kim, U. J., et al. (1992) Cloning and stable maintenance of 300-kilobase-pair fragments of human DNA in Escherichia coli using an F-factor-based vector. *Proc. Natl. Acad. Sci. USA* **89**, 8794–8797.
66. Ioannou, P. A., Amemiya, C. T., Garnes, J., et al. (1994) A new bacteriophage P1-derived vector for the propagation of large human DNA fragments. *Nature Genet.* **6**, 84–89.
67. Thomas, J. W., Prasad, A. B., Summers, T. J., et al. (2002) Parallel construction of orthologous sequence-ready clone contig maps in multiple species. *Genome Res.* **12**, 1277–1285.
68. Gregory, S. G., Sekhon, M., Schein, J., et al. (2002) A physical map of the mouse genome. *Nature* **418**, 743–750.
69. Marra, M., Kucaba, T., Sekhon, M., et al. (1999) A map for sequence analysis of the Arabidopsis thaliana genome. *Nature Genet.* **22**, 265–270.
70. Tao, Q., Chang, Y-L., Wang, J., et al. (2001) Bacterial artificial chromosome-based physical map of the rice genome constructed by restriction fingerprint analysis. *Genetics* **158**, 1711–1724.
71. Tilghman, S. M., Tiemeier, D. C., Seidman, J. G., et al. (1978) Intervening sequence of DNA identified in the structural portion of a mouse beta-globin gene. *Proc. Natl. Acad. Sci. USA* **75**, 725–729.
72. Gilbert, W. (1978) Why genes in pieces? *Nature* **271**, 501.
73. Moore, M. J. and Sharp, P. A. (1993) Evidence for two active sites in the spliceosome provided by stereochemistry of pre-mRNA splicing. *Nature* **365**, 364–368.
74. Kozak, M. (1987) An analysis of 5'-noncoding sequences from 699 vertebrate messenger RNAs. *Nucleic Acids Res.* **15**, 8125–8148.
75. Kessler, M. M., Beckendorf, R. C., Westhafer, M. A., and Nordstrom, J. L. (1986) Requirement of A-A-U-A-A-A and adjacent downstream sequences for SV40 early polyadenylation. *Nucleic Acids Res.* **14**, 4939–4952.
76. Beaudoing, E., Freier, S., Wyatt, J. R., Claverie, J. M., and Gautheret, D. (2000) Patterns of variant polyadenylation signal usage in human genes. *Genome Res.* 10, 1001–1010.
77. Uberbacher, E. C., Xu, Y., and Mural, R. J. (1996) Discovering and understanding genes in human DNA sequence using GRAIL. *Methods Enzymol.* **266**, 259–281.
78. Solovyev, V. V., Salamov, A. A., and Lawrence, C. B. (1994) Predicting internal exons by oligonucleotide composition and discriminant analysis of spliceable open reading frames. *Nucleic Acids Res.* **22**, 5156–5163.
79. Burge, C. and Karlin, S. (1997) Prediction of complete gene structures in human genomic DNA. *J. Mol. Biol.* **268**, 78–94.
80. Solovyev, V. V., Salamov, A. A., and Lawrence, C. B. (1995) Identification of human gene structure using linear discriminant functions and dynamic programming. *Proc. Int. Conf. Intel. Syst. Mol. Biol.* **3**, 367–375.
81. Guigo, R., Agarwal, P., Abril, J. F., Burset, M., and Fickett, J. W. (2000) An assessment of gene prediction accuracy in large DNA sequences. *Genome Res.* **10**, 1631–1642.
82. Antequera, F. and Bird, A. (1993) Number of CpG islands and genes in human and mouse. *Proc. Natl. Acad. Sci. USA* **90**, 11,995–11,999.
83. Fields, C., Adams, M. D., White, O. and Venter, J. C. (1994) How many genes in the human genome? *Nature Genet.* **7**, 345–346.
84. Roest-Crollius, H., Jaillon, O., Bernot, A., et al. (2000) Estimate of human gene number provided by genome-wide analysis using *Tetraodon nigroviridis* DNA sequence. *Nature Genet.* **25**, 235–238.
85. Liang, F., Holt, I., Pertea, G., Karamycheva, S., Salzberg, S. L., and Quackenbush, J. (2000) Gene index analysis of the human genome estimates approximately 120,000 genes. *Nature Genet.* **25**, 239–240.
86. Ewing, B. and Green, P. (2000) Analysis of expressed sequence tags indicates 35,000 human genes. *Nature Genet.* **25**, 232–234.
87. Thompson, J. D., Higgins, D. G., and Gibson, T. J. (1994) CLUSTAL W: improving the sensitivity of progressive multiple sequence alignment through sequence weighting, position-specific gap penalties and weight matrix choice. *Nucleic Acids Res.* **22**, 4673–4680.

88. Sonnhammer, E. L. and Durbin, R. (1995) A dot-matrix program with dynamic threshold control suited for genomic DNA and protein sequence analysis. *Gene* **167**, GC1–GC10.
89. Schwartz, S., Zhang, Z., Frazer, K. A., et al. (2000) PipMaker—A web server for aligning two genomic DNA sequences. *Genome Res.* **10**, 577–586.
90. Ning, Z., Cox, A. J., and Mullikin, J. C. (2001) SSAHA: a fast search method for large DNA databases. *Genome Res.* **11**, 1725–1729.
91. Kent, W. J. (2002) BLAT—the BLAST-like alignment tool. *Genome Res.* **12**, 656–664.
92. Altschul, S. F., Gish, W., Miller, W., Myers, E. W., and Lipman, D. J. (1990) Basic local alignment search tool. *J. Mol. Biol.* **215**, 403–410.
93. Altschul, S. F., Madden, T. L., Schaffer, A. A., et al. (1997) Gapped BLAST and PSI-BLAST: a new generation of protein database search programs. *Nucleic Acids Res.* **25**, 3389–3402.
94. Hardison, R., Slightom, J. L., Gumucio, D. L., Goodman, M., Stojanovic, N., and Miller, W. (1997) Locus control regions of mammalian beta-globin gene clusters: combining phylogenetic analyses and experimental results to gain functional insights. *Gene* **205**, 73–94.
95. Koop, B. F. and Hood, L. (1994) Striking sequence similarity over almost 100 kilobases of human and mouse T-cell receptor DNA. *Nature Genet.* **7**, 48–53.
96. Hardison, R., Slightom, J. L., Gumucio, D. L., Goodman, M., Stojanovic, N., and Miller, W. (1997) Locus control regions of mammalian beta-globin gene clusters: combining phylogenetic analyses and experimental results to gain functional insights. *Gene* **205**, 73–94.
97. Makalowski, W., Zhang, J., and Boguski, M. S. (1996) Comparative analysis of 1196 orthologous mouse and human full-length mRNA and protein sequences. *Genome Res.* 6, 846–857.
98. Ansari-Lari, M. A., Oeltjen, J. C., Schwartz, S., et al. (1998) Comparative sequence analysis of a gene-rich cluster at human chromosome 12p13 and its syntenic region in mouse chromosome 6. *Genome Res.* **8**, 29–40.
99. Jang, W., Hua, A., Spilson, S. V., Miller, W., Roe, B. A., and Meisler, M. H. (1999) Comparative sequence of human and mouse BAC clones from the mnd2 region of chromosome 2p13. *Genome Res.* **9**, 53–61.
100. Gottgens, B., Barton, L. M., Gilbert, J. G., et al. (2000) Analysis of vertebrate SCL loci identifies conserved enhancers. *Nature Biotechnol.* **18**, 181–186.
101. Pennacchio, L. A., Olivier, M., Hubacek, J. A., et al. (2001) An apolipoprotein influencing triglycerides in humans and mice revealed by comparative sequencing. *Science* **294**, 169–173.

12

Dideoxyfingerprinting for Mutation Detection

Ioannis Bossis, Antonios Voutetakis, and Constantine A. Stratakis

1. Introduction

Screening for mutations of the thousands of the sequence products provided by human genome analysis has proven to be a daunting task. The gold standard for identifying sequence alterations is direct sequencing. However, this method is labor- intensive and the least cost-effective. Since the mid-1980s, the need for rapid, high-throughput, accurate, and economical mutation analysis systems has lead to the development of several technologies, as an alternative to analysis by direct sequencing, which allowed detection of single mutations in long stretches of DNA (200–600 bp). These techniques include restriction endonuclease digestion of polymerase chain reaction (PCR) products (PCR-RFLP), denaturing gradient gel electrophoresis (DGGE), single-strand conformation polymorphism (SSCP), dideoxyfingerprinting (ddF), and heteroduplex mobility assay (HMA). Most of these methods utilize PCR for amplification of a region of the DNA, a physical or chemical treatment of amplified DNA (by restriction digestion or denaturation), separation of the amplicons by gel electrophoresis (by denaturing or nondenaturing), and visualization of the separated sequence strands (by autoradiography or fluorescence-based detection). Most recent modifications in some of these techniques allow the simultaneous separation and detection of DNA fragments with the use of sophisticated equipment such as high-performance liquid chromatography (HPLC) and capillary electrophoresis.

Originally described in 1989 *(1)*, SSCP analysis has been the most widely used method for the detection of mutations because of its simplicity and efficiency (see Chapter 7). In SSCP, DNA regions with potential mutations are first amplified by PCR in the presence of a radiolabeled dNTP. Single-stranded DNA fragments are then generated by denaturation of the PCR products and separated on a native polyacrylamide gel. As the denatured PCR product moves through the gel and away from the denaturant, it will regain a secondary structure that is sequence-dependent. The electrophoretic migration of single-stranded DNA is a function of its secondary structure. Therefore, PCR products that contain substitution differences, as well as insertions and deletions, will have different mobilities when compared with wild-type DNA.

Although SSCP is simple, rapid, and inexpensive, it also has some disadvantages. The major limitations of the technique are sequence-dependent-sensitivity and the inability to provide information about the location of a mutation in a DNA fragment. Efficacy studies, in which the technique was evaluated against a known mutation, showed that the sensitivity of SSCP can be highly variable in identifying sequence alterations *(2–5)*. This variability is seen not only between amplicons but also within the same amplicon if examined under different conditions. Overall, SSCP detects previously identified sequence changes in as many as 90% and as little as 60%

From: *Medical Biomethods Handbook*
Edited by: J. M. Walker and R. Rapley © Humana Press, Inc., Totowa, NJ

of the specimens, depending primarily on the sequence, the amplicon's size, and the location of the mutation.

To circumvent the disadvantages of SSCP, Sarkar and co-workers ***(6)*** proposed a new technique that became known as dideoxyfingerprinting (ddF). ddF is essentially a hybrid of SSCP and dideoxy sequencing in which primer extension products are generated in the presence of one dideoxynucleotide and subjected to chemical denaturation and electrophoresis on a nondenaturing polyacrylamide gel to exploit differences in secondary structure of single-stranded conformers. The resulting electrophoretic pattern resembles sequencing gels in which the mobility of extension products is determined by both size and conformation. In ddF, single-base and other sequence changes can result in the elimination of a normal band or appearance of an extra band (informative dideoxy component) and altered electrophoretic mobility of one or more sequence-termination products (informative SSCP component). The intensity of bands can also be used for dosage analysis, because detection of "half-loss" would indicate heterozygous mutation.

There are three major advantages of ddF over conventional SSCP: (1) The relative position of a mutation can be revealed as a result of the addition or deletion of a dideoxy-termination product; (2) the sensitivity is increased because, unlike SSCP, there are multiple bands that exhibit altered mobility, making it unlikely that a mutation will be missed (because the technique is based on the same extension principle as dideoxy sequencing, multiple DNA strands are generated that contain the sequence change); and (3) it permits large PCR fragments to be generated only once and analyzed in smaller subfragments using different primers. The originally described ddF could screen the same size of DNA as SSCP (250–350 bp) but with an astonishingly higher level of detection ***(6)***.

In recent years, modifications of the original procedure have resulted in new and improved variants of ddF. Bidirectional ddF (Bi-ddF), in which the dideoxy-termination reaction is performed simultaneously with two opposing primers, has allowed larger fragments (approx 600 bp) to be screened with almost 100% sensitivity ***(7)***. RNA ddF (R-ddF), in which RNA is used as starting material, enables identification of mutations that result in splicing errors and allows screening of genes with large intronic regions ***(2)***. Denaturing fingerprinting (dnF), a modification of Bi-ddF in which fingerprints are generated by performing denaturing gel electrophoresis on bidirectional "cycle-sequencing" reactions with each of two dideoxy terminators, has allowed screening of DNA regions with high GC content, avoiding the generation of "smearing" bands and, thus, increasing the sensitivity of the technique in identifying heterozygous mutations ***(8)***. Further modifications have streamlined the above procedures by adapting them to either automated fluorescent sequencers or capillary electrophoresis utilizing fluorescent dNTPs or primers (automated Bi-ddF, capillary electrophoresis-ddF) ***(9,10)***. Automated fingerprinting technology allows simultaneous analysis of a larger number of samples, higher reproducibility, faster data processing, and the ability to analyze longer sequences from a single reaction.

2. Principles of ddF and ddF Variants

2.1. The Method

In ddF, genomic DNA from control and test subjects (containing possible mutations) is PCR-amplified and the PCR products serve as the template for the subsequent Sanger sequencing reaction. The use of a proofreading polymerase in the PCR reaction, such as *pfu*, is recommended to avoid mutations introduced by *Taq* polymerase, which could be responsible for a false-positive outcome. Success of the PCR reaction is often verified by gel electrophoresis. The absence of nonspecific products is required in order to obtain meaningful results in the subsequent steps. Unincorporated dNTPs and unused primers should be removed by passing the PCR reaction through ultrafiltration columns or treatment with exonuclease I and shrimp alkaline phosphatase (SAP). Yield of the PCR-amplified DNA for each sample should be determined spectrophotometrically or by running a small amount through agarose gel electrophoresis and comparing it with known standards. When necessary, the volume is adjusted so that equivalent amounts of DNA template are added to the Sanger reaction for each different sample.

The Sanger sequencing reaction is basically a PCR reaction using one nested primer (linear amplification) and a combination of normal dNTPs and a single ddNTP (usually ddGTP or ddCTP) in excess. The primer is radiolabeled by the end-labeling technique using [^{32}P]ATP or [^{33}P]ATP and T4 polynucleotide kinase. [^{33}P]ATP is more expensive than [^{32}P]ATP but results in better autoradiographs (with sharper bands) and increased sensitivity. The primers can also be fluorescently labeled and the technique adapted to automated DNA analyzers. The Sanger reaction is typically performed in a total volume of 7–10 µL containing 20 µ*M* of each dNTP, 100 µ*M* ddGTP or ddCTP, 0.05 µ*M* of radiolabeled primer, 5–10 ng template DNA from the initial PCR reaction, and 0.4–1 U *Taq* polymerase. The ratio of ddGTP to dGTP often needs to be optimized for better results.

Upon completion of the PCR cycling protocol (usually 30 cycles), the reaction is stopped by adding 35–40 µL of stop/loading buffer (50% formamide, 7 *M* urea, 2 m*M* EDTA, 0.1% bromophenol blue, and 0.1% xylene cyanol) and denatured by heating to 80°C for 3–5 min. The samples are then snap-cooled and loaded into square wells of the gel (shark-tooth comps are not recommended). Originally, Sanger reactions were separated on 5.6% polyacrylamide gels at room temperature. However, 7.5% GeneAmp gels or 0.5xMDE gels provide superior results. Electrophoresis at 7–8°C is better than room temperature and further improves the quality of autoradiographs and efficiency in detecting mutations. After electrophoresis, the gel is transferred on chromatography paper, dried, and exposed to autoradiography film. The length of exposure should be determined empirically, and it is affected primarily by the specific activity of the radiolabeled primer, the amount of DNA template and number of PCR cycles in the Sanger reaction, and the sequence complexity of DNA. The resolution limit for detecting an abnormally migrating band is a change of a half-bandwidth on the upper part of the gel and a change of one-quarter bandwidth on the lower part.

A schematic representation of ddF is presented in **Fig. 1**, in which a control DNA (C) is compared with samples 1 and 2, each containing a single-base substitution. The DNA templates C, 1, and 2 are generated by PCR from genomic DNA and are subjected to the Sanger sequencing reaction using a single radiolabeled primer and ddGTP. In the example shown, the lane with sample 1 contains an extra band relative to the control lane as a result of T→G mutation and the use of ddGTP terminator in the Sanger reaction. In this case, the dideoxy component of ddF is informative. In addition, the SSCP component of ddF is informative. The top three bands are shifted because they have incorporated the mutation. In sample 2, a different mutation (C→T) is present. In this case, the dideoxy component is noninformative; however, the SSCP component is informative (the top two bands containing the mutation are shifted). From the example shown, it becomes apparent that certain mutations can be detected with much higher efficiency than others. For example, mutations that result in a dideoxy informative component (an extra or a missing band) are easier to detect.

From all possible single-base mutations (A→G, T→G, C→G, G→A, G→T, G→C, A→T, A→C, T→C, T→A, C→A, C→T), 50% will result in an informative dideoxy component. The efficiency in detecting mutations that result only in a SSCP-informative component is variable. The efficiency of the SSCP component for a given mutation depends on the number of abnormally migrating bands. Thus, mutations located far from the site complementary to the primer will be more difficult to detect. Typically, mutations as far as 250 bp from the primer-annealing site can be detected.

2.2. Bidirectional ddF

For screening larger DNA segments (up to 550 bp) with 100% sensitivity, bidirectional ddF (Bi-ddF) should be used *(7)*. In Bi-ddF, the Sanger termination reaction utilizes two opposite primers to simultaneously scan for mutations in both directions. In this way, dideoxy and SSCP components from both strands contribute to the sensitivity. If ddGTP is used in the termination

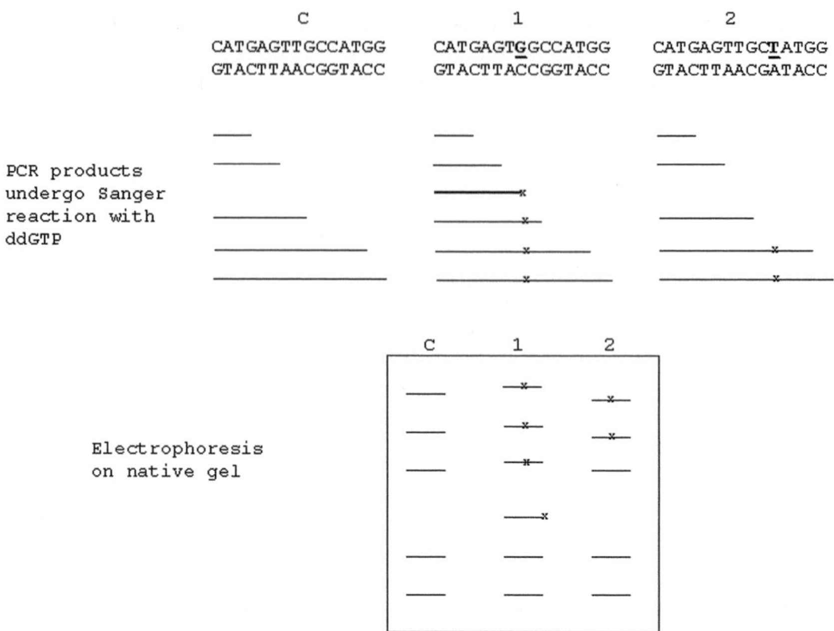

Fig. 1. Representative diagram of the principles underlying band detection in ddF. The upper panel shows control (c) DNA and two amplicons with two different point mutations. The dideoxy-sequencing bands bearing the mutations are shown along with those from the normal control. The lower panel shows the different migration pattern generated by the mutant bands in comparison to those from the control specimen when all specimens are run on a nondenaturing gel. The number of additional bands depends on the type and location of the mutation(s). For further details, *see* the text.

reaction, only A→T or T→A mutations will not have an informative component. As mentioned above, mutations that result in an informative dideoxy component in addition to an SSCP component are detected with greater sensitivity. In addition, the SSCP component in Bi-ddF is, on average, more informative because alterations of mobility can be detected in either the upstream or downstream direction. From a technical point of view, the only additional work required to adapt Bi-ddF over the original ddF is that the ratio of the downstream and upstream primer concentrations in the Sanger reaction needs to be optimized so that both termination products are of similar intensity. Bi-ddF can be also adapted to automated DNA analyzers by labeling the upstream and downstream primers with different fluorescent dyes *(9)*. Bi-ddF is ideally suited for mutation detection in hemizygous, homozygous, and cloned templates. Templates with heterozygous mutations are generally more difficult targets for Bi-ddF.

2.3. Denaturation Fingerprinting

Both conventional ddF and Bi-ddF perform extremely well in DNA sequences with an average G+C content. However, when these methods are applied to regions of high G + C-content (G+C>60%), extensive smearing of bands is observed decreasing the resolution and, therefore, the sensitivity of the technique. For these regions, denaturation fingerprinting (dnF) is the method of choice *(8)*. In dnF, the Sanger termination reactions are performed bidirectionally (like Bi-ddF) with two dideoxy terminators one of which is either ddATP or ddTTP and the other is ddCTP or ddGTP and the fingerprints are generated on denaturing gels to avoid smearing. Even though some secondary structure is still present, the use of denaturing gels in dnF diminishes contribution of the SSCP component in detecting mutations. There are two versions of dnF termed dnF_{2R} and dnF_{1R}. In dnF_{2R}, the two Sanger reactions are performed separately, electrophoresed on two separate lanes of a denaturing gel, and analyzed. If a bidirectional cycle sequencing reaction is performed with one ddNTP (e.g., ddCTP), only about 80% of mutations are detected based only on the dideoxy component. If the second bidirectional reaction is performed with ddATP or ddTTP, then almost 100% of mutations will be detected when both fingerprints are examined. In dnF_{1R}, only one fingerprint is generated by performing a single reaction with ddATP (or ddTTP) and a second chemically modified terminator (ROX-conjugated ddCTP or ROX-ddGTP), which retards the mobility of the same termination products by two nucleotides. It is not possible to use ddATP and ddCTP in the same reaction because certain mutations would result in simultaneous addition and deletion of a band at the same site. The use of a modified dideoxy nucleotide enables detection of all mutation types in a single Sanger reaction.

3. Clinical and Research Applications of ddF

ddF was initially proposed as a screening technique that could be beneficial for both molecular genetics and disease diagnostics. ddF was first used to examine genomic DNA from patients with hemophilia B. In the original report describing the technique, all single-base changes in a 1669-bp segment of the human *factor IX* gene were detected (100% sensitivity) with a very low rate of false-positive signals ***(6)***. Importantly, the approximate location of the sequence changes was determined, indicating that ddF is a not only a rapid screening method for the presence of mutations but also more efficient than SSCP.

Since the initial report, ddF has proven to be the method of choice for screening large genomic regions (containing multiple genes) where suspected mutations might be present. An elegant example for the successful application of the technique is the case of multiple endocrine neoplasia 1 (MEN 1). MEN 1 is an inherited cancer syndrome where affected individuals develop multiple parathyroid, enteropancreatic, and pituitary tumors. The MEN 1 approximate locus was originally identified by linkage analysis; ddF was then used to identify the gene harboring the disease-causing mutations. A very large region (2.8 Mb), containing a total of 15 candidate genes, was thoroughly evaluated in order to complete the task ***(11)***.

ddF can also be useful for the simultaneous screening of multiple candidate genes to determine the specific cause of a disease. Long QT syndrome is a cardiac disorder characterized by syncopes and sudden death. ddF was used to screen all exons in five candidate genes, encoding cardiac Na(+) and K(+) channels. In this case, analysis of the samples in both the sense and anti-sense direction (Bi-ddF) and the use of capillary array electrophoresis resulted in 100% sensitivity ***(12)***.

ddF is a relatively rapid and inexpensive molecular technique and seems appropriate when a large number of samples need assessment. Therefore, ddF is an ideal screening method for identifying potentially disease-contributing polymorphisms in a gene candidate for a complex disease or a susceptibility factor. For example, the technique was used to investigate whether polymorphisms in the pancreatic transcription factor neurogenin-3 (NGN3) contribute to the etiology of maturity-onset diabetes of the young or other forms of autosomal dominant diabetes ***(13)***. Ninety-one probands of families with autosomal dominant diabetes were screened for *NGN3* mutations. Three sequence differences were identified: a polymorphism not affecting the amino acid sequence (L75L), a CA insertion/deletion (ins/del) in intron 1 (–44 ins/del), and a C to T transition causing a serine to phenylalanine substitution (S199F). Similarly, ddF was used to investigate the 5' flanking region of the neurofibromatosis type 1 (*NF1*) gene in 380 neurofibromatosis type 1 patients ***(14)*** and the insulin receptor substrate 2 (*IRS-2*) gene in 14 pedigrees with early-onset autosomal dominant type 2 diabetes ***(15)***. In the same context, the technique has also been used in large-scale polymorphism studies in order to better understand the biological role of a protein. For example, Bi-ddF was used to screen for polymorphisms in the interleukin-3 (IL-3) enhancer, promoter, and gene coding regions in a group of 88 Caucasians. Analysis revealed the presence of several infrequent polymorphisms, which may affect IL-3 functionality ***(16)***. Similarly, ddF was used to screen the monoamine oxidase B (*MAO-B*)

Table 1
Examples of Successful Applications of ddF and Its Variants in Genetic Analysis of Human Disease

Disease	Gene	Ref.
Aniridia	*PAX6*	***30***
Hemolytic anemia	Glucose-6-phosphate Dehydrogenase (*G6PD*)	***36***
Hemophilia A	*factor VIII*	***33***
Hemophilia B	*factor IX*	***6,34***
Intermediate motor and sensory neuropathy	*myelin protein zero*	***29***
Isolated growth hormone deficiency	*growth hormone-1*	***31,32***
Long QT syndrome (LQTS)	*KCNQ1*, *KCNH2*	***12***
Multiple endocrine neoplasia 1 (MEN 1)	*MEN 1*	***11,21,22***
Schmid metaphyseal chondroplasia (SMCD)	*COL10A1*	***37***
Severe combined immunodeficiency (SCID)	Gamma chain of leukocyte cytokine receptors (*IL2RG*)	***35***

gene for polymorphisms in 100 alleles derived from patients with schizophrenia to identify sequence changes that might be associated with the disease ***(17)***. Interestingly, approximately 235 kb of genomic sequence was screened in this report in a short time, further underlining the value of ddF. The study concluded that mutations affecting the structure of the MAO-B protein are uncommon in the general population and are unlikely to contribute significantly to the genetic predisposition to schizophrenia. ddF has also been utilized to screen for schizophrenia-predisposing alterations within the dopamine D1 and D5 receptor (*DRD1* and *DRD5*) genes in a large number of patients; no association with the disease was found ***(18,19)***.

It has recently been realized that certain diseases can be caused by mutations or polymorphisms in promoter regions of certain genes. Such is the situation, for example, in the Langerhans cell histiocytosis (LCH) syndrome. In this condition, elevated levels of tumor necrosis factor-α (TNF-α), interferon-γ (INF-γ) and other cytokines is observed. In an elaborate study aimed to unravel the genetic cause of the disease ***(20)***, ddF was utilized to compare the TNF-α and IFN-γ promoter DNA sequences in LCH patients and normal individuals. ddF can be very useful in these cases because regulatory elements are often found far from transcription initiation sites and large pieces of DNA promoter regions need screening. Indeed, the method identified polymorphisms in the TNF-α promoter of LCH patients that result in increased production of the cytokine.

The method is also being routinely used for screening large tumor-related genes. Obviously, direct sequencing of these genes in all members of families with multiple cancer incidents is unrealistic. Therefore, ddF has been employed for screening genes such as *BRCA1* (breast/ovarian cancer), *p53* (follicular thyroid carcinoma and primary breast cancer), and *MEN 1* ***(21–27)***. Particularly, mutation analysis of the *BRCA1* gene is a challenge because of its size (22 exons encoding an 1863 amino acid protein) and wide distribution of mutations. It should be mentioned that alterations in the *BRCA1* sequence are responsible for approximately half of the cases of hereditary breast cancer, including the majority of familial breast and ovarian cancers. ddF has also been employed to unravel mutations in genes associated with other endocrine conditions such as *GNAS*, *GNAI2*, and the *GHRH receptor* gene ***(28)***. Therefore, ddF can help identify carriers of hereditary life-threatening diseases and optimize the management of those at risk (for proper genetic counseling).

In addition to the examples mentioned, ddF has aided in the identification of mutations in a great variety of human disorders (*see* **Table 1**). These include mutations in the *myelin protein*

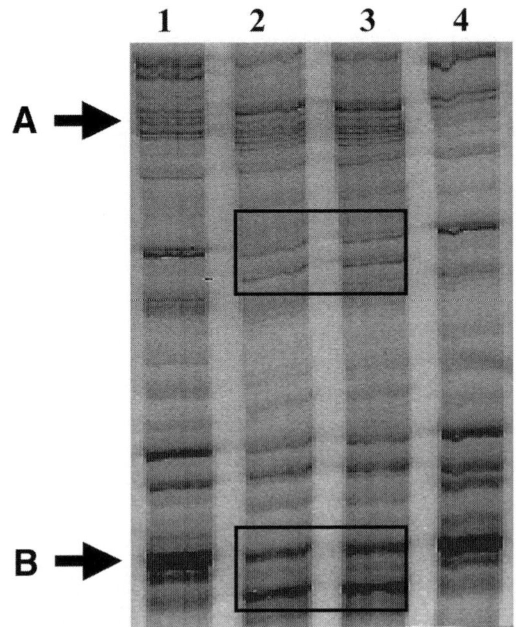

Fig. 2. Example of a ddF electrophoresis gel using DNA extracted from two siblings with Schmid metaphyseal chondroplasia (lanes 2 and 3, respectively) and a normal control (lane 4). All three individuals are heterozygotes for the same pathogenic substitution in nucleotide 2011 of the *COL10A1* gene (arrow A). Their mother (lane 1) is heterozygous for the T to C transition resulting in the appearance of two bands toward the top of the gel for each corresponding control band: one normal (same as the control in lane 4) and one with altered mobility because of the presence of the mutation. The two children (lanes 2 and 3) carry the same T to C transition in their maternal allele and an additional trinucleotide (CCC) deletion (3' untranslated region; downstream of nucleotide 2011) in their nonpathogenic, paternal allele. When only the CCC deletion is present in an amplicon (bottom part of the gel), two bands appear in lanes 2 and 3 (arrow B): a normal one from the maternal allele that moves along with the corresponding bands in lane 1 and 4 and a shifted one from the paternal allele containing the deletion. As we move toward the top of the gel, the DNA segments coming from the maternal allele start containing the C to T transition and are shifted in parallel with the band in lane 1 that has the same sequence. The end result in lanes 2 and 3 (arrow A) are two shifted bands (one containing the C to T transition and one containing the CCC deletion) for each corresponding control band.

zero gene in families with intermediate motor and sensory neuropathy ***(29)***, the developmental control gene *PAX6* in patients with aniridia ***(30)***, the *growth hormone-1* gene in sporadic cases with severe isolated growth hormone deficiency ***(31,32)***, *factor VIII* and *factor IX* gene in families with hemophilia A and B, respectively ***(33,34)***, mutations in the gamma chain of leukocyte cytokine receptors (*IL2RG*) gene in patients with severe combined immunodeficiency (SCID) ***(35)***, and in glucose-6-phosphate dehydrogenase–deficient patients ***(36)***.

Schmid metaphyseal chondroplasia (SMCD) is an autosomal dominant disorder of the skeleton. Patients with SMCD have mutations in the gene that codes for the α-1-chain of collagen X (*COL10A1*). Mutation analysis of the *COL10A1* gene is hampered by its size (exon 2 that encodes 95% of the α-1-chain contains 2940 bp); thus, ddF was employed in order to detect sequence changes in two SMCD siblings and their affected mother ***(37)***. This particularly complex case proved to be a challenge for the deciphering power of the ddF and is therefore described in detail. The above-mentioned three patients were heterozygotes for a pathogenic T to C transition (nucleotide 2011). In the ddF gel (*see* **Fig. 2**), all bands containing the mutation

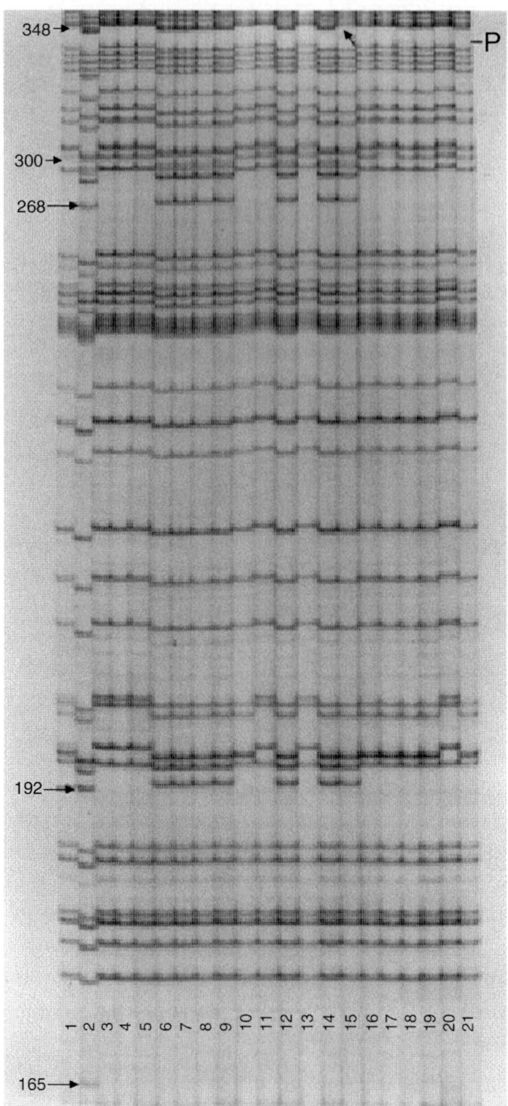

Fig. 3. Application of ddF in genotyping of the parasite *Schistosoma japonicum* (SJ). Schistosomiasis is caused by the blood fluke SJ and consists a major health problem in several areas around the world. There is considerable strain variation among SJ populations that correlates highly with pathogenicity of the micro-organism. In this analysis, low-level nucleotide variation in the *NADH dehydrogenase* mitochondrial gene in isolates of SJ is demonstrated. Notice a 6-base difference between lanes 2 and 3, a 5-base difference between lanes 1 and 2, a 4-base difference between lanes 9 and 10, a 3-base difference between lanes 10 and 11, a 2-base difference between lanes 17 and 19 and a single-base difference between lanes 14 and 15. (From **ref. *38***, with permission.)

should appear shifted in comparison to the corresponding normal bands (control lane). Because the patients had one physiological allele, the normally expected bands should also exist, resulting in the appearance of two bands (one normal and one shifted) for each band found in the control lane after nucleotide 2011. That was the case only for the affected mother. The two children also carried a polymorphism in their normal allele, inherited from their father (a trinucleotide [CCC] deletion in the 3' untranslated region downstream of nucleotide 2011) shifting their "normal" band as well. Because ddGTP was used in the termination reaction, an informative component (loss or gain of a band) was not observed.

The discriminative power of ddF can also be applied in the classification and/or identification of bacteria, parasites, and viruses. In such cases, screening of a reference gene can reveal strain-characteristic variations that can be thereafter used for quick and specific identification (*see* **Fig. 3**). Genotyping of microorganisms can be helpful in understanding the epidemiology and different clinical presentations of a contagious disease. Moreover, such knowledge can help physicians predict the course and severity of an infection and resistance to antibiotics prior to therapy initiation. For example, ddF was used to genotype a panel of *Borrelia burgdorferi* spirochete isolates from patients suffering from Lyme borreliosis. The most frequent and serious manifestation of disseminated Lyme borreliosis is neuroborreliosis. The goal of the study was to find a possible association of neuroborreliosis to certain *B. burgdorferi* genospecies, indicating species-dependent organotropism ***(39)***. In another study, ddF was used to detect mutations in the *rpoB* gene (known to confer drug resistance) in clinical isolates of mycobacteria causing tuberculosis. Such information can help physicians predict resistance to key components of therapeutic regimens and, therefore, choose the most appropriate treatment ***(40)***. ddF has also been used to distinguish between hepatitis virus (HCV) types and subtypes because there are genotype-specific clinical consequences in HCV infections, particularly in response to interferone treatment ***(38)***. In such cases, HCV RNA was reverse transcribed and PCR amplified prior to ddF analysis. The method proved rapid and less labor-intensive for routine genotyping than direct DNA sequence and thus of greater clinical applicability ***(38)***. ddF has been also used to display low-level nucleotide variation in mitochondrial genes of the human blood fluke *Schistosoma japonicum* and cestodes of the genus *Echinococcus* and in the ribosomal DNA of ascaridoid nematodes. In all cases, ddF proved very efficient for genotyping because parasites could be readily differentiated by their reproducible ddF profiles ***(41–43)***.

References

1. Orita, M., Iwahana, H., Kanazawa, H., Hayashi, K., and Sekiya, T. (1989) Detection of polymorphisms of human DNA by gel electrophoresis as single-strand conformation polymorphisms. *Proc. Natl. Acad. Sci. USA* **86**, 2766–2770.
2. Sarkar, G., Yoon, H. S., and Sommer, S. S. (1992) Screening for mutations by RNA single-strand conformation polymorphism (rSSCP):comparison with DNA-SSCP. *Nucleic Acids Res.* **20(4)**, 871–878.
3. Sheffield, V. C., Beck, J. S., Kwitek, A. E., Sandstrom, D. W., and Stone, E. M. (1993) The sensitivity of single-strand conformation polymorphism analysis for the detection of single base substitutions. *Genomics* **16(2)**, 325–332.
4. Hayashi, K. (1992) PCR-SSCP: a method for detection of mutations. *Genet. Anal. Tech. Appl.* **9(3)**, 73–79.
5. Spinardi, L., Mazars, R., and Theillet, C. (1991) Protocols for an improved detection of point mutations by SSCP. *Nucleic Acids Res.* **19(14)**, 4009.
6. Sarkar, G., Yoon, H., and Sommer, S. S. (1992) Dideoxy fingerprinting (ddF): a rapid and efficient screen for the presence of mutations. *Genomics* **13**, 441–443.
7. Qiang, L., Jinong, F., and Sommer, S. S. (1996) Bi-directional dideoxy fingerprinting (Bi-ddF): a rapid method for quantitative detection of mutations in genomic regions of 300–600 bp. *Hum. Mol. Genet.* **5(1)**, 107–114.
8. Qiang, L., Weinshenker, B. G., Wingerchuk, D. M., and Sommer, S. S. (1998) Denaturation fingerprinting: two related mutation detection methods especially advantageous for high G+C regions. *BioTechniques* **24**, 140–147.
9. Shevchenko, Y. O., Bale, S. J., and Compton, J. G. (1999) Mutation screening using automated bidirectional dideoxy fingerprinting. *BioTechniques* **28**, 134–138.
10. Larsen, L. A., Johnson, M., Brown, C., et al. (2001) Automated mutation screening using dideoxy fingerprinting and capillary array electrophoresis. *Hum. Mutat.* **18**, 451–457

11. Guru, S. C., Agarwal, S. K., Manickam, P., et al. (1997) A transcript map for the 2.8-Mb region containing the multiple endocrine neoplasia type 1 locus. *Genome Res.* **7(7)**, 725–735.
12. Larsen, L. A., Johnson, M., Brown, C., et al. (2001) Automated mutation screening using dideoxy fingerprinting and capillary array electrophoresis. *Hum. Mutat.* **18(5)**, 451–457.
13. Kim, S. H., Warram, J. H., Krolewski, A. S., and Doria, A. (2001) Mutation screening of the neurogenin-3 gene in autosomal dominant diabetes. *J. Clin. Endocrinol. Metab.* **86(5)**, 2320–2322.
14. Osborn, M., Cooper, D. N., and Upadhyaya, M. (2000) Molecular analysis of the 5'-flanking region of the neurofibromatosis type 1 (NF1) gene: identification of five sequence variants. *Clin. Genet.* **57(3)**, 221–224.
15. Bektas, A., Warram, J. H., White, M. F., Krolewski, A. S., and Doria, A. (1999) Exclusion of insulin receptor substrate 2 (IRS-2) as a major locus for early-onset autosomal dominant type 2 diabetes. *Diabetes* **48(3)**, 640–642.
16. Jeong, M. C., Navani, A., and Oksenberg, J. R. (1998) Limited allelic polymorphism in the human interleukin-3 gene. *Mol. Cell. Probes* **12(1)**, 49–53.
17. Sobell, J. L., Lind, T. J., Hebrink, D. D., Heston, L. L., and Sommer, S. S. (1997) Screening the monoamine oxidase B gene in 100 male patients with schizophrenia: a cluster of polymorphisms in African-Americans but lack of functionally significant sequence changes. *Am. J. Med. Genet.* **74(1)**, 44–49.
18. Liu, Q., Sobell, J. L., Heston, L. L., and Sommer, S. S. (1995) Screening the dopamine D1 receptor gene in 131 schizophrenics and eight alcoholics: identification of polymorphisms but lack of functionally significant sequence changes. *Am. J. Med. Genet.* **60(2)**, 165–171.
19. Sobell, J. L., Lind, T. J., Sigurdson, D. C., et al. (1995) The D5 dopamine receptor gene in schizophrenia: identification of a nonsense change and multiple missense changes but lack of association with disease. *Hum. Mol. Genet.* **4(4)**, 507–514.
20. Wu, W. S. and McClain, K. L. (1997) DNA polymorphisms and mutations of the tumor necrosis factor-alpha (TNF-alpha) promoter in Langerhans cell histiocytosis (LCH). *J. Interferon Cytokine Res.* **17(10)**, 631–635.
21. Prezant, T. R., Levine, J., and Melmed, S. (1998) Molecular characterization of the men1 tumor suppressor gene in sporadic pituitary tumors. *J. Clin. Endocrinol. Metab.* **83(4)**, 1388–1391.
22. Debelenko, L. V., Brambilla, E., Agarwal, S. K., et al. (1997) Identification of MEN1 gene mutations in sporadic carcinoid tumors of the lung. *Hum. Mol. Genet.* **6(13)**, 2285–2290.
23. Grebe, S. K., McIver, B., Hay, I. D., et al. (1997) Frequent loss of heterozygosity on chromosomes 3p and 17p without VHL or p53 mutations suggests involvement of unidentified tumor suppressor genes in follicular thyroid carcinoma. *J. Clin. Endocrinol. Metab.* **82(11)**, 3684–3691.
24. Lancaster, J. M., Berchuck, A., Futreal, P. A., and Wiseman, R. W. (1997) Dideoxy fingerprinting assay for BRCA1 mutation analysis. *Mol. Carcinog.* **19(3)**, 176–179.
25. Durocher, F., Tonin, P., Shattuck-Eidens, D., Skolnick, M., Narod, S. A., and Simard, J. (1996) Mutation analysis of the BRCA1 gene in 23 families with cases of cancer of the breast, ovary, and multiple other sites. *J. Med. Genet.* **33(10)**, 814–819.
26. Blaszyk, H., Hartmann, A., Schroeder, J. J., McGovern, R. M., Sommer, S. S., and Kovach, J. S. (1995) Rapid and efficient screening for p53 gene mutations by dideoxy fingerprinting. *Biotechniques* **18(2)**, 256–260.
27. Goebel SU, Heppner C, Burns AL, et al. (2000) Genotype/phenotype correlation of multiple endocrine neoplasia type 1 gene mutations in sporadic gastrinomas. *J. Clin. Endocrinol. Metab.* **85(1)**, 116–123.
28. Jorge, B. H., Agarwal, S. K., Lando, V. S., et al. (2001) Study of the multiple endocrine neoplasia type 1, growth hormone-releasing hormone receptor, Gs alpha, and Gi2 alpha genes in isolated familial acromegaly. *J. Clin. Endocrinol. Metab.* **86(2)**, 542–544.
29. Mastaglia, F. L., Nowak, K.J., Stell, R., et al. (1999) Novel mutation in the myelin protein zero gene in a family with intermediate hereditary motor and sensory neuropathy. *J. Neurol. Neurosurg. Psychiatry* **67(2)**, 174–179.
30. Gronskov, K., Rosenberg, T., Sand, A., and Brondum-Nielsen, K. (1999) Mutational analysis of PAX6: 16 novel mutations including 5 missense mutations with a mild aniridia phenotype. *Eur. J. Hum. Genet.* **7(3)**, 274–286.
31. Miyata, I., Eto, Y., Kamijo, T., Ogawa, M., Futrakul, A., and Phillips, J. A., 3rd. (1999) Screening for mutations in the GH-1 gene by dideoxy fingerprinting (ddF). *Endocr. J.* **46 (Suppl.)**, S71–S74.
32. Miyata, I., Cogan, J. D., Prince, M. A., Kamijo, T., Ogawa, M., and Phillips, J. A., 3rd. (1997) Detection of growth hormone gene defects by dideoxy fingerprinting (ddF). *Endocr. J.* **44(1)**, 149–154.

33. Lin, S. W., Lin, S. R., and Shen, M. C. (1993) Characterization of genetic defects of hemophilia A in patients of Chinese origin. *Genomics* **18(3)**, 496–504.
34. Thorland EC, Weinshenker BG, Liu JZ, et al. (1995) Molecular epidemiology of factor IX germline mutations in Mexican Hispanics: pattern of mutation and potential founder effects. *Thromb. Haemost.* **74(6)**, 1416–1422.
35. Puck, J. M., Pepper, A. E., Henthorn, P. S., et al. (1997) Mutation analysis of IL2RG in human X-linked severe combined immunodeficiency. *Blood* **89(6)**, 1968–1977.
36. Li, P., Thompson, J. N., Wang, X., and Song, L. (1998) Analysis of common mutations and associated haplotypes in Chinese patients with glucose-6-phosphate dehydrogenase deficiency. *Biochem. Mol. Biol. Int.* **46(6)**, 1135–1143.
37. Stratakis, C. A., Orban, Z., Burns, A. L., et al. (1996) Dideoxyfingerprinting (ddF) analysis of the type X collagen gene (COL10A1) and identification of a novel mutation (S671P) in a kindred with Schmid metaphyseal chondrodysplasia. *Biochem. Mol. Med.* **59(2)**, 112–117.
38. Fox, S. A., Lareu, R. R., and Swanson, N. R. (1995) Rapid genotyping of hepatitis C virus isolates by dideoxy fingerprinting. *J. Virol. Methods* **53(1)**, 1–9.
39. Lebech, A. M. (2002) Polymerase chain reaction in diagnosis of *Borrelia burgdorferi* infections and studies on taxonomic classification. *APMIS* **105 (Suppl.)**, 1–40.
40. Liu, Y. C., Huang, T. S., Huang, W. K., Chen, C. S., and Tu, H. Z. (1998) Dideoxy fingerprinting for rapid screening of rpoB gene mutations in clinical isolates of *Mycobacterium tuberculosis. J. Formos. Med. Assoc.* **97(6)**, 400–404.
41. Zhu, X., Bogh, H., and Gasser, R. B. (1999) Dideoxy fingerprinting of low-level nucleotide variation in mitochondrial DNA of the human blood fluke, *Schistosoma japonicum. Electrophoresis* **20(14)**, 2830–2833.
42. Gasser, R. B., Zhu, X., and McManus, D. P. (1998) Dideoxy fingerprinting: application to the genotyping of Echinococcus. *Int. J. Parasitol.* **28(11)**, 1775–1779.
43. Zhu, X. Q. and Gasser, R. B. (1998) Single-strand conformation polymorphism (SSCP)-based mutation scanning approaches to fingerprint sequence variation in ribosomal DNA of ascaridoid nematodes. *Electrophoresis* **19(8–9)**, 1366–1373.

13

Conformation-Sensitive Gel Electrophoresis

Marian Hill

1. Introduction

Conformation-sensitive gel electrophoresis (CSGE) was initially developed by Ganguly et al. in 1993 as a screening method to minimize the amount of nucleotide sequencing required when investigating large multiexon genes for mutations *(1)*. It is based on the ability to distinguish between homoduplexed and heteroduplexed DNA fragments by electrophoresis under partially denaturing conditions.

Conformation-sensitive gel electrophoresis is now a widely used screening method for mutation analysis in genetic disorders. It is ideal for the detection of single-base changes and small insertions or deletions, particularly in disorders where different unknown mutations are spread throughout large genes *(1–3)*. It has also become a valuable tool for the detection of single-nucleotide polymorphisms (SNPs) *(4,5)* (see Chapter 19).

1.1. CSGE Theory

DNA homoduplexes consist of double-stranded DNA fragments in which all of the bases are paired correctly with their complementary base on the opposite stand. Heteroduplex DNA contains mismatched bases, and in polymerase chain reaction (PCR) products that originate from a patient with a heterozygous mutation, they are formed when double-stranded DNA is allowed to dissociate then reanneal with the complementary strand originating from a different allele (*see* **Fig. 1**).

The presence of mismatched bases induces subtle conformational changes in the heteroduplex compared with the homoduplex, as the misaligned bases do not conform to the typical Watson–Crick base-pairing rules. In the presence of mildly denaturing solvents this conformational change is amplified and allows the differentiation of homoduplexes and heteroduplexes by polyacrylamide gel electrophoresis

2. Mechanism of CSGE

2.1. Sample Preparation for CSGE

The amplicon (DNA fragment generated by PCR) must be of good quality and sufficient concentration. Amplicon design is critical to the success of CSGE (*see* **Subheading 3.**). Hetetroduplex formation is carried out by heating the amplicon to separate the double-stranded DNA (typically at approx 98°C for 5 min), followed by an incubation step to allow reannealing of single-stranded DNA to the complementary strand (typically 65°C for 30 min). This results in the formation of both homoduplexes and heteroduplexes (*see* **Fig. 1**). In order to detect homozygous mutations, the amplicon must be mixed with an equivalent wild-type control for heteroduplex analysis.

From: *Medical Biomethods Handbook*
Edited by: J. M. Walker and R. Rapley © Humana Press, Inc., Totowa, NJ

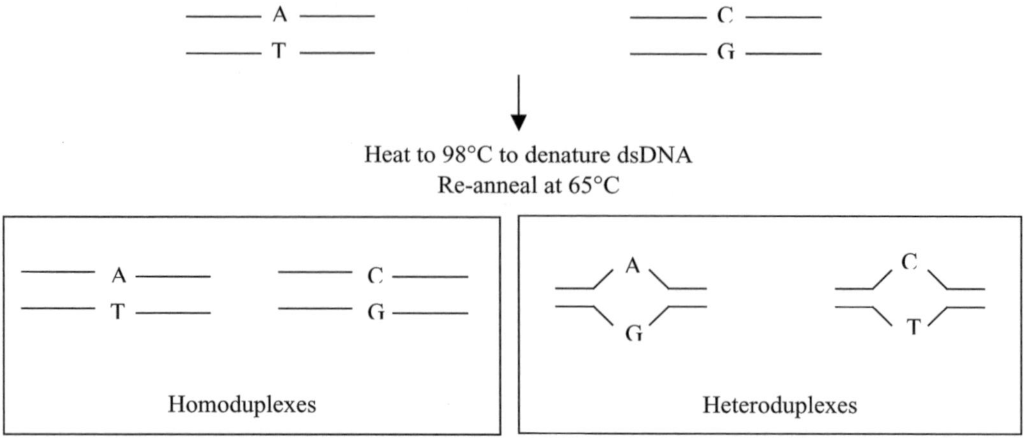

Fig. 1. Illustration of heteroduplex and homoduplex generation. A PCR product from a patient who is heterozygous (A/C) at a specific nucleotide position will contain two species of double-stranded DNA. Heating to 98°C will dissociate the double-stranded DNA, incubation at 65°C allows the strands to reanneal. Heteroduplexes are formed when a strand from one allele reanneals with the complementary strand from the other allele, and they will contain a mismatched basepair.

2.2. Gel Preparation

Conformation-sensitive gel electrophoresis involves the use of a nonproprietary polyacrylamide gel for electrophoretic separation. These gels are typically a manual sequencing format (e.g., 30 × 45 cm) and 1 mm thick consisting of 10% acrylamide with a 99 : 1 ratio of acrylamide to BAP [1,4-*bis*(acrolyl) piperazine]; this crosslinker was chosen because it increases conductivity and greatly improves gel strength. The gels are prepared in a Tris : taurine : EDTA buffer and also contain ethylene glycol (10%) and formamide (15%) as mild denaturants. Ammonium persulfate and TEMED (*N*, *N*, *N*, *N*'-tetramethylethylenediamine) are used to initiate polymerization.

2.3. Electrophoresis and Analysis

Samples are loaded onto the gel in a standard loading buffer. Electrophoresis is carried out (typically for 16 h at 400 V) in a Tris:taurine:EDTA buffer system, which is optimal for the separation of PCR products ***(1)***. Bands are usually visualized using a standard DNA stain such as ethidium bromide (Sigma) or Gelstar (BioWhittaker), although silver staining and ^{32}P-labeling have also been used in some laboratories ***(6,7)***.

Separated products are seen as bands under ultraviolet (UV) illumination. Homoduplexes are generally detected as a single band; one or more additional bands representing the co-migrating heteroduplexes might be seen if a mismatch is present. **Figure 2** illustrates the CSGE band patterns associated with two different mutations [a 1 bp deletion (–g) and a t→c substitution] in exon 9 of the *Factor XI* gene. Although the heteroduplex band/s are commonly seen above the homoduplex band, this type of pattern is not always seen, as illustrated in lane 3. In some cases, it might be possible to resolve up to four bands representing all variants. The pattern of bands is highly variable depending on the type of mutation and the sequence context, although insertions and deletions tend to produce the largest band separation, as they effect a larger conformation change. Because additional bands might also be seen as a result of a secondary structure, it is advisable to always include a wild-type control for comparison to avoid false-positive results.

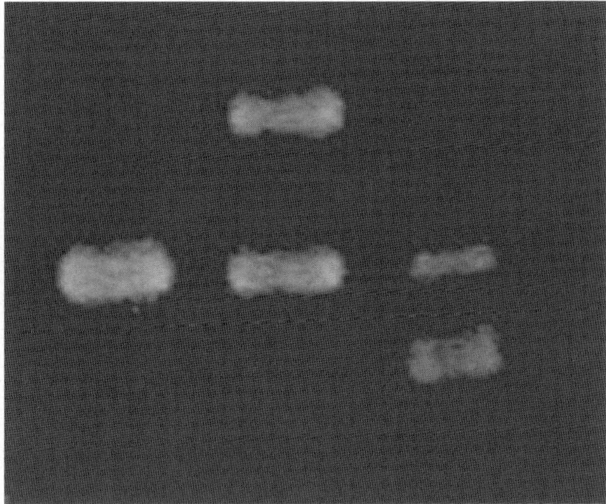

Fig. 2. Illustration of variation in band patterns seen in CSGE. *FXI* gene, exon 9. Lane 1: normal control, lane 2: patient with a single-base deletion (g); lane 3: patient with t → c substitution.

3. Factors Affecting CSGE Sensitivity

The exact nature of the mismatch, size of the amplicon, and location and sequence surrounding the mismatch (sequence context) will all affect the sensitivity of CSGE. The reported detection rate of this method has ranged from 60% of *BRCA1* mutations in a study coordinated by Eng et al. ***(8)*** to 100% in a number of studies, including those of Korkko et al., who recommended modified electrophoresis conditions and restricted amplicon size for optimum mutation detection in collagen genes ***(9)***.

3.1. Amplicon Size

Optimal amplicon size is 200–500 bp, although sequence mismatches have been detected in amplicons up to 800 bp in length. Size is limited by the inherent flexibility of DNA, which may mask any conformation change as a result of mismatch ***(10)***.

Korkko et al. ***(9)*** suggested that amplicons should be limited to below 450 bp for optimum sensitivity and demonstrated that a single-base polymorphism in the *COL1A2* gene, which could not be detected in a 755-bp amplicon, was clearly seen when PCR primers were redesigned to reduce the amplicon size to 276 bp.

3.2. Nature and Sequence Context

The band pattern seen is highly variable, as it is dependent on the nature of the mismatch and the surrounding nucleotide sequence (sequence context). In a recent update on CSGE, Ganguly et al. analyzed the ability of CSGE to distinguish specific mismatches within the same sequence context and found the following order of sensitivity ***(11)***:

$$G:G = G:T = T:G > G:A = A:G = T:T > A:A > C:T > C:C = C:A = A:C = T:C$$

In this study, four heteroduplexes (C:C, C:A, A:C, and T:C.) were indistinguishable from the homoduplex band. However, these mismatches are still detectable, as their complementary heteroduplexes (G:G, G:T, T:G, and A:G) resulted in a clearly defined additional band ***(11)***.

The importance of sequence context was also illustrated in this report, as a different migration pattern was seen with a C:T mismatch when the C was located in the sense or antisense strand ***(11)***.

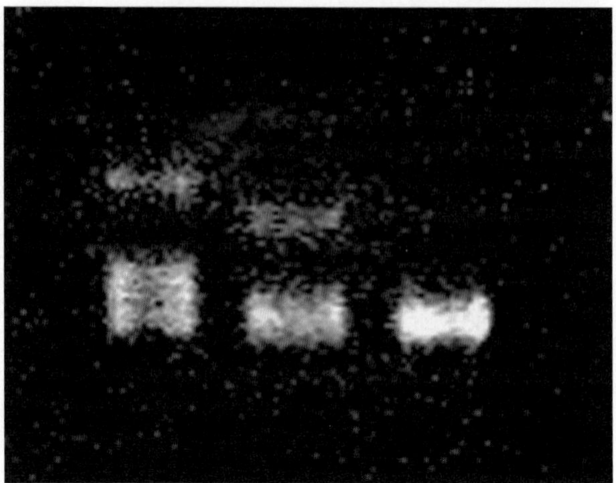

Fig. 3. Illustration of effect of mismatch position on ability on detection. *FXI* gene, exon 5. Primers were redesigned to position a c → a substitution at different points from the end of the amplicon. Size was maintained at 290–292 bp. Lane 1: mismatch is 174 bp from end; lane 2: mismatch is 70 bp from end; lane 3: mismatch is 42 bp from the end.

Initial studies with CSGE by Ganguly et al. ***(1)*** suggested that mismatches within high-temperature melting domains might be particularly difficult to resolve. However, the same mutations were detectable in later studies when primers were redesigned further away ***(9)***. It is, however, generally accepted that CSGE is more sensitive to mismatches within an AT-rich sequence context than a GC-rich region ***(11)***.

3.3. Location of Mismatch

Mismatches that are close to the end of the amplicon are less easily detected. Ganguly et al. ***(1)*** initially failed to detect a mismatch that was located 51 bp from one end of the PCR fragment. This became detectable when the primers were redesigned to position the mismatch 81 bp from the end. It is, therefore, advisable to allow for 50–100 bp of additional sequence at either end of the region of interest when designing primers for CSGE.

Figure 3 illustrates the effect of mismatch position on detection. Primers were redesigned to position a c → a substitution within exon 5 of the FXI gene at 174, 70, and 42bp from the end of the amplicon. The amplicon size was maintained at 290–292 bp. The ability to detect the mismatch is lost as the position of the mismatch approaches the end of the amplicon.

4. Limitations of CSGE

Conformation-sensitive gel electrophoresis is ideal for the detection of heterozygous single-base changes and small insertions or deletions; however, large deletions might go undetected. Because different homoduplexes are not generally resolved by CSGE, homozygous or hemizygous changes are not detectable unless the amplicon is mixed with an equivalent wild-type control to allow heteroduplex formation.

Changes in the band pattern can be subtle with some mismatches and experience is invaluable for interpretation of results and optimum amplicon design.

In common with most screening methods, CSGE provides limited information on the nature of a previously unknown sequence variation or its significance, and might not distinguish between two closely linked sequence variations. Although band patterns associated with specific mismatches within the same amplicon are reproducible, it is always advisable to sequence to confirm the nature of the mismatch.

Conformation-sensitive gel electrophoresis is particularly useful for genes that are not very polymorphic, such as *Factor VIII*. The sequencing load can increase significantly when highly polymorphic genes such as *BRCA1* are analyzed, owing to the detention of polymorphisms as false positives. However, it remains a powerful tool because it greatly reduces the amount of sequencing required during the investigation of such genes ***(12)***.

Conformation-sensitive gel electrophoresis has been compared with other screening methods in a number of studies. Markoff et al. reported that CSGE was better than single-stranded conformation polymorphism (SSCP) for the analysis of mutations in *BRCA1* ***(6)***, and although these two methods were comparable in a study by Eng et al. ***(8)***, they concluded that denaturing high-pressure liquid chromatography (dHPLC) was more sensitive. However, under optimum conditions, CSGE has been shown detect close to 100% of mutations ***(9)*** and has the advantage that it can be carried with a minimum of specialized equipment.

5. Applications and Modifications of CSGE

Conformation-sensitive gel electrophoresis is frequently used in the diagnosis and study of a range of genetic disorders, particularly those associated with different causative mutations that can be spread throughout the gene.

In hemophilia A, CSGE greatly reduces the sequencing required to identify mutations in the *Factor VIII* gene where it facilitates diagnosis and carrier analysis ***(13,14)***. CSGE has also been used in a number of other haemostatic disorders ***(15)***. CSGE was initially established to improve the analysis of multiple genes associated with collagen disorders ***(9)***. Korkko et al. reported that CSGE was a highly sensitive tool in the analysis of the *FBN1* gene in Marfan syndrome ***(2)***. Molecular diagnosis in this gene was useful for the confirmation of the diagnosis, which is valuable because of the extensive phenotypic variation in this disorder. Finnila et al. described the use of CSGE in the analysis of mitochondrial DNA in patients with occipital stroke ***(16)***.

Where a number of different fragments are to be screened, PCR reactions can be multiplexed and amplicons analysed simultaneously, providing that the products can be easily differentiated. Arancha et al. used this approach in a study of the *MEN 1* gene in endocrine neoplasia type 1 ***(17)***. Likewise multiple loading of suitable amplicons into the same lane is frequently carried out to increase throughput. CSGE has also been widely used in the analysis of sequence variations in cancer-susceptibility genes such as *BRCA1* and *BRCA2* ***(18,19)***.

The requirement for high-throughput testing, particularly for SNP analysis, has facilitated a number of modifications to the basic CSGE method to improve both speed and sensitivity. For example, CSGE was combined with restriction endonuclease fingerprinting (REF-CSGE) by Herzog et al. to reduce the number of confirmatory sequencing reactions required in the analysis of BRCA exon 11 mutations ***(20)***.

Leung et al. made modifications to the basic protocol to allow the use of CSGE as an inexpensive approach to high-throughput screening of SNPs in *TIGR/MYOC* genes in primary open-angle glaucoma ***(5)***. These included decreasing gel thickness, increasing the number of lanes per gel, and loading up to seven samples per lane. A two-stage screening strategy was used to minimize the sample mixing required for homozygote detection, and Sybr Gold DNA stain was used to increase sensitivity.

Higher throughput and sensitivity has also been achieved with the use of fluorescent labels and the automation of CSGE on genetic analyzers. Fluorescent-CSGE (F-CSGE) has been developed on both a gel and capillary format ***(4,15,19)***.

Gel composition might require modification to minimize fluorescent quenching, and fluorescent label is usually introduced by random incorporation of labeled dNTP during PCR. Bands are analyzed by instrument software, allowing greater sensitivity. This approach has been useful where the genes studied are highly polymorphic (e.g., *BRCA1* and *BRCA2*), as the band patterns resulting from known polymorphisms were more reproducible and therefore could easily be identified and dismissed ***(19)***.

Rozycka et al. described capillary-based CSGE in the analysis of SNPs and known sequence variations in the lipoprotein lipase and *BRCA2* genes. This was carried out using an ABI 310 genetic analyzer with a proprietary Genescan polymer *(4)*. Samples were loaded in 45% formamide, 30% ethylene glycol, and 10 m*M* EDTA, together with an internal size standard. The fluorescent label was introduced by the incorporation of labeled dNTPs. Although this method detected 7/7 short insertion or deletions, only 16/22 nucleotide substitutions in a series of mismatches in the lipoprotein lipase and *BRCA2* genes were detected. This was improved by limiting the amplicon size to below 350 bp and ensuring a 50 bp redundant sequence at each end of the amplicon. Although this method allows better resolution between closely migrating bands, transfer onto an automated platform might require modifications to ensure that sensitivity is maintained.

In addition to increased resolution, F-CSGE increases reproducibility by the inclusion of an internal size standard with each sample and throughput is particularly improved where four-color technology is available, as multiple fluorescent-labeled amplicons can be analyzed simultaneously. Although this approach is faster and more reproducible than the manual method, it does require specialized equipment and training, and therefore increased cost *(19)*.

6. Summary

Conformation-sensitive gel electrophoresis provides a highly sensitive and efficient screening method. It compares well in terms of sensitivity with other screening technologies and has the advantage of being simple to perform and cost-effective, as it can be carried out with minimal specialized equipment. CSGE also lends itself to automation and such modifications have been useful in increasing sample throughput, providing a technique that is valuable for both detection of unknown mutations in genetic disorders and in high-throughput screening for SNPs.

References

1. Ganguly, A., Rock, M. J., and Prockop, D. J. (1993) Conformation-sensitive gel electrophoresis for rapid detection of single-base differences in double-stranded PCR products and DNA fragments: evidence for solvent-induced bends in DNA heteroduplexes. *Proc. Natl. Acad. Sci. USA* **90**, 10,325–10,329.
2. Korkko, J., Kaitila, I., Lonnqvist, L., Peltonen, L., and Ala-Kokko, L. (2002) Sensitivity of conformation sensitive gel electrophoresis in detecting mutations in marfan syndrome and related conditions. *J. Med. Genet.* **39**, 34–41.
3. Ganguly, A. and Williams, C. (1997) Detection of mutations in multi-exon genes: comparison of conformation sensitive gel electrophoresis and sequencing strategies with respect to cost and time for finding mutations. *Hum. Genet.* **9**, 339–343.
4. Rozycka, M., Collins, N., Stratton, M. R., and Wooster, R. (2000) Rapid detection of DNA sequence variants by conformation sensitive capillary electrophoresis. *Genomics* **70(1)**, 34–40.
5. Leung, Y. F., Tam, PO-S., Tong W. C., et al. (2000) High through-put conformation-sensitive gel electrophoresis for discovery of SNPs. *Biotechniques* **30**, 334–340.
6. Markoff, A., Sormbroen, H., Bogdanova, D., et al. (1998) Comparison of conformation-sensitive gel electrophoresis and single-stranded conformation polymorphism analysis for detection of mutations in the BRCA1 gene using optimised conformation analysis protocols. *Eur. J. Hum. Genet.* **6(2)**, 145–150.
7. Lancaster, J. M., Wooster, R., Mangion, J., et al. (1996) BRCA2 mutations in primary breast and ovarian cancers. *Nature Genet.* **13**, 238–240.
8. Eng, C., Brody, L. C., Wagner, T. M., et al. and Steering Committee of the Breast Cancer Information Core (BIC) Consortium (2001) Interpreting epidemiological research: blinded comparison of methods used to estimate the prevalence of inherited mutations in BRCA1. *J. Med. Genet.* **38(12)**, 824–833.
9. Korkko J., Annunen S., Pihlajamaa T., Prockop D. J., and Ala-Kokko L. (1998) Conformation sensitive gel electrophoresis for simple and accurate detection of mutations: comparison with denaturing gradient gel electrophoresis and nucleotide sequencing. *Proc. Natl. Acad. Sci. USA* **95(4)**, 1681–1685.
10. Calladine, C. R., Collis, C. M., Drew, H. R., and Mott, M. R. (1991) A study of electrophoretic mobility of DNA in agarose and polyacrylamide gels. *J. Mol. Biol.* **221**, 981–1006.

11. Ganguly, A. (2002) An update on conformation sensitive gel electrophoresis. *Hum. Mutat.* **19**, 334–342.
12. Ganguly, A. and Williams, C. (1997) Detection of mutations in multi-exon genes: comparison of conformation sensitive gel electrophoresis and sequencing strategies with respect to cost and time for finding mutations. *Hum. Mutat.* **9(4)**, 339–343.
13. Hill, M. and Dolan, G. (2003) Mutation analysis in 32 families with haemophilia A. *Br. J. Haematol.* **121 (Suppl. 1)**, 64.
14. Williams, I. J., Abuzenadah, A., Winship, P., et al. (1998) Precise carrier diagnosis in families with haemophilia A: use of conformational sensitive gel electrophoresis for mutation screening and polymorphism analysis. *Thromb. Haemost. 79*, 723–726.
15. Hashemi, S. M. B., Hinks, J., Marsden, L., Peake, I. R., and Goodeve, A. C. (2003) Fluorescent conformation sensitive gel electrophoresis (F-CSGE) analysis of the vWF gene: A high throughput, sensitive mutation detection system. *J. Thromb. Haemost.* (abstract 1666).
16. Finnila, S., Hassinen, I. E., Majamaa, K. (2001) Phylogenetic analysis of mitochondrial DNA in patients with occipital stroke. Evaluation of mutations by using sequence of the entire coding region. *Mutat. Res.* **458(1–2)**, 31–39.
17. Arancha, C., Ruiz-Liorente, S., Cascon, A., et al. (2002) A rapid and easy method for multiplex endocrine neoplasia type 1 mutation detection using conformation-sensitive gel electrophoresis. *J Hum. Genet.* **47(4)**, 190–195.
18. Spitzer, E., Abbaszadegan, M. R., Schmidt, F., et al. (2000) Detection of BRCA1 and BRCA2 mutations in breast cancer families by a comprehensive two-stage screening procedure. *Int. J. Cancer* **85**, 474–481.
19. Ganguly, T., Dhulipala, R., Godmilow, L., and Ganguly, A. (1998) High-throughput fluorescence-based conformation-sensitive gel electrophoresis (F-CSGE) identifies six unique BRCA2 mutations and an overall low incidence of BRCA2 mutations in high risk BRCA1-negative breast cancer families. *Hum. Genet.* **102**, 549–556.
20. Herzog, J. S., Jancis, E. M., Liao, S., Somlo, G., and Weitzel, J. N. (2002) Restriction endonuclease fingerprinting enhanced conformation sensitive gel electrophoresis (REF-CSGE) in the analysis of BRCA1 exon 11 mutations in a high-risk breast cancer cohort. *Hum. Mutat.* **19(6)**, 656–663.

14

Amplification Refractory Mutation System and Molecular Diagnostics

Richard Kitching and Arun Seth

1. Introduction

Allele-specific polymerase chain reaction (PCR) was first described in 1989, with variations arising over the next few years such as allele-specific oligonucleotide PCR, mutant-allele-specific amplification (MASA), PCR amplification of specific alleles (PASA), and the amplification refractory mutation system (ARMS) ***(1–5)***. For convenience, the term ARMS will be used here when referring to them collectively.

These related methods utilize the difference in extension efficiency between primers with matched and mismatched 3' bases. The *Taq* DNA polymerase used in PCR reactions will extend primers with a mismatched 3' base 40- to 100-fold less efficiently than with a perfectly matched 3' base. A complete match between primer and template will yield a product, whereas one or several mismatches at the 3' end will inhibit extension. When used in combination with a common upstream primer, PCR amplification of either wild-type or mutant alleles produces DNA products of a predictable size, depending on which template is present. By detecting which primer pairs form the products, a sample's genotype can be determined. Product detection and identification are the most variable steps in an ARMS assay and the speed of product detection contributes to the throughput capacity of each particular method. The most straightforward is gel electrophoresis, whereby reaction products are separated from the oligonucleotide primers by size and detected by ultraviolet (UV) fluorescent DNA intercalating dye. Mass spectrometry detects and identifies the allele-specific products by molecular weight, DNA sequencing separates products by size and fluorescence labels ***(6,7)***. Other methods use fluorescence resonance energy transfer (FRET), fluorescence polarization (FP), luminescence, absorbance, and melting temperature ***(8–10)***. Most have been designed to be set up, performed, and analyzed in 96-well microplates.

The combination of ARMS with UV detection of DNA after electrophoresis is the most robust of these methods. Although cost-inefficient and time-consuming relative to high-throughput techniques, it avoids expensive instrumentation and costly probes. It is the most frequently used technique for the determination of single nucleotide polymorphisms (SNPs) outside of North America, western Europe, and Japan. SNPs have emerged as genetic marker of choice because of their high density and relatively even distribution in the human genome. About 1 in every 1000 bases along a human chromosome is estimated to differ between any two copies of that chromosome. Such a locus is referred to as polymorphic if the allele frequency of the most common variant is greater than 99% ***(11)***. Extensive sets of polymorphic sequences have been identified in the course of genome research, in studies of human genome diversity, and in the process of establishing mutation databases that eventually might include all common

From: *Medical Biomethods Handbook*
Edited by: J. M. Walker and R. Rapley © Humana Press, Inc., Totowa, NJ

variants of human genes *(12)*. A subset of these genetic markers are mutations that affect the structure of proteins or the activity of gene regulatory regions for genes that cause phenotypic disease. Fine mapping of disease loci and candidate gene association studies have identified many such mutations. The application of SNP analyses most suited to ARMS PCR is to survey the small numbers of sequence variants known to be associated with a specific disease.

2. Design of an ARMS Assay

Here we present an overview of the major steps required for design of an ARMS assay to detect a particular SNP in DNA samples from a human population such as was done for *BRCA1* *(5)*. SNPs are favored genetic markers because of their high density and relatively even distribution in the human genomes. Fine mapping of disease loci and candidate gene association studies by government and private companies have produced millions of nonredundant SNPs that are readily available for the design of ARMS primers (http://www.ncbi.nlm.nih.gov/SNP/ ; http://www.celera.com/). Typically, Pubmed searches for the gene of interest will result in one or more studies from which candidate SNPs can be extrapolated (see Chapter 19).

Multiple SNPs might be suitable for some applications, such as mutations that lead to truncation of the coding regions or inactivation of the promoter. These web-based or other software can be used to design oligonucleotides that have the SNP at the 3' end. Primers ending in 3'-A should be avoided if possible because of a generally lower extension efficiency, however, it might be possible to empirically define PCR conditions for such oligonucleotides in the absence of other candidates. Extension from the mutant or wild-type primer occurs only when the 3' end is complementary to the template DNA. In addition to the allele-specific primer, each ARMS reaction requires a second primer common to both reactions. The sequence chosen for the second primer will determine where the second primer hybridizes to the template DNA, allowing the designer to determine the size of the DNA fragment amplified in the PCR reaction. The second primer should be designed to amplify a DNA fragment large enough for the detection method to be employed. In the case of visual UV detection of reaction products after gel electrophoresis, a size of 200–600 nucleotides is suggested. If sequencing is to be used for confirmation, the amplified DNA should be suited to the limitations of the DNA sequencer. Be certain that the primer pair is entirely within a single exon of the gene. Coding regions for most genes are derived from cDNA sequences that lack introns and you will typically be using genomic template DNA, which might contain introns whose length exceeds the limits of the PCR reaction. Gene structures can be deduced by comparison of the full cDNA to human genomic DNA using the BLAST programs and databases available at the NCBI (www.ncbi.nih.nlm.gov) (see Chapter 28).

PCR conditions can have a significant effect on the specificity of DNA extension from a SNP-specific ARMS oligonucleotide. Although the quantitative dependence of PCR yields on target input conforms to a linear log–log relation at a low cycle number, it does not hold beyond about 1 n*M* of DNA *(13,14)*. Also of concern is the possibility of nonspecific or random hybridization to partially identical sequences in other regions of the genome. Thus, high-stringency conditions for hybridization and a low cycle number are the cautious approach to using complex genomic target DNA. Variable sample quality might affect the specificity of an individual reaction. Reference samples from unaffected patients are an important control for this. During the validation stage of assay development the presence of the mutation in DNA amplified using the mutant primer and the lack of the target mutation in DNA amplified by the wild-type primer should be confirmed by DNA sequencing. This can be done with samples whose genotype is determined by more laborious methods as described during the validation of our *BRCA1* ARMS assay *(5)*.

The presence of the SNP of interest is possible when amplification of the template DNA occurs using a primer whose 3' end is complementary to the SNP nucleotide. In this design, the presence of amplified DNA is detected by electrophoresis of the reaction products on a 1–2% agarose gel. Further confirmation can be done with the same sample in a second reaction to show the absence of amplification by a primer designed to hybridize with the wild-type

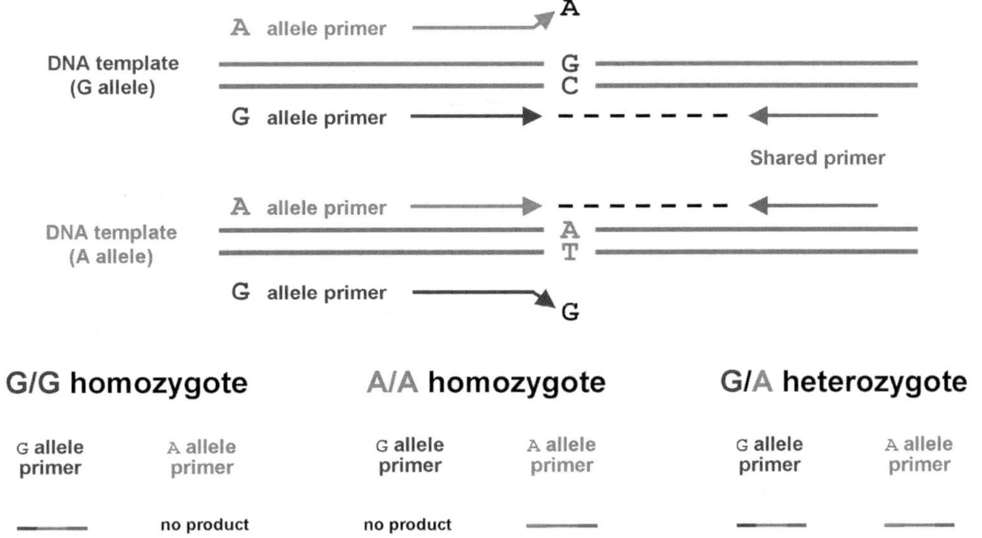

Fig. 1. Generalized diagram of ARMS PCR design and results. **Upper**: Separate PCR reactions are set up to allow extension from the mutant (red) or wild-type (blue) primer only when the 3' end of the primer is complementary to the 5' nucleotide of the template DNA. In addition to the allele-specific primer, each ARMS reaction requires a second primer common to both reactions (green). **Lower**: the presence of amplified DNA is detected by electrophoresis of the products from the separate wild-type and mutant reaction products; wild-type (blue) amplification without mutant (red) amplification indicates wild-type homozygosity; mutant (red) amplification without wild-type (blue) amplification indicates mutant homozygosity; and amplification products in both reactions indicates heterozygosity.

sequence (*see* **Fig. 1**). The final determination that the SNP was amplified from the sample DNA is by sequencing of DNA extracted from the agarose gel.

3. Automation and Detection Technologies

Most advances in the conversion of two-step PCR/electrophoresis ARMS have occurred in the automation of sample handling and detection of amplification products. Product detection and identification can be performed in many different ways. Mass spectrometry detects and identifies the allele-specific products by molecular weight; DNA sequencing separates products by size and fluorescence labels. Other methods use fluorescence resonance energy transfer (FRET), luminescence, absorbance, and melting temperature. Various combinations of these techniques with SNP-specific primer PCR have been tested and applied in recent years ***(9,10,15–18)***.

With the advent of robotic plate handlers and pipettors, automation of both amplification and detection can be done. Typically, microplates are set up with the oligonucleotide mixtures, sample template DNA, enzyme, and other reagents by using standard automated liquid-handling equipment, such that the only procedure not automated is the transfer of the plates to the PCR machine and, finally, the reader. This technology is fully compatible with various types of PCR machine and detection method. For example, microplate array diagonal gel electrophoresis (MADGE) is directly compatible with PCR microplates and has been used with ARMS PCR to generate gel images for semiautomatic computerized data analysis ***(15)***. It is expected that unifying robot arm technology will be more readily available in the future, allowing for fully automatic flowthrough.

The basic ARMS assay can easily be multiplexed with careful design of primer sets that produce distinct DNA product sizes for each SNP in a single reaction with each sample DNA.

This is demonstrated by the design of an assay for simultaneous identification of ten variants of the cytochrome P450 gene (*CYP2D6*) (debrisoquine 4-hydroxylase) ***(16)***. This liver enzyme metabolizes numerous drugs, including many antidepressants, neuroleptics, antiarrhythmics, and antihypertensive agents, and its variations can have profound effects on response to drug treatment. Variant alleles previously linked to poor drug metabolism in Caucasian and Asian populations were examined in DNA from 100 individuals previously genotyped by specific polymerase chain reaction–restriction fragment length polymorphism (PCR-RFLP) analysis. Four common primers and 10 allele-specific primers were used with each sample, and the size of the resulting DNA bands after electrophoresis were used to identify which alleles were present ***(16)***.

Elimination of gel electrophoresis entirely is possible with other detection systems, such as FRET probes ***(10,17)***. FRET is a physical phenomenon that occurs when two fluorescent groups are in proximity such that the "donor" fluorophore's emission spectrum overlaps the "acceptor's" excitation spectrum. FRET applications use this change in spectral character, detecting either quenching of donor emission or the increase in acceptor emission when the fluorophores are near each other, as happens when ARMS PCR produces an allele-specific product. Commercial descriptions of these reactions are available for both the Scorpion AS-PCR assay (http://www.dxsgenotyping.com/technology.htm), and the Amplifluor genotyping system (http://www.chemicon.com/Product/ProductDataSheet.asp?ProductItem=S7908). The main drawback of FRET with ARMS is the cost of labeling unique primers for each SNP; however, at least one method is described using universal energy transfer labeled primers ***(9)***. An alternative to expensive allele-specific labeled probes is to quantify the accumulation of allele-specific DNA amplicons by continuous fluorescence monitoring using the dye SYBR green I. This method cannot be multiplexed; however, automation allows one to include multiple samples and control reactions in each microplate. This approach has been used for human lymphocyte antigen (HLA) typing as well as for analysis of missense variants of the APC gene ***(18)***.

4. Applications

4.1. Genetic Disease

Linkage of candidate SNPs identified by ARMS has been used to investigate associations between individual SNPs in genetic disease, infectious disease susceptibility, oncogenes, tumor suppressor genes, HLA typing, forensics, and SNPs. The most popular of such studies is allele frequency determination in ethnic groups—most often β-thalassemia, sickle cell disease, and HLA typing. A typical research application investigates the frequency of candidate gene polymorphisms within an affected population, such as elderly women suffering from fractures received while falling or standing, an indicator of osteoporosis ***(19)***. The G to T polymorphism at a Sp1-binding site within intron 1 of the collagen I gene (*COL1A1*) has been considered to be a marker of low bone mineral density, with the consequent risk of bone fractures. DNA samples from 133 women at risk for osteoporosis were screened for a SNP that eliminates a DNA sequence that binds transcription factor proteins known to directly enhance expression of *COL1A1*. Results using the ARMS method were compared to those from a "mismatched" primer method, with DNA sequencing serving as the reference for comparison. ARMS was clearly superior to the mismatched primer method, with complete correspondence to the DNA sequencing results ***(19)***.

Genetic predisposition to thrombosis accounts for about 40–50% of episodes of venous thrombosis, and patients presenting with venous thrombosis could be candidates for mutation analysis. A single G to A base change at nucleotide 1691 of the *factor V* gene causes the arginine at position 506 to be replaced by glutamine at the site where protein C cleaves and inactivates *factor V*. The glutamine mutant Factor V protein is resistant to cleavage and inactivation by activated protein C at the mutated site. Factor V has both procoagulant properties after its activation by thrombin and an important role in the anticoagulant system as a cofactor to activated protein C. The high incidence of a single-base change in the codon for amino acid 506 has led to several applications using the ARMS technique *(6,20–23)*. All are directed to the G to A mutation and can be done using very small quantities of whole blood.

4.2. Infectious Disease

Strain identification for infectious diseases ranging from measles and mumps to drug-resistant tuberculosis and hepatitis also use ARMS *(24,25)*. Genetic characterization of wild-type viruses has become an invaluable tool for measles, mumps, and rubella control by identifying the transmission pathways of the viruses. Although serologically monotypic, at least 20 measles and 9 mumps genotypes have been recognized by sequencing of variable regions of their RNA genomes. RNA extracted from clinical specimens, usually oral swabs or other body fluids, is reverse-transcribed, and the resulting cDNA is amplified by PCR using primers specific to the viral genes of interest. These PCR amplicons are then cloned and sequenced for genotype. The ARMS version of measles and mumps genotyping uses a second "nested" PCR using genotype sequence-specific primers with the genomic cDNA amplicons resulting from the first round PCR *(24)*. As compared to the reverse transcriptase–PCR sequencing strategy, the "nested" ARMS method is faster and less expensive; however, it will not produce sequence data on novel genotypes.

A similar approach has been used to rapidly identify mutations in *Mycobacterium tuberculosis* *(25)*. Early diagnosis of multidrug-resistant tuberculosis is essential for efficient treatment and control of the disease. Rifampin resistance can be used as a surrogate marker for multidrug resistant tuberculosis, because more than 90% of them are also isoniazid resistant. In many nations and cities, public health agencies face limited budgets, with much of their infrastructure committed to culture methods that can take more than 1 mo, a timeframe and expense that appears to be unnecessarily long when compared to modern techniques such as genotyping with ARMS.

Genotyping offers the potential for rapid and inexpensive diagnosis of multidrug-resistant tuberculosis. Ninety-six percent of rifampin-resistant (Rif^r) *M. tuberculosis* strains possess genetic alterations in a small region of the *rpoB* gene. PCR single-strand conformational length polymorphism, dideoxyfingerprinting, heteroduplex analysis, and DNA sequencing can be used for analysis of *rpoB* gene mutations associated with rifampin resistance; however, these procedures are labor-intensive and time-consuming as compared to ARMS *(25)*.

4.3. HLA Typing

Some studies have evaluated ARMS methods for HLA class I typing for bone marrow and organ transplantation; however, it is most often used to provide background information for further studies in anthropology, organ transplantation, and major histocompatibility complex (MHC) disease associations in particular isolated regions of the world. Gene-longevity association studies of unrelated individuals, which search for nonrandom associations between polymorphisms at candidate loci and longevity, require that allele and genotype frequencies at polymorphic marker loci be compared between a long-lived group and a control group of randomly selected adults. A genetically homogeneous population is desirable in these studies, and one such investigation of Sardinian centenarians used ARMS as a rapid method for HLA typing that is minimally invasive, rapid, and inexpensive *(26)*. Although no associations were found between longevity and the 14 haplotypes, the ARMS results for HLA type were consistent and clear, validating its use *(26)*.

4.4. Forensics

Less clinically, ARMs has been used to forensically discriminate between species of cats in "tiger bone medicines," to identify plants used in drug preparations, to type the sex of morphologically similar birds, to identify somatic cattle clones, and to verify the quality of ancient DNA *(27–31)*. The chromohelicase DNA-binding gene has a sex-specific allele in most avian species, allowing ARMS to be more effective at distinguishing between male and female hatchlings of monomorphic birds ***(27)***. Several methods for chemical repair of low-quality ancient DNA exist, and sequence analysis of ARMS PCR products has been used to compare the success of these methods with medieval bone samples ***(30)***. Forensic component analysis by ARMS of plant and animal preparations used in traditional Asian medicines is an aid to regulation and investigation of their efficacy ***(28,29)***. In order to characterize the fate of *Bos indicus* donor cell mitochondria from a single genetic origin following nuclear transfer into cytoplasts of random genetic origin, ARMS was done with primers specific to either *Bos indicus* or *B. taurus* mtDNA haplotypes using fluorescently labeled primer pairs for six mitochondrial genes ***(31)***. Such combinations of multiplex primer pairs with high-technology detection systems represents the future of ARMS PCR wherein mutation-specific primer design remains the defining feature.

4.5. Carcinogenesis

As a research tool, ARMS methodology has a place in the surveyance of known mutations in proto-oncogenes and tumor suppressors. The *c-erbB-2* proto-oncogene (also known as *HER-2/neu*) encodes a transmembrane tyrosine kinase growth factor receptor, the proto-oncogene K-*ras* encodes a 21-kDa protein (known as p21) involved in the transduction of the external cell signals. *BRCA1* is a breast and ovarian cancer susceptibility gene mutated in 2–5% of breast cancer patients. Evaluation of candidate SNPs affecting oncogenes and tumor suppressor genes such as *BRCA1*, *ras*, and *c-erbB2* has led to the use of ARMS to screen populations for these and other gene mutations associated with early-onset cancers ***(32,33)***. Associations between clinical evidence of infection by particular human papilloma virus (HPV) strains have been compared to the genetic background of patients ***(33)***. *K-ras* mutations, *c-erbB-2* amplification, and the presence of HPV of high- and low-risk in abnormal cervical tissue samples infected with HPV-6, HPV-16, and HPV-18 were examined. HPV types 16 and 18 are "high-risk" strains associated with higher oncogenic potential, whereas HPV-6 and HPV-11 rarely result in invasive tumors. ARMS was used to survey these samples for *K-ras* codon 12 mutations. Although ARMS clearly identified which samples contained the *K-ras* mutation, no association was found with the strains examined ***(33)***. However, this study does indicate that ARMS is an appropriate tool for rapid identification of specific point mutations.

Several of the over 100 distinct *BRCA1* mutations that have been reported occur with greater frequency in particular subpopulations, such as women of Ashkenazi Jewish descent ***(34)***. A frequency of nearly 1% among the general population has been reported for a 40-bp deletion, 1294*del*40, within the large exon 11 of the *BRCA1* gene, which leads to translation termination at codon 397. The 185*del*AG frameshift mutation in exon 2 of *BRCA1* has a dinucleotide AG deletion at position 185 (codon 23), accounting for a significant number of breast cancer cases in Ashkenazi women. It is estimated that 1% of Jewish women carry this mutation and it also represents 13% of all *BRCA1* mutations characterized to date ***(35,36)***. Furthermore, this mutation accounts for an estimated 16% of breast and 39% of ovarian cancers in Ashkenazi women diagnosed before the age of 50 ***(37)***. In stark contrast to the 1/833 carrier frequency for *BRCA1* mutations in the general population, the frequency of 1 in 107 for the 185*del*AG mutation in Jewish women points to a strong demand for predictive genetic testing for breast cancer within this subpopulation.

We developed an inexpensive, single-step, PCR-based method that uses sequence specific oligonucleotide primers to distinguish between different alleles of *BRCA1* ***(5)***. Comparison of sample DNA amplified in separate mutant and wild-type primer pair reactions indicates which allele(s) are present (*see* **Fig. 2**). PCR with the wild-type (wt) primer pairs amplified the expected wt DNA fragments from a patient without *BRCA1* mutations and from a cell line (MCF-7) known to have wild-type *BRCA1*. Both normal and mutation-specific primers amplified DNA from two patients heterozygous for the 185*del*AG mutation. Two different primers specific to the 1294*del*40 mutation were also tested. Mutant primer I amplified both mutant and

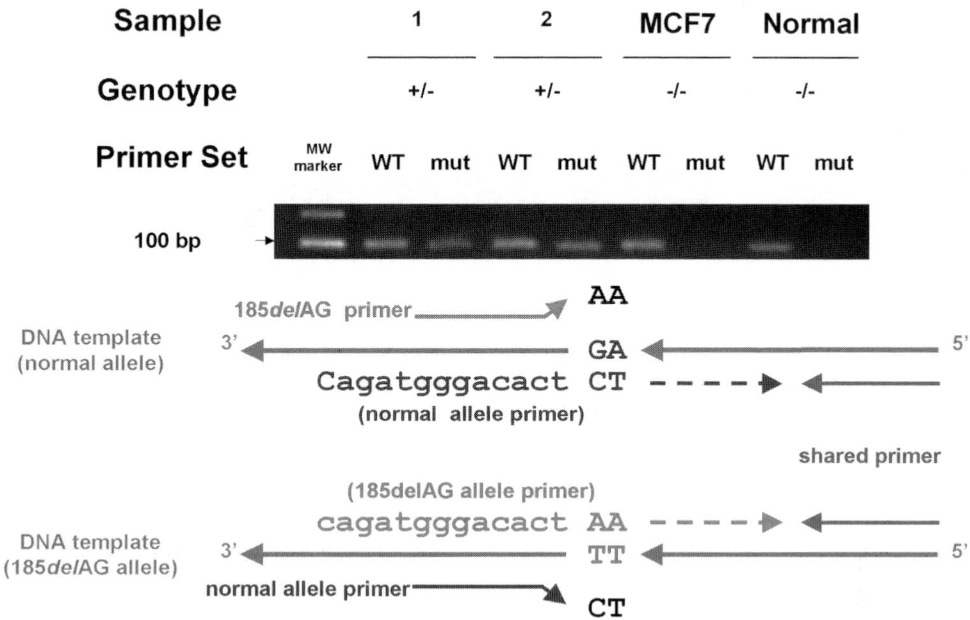

Fig. 2. Discrimination between normal and 185*del*AG alleles by ARMS PCR. WT: primer specific to wild-type BRCA1 sequence in PCR reaction; MUT: primer specific to 185*del*AG mutant BRCA1 sequence in PCR reaction; 1, 2, and Normal: patient template DNA sample; MCF-7: breast tumor cell line template DNA sample; ±: heterozygous for 185*del*AG BRCA1 mutation

wild-type products in a patient heterozygous for this deletion, whereas the 185*del*AG mutant primer II amplified only the mutant allele. PCR-amplified fragments were analyzed by DNA sequencing to confirm mutations detected by ASO PCR. Template DNA from the two patients heterozygous for 185*del*AG each produced one wild-type and one mutant allele-specific band of the same size (*see* **Fig. 2**). Samples found heterozygous for the 1294*del*40 mutation resulted in one wild-type and one mutant allele-specific band of different sizes as a result of the 40-bp deletion that characterizes this mutation (*see* **Fig. 3**). Heteroduplex analysis was used to confirm the ASO PCR results as well as the genotype of these patients, however, heteroduplex analysis cannot fully characterize the mutation in a single step as can our ASO method *(5)*.

The *RET* proto-oncogene is a receptor tyrosine kinase, a member of a family of cell surface proteins that transduce signals for cell growth and differentiation. The *RET* gene itself was found to be a classical oncogene, resulting in malignant transformation of cells in in vitro transfection assays. Mutations in the *RET* gene are associated with multiple endocrine neoplasia (MEN 2), Hirschsprung disease, and medullary thyroid carcinoma (MTC) *(38)*. Germline mutations in one of eight codons within *RET* cause the three subtypes of MEN 2; a somatic M918T mutation accounts for the largest proportion of *RET* mutations detected in medullary thyroid carcinomas. However, pheochromocytomas have a wider range of *RET* mutations and approx 25% of patients with Hirschsprung disease have germline mutations scattered throughout the length of *RET*. Thus, MEN2 and MTC are candidates for ARMS PCR designed to detect *RET* mutations that might underlie these familial cancers.

The *p53* tumor suppressor is a transcriptional factor with DNA sequence-specific binding and transactivation activity. Alteration or inactivation of *p53* by mutation or the interaction of *p53* with oncogene products of DNA tumor viruses can lead to cancer. The open reading frame of *p53* is 393 amino acids long, with the central region (consisting of amino acids from about

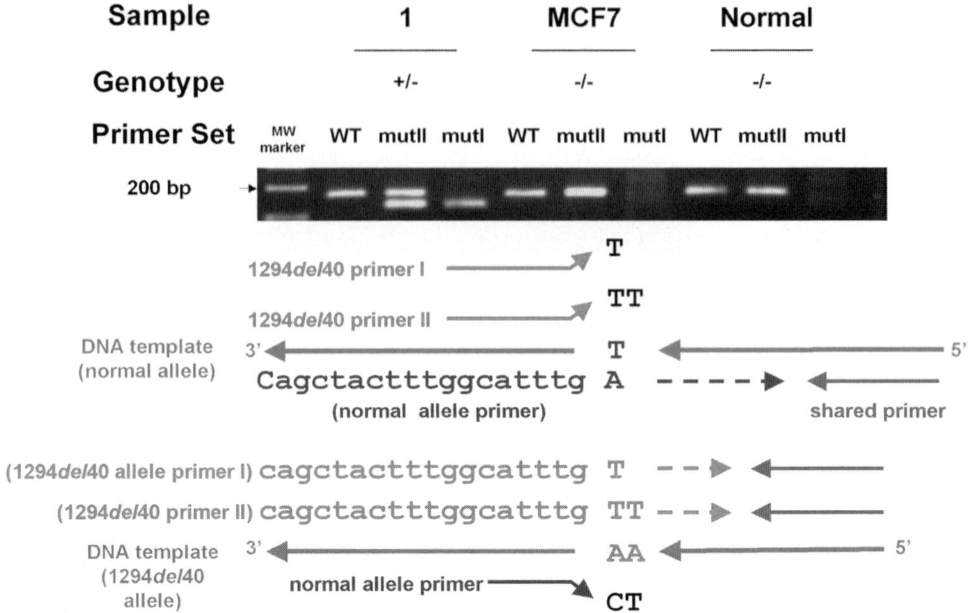

Fig. 3. Discrimination between normal and 1294*del*40 alleles by ARMS PCR. WT: primer specific to wild-type BRCA1 sequence in PCR reaction; MUT I and MUT II: primers specific to the 1294*del*40 mutant BRCA1 sequence in PCR reaction; 1 and Normal: patient template DNA sample; MCF-7: breast tumor cell line template DNA sample; ±: heterozygous for 1294*del*40 BRCA1 mutation.

100 to 300) containing the DNA-binding domain. The residues most frequently mutated in cancers are all at or near the protein–DNA interface, and over two-thirds of the missense mutations are in one of the three DNA loops.

Low et al. designed an ARMS PCR for the detection of *p53* mutations in DNA from tumor cells. with known *p53* point mutations at different sites ***(39,40)***. The technique was both sensitive and mutation-specific and cross-amplification of unmatched mutant *p53* DNA did not occur. Its application in the search for recurrent or minimal residual disease could improve the prognosis of patients by the early detection of tumors with known mutations.

Chronic contamination of food with the mycotoxin aflatoxin B1 is strongly associated with a G to T transversion in the third position mutation of codon 249 of the *p53* gene, resulting in substitution of serine for arginine. In Asian, sub-Saharan African, and other populations with significant aflatoxin B1 exposure, approximately half of hepatocellular carcinomas have this mutation whereas the codon 249 mutation, is rarely found in populations with negligible exposure ***(41–43)***. Furthermore, aflatoxin B1 has been shown to induce the G to T transversion in vitro ***(44–46)***. The combination of epidemiological association with a molecular mechanism supports the hypothesis that aflatoxin B1 has a causative and probably early role in hepatocarcinogenesis (reviewed in **ref.** ***47***).

The G to T transversion associated with aflatoxin B1-related hepatocellular cancer can be detected in the blood of potentially affected individuals; however, it is not sufficient to cause the disease. This suggests that screening for the codon 249 mutation in high aflatoxin areas is desirable in order to encourage avoidance of additional factors such as hepatitis B infection, cigaret smoking, or heavy alcohol consumption, which have synergistic effects on liver cancer risk ***(47)***. Such screening using rapid high-throughput technologies is unlikely to be supported by the developing economies of Asia and Africa. The ARMS method described here has a low equipment cost and robust chemistry that could be used as a public health measure against hepatocellular cancer in high-aflatoxin regions.

Acknowledgments

This work was supported by Canadian Breast Cancer Research Alliance grants to A. Seth.

References

1. Okayama, H., Curiel, D. T., Brantly, M. L., Holmes, M. D., and Crystal, R. G. (1989) Rapid, nonradioactive detection of mutations in the human genome by allele-specific amplification. *J. Lab. Clin. Med.* **114**, 105–113.
2. Newton, C. R., Graham, A., Heptinstall, L. E., et al. (1989) Analysis of any point mutation in DNA. The amplification refractory mutation system (ARMS). *Nucleic Acids Res.* **17**, 2503–2516.
3. Hayashi, N., Ito, I., Yanagisawa, A., et al. (1995) Genetic diagnosis of lymph-node metastasis in colorectal cancer. *Lancet* **345**, 1257–1259.
4. Sommer, S. S., Groszbach, A. R., and Bottema, C. D. (1992) PCR amplification of specific alleles (PASA) is a general method for rapidly detecting known single-base changes. *Biotechniques* **12**, 82–87.
5. Richter, S. and Seth, A. (1998) One step direct detection of recurrent mutations in the breast cancer susceptibility gene, BRCA1. *Int. J. Oncol.* **12**, 1263–1267.
6. Maher, C., Crowley, D., Cullen, C., Wall, C., Royston, D., and Fanning, S. (1999) Double fluorescent-amplification refractory mutation detection (dF-ARMS) of the factor V Leiden and prothrombin mutations. Thromb. Haemost. **81**, 76–80.
7. Humeny, A., Rodel, F., Rodel, C., et al. (2003) MDR1 single nucleotide polymorphism C3435T in normal colorectal tissue and colorectal carcinomas detected by MALDI-TOF mass spectrometry. *Anticancer Res.* **23**, 2735–2740.
8. Bengra, C., Mifflin, T. E., Khripin, Y., et al. (2002) Genotyping of essential hypertension single-nucleotide polymorphisms by a homogeneous PCR method with universal energy transfer primers. *Clin. Chem.* **48**, 2131–2140.
9. Myakishev, M. V., Khripin, Y., Hu, S., and Hamer, D. H. (2001) High-throughput SNP genotyping by allele-specific PCR with universal energy-transfer-labeled primers. *Genome Res.* **11**, 163–169.
10. Gelsthorpe, A. R., Wells, R. S., Lowe, A. P., Tonks, S., Bodmer, J. G., and Bodmer, W. F. (1999) High-throughput class I HLA genotyping using fluorescence resonance energy transfer (FRET) probes and sequence-specific primer-polymerase chain reaction (SSP-PCR). *Tissue Antigens* **54**, 603–614.
11. Graur, D. and Li, W.-H. (2000) *Fundamentals of Molecular Evolution*, 2nd ed., Sinauer Associates, Sunderland, MA, p XIV.
12. Cotton, R. G. and Horaitis, O. (2002) The HUGO Mutation Database Initiative. Human Genome Organization. *Pharmacogenom. J.*, **2**, 16–19.
13. Finn, G. K., Kurz, B. W., Cheng, R. Z., and Shmookler Reis, R. J. (1989) Homologous plasmid recombination is elevated in immortally transformed cells. *Mol. Cell Biol.* **9**, 4009–4017.
14. Ayyadevara, S., Thaden, J. J., and Shmookler Reis, R. J. (2000) Discrimination of primer 3'-nucleotide mismatch by taq DNA polymerase during polymerase chain reaction. *Anal. Biochem.* **284**, 11–18.
15. Ye, S., Dhillon, S., Ke, X., Collins, A. R., and Day, I. N. (2001) An efficient procedure for genotyping single nucleotide polymorphisms. *Nucleic Acids Res.* **29**, E88.
16. Roberts, R., Joyce, P., and Kennedy, M. A. (2000) Rapid and comprehensive determination of cytochrome P450 CYP2D6 poor metabolizer genotypes by multiplex polymerase chain reaction. *Hum. Mutat.* **16**, 77–85.
17. Albis-Camps, M. and Blasczyk, R. (1999) Fluorotyping of HLA-DRB by sequence-specific priming and fluorogenic probing. *Tissue Antigens* **53**, 301–307.
18. Bartlett, S., Straub, J., Tonks, S., Wells, R. S., Bodmer, J. G., and Bodmer, W. F. (2001) Alkaline-mediated differential interaction (AMDI): a simple automatable single-nucleotide polymorphism assay. *Proc. Natl. Acad. Sci. USA* **98**, 2694–2697.
19. Montanaro, L. and Arciola, C. R. (2002) Detection of the G→T polymorphism at the Sp1 binding site of the collagen type I alpha 1 gene by a novel ARMS-PCR method. *Genet. Test.* **6**, 53–57.
20. Thong, M. K., Law, H. Y., and Ng, I. S. (1996) Molecular heterogeneity of beta-thalassaemia in Malaysia: a practical approach to diagnosis. *Ann. Acad. Med. Singapore* **25**, 79–83.

21. Yandava, C. N., Zappulla, D. C., Korf, B. R., and Neufeld, E. J. (1996) ARMS test for diagnosis of factor VLeiden mutation, a common cause of inherited thrombotic tendency. *J. Clin. Lab. Anal.* **10**, 414–417.
22. Hezard, N., Cornillet, P., Droulle, C., Gillot, L., Potron, G., and Nguyen, P. (1997) Factor V Leiden: detection in whole blood by ASA PCR using an additional mismatch in antepenultimate position. *Thromb. Res.* **88**, 59–66.
23. Bathelier, C., Champenois, T., and Lucotte, G. (1998) ARMS test for diagnosis of factor V Leiden mutation and allele frequencies in France. *Mol. Cell. Probes* **12**, 121–123.
24. Samuel, D., Beard, S., Yang, H., Saunders, N., and Jin, L. (2003) Genotyping of measles and mumps virus strains using amplification refractory mutation system analysis combined with enzyme immunoassay: a simple method for outbreak investigations. *J. Med. Virol.* **69**, 279–285.
25. Fan, X. Y., Hu, Z. Y., Xu, F. H., Yan, Z. Q., Guo, S. Q., and Li, Z. M. (2003) Rapid detection of rpoB gene mutations in rifampin-resistant Mycobacterium tuberculosis isolates in shanghai by using the amplification refractory mutation system. *J. Clin. Microbiol.* **41**, 993–997.
26. Lio, D., Pes, G. M., Carru, C., et al. (2003) Association between the HLA-DR alleles and longevity: a study in Sardinian population. *Exp. Gerontol.* **38**, 313–317.
27. Ito, H., Sudo-Yamaji, A., Abe, M., Murase, T., and Tsubota, T. (2003) Sex identification by alternative polymerase chain reaction methods in falconiformes. *Zool. Sci.* **20**, 339–344.
28. Sasaki, Y., Fushimi, H., Cao, H., Cai, S. Q., and Komatsu, K. (2002) Sequence analysis of Chinese and Japanese curcuma drugs on the 18S rRNA gene and trnK gene and the application of amplification-refractory mutation system analysis for their authentication. *Biol. Pharm. Bull.* **25**, 1593–1599.
29. Wetton, J. H., Tsang, C. S., Roney, C. A., and Spriggs, A. C. (2002) An extremely sensitive species-specific ARMS PCR test for the presence of tiger bone DNA. *Forensic Sci. Int.* **126**, 137–144.
30. Pusch, C. M., Kayademir, T., Prangenberg, K., Conard, N. J., Czarnetzki, A., and Blin, N. (2002) Documenting ancient DNA quality via alpha satellite amplification and assessment of clone sequence diversity. *J. Appl. Genet.* **43**, 351–364.
31. Steinborn, R., Schinogl, P., Wells, D. N., Bergthaler, A., Muller, M., and Brem, G. (2002) Coexistence of Bos taurus and B. indicus mitochondrial DNAs in nuclear transfer-derived somatic cattle clones. *Genetics* **162**, 823–829.
32. Bahar, A. Y., Taylor, P. J., Andrews, L., et al. (2001) The frequency of founder mutations in the BRCA1, BRCA2, and APC genes in Australian Ashkenazi Jews: implications for the generality of U.S. population data. *Cancer* **92**, 440–445.
33. Mouron, S. A., Abba, M. C., Guerci, A., Gomez, M. A., Dulout, F. N., and Golijow, C. D. (2000) Association between activated K-ras and c-erbB-2 oncogenes with "high-risk" and "low-risk" human papilloma virus types in preinvasive cervical lesions. *Mutat. Res.* **469**, 127–134.
34. Friend, S., Borresen, A. L., Brody, L., et al. (1995) Breast cancer information on the web. *Nature Genet.* **11**, 238–239.
35. Struewing, J. P., Abeliovich, D., Peretz, T., et al. (1995) The carrier frequency of the BRCA1 185delAG mutation is approximately 1 percent in Ashkenazi Jewish individuals. *Nat. Genet.* **11**, 198–200.
36. Shattuck-Eidens, D., McClure, M., Simard, J., et al. (1995) A collaborative survey of 80 mutations in the BRCA1 breast and ovarian cancer susceptibility gene. Implications for presymptomatic testing and screening. JAMA **273**, 535—541.
37. Scully, R., Chen, J., Plug, A., et al. (1997) Association of BRCA1 with Rad51 in mitotic and meiotic cells. *Cell* **88**, 265–275.
38. Eng, C. and Mulligan, L. M. (1997) Mutations of the RET proto-oncogene in the multiple endocrine neoplasia type 2 syndromes, related sporadic tumours, and hirschsprung disease. *Hum. Mutat.* **9**, 97–109.
39. Low, E. O., Jones, A. M., Gibbins, J. R., and Walker, D. M. (2000) Analysis of the amplification refractory mutation allele-specific polymerase chain reaction system for sensitive and specific detection of p53 mutations in DNA. *J. Pathol.* **190**, 512–515.
40. Low, E. O., Gibbins, J. R., and Walker, D. M. (2000) In situ detection of specific p53 mutations in cultured cells using the amplification refractory mutation system polymerase chain reaction. *Diagn. Mol. Pathol.* **9**, 210–220.
41. Hsu, I. C., Metcalf, R. A., Sun, T., Welsh, J. A., Wang, N. J., and Harris, C. C. (1991) Mutational hotspot in the p53 gene in human hepatocellular carcinomas. *Nature* **350**, 427–428.
42. Bressac, B., Kew, M., Wands, J., and Ozturk, M. (1991) Selective G to T mutations of p53 gene in hepatocellular carcinoma from southern Africa. *Nature* **350**, 429–431.

43. Aguilar, F., Harris, C. C., Sun, T., Hollstein, M., and Cerutti, P. (1994) Geographic variation of p53 mutational profile in nonmalignant human liver. *Science* **264**, 1317–1319.
44. Puisieux, A., Lim, S., Groopman, J., and Ozturk, M. (1991) Selective targeting of p53 gene mutational hotspots in human cancers by etiologically defined carcinogens. *Cancer Res.* 51, 6185–6189.
45. Aguilar, F., Hussain, S. P., and Cerutti, P. (1993) Aflatoxin B1 induces the transversion of G→T in codon 249 of the p53 tumor suppressor gene in human hepatocytes. *Proc. Natl. Acad. Sci. USA* **90**, 8586–8590.
46. Cerutti, P., Hussain, P., Pourzand, C., and Aguilar, F. (1994) Mutagenesis of the H-ras protooncogene and the p53 tumor suppressor gene. *Cancer Res.* **54**, 1934s–1938s.
47. Staib, F., Hussain, S. P., Hofseth, L. J., Wang, X. W., and Harris, C. C. (2003) TP53 and liver carcinogenesis. *Hum. Mutat.* **21**, 201–216.

15

Ligase Chain Reaction

Carla Osiowy

1. Introduction

Nucleic acid amplification technologies have greatly facilitated medical diagnostics for genetic and infectious diseases through the exquisite sensitivity and specificity associated with these methods. Polymerase chain reaction (PCR) (*see* Chapter 6) ushered in these technologies and was soon accompanied by numerous newly developed amplification techniques, including ligase chain reaction (LCR). These nucleic acid amplification techniques result in the exponential increase of DNA such that the final product can be detected by nonisotopic means or without probe hybridization. Various techniques have been developed that amplify either the target DNA or the probes used to detect the specific target DNA. Ideally, any nucleic acid amplification technique used for diagnostic detection of DNA should incorporate high sensitivity and specificity and include effective discrimination of target DNA, low background values, ease of use, and the potential for automation. This chapter will describe the ligase chain reaction and highlight these qualities in light of its use as a diagnostic detection method.

1.1. Theory of Ligase Chain Reaction

Ligase chain reaction was initially reported by Barany in 1991 *(1,2)* as a method to amplify oligonucleotide probes or primers specific for a short DNA target sequence. LCR involves the use of two pairs of probes, each pair being complementary to a strand of the denatured target DNA. Each probe within a pair is designed to hybridize to adjacent stretches of DNA (*see* **Fig. 1**). If perfect hybridization occurs, particularly with the 3' end of the upstream primer, then ligation between the two probes will proceed. If a mismatch occurs, ligation is inhibited as a result of imperfect hybridization of the probe with target DNA, and, thus, the absence of a ligated product suggests the presence of at least a single-basepair mismatch within the probe sequence *(3)*.

Ligated probes can then serve as targets during subsequent rounds of denaturation and hybridization. In this manner, the ligated products accumulate exponentially when the reaction is carried out in a thermocycler. The use of a thermostable ligase, such as *Thermus thermophilus* ligase *(4)*, ensures that stringent conditions can be used during cyclical DNA amplification, including high-temperature strand denaturation and the use of high-melting-temperature probes. Such stringent conditions add to the increased specificity of LCR as a nucleic acid amplification technique *(2)*.

1.2. Detection of Point Mutations by LCR

As indicated in **Subheading 1.1.**, the presence of a single-basepair mismatch between the probe and target DNA at the junction of the two LCR probes will prevent ligation because of the lack of perfect hybridization. In this manner, LCR is well suited to detect point mutations,

From: *Medical Biomethods Handbook*
Edited by: J. M. Walker and R. Rapley © Humana Press, Inc., Totowa, NJ

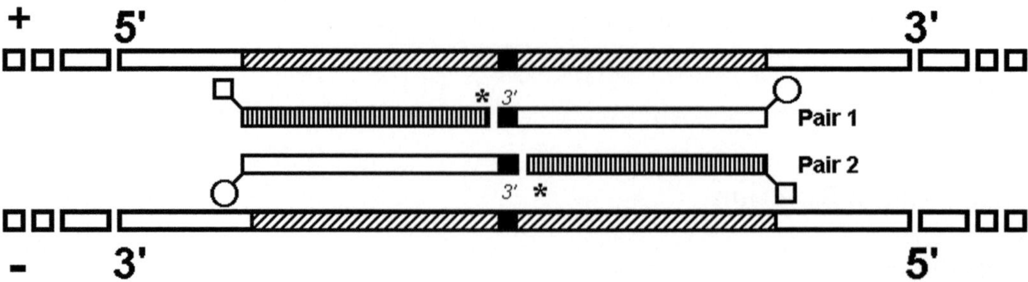

Fig. 1. Graphical representation of LCR probes and target DNA. The target region complementary to both pairs of LCR probes is shown (▨) on both the sense (+) and antisense (–) strand of the target DNA. Each pair of probes consists of a discriminating probe (□) and an adjacent probe (▥) which ligates to the discriminating probe. Sequence and probe features are indicated as follows: ■, point mutation nucleotide; 3', single-basepair overhang on the 3' end of the upstream probe; *, 5' phosphorylation on the ligating probe; ○, capture hapten; □, detection hapten.

such as those occurring in allelic genetic diseases, somatic mutations, or drug-resistance mutations in viruses or bacteria. Prior to the development of LCR, point mutations were usually detected by sequencing, allele-specific probe hybridization, or allele specific PCR ***(1,5,6)*** (*see* Chapter 25). These methods often lacked sensitivity or the ability to accurately discriminate the sequence of interest. Conversely, LCR demonstrates high sensitivity and is considered to have greater specificity for single-nucleotide polymorphisms than allele-specific PCR, as ligation is better able to discriminate mismatches when compared to primer extension ***(7)***.

Ligation-mediated detection of point mutations was originally developed using mesophilic ligases and a single probe set specific for one strand of the target DNA. Methods such as the oligonucleotide ligation assay (OLA) (see Chapter 22) ***(8)*** or ligation detection reaction (LDR) ***(5)*** were designed to detect point mutations in DNA targets normally first amplified by other means, such as PCR ***(8,9)***. OLA involves two adjacent oligonucleotide probes that upon ligation provide a positive result without subsequent amplification. On the other hand, LDR is cyclical and, thus, linear amplification of the two ligated probes is achieved following annealing to the complementary target ***(3)***. Subsequent methodology involving thermophilic ligases has led to the LCR and its associated advantages of point mutation detection and DNA amplification. Ligation-mediated detection of point mutations has been used to detect the human β-globin gene sickle cell mutation ***(1,10,11)***, the cystic fibrosis transmembrane conductance regulator gene mutation ***(12)***, colon tumor microsatellite sequence alterations ***(13)***, mutations in the *ras* family of oncogenes ***(14,15)***, the mutation responsible for bovine leukocyte adhesion deficiency ***(16)***, the mutation associated with type I Leber's hereditary optic neuropathy ***(17,18)***, extended spectrum β-lactamase resistance in bacteria ***(9,19)***, hepatitis B virus (HBV) mutants ***(20–22)***, AZT-resistant HIV mutants ***(23)***, and ganciclovir-resistant cytomegalovirus mutants ***(24)***.

Ligase chain reaction as a method for point mutation detection is not without certain disadvantages. It has been pointed out that this method is only able to detect known mutations and only a single mutation at a time ***(3,25,26)***. LCR requires that the entire target sequence complementary to the probes be known; therefore, exploratory analysis of unknown sequence mutations for phenotypic differences cannot be investigated by this method. Similarly, LCR is not easily used to detect multiple point mutations in a single assay, although this can be resolved through the optimization of multiplex LCR ***(3,12,13,27)***. Another issue is that ligation might be inhibited because of reasons other than the specific nucleotide mismatch of interest. For example, although a nucleotide mismatch at the 3' end of the upstream primer is the foremost reason for ligation not to proceed, lack of a ligation product may also result from other point mutations within the probe sequences close to the adjacent junction ***(21,23,28)*** or if a deletion or insertion occurs within the target sequence ***(29)***. In this way a negative result would not indicate the presence of a mismatch at the nucleotide of interest.

The ligation of probes not correctly hybridized to their complementary target results in target-independent ligation, which leads to false-positive results and poor specificity. This can occur with LCR; however, procedures to minimize this from occurring have been developed. Target-independent ligation has been minimized through the use of thermostable ligases ***(1)*** and probes having single-basepair overhangs, rather than blunt ends (*see* **Fig. 1**) ***(3)***. The addition of carrier DNA to the reaction also improves specificity ***(19)***, as does ensuring that only the probe ligating to the discriminatory probe is phosphorylated at its 5' end ***(2)***. It is equally important to use an appropriate negative template control, because the product of template-independent ligation is identical to a true positive ligation product ***(11,30)***. New developments in probe design, including padlock probes and Cleavase-modified LCR probes ***(31)***, can also increase assay specificity. Padlock probes have been reported to increase specificity and substantially reduce nonspecific binding during LCR ***(12,26)***. These probes consist of a single-stranded linear oligonucleotide of approx 90 bp that have 5' and 3' ends designed to hybridize adjacent to one another on the target sequence. Therefore, upon ligation, the probe is circularized and so it can withstand stringent wash conditions and can act as a template in a downstream rolling circle amplification reaction ***(26)***.

Because LCR is a DNA amplification technology similar to PCR, procedures to prevent carryover and contamination by environmental DNA must be followed ***(32)***. Unfortunately, ultraviolet (UV) decontamination of surfaces is not effective for LCR, as UV crosslinking is inefficient for short DNA products, such as those generated by LCR ***(30)***.

2. Ligase Chain Reaction Methodology

Ligase chain reaction proceeds by addition of target DNA to a reaction mixture containing a thermostable ligase, reaction buffer, and four oligonucleotide probes specific for the target. The oligonucleotide probes must be designed to be strictly target-specific, as the specificity of detection cannot be augmented by a posthybridization step ***(33)***. Ideally, the oligonucleotides are designed to all have similar melting temperatures and to be approx 20–25 nucleotides in length ***(1,28)***. To avoid template-independent, blunt-ended ligation, the discriminating probes are often designed to have 3' single-nucleotide overhangs ***(3)***, and the ligating probes are 5' phosphorylated (*see* **Fig. 1**). The ligase used is thermostable, such as *Tth* or *Pfu* ligases, and is therefore able to withstand the high temperatures required for DNA denaturation during cyclical reactions within a thermocycler. Another advantage of using *Tth* ligase is its increased fidelity over that observed with mesophilic ligases, such as T4 DNA ligase, leading to a significantly greater discrimination for mismatches occurring on the 3' side of the nick to be repaired ***(4,28)***.

The LCR reaction buffer normally consists of a Tris- or EPPS-buffered salt solution containing a number of enzyme cofactors and additives thought to enhance specificity. The reaction buffer is buffered to approx pH 7.5–8.3 using Tris at approx 20 m*M* ***(11,13,34)***, or EPPS at 50 m*M* ***(23,35)***. Potassium (20–100 m*M*) and magnesium (10–30 m*M*) salts are normally included in the reaction buffer; sodium salts are avoided as this ion is inhibitory to *Tth* DNA ligase ***(30)***. EDTA (1 m*M*), dithiothreitol (1–10 m*M*), and carrier DNA, such as herring sperm or salmon sperm DNA, have also been included in LCR reaction buffers ***(1,19,34)***. Inclusion of nonionic detergents, such as Triton X-100 or NP-40, although thought to improve the rate of ligation ***(3)***, could slightly increase nonspecific ligation. Depending on the thermostable ligase used in the reaction, ATP or NAD is added. Eubacterial ligases, such as *Tth* or *Taq* DNA ligase, require the coenzyme NAD, whereas archebacterial ligases, such as *Pfu* DNA ligase, require ATP as a reaction cofactor ***(28)***.

The number of cycles used for LCR amplification is often less than those used in a PCR reaction. Approximately 15–25 cycles have been found to be optimum, with increasing cycle number possibly producing a greater background without any improvement in product yield *(23,36)*, although this is dependent on the amount of target DNA initially present *(1)*. There are often only two temperature steps within each cycle: a denaturation step at approximately 95°C for approx 1 min, and an annealing step, at which temperature ligation also occurs with the thermostable ligase *(21,22,34)*.

Sensitivity of LCR is dependent on the sequence of the oligonucleotide probes precisely complementing the target sequence of interest, with a 50- to 500-fold increase in ligation products compared to those resulting from mismatched targets *(1)*. As well, the reduction of target-independent ligation products is crucial for increased sensitivity *(33)*, particularly for the detection of somatic mutations where a single-base change is present in a vast excess of wild-type sequence *(37,38)*. To overcome the obstacle of low mutant target levels within a sample, often the DNA target is first amplified by PCR prior to LCR *(20,34)*. LCR sensitivity has been found to match or exceed that of PCR *(33,35,39–42)*, with less than 10 targets detected in a background of 10^4–10^5 wild-type sequence targets or uninfected cells *(13,23,33)*.

2.1. Detection of Ligase Chain Reaction Products

Similar to PCR, gel electrophoresis of LCR products is one convenient method of detection. The electrophoretic gels used normally have a higher polyacrylamide content (approx 10–20%) to effectively separate the relatively short products (approx 40–50 nucleotides) from the reaction probes (approx 20–25 nucleotides). Products are detected in the gels by ethidium bromide staining or by autoradiography if a radiolabeled probe is used in the reaction *(21,34)*. Separation of LCR products by fast capillary electrophoresis coupled with laser-induced fluorescence analysis has also been described, thus allowing for extremely rapid detection of microquantities of LCR product *(18,43)*.

Alternatively, LCR products can be detected by capture on a solid phase. This method of detection has benefits both in terms of being nonisotopic and it could also slightly increase sensitivity. LCR lends itself well to detection in microtiter wells or on microparticles, as one probe might be haptenated with a capture molecule, such as biotin, while the other probe can be tagged with a label allowing fluorescent or colorimetric detection (*see* **Fig. 1**). The ligated products are then captured on a streptavidin-coated solid phase and detected through the presence of the label *(11,24,35)*. Microparticle capture and detection of LCR products has been automated and is available commercially through Abbott Laboratories *(44)*. Another method of detecting captured products involves matrix-assisted laser desorption/ionization time-of-flight mass spectrometry (MALDI-TOF-MS), which permits rapid and specific detection of femtomole amounts of DNA *(45)*.

2.2. Ligase Chain Reaction Modifications

Numerous modifications have been applied to the LCR to expand its utility, several of which will be described more fully in this subsection. Methods have been developed to increase LCR specificity (gap LCR, asymmetric gap LCR), sensitivity (nested LCR; (2)), and the number or type of targets detected by LCR (quantitative LCR, LCR of RNA targets, multiplex LCR) *(3)*.

The unique feature of gap LCR (gLCR) is the incorporation of a short gap of several nucleotides between the two adjacent probes complementary to each strand of the target DNA. This necessitates an initial extension of the upstream probe by a thermostable DNA polymerase prior to ligation occurring. This template-dependent extension is thought to increase the specificity of the reaction by reducing false-positive, blunt-ended ligation between the two probes *(2,3,23)*. The probes are designed to hybridize in a staggered fashion on both strands of DNA, such that only a single probe is discriminating for the mismatch of interest *(22)*. The DNA polymerase used must lack 5' to 3' exonuclease activity, and, normally, only the nucleotides required to fill the gap are added to the reaction *(22)*, although this effectively limits the gLCR reaction to detection of basepair changes of an A or T to a C or G or vice versa. Other basepair changes would necessitate addition of the discriminating nucleotide, although this has been shown not to adversely affect specificity ***(23)***. A method has also been developed (asymmetric gLCR) that permits specific amplification in the presence of all dNTPs ***(39)***.

Ligase chain reaction can be made semi-quantitative by including a standard curve of target molecules or titrated dilutions of cloned target DNA during an assay run. Known amounts or copies of target DNA can be processed in a similar fashion to the sample and the signal derived from each dilution plotted against the copy number or amount of DNA. Linear regression analysis of sample LCR signals then allow for approximate levels of sample target DNA to be determined. The strength of the LCR signal can be optimized such that it is directly related to the amount of target DNA in the samples ***(35)***. This type of analysis has been performed for HBV DNA and human immunodeficiency virus (HIV) RNA quantification using a microparticle-based enzyme immunoassay (MEIA) automated analyzer ***(35,46)***. Similarly, the approximate percentage of mutant DNA present in a background of wild-type DNA can be determined by comparison to a standard curve of mixed mutant and wild-type DNA. The mixture is composed of increasing ratios of cloned mutant and wild-type DNA, which is tested by LCR with mutant-specific probes. The percentage of mutant DNA present in a sample is then determined by comparison of the sample signal to signals plotted on the standard curve ***(20,22,24)***.

RNA targets, such as viral genomes or mRNA targets, can be detected by reverse transcribing the RNA prior to LCR detection. Hepatitis C virus (HCV) RNA has been detected by LCR of a cDNA target ***(39)*** and by reverse-transcription (RT)–PCR amplification prior to LCR ***(47)***. LCR with an initial RT-PCR step has also been used to detect GB virus C ***(48)***. RNA-templated DNA ligation has been described for direct analysis of mRNA transcripts ***(49)***. T4 DNA ligase is able to distinguish single-nucleotide variations within RNA sequences, however, with a substantially lower efficiency when compared to reactions having DNA templates. Directly analyzing RNA sequences rather than cDNA could provide a more accurate description of the relative abundance of specific mRNAs in cellular extracts ***(49)***.

3. Ligase Chain Reaction Applications

Ligase chain reaction methodology has been most useful in clinical diagnostics for detection of infectious disease agents and genetic polymorphisms leading to disease. Both DNA and RNA targets have been detected using LCR, and the technology has been automated and commercialized for the detection of specific microbes. LCR has also been used as a tool for genetic manipulations and investigations. There are several reports of LCR being used for in vitro synthesis of genes by ligation of short oligonucleotides followed by PCR amplification to generate the full-length gene sequence ***(50,51)***. Simultaneous site-directed mutagenesis of target DNA at multiple sites has also been accomplished using LCR ***(36,52)***. Genetic investigations involving LCR have included analysis of transcript splicing patterns within the human growth hormone gene cluster ***(53)*** and the study of rare intragenic crossover events in *Drosophila* genes ***(54)***.

3.1. Detection of Genetic Diseases

As referred to in **Subheading 1.2.**, LCR has been used for the detection of several genetic diseases of humans and animals, including cystic fibrosis, sickle cell anemia, type I Leber's hereditary optic neuropathy, hyperkalemic periodic paralysis ***(55)***, and bovine leukocyte adhesion deficiency. Although detection sensitivity for autosomal mutations is not as critical as somatic mutation detection, where a point mutation might be present in a vast excess of wild-type sequence, assay optimization is still imperative for sensitive and specific detection of genetic disease mutations. Internal controls are important for mutation detection, as they are for all LCR diagnostic assays, to control for amplification efficiency and the presence of inhibitors and to ensure that the same amount of template DNA is being investigated during multiple-sample diagnostics. Housekeeping genes, such as human growth hormone, can be coamplified at the time of mutant LCR testing of cellular samples *(11)*. Samples can also be spiked with nonspecific cloned DNA and simultaneously assayed with control probes to check the efficiency of amplification and for the presence of inhibitors. Detection of LCR products during genetic disease investigations has generally been accomplished by methods described in **Subheading 2.1.** In addition, each allele-specific probe can carry a unique tag to differentiate it from the wild-type allele or other mutations in a multiplex situation allowing for time-resolved fluorescent detection *(12)* or fluorescence resonance energy transfer (FRET) analysis *(28)*.

Concurrent detection of multiple genetic polymorphisms in a single sample has been performed using the oligonucleotide ligation assay (OLA), which involves three oligonucleotide probes. One probe is specific for the mutant allele, a second probe is specific for the wild-type allele, and the third probe acts as a common ligation probe for both *(28)*. Multiplexing of multiple genomic mutations is conveniently achieved by OLA, as each allele-specific probe can carry a different fluorescent tag or be of a different length, thus easily differentiating each product *(56)*. Multiplex OLA analysis has been used to simultaneously investigate 30 mutant loci throughout the cystic fibrosis gene and nine mutations in the 21-hydroxylase gene *(27,57)*. Similarly, LCR studies can employ two probe sets, one specific for wild-type sequence and the other specific for mutant sequence. The use of both sets is thought to reduce the incidence of clinical error and allow for confirmation of results *(17)*. Methods for multiplexed LCR have also been developed and have successfully been used for the detection of dystrophin gene deletions *(58)* and mutations associated with cystic fibrosis *(12)*.

3.2. Detection of Infectious Diseases

The detection and diagnosis of infectious disease agents has benefited enormously from the development of molecular detection technologies. Pathogenic bacterial and fungal species have been detected by LCR by taking advantage of the species-specific single-basepair differences resident within the 16S rDNA *(3,59,60)*. Numerous viral targets have also been investigated by LCR, both for diagnostic detection in clinical specimens and for detection of viral mutants. LCR has been used to detect HBV DNA in peripheral blood mononuclear cells *(35)* and cowpoxvirus DNA in viral isolate samples *(34)*. Viruses having RNA genomes, such as GB virus C *(48)*, HCV *(39,47)*, and HIV *(23,46)* have also been detected following RT-PCR or cDNA steps. Viral mutants, including HBV precore mutants *(20)* and HBV HBsAg mutants *(21,22)*, have been investigated. Aside from the great sensitivity afforded by PCR-coupled LCR, this method also has the advantage of specifically detecting host genome-integrated viral DNA much more reliably than PCR alone, as gene rearrangements can occur during integration leading to false-negative results with certain PCR primers *(33,35)*.

Ligase chain reaction is particularly well suited for detection of drug resistance in microorganisms, as the resistance mutation is often the result of a point mutation. Bacterial resistance to extended-spectrum β-lactam antibiotics has been investigated by PCR-coupled LDR and LCR *(9,19)*. Low levels of ganciclovir-resistant human cytomegalovirus (CMV) mutants and AZT-resistant HIV in high backgrounds of wild-type virus have both been sensitively detected by LCR analysis. CMV mutants, comprising approx 7.5% of an entire viral population *(24)*, and HIV mutants in a 10^4-fold excess of wild-type DNA *(23)* could be detected by LCR or gLCR. The obvious advantage of detecting drug-resistant mutants by LCR is that it allows for much faster detection of emerging resistant populations than would normally be detected by sequencing or isoelectric focusing analysis *(19)*.

3.2.1. Commercial Applications

Commercially available diagnostic kits using LCR technology have been developed by Abbott Laboratories (Abbott Park, IL) for numerous microorganisms. These kits have been well evaluated and will be discussed further in the following subsections. The Imx™ microparticle-based EIA system (Abbott Laboratories) has been used for detection of LCR products of various microorganisms, such as HIV *(23,46)*, HBV *(35)*, and GB virus C *(48)*.

Several life science companies market kits for LCR-based amplification of targets that include simple colorimetric detection of reaction products (LCR Kit by Stratagene, La Jolla, CA; AmpliTek by Bio-Rad Laboratories, Hercules, CA) ***(19)***. One advantage to performing LCR using commercial kits is the standardization of reaction components and methodology, such that presumably only the specific probes and sample DNA needs to be added.

The LCx system for semiautomated detection of microorganisms by LCR (Abbott Laboratories) is commercially available for *Mycobacterium tuberculosis*, *Chlamydia trachomatis*, and *Neisseria gonorrheae*. Specimen lysis and addition of DNA to LCR reaction tubes are performed manually, whereas detection of LCR products by MEIA is automated through use of the Abbott LCx Analyser. The LCx assay is generally very rapid, due to it being semiautomated, allowing for results to be obtained within 6–8 h ***(61,62)***. In general, each LCx reaction contains the components and probes necessary for a gap LCR reaction to amplify a target approx 40–60 nucleotides in length. Probe sets are commonly present in high concentrations in each reaction to increase the efficiency of hybridization, and the reaction is cycled approx 40 times to achieve up to a billion-fold amplification of the target sequence. MEIA detection involves four steps: (1) binding of LCR product to microparticle beads having capture molecules specific for a hapten present on one probe, (2) binding of alkaline phosphatase conjugated antibody specific for the hapten on a second, ligating probe, (3) production of a fluorescent signal following substrate utilization by alkaline phosphatase, and (4) fluorescent signal capture and value determination. As the automated system is entirely enclosed, the risk of amplicon contamination is decreased ***(63)***. False-positive results are further minimized by the inclusion of an amplicon-inactivation step that automatically follows the detection reaction in the LCx Analyser ***(64)***. The LCx kits have been found to be highly sensitive for the detection of these microorganisms. Specificity of detection is also observed to be very high, although false-negative results could occur as a result of gLCR inhibitors, and, as yet, the kits contain no internal control to check for amplification efficiency ***(65–67)***.

3.2.1.1. Mycobacterium tuberculosis

Traditionally, acid-fast staining and culture on solid and/or liquid media are the methods used for laboratory diagnosis of pulmonary tuberculosis, with culture accepted as the gold standard of detection because of its superior specificity ***(68,69)***. Although culture methods provide an accurate determination of *M. tuberculosis* (MTB), a more rapid and sensitive method of detection is required because cases of tuberculosis are increasing in certain populations ***(62,66)***. The MTB LCx assay is specific for a 44 bp segment of the single copy chromosomal gene encoding protein antigen b of MTB (64). The assay is validated only for pulmonary specimens, however non-pulmonary specimens have also been sensitively detected ***(63,65,66)***. LCx detection of MTB is highly specific, with 99% to 100% specificity compared to conventional procedures ***(41,61–64 ,66–68,70,71)***. Sensitivity of MTB LCx appears to be highly dependent on the acid-fast staining of the specimen, such that higher sensitivity occurs with smear-positive samples (96% to 100%) compared to smear-negative samples (57–72%) ***(61,63,64,72)***.

Specific advantages of the MTB LCx include the earlier and more specific detection of tuberculous bacilli in radiometric BACTEC culture-positive samples compared to confirmation by microscopic analysis ***(70,71)***. Earlier detection of MTB should prevent invasive testing normally used for confirmatory diagnostics in smear negative individuals, and hasten antituberculosis drug treatment in acutely infected patients ***(72)***.

As with other DNA amplification procedures, the presence of inhibitors in the sample often leads to false-negative results. It has been reported that the rate of LCR inhibition is approx 0.8% ***(73)***; thus, numerous specimens would be excluded in large-scale screening programs. Several studies have looked at how best to remove these inhibitors. Methods that have shown promise include pretreating the DNA specimen with sodium dodecyl sulfate (SDS) ***(68)*** or guanidinium isothiocyanate ***(66)*** followed by extensive washing ***(69)***. One study found that a major contributor to assay inhibition was a precipitate of calcium phosphate that acted as a source of inhibitory phosphate ions ***(73)***. In cases where spiking the sample with known amounts of DNA ensured that no inhibitors were present, false-negative results could still occur because of sampling problems caused by nonuniform distribution or clumping of very low levels of the tubercule bacilli within the sample ***(66,68)***. It has also been reported that LCx failed to detect a smear-positive MTB specimen because of a mutation resulting in deletion of the LCx target region ***(29)***.

Another important disadvantage of LCR detection compared to conventional MTB diagnostic procedures is the inability of the assay to differentiate between viable and non-viable organisms. Because DNA from nonviable organisms can be detected in patients following drug therapy, determination of the patient's progress cannot be made by LCR ***(61,63,66,68)***. Therefore, culture is still required to test for efficacy of antituberculosis drug therapy and MTB drug susceptibility, thus LCx can only be used as a complementary method for these diagnoses.

3.2.1.2. Neisseria gonorrheae

Conventional methods of diagnosing *N. gonorrheae* include the culture of endocervical or urethral swabs. The introduction of the LCx system for *N. gonorrheae* detection has enabled the use of urine specimens, which are more easily obtained and more resistant to variable transport and storage temperatures as compared to swab specimens ***(74–76)***. Urine specimens also have the advantage that they can be pooled without loss of sensitivity ***(75)***. The *N. gonorrheae* LCx target is the *opa* gene, of which *N. gonorrheae* has at least 11 copies per genome ***(75)***, thus increasing the sensitivity of detection. LCR detection of the *N. gonorrheae* pilin gene has also been performed to confirm positive LCx results or as a secondary test to resolve discrepancies between the swab culture and LCx ***(75,77,78)***. Highly sensitive LCx detection of *N. gonorrheae* has been observed with first-void urine specimens in comparison to swab culture ***(76)***; however, several studies have determined that LCx detection using swab specimens showed superior sensitivity compared to urine testing ***(79,80)***. Regardless, urine LCx testing provides a means of expanding *N. gonorrheae* screening programs to allow surveillance of low prevalence or asymptomatic populations that would not normally be conveniently screened by swab testing ***(74,81)***. An alternative to urine might be LCR testing of patient-obtained vaginal swabs, which are also obtained with greater compliance compared to endocervical swabs and demonstrate excellent sensitivity for *N. gonorrheae* detection ***(77)***.

3.2.1.3. Chlamydia trachomatis

Routine detection of *C. trachomatis* in urogenital swab specimens normally involves either the detection of chlamydial antigen by EIA, the staining of elementary bodies by direct fluorescent antibody (DFA) tests, or isolation of the microbe in tissue culture. Although for years chlamydial culture was considered the gold standard for detection ***(40,82)***, this has now changed because of the enhanced sensitivity yet comparable specificity of nucleic acid amplification tests such as PCR and LCR ***(42,83–86)***. Similar to culture, EIA and DFA tests demonstrate relatively low sensitivity compared to PCR and LCR, particularly with specimens from women and asymptomatic men ***(87)***. LCR detection of *C. trachomatis* by LCx has been carefully evaluated in numerous studies and is now accepted as an expanded or replacement gold standard of detection ***(42,85)***.

The LCx assay specifically targets the cryptic plasmid of *C. trachomatis*, which is present at approx 10 copies per organism ***(88)***. The target sequence is highly conserved among all serovars of *C. trachomatis*, yet is not found in other *Chlamydia* species ***(89)***, thus lending to the specificity of the assay. In studies confirming a positive LCx result or resolving discrepancies between LCx and other detection methods, LCR of the *C. trachomatis* major outer membrane protein gene was performed ***(87,90)***. This type of analysis, described as discrepant analysis, is often done when analyzing a new method such as the *C. trachomatis* LCx assay. It has been suggested that discrepant analysis introduces a bias in favor of the assay being evaluated, which has created a certain degree of controversy ***(91–93)***. However, more recent studies using an alternative to discrepant analysis have found similar results to those evaluations using discrepant analysis ***(83,94,95)***. Although the LCx assay is licensed for detection of *C. trachomatis* in female endocervical swab or urine specimens and male urethral swab or urine specimens, other specimens have been analyzed with this assay. Pharyngeal samples ***(96)*** and formalin-fixed paraffin-embedded tissues ***(97)*** have been assayed, as have synovial fluid cells for the diagnosis of reactive arthritis ***(98)***. The assay can also be modified to be quantitative ***(99)***.

The general consensus of numerous studies is that detection of *C. trachomatis* DNA by LCR is more sensitive and reproducible than culture, leading to it becoming a part of a new gold standard of detection ***(42,86,93)***. One of the greatest advantages of LCR over culture is its sensitive detection of *C. trachomatis* in urine specimens. Whereas previously urogenital swabs were required for detection of this organism, the present use of urine samples has allowed for wide-scale screening of *C. trachomatis*, particularly in low-prevalence settings or in asymptomatic groups ***(42,81,86,94,100)***. Diagnostic detection of *C. trachomatis* by LCR has also been shown to be more cost-effective in certain settings than conventional methods ***(84,101)***; however, for routine diagnostics of low-prevalence populations, it has been suggested that LCx be reserved for confirmation of positive EIAs because of the higher cost of LCx compared to EIA ***(102)***.

Despite the increased sensitivity of LCR, several concerns have been raised over certain limitations of the *C. trachomatis* LCx assay. There are differences in sensitivity between male and female specimens, particularly urine, such that the sensitivity of detection is much greater in male urine than female urine ***(76,83,103)***. This has led to suggestions that the diagnosis of *C. trachomatis* in women might require both urine and endocervical swab testing by LCx ***(85)***. There may even be sensitivity differences within the female population, such as pregnant vs nonpregnant women ***(103–106)***, menstruating vs nonmenstruating women ***(107)***, and younger vs older infected women, where younger women might have higher levels of the organism, thus allowing for easier detection ***(108)***. One of the most important concerns of the LCx assay is the occurrence of false-negative results during sample analysis. This most often occurs as a result of the presence of inhibitors, particularly in urine samples ***(104,109)***. Fortunately, these inhibitors appear to be easily minimized or removed by several days storage, freeze/thawing, or dilution ***(82,83,104,110)***. The failure to detect *C. trachomatis* might also occur as a result of loss or mutation within the *C. trachomatis* plasmid ***(99,109)***, by sampling variation caused by low inclusion counts within the sample ***(90,110,111)*** or by failure to preserve the specimen cold chain during transport and storage ***(100,112)***, although this storage requirement has been disputed ***(103)***. Other concerns include the often poor reproducibility of positive LCx results ***(111)*** and the need for an internal control within the assay to ensure that the DNA-extraction and DNA-amplification steps were performed correctly and that no inhibitors were present ***(87,98,104)***.

4. Future Prospects

Now that a greater understanding of the human genome is upon us, following its sequencing and characterization, the discovery of new genetic traits and disease markers is imminent. LCR will be an ideal method for research into these polymorphisms and mutations, with the advantage of future automation for wide-scale population screening. The development and commercialization of further LCR assays for other infectious disease agents or clinically important mutants of microorganisms is also a possibility for the future. For example, because of its high sensitivity, ease of use, and automation, LCR might be ideal for the screening of blood-borne viruses in transfusion settings ***(113)*** and for screening of drug-resistant bacterial or viral mutants ***(9,23,24)***. The development of high-density microarrays for multiplexing and detection of LCR products has already begun ***(7,27,28)*** and should allow for greater detection capability and automation for the future.

For any newly developed commercial LCR assay to succeed, it will be important to address the issue of amplification inhibitors and to include internal controls to ensure proper assay performance. As more and more laboratories adopt LCR technology and commercial kits for diagnostic purposes, it will also be necessary that appropriate quality assurance programs are instituted to maintain diagnostic efficiency *(42)*. With ever-increasing research and development into LCR methods and detection, LCR is poised to become an important detection technology for clinical- and discovery-based research in the future.

Acknowledgment

The author would like to thank Elizabeth Giles and Alan Smith for their helpful comments and suggestions during the preparation of this chapter.

References

1. Barany, F. (1991) Genetic disease detection and DNA amplification using cloned thermostable ligase. *Proc. Natl. Acad. Sci. USA* **88**, 189–193.
2. Barany, F. (1991) The ligase chain reaction in a PCR world. *PCR Methods Appl.* **1**, 5–16.
3. Wiedmann, M., Wilson,W. J., Czajka, J., Luo, J., Barany, F., and Batt, C. A. (1994) Ligase chain reaction (LCR)—overview and applications. *PCR Methods Appl.* **3**, S51–S64.
4. Luo, J., Bergstrom, D. E., and Barany, F. (1996) Improving the fidelity of Thermus thermophilus DNA ligase. *Nucleic Acids Res.* **24**, 3071–3078.
5. Landegren, U., Kaiser, R., Sanders, J., and Hood, L. (1988) A ligase-mediated gene detection technique. *Science* **241**, 1077–1080.
6. Bottema, C. D. and Sommer, S. S. (1993) PCR amplification of specific alleles: rapid detection of known mutations and polymorphisms. *Mutat. Res.* **288**, 93–102.
7. Schweitzer, B. and Kingsmore, S. (2001) Combining nucleic acid amplification and detection. Curr. Opin. Biotechnol. **12**, 21–27.
8. Nickerson, D. A., Kaiser, R., Lappin, S., Stewart, J., Hood, L., and Landegren, U. (1990) Automated DNA diagnostics using an ELISA-based oligonucleotide ligation assay. *Proc. Natl. Acad. Sci. USA* **87**, 8923–8927.
9. Niederhauser, C., Kaempf, L., and Heinzer, I. (2000) Use of the ligase detection reaction-polymerase chain reaction to identify point mutations in extended-spectrum beta-lactamases. *Eur. J. Clin. Microbiol. Infect. Dis.* **19**, 477–480.
10. Wu, D. Y. and Wallace, R. B. (1989) The ligation amplification reaction (LAR)—amplification of specific DNA sequences using sequential rounds of template-dependent ligation. *Genomics* **4**, 560–569.
11. Reyes, A. A., Carrera, P., Cardillo, E., et al. (1997) Ligase chain reaction assay for human mutations: the sickle cell by LCR assay. *Clin. Chem.* **43**, 40–44.
12. Landegren, U., Samiotaki, M., Nilsson, M., Malmgren, H., and Kwiatkowski, M. (1996) Detecting genes with ligases. *Methods* **9**, 84–90.
13. Zirvi, M., Nakayama, T., Newman, G., McCaffrey, T., Paty, P., and Barany, F. (1999) Ligase-based detection of mononucleotide repeat sequences. *Nucleic Acids Res.* **27**, e40i–e40viii.
14. Wilson, V. L., Wei, Q., Wade, K. R., et al. (1999) Needle-in-a-haystack detection and identification of base substitution mutations in human tissues. *Mutat. Res.* **406**, 79–100.
15. Martinez, A., Lehman, T. A., Modali, R., and Mulshine, J. L. (2003) Screening of mutations in the ras family of oncogenes by polymerase chain reaction-based ligase chain reaction. *Methods Mol. Biol.* **74**, 187–200.
16. Batt, C. A., Wagner, P., Wiedmann, M., Luo, J., and Gilbert, R. (1994) Detection of bovine leukocyte adhesion deficiency by nonisotopic ligase chain reaction. *Anim. Genet.* **25**, 95–98.
17. Zebala, J. A. and Barany, F. (1993) Implications for the ligase chain reaction in gastroenterology. *J. Clin. Gastroenterol.* **17**, 171–175.
18. Muth, J., Williams, P. M., Williams, S. J., Brown, M. D., Wallace, D. C., and Karger, B. L. (1996) Fast capillary electrophoresis-laser induced fluorescence analysis of ligase chain reaction products: human mitochondrial DNA point mutations causing Leber's hereditary optic neuropathy. *Electrophoresis* **17**, 1875–1883.
19. Kim, J. and Lee, H.-J. (2000) Rapid discriminatory detection of genes coding for SHV β-lactamases by ligase chain reaction. *Antimicrob. Agents Chemother.* **44**, 1860–1864.
20. Minamitani, S., Nishiguchi, S., Kuroki, T., Otani, S., and Monna, T. (1997) Detection by ligase chain reaction of precore mutant of Hepatitis B virus. *Hepatology* **25**, 216–222.

21. Devi Karthigesu, V., Mendy, M., Fortuin, M., Whittle, H. C., Howard, C. R., and Allison, L. M. C. (1995) The ligase chain reaction distinguishes hepatitis B virus S-gene variants. *FEMS Microbiol. Lett.* **131**, 127–132.
22. Osiowy, C. (2002) Sensitive detection of HBsAg mutants by a gap ligase chain reaction assay. *J. Clin. Microbiol.* **40**, 2566–2571.
23. Abravaya, K., Carrino, J. J., Muldoon, S., and Lee, H. H. (1995) Detection of point mutations with a modified ligase chain reaction (Gap-LCR). *Nucleic Acids Res.* **23**, 675–682.
24. Bourgeois, C., Sixt, N., Bour, J. B., and Pothier, P. (1997) Value of a ligase chain reaction assay for detection of ganciclovir resistance-related mutation 594 in UL97 gene of human cytomegalovirus. *J. Virol. Methods* **67**, 167–175.
25. Wolcott, M. J. (1992) Advances in nucleic acid-based detection methods. *Clin. Microbiol. Rev.* **5**, 370–386.
26. Andras, S. C., Power, J. B., Cocking, E. C., and Davey, M. R. (2001) Strategies for signal amplification in nucleic acid detection. *Mol. Biotechnol.* **19**, 29–44.
27. Winn-Deen, E. S. (1996) Multi-mutation screening using PCR and ligation—principles and applications. *Trends Biotechnol.* **14**, 112–114.
28. Jarvius, J., Nilsson, M., and Landegren, U. (2003) Oligonucleotide ligation assay. *Methods Mol. Biol.* **212**, 215–228.
29. Gilpin, C. M., Dawson, D. J., O'Kane, G., Armstrong, J. G., and Coulter, C. (2002) Failure of commercial ligase chain reaction to detect *Mycobacterium tuberculosis* DNA in sputum samples from a patient with smear-positive pulmonary tuberculosis due to a deletion of the target region. *J. Clin. Microbiol.* **40**, 2305–2307.
30. Shimer, G.H., Jr. and Backman, K. C. (1995) Ligase chain reaction. *Methods Mol. Biol.* **46**, 269–278.
31. Demchinskaya, A. V., Shilov, I. A., Karyagina, A. S., et al. (2001) A new approach for point mutation detection based on a ligase chain reaction. *J. Biochem. Biophys. Methods* **50**, 79–89.
32. Davies, P. O. and Ridgway, G. L. (1997) The role of polymerase chain reaction and ligase chain reaction for the detection of *Chlamydia trachomatis*. *Int. J. STD AIDS* **8**, 731–738.
33. Laffler, T., Carrino, J. J., and Marshall, R. L. (1993) The ligase chain reaction in DNA-based diagnosis. *Ann. Biol. Clin.* **50**, 821–826.
34. Pfeffer, M., Meyer, H., and Wiedmann, M. (1994) A ligase chain reaction targeting two adjacent nucleotides allows the differentiation of cowpox virus from other *Orthopoxvirus* species. *J. Virol. Methods* **49**, 353–360.
35. Trippler, M., Hampl, H., Goergen, B., et al. (1996) Ligase chain reaction (LCR) assay for semi-quantitative detection of HBV DNA in mononuclear leukocytes of patients with chronic hepatitis B. *J. Viral Hepat.* **3**, 267–272.
36. Rouwendal, G. J. A., Wolbert, E. J. H., Zwiers, L.-H., and Springer, J. (1996) Ligase chain reaction for site-directed in vitro mutagenesis. *Methods Mol. Biol.* **57**, 149–156.
37. Kalin, I., Shephard, S., and Candrian, U. (1992) Evaluation of the ligase chain reaction (LCR) for the detection of point mutations. *Mutat. Res.* **283**, 119–123.
38. Lee, H. H. (1996) Ligase chain reaction. *Biologicals* **24**, 197–199.
39. Marshall, R. L., Laffler, T., Cerney, M. B., Sustachek, J. C., Kratochvil, J., and Morgan, R. L. (1994) Detection of HCV RNA by the asymmetric gap ligase chain reaction. *PCR Methods Appl.* **4**, 80–84.
40. Schachter, J. (1997) DFA, EIA, PCR, LCR and other technologies: what tests should be used for diagnosis of Chlamydia infections? *Immunol. Invest.* **26**, 157–161.
41. Wang, S. X. and Tay, L. (1999) Evaluation of three nucleic acid amplification methods for direct detection of Mycobacterium tuberculosis complex in respiratory specimens. *J. Clin. Microbiol.* **37**, 1932–1934.
42. Black, C. M., Marrazzo, J., Johnson, R. E., et al. (2002) Head-to-head multicenter comparison of DNA probe and nucleic acid amplification tests for Chlamydia trachomatis infection in women performed with an improved reference standard. *J. Clin. Microbiol.* **40**, 3757–3763.
43. Cheng, J., Shoffner, M. A., Mitchelson, K. R., Kricka, L. J., and Wilding, P. (1996) Analysis of ligase chain reaction products amplified in a silicon-glass chip using capillary electrophoresis. *J. Chromatogr. A* **732**, 151–158.
44. Jungkind, D. (2001) Molecular testing for infectious disease. *Science* **294**, 1553–1555.
45. Jurinke, C., van den Boom, D., Jacob, A., Tang, K., Worl, R., and Koster, H. (1996) Analysis of ligase chain reaction products via matrix-assisted laser desorption/ionization time-of-flight mass spectrometry. *Anal. Biochem.* **237**, 174–181.

46. de Mendoza, C., Alcami, J., Sainz, M., Folgueira, D., and Soriano, V. (2002) Evaluation of the Abbott LCx quantitative assay for measurement of human immunodeficiency virus RNA in plasma. *J. Clin. Microbiol.* **40**, 1518–1521.
47. Crotty, P. L., Staggs, R. A., Porter, P. T., Killeen, A. A., and McGlennen, R. C. (1994) Quantitative analysis in molecular diagnostics. *Hum. Pathol.* **25**, 572–579.
48. Marshall, R. L., Cockerill, J., Friedman, P., et al. (1998) Detection of GB virus C by the RT-PCR LCx system. *J. Virol. Methods* **73**, 99–107.
49. Nilsson, M., Antson, D.-O., Barbany, G., and Landegren, U. (2001) RNA-templated DNA ligation for transcript analysis. Nucleic Acids Res. **29**, 578–581.
50. Au, L. C., Yang, F. Y., Yang, W. J., Lo, S. H., and Kao, C. F. (1998) Gene synthesis by a LCR-based approach: high-level production of leptin-L54 using synthetic gene in *Escherichia coli. Biochem. Biophys. Res. Commun.* **248**, 200–203.
51. Chalmers, F. M. and Curnow, K. M. (2001) Scaling up the ligase chain reaction-based approach to gene synthesis. *Biotechniques* **30**, 249–252.
52. Rouwendal, G. J. A., Wolbert, E. J. H., Zwiers, L.-H., and Springer, J. (1993) Simultaneous mutagenesis of multiple sites: application of the ligase chain reaction using PCR products instead of oligonucleotides. *Biotechniques* **15**, 68–76.
53. Boguszewski, C. L., Svensson, P. A., Jansson, T., Clark, R., Carlsson, L. M., and Carlsson, B. (1998) Cloning of two novel growth hormone transcripts expressed in human placenta. *J. Clin. Endocrinol. Metab.* **83**, 2878–2885.
54. Balles, J. and Pflugfelder, G. O. (1994) Facilitated isolation of rare recombinants by ligase chain reaction: selection for intragenic crossover events in the *Drosophila* optomotor-blind gene. *Mol. Gen. Genet.* **245**, 734–740.
55. Feero, W. G., Wang, J., Barany, F., et al. (1993) Hyperkalemic periodic paralysis: rapid molecular diagnosis and relationship of genotype to phenotype in 12 families. *Neurology* **43**, 668–673.
56. Shi, M. M. (2002) Technologies for individual genotyping: detection of genetic polymorphisms in drug targets and disease genes. *Am. J. Pharmacogenom.* **2**, 197–205.
57. Day, D. J., Speiser, P. W., White, P. C., and Barany, F. (1995) Detection of steroid 21-hydroxylase alleles using gene-specific PCR and a multiplexed ligation detection reaction. *Genomics* **29**, 152–162.
58. Jou, C., Rhoads, J., Bouma, S., et al. (1995) Deletion detection in the dystrophin gene by multiplex gap ligase chain reaction and immunochromatographic strip technology. *Hum. Mutat.* **5**, 86–93.
59. Wilson, W. J., Wiedmann, M., Dillard, H. R., and Batt, C. A. (1994) Identification of *Erwinia stewartii* by a ligase chain reaction assay. *Appl. Environ. Microbiol.* **60**, 278–284.
60. Hatziloukas, E., Tooley, P., and Carras, M. (1998) Ligase chain reaction-based detection of the potato pathogen *Phytophthora infestans*, in *Proc. COST 823: New Technologies to Improve Phytodiagnosis*, 12.
61. Moore, D. F. and Curry, J. I. (1998) Detection and identification of *Mycobacterium tuberculosis* directly from sputum sediments by ligase chain reaction. *J. Clin. Microbiol.* **36**, 1028–1031.
62. O'Connor, T. M., Sheehan, S., Cryan, B., Brennan, N., and Bredin, C. P. (2000) The ligase chain reaction as a primary screening tool for the detection of culture positive tuberculosis. *Thorax* **55**, 955–957.
63. Tortoli, E., Lavinia, F., and Simonetti, M. T. (1997) Evaluation of a commercial ligase chain reaction kit (Abbott LCx) for direct detection of *Mycobacterium tuberculosis* in pulmonary and extrapulmonary specimens. *J. Clin. Microbiol.* **35**, 2424–2426.
64. Lindbrathen, A., Gaustad, P., Hovig, B., and Tonjum, T. (1997) Direct detection of Mycobacterium tuberculosis complex in clinical samples from patients in Norway by ligase chain reaction. *J. Clin. Microbiol.* **35**, 3248–3253.
65. Palacios, J. J., Ferro, J., Ruiz Palma, N., et al. (1998) Comparison of the ligase chain reaction with solid and liquid culture media for routine detection of *Mycobacterium tuberculosis* in nonrespiratory specimens. *Eur. J. Clin. Microbiol. Infect. Dis.* **17**, 767–772.
66. Gamboa, F., Dominguez, J., Padilla, E., et al. (1998) Rapid diagnosis of extrapulmonary tuberculosis by ligase chain reaction amplification. *J. Clin. Microbiol.* **36**, 1324–1329.
67. Lumb, R., Davies, K., Dawson, D., et al. (1999) Multicenter evaluation of the Abbott LCx Mycobacterium tuberculosis ligase chain reaction assay. *J. Clin. Microbiol.* **37**, 3102–3107.
68. Ausina, V., Gamboa, F., Gazapo, E., et al. (1997) Evaluation of the semiautomated Abbott LCx *Mycobacterium tuberculosis* assay for direct detection of *Mycobacterium tuberculosis* in respiratory specimens. *J. Clin. Microbiol.* **35**, 1996–2002.

69. Ruiz-Serrano, M. J., Albadalejo, J., Martinez-Sanchez, L., and Bouza, E. (1998) LCx: A diagnostic alternative for the early detection of *Mycobacterium tuberculosis* complex. *Diagn. Microbiol. Infect. Dis.* **32**, 259–264.
70. Tortoli, E., Lavinia, F., and Simonetti, M. T. (1998) Early detection of *Mycobacterium tuberculosis* in BACTEC cultures by ligase chain reaction. *J. Clin. Microbiol.* **36**, 2791–2792.
71. Wang, L. and Tay, L. (2002) Early identification of *Mycobacterium tuberculosis* complex in BACTEC cultures by ligase chain reaction. *J. Med. Microbiol.* **51**, 710–712.
72. Jouveshomme, S., Cambau, E., Trystram, D., et al. (1998) Clinical utility of an amplification test based on ligase chain reaction in pulmonary tuberculosis. *Am. J. Respir. Crit. Care Med.* **158**, 1096–1101.
73. Leckie, G. W., Erickson, D. D., He, Q., et al. (1998) Method for reduction of inhibition in a *Mycobacterium tuberculosis*-specific ligase chain reaction DNA amplification assay. *J. Clin. Microbiol.* **36**, 764–767.
74. Smith, K. R., Ching, S., Lee, H., et al. (1995) Evaluation of ligase chain reaction for use with urine for identification of *Neisseria gonorrhoeae* in females attending a sexually transmitted disease clinic. *J. Clin. Microbiol.* **33**, 455–457.
75. Kacena, K. A., Quinn, S. B., Hartman, S. C., Quinn, T. C., and Gaydos, C. A. (1998) Pooling of urine samples for screening for *Neisseria gonorrhoeae* by ligase chain reaction: accuracy and application. *J. Clin. Microbiol.* **36**, 3624–3628.
76. Stary, A. (1999) Correct samples for diagnostic tests in sexually transmitted diseases: which sample for which test? *FEMS Immunol. Med. Microbiol.* **24**, 455–459.
77. Hook, E. W., III, Ching, S. F., Stephens, J., Hardy, K. F., Smith, K. R., and Lee, H. H. (1997) Diagnosis of *Neisseria gonorrhoeae* infections in women by using the ligase chain reaction on patient-obtained vaginal swabs. *J. Clin. Microbiol.* **35**, 2129–2132.
78. Kehl, S.C., Georgakas, K., Swain, G. R., et al. (1998) Evaluation of the Abbott LCx assay for detection of *Neisseria gonorrhoeae* in endocervical swab specimens from females. *J. Clin. Microbiol.* **36**, 3549–3551.
79. Buimer, M., Van Doornum, G. J. J., Ching, S., et al. (1996) Detection of Chlamydia trachomatis and *Neisseria gonorrhoeae* by ligase chain reaction-based assays with clinical specimens from various sites: Implications for diagnostic testing and screening. *J. Clin. Microbiol.* **34**, 2395–2400.
80. Carroll, K. C., Aldeen, W. E., Morrison, M., Anderson, R., Lee, D., and Mottice, S. (1998) Evaluation of the Abbott LCx ligase chain reaction assay for detection of *Chlamydia trachomatis* and *Neisseria gonorrhoeae* in urine and genital swab specimens from a sexually transmitted disease clinic population. *J. Clin. Microbiol.* **36**, 1630–1633.
81. Brodine, S. K., Shafer, M., Shaffer, R. A., et al. (1998) Asymptomatic sexually transmitted disease prevalence in four military populations: application of DNA amplification assays for Chlamydia and gonorrhea screening. *J. Infect. Dis.* **178**, 1202–1204.
82. de Barbeyrac, B., Rodriguez, P., Dutilh, B., Le Roux, P., and Bebear, C. (1995) Detection of *Chlamydia trachomatis* by ligase chain reaction compared with polymerase chain reaction and cell culture in urogenital specimens. *Genitourin. Med.* **71**, 382–386.
83. Puolakkainen, M., Hiltunen-Back, E., Reunala, T., et al. (1998) Comparison of performances of two commercially available tests, a PCR assay and a ligase chain reaction test, in detection of urogenital *Chlamydia trachomatis* infection. *J. Clin. Microbiol.* **36**, 1489–1493.
84. Waites, K. B., Smith, K. R., Crum, M. A., Hockett, R. D., Wells, A. H., and Hook, E.W., III. (1999) Detection of *Chlamydia trachomatis* endocervical infections by ligase chain reaction versus ACCESS Chlamydia antigen assay. *J. Clin. Microbiol.* **37**, 3072–3073.
85. Rabenau, H. F., Kohler, E., Peters, M., Doerr, H. W., and Weber, B. (2000) Low correlation of serology with detection of *Chlamydia trachomatis* by ligase chain reaction and antigen EIA. *Infection* **28**, 97–102.
86. Watson, E. J., Templeton, A., Russell, I., et al. (2002) The accuracy and efficacy of screening tests for *Chlamydia trachomatis*: a systematic review. *J. Med. Microbiol.* **51**, 1021–1031.
87. Rumpianesi, F., Donati, M., La Placa, M., Negosanti, M., D'Antuono, A., and Cevenini, R. (1996) Use of the ligase chain reaction on urine of men and their female sexual partners for detection of genital *Chlamydia trachomatis* infection. *Clin. Microbiol. Infect.* **2**, 123–126.
88. Palmer, L. and Falkow, S. (1986) A common plasmid of *Chlamydia trachomatis*. *Plasmid* **16**, 52–62.
89. Joseph, T., Nano, F. E., Garon, C. F., and Caldwell, H. D. (1986) Molecular characterization of *Chlamydia trachomatis* and *Chlamydia psittaci* plasmids. *Infect. Immun.* **51**, 699–703.
90. Bassiri, M., Hu, H.-Y., Domeika, M.A., et al. (1995) Detection of Chlamydia trachomatis in urine specimens from women by ligase chain reaction. *J. Clin. Microbiol.* **33**, 898–900.

91. Hadgu, A. (1997) Bias in the evaluation of DNA-amplification tests for detecting Chlamydia trachomatis. *Statist. Med.* **16**, 1391–1399.
92. Chernesky, M., Sellors, J., and Mahony, J. (1998) Bias in the evaluation of DNA-amplification tests for detecting *Chlamydia trachomatis*: letter to the editor. *Statist. Med.* **17**, 1055–1066.
93. Schachter, J. (1998) Bias in the evaluation of DNA-amplification tests for detecting *Chlamydia trachomatis*: letter to the editor. *Statist. Med.* **17**, 1527–1530.
94. Johnson, R. E., Green, T. A., Schachter, J., et al. (2000) Evaluation of nucleic acid amplification tests as reference tests for *Chlamydia trachomatis* infections in asymptomatic men. *J. Clin. Microbiol.* **38**, 4382–4386.
95. Van Dyck, E., Ieven, M., Pattyn, S., Van Damme, L., and Laga, M. (2001) Detection of *Chlamydia trachomatis* and *Neisseria gonorrhoeae* by enzyme immunoassay, culture, and three nucleic acid amplification tests. *J. Clin. Microbiol.* **39**, 1751–1756.
96. Winter, A. J., Gilleran, G., Eastick, K., and Ross, J. D. C. (2000) Comparison of a ligase chain reaction-based assay and cell culture for detection of pharyngeal carriage of *Chlamydia trachomatis*. *J. Clin. Microbiol.* **38**, 3502–3504.
97. Noguchi, Y., Yabushita, H., Noguchi, M., Fujita, M., Asai, M., and Del Carpio, C. A. (2002) Detection of *Chlamydia trachomatis* infection with DNA extracted from formalin-fixed paraffin-embedded tissues. *Diagn. Microbiol. Infect. Dis.* **43**, 1–6.
98. Nikkari, S., Puolakkainen, M., Yli-Kerttula, U., Luukkainen, R., Lehtonen, O.-P., and Toivanen, P. (1997) Ligase chain reaction in detection of *Chlamydia* DNA in synovial fluid cells. *Br. J. Rheumatol.* **36**, 763–765.
99. Blocker, M. E., Krysiak, R. G., Behets, F., Cohen, M. S., and Hobbs, M. M. (2002) Quantification of *Chlamydia trachomatis* elementary bodies in urine by ligase chain reaction. *J. Clin. Microbiol.* **40**, 3631–3634.
100. Neu, N., Grumet, S., McNees, A., et al. (1999) Screening for *Chlamydia trachomatis* in young men by ligase chain reaction. *Pediatr. Infect. Dis. J.* **18**, 649–650.
101. Battle, T. J., Golden, M. R., Suchland, K. L., et al. (2001) Evaluation of laboratory testing methods for *Chlamydia trachomatis* infection in the era of nucleic acid amplification. *J. Clin. Microbiol.* **39**, 2924–2927.
102. Dean, D., Ferrero, D., and McCarthy, M. (1998) Comparison of performance and cost-effectiveness of direct fluorescent-antibody, ligase chain reaction, and PCR assays for verification of chlamydia enzyme immunoassay results for populations with a low to moderate prevalence of *Chlamydia trachomatis* infection. *J. Clin. Microbiol.* **36**, 94–99.
103. Clad, A., Prillwitz, J., Hintz, K. C., et al. (2001) Discordant prevalence of *Chlamydia trachomatis* in asymptomatic couples screened using urine ligase chain reaction. *Eur. J. Clin. Microbiol. Infect. Dis.* **20**, 324–328.
104. Mahony, J., Chong, S., Jang, D., et al. (1998) Urine specimens from pregnant and nonpregnant women inhibitory to amplification of *Chlamydia trachomatis* nucleic acid by PCR, ligase chain reaction, and transcription-mediated amplification: identification of urinary substances associated with inhibition and removal of inhibitory activity. *J. Clin. Microbiol.* **36**, 3122–3126.
105. Jensen, I. P., Thorson, P., and Moller, B. R. (1997) Sensitivity of ligase chain reaction assay of urine from pregnant women for *Chlamydia trachomatis*. *Lancet* **349**, 329–330.
106. Gaydos, C. A., Howell, M. R., Quinn, T. C., Gaydos, J. C., and McKee, K. T., Jr. (1998) Use of ligase chain reaction with urine versus cervical culture for detection of *Chlamydia trachomatis* in an asymptomatic military population of pregnant and nonpregnant females attending Papanicolaou smear clinics. *J. Clin. Microbiol.* **36**, 1300–1304.
107. Horner, P. J., Crowley, T., Leece, J., Hughes, A., Smith, G. D., and Caul, E. O. (1998) *Chlamydia trachomatis* detection and the menstrual cycle. *Lancet* **351**, 341–342.
108. Webster Dicker, L., Mosure, D. J., Levine, W. C., Black, C. M., and Berman, S. M. (2000) Impact of switching laboratory tests on reported trends in *Chlamydia trachomatis* infections. *Am. J. Epidemiol.* **151**, 430–435.
109. Notomi, T., Ikeda, Y., Okadome, A., and Nagayama, A. (1998) The inhibitory effect of phosphate on the ligase chain reaction used for detecting *Chlamydia trachomatis*. *J. Clin. Pathol.* **51**, 306–308.
110. Thomas, B., Pierpoint, T., Taylor-Robinson, D., and Renton, A. (2001) Qualitative and quantitative aspects of the ligase chain reaction assay for *Chlamydia trachomatis* in genital tract samples and urines. *Int. J. STD AIDS* **12**, 589–594.

111. Castriciano, S., Luinstra, K., Jang, D., et al. (2002) Accuracy of results obtained by performing a second ligase chain reaction assay and PCR analysis on urine samples with positive or near-cutoff results in the LCx test for *Chlamydia trachomatis*. *J. Clin. Microbiol.* **40**, 2632–2634.
112. Thomas, B., Pierpoint, T., Taylor-Robinson, D., and Renton, A. (2001) Reduced detection of *Chlamydia trachomatis* by the ligase chain reaction assay due to suboptimal storage of urine. *Eur. J. Clin. Microbiol. Infect. Dis.* **20**, 581–583.
113. Allain, J.-P. (2000) Genomic screening for blood-borne viruses in transfusion settings. *Clin. Lab. Haem.* **22**, 1–10.

16

Chemical Cleavage of Mismatch

Theory and Clinical Applications

Neil V. Whittock and Louise Izatt

1. Introduction

Direct nucleotide sequencing is the reference standard critical to all molecular biology whether it is used for the elucidation of an entire genome or the characterization of a specific mutation. Sequencing protocols were initially developed using either dideoxy nucleotides ***(1)*** or chemicals ***(2)***, although the latter became less attractive because of the toxic nature of the chemicals used. It is obvious that if sequencing was less expensive, then there would be no need for mutation-scanning techniques. However, although fluorescent technology is improving and capillary electrophoresis is replacing polyacrylamide gel electrophoresis, the cost of consumables is rising and laboratories are, therefore, forced to use other techniques to detect mutations. Mutation-scanning techniques need to detect new mutations within the entire coding region of a gene and there are several methods available to scan for sequence changes in either cellular RNA or genomic DNA. These include denaturing gradient gel electrophoresis (DGGE) ***(3)*** (see Chapter 8), chemical cleavage of mismatch (CCM) ***(4)***, enzyme mismatch cleavage (EMC) ***(5,6)***, single-stranded conformation polymorphism (SSCP) (see Chapter 7) ***(7)***, heteroduplex analysis (HA) ***(8)***, conformation-sensitive gel electrophoresis (CSGE) ***(9)***, the protein truncation test (PTT) ***(10)***; and, more recently, denaturing high-performance liquid chromatography (DHPLC) ***(11,12)*** (see Chapter 24). The most critical factor that determines the success of any gene screening protocol is the sensitivity of the detection technique. The sensitivities of these methods vary greatly depending on the size of DNA/RNA template screened. For example, SSCP has a sensitivity of >95% for fragments of 155 bp, but this is reduced to only 3% for 600 bp ***(13)***. Once optimised, DGGE has a sensitivity of approx 99% for fragments of up to 500 bp ***(14)***, and CSGE has a sensitivity of 90–100% for fragments of up to 450 bp ***(15)***. CCM and EMC ,on the other hand, have sensitivities of 95–100% for fragments >1.5 kb in size ***(16,17)*** and are ideal for screening compact genes where more than one exon can be amplified together using genomic DNA as the template. All of these techniques detect sequence changes such as nonsense, frame shift, splice site, and missense mutations, as well as polymorphisms, however, the PTT screens only for truncating mutations and is predicted to have a sensitivity of >95% and can be used for RNA or DNA fragments in excess of 3 kb.

2. Chemical Cleavage of Mismatch

Several studies have demonstrated that mismatched bases become reactive with some chemical compounds. For example, mismatched guanine and thymine bases react with carbodiimide ***(18)***, mismatched cytosine bases react with hydroxylamine (NH_2OH), and mismatched thymine bases react with osmium tetroxide (OsO_4) ***(4)***. In addition, it has also been shown that

From: *Medical Biomethods Handbook*
Edited by: J. M. Walker and R. Rapley © Humana Press, Inc., Totowa, NJ

potassium permangenate ($KMNO_4$) in association with tetramethylammonium chloride (TMAC) ***(19)***, or potassium permangenate ($KMNO_4$) in association with tetraethylammonium chloride (TEAC) ***(20)*** can react with mismatched thymine bases. These reactivities have subsequently been exploited in the development of two mismatch-detection techniques, namely, the carbodiimide method ***(18)*** and the more popular chemical cleavage of mismatch ***(4)***. Chemical cleavage of mismatch (CCM), also known as chemical mismatch cleavage (CMC), was initially developed and applied to the detection of single-base mismatches in cloned fragments of DNA, but it was subsequently applied to screening genomic DNA and mRNA by the use of polymerase chain reaction (PCR) amplification ***(4,21,22)***. The procedure is based on the detection of mismatches in DNA hybrids called heteroduplexes generated by mixing probe (wild-type) and target (patient) DNA. The normal Watson–Crick basepairing is disrupted at the point of sequence change and the resultant structure is then reacted with chemicals that recognize the specific mismatches. These mismatches are treated with piperidine, which results in the cleavage of the DNA strand at or near the mismatch. The resultant cleavage fragments are subsequently analyzed by electrophoresis through denaturing polyacrylamide gels (*see* **Figs. 1** and **2**). Depending on available laboratory facilities, the probe/target DNA can either be radiolabeled or fluorescently labeled, with the cleavage products being detected using either autoradiography or an automated fluorescent analyzer. Because CCM can detect all types of mismatch, it is ideally suited for screening genes with missense, nonsense, frame-shift, or splice-site mutations. The CCM procedure is technically more involved than other scanning techniques such as SSCP or HA and can include the use of toxic chemicals. However, in contrast to SSCP or HA, CCM can be used to screen relatively large fragments in excess of 1.5 kb and will accurately pinpoint the position of the sequence change.

2.1 Template for PCR

If gene organization is known, then mutation-scanning techniques using patient genomic DNA as the template can be designed. Alternatively, an RNA-based method can be employed using gene transcripts that have been isolated from either specific tissues such as skin or ectopic transcripts from peripheral blood lymphocytes. Where genomic DNA is used as the template, approx 25–35 cycles of PCR will be required. However, when total RNA is used as the template, the relative quantities of the transcript can be very low and perhaps a nested PCR protocol will be required to obtain enough template for the CCM procedure. If differential mRNA expression of each mutant allele occurs, the sensitivity of CCM enables detection of as little as 10% of the mutant transcript ***(23)***.

2.2. Labeling of RNA/DNA

Since its development, CCM ***(4)*** has improved in terms of both sensitivity and safety. The original procedure used DNA probes that had been radioactively end labeled and this strategy was successfully applied to the detection of mutations in patients with the X-linked disorder human ornithine transcarbamoylase deficiency ***(24)***. Although the technique was suggested to detect close to 100% of single-base substitutions ***(25)***, the use of radioactivity with toxic chemicals was far from ideal. Although other nonisotopic methods have been reported ***(26)***, the introduction of fluorescent-based gel technology has greatly improved the safety aspect of the procedure. Fluorescent CCM (FCCM) is also known as fluorescence-assisted mismatch analysis (FAMA). The DNA strand can now either be end labeled using fluorescently labeled primers ***(23)*** or internally labelled using fluorescently labeled dUTPs ***(16)***. Analysis of cleavage products gives a precise indication of the position from either end of where the mismatch is positioned. For example, a product of 1200 bp could yield products of 400 bp and 800 bp. However, the user will not know whether the mutation is 400 bp from the 5' end or 400 bp from the 3' end of the target DNA, and two sets of sequencing reactions will have to be performed to pinpoint the sequence change. However, procedures that utilize different fluorescent labels on the 5' and 3' primers will allow polarization of the target DNA and will result in only one set of

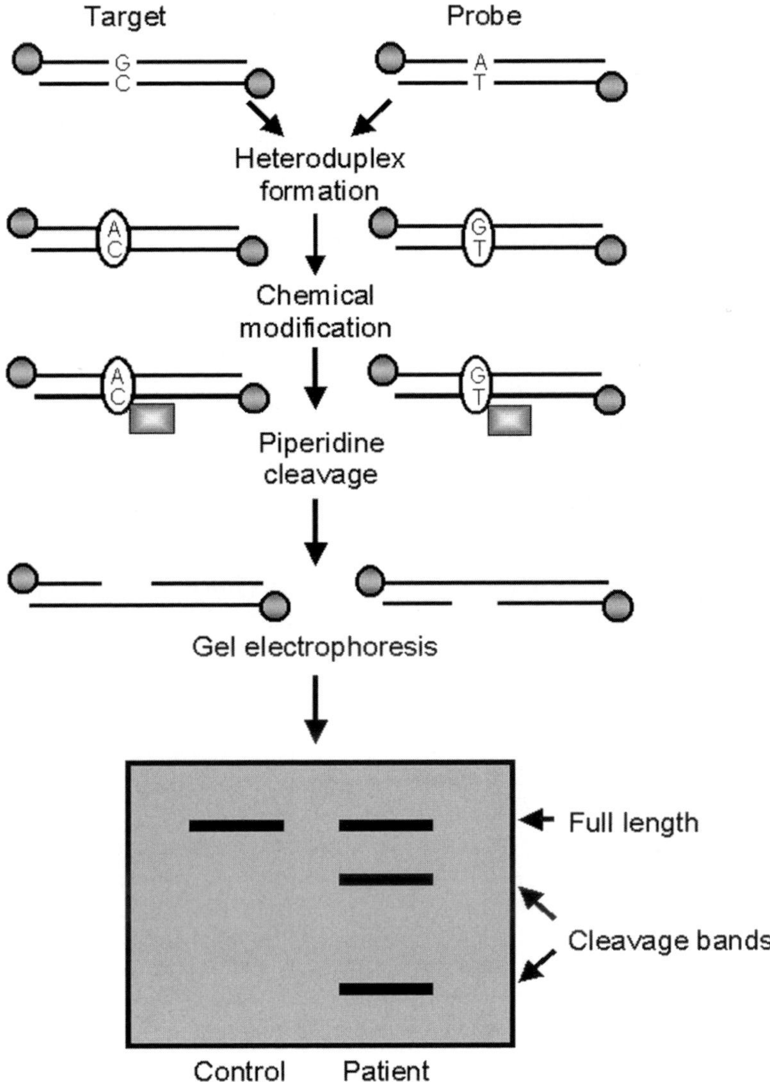

Fig. 1. Principles of the CCM mutation-scanning technique using labeled target and probe. Target and probe DNA is hybridized to form two heteroduplexes (A-C and G-T). The mismatches are chemically modified using either NH_2OH, OsO_4, or $KMNO_4$/TEAC, and the DNA is cleaved at the mismatch site by the use of piperidine. Cleavage products are electrophoresed through a polyacrylamide gel and detected using either autoradiography or fluorescence. Circles represent end labeling, and rectangles represent the chemical modification.

sequencing reactions to pinpoint the mutation. Whether internally labeled or end-labeled procedures are used, both will achieve the same sensitivity of detection. The overall efficiency of CCM depends on whether both the probe and/or target DNA are labeled. Studies have shown that some thymine–guanine mismatches are resistant to modification by osmium tetroxide *(25,27)*. However, the complementary DNA strand heteroduplex will have an adenine–cytosine mismatch that will be modified by hydroxylamine. Therefore, to yield a theoretical 100% detection efficiency, it will be necessary for both probe and target to be labeled.

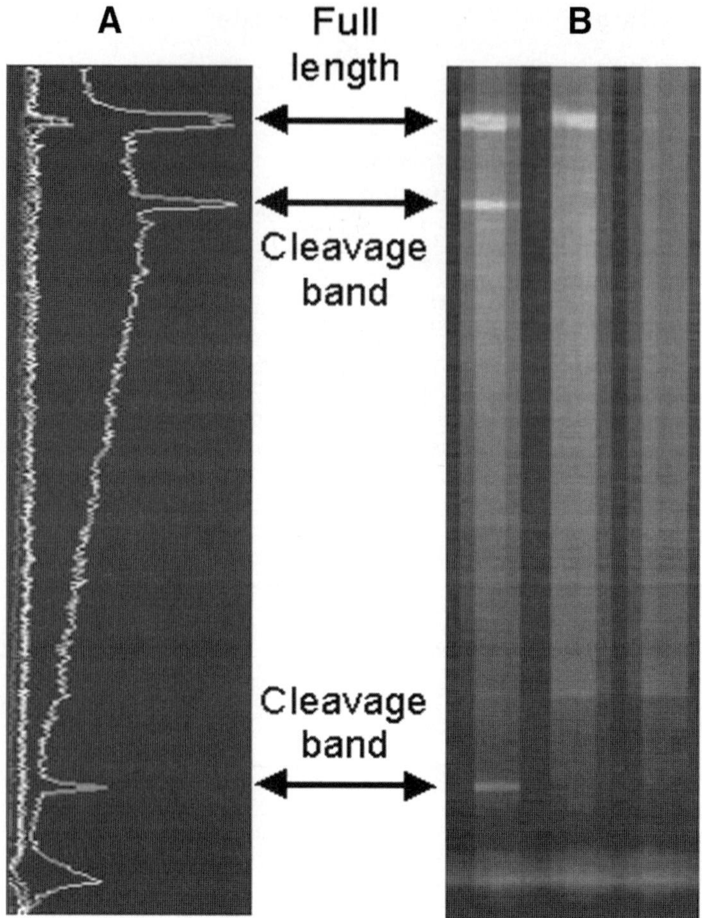

Fig. 2. Fluorescent CCM gel and analysis. **(A)** The electropherogram is generated by analyzing the fluorescence of the individual lanes of a fluorescent polyacrylamide gel. **(B)** Both cleavage products and full-size uncleaved products are arrowed.

2.3. Heteroduplex Formation

The efficient formation of a heteroduplex is critical for chemical modification. However, correct heteroduplex formation can be inhibited by the size and/or the sequence context of the fragment being studied *(28)*. In diseases with autosomal dominant inheritance, only one allele will carry a mutation and, thus, the patients DNA will be heterozygous and can naturally form a heteroduplex. However, in autosomal recessive disorders, the patient DNA might be homozygous and will, therefore, not form a heteroduplex. In this case or in cases where the transmission is not known, then the patient DNA must be mixed with an equal amount of control DNA. Heteroduplex formation is performed as in other techniques by denaturation of the DNA followed by controlled cooling to allow mispairing of the target and probe DNA.

2.4. Chemical Modification

Many studies have utilized hydroxylamine and/or osmium tetroxide protocols, but in the last few years, the osmium tetroxide procedures have been replaced by the use of potassium permanganate/tetraethylamine chloride ***(20,29,30)***. This overcomes the lack of reactivity of some thymine–guanine mismatches ***(25,27)*** as well as avoiding the use of potentially dangerous chemicals. In addition, a combined single-tube method—single tube chemical cleavage of mismatch method—has been developed for the treatment of heteroduplexes by potassium permanganate/tetraethylamine chloride and hydroxylamine without the requirement for intermediate purification steps and has been shown to detect all thymidine and cytidine mismatches ***(29)***.

2.5. Solid-Phase CCM

The original protocol ***(4)*** required several laborious ethanol precipitation steps that severely limited the number of reactions performed in a single day. However, the use of biotinylated primers, in conjunction with streptavidin-coated paramagnetic beads to yield a solid-phase CCM protocol, allows simple and effective washing of the cleavage products, permitting high-throughput screening using microtiter plate technology ***(16)***. Using excess target DNA favors maximum heteroduplex formation between the probe and target. Any excess target homoduplexes formed are removed in the supernatant after binding to streptavidin beads because they are nonbiotinylated. Very recently, a further development of the solid-phase CCM protocol has occurred whereby the DNA is immobilized on solid silica supports. This newer version of solid-phase CCM is relatively fast, cost-effective, and sensitive for detection of mismatches and mutations ***(31,32)***.

2.6. Mismatch Detection

As in the original protocols, radioactively end-labeled cleavage products can be separated using standard denaturing polyacrylamide gel electrophoresis (PAGE) apparatus and detected using standard X-ray film. However, if fluorescent technology is available, then PAGE can be performed on apparatus such as an Applied Biosystems ABI373 or ABI377, and the cleavage products can be accurately sized by the software programs available. In line with the switch from gel electrophoresis to capillary electrophoresis, the CCM procedure has now been applied to the latter technique that will allow greater throughput on capillary-based detection platforms ***(33)***.

3. Clinical Applications

Fluorescent CCM has been applied to the screening of many diseases, including Wilms' tumor patients (*WT1*) ***(34)***, Fanconi anemia (*FANCA* and *FANCAC*) ***(35,36)***, Fabry disease (*α -galactosidase A*) ***(37)***, hemophilia A (*factor VIII*) ***(38,39)***, hereditary breast and ovarian cancer (*BRCA1*) ***(40)***, and has also been applied in the detection of rhodopsin mutations in the prenatal diagnosis of retinitis pigmentosa ***(41)***. In addition, CCM has been applied to detecting different isolates of the dengue virus type 2, indicating its use in epidemiological studies ***(42)*** and in the detection of mutations associated with rifampin resistance of *M. tuberculosis* ***(43)***. Here, we give examples of the use of CCM in the diagnosis of two disorders. The first uses a genomic PCR template strategy in the detection of mutations in dystrophic epidermolysis bullosa ***(44)***, and the second uses an RNA nested PCR template strategy in the detection of mutations in ataxia telangiectasia ***(45)***.

3.1. Clinical Application I. Mutation Detection in Dystrophic Epidermolysis Bullosa

Epidermolysis bullosa (EB) is a term given to a group of mechanobullous disorders characterized by fragility of the skin and mucous membranes following mechanical trauma. There are over 20 types of EB, but they can be grouped into three main subtypes based on the level of cleavage within the skin ***(46)***, namely, EB simplex, junctional EB, or dystrophic EB. Dystrophic EB (DEB) can be inherited either as a severe autosomal recessive or milder autosomal dominant disorder, both of which lead to scarring of affected tissues. In the severe recessive Hallopeau–Siemens form (MIM 226600), blistering occurs over areas subjected to everyday trauma such as the hands, feet, elbows, and knees, as well as erosions and scarring of the mouth, oropharynx, supraglottic structures, vocal chords, and trachea. Autosomal dominant forms of DEB include pretibial (MIM 131850), Pasini (MIM 131750), Bart's syndrome ***(47)*** (MIM

132000), and transient bullous dermolysis of the newborn (TBDN) (MIM 131705), all of which have a much milder phenotype than recessive DEB. A distinct form of DEB, EB pruriginosa (MIM 604129), which can be inherited in either an autosomal dominant or recessive manner ***(48–50)***, is characterized by marked itching and presence of prurigo-like or lichenoid features, with scarring on the limbs ***(51,52)***.

The gene mutated in all forms of DEB, *COL7A1* located on 3p21.1, comprises 118 exons, spanning over 32 kb ***(53)***. The 9-kb transcript contains an open reading frame of 8832 bp, encoding a protein of 2944 amino acids ***(53–55)***. The type VII collagen protein consists of a central collagenous domain consisting of gly-X-Y repeats and noncollagenous domains at the amino (NC-1) and carboxyl (NC-2) ends. The NC-1 domain consists of subdomains with homology to known adhesive proteins, including cartilidge matrix protein, nine fibronectin type III-like repeats (FNIII), and the A domain of von Willebrand factor ***(53,54,56,57)***. In vivo, three pro-α molecules form parallel homotrimers [α1(VII)], which then form antiparallel dimers via disulfide bridges ***(58)*** with the subsequent removal of part of the NC-2 domains ***(59,60)***. These antiparallel dimers further aggregate laterally to form anchoring fibrils ***(58,61)***. To date, over 200 different *COL7A1* lesions have been described that consist of missense, nonsense, splice-site, or frame-shift mutations. Most cases of dominant DEB are caused by missense mutations within the collagenous domain and consist mainly of glycine substitutions that cause destabilization of the gly-X-Y repeats necessary for the integrity of the triple helix. With a few exceptions, the severe autosomal recessive forms of DEB are caused by truncating mutations (i.e., nonsense, frame-shift, and splice-site mutations) on both alleles either in a homozygous or, more commonly, a compound heterozygous state. These mutations should lead to type VII collagen molecules lacking certain domains that are necessary for anchoring fibril formation. However, before translation can occur in vivo, the majority of the transcripts are degraded via nonsense-mediated mRNA decay ***(62)*** and hence no anchoring fibrils are formed. The intermediate forms of recessive DEB are caused by a range of different types of *COL7A1* mutations, including a truncating mutation on one allele and a missense mutation on the other allele, splice-site mutations, or a delayed termination codon. In addition, not all glycine substitutions result in dominant negative interference: some might be inherited recessively and only contribute to a DEB phenotype when inherited *in trans* with a *COL7A1* mutation on the other allele ***(63)***. In the past, *COL7A1*-mutation-screening studies have used either DNA or RNA as the template and have used detection techniques such as SSCP ***(64,65)***, direct sequencing ***(66,67)***, DGGE ***(68,69)***, HA ***(70)*** or CSGE ***(71–73)***. More recently, a protocol has been published ***(74)*** that permits screening of the whole *COL7A1* coding region and its splice sites via the amplification of all 118 exons from genomic DNA using 72 separate PCR assays, followed by scanning of sequence changes using CSGE. Despite the relative technical simplicity of this method, the considerable number of PCR assays required reduces the practicability of using this mutation detection strategy for a large number of samples. Therefore, because of its compactness, the *COL7A1* gene is an ideal candidate for a glenomic PCR-based CCM strategy. Although an RNA-based strategy using ectopic blood lymphocyte total RNA would further reduce the number of assays required to screen the coding region, this was rejected because of alternately spliced *COL7A1* transcripts found in lymphocyte RNA ***(44)***. A solid-phase fluorescent CCM procedure was developed that used 21 individual PCR assays ranging between 620 bp and 1604 bp in size ***(44)*** (*see* **Fig. 3**). These fragments encompass all exons, splice sites, and > 80% of introns. Only probe DNA was internally labeled using fluorescently nucleotides. In a comparative study on 91 known mutations, the FCCM procedure detected 81% of *COL7A1* mutations, whereas CSGE detected 75%, and in a blind study FCCM detected 90% of the 21 predicted mutations ***(44)***. The lack of 100% detection of mutations can be explained by the fact that target DNA was not labeled. This would mean that some thymine–guanine mismatches would remain unmodified and subsequently not cleaved ***(25,27)***. Although the FCCM protocol might involve a few more stages than CSGE, this is easily outweighed by its advantages. The number of primers required for the FCCM PCR assays is only 84 (42 standard and 42 biotinylated), compared with 144 for CSGE. The fluorescent analyzer software can locate the position of the

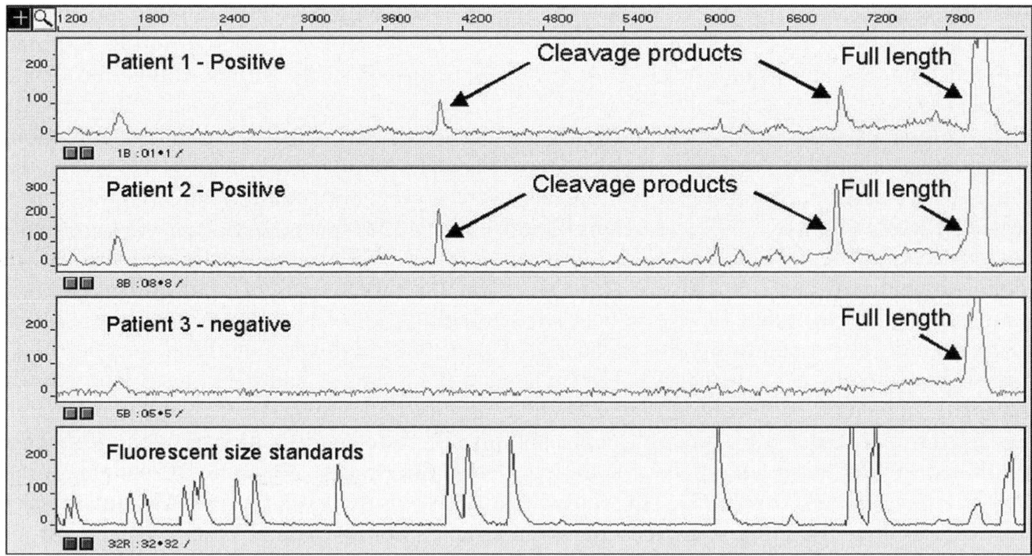

Fig. 3. Fluorescent CCM analysis of *COL7A1* demonstrating the mutation 7786delG. These electropherograms are generated by the Genescan software (ABI) of fragment 19 (exons 103–108) for two patients heterozygous for the common British mutation 7786delG (lanes 1 and 2) and a negative control lane 3. Lane 4 contains the size standards. Cleavage peaks can be seen for the 7786delG mutation (458 bp + 1082 bp) in lanes 1 and 2, but are absent for the negative control. The full-size uncleaved DNA peak is 1540 bp.

sequence change compared with CSGE, which does not give any indication of where the change is within the fragment. This permits the sequencing of a very small region of DNA to allow for the characterization of the mutation. Based on running three fluorescent gels per day, this FCCM protocol can be used to screen the entire *COL7A1* gene for mutations in six patients per day. However, if the three different fluorescent dUTPs (i.e., blue, green, and yellow) are used to create a multiplex in each well of the gel, this permits the use of an exceedingly high-throughput screening procedure in which the entire *COL7A1* gene can be screened for 18 patients per day (540 kb).

3.2. Clinical Application II: Mutation Screening in Ataxia Telangiectasia

Ataxia telangiectasia (A-T) is an autosomal recessive disease affecting between 1/300,000 and 1/40,000 individuals worldwide ***(75,76)*** and is characterized by progressive cerebellar degeneration, oculocutaneous telangiectasia (red vascular malformations of the skin and eyes), oculomotor apraxia (incoordination of eye movements), immunodeficiency, raised serum α-fetoprotein (AFP), chromosomal instability, radiosensitivity, and a 100-fold increase in cancer susceptibility, with lymphoreticular malignancies predominating ***(77)***. The A-T gene, *ATM* (ataxia telangiectasia mutated), located on chromosome 11q22-23 ***(78)*** comprises 66 exons, spanning over 150 kb, with the start codon in exon 4 ***(79,80)***. The 13-kb transcript contains an open reading frame of 9168 bp, encoding a protein of 3056 amino acids ***(81)*** that is principally located within the nucleus of normal cells ***(82)*** and demonstrates homology at its carboxy terminus to phosphatidylinositol (PI) 3-kinase. To date, over 300 *ATM* mutations have been reported that span the entire coding region ***(83)*** with no mutational hotspots. The majority of mutation studies have employed scanning methods using cDNA to identify *ATM* mutations, including PTT ***(84)***, SSCP ***(85)***, RNAse cleavage ***(86)***, and restriction endonuclease fingerprinting (REF) ***(87)***. However, some studies have used direct sequencing of cDNA ***(88)***. The overall mutation detection rate has been high (approx 70%), with base substitutions, small deletions, and insertions that result in truncation of the *ATM* protein ***(83)***, in addition to in-frame deletions and missense mutations, which are predicted to affect the structure/function of the protein ***(89)***.

Therefore, because solid-phase FCCM offers efficient and sensitive mutation detection, without bias toward any particular mutation type, this methodology was employed to screen the entire *ATM* transcript. Ectopic *ATM* transcripts were used as the template for RT-PCR, and a nested PCR strategy was employed to amplify the entire *ATM* open reading frame in eight first-round reactions, themselves serving as templates for nine overlapping reactions in the second round ranging between 0.9 and 1.7 kb in size. For the target (patient) and probe (control) samples, the second-round PCR products were internally labelled with the incorporation of fluorescent dUTPs. In addition, the probe cDNA was amplified using modified second round primer sets whereby each primer was biotinylated at the 5' end. A sample of each biotinylated internally labeled probe was combined with internally labeled target. Here, chemical mismatch was performed using hydroxylamine and osmium tetroxide modification sequentially. This optimized FCCM technique allows the entire *ATM* coding region for up to 10 patients to be analyzed on each FCCM gel ***(45)***. This study identified 7 out of a possible 10 *ATM* mutations in five A-T families, giving a sensitivity of 70%, which is comparable to other *ATM*-mutation-detection methods used [e.g., PTT ***(84)*** and REF ***(87)***]. However, assuming all A-T individuals are compound heterozygotes, three *ATM* mutant alleles were not detected, despite extending the scanning method to include the *ATM/E14* bidirectional promoter region. The undetected *ATM* mutations can, therefore, be regulatory mutations in remote promoter elements, as reported for other genes (e.g., *BRCA1* ***[90]***) or can destabilize the mRNA so that only a single copy of the transcript is amplified ***(91)***. Other possible mutation mechanisms include gross deletions/insertions of the gene, RNA editing, position effect variegation ***(92)***, or gene modifiers altering expression or function of *ATM* ***(93)***.

4. Summary

Since its development, chemical cleavage of mismatch has proved to be a very efficient method for the identification of sequence changes in DNA and RNA. Throughout its usage in different studies and its optimization to solid-phase fluorescent platforms, it has demonstrated its value to the molecular biology community and, to date, remains the fastest, most accurate, and most sensitive fluorescent mutation-scanning technique available.

References

1. Sanger, F., Nicklen, S., and Coulson, A. R. (1977) DNA sequencing with chain-terminating inhibitors. *Proc. Natl. Acad. Sci. USA* **74**, 5463–5467.
2. Maxam, A. M. and Gilbert, W. (1977) A new method for sequencing DNA. *Proc. Natl. Acad. Sci. USA* **74**, 560–564.
3. Myers, R. M., Maniatis, T., and Lerman, L. S. (1987) Detection and localization of single base changes by denaturing gradient gel electrophoresis. *Methods Enzymol.* **155**, 501–527.
4. Cotton, R. G., Rodrigues, N. R., and Campbell, R. D. (1988) Reactivity of cytosine and thymine in single-base-pair mismatches with hydroxylamine and osmium tetroxide and its application to the study of mutations. *Proc. Natl. Acad. Sci. USA* **85**, 4397–4401.
5. Mashal, R. D., Koontz, J., and Sklar, J. (1995) Detection of mutations by cleavage of DNA heteroduplexes with bacteriophage resolvases. *Nature Genet.* **9**, 177–183.
6. Youil, R., Kemper, B. W., and Cotton, R. G. (1995) Screening for mutations by enzyme mismatch cleavage with T4 endonuclease VII. *Proc. Natl. Acad. Sci. USA* **92**, 87–91.
7. Orita, M., Iwahana, H., Kanazawa, H., Hayashi, K., and Sekiya, T. (1989) Detection of polymorphisms of human DNA by gel electrophoresis as single-strand conformation polymorphisms. *Proc. Natl. Acad. Sci. USA* **86**, 2766–2770.
8. White, M. B., Carvalho, M., Derse, D., O'Brien, S. J., and Dean, M. (1992) Detecting single base substitutions as heteroduplex polymorphisms. *Genomics* **12**, 301–306.

9. Ganguly, A., Rock, M. J., and Prockop, D. J. (1993) Conformation-sensitive gel electrophoresis for rapid detection of single-base differences in double-stranded PCR products and DNA fragments: evidence for solvent-induced bends in DNA heteroduplexes. *Proc. Natl. Acad. Sci. USA* **90**, 10,325–10,329.
10. Roest, P. A., Roberts, R. G., Sugino, S., van Ommen, G. J., and den Dunnen, J. T. (1993) Protein truncation test (PTT) for rapid detection of translation-terminating mutations. *Hum. Mol. Genet.* **2**, 1719–1721.
11. Oefner, P. J. and Underhill, P. A. (1998) DNA mutation detection using denaturing high-performance liquid chromatography (DHPLC). *Curr. Protocols Hum. Genet.* **19(Suppl)**, 7101–7102.
12. Oefner, P. J. and Underhill, P. A. (1995) Comparative DNA sequence by denaturing high performance liquid chromatography (DHPLC). *Am. J. Hum. Genet.* **57**, A266.
13. Sheffield, V. C., Beck, J. S., Kwitek, A. E., Sandstrom, D. W., and Stone, E. M. (1993) The sensitivity of single-strand conformation polymorphism analysis for the detection of single base substitutions. *Genomics* **16**, 325–332.
14. Fodde, R. and Losekoot, M. (1994) Mutation detection by denaturing gradient gel electrophoresis (DGGE). *Hum. Mutat.* **3**, 83–94.
15. Korkko, J., Annunen, S., Pihlajamaa, T., Prockop, D. J., and Ala-Kokko, L. (1998) Conformation sensitive gel electrophoresis for simple and accurate detection of mutations: comparison with denaturing gradient gel electrophoresis and nucleotide sequencing. *Proc. Natl. Acad. Sci. USA* **95**, 1681–1685.
16. Rowley, G., Saad, S., Giannelli, F., and Green, P. M. (1995) Ultrarapid mutation detection by multiplex, solid-phase chemical cleavage. *Genomics* **30**, 574–582.
17. Babon, J. J., McKenzie, M., and Cotton, R. G. (2003) The use of resolvases T4 endonuclease VII and T7 endonuclease I in mutation detection. *Mol. Biotechnol.* **23**, 73–81.
18. Novack, D. F., Casna, N. J., Fischer, S. G., and Ford, J. P. (1986) Detection of single base-pair mismatches in DNA by chemical modification followed by electrophoresis in 15% polyacrylamide gel. *Proc. Natl. Acad. Sci. USA* **83**, 586–590.
19. Gogos, J. A., Karayiorgou, M., Aburatani, H., and Kafatos, F. C. (1990) Detection of single base mismatches of thymine and cytosine residues by potassium permanganate and hydroxylamine in the presence of tetralkylammonium salts. *Nucleic Acids Res.* **18**, 6807–6814.
20. Roberts, E., Deeble, V. J., Woods, C. G., and Taylor, G. R. (1997) Potassium permanganate and tetraethylammonium chloride are a safe and effective substitute for osmium tetroxide in solid-phase fluorescent chemical cleavage of mismatch. *Nucleic Acids Res.* **25**, 3377–3378.
21. Montandon, A. J., Green, P. M., Giannelli, F., and Bentley, D. R. (1989) Direct detection of point mutations by mismatch analysis: application to haemophilia B. *Nucleic Acids Res.* **17**, 3347–3358.
22. Dahl, H. H., Lamande, S. R., Cotton, R. G., and Bateman, J. F. (1989) Detection and localization of base changes in RNA using a chemical cleavage method. *Anal. Biochem.* **183**, 263–268.
23. Verpy, E., Biasotto, M., Meo, T., and Tosi, M. (1994) Efficient detection of point mutations on color-coded strands of target DNA. *Proc. Natl. Acad. Sci. USA* **91**, 1873–1877.
24. Grompe, M., Muzny, D. M., and Caskey, C. T. (1989) Scanning detection of mutations in human ornithine transcarbamoylase by chemical mismatch cleavage. *Proc. Natl. Acad. Sci. USA* **86**, 5888–5892.
25. Forrest, S. M., Dahl, H. H., Howells, D. W., Dianzani, I., and Cotton, R. G. (1991) Mutation detection in phenylketonuria by using chemical cleavage of mismatch: importance of using probes from both normal and patient samples. *Am. J. Hum. Genet.* **49**, 175–183.
26. Saleeba, J. A. and Cotton, R. G. (1993) Chemical cleavage of mismatch to detect mutations. *Methods Enzymol.* **217**, 286–295.
27. Cotton, R. G. (1993) Current methods of mutation detection. *Mutat. Res.* **285**, 125–144.
28. Smith, M. J., Humphrey, K. E., Cappai, R., Beyreuther, K., Masters, C. L., and Cotton, R. G. (2000) Correct heteroduplex formation for mutation detection analysis. *Mol. Diagn.* **5**, 67–73.
29. Lambrinakos, A., Humphrey, K. E., Babon, J. J., Ellis, T. P., and Cotton, R. G. (1999) Reactivity of potassium permanganate and tetraethylammonium chloride with mismatched bases and a simple mutation detection protocol. *Nucleic Acids Res.* **27**, 1866–1874.
30. Deeble, V. J., Roberts, E., Robinson, M. D., Woods, C. G., Bishop, D. T., and Taylor, G. R. (1997) Comparison of enzyme mismatch cleavage and chemical cleavage of mismatch on a defined set of heteroduplexes. *Genet Test.* **1**, 253–259.
31. Bui, C. T., Lambrinakos, A., Babon, J. J., and Cotton, R. G. (2003) Chemical cleavage reactions of DNA on solid support: application in mutation detection. *BMC Chem. Biol.* **3**, 1.

32. Bui, C. T., Babon, J. J., Lambrinakos, A., and Cotton, R. G. (2003) Detection of mutations in DNA by solid-phase chemical cleavage method. A simplified assay. *Methods Mol. Biol.* **212**, 59–70.
33. Ren, J. (2001) Chemical mismatch cleavage analysis by capillary electrophoresis with laser-induced fluorescence detection. *Methods Mol. Biol.* **163**, 231–239.
34. Little, M. H., Prosser, J., Condie, A., Smith, P. J., Van Heyningen, V., and Hastie, N. D. (1992) Zinc finger point mutations within the WT1 gene in Wilms tumor patients. *Proc. Natl. Acad. Sci. USA* **89**, 4791–4795.
35. Gibson, R. A., Morgan, N. V., Goldstein, L. H., et al. (1996) Novel mutations and polymorphisms in the Fanconi anemia group C gene. *Hum. Mutat.* **8**, 140–148.
36. Tipping, A. J., Pearson, T., Morgan, N. V., et al. (2001) Molecular and genealogical evidence for a founder effect in Fanconi anemia families of the Afrikaner population of South Africa. *Proc. Natl. Acad. Sci. USA* **98**, 5734–5739.
37. Germain, D., Biasotto, M., Tosi, M., Meo, T., Kahn, A., and Poenaru, L. (1996) Fluorescence-assisted mismatch analysis (FAMA) for exhaustive screening of the alpha-galactosidase A gene and detection of carriers in Fabry disease. *Hum. Genet.* **98**, 719–726.
38. Schwaab, R., Oldenburg, J., Lalloz, M. R., et al. (1997) Factor VIII gene mutations found by a comparative study of SSCP, DGGE and CMC and their analysis on a molecular model of factor VIII protein. *Hum. Genet.* **101**, 323–332.
39. Waseem, N. H., Bagnall, R., Green, P. M., and Giannelli, F. (1999) Start of UK confidential haemophilia A database: analysis of 142 patients by solid phase fluorescent chemical cleavage of mismatch. Haemophilia Centres. *Thromb. Haemost.* **81**, 900–905.
40. Greenman, J., Mohammed, S., Ellis, D., et al. (1998) Identification of missense and truncating mutations in the BRCA1 gene in sporadic and familial breast and ovarian cancer. *Genes Chromosomes Cancer.* **21**, 244–249.
41. Tessitore, A., Toniato, E., Gulino, A., et al. (2002) Prenatal diagnosis of a rhodopsin mutation using chemical cleavage of the mismatch. *Prenat. Diagn.* **22**, 380–384.
42. Lin, B., Cotton, R. G., Trent, D. W., and Wright, P. J. (1992) Geographical clusters of dengue virus type 2 isolates based on analysis of infected cell RNA by the chemical cleavage at mismatch method. *J. Virol. Methods.* **40**, 205–218.
43. Bahrmand, A. R., Marashi, S. M., Bakayeva, T. G., and Bakayev, V. V. (2000) Chemical cleavage of mismatches in heteroduplexes of the rpoB gene for detection of mutations associated with resistance of Mycobacterium tuberculosis to rifampin. *Scand. J. Infect. Dis.* **32**, 395–398.
44. Whittock, N. V., Ashton, G. H., Mohammedi, R., Mellerio, J. E., et al. (1999) Comparative mutation detection screening of the type VII collagen gene (COL7A1) using the protein truncation test, fluorescent chemical cleavage of mismatch, and conformation sensitive gel electrophoresis. *J. Invest. Dermatol.* **113**, 673–686.
45. Izatt, L., Vessey, C., Hodgson, S. V., and Solomon, E. (1999) Rapid and efficient ATM mutation detection by fluorescent chemical cleavage of mismatch: identification of four novel mutations. *Eur. J. Hum. Genet.* **7**, 310–320.
46. Fine, J. D., Eady, R. A., Bauer, E. A., et al. (2000) Revised classification system for inherited epidermolysis bullosa: report of the Second International Consensus Meeting on diagnosis and classification of epidermolysis bullosa. *J. Am. Acad. Dermatol.* **42**, 1051–1066.
47. Bart, B. J., Gorlin, R. J., Anderson, V. E., and Lynch, F. W. (1966) Congenital localized absence of skin and associated abnormalities resembling epidermolysis bullosa. A new syndrome. *Arch. Dermatol.* **93**, 296–304.
48. Lee, J. Y., Chen, H. C., and Lin, S. J. (1993) Pretibial epidermolysis bullosa: a clinicopathologic study. *J. Am. Acad. Dermatol.* **29**, 974–981.
49. Lee, J. Y., Pulkkinen L., Liu, H. S., Chen, Y. F., and Uitto, J. (1997) A glycine-to-arginine substitution in the triple-helical domain of type VII collagen in a family with dominant dystrophic epidermolysis bullosa pruriginosa. *J. Invest. Dermatol.* **108**, 947–949.
50. Mellerio, J. E., Ashton, G. H., Mohammedi, R., et al. (1999) Allelic heterogeneity of dominant and recessive COL7A1 mutations underlying epidermolysis bullosa pruriginosa. *J. Invest. Dermatol.* **112**, 984–987.
51. McGrath J. A., Schofield O. M., and Eady R. A. (1994) Epidermolysis bullosa pruriginosa: dystrophic epidermolysis bullosa with distinctive clinicopathological features. *Br. J. Dermatol.* **130**, 617–625.
52. Cambiaghi, S., Brusasco, A., Restano, L., Cavalli, R., and Tadini G. (1997) Epidermolysis bullosa pruriginosa. *Dermatology* **195**, 65–68.

53. Christiano, A. M., Hoffman, G. G., Chung-Honet, L. C., et al. (1994) Structural organization of the human type VII collagen gene (COL7A1), composed of more exons than any previously characterized gene. *Genomics* **21**, 169–179.
54. Parente, M. G., Chung, L. C., Ryynanen, J., et al. (1991) Human type VII collagen: cDNA cloning and chromosomal mapping of the gene. *Proc. Natl. Acad. Sci. USA* **88**, 6931–6935.
55. Christiano, A. M., Greenspan, D. S., Lee, S., and Uitto, J. (1994) Cloning of the human type VII collagen. Complete primary sequence of the alpha1(VII) chain and identification of intragenic polymorphisms. *J. Biol. Chem.* **269**, 20,256–20,262.
56. Christiano, A. M., Rosenbaum, L. M., Chung-Honet, L. C., et al. (1992) The large non-collagenous domain (NC-1) of type VII collagen is amino-terminal and chimeric. Homology to cartilage matrix protein, the type III domains of fibronectin and the A domains of von Willebrand factor. *Hum. Mol. Genet.* **1**, 475–481.
57. Gammon, W. R., Abernethy, M. L., Padilla, K. M., et al. (1992) Noncollagenous (NC1) domain of collagen VII resembles multidomain adhesion proteins involved in tissue-specific organization of extracellular matrix. *J. Invest. Dermatol.* **99**, 691–696.
58. Bachinger, H. P., Morris, N. P., Lunstrum, G. P., et al. (1990) The relationship of the biophysical and biochemical characteristics of type VII collagen to the function of anchoring fibrils. *J. Biol. Chem.* **265**, 10,095–10,101.
59. Bruckner-Tuderman, L., Nilssen, O., Zimmermann, D. R., et al. (1995) Immunohistochemical and mutation analyses demonstrate that procollagen VII is processed to collagen VII through removal of the NC-2 domain. *J. Cell. Biol.* **131**, 551–559.
60. Lunstrum, G. P., Kuo, H. J., Rosenbaum, L. M., et al. (1987) Anchoring fibrils contain the carboxyl-terminal globular domain of type VII procollagen, but lack the amino-terminal globular domain. *J. Biol. Chem.* **262**, 13,706–13,712.
61. Burgeson, R. E. (1993) Type VII collagen, anchoring fibrils, and epidermolysis bullosa. *J. Invest. Dermatol.* **101**, 252–255.
62. Christiano, A. M., Amano, S., Eichenfield, L. F., Burgeson, R. E., and Uitto, J. (1997) Premature termination codon mutations in the type VII collagen gene in recessive dystrophic epidermolysis bullosa result in nonsense-mediated mRNA decay and absence of functional protein. *J. Invest. Dermatol.* **109**, 390–394.
63. Christiano, A. M., McGrath, J. A., and Uitto, J. (1996) Influence of the second COL7A1 mutation in determining the phenotypic severity of recessive dystrophic epidermolysis bullosa. *J. Invest. Dermatol.* **106**, 766–770.
64. Christiano, A. M., Greenspan, D. S., Hoffman, G. G., et al. (1993) A missense mutation in type VII collagen in two affected siblings with recessive dystrophic epidermolysis bullosa. *Nature Genet.* **4**, 62–66.
65. Dunnill, M. G., McGrath, J. A., Richards, A. J., et al. (1996) Clinicopathological correlations of compound heterozygous COL7A1 mutations in recessive dystrophic epidermolysis bullosa. *J. Invest. Dermatol.* **107**, 171–177.
66. Gardella, R., Belletti, L., Zoppi, N., Marini, D., Barlati, S., and Colombi, M. (1996) Identification of two splicing mutations in the collagen type VII gene (COL7A1) of a patient affected by the localisata variant of recessive dystrophic epidermolysis bullosa. *Am. J. Hum. Genet.* **59**, 292–300.
67. Winberg, J. O., Hammami-Hauasli, N., Nilssen, O., et al. (1997) Modulation of disease severity of dystrophic epidermolysis bullosa by a splice site mutation in combination with a missense mutation in the COL7A1 gene. *Hum. Mol. Genet.* **6**, 1125–1135.
68. Hovnanian, A., Hilal, L., Blanchet-Bardon, C., et al. (1994) Recurrent nonsense mutations within the type VII collagen gene in patients with severe recessive dystrophic epidermolysis bullosa. *Am. J. Hum. Genet.* **55**, 289–296.
69. Hilal, L., Rochat, A., Duquesnoy, P., et al. (1993) A homozygous insertion–deletion in the type VII collagen gene (COL7A1) in Hallopeau-Siemens dystrophic epidermolysis bullosa. *Nature Genet.* **5**, 287–293.
70. Christiano, A. M., D'Alessio, M., Paradisi, M., et al. (1996) A common insertion mutation in COL7A1 in two Italian families with recessive dystrophic epidermolysis bullosa. *J. Invest. Dermatol.* **106**, 679–684.
71. Christiano, A. M., LaForgia, S., Paller, A. S., McGuire, J., Shimizu, H., and Uitto, J. (1996) Prenatal diagnosis for recessive dystrophic epidermolysis bullosa in 10 families by mutation and haplotype analysis in the type VII collagen gene (COL7A1). *Mol. Med.* **2**, 59–76.

72. Hammami-Hauasli, N., Kalinke, D. U., Schumann, H., et al. (1997) A combination of a common splice site mutation and a frameshift mutation in the COL7A1 gene: absence of functional collagen VII in keratinocytes and skin. *J. Invest. Dermatol.* **109**, 384–389.
73. Mellerio, J. E., Dunnill, M. G., Allison, W., et al. A. (1997) Recurrent mutations in the type VII collagen gene (COL7A1) in patients with recessive dystrophic epidermolysis bullosa. *J. Invest. Dermatol.* **109**, 246–249.
74. Christiano, A. M., Hoffman, G. G., Zhang, X., et al. (1997) Strategy for identification of sequence variants in COL7A1 and a novel 2-bp deletion mutation in recessive dystrophic epidermolysis bullosa. *Hum. Mutat.* **10**, 408–414.
75. Swift, M., Morrell, D., Cromartie, E., Chamberlin, A. R., Skolnick, M. H., and Bishop, D. T. (1986) The incidence and gene frequency of ataxia-telangiectasia in the United States. *Am. J. Hum. Genet.* **39**, 573–583.
76. Woods, C. G., Bundey, S. E., and Taylor, A. M. (1990) Unusual features in the inheritance of ataxia telangiectasia. *Hum. Genet.* **84**, 555–562.
77. Lavin, M. F. and Shiloh, Y. (1997) The genetic defect in ataxia-telangiectasia. *Annu. Rev. Immunol.* **15**, 177–202.
78. Gatti, R. A., Berkel, I., Boder, E., et al. (1988) Localization of an ataxia-telangiectasia gene to chromosome 11q22-23. *Nature* **336**, 577–580.
79. Uziel, T., Savitsky, K., Platzer, M., et al. (1996) Genomic organization of the ATM gene. *Genomics* **33**, 317–320.
80. Platzer, M., Rotman, G., Bauer, D., et al. (1997) Ataxia-telangiectasia locus: sequence analysis of 184 kb of human genomic DNA containing the entire ATM gene. *Genome Res.* **7**, 592–605.
81. Savitsky, K., Sfez, S., Tagle, D. A., et al. (1995) The complete sequence of the coding region of the ATM gene reveals similarity to cell cycle regulators in different species. *Hum. Mol. Genet.* **4**, 2025–2032.
82. Lakin, N. D., Weber, P., Stankovic, T., Rottinghaus, S. T., Taylor, A. M., and Jackson, S. P. (1996) Analysis of the ATM protein in wild-type and ataxia telangiectasia cells. *Oncogene* **13**, 2707–2716.
83. Concannon, P. and Gatti, R. A. (1997) Diversity of ATM gene mutations detected in patients with ataxia- telangiectasia. *Hum. Mutat.* **10**, 100–107.
84. Telatar, M., Wang, Z., Udar, N., et al. (1996) Ataxia-telangiectasia: mutations in ATM cDNA detected by protein-truncation screening. *Am. J. Hum. Genet.* **59**, 40–44.
85. Wright, J., Teraoka, S., Onengut, S., et al. (1996) A high frequency of distinct ATM gene mutations in ataxia-telangiectasia. *Am. J. Hum. Genet.* **59**, 839–846.
86. Toyoshima, M., Hara, T., Zhang, H., et al. (1998) Ataxia-telangiectasia without immunodeficiency: novel point mutations within and adjacent to the phosphatidylinositol 3-kinase-like domain. *Am. J. Med. Genet.* **75**, 141–144.
87. Gilad, S., Khosravi, R., Harnik, R., et al. (1998) Identification of ATM mutations using extended RT-PCR and restriction endonuclease fingerprinting, and elucidation of the repertoire of A-T mutations in Israel. *Hum. Mutat.* **11**, 69–75.
88. van Belzen, M. J., Hiel, J. A., Weemaes, C. M., et al. (1998) A double missense mutation in the ATM gene of a Dutch family with ataxia telangiectasia. *Hum. Genet.* **102**, 187–191.
89. Stankovic, T., Kidd, A. M., Sutcliffe, A., et al. (1998) ATM mutations and phenotypes in ataxia-telangiectasia families in the British Isles: expression of mutant ATM and the risk of leukemia, lymphoma, and breast cancer. *Am. J. Hum. Genet.* **62**, 334–345.
90. Xu, C. F., Chambers, J. A., Nicolai, H., et a;. (1997) Mutations and alternative splicing of the BRCA1 gene in UK breast/ovarian cancer families. *Genes. Chromosomes Cancer* **18**, 102–110.
91. Brown, M. A. (1997) Tumor suppressor genes and human cancer. *Adv. Genet.* **36**, 45–135.
92. Vorechovsky, I., Luo, L., Lindblom, A., et al. (1996) ATM mutations in cancer families. *Cancer Res.* **56**, 4130–4133.
93. Roy, N., Laflamme, G., and Raymond, V. (1992) 5' untranslated sequences modulate rapid mRNA degradation mediated by 3' AU-rich element in v-/c-fos recombinants. *Nucleic Acids Res.* **20**, 5753–5762.

17

The Protein Truncation Test

Johan T. den Dunnen

1. Introduction

Many techniques are available to efficiently pick up (point) mutations, all with their own strong and weak points and limitations. The most powerful qualitative techniques include denaturing gradient gel electrophoresis (DGGE) (see Chapter 8), chemical cleavage of mismatches (CCM) (see Chapter 7), and denaturing high-performance liquid chromatography (DHPLC). Of the quantitative techniques, Southern blotting, quantitative polymerase chain reaction (qPCR) (see Chapter 25) and multiplex ligation-dependent probe amplification (MLPA) are most popular. The protein truncation test (PTT) *(1)*, also known as the in vitro synthesized protein assay (IVSP) *(2)*, is rather exceptional and has several unique features. The most prominent of these are that it uses an RNA template, scans for mutations at the protein level, has a high sensitivity, and hardly gives false positives. In addition, PTT detects truncating mutations only and the large majority of these are disease causing and thus directly clinically relevant. No PTT-detected alterations have been reported that could not be confirmed by sequence analysis; PTT analysis could thus be used directly for diagnostic purposes.

The protein truncation test was first applied to detect mutations in the human *DMD* gene *(1)*. Mutations in this gene cause Duchenne and Becker muscular dystrophy (DMD/BMD) *(4)*. Two-thirds of patients have large (intragenic) deletions and duplications; the rest are mainly point mutations. Because the *DMD* gene is the largest gene found in nature to date, with 79 exons spread over 2.4 Mb *(6,7)*, detection of the remaining mutations is a daunting task. In addition, the gene contains many polymorphic sites (i.e., nonpathogenic nucleotide changes). Mutation detection using standard DNA-based technology will pick up these as "false positive" signals. The first step around these problems was to focus mutation analysis on the transcribed coding region, thereby condensing 79 individual exons into one 11-kb molecule, the DMD mRNA. Currently RNA-based PTT requires analysis of only 5–12 segments *(8,9)*. The second step focused on the expected type of mutations; that is, most mutations identified cause premature translation termination. Therefore, a method was designed that translated the mRNA (cDNA) into protein and focused on the size of the translation products; truncating mutations should yield shorter products that should be easy to detect. Inherent to this approach was the fact that all polymorphisms, mostly yielding amino acid substitutions or no change at all, would not be detected. Consequently, the PTT assay as designed specifically zooms in on pathogenic changes only.

2. The Method

A PTT assay (*see* **Fig. 1A**) starts with the isolation of an appropriate PCR template, usually RNA from freshly drawn blood of a patient, although, in some instances, DNA can be used. Next, the messenger RNA (mRNA) is reverse transcribed to make a DNA copy (cDNA) of the

From: *Medical Biomethods Handbook*
Edited by: J. M. Walker and R. Rapley © Humana Press, Inc., Totowa, NJ

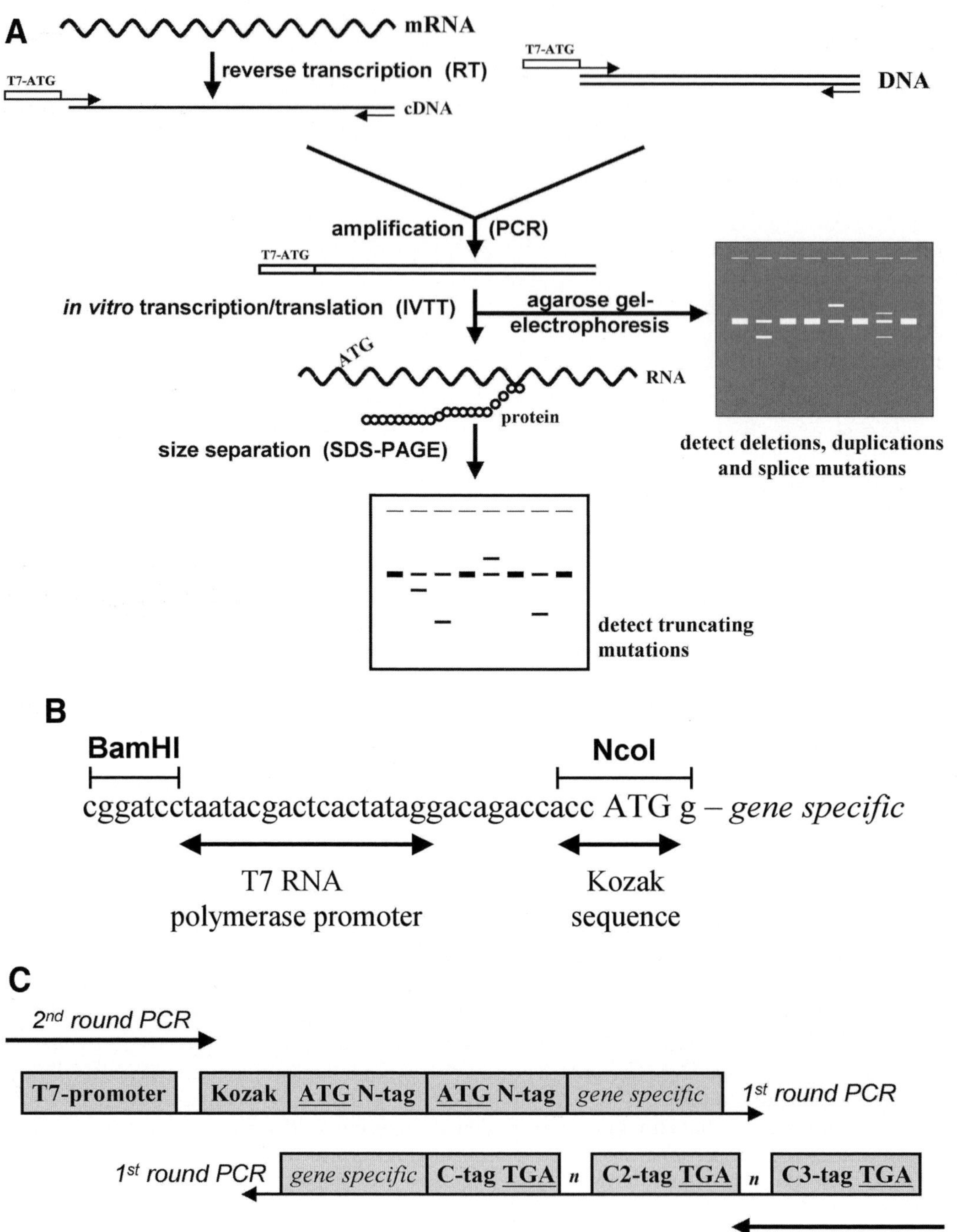

Fig. 1. The protein truncation test (PTT). **(A)** schematic representation showing the individual steps of a PTT assay. **(B)** Example sequence of a tailed PTT primer used to introduce sequences facilitating subsequent transcription/translation. The basic elements include a T7 RNA polymerase promoter sequence (T7), a spacer, and a translation initiation (Kozak) sequence with ATG codon. A 5' *Bam*HI site and an *Nco*I site (covering the ATG codon) facilitate cloning. **(C)** N- and C-terminal protein tagging in PTT. To ensure effective translation initiation, the forward primer contains two N-terminal tags in-frame with the gene-specific sequence.

message. Then, two rounds of PCR are performed, both to amplify the target sequence and to introduce sequences that facilitate subsequent in vitro transcription/translation. The latter is achieved using tailed PCR primers introducing an RNA polymerase promoter and a translation initiation sequence with an in-frame ATG codon (Kozak sequence) directly upstream of the sequence to be translated (*see* **Fig. 1B**). After first- and second-round PCR, DNA fragments are electrophoresed to check their quality and quantity as well as to determine their size. The PCR products are then transcribed using RNA polymerase (usually T7) and the transcripts are translated using an in vitro translation system. Finally, protein gel electrophoresis (sodium-dodecyl sulfate–polyacrylamide gel electrophoresis [SDS-PAGE]) is used for size separation of the translation products. Depending on the labeling method used, the gel is either analyzed directly or after membrane transfer of the translation products (Western blotting).

Mutations present in the sample analyzed emerge as translation products of altered size, usually shorter products derived from premature translation termination caused by nonsense and frame-shift mutations. The size of the translation product is used to deduce the position of the truncating mutation, and the respective segment of the PCR product (cDNA) is sequenced to identify the change at nucleotide level. Finally, when a mutation is found, this is confirmed on genomic DNA by sequence analysis of the respective gene segment.

3. Practical Considerations

3.1. The Template

In a standard PTT assay, RNA is used as a starting template. In rare cases (i.e., when the gene contains large exons), DNA can also be used ***(1,10,11)***. The majority of genes, however, do not contain such a favorable structure, but they are split into many small exons, leaving RNA as the only workable template.

Preferably, RNA should be isolated from a tissue where the gene of interest is expressed at significant levels. In neuromuscular disorders, this source would be muscle (i.e., RNA isolated from a muscle biopsy). In many cases, such a source is not readily available and RNA is isolated from blood. Alternative sources could be buccal swabs ***(12)***, excrements ***(13)***, cultured cells (e.g., fibroblast cell lines obtained from skin biopsies), or chorionic and amniotic fluid cells ***(8,12)***. During harvesting of the cells, one should realize that any step in the protocol can influence gene expression, potentially introducing undesired artefacts. A well-known example is the effect of a "cold shock," which could be shown to promote the insertion of a cryptic exon in the NF1 mRNA ***(14)***, initially erroneously reported as related to pathogenicity.

Detection of truncating mutations from RNA samples is complicated by a cellular process designated nonsense-mediated mRNA decay (NMD) ***(15)***. This process, believed to act as a surveillance and defense mechanism against the production of deleterious out-of-frame protein sequences, specifically degrades RNA molecules that contain premature translation termination mutations. As a result, in recessive diseases, NMD will reduce the expression level of the mRNA transcribed from the mutated allele (i.e., the mRNA containing the mutation to detect). Consequently, the mutation might be difficult to detect and it is essential to verify whether RNA from both alleles was amplified and thus scanned for the presence of mutations. This

---

Fig. 1 *(continued from facing page)* The reverse primer encodes a C-terminal protein tag (C1), followed by a stop codon (TGA). To ensure detection of frame-shifting mutations, which do not encounter a stop codon before the end of the translated fragment, two additional out-of-frame protein tags (C2), one for each alternative frame, are included as well. Antibodies directed against the tags (N, C1, and C2) can be used for various purification and detection protocols. For a segmented PTT assay (*see* **Fig. 2**), when the primers indicated are used for first-round PCR, second-round PCR can be performed using general primers. PCR: polymerase chain reaction; RT: reverse transcription.

verification can be performed most effectively using polymorphisms present in the transcribed region *(12)*. Using cultured cells, the NMD process can be reduced by culturing the cells in the presence of cycloheximide, an inhibitor of protein translation *(16)*.

3.2. Target Amplification

Amplification starts with a reverse transcription (RT) reaction, usually followed by two rounds of PCR amplification. During these PCR reactions, specifically designed tailed primers are used to introduce sequences that facilitate subsequent transcription/translation (*see* **Subheading 4.**). The RT reaction can be initiated using two different approaches, random *(8,12)* or gene-specific priming *(9)*. Depending on the level of expression of the gene in the sample used, it is often necessary to perform a nested PCR reaction to amplify the target sequences to a detectable level. This approach is very powerful and allows the amplification of nearly any gene from, for example, blood samples, based on the minute levels of "ectopic" ("illegitimate") transcription that are present for most genes in any given tissue *(17)*. When DNA is used as a template or when RNA expression in the target sample is high *(11)*, nested PCR is not necessary and the tailed primers can be used directly in the first-round PCR reaction.

Problems with a PTT assay are most frequently caused by a failing RT reaction, the first and most critical step of the entire procedure. The RT step is very sensitive, failing most frequently from poor quality RNA. Because nested PCR is used, the PTT assay is very sensitive to contamination of the RT-PCR sample and reagents with trace amounts of cDNA or cloned genomic DNA. Consequently, a "pre-PCR" laboratory should be established to keep the RNA sample and primers physically separated from the RT-PCR and subsequent steps of the procedure. A control reaction (no RNA and/or no enzyme added) should always be part of the assay to exclude possible contaminations.

3.3. Transcription/Translation

Transcription and translation are performed using in vitro reactions, either coupled (one-tube reaction) or separate *(12)*. The advantage of performing separate reactions is that every step can be controlled and that RNA yield can be determined before translation. A coupled one-tube in vitro transcription/translation gives a reduced chance of contamination and is simpler, but the reactions are more difficult to control. Most translation systems use a T7 RNA polymerase to generate RNA transcripts; some use T3 or SP6. In vitro translation is usually performed using a rabbit reticulocyte lysate *(12)*, sometimes using a wheat germ extract *(18)*. The method to detect the translation products is decided at the level of translation. Recently, several new methodologies have been reported *(13,19–22)*. Initially, radioactively labeled amino acids (mostly [^{3}H]leucine or [^{35}S]methionine) were incorporated and translation products were detected using autoradiography after SDS-PAGE separation *(1)*. Currently, nonradioactive methods are used and either biotin-labeled lysine is incorporated or protein-tag sequences are added N or C terminally (*see* **Subheading 4.**). Detection is accomplished through immunological detection of chemiluminescently labeled antibodies directed at the biotin-labeled amino acids or to the incorporated protein tag. Gite et al. *(21)* recently used incorporation of fluorescently labeled lysine facilitating direct in-gel detection of the translation products after SDS-PAGE. This methodology is attractive because it obviates the time-consuming Western blotting step.

The quality of the in vitro translation system, which varies from batch to batch, can be a critical parameter in the PTT assay—in particular, because translation of certain cDNA segments can be very difficult *(12)*. Although translation for most cDNA segments with moderate-quality translation batches is sufficient, translation of specific fragments will fail totally.

3.4. Detection of Mutations

After size separation and signal detection, analysis of the translation products should reveal clear bands and the identification of mutations affecting the reading frame should be rather

Table 1
Effect of Different Mutations on the Translational Reading Frame in Relation to the Altered Products Observed after PTT Analysis

Cause	Shorter translation product	Larger translation product
Nonsense mutation	Yes	No
Frame-shift mutation	yes	Rarely[a]
Mutation translation initiation site	No (yes when downstream new initiation site[a])	Rarely (yes when upstream new initiation site[a])
Mutation translation termination site	No	Yes[a]
Splice mutation	Yes	Rarely (in frame intronic insertion)
Genomic deletion	Yes	No
Genomic duplication	Yes (when out of frame)	Yes (when in frame)

[a]Detectable only when the PTT assay was designed to pick up this type of change.

straightforward (*see* **Table 1**). Detection of a truncated fragment should always be verified by a second, completely independent assay starting from a new RT reaction. It might happen that truncated fragments cannot be reproduced *(13)*. All translation products that do not run at the expected size point to the presence of mutations. In general, mutations are truncating and appear as faster-migrating translation products, but sometimes larger translation products can be found (*see* **Table 1**). The data obtained from protein analysis should be compared with the size of the RT-PCR product; when mutations affecting splicing are involved, the size of the RT-PCR product will be altered.

The most frequently observed changes are shorter translation fragments (*see* **Table 1**), caused by mutations that either directly introduce a stop codon (nonsense change) or cause a shift in the normal reading frame, generating a premature stop codon (frame-shift mutation). Because PTT is an RNA-based technology, mutations affecting RNA processing (in particular, splicing) will also be detected. Skipped exons give shorter translation products, whereas shifts of the splice site, either into the exon or into the intron, generate shorter or larger translation products, respectively. In addition to RT-PCR products of altered size, most splice mutations also yield out-of-frame RNAs and, thus, truncated translation products (*see* **Table 1**). Analysis of the RT-PCR products will not only reveal splice mutations but also the presence of genetic rearrangements: decreased sizes indicating genomic deletions and increased sizes duplications or insertions.

Sometimes, larger translation products are detected. These derive most often from in-frame duplications or intronic mutations that incorporate intron sequences in the mature transcript. In rare cases, the natural stop codon in the gene is mutated *(23)* or frame-shift mutations do not encounter a stop codon in the shifted reading frame, generating a protein of increased length. Note that such changes will only be detected when the PTT assay is designed to pick up such changes (*see* **Subheading 4.**).

Mutation detection using transcription/translation can give surprising results. Chang et al. *(24)* reported the identification of a patient with a peroxisome-biogenesis disorder presenting a relatively mild clinical phenotype for two seemingly severe (early) truncating *PEX12* mutations. In vitro transcription/translation of the cDNA segment expected to direct the synthesis of an eight-amino-acid protein surprisingly produced a 29-kDa product. Apparently, the mutation activated translation initiation at an internal ATG codon, yielding a partially functional protein and a mild disease phenotype. Note that, in such cases, the length of the truncated product does point to the site of the mutation, but not counted from the C-terminal end.

3.5. Sequence Analysis

The cause of a detected change should be confirmed first by sequencing the RT-PCR product. Unlike other mutation-detection methods, excluding chemical mismatch cleavage, PTT pinpoints the site of the mutation by the length of the truncated translation product. This characteristic facilitates focused sequence analysis to reveal the underlying mutation. Identification of the sequence change in the cDNA should be straightforward and the effect of change should match the expected effect on protein translation. Finally, the genomic DNA of the patient should be sequenced to confirm the presence of the change at the genomic level.

Especially when tagged translation products are used, a PTT assay is usually much more sensitive than a sequence analysis. Low expression of the mutated allele (10% or lower) (e.g., as a consequence of nonsense-mediated mRNA decay) might give a clearly visible truncated translation product, whereas sequence analysis of the cDNA will not reveal the causative change ***(13,20)***. In general, identification of the change in genomic DNA should be easier because here, the mutation should be present at 50% (one of two alleles present). Exceptions to this rule are samples derived from tumors that are often not clonal and/or contain a mix of cells from different tissues.

Laken et al. ***(25)*** found truncated translation products from a segment of the *APC* gene that could not be confirmed by changes in genomic DNA. The only nucleotide change detected would result in an amino acid substitution. However, upon sequence analysis of individually cloned cDNA fragments, a range of different mutations was identified in close proximity to the original change, which all caused premature translation termination. Rather than altering the function of the encoded protein, this mutation somehow created a hypermutable region in the gene, indirectly causing colon cancer predisposition.

4. Design of a PTT-Assay

Before a PTT-based mutation analysis is initiated, one should check whether the RNA template to be analyzed can be amplified from the template available, i.e., whether it is expressed. In addition, one needs to check whether that transcript has the same structure as the transcript present in the affected tissue. In, for example, Duchenne muscular dystrophy (DMD), it has been reported that muscle RNA, in contrast to lymphocyte RNA, might contain an additional exon "X" spliced between exons 1 and 2 ***(26)***. When RNA templates that are not derived from a disease-affected tissue are used, mutations influencing only tissue-specific expression, including tissue-specific promoters and tissue-specific splicing, will not be detected. RNA samples derived from peripheral blood lymphocytes are usually suitable for amplifying any gene of interest. The author has experienced only one exception to this rule; amplification of the ornithine transcarbamylase (*OTC*) nRNA always failed ***(12)***.

In most cases, the gene of interest is too large to permit PTT analysis from one fragment. Consequently, the gene has to be split into several partly overlapping segments of 1–2 kb in length together spanning the complete coding region (*see* **Fig. 2**). Although proteins can be translated from templates of up to 4 kb, it is rather difficult to amplify such fragments reproducibly unless specific precautions are taken ***(9)***. In practice, the ability to detect small mobility shifts of the translation products in SDS-PAGE determines the most favorable lengths, with fragments of 1–2 kb usually giving most optimal results.

Because expression of the target gene in the (blood) sample of the patient is usually rather low, RT-PCR amplification is performed using two rounds of PCR: a first-round PCR and a second-round nested PCR. To allow subsequent in vitro transcription and translation, PCR needs to be performed with specifically designed tailed oligonucleotide primers (*see* **Fig. 1B**). The 5' tail of the forward PCR primer is used to introduce an RNA promoter and an in-frame translation initiation sequence ***(1,11,12)***. These primers contain an 18-bp bacteriophage T7

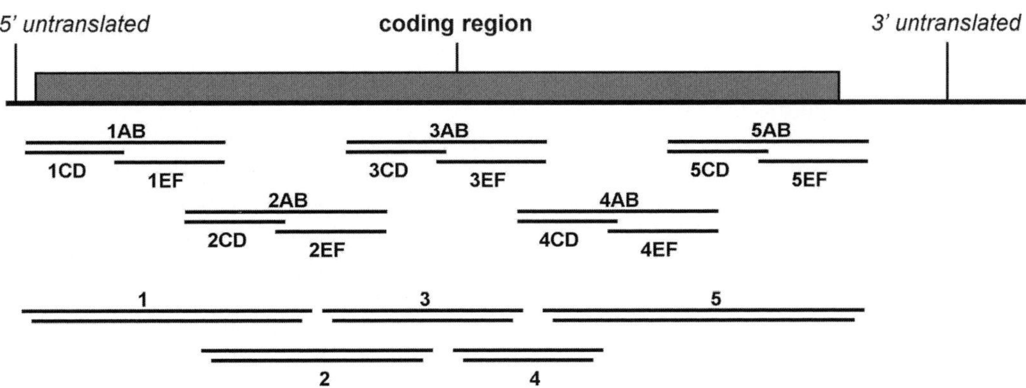

Fig. 2. A segmented PTT-assay. For PTT analysis of the human dystrophin (*DMD*) gene, it is split into several overlapping segments. **Top**:10-segment set used by Roest et al. *(8)*; **bottom**: 5-segment set used by Whittock et al. *(9)*.

promoter and an 8-bp eukaryotic translation initiation signal, including an ATG translation initiation codon (Kozak sequence). Although in most reports translation products are generated by runoff translation from the amplified segments, it has been suggested that it is advantageous to add a specific translation termination signal to the reverse primer (see below). Because pathogenic mutations have been found that affect the natural stop codon, thereby extending the normal reading frame *(23)*, the primer used to amplify the last segment of the target gene should be well 3' of the stop codon. Primers for PTT assays are rather long and using high-quality full-length primers is essential for the assay to work, especially because synthesis of such primers is not trivial.

Construction of a segmented set of sequences covering the target gene requires careful attention, and choice of primers is critical to the ultimate sensitivity of the assay. To construct a good set of overlapping segments, the following points should be addressed: (1) the ATG translation initiation codon in the tailed sense primer should be in frame with the coding sequence; (2) flanking segments should contain an overlap ensuring detection of mutations close to their ends—in general, 150- to 200-bp overlaps (equivalent to 5–8 kDa after translation) for 1- to 2-kb segments should be sufficient; (3) the reverse primer should not be selected from near the end of a region where a large open reading frame is present in one, or both, of the alternative reading frames—in such cases, a frame-shift mutation will not cause premature translation termination; (4) primers of flanking segments should be located in different exons to minimize the risk of not amplifying an allele in which the primer-binding site lies in a deleted exon; (5) primers spanning bordering exons (i.e., sequences genomically interrupted by introns) have the advantage of reducing background (amplification of alternatively spliced transcripts or contaminating genomic DNA) at the cost of being more sensitive to amplification failure when a mutation is present that influences splicing of either or both exons.

The boundaries of the translated fragments provide the most critical region to overlook a mutation. N-Terminal (early) mutations result in a product that is too small to detect because no or little label is incorporated or because the product might migrate too far and be lost from the gel. C-Terminal (late) mutations result in a size difference that is not resolved near the top of the gel. In genes analyzed in segments, these problems can be circumvented by proper selection of the overlapping regions. This safety check is not possible for the translation initiation and translation termination sites of a gene. Here, special care should be taken not to miss a mutation. The emerin gene, involved in Emery–Dreifuss muscular dystrophy (EMD), is an example in which extreme N-terminal mutations occur, even including the ATG-translation initiation codon itself ***(27,28)***. Extreme C-terminal mutations have also been reported several times ***(29)***, some affecting the natural stop codon directly, thereby extending the normal reading frame ***(23)***. Note that the disease-causing potential of mutations near the extreme end of a protein should always be questioned ***(3)***.

Recently, PTT assays have been designed such that N-terminal ***(12,19)*** and/or C-terminal ***(13,20)*** protein sequences, so-called protein tags, are incorporated (*see* **Fig. 1C**). This is achieved by using primers that encode the tag sequence and fuse it directly in-frame with the normal protein sequence (*see* **Fig. 1C**). The inclusion of these tags has several advantages: (1) One has more possibilities for detection and purification of the translation products. This possibility can be used to specifically detect very low amounts of truncated products ***(13,20)***; (2) The background derived from undesired secondary translation initiation products, which for specific segments can be rather prominent ***(12)***, will not be visible when tagged and, thus, correctly initiated translation products are detected. Although, in general, the translation pattern is fairly constant and changes are easy to detect, these background products might disguise truncated products; (3) with one tag per translation product, the signal strength is equal for all products and not variable based on the amount of label incorporated (i.e., larger fragments giving stronger signals). (4) There is no chance that the labeled amino acid used for detection is not present in the translated segment; this is critical when mutations produce short truncated products. (5) Incorporation of C-terminal protein tags, both in-frame and out-of-frame, facilitate the analysis of extremely C-terminal mutations (*see* **Fig. 1C**). In addition, they facilitate readout using specific protocols, either enzyme-linked immunosorbent assay based ***(21,30)*** or using multiple colors ***(13)***.

Many different tags have been used successfully ***(12,13,19–21)***, the main characteristic being the availability of a good antibody for their detection. The system we currently use adds the tag-encoding sequence during the first-round PCR, facilitating second-round PCR with uniform primers (*see* **Fig. 1C**) ***(12)***. Note that mutations in the translation initiating ATG cannot be detected when tag sequences are added. In such cases, priming in the natural 5'-untranslated region of the gene is required ***(28)***.

The protein truncation test is a rather new technique that is still in its infancy ***(31)***, especially with respect to the last step—analysis of the translation products using SDS-PAGE. There are several possibilities to increase the sensitivity of PTT and expand its detection range. So far, most impressive modifications have been obtained using protein tagging, especially regarding the reduction of undesired background products (derived from secondary translation initiation) as well as the overall sensitivity of the assay ***(12,13,19–21)***. Protein tagging has many potential applications and, as shown by Traverso et al. ***(13)***, can even be performed in a multicolor format, opening several new designs.

Concluding Remarks

In the standard PTT assay, only the length of translation products is used as an analytical tool to detect mutations. Other methods of protein separation that extend the detection range to other mutation types are available. For example, many thalassemia mutations can be detected by mobility shifts after electrophoresis of the α-globulin and β-globin proteins. Isoelectric focusing is an obvious candidate for expanding the level of detection to include amino acid substitutions and small in-frame insertions and deletions ***(31)***. Garvin et al. ***(22)*** reported the analysis of translation products using mass spectrometry. This approach identifies any change in the protein sequence, including amino acid substitutions, and facilitates a high-throughput analysis. It should be noted that using this approach, one of the most prominent features of the PTT, zooming in on truncating disease-causing mutations only, is lost.

Finally, for genes containing small coding regions, PTT enables translation of the entire gene product and, thus, facilitates direct functional assays of the protein (e.g., measurement of enzymatic activity). In this respect, it should be realized that even when a mutation can be detected at the DNA level, its disease-causing phenotype can only be verified with a functional test (i.e., at the protein level). For this, RT-PCR and in vitro transcription/translation can provide the ultimate analytical tool.

5. Applications

The protein truncation test can be applied as a mutation analysis tool for any gene where truncating mutations constitute a significant fraction of the pathogenic mutations *(31)*. The number of disease genes for which a major fraction of patients carry nonsense and frame-shift mutations is increasing continuously. Among these are a range of tumor-related genes, including breast cancer *BRCA1* *(11)* and *BRCA2* *(32)*, colon cancer related to familial adenomatous polyposis (APC) *(2,10,13)* and hereditary nonpolyposis colon cancer (HNPCC) *(33,34)*, and neurofibromatosis *NF1* *(35)* and *NF2* *(36)*. Other examples include Duchenne muscular dystrophy *(1,8,9)*, polycystic kidney disease *(37)*, and even cystic fibrosis *(38)*. For a recent review, *see* **ref. *31***.

The basis of the PTT assay, (i.e., generating [segments of] proteins using in vitro transcription/translation) can be applied for many purposes. An interesting PTT-specific application lays in the identification of new disease genes. When only fragments of the sequence of a specific candidate disease gene are known tailed primers can be devised and used directly to scan large numbers of patient samples for the potential presence of truncating mutations. A first example of this approach was the identification of the gene involved in Rubinstein–Taybi syndrome *(39)*.

References

1. Roest, P. A. M., Roberts, R. G., Sugino, S., Van Ommen, G. J. B., and Den Dunnen, J. T. (1993) Protein truncation test (PTT) for rapid detection of translation-terminating mutations. *Hum. Mol. Genet.* **2**, 1719–1721.
2. Powell, S. M., Petersen, G. M., Krush, A. A. J., et al. (1993) Molecular Diagnosis of familial adenomatous polyposis. *N. Engl. J. Med.* **329**, 1982–1987.
3. Mazoyer, S., Dunning, A. M., Serova, O., et al. (1996) A polymorphic stop codon in Brca2. *Nature Genet.* **14**, 253–254.
4. Koenig, M., Hoffman, E. P., Bertelson, C. J., Monaco, A. P., Feener, C. A., and Kunkel, L. M. (1987) Complete cloning of the Duchenne muscular dystrophy (DMD) cDNA and preliminary genomic organization of the DMD gene in normal and affected individuals. *Cell* **50**, 509–517.
5. Steimle, V., Otten, L. A., Zufferey, M., and Mach, B. (1993) Complementation cloning of an MHC class II transactivator mutated in hereditary MHC class II deficiency (or bare lymphocyte syndrome). *Cell* **75**, 135–146.
6. Den Dunnen, J. T., Grootscholten, P. M., Bakker, E., et al. (1989) Topography of the DMD gene: FIGE and cDNA analysis of 194 cases reveals 115 deletions and 13 duplications. *Am. J. Hum. Genet.* **45**, 835–847.
7. Roberts, R. G., Coffey, A. J., Bobrow, M., and Bentley, D. R. (1992) Determination of the exon structure of the distal portion of the dystrophin gene by vectorette PCR. *Genomics* **13**, 942–950.
8. Roest, P. A. M., Bakker, E., Fallaux, F. J., Verellen-Dumoulin, C., Murry, C. E., and Den Dunnen, J. T. (1999) New possibilities for prenatal diagnosis of muscular dystrophies: forced myogenesis with an adenoviral MyoD-vector. *Lancet* **353**, 727–728.
9. Whittock, N. V., Roberts, R. G., Mathew, C. G., and Abbs, S. J. (1997) Dystrophin point mutation screening using a multiplexed protein truncation test. *Genet. Test.* **1**, 115–123.
10. Van Der Luijt, R. B., Meera Kahn, P., Vasen, H., et al. (1994) Rapid detection of translation-terminating mutations at the adenomatous polyposis coli (APC) gene by direct protein truncation test. *Genomics* **20**, 1–4.
11. Hogervorst, F. B. L., Cornelis, R. S., Bout, M., et al. (1995) Rapid detection of BRCA1 mutations by the protein truncation test. *Nature Genet.* **10**, 208–212.
12. Den Dunnen, J. T. (1996) Protein truncation test, in *Current Protocols In Human Genetics* (Dracopoli, N. C., Haines, J. L., Korf, B. R., et al., eds.), Wiley, New York.
13. Traverso, G., Diehl, F., Hurst, R., et al. (2003) Multicolor in vitro translation. *Nature Biotechnol.* **21**, 1093-1097.

14. Ars, E., Serra, E., De La Luna, S., Estivill, X., and Lazaro, C. (2000) Cold shock induces the insertion of a cryptic exon in the neurofibromatosis type 1 (NF1) mRNA. *Nucleic Acids Res.* **28**, 1307–1312.
15. Culbertson, M. R. (1999) RNA surveillance. Unforeseen consequences for gene expression, inherited genetic disorders and cancer. *Trends. Genet.* **15**, 74–80.
16. Lamande, S. R., Bateman, J. F., Hutchison, W., et al. (1998) Reduced collagen VI causes Bethlem myopathy: a heterozygous COL6A1 nonsense mutation results in mRNA decay and functional haploinsufficiency. *Hum. Mol. Genet.* **7**, 981–989.
17. Chelly, J., Concordet, J., Kaplan, J. C., and Kahn, A. (1989) Illegitimate Transcription Of Any Gene In Any Cell Type. *Proc. Natl. Acad. Sci. USA* **86**, 2617–2621.
18. Hope, I. A. and Struhl, K. (1985) GCN4 protein synthesized in vitro, binds HIS3 regulatory sequences: implications for general control of amino acid biosynthetic genes in yeast. *Cell* **43**, 177–188.
19. Rowan, A. J. and Bodmer, W. F. (1997) Introduction of a myc reporter tag to improve the quality of mutation detection using the protein truncation test. *Hum. Mutat.* **9**, 172–176.
20. Kahmann, S., Herter, P., Kuhnen, C., et al. (2002) A non-radioactive protein truncation test for the sensitive detection of all stop and frameshift mutations. *Hum. Mutat.* **19**, 165–172.
21. Gite, S., Lim, M., Carlson, R., Olejnik, J., Zehnbauer, B., and Rothschild, K. (2003) A high-throughput nonisotopic protein truncation test. *Nature Biotechnol.* **21**, 194–197.
22. Garvin, A. M., Parker, K. C., and Haff, L. (2000) Maldi-tof based mutation detection using tagged in vitro synthesized peptides. *Nature Biotechnol.* **18**, 95–97.
23. Martinez-Gimeno, M., Gamundi, M. J., Hernan, I., et al. (2003) Mutations in the pre-mRNA splicing-factor genes PRPF3, PRPF8, and PRPF31 in Spanish families with autosomal dominant retinitis pigmentosa. *Invest. Ophthalmol. Vis. Sci.* **44**, 2171–2177.
24. Chang, C. C. and Gould, S. J. (1998) Phenotype–genotype relationships in complementation group 3 of the peroxisome-biogenesis disorders. *Am. J. Hum. Genet.* **63**, 1294–1306.
25. Laken, S. J., Petersen, G. M., Gruber, S. B., et al. (1997) Familial colorectal cancer in ashkenazim due to a hypermutable tract in APC. *Nature Genet.* **17**, 79–83.
26. Roberts, R. G., Bentley, D. R., and Bobrow, M. (1993) Infidelity in the structure of ectopic transcripts: a novel exon in lymphocyte dystrophin transcripts. *Hum. Mutat.* **2**, 293–299.
27. Yates, J., Aksmanovic, V., Mcmahon, R., Bione, S., and Toniolo, D. (1996) Mutation analysis in emery-dreifuss muscular dystrophy. *Eur. J. Hum. Genet.* **4**, 62–63.
28. De Koning Gans, P. A. M., Ginjaar, H. B., Bakker, E., Yates, J. R. W., and Den Dunnen, J. T. (1999) A protein truncation test for emery-dreifuss muscular dystrophy (EMD): detection of N-terminal truncating mutations. *Neuromusc. Disord.* **9**, 247–250.
29. Shattuck-Eidens, D., Mcclure, M., Simard, J., et al. (1995) A collaborative survey of 80 mutations in the brca1 breast and ovarian cancer susceptibility gene. *J. Am. Med. Assoc.* **273**, 535–541.
30. Van Ommen, G. J. B., Bakker, E., and Den Dunnen, J. T. (1999) The human genome project and the future of diagnostics, treatment, and prevention. *Lancet* **354**, Si5–Si10.
31. Den Dunnen, J. T. and Van Ommen, G. J. B. (1999) The protein truncation test: a review. *Hum. Mutat.* **14**, 95–102.
32. Lancaster, J. M., Wooster, R., Mangion, J., et al. (1996) BRCA2 mutations in primary breast and ovarian cancers. *Nature Genet.* **13**, 238–240.
33. Farrington, S. M., Lin-Goerke, J., Ling, J., et al. (1998) Systematic analysis of hMSH2 and hMLH1 in young colon cancer patients and controls. *Am. J. Hum. Genet.* **63**, 749–759.
34. Kohonen-Corish, M., Ross, V. L., Doe, W. F., et al. (1996) RNA-based mutation screening in hereditary nonpolyposis colorectal cancer. *Am. J. Hum. Genet.* **59**, 818–824.
35. Heim, R. A., Silverman, L. M., Farber, R. A., Kam-Morgan, L. N. W., And Luce, M. C. (1995) Screening for truncated NF1 proteins. *Nature Genet.* **8**, 218–219.
36. Parry, D. M., Maccollin, M. M., Kaiser Kupfer, M. I., et al. (1996) Germ-line mutations in the neurofibromatosis 2 gene: correlations with disease severity and retinal abnormalities. *Am. J. Hum. Genet.* **59**, 529–539.
37. Roelfsema, J. H., Spruit, L., Saris, J. J., et al. (1997) Mutation detection in the repeated part of the PKD1 gene. *Am. J. Hum. Genet.* **61**, 1044–1052.
38. Romey, M. C., Tuffery, S., Desgeorges, M., Bienvenu, T., Demaille, J., and Claustres, M. (1996) Transcript analysis of CFTR frameshift mutations in lymphocytes using the revesre transription-polymerase chain reaction technique and the protein truncation test. *Hum. Genet.* **98**, 328–332.
39. Petrij, F., Giles, R. H., Dauwerse, J. G., et al. (1995) Rubinstein-Taybi syndrome caused by mutations in the transcriptional co-activator CBP. *Nature* **376**, 348–351.

18

Linkage, Allele Sharing, and Association

Mara Giordano

1. Human Disease Gene Mapping

It has been estimated that the human genome contains 25,000–30,000 genes that are distributed on the 24 chromosomes. Therefore, the identification and localization of any particular gene responsible for a given trait or disease has always been a difficult task.

The most successful mapping approach that led to the identification of hundreds of genes involved in human disease is positional cloning (i.e., the isolation of a gene solely on the basis of its chromosomal location with no information about its biochemical function). In this approach, genetic markers (i.e., polymorphic loci with a well-defined chromosomal position) with no relation with the disease are tested in affected families to analyze their cosegregation with the disease. If a marker cosegregates with the disease, this means that the disease gene is located close to the marker locus ("linked"). Thus, this procedure allows one to find the position of the gene causing the disease and, after that, to isolate it to study its function and its role in the disease. This approach is equally applicable to any kind of disease for which the biochemical nature is unknown. One of the first genes to be identified by positional cloning was the gene involved in cystic fibrosis, named CFTR (cystic fibrosis transmembrane conductance regulator) in 1989 *(1)*. Nothing was initially known about the function of the mutated protein or about the physiological role of the normal product. After the position in the genome was well established on chromosome 7, the gene was cloned and sequenced. Then, it was demonstrated that the protein was a regulated Cl^- channel located in the apical membrane of the epithelial cells affected in the disease. During the course of the last 15 yr thanks to the identification of thousands of genetics markers distributed all over the genome, this strategy allowed to map more than 1000 monofactorial disorders.

The localization of a disease gene with respect to specific genetic markers is investigated through linkage analysis. Linkage analysis utilizes the phenomenon whereby alleles at neighboring loci on the same chromosome will be transmitted together more frequently than loci located on different chromosomes. The final goal is to define the position of disease genes in relation to that of well mapped loci.

2. Independent and Linked Loci

If two genes lie on different chromosomes, they will segregate independently at meiosis (Mendel II law): an allele at one locus on one chromosome is transmitted together with a given allele at another locus on another chromosome with 50% probability. Assuming a pair of loci A and B (*see* **Fig. 1**), each with two alleles (A,a and B,b), a double-heterozygous individual (AaBb) can produce four types of gamets. Two of these reconstitute the parental gametes and are referred as "parentals" (identical to those received from his parents). The other two types of

From: *Medical Biomethods Handbook*
Edited by: J. M. Walker and R. Rapley © Humana Press, Inc., Totowa, NJ

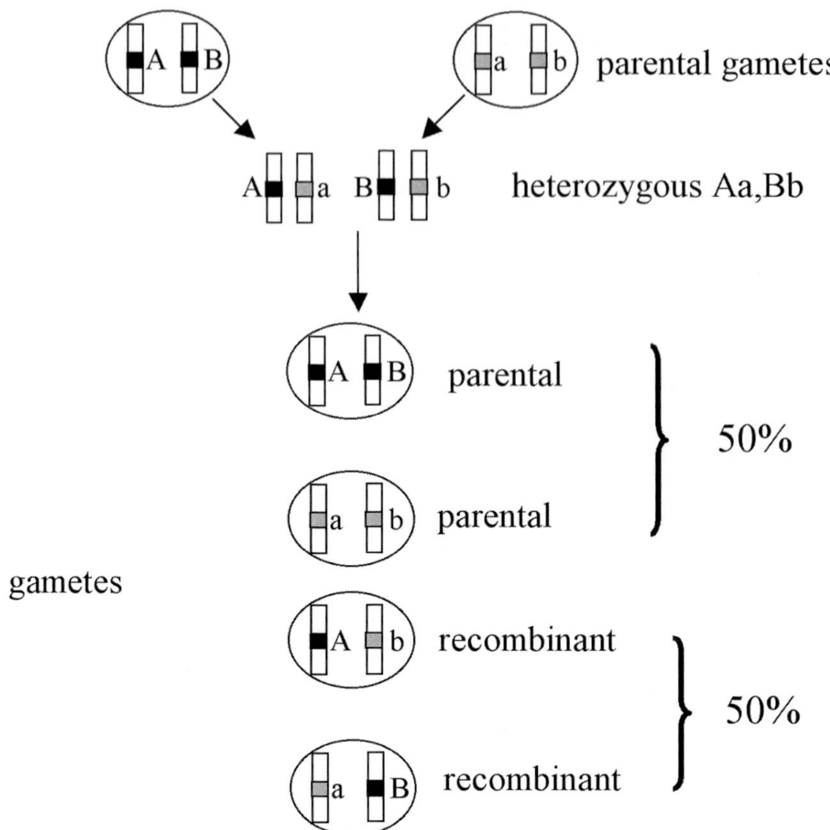

Fig. 1. Gametic combinations generated by a double-heterozygous individual for two independent loci A (with alleles A and a) and B (with alleles B and b). Fifty percent of the combinations are identical to the parental combinations and 50% are recombinant.

gamets are generated by the independent assortment of the parental alleles and are referred to as "recombinants." For two independent loci, the ratio parentals: recombinants is 50 : 50.

If two loci are syntenic (i.e., they lie on the same chromosome), they do not segregate independently. Genetic linkage is the phenomenon whereby alleles at loci close together on the same chromosome do not assort independently at meiosis, but they will tend to be inherited together. If the linkage between two loci is complete, an individual heterozygous at both loci (AaBb) will produce only two types of gamete. In **Fig. 2**, the loci A and B are closely linked: All of the gametes will carry the parental combination (i.e., 50% AB and 50% ab).

An example of independent and linked loci segregating in the same family is reported in **Fig 3**, in which is shown a large pedigree with some members affected by a rare autosomal dominant disease, in which each affected individual has inherited a mutant copy of the disease gene (D) and is Dd heterozygous, whereas unaffected subjects have two normal alleles at the disease locus (dd genotype). All of the members of the family have been tested with two markers: marker A on chromosome 2, with alleles A_1 and A_2, and marker B on chromosome 5, with alleles B_1 and B_2. Following the inheritance of marker A alleles and the inheritance of the disease, the affected members of the pedigree transmit either the marker allele A_1 or A_2 to affected offspring with equal likelihood, as expected by random segregation of the disease and the marker. Instead, when comparing the inheritance of marker B alleles and the disease, all affected members of the pedigree transmit the allele B_1 to the affected offspring and none of the

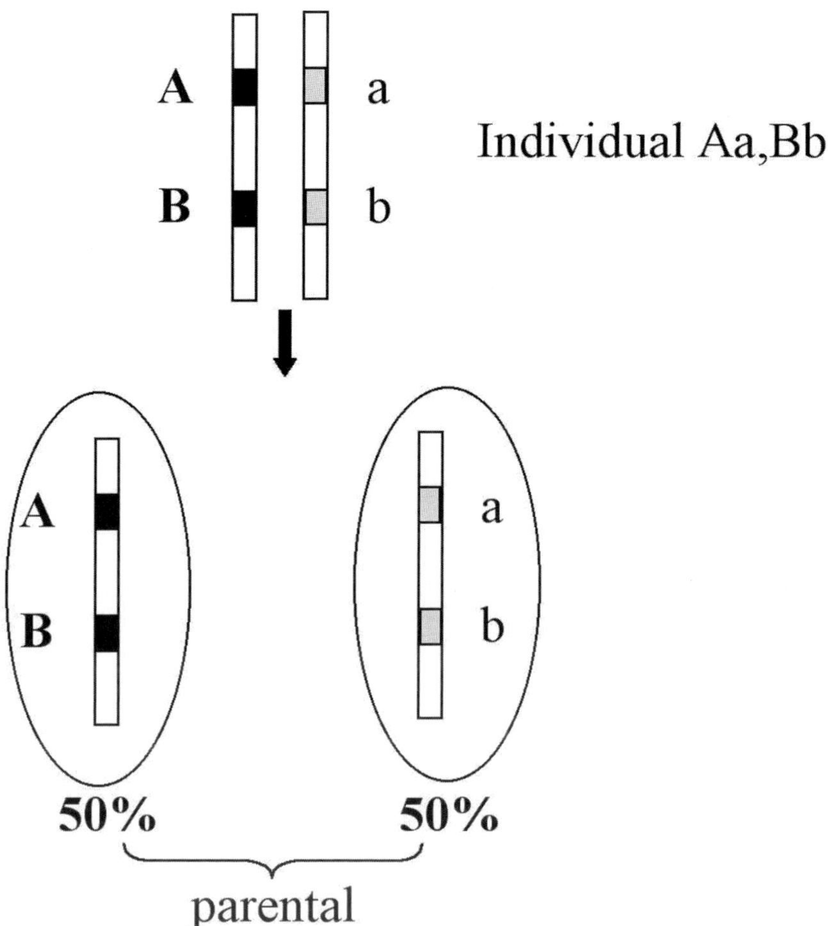

Fig. 2. Gametic combinations generated by a double-heterozygous individual for two linked loci A and B. All of the gametes carry the parental combinations.

unaffected members of this pedigree has inherited this same allele. The precise cosegregation of B_1 and D alleles allows us to hypothesize that the marker locus B and the disease locus are physically linked on chromosome 5.

In geneticist terminology, two alleles at linked loci are said to be in coupling (or in cis) when they are on the same parental chromosome (e.g., B_1 and D in **Fig. 3**) and in repulsion (or in trans) when they are on two different chromosomes (e.g., B_1 and d). The only consequence of different phases is that the gametes that are called recombinant in one case are nonrecombinant in the other.

3. Recombination Fraction

The family reported in **Fig. 4** shows the segregation of the nail–patella syndrome, a rare autosomal dominant disorder, involving nail and skeletal deformities. The affected individuals are heterozygotes Dd for the disease allele, whereas the non affected are homozygotes dd. All of the family members were genotyped for the ABO blood group determined by three different alleles, A and B codominant and each dominant to O. Individual II-1 inherits the D-A allelic combination from her father (I-1) and the allelic combination d-B from her mother (I-2). She transmits to her eight children the combination D-A four times and the combination d-B three

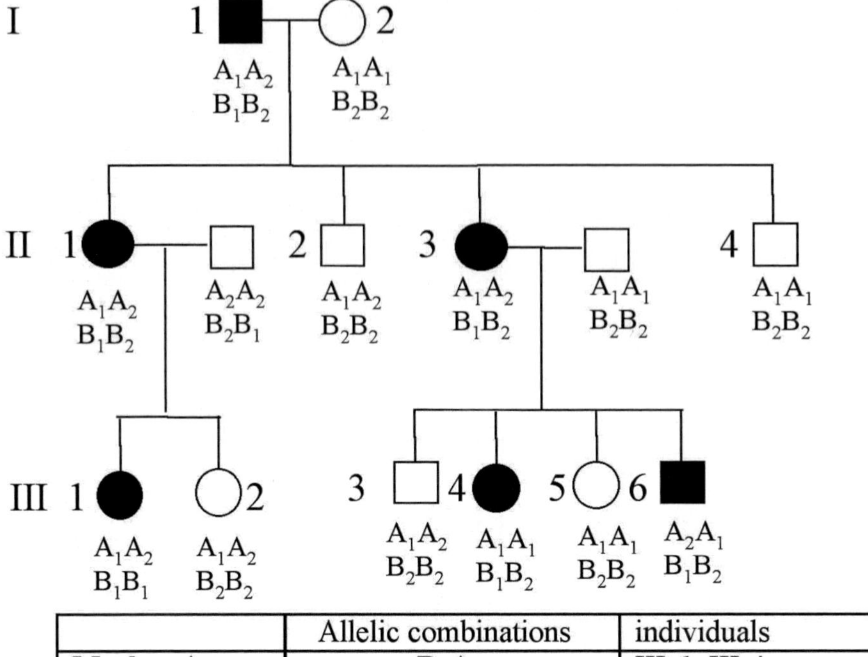

	Allelic combinations	individuals
Marker A	D-A_1	III-1, III-4
	D-A_2	II-1, II-3, III-6
	d-A_1	II-4, III-2, III-5
	d-A_2	II-2, III-3
Marker B	**D-B_1**	II-1, II-3, III-1, III-4, III-6
	D-B_2	
	d-B_1	
	d-B_2	II-2, II-4, III-2, III-3, III-5

Fig. 3. Pedigree with members affected by an autosomal dominant disease. All the individuals were genotyped for two markers A (alleles A_1 and A_2) and B (alleles B_1 and B_2). Individuals carrying the various gametic combinations of the disease locus alleles with each of the two markers are reported in the table. For marker A, all of the possible allelic combinations are detected, as expected for independent loci. For marker B, only two allelic combinations are present (D-B_1 and d-B_2), suggesting that the disease and marker B locus are linked and that the disease allele D is carried in cis with marker allele B_1.

times, and in just one case, she transmits the recombinant combination d-A (II-3). Thus, the parental combinations D-A and d-B are cotransmitted to members of the third generation more frequently than expected by chance. This suggests that the nail–patella and the ABO group loci are linked.

As stated above, most but not all the individuals inherit the parental allelic combination. Individual III-3 received the marker allele A from her mother but not the disease-allele D. This is a recombinant, originated by a crossing-over event between the loci ABO and nail–patella which occurred during the maternal meiosis.

Crossing-over is the reciprocal exchange of corresponding segments between the nonsister chromatids of two homologous chromosomes. It is assumed to be random along the length of the

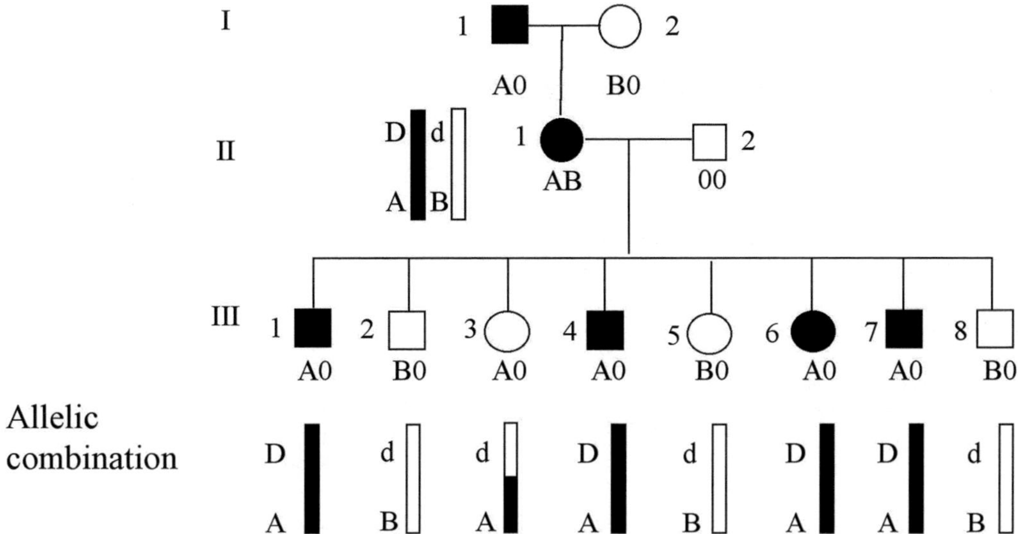

Fig. 4. Pedigree with members affected by Nail–Patella syndrome. The genotype at the ABO locus is reported for each individual. The parental combinations of II-1 (D-A and d-B) and the gametes transmitted to her eight children are reported. All of the children but III-3 (recombinant) received a parental combination.

chromosome. Thus, the more distant the loci, the greater is the probability for a crossing-over to occur between them somewhere along the length of the chromosome. As a consequence alleles at loci on the same chromosome co-segregate at a rate related to the distance between them. The distance separating two loci is measured by the recombination fraction (ϑ), which corresponds to the probability that a recombination event occurs between the two loci (*see* **Fig. 5**).

Because the frequency of recombinant gametes is a function of the distance between two genes, it is used as a measure of the so-called "genetic distance" between the two genes. One percent recombination is referred to as one map unit, or 1 centimorgan (cM), distance. The human genome has been estimated by recombination studies to be about 3000 cM. As the physical length of the haploid genome is 3000 Mb, 1 cM can be approximated to 1 Mb (1 million bases).

The maximum measurable genetic distance between two synthenic loci is 50 cM (50% recombination). This is explained by the fact that if a single crossing-over occurs, only two of the four chromatids are involved (*see* **Fig. 6A**). Considering two synthenic loci, a single crossover event occurring between them generates two recombinant gametes out of four. Therefore, 100% probability of a crossover event results in 50% recombinant gametes. However, two loci far apart can be separated by more than one crossing-over event. Double crossovers can involve two, three, or four chromatids (*see* **Fig. 6B**). As shown in **Fig. 6B**, because the three types of double crossing-over occur randomly, the overall effects of this double recombination event give, on average, 50% recombinants. Consequently, the proportion of the parental and recombinant gametic combinations when two loci are far apart on the same chromosome approximates to 50%, exactly as is seen for loci located on different chromosomes. Thus, when $\theta > 0.5$, the two loci are not linked, because they are either on different chromosomes or far apart on the same chromosome.

4. Informative Matings

Linkage analysis between two loci can be performed only when it is possible to distinguish individuals who inherit recombinant allelic combinations from individuals who inherit nonre-

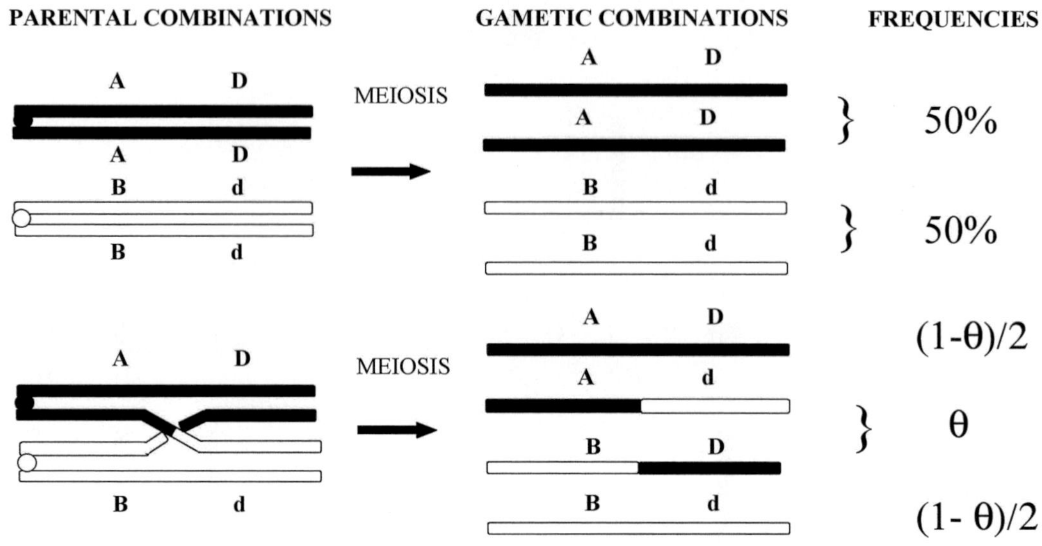

Fig. 5. Gametic combinations for two linked loci: (1) closely linked loci: no crossing-over occurs between the two loci. All of the gametic combinations are parental. (2) Crossing-over between the two loci occurs at a rate related to the distance separating the two loci. The proportion of recombinant gametes is equal to the recombination fraction (ϑ) and the proportion of nonrecombinant gametes is $1-\vartheta$, each type being $(1-\vartheta)/2$.

combinant (parental) combinations. Recombination events can be detected only when the individual is heterozygous at both loci and the phase of the parental alleles is known. The phase can be established for sure only in three-generation pedigrees with available grandparents. In the family in **Fig. 7A**, it is not possible to count recombinants and nonrecombinants because the phase of the parental gametes in unknown. The same family is reported with the first-generation genotypes available (*see* **Fig. 7B**). Now, the phase can be determined (B-D) and it is possible to distinguish whether a gamete passed to the third generation is the product of a recombination event or is a parental (nonrecombinant) gamete and to count recombinants and nonrecombinants in the offspring. In some pedigrees, even when the grandparents are accessible, the phase remains undetectable. This occurs when the parent useful for linkage analysis is homozygous (*see* **Fig. 7C**) or when the parents, although heterozygous, share the same marker alleles and their children are also heterozygous (*see* **Fig. 7D**).

5. Markers Used For Linkage Analysis

As stated above, the aim of linkage analysis is to find the rough location of the gene of interest relative to another DNA sequence (another gene or an anonymous DNA sequence), called a genetic marker, for which distinguishable alleles are known and with a precise chromosomal localization. Genetic markers are sequences that show polymorphism (variations in sequence or size) in the population. Their alleles can be identified either by a recognizable phenotypic effect or directly by molecular analysis of the DNA. In the past, genes with polymorphic alleles that could be scored at the phenotypic level were used for mapping (e.g., blood groups and electrophoretically separable isozymes). In the last two decades, the important advances in molecular biology have made possible the rapid development of new technologies leading to the identification of numerous markers at the DNA level.

As the identification of recombinants requires double heterozygotes for both the marker allele and the disease loci, a crucial feature of the marker is that it should have a high heterozygosity index so that a randomly selected individual has a high probability of being heterozy-

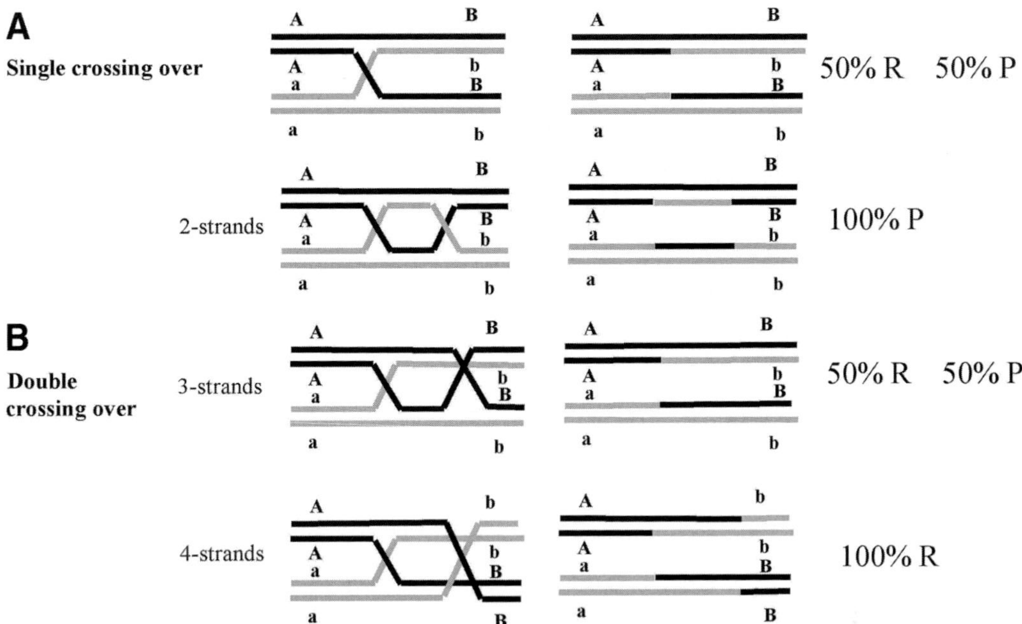

Fig. 6. Each crossing-over event involves two of the four sister chromatids: **(A)** a single crossing-over generates two recombinant and two nonrecombinant chromatids; **(B)** a double crossing-over generates only non recombinant chromatids if it involves two strands; a threestrand double crossing-over generates two recombinants and two nonrecombinants; a four-strand double crossing-over generates only recombinants. The three types of double crossing-over occur randomly, so the average effect is 50% recombinants.

gous. The heterozygosity of a marker corresponds to the chance that a randomly selected person will be heterozygous for that marker. This depends on the number of alleles and their relative frequency. If the marker alleles are A_1, A_2, A_3,…, and their frequencies are p_1, p_2, p_3,…, the heterozygosity is

$$H = 1 - \sum p_i^2$$

where *pi* corresponds to the frequency of the p_i-th allele, and p^2 is the fraction of homozygotes. For biallelic markers, the maximum heterozygosity is 0.5, corresponding to the presence of two alleles with equal frequency [$H = 1 - (0.5^2 + 0.5^2) = 0.5$].

The first generation of DNA markers was represented by RFLPs (restriction fragment length polymorphisms) corresponding to DNA variations located within a restriction enzyme recognition site. The analysis of RFLPs required isolation of a large amount of DNA, restriction digests, Southern blotting, and hybridization with radiolabeled probes. The major limitation of RFLPs was their uninformativeness, as they are biallelic markers with the maximum heterozygosity of 0.5.

A great improvement was given by minisatellite VNTR (variable number tandem repeat)—tandemly repeated DNA sequences of 100–1000 bp that show an elevated number of alleles in the population and, thus, a high level of heterozygosity. However, they are not uniformly distributed all over the genome (they tend to cluster near telomeres) and have the drawback of requiring typing by Southern blotting and radioactive detection.

The last generation of highly informative polymorphic markers is represented by microsatellites, which consist of particular DNA sequence motifs ranging from 1 to 6 basepair

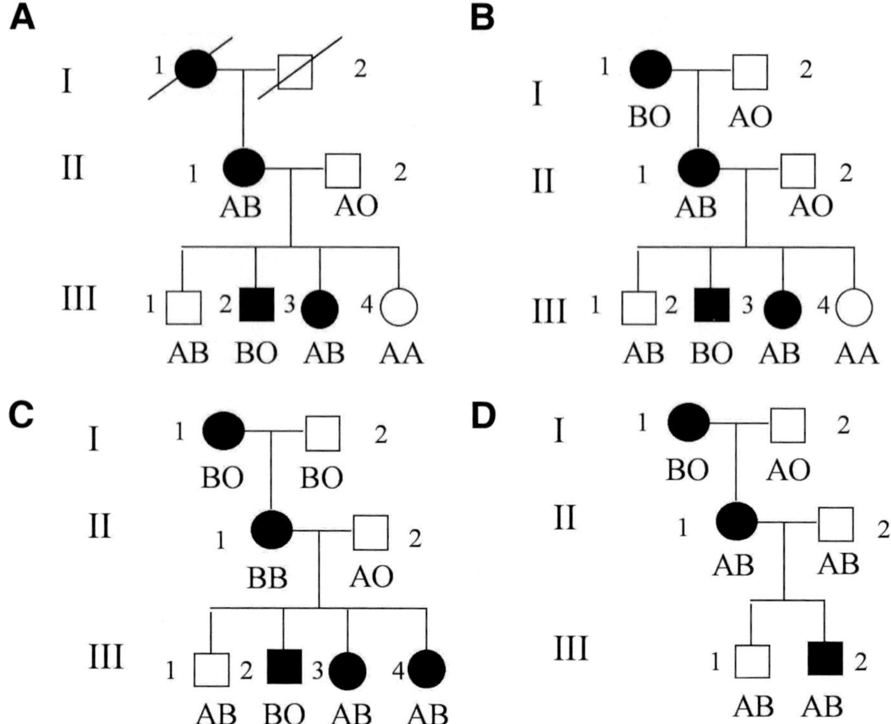

Fig. 7. Informative and uninformative meioses for linkage analysis. **(A)** Uninformative meioses: the grandparents are not available and the phase of II-1 (useful for linkage analysis) is unknown. **(B)** Informative meioses: the genotypes of the first generation are now available and the phase can be determined (B-D and A-D). **(C)** Uninformative meioses: II-1 is homozygous at the ABO locus and it is not possible to distinguish which alleles she transmits to the offspring. **(D)** Uninformative meioses: the two parents II-1 and II-2 are heterozygous for the same marker alleles; the children could have inherited A from the mother and B from the father, or vice versa.

(bp) tandem repetitions. They usually have multiple alleles differing in the repetition number and show a high level of heterozygosity. The presence of numerous different alleles decreases the probability that the two parents are heterozygous for the same alleles (as the example of **Fig. 7C**), thus decreasing the probability of uninformative meioses. Microsatellites are uniformly distributed all over the genome and are easily genotyped by polymerase chain reaction (PCR). These markers became the most popular genetic markers and a high-density microsatellite map including more than 5000 microsatellites has been developed *(2)*.

6. The LOD Score Analysis

When completely informative pedigrees are available, it is possible to detect unambiguously recombinants and nonrecombinants. Crossing-over events can be identified for sure in three-generation pedigrees, where the phase of the parental gametes can be defined. This is the case of the family reported in **Fig. 4,** where it is possible to establish that the AB0 locus is linked to the nail–patella locus, as the A allele shows a strong cosegregation with the disease allele. We can directly identify the recombinants and nonrecombinants unambiguously because the phase of the double-heterozygous individual II-1, useful for the linkage analysis, is known. We observe one recombinant (III-3) in eight meioses with an estimated recombination fraction of 1/8 (12.5%). We then need a statistical test to evaluate whether the observed ratio 1 : 8 is significantly different from 1 : 1, the ratio expected in the null hypothesis of no linkage (see below).

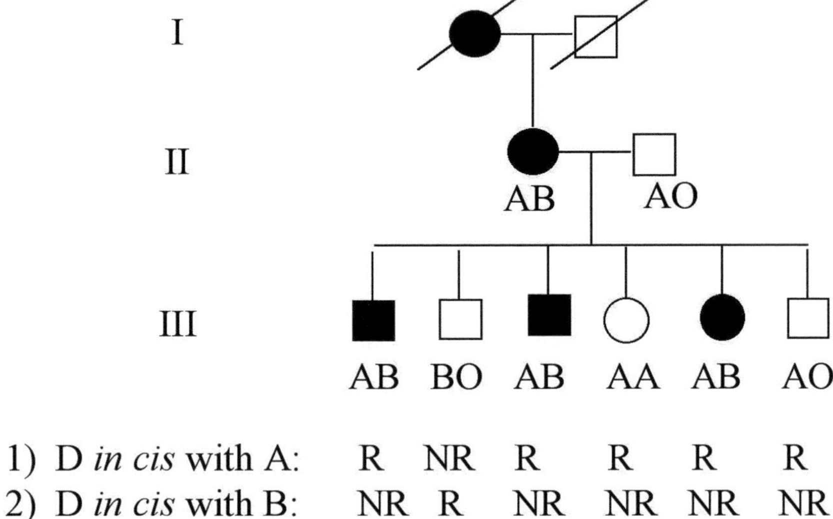

Fig. 8. Family with Nail–Patella syndrome, typed for the ABO locus. II-1 is a double heterozygous (AB, Dd) individual but the phase is unknown. If the phase is D-A (1), there are five recombinants and one nonrecombinant. Alternatively, if the phase is D-B (2), there are one recombinant and five nonrecombinants. R = recombinant; NR = nonrecobinant.

In most human pedigrees, because of the small family size and the difficulty in collecting grandparent DNA, it is impossible to determine the allele phase and to directly count the recombinants. In **Fig. 8**, individual II-1 is again double heterozygous, but the phase is unknown. Therefore, among her children, there can be either five recombinants and one nonrecombinant or there are five nonrecombinants and one recombinant. In this case, we cannot identify unambiguously recombinants and nonrecombinants.

The test used to calculate the significance of the observations indicating that two loci are linked at a certain genetic distance (ranging from $\vartheta = 0$ to $\vartheta = 0.5$) is the LOD score test. It evaluates the overall likelihood of the pedigree, on the alternative assumption that the loci are linked ($\vartheta < 0.5$) or not linked ($\vartheta = 0.5$). The ratio of these two likelihoods gives the odds of linkage and it is expressed as the log of this ratio (LOD stands for "log of odds"). Being expressed as a logarithm, the LOD score has the useful feature that scores from different matings for which the same loci are analyzed can be added, hence providing a cumulative set of data either supporting or not supporting some particular linkage value.

When performing a LOD score test we ask "what is the likelihood that the detected meioses (births) come from linked loci with a certain percentage of recombination ($0 < \vartheta < 0.5$) as compared with the probability that the same sequence would have been produced by independent loci ($\vartheta = 0.5$)?" In each family, the probability for an individual being recombinant is ϑ and the probability of being nonrecombinant is $1–\vartheta$. To determine the LOD score, different values of ϑ are chosen. The Z value (LOD score) is calculated for each, and then the ϑ at which the maximum LOD score (Z_{max}) is obtained gives the best estimate (maximum likelihood estimate, MLE) of the recombination fraction between the two loci. This value of ϑ is the estimate of the map distance between the marker and the disease locus.

6.1. Example of Lod Score Calculation When the Phase is Known

In the family of **Fig. 4**, we can establish that among the children, there are one recombinant (*a priori* probability ϑ) and seven nonrecombinants (*a priori* probability $1\text{-}\vartheta$) because the phase in individual II-1 is known (A is in coupling with D). The overall likelihood given linkage is

$(1-\vartheta)^7\vartheta$. The likelihood given no linkage is $(0.5)^8$. The linkage/nonlinkage likelihood ratio is $(1-\vartheta)^7\vartheta / (0.5)^8$ and the LOD score (logarithm of the ratio) is

$$Z(\vartheta) = \log[(1-\vartheta)^7 \vartheta / (0.5)^8].$$

We can calculate the Z values for different ϑ values. For example, determine Z at $\vartheta = 0.2$:

$$\text{Probability given linkage} = (1-0.2)^7 (0.2) = 0.0419,$$

$$\text{Probability with no linkage} = 0.5^8 = 0.0039,$$

$$\text{LOD score} = \log(0.0419 / 0.0039) = 1.031.$$

The calculation is repeated with other ϑ values, including 0.125, the recombination fraction value calculated by directly counting the recombinant and nonrecombinants, giving the following results:

ϑ	0	0.1	0.125	0.2	0.3	0.4	0.5
Z	$-\infty$	1.08	1.1163	1.031	0.801	0.458	0

The value of ϑ for which Z reaches a maximum positive value (Z_{max}) is 1.1163 at a ϑ value of 0.125. However, this would not be considered a significant LOD value. In fact, the linkage hypothesis is conventionally considered statistically significant when it is 1000-fold (LOD score = 3) more likely than the independence hypothesis. To obtain such high a value, even if the linkage hypothesis is correct, larger sample sizes are needed. This could be done either by extending the analysis to other family members (if they are available) or by adding the Z value calculated from other pedigrees.

6.1.1. Example of LOD Score Calculation When the Phase is Unknown

In the family reported in **Fig. 8**, the phase of the parental gamete is unknown. This represents the situation more frequently encountered in human pedigrees. There is no way to know for certain if the D allele at the nail–patella locus is in cis with the A allele at the ABO locus (D-A; prior probability 1/2) or with the B allele (D-B; prior probability 1/2). If the correct phase is D-A, there are five nonrecombinants and one recombinant. The overall likelihood of the observed sequence of meioses is $1/2\ (1-\vartheta)^5\vartheta$. If the correct phase is D-B, there is one recombinant and five nonrecombinants. The overall likelihood of the latter sequence of meioses is $1/2\ (1-\vartheta)\vartheta^5$.

The overall likelihood is then calculated as

$$Z(\vartheta) = \log\left\{\left[\frac{1}{2}(1-\vartheta)^5\vartheta + \frac{1}{2}(1-\vartheta)\vartheta^5\right]/(0.5)^6\right\}$$

This allows for either phases with equal prior probability. The calculation is performed at different ϑ values giving the following results:

J	0	0.1	0.2	0.3	0.4	0.5
Z	$-\infty$	0.276	0.323	0.222	0.076	0

Many families can be scored and the Z values obtained can be added to reach a significant result demonstrating linkage between the two loci. The total results can be displayed graphically (*see* **Fig. 9**). The graph was obtained from the LOD scores calculated in different families including the pedigrees of **Figs. 4** and **8**. In this example, the LOD score at $\vartheta = 0$ tends to $-\infty$. This is generally the case if there has been at least one recombination event between the loci, where it is clear that the two loci cannot be zero distance apart. In the graph, the maximum value of Z occurs at a recombination fraction of 0.1 and it is higher than 3. Therefore, we can feel confident that the two loci are linked and the best estimate of their genetic distance is ϑ =0.1, which corresponds to 10 cM.

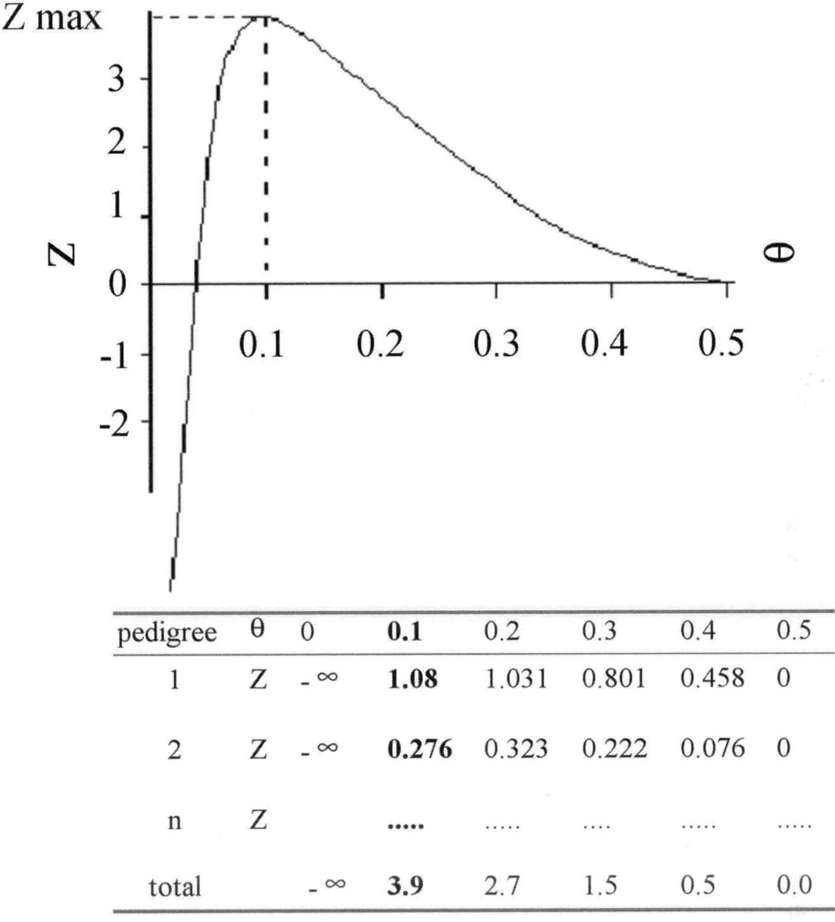

pedigree	θ	0	**0.1**	0.2	0.3	0.4	0.5
1	Z	- ∞	**1.08**	1.031	0.801	0.458	0
2	Z	- ∞	**0.276**	0.323	0.222	0.076	0
n	Z		.....	.....	....	.....	.....
total		- ∞	**3.9**	2.7	1.5	0.5	0.0

Fig. 9. Graph of LOD score against recombination fraction from a hypothetical set of families. LOD score values are reported for pedigrees 1 and 2 corresponding to the families of Figs. 4 and 8, respectively. All of the LOD score values from *n* families are added and the total LOD scores are reported for different ϑ values. The maximum value of Z ($Z_{max} = 3.9$) occurs at a recombination fraction of 0.1, suggesting that the disease locus and the ABO locus are 10 cM apart.

Linkage is conventionally excluded when $Z < -2$ which corresponds to 100 : 1 odds against linkage. Values of Z between –2 and 3 are defined inconclusive. In the examples of nail–patella affected pedigrees reported above, the LOD scores could be easily calculated manually. However in many monofactorial diseases several factors can complicate linkage analysis. One of these is the incomplete penetrance. Nail–patella syndrome is a dominant disease with complete penetrance; that is, all of the individuals carrying the Dd genotype manifest the clinical symptoms typical of the syndrome. In several dominant diseases, the penetrance is not 100%, but it is incomplete (i.e., not all of the individuals carrying the Dd genotype are affected), thus the genotype of unaffected family members cannot be known for certain. Another complicating factor is genetic heterogeneity, which means that mutations in different genes are responsible for the disease. Consequently, in different families, the disease will be coinherited with different marker loci.

Therefore, several sophisticated software packages have been developed (such as the LINKAGE package) to determine the LOD score values for a continuous range of ϑ from very complex pedigree data, considering factors such as the incomplete penetrance and the genetic heterogeneity.

6.1.1.1. Multipoint Linkage Analysis

Up to now, we have only considered linkage analysis with two loci (the disease locus and one marker locus). However, by performing two-point linkage analysis, we have no information about the order of the loci. In the example of the nail-patella locus, we know that it is at about 10 cM from the ABO locus, but it is impossible to determine on which side with respect to the marker locus it is located.

In multipoint analysis, the inheritance of multiple loci in an interval is considered simultaneously. Instead of calculating the recombination fraction between a single marker and the disease locus "D," the recombination fraction is determined between the disease locus and several markers at the same time. In this approach, a fixed map of known markers is used to establish the position of an unknown disease locus. The locus D is moved across the map and the LOD score values are calculated for each possible point of the map considering all the markers of the map simultaneously. A great advantage of multipoint linkage analysis is that it gives a higher LOD score because the LOD-scores obtained at the multiple points of the map can be added. Programs such as LINKMAP (part of the LINKAGE package) and GENHUNTER are commonly used to compute the overall likelihood of the pedigree at each position.

The basic approach currently used in genetic mapping of human diseases involves the analysis of the segregation of many markers distributed all over the genome. To this aim, a map of about 300 microsatellites spaced, on average, 10 cM is used. In affected families, one or few microsatellites will be eventually identified that cosegregate with the disease, giving significant LOD score values.

The next step is collecting as many families as possible and using a denser marker map to refine the chromosomal position in order to identify the shorter interval containing the disease gene.

7. Genetic Dissection of Complex Diseases

Many common diseases such as diabetes, multiple sclerosis, asthma, and cardiovascular diseases are referred to as complex disorders because, although they show a tendency to cluster within families, they do not follow the simple Mendelian rules of inheritance. The relationship between the genotype and the phenotype in complex diseases is complicated by the following factors:

1. More than one gene is involved (i.e., disease susceptibility might require the simultaneous presence of mutations in multiple genes [polygenic inheritance]). These genes might contribute together to the disease in an additive fashion or by their product interaction (multiplicative model).
2. Variations in different genes might result in identical phenotypes (genetic heterogeneity). This means that affected individuals might possess disease-contributing variations in different genes.
3. Variants that contribute to the disease will probably have much more subtle effects on the activity of gene products than those observed in Mendelian disorders. Nucleotide variations involved in such diseases are likely not pathogenic mutations but rather polymorphisms also present in the normal population. Consequently, most susceptibility disease alleles only confer a modest risk to disease susceptibility and individuals carrying predisposing allele might not manifest the disease (incomplete penetrance).
4. The disease is determined by a complex interaction between the susceptibility genes and the environment (multifactorial inheritance). For some diseases, the environmental factors are known. For example, in obesity, the environmental factor is food intake. For many other complex diseases, such as mental illness and multiple sclerosis, there is only a slight idea of what the environmental factors are.

5. In some cases, the disease is triggered by environmental factors in the absence of a predisposing genotype (phenocopies).

For all of these reasons, the identification of genes involved in multifactorial diseases is much more complicated than Mendelian disease genes. Parametric linkage analysis, which requires a precise genetic model that specifies the modality of transmission and the penetrance of each genotype, is not suitable for mapping genes involved in complex diseases. The analysis of a complex disease is performed through methods that do not need to specify the transmission model, and these methods are termed non-parametric or "model free." Nonparametric analysis ignores the unaffected people, thereby eliminating the problem of nonpenetrant carriers of disease-predisposing alleles. Such tests look for alleles or chromosomal segments that are shared between affected people by searching for markers shared by the affected members either within families (linkage) or in the whole population (association).

Nonparametric linkage analysis investigates the coinheritance of the disease and of marker alleles in affected relatives. If an allele at a marker locus tends to be coinherited by the affected family members more frequently than expected by the Mendelian random segregation, then the allele might be linked to a disease-susceptibility gene.

Association relies on the detection of markers shared by affected people in the population through the identification of allele frequencies that differ in patients and controls (association). Association studies test whether a particular allele, genotype, or haplotype is overrepresented (positive association) or underrepresented (negative association) in affected unrelated individuals compared with unaffected controls.

In principle, searching for association (i.e., counting how many times the marker and the disease are present together in the population) is the same operation as that of looking at linkage in families (i.e., counting how many times the marker and the disease are inherited together by relatives). The difference is that all affected individuals in a population are now treated as distant relatives and one tests the frequency by which they have inherited jointly the disease and the marker from a common ancestor. Thus, one looks not only to the meioses in one or few generations as in linkage analysis but also to all meioses in the history of the population.

Linkage and association provide different types of information. Each has its limitations and advantages. Linkage carries the advantage over association studies that, in general, it extends over greater genetic distances and can therefore be detected using fewer markers. Conversely, association has the advantage that it is considerably more powerful than linkage and therefore requires smaller study populations. One adopted strategy is first to determine by whole-genome linkage analysis a chromosomal region where the susceptibility gene is located (positional cloning) and then to refine the map by association analysis.

8. Identical-by-State vs Identical-by-Descent Alleles

The analysis of chromosomal regions shared by affected individuals in the same family needs to distinguish alleles that are identical-by-state (IBS) from alleles identical-by-descent (IBD). Any two identical copies of an alleles are said to be IBS. If these allele are inherited from the same ancestor, then they are also IBD. It is possible to determine if two individuals of the same pedigree have the same allele at a marker locus just by typing them. For example, in **Fig. 10A**, the two affected children were typed for the marker A and they both have allele A_1 and differ for the other allele. Therefore, the number of IBS allele is 1. In order to establish if this allele is also IBD, the typing of additional family members is required. The genotypes of the parents reveal that alleles A1 are not IBD: one child inherits A_1 from the mother and the other from the father (*see* **Fig. 10B**). In the family reported in **Fig. 10C**, again both the two affected children have the allele A_1. In this case, because they inherit it from the heterozygous A_1A_2 father, the allele is IBD. In some cases although both parents have been genotyped, it is not possible to infer if the identical allele is IBD. This situation is encountered when one or both parents are homozygous (*see* **Fig. 10D**) and when both parents and the children are heterozygous for the same genotype (*see* **Fig. 10E**). Clearly, IBD implies IBS, but not vice versa.

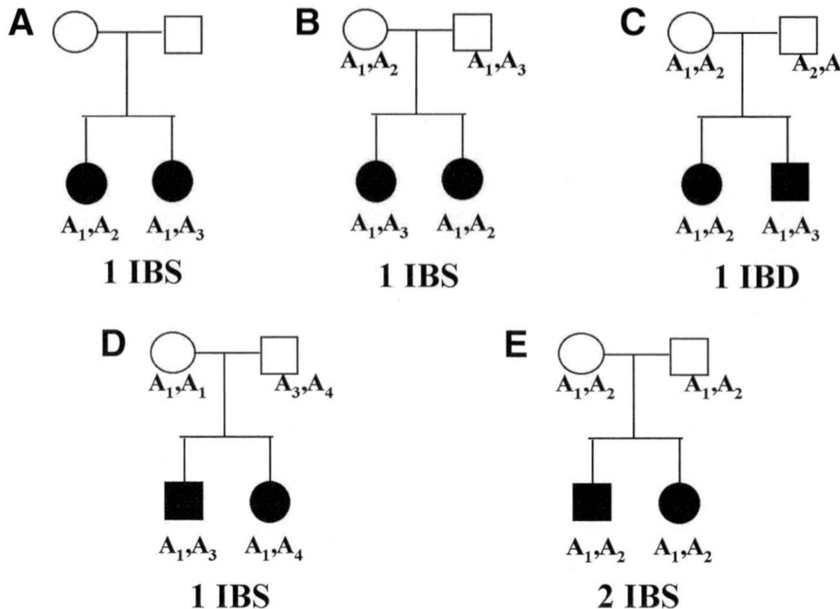

Fig. 10. Identical-by-state (IBS) and identitical-by-descent (IBD) alleles. **(A)** The two sibs have one identical allele (A_1). As the parents' genotypes are not available, we do not know if A_1 is IBS or IBD. The genotypes of the parents reveal that A_1 is IBS **(B)**. **(C)** The two children share one IBD; they inherited it from the heterozygous mother. **(D)** Both of the two children inherited A1 from their homozygous mother. The allele is IBS for sure, but we cannot distinguish if it is also IBD. **(E)** Parents and children are heterozygous for the same genotype; the children could have inherited A_1 from the mother and B_1 from the father, or vice versa. They have two IBS alleles; we cannot distinguish if they are also IBD.

9. Affected Sib Pair Analysis

Affected sib pair (ASP) analysis is the simplest form of interfamilial allele sharing methods. It relies on the fact that if a marker is linked to a disease locus, then the same marker allele will be inherited by two affected siblings more often than would be expected by chance. In this approach, families with two affected children are collected and genotyped. The probability that, for a given locus, affected siblings will share by chance neither, one, or two IBD alleles at a given locus is 0.25, 0.50, and 0.25, respectively (*see* **Fig. 11**). Linkage is present if that distribution is significantly distorted in favor of one or two sharing. Excess allele sharing is calculated using the chi-square (χ^2) statistic.

In **Table 1** are shown the results from a study on sib pairs affected by type 1 diabetes (T1DM), with markers positioned on different chromosomes. For each marker the transmission from one parent to the two affected sibs is reported. If the locus is not linked to the disease, 50% of the sib pairs' two children should share the same allele. On the other hand, if the locus is linked to the disease, the proportion of sib pairs sharing the same allele should exceed 50%. A strong deviation of the observed IBD sharing percentage from the expected values is observed for a marker on chromosome 6 mapping within the human leukocyte antigen (HLA) region. This distortion is statistically significant, indicating that the HLA region is linked to T1DM.

9.1. Whole-Genome Screening

The usual practice in mapping complex traits by ASP analysis involves genotyping multiplex families with two affected children (multiplex families) for hundreds of microsatellites

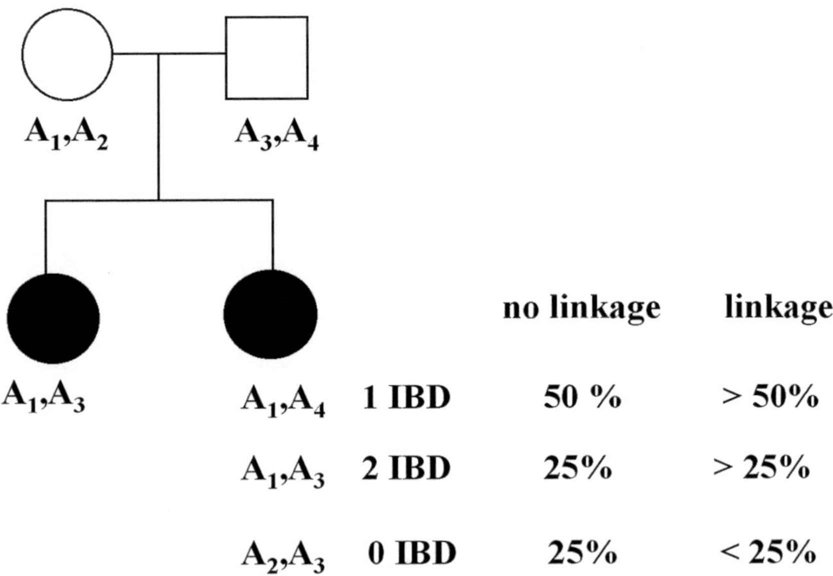

Fig. 11. Affected sib pair linkage analysis. By random segregation, sib pairs share none, one, or two IBD alleles 25%, 50%, or 25% of the times, respectively. If linkage is present, the siblings will tend to share one or two IBD alleles more often than 50% and 25%, respectively.

Table 1
Affected Sib Pair Analysis in Type 1 Diabetes

Chromosome	Marker locus	No. of informative meiosis	No. of ASP who inherited from one parent: Same allele (1 IBD)	No. of ASP who inherited from one parent: Different alleles (0 IBD)	Chi-square
1	D1S234	142	73 (51%)	69	0.062
1	D1S238	147	75 (51%)	72	0.061
2	D2S131	138	67 (49%)	71	0.116
2	D2S125	147	68 (46%)	79	0.823
3	D3S1297	125	60 (48%)	65	0.2
3	D3S1265	139	70 (50%)	69	0.007
6	**D6S258**	**117**	**90 (77%)**	**27**	**33.923**
6	DSS291	102	54 (53%)	48	0.353
7	D7S503	138	71 (51%)	67	0.116
7	D7S684	140	68 (48%)	72	0.114
9	D9S144	123	61 (50%)	62	0.008
9	D9S118	138	68 (53%)	70	0.029

markers (300–400) throughout the genome (whole-genome screening). When an elevated degree of allele sharing is found with a marker, it means that a susceptibility gene might be located nearby. As in the case of a parametric LOD score, it is crucial to evaluate the strength of findings of linkage at different sites in the genome. This is usually reported as the LOD score or the *p* value. A maximum LOD score ratio is used to test whether the degree of sharing is significantly distorted from the expected sharing under the hypothesis of no linkage, just like the parametric LOD score is used to test whether a recombination frequency less than 0.5 is signifi-

Table 2
Lander and Kruglyak Significance Criteria for Mapping Loci Involved in Complex Diseases by Whole-Genome Scans in Affected Sib Pairs

Category	Range of p-values	Range of LOD scores
No linkage	1.00–0.0008	0–2.1
Suggestive linkage	0.0007–0.00003	2.2–3.5
Significant linkage	0.00002–0.0000004	3.6–5.4
Highly significant linkage	≤ 0.0000003	≥5.4
Confirmed linkage	Significant linkage in an initial study, confirmed in an independent sample	

cant. The *p* value corresponds to the probability that the observed deviation in allele sharing has arisen by chance under the independent assortment of the alleles. Several geneticists have suggested some guidelines for interpreting the significance of evidence of linkage data obtained from whole-genome screening of ASP to avoid false-positive results. At the same time, caution is needed not to run the risk of missing true hints of linkage. The significance of a positive linkage result depends on how often such deviation would arise by chance in a whole-genome search. Lander and Kruglyak *(3)* proposed some significance criteria for mapping loci involved in complex diseases by whole genome scans categorizing the significant *p* value, as shown in **Table 2**. In their guidelines, the authors suggested that to reach a significant linkage, a threshold of LOD ≥ 3.6 must be imposed. Moreover, a linkage study that gave a significant LOD score must be replicated in a second independent set of families (replication study) to be credible and therefore confirmed.

The most widely used software for the nonparametric LOD score is the GENEHUNTER program, which performs multipoint analysis in affected sib pairs. The results can be graphically shown as curves reporting the LOD score values (*y*-axis) for each chromosome marker (*x*-axis). In **Fig. 12** is presented a graph with a linkage curve for chromosome 6 obtained from a genomewide scan on sib pairs affected by celiac disease (CD) *(4)*. A highly significant LOD score is reached in correspondence of the HLA region, which is known to play an important role in CD susceptibility.

10. The Example of Type 1 Diabetes

Type 1 diabetes mellitus is an example of complex disease for which several genome-wide linkage scans in affected sib pairs were conducted. It is caused by the autoimmune destruction of pancreatic B-cells affecting young people and requires lifelong insulin treatment. As in the case of many other autoimmune diseases, linkage to HLA region (called IDDM1 locus, insulin-dependent diabetes mellitus locus) on chromosome 6p21 has been consistently recognized across multiple studies. It has been estimated that HLA-conferred susceptibility might account for less than 50% of the genetic risk. This is clearly demonstrated by twin studies in which the concordance between dizygotic twins that share HLA haplotypes is much lower than concordance in monozygotic twins. This suggests that there is a substantial role for non-HLA genes. Using the affected sib pair non-parametric linkage analysis, several groups have identified a number of candidate susceptibility loci for T1DM, many of which have received a specific name (IDDM1 through 18; *see* **Table 3**) as a temporary operational designation. In **Table 3** are reported the results for a whole-genome scan on affected sib pairs performed for the US population *(5)*. In marked contrast with the strong effect of HLA, non-HLA loci likely provide only small additional contributions to disease susceptibility. Only two loci besides HLA have been replicated in independent studies. IDDM2 on chromosome 11p15 maps to a VNTR sequence located 596 bp upstream of the insulin gene translation start site. It has been suggested that the length of this VNTR affects insulin expression and T-cell tolerance to the protein. A region on

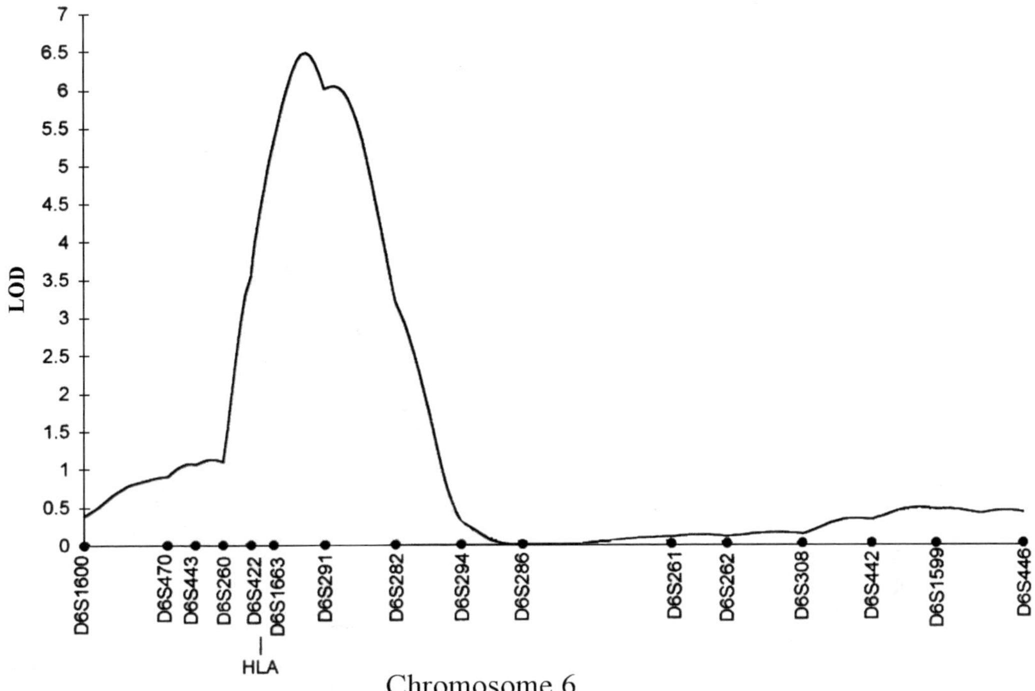

Fig. 12. Affected sib pair linkage analysis in coliac disease with markers on chromosome 6 ***(4)***. Multipoint LOD scores are plotted relatively to markers along the chromosome. A highly significant peak of linkage is detected in correspondence of markers mapping within the HLA region.

chromosome 2q34 (IDDM12) has been linked to type 1 diabetes by several groups ***(6–9)*** and it maps near the CTLA4/CD28 gene complex. These two genes are both expressed by T-cells: CD28 mediates T-cell activation, whereas CTLA4 induces apoptosis. A polymorphism in the first exon and a microsatellite in the 3' untranslated region of the CTLA-4 gene have been found to be associated to T1DM in independent studies ***(10–12)***.

Among the remaining loci, replication of linkage evidence was reported in different data sets for the susceptibility intervals IDDM4 (11q13), IDDM5 (6q25), IDDM6 (18q21), IDDM8 (6q27) although there is not as yet an indication for the likely susceptibility gene in these regions.

It is often difficult to detect and replicate linkage data and considerable heterogeneity is present among data sets. This can be attributed both to statistical errors (including false positives and false negatives) and to biological causes such as the large number of involved loci, genetic heterogeneity, variable disease gene frequencies and environmental factors. Given this situation, the study of many hundreds of affected sib-pairs of the order of thousands would be required to establish linkage and this material is not easy to collect.

Sib pair analysis has been used to investigate many other complex diseases such as asthma, diabetes, multiple sclerosis, schizophrenia, and celiac disease. All of these studies have been successful to a limited extent and progress has been extremely slow. Only a few genes and some genetic regions outside the HLA involved in complex diseases have been identified and replicated in different studies. Linkage analysis might be ineffective for mapping complex disease loci with low phenotypic effect and it can succeed only with moderate to large effect variations. Association studies provide a more powerful approach to detect disease susceptibility genes with a relatively modest effect.

Table 3
Reported Susceptibility Loci for Type 1 Diabetes

Locus	Chromosome	Lod score
IDDM1	6p21.3	65.8
IDDM2	11p15.5	4.28
IDDM3	15q26	0.01
IDDM4	11q13	1.37
IDDM5	6q25-q27	1.60
IDDM6	18q21	0.01
IDDM7	2q31-q33	2.62
IDDM8	6q25-q27	0.94
IDDM9	3q22-q25	0.50
IDDM10	10p11-q11	2.80
IDDM11	14q24.3-q31	1.22
IDDM12	2q33	0.77
IDDM13	2q34	0.23
IDDM15	6q21	2.36
IDDM17	10q25	1.38
IDDM18	5q33-34	0.05
D1S617	1q42	2.27
D16S3098	16q22-24	4.13

Table 4
Some Examples of HLA-Associated Diseases

		% Individuals carrying the associated allele	
Disease	Allele	Patients	Controls
Ankylosing spondylitis	B27	90%	9%
Type 1 diabetes	DR3	52%	23%
	DR4	74%	24%
	DR3 or DR4	93%	43%
Multiple sclerosis	DR2	60%	30%
Rheumatoid arthritis	DR4	81%	24%
Coliac disease	DQ2	95%	28%
Grave's disease	DR3	65%	27%
Narcolepsy	DR2	95%	33%

11. Association Studies

In general, an allele is said to be positively associated to a disease when its frequency is significantly higher in patients than in controls. Some examples of association of HLA alleles with autoimmune diseases are illustrated in **Table 4**.

A positive association with the disease could imply that the tested allelic variation is directly involved in the etiology of the disease. This might be the case of deleterious amino acid changes or promoter variations that alter gene expression.

Alternatively, a positive association could be the result of linkage disequilibrium (LD) between a neutral marker allele and the disease-predisposing allele, implying a close physical linkage of the marker and the disease involved gene. LD usually reflects the historical origin of a mutation on a specific chromosome bearing a characteristic set of allelic combinations. If the marker and the disease loci are very close to each other, the combination of the alleles at the two loci that was present in the ancestor might have been preserved through many generations. (This conservation is referred to as "linkage disequilibrium.") Thus, the particular allele combination would be found more often than expected by chance, i.e. considering the frequencies of the two alleles). The combination of alleles between more distant loci would be lost because of the increased probability of crossing-over between the two. As a consequence, when it does not involve the causal variations, association depends on the presence of linkage disequilibrium between the tested markers and the disease gene and it can be observed only using markers located very near to it. Conversely, detecting association of a marker with the disease indicates that the marker is located very close to the disease gene. An international project called the HapMap project has been launched to characterize the LD pattern throughout the genome, in the hope of facilitating association studies. It has been discovered that LD is highly structured into discrete islands, called haplotype blocks *(13)*. Each block comes only in a few common patterns and few markers can represent a haplotype block. Thus, when the haplotypic structure of the human genome is completed, the search for genes underlying common diseases will require testing only few single-nucleotide polymorphisms marking different versions of a given chromosome region. In the future, whole-genome association studies can be envisaged.

12. Markers Used in Association Study

The markers of choice in association studies are represented by single-nucleotide polymorphisms (SNPs; see Chapter 19), which include not only the traditional RFLPs but also nucleotide substitutions that do not alter or create a restriction enzyme site. With respect to microsatellites, they are less informative because they are mostly biallelic, whereas microsatellites have many alleles. However, SNPs are more frequent, occurring approx 1 per 500–1000 bp, leading to more than 3 million in the genome and they are also present within all genes. Moreover, they are less mutating than microsatellites and, thus, more stable in the population. SNPs also have a further advantage over microsatellites: whereas microsatellites themselves rarely contribute to the disease state and are usually located outside the coding regions, SNPs can also have a functional relevance if they occur in the coding or regulatory regions of a gene.

A map of the human genome sequence variation containing about 1.42 million SNPs has been developed recently *(14)*. This high-density SNP map provides, on average, 1 SNP every 1.9 Kb, with about 60,000 SNPs falling in coding sequences.

13. Relative Risk and Odds Ratio

The strength of the association between a marker and a disease is usually described by the relative risk (RR) or the odds ratio (OR). These parameters are widely used in epidemiology to asseses the risk conferred by the exposition to a risk factor that in genetics is the presence or the absence of a certain marker allele at a specific locus. The RR is used in prospective studies, in which one of the random sample consists of subjects who possess the risk factor (a marker allele) and the other random sample consists of subjects who do not possess the risk factor. The individuals of each sample are then followed in time to test which subjects eventually develop a certain disease and which subjects will not.

The scheme of the study can be represented as a 2×2 contingency table in the following form:

	Disease		
Marker	Present	Absent	Total
Present	a	b	$a + b$
Absent	c	d	$c + d$
Total	$a + c$	$b + d$	

The proportion of subjects with the risk factor who develop the disease is $a/(a+b)$ and it can be considered as a measure of the risk of developing the disease in the presence of the risk factor. Similarly, the risk of developing the disease in the absence of the risk factor is $c/(c+d)$.

The RR compares the risk of developing the disease in presence of the allele with the risk in absence and it is defined by the ratio of the two risks:

$$\mathrm{RR} = \frac{a/(a+b)}{c/(c+d)} = \frac{a(c+d)}{c(a+b)}$$

The odds ratio is, by definition, the ratio of two odds for an event (i.e., the ratio between the probability of one event and the probability that the event does not occur). The OR is used in retrospective studies in which one of the random sample consists of individuals who developed the disease (cases) and the other random samples consists of unaffected individuals (controls).

In this kind of approach, the number of affected and unaffected is fixed by the researcher and, thus, it makes no sense to estimate the risk of developing the disease in presence or in absence of the risk factor. In this case, a comparison of the presence of a risk factor for disease in a sample of affected subjects and nonunaffected controls is made: the number of individuals with disease who possess the allele over those with disease who do not possess it (*a*/*b*) divided by those without disease who were exposed over those without disease who were not exposed (*c*/*d*). Thus,

$$\mathrm{OR} = \frac{a/b}{c/d} = \frac{ad}{bc}$$

This measure should be used for association case-control studies where we retrospectively look at a disease allele in patients and controls. For example, in one study of 91 patients affected by late-onset Alzheimer disease and 71 healthy controls, a particular allele (ε4) of the apoliprotein E (ApoE) was found in 58 patients and 16 controls:

	Alzheimer's disease		
Allele ε4	Present (patients)	Absent (controls)	Total
Present (ε4+)	58	16	74
Absent (ε4–)	33	55	88
Total	91	71	

The OR is then (58×55)/(16×33) = 6. This mean that the odds of developing Alzheimer's disease are six time higher in individuals who carry the ApoE-ε4 allele with respect to those individuals who do not carry this polymorphism.

14. Transmission Disequilibrium Test

Case-control association studies are largely prone to detect spurious associations between markers and diseases resulting from population admixture. If population admixture is present or has occurred in the recent past, then this can give rise to an association between alleles that is not the result of linkage disequilibrium between loci. In mixed populations, any trait present more frequently in a given ethnic group will be apparently associated with any allele more common in that group. To take a trivial example, in the San Francisco Bay area, a positive association is found between the ability to use chopsticks and the HLA-A1 allele. This association is a consequence of population admixture because in the Chinese population (with many members living in San Francisco), the allele HLA-1 is frequent and it is rare in other ethnic groups. A crucial point in association studies is, thus, the case-control sample selection.

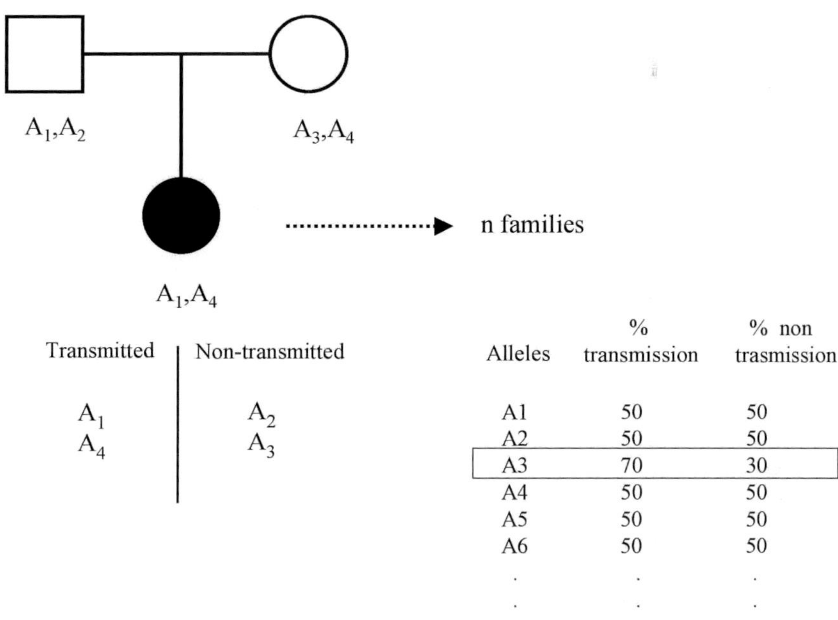

Alleles	% transmission	% non trasmission
A1	50	50
A2	50	50
A3	70	30
A4	50	50
A5	50	50
A6	50	50
.	.	.
.	.	.
.	.	.

Fig. 13. The transmission disequilibrium test is performed on families with an affected child (simplex). *n* families are genotyped with a microsatellite marker with multiple alleles. The allele transmitted from heterozygous parents to the affected child are counted. Each allele has a 50% chance of being transmitted. If an allele is involved in the disease, the chance of being transmitted will be higher than 50%. Allele A3 is transmitted 70% of the times and, thus, it might be in linkage disequilibrium with a disease-susceptibility allele.

The problem of population stratification can be reduced to some extent by using genetically isolated populations for case-control studies, for instance, the Finnish or the Sardinian populations. In a mixed population, a solution comes from family-based association tests as the transmission disequilibrium test (TDT) ***(15)***.

The transmission disequilibrium test is performed on nuclear families with one affected offspring and the two parents (simplex families; *see* **Fig. 13**). TDT compares the number of times that each parent transmits a given marker allele to an affected offspring with the number of times that he does not transmit it. By combining the data of many families, each allele has a 50% chance of being transmitted and a 50% of not being transmitted. If an allele at a locus is responsible for the disease or if it is very close to the disease locus and in linkage disequilibrium with it, the chance of being transmitted will be higher than 50%. This method detects true association, as a result of linkage disequilibrium, eliminating the possibility of spurious association resulting from population stratification. As an example, in a set of Type 1 diabetes families ascertained in the United States ***(4)***, the DQB1*0302 allele is preferentially transmitted to affected offspring (*see* **Table 5**)

A limitation of the TDT is that it can be performed only when the parents are heterozygous for the marker locus. This creates a loss of statistical power because the number of individuals useful for the analysis is confined to heterozygotes. Moreover, because it requires the parental genotypes to be specified, it is not feasible when late-onset diseases are studied, such as Alzheimer's disease. Consequently, although insensitive to population stratification, the family-based methodologies can be difficult to implement and might require significantly more patients and family collections than case-control studies.

Table 5
TDT Result for the HLA-DQB1*0302 Allele in Type 1 Diabetes Families

	No. of times the allele is		
	Transmitted	Non transmitted	$\chi 2$
Observed	393 (80.3%)	96	180
Expected	244.5 (50%)	244.5	

Acknowledgments

The author thanks Professor Patricia Momigliano-Richiardi for her comments and critically reviewing this chapter.

References

1. Rommens, J. M., Iannuzzi, M. C., Kerem, B., et al. (1989) Identification of the cystic fibrosis gene: chromosome walking and jumping. *Science* **245**, 1059–1065.
2. Dib C., Fauré S., Fizames C., et al. (1996) A comprehensive genetic map of the human genome based on 5,264 microsatellites. *Nature* **380**, 152–154.
3. Lander, E. and Kruglyak, L. (1995) Genetic dissection of complex traits: guidelines for interpreting and reporting linkage results. *Nature Genet.* **11**, 241–247.
4. Greco, L., Corazza, G., Babron, M. C., et al. (1998) Genome search in celiac disease *Am. J. Hum. Genet.* **62**, 669–675.
5. Cox, N. J., Wapelhorst, B., Morrison, V. A., et al. (2001) Seven regions of the genome show evidence of linkage to type 1 diabetes in a consensus analysis of 767 multiplex families. *Am. J. Hum. Genet.* **69**, 820–830.
6. Davies, J. L., Kawaguchi, Y., Bennett, S. T., et al (1994). A genome-wide search for human type 1 diabetes susceptibility genes. *Nature* **371**, 130–136.
7. Morahan, G., Huang, D., Tait, B. D. Colman, P. G., Harrison, L. C. (1996) Markers on distal chromosome 2q linked to insulin-dependent diabetes mellitus. *Science* **272**, 1811–1813.
8. Esposito, L., Hill, N. J., Pritchard, L. E., et al. (1998) Genetic analysis of chromosome 2 in type 1 diabetes: analysis of putative loci IDDM7, IDDM12, and IDDM13 and candidate genes NRAMP1 and IA-2 and the interleukin-1 gene cluster. *Diabetes* **47**, 1797–1799.
9. Mein C. A., Esposito L., Dunn M. G., et al. (1998). A search for type 1 diabetes susceptibility genes in families from the United Kingdom. *Nature Genet.* **19**, 297–300.
10. Nisticò, L., Buzzetti, R., Pritchard, L. E., et al. (1996). The CTLA-4 gene region of chromosome 2q33 is linked to, and associated with, type 1 diabetes. Belgian Diabetes Registry. *Hum. Mol. Genet.* **5**, 1075–1080.
11. Marron, M. P., Raffel, L. J., Garchon, H. J., et al. (1997) Insulin-dependent diabetes mellitus (IDDM) is associated with CTLA4 polymorphisms in multiple ethnic groups. *Hum. Mol. Genet.* **6**, 1275–1282.
12. Ueda, H., Howson, J. M., Esposito, L., et al. (2003) Association of the T-cell regulatory gene CTLA4 with susceptibility to autoimmune disease. *Nature* **423**, 506–511.
13. Wall, J. D. and Pritchard, J. K. (2003) Haplotype blocks and linkage disequilibrium in the human genome. *Nature Rev. Genet.* **4**, 587–597.
14. Sachidanandam, R., Weissman, D., Schmidt, S. C., et al. (2001) A map of human genome sequence variation containing 1.42 million single nucleotide polymorphisms. *Nature* **409**, 928–933.
15. Spielman, R. S., McGinnis, R. E., and Ewens, W. J. (1993) Transmission test for linkage disequilibrium: the insulin gene region and insulin-dependent diabetes mellitus (IDDM). *Am. J. Hum. Genet.* **52**, 506–516.

19

Single-Nucleotide Polymorphisms

Technology and Applications

Cyril Mamotte, Frank Christiansen, and Lyle J. Palmer

1. Introduction

The genomics revolution is transforming epidemiology, medicine, and drug discovery ***(1–7)*** and there is an ongoing refocusing of effort away from family-based linkage studies toward population-based genetic association studies for complex phenotypes ***(3,8–11)***. The generation of new genomic knowledge and its integration into epidemiological and clinical research projects in industry and academia are exponentially increasing trends. The genetic basis of disease susceptibility, disease progression and severity, and response to therapy for many complex conditions has been increasingly emphasized in medical research, with the ultimate goal of improving preventive strategies, diagnostic tools, and therapies ***(4,5,12,13)***. Enormous effort in both academia and industry has been expended in genetic studies of complex human diseases over the last decade. Concomitant technical developments in molecular genetics and in the use of polymorphism directly derived from DNA sequence have occurred, and extensive catalogs of DNA sequence variants across the human genome have been constructed ***(14–16)*** (see the Appendix, p. 247). The completion of the human genome project and the application of high-throughput technologies for polymorphism detection to the investigation of complex disease genetics has created unprecedented opportunities for understanding the pathogenic basis of common human diseases ***(1,16)***.

This chapter provides a technical review of the platforms currently available for genotyping single-nucleotide polymorphisms (SNPs) and summarizes the current and potential future applications of SNPs to our understanding of the pathophysiology of complex human diseases. Although the genomics revolution and the generation of high-density SNP maps has also greatly benefited investigations of Mendelian diseases, we restrict our discussion here to common complex human conditions such as obesity and cardiovascular disease that are determined by multiple genetic and environmental factors; such diseases constitute the principal health burden in the developed nations ***(1,4,5,17)***.

2. Genetic Polymorphism

Genetic polymorphism arises from mutation. The simplest class of polymorphism derives from a single-base mutation that substitutes one nucleotide for another, termed a "single-nucleotide polymorphism" or SNP (pronounced "snip") ***(18)***. SNPs are recognized through a variety of techniques that exploit the known DNA sequence variant ***(18)***. Previous nomenclature was based on the method used to detect a particular SNP (e.g., SNPs detected via the identification of restriction enzyme sites were called "restriction fragment length polymorphisms" [RFLPs])

From: *Medical Biomethods Handbook*
Edited by: J. M. Walker and R. Rapley © Humana Press, Inc., Totowa, NJ

(19). In addition to RFLPs, other types of SNP are detectable using an increasingly wide range of genotyping techniques (*see* **Subheading 3.1.**), or by direct sequencing ***(18)***.

The frequency of SNPs across the human genome is higher than for any other type of polymorphism ***(20)***. Precise estimates of SNP frequency are difficult to determine and often vary across different populations and genomic regions. On average, the genome of any two humans will differ from one another by around 0.1% of nucleotide sites (i.e., 1 SNP for every 1 kilobase [kb] ***(20,21)***. SNPs can be found in coding or regulatory regions of a gene and, thus, can directly affect gene function or expression. Most SNPs are likely to arise from a single historical mutational event, as the mutation rate is relatively low (around 10^{-8} per site per generation) compared to the most recent common ancestor of any two humans (around 10^4 generations) ***(22)***.

Because of their potential biological importance, the common SNPs in the human genome increasingly have been the subject of large-scale cataloging projects funded by both government and industry groups ***(14,15,23)***. It has been estimated that there are likely to be around 10 million common (≥1% frequency) SNPs in the human genome and that these common SNPs account for around 90% of human genetic variation ***(15,24)***. Worldwide public efforts have so far identified over 7 million SNPs, over 3 million of which are validated (www.ncbi.nlm.nih.gov/SNP/index.html). Many remaining SNPs are likely to be discovered in the near future through increased efforts by industry and academic research groups, aided by improvements in sequencing technology and capacity and the increased availability of high-efficiency mutation scanning techniques. Although these enterprises are constructing large SNP databases that are constantly updated and growing rapidly (see the Appendix), this process is not yet complete ***(15,24)***. The information contained in these databases is far from infallible and, as yet, there have been few systematic reviews of the accuracy of these data—an indeterminate proportion of which will be the result of sequencing errors ***(14,15)***. Limitations related to cost and the current incomplete status of SNP databases has meant that the association analysis of SNPs in complex disease genetics has, so far, generally been limited to polymorphisms within biologically plausible candidate loci. Many investigators interested in a specific genes or pathways have independently sought to identify sequence variants by primary resequencing ***(24,25)***.

In the context of medical research, SNPs are finding widespread use in fine mapping of genetic disorders, in the delineation of genetic influences in multifactorial diseases such as breast cancer, cardiovascular disease, and asthma, in haplotype mapping, and as genetic markers to predict responses to drugs and adverse drug reactions ***(17)***. These applications will involve SNP analyses on an unprecedented scale ***(26)***. Thus, there are dual imperatives to develop advanced technologies to detect SNPs for research and clinical applications and to concomitantly improve statistical approaches and study designs to enable the effective incorporation of SNP data into epidemiology and clinical medicine. To that end, considerable efforts are now under way by industry and research communities to establish cost-effective, high-throughput, and accurate SNP typing technologies and improved statistical methodologies ***(26)***.

3. SNP Discovery and Genotyping

The last decade has seen a dramatic increase in molecular genetic technologies that can potentially be used to understand the biological basis of complex human diseases. The current focus is on SNP discovery leading to the creation of SNP catalogs and on improving technologies for high-throughput and sensitive SNP genotyping ***(26)***. However, genotyping technologies are changing very rapidly, and it remains unclear as to the optimal genotyping platform and strategy to adopt. The key features of a useful genotyping assay include the ability to interrogate a unique genomic locus, accurate allele discrimination, the capacity for scaling assays to high-throughput (with high completion rates), the use of minimal amounts of DNA, low cost, and ease of implementing quality control measures.

The following subsections summarize, in broad terms, some of the current and emerging techniques for SNP genotyping. Most of the techniques described are polymerase chain reaction (PCR) based. Although such assays constitute the bulk of SNP detection technologies, there are SNP assays that are not PCR based, the Invader® (or flap probe cleavage) assay being a notable example. For the latter, and for greater technical detail on many of the techniques described, the reader is directed to numerous recent reviews ***(26–31)***. Although not addressed in this chapter, we note that DNA pooling is a new technique with considerable potential to both enhance rational SNP choice and to reduce genotyping cost for large-scale association studies ***(32,33)***.

3.1. SNP Detection Chemistries (Fig. 1)

3.1.1. Enzyme Digestion

Restriction enzyme digestion of PCR-amplified DNA followed by electrophoresis and ethidium bromide staining has been used extensively for SNP genotyping. Distinction of alleles is based on the selectivity of restriction endonucleases for short and specific DNA sequences, which are often disrupted by a SNP (*see* **Fig. 1**). Even when a SNP does not result in the creation or abolition of a restriction site, it is often possible to introduce artificial restriction sites by using mutagenic PCR primers ***(34)***. However, in the context of high-throughput genotyping, restriction enzyme analysis has several limitations. The processes, gel electrophoresis in particular, are not readily automated; some restriction enzymes are very expensive, and if a SNP does not affect a restriction site, it is not always possible to introduce an artificial restriction site; optimal incubation conditions for restriction enzymes can vary substantially, and if many SNPs are being examined, storage of a large catalog of enzymes is required. Nevertheless, the method is very simple and requires very little in terms of instrumentation and remains in use in many laboratories conducting SNP typing on a small scale. The use of 96-well gels, such as the MADGE system ***(35)*** or Invitrogen's E-gels (www.invitrogen.com) can be helpful in scaling up to higher throughput.

3.1.2. Allele-Specific Amplification

Allele-specific amplification, also known as the amplification refractory mutation system (ARMS), is also a commonly used method for SNP detection ***(36)*** (see Chapter 14). In this approach, the presence or absence of an amplification product is the basis for SNP typing. The method is based on the observation that under carefully controlled conditions, extension of PCR primers, and therefore PCR amplification, only occurs if nucleotides near the 3' end of the PCR primer are exactly complementary to the primer-binding site on the target sequence (*see* **Fig. 1**). In general, one of the pairs of PCR primers is positioned such that its 3' nucleotide sits over the polymorphic SNP site. In early embodiments, the result was determined by gel electrophoresis and ethidium bromide staining, an approach that was very amenable for low-cost, small-scale applications and is still in use today. A very simple but useful variation of the technique, termed mutagenically separated PCR (MS-PCR), uses two ARMS primers of different lengths for each SNP site in a PCR mix, one for each allele of a biallelic SNP, resulting in amplicons of different lengths for the two alleles ***(37)***. Subsequent gel electrophoresis shows at least one of the two allelic products, therefore excluding the possibility of false-negative results resulting from nonamplification. This approach has been adapted for very high throughput by Qiagen Genomics Inc. (www.qiagen.com). Their Masscode™ system, now available through Bioserve Biotechnologies Ltd. (www.bioserve.com), uses small molecular-weight tags, or mass tags that are covalently attached through a photo-cleavable linker to the ARMS primer, with each ARMS primer labeled with a tag of different molecular weight ***(38)***. Currently, there are some 48 different tags that can be used to simultaneously multiplex 24 SNPs (two tags per SNP) in a single PCR. For any one SNP, genotyping is based on comparison of the relative abundance of the two relevant mass tags by mass spectrometry. It is important to note that in this assay, accuracy is still based solely on differential amplification, which must be carefully optimized for each polymorphic site to be tested.

A more robust approach for allele-specific amplification is to use the oligonucleotide ligation assay (OLA) (see Chapter 22). The method is based on the principle that the ligation, or joining, of two adjacent oligonucleotides hybridized to a target DNA template by DNA ligases

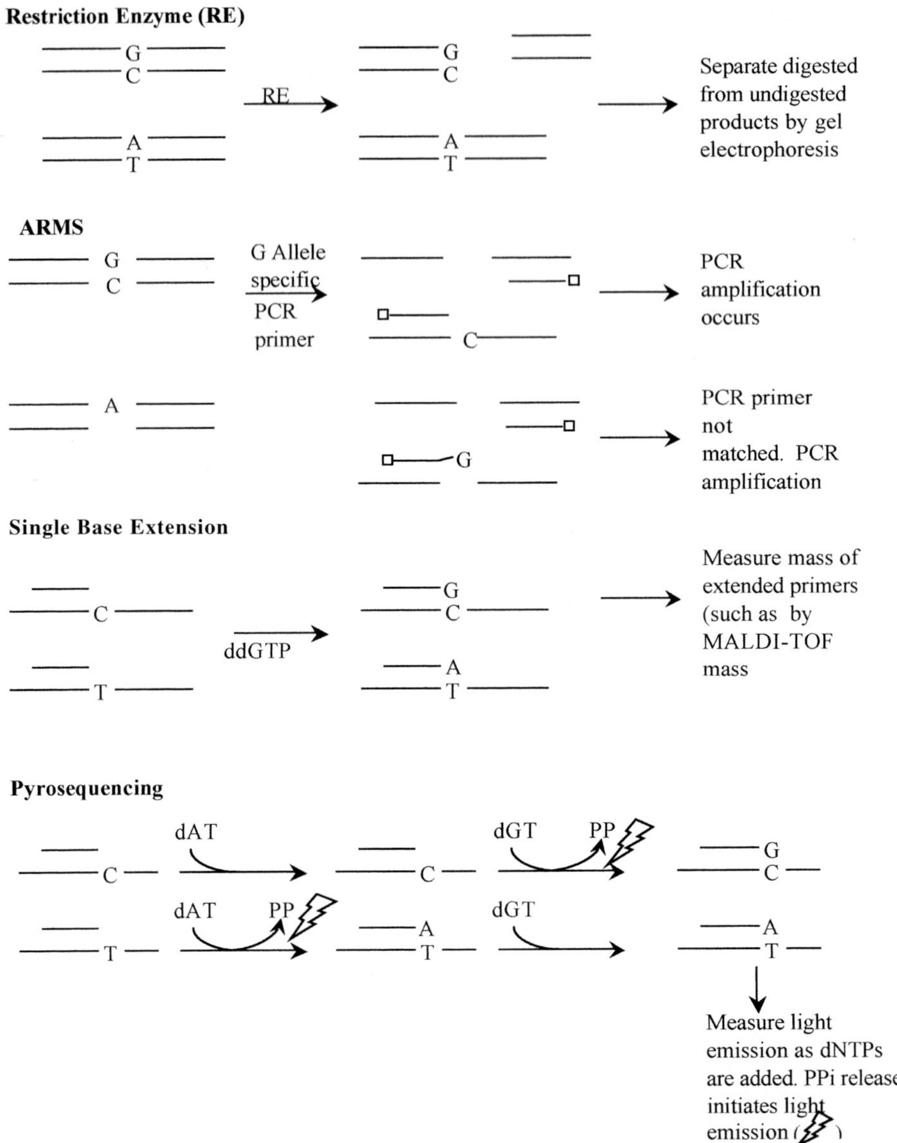

Fig. 1. Schematic diagram of several SNP typing technologies for G→A substitution. PPi, pyrophosphate; ARMS, amplification refractory mutation system. Additional configurations for single-base extension and ARMS are described in the text.

is very sensitive to mismatches at the ligation site. In OLA, a region spanning the SNP is first amplified by PCR. An aliquot of the amplicon is then added to a mix containing a thermostable DNA ligase and three oligonucleotides. Two of the oligonucleotides are allele-specific oligonucleotides (ASOs) and have their 3' nucleotides complementary to one or other of the two nucleotides defining a biallelic SNP, whereas the third binds immediately downstream to the ASOs. The mixture is taken through several cycles of denaturation, annealing, and ligation in a thermocycler, resulting in linear accumulation of ligated products. Numerous approaches can be used for the detection of the ligated products. For example, the ASOs can be differentially labeled with fluorescent or hapten labels and ligated products detected by fluorimetry or colorimetric enzyme-linked immunosorbent assay (ELISA), respectively ***(39,40)***. For electrophoresis-based systems, use of mobility modifier tags or variation in probe lengths coupled with fluorescence detection enables the multiplex analysis of several SNPs in a single tube ***(41,42)***. When used on arrays, ASOs can be spotted at specific locations or addresses on a chip, PCR-amplified DNA added and ligation to labeled oligonucleotides at specific addresses on the array measured ***(43)***.

Whereas most of the above techniques involve PCR amplification and then DNA ligation, this is reversed in the multiplex ligation-dependent probe amplification assay (MLPA). For any particular SNP using this approach, the two ASO probes and the third common probe are hybridized to genomic DNA, and after allele-specific ligation by DNA ligase, the joined probes are PCR amplified. SNP alleles can be differentiated by varying the length of the two ASOs and fragment size analysis on a capillary sequencer. Likewise, multiplexing can be achieved by varying the lengths of the ASOs for different SNP loci and, finally, tagging of the 5' tails of the ligation oligonucleotides with common PCR primer binding sites enables use of a single set of PCR primers for amplification of all ligated products. Using this principle, it is possible to multiplex amplification of numerous loci, and in the original report, 40 loci were amplified simultaneously ***(44)***. Applied Biosystems SNPlex™ assay (www.appliedbiosystems.com), which also uses multiplex amplification of ligated probes and analysis on a capillary sequencer, is a variation on this general approach.

The concept of amplifying hybridized probes has also been applied to microbead arrays using fluorescent detection. Illumina's GoldenGate assay (www.illumina.com) uses allele-specific primer extension, similar to ARMS primers, for probe synthesis and subsequent amplification of the synthesized probes for all SNP loci probes with common PCR primers. Three PCR primers are used; one is common and two are labeled with different fluorescent dyes and bind to sequence tags on one or other of the allele-specific primers used for probe synthesis. The concept of multiplex probe amplification based on the use of common PCR primer site tags is somewhat similar to MLPA and the SNPlex assay, but probe synthesis is by allele extension and the microbead array format affords a much higher order of multiplexing, greater than 1000-fold, corresponding to in excess of 100,000 genotypes per array matrix.

3.1.3. Single-Base Extension

Single-base extension or minisequencing (see Chapter 9) involves the annealing of an oligonucleotide primer to the single strand of a PCR amplicon, with the 3' end of the primer binding one base short of the SNP site (*see* **Fig. 1**). DNA polymerase then catalyzes extension of the primer in the presence of chain-terminating dideoxynucleotides. The dideoxynucleotide incorporated is complementary to the base at the SNP site. Accuracy comes from the specificity of dideoxynucleotide incorporation by DNA polymerase. This general approach has numerous advantages. The specificity of dideoxynucleotide incorporation by DNA polymerase makes the method specific and robust; incorporation of a single dideoxynucelotide can be monitored relatively easily using a number of chemistries and platforms suitable for small-scale or high-throughput genotyping and the binary nature of the results makes data analyses for many of the numerous embodiments very simple. The different chemistries and platforms include the following: use of hapten-labeled dideoxynucleotides, followed by colorimetric detection using enzyme-labeled antibodies and chromogenic substrates (Orchid's 96-Well colorimetric SNP-IT™ assay; www.orchid.com); differential labeling of the four ddNTPs and fluorescent detection of the incorporated ddNTPs using solid-phase capture on bead arrays, such as color-coded microbeads (Luminex; www.luminex.com); two-dimensional microarrays such as Orchid's SNPstream UHT system and as described by Fan et al. ***(45)***, or by electrophoresis on a sequencing gel ***(46)***, fluorescence polarization (*see* **Subheading 3.1.6.**); and mass spectrometry. The latter does not require labeling of ddNTPs; because of the intrinsic difference in molecular weight of the four dideoxynucleotides, genotypes can be assigned on the basis of the mass of the extended primer, most commonly using matrix-assisted laser desorption/ionization time-of

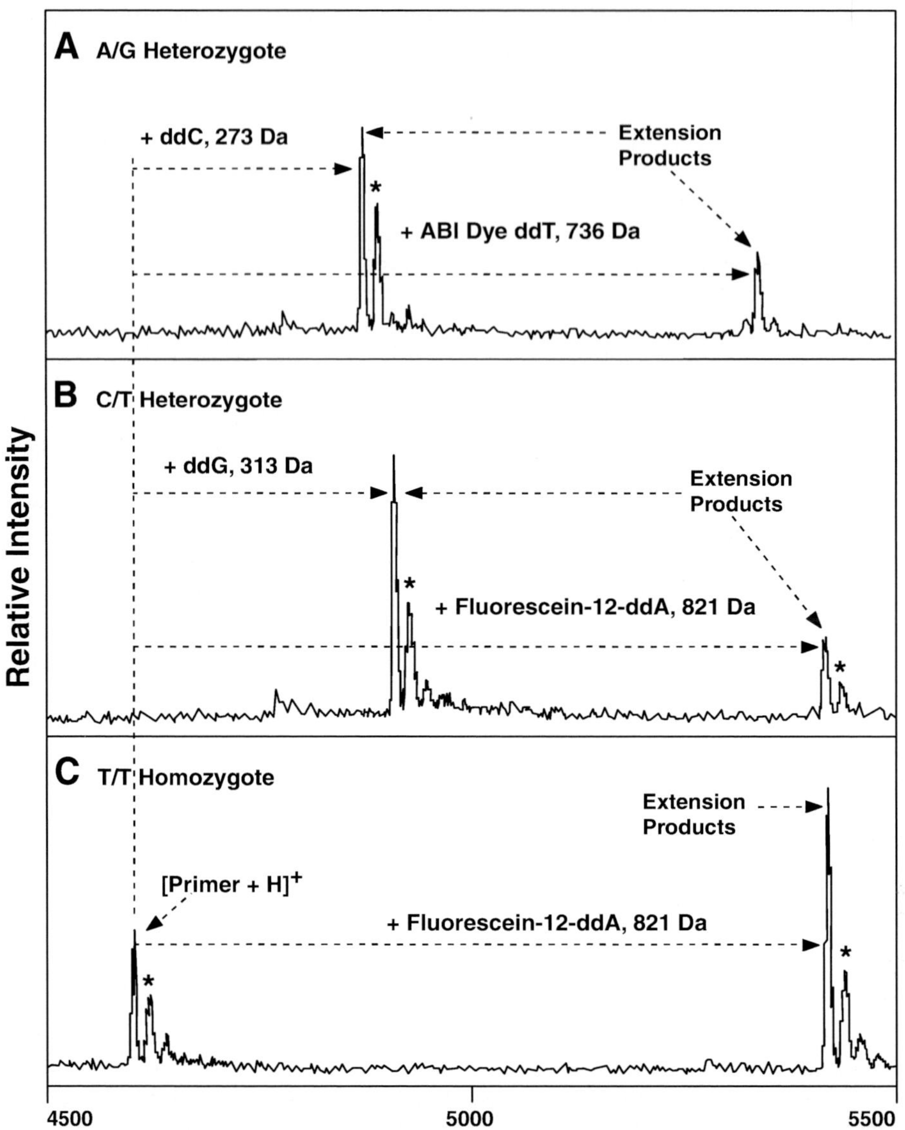

Fig. 2. Genotyping using mass spectrometry on a Sequenom system. SNP alleles are interrogated by mass spectrometry time-of-flight (MALDI-TOF). Polymorphism is tagged by nucleotide mass.

flight mass spectrometry (MALDI-TOF) *(47)*. Analysis by MALDI-TOF is very rapid and forms the basis for Sequenom's Mass Array technology (www.sequenom.com) (*see* **Fig. 2**). Multiplexing is possible on many of these platforms. For example, by using sequencing primers of different lengths, it is possible to interrogate multiple SNPs with the dideoxynucleotides simultaneously by MALDI-TOF or capillary electrophoresis, enabling high-throughput genotyping and efficient use of genomic DNA. Even higher orders of multiplexing are achievable on two-dimensional microarrays or bead arrays (e.g., Affymetrix; www.affymetrix.com).

3.1.4. Allele-Specific Oligonucleotide Hybridization Probes

Single-nucleotide polymorphisms can be typed by hybridization of ASOs. Two oligonucleotides, one complementary to each of the two variants of a biallelic SNP, are hybridized to PCR-amplified DNA spanning the mutation site. The basic requirement is that binding of ASO hybridization probes is highly specific. Several configurations are possible. In its simplest and earliest embodiments, PCR products were immobilized onto a solid surface and hybridization of radiolabeled ASO probes measured. Alternatively, one can immobilize the allele-specific oligonucleotides onto a solid surface and measure binding of PCR products containing a quantifiable label (e.g., biotin or fluorescent labels). The use of microarrays comprised of hundreds of ASOs immobilized onto solid surfaces to form an array of ASOs is a common approach for large-scale genotyping of SNPs and large-scale mutation screening (e.g., Affymetrix GeneChip Mapping 10K Array), an approach that is highly amenable to automation and high throughput. However, it can often be difficult to unequivocally distinguish between homozygous and heterozygous genotypes on the basis of differential hybridization to ASOs. Hybridization and washing conditions need to be strictly controlled and specific for any particular SNP/ASO combination. This presents a problem for standardization of conditions, particularly when developing methods for hundreds if not many tens of thousands of SNP assays in a microarray format. For that reason, microarrays often use a panel of ASOs for each SNP, and genotyping results are on the basis of assessing the results from all probes using sophisticated algorithms. However, even when using a high density or a highly redundant set of ASOs, correct assignment of genotypes can be far from satisfactory, with reported error rates as high as 30% ***(48)***.

3.1.5. Homogeneous Methods

Most of the chemistries described above require postamplification processing, such as PCR probe hybridization, washing steps, or probe extension reactions, depending on the exact chemistry. As a result, they also require a high degree of automation and require liquid-handling systems to achieve high throughput. In homogeneous or closed-tube assays, however, genomic DNA and all of the reagents required for amplification and genotyping or SNP typing are added simultaneously. Measurement of amplification and SNP typing is usually based on fluorescence measurements and can be achieved without the need for any postamplification additions, separations, or processing. This has some important benefits. The time taken for analysis is greatly reduced; the absence of any post-PCR processing minimizes the risk of contamination of work areas with amplified products, and there is a decreased requirement for automation and wet areas in the laboratory. An example of this approach is the 5' fluorogenic nuclease or TaqMan® assay ***(49)*** (www.appliedbiosystems.com). This assay uses two dual-labeled ASO hybridization probes, one for each allele. Both are labeled with a common quencher dye but a different fluorescent reporter dye. They have identical sequences except for a single base corresponding to the SNP site. Genotyping is determined by measurement of the signal intensity of the two allele-specific reporter dyes after PCR amplification (*see* **Fig. 3**). The assay is usually carried out in a 96-well formats, but availability of a 384-well version of Applied Biosystems' AB7900HT real-time thermocycler with a capacity of 100,000 SNP typings per day enables very high throughput. The specificity of the approach is based on differential hybridization of the two probes to their corresponding target sequences. As with all systems based on hybridization, it can sometimes be difficult to achieve very specific binding of probes to one or other allele. Applied Biosystems has developed assays for over 120,000 SNPs in the human genome. A design and manufacture service is also available for any additional SNPs required (www.appliedbiosystems.com).

There are several alternative homogeneous PCR- and hybridization-based techniques, including, among others, molecular beacons ***(50,51)***, scorpion probes ***(52,53)***, and melting-curve analysis of hybridization probes. The last of these techniques has become a commonly used approach, particularly for the genotyping of clinically important mutations, including the Factor V and the hemochromatosis C282Y mutations ***(54)*** (*see* **Fig. 4A**). Typically, the approach

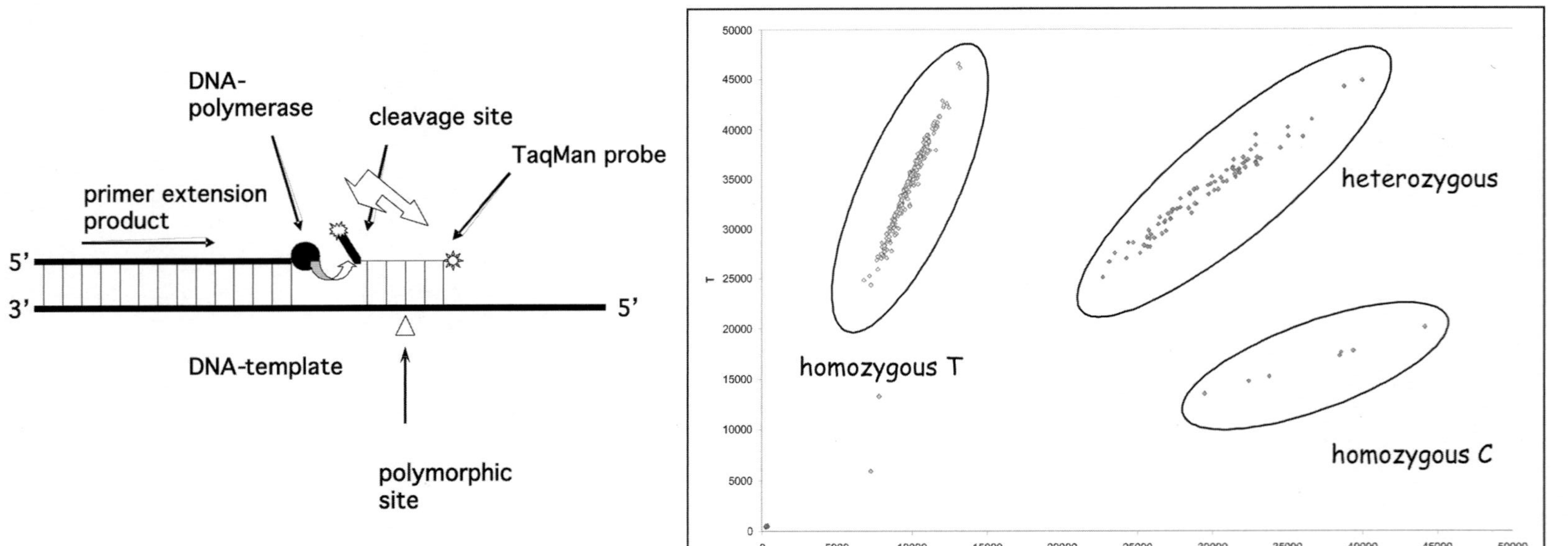

Fig. 3. Allele discrimination using TaqMan technology and end-point fluorescence reading. The graph shows readings using a SNP assay-on demand probe, located in the *SVAT* gene. Genotyping is determined by measurement of the fluorescence resulting from the two allele-specifi reporter dyes after all PCR cycles have been completed. Genotyping results fall in three clusters for the three possible genotypes. For the T genotype, fluorescence with the T allele-specific probe; for the CC genotype, fluorescence with the A allele-specific probe; and for TC heterozygote fluorescence with both probes.

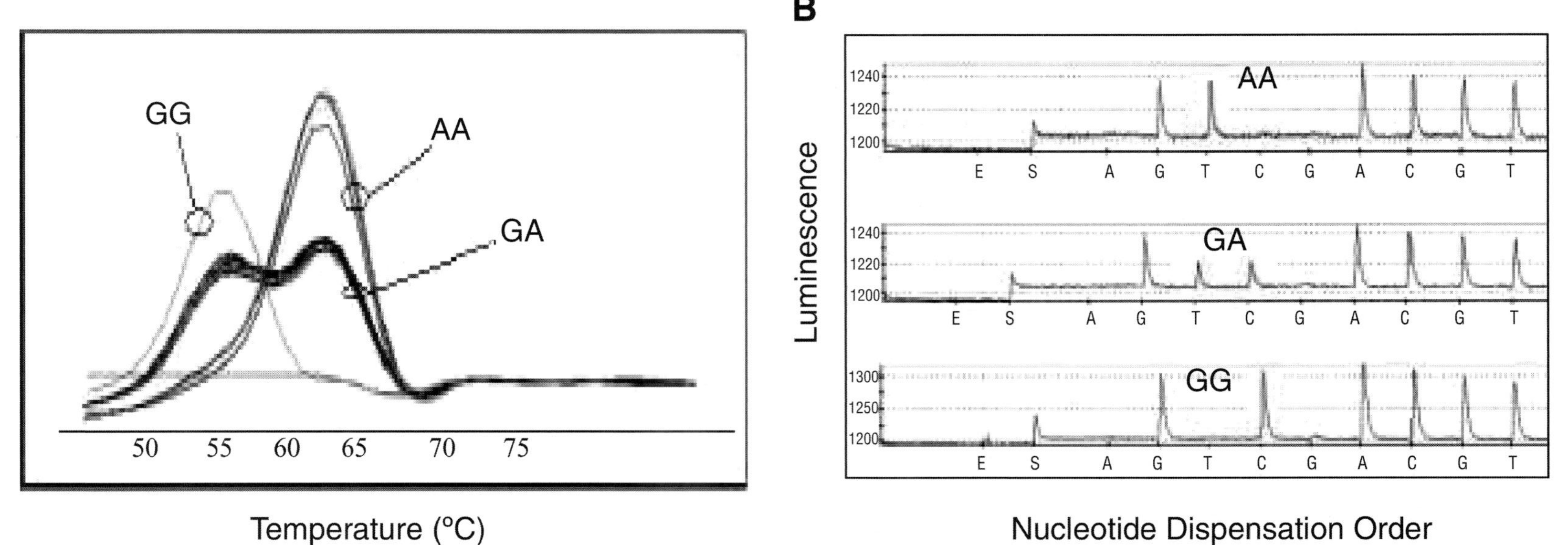

Fig. 4. Genotyping for the hemochromatosis G845A mutation using FRET probes and melting-curve analysis (A) and Pyrosequencing® (B). AA, G845A heterozygotes; GG wild-type homozygotes. For melting-curve analysis (A), three different melting curves are shown for the three possible genotypes. These are plotted as –d(fluorescence)/d(temperature), or –dF/dT, vs temperature (T) and represent changes in fluorescence of the FRET complex as it is heated through its melting temperature at the end of PCR amplification. The apex of the curves represents the melting point for the fluorescent complexes. Although the FRET probes bind to both alleles to form a fluorescent complex, they are complementary to the A allele but mismatched to the G allele by one base. Consequently, the melting temperature of the fluorescent complex is higher for the A allele than the G allele. Heterozygotes have two apexes representing both alleles. For Pyrosequencing (B), an oligonucleotide primer is hybridized to a single strand of PCR-amplified DNA, one base from the SNP site. dNTPs are then added in order as shown on the x-axis (A, G, T, etc.). Incorporation of a nucleotide by DNA polymerase results in release of pyrophosphate, which, in turn, initiates a luciferase-catalyzed reaction resulting in light emission, shown as peaks in the diagram. If a nucleotide is not incorporated, because of the absence of a complementary base such as the first T added for the GG genotype, no light is emitted.

involves the inclusion of PCR primers as well as two oligonucleotide probes, one of which is an ASO that binds across the SNP site and the other is an anchor probe that binds to an adjacent site. The adjacent 3' and 5' ends of the two probes are labeled with fluorescent dyes; one dye, the donor fluorophore (dye A), can be directly excited by a light source, and the other, an acceptor dye (dye B), fluoresces in response to fluorescent resonance energy transfer (FRET) from the donor fluorophore. Dye B only fluoresces when the two probes are hybridized to adjacent positions. Once PCR amplification is complete, heating of the sample through the melting temperature of the ASO yields a fluorescence–temperature curve. A single-nucleotide change causes the probe to dissociate or melt at a lower temperature than if the ASO is completely complementary. For melting-curve analysis, the readout, a melting curve, is graphical. This makes the approach less amenable to automatic SNP scoring than many of the other technologies described, which have a binary readout. OLA (*see* **Subheading 3.1.2.**) can also be performed in a homogenous system by use of FRET probes *(55)*.

3.1.6. Fluorescent Polarization

When fluorescent molecules are excited by plane-polarized light, the fluorescent light emitted is also polarized. The extent of fluorescence polarization (FP), the degree to which the emitted light remains polarized in a particular plane, is proportional to the speed at which the molecules rotate and tumble in solution, which, under constant pressure, temperature, and viscosity, is directly proportional to the molecular weight. Hence, when a small fluorescent molecule is incorporated into a larger molecule, there is an increase in FP (*see* **Fig. 5**). A high signal can only be achieved when the concentration of unreacted fluorescent labels is low, which can be effected by ensuring these are almost totally consumed or removed. The technique can be applied to a number of the chemistries already described. The FP template-directed dye-terminator incorporation assay (FP-TDI), for example, is a minisequencing method *(55)*. Following PCR amplification, unincorporated primers and nucleotides are degraded enzymatically, the enzymes heat inactivated, and a minisequencing reaction using DNA polymerase and fluorescent-labeled dideoxynucleotides performed. FP is measured as the ratio of fluorescence in the vertical and horizontal planes, typically using black 96- or 384-well plates on an FP reader, such as the Perkin-Elmer Victor, the Tecan Ultra, or Molecular Devices' Analyst™. This general principle forms the basis of Perkin-Elmer's AcycloPrime™-FP SNP Detection Kit (www.perkinelmer.com). The 5' nuclease assay described previously, where an oligonucleotide probe is digested to a lower-molecular-weight species, is also amenable to FP *(56)*, but with the added benefit of not requiring a quencher.

3.1.7. Pyrosequencing

Pyrosequencing™ is a homogeneous rapid-sequencing technique particularly suitable for sequencing short segments of PCR-amplified DNA. Unlike traditional sequencing, it does not require any electrophoresis. It is based on the generation of pyrophosphate (PPi) whenever a deoxynucleotide is incorporated during extension of the sequencing primer, as either dGTP, dCTP, cTTP, and a dATP analogs are added in a predetermined order, one at a time (**Fig. 1**). The generation of pyrophosphate is coupled to a luciferase-catalyzed reaction resulting in light emission if the particular dNTP added is incorporated, yielding a quantitative and distinctive pyrogram (*see* **Fig. 3**) (www.pyrosequencing.com). Results are automatically assigned by pattern recognition. Pyrosequencing is particularly suitable for SNP analysis; sequencing usually occurs over an 8-bp region. The approach has several advantages. The setup costs for any one SNP is very low, the quality of the results is high with very clear distinction between genotypes, the approach does not require a high level of technical expertise, and the quantitative nature of the signals produced makes it suitable for gene-dosage studies, including estimation of gene frequencies on pooled DNA. Some of these features make pyrosequencing highly amenable for numerous clinical applications. Its major disadvantage is the cost of the reagents and a limited capacity for multiplexing.

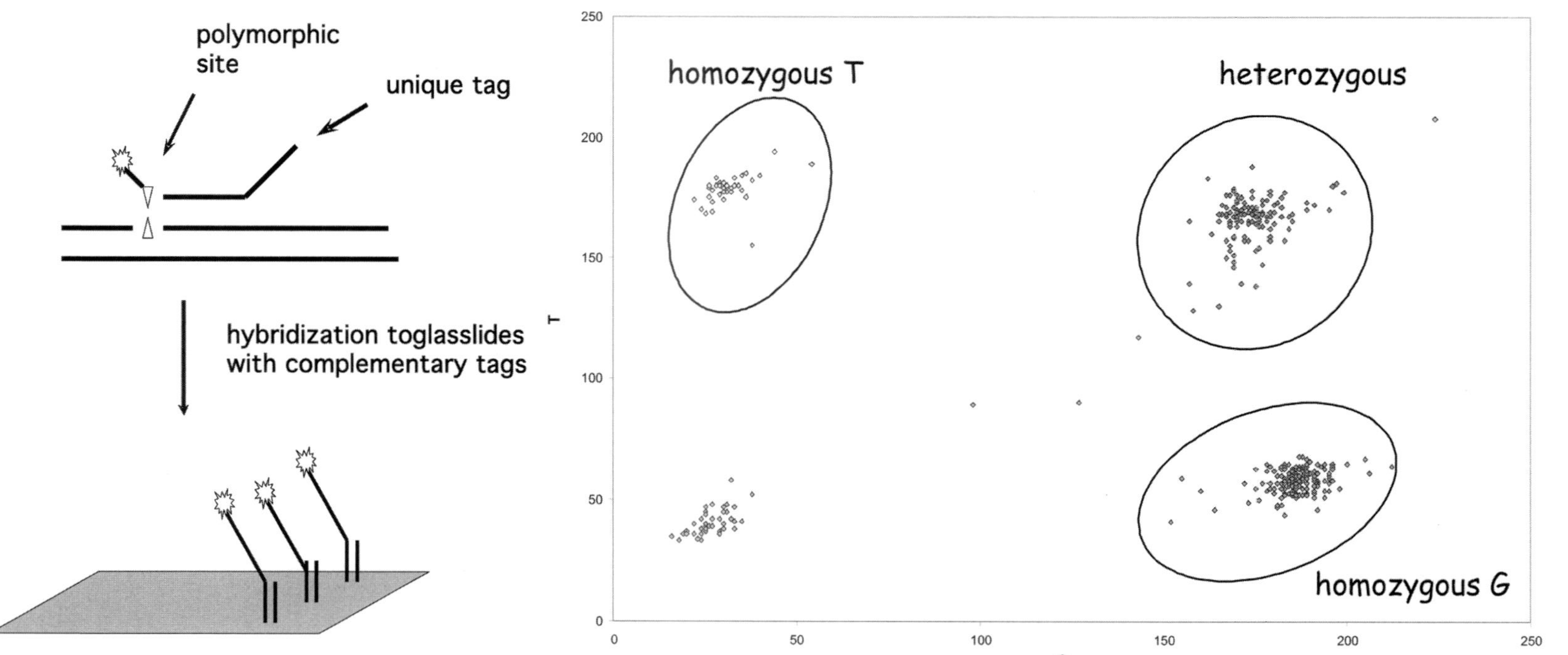

Fig. 5. Allele discrimination using template directed dye terminator incorporation technology and fluorescence polarization.

3.2. Choice of Genotyping Platforms

The choice of genotyping platform requires numerous considerations, including scale, cost, compatibility with other instruments in the laboratory, requirements for liquid-handling systems, technical support from manufacturers, the technical expertise available on site, throughput, and whether SNP genotyping will be in the context of diagnostic or research applications. Currently, unless the required throughput is very high (e.g., numbering in the hundreds of thousands per day), the use of homogeneous chemistries coupled with fluorescent plate readers have much to recommend them. They have some clear advantages: they dispense with the need for any post-PCR processing, thus analysis time is greatly reduced; and the risk of contamination of work areas with PCR products is minimized, because they are not technically demanding, and there are minimal requirements for robotics or liquid-handling systems. Throughput on some platforms can be reasonably high (e.g., approx 100,000 per day for Applied Biosystems' 7900HT). Homogeneous methods are particularly useful where rapid analysis might be required (e.g., in diagnosis). This aspect will be particularly important in pharmacogenetics (i.e., to determine a patients pharmacogenetic profile prior to prescription of drugs). However, the requirement for fluorescent-labeled probes can result in high initial setup costs, probe design can be problematic, and it is not always possible to clearly distinguish genotypes. In this context, availability of a probe design and manufacture service or use of validated assays is useful. Finally, on the basis of current technology, homogeneous assays have less scope for multiplexing, and as the number of samples requiring analysis and the size of SNP panels increases, reagent costs could become prohibitive. Analysis by MALDI-TOF analysis, a popular choice for high-throughput SNP typing, requires a larger investment in capital expenditure and technical expertise, but has a lower reagent cost and is more amenable to multiplexing. There are numerous steps following amplification and a greater requirement for liquid-handling systems, but recent developments, such as the GOOD assay, have done much to both simplify the process and reduce the cost ***(57,58)***. As in pyrosequencing, the signal output by MALDI-TOF is quantitative and includes data such as relative allele abundance, which makes it highly amenable for estimating allele frequencies for SNPs on pooled DNA samples, which is important to validate SNP frequencies and prioritize SNP assays. However, as the number of SNPs being examined expands further, higher orders of multiplexing than are currently possible by MALDI-TOF will be required to make more efficient used of genomic DNA and to reduce the cost of reagents, including PCR primers and enzymes, and disposables. Very high orders of multiplexing are currently only possible with high-density arrays, either in two-dimensional or microbead format.

Finally, we note that it is increasingly the case that SNP genotyping is being "industrialized," with genotyping services being provided to individual research groups by very high-throughput, cost-effective, and centralized genotyping core facilities provided by academic institutions, national governments, or industry (e.g., www.sanger.ac.uk/; www.genomequebec.com/index_e.asp).

4. Applications of SNP to Complex Human Diseases

4.1. Genetic Linkage and Association Studies of Complex Disease

Two types of study have been widely employed to attempt to identify the genetic determinants of complex diseases: positional cloning and candidate gene association studies. Positional cloning begins with the identification of a chromosomal region that is transmitted within families along with the disease phenotype of interest. This phenomenon is described as genetic linkage. Positional cloning has been extremely useful in the identification of genes responsible for diseases with simple Mendelian inheritance ***(17)***, such as cystic fibrosis ***(59)***. The application of linkage analysis to complex disorders without obvious Mendelian inheritance has been much less successful to date ***(60,61)***.

The growing recognition of the limitations of linkage analysis in complex human diseases has seen a shift in emphasis away from linkage analysis and microsatellite markers toward SNP genotyping and different analytical strategies based on association and haplotype analysis ***(8,9,23,60,62)***. Association analyses are now recognized as essential for localizing susceptibility loci, and they are intrinsically more powerful than linkage analyses in detecting weak genetic effects ***(8,63,64)***. A small but increasing number of genes associated with complex diseases have been discovered by association-based linkage disequilibrium (LD) mapping ***(22,64)***.

Association studies rely on the detection of polymorphisms in candidate genes or regions and the demonstration that particular alleles are associated with one or more phenotypic traits. However, analyses of specific alleles suggesting a statistical association between an allele and a phenotypic trait will be the result of one of three situations ***(65)***. First, the finding could be the result of chance or artifact (e.g., confounding or selection bias). Second, the allele can be in LD with an allele at another locus that directly affects the expression of the phenotype. Third, the allele itself can be functional and directly affect the expression of the phenotype.

Importantly, most SNPs are not likely to alter gene structure or function in any way and, therefore, might not be directly associated with any change in phenotype ***(66)***. Thus, it is important to ascertain whether the DNA sequence variant under consideration is potentially functional (i.e., could lead to the observed biology) or is a marker in LD with another DNA sequence variant that is the actual cause of the phenotype of interest. Because functional data are rarely available for a given SNP, tests of genetic association using SNPs are generally based on LD. LD arises in a population from the coinheritance of alleles at loci that are in close physical proximity on an individual chromosome (i.e., a haplotype). Many factors can influence LD, including genetic drift, population growth, admixture, population structure, natural selection, variable recombination and mutation rates, and gene conversion ***(67,68)***. A project currently underway to describe the disequilibrium patterns across the human genome in at least some ethnic groups will shed important new light on the ultimate utility of SNPs for the disequilibrium mapping of disease genes ***(22)*** (*see* **Subheading 4.4.3.**).

4.2. SNP Analysis and Complex Human Disease

There are a number of important potential advantages to using SNPs to investigate the genetic determinants of complex human diseases compared to other types of genetic polymorphism ***(8,69,70)***. First, SNPs are plentiful throughout the human genome, being found in exons, introns, promotors, enhancers, and intergenic regions, allowing them to be used as markers in dense positional cloning investigations using both randomly distributed markers and markers clustered within genes ***(71,72)***. Furthermore, the abundance of SNPs makes it likely that alleles at some of these polymorphisms are themselves functional. Second, groups of adjacent SNPs can exhibit patterns of LD and haplotypic diversity that could be used to enhance gene mapping ***(73)*** and can highlight recombination "hot spots" ***(74)***. Third, interpopulation differences in SNP frequencies can be used in population-based genetic studies ***(75,76)***. Finally, there is good evidence that SNPs are less mutable than other types of polymorphism ***(77,78)***. The resultant greater stability can allow more consistent estimates of LD and gene–phenotype associations.

There are five primary areas of potential application for SNP technologies in improving our understanding of complex disease pathophysiology: gene discovery and mapping; association-based candidate polymorphism testing; pharmacogenetics, diagnostics, and risk profiling; prediction of response to nonpharmacological environmental stimuli; and homogeneity testing and epidemiological study design ***(9)***. Although only a few of these areas are currently areas of active research, it is likely that each of these areas will become increasingly relevant to many complex human diseases. We discuss two of these areas in more detail below.

4.2.1. Candidate Gene Polymorphism Testing

Single-nucleotide polymorphisms for use in genetic applications and research are derived both from public and proprietary SNP databases (see the Appendix, p. 247) and from targeted SNP discovery by mutation detection ***(79)*** or primary resequencing in candidate genes or regions ***(24,80)***.

Over 1200 genes in the human genome have been shown to be causally associated with Mendelian diseases ***(17)***, and SNPs in a number of these genes are routinely tested for clinical purposes. The HFE C282Y (hemochromatosis), apo E4 (Alzheimer's), Factor V Leiden (thrombophilia), and common CFTR (cystic fibrosis) SNPs are a few of many examples (www.ncbi.nlm.nih.gov/entrez/query.fcgi?db=OMIM). Although the numbers are much smaller, there is also an increasing number of genes that have been definitively associated with common diseases such as type 1 diabetes ***(81)***, asthma ***(82)***, and inflammatory bowel disease ***(83,84)***. As more and more genetic factors of relevance to complex disease and pharmacotherapy are identified, very large panels of SNPs are likely to be employed routinely in clinical applications ***(4,5,12,13)***.

The number of biologically plausible candidate genes that might be involved in the determination of any complex human traits is generally very large. There is now an extensive and growing list of candidate genes investigated for association with complex traits. Obesity provides an instructive example; there has been an enormous amount of work directed toward understanding the genetics of this condition ***(85)***. As of October 2003, 272 studies have reported positive associations with 90 candidate genes; 15 of these candidate genes have been supported by at least five positive studies. One problem for investigations of common complex diseases is that the number of biologically plausible candidate genes will often number in the hundreds (or even thousands) for any given pathogenic pathway ***(86)***.

4.2.2. Pharmacogenetics

An expanding area of interest in the application of SNPs to investigations of disease pathophysiology is the stratification of populations by their genetically determined response to therapeutic drugs ("pharmacogenetics"). Ideally, we would be able to stratify a population into responders, nonresponders, and those with adverse side effects ***(87)***. The ultimate goal of such a stratification is to improve the efficacy of drug-based interventions and to expedite targeted drug discovery and development. Pharmacogenetic initiatives are currently an area of very active research in many complex human diseases ***(13,88,89)***.

An excellent example of the increasing integration of pharmacogenetic information into clinical practice comes from western Australia, where the screening of human leukocyte antigen (HLA) polymorphisms in newly incident human immunodeficiency virus (HIV) cases has been used to virtually eliminate a hypersensitivity reaction to a commonly used retroviral therapy ***(90)***. Applications such as this one could mark the beginning of the clinical use of genotyping at an individual level as an adjunct to pharmacotherapy for HIV and many other disorders.

4.3. Disease Association Studies: Methodological and Study Design Issues

Limitations in SNP association analyses of complex human diseases arise from technical issues in genotyping (*see* **Subheading 3.**) and from the now well-described general limitations of our understanding of LD ***(68)*** and of investigating gene–phenotype associations involving multiple interacting genetic and environmental factors ***(61,70,91)***. The growing focus on SNP genotyping has made it clear that concomitant statistical advances in the LD mapping of complex traits will also be required ***(61,92–94)***. The SNP genotyping effort has caused a broad reexamination of mapping methodologies and study designs in complex human disease ***(10,61 ,67,95,96)***. The testing of large numbers of SNPs for association with one or more traits raises important statistical issues regarding the appropriate false-positive rate of the tests and the level of statistical significance to be adopted given the multiple testing involved ***(8,64,96)***. The required methodological development in genetic statistics is nontrivial given the complexity of common human diseases. The fundamental issue of how to deal with the sheer volume of data being produced has only just begun to be addressed, and it is clear that methodological development in biostatistics is lagging behind the technical capacity to generate SNP genotypes ***(97,98)***. The optimal strategies for the detection and application of SNPs to our understanding of the genetic epidemiology of common diseases thus remain unclear.

A number of recent articles have addressed the features of a "good" genetic association study ***(61,64,65,99,100)***. The increasing focus on study design has resulted from the growing realization that genetic association studies of complex phenotypes have tended to either fail to discover susceptibility loci or have failed to replicate across studies ***(91,100–103)***. Potential reasons for the nonreplication of true positive association results include interinvestigator and interpopulation heterogeneity in study design, analytic method, phenotype definition, genetic structure, environmental exposures, and markers typed. It is increasingly clear that large sample sizes, rigorous *p*-value thresholds, and replication in multiple independent datasets are all necessary for reliable results in genetic association studies of complex human diseases ***(4,64,91 ,99,103)***. In general, lack of ability to define modest genetic associations and to replicate results have often involved poor study design and execution, in particular a lack of appreciation for the sample sizes required to detect modest genetic effects and overinterpretation of marginal results ***(10)***. For most complex human diseases, the reality of multiple disease-predisposing genes of modest individual effect, gene–gene interactions, gene–environment interactions, interpopulation heterogeneity of both genetic and environmental determinants of disease, and the concomitant low statistical power mean that both initial detection and replication will likely be very difficult ***(61,94,96)***.

There are several important statistical issues related to the disequilibrium mapping of complex disease genes ***(94,104)***. These include multiple testing and statistical power, the potentially biasing effects of population stratification, and choice of analytic strategy in the measurement of LD and the imputation of haplotypes.

4.3.1. Statistical Power and Multiple Testing

Growing experience with complex disease genetics has made clear the need to restrict types I and II error in genetic studies ***(8,61,105)***; most complex human phenotypes are likely to be heterogeneous and to involve multiple genes of small individual effect. Power for studies of allelic association will depend on sample size, effect size of the susceptibility locus, strength of LD with a marker, and frequencies of susceptibility and marker alleles ***(64)***. **Figure 6** shows some simple estimations of required sample sizes of cases needed to detect a true odds ratio (OR) of 1.5 with 80% power and type I error probability (α) of either 0.05 or 0.005. Even for the "best-case scenario," a common SNP acting in a dominant fashion, a relatively large sample size of more than 800 subjects is required at an α of 0.05 (*see* **Fig. 6**).

Multiple testing issues are likely to be an issue in many genetic association studies of candidate loci where either multiple SNPs in one gene or multiple SNPs in several loci are tested, or both ***(106)***, suggesting that an α of 0.005 is probably more realistic than an α of 0.05. Using the more realistic α of 0.005 or assuming an uncommon SNP that acts in a recessive fashion leads to the need for very large (in some cases logistically improbable) sample sizes. Finally, **Fig. 6** assumes an effect size (OR=1.5) that, in the context of a common, multifactorial disease such as asthma or diabetes, can be quite large. Assuming a smaller effect might be more realistic for many genes and would lead to concomitantly higher required sample sizes. Simulation studies have also suggested that genes of small effect are not likely to be detectable by association studies in sample sizes of less than 500 ***(93)***.

Although these power calculations are simple and make a number of conservative assumptions, they clearly demonstrate that the sample sizes used in many of the small case-control association studies of complex phenotypes conducted to date will have had insufficient power to detect even quite a large effect of a SNP. This suggests that genetic association studies have generally been underpowered to date ***(4,10,86,94,96)***, that larger-scale studies than those currently being performed by many groups will be needed in future, and that very large cohort studies are needed for genetic epidemiological investigations of common diseases ***(11,107)***.

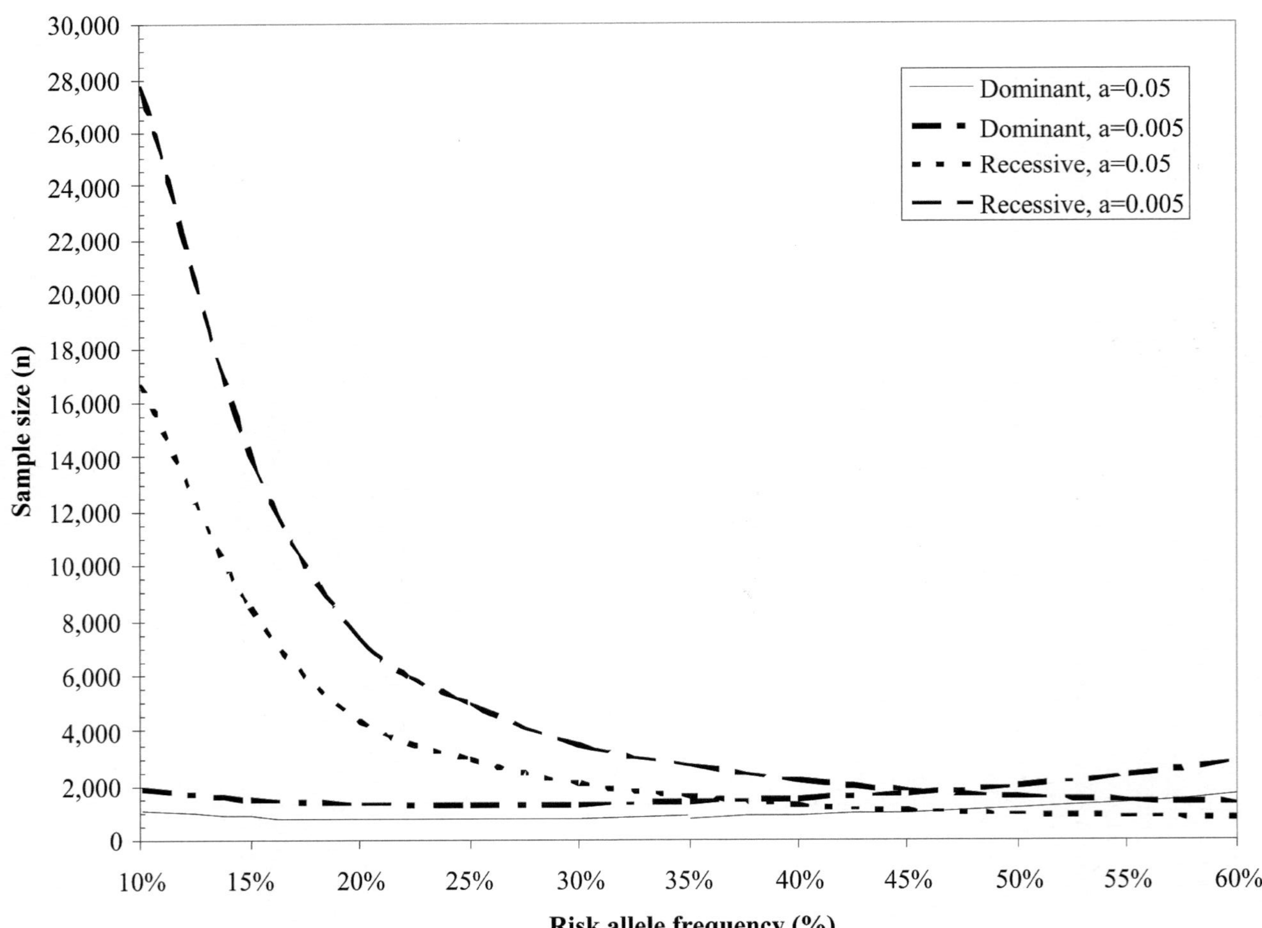

Fig. 6. Sample size estimates (cases plus controls) for case-control analyses of SNPs (1 control per case; detectable difference of OR ≥1.5; power=80%) as a function of genetic model, frequency of risk-increasing allele in general population, and type I error rate (α).

The testing of large numbers of SNPs for association with one or more traits raises important statistical issues regarding the appropriate false-positive rate of the tests and the level of statistical significance to be adopted given the multiple testing involved ***(96)***. Post hoc corrections tend to be too conservative, especially as many [such as a simple Bonferoni correction ***(108)***] do not take proper account of the nonindependence of SNPs in LD with each other. Although sophisticated statistical techniques are available to attempt to correct for multiple comparisons ***(109)***, replication of genetic association findings in multiple independent population samples has become the gold standard for complex disease genetics ***(8,65,110)***.

4.3.2. Population Stratification

Genetic heterogeneity is a major issue complicating gene discovery in complex diseases ***(111–114)***. Heterogeneity testing can be used to test explicitly for population stratification in association analyses ***(115)***, and to assess the potential generalizability of SNP–phenotype associations. In addition to variation in allele frequencies, there is also a high degree of variation in LD strength between populations of different origins ***(114,116)*** and also between different genomic regions ***(117,118)***.

For association studies of many complex diseases, case-control designs have become the approach of choice. Case-control association analyses are now recognized as well suited for localizing susceptibility loci ***(65)***, and they are intrinsically more powerful and efficient than family-based linkage and association analyses in detecting weak genetic effects ***(63,96,119)***. Furthermore, nested case-control studies can take advantage of large cohort studies such as the Busselton Health Study ***(120)***, in which extensive longitudinal data are available on a large, population-based sample of subjects.

However, the single largest criticism of genetic case-control studies to date has been potentially undetected population stratification: Spurious association can arise in a case-control study when allelic frequencies vary across subpopulations (e.g. subjects from different ethnic groups) ***(121)***. Such population stratification can result from recent admixture or from poorly matched cases and controls. Genotyping of random panels of SNPs can be used to partition study populations into genetically homogeneous groups, and there are now promising methods for assessing and, if necessary, correcting for population stratification ***(122–128)***. Furthermore, there is growing empirical and theoretical evidence suggesting that well-designed, well-conducted, and appropriately analyzed and interpreted population-based studies with unrelated controls are robust to bias from population stratification ***(119)***. It is becoming apparent that the potential for bias related to cryptic population substructure in population-based studies has been greatly exaggerated in the literature.

4.3.3. Haplotyping

Haplotypes are linear arrangements of closely linked alleles on individual chromosomes, which are usually inherited as a unit. With sufficient LD, haplotypes can be used in association studies to map common alleles that might influence the susceptibility to common diseases, as well as for reconstructing genomic evolution ***(129)***. In complex diseases, where multiple variant loci can contribute to disease susceptibility, haplotypes are potentially important because different combinations of particular syntenic alleles in the same gene can act as a "meta-allele" and have different effects on the protein product and on transcriptional regulation. In addition, haplotypes can offer advantages in terms of statistical power to detect a true association with a given sample size compared with analyses based on single SNP or combinations of SNPs ***(17,98,130)***. Haplotypes are useful both for the mapping of susceptibility loci and the evaluation of sequence variation at multiple sites along a chromosome (*see* **Subheading 4.4.4.**).

In nearly all large population-based SNP association studies, genotypic data are phase-unknown, so haplotypes for double heterozygotes are unknown and must be imputed ***(131)***. There has been substantial progress over the past few years in the development and implementation of appropriate statistical methods, and the imputation of haplotypes from genotypic data in large population-based samples or in family data is surprisingly efficient ***(131,132)***. Probably the most widely utilized approach to estimate haplotype frequencies in unphased genotypic data has been the expectation maximization (EM) algorithm, as implemented in software such as ARLEQUIN ***(133)***. New maximum-likelihood methods have recently been developed to enable the testing of statistical association between haplotypes and a wide variety of traits, including binary, ordinal, and quantitative traits. These methods, based on score tests, also allow adjustment for nongenetic covariates ***(134)***. Haplotype imputation algorithms based on Markov chain Monte Carlo methods have also recently been developed. Useful implementations include the software programs HAPLOTYPER ***(135)*** and PHASE ***(136)***.

General problems for haplotype analysis that have not yet been fully resolved include methods for dealing appropriately with missing data—results from sequencing or genotyping any given SNP are rarely 100% complete—and a general issue with multiple comparisons and haplotype choice. The latter problem can be ameliorated by the growing utilization of phylogenetic approaches derived from population genetics in human gene discovery investigations ***(137)***.

4.4. Future Directions

We discuss in the following subsections some future directions for SNP-based association studies. This is a small sample of the many possible future uses for SNP-based technologies. For instance, we do not discuss here the vast and largely untapped potential of comparative genomics to inform us both as to the regulatory sequences that control the activity of human genes and as to the evolutionary forces that have shaped the human genome ***(97,138,139)***.

4.4.1. Diagnostics and Risk Profiling

Following identification of a SNP or SNP-based haplotype that is closely associated with a disease or associated trait, it might be possible to use this information to develop diagnostic tests. The ability to determine risk of disease prior to the onset of symptoms would be potentially of great benefit in many conditions. As for all diagnostic genetic tests, the utility and ultimate success of diagnostic testing for disease susceptibility using SNPs in a particular population would depend on the extent and nature of disease heterogeneity, the frequency of the high-risk allele and the concomitant attributable risk, the penetrance of a specific allele, and the ability to define a useful risk model including other genetic factors, important environmental risk factors, and interactions between the SNP and factors such as age and gender ***(9,140)***. In addition, there are both technical problems with routine genetic testing, largely related to false negatives, and important ethical and psychosocial concerns that remain unresolved ***(140–142)***.

Understanding how etiological factors act at a population level will a critical step for the clinical utilization of new genomic knowledge and tools to improve health outcomes ***(4,143, 144)***. Ultimately, genetic knowledge will only become useful in the clinical arena if placed back into an epidemiological and public health context ***(5–7,12,145)***. It is, therefore, clear that very large, longitudinal, well-characterized, population-based studies originally established for epidemiological purposes, such as the Nurses Health Study ***(146)*** and the Busselton Health Study ***(120)***, will be critical to the future implementation of diagnostic SNP-based tests for complex phenotypes in general populations ***(100)***.

4.4.2. Gene-Environment Interaction

The environment has largely been ignored in genetic investigations of complex human diseases, despite a growing consensus that gene–environment interactions are likely to play a critical role in many human diseases ***(6,7,11,147)***. In addition to pharmacogenetic applications, the identification of groups of individuals likely to be affected by other environmental exposures because of their genetic susceptibility might also be beneficial to our future understanding and treatment of many diseases ***(148)***. Examples of potentially important environmental factors that might interact with underlying genetic susceptibilities include cigaret smoke exposure, exposure to common pollutants and infections, housing and lifestyle factors, *in utero* factors acting during pregnancy, and diet ***(7)***. Prediction of response to environmental factors in individuals genetically predisposed to disease is potentially of major public health and economic significance ***(6,7,11)***. The incorporation of SNP-based genetic data into public health policy and practice could, therefore, become an increasingly important factor in preventive medicine. The ongoing integration of genetics with mainstream epidemiology will be the critical enabling factor for the future of public health genetics.

"Epigenetic" effects on gene expression arise during development through mechanisms such as DNA methylation or histone modifications that do not involve DNA mutations ***(149–153)***. Epigenetic regulation is one of the fundamental mechanisms by which the human genome integrates environmental and intrinsic influences. In addition to the study of gene–environment interaction, epigenetic epidemiology is likely to be an area of great future importance to our understanding of the pathogenesis and treatment of many complex diseases.

4.4.3. The HapMap Project

Linkage disequilibrium mapping relies on gene–phenotype associations at the level of population ***(152)***, and requires a dense map of markers ***(96)***. LD mapping can also be enhanced using haplotype analysis; although haplotype analysis in practice has proven difficult ***(153)***, it is likely to be more powerful than focusing on a single SNP locus.

A number of studies have now shown that the human genome can be parsed into "haplotype blocks" of limited or no recombination separated by smaller (typically < 1 kb) regions in which recombination tends to occur ***(154–156)***. Haplotype blocks are regions in which there is a relatively high and consistent LD between SNPs. Block size is highly variable across the human genome and can range from < 1 kb to > 100 kb in length, depending on the genomic region and ethnic background of the sample ***(154,157)***. They can represent a fundamental unit of human heredity, being transmitted unchanged (apart from relatively rare mutational and recombinational events) from generation to generation. The human genome is characterized by limited haplotypic diversity ***(156)***.

Defining and genotyping a relatively small number of "haplotype-tagging" SNPs within a haplotype block can allow unambiguous determination of the specific haplotypes that each individual posseses, and "captures" all or most of the LD within that block ***(98,156,158–163)***. By this means, SNP–phenotype association studies can be performed relatively efficiently in contrast to identifying and genotyping all sites of variation within a block ***(158)***. Given a set of haplotypes, the minimal set of SNPs needed to unambiguously assign a haplotype to an individual ("haplotype-tagging SNPs") can be determined using various statistical approaches ***(98,159–164)***. For European samples, the use of haplotype-tagging SNPs generally identifies a subset of SNPs that should be genotyped that is between a tenth and a third of the size of the entire set of SNPs ***(25)***.

The international HapMap project ***(165)*** aims to enable the use of haplotype blocks in the human genome for gene mapping by determining the common patterns of DNA sequence variation in the human genome and making these data publicly available ***(22)***. The international HapMap consortium is developing a map of these patterns across the genome by determining the genotypes of 1 million or more sequence variants, their frequencies, and the degree of association between them in DNA samples from populations with ancestry from parts of Africa, Asia, and Europe. If successful, the HapMap will facilitate the discovery of SNPs that affect common disease and the utilization of SNPs in diagnostic and therapeutic applications ***(22)***.

Extensive resequencing and large-scale SNP genotyping of entire human chromosomes and large genomic regions has, however, raised some doubts as to the feasibility and ultimate utility of the HapMap project. The definitions and underlying causes of a "block" have been controversial ***(98,100,166,167)***; block definition appears to be somewhat arbitrary and will greatly depend on the context of the specific application, the genomic region being investigated, the scale (i.e., marker density) at which a genomic region is typed, and the demographic history of the population under study ***(98,167–169)***. Although the public databases are reasonably complete for European populations, this does not yet appear to be true for other ethnic groups ***(24)***. The ultimate utility of the HapMap project for our understanding of the genetic epidemiology of complex diseases has yet to be determined ***(98,100,167,170)***.

4.4.4. Whole-Genome SNP Association

The growing density of SNP maps together with the identification of genes associated with the human genome project ***(171)*** has made genomewide association analyses technically feasible for many conditions ***(172)***. However, testing all of the common SNPs for association is currently impractical. There are two strategies for reducing the potential number of SNPs that need to be typed in any given gene or region or in the whole genome without loss of power ***(72)***—one based on the use of haplotype-tagging SNPs across the genome (*map based*) and one based on the genotyping of all potentially functional SNPs across the genome (*sequence based*) ***(17,173)***. However, trade-offs in power to detect modest genetic effects through association rather than linkage ***(96,174)*** are likely to be offset by the need for very large sample sizes and a substantial statistical penalty necessary to correct for multiple comparisons. Further limitations arise from the cost of typing the very large number of markers (suggested to be at least 300,000 and up to 1 million tagged SNPs in a general outbred population) required for a genomewide association analysis ***(154,156,174,175)***, and the uncertain properties of LD between alleles of tightly linked SNPs across the genome and between different ethnic groups ***(98,100,114,176, 177)***. The latter issue leads to uncertainty regarding optimal SNP marker choice for a genomewide scan, particularly with regard to the detection of uncommon susceptibility loci and when investigating non-European populations ***(24,98)***. The feasibility of whole-genome association scans will critically depend on knowledge of the haplotype structure of the human genome ***(156,157)*** (*see* **Subheading 4.4.3.**). In addition, the optimal statistical approaches to analyze a whole-genome scan for association with complex phenotypes remain unclear ***(96,98, 178,179)***.

Although SNP mapping poses multiple and serious problems if used in genomewide strategies, these problems might become more tractable if the HapMap project fulfilss its stated goals ***(22,24 ,98)*** (*see* **Subheading 4.4.3.**). Alternatively, it can become apparent that primary resequencing of candidate genes and regions of interest by individual research groups will continue to be the approach of choice ***(25)***.

5. Conclusions

There are numerous chemistries and platforms available for SNP genotyping and the choice of platforms requires many considerations. Speed and simplicity of analysis, even in high-throughput settings, often has to be weighed against lower reagent costs. Genotyping accuracy is important for all applications, but is absolutely critical in clinical or diagnostic settings. Several manufacturers offer genomewide SNP panels, particular in microarray format; these are currently the most efficient approaches in terms of cost per SNP, but they do not cover all SNPs nor suit all needs. It is clear that no one platform can yet meet these many varied requirements. It is, therefore, not uncommon for a single group to use a number of different chemistries and platforms. Nevertheless, although many challenges remain, it is clear that the chemistries and platforms currently available make far more efficient use of limited funds and, arguably, an even more precious resource, genomic DNA from many thousands of subjects, than previously possible. The pace of current developments in SNP genotyping technology with regard to throughput and sensitivity gives cause for further optimism that substantially greater efficiencies will be possible in the very near future.

The technology for detecting SNPs has thus undergone rapid development, extensive catalogs of SNPs across the genome have been constructed, and SNPs have been increasingly used as a means for investigating the genetic etiology of complex human diseases. The potential areas of application for SNP technology in improving our understanding of disease pathophysiology include the following: gene discovery and mapping; association-based candidate polymorphism testing; pharmacogenetics; diagnostics and risk profiling; prediction of response to nonpharmacological environmental stimuli; homogeneity testing and epidemiological study design. It is likely that all of these areas will become increasingly important in investigations of the genetic susceptibility to common, complex human diseases. However, there are technical, statistical, ethical, and psychosocial issues that remain unresolved in the use of SNP technology to investigate these aspects of the pathophysiology of complex human diseases. The complexity of local haplotype structure in the human genome and the extent of LD remain poorly defined.

Genetic approaches to complex diseases offer great potential to improve our understanding of common disease pathophysiology, but they also offer significant challenges. Despite much progress in defining the genetic basis of many diseases in the last decade, accompanied by rapid technical progress in SNP genotyping technologies, further methodological and applied research is required. However, a large number of groups are currently active in addressing methodological problems in SNP genotyping and statistical genetic, and our growing understanding of the architecture of the human genome—aided by projects such as HapMap—will likely accelerate our understanding of the pathophysiology of many conditions. The understanding of disease pathophysiology can then enter into the realm of clinical and population genetics.

Acknowledgments

We thank Professor Luba Kalaydjieva for helpful comments on this chapter and Dr. Sibylle Schwab for assistance with some of the genotyping figures.

Appendix: Selected Web Sites

SNP databases

- dbSNP Polymorphism Repository [http://www.ncbi.nlm.nih.gov/SNP/]
- Cancer Genome Anatomy project [http://cgap.nci.nih.gov/]
- Génome Québec [http://www.genomequebec.com/index_e.asp]
- The Golden Path [http://genome.ucsc.edu]
- The Human Genome Variation Society [http://www.genomic.unimelb.edu.au/mdi/]
- Human SNP Database [http://www-genome.wi.mit.edu/SNP/human/index.html]
- The International HapMap Project [http://www.hapmap.org/]
- LocusLink [http://www.ncbi.nih.gov/LocusLink/]
- NHLBI Programs for Genomic Applications Resources [http://pga.lbl.gov/PGA/PGA_inventory.html]
- OMIM™—Online Mendelian Inheritance in Man™ [http://www.ncbi.nlm.nih.gov/entrez/query.fcgi?db=OMIM]
- SNP Consortium [http://snp.cshl.org/].
- SNP View [http://snp.gnf.org/].
- The Sanger Centre [http://www.sanger.ac.uk/].
- The Whitehead Institute [http://www.wi.mit.edu/home.html].

Software

- An extensive list of genetic analysis software [http://linkage.rockefeller.edu/soft/list.html].

References

1. Khoury, M. J. (1997) Genetic epidemiology and the future of disease prevention and public health. *Epidemiol. Rev.* **19**, 175–180.
2. Nagy, A., Perrimon, N., Sandmeyer, S., and Plasterk, R. (2003) Tailoring the genome: the power of genetic approaches. *Nature Genet.* **33 (Suppl.)**, 276–284.
3. Zerhouni, E. (2003) Medicine. The NIH roadmap. *Science* **302**, 63–72.
4. Goldstein, D. B., Tate, S. K., and Sisodiya, S. M. (2003) Pharmacogenetics goes genomic. *Nature Rev. Genet.* **4**, 937–947.
5. Merikangas, K. R. and Risch, N. (2003) Genomic priorities and public health. *Science* **302**, 599–601.

6. Kelada, S. N., Eaton, D. L., Wang, S. S., Rothman, N. R., and Khoury, M. J. (2003) The role of genetic polymorphisms in environmental health. *Environ. Health Perspect.* **111**, 1055–1064.
7. Shostak, S. (2003) Locating gene-environment interaction: at the intersections of genetics and public health. *Soc. Sci. Med.* **56**, 2327–2342.
8. Risch, N. J. (2000) Searching for genetic determinants in the new millennium. *Nature* **405**, 847–856.
9. Schork, N. J., Fallin, D., and Lanchbury, J. S. (2000) Single nucleotide polymorphisms and the future of genetic epidemiology. *Clin. Genet.* **58**, 250–264.
10. Cardon, L. R. and Palmer, L. J. (2003) Population stratification and spurious allelic association. *Lancet* **361**, 598–604.
11. Wright, A. F., Carothers, A. D., and Campbell, H. (2002) Gene–environment interactions—the BioBank UK study. *Pharmacogenomics J.* **2**, 75–82.
12. Burke, W. (2003) Genomics as a probe for disease biology. *N. Engl. J. Med.* **349**, 969–974.
13. Johnson, J. A. (2003) Pharmacogenetics: potential for individualized drug therapy through genetics. *Trends Genet.* **19**, 660–666.
14. Varmus, H. (2003) Genomic empowerment: the importance of public databases. *Nature Genet.* **35 (Suppl. 1)**, 3.
15. Reich, D. E., Gabriel, S. B., and Altshuler, D. (2003) Quality and completeness of SNP databases. *Nature Genet.* **33**, 457–458.
16. Venter, J. C., Levy, S., Stockwell, T., Remington, K., and Halpern, A. (2003) Massive parallelism, randomness and genomic advances. *Nature Genet.* **33 (Suppl.)**, 219–227.
17. Botstein, D. and Risch, N. (2003) Discovering genotypes underlying human phenotypes: past successes for mendelian disease, future approaches for complex disease. *Nature Genet.* **33 (Suppl)**, 228–237.
18. Marth, G. T., Korf, I., Yandell, M. D., et al. (1999) A general approach to single-nucleotide polymorphism discovery. *Nature Genet.* **23**, 452–456.
19. Botstein, D., White, R. L., Skolnick, M., and Davis, R. W. (1980) Construction of a genetic linkage map in man using restriction fragment length polymorphisms. *Am. J. Hum. Genet.* **32**, 314–331.
20. Wang, D. G., Fan, J. B., Siao, C. J., et al. (1998) Large-scale identification, mapping, and genotyping of single- nucleotide polymorphisms in the human genome. *Science* **280**, 1077–1082.
21. Cargill, M., Altshuler, D., Ireland, J., et al. (1999) Characterization of single-nucleotide polymorphisms in coding regions of human genes. *Nature Genet.* **22**, 231–238.
22. The International HapMap Project (2003) The International HapMap Project. *Nature* **426**, 789–796.
23. Gray, I. C., Campbell, D. A., and Spurr, N. K. (2000) Single nucleotide polymorphisms as tools in human genetics. *Hum. Mol. Genet.* **9**, 2403–2408.
24. Carlson, C. S., Eberle, M. A., Rieder, M. J., Smith, J. D., Kruglyak, L., and Nickerson, D. A. (2003) Additional SNPs and linkage-disequilibrium analyses are necessary for whole-genome association studies in humans. *Nature Genet.* **33**, 518–521.
25. Lazarus, R., Vercelli, D., Palmer, L. J., et al. (2002) Single nucleotide polymorphisms in innate immunity genes: abundant variation and potential role in complex human disease. *Immunol. Rev.* **190**, 9–25.
26. Kirk, B. W., Feinsod, M., Favis, R., Kliman, R. M., and Barany, F. (2002) Single nucleotide polymorphism seeking long term association with complex disease. *Nucleic Acids Res.* **30**, 3295–3311.
27. Shi, M. M. (2001) Enabling large-scale pharmacogenetic studies by high-throughput mutation detection and genotyping technologies. *Clin. Chem.* **47**, 164–172.
28. Kwok, P. Y. (2000) High-throughput genotyping assay approaches. *Pharmacogenomics* **1**, 95–100.
29. Kwok, P. Y. (2001) Methods for genotyping single nucleotide polymorphisms. *Annu. Rev. Genomics Hum. Genet.* **2**, 235–258.
30. Gut, I. G. (2001) Automation in genotyping of single nucleotide polymorphisms. *Hum. Mutat.* **17**, 475–492.
31. Landegren, U., Nilsson, M., and Kwok, P. Y. (1998) Reading bits of genetic information: methods for single-nucleotide polymorphism analysis. *Genome Res.* **8**, 769–776.
32. Wang, S., Kidd, K. K., and Zhao, H. (2003) On the use of DNA pooling to estimate haplotype frequencies. *Genet. Epidemiol.* **24**, 74–82.
33. Sham, P., Bader, J. S., Craig, I., O'Donovan, M., and Owen, M. (2002) DNA pooling: a tool for large-scale association studies. *Nature Rev. Genet.* **3**, 862–871.
34. Mamotte, C. D. and van Bockxmeer, F. M. (1993) A robust strategy for screening and confirmation of familial defective apolipoprotein B-100. *Clin. Chem.* **39**, 118–121.

35. Day, I. N. and Humphries, S. E. (1994) Electrophoresis for genotyping: microtiter array diagonal gel electrophoresis on horizontal polyacrylamide gels, hydrolink, or agarose. *Anal. Biochem.* **222**, 389–395.
36. Newton, C. R., Summers, C., Heptinstall, L. E., et al. (1991) Genetic analysis in cystic fibrosis using the amplification refractory mutation system (ARMS): the J3.11 MspI polymorphism. *J. Med. Genet.* **28**, 248–251.
37. Rust, S., Funke, H., and Assmann, G. (1993) Mutagenically separated PCR (MS-PCR): a highly specific one step procedure for easy mutation detection. *Nucleic Acids Res.* **21**, 3623–3629.
38. Kokoris, M., Dix, K., Moynihan, K., et al. (2000) High-throughput SNP genotyping with the Masscode system. *Mol. Diagn.* **5**, 329–340.
39. Tobe, V. O., Taylor, S. L., and Nickerson, D. A. (1996) Single-well genotyping of diallelic sequence variations by a two-color ELISA-based oligonucleotide ligation assay. *Nucleic Acids Res.* **24**, 3728–3732.
40. Samiotaki, M., Kwiatkowski, M., Parik, J., and Landegren, U. (1994) Dual-color detection of DNA sequence variants by ligase-mediated analysis. *Genomics* **20**, 238–242.
41. Grossman, P. D., Bloch, W., Brinson, E., et al. (1994) High-density multiplex detection of nucleic acid sequences: oligonucleotide ligation assay and sequence-coded separation. *Nucleic Acids Res.* **22**, 4527–4534.
42. Baron, H., Fung, S., Aydin, A., et al. (1997) Oligonucleotide ligation assay for detection of apolipoprotein E polymorphisms. *Clin. Chem.* **43**, 1984–1986.
43. Zhong, X. B., Reynolds, R., Kidd, J. R., et al. (2003) Single-nucleotide polymorphism genotyping on optical thin-film biosensor chips. *Proc. Natl. Acad. Sci. USA* **100**, 11,559–11,564.
44. Schouten, J. P., McElgunn, C. J., Waaijer, R., Zwijnenburg, D., Diepvens, F., and Pals, G. (2002) Relative quantification of 40 nucleic acid sequences by multiplex ligation-dependent probe amplification. *Nucleic Acids Res.* **30**, e57.
45. Fan, J. B., Chen, X., Halushka, M. K., et al. (2000) Parallel genotyping of human SNPs using generic high-density oligonucleotide tag arrays. *Genome Res.* **10**, 853–860.
46. Pastinen, T., Partanen, J., and Syvanen, A. C. (1996) Multiplex, fluorescent, solid-phase minisequencing for efficient screening of DNA sequence variation. *Clin. Chem.* **42**, 1391–1397.
47. Li, J., Butler, J. M., Tan, Y., et al. (1999) Single nucleotide polymorphism determination using primer extension and time-of-flight mass spectrometry. *Electrophoresis* **20**, 1258–1265.
48. Pastinen, T., Raitio, M., Lindroos, K., Tainola, P., Peltonen, L., and Syvanen, A. C. (2000) A system for specific, high-throughput genotyping by allele-specific primer extension on microarrays. *Genome Res.* **10**, 1031–1042.
49. Livak, K. J. (1999) Allelic discrimination using fluorogenic probes and the 5' nuclease assay. *Genet. Anal.* **14**, 143–149.
50. Tyagi, S., Bratu, D. P., and Kramer, F. R. (1998) Multicolor molecular beacons for allele discrimination. *Nature Biotechnol.* **16**, 49–53.
51. Tyagi, S. and Kramer, F. R. (1996) Molecular beacons: probes that fluoresce upon hybridization. *Nature Biotechnol.* **14**, 303–308.
52. Thelwell, N., Millington, S., Solinas, A., Booth, J., and Brown, T. (2000) Mode of action and application of Scorpion primers to mutation detection. *Nucleic Acids Res.* **28**, 3752–3761.
53. Solinas, A., Brown, L. J., McKeen, C., et al. (2001) Duplex Scorpion primers in SNP analysis and FRET applications. *Nucleic Acids Res.* **29**, E96.
54. von Ahsen, N., Schutz, E., Armstrong, V. W., and Oellerich, M. (1999) Rapid detection of prothrombotic mutations of prothrombin (G20210A), factor V (G1691A), and methylenetetrahydrofolate reductase (C677T) by real-time fluorescence PCR with the LightCycler. *Clin. Chem.* **45**, 694–696.
55. Chen, X., Levine, L., and Kwok, P. Y. (1999) Fluorescence polarization in homogeneous nucleic acid analysis. *Genome Res.* **9**, 492–498.
56. Latif, S., Bauer-Sardina, I., Ranade, K., Livak, K. J., and Kwok, P. Y. (2001) Fluorescence polarization in homogeneous nucleic acid analysis II: 5'-nuclease assay. *Genome Res.* **11**, 436–440.
57. Sauer, S., Lechner, D., Berlin, K., et al. (2000) A novel procedure for efficient genotyping of single nucleotide polymorphisms. *Nucleic Acids Res.* **28**, E13.
58. Sauer, S., Lechner, D., Berlin, K., et al. (2000) Full flexibility genotyping of single nucleotide polymorphisms by the GOOD assay. *Nucleic Acids Res.* **28**, E100.
59. Zielenski, J. and Tsui, L. (1995) Cystic fibrosis—genotypic and phenotypic variations. *Annu. Rev. Genet.* **29**, 777–807.

60. Altmuller, J., Palmer, L. J., Fischer, G., Scherb, H., and Wjst, M. (2001) Genomewide scans of complex human diseases: true linkage is hard to find. *Am. J. Hum. Genet.* **69**, 93–950.
61. Cardon, L. R. and Bell, J. I. (2001) Association study designs for complex diseases. Nat. Rev. Genet. 2, 91–99.
62. Keavney, B. (2000) Genetic association studies in complex diseases. *J. Hum. Hypertens.* **14**, 361–367.
63. Elston, R. (1995) The genetic dissection of multifactorial traits. *Clin. Exp. Allergy* **2**, 103–106.
64. Zondervan, K. T. and Cardon, L. R. (2004) The complex interplay among factors that influence allelic association. *Nature Rev. Genet.* **5**, 89–100.
65. Silverman, E. K. and Palmer, L. J. (2000) Case-control association studies for the genetics of complex respiratory diseases. *Am. J. Respir. Cell. Mol. Biol.* **22**, 645–648.
66. Collins, A., Lonjou, C., and Morton, N. E. (1999) Genetic epidemiology of single-nucleotide polymorphisms. *Proc. Natl. Acad. Sci. USA* **96**, 15,173–15,177.
67. Weeks, D. and Lathrop, G. (1995) Polygenic disease: methods for mapping complex disease traits. *TIG* **11**, 513–519.
68. Ardlie, K. G., Kruglyak, L., and Seielstad, M. (2002) Patterns of linkage disequilibrium in the human genome. *Nature Rev. Genet.* **3**, 299–309.
69. Collins, F. S., Patrinos, A., Jordan, E., Chakravarti, A., Gesteland, R., and Walters, L. (1998) New goals for the U.S. Human Genome Project: 1998–2003. **Science** *282*, 682–689.
70. Palmer, L. J. and Cookson, W. O. C. M. (2001) Using single nucleotide polymorphisms (SNPs) as a means to understanding the pathophysiology of asthma. *Respir. Res.* **2**, 102–112.
71. Kruglyak, L. (1997) The use of a genetic map of biallelic markers in linkage studies. *Nature Genet.* **17**, 21–24.
72. Collins, F. S., Guyer, M. S., and Charkravarti, A. (1997) Variations on a theme: cataloging human DNA sequence variation. *Science* **278**, 1580–1581.
73. Nickerson, D. A., Whitehurst, C., Boysen, C., Charmley, P., Kaiser, R., and Hood, L. (1992) Identification of clusters of biallelic polymorphic sequence-tagged sites (pSTSs) that generate highly informative and automatable markers for genetic linkage mapping. *Genomics* **12**, 377–387.
74. Chakravarti, A. (1998) It's raining SNPs, hallelujah? [news]. *Nature Genet.* **19**, 216–217.
75. McKeigue, P. M. (1998) Mapping genes that underlie ethnic differences in disease risk: methods for detecting linkage in admixed populations, by conditioning on parental admixture. *Am. J. Hum. Genet.* **63**, 241–251.
76. Kuhner, M. K., Beerli, P., Yamato, J., and Felsenstein, J. (2000) Usefulness of single nucleotide polymorphism data for estimating population parameters. *Genetics* **156**, 439–447.
77. Stallings, R. L., Ford, A. F., Nelson, D., Torney, D. C., Hildebrand, C. E., and Moyzis, R. K. (1991) Evolution and distribution of (GT)n repetitive sequences in mammalian genomes. *Genomics* **10**, 807–815.
78. Brookes, A. J. (1999) The essence of SNPs. *Gene* **8**, 177–186.
79. Edwards, J. and Bartlett, J. M. (2003) Mutation and polymorphism detection: a technical overview. *Methods Mol. Biol.* **226**, 287–294.
80. Kruglyak, L. and Nickerson, D. A. (2001) Variation is the spice of life. *Nature Genet.* **27**, 234–236.
81. Dorman, J. S., LaPorte, R. E., Stone, R. A., and Trucco, M. (1990) Worldwide differences in the incidence of type I diabetes are associated with amino acid variation at position 57 of the HLA-DQ beta chain. *Proc. Natl. Acad. Sci. USA* **87**, 7370–7374.
82. Allen, M., Heinzmann, A., Noguchi, E., et al. (2003) Positional cloning of a novel gene influencing asthma from chromosome 2q14. *Nature Genet.* **35**, 258–263.
83. Hugot, J. P., Chamaillard, M., Zouali, H., et al. (2001) Association of NOD2 leucine-rich repeat variants with susceptibility to Crohn's disease. *Nature* **411**, 599–603.
84. Ogura, Y., Bonen, D. K., Inohara, N., et al. (2001) A frameshift mutation in NOD2 associated with susceptibility to Crohn's disease. *Nature* **411**, 603–606.
85. Chagnon, Y. C., Rankinen, T., Snyder, E. E., Weisnagel, S. J., Perusse, L., and Bouchard, C. (2003) The human obesity gene map: the 2002 update. *Obes. Res.* **11**, 313–367.
86. Palmer, L. and Cookson, W. (2001) Using single nucleotide polymorphisms as a means to understanding the pathophysiology of asthma. *Respir. Res.* **2**, 102–112.
87. Stephens, J. C. (1999) Single-nucleotide polymorphisms, haplotypes, and their relevance to pharmacogenetics. *Mol. Diagn.* **4**, 309–317.
88. Rose, C. M., Marsh, S., Ameyaw, M. M., and McLeod, H. L. (2003) Pharmacogenetic analysis of clinically relevant genetic polymorphisms. *Methods Mol. Med.* **85**, 225–237.

89. Lele, R. D. (2003) The human genome project: its implications in clinical medicine. *J. Assoc. Physicians India* **51**, 373–380.
90. Mallal, S., Nolan, D., Witt, C., et al. (2002) Association between presence of HLA-B*5701, HLA-DR7, and HLA-DQ3 and hypersensitivity to HIV-1 reverse-transcriptase inhibitor abacavir. *Lancet* **359**, 727–732.
91. Weiss, K. M. and Terwilliger, J. D. (2000) How many diseases does it take to map a gene with SNPs? *Nature Genet.* **26**, 151–157.
92. Zhao, L.P., Aragaki, C., Hsu, L., and Quiaoit, F. (1998) Mapping of complex traits by single-nucleotide polymorphisms. *Am. J. Hum. Genet.* **63**, 225–240.
93. Long, A. D. and Langley, C. H. (1999) The power of association studies to detect the contribution of candidate genetic loci to variation in complex traits. *Genome Res.* **9**, 720–731.
94. Terwilliger, J. D. and Goring, H. H. (2000) Gene mapping in the 20th and 21st centuries: statistical methods, data analysis, and experimental design. *Hum. Biol.* **72**, 63–132.
95. Lander, E. and Schork, N. (1994) Genetic dissection of complex traits. *Science* **265**, 2037–2048.
96. Risch, N. and Merikangas, K. (1996) The future of genetic studies of complex human diseases. *Science* **273**, 1516–1517.
97. Wolfe, K. H. and Li, W. H. (2003) Molecular evolution meets the genomics revolution. *Nature Genet.* **33 (Suppl.)**, 255–265.
98. Cardon, L. R. and Abecasis, G. R. (2003) Using haplotype blocks to map human complex trait loci. *Trends Genet.* **19**, 135–140.
99. Dahlman, I., Eaves, I. A., Kosoy, R., et al. (2002) Parameters for reliable results in genetic association studies in common disease. *Nature Genet.* **30**, 149–150.
100. Goldstein, D. B., Ahmadi, K. R., Weale, M. E., and Wood, N. W. (2003) Genome scans and candidate gene approaches in the study of common diseases and variable drug responses. *Trends Genet.* **19**, 615–622.
101. Ioannidis, J. P., Ntzani, E. E., Trikalinos, T. A., and Contopoulos-Ioannidis, D. G. (2001) Replication validity of genetic association studies. *Nature Genet.* **29**, 30–309.
102. Lohmueller, K. E., Pearce, C. L., Pike, M., Lander, E. S., and Hirschhorn, J. N. (2003) Meta-analysis of genetic association studies supports a contribution of common variants to susceptibility to common disease. *Nature Genet.* **33**, 177–182.
103. Tabor, H. K., Risch, N. J., and Myers, R. M. (2002) Opinion: candidate-gene approaches for studying complex genetic traits: practical considerations. *Nature Rev. Genet.* **3**, 391–397.
104. Olson, J. M., Witte, J. S., and Elston, R. C. (1999) Genetic mapping of complex traits. *Stat. Med.* **18**, 2961–2981.
105. Lander, E. and Kruglyak, L. (1995) Genetic dissection of complex traits: guidelines for interpreting and reporting linkage results. *Nature Genet.* **11**, 241–247.
106. Witte, J. S., Elston, R. C., and Cardon, L. R. (2000) On the relative sample size required for multiple comparisons. *Statist. Med.* **19**, 369–372.
107. Austin, M. A., Harding, S., and McElroy, C. (2003) Genebanks: a comparison of eight proposed international genetic databases. *Community Genet.* **6**, 37–45.
108. Rosner, B. *Fundamental of Biostatistics*. 3rd ed., PWS-Kent, Boston, MA.
109. Lee, W. C. (2002) Testing for candidate gene linkage disequilibrium using a dense array of single nucleotide polymorphisms in case-parents studies. *Epidemiology* **13**, 545–551.
110. Weiss, S. T., Silverman, E. K., and Palmer, L. J. (2001) Case-control association studies in pharmacogenetics. *Pharmacogenomics J.* **1**, 157–158.
111. Palmer, L. J. and Cookson, W. O. C. M. (2000) Genomic approaches to understanding asthma. *Genome Res.* **10**, 1280–1287.
112. Feldman, M. W., Lewontin, R. C., and King, M. C. (2003) Race: a genetic melting-pot. *Nature* **424**, 374.
113. Risch, N., Burchard, E., Ziv, E., and Tang, H. (2002) Categorization of humans in biomedical research: genes, race and disease. *Genome Biol.* **3**, 2007.
114. Shifman, S., Kuypers, J., Kokoris, M., Yakir, B., and Darvasi, A. (2003) Linkage disequilibrium patterns of the human genome across populations. *Hum. Mol. Genet.* **12**, 771–776.
115. Roewer, L., Kayser, M., de Knijff, P., et al. (2000) A new method for the evaluation of matches in non-recombining genomes: application to Y-chromosomal short tandem repeat (STR) haplotypes in European males. *Forensic Sci. Int.* **114**, 31–43.

116. Zavattari, P., Deidda, E., Whalen, M., et al. (2000) Major factors influencing linkage disequilibrium by analysis of different chromosome regions in distinct populations: demography, chromosome recombination frequency and selection. *Hum. Mol. Genet.* **9**, 2947–2957.
117. Watkins, W. S., Zenger, R., O'Brien, E., et al. (1994) Linkage disequilibrium patterns vary with chromosomal location: a case study from the von Willebrand factor region. *Am. J. Hum. Genet.* **55**, 348–355.
118. Jorde, L. B., Watkins, W. S., Carlson, M., et al. (1994) Linkage disequilibrium predicts physical distance in the adenomatous polyposis coli region. *Am. J. Hum. Genet.* **54**, 884–898.
119. Cardon, L. R. and Palmer, L. J. (2003) Population stratification and spurious allelic association. *Lancet* **361**, 598–604.
120. Welborn, T. (1998) *The Busselton Study: Mapping Population Health.* Australasian Medical Publishing, Sydney.
121. Ewens, W. and Spielman, R. (1995) The transmission/disequilibrium test: history, subdivision, and admixture. *Am. J. Hum. Genet.* **57**, 455–464.
122. Satten, G. A., Flanders, W. D., and Yang, Q. (2001) Accounting for unmeasured population substructure in case-control studies of genetic association using a novel latent-class model. *Am. J. Hum. Genet.* **68**, 466–477.
123. Devlin, B., Roeder, K., and Wasserman, L. (2001) Genomic control, a new approach to genetic-based association studies. *Theor. Popul. Biol.* **60**, 155–166.
124. Devlin, B., Roeder, K., and Bacanu, S. A. (2001) Unbiased methods for population-based association studies. *Genet. Epidemiol.* **21**, 273–284.
125. Overall, A. D. and Nichols, R. A. (2001) A method for distinguishing consanguinity and population substructure using multilocus genotype data. *Mol. Biol. Evol.* **18**, 2048–2056.
126. Bacanu, S. A., Devlin, B., and Roeder, K. (2002) Association studies for quantitative traits in structured populations. *Genet. Epidemiol.* **22**, 78–93.
127. Pritchard, J. K. and Rosenberg, N. A. (1999) Use of unlinked genetic markers to detect population stratification in association studies. *Am. J. Hum. Genet.* **65**, 220–228.
128. Pritchard, J. K., Stephens, M., Rosenberg, N. A., and Donnelly, P. (2000) Association mapping in structured populations. *Am. J. Hum. Genet.* **67**, 170–181.
129. Weiss, K. M. and Clark, A. G. (2002) Linkage disequilibrium and the mapping of complex human traits. *Trends Genet.* **18**, 19–24.
130. Fallin, D., Cohen, A., Essioux, L., et al. (2001) Genetic analysis of case/control data using estimated haplotype frequencies: application to APOE locus variation and Alzheimer's disease. *Genome Res.* **11**, 143–151.
131. Thomas, S., Porteous, D., and Visscher, P.M. (2004) Power of direct vs. indirect haplotyping in association studies. *Genet. Epidemiol.* **26**, 116–124.
132. Schaid, D. J. (2002) Relative efficiency of ambiguous vs. directly measured haplotype frequencies. *Genet. Epidemiol.* **23**, 426–443.
133. Excoffier, L., Smouse, P. E., and Quattro, J. M. (1992) Analysis of molecular variance inferred from metric distances among DNA haplotypes: application to human mitochondrial DNA restriction data. *Genetics* **131**, 479–491.
134. Schaid, D. J., Rowland, C. M., Tines, D. E., Jacobson, R. M., and Poland, G. A. (2002) Score tests for association between traits and haplotypes when linkage phase is ambiguous. *Am. J. Hum. Genet.* **70**, 425–434.
135. Niu, T., Qin, Z. S., Xu, X., and Liu, J. S. (2002) Bayesian haplotype inference for multiple linked single-nucleotide polymorphisms. *Am. J. Hum. Genet.* **70**, 157–169.
136. Stephens, M., Smith, N. J., and Donnelly, P. (2001) A new statistical method for haplotype reconstruction from population data. *Am. J. Hum. Genet.* **68**, 978–89.
137. Templeton, A. R. (1996) Cladistic approaches to identifying determinants of variability in multifactorial phenotypes and the evolutionary significance of variation in the human genome. *Ciba Found. Symp.* **197**, 259–277.
138. Editorial (2003) Compare and contrast. *Nature* **426**, 750–751.
139. Castillo-Davis, C. I. and Hartl, D. L. (2003) Conservation, relocation and duplication in genome evolution. *Trends Genet.* **19**, 593–597.
140. Yan, H., Kinzler, K. W., and Vogelstein, B. (2000) Tech.sight. Genetic testing—present and future. *Science* **289**, 1890–1892.
141. van Ommen, G. J., Bakker, E., and den Dunnen, J. T. (1999) The human genome project and the future of diagnostics, treatment, and prevention. *Lancet* **354(Suppl. 1)**, SI5–S10.

142. Ross, L. F. and Moon, M. R. (2000) Ethical issues in genetic testing of children. *Arch. Pediatr. Adolesc. Med.* **154**, 873–8739.
143. Ohlstein, E. H., Ruffolo, R. R., Jr., and Elliott, J. D. (2000) Drug discovery in the next millennium. *Annu. Rev. Pharmacol. Toxicol.* **40**, 177–191.
144. Chanda, S. K. and Caldwell, J. S. (2003) Fulfilling the promise: drug discovery in the post-genomic era. *Drug Discov. Today* **8**, 168–174.
145. Khoury, M. J., McCabe, L. L., and McCabe, E. R. (2003) Population screening in the age of genomic medicine. *N. Engl. J. Med.* **348**, 50–58.
146. Colditz, G. A., Manson, J. E., and Hankinson, S. E. (1997) The Nurses' Health Study: 20-year contribution to the understanding of health among women. *J. Womens Health* **6**, 49–62.
147. Clayton, D. and McKeigue, P. M. (2001) Epidemiological methods for studying genes and environmental factors in complex diseases. *Lancet* **358**, 1356–1360.
148. Davey Smith, G. and Ebrahim, S. (2003) "Mendelian randomization": can genetic epidemiology contribute to understanding environmental determinants of disease? *Int. J. Epidemiol.* **32**, 1–22.
149. Kopelovich, L., Crowell, J. A., and Fay, J. R. (2003) The epigenome as a target for cancer chemoprevention. *J. Natl. Cancer Inst.* **95**, 1747–1757.
150. Andersen, A. A. and Panning, B. (2003) Epigenetic gene regulation by noncoding RNAs. *Curr. Opin. Cell Biol.* **15**, 281–289.
151. Jaenisch, R. and Bird, A. (2003) Epigenetic regulation of gene expression: how the genome integrates intrinsic and environmental signals. *Nature Genet.* **33 (Suppl.)**, 245–254.
152. Jorde, L. (1995) Linkage disequilibrium as a gene-mapping tool [editorial; comment]. *Am. J. Hum. Genet.* **56**, 11–14.
153. Toivonen, H. T., Onkamo, P., Vasko, K., et al. (2000) Data mining applied to linkage disequilibrium mapping. *Am. J. Hum. Genet.* **67**, 133–145.
154. Gabriel, S. B., Schaffner, S. F., Nguyen, H., et al. (2002) The structure of haplotype blocks in the human genome. *Science* **296**, 2225–2229.
155. Daly, M. J., Rioux, J. D., Schaffner, S. F., Hudson, T. J., and Lander, E. S. (2001) High-resolution haplotype structure in the human genome. *Nature Genet.* **29**, 229–232.
156. Patil, N., Berno, A. J., Hinds, D. A., et al. (2001) Blocks of limited haplotype diversity revealed by high-resolution scanning of human chromosome 21. *Science* **294**, 1719–1723.
157. Tsui, C., Coleman, L. E., Griffith, J. L., et al. (2003) Single nucleotide polymorphisms (SNPs) that map to gaps in the human SNP map. *Nucleic Acids Res.* **31**, 4910–4916.
158. Johnson, G. C., Esposito, L., Barratt, B. J., et al. (2001) Haplotype tagging for the identification of common disease genes. *Nature Genet.* **29**, 233–237.
159. Sebastiani, P., Lazarus, R., Weiss, S. T., Kunkel, L. M., Kohane, I. S., and Ramoni, M. F. (2003) Minimal haplotype tagging. *Proc. Natl. Acad. Sci. USA* **100**, 9900–9905.
160. Schulze, T. G., Zhang, K., Chen, Y. S., Akula, N., Sun, F., and McMahon, F. J. (2004) Defining haplotype blocks and tag single-nucleotide polymorphisms in the human genome. *Hum. Mol. Genet.* **13**, 335–342.
161. Chapman, J. M., Cooper, J. D., Todd, J. A., and Clayton, D. G. (2003) Detecting disease associations due to linkage disequilibrium using haplotype tags: a class of tests and the determinants of statistical power. *Hum. Heredity* **56**, 18–31.
162. Zhang, K., Calabrese, P., Nordborg, M., and Sun, F. (2002) Haplotype block structure and its applications to association studies: power and study designs. *Am. J. Hum. Genet.* **71**, 1386–1394.
163. Ke, X. and Cardon, L. R. (2003) Efficient selective screening of haplotype tag SNPs. *Bioinformatics* **19**, 287–288.
164. Wiuf, C., Laidlaw, Z., and Stumpf, M. P. (2003) Some notes on the combinatorial properties of haplotype tagging. *Math. Biosci.* **185**, 205–216.
165. Couzin, J. (2002) Human genome. HapMap launched with pledges of $100 million. *Science* **298**, 941–942.
166. Phillips, M. S., Lawrence, R., Sachidanandam, R., et al. (2003) Chromosome-wide distribution of haplotype blocks and the role of recombination hot spots. *Nature Genet.* **33**, 382–387.
167. Wall, J. D. and Pritchard, J. K. (2003) Haplotype blocks and linkage disequilibrium in the human genome. *Nature Rev. Genet.* **4**, 587–597.
168. Goldstein, D. B. and Weale, M. E. (2001) Population genomics: linkage disequilibrium holds the key. *Curr. Biol.* **11**, R576–R579.
169. Tishkoff, S. A. and Verrelli, B. C. (2003) Role of evolutionary history on haplotype block structure in the human genome: implications for disease mapping. *Curr. Opin. Genet. Dev.* **13**, 569–575.

170. Clark, A. G. (2003) Finding genes underlying risk of complex disease by linkage disequilibrium mapping. *Curr. Opin. Genet. Dev.* **13**, 296–302.
171. Fields, S. (1997) The future is function. *Nature Genet.* **15**, 325–327.
172. Matsuzaki, H., Loi, H., Dong, S., et al. (2004) Parallel genotyping of over 10,000 SNPs using a one-primer assay on a high-density oligonucleotide array. *Genome Res.* **14**, 414–425.
173. Peltonen, L. and McKusick, V. A. (2001) Genomics and medicine. Dissecting human disease in the postgenomic era. *Science* **291**, 1224–1229.
174. Kruglyak, L. (1999) Prospects for whole-genome linkage disequilibrium mapping of common disease genes. *Nature Genet.* **22**, 139–144.
175. Judson, R., Salisbury, B., Schneider, J., Windemuth, A., and Stephens, J. C. (2002) How many SNPs does a genome-wide haplotype map require? *Pharmacogenomics* **3**, 379–391.
176. Terwilliger, J. D. and Weiss, K. M. (1998) Linkage disequilibrium mapping of complex disease: fantasy or reality? *Curr. Opin. Biotechnol.* **9**, 578–594.
177. Abecasis, G. R., Noguchi, E., Heinzmann, A., et al. (2001) Extent and distribution of linkage disequilibrium in three genomic regions. *Am. J. Hum. Genet.* **68**, 191–197.
178. Hoh, J. and Ott, J. (2003) Mathematical multi-locus approaches to localizing complex human trait genes. *Nature Rev. Genet.* **4**, 701–709.
179. Hoh, J., Wille, A., and Ott, J. (2001) Trimming, weighting, and grouping SNPs in human case-control association studies. *Genome Res.* **11**, 2115–2119.

20

cDNA Microarrays

Phillip G. Febbo

1. Background and Theory of cDNA Microarrays

1.1. Overview

Clinicians and scientists are limited in their ability to understand human disease and cellular biology by the technologies available to measure the state of the organism or cell. For millennia, scientists have studied human biology and disease based on anatomical observations; for centuries, decisions have been based on microscopic observations, and over the past several decades, decisions have been based on the status of specific genes associated with disease. In the mid-1990s, cDNA microarray technology emerged that simultaneously measured the expression of thousands of genes ***(1–5)***. These expression microarrays have rapidly evolved to cover most of the 34,000 genes in the human genome ***(6)*** and now offer clinicians and scientists an unprecedented level of detail through which they can observe human disease and cellular biology.

The central position of gene expression in cellular homeostasis makes it a provocative window through which to view cellular biology ***(7)***. From the genetic and epigenetic events that cause disease come profound changes in the cell biology that are reflected in the global pattern of gene expression ***(8)***. These changes can then be analyzed to identify specific target genes of particular interest ***(9)*** or to implicate cellular processes with mechanistic import ***(10)***.

Scientists and clinicians have seized on the new modality of cDNA microarrays to assay and observe cellular biology. Investigators began applying microarrays to single-cell organisms and have advanced up the phylogenetic tree. They began with very simple experimental designs in simple model systems and have progressed to large, multifaceted experiments in complex models. As the technology has developed, so too have the methods by which to analyze and interpret data of increasing complexity generated by microarrays.

1.2. RNA Expression

RNA stands between DNA and protein in the central dogma of life proposed by Francis Crick ***(11)***. Although the majority of messenger RNA species have no direct role in cellular metabolism, the relative abundance of each gene's transcript can reflect the identity and state of each cell. Thus, accurately measuring the pattern of expression across all genes [herein referred to as the "transcriptome" ***(12)***] is a potential mechanism by which to assay cellular biology.

Messenger RNA (mRNA) comprises approx 1% of total cellular RNA but is the template for protein translation and the major focus of expression analysis. The posttranscriptional addition of a poly-A tail (3' polyadenylation) to the majority of mature mRNA species is used opportunistically to process samples for microarrays, thereby enabling a broad representation of the transcriptome. The product resulting from mRNA processing for microarrays (henceforth

From: *Medical Biomethods Handbook*
Edited by: J. M. Walker and R. Rapley © Humana Press, Inc., Totowa, NJ

referred to as target) depends on the microarray platform with the two most common forms being single-stranded complementary DNA (cDNA) following reverse transcription of mRNA and antisense RNA (aRNA) resulting from in vitro transcription of a double-stranded cDNA template.

Two key features of target synthesis are critical to the successful microarray analysis. First, the resulting target must be in a form that can hybridize to its complementary sequence. In general, the target is single stranded and not too heavily laden with large macromolecules so as to interfere with complementary basepair binding. In addition, the final target is often fragmented to shorter lengths to minimize secondary structure that can significantly affect complementary binding. Second, the target must be detectable when bound to its complementary sequence. Although this is achieved by incorporating radioactive isotopes or fluorescent-dye-labeled nucleotides into the final synthesis of target, a balance must be struck between successful detection and steric hindrance of complementary binding. However, through the judicious use of radioactive labeling or methods of fluorescent signal amplification, successful labeling of the target results in sensitivities in the attomolar range.

1.3. cDNA Microarrays

Once a labeled target is synthesized from mRNA, microarrays are used to detect the relative abundance of each transcript. The platforms upon which DNA arrays have been made range from the earliest having cDNA clones on Nylon filters to silicon wafers that use photolithography to create oligonucleotides of known sequence. In general, microarray detection of a labeled target differs from previous broad surveys of gene expression [e.g., SAGE ***(13)***] and differential display ***(14)*** in that the curated platform of the microarray allows immediate identification of genes differentially expressed whereas previous techniques required intensive cloning and sequencing efforts.

The two major platforms currently in use for large-scale expression analysis are spotted microarrays and synthesized microarrays. Spotted microarrays are created by adhering individual species from curated cDNA libraries onto a glass slide. Each spotted microarray element is a known cDNA assigned to a specific location on a two-dimentional surface. Synthesized oligonucleotide microarrays use chemistry to create a grid of unique oligonucleotides (called features) complementary to known genes ***(5)***.

Each platform has advantages and disadvantages, but all use the basic chemistry of sequence-specific hybridization. After a labeled target is hybridized to the microarray, the amount of radioactivity or fluorescence at each location on the microarray (signal) is determined. Because the intensity of signal at any location is dependent on many variables, microarray determination of expression is relative not absolute. Although this is obvious with the dual-hybridization approach applied to most spotted arrays, it is equally true for synthesized arrays. Thus, detection of a gene on a microarray is almost meaningless unless evaluated in the context of a well-designed experiment or previously subjected to rigorous analysis to understand the correlation between detected expression and absolute expression.

1.4. Microarray Data Analysis

With the high-density organization of spotted elements or features on microarrays, the expression data representing the transcriptome is sizable and complex. Successful analysis of data generated from microarrays involves (1) an assessment of microarray staining quality, (2) normalization of expression across microarrays in an experiment to minimize technically introduced variation, (3) exclusion of genes not detected as being expressed or without expression variation, and (4) expression analysis. Here, expression analysis is defined as the application of computational algorithms in order to find structure within complex expression patterns.

Although the computational approaches applied to expression analysis have ranged from the very simple (e.g., fold difference in mean expression of genes between classes of samples) to the very complex (e.g., support vector machines), there are two basic approaches to expression analysis: supervised and unsupervised analysis. During supervised analysis, sample identifiers are used to assess the correlation between gene expression and sample characteristics. In unsupervised analysis, no sample identification is provided during analysis.

The computational tools for microarray analysis have advanced along with the technology to measure gene expression. The computational demands engendered by technologies assaying a cell's transcriptome required the assembly of collaborative investigative teams to successfully design, implement, and analyze data generated from expression arrays. These multidisciplinary teams create an exciting environment within which important discoveries are being made and represent a paradigm for the future application of other genomic technologies.

2. Practical Steps Involved in cDNA Microarrays

As outlined above, cDNA microarray analysis involves study design, RNA isolation, target synthesis, microarray hybridization and scanning, data processing, and data analysis. This section discusses the practical steps of cDNA microarray analysis and highlights areas of particular importance to foster a strong conceptual foundation. The specifics of each step involved in cDNA microarray analysis will evolve along with the continued advancement of technology platforms. However, many of the challenges involved in effective study design, such as RNA isolation, target synthesis, microarray development, microarray hybridization and scanning, and analysis will remain consistent and these are discussed below.

2.1. Study Design

Although there are no hard and fast rules for the successful design of microarray studies, some general guidelines should be kept in mind when planning an experiment. First, microarrays include probes for tens of thousands of genes and, thus, create experiments where there are significantly more variables (i.e., genes) than experimental samples. This challenges standard methods of evaluation and interpretation and requires specific approaches during the experimental design ***(15)***. Additionally, study design also has to anticipate the significant biologic and technical variation inherent to most experimental systems and RNA processing. While this topic has been reviewed in detail elsewhere ***(16)***, the important elements for study design are discussed below.

In general, the number of samples required for expression analysis will be dependent on the specific question being addressed and the experimental "system" being used ***(17)***. As the anticipated difference in gene expression across an experiment increases, the number of samples required decreases. Similarly, for well-controlled experimental systems where biological variability can be minimized, fewer samples are required than for experiments using primary clinical samples. Other significant factors influencing study design are the specific array platform to be used and the anticipated methods of analysis ***(17)***.

Although methods have been proposed for sample size calculation in classification models ***(18,19)***, there is generally insufficient *a priori* knowledge upon which to base accurate sample size estimates. Practically, it is often most helpful to review the literature to determine how many samples were required to perform a successful microarray experiment in a similar system.

For in vitro cell-based experiments, triplicate samples for each experimental condition are generally sufficient. Collection of samples over a time-course can significantly help identify genes with small but reproducible expression changes in response to a stimulus but late time-point untreated (or vehicle-treated) controls are necessary. The best experiments have internal control genes that are followed in parallel on the same cells for which expression will be measured through a technique independent of microarrays.

For clinical samples, it is often impractical to perform duplicate or triplicate analysis on each sample. In addition, the effects of tissue heterogeneity, differences in specimen handling, and biological variability are likely to affect final expression patterns more than technically introduced variation during target preparation, and resources are better spent obtaining additional unique samples than duplicating samples. All of the factors mentioned can obfuscate expression data in clinical samples and, in general, mandate larger sample sizes for experiments using primary samples than those using cell-culture systems.

A critical part of the design of spotted arrays is choosing the appropriate control sample that serves as a common reference for all of the test samples. There are alternative design schemes for dual-hybridization techniques, but the use of a single standard reference is most common. In general, an RNA sample pooled from many cell lines or samples is used to provide optimal coverage of genes present on the array. Alternatively, it has been suggested that pooling a small number samples with diverse expression could result in higher-quality expression data ***(20)***.

2.2. RNA Isolation

cDNA microarray analysis begins with RNA isolation. Currently, fresh cells or rapidly frozen tissue are required for genomewide expression analysis. Great care must be taken to minimize RNA degradation, as RNase activity is ubiquitous and introduces uninformative changes in the measured transcriptome.

Many methods of RNA isolation are compatible with microarray analysis. The best methods preserve mRNA transcript length, minimize DNA contamination, and result in relatively concentrated RNA in solution without excessive salt or protein contaminants. Prior to target synthesis, if an investigator is relatively new to RNA isolation techniques, it is best to view the RNA (approx 0.5 µg) after separation on a 1% denaturing agarose gel. Specifically, high-quality RNA samples should have clear and sharp 18S and 28S ribosomal RNA bands and a diffuse smear representing the less abundant mRNA species of variable lengths. Diffuse ribosomal bands or heavy smears close to the gel-loading wells can indicate RNA degradation or sample contamination, respectively, and the RNA from these samples is likely to yield poor quality expression data and should not be used. Although there are alternative means to assess RNA quality, there is no gold standard and the simple method mentioned here is generally sufficient.

Standardizing the amount of input RNA across experiments is important regardless of the microarray platform subsequently used. Additionally, more RNA is not always better. Excessive starting RNA can saturate the target preparation reactions and negatively affect yield. The amount of total RNA required depends on the specific protocol but ranges from as low as approx 50 ng ***(21)*** to as much as 15 µg ***(22)***.

There is great interest in applying microarrays to paraffin-embedded tissue samples because of the relative scarcity of frozen tissue compared to paraffin-embedded tissue. Formalin fixation of tissue prior to paraffin embedding, routinely used to process surgical specimens, crosslinks and fragments RNA. As a result, RNA isolated from paraffin-embedded specimens tends to be of very short length and would not pass the quality assessment discussed above. Standard techniques for labeling RNA prior to microarray analysis are ineffective and resultant expression data quality is poor. Novel approaches, including targeted multiplexed reverse transcription–polymerase chain reaction (RT-PCR), redesign of microarrays to bias probes to the 3' end of gene transcripts, and new enzymological approaches, will be tested rigorously over the next few years. If these prove successful in generating high-quality expression data, they are likely to rapidly replace existing techniques for tissue-based projects.

In the end, prior to starting target synthesis, the required amount of RNA should be in a standardized volume of solution. Excessive vacuum concentration should be avoided and an RNA pellet should never be allowed to fully air-dry. Although it might seem trivial, beginning with high-quality, contaminant-free RNA is a critical step in microarray analysis and its importance cannot be overstated.

2.3. Target Synthesis

There are a myriad of specific protocols for target preparation. The specific protocol used will depend on the amount of starting RNA and the microarray platform used. When relatively unlimited RNA is available, a target for spotted arrays is created by directly incorporating aminoallyl into the first cDNA strand synthesis from approx 2 µg of mRNA or approx 100 µg of total RNA *(23)*. The fluorescent dyes (Cy3 or Cy5) are subsequently covalently bound to the aminoallyl linkers to label the product. Oligo-dT or random hexamer primers are used to prime cDNA synthesis and second- or third-generation reverse transcriptases are used to allow cDNA synthesis at higher temperatures that minimize secondary RNA structure and maximize cDNA length. The resulting product is processed and column purified to remove RNA and unbound dye and concentrate the labeled cDNA product. A detailed description of this protocol can be found at http://cmgm.stanford.edu/pbrown/protocols/index.html .

Synthesized arrays take advantage of in vitro transcription to amplify the RNA transcript number and to create a labeled product of the correct complementation to the designed features on the arrays. Briefly, single-stranded cDNA is synthesized from RNA using a oligo-dT primer that also includes a T7 polymerase site. Double-stranded cDNA (dscDNA) is created using a combination of RNase H to digest the RNA from the RNA/DNA duplex, DNA polymerase to create the second strand, and DNA ligase to complete the phosphate backbone of the second strand. The dscDNA can subsequently be used with T7 mediated in vitro transcription (IVT) to create antisense RNA (aRNA). During IVT, biotin-labeled nucleotides (bio-UTP and bio-CTP) are incorporated into the aRNA for eventual detection on the microarrays. Specifics for this protocol can be found at http://www.affymetrix.com/support/technical/other/cdna_protocol_manual.pdf .

When RNA is of very limited quantity, methods of RNA amplification can be applied to obtain a sufficient target for expression analysis. Again, there are a number of methods commercially available and more are in development. The most common method of RNA amplification involves sequential rounds of dscDNA synthesis and IVT *(24)*. For this protocol, the first round of cDNA synthesis and IVT proceeds similar to that outlined above but biotin-labeled nucleotides are not incorporated into the aRNA. Instead, the aRNA is used as a template for an additional round of cDNA synthesis. Because of the antisense orientation of the aRNA, random hexamers are used to prime first-strand cDNA synthesis, and the oligo-dT primer with the T7 polymerase start site is used to create the dscDNA. This sequential method significantly increases the target yield and makes it feasible to obtain expression data from only 50 ng of total RNA. There is some cost to amplification; the lengths of the amplified target decrease with each round and the signal intensity diminishes. This cost has to be balanced with the practicality of obtaining increased amounts of RNA for an experiment.

2.4. Microarray Design

2.4.1. Spotted Arrays

There are significant but surmountable technical challenges to the reproducible creation of cDNA arrays because of sequence specific differences between the individual cDNAs, differences in the lengths of cDNA species, and the chemistry of fixing cDNA to surfaces. These technical challenges necessitate a dual-hybridization approach (test sample and control sample) in order to obtain reproducible expression data from cDNA arrays (*see* below).

Most recently, investigators have switched to spotting short oligonucleotides of similar length onto glass slides. Spotted oligonucleotide arrays have greater specificity and sensitivity *(25)* than the cDNA arrays but still require a dual-hybridization approach.

2.4.2. Synthesized Arrays

The quaternary genetic code allows for the development of small oligonucleotide probes that are highly specific to genes of interest. Using complemenary sequence hybridization is at the backbone of this technology. As such, the chemistry around minimizing secondary structure while maintaining strand integrity and minimizing mutational events is critical to cDNA array technology. Additionally, incorporating sufficient controls so as to account for nonspecific hybridization ("cross-hybridization") is also critical for success.

Synthesized oligonucleotide microarrays (i.e., Affymetrix) do not require a dual-hybridization approach but are only available commercially. In this process, chemical masking is combined with nucleic acid synthesis to create a grid of unique 25-mer oligonucleotides *(5)*. The chemical precision of the process (adapted from the synthetic process of microprocessors) abrogates the need for the hybridization of a control and a test sample on each microarray. However, as the detection of each gene is dependent on the success of the synthesized probes and primary structure of the gene, expression should still be viewed as relative rather than absolute. Although test and control samples are not required for the same array, understanding the expression of any single gene will require that results from several samples to be compared.

2.5. Microarray Hybridization and Scanning

The labeled target is hybridized to either spotted or synthesized microarrays and scanned to generate the expression data. Hybridization occurs at temperatures around 60°C in a buffer that minimizes nonspecific binding of target to probe and maximizes signal intensity for true hybridization. Hybridization buffer contains a combination of competing unlabeled DNA, salts, and serum that minimizes secondary structure of nucleic acids and saturates sites likely to bind nonspecifically with nucleic acids. After several hours of hybridization, the arrays are washed with buffers of variable stringency to remove excess, unbound target.

A laser is used to excite and measure the emission of the fluorescent moieties incorporated into the target to determine the amount of target bound to each microarray element. For spotted arrays, two-color excitation and emission is measured, whereas with synthetic microarrays, only a single emission is measured. The end result for both array platforms is a data file with fluorescence intensity at each coordinate on the grid of the microarray. These fluorescence intensities are subsequently used to calculate expression for each gene represented on the array.

2.6. Data Processing

The first step in data analysis is calculating gene expression from the emission intensity for each element on the microarray. For spotted arrays, the intensity of the test sample dye is divided by the intensity of the reference sample. A log2(ratio) value reflects the fold difference in expression for the gene in the test sample compared to the reference sample. Because these values can be dependent on fluorescence intensity, log2(ratio) are often normalized using a locally weighted linear regression (lowess) *(26)*.

For synthetic arrays, dual hybridization is not performed and gene expression is determined by the difference in staining intensity between matched and mismatched probes. The matched probes are perfectly complementary to the gene of interest, whereas the mismatched probe differs by one nucleic acid in the 13th position. Staining intensity at the matched probe is used to represent specific hybridization and staining at the mismatched probe is considered nonspecific. Currently, the MAS5 software provided with Affymetrix microarrays determines gene expression by performing a ranking statistic for the difference in hybridization between each of the 11 pairs of matched and mismatched probe sets for each gene. Based on the values and consistency between the 11 sets of probes, the software will assign a confidence call (Present, Absent, Moderate) and a value. Alternative methods have been proposed to obtain expression values from synthetic arrays *(27)*, and this remains an active area of research.

Once expression values are assigned for each gene on the array, all of the arrays from the experiment are brought together for analysis. Array-to-array differences in overall fluorescence intensity are somewhat unavoidable during the staining, washing, and scanning of arrays. Because of this, microarrays are often scaled together (also referred to as normalized) to minimize the effect of this variation on subsequent analysis (reviewed in **ref.** ***28***). Most approaches to array normalization assume that overall microarray intensity should be equal across an experiment and use linear adjustments to make equal median or mean expression for each array in an experiment. A few examples of alternative strategies to normalize microarray data include intensity-based *(26)*, nonlinear *(29)*, and rank-invariant *(30)* normalization techniques. Although this process might appear trivial, there can be profound effects on subsequent analysis *(31)*. Simple linear scaling can decrease true differential expression between samples if cells within an experiment differ significantly with respect to their global transcriptional activity. Scaling can also artificially accentuate variations in gene expression if an array of poor quality and poor overall staining is included because of the high scaling factors required *(32)*. Thus, integral to the scaling together of microarrays in an experiment is quality assessment *(33)* Again, there are multiple methods with which to assess the quality of each individual microarray *(30,34,35)*. The critical element of quality assessment is identifying microarrays (or regions of microarrays) that are of poor quality and need to be excluded from subsequent analysis. The specific approach and execution of quality assessment is dependent on platform and experimental design.

2.7. Expression Analysis

Expression analysis applies computational methods to find informative structure within the data generated from microarrays. One of Albert Einstein's quotes anticipated the basic approach of expression analysis: "Out of clutter, find Simplicity, from discord, find Harmony. In the middle of difficulty lies opportunity." However research in the field should keep another important quote from Einstein in mind: "Make everything as simple as possible but no simpler." Although some experiments can be successfully analyzed with very simple analysis, the complexity of the data generated by microarrays often mandate sophisticated analytic approaches. Because of this, microarray technology has forged strong collaborations between biologists and computational analysts and has helped foster a new field of computational biology.

As mentioned, there are two basic approaches to expression analysis, supervised and unsupervised analysis. Each has its strengths and weaknesses and these will be discussed along with specific examples.

2.7.1. Unsupervised Analysis

Unsupervised analysis imports no knowledge about the samples being analyzed other than the expression data. This type of analysis is least biased and most likely to define structure in a dataset that is novel and not based on *a priori* knowledge or assumptions about the question being addressed. Examples of unsupervised analysis include hierarchical clustering *(36)* and self-organized maps *(37)*. Although the least biased, unsupervised methods of analysis are the most susceptible to technical variation. Global differences between in vitro transcription batches or differences in the cellular components comprising clinical samples can dominate the structure detected by unsupervised means. Thus, more subtle phenotypes are often lost during unsupervised analysis because of competing expression structure.

Initially, most unsupervised analytic techniques lacked a measure against which to determine how different the actual results are compared to that expected by chance alone. One approach to determining the significance of discovered clusters is to annotate each gene on the microarray with characteristics (e.g., gene ontology [GO] information) and then determine the likelihood of having a number of genes sharing the same GO terms within any specific subcluster *(38)*.

2.7.2. Supervised Analysis

Supervised analysis imports information associated with each sample that is used to analyze microarray data. This analysis is less sensitive to technical artifact provided the samples were processed together or in a randomized manner. However, these methods are also sensitive to overfitting because of the large number of variables (genes). Basically, the sample information is used to identify genes with correlating expression patterns. Many measures of correlation have been applied, including fold change, *t*-tests, Z-tests, Pearson's correlation coefficient, and a variant of a signal-to-noise (S2N) metric *(22)*, just to name a few. While most of these measures of correlation to test for statistical significance, alternative means of assessing significance are generally required because of multiple hypothesis testing and the lack of independence of gene expression between some genes on the microarray.

Permutation testing is one means by which investigators have started to assess the significance of associations between gene expression and sample characteristics. In this method, after the degree of correlation between gene expression and the sample characteristic of interest is established, sample identification is randomized. Such randomization breaks any true structure between gene expression and the characteristic of choice. The same methods are then applied to the data after randomized identifications and the degree of correlation between gene expression and the same characteristic is again calculated. This process is repeated *n* times. The results are then combined and the experimental data are compared to the aggregate results from *n* random permutations. If the experimental correlation is stronger than that found with the randomized data at a frequency of 0.05 (or another predetermined level of significance), then statistical significance is inferred.

This is one example of a commonly used methodology; alternatives exist and there will be continued evolution in the statistical methods used to assess the significance between gene expression and sample characteristics to make discovery. Whereas the specifics are less important, it is critical to understand that sophisticated methods are often required in order to properly account for the multiple-hypothesis testing problem. However, the best test of a finding during supervised (or unsupervised) analysis is to determine if the same structure discovered in an initial dataset is present in an independent dataset, this is called validation (*see* **Subheading 2.8.**).

Both unsupervised and supervised methods continue to be developed. There are very interesting methods being applied to microarray data in order to identify gene copy number changes ***(39)***, biological pathway activity ***(10,39)***, drug sensitivity ***(40)***, and many other applications. Many recent computational approaches use the expression of sets of genes rather than single genes in order to limit the effects of biological and technical variability and such approaches seem to hold great promise ***(41)***. As computational approaches to microarray data continue to evolve, it will be very interesting to see if this field continues to diversify into a practically limitless number of options or into a relatively set number of accepted standards.

2.8. Validation

Once a complex genomic pattern is associated with disease behavior, validating the model becomes of prime importance ***(42)***. If the goal of an experiment is to identify specific marker genes associated with an experimental or clinical feature, validation confirms differential gene expression independent of the microarray analysis. Although there is no "gold standard," quantitative RT-PCR has been widely adopted as the most appropriate method to validate differential expression of a single gene. Alternatively, some investigators look at protein expression to validate genes identified by microarrays. Although not directly confirming the microarray results, finding changes in protein expression for a marker gene lends additional support to importance of the identified gene.

If expression analysis develops a model from gene expression predicting phenotype or clinical outcome, validation demands that a model, developed in a training set, be applied without changes to an independent set of tumors (none of which were used during the training of the model) and accurately reproduce the preliminary findings ***(43)***. Although this represents the ideal, investigators often resort to alternative means of validation because of limited access to clinical specimens and a desire to maximize the information obtained from those that are available ***(44)***.

Cross-validation, leave-one-out, and bootstrapping are different mechanisms to perform resubstitution estimates of a model's error rate and represent internal validation ***(45–47)***. Routinely used, these methods maximize disease sampling by allowing investigators to use all available samples to train and test a model ***(47–51)***. While decreasing the risk of overfitting, resubstitution approaches tend to underestimate the true error rate of a model ***(44)*** and further validation on independent samples is required prior to clinical application.

Applying internally validated models on independent samples obtained from different experiments, laboratories, or locations is a more stringent validation test. The application of genomic models across laboratories or clinics help determine if developed models remain accurate despite differences in tissue collection, tissue processing, and data acquisition. In addition, with many datasets becoming publicly available, investigators can often use these datasets to determine if genes associated with a specific phenotype are similarly correlated in an independent dataset.

2.9. Complementary Analysis

Expression analysis is one window through which to view the biology of cells. There is intense investigation ongoing to combine the complex data from microarrays from that generated using other platforms. Data from methods to broadly assay the DNA alterations in an individual (genomic) or in disease (somatic), including single-nucleotide polymorphism *(52)* and comparative genomic hybridization microarrays *(8)*, can complement RNA expression data and result in novel discoveries. With the evolution and maturation of proteomics, certainly combining serum- or tissue-based patterns of protein expression with RNA expression holds promise. Finally, other rich sources of complex data such as the literature can be used to complement our analysis of microarray data *(39)*. These analyses face significant challenges with respect to gene annotation and statistical interpretation, but such combined approaches are likely to typify the next wave of array-based discoveries.

In conclusion, from start to finish, expression analysis works to identify marker genes or gene patterns correlating with cellular biology or disease characteristics. Like other methods, microarray analysis requires scientific experimental design and rigorous execution. The complexity of data generated often mandates sophisticated computational analysis previously outside the realm of the biological sciences. To handle the complexity of analysis and facilitate reproducibility, standards of laboratory practice need to be established and some have already been proposed, if not widely accepted *(53)*. However, as with past technological advances, the community of scientific investigators has risen to meet these challenges. As the challenges of microarray analysis discussed above have been met, this powerful investigational tool has been focused on addressing basic and clinical questions examples of which are discussed below.

3. Applications of cDNA Microarrays (Clinical/Medical)

Microarrays have contributed significantly to the genomics revolution over the past decade and will continue to play an important role in our understanding of cellular biology. In some settings, cDNA microarrays will be remain a tool for basic discovery; in others, microarrays will be directly applied to patient care. At this point, it is easy to forecast a lasting role for microarrays but impossible to accurately anticipate specific applications. Although a complete review of the important literature demonstrating the use of cDNA microarrays in biomedical and clinical research is not possible given space limitations, here we highlight recent examples of microarrays applications within biomedical and clinical research and discuss future directions.

3.1. Biological Discovery

The ability to simultaneously measure tens of thousands of genes in cellular systems offers an unbiased approach to improving our understanding of the basic biology underlying important cellular functions or phenotypes. Expression analysis has improved our understanding of the molecular mechanisms that are necessary and/or sufficient for specific, well-defined phenotypes. Early studies applied microarray analysis to describe the global transcriptional activity during replication in yeast *(54)*, synchronized proliferation of fibroblasts *(55)*, and controlled hematopoietic differentiation *(56)* and identified sets of genes whose coordinate expression were associated with basic cellular processes such as mitosis and differentiation. These and many additional studies have demonstrated that systems with robust biology, profound domain knowledge, and strong in vitro models can be interrogated and analyzed using cDNA microarrays in order to make new discoveries.

Although rigorous scientific methods are required to generate informative expression patterns, microarray data are often viewed as hypothesis generating and significant work goes into validating observations and definitively testing novel hypotheses with more standard laboratory methodologies ***(42)***. Functional validation of genes identified via microarray experiments has become commonplace in laboratories and there are thousands of manuscripts reporting the use of microarrays to identify candidate genes in a biological system.

Because of the rigor required to validate even simple associations, many unexpected yet intriguing findings lack true validation. Often, the optimistic experimental goals of mapping all of the important pathways associated with a complex cellular trait are reduced to the identification and further exploration of one or a few genes not previously known to be involved with the studied phenotype. Initial aspirations of using microarrays to comprehensively map the molecular pathways underlying complex cellular phenotypes was not immediately realized and continues to be a challenge ***(57)***.

As cDNA microarray technology and associated computational analysis continue to improve, investigators have focused on more complex systems including multicellular organisms. Developmental transcriptional maps of *Caenorhabditis elegans* ***(58,59)***, *Drosophila melanogaster* ***(60–62)***, and *Danio rerio* (zebrafish) ***(63)*** have been derived using microarrays in order to understand the temporal relationships of gene expression in organisms. Data from organism-based studies have subsequently been combined to identify interspecies transcriptional similarities. For example, Stuart et al. used data from 3182 microarrays to identify genes with coordinate RNA expression conserved across species and successfully inferred gene function based on shared expression ***(64)***. In a similar approach, McCarroll et al. identified sets of genes with expression commonly associated with aging in both worms and flies ***(65)***.

Gene expression patterns have also been interrogated to implicate drug mechanism. In a seminal paper, Hughes et al. used microarrays to interrogate the expression patterns from 300 yeast experiments involving either gene mutation or chemical treatment ***(66)***. In this work, the function of previously uncharacterized genes could be inferred from the expression similarity between yeasts with these genes mutated and other yeast cultures harboring mutations in functionally characterized genes and or exposed to chemicals affecting specific pathways ***(66)***. The use of global gene expression as a phenotype with which to infer function for genes or chemicals is an approach that holds great promise ***(67)***. In a recent example underscoring the potential of this approach, Stegmaier et al. used microarrays to identify a signature for therapy-induced differentiation in human leukemia cells and then used the signature to identify novel differentiation-causing agents from a screen of 1739 compounds ***(68)***.

3.2. Medical/Clinical Applications

Microarrays are now a frequent element of laboratory-based biomedical investigation. However, along with biomedical discovery, there is great hope that microarrays will improve medical care. Microarrays have been applied to clinical medicine in order to better understand the underlying biology and physiology, to identify marker genes for specific disease behavior, and to improve disease prognosis and treatment response prediction. Here, we will review some of the early applications of microarrays to clinical disease. It is worth noting that the application of microarrays to cancer currently dominates the literature because of the ability to obtain tumor tissue for analysis. However, other fields of medicine are currently applying microarray analysis to address fundamental biological, pathological, and clinical questions and a more diverse literature will be forthcoming.

3.2.1. Diagnosis

cDNA arrays have been applied to primary human samples of complex phenotypes in order to identify candidate marker genes for disease, to discover molecular classes of diseases, and to molecularly describe clinical behavior. Early on, it became clear that tissue- and cancer-specific patterns of expression existed. In an early seminal article, Golub et al. found strong expression differences between the two major forms of leukemia, acute myeloblastic leukemia (AML) and acute lymphoblastic leukemia (ALL), using microarrays ***(22)***. In that article, the investigators could accurately predict the identity of an unknown sample (either AML and ALL) based on gene expression alone. This work has been extended in ALL and distinct expression patterns associated with the common chromosomal abnormalities underlying leukemia have been identified ***(69)***. Thus, global expression patterns were found to underly both histological and genetic distinctions in leukemia.

Large differences in global gene expression also exist between different types of solid tumor and lymphoma. For solid tumors, gene expression can re-establish histological differences between solid tumors from different organs with an accuracy of approx 80% ***(70,71)***. The genes found to best discriminate between tumor types were often tissue-specific rather than tumor-specific and poorly differentiated tumors continue to be difficult to classify ***(71)***.

Gene expression can easily classify follicular lymphomas from diffuse large B-cell lymphomas (DLBCL) ***(72)*** and can further separate diffuse B-cell lymphomas into two classes likely based on their origin within lymph nodes: germinal center B-like and activated B-like DLBCL ***(73)***.

Investigators also applied microarrays to detect gene expression differences between normal tissues and cancers derived from those tissues in order to identify novel markers for cancer detection or diagnosis. Profound differences in gene expression have been found between normal and tumor tissue for breast, prostate, ovarian, lung, brain, gastric, esophageal, and most other forms of cancer ***(47,74–75)***. In a specific example, two groups simultaneously used microarrays to identified a gene over expressed in prostate tumors compared to normal tumors called α-methylacyl coenzyme A racemase (*AMACR*) and validated their results using immunohistochemistry ***(79,80)***. Some clinical pathologists now use *AMACR* to help diagnose cancer in prostate biopsies, representing a successful evolution from microarray discovery to clinical application.

Although other examples exist for microarray-derived diagnostic tissue-based tests, there have been no novel serological tests yet introduced into the clinic based on a gene expression experiment. For any gene, there are many biochemical and cellular processes between mRNA expression in a cell and protein circulating in the blood. This has likely contributed to the difficulty in translating microarray data to novel serological tests but work is ongoing and it is likely that some candidates will be successfully translated to the clinic as blood-based diagnostic tests.

Investigators have also used microarrays to detect gene expression changes associated with the genetics of cancer. As mentioned above, gene expression changes have been found that correlate with the major chromosomal abnormalities in ALL ***(69)***. Similarly, breast cancer tumors harboring mutations in the *BRCA1* and *BRCA2* cancer predisposition genes also have distinct gene expression signatures measured by microarrays ***(81)***.

In addition to recapitulating known histological or genetic traits of disease, microarrays have identified previously unrecognized molecular subclasses for some solid tumors. Gene expression not only separates the different major types of lung cancer (e.g., small cell vs adenocarcinoma) but can also subclassify the most common form of lung cancer, adenocarcinoma of the lung ***(82,83)***. Specifically, gene expression analysis identified a subgroup within adenocarcinomas that expressed neuroendocrine marker genes and had worse survival compared to the other groups ***(83)***. In a similar example, gene expression identified a new subgroup of tumors (MLL) from within cases of leukemia with very similar morphology but remarkably different clinical behavior ***(84)***.

There are also very good examples of investigators using microarrays to assess gene expression differences between normal and diseased tissues that do not focus on cancer. Investigators have used microarrays to identify gene expression changes associated with heart failure ***(85)***, diabetes ***(10)***, inflammatory diseases mellitus ***(86)***, diabetic nephropathy ***(87)***, multiple sclerosis ***(88,89)***, and many others. Most of these studies apply microarrays in order to better understand the biology underlying disease and to identify potential diagnostic markers of disease. Often, a major challenge is determining the most appropriate "normal" tissue with which to compare the diseased tissue. For example, whereas the robust immune response causes damage to normal tissues in some inflammatory diseases, the gene expression changes caused by the inflammation is likely to overwhelm expression changes because of the root cause of the disease. Similarly, the cellular archictecture of diseased tissues is often different than the corresponding normal tissue (i.e., atherosclerotic plaques vs normal arterial intima) and expression changes might be more related to the tissue composition than disease pathogenesis. In analyzing this literature, it is very important to keep these points in mind.

Mootha et al. combined excellent design with novel computation in order to investigate the metabolic implications of diabetes mellitus ***(10)***. These investigators processed skeletal muscle biopsies from normal and diabetic patients for microarray analysis. On the first analysis, there were little differences between the sets of muscle biopsies with respect to histology and supervised analysis of gene expression. Specifically, there were no individual genes with statistically significant expression differences between normal or diabetic muscle. However, when a novel computational method was applied called gene set enrichment analysis (GSEA), there was a significant decrease in the expression of genes involved in oxidative phosphorylation which are also high at sites of glucose disposal.

Be it cancer or other human disease, microarray analysis can redefine diseases from clinical and pathological collections of findings to molecular entities. Thus, differentiating disease diagnosis based on organ site might start to hold less weight than common underlying biology. The potential for this approach might be realized in anticipating disease outcome and choosing therapy. Diagnosing cancers based on specific pathway activation rather than tissue of origin might allow more effective prognosis and treatment than the diagnostic classifications used currently. While provocative, such musing is preliminary and what has been demonstrated with respect to disease prognosis and treatment choice using microarrays will be discussed below.

3.2.2. Prognosis

Diagnosis and prognosis often are related in medicine. However, whereas diagnosis focuses on the current disease state, prognosis focuses on the future behavior of disease in the context of the individual. Investigators have applied microarray analysis to assess if gene expression patterns correlate with disease outcome and have met with some success.

Oncologists remain optimistic that by understanding the biology and genetics of an individual's tumor, they will be able to accurately predict outcome. Microarrays have identified gene expression patterns that are associated with disease progression and/or patient survival in breast cancer ***(90–93)***, prostate cancer ***(47,94)***, lymphoma ***(95,96)***, lung cancer ***(83)***, and some brain tumors ***(50)***, among others. Outcome can be defined as recurrence following definitive surgery, development of metastasis, or death from disease. Regardless, the preliminary success of the studies mentioned above suggests that gene expression changes within localized tumors can anticipate recurrence, metastasis, and possibly death from disease.

Interestingly, although most studies focused on specific types of cancer, expression differences between local and metastatic tumors (of multiple cancer types) were also used to predict outcome ***(97)***. In this report, Ramaswamy et al. applied a gene expression signature comprised of genes differentially expressed between local and metastatic tumors to successfully predict outcome in breast, prostate, and a type of brain tumor but not in lymphoma, again supporting the idea that some local tumors are preprogrammed for recurrence or progression following local therapy and that microarray analysis can measure this programming and anticipate outcome.

Breast cancer is the disease furthest along with respect to the clinical application of microarrays. As touched upon above, two independent groups have found that microarray analysis can predict the development of metastasis and survival in women initially diagnosed with localized disease ***(90,92)*** and one group has validated their results in a larger cohort of women with disease ***(93)***. Today, clinical trials are ongoing that test how these predictive models derived from microarrays compare to standard risk stratification using clinical and pathological features of breast cancer. Even without clinical trials demonstrating the usefulness of these genomic tests, at least two companies are now offering microarray-based testing for women diagnosed with localized breast cancer (http://www.genomichealth.com). These tests are expensive and need to be compared to standard prognostic methods before they should be used broadly, but they have the potential to significantly improve our ability to identify patients with high- or low-risk disease so as to optimize management.

Noncancer diseases have been studied with respect to the ability of microarrays to anticipate the disease course, although this area of investigation remains less well developed. There is some evidence that expression analysis can model the progression of heart failure ***(98,99)***, Parkinson's disease ***(100)***, kidney disease ***(101)***, and lung injury ***(102)***, among others, but these studies largely involve laboratory models of disease and little has been directly tested on human disease. That being said, strong expression patterns associated with a variety of diseases are being identified and subsequently tested on specimens collected from patients in clinical trials.

3.2.3. Treatment Choice

For a patient diagnosed with a disease, it is helpful for physicians to be able to anticipate the severity and natural history of the disease. However, for patients found to have aggressive disease by conventional or molecular means, it becomes critical for investigators to optimize therapy. Microarrays, by assaying the underlying biology of disease processes, have the potential to anticipate response to specific therapies.

The best examples to date are also in oncology. As cancer chemotherapy is toxic and has profound side effects, only those patients likely to respond would ideally be treated. Breast cancer expression patterns were found to correlate with sensitivity to two different types of chemotherapy: a taxane ***(103)*** and anthracyclines ***(104)***. The subclass of leukemia detected by microarrays (MLL) was subsequently found to be sensitive to agents inhibiting one of the genes (FLT3) detected as over expressed by microarrays ***(105)***.

In the future, it is possible that treatment choice will be determined by the molecular biology of the disease as measured by microarrays and other genomic techniques rather than clinical or histopathological features. It is premature to feel confident that microarray-based tests will be used to determine individualized treatment but the work performed so far supports this possibility.

4. Future Directions

As cDNA microarray technology and analysis continue to improve, it is important to understand the critical steps that currently limit our ability to map complex cellular phenotypes using expression analysis and apply microarrays clinically as biomedical assays. Initially, the number of genes analyzed was thought to be critical, with more genes denoting better microarrays. However, current cDNA technologies are very close to comprehensively covering the approx 25,000–30,000 genes in the human genome and current technologies likely represent a sufficient sampling of the human transcriptome. Also, the ability to measure thousands of genes simultaneously initially outpaced our analytic tools. However, after recruiting scientists from fields outside of biology, the development of tools for expression analysis is now in step with microarray technology.

It is more likely that our inability to use microarrays for disease diagnosis, prognosis, and treatment has more to do with sampling size, gene annotation, and the onerous work of functional validation than with any single technical limitation. Microarray experiments remain expensive and, as such, experimental designs have to be limited. Access to less expensive technologies and the ability to combine data from different microarray platforms will add significantly to our ability to identify important mechanisms active in complex phenotypes.

However, even with great annotation and larger experiments, RNA expression alone might not sufficiently assay a disease state to allow accurate disease prediction. For a comprehensive picture of disease, it is likely that multiple genomic technologies will have to be brought together. It is clear that the developing proteomic technologies together with the DNA-based arrays measuring allelic loss or gain will provide complementary information to expression analysis. Data from disease tissues (i.e., tumors) as well has the host will be combined in the future in order to obtain a full description of disease state.

Microarrays are now a fundamental part of basic biomedical laboratory practice. Early evidence suggests they have a role in clinical medicine, and work over the next decade will test the full potential of this novel technique to influence how we treat patients.

References

1. Schena, M., et al. (1995) Quantitative monitoring of gene expression patterns with a complementary DNA microarray. *Science* **270(5235)**, 467–470.
2. Lipshutz, R. J., et al. (1995) Using oligonucleotide probe arrays to access genetic diversity. *Biotechniques* **19(3)**, 442–447.
3. Chu, S., et al. (1998) The transcriptional program of sporulation in budding yeast. *Science* **282(5389)**, 699–705.
4. Wodicka, L., et al. (1997) Genome-wide expression monitoring in *Saccharomyces cerevisiae. Nature Biotechnol.* **15(13)**, 1359–1367.
5. Lockhart, D. J., et al. (1996) Expression monitoring by hybridization to high-density oligonucleotide arrays. *Nature Biotechnol.* **14(13)**, 1675–1680.
6. Baltimore, D. (2001) Our genome unveiled. *Nature* **409**, 814–816.
7. Hanahan, D. and Weinberg, R. A. (2000)The hallmarks of cancer. *Cell* **100(1)**, 57–70.
8. Pollack, J. R., et al. (2002) Microarray analysis reveals a major direct role of DNA copy number alteration in the transcriptional program of human breast tumors. *Proc. Natl. Acad. Sci .USA* **99(20)**, 12,963–12,968.
9. Varambally, S., et al. (2002) The polycomb group protein EZH2 is involved in progression of prostate cancer. *Nature* **419(6907)**, 624–629.
10. Mootha, V. K., et al. (2003) PGC-1alpha-responsive genes involved in oxidative phosphorylation are coordinately downregulated in human diabetes. *Nature Genet.* **34(3)**, 267–273.
11. Crick, F. (1970) Central dogma of molecular biology. *Nature* **227(258)**, 561–563.
12. Velculescu, V. E., et al. (1997) Characterization of the yeast transcriptome. *Cell* **88(2)**, 243–251.
13. Velculescu, V. E., et al. (1995) Serial analysis of gene expression. *Science* **270(5235)**, 484–487.
14. Liang, P. and Pardee, A. B. (1992) Differential display of eukaryotic messenger RNA by means of the polymerase chain reaction. *Science* **257(5072)**, 967–971.
15. Kerr, M. K. and Churchill, G. A. (2001) Statistical design and the analysis of gene expression microarray data. *Genet. Res.* **77(2)**, 123–128.
16. Churchill, G. A. (2002) Fundamentals of experimental design for cDNA microarrays. *Nature Genet.* **32 (Suppl.)**, 490–495.
17. Simon, R., Radmacher, M. D. and Dobbin, K. (2002) Design of studies using DNA microarrays. *Genet. Epidemiol.* **23(1)**, 21–36.
18. Mukherjee, S., et al. (2003) Estimating dataset size requirements for classifying DNA microarray data. *J. Comput. Biol.* **10(2)**, 119–142.
19. Van Der Laan, M.J. and J. Bryan, Gene expression analysis with the parametric bootstrap. Biostatistics, 2001. **2**(4): p. 445-61.
20. Yang, I.V., et al., Within the fold: assessing differential expression measures and reproducibility in microarray assays. Genome Biology, 2002. **3**: p. 62.1–62.12.
21. Ma, X. J., et al. (2003) Gene expression profiles of human breast cancer progression. *Proc. Natl. Acad. Sci. USA* **100(10)**, 5974–5979.
22. Golub, T. R., et al. (1999) Molecular classification of cancer: class discovery and class prediction by gene expression monitoring. *Science* **286(5439)**, 531–537.
23. Perou, C. M., et al. (2000) Molecular portraits of human breast tumours. *Nature* **406(6797)**, 747–752.
24. Baugh, L. R., et al. (2001) Quantitative analysis of mRNA amplification by in vitro transcription. *Nucleic Acids Res.* **29(5)**, E29.
25. Barrett, J. C. and Kawasaki, E. S. (2003) Microarrays: the use of oligonucleotides and cDNA for the analysis of gene expression. *Drug Discov. Today* **8(3)**, 134–141.
26. Yang, Y. H., et al. (2002) Normalization for cDNA microarray data: a robust composite method addressing single and multiple slide systemic variation. *Nucelic Acids Res.* **30(4)**, e15.

27. Li, C. and Wong, W. H. (2001) Model-based analysis of oligonucleotide arrays: expression index computation and outlier detection. *Proc. Natl. Acad. Sci. USA* **98(1)**, 31–36.
28. Quackenbush, J. (2002) Microarray data normalization and transformation. *Nature Genet.* **32 (Suppl.)**, 496–501.
29. Workman, C., et al. (2002) A new non-linear normalization method for reducing variability in DNA microarray experiments. *Genome Biol.* **3(9)**, research0048.
30. Tseng, G. C., et al. (2001) Issues in cDNA microarray analysis: quality filtering, channel normalization, models of variations and assessment of gene effects. *Nucleic Acids Res.* **29(12)**, 2549–2557.
31. Hoffmann, R., Seidl, T., and Dugas, M. (2002) Profound effect of normalization on detection of differentially expressed genes in oligonucleotide microarray data analysis. *Genome Biol.* **3(7)**, p. research0033.
32. Hautaniemi, S., et al. (2003) A novel strategy for microarray quality control using Bayesian networks. *Bioinformatics* **19(16)**: 2031–2038.
33. Smyth, G. K., Yang, Y. H. and Speed, T. (2003) Statistical issues in cDNA microarray data analysis. *Methods Mol. Biol.* **224**, 111–136.
34. Jenssen, T. K., et al. (2002) Analysis of repeatability in spotted cDNA microarrays. *Nucleic Acids Res.* **30(14)**, 3235–3244.
35. Gollub, J., et al. (2003) The Stanford Microarray Database: data access and quality assessment tools. *Nucleic Acids Res.* **31(1)**, 94–96.
36. Eisen, M. B., et al. (1998) Cluster analysis and display of genome-wide expression patterns. *Proc. Natl. Acad. Sci. USA* **95(25)**, 14,863–14,868.
37. Tamayo, P., et al. (1999) Interpreting patterns of gene expression with self-organizing maps: methods and application to hematopoietic differentiation. *Proc. Natl. Acad. Sci. USA* **96(6)**, 2907–2912.
38. Zhong, S., Li, C., and Wong, W. H. (2003) ChipInfo: Software for extracting gene annotation and gene ontology information for microarray analysis. *Nucleic Acids Res.* **31(13)**, 3483–3486.
39. Jenssen, T. K., et al. (2001) A literature network of human genes for high-throughput analysis of gene expression. *Nature Genet.* **28(1)**, 21–28.
40. Butte, A. J., et al. (2000) Discovering functional relationships between RNA expression and chemotherapeutic susceptibility using relevance networks. *Proc. Natl. Acad. Sci. USA* **97(22)**, 12,182–12,186.
41. Nevins, J. R., et al. (2003) Towards integrated clinico-genomic models for personalized medicine: combining gene expression signatures and clinical factors in breast cancer outcomes prediction. *Hum. Mol. Genet.* **12**, (Epub 8/19/03).
42. Chuaqui, R. F., et al. (2002) Post-analysis follow-up and validation of microarray experiments. *Nature Genet.* **32 (Suppl.)**, 509–514.
43. Slonim, D. K. (2002) From patterns to pathways: gene expression data analysis comes of age. *Nature Genet.* **32 (Suppl.)**, 502–508.
44. Dougherty, E. (2001) Small sample issues for microarray-based classification. *Comp. Function. Genom.* **2**, 28–34.
45. Kerr, M. K. and Churchill, G. A. (2001) Bootstrapping cluster analysis: assessing the reliability of conclusions from microarray experiments. *Proc. Natl. Acad. Sci. USA* **98(16)**, 8961–8965.
46. Azuaje, F. (2003) Genomic data sampling and its effect on classification performance assessment. *BMC Bioinformatics* **4(1)**, 5.
47. Singh, D., et al. (2002) Gene expression correlates of clinical prostate cancer behavior. *Cancer Cell* **1(2)**, 203–209.
48. Dyrskjot, L., et al. (2003) Identifying distinct classes of bladder carcinoma using microarrays. *Nature Genet.* **33(1)**, 90–96.
49. Nutt, C. L., et al. (2003) Gene expression-based classification of malignant gliomas correlates better with survival than histological classification. *Cancer Res.* **63(7)**, 1602–1607.
50. Pomeroy, S. L., et al. (2002) Prediction of central nervous system embryonal tumour outcome based on gene expression. *Nature* **415(6870)**: 436–442.
51. Shipp, M. A., et al. (2002) Diffuse large B-cell lymphoma outcome prediction by gene-expression profiling and supervised machine learning. *Nature Med.* **8(1)**, 68–74.
52. Stickney, H. L., et al. (2002) Rapid mapping of zebrafish mutations with SNPs and oligonucleotide microarrays. *Genome Res.* **12(12)**, 1929–1934.
53. Brazma, A., et al. (2001) Minimum information about a microarray experiment (MIAME)-toward standards for microarray data. *Nature Genet.* **29(4)**, 365–371.

54. Spellman, P. T., et al. (1998) Comprehensive identification of cell cycle-regulated genes of the yeast *Saccharomyces cerevisiae* by microarray hybridization. *Mol. Biol. Cell* **9(12)**, 3273–3297.
55. Iyer, V. R., et al. (1999) The transcriptional program in the response of human fibroblasts to serum. *Science* **283(5398)**, 83–87.
56. Le Naour, F., et al. (2001) Profiling changes in gene expression during differentiation and maturation of monocyte-derived dendritic cells using both oligonucleotide microarrays and proteomics. *J. Biol. Chem.* **276(21)**, 17920–17931.
57. Quackenbush, J. (2003) Genomics. Microarrays—guilt by association. *Science* **302(5643)**, 240–241.
58. Kim, S. K., et al. (2001) A gene expression map for *Caenorhabditis elegans*. *Science* **293(5537)**, 2087–2092.
59. Walhout, A. J., et al. (2002) Integrating interactome, phenome, and transcriptome mapping data for the *C. elegans* germline. *Curr. Biol.* **12(22)**, 1952–1958.
60. White, K. P., et al. (1999) Microarray analysis of *Drosophila* development during metamorphosis. *Science* **286(5447)**, 2179–2184.
61. Jin, W., et al. (2001) The contributions of sex, genotype and age to transcriptional variance in *Drosophila melanogaster*. *Nature Genet.* **29(4)**, 389–395.
62. Arbeitman, M. N., et al. (2002) Gene expression during the life cycle of *Drosophila melanogaster*. *Science* **297(5590)**, 2270–2275.
63. Ton, C., et al., Construction of a zebrafish cDNA microarray: gene expression profiling of the zebrafish during development. Biochem Biophys Res Commun, 2002. **296**(5): p. 1134-42.
64. Stuart, J. M., et al. (2003) A gene-coexpression network for global discovery of conserved genetic modules. *Science* **302(5643)**, 249–255.
65. McCarroll, S. A., et al. (2004) Comparing genomic expression patterns across species identifies shared transcriptional profile in aging. *Nature Genet.* **36(2)**, 197–204.
66. Hughes, T. R., et al. (2000) Functional discovery via a compendium of expression profiles. *Cell* **102(1)**, 109–126.
67. Schadt, E. E., Monks, S. A., and Friend, S. H. (2003) A new paradigm for drug discovery: integrating clinical, genetic, genomic and molecular phenotype data to identify drug targets. *Biochem. Soc. Trans.* **31(2)**, 437–443.
68. Stegmaier, K., et al. (2004) Gene expression-based high-throughput screening (GE-HTS) and application to leukemia differentiation. *Nature Genet.* **36(3)**, 257–263.
69. Yeoh, E. J., et al. (2002) Classification, subtype discovery, and prediction of outcome in pediatric acute lymphoblastic leukemia by gene expression profiling. *Cancer Cell* **1(2)**, 133–143.
70. Su, A. I., et al. (2001) Molecular classification of human carcinomas by use of gene expression signatures. *Cancer Res.* **61(20)**, 7388–7393.
71. Ramaswamy, S., et al. (2001) Multiclass cancer diagnosis using tumor gene expression signatures. *Proc. Natl. Acad. Sci .USA* **98(26)**, 15,149–15,154.
72. Chan, W. C. and Huang, J. Z. (2001) Gene expression analysis in aggressive NHL. *Ann. Hematol.* **80 (Suppl. 3)**, B38–B41.
73. Alizadeh, A. A., et al. (2000) Distinct types of diffuse large B-cell lymphoma identified by gene expression profiling. *Nature* **403(6769)**, 503–511.
74. Ono, K., et al. (2000) Identification by cDNA microarray of genes involved in ovarian carcinogenesis. *Cancer Res.* **60(18)**, 5007–5011.
75. Sallinen, S. L., et al. (2000) Identification of differentially expressed genes in human gliomas by DNA microarray and tissue chip techniques. *Cancer Res.* **60(23)**, 6617–6622.
76. Lu, J., et al. (2001) Gene expression profile changes in initiation and progression of squamous cell carcinoma of esophagus. *Int. J. Cancer* **91(3)**, 288–294.
77. Mori, M., et al. (2002) Analysis of the gene-expression profile regarding the progression of human gastric carcinoma. *Surgery* **131(1 Suppl.)**, S39–S47.
78. Jiang, Y., et al. (2002) Discovery of differentially expressed genes in human breast cancer using subtracted cDNA libraries and cDNA microarrays. *Oncogene* **21(14)**, 2270–2282.
79. Rubin, M. A., et al. (2002) Alpha-Methylacyl coenzyme A racemase as a tissue biomarker for prostate cancer. *JAMA* **287(13)**, 1662–1670.
80. Luo, J., et al. (2002) Alpha-methylacyl-CoA racemase: a new molecular marker for prostate cancer. *Cancer Res.* **62(8)**, 2220–2226.
81. Hedenfalk, I., et al. (2001) Gene-expression profiles in hereditary breast cancer. *N. Engl. J. Med.* **344(8)**, 539–548.

82. Garber, M. E., et al. (2001) Diversity of gene expression in adenocarcinoma of the lung. *Proc. Natl. Acad. Sci. USA* **98(24)**, 13,784–13,789.
83. Bhattacharjee, A., et al. (2001) Classification of human lung carcinomas by mRNA expression profiling reveals distinct adenocarcinoma subclasses. *Proc. Natl. Acad. Sci. USA* **98(24)**, 13,790–13,795.
84. Armstrong, S. A., et al. (2002) MLL translocations specify a distinct gene expression profile that distinguishes a unique leukemia. *Nature Genet.* **30(1)**, 41–47.
85. Yussman, M. G., et al. (2002) Mitochondrial death protein Nix is induced in cardiac hypertrophy and triggers apoptotic cardiomyopathy. *Nature Med.* **8(7)**, 725–730.
86. Heller, R. A., et al. (1997) Discovery and analysis of inflammatory disease-related genes using cDNA microarrays. *Proc. Natl. Acad. Sci. USA* **94(6)**, 2150–2155.
87. Baelde, H. J., et al. (2004) Gene expression profiling in glomeruli from human kidneys with diabetic nephropathy. *Am. J. Kidney Dis.* **43(4)**, 636–650.
88. Whitney, L. W., et al. (1999) Analysis of gene expression in mutiple sclerosis lesions using cDNA microarrays. *Ann. Neurol.* **46(3)**, 425–428.
89. Chabas, D., et al. (2001) The influence of the proinflammatory cytokine, osteopontin, on autoimmune demyelinating disease. *Science* **294(5547)**, 1731–1735.
90. van ‘t Veer, L. J., et al. (2002) Gene expression profiling predicts clinical outcome of breast cancer. *Nature* **415(6871)**, 530–536.
91. West, M., et al. (2001) Predicting the clinical status of human breast cancer by using gene expression profiles. *Proc. Natl. Acad. Sci. USA* **98(20)**, 11,462–11,467.
92. Huang, E., et al. (2003) Gene expression predictors of breast cancer outcomes. *Lancet* **361(9369)**, 1590–1596.
93. van de Vijver, M. J., et al. (2002) A gene-expression signature as a predictor of survival in breast cancer. *N. Engl. J. Med.* **347(25)**, 1999–2009.
94. Henshall, S. M., et al. (2003) Survival analysis of genome-wide gene expression profiles of prostate cancers identifies new prognostic targets of disease relapse. *Cancer Res.* **63(14)**, 4196–4203.
95. Li, S., et al. (2001) Comparative genome-scale analysis of gene expression profiles in T cell lymphoma cells during malignant progression using a complementary DNA microarray. *Am. J. Pathol.* **158(4)**, 1231–1237.
96. Wright, G., et al. (2003) A gene expression-based method to diagnose clinically distinct subgroups of diffuse large B cell lymphoma. *Proc. Natl. Acad. Sci. USA* **100(17)**, 9991–9996.
97. Ramaswamy, S., et al. (2003) A molecular signature of metastasis in primary solid tumors. *Nature Genet.* **33(1)**, 49–54.
98. Blaxall, B. C., et al. (2003) Differential myocardial gene expression in the development and rescue of murine heart failure. *Physiol. Genom.* **15(2)**, 105–114.
99. Ueno, S., et al. (2003) DNA microarray analysis of in vivo progression mechanism of heart failure. *Biochem. Biophys. Res. Commun.* **307(4)**, 771–777.
100. Youdim, M. B., et al. (2002) Early and late molecular events in neurodegeneration and neuroprotection in Parkinson's disease MPTP model as assessed by cDNA microarray; the role of iron. *Neurotox. Res.* **4(7–8)**, 679–689.
101. Eikmans, M., et al. (2002) RNA expression profiling as prognostic tool in renal patients: toward nephrogenomics. *Kidney Int.* **62(4)**, 1125–1135.
102. Leikauf, G. D., et al. (2002) Acute lung injury: functional genomics and genetic susceptibility. *Chest* **121(3 Suppl.)**: 70S–75S.
103. Chang, J. C., et al. (2003) Gene expression profiling for the prediction of therapeutic response to docetaxel in patients with breast cancer. *Lancet* **362(9381)**, 362–369.
104. Faneyte, I. F., et al. (2003) Breast cancer response to neoadjuvant chemotherapy: predictive markers and relation with outcome. *Br. J. Cancer* **88(3)**, 406–412.
105. Armstrong, S. A., et al. (2003) Inhibition of FLT3 in MLL. Validation of a therapeutic target identified by gene expression based classification. *Cancer Cell* **3(2)**, 173–183.

21

Nucleic Acid Sequence-Based Amplification

Katherine Loens, D. Ursi, H. Goossens, and M. Ieven

1. Introduction

Nucleic acid sequence-based amplification (NASBA; bioMérieux, Boxtel, The Netherlands) is a commercially available amplification procedure that uses RNA as the target. It makes use of the simultaneous enzymatic activities of avian myeloblastosis virus reverse transcriptase (AMV-RT), RNase H, and T7 RNA polymerase, under isothermal conditions. The constant temperature maintained throughout the amplification reaction allows each step of the reaction to proceed as soon as an amplification intermediate becomes available. Products of NASBA are single stranded and, thus, can be applied to detection formats using probe hybridization without any denaturation step.

A nucleic acid detection assay is generally composed of three components: nucleic acid extraction, amplification, and detection. The quality of the result is highly dependent on the efficiency and reproducibility of each of the components.

This chapter describes the various steps involved, comprising quantitative possibilities, models for RNA and DNA NASBA, and the advantages and disadvantages of NASBA. Finally, an overview of published NASBA applications is presented and a number of NASBA tests are described in more detail.

2. Nucleic Acid Extraction

The NASBA procedure makes use of the Boom nucleic acid extraction procedure for isolation of nucleic acids from clinical specimens *(1)*. The method is based on the lysing and nuclease-inactivating properties of guanidinium thiocyanate, a chaotropic agent, together with the nucleic-acid-binding properties of silica particles. One hundred microliters of a clinical sample is added to a lysis buffer, lysis of the material takes place, and the released nucleic acid binds to silica particles, forming complexes that can be rapidly sedimented by centrifugation. These complexes are then washed twice with guanidine thiocyanate containing washing buffer, twice with 70% ethanol, and once with acetone. The complexes are dried, and nucleic acids are subsequently eluted in water or in an aqueous low-salt buffer in the initial reaction vessel. The isolation can either be performed manually or by an automated isolation system called the "NucliSens Extractor" (bioMérieux) *(2–4)*.

Pretreatment of clinical samples and modifications of the extraction procedure have also been described: protease treatment *(5)*, proteinase K and lysozyme treatment *(6)*, mechanical disruption *(7)*, and the use of a lysis buffer with a lower pH *(5)*.

From: *Medical Biomethods Handbook*
Edited by: J. M. Walker and R. Rapley © Humana Press, Inc., Totowa, NJ

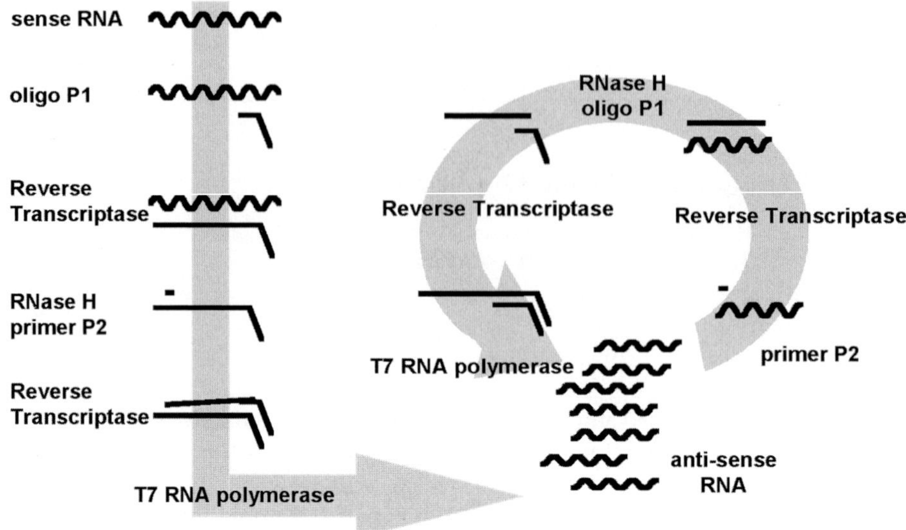

Fig. 1. Schematic representation of NASBA amplification. The straight arrow represents the initiation phase and the circular arrow represents the cyclic phase. The activities of reverse transcriptase, RNase H, T7 RNA polymerase, and the primer-binding activities are indicated. The polarity of the T7 RNA polymerase product, amplified in the cyclic phase, is complementary to that of the target nucleic acid.

3. Nucleic Acid Amplification

3.1. RNA Amplification

Using an RNA template, a NASBA reaction comprises three enzymes, two specific primers, nucleoside triphosphates, and appropriate buffer components. The three enzymes involved are T7 RNA polymerase, avian myeloblastosis virus reverse transcriptase, and RNase H, each acting continuously on its appropriate substrates. The reaction starts with hybridization of an oligonucleotide primer that contains a T7 RNA polymerase-binding site to the target RNA. Reverse transcriptase elongates the primer, creating a cDNA copy of the RNA template and forming a RNA/cDNA hybrid. RNase H recognizes this hybrid as the substrate and hydrolyzes the RNA portion of the hybrid, leaving single-stranded cDNA. A second oligonucleotide primer anneals to the cDNA strand, again forming a substrate suitable for reverse transcriptase extension. This extension finally renders the promotor portion of the nucleic acid sequence double stranded and transcriptionally active. Recognizing the now functional promotor, T7 RNA polymerase produces multiple copies of antisense RNA transcripts of the original target sequence. Each new antisense RNA molecule can, in its turn, again be converted to a T7 promotor containing double-stranded cDNA in a similar way, except that primer annealing and extension occur in reverse order because the newly generated RNA template is opposite in orientation to the original target. Again, many copies are generated from each RNA target that re-enters the reaction resulting in exponential amplification (*see* **Fig. 1**). Amplification of approx 10^6 to 10^9-fold is obtained within 90 min. An excellent overview of primer and probe design rules is given by Deiman et al. *(8)* (*see* **Tables 1** and **2**).

3.2. DNA Amplification

Amplification of DNA by NASBA has also been described ***(9–11)***. Yates et al. ***(11)*** showed that with modifications like primer design, sample extraction method, and template denaturation, the NASBA technique can be made suitable for amplification of a DNA template result-

Table 1
Primer Design Rules

The distance between the binding site of the forward primer (P1, complementary to the target sequence) and that of the reverse primer (P2, identical to the target sequence) should be about 80–200 nucleotides, resulting in amplicons with a length of 120–150 nucleotides.
If the target is obtained from different sources, a sequence alignment should be performed. The primers should be directed against the conserved regions.
The P1 contains a 5' T7 promoter sequence consisting of 25 nucleotides: 5' AAT TCT AAT ACG ACT CAC TAT AGG G 3'.
A purine-rich region of 6–10 nucleotides directly downstream of the 5' promoter sequence and upstream of the 3' hybridizing sequence of P1 could improve amplification. If the 5' part of the hybridizing sequence is purine rich, an additional purine stretch is not required.
Pyrimidine stretches in the first 10 nucleotides downstream of the T7 promoter sequence should be avoided.
If the Basic Kit is used, a defined nonhybridizing sequence consisting of 20 nucleotides: 5' GAT GCA AGG TCG CATATG AG 3' can be added to the 5' end of P2 for generic ECL detection.
The hybridizing parts of both primers should be 20–30 nucleotides.
The G/C content of the hybridizing part should be 40–60%.
The final nucleotide at the 3' end of the hybridizing sequence is preferably an A-residue.
A G/C-rich region at the 5' end of the hybridizing part and an A/T-rich region at the 3' end of the hybridizing part is preferred.
Full-length primers should be used. Primers purified by polyacrylamide gel electrophoresis or HPLC are recommended.
Tracks of four or more of the same nucleotides in the primer sequence should be avoided.
Both primer sequences should be screened for undesired matches with other nucleic acid sequences by using a DNA/RNA sequence databank.
Both primer sequences should be checked for internal secondary structures by using a DNA-folding program.
The secondary structure of the target sequence and amplicon sequence could be predicted with an RNA-folding program. Heavily structured target and/or amplicon sequences should be avoided.

Source: **ref.** ***8***.

ing in RNA amplicons: A primer pair was chosen to amplify a single-stranded region of the hepatitis B virus (HBV) genome. Furthermore, the addition of α-casein to the lysis buffer improved detection by HBV DNA real-time NASBA 10-fold. The NASBA reaction mixtures were incubated at 95°C instead of the standard 65°C for 5 min for denaturation, followed by incubation at 41°C for 5 min before adding the enzyme mixture. The assay has a detection range of 10^3–10^9 HBV DNA copies/mL of plasma or serum, with good reproducibility and precision.

3.3. DNA Amplification Under RNA NASBA Reaction Conditions

DNA can be used as a template in the absence of RNA in some RNA-based NASBA systems. However, the sensitivity decreases when using DNA as the only target when compared with the corresponding RNA ***(12)***, indicating that the RNA will be amplified preferentially. This is illustrated by a NASBA test designed to detect mRNA of the human cytomegalovirus ***(13)***. The specificity for mRNA in the presence of viral DNA was shown in human cytomegalovirus (HCMV)-infected cells. The results show that NASBA is highly specific for RNA, and only in the absence of target RNA or in case of an enormous excess of target DNA over RNA (more than 1000-fold) can DNA be amplified by NASBA. Heim et al. ***(14)*** demonstrated that

Table 2
Probe Design Rules

The hybridizing sequence of the probe is identical to the target sequence and, thus, complementary to the amplicon sequence.
The hybridizing sequence should be 20–25 nucleotides in length.
The G/C content should be 40–60%.
The hybridizing sequence should not overlap with the P1 or P2 binding site.
The melting temperature of the hybridizing sequence should be at least 7–10°C higher than the NASBA annealing temperature (41°C).
Internal structures in the hybridizing sequence and dimerization or interactions with the primers should be avoided.
ECL detection
Biotin is attached to the 5' end of the capture probe to immobilize it on streptavidin-coated paramagnetic beads.
In the Basic Kit, the hybridizing sequence of the generic detection probe is identical to the 5' tail of the P2.
Molecular beacons
Two complementary stem sequences should be added, one at each end of the hybridizing sequence (loop).
A G-residue can act as a quencher. A C-residue is therefore recommended directly adjacent to the fluorophore.
The stem region should be 6–7 basepairs (bp) in length, dependent on the length of the loop: A 6-bp stem is recommended for a loop of 20 nucleotides and a 7-bp stem for a loop of 25 nucleotides.
The G/C content of the stem region should be 70–80%.
The melting temperature of a 6-bp stem should be approx 60–65°C; of a 7-bp stem, it should be approx 65–70%.
A DNA-folding program could be used to predict the structure and melting temperature of the beacon, although the latter should preferably be determined by means of a melting-curve analysis.
Beacons with multiple structures predicted should be avoided.
A free energy (ΔG) of –3 ± 0.5 kcal/mol is recommended.

Source: **ref. *8***.

10 copies of mRNA for the intronless human interferon-β gene could be detected by NASBA, whereas 100 ng of genomic DNA gave negative NASBA results. The results suggest that amplification of DNA in NASBA systems originally designed to amplify RNA is only possible for primers directed against regions in the DNA that are easily accessible (e.g., low-melting-point areas or single-stranded regions) ***(8)***. In fact, NASBA can amplify single-stranded RNA targets selectively in the presence of double-stranded DNA of identical sequence.

4. Amplicon Detection

Amplified products of a NASBA reaction can, in principle, be detected on ethidium bromide–stained gels, but poor resolution of short RNA fragments often hampers interpretation of results. However, because the product of a NASBA reaction is single-stranded RNA, it can be easily detected by hybridization with sequence-specific probes. Different techniques can be applied: enzyme-linked gel assay ***(5,15,16)***, electrochemiluminescence ***(17–22)***, Northern blot ***(23–25)***, enzyme-linked immunosorbent assay (ELISA) ***(26)***, and real-time NASBA ***(11,27,28)***.

4.1. Enzyme-Linked Gel Assay

Because radioactive detection has major drawbacks, a nonradioactive hybridization method for identification of NASBA products, enzyme-linked gel assay (ELGA), was used in the past.

Horseradish peroxidase-conjugated oligonucleotide probes (HRP probes) allow for liquid hybridization. Subsequent electrophoresis of the hybridization reaction discriminates between a free HRP probe and a HRP probe that has specifically hybridized to the NASBA product because the latter will migrate slower into the gel than the unbound probe. By soaking the gels in the appropriate substrate solution, the hybridization reaction products can be visualized.

4.2. Electrochemiluminescence

This is accomplished in an electrochemiluminescence (ECL) reader, which employs a magnetic bead capture format. An oligonucleotide capture probe (specific for the amplification product) is synthesized with a biotin molecule at the 5' terminus. This capture probe is then bound to streptavidin-coated magnetic beads. Amplification reaction products are captured onto the surface of magnetic beads by a standard hybridization reaction. Detection of the captured product is achieved by the hybridization of a second detection probe to a region on the amplification product that is distinct from the capture probe site. The detector probe is labeled at the 5' terminus with ruthenium, the compound responsible for the subsequently produced ECL signal. Once this sandwich hybridization is completed, the material is loaded into the reader and the amount of induced ECL signal is quantitated by a photomultiplier tube (*see* **Fig. 2**) *(21)*.

4.3. Real-Time Detection Using Molecular Beacons

Recently, a novel nucleic acid detection technology, based on probes (molecular beacons) that fluoresce only upon hybridization with their target, has been introduced. Molecular beacons are single-stranded oligonucleotides with a stem–loop structure. The loop portions contain the sequence complementary to the target nucleic acid, whereas the stem is unrelated to the target and has a double-stranded structure. One arm of the stem is labeled with a fluorescent dye, and the other arm is labeled with a nonfluorescent quencher. In this stem–loop state, the probe does not produce fluorescence because the energy is transferred to the quencher and released as heat. When the molecular beacon hybridizes to its target, it undergoes a conformational change that separates the fluorophore and the quencher, and the bound probe fluoresces brightly (*see* **Fig. 3**) *(29)*.

When molecular beacons are used as a detection and quantification system for NASBA to monitor the amplification, it is crucial for the molecular beacon to unfold at a temperature of 41°C when the target sequence is present.

Molecular beacons enable the performance of a one-tube assay, suitable for high-throughput applications. In addition, they allow real-time monitoring of the NASBA amplification reaction because hybridization occurs simultaneously with the production of the amplicons, in contrast to postamplification gel and ECL detection. Combination of the amplification and detection procedures into real-time NASBA considerably reduces the assay time; it lowers hands-on time and limits considerably the potential contamination between samples.

Multiplex reactions are possible by the introduction in the NASBA reaction of different primer sets and molecular beacons with different fluorescent labels *(27,30,31)*.

5. Multiplex NASBA

Duplex NASBAs have been described: a hepatitis A virus and rotavirus NASBA with detection by Northern blotting *(32)*, a combination of potato leafroll virus and potato virus Y in potato tubers by real-time NASBA *(33)*, and a combination of tissue factor and CD14 mRNA with ECL detection *(34)*.

Greijer et al. designed a multiplex real-time NASBA to quantify HCMV IE1 mRNA by competitive coamplification of wild-type and calibrator RNA with simultaneous detection of late pp67mRNA in whole-blood samples of CMV-infected lung transplant patients *(27)*. Three different fluorophores were used to label the molecular beacon probes: FAM, ROX, and Cascade blue. The assay was evaluated using RNA from infected cells and sequential whole-blood samples of CMV-infected lung transplant recipients. Despite the somewhat lower sensitivity of

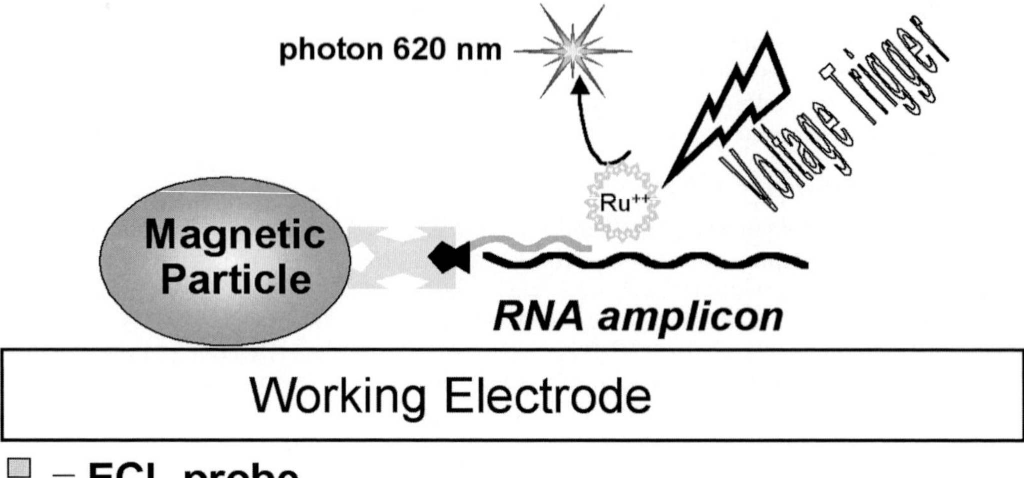

Fig. 2. ECL detection. Amplicons are hybridized to target-specific probes (an electrochemiluminescent probe and a second probe coupled to paramagnetic beads). Following hybridization, the bead/amplicon/ECL probe complexes are captured at the magnet electrode of the automated ECL reader. Subsequently, a voltage pulse triggers the ECL reaction and counts are measured.

the real-time NASBA compared with the conventional NASBA, the simultaneous quantification of IE1 and detection of pp67 RNA was reproducible and accurate.

Loens et al. developed a multiplex real-time NASBA, using FAM, ROX, and Cy5 as fluorophores, for the simultaneous detection of *Mycoplasma pneumoniae*, *Chlamydia pneumoniae*, and *Legionella pneumophila* RNA in respiratory specimens ***(31)***. They found a slightly lower sensitivity of the multiplex real-time assay compared to the conventional ECL detection on spiked respiratory samples, especially in samples with low ECL counts (own unpublished results).

6. Quantitative Assays

Quantification of RNA by NASBA is based on the coamplification of the target with three internal standard RNAs within a single amplification reaction. Three independent constructs to be used for production of internal standards, each with a unique probe detection sequence of 20 nucleotides, are generated. The probe sites are randomized in each construct so that the A, C, T, and G contents are maintained but with a different base order. Therefore, RNA produced from each of these constructs will be amplified by NASBA with kinetics equal to that of the original wild-type (wt) RNA target. Moreover, specific ECL probes complementary to the randomized regions of the calibrators to be used in hybridization analysis to distinguish between amplified wild type wt and calibrator RNA will have the same melting temperature. Before extraction, specific quantities of each calibrator are spiked into the biological sample lysate. Consequently, the calibrators serve as an independent set of internal controls for each individual sample. By including the calibrators in the assay from the time of sample lysis, any loss of wt analyte in the sample is normalized against a corresponding degree of loss of the calibrator RNAs. After coextraction of the wt and calibrator RNAs, the material is coamplified in a single tube. For ECL detection, the reaction product is then divided into four separate detection assays, with each of the four detector probes. The corresponding ECL signal is then determined for each of

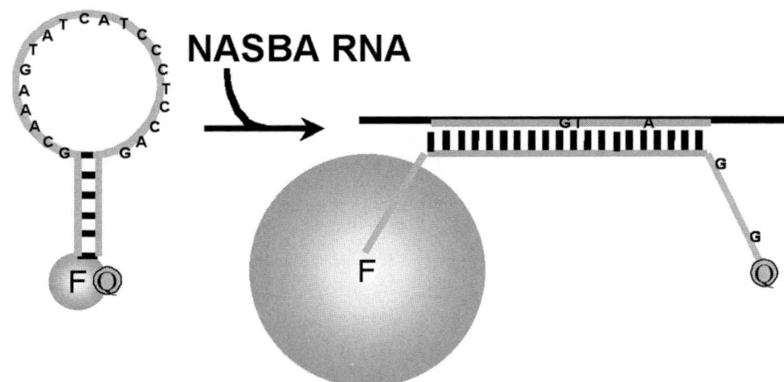

Fig. 3. Real-time detection; molecular beacon technology. In the absence of a complementary sequence, the stem of the molecular beacon hairpin structure is closed and the quencher (Q) prevents the detection of the fluorescent signal emitted by the fluorophore (F). By hybridization of the loop sequence of the beacon with an amplicon, the stem region is forced open and the fluorescent signal becomes detectable.

the four hybridizations. Because the original input quantities for the three calibrators are known, the amount of wt RNA present in the original sample can be calculated by determining the ratio of wt ECL signal to the signal of each calibrator in regression analysis ***(35)***.

When molecular beacons are used as a quantification system for NASBA, only one calibrator with a fixed concentration is included to quantify the target input. Amplification and detection take place in the same test tube and the relation between the kinetics of the fluorescence results obtained from the calibrator beacon and that of the target nucleic acid beacon can be used for quantification.

7. Advantages and Disadvantages of NASBA

One advantage of NASBA compared with polymerase chain reaction (PCR) is that it is a continuous, isothermal process, not requiring a thermocycler. The constant temperature maintained throughout the amplification reaction allows each step of the reaction to proceed as soon as an amplification intermediate becomes available. Thus, the exponential kinetic of the NASBA process, which is caused by multiple transcription of RNA copies from a given DNA product, is intrinsically more efficient than DNA-amplification methods limited to binary increases per cycle ***(36)***.

Products of NASBA are single stranded and, thus, can be applied to detection formats using probe hybridization without any denaturation step.

RNA being the genomic material of many RNA viruses, an RNA-based amplification technique in contrast to PCR avoids an additional reverse transcription (RT) step, thus reducing the risk of contamination and lowering hands-on time.

Nucleic acid sequence-based amplification can be targeted at ribosomal RNA. The detection of microorganisms by a ribosomal RNA-based amplification technique might be more sensitive than PCR because of the presence of multiple ribosomal RNA copies, implying at the same time biological activity. Therefore, it could be used in viability studies.

There are also some disadvantages of NASBA. RNA integrity is the main cause of concern for NASBA, as for RT-PCR and other RNA amplification procedures. Furthermore, because the specificity of the reactions is dependent on thermolabile enzymes, the reaction temperature cannot exceed 42°C without compromising it. However, the specificity is guaranteed by the addition of dimethyl sulfoxide in the amplification reaction and additional hybridization with target specific probes by ELGA, ECL detection, or real-time detection. Finally, the length of the amplified RNA target sequence should be in the range of 120–250 nucleotides, shorter or longer sequences being amplified less efficiently.

8. RNA Stability

The stability of RNA can be affected during collection, processing, and storage of specimens prior to isolation. The addition of RNase inhibitors to the clinical specimens, such as guanidine thiocyanate, is required to preserve RNA integrity.

Most RNA degradation is likely to occur during transportation of the specimens, resulting in false negatives in both NASBA and RT-PCR. Furthermore, although the vast majority of samples are tested within 1 wk of collection, samples to be analyzed for research purposes are often held for longer periods of time and are batch tested later. For such delayed testing, RNA stability over extended periods of time is a critical issue. Two considerations are relevant: First, RNA levels might be slightly reduced with one cycle of freeze–thaw; second, storage over time also generally results in loss of RNA. Although both components can independently result in insignificant changes in detectable RNA copies, the combination of both factors could result in a significant reduction of RNA. This was clearly demonstrated by a NASBA QT with heparinized plasma samples containing human immunodeficiency virus-1 (HIV-1) RNA, where a significant decrease in RNA levels resulted from a combination of losses from both one freeze–thaw cycle and storage over time ***(37)***. Studying the stability of HIV-1 RNA by a NASBA QT in whole blood, plasma, and serum before and after addition of lysis buffer, Bruisten et al. (***38***) concluded that the specimens could be kept at 4°C in lysis buffer, provided transportation time was as short as possible. Under these conditions, the HIV load did not decline up to 14 d. At 30°C, however, the load declined significantly after 2 d. Furthermore, the authors suggested that specimens should be processed on the day of sampling or stored at –70°C in lysis buffer, thus stabilizing the RNA for at least 6 mo. On the other hand, Griffith et al. ***(39)*** subjected serum from HIV-infected patients to multiple freeze–thaw cycles and found no significant loss of detectable HIV RNA.

RNA degradation in lysis buffer produced with guanidine thiocyanate (GuSCN) from two different sources was monitored by Loens et al. ***(5)***. Lysis buffers were prepared using either GuSCN from Sigma (Sigma-Aldrich NV, Bornem, Belgium), or from ICN (ICN Biomedicals NV, Asse, Belgium). *M. pneumoniae* cells PI 1428 were added to both versions of the lysis buffer and the resulting specimens were stored at –80°C or at room temperature for different periods of time prior to extraction. GuSCN from ICN was qualified as unsuitable for RNA extraction. Samuelson et al. ***(40)*** reported problems with an unexplained inhibitory factor using the RNAce Purification system (Bioline, London, UK), a commercially available extraction kit based on the same chemistry as the Boom procedure, when low dilutions of the eluate were processed. These studies ***(38,40)*** suggest that quality control of buffers for storage of clinical material is critical, particularly when RNA is to be analyzed.

To overcome part of this problems and to improve the reproducibility of the in-house-developed NASBAs, standardized reagents "NucliSens Basic Kit" are now commercially available. They contain all necessary reagents for (1) nucleic acid release and inactivation of RNases and DNases, (2) silica-based extraction of nucleic acids, (3) NASBA reagents, and (4) reagents for ECL detection, including the ECL probe, and paramagnetic particles to link the capture probe for detection. Primers and a target-specific biotinylated capture probe are to be synthesized for each target. The P1 primer should contain a 5' terminal T7 RNA polymerase promotor sequence plus a sequence complementary to the target RNA, whereas P2 contains a short sequence identical to the target RNA and a 5' stretch of 20 nucleotides as the ECL tail, unrelated to the target RNA ***(17,19,20,41,42,43)***.

9. NASBA Applications

9.1. NASBA Applications for the Detection of Infectious Agents

The technique has been employed for the detection of many classes of infectious agents in different types of specimens: blood ***(44)***, serum ***(39,45)***, plasma ***(39)***, seminal plasma ***(46)***, semen ***(47)***, respiratory tract specimens ***(5)***, nasopharyngeal biopsies ***(48)***, genital tract specimens ***(25,41)***, urine ***(49)***, skin biopsies ***(25)***, saliva ***(50)***, cerebrospinal fluid ***(51,52)***, stools ***(17)***, breast milk ***(53)***, cervical–vaginal lavage fluid ***(53)***, amniotic fluid ***(54)***, sewage-treatment efflux ***(26)***, potato tubers ***(55,56)***, liquid whole eggs ***(42)***, and poultry ***(57)***.

9.2. Bloodborne Pathogens

9.2.1. Human Immunodeficiency Virus Type 1 Genomic RNA

After an HIV-1 infection, an acute phase with extensive viremia is followed by an asymptomatic carrier state. This latent phase is characterized by low but constant amounts of virus and infected cells in the circulation and can last for several years, ultimately progressing to full-blown autoimmune deficiency syndrome (AIDS). Nucleic acid amplification tests (NAATs) are used to detect the presence of HIV-1 RNA in the blood of blood donors, to quantitate the amount of HIV-1 RNA present in plasma, and to monitor the antiviral therapy. Since the first HIV-1 NASBA was described in 1991 by Kievits et al. ***(23)***, the assay has been improved considerably. Since 1995, a quantitative NASBA test for HIV-1 viral load determination based on ECL detection is commercially available (NASBA HIV-1 RNA QT, bioMérieux). The primers and probes are targeted at the *gag* region of the viral genome ***(21)***. This NASBA assay, as well as other commercially available amplification tests, is unable to detect all HIV-1 subtypes or quantitate them reliably ***(58)***.

The more sensitive second-generation assay, NucliSens HIV-1 QT (bioMérieux) ***(3,53,59,60)***, can detect a range of HIV-1 RNA between 25 and >5×10^6 copies/mL and has a 95% detection rate of 176 copies/mL, using 1 mL of EDTA, citrate, or heparin plasma. Quantification in the NucliSens HIV-1 QT assay is achieved by coamplification of the HIV-1 RNA in the sample, together with three internal calibrators (Qa, Qb, and Qc). Detection is done by ECL ***(59)***. When Murphy et al. ***(3)*** compared the Roche COBAS AMPLICOR MONITOR version 1.5, the NucliSens HIV-1 QT with extractor, and the Bayer Quantiplex Version 3.0 for quantification of HIV-1 RNA in 460 plasma specimens from HIV-1-infected patients, the NucliSens HIV-1 QT showed 100% specificity and the best sensitivity with an input of 2 mL. In another recent study, the same specificity was displayed by the NucliSens HIV-1 QT assay ***(60)***. Comparison of the NucliSens HIV-1 QT and the ultrasensitive AMPLICOR HIV-1 Monitor version 1.0 assays showed both assays to be equivalent in detecting HIV-1 RNA in concentrations of 10^3–10^5 copies/mL. The NucliSens assay was more sensitive at lower RNA concentrations than the AMPLICOR assay.

When Spearman et al. ***(61)*** compared the Roche MONITOR and the NucliSens HIV-1 QT assays for the quantitative detection of HIV-1 RNA in cerebrospinal fluid, both assays reliably quantified the HIV-1 RNA.

Recently, a real-time NASBA was described for the quantification of HIV-1 isolates ***(28)*** based on the amplification of a long terminal repeat region of the HIV-1 genome ***(62)***. All HIV-1 groups, subtypes, and circulating recombinant forms (CRFs) could be detected and quantitated with equal efficiency, including the group N isolate YBF30 and the group O isolate ANT70. The throughput of the assay was high, with 96 samples amplified and detected within 60 min.

9.2.2. Hepatitis C Virus Genomic RNA

Hepatitis C virus (HCV) is the etiologic agent of most posttransfusion non-A, non-B hepatitis cases. In 1994, a first NASBA assay, with ELGA detection, for the amplification of the HCV 5' noncoding region RNA (5'NCR) in serum was described ***(63)***. When Lunel et al. ***(64)***

compared this NASBA with an in-house PCR, an in-house nested PCR, Amplicor (Roche), and, finally, branched DNA (bDNA, Quantiplex 1.0, Chiron diagnostics) for the detection of HCV RNA in serum from HCV-positive and negative individuals; the assays with the highest sensitivity were Amplicor, NASBA, and nested in-house PCR, followed by in-house PCR and then branched DNA. False-positive results were obtained with Amplicor and in-house nested PCR.

The conserved 5'NCR has been used as a target for the further development of qualitative and quantitative ECL-based NASBA assays for genotypes 1a, 1b, 2, 3, 4, and 5 ***(65,66)***. Again, quantification was done in the presence of three calibrators. The HCV NASBA QT assay was over 10 times more sensitive than the bDNA assay, whereas the quantitative results of both assays were highly concordant. The HCV NASBA QT assay was comparable in sensitivity with the HCV MONITOR assay (Roche), but the latter yielded consistently lower values ***(62)***. Similar conclusions were made by Lunel et al. ***(63)***.

9.2.3. Human Cytomegalovirus mRNA

Primary infection of immunocompetent individuals with CMV, a DNA virus, usually does not result in complications. In contrast, immunocompromised patients, like AIDS and transplant patients, can suffer severe complications during primary infection and, to a lesser degree, during secondary infection. Detection of CMV at an early stage of infection is a prerequisite for effective pre-emptive antiviral therapy. The expression of some mRNAs can be used as a specific indicator for viral replication. The first described NASBA, a qualitative assay, targeted the late pp67 (UL65) mRNA ***(67)***; it was applied on mRNA isolated from whole blood from renal transplant patients. NASBA results were compared to results of the pp65 antigenemia assay and virus isolation on cell culture. The sensitivity of NASBA proved to be higher than that of the antigenemia assay, whereas the sensitivities of cell culture and NASBA were comparable. NASBA detected the onset of CMV infection simultaneously with cell culture and the antigenemia assay. The NucliSens CMV pp67 test (bioMérieux, The Netherlands) successfully monitored CMV disease in HIV-infected individuals ***(68)***, solid-organ transplant recipients ***(69)***, and heart, lung, and bone marrow transplant recipients ***(70)*** and the assay could be used as a virologic marker for initiation of pre-emptive antiviral CMV therapy ***(70)***. Although the assay is not the most sensitive available, it specifically detects the clinical relevant stages of infection and has the best specificity and positive predictive value for disease development ***(52,68,71)***. The NucliSens CMV pp67 applied on cerebrospinal fluid samples from HIV-infected patients with a postmortem histopathological diagnosis of CMV encephalitis ***(72)*** and from HIV-1-infected patients with an active CMV central nervous system (CNS) disease ***(52)*** proved to be useful to diagnose active CMV CNS disease.

Other CMV target mRNAs for NASBA amplification and detection include the β2.7 transcript ***(73,74)***, and the immediate early (IE) 1 mRNA (UL123) ***(75–77)***, which plays an essential role in viral replication.

Goossens et al. ***(71)*** compared the immediate early 1 and late pp67 mRNA detection by NASBA in renal and liver transplant patients. The first was found to be the most sensitive test, making it an attractive screening test for the early detection of CMV infection. Comparable results were reported by Gerna et al. ***(77)*** in bone marrow transplant recipients.

Recently, quantitative NASBA assays for the detection of IE 1, and pp67 mRNA expression ***(78)*** in lung transplant recipients and for the quantitative detection of IE 1, pp67, and immune evasion genes *US3*, *US6*, and *US11* in cells infected with CMV ***(78)*** have been described. It was concluded that effective antiviral treatment was reflected by a rapid disappearance of pp67 mRNA confirming the clinical utility of the assay, whereas IE1 mRNA remained detectable for longer periods and should be further validated in prospective studies ***(79)***. Additionally, a multiplex real-time NASBA was described for the simultaneous detection and quantification of CMV encoded IE 1 and pp67 mRNA ***(27)***.

9.3. Respiratory Pathogens

Rapid diagnostic techniques for respiratory pathogens are not only important for clinical-epidemiological reasons but are also useful to initiate appropriate treatment within the first 24 h or interrupt empyrical treatment when the disease is caused by a microbial species requiring a different treatment.

9.3.1. Mycoplasma pneumoniae

Mycoplasma pneumoniae is a common etiologic agent of respiratory tract infections in humans, responsible for 15–20% of all cases of pneumonia ***(80)*** and a wide range of mild to serious extrapulmonary complications ***(81,82)***. The specificity and sensitivity of the NASBA assay were evaluated by Loens et al. ***(5)***. *M. pneumoniae* RNA prepared from a plasmid construct was used to assess the sensitivity of the assay, and an internal control for the detection of inhibitors was constructed. Using primers and probes targeting the *M. pneumoniae* 16S rRNA, positive results were obtained with nucleic acid extracts from *M. pneumoniae* types 1 and 2 and *M. genitalium*, but from none of the other *Mycoplasma* spp., respiratory pathogens, or commensal flora. The sensitivity of the NASBA assay was 10 molecules of in vitro-generated wild-type *M. pneumoniae* RNA and five color changing units (CCU) of *M. pneumoniae*. In protease-treated spiked throat swabs, nasopharyngeal aspirates, bronchoalveolar lavages, and sputum, the sensitivity of the NASBA assay, in the presence of the internal control, was $2{\times}10^4$ molecules of in vitro-generated RNA or 5 CCU of *M. pneumoniae*. The sensitivity of the NASBA assay was comparable to that of a PCR targeted at the P1 adhesin gene used in the study by Ieven et al. ***(83)***. Fifteen clinical specimens positive for *M. pneumoniae* by PCR were also positive by NASBA. Typing of *M. pneumoniae* by NASBA was described by Ovyn et al. ***(15)***. The commercially available NASBA NucliSens Basic Kit assay for the detection of *M. pneumoniae* 16S rRNA in respiratory specimens was developed and compared to standard NASBA and PCR assays previously described by the authors (20). The specificity and sensitivity of the NucliSens Basic Kit assay were comparable to the specificity and sensitivity of the corresponding standard NASBA assay. The most important advantage of the NucliSens Basic Kit is the provision of standardized reagents, improving the reproducibility compared with home-made assays. Additionally, ECL detection is less time-consuming than the previously used ELGA detection system ***(5,15)*** and the interpretation of the results is more objective. A real-time NASBA for the detection of *M. pneumoniae* was developed by the same group ***(84)***. The authors concluded that real-time and conventional NASBA show high concordance in sensitivity and specificity, with a clear advantage for the real-time technology regarding handling, speed, and number of samples that can easily be tested in a single run. Moreover, the real-time NASBA assay requiring fewer manipulations and producing results without postamplification processing reduces the potential risk for product carryover.

9.3.2. Other Respiratory Pathogens

A conventional NASBA assay for the detection of the 16S rRNA of *Bordetella pertussis* ***(85)*** and real-time NASBA assays for the detection of the 16S rRNA of *Chlamydia pneumoniae* ***(86)*** and *Legionella pneumophila* ***(87)*** in respiratory specimens have been developed. Finally, a multiplex real-time NASBA for the simultaneous detection of *M. pneumoniae, C. pneumoniae*, and *L. pneumophila* 16S rRNA was developed ***(34)***.

9.3.3. Mycobacterium tuberculosis

Van der Vliet et al. designed a NASBA assay based on a highly conserved region of the 16S rRNA of *Mycobacteria* ***(16)***. A species-specific probe for the detection of *M. tuberculosis* amplicons was used in an ELGA-based detection system. Nucleic acids derived from 200 bacteria resulted in a positive signal. The assay correctly identified 32 *M. tuberculosis* strains isolated from different parts of the world.

9.3.4. Picornaviridae

Human rhinoviruses (HRVs) and enteroviruses belong to the family of the picornaviridae, small nonenveloped viruses containing a positive-sense, single-stranded RNA genome. Whereas rhinoviruses are the major cause of the common cold and are implicated in acute asthma exacerbations in children and adults, enteroviruses are important pathogens, with a wide range of clinical manifestations ranging from severe CNS disease with paralysis to mild respiratory symptoms. Targeting the HRV 5'NCR, two NASBAs for HRV typing were developed ***(88)***. Another assay with the 5'NCR and viral protein 4 was developed for the detection of HRV in respiratory specimens. The assay proved to be specific and sensitive and was succesfully applied to analysis of HRV sequences in respiratory specimens ***(40)***.

Three groups applied NASBA for the detection of enteroviruses in respiratory samples ***(17)***, cerebrospinal fluid ***(17,51,89)***, and stools ***(17)***.

Nucleic acid sequence-based amplification assays have also been developed for the detection of other respiratory viruses such as respiratory syncytial virus ***(90)***, parainfluenzaviruses ***(91)***, and influenza viruses ***(92)***.

9.4. Other NASBA Applications for Detection of Infectious Agents

Various other NASBA applications have been described: for malaria parasites ***(44,93)***, *Cryptosporidium parvum* ***(94)***, *Campylobacter* species ***(6,57,95)***, *Chlamydia trachomatis* ***(49)***, *Escherichia coli* ***(96)***, *Listeria monocytogenes* ***(6,9)***, *Mycobacterium leprae* ***(7)***, *Neisseria gonorrhoeae* ***(41)***, *Salmonella enterica* ***(42,97)***, *Clavibacter michiganensis* subsp. *sepedonicus* ***(56)***, *Ralstonia solanacearum* ***(98)***, *Aspergillus* ***(19)***, *Candida* species ***(99–101)***, including a wide range of viruses such as rotavirus ***(26)***, hepatitis A virus ***(102)***, human papillomavirus 16 ***(25)***, varicella zoster virus ***(103)***, Epstein–Barr virus ***(17)***, dengue virus ***(22)***, West Nile virus ***(18)***, St Louis encephalitis virus ***(18)***, rabies ***(50)***, simian immunodeficiency virus ***(104)***, feline immunodeficiency virus ***(105)***, avian influenza subtype H5 viruses ***(106)***, potato leafroll virus ***(55)***, potato virus Y ***(30)***, citrus tristeza virus ***(107)***, and apple stem pitting virus ***(108)***.

9.5. Viability Studies

Van der Vliet et al. and Morré et al. ***(109,110)*** applied NASBA to study the influence of antibiotics on bacterial viability. Van der Vliet et al. ***(109)*** exposed *M. smegmatis* to various concentrations of rifampicin and ofloxacin suspended in broth for different periods of time. During the drug exposure, the viability of the *Mycobacteria*, expressed as the number of colony-forming units (CFUs) was compared with the presence of 16S rRNA as determined by NASBA and with the presence of DNA coding for the 16S rRNA as determined by PCR. Exposure of the *Mycobacteria* to increasing concentrations of drug resulted in a decreasing number of CFUs and a decrease in NASBA signal. Drug exposure of *M. smegmatis* resulted in a negative NASBA signal after 7 d. The number of CFUs did decrease, whereas the amount of DNA as detected by PCR did not. This implies that unlike degradation of DNA, decay of 16S rRNA in an in vitro system occurs rapidly after cell death. Morré et al. ***(110)*** investigated the value of RNA detection by NASBA for the monitoring of *Chlamydia trachomatis* infections after antibiotic treatment. *C. trachomatis* RNA was found in 24 smears before antibiotic treatment and in two smears taken 1 wk after treatment; no *C. trachomatis* RNA was detected after 2 wk or more. In contrast, *C. trachomatis* DNA was found in all specimens before treatment, but five smears were still positive after 3 wk of antibiotic treatment. *C. trachomatis* DNA and RNA were also found in 15 urine specimens before treatment, whereas 1 wk after treatment, 4/15 were *C. trachomatis* DNA positive and 1/15 was RNA positive.

Simpkins et al. found consistent and highly significant differences between the mRNA amplifications extracted from viable and heat-killed *Salmonella enterica* cells, whereas PCR amplification of both kind of samples was unaffected ***(97)***. On the other hand, when Birch et al. ***(111)*** assessed three amplification techniques (mRNA NASBA, mRNA RT-PCR, and DNA

PCR) for their ability to detect nucleic acid persistence in an *E. coli* strain following heat-killing, NASBA offered the greatest sensitivity to detect a decrease in viability although residual DNA and mRNA could be detected by PCR and NASBA. The same conclusion was reached by Uyttendaele et al. ***(112)*** studying the difference between viable and nonviable *C. jejuni*. They showed that 16S rRNA, or at least the sequence enclosed by the primer set applied, is fairly stable and resistant to heating at 100°C. Thus, the correlation between cell viability and persistence of nucleic acids must be well characterized for particular situations before molecular techniques can be substituted for traditional culture methods.

9.6. NASBA Application for the Detection of Noninfectious Targets

BCR-ABL mRNA in Ph+ chronic myeloid leukemia ***(24)***, factor V Leiden ***(113)***, circulating breast cancer cells ***(114)***, macrophage-derived chemokine gene expression ***(115)***, cytokine production ***(116)***, in vivo kinetics of tissue factor messenger RNA ***(117)***, tumor necrosis factor (TNF)-α mRNA ***(118)***, and circulating tumor cells ***(119)*** have all been more or less successfully detected by the NASBA procedure.

10. Conclusions

Industrially manufactured NASBA-based assays containing the necessary components required for isolation, amplification, and detection, such as the NucliSens Basic Kit (bioMérieux), are becoming increasingly available (e.g., HIV-1 genomic RNA and CMV mRNA assays). They are user-friendly and specific primers and probes for various other applications ***(17,19,20,41)*** have been designed that can be used in combination with the NucliSens Basic Kit (bioMérieux). These assays need further evaluation in clinical settings.

At present, ECL is the most sensitive method for the detection of the NASBA amplicons. Real-time applications are increasingly being evaluated, including multiplex formats ***(27,31)***. A major advantage of the use of molecular beacons is the ability to carry out real-time detection during amplification, which reduces the overall time of the assay, the risk of contamination, and provides information about the kinetics of the reaction.

Because NASBA has been shown not to suffer from sensitivity and specificity problems associated with other isothermal techniques, these unique features make it very suitable for wide applications in research and diagnosis. As the processes become more refined and standardized, NASBA can replace some current methods for rapid clinical laboratory diagnosis.

Acknowledgments

The financial support from the European Commission QLK2-CT-2000-00294 is gratefully acknowledged. The authors wish to thank P. van Aarle (bioMérieux, Boxtel, The Netherlands) for providing the figures.

References

1. Boom, R., Sol, C. J. A., Salimans, M. M. M., Jansen, C. L., Wertheim-van Dillen, P. M. E., and van der Noordaa, J. (1990) Rapid and simple method for purification of nucleic acids. *J. Clin. Microbiol.* **28**, 495–503.
2. van Buul, C., Cuypers, H., Lelie, P., et al. (1998) The NucliSens Extractor for automated nucleic acid isolation. *Infusionsther. Transfusionsmed.* **25**, 147–151.
3. Murphy, D. G., Côté, L., Fauvel, M., René, P., and Vincelette, J. (2000) Multicenter comparison of Roche COBAS AMPLICOR MONITOR version 1.5, Organon Teknika NucliSens QT with extractor, and Bayer Quantiplex version 3.0 for quantification of human immunodeficiency virus type 1 RNA in plasma. *J. Clin. Microbiol.* **38**, 4034–4041.
4. Gobbers, E., Oosterlaken, T. A. M., van Bussel, M. J. A. W. M., Melsert, R., Kroes, A. C. M., and Claas, E. C. J. (2001) Efficient extraction of virus DNA by NucliSens extractor allows sensitive detection of hepatitis B virus by PCR. *J. Clin. Microbiol.* **39**, 4339–4343.
5. Loens, K., Ursi, D., Ieven, M., et al. (2002) Detection of *Mycoplasma pneumoniae* in spiked clinical samples by nucleic acid sequence-based amplification. *J. Clin. Microbiol.* **40**, 1339–1345.

6. Uyttendaele, M., Schukkink, R., Van Gemen, B., and Debevere, J. (1995) Development of NASBA, a nucleic acid amplification system, for identification of *Listeria monocytogenes* and comparison to ELISA and a modified FDA method. *Int. J. Food Microbiol.* **27**, 77–89.
7. van der Vliet, G. M. E., Cho, S.-N., Kampirapap; K., et al. (1996) Use of NASBA RNA amplification for detection of *Mycobacterium leprae* in skin biopsies from untreated and treated leprosy patients. *Int. J. Leprosy* **64**, 396–403.
8. Deiman, B., van Aarle, P., and Sillekens, P. (2002) Characteristics and applications of nucleic acid sequence-based amplification (NASBA). *Mol. Biotechnol.* **20**, 163–179.
9. Blais, W. L., Turner, G., Sooknanan, R., Malek, L. T., and Phillippe, L. M. (1996) A nucleic acid sequence-based amplification (NASBA) system for Listeria monocytogenes and simple method for detection of amplimers. *Biotechnol. Tech.* **10**, 189–194.
10. Sooknanan, R., Van Gemen, B., and Malek, L. (1995) *Nucleic Acid Sequence Based Amplification. Molecular Methods for Virus Detection*, Academic, London, pp. 261–285.
11. Yates, S., Penning, M., Goudsmit, J., et al. (2001) Quantitative detection of hepatitis B virus DNA by real-time nucleic acid sequence based amplification with molecular beacon detection. *J. Clin. Microbiol.* **39**, 3656–3665.
12. Voisset, C., Mandrand, B., and Paranhos-Baccala, G. (2000) RNA amplification technique, NASBA, also amplifies homologuous plasmid DNA in non-denaturing conditions. *BioTechniques.* **29**, 236–240.
13. Greijer, A. E., Dekkers, C. A. J., and Middeldorp, J. M. (2000) Human cytomegalovirus virions differentially incorporate viral and host cell RNA during the assembly process. *J. Virol.* **74**, 9078–9082.
14. Heim, A., Grumbach, I. M., Zeuke, S., and Top, B. (1998) Highly sensitive detection of gene expression of an intronless gene: amplification of mRNA, but not genomic DNA by nucleic acid sequence-based amplification. *Nucleic Acids Res.* **26**, 2250–2251.
15. Ovyn, C., Van Strijp, D., Ieven, M., Ursi, D., Van Gemen, B., and Goossens, H. (1996) Typing of *Mycoplasma pneumoniae* by nucleic acid sequence based amplification, NASBA. *Mol. Cell. Probes* **10**, 319–324.
16. van der Vliet, G. M. E., Schukkink, R. A. F., van Gemen, B., Schepers, P., and Klatser, P. R. (1993) Nucleic acid sequence based amplification (NASBA) for the identification of *Mycobacteria*. *J. Gen. Microbiol.* **139**, 2423–2429.
17. Fox, J. D., Han, S., Samuelson, A., Zhang, Y., Neale, M. L., and Westmoreland, D. (2002) Development and evaluation of nucleic acid sequence-based amplification (NASBA) for diagnosis of enterovirus infections using the NucliSens Basic Kit. *J. Clin. Virol.* **24**, 117–130.
18. Lanciotti, R. S. and Kerst, A. J. (2001) Nucleic acid sequence based amplification assays for rapid detection of West Nile and St. Louis encephalitis viruses. *J. Clin. Microbiol.* **39**, 4506–4513.
19. Loeffler J., Hebart, H., Cox, P., Flues, N., Schumacher, U., and Einsele, H. (2001) Nucleic acid sequence based amplification of *Aspergillus* RNA in blood samples. *J. Clin. Microbiol.* **39**, 1626–1629.
20. Loens, K., Ieven, M., Ursi, D., Foolen, H., Sillekens, P. and Goossens, H. (2003) Application of NucliSens Basic Kit for the detection of *Mycoplasma pneumoniae* in respiratory specimens. *J. Microbiol. Methods* **54**, 127–130.
21. van Gemen, B., Van Beuningen, R., Nabbe, A., et al. (1994) A one tube quantitative HIV-1 RNA NASBA nucleic acid amplification assay using electrochemiluminescent (ECL) labelled probes. *J. Virol. Methods* **49**, 157–168.
22. Wu, S.-J., Lee, E. M., Putvatana, R., et al. (2001) Detection of dengue viral RNA using a nucleic acid sequence based amplification assay. *J. Clin. Microbiol.* **39**, 2794–2798.
23. Kievits, T., van Gemen, B., van Strijp, D., et al. (1991) NASBA™ isothermal enzymatic in vitro nucleic acid amplification optimized for the diagnosis of HIV-1 infection. *J. Virol. Methods.* **35**, 273–286.
24. Sooknanan, R., Malek, L., Wang, X.-H., Siebert, T., and Keating A. (1993) Detection and direct sequence identification of BCR-ABL mRNA in Ph+ chronic myeloid leukemia. *Exp. Hematol.* **21**, 1719–1724.
25. Smits, H. L., van Gemen, B., Schukkink, R., et al. (1995) Application of the NASBA nucleic acid amplification method for the detection of human papillomavirus type 16 E6-E7 transcripts. *J. Virol. Methods* **54**, 75–81.

26. Jean, J., Blais, B., Darveau, A., and Fliss, I. (2002) Rapid detection of human rotavirus using colorimetric nucleic acid sequence based amplification (NASBA) enzyme linked immunosorbent assay in sewage treatment effluent. *FEMS Microbiol. Lett.* **210**, 143–147.
27. Greijer, A. E., Adriaanse, H. M. A., Dekkers, C. A. J., and Middeldorp, J. M. (2002) Multiplex real-time NASBA for monitoring expression dynamics of human cytomegalovirus encoded IE1 and pp67 RNA. *J. Clin. Virol.* **24**, 57–66.
28. de Baar, M. P., Van Dooren, M. W., De Rooij, E., et al. (2001) Single rapid real-time monitored isothermal RNA amplification assay for quantification of human immunodeficiency virus type 1 isolates from groups M, N, and O. *J. Clin. Microbiol.* **39**, 1378–1384.
29. Leone G., van Schijndel, H., van Gemen, B., Russel Kramer, F., and Schoen, C. D. (1998) Molecular beacon probes combined with amplification by NASBA enable homogeneous, real-time detection of RNA. *Nucleic Acids Res.* **26**, 2150–2155.
30. Szemes, M., Klerks, M. M., Van den Heuvel, J. F. J. M., and Schoen, C. D. (2002) Development of a multiplex AmpliDet RNA assay for simultaneous detection and typing of potato virus Y isolates. *J. Virol. Methods* **100**, 83–96.
31. Loens, K., Beck, T., Sillekens, P., et al. (2003) Comparison of mono and multiplex real-time NASBA for the detection of *M. pneumoniae, C. pneumoniae,* and *L. pneumophila* in respiratory specimens. 13th European Congress of Clinical Microbiology and Infectious Diseases, abstract O210.
32. Jean, J., Blais, B., Darveau, A., and Fliss, I. (2002) Simultaneous detection and identification of hepatitis A virus and rotavirus by multiplex nucleic acid sequence-based amplification (NASBA) and microtitre plate hybridization system. *J. Virol. Methods* **105**, 123–132.
33. Klerks, M. M., Leone, G. O. M., Verbeek, M., van den Heuvel, J. F. J. M., and Schoen, C. D.. (2001) Development of a multiplex AmpliDet RNA for the simultaneous detection of potato leafroll virus and potato virus Y in potato tubers. *J. Virol. Methods* **93**, 115–125.
34. van Deursen, P., Gunther, A. W., Spaargaren-van Riel, C. C., et al. (1999) A novel quantitative multiplex NASBA method: application to measuring tissue factor and CD14 mRNA levels in human monocytes. *Nucleic Acids Res.* **27**, e15:i–e15:vi.
35. van Gemen, B., Van de Wiel, P., van Beuningen, R., et al. (1995) The one-tube quantitative HIV-1 RNA NASBA: precision, accuracy, and application. *PCR Methods Applic.* **4**, s177–s184.
36. Sooknanan, R. and Malek, L. T. (1995) NASBA a detection and amplification system uniquely suited for RNA. *Bio/Technology* **13**, 563–564.
37. Ginocchio, C. C., Wang, X.-P., Kaplan, M. H., et al. (1997) Effects of specimen collection, processing, and storage conditions on stability of human immunodeficiency virus type 1 RNA levels in plasma. *J. Clin. Microbiol.* **35**, 2886–2893.
38. Bruisten, S. M., Oudshoorn, P., van Swieten, P., et al. (1997) Stability of HIV-1 RNA in blood during specimen handling and storage prior to amplification by NASBA-QT. *J. Virol. Methods* **67**, 199–207.
39. Griffith, B. P., Rigsby, M. O., Garner, R. B., Gordon, M. M., and Chacko, T. M. (1997) Comparison of the Amplicor HIV-1 monitor test and the nucleic acid sequence-based amplification assay for quantitation of human immunodeficiency virus RNA in plasma, serum and plasma subjected to freeze–thaw cycles. *J. Clin. Microbiol.* **35**, 3288–3291.
40. Samuelson, A., Westmoreland, D., Eccles, R., and Fox, J. D. (1998) Development and application of a new method for amplification and detection of human rhinovirus RNA. *J. Virol. Methods* **71**, 197–209.
41. Mahony, J. B., Song, X., Chong, S., Faught, M., Salonga, T., and Kapala, J. (2001) Evaluation of the NucliSens Basic Kit for detection of *Chlamydia trachomatis* and *Neisseria gonorrhoeae* in genital tract specimens using nucleic acid sequence based amplification of 16S rRNA. *J. Clin. Microbiol.* **39**, 1429–1435.
42. Cook, N., Ellison, J., Kurdziel, A. S., Simpkins, S., and Hays, J. P. (2002) A nucleic acid sequence-based amplification method to detect Salmonella entereica serotype enteritidis strain PT4 in liquid whole egg. *J. Food Protect.* **7**, 1177–1178.
43. Tetali, S., Lee, E. M., Kaplan, M. H., Romano, J. W., and Ginocchio, C. C. (2001) Chemokine receptor CCR5 Δ32 genetic analysis using multiple specimen types and the NucliSens Basic Kit. *Clin. Diagn. Lab. Immunol.* **8**, 965–971.
44. Smits, H. L., Gussenhoven, G. C., Terpstra, W., Schukkink, R. A. F., van Gemen, B., and van Gool, T. (1997) Detection, identification and semi-quantification of malaria parasites by NASBA amplification of small subunit ribosomal RNA sequences. *J. Microbiol. Meth.* **28**, 65–75.

45. Hollingsworth, R. C., Sillekens, P., van Deursen, P., Neal, K. R., Irving, W. L., on behalf of the Trent HCV Study Group (1996) Serum HCV RNA levels assessed by quantitative NASBA: stability of viral load over time and lack of correlation with liver disease. *J. Hepatol.* **25**, 301–306.
46. Dyer, J. R., Gilliam, B. L., Eron, J. J., Jr., Grosso, L., Cohen, M. S., and Fiscus, S. A. (1996) Quantitation of human immunodeficiency virus type 1 RNA in cell free seminal plasma: comparison of NASBA™ with Amplicor™ reverse transcription–PCR amplification and correlation with quantitative culture. *J. Virol. Methods* **60**, 161–170.
47. Christie, I. L., Mullen, J. E., Braude, P. R., et al. (1998) Assisted conception in HIV discordant couples: evaluation of semen processing techniques in reducing HIV viral load. *J. Reprod. Immunol.* **41**, 301–306.
48. Brink, A. A. T. P., Vervoort, M. B. H. J., Middeldorp, J. M., Meijer, C. J. L. M., and Van den Brule, A. J. C. (1998) Nucleic acid sequence-based amplification, a new method for analysis of spliced and unspliced Epstein–Barr virus latent transcripts, and its comparison with reverse transcriptase PCR. *J. Clin. Microbiol.* **36**, 3164–3169.
49. Morré, S. A., Sillekens, P., Jacobs, M. V., et al. (1996) RNA amplification by nucleic acid sequence based amplification with an internal standard enables reliable detection of *Chlamydia trachomatis* in cervical scrapings and urine samples. *J. Clin. Microbiol.* **34**, 3108–3114.
50. Wacharapluesadee, S. and Hemachudha, T. (2001) Nucleic acid sequence based amplification in the rapid diagnosis of rabies. *Lancet* **358**, 892–893.
51. Heim, A. and Schumann, J. (2002) Development and evaluation of a nucleic acid sequence-based amplification (NASBA) protocol for the detection of enterovirus RNA in cerebrospinal fluid samples. *J. Virol. Methods* **103**, 101–107.
52. Zhang, F., Tetali, S., Wang, X. P., Kaplan, M. H., Cromme, F. V., and Ginocchio, C. C. (2000) Detection of human cytomegalovirus pp67 late gene transcripts in cerebrospinal fluid of human immunodeficiency virus type 1-infected patients by nucleic acid sequence based amplification. *J. Clin. Microbiol.* **38**, 1920–1925.
53. Shepard, R. N., Schock, J., Robertson, K., et al. (2000) Quantitation of human immunodeficiency virus type 1 RNA in different biological compartments. *J. Clin. Microbiol.* **38**, 1414–1418.
54. Revello, M. G., Lilleri, D., Zavattoni, M., Furione, M., Middeldorp, J., and Gerna, G. (2003) Prenatal diagnosis of congenital human cytomegalovirus infection in amniotic fluid by nucleic acid sequence-based amplification assay. *J. Clin. Microbiol.* **41**, 1772–1774.
55. Leone, G., van Schijndel, H. B., van Gemen, B., and Schoen, C. D. (1997) Direct detection of potato leafroll virus in potato tubers by immunocapture and the isothermal nucleic acid sequence based amplification method NASBA. *J. Virol. Methods* **66**, 19–27.
56. van Beckhoven, J. R. C. M., Stead, D. E., and van der Wolf, J. M. (2002) Detection of *Clavibacter michiganensis* subsp. *sepedonicus* by AmpliDet RNA, a new technology based on real-time monitoring of NASBA amplicons with a molecular beacon. *J. Appl. Microbiol.* **93**, 840–849.
57. Uyttendaele, M., Schukkink, R., van Gemen, B., and Debevere, J. (1996) Comparison of the nucleic acid amplification system NASBA and agar isolation for detection of pathogenic *Campylobacters* in naturally conatminated poultry. *J. Food Protect.* **59**, 683–687.
58. Gobbers, E., Fransen, K., Oosterlaken, T., et al. (1997) Reactivity and amplification efficiency of the NASBA HIV-1 RNA amplification system with regard to different HIV-1 subtypes. *J. Virol. Methods* **66,** 293–301.
59. Notermans, D. W., De Wolf, F., Oudshoorn, P., et al. (2000) Evaluation of a second-generation nucleic acid sequence based amplification assay for quantification of HIV type 1 RNA and the use of ultrasensitive protocol adaptations. *AIDS Res. Hum. Retrov.* **16**, 1507–1517.
60. Ginocchio, C. C., Kemper, M., Stellrecht, K. A., and Witt, D. J. (2003) Multicenter evaluation of the performance characteristics of the NucliSens HIV-1 QT assay used for quantitation of human immunodeficiency virus type 1 RNA. *J. Clin. Microbiol.* **41,** 164–173.
61. Spearman, P., Fiscus, S. A., Smit J. R. M., et al. (2001) Comparison of Roche MONITOR and Organon Teknika NucliSens assays to quantify human immunodeficiency virus type 1 RNA in cerebrospinal fluid. *J. Clin. Microbiol.* **39**, 1612–1614.
62. de Baar, M. P., van der Schoot, A. M., Goudsmit, J., et al. (1999) Design and evaluation of a human immunodeficiency virus type 1 RNA assay using nucleic acid sequence-based amplification technology able to quantify both group M and O viruses by using the long terminal repeat as target. *J. Clin. Microbiol.* **37**, 1813–1818.
63. Sillekens, P., Kok, W., van Gemen, B., et al. (1994) *Specific Detection of HCV RNA Using NASBA™ as a Diagnostic Tool. Hepatitis C Virus GEMHEP*, John Libbey Eurotext, Paris, pp. 71–82.

64. Lunel, F., Mariotti, M., Cresta, P., De La Croix, I., Huraux, J.–M., and Lefrère J.-J. (1995) Comparative study of conventional and novel strategies for the detection of hepatitis C virus RNA in serum: amplicor, branched-DNA, NASBA and in-house PCR. *J. Virol. Methods* **54**, 159–171.
65. Damen, M., Sillekens, P., Cuypers, H. T. M., Frantzen, I., and Melsert R. (1999) Characterization of the quantitative HCV NASBA assay. *J. Virol. Methods* **82**, 45–54.
66. Lunel, F., Cresta, P., Vitour, D., et al. (1999) Comparative evaluation of hepatitis C virus RNA quantitation by branched DNA, NASBA, and monitor assay. *Hepatology* **29**, 528–535.
67. Blok, M. J., Goossens, V. J., Vanherle, S. J. V., et al. (1998) Diagnostic value of monitoring human cytomegalovirus late pp67 mRNA expression in renal-allograft recipients by nucleic acid sequence-based amplification. *J. Clin. Microbiol.* **36**, 1341–1346.
68. Blank, B. S. N., Meenhorst, P. L., Pauw, W., et al. (2002) Detection of late pp67-mRNA by NASBA in peripheral blood for diagnosis of human cytomegalovirus disease in AIDS patients. *J. Clin. Virol.* **25**, 29–38.
69. Witt, D. J., Kemper, M., Stead, A., et al. (2000) Analytical performance and clinical utility of a nucleic acid sequence based amplification assay for detection of cytomegalovirus infection. *J. Clin. Microbiol.* **38**, 3994–3999.
70. Gerna G., Baldanti, F., Middeldorp, J. M., et al. (1999) Clinical significance of expression of human cytomegalovirus pp 67 late transcript in heart, lung and bone marrow transplant recipients, as determined by nucleic acid sequence-based amplification. *J. Clin. Microbiol.* **37**, 902–911.
71. Goossens, V. J., Blok, M. J., Christiaans, M. H. L., et al. (1999) Diagnostic value of nucleic acid sequence-based amplification for the detection of cytomegalovirus infection in renal and liver transplant recipients. *Intervirology* **42**, 373–381.
72. Bestetti, A., Pierotti, C., Terreni, M., et al. (2001) Comparison of three nucleic acid amplification assays of cerebrospinal fluid for diagnosis of cytomegalovirus encephalitis. *J. Clin. Microbiol.* **39**, 1148–1151.
73. Aono, T., Kondo, K., Miyoshi, H., et al. (1998) Monitoring of human cytomegalovirus infections in pediatric bone marrow transplant recipients by nucleic acid sequence-based amplification. *J. Infect. Dis.* **178**, 1244–1249.
74. Oldenburg, N., Lam, K. M. C., Khan, M. A., et al. (2002) Evaluation of human cytomegalovirus gene expression in thoracic organ transplant recipients using nucleic acid sequence-based amplification. *Transplantation* **70**, 1209–1215.
75. Blok, M. J., Christiaans, M. H. L., Goossens, V. J., et al. (1998) Evaluation of a new method for early detection of active cytomegalovirus infections. A study in kidney transplant recipients. *Transplant Int.* **11 (Suppl. 1)**, S107–S109.
76. Blok, M. J., Christiaans, M. H. L., Goossens, V. J., et al. (1999) Early detection of human cytomegalovirus infection after kidney transplantation by nucleic acid sequence-based amplification. *Transplantation* **67**, 1274–1277.
77. Gerna, G., Baldanti, F., Lilleri, D., et al. (2000) Human cytomegalovirus immediate-erly mRNA detection by nucleic acid sequence-based amplification as a new parameter for preemptive therapy in bone marrow transplant recipients. *J. Clin. Microbiol.* **38**, 1845–1853.
78. Greijer, A. E., Adriaanse, H. M. A., Kahl,, M., et al. (2001) Quantitative competitive NASBA for measuring mRNA expression levels of the immediate early 1, late pp67, and immune evasion genes US3, US6, and US11 in cells infected with human cytomegalovirus. *J. Virol. Methods* **96**, 133–147.
79. Greijer, A. E., Verschuuren, E. A. M., Harmsen, M. C., et al. (2001) Direct quantification of human cytomegalovirus immediate-early and late mRNA levels in blood of lung transplant recipients by competitive nucleic acid sequence-based amplification. *J. Clin. Microbiol.* **39**, 251–259.
80. Foy, H. M. (1993) Infections caused by *Mycoplasma pneumoniae* and possible carrier state in different populations of patients. *Clin. Infect. Dis.* **17(Suppl. 1),** S37–S46.
81. Clyde, W. A., Jr. (1993) Clinical overview of typical *Mycoplasma pneumoniae* infections. *Clin. Infect. Dis.* **17(Suppl. 1)**, S32–S36.
82. Ieven, M., Demey, H., Ursi, D., van Goethem, G., Cras, P., and Goossens, H. (1998) Fatal encephalitis caused by *Mycoplasma pneumoniae* diagnosed by the polymerase chain reaction. *Clin. Infect. Dis.* **27**, 1552–1553.
83. Ieven, M., Ursi, D., Van Bever, H., Quint, W., Niesters, H. G. M., and Goossens H. (1996) Detection of *Mycoplasma pneumoniae* by two polymerase chain reactions and role of *M. pneumoniae* in acute respiratory tract infections in pediatric patients. *J. Infect. Dis.* **173**, 1445–1452.
84. Loens, K., Ieven, M., Ursi, D., et al. (2003) Detection of *Mycoplasma pneumoniae* by real-time NASBA. *J. Clin. Microbiol.* **41**, 4448–4450.

85. Loens, K., Beck, T., Sillekens, P., et al. (2003) Development of conventional NASBA for the detection of *Bordetella pertussis* in respiratory specimens, 13th European Congress of Clinical Microbiology and Infectious Diseases, abstract P1369.
86. Coombes, B. K. and Mahony, J. B. (2000) Nucleic acid sequence-based amplification (NASBA) of *Chlamydia pneumoniae* major outer membrane protein (ompA) mRNA with bioluminescent detection. *Comb. Chem. High Throughput Screen.* **3**, 315–327.
87. Sillekens, P., Overdijk, M., Roosmalen, I., et al. (2002) Molecular diagnosis of *Legionella* infections using NASBA with real-time detection. *Clin. Microbiol. Infect.* **8 (Suppl. 1)**, 147.
88. Loens, K., Ieven, M., Ursi, D., et al. (2003) Improved detection of rhinoviruses by NASBA® after nucleotide sequence determination of the 5'NCR of additional rhinovirus strains. *J. Clin. Microbiol.* **41**, 1971–1976.
89. Landry, M. L., Garner, R., and Ferguson, D. (2003) Rapid enterovirus RNA detection in clinical speciemns by using nucleic acid sequence-based amplification. *J. Clin. Microbiol.* **41**, 346–350.
90. Rahman, A., John, R., Westmoreland, D., and Fox, J. D. (2002) Clinical evaluation of molecular amplification assays based on NASBA and RT-PCR for diagnosis of respiratory syncytial virus infection, Pan American Society for Clinical Virology, Florida.
91. Hibbitts, S., Rahman, A., Westmoreland, D., and Fox, J. D. (2002) Diagnosis of parainfluenza virus infections using NucliSens® Basic Kit NASBA with "end point" and "real-time" detection, 12th European Congress of Clinical Microbiology and Infectious Diseases.
92. Rahman, A., Westmoreland, D., and, Fox J. D. (2002) Evaluation of "end point" and "real time" molecular-based assays for amplification and detection of influenza viruses, 12th European Congress of Clinical Microbiology and Infectious Diseases.
93. Schoone, G. J., Oskam, L., Kroon, N. C. M., Schallig, H. D. F. H., and Omar, S. A. (2000) Detection and quantification of *Plasmodium falciparum* in blood samples using quantitative nucleic acid sequence based amplification. *J. Clin. Microbiol.* **38**, 4072–4075.
94. Baeumner, A. J., Humiston, M. C., Montagna R. A., and Durst R. A. (2001) Detection of viable oocysts of *Cryptosporidium parvum* following nucleic acid sequence-based amplification. *Anal. Chem.* **73**, 1176–1180.
95. Uyttendaele, M., Schukkink, R., van Gemen, B., and Debevere, J. (1994) Identification of *Camplylobacter jejuni*, *Campylobacter coli* and *Camplylobacter lari* by the nucleic acid sequence based amplification system NASBA™. *J. Appl. Bacteriol.* **77**, 694–701.
96. Min, J. and Baeumner, A. J. (2002) Highly sensitive and specific detection of viable *Escherichia coli* in drinking water. *Anal. Biochem.* **303**, 186–193.
97. Simpkins, S. A., Chan, A. B., Hays, J., Pôpping, B., and Cook, N. (2000) An RNA transcription based amplification technique (NASBA) for the detection of viable *Salmonella enterica*. *Lett. Appl. Microbiol.* **30**, 75–79.
98. Bentsink, L., Leone, G. O. M., van Beckhoven, J. R. C. M., Van Schijndel, H. B., Van Gemen, B., and Van der Wolf, J. M. (2002) Amplification of RNA by NASBA allows direct detection of viable cells of *Ralstonia solanacearum* in potato. *J. Appl. Microbiol.* **93**, 647–655.
99. Widjojoatmodjo, M. N., Borst, A., Schukkink, R. A. F., et al. (1999) Nucleic acid sequence based amplification (NASBA) detection of medically important *Candida* species. *J. Microbiol. Methods* **38**, 81–90.
100. Borst, A., Leverstein-Van Hall, M. A., Verhoef, J., and Fluit, A. C. (2001) Detection of *Candida* spp. in blood cultures using nucleic acid sequence-based amplification (NASBA). *Diagn. Microbiol. Infect. Dis.* **39**, 155–160.
101. Borst, A., Verhoef, J., Boel, E., and Fluit, A. C. (2002) Clinical evaluation of a NASBA-based assay for detection of *Candida* spp. in blood and blood cultures. *Clin. Lab.* **48**, 487–492.
102. Jean, J., Blais, B., Darveau, A., and Fliss, I. (2001) Detection of hepatitis A virus by the nucleic acid sequence based amplification technique and comparison with reverse trasncription PCR. *Appl. Environm. Microbiol.* **67**, 5593–5600.
103. Mainka, C., Fuss, B., Geiger, H., Höfelmayr, H., and Wolff, M. H. (1998) Characterization of viremia at different stages of varicella zoster virus infection. *J. Med. Virol.* **56**, 91–98.
104. Romano, J. W., Shurtliff, R. N., Dobratz, E., et al. (2000) Quantitative evaluation of simian immunodeficiency virus infection using NASBA technology. *J. Virol. Methods* **86**, 61–70.
105. Jordan, H. J., Scappino, L. A., Moscardini, M., and Pistello, M. (2002) Detection of feline immunodeficiency virus RNA by two nucleic acid sequence based amplification (NASBA) formats. *J. Virol. Methods* **103**, 1-13.

106. Collins, R. A., Ko, L.-S., So, K.-L., Ellis, T., Lau, L.-T., and Yu, A. C. H.. (2002) Detection of highly pathogenic and low pathogenic avian influenza subtype H5 (Eurasian lineage) using NASBA. *J. Virol. Methods* **103**, 213–225.

107. Lair, S. V., Mirkow, T. E., Dodds, J. A., and Murphy, M. F. (1993) A single temperature amplification technique applied to the detection of citrus tristeza viral RNA in plant nucleic acid extracts. *J. Virol. Methods* **47**, 141–152.

108. Klerks, M. M., Leone, G., Linder, J. L., Schoen, C. D., and van den Heuvel, J. F. J. M. (2001) Rapid and sensitive detection of apple stem pitting virus in apple trees through RNA amplification and probing with fluorescent molecular beacons. *Phytopathology* **91**, 1085–1091.

109. van der Vliet, G. M. E., Schepers, P., Schukkink, R. A. F., van Gemen, B., and Klatser, P. R. (1994) Assessment of mycobacterial viability by RNA amplification. *Antimicrob. Agents Chemother.* **38**, 1959–1965.

110. Morré, S. A., Sillekens, P. T. G., Jacobs, M. V., et al. (1998) Monitoring of *Chlamydia trachomatis* infections after antibiotic treatment using RNA detection by nucleic acid sequence based amplification. *J. Clin. Pathol. Mol. Pathol.* **51**, 149–154.

111. Birch, L., Dawson, C. E., Cornett, J. H., and Keer, J. T. (2001) A comparison of nucleic acid amplification techniques for the assessment of bacterial viability. *Lett. Appl. Microbiol.* **33,** 296–301.

112. Uyttendaele, M., Bastiaans, A., and Debevere, J. (1997) Evaluation of the NASBA nucleic acid amplification system for assessment of the viability of *Campylobacter jejuni. Int. J. Food Microbiol.* **37**, 13–20.

113. Reitsma, P. H., van der Velden, P. A., Vogels, E., et al. (1996) Use of the direct RNA amplification technique NASBA to detect factor V Leiden, a point mutation associated with APC resistance. *Blood Coagul. Fibrin.* **7**, 659–663.

114. Lambrechts, A. C., Bosma, A. J., Klaver, S. G., et al. (1999) Comparison of immunocytochemistry, reverse transcriptase polymerase chain reaction, and nucleic acid sequence based amplification for the detection of circulating breast cancer cells. *Breast Cancer Res. Treat.* **56**, 219–231.

115. Romano, J. W., Shurtliff, R. N., Grace, M., et al. (2001) Macrophage-derived chemokine gene expression in human and macaque cells: mRNA quantification using NASBA technology. *Cytokine* **13**, 325–333.

116. Heim, A., Zeuke, S., Weiss, S., Ruschewski, W., and Grumbach, I. M. (2000) Transient induction of cytokine production in human myocardial fibroblasts by Coxsackievirus B3. *Circ. Res.* **86**, 753–759.

117. Franco, R. F., De Jonge, E., Dekkers, P. E. P., et al. (2000) The in vivo kinetics of tissue factor messenger RNA expression during human endotoxemia: Relationship with activation of coagulation. *Blood* **96**, 554–559.

118. Darke, B. M., Jackson, S. K., Hanna, S. M., and Fox, J. D. (1998) Detection of human TNF-α mRNA by NASBA. *J. Immunol. Methods* **212**, 19–28.

119. Burchill, S. A., Perebolte, L., Johnston, C., Top, B., and Selby, P. (2002) Comparison of the RNA-amplification based methods RT-PCR and NASBA for the detection of circulating timour cells. *Br. J. Cancer* **86**, 102–109.

22

Oligonucleotide Ligation Assay

Faye A. Eggerding

1. Introduction

Advances in molecular genetics and biotechnology are changing the practice of medicine. Completion of the human genome sequence and development of high-throughput semiautomated DNA analysis techniques have resulted in an exponential increase in the rate of discovery of new disease genes. The challenge of understanding complex, multigenic diseases, and the role of genetic variability in disease risk is now within reach. Identification of genetic mutation is assuming increasing importance in cancer diagnosis and in the detection and characterization of infectious agents. The emerging field of molecular genetic medicine has resulted in a need for new technologies capable of probing large numbers of gene sequences with sufficient selectivity to distinguish single-nucleotide differences.

Current methods for gene analysis fall into two broad categories referred to as scanning methods to detect mutations at unspecified DNA sites and diagnostic methods to detect previously identified mutations *(1–3)*. Single-nucleotide substitutions and small deletions or insertions are the most common disease-producing mutations in the human genome. Although all DNA sequences are subject to mutation, different genes have distinct patterns of mutational inactivation. For example, in some genes, single-base substitutions (point mutations) occur at known positions within the gene; in other genes, mutations occur *de novo* at scattered sites throughout the gene. DNA sequencing is considered the definitive procedure both for detection of known mutations and identification of unknown mutations. However, the expense and labor-intensive nature of DNA sequencing necessitated development of alternative strategies for screening DNA for mutation. Mutation-scanning methods, including denaturing gradient gel electrophoresis (DGGE) (see Chapter 8), single-strand conformational polymorphisms (SSCPs) (see Chapter 7), and ribonuclease (RNase) cleavage, are designed to survey stretches of DNA for mismatched nucleotides, but they give no information as to the nature of the mutated base. Diagnostic methods, including allele-specific oligonucleotide hybridization, allele-specific amplification, and oligonucleotide ligation, are more stringent than scanning methods and unequivocally detect known genetic mutations. This chapter focuses on oligonucleotide ligation methods and their applications in diagnostic genetic analysis.

1.1. Principle of Oligonucleotide Ligation

Oligonucleotide ligation is a robust diagnostic strategy for discrimination of known DNA sequence variants that combines polymerase chain reaction (PCR) with an oligonucleotide ligation assay (OLA) *(4,5)*. In the oligonucleotide ligation reaction, two oligonucleotides or probes (usually 20-mers in size) are hybridized to a DNA target sequence (usually PCR amplicons of the gene of interest). The two oligonucleotide probes abut each other when

From: *Medical Biomethods Handbook*
Edited by: J. M. Walker and R. Rapley © Humana Press, Inc., Totowa, NJ

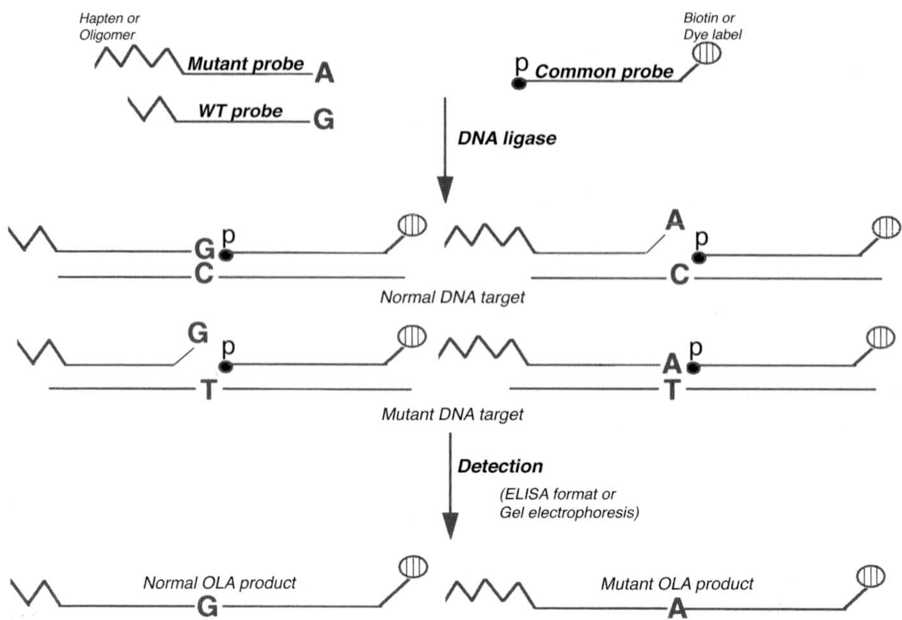

Fig. 1. Diagram of the ligation reaction. The OLA relies on PCR amplification to replicate DNA target sequences and probe ligation to distinguish sequence variants (G→A transition, as illustrated). Probes hybridizing in juxtaposition to target DNA are joined by DNA ligase; probes mismatched at their junction are not ligated. Allelic probes are modified at their 5' ends by the addition of either specific hapten groups (e.g., digoxigenin or fluorescein) for enzyme-linked immunosorbent assay (ELISA) detection or oligomers of differing size for detection by electophoresis. The downstream, common probe is 5'-phosphorylated (p) and labeled with biotin or a fluorescent dye at its 3' end. Ligation products are detected colorimetrically in an ELISA format or by denaturing polyacrylamide gel electrophoresis. Biotin-labeled ligation products are captured on streptavidin-coated microtiter plates and, after washing to remove unligated probe, an ELISA reaction is performed with anithapten antibodies labeled with different enzyme reporters (e.g., alkaline phosphate and horseradish peroxidase) to detect allele-specific haptens. After the addition of chromogenic substrates for each reporter enzyme, the appearance of color indicates that ligation has occurred between and allele-specific probe and the common, biotinylated probe. Ligation of allele-specific probes with unique mobility modifiers [e.g., poly(A) oligomers] to a fluorescent dye-tagged common probe gives each ligation product a unique electrophoretic signature when detected using an automated DNA sequencer.

hybridized to target DNA such that the 3' end of the upstream oligonucleotide is immediately adjacent to the 5' end of the downstream oligonucleotide. DNA ligase covalently joins the two juxtaposed oligonucleotides only if the nucleotides at their junction are perfectly base-paired with the target DNA ***(6–9)***. The terminal and penultimate nucleotides on both sides of the junction of the two probes must be correctly base-paired for the ligase enzyme to covalently join the probes. The principle of the reaction is illustrated in **Fig. 1**.

Oligonucleotide probes of known sequence can be synthesized to examine allelic variants at any given locus. For example, genotyping a normal and mutant allele requires synthesis of two allelic- or target-specific probes, representing the normal and mutant allele, and one common probe complementary to both alleles. The normal and mutant allelic probes are designed such that the distinguishing allelic nucleotide is located at the 3' end of the probe; otherwise, the

Table 1
Advantages and Limitations of Oligonucleotide Ligation

Advantages
Accurate distinction of any base substitution or insertion/deletion mutation
Specificity (chance of a false positive essentially zero)
Sensitivity (detects rare sequence variants in background of normal DNA)
Rapid, economical, and easy to perform (one-step reaction)
Adapts well to multiplexing
Minimal sample preparation required
Results simple to interpret
Tolerates some degree of sequence polymorphism
Nonradioactive and nontoxic reagents
Compatible with multiple detection formats
Amenable to automation and computerized genotyping
Limitation
Detects only known mutations
Special equipment might be required
Sequence polymorphisms near ligation site might affect ligation
G→T mismatches least well discriminated

probes are identical. Mismatches at the 3' end of a target-specific probe disrupt ligation of the target and common probes. The common oligonucleotide probe is complementary to sequences adjacent to and downstream from the allelic probes and contains a phosphate group at its 5' end to enable formation of a phosphodiester bond in the presence of ligase.

Variants of the ligation method have been developed. The ligase chain reaction is a less robust variant in which complementary pairs of oligonucleotide probes are used, allowing exponential amplification of the ligation signal ***(10,11)***. Another modification uses a padlock probe, which is an oligonucleotide complementary to the target sequence at its 5' and 3' ends but not in its middle portion ***(12,13)***. When perfectly paired to a target sequence and treated with ligase, the probe circularizes and can be amplified by rolling circle replication ***(14)***.

2. Ligation Protocol

There are a number of stringent requirements that must be met in the development of a clinical molecular genetic test. The test must be highly specific in discriminating allelic variants and sensitive enough to detect a sequence variant against a background of normal sequence. In addition, the test protocol should require use of nontoxic and nonradioactive reagents, and it should be simple enough so that it can be performed robustly and inexpensively on a routine basis. Features of the OLA address these requirements and make it ideally suited for typing single-base substitutions and small deletion or insertion mutations. Some advantages and disadvantages of the method are listed in **Table 1**.

Specificity of the reaction is ensured because ligation is dependent on both probe hybridization and the specificity of ligase enzyme in distinguishing single-basepair mismatches. Because ligation depends only on perfect base-pairing of nucleotides at the probe junction, nonstringent annealing conditions can be used without compromising the specificity of the ligation reaction, thus enabling the typing of multiple nucleotide substitutions in one assay. The oligonucleotide probes can be labeled with reporter molecules or fluorescent dyes and mobility modifiers making the results of the assay simple to interpret. **Figure 2** outlines the steps of the OLA, which include an initial PCR amplification of the target sequence, ligase-mediated mutation analysis, followed by detection of the ligation products.

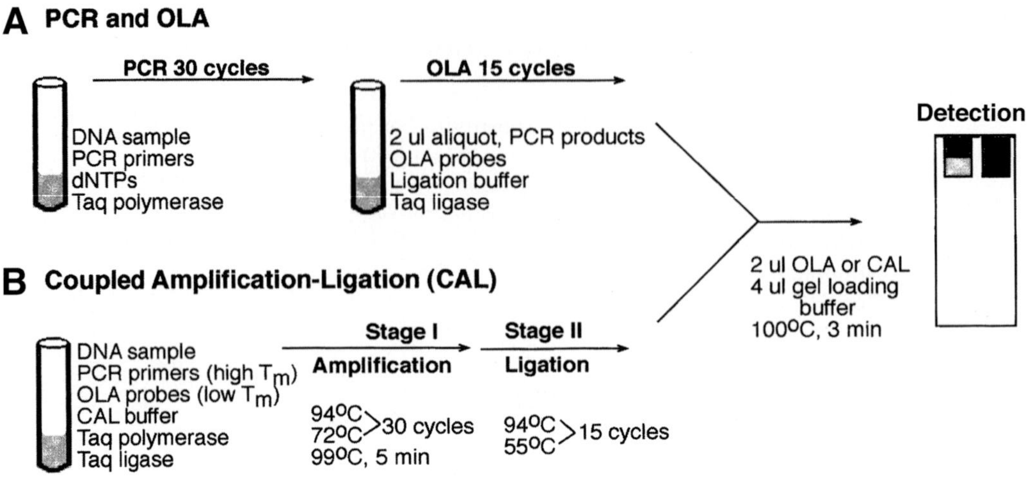

Fig. 2. Outline of the laboratory procedures for the OLA. The assay can be performed in two sequential steps **(A)** or in a one-step, single-tube format **(B)**. Use of high-T_m PCR primers enables amplification in stage I; genotyping by oligonucleotide ligation occurs in stage II simply by lowering the temperature to anneal low-T_m ligation probes. The reaction is monitored by electrophoresis as shown or ligation products can be detected spectrophotometrically using a microtiter plate reader.

2.1. Sample Preparation and PCR Amplification

Target DNA samples to be examined for sequence variants are prepared for analysis by amplification of genomic or plasmid DNAs or cDNAs using PCR. Amplifications can be performed in a multiplex format. Initial samples may be obtained from sources including whole blood, plasma, saliva, tissue sections, dried blood spots on Guthrie cards, bacterial DNA, and viral DNA or RNA. Rigorous DNA purification is not required for the precision of the assay. Following PCR, the sample is subjected to a heating step at 99°C for 5 mins to inactivate *Taq* DNA polymerase. PCR amplicons require no futher purification or removal of the PCR primers and can be used directly in the ligation reaction.

2.2. Oligonucleotide Ligation Step

After amplification, a ligation mix is added directly to each PCR sample in microtiter wells or in individual PCR tubes. The ligation mix contains the oligonucleotide ligation probes, thermostable DNA ligase, and NAD^+ as a reaction cofactor. A biallelic locus requires three oligonucleotide probes or primers for ligation analysis (one common and two allelic- or target-specific probes). During each ligation cycle, the target amplicons denature and anneal to the appropriate oligonucleotide probes. At each locus, one allelic and one common probe anneal to the target amplicon, and if perfectly matched at their junction, the ligase enzyme covalently joins them. Linear amplification of ligation products is achieved by using a thermostable DNA ligase and a thermocycling profile of initial denaturation at 98°C for 3 min followed by 15–20 cycles of denaturation at 95°C and anneal–ligation at 50–55°C. Specificity of ligation can be enhanced by performing ligations at or close to the melting temperature of the OLA probes to destabilize hybridization of imperfectly matched probes.

2.3. Detection of Ligation Products

Numerous detection schemes have been developed ***(15–19)***. In one format, the common ligation probe is 3' end labeled with biotin, and allelic probes are marked with different hapten reporter groups (i.e., digoxigenin and fluorescein). Ligation products are captured in streptavidin coated microtiter plates and wild-type and mutant genotypes are detected colorimetrically ***(15)***. An enzyme-linked immunosorbent assay (ELISA) is performed with hapten-specific antibodies labeled with different enzyme reporters (i.e., alkaline phosphatase and horseradish peroxidase) to identify if either or both of the allele-specific probes have been ligated to the biotin-labeled common probe. Results are easily interpreted by measurement of spectrophotometric absorbance using an ELISA plate reader.

An alternative method of detection relies on high-resolution electrophoretic separation using a fluorescent DNA sequencer ***(17–19)***. In this format, common ligation probes contain different fluorescent tags and allelic probes contain 5' end variable-length, oligomeric mobility modifiers. Mobility modifiers can be nucleotides in the form of noncomplementary 5'-poly(A) extensions or non-nucleotide molecules (e.g., pentaethylene oxide [PEO] oligomers). PEO molecules are hydrophilic and negatively charged and can be added to the allelic probes during oligonucleotide synthesis using standard automated phosphoramidite chemistry. Ligated probes are separated in denaturing sequencing gels using an automated fluorescent DNA sequencer. This combination of fluorescent tags with mobility modifiers gives each ligation product a unique electrophoretic signature and enables extensive sample multiplexing.

2.4. Coupled Amplification and Ligation

A modification of the PCR and OLA procedure allows simultaneous multiplex amplification and genotyping by ligation in a single-tube, one-step format (*see* **Fig. 2**) ***(20)***. Amplification and genotyping can be sustained in one reaction as a result of differences in melting temperatures (T_m) of hybrids formed among PCR primers, oligonucleotide probes, and template DNA. Long, high-T_m PCR primers enable amplification and elongation to be carried out at 72°C; allelic discrimination occurs in the second phase of the reaction simply by lowering the temperature to facilitate annealing and ligation of low T_m allelic and common OLA probes to the amplified target sequences. Ligation products are analyzed using a fluorescent DNA sequencer.

3. Applications of Oligonucleotide Ligation

The ligation assay was originally developed and applied to analysis of β-globin alleles in sickle cell disease using radioactive detection ***(21)***. Use of fluorescent labeling and simplification and automation of the detection of ligation products have resulted in numerous clinical applications for this technology. **Table 2** presents representative applications of oligonucleotide ligation for genetic analysis. The applications fall into three main categories: (1) analysis of disease genes for known mutations, (2) analysis of viral and bacterial pathogens, and (3) research applications.

3.1. Disease Genes

A clinical test for cystic fibrosis capable of analyzing 60 different alleles is commercially available ***(19,20)***. The test combines multiplex PCR and OLA with detection by sequence-coded separation based on fluorescence and unique electrophoretic mobility. A similar strategy has recently been used for prenatal diagnosis of sickle cell disease ***(22)***. Allelic probes specific for normal and mutant β-globin contain both fluorescent and mobility tags at their 5' ends for detection by capillary electrophoresis. The authors note that the test requires minute amounts of blood or amniotic fluid (0.2 μL) and minimal DNA purification such that diagnostic results can be reported the same day the specimen is received. Chakravarty et al. reported use of oligonucleotide ligation as a high-throughput and precise method for the detection of the factor V Leiden mutation (R506Q) ***(23)***. Ligation products are captured in microtiter plates containing streptavidin to bind biotin-labeled common oligonucleotide probes; allelic probes carry lanthanide labels for sensitive detection using a fluorescent microplate reader.

The analysis of mutations associated with cancer presents complexity that is not encountered in studies of inherited or germline disease mutations. The challenge is to identify somatic

Table 2
Applications of Oligonucleotide Ligation

Application	Assay Format	Ref.
Sickle cell disease (β-globin)	PCR/OLA, electrophoresis	22
Cystic fibrosis (CFTR)	PCR/OLA, CAL, sequence-codedelectrophoresis	20 20 19
Factor V Leiden (R506Q)	PCR/OLA, enhanced lanthanide fluroescence	23
Ras oncogenes	PCR/OLA, CAL, gel electrophoresis	24
HIV-1 pol gene	RT-PCR/OLA, colorimetric microtiter plate	25
HIV-1 protease gene	RT-PCR/OLA, colorimetric microtiter plate	26
S. pneumoniae (topoisomerase genes)	PCR/OLA, colorimetric microtiter plate	27
H. pylori (23S rRNA gene)	PCR/OLA, colorimetric microtiter plate	28
SNPs (single-nucleotide polymorphisms)	PCR/OLA, colorimetric microtiter plate	30 29
	PCR/OLA, sequence tag fluorescent microspheres	32
	PCR/OLA, real-time fluorescence resonance energy transfer	31
Human centromere DNA	Padlock OLA probes, *in situ* detection	13

mutations in tumor cells in a background of normal cells. The application of oligonucleotide ligation to analysis of K-ras mutations in pancreatic cancer specimens is shown in **Fig. 3** ***(24)***. The presence of K-ras mutation is a biomarker for pancreactic and colorectal cancer. Ras genes acquire oncogenic potential through point mutation in codons 12, 13, or 61, producing amino acid substitutions that lock the protein in an active, growth-promoting conformation. Sequences and design of oligonucleotide probes used for detection of K-ras codon 12, 13, and 61 mutations are shown in **Table 3**. A signal indicating the presence of a G→A transition in codon 12 of the *K-ras* gene producing a Gly12Asp substitution is clearly evident in the ligation assay (*see* **Fig. 3E**). Sequencing electrophoretograms of the same DNA sample fail to detect the mutant base identified by OLA (*see* **Fig. 3E**, inset). The sensitivity of the ligation assay in detecting minority mutant ras genotypes is shown in **Fig. 4**. The assay can detect ras mutations in tumor cells making up less than 5% of the sample population; in comparison, mutant sequences must make up at least 25% of the sample for detection by fluorescent-based DNA sequencing.

3.2. Infectious Diseases

Strains of bacteria and viral pathogens are often distinguished by variations in single-nucleotide sequences. Genetic analysis of infectious disease (e.g., latent viral infection), similar to cancer, requires detection of sequence variants present in low abundance against a background of normal sequence. Edelstein et al. and Beck et al. described the application of oligonucletotide ligation to detection of HIV-1 *pol* and *protease* gene mutations associated with resistance to antiretroviral drugs ***(25,26)***. Plasma HIV-1 RNA from clinical samples was amplified by reverse transcription–PCR and genotyped by oligonucleotide ligation. In an anlaysis of 700 *pol* codons from 175 patient specimens, only 15 failed to be genotyped by OLA (2.1%) ***(25)***. Indeterminate

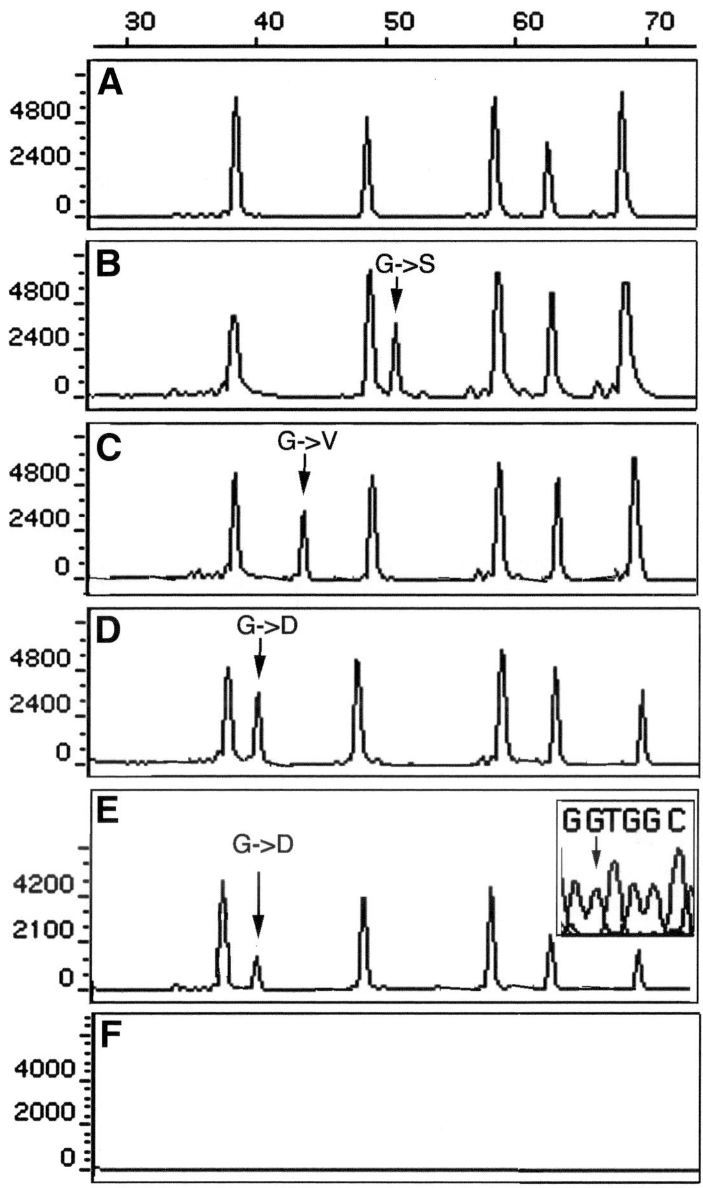

Fig. 3. Analysis of *K-ras* oncogene mutations in pancreatic cancer patients. DNA extracted from tumor specimens was amplified by PCR and analyzed by multiplex oligonucleotide ligation for mutation in *K-ras* codons 12, 13 and 61 using the probes shown in Table 3. Ligation products were separated in 8% polyacrylamide gels in a fluorescent DNA sequencer. Electrophoretogram tracings representing real-time fluorescence detection of ligation products are shown. *y*-axes display peak heights measured by fluorescence intensity in arbitrary units, and *x*-axes display the computed sizes of the ligation products in bases. The top panel **(A)** is a normal DNA control, **panels C** through **E** are from pancreatic tumor DNA, and **panel F** is a negative, no DNA control. Note in **panel E** that sequence analysis failed to detect a G→A mutation that was evident by OLA analysis (GGT→GAT, G→D).

Table 3
Oligonucleotide Probe Design for K-ras Mutation Analysis

Codon[a]	Amino acid	5'-Oligonucleotide (allelic probe)[b]	3'-Oligonucleotide (common probe)[c]	Size (bp)[d]
12-1	G (wt)	aaacttgtggtagttggagct**g**	gtggcgtaggcaagagtgc	49 (wt)
	S	aaacttgtggtagttggagct**a**		51
	R	aaacttgtggtagttggagct**c**		53
	C	aaacttgtggtagttggagct**t**		55
12-2	G (wt)	ttgtggtagttggagctg**g**	tggcgtaggcaagagtgcc	38 (wt)
	D	aacttgtggtagttggagctg**a**		41
	V	aaacttgtggtagttggagctg**t**		44
	A	ttgtggtagttggagctg**c**		46
13-1	G (wt)	gtggtagttggagctggt**g**	gcgtaggcaagagtgcctt	62 (wt)
	C	tgtggtagttggagctggt**t**		65
13-2	G (wt)	tggtagttggagctggtg**g**	cgtaggcaagagtgccttg	58 (wt)
	D	gtggtagttggagctggtg**a**		60
61-3[e]	Q (wt)	tcattgcactgtactcctc**t**	tgacctgctgtgtcgagaat	68 (wt)
	H	tcattgcactgtactcctc**a**		73

[a]Codon number and mutation position in the codon.
[b]Synthesized with 5' mobility modifiers (oligomers of PEO).
[c]Synthesized with a 5' phosphate group and 3' fluorescent label.
[d]Size in basepairs of ligation products.
[e]Synthesized in antisense orientation.

OLA results were the result of alternative or polymorphic sequences located close to the ligation site. These studies both confirm the greater sensitivity of OLA than DNA sequencing in detecting minor populations of mutant virus present in clinical samples. Because the incidence of HIV-1 drug-resistant variants is increasing because of widespread use of antiretrovirals, sensitive methods for screening patients are needed to guide selection of appropriate therapeutic agents.

Genetic tests are particularly useful in identification of bacterial pathogens because they overcome the need for bacterial growth in culture, which is slow and costly. Accurate and early diagnosis of bacterial disease can be life saving because appropriate treatments can be applied in a timely manner before infection becomes advanced. Recent studies have validated the PCR-OLA method as a rapid and accurate method for identifying quinolone-resistant strains of *Streptococcus pneumoniae*, a major cause of meningitis, bronchitis, and pneumonia ***(27)***. The technique has also been applied to *Helicobacter pylori* DNA isolated directly from gastric biopsy material for rapid and accurate identification of antibiotic resistance markers in ribosomal RNA genes of *H. pylori* ***(28)***.

3.3. Research Applications

Single-nucleotide polymorphisms (SNPs), which make up most of the genetic variation in the human genome, are single-nucleotide substitutions occurring on, average, 1 every 1000 basepairs along the human genome (see Chapter 19). Much research is now being directed toward association of specific SNPs with susceptibility to common diseases with the goal of predicting disease risk and individual response to therapy. These studies will require genotyping an extensive catalog of SNPs in large populations to identify disease associations. With the demand for SNP genotyping exploding, innovative technologies that facilitate accurate, simple, and high-throughput genotyping are essential. Oligonucleotide ligation has significant advantages in this regard. The method has successfully been applied to SNP genotyping using a semiautomated robotic workstation capable of processing 1200 samples per day, and it is esti-

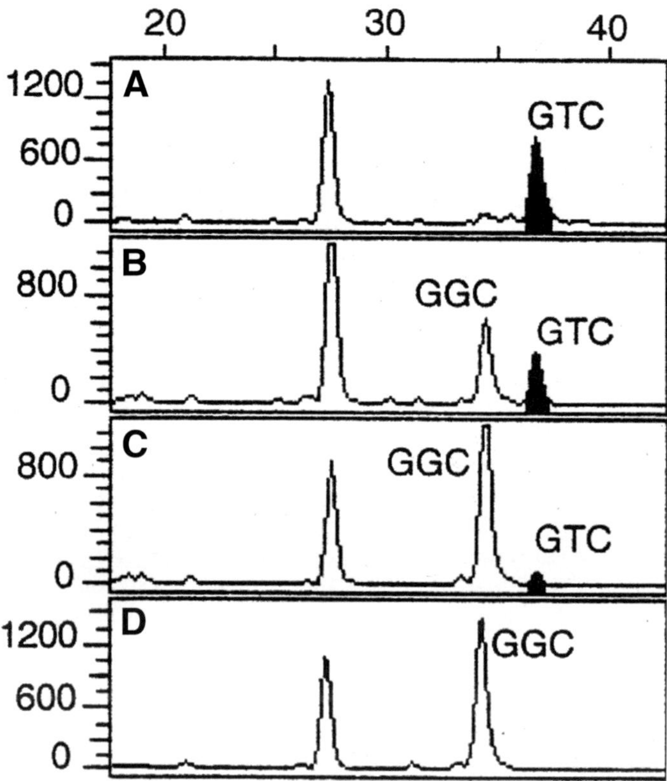

Fig. 4. Sensitivity of the ligation reaction. Genomic DNA extracted from T24 human bladder cancer cells was diluted with normal DNA and analyzed for *H-ras* mutation in multiplex amplification–ligation (CAL) reactions. **Panel A**: Undiluted T24 cell DNA, homozygous for a G→T transversion (GGC→GTC). T24 and wt DNA in a 1 : 10 ratio (**Panel B**), a 1 : 100 ratio (**Panel C**) and a 1 : 200 ratio (**Panel D**).

mated with the use of higher-density microtiter plates and multiplexing, the capacity could be increased to almost 10,000 assays per day ***(29,30)***.

3.4. The Future

The unique sensitivity and specificity of oligonucleotide ligation in discriminating single-nucleotide differences combined with its simplicity and flexibility make it likely that this technology will continue to develop new adaptations to keep pace with the ever-increasing need for gene-specific analytic strategies. Promising new applications include the use of padlock probes, oligonucleotide probes that circularize upon ligation, for *in situ* genotyping coupled with isothermal rolling circle amplification of these probes, thus circumventing the need for PCR amplification ***(13)***. A spectrum of new formats for detection of ligation products is also emerging that will provide increased sample throughput and rapid screening for mutation in large numbers of genes ***(31,32)***. Such developments include the addition of unique identifier sequences to ligation probes for detection by hybridization to oligonucleotide microarrays or fluoresecent microspheres containing complementary sequences ***(32)***. The pace of new innovations in gene-probing techniques will guarantee continued improvement in our ability to diagnose and treat disease as well as the elucidation of the molecular mechanisms underlying diseases.

References

1. Cotton, R. (1997) *Mutation Detection*, Oxford University Press, New York.
2. Mashal, R. and Sklar, J. (1996) Practical methods of mutation detection. *Curr. Opin. Genet. Dev.* **6**, 275–280.
3. Landegren, U. (1996) *Laboratory Protocols for Mutation Detection*. Oxford University Press, New York.
4. Landegren, U. (1993) Ligation-based DNA diagnostics. *BioEssays* **15**, 761–764.
5. Jarvius, J., Nilsson, M., and Landegren, U. (2003) Oligonucleotide ligation assay, in *Single Nucleotide Polymorphisms: Methods and Protocols* (Kwok, P.-Y., ed.), Humana, Totowa, NJ, pp. 215–228.
6. Barany, F. (1991) Genetic disease detection and DNA amplification using cloned thermostable ligase. *Proc. Natl. Acad. Sci. USA* **88**, 189–193.
7. Luo, J, Bergstrom, D. and Barany, F. (1996) Improving the fidelity of *Thermus thermophilus* DNA ligase. *Nucleic Acids Res.* **24**, 3071–3087.
8. Luo, J. and Barany, F. (1996) Identification of essential residues in *Thermus thermoophilus* DNA ligase. *Nucleic Acids Res.* **24**, 3079–3085.
9. Pritchard, C. and Southern, E. (1997) Effects of base mismatches on joining of short oligodeoxynucleotides by DNA ligases. *Nucleic Acids Res.* **25**, 3403–3407.
10. Wu, D. and Wallace, R. (1989) The ligation amplification reaction (LAR)-amplification of specific DNA sequences using sequential rounds of template-dependent ligation. *Genomics* **5**, 560–569.
11. Wiedmann, M., Wilson, W., Czajka, J., Luo, J., Barany, F., and Batt, C. (1994) Ligase chain reaction (LCR)-overview and applications. *PCR Methods Applic.* **3**, S51–S64.
12. Nilsson, M., Malmgre, H., Samiotaki, M., Kwiatkowski, M., Chowdhary, B., and Landegren, U. (1994) Padlock probes: circularizing oligonucleotides for localized DNA detection. *Science* **265**, 2085–2088.
13. Nilsson, M., Krejci, K., Koch, J., Kwiatkowski, M., Gustavsson, P., and Landegren, U. (1997) Padlock probes reveal single-nucleotide differences, parent of origin and *in situ* distribution of centromeric sequences in human chromosomes 13 and 21. *Nature Genet.* **16**, 252–255.
14. Baner, J., Nilsson, M., Mendel-Hartvig, M., and Landegren, U. (1998) Signal amplification of padlock probes by rolling circle replication. *Nucleic Acids Res.* **26**, 5073–5078.
15. Nickerson, D., Kaiser, R., Lappin, S., Stewart, J., Hood, L., and Landegren, U. (1990) Automated DNA diagnostics using an ELISA-based oligonucleotide ligation assay. *Proc. Natl. Acad. Sci. USA* **87**, 8923–8927.
16. Samiotaki, M., Kwiatkowski, M., Parik, J., and Landegren, U. (1994) Dual-color detection of DNA sequence variants through ligase-mediated analysis. *Genomics* **20**, 238–242.
17. Grossman, P., Bloch, W., Brinson, E., et al. (1994) High-density multiplex detection of nucleic acid sequences: oligonucleotide ligation assay and sequence-coded separation. *Nucleic Acids Res.* **22**, 4527–4534.
18. Eggerding, F., Iovannisci, D., Brinson, E., Grossman, P., and Winn-Deen, E. (1995) Fluorescence-based oligonucleotide ligation assay for analysis of cystic fibrosis transmembrane conductance regulator gene mutations. *Hum. Mutat.* **5**, 153–165.
19. Brinson, E., Adriano, T., Bloch, W., et al. (1997) introduction to PCR/OLA/SCS, a multiplex DNA test, and its application to cystic fibrosis. *Genet. Test.* **1**, 61–68.
20. Eggerding, F. (1995) A one-step coupled amplification and oligonucleotide ligation procedure for multiplex genetic typing. *PCR Methods Applic.* **4**, 337–345.
21. Landegren, U., Kaiser, R., Sanders, J., and Hood, L. (1988) A ligase-mediated gene detection technique. *Science* **241**, 1077–1080.
22. Day, N., Tadin, M., Christiano, A., Lanzano, P, Piomelli, S., and Brown, S. (2002) Rapid prenatal diagnosis of sickle cell diseases using oligonucleotide ligation assay coupled with laser-induced capillary fluorescence detection. *Prenat. Diagn.* **22**, 686–691.
23. Chakravarty, A., Hansen, T., Horder, M., and Kristensen, S. (1997) A fast and robust dual-label nonradioactive oligonucleotide ligation assay for detection of factor V Leiden. *Thromb. Haemost.* **78**, 1234–1236.
24. Eggerding, F. (2000) Fluorescent oligonucleotide ligation technology for identification of *ras* oncogene mutations. *Mol. Biotechnol.* **14**, 223–233.
25. Edelstein, R., Nickerson, D., Tobe, V., Manns-Arcuino, L., and Frenkel, L. (1998) Oligonucleotide ligation assay for detecting mutations in the human immunodeficiency virus type 1 *pol* gene that are associated with resistance to zidovudine, didanosine, and lamivudine. *J. Clin. Microbiol.* **36**, 569–572.

26. Beck, I., Mahalanabis, M., Pepper, G., et al. (2002) Rapid and sensitive oligonucleotide ligation assay for detection of mutations in human immunodeficiency virus type 1 associated with high-level resistance to protease inhibitors. *J. Clin. Microbiol.* **40**, 1413–1419.
27. Bui, M-H., Stone, G., Nilius, A., Almer, L., and Flamm, R. (2003) PCR-oligonucleotide ligation assay for detection of point mutations associated with quinolone resistance in *Streptococcus pneumoniae. Antimicrob. Agents Chemother.* **47**, 1456–1459.
28. Stone, G., Shortridge, D., Versalovic, J., et al. (1997) A PCR-oligonucleotide ligation assay to determine the prevalence of 23S rRNA gene mutations in clarithromycin-resistant *Helicobacter pylori. Antimicrob. Agents Chnemother.* **41**, 712–714.
29. Delahunty, C., Ankener, J. W., Deng, Q., Eng, J., and Nickerson, D. (1996) Testing the feasibility of DNA typing for human identification by PCR and an oligonucleotide ligation assay. *Am. J. Hum. Genet.* **58**, 1239–1246.
30. Tobe, V., Taylor, S., and Nickerson D. (1996) Single-well genotyping of diallelic sequence variations by a two-color ELISA-based oligonucleotide ligation assay. *Nucleic Acids Res.* **24**, 3728–3732.
31. Chen, X., Livak, K., and Kwok, P.-Y. (1998) A homogeneous, ligase-mediated DNA diagnostic test. *Genome Res.* **8**, 549–556.
32. Iannone, M., Taylor J., Chen, J., et al. (2000) Multiplexed single nucleotide polymorphism genotyping by oligonucleotide ligation and flow cytometry. *Cytometry* **39**, 131–140.

23

Quantitative TaqMan Real-Time PCR

Diagnostic and Scientific Applications

Jörg Dötsch, Ellen Schoof, and Wolfgang Rascher

1. Introduction

The invention of real-time polymerase chair reaction (PCR) has revolutionized the quantification of gene expression and DNA copy number measurements. However, after the first documentation of real-time PCR in 1993 *(1)*, it took several years for this method to become a mainstream tool. PCR generates DNA copies in an exponential way. As soon as resources are exhausted, however, the so-called plateau phase of PCR reaction is reached, making quantification very unreliable. Therefore, quantification appears most reliable in the early exponential phase of PCR (i.e., in a "real-time" fashion). To ensure measurements in this phase of the PCR cycle, real-time PCR measures as soon as the threshold of detection is definitely reached. The cycle of PCR at which this occurs is then named the threshold cycle *(2)* (*see* **Fig. 1**).

It is the objective of this chapter to describe the possibilities of TaqMan real-time PCR for mRNA and DNA quantification and to discuss pitfalls and alternatives.

2. Principles of TaqMan Real-Time PCR

The use of the TaqMan reaction has been described in a number of original and review articles *(3–5)*. This approach makes use of the 5' exonuclease activity of the DNA polymerase (AmpliTaq Gold). Briefly, within the amplicon defined by a gene-specific PCR primer pair, an oligonucleotide probe labeled with two fluorescent dyes is created, designated as the TaqMan probe. As long as the probe is intact, the emission of the reporter dye (i.e., 6-carboxy-fluorescein, FAM) at the 5' end is quenched by the second fluorescence dye (6-carboxy-tetramethyl-rhodamine, TAMRA) at the 3' end. During the extension phase of PCR, the polymerase cleaves the TaqMan probe, resulting in a release of reporter dye. The increasing amount of reporter dye emission is detected by an automated sequence detector combined with a dedicated software (ABI Prism 7700 Sequence Detection System, Perkin-Elmer, Foster City, CA). The algorithm normalizes the reporter signal (Rn) to a passive reference. Next, the algorithm multiplies the standard deviation of the background Rn in the first few cycles (in most PCR systems, cycles 3–15, respectively) by a default factor of 10 to determine a threshold. The cycle at which this baseline level is exceeded is defined as the threshold cycle (Ct) (*see* **Fig. 1**). Ct has a linear relation with the logarithm of the initial template copy number. Its absolute value additionally depends on the efficiency of both DNA amplification and cleavage of the TaqMan probe. The Ct values of the samples are interpolated to an external reference curve constructed by plotting the relative or absolute amounts of a serial dilution of a known template vs the corresponding Ct values.

From: *Medical Biomethods Handbook*
Edited by: J. M. Walker and R. Rapley © Humana Press, Inc., Totowa, NJ

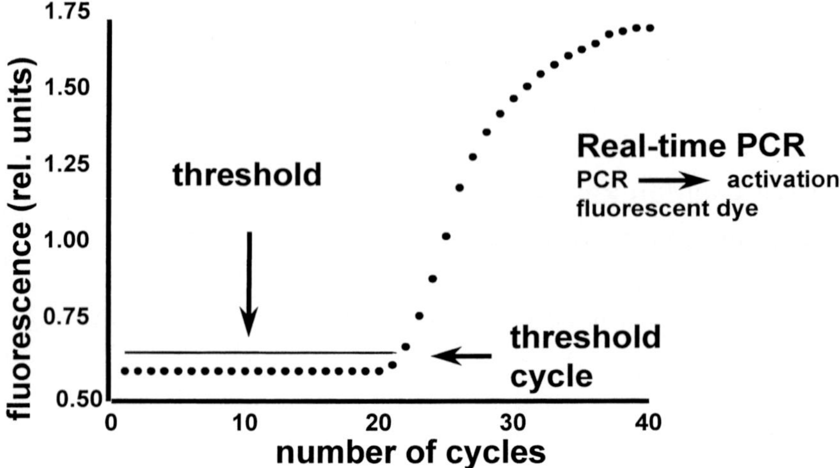

Fig. 1. Sketch of the principle of TaqMan PCR for the quantification of gene expression. By measuring the amplicon concentration in the early exponential phase of the PCR reaction, the exhaustion of reagents is avoided. (Modified from **ref. *3***.)

The oligonucleotides of each target of interest can be designed by the Primer Express software (Perkin-Elmer) using uniform selection parameters.

3. Reliability and Validation of TaqMan Real-Time PCR

The use of TaqMan real-time PCR for the quantification of gene expression has been shown to be at least as reliable as the application other quantitative PCR techniques, like competitive PCR ***(6,7)***. Whereas the expression of highly expressed genes like the housekeeping gene glyceraldehyde-3-phosphate is well correlated between the two methods, for the determination of genes with lower expression like the neuropeptide Y TaqMan PCR is much more sensitive than competitive PCR ***(6)*** (*see* **Fig. 2**). In addition, the spectrum of linear measurements for real-time PCR is in the range of 10^6, in contrast to 10^2 in competitive RT-PCR. Finally, a considerably higher number of samples per day can be measured by real-time PCR (up to 400 measurements). In comparison to the Northern blot assessment, only a minimal fraction of mRNA is necessary to quantify gene expression by real-time PCR ***(8,9)***.

RNA can be extracted using standard techniques like commercial RNA isolation kits (e.g., RNAzol-B isolation kit; WAK-Chemie Medical GmbH, Bad Homburg, Germany) or conventional phenol–chloroform extraction for DNA ***(10)***.

4. Applications for TaqMan Real-Time PCR

A number of applications for TaqMan real-time PCR have been introduced in the last few years, the most important being the quantification of gene expression. Some of the most important applications and potential applications will be discussed below.

4.1. Quantification of Gene Expression

Quantification of gene expression has been facilitated to a considerable degree by the use of real-time techniques such as TaqMan PCR. This technique has practically replaced less sensitive and more time-consuming methods such as Northern blot or RNAse protection assay. Quantitative competitive PCR ***(7)*** is less effective and less sensitive as well ***(6)***.

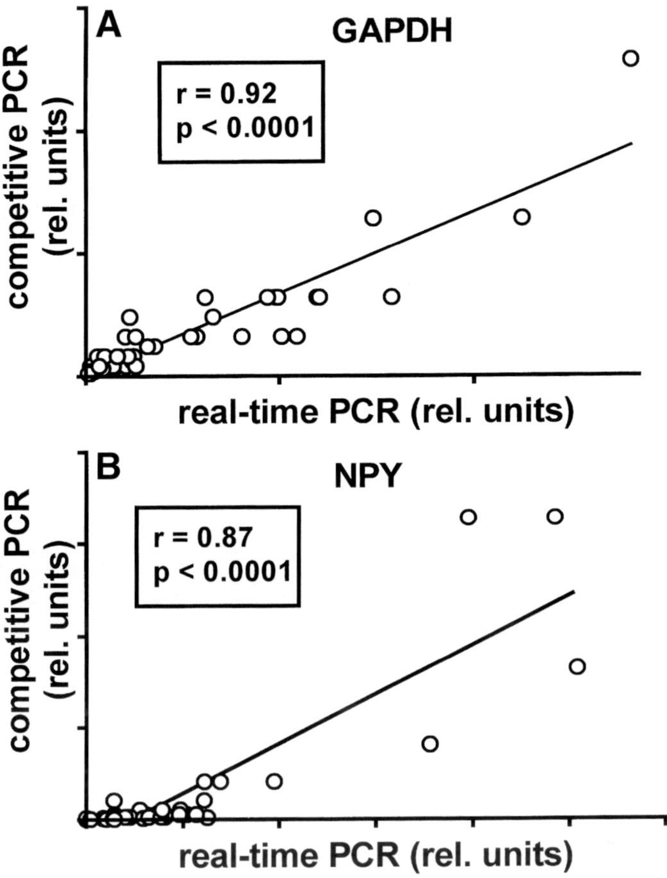

Fig. 2. Relation of gene expression as assessed by competitive quantitative and TaqMan real-time PCR. **(A)** mRNA expression of GAPDH. $r = 0.92$, $p < 0.001$. **(B)** mRNA expression of NPY. $r = 0.87$, $p < 0.001$ (Adapted from **ref. *6*.**)

Crucial aspects in the measurements of gene expression by real-time PCR are the preparation of RNA especially in samples that only contain small amounts of the specific mRNA, such as single-cell picking or microdissection (*see* **Subheading 7.**). On the other hand, quantification can be difficult; in general, housekeeping genes can be applied (e.g., by using duplex approaches). Alternatively, external standards can be considered ***(11)***.

In most cases of mRNA quantification, gene expression has to be related to housekeeping genes that are expressed relatively stable throughout the cells. In the studies performed on gene expression in neuroblastomas so far, three different housekeeping genes have been assessed: the more traditional genes glyceraldehyde-3-phosphate (GAPDH) ***(6)*** and β-actin ***(10)*** and the neuronal marker protein gene product 9.5 (PGP9.5) ***(6)***. One major general difficulty in the use of housekeeping genes for the determination of mRNA transcript ratios is the possibility that their expression might also be altered by coexpression of pseudogenes and environmental changes (e.g., by hypoxia in case of GAPDH) ***(12)***. Pseudogenes can be eliminated using primer combinations that are intron spanning. However, unidentified influences cannot be dealt with as easily. Therefore, we used at least two housekeeping genes for quantification of mRNA expression. However, this aspect clearly needs further evaluation. In our group, a number of primers and TaqMan probes have been used for housekeeping gene amplification ***(6,13)*** (**Table 1**).

Table 1
Primers and TaqMan Probes Used for the Quantification of Human Housekeeeping Gene Expression

β-Actin	Probe	CCAGCCATGTACGTTGCTATCCAGGC
	Forward	GCGAGAAGATGACCCAGGATC
	Reverse	CCAGTGGTACGGCCAGAGG
GAPDH	Probe	CCTCAACTACATGGTTTACATGTTCCAATATGATTCCAC
	Forward	GCCATCAATGACCCCTTCATT
	Reverse	TTGACGGTGCCATGGAATTT
PBGD	Probe	CTTCGCTGCATCGCTGAAAGGGC
	Forward	TGTGCTGCACGATCCCG
	Reverse	ACACTGCAGCCTCCTTCCAG
HPRT	Probe	CGCAGCCCTGGCGTCGTGATTA
	Forward	CCGGCTCCGTTATGGC
	Reverse	GGTCATAACCTGGTTCATCATCA
β2MG	Probe	TGATGCTGCTTACATGTCTCGATCCCA
	Forward	TGACTTTGTCACAGCCCAAGATA
	Reverse	CCAAATGCGGCATCTTC

Abbreviations: GAPDH: glyceraldehyd-3-phosphate dehydrogenase, PBGD: porphobilinogen deaminase, HPRT: hypoxanthine-guanine-phosphoribosyl-transferase, β2MG: β2-microglobulin.

Source: Data from **refs. *6*** and ***13–15***.

4.2. Quantification of Gene Copy Number, Determination of Minimal Residual Disease, and Allelic Discrimination in Malignant Tumors

4.2.1. MYCN Detection by TaqMan PCR in Neuroblastoma Tissue

DNA copy number, MYCN amplification in neuroblastomas, is of great potential interest with regard to the prognosis of disease. In fact, in clinical practice, MYCN gene expression correlates with both advanced disease stage ***(16)*** and rapid tumor progression ***(17)***.

Several methods have been used for the detection of MYCN detection mainly based on Southern or dot blot ***(18,19)***, on quantitative PCR ***(20)***, and on fluorescent *in situ* hybridization techniques ***(21)***. The use of most PCR methods is restricted, however, by the fact that end-point measurements are used for quantification. Therefore, Raggi and co-workers ***(10)*** introduced a TaqMan real-time base method for the determination of MYCN amplification in neuroblastomas. The authors demonstrate a precise assay with an interassay coefficient of variation of 13% and an intraassay coefficient of variation of 11%. The threshold cycle for the detection of MYCN correlates in an inverse linear way with the logarithm of the input of genomic DNA molecules. There is a good linear relationship between the MYCN amplification measured by TaqMan real-time PCR and competitive PCR. Using Kaplan–Meier survival curves, the authors showed that the amplification of MYCN as assessed by TaqMan real-time PCR is closely linked to cumulative survival, as this had already been demonstrated with several other techniques for the quantification of MYCN amplification.

4.2.2. Minimal Residual Disease

Another important aspect of real-time PCR in the field of oncology is the detection of minimal residual disease that is 100- to 1000-fold more sensitive than traditional methods. As few as five copies can be detected in one reaction ***(11)***, but the maximal input of DNA during sample preparation is the limiting step. This limitation must, therefore, be considered when looking for minimal residual disease in malignant diseases such as childhood or adulthood acute lymphoblastic leukemia ***(22,23)***.

4.2.3. Allelic Discrimination and Haploinsufficiency

Various methods have been used to study tumor cytogenetic aberrations in malignant tumors for clinical decision-making. Initially, conventional cytogenetic techniques were applied. Karyotyping by conventional cytogenetics, however, depends on dividing cells, and successful evaluations are often hampered by inferior metaphase quality. Following this, restriction fragment length polymorphisms (RFLPs), PCR-based microsatellite, and fluorescence *in situ* hybridization (FISH) analyses have been applied to overcome the restrictions of conventional cytogenetics. However, RFLP- and PCR-based microsatellite analyses depend on informative loci in the region of interest and the need for normal reference DNA of the respective patient, whereas FISH analyses are sometimes hampered by inferior tissue quality and hybridization probe availability. The latest technique used for routine detection of cytogenetic aberrations is comparative genomic hybridization CGH *(24)*. CGH has proved to be consecutively applicable to tumor specimens for the detection of most aberrations known from conventional cytogenetics, but deletions of smaller DNA regions might be undetectable. Thus, this technique should be used to prescreen tumor samples for gross cytogenetic aberrations and be supplemented by a TaqMan PCR-based approach to detect loss or gain of DNA on the single-gene level.

4.3. Pathogene Detection and Quantification

Real-time PCR can be of great benefit in the detection and quantification of pathogens such as viruses, bacteria, and fungi. The method proves to be useful in the use for environmental detection *(25)* and in infected patients.

Several difficulties, however, can occur *(11)*. In contrast to traditional methods of microbiology, vital and dead microorganisms are both detected. Second, most infectious organisms are characterized by a high mutation rate, which might influence the estimation of viral or bacterial load dramatically *(26)*. Finally, quantification necessitates the use of reliable standards. This can be achieved by using duplex or multiplex assays *(27)*. To assure permanent quality and the possibility of comparing results, international standardization will have to be obtained in the future.

5. Limitations and Pitfalls in the Use of TaqMan Real-Time PCR

The use of TaqMan PCR can be particularly difficult if gene expression at a low level is to be quantified. One major pitfall in this context is the accidental determination of genomic DNA when RT-PCR is intended. There are various approaches to meet this problem: It is always useful to select primer combinations that are intron spanning *(2)*. If there is no possibility to select intron-spanning primers, RNA samples can be pretreated with DNAse. However, this measure should not be chosen routinely and can also be deleterious if only small amounts of RNA are present that are partially destroyed as well *(28)*.

One other difficulty is the high degree of technical expertise that is required to achieve as low a variation coefficient and as sensitive a measurement as possible. It could be shown that the degree of technical expertise can alter the gene level that is measured by up to 1000-fold *(28)*.

Because real-time PCR is highly sensitive, the risk of having interference with minor contamination is quite considerable. On the other hand, the risk of false-negative results must not be underestimated because post-hoc PCR steps are not "visible" to the degree that is provided by the more traditional methods for gene quantification *(11)*.

6. Alternative Real-Time PCR Methods

There are other methods for real-time PCR not relying on exonuclease cleavage of a specific probe to generate a fluorescence signal. One of them, the LightCycler System (Roche Molecular), makes use of so-called "hybridization probes" *(29 , 30)*. Like exonuclease probes, hybridization probes are used in addition to the PCR primers. However, unlike the first, hybridization probes combine two different fluorescent labels to allow resonance energy transfer. One of them is activated by external light. When both probes bind very closely at the DNA molecules generated by the PCR-amplification process, the emitted light from the first dye activates the fluorescent dye of the second probe. This second dye emits light with a longer wavelength, which is measured every cycle. Thus, the fluorescence intensity is directly correlated to the extent of probe hybridization and, subsequently, directly related to the amount of PCR product. A major advantage of the LightCycler is the very short time of the PCR run. However, the TaqMan system allows one to analyze a higher number of samples at one time and, at least theoretically, might be more accurate, as there is an internal reference dye to monitor minor variations in sample preparation.

Apart from exonuclease and hybridization methods for real-time PCR, there are other options, including hairpin probes, hairpin primers, and intercalating fluorescent dyes. Hairpin probes, also known as molecular beacons, contain reverse complement sequences at both ends binding together while the rest of the strand remains single stranded, creating a panhandlelike structure. In addition, there are fluorescent dyes at both ends of the molecule: a reporter and a quencher similar to the TaqMan probes. In the panhandlelike conformation, there is no fluorescence, as the fluorescent reporter at one end and the quencher at the other end of the probe are very close to one another ***(31)***. As the central part of a molecular beacon consists of a target-specific sequence, both ends are separated from each other when this part of the molecule is bound to the PCR product and a fluorescence signal can be emitted from the reporter dye. Hairpin primers, also named "amplifluor primers," are similar to molecular beacons, but fluorescence is generated as they become incorporated into the double-stranded PCR product during amplification. Another very simple technique for monitoring the generation of PCR product in a real-time fashion is the use of intercalating dyes, such as EtBr and SYBR green I, which do not bind to single-stranded DNA but to the double-stranded PCR product ***(30,32)***. However, hairpin primers and intercalating dyes do not offer the high specificity of the probe-based techniques and a positive signal might even be generated by primer dimers.

7. Future Developments

From a *diagnostic point of view*, one interesting aspect for the future might be the use of semiautomated or automated real-time devices for the assessment of gene expression and amplification, putting into consideration the relatively easy and low time-consuming method of measurement. For diseases such as neuroblastoma, real-time PCR might help to quantify more prognostic markers like the nerve growth factor receptor (TRKA gene) ***(33)***, the expression of genes involved in multidrug resistance (*MDR1* and *MRP*) ***(34,35)***, and genes related to tumor invasion and metastasis (*nm23* and *CD44*) ***(36,37)***. There are first reports on the use of multiplex real-time PCR, a development that is certainly going to facilitate diagnostic procedure in the next 5 yr ***(38,39)***.

From a *research point of view*, real-time PCR might facilitate the identification of new prognostic markers, because a large number of samples can be processed in a relatively short time ***(40,41)***. Another future application of TaqMan PCR is the confirmation of results obtained by cDNA microarrays, which will be abundantly used for cDNA screening. Of particular interest might be the chance to determine gene expression in very few cells using single-cell picking ***(42,43)***. Using this approach, not only can minimal involvement of tumor cells be visualized but also nonhomogenous distributions in malignant tumor might be monitored with respect to essential prognostic markers. It is of importance that *in situ* gene expression after laser capture microdissection might not only be performed in frozen sections but also from formalin-fixed and paraffin-embedded biopsies ***(44)***.

8. Conclusions

TaqMan real-time PCR provides a reliable technology for the quantification of gene expression. However, a number of preconditions have to be met for each new marker. First, the system parameters reflecting amplification efficiency (slope) and linearity should match the minimal requirements. This should include the calculation of the intra-assay and interassay coefficient of variation with regard to the threshold cycle at which the amplification signal is detected. Second, the assay itself should be carefully evaluated with regard to a linear relationship between threshold cycle and the logarithm of a serial dilution of a reference sample. Third, real-time PCR results should be compared with a second, independent method for quantification, like quantitative competitive PCR. Finally, it should be assessed whether the results obtained with regard to clinical outcome represent the experiences obtained with other methods for gene detection.

Apart from a high degree of precision, practical advantages of real-time PCR are easy handling, rapid measurements, and a broad linear range for the measurements. Whereas competitive PCR only allows for the determination of few samples in one assay, TaqMan real-time PCR provides the opportunity to measure more than 80 samples at one time in a 96-well plate together with the control reactions needed.

References

1. Higuchi, R., Fockler, C., Dollinger, G., and Watson R. (1993) Kinetic PCR analysis: real-time monitoring of DNA amplification reactions. *Biotechnology* **11**, 1026–1030.
2. Ginzinger, D. G. (1993) Gene quantification using real-time quantitative PCR: an emerging technology hits the mainstream. *Exp. Hematol.* **30**, 503–512.
3. Dötsch, J., Repp, R., Rascher, W., and Christiansen, H. (2001) Diagnostic and scientific applications of TaqMan real-time PCR in neuroblastomas. *Expert Rev. Mol. Diagn.* **1**, 233–238.
4. Dötsch, J., Nüsken, K. D., Knerr, I., Kirschbaum, M., Repp, R., and Rascher, W. (1999) Leptin and neuropeptide Y gene expression in human placenta: ontogeny and evidence for similarities to hypothalamic regulation. *J. Clin. Endocrinol. Metab.* **84**, 2755–2758.
5. Orlando, C., Pinzani, P., and Pazzaggli, M. (1998) Developments in quantitative PCR. *Clin. Chem. Lab. Med.* **36**, 255–269.
6. Dötsch, J., Harmjanz, A., Christiansen, H., Hänze, J., Lampert, F., and Rascher, W. (2000) Gene expression of neuronal nitric oxide synthase and adrenomedullin in human neuroblastoma using real-time PCR. *Int. J. Cancer* **88,** 172–175.
7. Förster, E. (1994) Rapid generation of internal standards for competitive PCR by low-stringency primer annealing. *BioTechniques* **16**, 1006–1008.
8. Heid, C. A., Stevens, J., Livak, K. J., and Williams, P. M. (1996) Real-time quantitative PCR. *Genome Res.* **6**, 986–994.
9. Gibson, U. E., Heid, C. A., and Williams, P. M. (1996) A novel method for real-time quantitative RT-PCR. *Genome Res.* **6**, 995–1001.
10. Raggi, C. C., Bagnoni, M. L., Tonini, G. P., et al. (1999) Real-time quantitative PCR for the measurement of MYCN amplification in human neuroblastoma with the TaqMan detection system. *Clin. Chem.* **45**, 1918–1924.
11. Klein, D. (2002) Quantification using real-time PCR technology: applications and limitations. *Trends Mol. Med.* **8**, 257–260.
12. Bustin, S. A. (2000) Absolute quantification of mRNA using real-time reverse transcription polymerase chain reaction assays. *J. Mol. Endocrinol.* **25,** 16–193.
13. Schoof, E., Girstl, M., Frobenius, W., et al. (2001) Decreased gene expression of 11beta-hydroxysteroid dehydrogenase type 2 and 15-hydroxyprostaglandin dehydrogenase in human placenta of patients with preeclampsia. *J. Clin. Endocrinol. Metab.* **86**, 1313–1317.
14. Schoof, E., Stuppy, A., Harig, F., et al. (2003) No influence of surgical stress on postoperative leptin gene expression in different adipose tissues and soluble leptin receptor plasma levels. *Horm. Res.* **59**, 184–190.
15. Trollmann, R., Amann, K., Schoof, E., et al. (2003) Hypoxia activates the human placental vascular endothelial growth factor system in vitro and in vivo: up-regulation of vascular endothelial growth factor in clinically relevant hypoxic ischemia in birth asphyxia. *Am. J. Obstet. Gynecol.* **188**, 517–523.
16. Brodeur, G. M., Seeger, R. C., Schwab, M., Varmus, H. E., and Bishop, J. M. (1984) Amplification of N-myc in untreated human neuroblastomas correlates with advanced disease stage. *Science* **224**, 1121–1124.
17. Seeger, R. C., Brodeur, G. M., Sather, H., et al. (1985) Association of multiple copies of the N-myc oncogene with rapid progression of neuroblastomas. *N. Engl. J. Med.* **313**, 1111–1116.

18. Pession, A., Trere, D., Perri, P., et al. (1997) N-myc amplification and cell proliferation rate in human neuroblastoma. *J. Pathol.* **183**, 339–344.
19. De Cremoux, P., Thioux, M., Peter, M., et al. (1997) Polymerase chain reaction compared with dot blotting for the determination of N-myc gene amplification in neuroblastoma. *Int. J. Cancer* **72,** 518–521.
20. Sestini, R., Orlando, C., Zentilin, L., et al. (1995) Gene amplification for c-erbB-2, c-myc, epidermal growth factor receptor, int-2, and N-myc measured by quantitative PCR with a multiple competitor template. *Clin. Chem.* **41**, 826–832.
21. Shapiro, D. N., Valentine, M. B., Rowe, S. T., et al. (1993) Detection of N-myc gene amplification by fluorescence in situ hybridization. Diagnostic utility for neuroblastoma. *Am. J. Pathol.* **142,** 1339–1346.
22. Kwan, E., Norris, M. D., Zhu, L., Ferrara, D., Marshall, G. M., and Haber, M. (2000) Simultaneous detection and quantification of minimal residual disease in childhood acute lymphoblastic leukaemia using real-time polymerase chain reaction. *Br. J. Haematol.* **109**, 430–434.
23. Cilloni, D., Gottardi, E., De Micheli, D., et al. (2002) Quantitative assessment of WT1 expression by real-time quantitative PCR may be a useful tool for monitoring minimal residual disease in acute leukemia patients. *Leukemia* **16,** 2115–2121.
24. Brinkschmidt, C., Christiansen, H., Terpe, H. J., et al. (1997) Comparative genomic hybridization (CGH) analysis of neuroblastomas—an important methodological approach in pediatric tumour pathology. *J. Pathol.* **181**, 394–400.
25. Brunk, C. F., Li, J., Avaniss-Aghajani, E. (2002) Analysis of specific bacteria from environmental samples using a quantitative polymerase chain reaction. *Curr Issues Mol Bio*l. **4,** 13—18.
26. Klein, D., Leutenegger, C. M., Bahula, C., et al. (2001) Influence of preassay and sequence variations on viral load determination by a multiplex real-time reverse transcriptase-polymerase chain reaction for feline immunodeficiency virus. *J. Acquired Immune Defic. Syndr.* **26**, 8–20.
27. Gruber, F., Falkner, F. G., Dorner, F., and Hammerle, T. (2001) Quantitation of viral DNA by real-time PCR applying duplex amplification, internal standardization, and two-color fluorescence detection. *Appl. Environ. Microbiol.* **67**, 2837–2839.
28. Bustin S. A. (2002) Quantification of mRNA using real-time reverse transcription PCR (RT-PCR): trends and problems. *J. Mol. Endocrinol.* **29**, 23–39.
29. Wittwer, C., Herrmann, M., Moss, A., and Rasmussen, R. (1997) Continuous fluorescence monitoring of rapid cycle DNA amplification. *BioTechniques* **22**, 130–138.
30. Giulietti, A., Overbergh, L., Valckx, D., Decallonne, B., Bouillon, R., and Mathieu, C. (2001) An overview of real-time quantitative PCR: applications to quantify cytokine gene expression. *Methods* **25**, 386–401.
31. Tyagi, S. and Kramer, R. (1996) Molecular beacons: probes that fluorescence upon hybridization. *Nature Biotechnol.* **14**, 303–308.
32. Zubritsky, E. (1999) Widespread interest in gene quantification and high-throughput assays are putting quantitative PCR back in the spotlight. *Anal. Chem.* **71**, 191A–195A.
33. Nakagawara, A., Arima-Nakagawara, M., Scavarda, N. J., Azar, C. G., Cantor, A. B., and Brodeur, G. M. (1993) Association between high levels of expression of the TRK gene and favorable outcome in human neuroblastoma. *N. Engl. J. Med.* **328**, 847–854.
34. Chan, H. S., Haddad, G., Thorner, P. S., et al. (1991) P-Glycoprotein expression as a predictor of the outcome of therapy for neuroblastoma. *N. Engl. J. Med.* **325**, 1608–1614.
35. Norris, M. D., Bordow, S. B., Marshall, G. M., Haber, P. S., Cohn, S. L., and Haber, M. (1996) Expression of the gene for multidrug-resistance-associated protein and outcome in patients with neuroblastoma. *N. Engl. J. Med.* **334**, 231–238.
36. Leone, A., Seeger, R. C., Hong, C. M., et al. (1993) Evidence for nm23 RNA overexpression, DNA amplification and mutation in aggressive childhood neuroblastomas. *Oncogene* **8**, 855–865.
37. Favrot, M. C., Combaret, V., and Lasset, C. (1993) CD44—a new prognostic marker for neuroblastoma. *N. Engl. J. Med.* **329**, 1965.
38. Hahn, S., Zhong, X. Y., Burk, M. R., Troeger, C., and Holzgreve, W. (2000) Multiplex and real-time quantitative PCR on fetal DNA in maternal plasma. A comparison with fetal cells isolated from maternal blood. *Ann. NY Acad. Sci.* **906**, 148–152.
39. Von Ahsen, N., Oellerich, M., and Schutz, E. (2000) A method for homogeneous color-compensated genotyping of factor V (G1691A) and methylenetetrahydrofolate reductase (C677T) mutations using real-time multiplex fluorescence PCR. *Clin. Biochem.* **33**, 535–539.

40. Norris, M. D., Burkhart, C. A., Marshall, G. M., Weiss, W. A., and Haber, M. (2000) Expression of N-myc and MRP genes and their relationship to N-myc gene dosage and tumor formation in a murine neuroblastoma model. *Med. Pediatr. Oncol.* **35**, 585–589.
41. Kawamoto, T., Shishikura, T., Ohira, M., et al. (2000) Association between favorable neuroblastoma and high expression of the novel metalloproteinase gene, nbla3145/XCE, cloned by differential screening of the full-length-enriched oligo-capping neuroblastoma cDNA libraries. *Med. Pediatr. Oncol.* **35**, 628–631.
42. Fink, L., Seeger, W., Ermert, L., et al. (1998) Real-time quantitative RT-PCR after laser-assisted cell picking. *Nature Med.* **4**, 1329–1333.
43. von der Hardt, K., Kandler, M. A., Fink, L., et al. Laser-assisted microdissection and real-time PCR detect anti-inflammatory effect of perfluorocarbon. *Am. J. Physiol. Lung Cell Mol. Physiol.* **285**, L55–L62.
44. Lehmann, U. and Kreipe, H. (2001) Real-time PCR analysis of DNA and RNA extracted from formalin-fixed and paraffin-embedded biopsies. *Methods* **2**, 409–418.

24

Use of Denaturing High-Performance Liquid Chromatography in Molecular Medicine

Elizabeth L. Rugg and Gareth J. Magee

1. Introduction

The molecular diagnosis of hereditary and somatic disorders is a rapidly expanding field in modern medicine. Mutations in more than 1500 genes have been found to be associated with human diseases,* and this figure is likely to rise significantly over the next few years as the genetic basis of more conditions becomes known. DNA sequencing is still probably the most widely used and reliable method of detecting gene mutations and sequence variations. However, it is expensive and time-consuming, and, increasingly, other methods are employed to screen DNA fragments for sequence variations. For a screening method to be useful, it must be fast, inexpensive and applicable to most genes. Ideally, it needs to provide information about the nature and position of a mutation and to minimize exposure of laboratory staff to hazardous reagents. Many of the protocols currently used fit some of these criteria, but there are few, if any, that fulfill all.

Mutation detection can involve the confirmation of predetermined mutations or the identification of unknown genetic lesions. For some conditions, a small number of mutations are responsible for the majority of disease cases and screens are often used to identify these common variations. For other conditions, most mutations might be specific to each case and it might be necessary to screen multiple genes to be sure of identifying the pathogenic lesion. The type of mutation is an important consideration in deciding the best method of mutation detection. For example, some congenital disorders and cancers are caused by large chromosomal deletions or rearrangements and techniques such as fluorescence *in situ* hybridization would be the method of choice rather than DNA sequencing. Many inherited disorders result from single-nucleotide substitutions or small insertions/deletions and methods employed to identify these must be sensitive enough to detect these small alterations.

Denaturing high-performance liquid chromatography (DHPLC) is one of many techniques that have been developed to screen genes for sequence variations. Depending on the mode of operation, DHPLC can be used to detect single-nucleotide substitutions or small insertions or deletions in double-stranded DNA fragments as well as analyze and purify single-stranded nucleic acids [reviewed by Xiao and Oefner, 2001 ***(1)***]. Although DHPLC can identify DNA fragments that contain sequence variations, additional methods (usually DNA sequencing) are required to confirm the precise nature of the mutation.

Denaturing HPLC is a form of ion-pair reversed-phased chromatography in which nucleic acids can be bound to a hydrophobic column [usually poly(styrene-divinylbenzene particles)]

---

*Estimated from information available at http://www.ncbi.nlm.nih.gov/LocusLink.

From: *Medical Biomethods Handbook*
Edited by: J. M. Walker and R. Rapley © Humana Press, Inc., Totowa, NJ

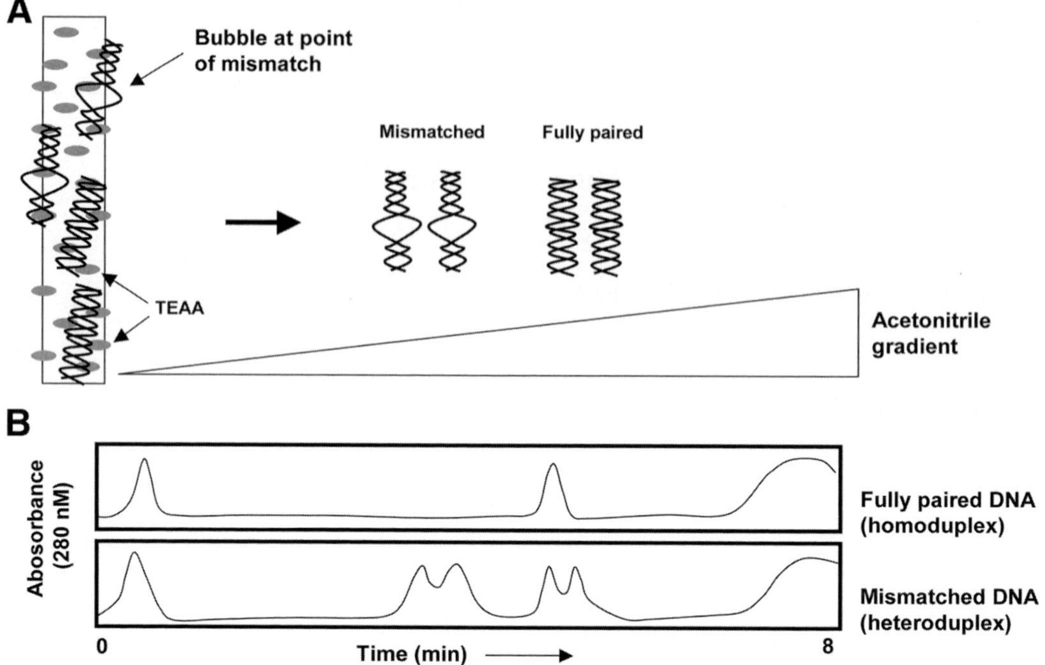

Fig. 1. Schematic diagram of the DHPLC analysis. **(A)** The double-stranded DNA is bound to a hydrophobic column through an intermediary such as TEAA. Fully paired DNA binds more efficiently than DNA containing a mismatch. As an increasing gradient of an organic solvent such as acetonitrile is flushed over the column, the TEAA is removed and the DNA eluted. DNA containing the mismatch is eluted earlier in the gradient than fully paired DNA. **(B)** Schematic diagram of the DHPLC profile generated by a fully paired sample (*top*) that elutes as a single peak and a fully resolved sample containing a DNA mismatch (*bottom*). In the lower trace, the two mismatched alleles can be seen to the left as a double peak, with the two homoduplex species resolving as a doublet to the right.

via an intermediary such as triethylammonium acetate (TEAA). The affinity of this interaction is dependent on size, nucleotide composition, and column temperature. An increasing gradient of an organic solvent, usually acetonitrile, is passed over the column and gradually removes the TEAA, leading to the release of bound DNA (or RNA) from the column. After being eluted, the DNA can be detected via ultraviolet (UV) absorbance or, if the DNA has been labeled, fluorescence. Under nondenaturing conditions (i.e., column temperatures at which double-stranded DNA remains fully paired), the interaction is almost completely dependent on fragment size. Working in this mode, it is possible to accurately size and quantify PCR fragments *(2,3)*. At high column temperatures, double-stranded DNA is completely denatured into single strands and elution from the column is dependent on both size and sequence. Oefner demonstrated that for short fragments (< 30 nucleotides), it is possible to separate all oligonucleotides differing by a single base and that for larger fragments (up to at least 60 nucleotides), all possible substitutions can be detected with the exception of a C to G transversion *(4)*. For most purposes, DHPLC is performed under partially denaturing conditions. In this mode, a temperature is chosen at which double-stranded DNA containing a mismatch starts to dissociate while completely matched DNA remains as a double strand. DNA containing a mismatch has a slightly lower affinity for the column because it makes fewer ion-paring bonds with the hydrophobic column and will elute before the fully paired DNA (*see* **Fig. 1**). Typically, single-nucleotide

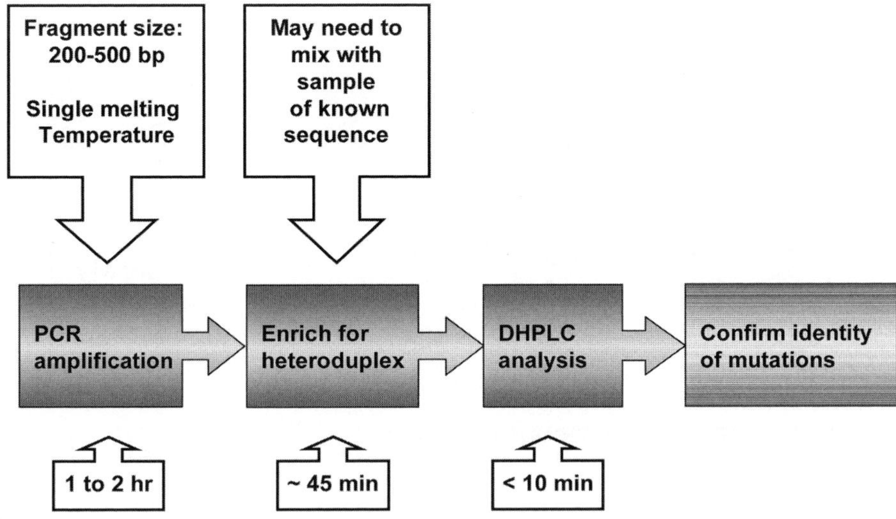

Fig. 2. The steps involved in DHPLC analysis of a gene fragment. The time for analysis varies depending on the sample, but the complete process, including PCR amplification, normally takes about 3 h with the actual DHPLC analysis taking less than 10 min per sample. The identity of a mutation is usually determined by DNA sequencing.

mismatches can be reliably detected in DNA fragments of 300–400 nucleotides *(5,6)*, although much higher detection limits (up to 1500 bp) have been reported *(7)*.

2. DHPLC Analysis: Practical Considerations

2.1. Amplification of DNA Fragments

The basic steps involved in DHPLC analysis are outlined in **Fig. 2**. The first step is polymerase chain reaction (PCR) amplification of DNA fragments from the gene of interest (see Chapter 6). The sensitivity of DHPLC is dependent on fragment size and nucleotide composition. Ideally, the fragments should be between 200 and 500 basepairs (bp) and have a single-domain melting temperature. The melting temperature is the temperature at which double-stranded DNA dissociates into single strands. DHPLC analysis relies on the fact that the melting temperature of mismatched (heteroduplex) DNA is lower than correctly paired (homoduplex) DNA. The temperature for the analysis should be such that the DNA melts at the point of the mismatch to produce a single-stranded "bubble," but the temperature is below that at which the whole DNA fragment melts (*see* **Fig. 1**). The melting temperature for a particular fragment can be predicted using software applications such Wavemaker™ (Transgenomic Ltd, Crewe, UK) or Melt (http://insertion.stanford.edu./melt.html). Empirical testing shows that not all mutations are resolved by analysis at the predicted melting temperature. However, a number of studies suggest that all mutations can be detected by repeat analysis at 2°C above and below the predicted melting temperature *(8,9)*. In practice, this means that samples that elute as a single peak should be reanalyzed at different temperatures to be sure of identifying all sequence variations.

The choice of the PCR enzyme is important in order to minimize the introduction of sequence variations during PCR amplification, and a polymerase with proofreading capabilities should always be used. In addition, DHPLC columns are expensive and some PCR buffer additives can be detrimental and reduce the life-span of the column. Polymerases such as Optimase™

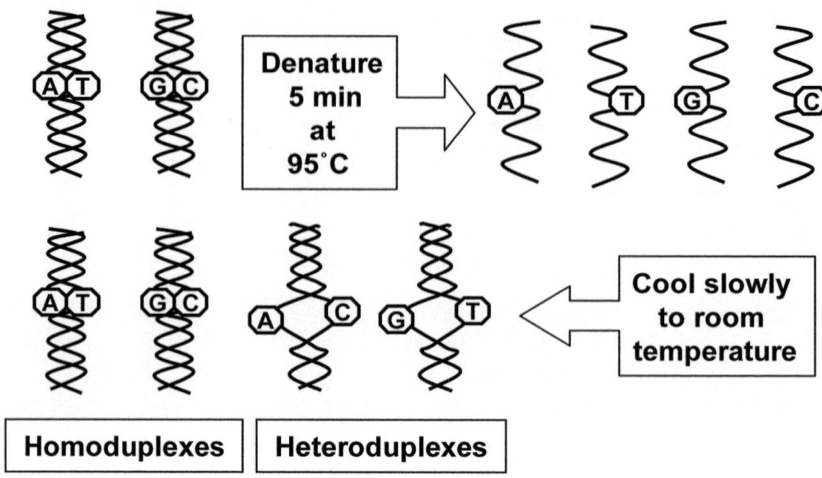

Fig. 3. A schematic diagram showing the process of heteroduplex enrichment. This step is important to maximize the chances of detecting sequence variations. PCR amplification generates homoduplexes and it is important that these be completely dissociated and heteroduplexes allowed to form.

(Transgenomic Ltd, Crewe, UK), Amplitaq Gold (Applied Biosystems, Warrington, UK), and Hot StarTaq (Qiagen, Crawley, UK) are compatible with the Wave™ system. Oil should never be used with samples destined for DHPLC analysis.

2.2. Heteroduplex Enrichment

To maximize the chances of detecting a sequence variation, it is necessary to enrich the sample for heteroduplex DNA. This is achieved by completely denaturing the PCR fragment and then allowing it to slowly reanneal (*see* **Fig. 3**). The sample will then comprise equal amounts of four species; two homoduplexes and two heteroduplexes. In cases of recessive disease, where the sample is expected to be homozygous for a mutation, heteroduplexes can be artificially produced by mixing the PCR fragment from the patient sample with an equal amount of PCR fragment generated from wild-type DNA.

2.3. Equipment and Running Costs

The WAVE system (Transgenomic, Ltd, Crewe, UK) is currently the only commercial system available that has been specifically developed for DHPLC analysis. After an initial outlay for the purchase of the equipment, there are few running costs, and, importantly, no further treatment of PCR-amplified DNA fragments is necessary before analysis, other than enriching for heteroduplexes. It has been estimated that DHPLC analysis of a PCR fragment is around 8 times faster and 10 times less expensive than direct sequencing *(5)*.

2.4. DHPLC Analysis of "Nonideal" Fragments

Although one should try to meet the optimum conditions for DHPLC, this is not always possible. For example, it might not be possible to generate a PCR fragment of ideal size or with a single melting domain. In such cases, empirical testing is required to determine if DHPLC is suitable for a particular analysis. Fragments with multiple melting domains can be analyzed but often require DHPLC analysis at more than one temperature. An example of the successful analysis of a "difficult" sample is illustrated in **Subheading 2.5.**

2.5. A Case Study: DHPLC Identification of Sequence Variations in KRT 14

Mutations in *KRT14*, the gene encoding keratin 14, cause the inherited blistering disorder *epidermolysis bullosa simplex*. This can be a severe condition and there is demand from patients for prenatal diagnosis. Approximately 25% of mutations causing the severest disease are single-nucleotide substitutions in exon 1 of *KRT14*, with the remaining pathogenic mutations occurring in other regions of *KRT14* or another keratin gene *KRT5*. A screen that is able to rapidly identify sequence variation in exon 1 of *KRT14* would speed up the identification of pathogenic mutations and reduce the amount of DNA sequencing required. Specific amplification of exon 1 of *KRT14* is complicated by the existence of a pseudogene with high sequence homology to the expressed gene, which restricts the choice of PCR primers. Previous studies have demonstrated that a 967-bp fragment can specifically be amplified without amplification of the pseudogene; however, this fragment contains a number of polymorphic sequence variations that preclude screening by established methods such as a single-strand conformation polymorphism analysis ***(10)***. In addition, analysis of the DNA sequence of this fragment using Wavemaker™ software suggests that there are at least three different melting temperature domains of approximately 60°C, 63°C, and 65°C within this fragment (*see* **Fig. 4A**). Despite the fact that the data suggest that this fragment of the *KRT14* is not suitable for DHPLC analysis, empirical testing revealed that reproducible DHPLC profiles could be obtained that distinguished between all sequence compositions tested (see **Fig. 4B–D**). Analysis of the fragment at a single temperature of 65°C was sufficient to discriminate between different polymorphic variants and identify the new mutations against a polymorphic background.

3. Clinical Applications of DHPLC

The screening of genes for sequence variations can be carried out for a number of reasons. There is an increased demand from individuals with identified inherited disorders to know the exact nature of their genetic lesion. This might be simply for personal information and reassurance or because they wish to take steps to avoid passing on the affected gene to future generations by having preimplantation or prenatal testing. Individuals with no disease but with a family history of a recessive disorder might wish to determine if they carry a copy of a disease gene. Genetic testing is also used by the clinician for differential molecular diagnosis of conditions that present with similar phenotypes, to predict the likelihood of individuals developing adult-onset disorders such as Huntington's disease or Alzheimer's disease, or to assess the risk of developing adult-onset cancers. The remainder of this chapter gives examples of some of the many applications of DHPLC in molecular medicine. Additional applications in genotyping and mutation detection are described in recent reviews ***(1,11)***.

3.1. Diagnosis of Inherited Disorders

Denaturing HPLC is most widely used to screen for pathogenic mutations causing inherited disorders and has become a routine method in many clinical genetics laboratories. Its success lies in its relative simplicity and the fact that the same platform can be used to detect many different types of mutation. A direct comparison between DHPLC and fluorescent single-strand conformation polymorphism (F-SSCP) analysis found the two methods to have similar specificity, and although DHPLC was marginally less sensitive, this was offset by the high throughput of DHPLC ***(12)***.

The first gene to be analyzed for mutations using DHPLC was *CACNA1A*. Mutations in this gene cause familial hemiplegic migraine and episodic ataxia type-2 ***(13)***. Since then, DHPLC has been used to screen more than 300 genes for sequence variations associated with a wide range of inherited disorders. These include many common conditions such as hemophilia ***(14–17)***, cystic fibrosis ***(8,18–23)***, and deafness ***(9,24–27)***, as well as rarer disorders such as hereditary neuropathies ***(28)*** and a number of skin diseases ***(10,24,29–33)***. A full list of genes and disorders analyzed by DHPLC to date can be found at http://insertion.stanford.edu/human_genes_DHPLC.

Hearing loss is the most common sensory disability in humans and a significant proportion of cases (>50% in the developed world) are genetic in origin. More than 100 genes are believed to be involved in various forms of deafness, and although only some of the genes have been

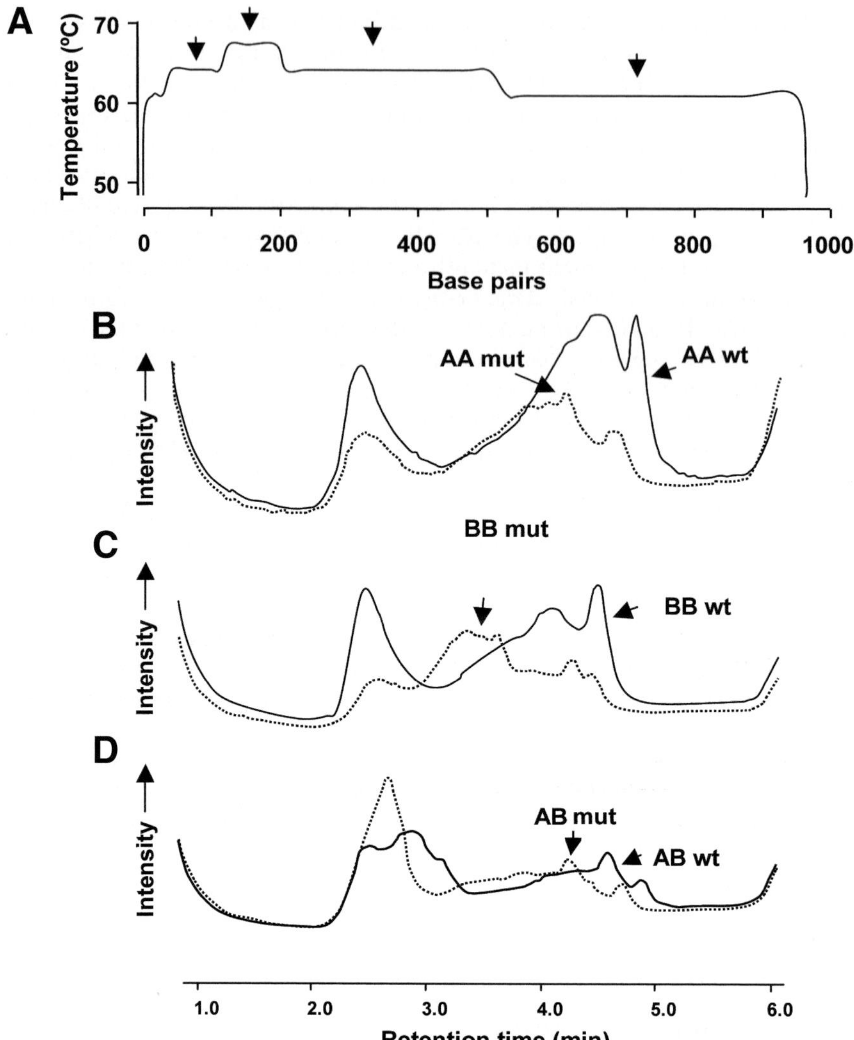

Fig. 4. (**A**) Melting temperature profile of a fragment of *KRT14* determined using the Wavemaker program (Transgenomic). Note the different melting domains indicated by the arrows. This fragment is predicted to melt at three different temperatures. (**B–D**) DHPLC analysis of a fragment of *KRT14* that is frequently mutated in patients with epidermolysis bullosa simplex. This fragment of DNA is contains seven polymorphisms, and in the UK population, these segregate into two common haplotypes (A and B). The solid lines show DNA amplified from a normal individual and the dashed lines are the profiles from patients with known pathogenic mutations in this fragment of *KRT14*. (**B**) The profiles that result from individuals homozygous for the A allele; (**C**) individuals homozygous for the B allele; (**D**) individuals heterozygous for A and B. The DHPLC profiles are different in each case and normal samples with the haplotypes AA, AB, and BB could easily be distinguished from each other. These profiles are reproducible from run to run and between different individuals. The profiles are altered when a mutation is present in the samples and the new profiles are different from the existing profiles. These results demonstrate the sensitivity of DHPLC to sequence variations and the ability to discriminate between single-nucleotide variations on a polymorphic background.

identified, mutation screening is becoming an important method of diagnosis and carrier detection. The identification of the underlying genetic lesions is important in order that appropriate genetic counselling can be provided. Approximately 50% of recessive nonsyndromic hereditary deafness is caused by mutations in the *GJB2* gene, which encodes the gap-junction protein connexin 26 ***(34)***. *GJB2* is a small gene that is relatively easy to sequence; however, because of the time and expense of DNA sequencing, it is usually more cost-effective and labor-efficient to identify samples with sequence variations prior to sequencing. A variety of methods have been employed to screen for *GJB2* mutations, including single-strand conformation polymorphism analysis, denaturing gradient gel electrophoresis, and various allele-specific PCR and oligonucleotide-binding protocols, but all of these methods are either limited in sensitivity or are so specific that a separate assay has to be developed for each mutation. The suitability of DHPLC to screen DNA from patients with hearing impairment for mutations in *GJB2* has been assessed in two recent studies ***(9,26)***. Both studies concluded that DHPLC provided a rapid and reliable method of scanning for mutations in this gene.

Many inherited disorders are genetically heterogeneous, and DHPLC is a particularly useful method for screening samples when mutations in different genes can cause the same condition. For example, Charcot–Marie–Tooth neuropathy is an inherited neuropathology, which can be caused by mutations in at least nine genes. Although there is some phenotype/genotype correlation, many patients present with a range of overlapping symptoms that are often difficult to distinguish from acquired neuropathies. The identification of the underlying genetic lesion is extremely useful for differential diagnosis of these various disorders. Takashima and colleagues used DHPLC to screen four of the genes (*PM22, MPZ, GJB1,* and *ERG2*) that are frequently mutated in this condition ***(28)***. They directly compared DHPLC analysis with DNA sequencing and found that, under optimal conditions, all fragments that contained sequence variations identified by DNA sequencing were also identified by DHPLC. In addition, they found two mutations by DHPLC that were not identified by the automated sequencing analysis program, although visual analysis of the sequence traces suggested their presence. These mutations were confirmed by restriction enzyme digestion, demonstrating that in some situations, DHPLC is more sensitive than sequencing ***(28)***.

Denaturing HPLC has been applied to the identification of mutations underlying several skin disorders. Again, this technique has proved particularly valuable where there is a need to screen several genes. For example, *epidermolysis bullosa* (EB) is a group of related skin fragility conditions caused by mutations in 10 different genes. Although detailed phenotypic analysis of the patients often gives an indication of which particular gene is affected, some forms of EB result from mutations in more than one gene, necessitating analysis of multiple genes. Even when the condition can be easily ascribed to a single gene, as is usually the case for dystrophic *epidermolysis bullosa*, screening is required because of the large size of the *COL7A* gene. DHPLC provides a simple, rapid, and accurate means of achieving this screen ***(31)***. DHPLC has also been used to screen patients for mutations in *KRT9* which cause *epidermolytic palmoplantar keratoderma* ***(10)***. In this case, identification of the mutations was used to provide molecular confirmation of the disorder and differential diagnosis from the more serious condition, tylosis, which is associated with esophageal cancer.

As well as identifying pathogenic mutations, DHPLC can also be used to confirm that an identified novel sequence variation is likely to be the pathogenic mutation and not a polymorphic variation. Common polymorphisms are defined as sequence variations that occur at a frequency of 5% or more in the population. To exclude the possibility that a sequence variation is a common polymorphism, it is necessary to screen at least 100 chromosomes (50 DNA samples) from unaffected, unrelated individuals from the same genetic background as the affected individual. Many methods have been employed for screening and often they are specific for the particular mutation. For example, if a mutation alters a restriction enzyme site, then it is possible to screen samples for the presence or absence of the site. In many cases, no restriction site is altered and it is necessary to use other screening methods such as allele-specific PCR or

PCR-mediated site-directed mutagenesis. Development of these screens can be time-consuming and expensive. DHPLC offers a rapid and reliable alternative method of screening ***(10)***.

3.2. DHPLC for the Detection of Mosaicism

Mosaicism arises when a mutation occurs early in development and the cells arising from the mutated cell carry the mutation. The degree to which mosaicism is evident in an individual depends on the disease phenotype and might not be obvious. If cells giving rise to sperm or ova are affected, it is possible that offspring might have the full-blown condition and this has implications for genetic counseling. It can be difficult to detect the genetic lesion in individuals with mosaicism, even when the disease is apparent, because only a proportion of the cells carry the mutation. In these cases, DNA sequencing invariably results in a normal sequence. Studies in which DNA samples containing known sequence variations have been mixed in varying proportions suggest that DHPLC is capable of detecting mutant alleles when they are as low as 1 : 30 (3.25%) of the total DNA ***(35)***. Jones and colleagues demonstrated in three cases of tuberous sclerosis that abnormal DHPLC profiles were the result of low-level mosaicism in *TSC1* or *TSC2* ***(36)***. In two of these cases, the proportion of mutant allele in the sample was less than 10%, demonstrating the sensitivity of DHPLC to the presence of heteroduplexes. In all three cases, DNA sequencing of PCR fragments failed to detect any abnormality and the mutation could only be identified following analysis of cloned fragments. DHPLC traces should be considered as indicative of sequence variations; these can be caused by PCR artefacts, but could also indicate mosaicism.

3.3. Cancer

The ability of DHPLC to detect the mutation's sequence variations against a varying background of the normal alleles has applications for the early detection of cancers. It is well recognized that an early diagnosis of a tumor is associated with a better long-term outcome. Some genes, such as the tumor suppressor gene *p53*, are mutated in a high proportion of tumors and there is evidence that mutations in this gene are associated with a poor prognosis. DHPLC has been used to screen a range of tissues for somatic *p53* mutations, including ovarian tumors ***(37)***, colorectal carcinomas ***(38)***, lymphomas ***(39)***, and leukemia ***(35)***. Keller et al. determined the sensitivity and specificity to be in the range of 95% for detecting *p53* mutations in colorectal tumors ***(38)***. In a recent study that directly compared mutation detection by DHPLC with denaturing gradient gel electrophoresis, DHPLC was found to be slightly less sensitive for detecting *TP53* mutations than DGGE in a range of tumor samples. However, it was still more sensitive than sequencing and faster and far less hazardous than DGGE ***(40)***.

Denaturing HPLC has also been used to screen genes for sequence variations that predispose individuals toward developing cancers and other late-onset conditions (*see* http://insertion.stanford.edu/human_genes_DHPLC for a complete list of publications). Screening for such predisposing gene mutations often involves a large number of samples, and techniques such as DHPLC are particularly attractive because of their high throughput and low cost. For example, heterozygous mutations in the *ATM* gene are associated with a higher risk of developing breast cancer in women ***(41)***. DHPLC had been used by many laboratories to screen for mutations in this gene and, recently, Bernstein and colleagues carried out a study demonstrating the feasibility of using DHPLC for multicenter screening ***(42)***.

3.4. Mitochondrial DNA

In recent years, it has become apparent that mutations in mitochondrial DNA (mtDNA) as well as nuclear DNA can result in clinical disorders ***(43)***. Mitochondrial disorders are associated with a wide range of phenotypes that can affect any of the body's systems and can vary markedly both between and within families. Each cell in the human body contains between 100 and 10,000 copies of mtDNA and individuals with pathogenic mutations in mtDNA are usually heteroplasmic (i.e., they have a mixture of wild-type and mutant mtDNA). The proportion of wild-type and mutant mtDNA can vary across tissues and between cells within the same tissue, making it difficult to screen for mutations. The sensitivity and specificity of DHPLC to screen for mutations in mtDNA was first validated by van den Bosch et al., who demonstrated that known mutations could be detected in samples of varying heteroplasmy with a sensitivity as low as 0.5% *(44)*. The same study also demonstrated the feasibility of using DHPLC to screen the entire mitochondrial genome of patients suspected of mitochondrial disease and identified the underlying mutations in three out of six cases.

3.5. Forensic Medicine

There are a small number of articles describing the use of DHPLC for forensic identification and it is likely that as DHPLC becomes more established, its application in this area will increase. Shinka and colleagues have used DHPLC to analyze the amelogenin gene and have developed a rapid assay to determine the presence of X and Y chromosomal by DHPLC. This could be applied to many different areas, including forensic medicine, prenatal diagnosis, animal breeding, and anthropology *(45)*.

Analysis of hypervariable regions of mtDNA is frequently used in forensic genotyping. Mitochondrial DNA tends to be more stable than nuclear DNA, and because of its high copy number per cell, it is a particularly useful target for analysis when samples are limited. However, forensic DNA samples are often a mixture from two or more individuals and DNA sequencing often gives ambiguous results. In a recent study, LaBerge and colleagues demonstrated that DHPLC is a reliable method for detecting mtDNA variations in forensic samples *(46)*. They also presented evidence that it is possible to separate the homoduplexes and heteroduplexes allowing accurate sequencing of the individual products. It remains to be established if this will be reliable enough to meet the high standards required for forensic science.

4. Summary

Denaturing HPLC is a low-cost, rapid, specific, and sensitive method for mutation detection. It can be applied to a wide range of conditions and has become established as a method of choice for the identification of DNA mutations in many inherited disorders. An emerging feature of DHPLC analysis is the ability to detect sequence variations in samples where the stochiometry of the alleles is not 1 : 1, such as in somatic tumors, mosaicism, and mitochondrial disease. In these cases, DHPLC is frequently more sensitive than DNA sequencing.

References

1. Xiao, W. and Oefner, P. J. (2001) Denaturing high-performance liquid chromatography: a review. *Hum. Mutat.* **17**, 439–474.
2. Huber, C. G., Oefner, P. J., and Bonn, G. K. (1995) Rapid and accurate sizing of DNA fragments by ion-pair chromatography on alkylated nonporous poly(styrene-divinylbenzene) particles. *Anal. Chem.* **67**, 578–585.
3. Doris, P. A., Oefner, P. J., Chilton, B. S., and Hayward-Lester, A. (1998) Quantitative analysis of gene expression by ion-pair high-performance liquid chromatography. *J. Chromatogr. A* **8**, 47–60.
4. Oefner, P. J. (2000) Allelic discrimination by denaturing high-performance liquid chromatography. *J. Chromatogr. B: Biomed. Sci. Applic.* **739**, 345–355.
5. Arnold, N., Gross, E., Schwarz-Boeger, U., et al. (1999) A highly sensitive, fast, and economical technique for mutation analysis in hereditary breast and ovarian cancers. *Hum. Mutat.* **14**, 333–339.
6. Choy, Y. S., Dabora, S. L., Hall, F., et al. (1999) Superiority of denaturing high performance liquid chromatography over single-stranded conformation and conformation-sensitive gel electrophoresis for mutation detection in TSC2. *Ann. Hum. Genet.* **63(Pt. 5)**, 383–391.
7. O'Donovan, M. C., Oefner, P. J., Roberts, S. C., et al. (1998) Blind analysis of denaturing high-performance liquid chromatography as a tool for mutation detection. *Genomics* **52**, 44–49.
8. Jones, A. C., Austin, J., Hansen, N., et al. (1999) Optimal temperature selection for mutation detection by denaturing HPLC and comparison to single-stranded conformation polymorphism and heteroduplex analysis. *Clin. Chem.* **45**, 1133–1140.

9. Lin, D., Goldstein, J. A., Mhatre, A. N., Lustig, L. R., Pfister, M., and Lalwani, A. K. (2001) Assessment of denaturing high-performance liquid chromatography (DHPLC) in screening for mutations in connexin 26 (GJB2). *Hum. Mutat.* **18**, 42–51.
10. Rugg, E. L., Common, J. E., Wilgoss, A., et al. (2002) Diagnosis and confirmation of epidermolytic palmoplantar keratoderma by the identification of mutations in keratin 9 using denaturing high-performance liquid chromatography. *Br. J. Dermatol.* **146**, 952–957.
11. Frueh, F. W. and Noyer-Weidner, M. (2003) The use of denaturing high-performance liquid chromatography (DHPLC) for the analysis of genetic variations: impact for diagnostics and pharmacogenetics. *Clin. Chem. Lab. Med.* **41**, 452–461.
12. Ellis, L. A., Taylor, C. F., and Taylor, G. R. (2000) A comparison of fluorescent SSCP and denaturing HPLC for high throughput mutation scanning. *Hum. Mutat.* **15**, 556–564.
13. Ophoff, R. A., Terwindt, G. M., Vergouwe, M. N., et al. (1996) Familial hemiplegic migraine and episodic ataxia type-2 are caused by mutations in the Ca^{2+} channel gene CACNL1A4. *Cell* **87**, 543–552.
14. Oldenburg, J., Ivaskevicius, V., Rost, S., et al. (2001) Evaluation of DHPLC in the analysis of hemophilia A. *J. Biochem. Biophys. Methods* **47**, 39–51.
15. Frusconi, S., Passerini, I., Girolami, F., et al. (2002) Identification of seven novel mutations of F8C by DHPLC. *Hum. Mutat.* **20**, 231–232.
16. Bogdanova, N., Markoff, A., Pollmann, H., et al. (2002) Prevalence of small rearrangements in the factor VIII gene F8C among patients with severe hemophilia A. *Hum. Mutat.* **20**, 236–237.
17. Castaldo, G., Nardiello, P., Bellitti, F., et al. (2003) Haemophilia B: from molecular diagnosis to gene therapy. *Clin. Chem. Lab. Med.* 41, 445–451.
18. Liu, W., Smith, D. I., Rechtzigel, K. J., Thibodeau, S. N., and James, C. D. (1998) Denaturing high performance liquid chromatography (DHPLC) used in the detection of germline and somatic mutations. *Nucleic Acids Res.* **26**, 1396–1400.
19. Le Marechal, C., Audrezet, M. P., Quere, I., Raguenes, O., Langonne, S., and Ferec, C. (2001) Complete and rapid scanning of the cystic fibrosis transmembrane conductance regulator (CFTR) gene by denaturing high-performance liquid chromatography (D-HPLC): major implications for genetic counselling. *Hum. Genet.* **108**, 290–298.
20. Audrezet, M. P., Chen, J. M., Le Marechal, C., et al. (2002) Determination of the relative contribution of three genes—the cystic fibrosis transmembrane conductance regulator gene, the cationic trypsinogen gene, and the pancreatic secretory trypsin inhibitor gene—to the etiology of idiopathic chronic pancreatitis. *Eur. J. Hum. Genet.* **10**, 100–106.
21. Ravnik-Glavac, M., Atkinson, A., Glavac, D., and Dean, M. (2002) DHPLC screening of cystic fibrosis gene mutations. *Hum. Mutat.* **19**, 374–383.
22. Scotet, V., Gillet, D., Dugueperoux, I., et al. (2002) Spatial and temporal distribution of cystic fibrosis and of its mutations in Brittany, France: a retrospective study from 1960. *Hum. Genet.* **111**, 247–254.
23. Girardet, A., Cathala, P., and Claustres, M. (2003) Rapid detection of the deltaF508 mutation in single cells using DHPLC: implications for preimplantation genetic diagnosis. *J. Assist. Reprod. Genet.* **20**, 153–156.
24. Kelsell, D. P., Wilgoss, A. L., Richard, G., Stevens, H. P., Munro, C. S., and Leigh, I. M. (2000) Connexin mutations associated with palmoplantar keratoderma and profound deafness in a single family. *Eur. J. Hum. Genet.* **8**, 141–144.
25. Weigell-Weber, M., Schinzel, A., and Hergersberg, M. (2000) Hereditary hearing loss due to mutations in the connexin-26 gene. *Schweiz. Med. Wochenschr.* **130**, 1072–1077.
26. Pallares-Ruiz, N., Blanchet, P., Mondain, M., et al. (2001) Evaluation of dHPLC for CX26 mutation screening in patients from southern France with sensorineural deafness. *Genet. Test.* **5**, 339–343.
27. Mhatre, A. N., Weld, E., and Lalwani, A. K. (2003) Mutation analysis of Connexin 31 (GJB3) in sporadic non-syndromic hearing impairment. *Clin. Genet.* **63**, 154–159.
28. Takashima, H., Boerkoel, C. F., and Lupski, J. R. (2001) Screening for mutations in a genetically heterogeneous disorder: DHPLC versus DNA sequence for mutation detection in multiple genes causing Charcot–Marie–Tooth neuropathy. *Genet. Med.* **3**, 335–342.
29. Jacobsen, N. J., Lyons, I., Hoogendoorn, B., et al. (1999) ATP2A2 mutations in Darier's disease and their relationship to neuropsychiatric phenotypes. *Hum. Mol. Genet.* **8**, 1631–1636.
30. Bitoun, E., Chavanas, S., Irvine, A. D., et al. (2002) Netherton syndrome: disease expression and spectrum of SPINK5 mutations in 21 families. *J. Invest. Dermatol.* **118**, 352–361.

31. Pfendner, E. G., Nakano, A., Pulkkinen, L., Christiano, A. M., and Uitto, J. (2003) Prenatal diagnosis for epidermolysis bullosa: a study of 144 consecutive pregnancies at risk. *Prenat. Diagn.* **23**, 447–456.
32. Wilgoss, A., Leigh, I. M., Barnes, M. R., et al. (1999) Identification of a novel mutation R42P in the gap junction protein beta-3 associated with autosomal dominant erythrokeratoderma variabilis. *J. Invest. Dermatol.* **113**, 1119–1122.
33. Terron-Kwiatkowski, A., Paller, A. S., Compton, J., Atherton, D. J., McLean, W. H., and Irvine, A. D. (2002) Two cases of primarily palmoplantar keratoderma associated with novel mutations in keratin 1. *J. Invest. Dermatol.* **119**, 966–971.
34. Rabionet, R., Gasparini, P., and Estivill, X. (2000) Molecular genetics of hearing impairment due to mutations in gap junction genes encoding beta connexins. *Hum. Mutat.* **16**, 190–202.
35. Leonard, D. G., Travis, L. B., Addya, K., et al. (2002) p53 mutations in leukemia and myelodysplastic syndrome after ovarian cancer. *Clin. Cancer Res.* **8**, 973–985.
36. Jones, A. C., Sampson, J. R., and Cheadle, J. P. (2001) Low level mosaicism detectable by DHPLC but not by direct sequencing. *Hum. Mutat.* **17**, 233–234.
37. Gross, E., Kiechle, M., and Arnold, N. (2001) Mutation analysis of p53 in ovarian tumors by DHPLC. *J. Biochem. Biophys. Methods* **47**, 73–81.
38. Keller, G., Hartmann, A., Mueller, J., and Hofler, H. (2001) Denaturing high pressure liquid chromatography (DHPLC) for the analysis of somatic p53 mutations. *Lab. Invest.* **81**, 1735–1737.
39. Quintanilla-Martinez, L., Kremer, M., Keller, G., et al. (2001) p53 Mutations in nasal natural killer/T-cell lymphoma from Mexico: association with large cell morphology and advanced disease. *Am. J. Pathol.* **159**, 2095–2105.
40. Breton, J., Sichel, F., Abbas, A., Marnay, J., Arsene, D., and Lechevrel, M. (2003) Simultaneous use of DGGE and DHPLC to screen TP53 mutations in cancers of the esophagus and cardia from a European high incidence area (Lower Normandy, France). *Mutagenesis* **18**, 299–306.
41. Athma, P., Rappaport, R., and Swift, M. (1996) Molecular genotyping shows that ataxia-telangiectasia heterozygotes are predisposed to breast cancer. *Cancer Genet. Cytogenet.* **92**, 130–134.
42. Bernstein, J. L., Teraoka, S., Haile, R. W., et al. (2003) Designing and implementing quality control for multi-center screening of mutations in the ATM gene among women with breast cancer. *Hum. Mutat.* **21**, 542–550.
43. Chinnery, P. F., Howell, N., Andrews, R. M., and Turnbull, D. M. (1999) Clinical mitochondrial genetics. *J. Med. Genet.* **36**, 425–436.
44. van Den Bosch, B. J., de Coo, R. F., Scholte, H. R., et al. (2000) Mutation analysis of the entire mitochondrial genome using denaturing high performance liquid chromatography. *Nucleic Acids Res.* **28**, E89.
45. Shinka, T., Naroda, T., Tamura, T., Sasahara, K., and Nakahori, Y. (2001) A rapid and simple method for sex identification by heteroduplex analysis, using denaturing high-performance liquid chromatography (DHPLC). *J. Hum. Genet.* **46**, 263–266.
46. LaBerge, G. S., Shelton, R. J., and Danielson, P. B. (2003) Forensic utility of mitochondrial DNA analysis based on denaturing high-performance liquid chromatography. *Croat. Med. J.* **44**, 281–288.

25

Quantitative PCR

David Sugden

1. Introduction

Commonly used methods to quantify RNA and DNA include Northern and Southern blotting, RNase protection assays, and *in situ* hybridization (see Chapter 29). Because these methods analyze nonamplified RNA or DNA, they are of low sensitivity and require relatively large amounts of nucleic acid. Another method, thousands of times more sensitive than these traditional techniques, combines reverse transcription (RT) and the polymerase chain reaction (PCR). Although RT-PCR is an exquisitely sensitive and specific technique, obtaining quantitative data presents a difficult challenge ***(1–4)***.

The goal of all quantitative PCR methods is to determine the initial number of molecules of a given target from the amount of product generated during PCR. A major obstacle to achieving this goal is the exponential nature of PCR itself. Under ideal conditions, when the reaction efficiency is 100% (i.e., E=1), the amount of product generated increases exponentially, doubling with each cycle of PCR. In practice, the efficiency of amplification can be considerably less than this and can vary substantially. Reaction efficiency depends on many factors, including the primer sequences, the length of the amplicon and its GC content, and sample impurities ***(5)***. These factors affect primer binding, the melting point of the target sequence, and the processivity of the *Taq* DNA polymerase. Importantly, amplification of the same sequence in replicate tubes using the same PCR program can give different efficiency values (e.g., 0.8–0.99) even when a master mix of reaction components is used ***(6)***. This occurs because of small sample-to-sample differences in cycling conditions, which lead to small variations in reaction efficiency. Because of the exponential nature of PCR, this results in substantial differences in product yield as the reaction progresses. Tube-to-tube variation in reaction efficiency can be significant and unpredictable. A difference in efficiency of as little as 5% between two samples with the same initial copy number can result in one sample having twice as much product after 26 cycles of PCR ***(7)***.

Another difficulty in obtaining quantitative data is that there is a linear relationship between the number of target molecules present in a sample at the start and at the end of the PCR reaction *only* during the exponential phase of PCR. During PCR, product accumulates exponentially initially, but then slows as the concentration increases to such an extent that reassociation of sense and antisense product strands competes with primers for further annealing and extension. Additional factors that can contribute to this "plateau" effect include the accumulation of polymerase inhibitors and a loss of *Taq* activity ***(8)***.

From: *Medical Biomethods Handbook*
Edited by: J. M. Walker and R. Rapley © Humana Press, Inc., Totowa, NJ

2. RT-PCR Quantification Strategies

Quantitative PCR methods have been devised that use equipment and techniques common to most molecular biology laboratories (a PCR block, agarose gel electrophoresis, densitometry analysis). These methods endeavour to overcome the difficulties of tube-to-tube variation in efficiency and the limitations imposed by the need for measurement during the exponential phase of PCR. Often, these "end-point" methods separate the amplicon from other reaction components by agarose gel electrophoresis and quantitate by staining with ethidium bromide (EtBr) ***(9)*** or another intercalating fluorescent dye such as SYBR green I ***(10)***. Measurement of incorporated radiolabeled nucleotides or primers followed by autoradiography or phosphoimaging, and hybridization-based strategies such as Southern blotting using radiolabeled amplicon-specific probes have also been used. In order to ensure that quantitation occurs during the exponential phase, it is necessary to sample and analyze the product every cycle or to run multiple serial dilutions of each cDNA.

Some assays have been developed that coamplify the target gene and an endogenous gene as an internal standard. Common endogenous standards include housekeeping gene mRNAs such as β-actin or glyceraldehyde-3-phosphate dehydrogenase (GAPDH) ***(11–13)*** and 28S ribosomal RNA ***(14)***. The endogenous standard is amplified using a second pair of gene-specific primers. Reactions can be run in separate tubes, but tube-to-tube variation remains a problem, or in the same tube, in which case it is essential that the primer pairs function truly independently—a condition that can be difficult to achieve. Because the endogenous standard and target RNA in each sample are processed together throughout the experiment, differences in RNA and cDNA synthesis yield are minimized. The ratio of target to endogenous standard produced can be compared between samples in order to measure relative changes in gene expression. To be reliable, the expression level of the endogenous control should not vary between samples. Unfortunately, few genes are expressed in a strictly constitutive manner, including β-actin ***(15)*** and GAPDH ***(16)***. In addition, it is often difficult to ensure that the reaction is analyzed during the exponential phase, as the mRNA for many endogenous standards is abundant and accumulation of the endogenous standard amplicon can reach a plateau in relatively few cycles, perhaps before the target amplicon can even be detected.

3. Competitive RT-PCR

Another approach is competitive RT-PCR, which has the important advantage that it is not necessary to analyze products only during the exponential phase of PCR ***(17)***. In this method, an internal standard that shares the same primer sequences as the target is amplified in the same tube, leading to competition for reagents. The best internal standard is an exogenous RNA that is spiked into the tissue RNA. Internal standard and target RNAs are then reverse transcribed in the same reaction, allowing a control for varying RT efficiency. Great care must be taken with RNA standards because of the well-known susceptibility of RNA to degradation. Alternatively, an exogenous DNA added prior to PCR can be used as an internal standard. Such internal DNA standards can be homologous or heterologous. A homologous competitor is designed to have the same sequence as the target except for a unique restriction site or a small deletion (or insertion), thus allowing target and competitor amplicons to be easily separated by agarose gel electrophoresis and quantified after PCR. One problem with this type of competitor is that during the later stages of PCR when the concentrations of target and competitor products are high, heteroduplexes can form between target and competitor strands. These heteroduplexes run on an agarose gel at an intermediate size and can complicate quantitation of genuine target and competitor bands. A heterologous competitor is one that shares the target primer sequences but contains a completely different intervening sequence. Heterologous competitors can be made easily for any primer pair simply by low-stringency amplification of DNA of an evolutionarily distantly related species ***(18)***. The resulting multiple products are separated by agarose gel electrophoresis and a band differing in size from the target by approx 25% is selected and purified. It is important to confirm that the putative competitor DNA only amplifies when both forward and reverse primers are present. The quantity of competitor cut from the gel and purified can be assessed by densitometry by running an aliquot(s) alongside markers of known size and amount. Products of suitable size that are easily visible by EtBr staining will contain more than enough copies of competitor for many hundreds of assays (e.g., 10 ng of a 300-bp competitor is approx 3×10^{10} copies). Competitors can also be cloned to generate even larger amounts that can be quantitated by spectrophotometry. Because the internal standard and target use the same primer sequences and generate amplicons of similar size and composition, they should be amplified with the same efficiency. This should be checked in preliminary experiments by sampling the reaction during the exponential phase and analyzing products by EtBr-agarose gel electrophoresis.

In competitive PCR, a series of tubes containing a fixed amount of tissue or cell cDNA are set up with varying amounts of the competitor (*see* **Fig. 1**). The greater the initial concentration of competitor, the more likely it is that the primers will bind to and amplify it rather than the target cDNA. Gel electrophoresis separates target and competitor products at the end of the reaction and the intensity of each is quantified. Because target and competitor have been amplified in the same tube, any potential tube-to-tube variations in reaction conditions are controlled. A graph of the logarithm of the ratio of target amplicon intensity/competitor amplicon intensity vs concentration of competitor spiked into the reaction is linear. The concentration of target can be determined from this graph, as the logarithm of the ratio of target amplicon intensity/competitor amplicon intensity is zero when the concentration of target and competitor are equal at the start of PCR (*see* **Fig. 2**).

Competitive RT-PCR ingeniously circumvents some of the substantial problems in making PCR quantitative. However, substantial practical problems remain. First, all competitive PCR assays are very labor intensive, unsuited to high throughput, and expensive. Several PCR tubes must be set up with varying concentrations of competitor (five to six is typical) for each sample to be measured, and post-PCR processing is required for all reactions. Second, densitometry of EtBr-stained gels typically gives an accurate measure of the target/competitor amplicon ratio between 0.1 and 10, allowing a limited dynamic range to the assay (maximum 100-fold). If the concentration range of competitor selected is inappropriate for a given target sample, a new series of reactions must be run. Third, although RNA competitors control for variations in RT efficiency between samples, they are labile and not suited for long-term storage, and if the concentrations that are spiked into the target sample prior to RT are inappropriate (see above), then new tissue or cell samples must be collected. DNA competitors added prior to PCR have the advantage that the same cell or tissue cDNA sample can be used to measure the expression of multiple genes.

4. Real-Time PCR

Techniques like competitive RT-PCR, which rely on "end-point" analysis of amplicon quantity have largely been superseded in the last few years by the development of novel PCR instruments that combine amplification with fluorescence detection and quantitation of product. Such instruments provide cycle-by-cycle measurement of accumulating product in real time, allowing the entire course of the PCR process to be accurately defined, enabling reproducible determination of starting template concentration. Real-time PCR assays have many advantages. These include a very large dynamic range (up to 8 log units), high throughput, and a very high sensitivity, with a typical assay able to measure as little as 10 copies of an amplicon. Furthermore, real-time PCR assays have high precision—interassay and intraassay coefficients of variation are < 10% when measuring only 100 copies. Assays are run in a closed-tube system and no handling or manipulation of PCR products is required, minimizing the risk of cross-contamination between samples. A variety of fluorescence detection strategies have been developed, including sequence-specific methods, which have the potential for multiplex assays to measure two or more products in a single tube.

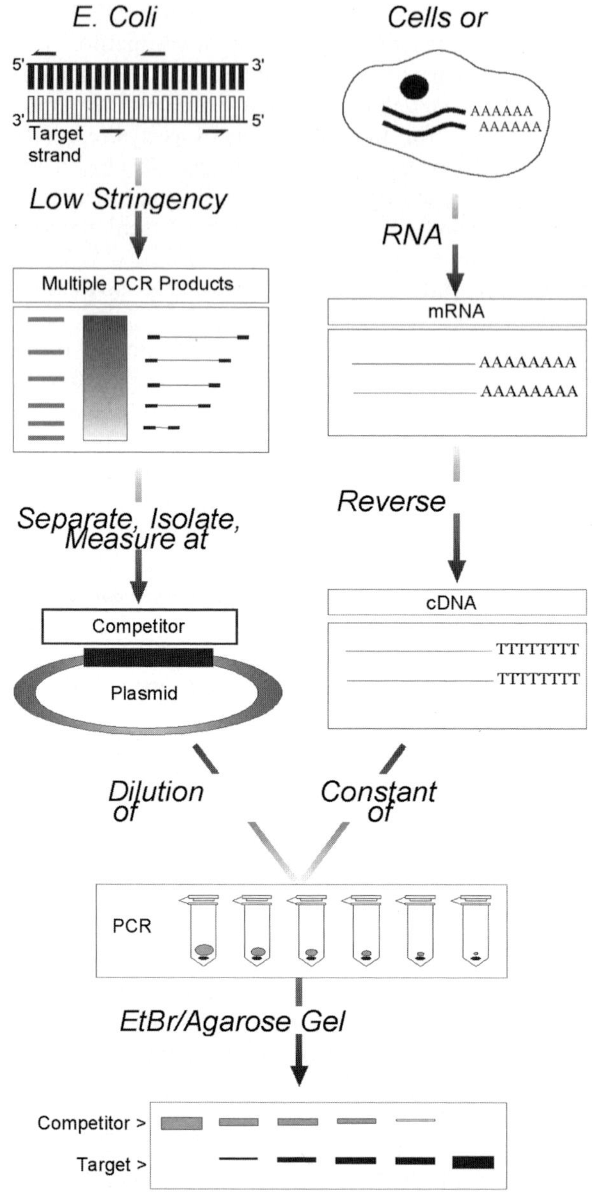

Fig. 1. Diagram showing the competitive RT-PCR procedure.

5. Detection Chemistries

5.1. SYBR Green

SYBR green I is a minor groove DNA-binding dye ***(19)***. In solution (i.e., not bound to DNA), its fluorescence is low; on binding to double-stranded DNA, fluorescence increases (*see* **Fig. 3**). Of the detection chemistries available for real-time PCR, it is the least expensive, not requiring the synthesis of a target-specific probe, and can be used with any pair of primers. Thus, it is particularly useful in developing a real-time quantitative assay when primers are already available that are known to generate a single product with high yield. In such circumstances, we have found it generally possible to develop a real-time assay very quickly, using

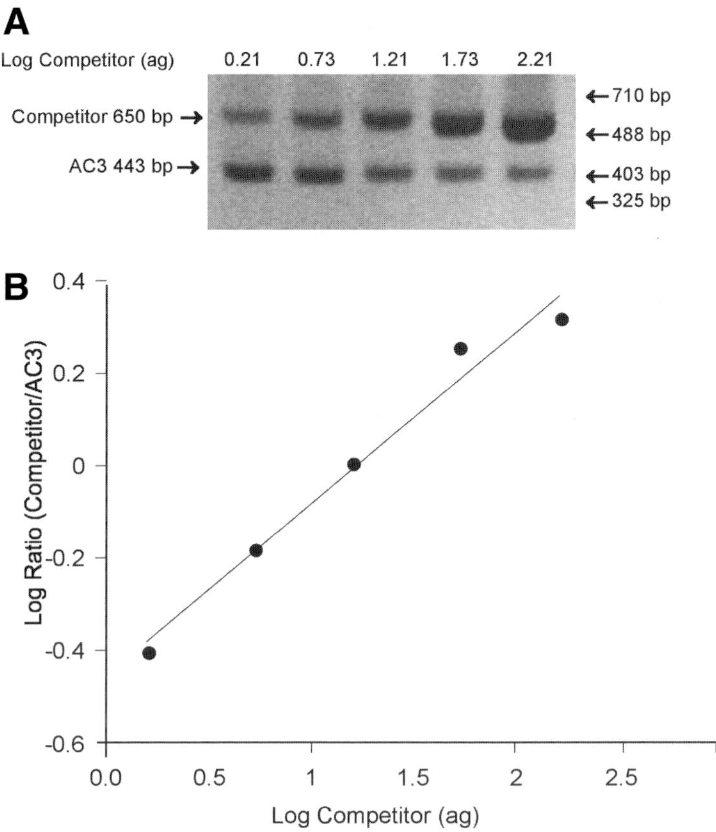

Fig. 2. Example of competitive RT-PCR assay. (**A**) EtBr-stained agarose gel showing competitor (650 bp) and target (rat adenylate cyclase 3, AC3; 443 bp). A constant amount of cDNA (equivalent to approx 2000 cells) was coamplified with a threefold serial dilution of known amounts of competitor (1.6, 5.4, 16.2, 53.7, 162.2 ng) for 40 cycles using AC3-specific primers. (**B**) The ratio of competitor product/AC3 target product was determined by densitometry and the linear regression line drawn (r^2=0.979, $p < 0.005$). The amount of AC3 present in the cDNA sample can be calculated from the regression equation and is equal to the concentration of competitor giving a ratio of target/competitor product = 1 (log=0).

the annealing temperature already optimized in "regular" PCR and the reagents provided in various SYBR green I kits. Recent kits include a hot-start *Taq*, which must be activated in a preliminary step (e.g., QuantiTect, Qiagen; FastStart, Roche Applied Science), allowing reaction setup on the bench, and a reaction buffer formulation that dispenses with the need to optimize the concentration of Mg^{2+} in the reaction. As the PCR progresses, increasing amounts of SYBR green I bind to the double-stranded amplicon, giving an increase in fluorescence that is proportional to the concentration of the product. As SYBR green I fluoresces when bound to double-stranded DNA, fluorescence is measured once each cycle at the end of each elongation step. Real-time assays using SYBR green I are as sensitive as those using any of the sequence-specific fluorescent detection strategies.

One potential drawback to using SYBR green I is that it will detect not only the specific target but also nonspecific products, including primer–dimers. However, in our hands this has not proven to be a difficulty if sufficient care is taken in designing specific primers (as it should be in any PCR). In any case, it is possible to achieve additional specificity and to verify the

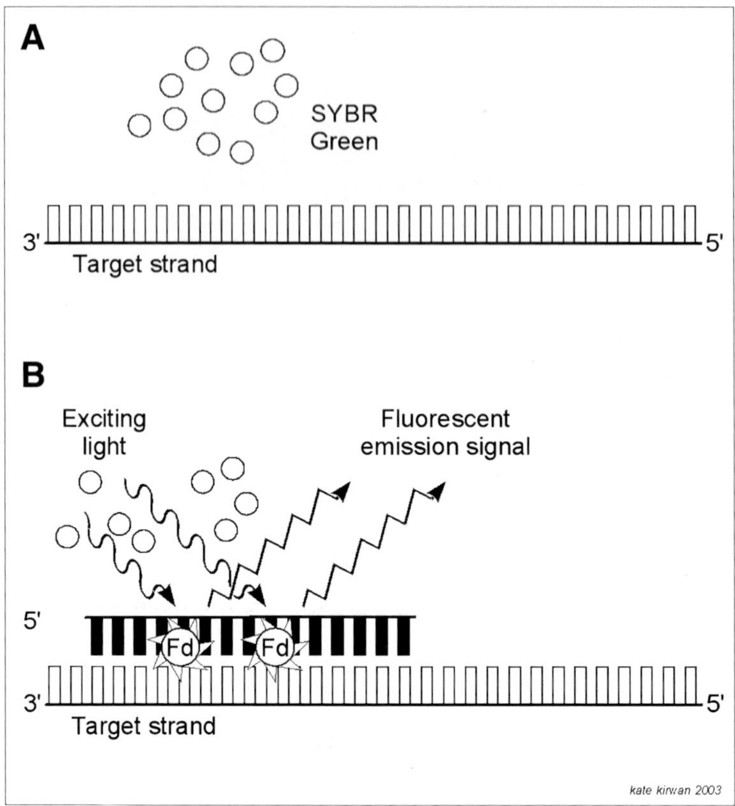

Fig. 3. Use of SYBR green I to detect accumulating amplicons in real-time quantitative PCR. SYBR green I shows a large increase in fluorescence on binding to the double-stranded amplicon.

identity of the product in each amplified sample by using the melting analysis function available on many real-time instruments. At the end of amplification, the temperature is lowered to 5°C above the annealing temperature. This allows all double-stranded products to anneal and SYBR green I to bind, giving maximum fluorescence. The instrument is then programmed to increase temperature slowly (0.1°C/s) while continuously monitoring the fluorescence signal. If a single product has been generated during PCR, fluorescence will fall dramatically, as all the identical amplicon molecules denature as their melting temperature is reached. The first negative derivative ($-dF/dT$) is plotted against temperature by the instrument software, and a single peak is generated for a single-amplicon species. If primer–dimers have been generated, these will typically melt at a lower temperature, as they are much shorter than the genuine product. Having established the melting temperature (T_m) of the genuine amplicon, the amplification step of the real-time assay program can then be modified to include an additional heating step to a temperature 2–3°C below the product T_m before acquiring fluorescence. In this way, fluorescence as a result of primer–dimers is eliminated, whereas that produced by the genuine amplicon is collected.

5.2. Hydrolysis or TaqMan Probes

Hydrolysis probes (also known as TaqMan probes) exploit the 5'→3' exonuclease activity of the *Taq* DNA polymerase used for amplification of the template ***(20)***. In addition to the specific sense and antisense primers that define the ends of the amplicon, a third target-specific oligo-

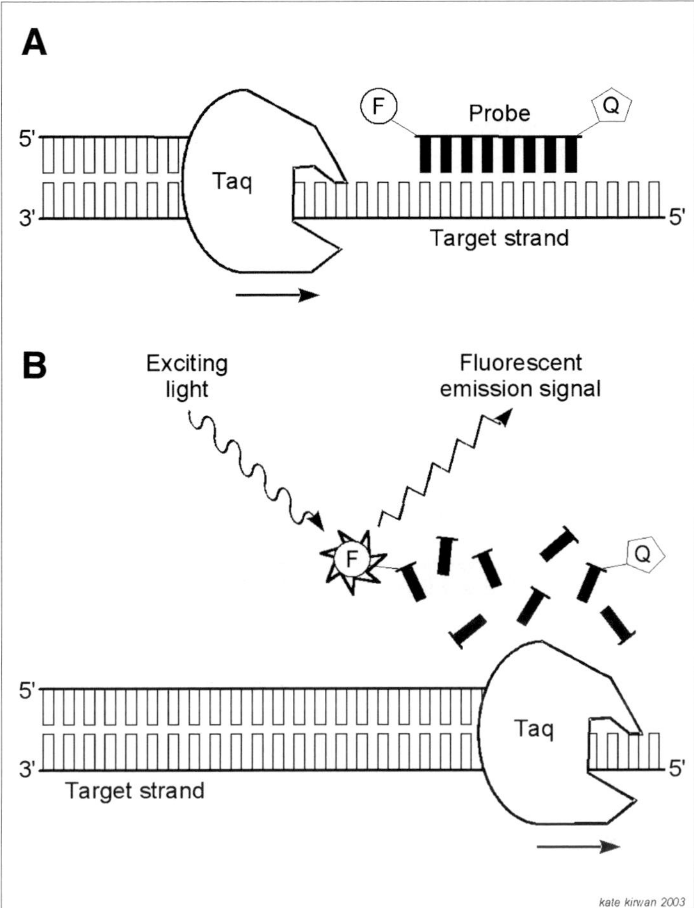

Fig. 4. Principle of hydrolysis or TaqMan probes. A probe complementary to the specific target is synthesised with a 5' fluophore (F) and a 3' quencher dye (Q). The probe hybridizes to the target amplicon, but in the intact probe, the quencher inhibits fluorescence emission (**a**). During the extension phase of PCR (**b**), the 5' → 3'-exonuclease activity of *Taq* polymerase cleaves the probe, releasing the fluophore, allowing it to emit a fluorescent signal.

nucleotide is included in the reaction mix. This oligo has a 5'-fluorescent reporter dye, such as FAM (6-carboxyfluorescein) and a quencher dye such as TAMRA (6-carboxytetramethylrhodamine) bound to the 3' end of the probe through a linker. In the absence of the specific amplicon, the fluorescence emission of the 5' reporter dye is quenched by the 3' dye. When the specific amplicon is generated, the probe anneals after the denaturation step and remains hybridized while the polymerase extends the primer until the enzyme reaches the hybridized probe. As the 5' exonuclease activity of the *Taq* is double-strand-specific, the polymerase hydrolyzes the 5' reporter, freeing it from the quenching effect of the 3' dye (*see* **Fig. 4**). Thus, the fluorescence emission of the reporter dye increases with each PCR cycle, reflecting the accumulation of the specific product. As the labeled probes have a T_m of around 70°C, a combined annealing and polymerization step at 60–62°C is recommended to ensure that the probe remains bound to its target during primer extension and to ensure optimal 5'→3'-exonuclease activity of the *Taq* ***(21)***.

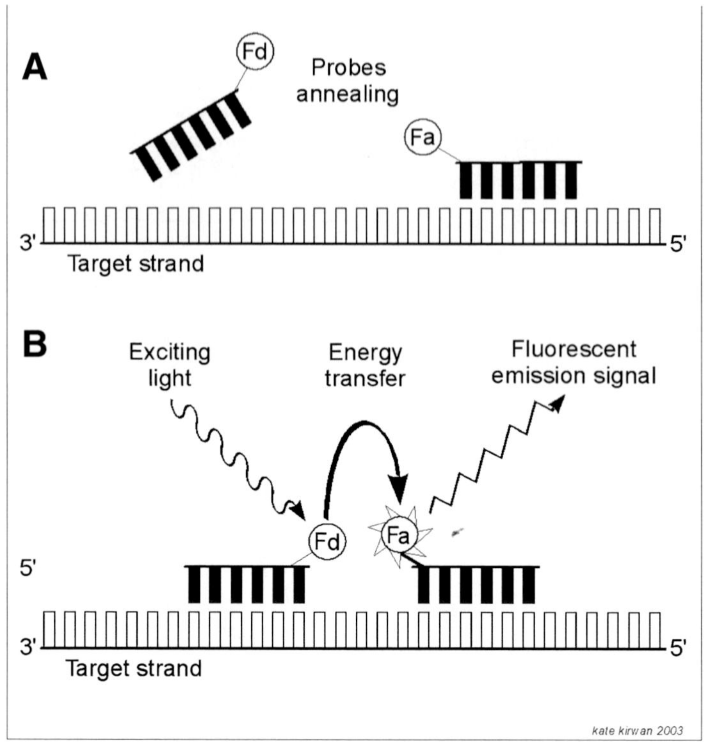

Fig. 5. Principle of the hybridization or FRET probe procedure. (**A**) Two sequence-specific probes are designed to anneal to the target in close proximity (a 1- to 5-base gap is optimal). One probe is labeled with a 3'-fluorescent donor (Fd; fluoroscein); the other has a 5'-acceptor (Fa; LC Red 640 or 705). On hybridization to the specific target amplicon in a head-to-tail fashion (**B**), the energy absorbed by the donor fluophore is transferred to the acceptor fluophore, which then emits fluorescence (FRET).

5.3. FRET Probes

Another detection strategy makes use of two target-specific hybridization probes designed to anneal head-to-tail on the target amplicon. One probe carries a 3'-fluorescein donor that emits green light when excited; the second probe has a 5'-acceptor fluorophore whose excitation spectrum overlaps the 3'-fluorescein donor. On annealing to the target amplicon during PCR, excitation of the donor results in highly efficient fluorescence resonance energy transfer (FRET) to the acceptor, which then emits red fluorescent light (*see* **Fig. 5**). The intensity of light emitted at the longer wavelength from the second dye is proportional to the amount of PCR product synthesized. Background fluorescence is low because emission by the acceptor fluophore can only occur when the two probes are brought into close proximity upon annealing to the genuine amplicon. Because the probes are not hydrolyzed, fluorescence is reversible, allowing melting-curve analysis (see above), a useful tool enabling confirmation of product identity in each amplified sample.

5.4. Other Probe Strategies

Other probe-based detection systems have been described, including molecular beacons *(22)*, Scorpion primers *(23)*, and LUX (light upon extension) primers *(24)*.

A molecular beacon probe consists of a target-specific oligonucleotide probe flanked on each side by a nontarget-specific complementary sequence of six to eight nucleotides, with a

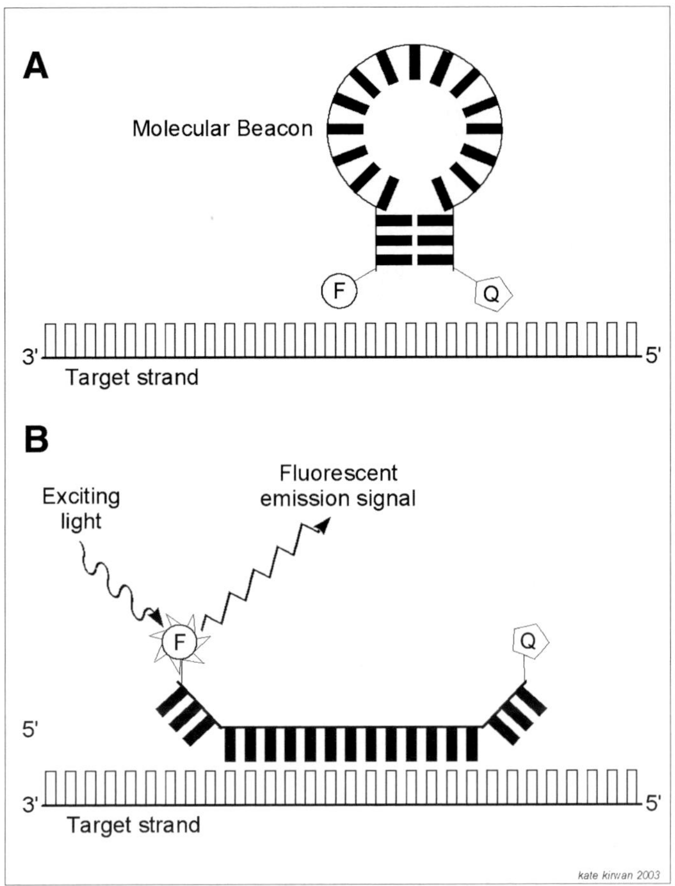

Fig. 6. Use of molecular beacon (or hairpin) probes in real-time quantitative PCR. In the absence of a specific target, the complementary 5' and 3' arms of the probe force it to adopt a hairpin structure (**A**) bringing the fluophore (F) and quencher (Q) into close proximity, eliminating fluorescence emission. When a specific target amplicon is present, the probe changes conformation, allowing the target-specific internal region to hybridize, separating quencher from fluophore, which then emits a fluorescence signal (**B**).

fluorescent marker attached to one arm and a quencher linked to the other arm (*see* **Fig. 6**). In solution, the beacon adopts a hairpin structure, bringing the fluophore and quencher close together and efficiently quenching fluorescence emission. On annealing to the complementary sequence of a target amplicon, fluophore and quencher are forced apart, allowing fluorescence to increase.

Scorpion primers combine both target-specific primer and probe sequence in a single molecule. The probe sequence is linked to the 5' end of a primer via a nonamplifiable "stopper" moiety and comprises complementary stem sequences flanking a target-specific probe sequence (*see* **Fig. 7**). As with molecular beacons, the 5' fluophore is quenched by a 3' quencher as the probe is held in a hairpin loop structure in the unhybridized state. During the extension phase of PCR, the probe sequence of the Scorpion is able to bind to its complementary sequence in the newly formed strand. An advantage of Scorpions is that unimolecular hybridization is kinetically more favorable, as it does not require the chance meeting of probe and amplicon, allowing more rapid cycling and enhanced signal strength ***(25)***.

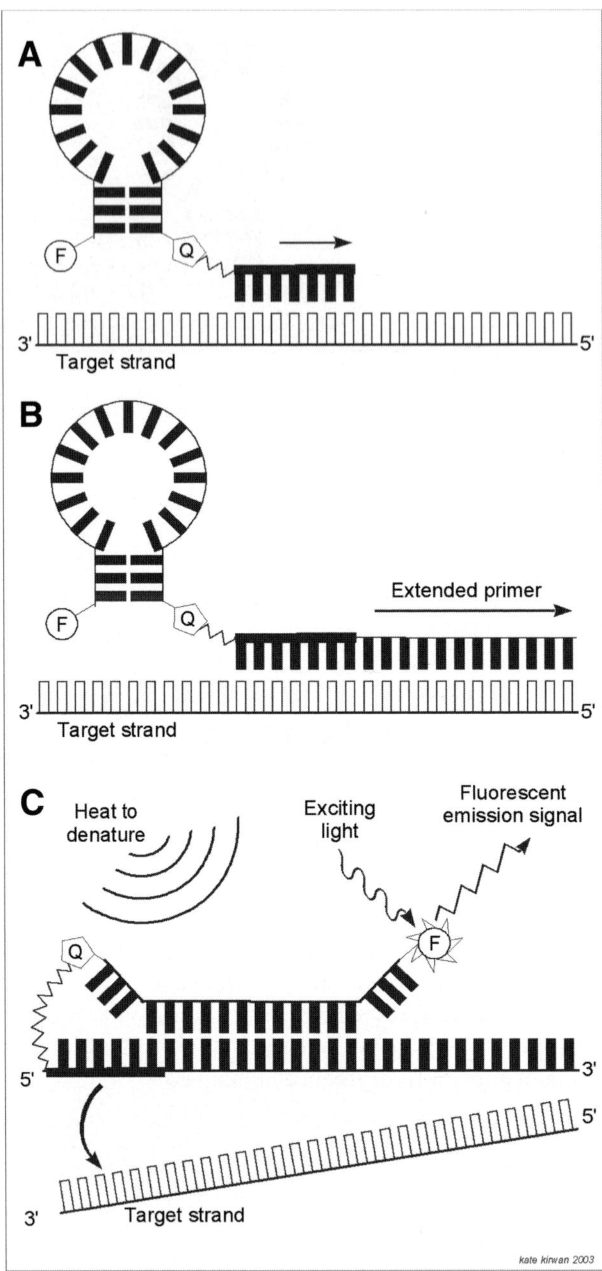

Fig. 7. Mechanism of action of Scorpion primers. Scorpion primers combine target-specific primer and probe in a single sequence. The primer is separated from the probe by nonamplifiable "stopper" moiety (/\/\/). As in molecular beacons, complementary arms flanking a target-specific sequence have fluophore (F) and quencher (Q) dyes. In the absence of the specific amplicon, the complementary probe arms anneal, bringing fluophore and quencher into close proximity to eliminate fluorescence (**A**). As the primer extends, the Scorpion probe remains quenched (**B**), but on heating, the two strands of the amplicon and the complementary arms of the probe dissociate. This allows the target specific sequence to anneal to the amplicon, separating fluophore and quencher, thus restoring fluorescence emission (**C**).

The LUX primers also utilize a single fluorogenic primer ***(26)***. However, instead of containing a quencher moiety, LUX primers are designed to be self-quenched until the primer is incorporated into a double-stranded PCR product. Research on the effects of primary and secondary oligonucleotide structure on the emission properties of conjugated fluophores ***(27)*** provided an understanding of the design factors important for efficient self-quenching, allowing this phenomenon to be exploited for quantitative PCR. LUX primers are designed to have a G or C 3'-terminal nucleotide, include a fluophore attached to the second or third base (T) from the 3' end, and have a five to seven nucleotide 5' tail that is complementary to the 3' end of the primer. The primer forms a blunt-end hairpin structure at temperatures below its T_m which has low fluorescence. Various fluorescent dyes can be used, allowing the potential for multiplex assays to simultaneously quantitate multiple genes.

6. Assay Design, Analysis, and Normalization

The optimal amplicon length for a real-time assay is around 100 bp. In general, shorter products amplify more efficiently than longer ones and are more tolerant of reaction conditions; they also allow shorter extension times and more rapid sample throughput. However, we have, on occasion, made use of primers already in use in our laboratory for qualitative PCR, which generate specific, if rather long, products to establish workable real-time quantitative assays. The longest of these generated an amplicon of 915 bp. In such cases, assay sensitivity is reduced, typically to approx 100 copies, rather than 10 copies usually achievable with shorter amplicons. For detection strategies making use of hybridization probes, the amplicon length has no influence on sensitivity. With SYBR green I detection, longer products should generate increased fluorescence, but, in practice, the loss in amplification efficiency observed negates any gain in fluorescence signal.

Primer design is crucial to establishing a specific and sensitive real-time assay. When measuring gene expression, primers should be designed to anneal to separate exons to avoid amplification of contaminating genomic DNA. If intron/exon boundaries are not known, RNA samples should be treated with RNase-free DNase prior to reverse transcription. Use of SYBR green I detection necessarily requires that a single amplicon be generated. Although hybridization detection strategies might at first sight appear to be more forgiving of amplification of nonspecific products, in practice such misamplification or primer dimerization will reduce assay sensitivity and/or reliability. Time devoted to careful choice of primers and to optimizing assay conditions is always time well spent.

The methods used for analysis of real-time PCR assays give data that are either absolute or relative. Absolute quantification requires that standards of known copy number be amplified in the same run, whereas relative quantification allows the fold change between samples to be calculated. Analysis software is provided by the manufacturers of real-time PCR instruments, enabling the user to determine Ct values for each tube (i.e., the fractional cycle when the fluorescence signal reaches a threshold set by the user within the linear phase of the reaction). Quantitation is based on the fact that there is a linear relationship between Ct and the logarithm of the initial copy number. Absolute quantitation requires that the standards and sample unknowns amplify with equal efficiency. Recent publications have discussed the merits of relative and absolute quantitation ***(28)*** and described the use of new algorithms allowing automated detection and characterisation of the exponential phase of amplification curves to give increased precision in the determination of Ct values ***(29)***.

It is necessary to normalize quantitative PCR data to account for variations in starting material, mRNA (or DNA) extraction, and differences in reverse-transcription efficiency between samples. To do this, an internal reference gene is quantitated in the same samples. Ideally, this internal standard should be expressed at a constant level in different tissues and should not vary with experimental treatment. The issue of which gene is most appropriate for normalization has been much discussed (*see* **refs. *30*** and ***31***), but not satisfactorily resolved. Housekeeping genes such as GAPDH and β-actin are often used, although there is evidence that the expression of these genes can vary substantially. Data showing that normalization is best accomplished using the geometric mean of several housekeeping genes ***(32)*** might not offer a practicable solution for most investigators. Ribosomal RNA, which constitutes 85–90% of the total cellular RNA has also been used and validated in recent studies ***(33,34)***, although its suitability has been questioned ***(35)***. Normalization to total RNA quantity has been advocated as the least unreliable method ***(30)***. Total RNA content can be determined by spectrophotometry ($A_{260\ nm}$) although this method is relatively insensitive and prone to interference from protein and free-nucleotide contaminants. An alternative, sensitive (as low as 5 ng/mL), and accurate method uses the dye RiboGreen, which shows significant fluorescence on binding to nucleic acid ***(36)***.

7. Applications of Quantitative Real-Time PCR

The use of quantitative real-time assays has increased enormously in the last 5 yr. A PubMed search using the key word "real-time PCR" gave 45 citations for 1998–1999, and 2662 for 2002 (January–August). This reflects the increasing availability of real-time cyclers and associated primers, probes, and kits from established manufacturers, the development of additional fluorescence detection strategies, and a growing realization and acceptance that rapid, reliable quantitative measurements are practicable. Another important factor is the remarkable utility of the technique itself, which has found applications in a very wide variety of research and clinical areas. Some of these applications are highlighted.

7.1. Genotyping

Real-time assays using many of the established fluorescence detection strategies have been described for detecting small germline mutations/polymorphisms in the genes that cause common inherited diseases. These include cystic fibrosis [the cystic fibrosis transconductance regulator gene, *CFTR* ***(37)***], emphysema [α_1-antitrypsin gene ***(38)***], venous thrombosis [factor V Leiden gene ***(39)***], hypercholesterolemia [the *ApoE* gene ***(40)***], hemochromatosis [the hemochromatosis gene ***(41)***], and inherited metabolic disorders [glycogen storage disease type Ia and acyl-CoA dehydrogenase deficiency ***(42)***]. Assays to detect polymorphisms in drug-metabolizing enzymes ***(43)*** and human leukocyte antigen (HLA) alleles ***(44)*** have also been described. Allele discrimination depends on allele-specific primers or probes and postamplification melting-curve analysis.

Recent work has shown the utility of high-resolution analysis of product melting curves for genotyping ***(45,46)***. Rather than using a specific fluorescently labeled probe for each gene of interest, a generic double-stranded DNA dye, LCGreen, is used. The method can distinguish a single-nucleotide polymorphism within a 544-bp PCR product and shows promise as a mutation screening tool for identifying unknown sequence variants.

7.2. Detection of Pathogens

The evolution of molecular diagnostics has had a great impact in the area of infectious disease. Many clinical laboratories offer molecular-based testing for pathogens that include both amplification-based and non-amplification-based methods. Increasing emphasis is being placed on quantitative assays for rapid detection of infectious agents, including many pathogenic viruses, bacteria, and yeast, and identification of drug-resistance markers.

7.2.1 Viral Load

Viral genome quantification has made a significant contribution to the diagnosis and management of a number of viral infections ***(47,48)***. Real-time PCR is rapid and sensitive, has an enormous dynamic range, has low intra- and interassay variability, and can utilize templates from a variety of samples ***(49)***. The technique has been used to indicate the extent of active infection, virus–host interactions, and the response to antiviral therapy and to monitor disease progression and viral reactivation and persistence in chronic disease. For example, viral load testing has revolutionized the management of antiretroviral therapy in human immunodeficiency virus-1 (HIV-1)-infected individuals ***(50)***. Quantitative PCR is valuable in determining cytomegalovirus (CMV) load in patients following solid-organ or bone marrow transplantation. In such patients, monitoring viral load can predict CMV disease and relapse and serve as a guide for pre-emptive antiviral therapy ***(51)***. The role of viral genome quantification in the clinical management of patients infected with HIV, hepatitis B and C virus, and CMV ***(52)*** and in typing influenza strains ***(53)*** has been recently reviewed.

7.2.2. Microbial Pathogens

Real-time PCR assays for a number of important microbial pathogens have been described. These include assays able to distinguish *Enterococcus faecium* from *E. faecalis*, to detect vancomycin resistance, to discriminate *Staphylococcus aureus* and coagulase negative *Staphylococci* and analyze methicillin resistance, and to quantitate clinically important species of *Legionella* in respiratory samples ***(54–56)***. Another assay using pan-fungal 18S rRNA primers and a specific probe can measure the seven most common pathogenic species of *Candida*, responsible for approx 80% of systemic fungal infections and a major cause of morbidity and mortality in immunocompromised patients ***(57)***.

7.2.3. Monitoring Food Safety

Contamination of food continues to be a significant public health problem. Rapid, accurate, and sensitive analysis is necessary to ensure food quality and trace outbreaks of bacterial pathogens within the food supply. Real-time PCR has been used to diagnose outbreaks of food-borne disease. For example, Chen et al. ***(58)*** established a sensitive real-time assay using a hydrolysis probe to monitor expression of a *Salmonella* gene (*inv*A gene), which had >98% correlation with a culture method for detecting the pathogen. Another real-time assay targets amplification of a 122-bp fragment of the *Salmonella him*A gene ***(59)***, this time using molecular beacons for detection and quantitation of product. Simultaneous detection of three pathogens (*Listeria monocytogenes*, *Salmonella* strains, and *Escherichia coli* O157:H7) using specific primer sets has recently been described using SYBR green I and melting-curve analysis ***(60)***.

7.3. Oncology

Increasingly, molecular techniques are being used to understand, characterize, monitor, and treat cancers. As expression microarray methods and comparative genomic hybridization identify the genes responsible for driving malignant cell proliferation and metastasis, cell division, repair, apoptosis, and angiogenesis in tumors, real-time PCR will play an increasingly valuable role in clinical testing. Real-time PCR is an attractive technique in this regard, as it is robust, rapid, versatile, and cost-effective and requires very small amounts of tissue. Assays can be designed to provide information about gene expression, gene amplification, or loss and can detect point mutations.

7.3.1. Cancer Diagnostics

Quantitative real-time PCR, with its ability to determine gene duplications and deletions and identify small mutations, is widely used for initial diagnosis and follow-up ***(61)***. For example, a real-time PCR assay using hydrolysis probes successfully detected the reciprocal chromosomal translocation t(14:18)(q32:21), involving the *bcl-2* gene on chromosome 18q21 and the immunoglobulin heavy-chain gene in 14q32, found in 90% of follicular lymphomas ***(62)***. Samples diluted to six to eight copies of the target DNA were consistently detected.

7.3.2. Monitoring Response to Therapy

Quantitative PCR can be a valuable tool for monitoring the progress of hematological malignancies, refining treatment regimens, and detecting disease recurrence. Specific translocation or expression markers are currently available for 70–90% of all acute myeloid leukemias. For most markers, a lack of decline in transcript levels by less than 2 log units after chemotherapy has been established as a poor prognostic sign, and an increase is almost invariably associated with relapse ***(61)***.

In one example, FRET probe assays were established to measure the level of mRNA encoding the BCL-ABL fusion transcript and subunit 2c of the interferon-α (IFN-αR2c) receptor in the blood of chronic myeloid leukemia (CML) patients ***(63)***. It was found that the response of patients to IFN-α did not correlate with BCR-ABL level at diagnosis but was significantly associated with IFN-αR2c mRNA level, thus allowing early identification of IFN-α responsive and unresponsive patients and improving CML management.

7.3.3. Quantitation of Minimal Residual Disease

Many patients with leukemia or lymphoma achieve a complete clinical remission but eventually relapse because residual tumor cells are not detected by conventional staging procedures. Quantitation of molecular disease markers by real-time PCR can offer a more sensitive means of detecting minimal residual disease (MRD) in such patients ***(64,65)***. A number of studies on hematological malignancies support the clinical utility of quantitative PCR for detecting MRD, allowing refinement of an individual patient's treatment regimen. (*see* **ref. *66*** for review). The clinical utility of quantitative PCR for detecting solid-tumor MRD is less well established. Solid tumors are rarely characterized by specific chromosomal translocations. Furthermore, analysis of transcripts is made difficult because of insufficient specificity of most markers and the basal ("illegitimate") background transcription, which can be revealed by a technique as sensitive as quantitative RT-PCR. The potential of quantitative PCR as a tool for detecting circulating tumor cells and micrometastases in blood, bone marrow, lymph nodes, urine, and stools has recently been reviewed ***(67)***.

7.4. Genetically Modified Organisms

In the United States in 1999, almost half of the acres planted with corn, cotton, and soybean used genetically modified varieties, and 60% of food products in US supermarkets contained genetically modified organisms (GMOs) ***(68,69)***. Introducing genes conferring desirable properties such as herbicide and insect tolerance into crops such as maize, soybean, and cotton has caused continuing debate and controversy, particularly in the United Kingdom and the European Union, and has led to legislation requiring the labeling of GM foods and food ingredients. In turn, this has required the development of accurate, quantitative testing methods for GMOs in crops, foods, and food ingredients to ensure compliance with regulations governing labeling and transfer of GM seeds, foods, and food ingredients. Quantitation is a crucial aspect, as EU legislation established a threshold of 1% of GMOs in foods as the basis for mandatory labeling. Testing methods include those that detect and quantitate the protein(s) expressed by the novel gene and also methods detecting the introduced DNA itself. Real-time PCR using TaqMan probes has been successfully used to detect GM soy and maize present in a variety of food products ***(70)***. Multiplex assays to quantitate GMO transgenes in grains have been described that can readily measure as little as 0.1% of transgenic DNA ***(71)***. Quantitative real-time assay kits using TaqMan and FRET probes are commercially available that can measure the transgenes found in more than 95% of the presently available GMO crops.

7.5. Bioterrorism

The concerns of many governments in recent years about the potential use of biological warfare agents has focused efforts to develop sensitive assays able to rapidly identify specific microorganisms thought likely to be used by terrorists ***(72)***. Rapid and reliable testing is vital for diagnosing cases of infection with microorganisms such as *Bacillus anthracis* when speedy therapeutic intervention is essential. Real-time PCR assays have been developed to specifically identify unique targets on *B. anthracis* virulence plasmids using either SYBR green I or FRET probes ***(73,74)***. The assay discriminates *B. anthracis* from other *Bacillus* species and other bacterial strains and can be completed in 1 h ***(74)***. Real-time PCR assays for other potential agents include hydrolysis probe assays targeting the hemagglutinin gene of the smallpox virus *(75)*, the plasminogen activator gene (*pla*) gene of *Yersinia pestis* (plague) *(76)*, and the outer membrane protein (*fop*) gene of *Francisella tularensis* (tularemia) *(77)*.

Viral hemorrhagic fever (VHF) is caused by a number of RNA viruses (e.g., Ebola, Marburg, Lassa, Dengue, Yellow fever), many of which are endemic in tropical or subtropical regions and have been considered to be potential biological weapons *(72)*. Real-time assays using SYBR green I or hydrolysis probes have been established for quantification of six important VHF agents. The detection limit is as low as 10 copies per reaction using primers/probes designed against regions of gene targets that are well conserved in multiple isolates of each virus *(78)*.

Hardened instruments able to withstand the rigors of operation in battlefield conditions have been designed (R.A.P.I.D., Idaho Technology), and a hand-held, portable real-time thermal cycler weighing under 1 kg has been successfully used to detect DNA extracted from *B. anthracis* isolated during the bioterrorism incident in Washington, DC in October 2001 *(79)*.

7.6. Other Applications

Real-time PCR can be applied to detect and quantitate plant and seed contamination by fungal pathogens and to study the development of fungicide resistance *(80)*. The speed and sensitivity of quantitative PCR has the potential to improve seed health testing and disease control by informing decisions about the use of fungicides.

In forensic science, quantitative assays targeting a fragment of the mitochondrial gene, *cytochrome-b*, might be useful as a rapid means of species identification. A similar assay, which can detect very small traces of tiger DNA in complex mixtures, could be a valuable tool in countering the thriving illegal global trade in traditional Chinese medicines prepared from this highly endangered species *(81)*.

Recent work has investigated the utility of cell-free analysis of fetal DNA in samples of maternal blood, a noninvasive approach to prenatal diagnosis that could avoid the small, but significant, risk of fetal loss associated with procedures such as amniocentesis and chorionic villus sampling *(82)*. Real-time quantitative PCR can be used to detect fetal loci absent from the maternal genome, such as Y-chromosome-specific sequences to determine fetal sex, and fetal rhesus D status where the mother is negative for this blood group antigen, and it has potential application in the prenatal diagnosis of sex-linked disorders *(83,84)*. Quantitative abnormalities in fetal DNA in maternal serum have also been reported in pre-eclampsia, which might have diagnostic importance *(85)*. The recent finding that fetal mRNA derived from the placenta can be measured in maternal plasma suggests that a gender-independent approach could be devised for noninvasive prenatal screening and diagnosis *(86)*.

Finally, quantitative PCR has numerous applications in basic biological research. The transcription of many genes can change in response to a wide variety of extracellular signals during development and differentiation, as part of normal physiological function, and during disease. Thus, analysis of steady-state levels of specific mRNAs is important in a broad range of research areas. The specificity, sensitivity, accuracy, and speed of real-time quantitative PCR make it the method of choice for such studies.

References

1. Wang, A. M., Doyle, M. V., and Mark, D. F. (1989) Quantitation of mRNA by the polymerase chain reaction. *Proc. Natl. Acad. Sci. USA* **86,** 9717–9721.
2. Foley, K. P., Leonard, M. W., and Engel, J. D. (1993) Quantitation of RNA using the polymerase chain reaction. *Trends Genet.* **9,** 380–385.
3. Eidne, K.A. (1991) The polymerase reaction and its uses in endocrinology. *Trends Endocr. Med.* **2,** 69–175.
4. Becker-Andre, M. and Hahlbrock, K. (1989) Absolute mRNA quantification using the polymerase chain reaction (PCR). A novel approach by a PCR aided transcript titration assay (PATTY). *Nucleic Acids Res.* **17,** 9437–9446.

5. McDowell, D. G., Burns, N. A., and Parkes, H. C. (1998) Localised sequence regions possessing high melting temperatures prevent the amplification of a DNA mimic in competitive PCR. *Nucleic Acids Res.* **26,** 3340–3347.
6. Wiesner, R. J. (1992) Direct quantification of picomolar concentrations of mRNAs by mathematical analysis of a reverse transcription/exponential polymerase chain reaction assay. *Nucleic Acids Res.* **20,** 5863–5864.
7. Freeman, W. M., Walker, S. J., and Vrana, K. E. (1999) Quantitative RT-PCR: pitfalls and potential. *Biotechniques* **26,** 112–122, 124–125.
8. Kainz, P. (2000) The PCR plateau phase—towards an understanding of its limitations. *Biochim. Biophys. Acta* **1494,** 23–27.
9. Ririe, K. M., Rasmussen, R. P., and Wittwer, C. T. (1997) Product differentiation by analysis of DNA melting curves during the polymerase chain reaction. *Anal. Biochem.* **245,** 154–160.
10. Schneeberger, C., Speiser, P., Kury, F., and Zeillinger, R. (1995) Quantitative detection of reverse transcriptase–PCR products by means of a novel and sensitive DNA stain. *PCR Methods Appl.* **4,** 234–238.
11. Noonan, K. E., Beck, C., Holzmayer, T. A., et al. (1990) Quantitative analysis of MDR1 (multidrug resistance) gene expression in human tumors by polymerase chain reaction. *Proc. Natl. Acad. Sci. USA* **87,** 7160–7164.
12. Murphy, L. D., Herzog, C. E., Rudick, J. B., Fojo, A. T., and Bates, S. E. (1990) Use of the polymerase chain reaction in the quantitation of mdr-1 gene expression. *Biochemistry* **29,** 10,351–10,356.
13. Kinoshita, T., Imamura, J., Nagai, H., and Shimotohno, K. (1992) Quantification of gene expression over a wide range by the polymerase chain reaction. *Anal. Biochem.* **206,** 231–235.
14. Khan, I., Tabb, T., Garfield, R. E., and Grover, A. K. (1992) Polymerase chain reaction assay of mRNA using 28S rRNA as internal standard. *Neurosci. Lett.* **147,** 114–117.
15. Siebert, P. D. and Fukuda, M. (1985) Induction of cytoskeletal vimentin and actin gene expression by a tumor-promoting phorbol ester in the human leukemic cell line K562. *J. Biol. Chem.* **260,** 3868–3874.
16. Shinohara, M. L., Loros, J. J., and Dunlap, J. C. (1998) Glyceraldehyde-3-phosphate dehydrogenase is regulated on a daily basis by the circadian clock. *J. Biol. Chem.* **273,** 446–452.
17. Siebert, P. D. and Larrick, J. W. (1992) Competitive PCR. *Nature* **359,** 557–558.
18. Uberla, K., Platzer, C., Diamantstein, T., and Blankenstein, T. (1991) Generation of competitor DNA fragments for quantitative PCR. *PCR Methods Applic.* **1,** 136–139.
19. Morrison, T. B., Weis, J. J., and Wittwer, C. T. (1998) Quantification of low-copy transcripts by continuous SYBR Green I monitoring during amplification. *Biotechniques* **24,** 954–958, 960, 962.
20. Holland, P. M., Abramson, R. D., Watson, R., and Gelfand, D. H. (1991) Detection of specific polymerase chain reaction product by utilizing the 5'→3'-exonuclease activity of *Thermus aquaticus* DNA polymerase. *Proc. Natl. Acad. Sci. USA* **88,** 7276–7280.
21. Tombline, G., Bellizzi, D., and Sgaramella, V. (1996) Heterogeneity of primer extension products in asymmetric PCR is due both to cleavage by a structure-specific exo/endonuclease activity of DNA polymerases and to premature stops. *Proc. Natl. Acad. Sci. USA* **93,** 2724–2728.
22. Tyagi, S. and Kramer, F. R. (1996) Molecular beacons: probes that fluoresce upon hybridization. *Nature Biotechnol.* **14,** 303–308.
23. Whitcombe, D., Theaker, J., Guy, S. P., Brown, T., and Little, S. (1999) Detection of PCR products using self-probing amplicons and fluorescence. *Nature Biotechnol.* **17,** 804–807.
24. Lowe, B., Avila, H. A., Bloom, F. R., Gleeson, M., and Kusser, W. (2003) Quantitation of gene expression in neural precursors by reverse-transcription polymerase chain reaction using self-quenched, fluorogenic primers. *Anal. Biochem.* **315,** 95–105.
25. Thelwell, N., Millington, S., Solinas, A., Booth, J., and Brown, T. (2000) Mode of action and application of Scorpion primers to mutation detection. *Nucleic Acids Res.* **28,** 3752–3761.
26. Nazarenko, I., Lowe, B., Darfler, M., Ikonomi, P., Schuster, D., and Rashtchian, A. (2002) Multiplex quantitative PCR using self-quenched primers labeled with a single fluorophore. *Nucleic Acids Res.* **30,** e37.
27. Nazarenko, I., Pires, R., Lowe, B., Obaidy, M., and Rashtchian, A. (2002) Effect of primary and secondary structure of oligodeoxyribonucleotides on the fluorescent properties of conjugated dyes. *Nucleic Acids Res.* **30,** 2089–2195.
28. Peirson, S. N., Butler, J. N., and Foster, R. G. (2003) Experimental validation of novel and conventional approaches to quantitative real-time PCR data analysis. *Nucleic Acids Res.* **31,** e73.

29. Wilhelm, J., Pingoud, A., and Hahn, M. (2003) Validation of an algorithm for automatic quantification of nucleic acid copy numbers by real-time polymerase chain reaction. *Anal. Biochem.* **317,** 218–225.
30. Bustin, S. A. (2000) Absolute quantification of mRNA using real-time reverse transcription polymerase chain reaction assays. *J. Mol. Endocrinol.* **25,** 169–193.
31. Bustin, S. A. (2002) Quantification of mRNA using real-time reverse transcription PCR (RT-PCR): trends and problems. *J. Mol. Endocrinol.* **29,** 23–39.
32. Vandesompele, J., De Preter, K., Pattyn, F., et al. (2002) Accurate normalization of real-time quantitative RT-PCR data by geometric averaging of multiple internal control genes. *Genome Biol.* **3,** RESEARCH0034.
33. Goidin, D., Mamessier, A., Staquet, M. J., Schmitt, D., and Berthier-Vergnes, O. (2001) Ribosomal 18S RNA prevails over glyceraldehyde-3-phosphate dehydrogenase and beta-actin genes as internal standard for quantitative comparison of mRNA levels in invasive and noninvasive human melanoma cell subpopulations. *Anal. Biochem.* **295,** 17–21.
34. Schmittgen T. D. and Zakrajsek, B. A. (2000) Effect of experimental treatment on housekeeping gene expression: validation by real-time, quantitative RT-PCR. *J. Biochem. Biophys. Methods* **46,** 69–81.
35. Solanas, M., Moral, R., and Escrich, E. (2001) Unsuitability of using ribosomal RNA as loading control for Northern blot analyses related to the imbalance between messenger and ribosomal RNA content in rat mammary tumors. *Anal. Biochem.* **288,** 99–102.
36. Jones, L. J., Yue, S. T., Cheung, C. Y., and Singer, V. L. (1998) RNA quantitation by fluorescence-based solution assay: RiboGreen reagent characterization. *Anal. Biochem.* **265,** 368–374.
37. Gundry, C. N., Bernard, P. S., Herrmann, M. G., Reed, G. H., and Wittwer, C. T. (1999) Rapid F508del and F508C assay using fluorescent hybridization probes. *Genet. Test.* **3,** 365–370.
38. von Ahsen, N., Oellerich, M., and Schutz, E. (2000) Use of two reporter dyes without interference in a single-tube rapid-cycle PCR: alpha(1)-antitrypsin genotyping by multiplex real-time fluorescence PCR with the LightCycler. *Clin. Chem.* **46,** 156–161.
39. Nauck, M., Marz, W., and Wieland, H. (2000) Evaluation of the Roche diagnostics LightCycler-Factor V Leiden Mutation Detection Kit and the LightCycler-Prothrombin Mutation Detection Kit. *Clin. Biochem.* **33,** 213–216.
40. Nauck, M., Hoffmann, M. M., Wieland, H., and Marz, W. (2000) Evaluation of the *apo E* genotyping kit on the LightCycler. *Clin. Chem.* **46,** 722–724.
41. Mangasser-Stephan, K., Tag, C., Reiser, A. and Gressner, A.M. (1999) Rapid genotyping of hemochromatosis gene mutations on the LightCycler with fluorescent hybridization probes. *Clin. Chem.* **45,** 1875–1878.
42. Fujii, K., Matsubara, Y., Akanuma, J., et al. (2000) Mutation detection by TaqMan-allele specific amplification: application to molecular diagnosis of glycogen storage disease type Ia and medium-chain acyl-CoA dehydrogenase deficiency. *Hum. Mutat.* **15,** 189–196.
43. Hiratsuka, M., Agatsuma, Y., Omori, F., et al. (2000) High throughput detection of drug-metabolizing enzyme polymorphisms by allele-specific fluorogenic 5' nuclease chain reaction assay. *Biol. Pharm. Bull.* **23,** 1131–1135.
44. Bon, M. A., van Oeveren-Dybicz, A., and van den Bergh, F. A. (2000) Genotyping of HLA-B27 by real-time PCR without hybridization probes. *Clin. Chem.* **46,** 1000–1002.
45. Wittwer, C. T., Reed, G. H., Gundry, C. N., Vandersteen, J. G., and Pryor, R. J. (2003) High-resolution genotyping by amplicon melting analysis using LCGreen. *Clin. Chem.* **49,** 853–860.
46. Gundry, C. N., Vandersteen, J. G., Reed, G. H., Pryor, R. J., Chen, J., and Wittwer, C.T. (2003) Amplicon melting analysis with labeled primers: a closed-tube method for differentiating homozygotes and heterozygotes. *Clin. Chem.* **49,** 396–406.
47. Schutten, M. and Niesters H. G. (2001) Clinical utility of viral quantification as a tool for disease monitoring. *Expert Rev. Mol. Diagn.* **1,** 53–62.
48. Niesters, H. G. (2002) Clinical virology in real time. *J. Clin. Virol.* **25(Suppl 3),** S3–12.
49. Mackay, I. M., Arden, K. E., and Nitsche, A. (2002) Real-time PCR in virology. *Nucleic Acids Res.* **30,** 1292–1305.
50. Carpenter, C. C., Cooper, D. A., Fischl, M. A., et al. (2000) Antiretroviral therapy in adults: updated recommendations of the International AIDS Society–USA Panel. *JAMA* **283,** 381–390.
51. Kelley, V. A. and Caliendo, A. M. (2001) Successful testing protocols in virology. *Clin. Chem.* **47,** 1559–1562.

52. Berger, A. and Preiser, W. (2002) Viral genome quantification as a tool for improving patient management: the example of HIV, HBV, HCV and CMV. *J. Antimicrob. Chemother.* **49,** 713–721.
53. Schweiger, B., Zadow, I., Heckler, R., Timm, H., and Pauli, G. (2000) Application of a fluorogenic PCR assay for typing and subtyping of influenza viruses in respiratory samples. *J. Clin. Microbiol.* **38,** 1552–1558.
54. Rantakokko-Jalava, K. and Jalava, J. (2001) Development of conventional and real-time PCR assays for detection of *Legionella* DNA in respiratory specimens. *J. Clin. Microbiol.* **39,** 2904–2910.
55. Elsayed, S., Chow, B. L., Hamilton, N. L, Gregson, D. B., Pitout, J. D., and Church, D. L. (2003) Development and validation of a molecular beacon probe-based real-time polymerase chain reaction assay for rapid detection of methicillin resistance in *Staphylococcus aureus*. *Arch. Pathol. Lab. Med.* **127,** 845–849.
56. Palladino, S., Kay, I. D., Flexman, J. P., et al. (2003) Rapid detection of *vanA* and *vanB* genes directly from clinical specimens and enrichment broths by real-time multiplex PCR assay. *J. Clin. Microbiol.* **41,** 2483–2486.
57. White, P. L., Shetty, A., and Barnes, R. A. (2003) Detection of seven *Candida* species using the Light-Cycler system. *J. Med. Microbiol.* **52,** 229–238.
58. Chen, S., Yee, A., Griffiths, M., et al. (1997) The evaluation of a fluorogenic polymerase chain reaction assay for the detection of *Salmonella* species in food commodities. *Int. J. Food Microbiol.* **35,** 239–250.
59. Chen, W., Martinez, G., and Mulchandani, A. (2000) Molecular beacons: a real-time polymerase chain reaction assay for detecting *Salmonella*. *Anal. Biochem.* **280,** 166–172.
60. Bhagwat, A. A. (2003) Simultaneous detection of *Escherichia coli* O157:H7, *Listeria monocytogenes* and *Salmonella* strains by real-time PCR. *Int. J. Food Microbiol.* **84,** 217–224.
61. Jaeger, U. and Kainz, B. (2003) Monitoring minimal residual disease in AML: the right time for real time. *Ann. Hematol.* **82,** 139–147.
62. Estalilla, O. C., Medeiros, L. J., Manning, J. T. Jr., and Luthra, R. (2000) 5'→3' exonuclease-based real-time PCR assays for detecting the t(14;18)(q32;21): a survey of 162 malignant lymphomas and reactive specimens. *Mod. Pathol.* **13,** 661–666.
63. Barthe, C., Mahon, F. X., Gharbi, M. J., et al. (2001) Expression of interferon-alpha (IFN-alpha) receptor 2c at diagnosis is associated with cytogenetic response in IFN-alpha-treated chronic myeloid leukemia. *Blood* **97,** 3568–3573.
64. Ginzinger, D. G. (2002) Gene quantification using real-time quantitative PCR: an emerging technology hits the mainstream. *Exp. Hematol.* **30,** 503–512.
65. van der Velden, V. H., Hochhaus, A., Cazzaniga, G., Szczepanski, T., Gabert, J., and van Dongen, J. J. (2003) Detection of minimal residual disease in hematologic malignancies by real-time quantitative PCR: principles, approaches, and laboratory aspects. *Leukemia* **17,** 1013–1034.
66. Mocellin, S., Rossi, C. R., Pilati, P., Nitti, D., and Marincola, F.M. (2003) Quantitative real-time PCR: a powerful ally in cancer research. *Trends Mol. Med.* **9,** 189–195.
67. Bockmann, B., Grill, H. J., and Giesing, M. (2001) Molecular characterization of minimal residual cancer cells in patients with solid tumors. *Biomol. Eng.* **17,** 95–111.
68. Ahmed, F. E. (2002) Detection of genetically modified organisms in foods. *Trends Biotechnol.* **20,** 215–223.
69. Beachy, R. N. (1999) Facing fear of biotechnology. *Science* **285,** 335.
70. Vaitilingom, M., Pijnenburg, H., Gendre, F., and Brignon, P. (1999) Real-time quantitative PCR detection of genetically modified Maximizer maize and Roundup Ready soybean in some representative foods. *J. Agric. Food Chem.* **47,** 5261–5266.
71. Permingeat, H. R., Reggiardo, M. I., and Vallejos, R. H. (2002) Detection and quantification of transgenes in grains by multiplex and real-time PCR. *J. Agric. Food Chem.* **50,** 4431–4436.
72. Broussard, L. A. (2001) Biological agents: weapons of warfare and bioterrorism. *Mol. Diagn.* **6,** 323–333.
73. Lee, M. A., Brightwell, G., Leslie, D., Bird, H., and Hamilton, A. (1999) Fluorescent detection techniques for real-time multiplex strand specific detection of *Bacillus anthracis* using rapid PCR. *J. Appl. Microbiol.* **87,** 218–223.
74. Bell, C. A., Uhl, J. R., Hadfield, T. L., et al. (2002) Detection of *Bacillus anthracis* DNA by LightCycler PCR. *J. Clin. Microbiol.* **40,** 2897–2902.
75. Ibrahim, M. S., Kulesh, D. A., Saleh, S. S., et al. (2003) Real-time PCR assay to detect smallpox virus. *J. Clin. Microbiol.* **41,** 3835–3839.

76. Higgins, J. A., Ezzell, J., Hinnebusch, B. J., Shipley, M., Henchal, E. A., and Ibrahim, M. S. (1998) 5' nuclease PCR assay to detect *Yersinia pestis*. *J. Clin. Microbiol.* **36,** 2284–2288.
77. Higgins, J. A., Hubalek, Z., Halouzka, J., et al. (2000) Detection of *Francisella tularensis* in infected mammals and vectors using a probe-based polymerase chain reaction. *Am. J. Trop. Med. Hyg.* **62,** 310–318.
78. Drosten, C., Gottig, S., Schilling, S., et al. (2002) Rapid detection and quantification of RNA of Ebola and Marburg viruses, Lassa virus, Crimean–Congo hemorrhagic fever virus, Rift Valley fever virus, dengue virus, and yellow fever virus by real-time reverse transcription–PCR. *J. Clin. Microbiol.* **40,** 2323–2330
79. Higgins, J. A., Nasarabadi, S., Karns, J. S., et al. (2003) A handheld real time thermal cycler for bacterial pathogen detection. *Biosens. Bioelectron.* **18,** 1115–1123.
80. McCartney, H. A., Foster, S. J., Fraaije, B. A., and Ward, E. (2003) Molecular diagnostics for fungal plant pathogens. *Pest Manag. Sci.* **59,** 129–142.
81. Wetton, J. H., Tsang, C. S., Roney, C. A., and Spriggs, A. C. (2002) An extremely sensitive species-specific ARMS PCR test for the presence of tiger bone DNA. *Forensic Sci. Int.* **126,** 137–144.
82. Siva, S. C., Johnson, S. I., McCracken, S. A., and Morris, J. M. (2003) Evaluation of clinical usefulness of isolation of fetal DNA from the maternal circulation. *Aust. NZ J. Obstet. Gyneacol.* **43,** 10–15
83. Costa, J. M., Giovangrandi, Y., Ernault, P., et al. (2002) Fetal RHD genotyping in maternal serum during the first trimester of pregnancy. *Br. J. Haematol.* **119,** 255–260.
84. Costa, J. M., Benachi, A., Olivi, M., Dumez, Y., Vidaud, M., and Gautier, E. (2003) Fetal expressed gene analysis in maternal blood: a new tool for noninvasive study of the fetus. *Clin. Chem.* **49,** 981–983.
85. Lo, Y. M. D., Leung, T. N., Tein, M. S. C., et al. (1999) Quantitative abnormalities of fetal DNA in maternal serum in preeclampsia. *Clin. Chem.* **45,** 184–188.
86. Ng, E. K., Tsui, N. B., Lau, T. K., et al. (2003) mRNA of placental origin is readily detectable in maternal plasma. *Proc. Natl. Acad. Sci. USA* **100,** 4748–4753.

26

Liquid Chromatography–Mass Spectrometry of Nucleic Acids

Herbert Oberacher and Walther Parson

1. Introduction

1.1. Ion-Pair Reversed-Phase Liquid Chromatography of Nucleic Acids

By virtue of its high-resolving capability, short analysis time, and advanced instrumentation high-performance liquid chromatography (HPLC) has become a versatile technique for the separation and characterization of nucleic acids. Among the various chromatographic modes, which have been summarized in several reviews *(1–6)*, ion-pair reversed-phase HPLC (IP-RP-HPLC) represents the most popular chromatographic technique applicable to the separation of single- and double-stranded DNA molecules.

In IP-RP-HPLC, the phase system comprises a hydrophobic stationary phase and a hydro-organic mobile phase modified with an ion-pair reagent that consists of an amphiphilic ion carrying both charged and hydrophobic groups and a small, hydrophilic counterion. Among the various trialkyl- and tetraalkylammonium salts, triethylammonium acetate represents the most commonly applied ion-pair reagent. According to the electrostatic retention model developed by Ståhlberg *(7)*, positively charged, hydrophobic triethylammonium ions are adsorbed onto the nonpolar surface of the stationary phase, resulting in the formation of an electric double layer having an excess of positive charges near the surface, which can interact with the negatively charged nucleic acid molecules (*see* **Fig. 1**).

The magnitude of electrostatic interaction and, thus, retention is determined by several factors including hydrophobicity of the column packing, charge, hydrophobicity and concentration of the pairing ion, ionic strength, temperature and dielectric constant of the mobile phase, concentration of the organic modifier [see Eq. (10) in **ref. 7**], and charge and size of the nucleic acid molecule. Elution of the adsorbed nucleic acids is effected by a decrease in the surface potential resulting from desorption of the amphiphilic ions from the stationary phase with a gradient of increasing organic modifier concentration. Because the number of charges uniformly increases with size, double-stranded DNA molecules are separated according to chain length in IP-RP-HPLC *(8)*. However, additional solvophobic interactions between hydrophobic regions of the solutes (e.g., nucleobases) and the hydrophobic surface of the stationary phase are possible and contribute considerably to the sequence dependence of retention with single-stranded oligodeoxynucleotides *(9,10)*.

1.2. Principles of Electrospray Ionization–Mass Spectrometry of Nucleic Acids

Mass spectrometry (MS) is one of the most important physical methods in today's analytical chemistry. A particular advantage of MS, compared to other molecular spectroscopic methods, is its high sensitivity down to the zeptomole level *(11)*, so that it provides one of the few methods that is entirely suitable for the identification of trace amounts of biological substances.

From: *Medical Biomethods Handbook*
Edited by: J. M. Walker and R. Rapley © Humana Press, Inc., Totowa, NJ

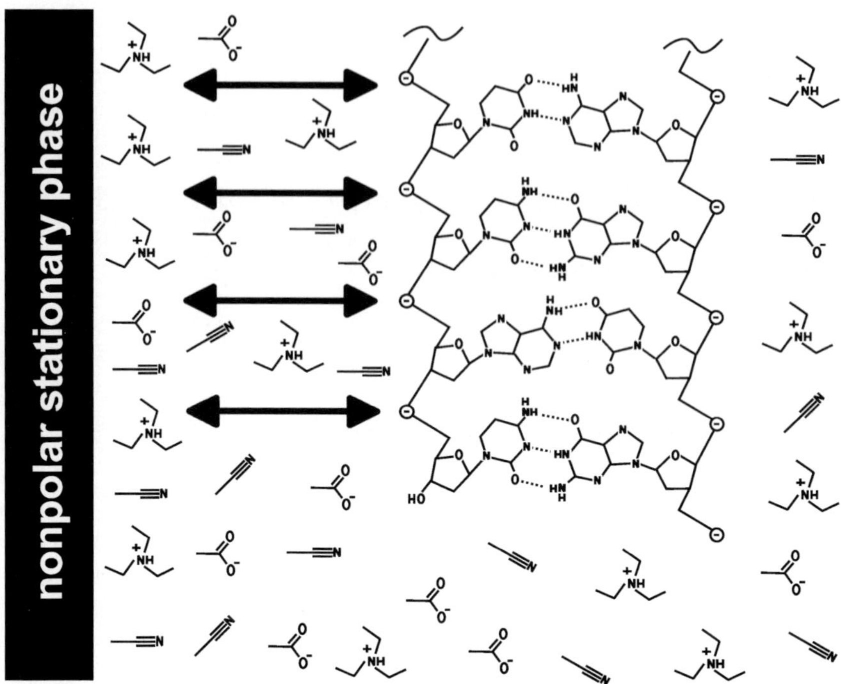

Fig. 1. Schematic illustration of the mechanism explaining the chromatographic retention of nucleic acids in ion-pair reversed-phase high-performance liquid chromatography.

Anyhow, the application of MS to the analysis of nucleic acids has long been limited by experimental problems related to their very high polarity, polyanionic nature, and high molecular mass. The ability to produce intact ions in the gas phase from highly polar molecules gained by the introduction of electrospray ionization (ESI) ***(12)*** at the end of the 1980s opened a new chapter in MS nucleic acid research. The further development of the associated instrumentation has led to a dramatic increase in the accessible mass range, resolution, and sensitivity, making it possible to routinely mass analyze higher-molecular-mass nucleic acids ***(13–19)*** and even genomic DNA with a molecular mass of several million ***(20,21)***.

Electrospray ionization is known as a soft ionization technique that allows the production of gas-phase ions from electrolyte ions in solution. The details of these processes have been investigated extensively ***(22)*** and a brief summary is presented in **Fig. 2**. The ionization process can be divided into three major steps, which include the production of charged droplets at the electrospray capillary tip kept at high voltage (*see* **Fig. 2A**), the shrinkage of the charged droplets by solvent evaporation and repeated droplet disintegrations (*see* **Fig. 2B**), and the release of ions into the gas phase (*see* **Fig. 2C**).

The high voltage applied to the capillary tip of the electrospray ion source results in a high electric field that penetrates the solution at the capillary tip. When the electrospray tip is the negative electrode, positive charge is removed by electrolysis generating an excess of negative ions in the solution. The negative ions drift downfield of the solution (i.e., away from the tip) toward the meniscus of the solution. Because of repulsion between the negative ions, the surface begins to expand, a Taylor cone forms, and if the applied field is sufficiently high, a fine jet emerges from the cone tip, which breaks up into small, charged droplets. The charged droplets produced by the spray shrink owing to solvent evaporation while the charge remains constant. The decrease of droplet radius at constant droplet charge leads to an increase in the electrostatic repulsion of the charges at the surface until the droplets surpass the Raleigh stabil-

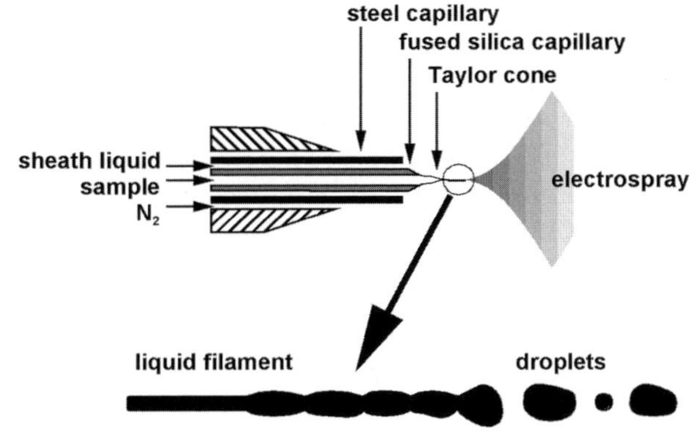

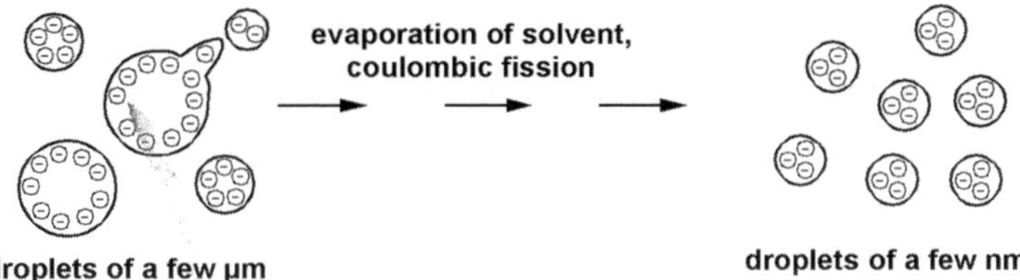

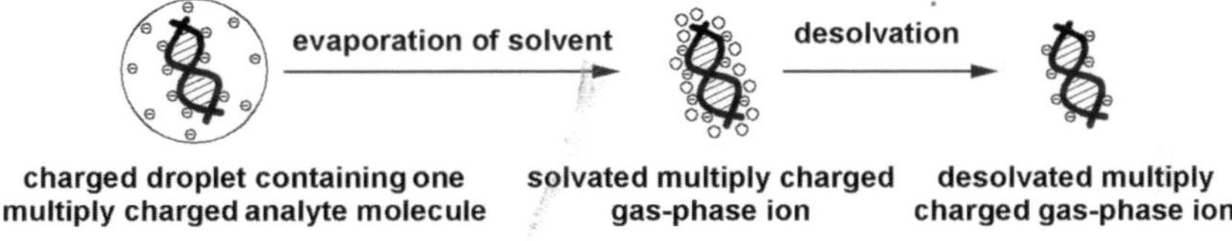

Fig. 2. Major processes involved in the formation of gas-phase ions by ESI operated in the negative ion mode. (**A**) Enrichment of the liquid surface by negative charge upon electrolysis leads to the formation of a cone and jet emitting droplets with an excess of negative ions. (**B**) Charged droplets shrink by evaporation and split into smaller droplets because of coulombic fission. (**C**) Repeated evaporation and coulombic fission finally result in the formation of desolvated, multiply-charged gas-phase ions, either by the formation of extremely small droplets that contain only a single ion [charged residue mechanism ***(23)***] or by the emission of ions from very small and highly charged droplets [ion evaporation mechanism ***(24)***].

ity limit, after which the droplets undergo fragmentation, generally referred to as coulombic fission. Repeated evaporation and coulombic fission finally result in the formation of desolvated gas-phase ions, either by the formation of extremely small droplets that contain only a single ion [charged residue mechanism ***(23)***] or by the emission of ions from very small and highly charged droplets [ion evaporation mechanism ***(24)***]. The produced ions are collected and guided by an ion transfer optic from the ion-source region to the central part of a MS, the mass analyzer. The analyzer uses dispersion or filtering to sort ions according to their mass-to-charge (m/z) ratios. The most commonly used mass analyzers so far are linear quadrupole and quadrupole ion trap mass analyzers, although sectorfield, time-of-flight, and Fourier transform ion cyclotron resonance mass analyzers are also applicable in principle.

One notable feature of ESI is its ability to produce multiply-charged gas-phase ions of nucleic acids by the removal of several protons from the acidic sugar–phosphate backbone. A typical negative ion spectrum of an oligomeric nucleic acid molecule consists of a series of peaks, each of which represents a multiply charged ion of the intact molecule having a specific number of protons removed from the phosphodiester groups (*see* **Fig. 3B**). The *m/z* values for the ions have the general form $(M–z\mathrm{H})/z$, where *z* equals the number of protons removed. It follows that the molecular mass can be readily calculated from two measured adjacent *m/z* values, given the additional information that the difference in their charge states equals 1 ***(25)***. Once *M* and *z* are determined for one pair of signals, all other *m/z* signals can be deconvoluted into one peak on a real mass scale, yielding the molecular mass with a mass accuracy of better than 0.01% ***(26–28)***. Because of this, MS can detect and identify sequence variations with high sensitivity. Moreover, whereas other indicators like retention or migration times are influenced by experimental conditions, the molecular mass is an intrinsic property and, therefore, independent from the physical environment, which makes MS more reliable than any other method for the characterization of nucleic acids.

However, a significant problem in ESI-MS of nucleic acids is the partial substitution of protons in the sugar–phosphate backbone by variable numbers of other cations, such as sodium, potassium, magnesium, or iron. Because of this cation adduction, the pseudomolecular ions are dispersed among several different species of characteristic *m/z* ratios, resulting in highly complex spectra and decreased sensitivity ***(29,30)***. **Figure 3A** illustrates the mass spectrum of the sodium salt of a pentadecameric oligodeoxythymidylic acid, where extensive cation adduction is obviously predominant in the signals representing the lower charge states of the intact molecule (2– charge state around *m/z* 2250, 3– charge state around *m/z* 1500). Accurate mass measurements are severely hampered by such cation adduction, particularly with large nucleic acids, because the numerous signals for the higher charge states of adducted species merge into one unresolved peak shifted to higher mass relative to the fully protonated species. Effective removal of cations is, therefore, a prime requisite to obtain mass spectra of high quality from nucleic acids and accurate molecular masses. Commonly applied off- and on-line sample preparation techniques include multiple ethanol precipitation ***(31)***, solid-phase extraction ***(32)***, microdialysis ***(33)***, affinity purification ***(34)***, cation exchange ***(35)***, liquid chromatography ***(36–40)***, and combinations thereof ***(18,41,42)***.

2. Hyphenation of Liquid Chromatography and Electrospray Ionization Mass Spectrometry

2.1. Instrumental Aspects

Although IP-RP-HPLC ***(5,6)*** and ESI-MS ***(13,43)*** on their own do represent powerful analytical tools for the characterization of nucleic acids, the on-line hyphenation of the chromatographic separation and MS structure investigation enhances the productivity of both analytical methods to characterize nucleic acids in complex mixtures of biological origin ***(39)***. Moreover, the application of the combined method allows the investigation of problems that could not be solved by one method alone. However, for the optimum performance of ion-pair reversed-phase high-performance liquid chromatography hyphenated to electrospray ionization–mass spectrometry (ICEMS), the flow of the sample solution introduced into the ion source of the MS should be in the low-microliter or nanoliter per minute range ***(44)***. Unfortunately, conventional HPLC applying columns with an inner diameter of 4.6 mm is performed at flow rates larger than 100 μL/min. Therefore, either postcolumn splitting of the column effluent ***(36)*** or the miniaturization of the chromatographic separation system by using columns in the capillary format is recommended ***(45–47)***. Currently, because of its approved advantages like increased mass sensitivity with concentration-sensitive detectors ***(48)***, higher separation efficiency and better resolving power ***(49,50)***, and low consumption of mobile and stationary phases, miniaturization represents the common way of adapting the flow rate.

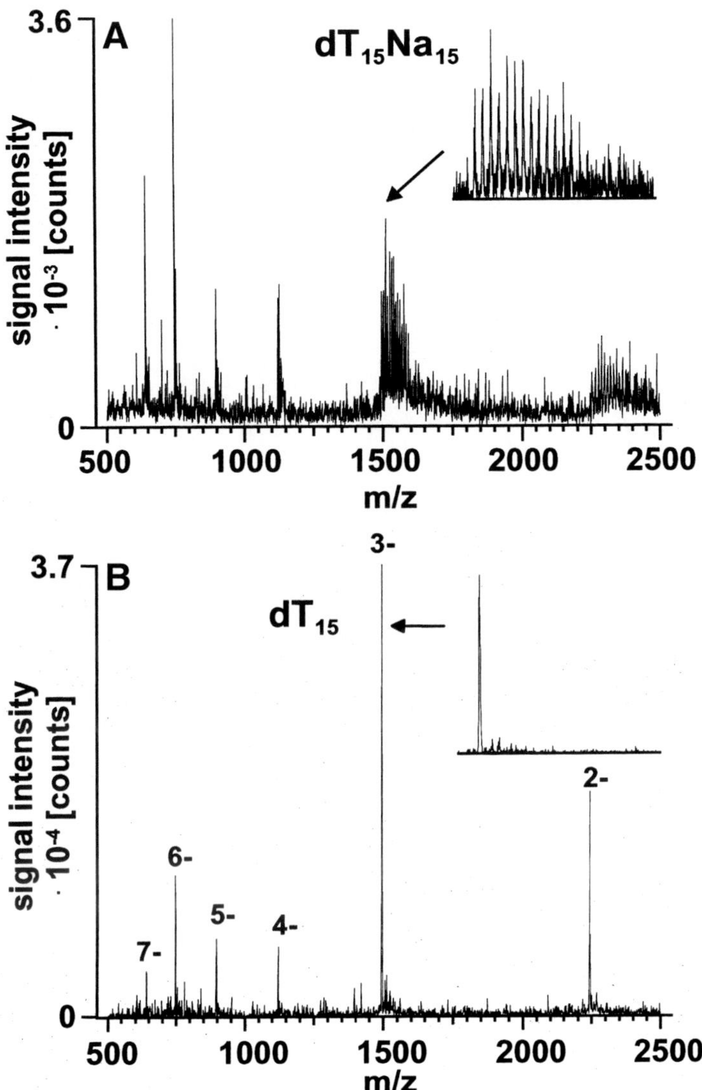

Fig. 3. Mass spectra of $(dT)_{15}$ (**A**) without and (**B**) with online cation-exchange sample preparation. Insets, expanded views of the 3^- charge state. Cation-exchange microcolumn, ROMP-$(COOH)_2$, 45 × 0.8 mm; sheath gas, nitrogen; flow injection into 10 m*M* triethylamine in 50% water–50% acetonitrile (v/v), 3 µL/min; injection volume, 5 µL; scan, 500–2500 amu; sample, 400 pmol $dT_{15}Na_{15}$. (Reproduced from **ref. *35*** by permission of the American Chemical Society [copyright 1998].)

However, the large-scale transition to miniaturized HPLC systems was only possible because of the commercially availability of reliable instrumentation and column technology. Microcolumn HPLC systems need to be designed and operated with the utmost attention to eliminating extracolumn band dispersion attributable to the sampling volume, detection volume, connecting tubing, and system time constant ***(51,52)***. Introduction of small sample volumes and amounts into microcolumns with microinjectors is mandatory for preventing column overloading and minimizing peak variance. Because the gradient delay volume must be kept at

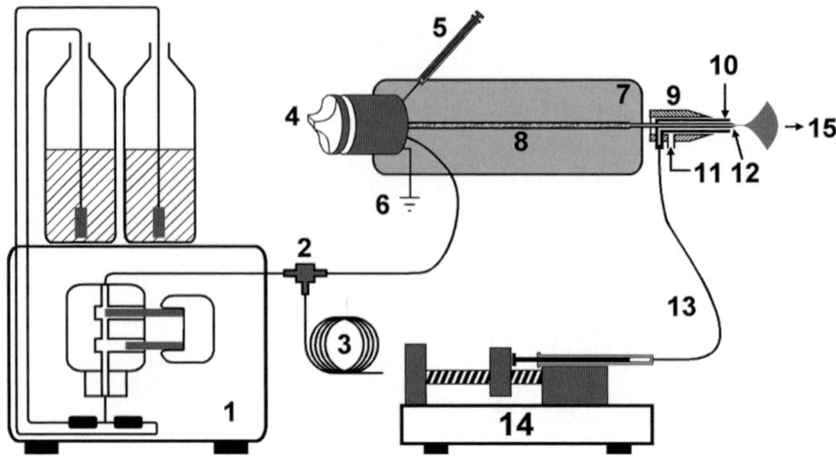

Fig. 4. Generic illustration of an instrumental configuration suitable for nucleic acid analysis by liquid chromatography–mass spectrometry. **1**, low-pressure binary gradient micropump; **2**, splitting tee; **3**, restriction capillary; **4**, 20- to 500-nL microinjector; **5**, syringe for sample introduction; **6**, ground; **7**, column thermostat; **8**, capillary column; **9**, triaxial electrospray ion source with connections for sheath gas and sheath liquid; **10**, metal needle at high voltage; **11**, sheath gas line; **12**, fused silica spray capillary; **13**, sheath liquid line; **14**, syringe pump; **15**, mass analyzer. (Reproduced from **ref. *39*** by permission of Wiley–VCH [copyright 2001].)

a minimum, carrying out gradient elution in miniaturized HPLC is more complicated than using conventional solvent delivery systems. Some modifications of commercially available solvent delivery systems include stepwise gradients ***(53)***, split-flow operation ***(54)***, preformed gradients ***(55)***, and miniaturized diluting chambers ***(49,56)***. To date, commercially available micro-HPLC instrumentation with micromixing chambers is capable of performing reproducible gradients at flow rates as low as 5–10 μL/min without solvent splitting, whereas gradients at flow rates smaller than 5 μL/min are regularly generated using the splitting technique.

Figure 4 outlines an instrumental setup suitable for the separation and mass analysis of nucleic acids by ICEMS. Separation columns can either be fused silica capillaries packed with suitable stationary-phase particles, such as poly(styrene/divinylbenzene) beads chemically modified with octadecyl groups (PS/DVB-C18) ***(57)*** or capillary monoliths comprising a single piece of a porous polymer, such as monolithic poly(styrene/divinylbenzene) ***(58)***. The optimal flow rate for capillary columns of 200 μm inside diameter (i.d.) lies between 1 and 3 μL/min, and flow rates for other column diameters can be calculated using the relation (flow rate$_1$) : (flow rate$_2$) = (column i.d.$_1$)^2 : (column i.d.$_2$)^2.

The HPLC system comprises a low-pressure gradient micropump capable of forming reproducible binary gradients at a flow rate of 50 μL/min and higher. A primary flow of 150 μL/min can be split by a ratio of 1 : 50 to 1 : 150 to 1.0–3.0 μL/min with the help of a tee piece and a 50-μm-i.d. fused silica restriction capillary. Samples are injected by means of a microinjector with a 20- to 500-nL sample loop. Because elevated temperatures can be used to achieve a partial or complete denaturing of nucleic acids ***(10,59)***, a miniaturized column thermostat made from 3.3-mm-outside diameter (o.d.) copper tubing and heated by means of a circulating water bath is used to keep the column temperature at 20–80°C. The exit of the capillary column is directly connected to the ESI spray capillary (fused silica, 90 μm o.d., 20 μm i.d.) by means of a microtight union.

2.2. Optimization of Chromatographic Conditions

An important issue in the conjugation of HPLC and ESI-MS is the proper choice of a suitable chromatographic phase system, which allows the efficient MS detection of highly resolved analytes. Generally, the choice of mobile-phase compositions compatible with ESI-MS detection is restricted to components of high volatility, low surface tension, and low conductivity ***(44,60–62)***. Although only volatile mobile-phase components, namely water, acetonitrile, and ion pair reagents like triethylammonium acetate, are employed in IP-RP-HPLC, which, in principle, would allow the direct introduction of the column effluent into the electrospray ion source. Bleicher and Bayer ***(36)*** as well as Apffel et al. ***(37,63)*** reported that the ordinary used solvent, 100 m*M* triethylammonium acetate, is not a suitable ion-pair reagent for ICEMS because of the poor detectability of the eluted oligonucleotides by ESI-MS. The investigations of these two groups revealed that the major factors responsible for the observed low intensities of oligonucleotide signals were high surface tension because of the low concentration of the organic solvent and high conductivity because of the relative high concentration of the ion-pair reagent. Therefore, Apffel et al. ***(37,63)*** recommended 400 m*M* hexafluoroisopropanol/2.2 m*M* triethylamine as the ion-pair reagent and methanol as the organic modifier to improve the sensitivity of ICEMS. Even though triethylammonium hexafluoroisoproanolate enabled the analysis of single-stranded oligonucleotides up to 75-mers, separation efficiency for larger oligomers and double-stranded DNA was notably impaired compared to that with 100 m*M* triethylammonium acetate, presumably because of the relatively low concentration of triethylammonium ions in the mobile phase. This example clearly demonstrates that solvent systems for chromatography and MS cannot be optimized independently, and, thus, a compromise has to be found to enable the efficient on-line coupling of both techniques.

In search for alternative mobile-phase compositions, Huber and Krajete ***(46)*** investigated the influence of different triethylammonium bicarbonate concentrations on the chromatographic and MS performance. At concentrations of 100 m*M*, triethylammonium bicarbonate gave excellent chromatographic efficiencies, but ESI-MS detection sensitivity was still poor because of the high ionic strength in the electrosprayed eluent. A reduction in ion-pair concentration resulted in a moderate loss of chromatographic resolving power (*see* **Fig. 5A**) but facilitated a considerable increase in MS signal intensity (*see* **Fig. 5B**). Hence, the use of 25–50 m*M* triethylammonium bicarbonate as ion-pair reagent for ICEMS was suggested. By applying 50 m*M* triethylammonium bicarbonate, ESI-MS detection limits in a quadrupole ion-trap MS with full scan and selected ion monitoring mode were determined to be in the range of 104 fmol (signal-to-noise ratio of 3 : 1) and 710 amol (signal-to-noise ratio of 3 : 1), respectively, for a $(dT)_{16}$ oligonucleotide ***(64)***. Moreover, 12 fmol of double-stranded DNA restriction fragments up to a size of 434 base pairs (bp) were detected with 25 m*M* triethylammonium bicarbonate as the ion-pair reagent ***(58,65)***.

A further improvement in detection limits was possible by the use of butyldimethylammonium bicarbonate as the ion-pair reagent instead of triethylammonium bicarbonate ***(27,66)***. The analyses of 500 and 100 fmol of the reverse and forward strands (identified by the subscripts r and f) of a 83-bp polymerase chain reaction (PCR) product utilizing 25 m*M* triethylammonium bicarbonate or butyldimethylammonium bicarbonate as the ion-pair reagent are illustrated in **Fig. 6**. Gradients of 2–12% (*see* **Fig. 6A**) and 6–28% acetonitrile (*see* **Fig. 6B**) were required to elute the DNA from the column with triethylammonium bicarbonate and butyldimethylammonium bicarbonate, respectively, representing a more than twofold increase in applicable acetonitrile concentration with the more hydrophobic ion-pair reagent. Compared to the chromatogram shown in **Fig. 6A**, the signal-to-noise ratio in **Fig. 6B** was enhanced by a factor of approx 2, which, together with the fivefold decrease in sample amount, corresponds to an improvement in detection sensitivity by a factor of 10. Therefore, detection limits in the attomol range were reached for double-stranded DNA fragments, and mass spectra, from which the intact molecular masses could be calculated, were obtained for 753 amol of a 331-mer and 15 fmol of a 501-mer ***(66)***.

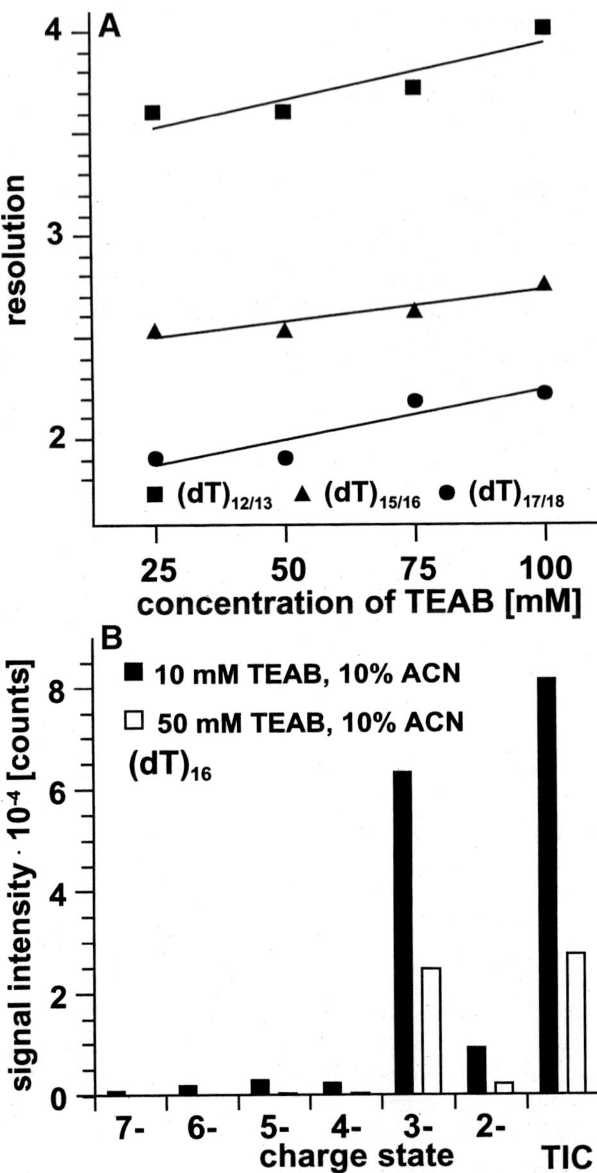

Fig 5. Influence of ion-pair reagent concentration on (**A**) chromatographic resolution and (**B**) signal intensity in ion-pair reversed-phase HPLC-ESI-MS of oligodeoxynucleotides. Conditions in (**A**): column, PS-DVB-C18, 2.1 µm, 60 × 4.0 mm i.d.; mobile phase; (A) 25–100 m*M* triethylammonium bicarbonate, pH 7.50, (B) 25–100 m*M* triethylammonium bicarbonate, pH 7.50, 10% acetonitrile; linear gradient, 50–100% B in 15 min; flow rate, 550 µL/min; temperature, 50°C; detection, ultraviolet (UV), 254 nm; sample, $(dT)_{12-18}$, 36 ng each. Conditions in (**B**): cation-exchange microcolumn, DOWEX 50 WX8, 20 × 0.50 mm; sheath gas, nitrogen; scan, 200–2500 amu; direct infusion of 0.24 mg/mL $(dT)_{16}$ in 10 m*M* and 50 m*M* triethylammonium bicarbonate, pH 8.90, respectively, 10% acetonitrile; flow rate, 3.0 µL/min. (Reproduced from **ref. *46*** by permission of the American Chemical Society [copyright 1999].)

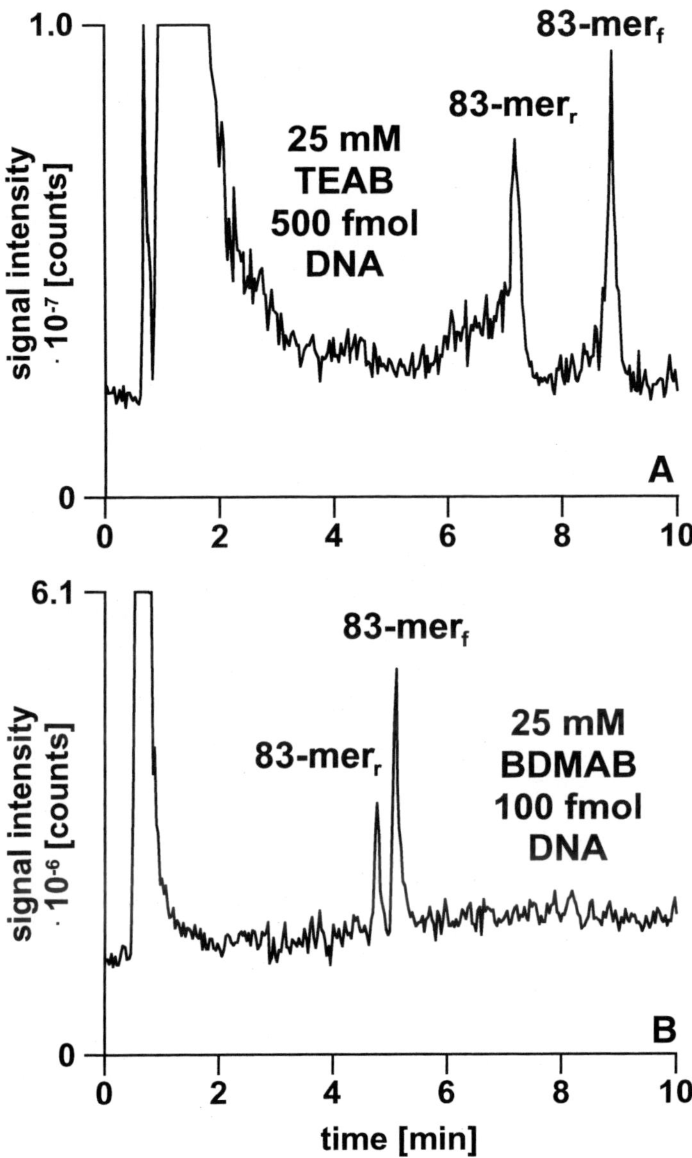

Fig 6. Comparison of detection sensitivity using (**A**) triethyammonium bicarbonate and (**B**) butyldimethylammonium bicarbonate as the ion-pair reagent. Column, PS-DVB monolith, 60 × 0.20 mm i.d.; mobile phase in (**A**): (A) 25 m*M* triethyammonium bicarbonate, pH 8.40, (B) 25 m*M* triethyammonium bicarbonate, pH 8.40, 20% acetonitrile; linear gradient, 10–60% B in 10 min; flow rate, 3.0 µL/min; temperature, 50°C; mobile phase in (**B**): (A) 25 m*M* butyldimethylammonium bicarbonate, pH 8.40, (B) 25 m*M* butyldimethylammonium bicarbonate, pH 8.40, 40% acetonitrile; linear gradient, 15–70% B in 10 min; flow rate, 3.0 µL/min; temperature, 70°C; sheath liquid, 3.0 µL/min acetonitrile; sheath gas, nitrogen; scan, 500-2000 amu; sample, 83-bp PCR product, approx 500 (**A**) and 100 (**B**) fmol. (Reproduced from **ref. *39*** by permission of Wiley–VCH [copyright 2001].)

2.3. Effect of Sheath Liquid on MS Sensitivity

In order to decrease the surface tension and to increase the volatility of the electrosprayed solution, the postcolumn addition of organic solvents to the column effluent was suggested ***(46,64,67)***. Huber and Krajete ***(46,64)*** investigated the usefulness of different sheath liquids, which can be added directly to the column effluent via a tee piece or through a coaxial electrospray probe (*see* **Fig. 4**). They could demonstrate that the total ion current-to-noise ratios for $(dT)_{24}$, which was directly infused into the mass spectrometer, increased in the order no sheath liquid<hexafluoroisopropanol<isopropanol<50 m*M* triethylamine in acetonitrile<methanol<acetonitrile. Interestingly, the simple addition of acetonitrile resulted in a more than sevenfold increase in signal intensity compared to the detection without sheath liquid.

2.4. Influence of Instrument Tuning on the Detectability of Nucleic Acids

Proper instrument tuning is a prime prerequisite for successful ESI-MS of biological macromolecules, where the low amounts of available sample usually necessitate operation with utmost detection sensitivity. Therefore, modern mass spectrometers are equipped with automatic tuning routines, which perform a fully automated optimization of all parameters of the ion transfer optics and mass analyzer for maximum ion transmission and detection. Under autotune operation, a defined solution of a tuning substance or a mixture of tuning substances is introduced into the mass spectrometer at a steady rate, whereas the instrument is calibrated and tuned to ensure its optimum performance. Common tuning compounds for ESI-MS are cluster ions of water ***(68)***, polyethylene glycol ***(69)***, alkali metal salt clusters ***(70)***, Ultramark 1621 ***(71)***, alkali metal trifluoroacetate clusters ***(72)***, and proteins ***(73)***. However, for nucleic acid analysis the autotune settings are usually refined by using oligonucleotides as tuning substances in order to achieve mass spectra of highest quality and to obtain lowest limits of detection. In this context, it was recently shown that highly charged oligodeoxynucleotide anions were only detectable with high sensitivity upon tuning of the MS parameters with an oligodeoxynucleotide having approximately the same size and charge state ***(74)***. For example, the poor quality mass spectrum obtained for an 104-mer oligodeoxynucleotide on a quadrupole ion-trap mass spectrometer tuned upon the signal of the 4 charge state of $(dT)_{24}$ (*see* **Fig. 7A**; low charge state tuning) could be drastically improved after tuning of the instrument with the 24 charge state of an 80-mer (*see* **Fig. 7B**, high charge state tuning). Obviously, only by using the high charge state tuning, a nearly Gaussian distribution of multiply-charged ions covering charge states from 16 to 49 was observed and deconvoluted into a molecular mass of 31,749, which differed from the theoretical value by only 0.015%.

3. Applications

3.1. Characterization of Synthetic Oligonucleotides

The core technology that drives the most widely used automated oligonucleotide synthesis devices is the phosphoramidite method of DNA synthesis, as developed by Caruthers et al. at the begin of the 1980s ***(75,76)***. By applying this technique, the fully automated synthesis of oligonucleotides of 100 or more bases is feasible within hours. This ready access to synthetic oligonucleotides has revolutionized the fields of molecular biology and biochemistry, facilitating the development of PCR, terminatory DNA sequencing, recombinant DNA technology, gene construction, chromosome mapping, site-directed mutagenesis, and numerous other powerful techniques. As a result, synthetic oligonucleotides have helped drive the development of biotechnology, nucleic acid therapeutics, and genomic sequencing efforts, facilitated target identification for traditional pharmaceutical development, and improved the sensitivity and scope of diagnostics.

However, neither the coupling reaction producing the growing polymer chain nor the subsequent deprotection of the full-length oligonucleotide occurs with 100% efficiency. Assuming a

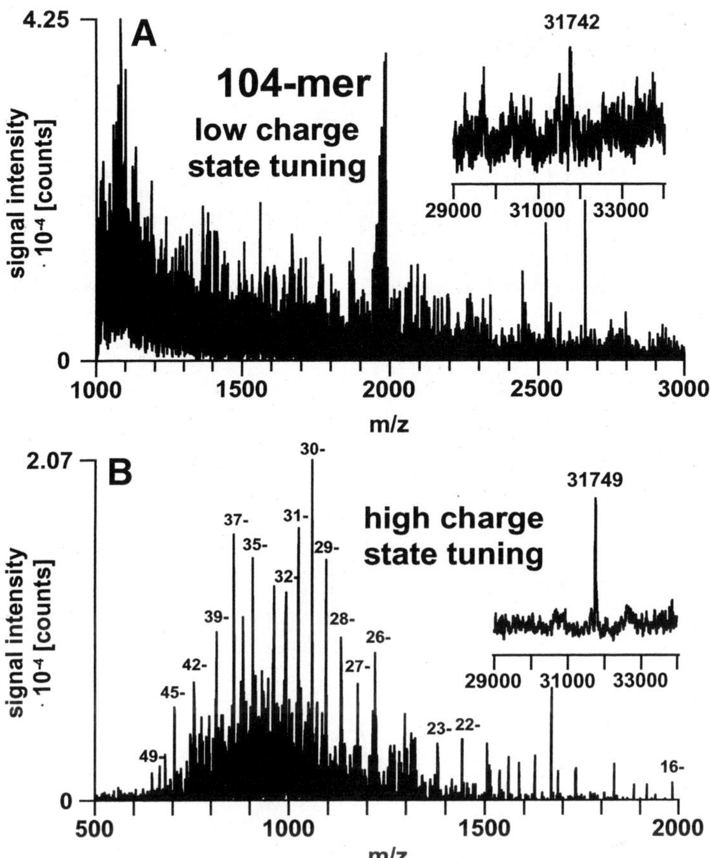

Fig. 7. Comparison of the mass spectra of an 104-mer extracted from the total ion chromatograms of the raw synthetic product acquired with (**A**) low and (**B**) high charge state tuning. Column, PS-DVB monolith, 60 × 0.20 mm i.d.; mobile phase, (A) 25 m*M* triethylammonium bicarbonate, pH 8.40, (B) 25 m*M* triethylammonium bicarbonate, pH 8.40, 20% acetonitrile; linear gradient, 5–50% B in 10.0 min; flow rate, 2.0 μL/min; temperature, 50°C; instrument fine-tuned with (**A**) $(dT)_{24}$, (**B**) an 80-mer; scan, (**A**) 1000–3000, (**B**) 500–2000, sheath liquid, 3.0 μL/min acetonitrile; sample, 500 ng of an raw 104-mer. (Reproduced from **ref.** ***74*** by permission of Wiley–VCH [copyright 2003].)

coupling efficiency of 98–99% per synthesis cycle, the maximum yield of a 32-mer oligodeoxynucleotide will be only 52–72%. Therefore, the contamination of the target sequence with a number of failure sequences or partially deprotected sequences is generally observed ***(46,77,78)***. Thus, the purification of synthetic oligonucleotides by solid-phase extraction, polyacrylamide gel electrophoresis, or HPLC is obligatory for most applications. Moreover, simple, reproducible, and sensitive analytical methods like ICEMS are of utmost importance for purity control of the fabricated oligonucleotides.

Figure 8 illustrates the analysis of 65 ng of a crude 32-mer oligodeoxynucleotide by ICEMS. The target component eluted at 6.9 min and represented approx 65% of all components in the reaction mixture. Extraction and analysis of the mass spectra allowed the identification of as many as 33 by-products on the basis of their molecular masses (*see* **Table 1**). Failure sequences ranging from the 10-mer to the 31-mer eluted before the target product, whereas partially protected sequences containing one to three isobutyryl-protecting groups (ibu) eluted after the

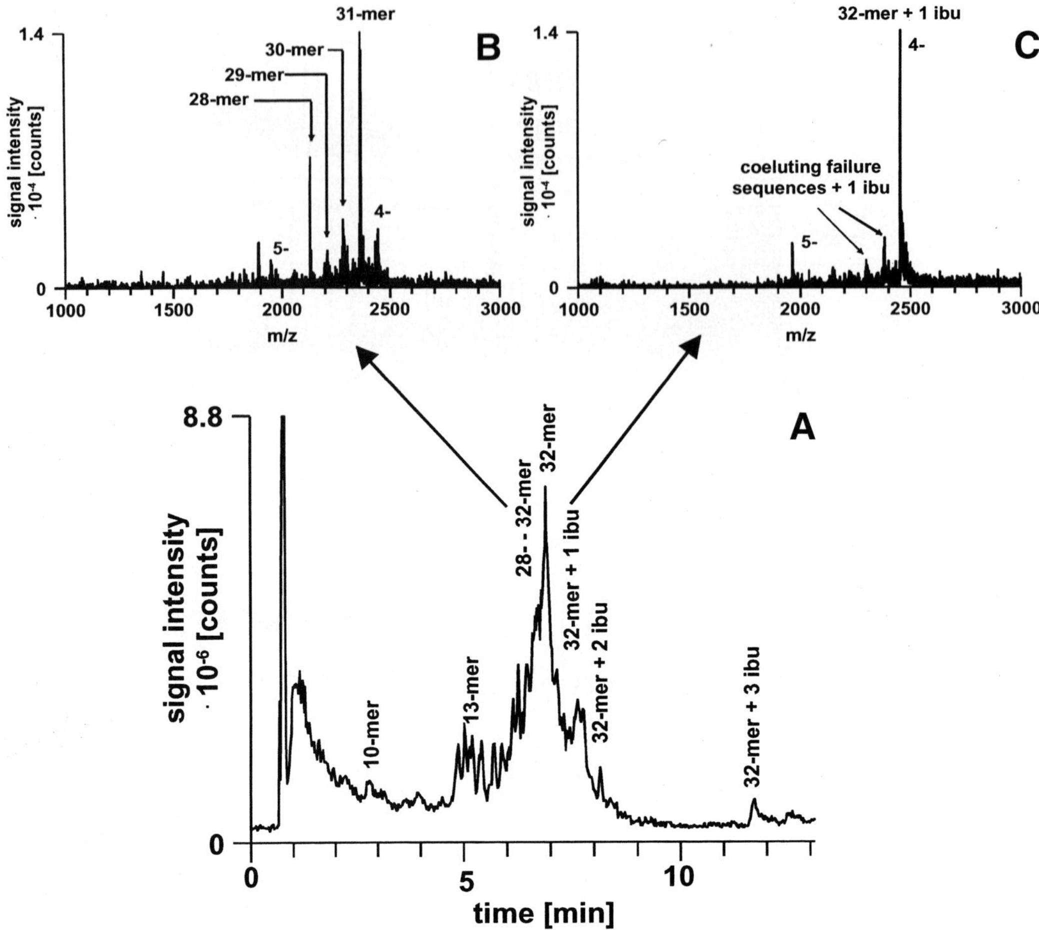

Fig. 8. Reconstructed total ion current chromatogram and mass spectra of byproducts from a synthetic 32-mer oligonucleotide analyzed by ICEMS. Column, PS/DVB monolith, 60 × 0.2 mm i.d.; mobile phase, (A) 25 m*M* triethylammonium bicarbonate, pH 8.40, (B) 25 m*M* triethylammonium bicarbonate, pH 8.40, 20% acetonitrile; linear gradient, 10–60% B in 10 min; flow rate, 2.0 µL/min; temperature, 50°C; sheath liquid, 2.0 µL/min acetonitrile; sheath gas, nitrogen; scan, 1000–3000 amu; sample, 65 ng raw product. (Reproduced from **ref. *39*** by permission of Wiley–VCH [copyright 2001].)

target product. The ibu is used during solid-phase synthesis to protect the amino group of deoxyguanosine and has been shown to be difficult to hydrolyze during the final deprotection step *(46,78)*. **Figure 8B,C** illustrates examples for mass spectra of failure sequences and the target sequence protected with one ibu. With analysis times of less than 15 min and its high information content, ICEMS analysis is eminently suited to monitor the effectiveness of chemical synthesis of nucleic acids, to estimate the synthesis yield on the basis of relative peak areas, and to spot major by-products resulting from failure of chain elongation or partial deprotection.

3.2. Length Determination of Double-Stranded DNA Fragments

The length of a double-stranded DNA fragment, usually expressed as the number of bp, is basic information to characterize a DNA molecule. It not only can be used for the identification of nucleic acids generated by DNA restriction digestion or PCR but also for the detection of

Table 1
Identification of the Components in the Reaction Mixture from the Solid-Phase Synthesis of a 32-mer

Retention time (min)	m/z	Charge	Calculated mass	Identification	Theoretical mass
2.82	1505.1	2–	3012.2	10-mer	3013
4.86	1657.5	2–	3317.0	11-mer	3317
5.01	1201.5	3–	3606.3	12-mer	3606
5.18	1310.6	3–	3934.8	13-mer	3936
5.18	1511.4	3–	4537.2	15-mer	4538
5.41	1414.9	3–	4247.7	14-mer	4249
5.41	1620.8	3–	4865.4	16-mer	4867
5.66	1722.5	3–	5170.5	17-mer	5172
5.96	1824.0	3–	5475.0	18-mer	5476
6.20	1920.4	3–	5764.2	19-mer	5765
6.44	2120.8	3–	6365.4	21-mer	6367
6.64	1666.9	4–	6671.6	22-mer	6671
6.64	1831.8	4–	7331.2	24-mer	7330
6.64	1913.4	4–	7657.6	25-mer	7659
6.64	1985.8	4–	7947.2	26-mer	7948
6.64	2024.6	3–	6076.8	20-mer	6077
6.73	1666.5	4–	6670.0	22-mer	6671
6.73	1749.1	4–	7000.4	23-mer	8527
6.73	1830.1	4–	7324.4	24-mer	7330
6.73	2058.4	4–	8237.6	27-mer	8237
6.78	2130.4	4–	8525.6	28-mer	8527
6.78	2206.1	4–	8828.4	29-mer	8831
6.78	2285.1	4–	9144.4	30-mer	9144
6.78	2366.8	4–	9471.2	31-mer	9473
6.89	2130.4	4–	8525.6	28-mer	8527
6.89	2206.1	4–	8828.4	29-mer	8831
6.89	2285.1	4–	9144.4	30-mer	9144
6.89	2366.8	4–	9471.2	31-mer	9473
6.89	2439.1	4–	9760.4	32-mer	9762
7.61	2302.1	4–	9212.4	30-mer + ibu[a]	9215
7.61	2384.0	4–	9540.0	31-mer + ibu	9544
7.61	2456.6	4–	9830.4	32-mer + ibu	9834
8.36	2474.1	4–	9900.4	32-mer + 2 ibu	9905
11.70	2491.0	4–	9968.0	32-mer + 3 ibu	9976

[a]ibu = isobutyryl protecting group.
Source: From **ref. *39*** by permission of Wiley-VCH (copyright 2001).

insertions and deletions in DNA sequences. Gel electrophoresis can be applied to determine the size of DNA fragments by using the reciprocal relationship between fragment length and electrophoretic mobility ***(79)***. However, because the electrophoretic mobility of a DNA fragment is affected by its sequence-dependent secondary structure, local sizing accuracy will be compromised by standard fragments with anomalous mobility. Similarly, the influence of AT content on retention of DNA fragments resulting in occasional inversions of eluted peaks as a function of molecular size prevented the employment of anion-exchange chromatography for size-accurate fragment identification ***(80)***. Consequently, IP-RP-HPLC was shown to be the most accurate separation method for the determination of the size of double-stranded DNA fragments, which allowed the calculation of fragment size from retention data with an accuracy of better than 3.2% *(8)*.

A new era in the sizing of DNA in terms of sizing accuracy has begun with the introduction of ICEMS *(58,65,66,81)*. To illustrate this, the molecular masses of DNA restriction fragments from four different commercially available DNA sizing standards were determined by ICEMS with a relative mass deviation smaller than 0.12% (average 0.049%) (*see* **Table 2**). The corresponding lengths of each of the 40 fragments (in basepairs) were calculated from the molecular mass by subtracting the mass of H_2O from the molecular mass and dividing by the average mass of an AT and GC basepair [mass of (AT-H_2O) = 617.4; mass of (GC-H_2O) = 618.4; average = 617.9). Because the mass of an AT basepair differs from that of a GC base pair only by one mass unit, the length of a double-stranded DNA fragment in basepairs is essentially independent of base composition. It can be deduced from the last column of **Table 2** that the measured lengths were in excellent agreement with the actual lengths (maximum relative deviation of 0.13%) and that the sizing accuracy is more than one order of magnitude better than that of mass measurements based on chromatographic retention data *(8)*.

The observed high accuracy of mass and length determinations by ICEMS readily facilitates the detection of sequence variations that cause a change in size of double-stranded nucleic acids, namely insertions and deletions. In this context, the applicability of ICEMS was evaluated through the characterization of restriction fragments of the pUC18 cloning vector *(66)* (**Fig. 9**). After digestion with *Hae*III, all fragment masses were determined with a maximum mass deviation of ±0.09%. The only exemption was the 257-bp fragment showing a mass deviation of –0.39%, which differed significantly from the average value for the mass deviation given above. The measured mass difference of 620 mass units (measured mass: 158,220; theoretical mass of the 257-mer: 158,840; *see* **Fig. 10**) corresponded almost exactly to the mass of 1 bp. In fact, Sanger sequencing revealed the deletion of a C-G basepair at position 184, proving that the fragment actually was a 256-mer (theoretical mass: 158,222; *see* **Fig. 10**).

Another application of ICEMS involved the size determination of short tandem repeat (STR) loci *(27)*. STRs are DNA segments typically found in noncoding regions, which are composed of repeating units of dinucleotide to hexanucleotide motifs. Because the number of repeat units is polymorphic in human populations, STRs are very useful for identity testing and genetic mapping *(82)*. In this context, ICEMS was applied to the genotyping of polymorphic STR loci from the human tyrosine hydroxylase gene (humTH01). In all cases, the ICEMS results were in full accordance with the alleles identified by conventional capillary electrophoretic sizing, although as a result of the use of the molecular masses of the PCR amplicons for the allelic discrimination, no DNA sizing standard or allelic ladder as internal standard was necessary for the reliable genotyping of the STR markers.

3.3. Genotyping of Single-Nucleotide Polymorphisms by ICEMS

3.3.1. Introduction

After completion of the human genome sequence by the year 2001 *(83,84)*, the discovery of small dissimilarities in DNA sequences of different individuals, so-called single-nucleotide polymorphisms (SNPs), will gain growing significance *(85,86)*. Information about genetic diversity facilitates valuable insights into inherited disposition to disease as well as in human origin and gene migration. The gold standard for the determination of DNA sequences is the fully automated Sanger sequencing method employing multiplexed capillary electrophoretic analysis *(87)*. Because polymorphisms between two randomly chosen chromosomes occur only at a frequency of 1 in 800–62000 bp, their discovery by Sanger sequencing would dissipate a lot of time and experimental effort for the determination of already known sequences. In due consequence, a high number of different methodologies for screening for polymorphisms and for determining individual allelic states have been developed, which have been reviewed recently in **refs.** *88–90*. In this context, ICEMS was proven to be a powerful analytical tool *(28,39,66,81,91–94)*.

Table 2
Molecular Masses and Deduced Fragment Sizes of Double-Stranded DNA Fragments

Fragment length [bp][a]	Source[b]	Theoretical molecular mass	Measured molecular mass	Relative deviation in mass (%)	Calculated length (bp)	Relative deviation in size (%)
46	3	28,501.7	28476	–0.090	46.03	0.06
49	3	30,316.8	30,299	–0.059	48.98	–0.05
51	4	31,559.6	31,565	0.017	51.03	0.05
57	4	35,263.0	35,252	–0.031	56.99	–0.01
57	3	35,260.1	35,240	–0.057	56.97	–0.05
63	3	38,964.5	39,006	0.106	63.07	0.11
64	4	39,592.8	39,573	–0.050	63.99	–0.02
67	2	41,438.1	41,423	–0.036	66.98	–0.03
80	1	49,475.4	49,535	0.121	80.11	0.13
80	4	49,475.4	49,494	0.038	80.04	0.05
89	4	55,039.0	55,058	0.035	89.05	0.05
90	3	55,658.4	55,690	0.057	90.07	0.08
100	3	61,825.5	61,803	–0.036	99.96	–0.04
102	1	63,067.3	63,123	0.088	102.10	0.10
104	4	64,313.0	64,391	0.121	104.15	0.14
110	2	68,005.6	67,977	–0.042	109.95	–0.04
111	2	68,623.0	68,623	0.000	111.00	0.00
123	4	76,045.8	76,059	0.017	123.03	0.03
124	4	76,675.1	76,731	0.073	124.12	0.10
147	2	90,883.4	90,862	–0.024	146.99	–0.01
174	1	107,566.3	107,640	0.068	174.14	0.08
184	4	113,747.4	113,802	0.048	184.12	0.06
190	2	117,438.0	117,424	–0.012	189.98	–0.01
192	4	118,668.8	118,722	0.045	192.08	0.04
213	4	131,674.0	131,733	0.045	213.13	0.06
226	3	139,688.5	139,718	0.021	226.06	0.03
234	4	144,646.6	144,708	0.042	234.13	0.06
242	2	149,566.0	149,596	0.020	242.04	0.02
256	1	158,221.7	158,220	–0.001	256.00	0.00
257	3	158,848.0	158,849	0.001	257.02	0.01
267	1	165,018.1	165,221	0.123	267.33	0.12
267	4	165,019.1	165,091	0.044	267.12	0.05
281	3	173,693.5	173,522	–0.099	280.76	–0.08
298	1	184,198.4	184,262	0.035	298.15	0.05
331	2	204,588.8	204,611	0.011	331.08	0.02
403	3	249,099.7	248,874	–0.091	402.71	–0.07
404	2	249,643.9	249,750	0.043	404.13	0.03
434	1	268,240.4	268,340	0.037	434.22	0.05
489	2	302,222.7	302,391	0.056	489.32	0.07
501	2	309,574.2	309,567	–0.002	500.94	–0.01

[a]Actual length derived from the sequence.

[b]1, pUC18 DNA–*Hae*III digest; 2, pUC19 DNA–*Msp*I digest; 3, pBR322 DNA-*AluI* digest; 4, pBR322DNA–*Hae*III digest.

Source: From **ref. *81*** by permission of Elsevier Science (copyright 2002).

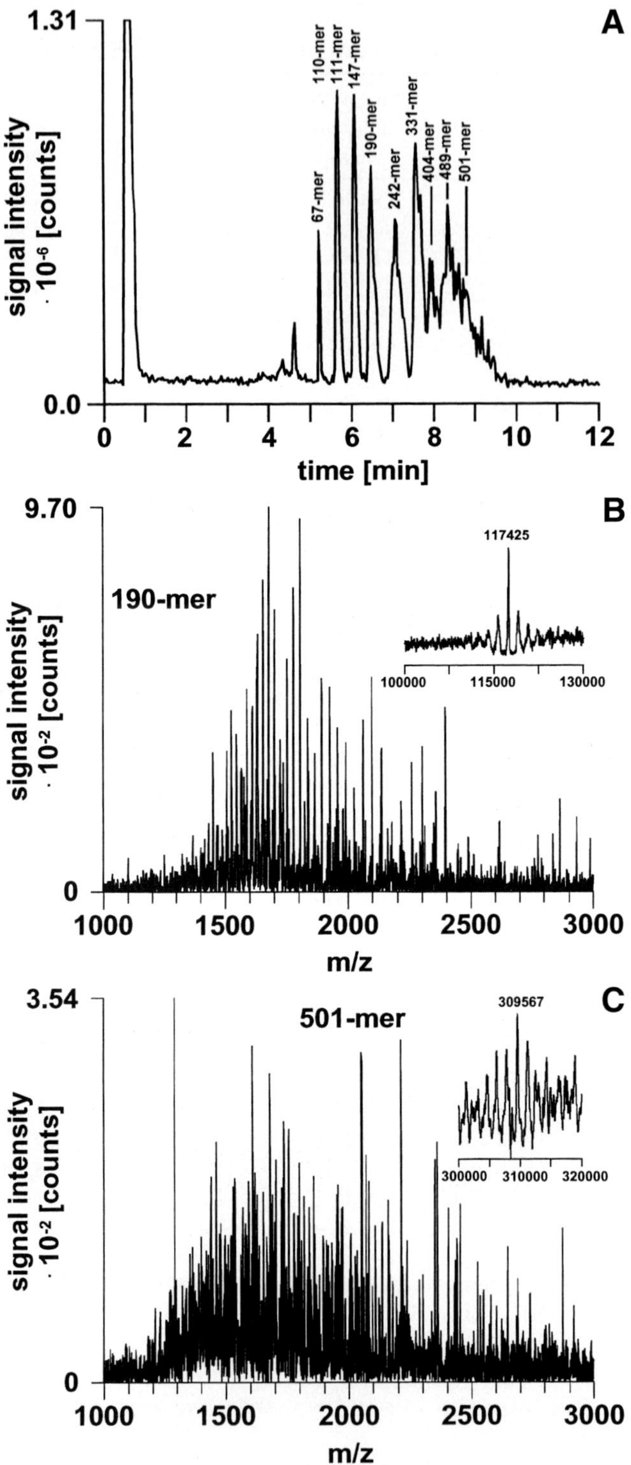

Fig. 9. Chromatographic separation and mass analysis of the double-stranded restriction fragments of a pUC19 *Msp*I digest: (**A**) reconstructed total ion current chromatogram; (**B**) mass spectrum of the double-stranded 190-mer; (**C**) mass spectrum of the double-stranded 501-mer. Column: PS/DVB monolith, 60 × 0.2 mm i.d.; mobile phase: (A) 25 m*M* butyldimethylammonium

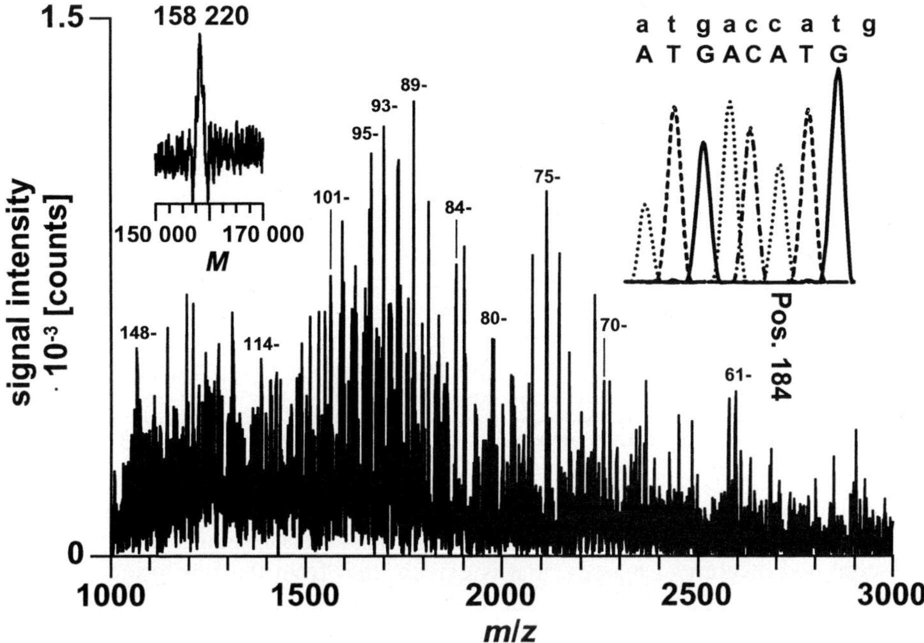

Fig. 10. Mass spectrum for the detection of a deletion in a double-stranded DNA fragment from the cloning vector pUC18 by means of ICEMS. Column: PS/DVB monolith, 60 × 0.2 mm i.d.; mobile phase: (A) 25 m*M* triethylammonium bicarbonate, pH 8.4, (B) 25 m*M* triethylammonium bicarbonate, pH 8.4, 20% acetonitrile; linear gradient: 5–32% B in 3 min, 32–38% B in 7 min, 38–43% B in 12 min; flow rate: 3.0 µL/min; temperature: 25°C; sheath gas, nitrogen; sheath liquid, 3.0 µL/min acetonitrile; scan: 1000–3000 amu; sample, 12 fmol pUC18 DNA–*Hae*III digest. The inset on the left side shows the deconvoluted mass spectrum; the one on the right side shows the sequence determined by Sanger DNA sequencing (in uppercase letters), which confirms the deletion suspected by mass spectrometry as a missing G-C basepair at position 184 in the published sequence (in lowercase letters). (Reproduced from **ref. *66*** by permission of Wiley–VCH [copyright 2001].)

3.3.2. Mutation Detection by ICEMS

In general, for the detection of sequence variations in PCR-amplified sequences, ICEMS is performed under completely denaturing conditions by applying elevated temperatures. Because the mass of an AT basepair (617.4 Dalton) differs from that of a GC basepair (618.4 Dalton) only by one mass unit, base changes in double-stranded chromosomal fragments would go undetected because they do not cause a sufficient difference in mass. In contrast, by measuring the masses of the single strands, base substitutions both in heterozygous and homozygous samples can be unequivocally identified in DNA fragments with lengths of more than 80 bp by taking advantage of mass differences of at least 9 Daltons (A>T) up to 40 Daltons (G>C).

Fig. 9. *(continued)* bicarbonate, pH 8.4, (B) 25 m*M* butyldimethylammonium bicarbonate, pH 8.4, 40% acetonitrile; linear gradient: 10–40% B in 3 min, 40–50% B in 12 min; flow rate: 3.0 µL/min; temperature: 25°C; sheath gas: nitrogen; sheath liquid: 3.0 µL/min acetonitrile; scan, 1000–3000 amu; sample: 15 fmol digest. (Reproduced from **ref. *39*** by permission of Wiley–VCH [copyright 2001].)

Table 3
Allele Identification Though Comparison of the Eight Theoretically Possible with the Measured Masses

Allele	Theoretical molecular mass	Measured molecular mass
A+dA_f	15,893.54	15,890
A_r	15,806.42	15,806
T+dA_f	15,884.52	—
T_r	15,797.40	—
G+dA_f	15,909.54	—
G_r	15,871.47	—
C+dA_f	15,869.51	15,869
C_r	15,831.44	15,831

Source: From **ref. *66*** by permission of Wiley-VCH (copyright 2001).

An example for the recognition of an A>C polymorphism in a heterozygous individual is illustrated in **Table 3**. Based on the high accuracy of mass measurements and a comparison of the measured and theoretical masses, it was evident that from the four principally possible alleles only the alleles A and C fit the experimental data. As expected, the masses of the forward strands were shifted by 312 mass units to higher mass resulting from the nontemplate addition of an additional deoxyadenosine by *Taq* polymerase ***(27,66,95)***.

3.3.3. Resequencing of Multiple SNPs

The information content of SNPs increases considerably when several SNPs are combined to form haplotypes. Haplotypes are usually inferred from individual unphased SNP genotypes. Although accuracy of such inferences tends to be excellent ***(96)***, experimental confirmation is still desirable.

In this context, ICEMS was shown to be an inexpensive and rapid alternative ***(92)*** to the traditionally applied techniques like cloning or allele-specific PCR, followed by Sanger sequencing ***(88)***. For example, the identification of SNPs in a 76-bp amplicon containing two polymorphic sites, namely A>T and A>G, located at positions 26 and 40, respectively, from the 5' end of the forward primer, is depicted in **Fig. 11**. The reverse and forward strands of the double-stranded PCR product were partly separated. Although it is feasible to resolve the two strands completely, as chromatography on a poly(styrene/divinylbenzene) matrix separates single-stranded DNA fragments as a function of base composition rather than size ***(10)***, this is not necessary because the masses of the eluting single-stranded DNA fragments can be determined unambiguously even if eluted in a single peak. Mass spectra for both single strands were extracted from the reconstructed ion chromatogram (*see* **Fig. 11A**) and deconvoluted to yield intact molecular masses of 23915 and 23218, respectively (*see* **Fig. 11B and C**). The measured molecular masses corresponded excellently with the theoretical masses inferred from the forward and reverse sequences of the A,A allele and, thus, clearly identified the sample as a homozygous A,A haplotype.

3.3.4. High-Throughput Analysis of PCR Products

By genotyping of the Y-chromosomal SNP M9, the high reliability and speed of ICEMS was approved ***(91)***. Moreover, the suitability of ICEMS for forensic SNP genotyping was demonstrated. The M9 locus, a C>G transversion, was initially detected by denaturing HPLC ***(97)*** and defines an ancestral lineage that is found in all geographical regions except Africa. The two alleles of the M9 locus were differentiated in 62-bp-long amplicons on the basis of intact molecular mass measurements. Because sample purification by liquid chromatography usually remains the time-limiting step during the analysis, the duty cycle for a single run was reduced

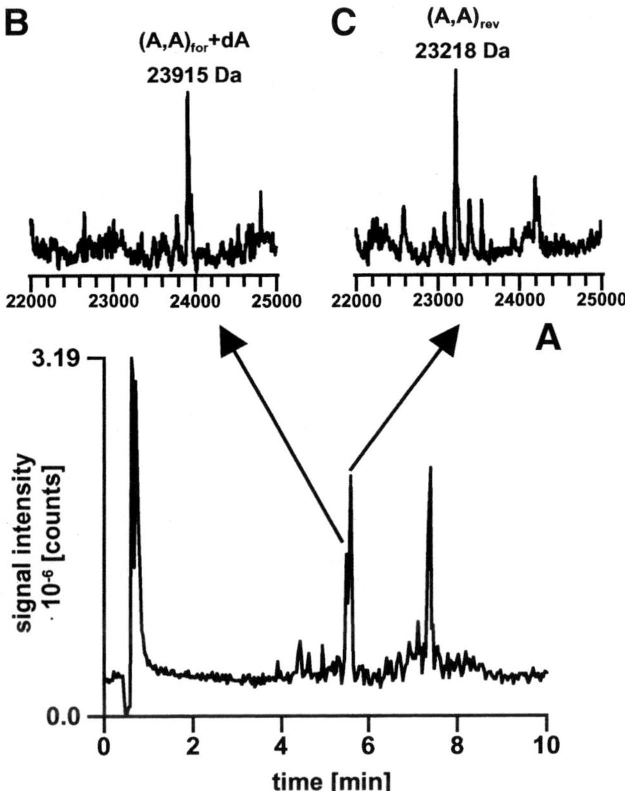

Fig. 11. Determination of multiple SNPs in a homozygous 76-bp amplicon of STS G-109954 by ICEMS. Column, PS-DVB monolith, 60 × 0.20 mm i.d.; mobile phase, (A) 25 m*M* butyldimethylammonium bicarbonate, pH 8.40, (B) 25 m*M* butyldimethylammonium bicarbonate, pH 8.40, 40% acetonitrile; linear gradient, 5–70% B in 10 min, flow rate, 2.0 µL/min; temperature, 70°C; scan, 500–2000 amu; sheath gas, nitrogen; postcolumn addition of 3.0 µL/min acetonitrile; sample, 500 nL PCR product. (Reproduced from **ref. *92*** by permission of Oxford University Press [copyright 2002].)

to 2 min by applying a separation time of 45 s, a column regeneration time of 45 s, and column equilibration time of 30 s (*see* **Fig. 12**), which, in principle, enables the analysis of more than 700 samples per day on one system. Moreover, in a blind study, 90 different DNA samples from unrelated Caucasian males were analyzed using both ICEMS and a restriction fragment length polymorphism assay employing capillary gel electrophoretic analysis. Of these, two samples could successfully be genotyped only by ICEMS, whereas no signals were detected by the other method. This observation was attributed to the low amount of DNA present in these two samples. All other samples were assigned correctly by both methods and the genotypes determined correlated 100%.

3.3.5. Detection of Alleles Predisposing to Ovarian and Breast Cancer

Most ovarian cancers are considered sporadic, but about 5–10% of cases present with a family history of ovarian and/or breast cancer that suggests inheritance of predisposing factors. The first gene predisposing to ovarian and breast cancer, named *BRCA1*, was cloned by Miki et al. in 1994 ***(98)***. Mutations in *BRCA1* are found in more than 80% of families with a history of

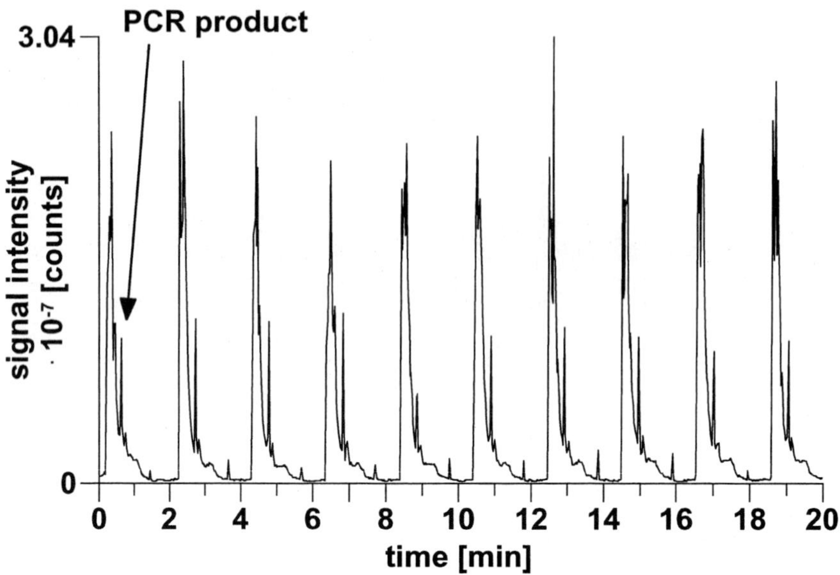

Fig. 12. High-speed, repetitive analysis of a 61-bp PCR amplicon of the Y-chromosomal SNP locus M9. Column, monolithic PS-DVB, 60 × 0.20 mm i.d.; mobile phase, (A) 25 m*M* butyldimethylammonium bicarbonate, pH 8.40, (B) 25 m*M* butyldimethylammonium bicarbonate, pH 8.40, 40% acetonitrile; linear gradient, 5–40% B in 45 s, 100% B for 30 s, 5% B for 45 s; flow rate, 2.0 μL/min; temperature, 70°C; scan, 500–2000 amu; sheath gas, nitrogen; sheath liquid, 3 μL/min acetonitrile; sample, M9 allele G, approx 100 fmol. (Reproduced from **ref. *91*** by permission of Elsevier Science [copyright 2002].)

both breast and ovarian cancer ***(99)***. Therefore, a fast and reliable screening tool for the detection of sequence variations in *BRCA1* is of utmost importance in presymptomatic tumor diagnostics.

In this context, partially denaturing HPLC (DHPLC) ***(94,100,101)*** has emerged as one of the most sensitive physical mutation-screening methods. However, a recent study reported difficulties in the detection of two mutations in exon 5 of *BRCA1*, namely 286A>G and 300T>G ***(102)***. This led to the recommendation to sequence exon 5 in all cases, in which no disease-causing mutation could be detected in any of the BRCA1 exons. Although rare, such observations have underscored the need for developing alternative screening methods like ICEMS that are independent of experimental conditions like temperature or denaturant concentration for the detection of a sequence variation in a DNA fragment of interest.

The outstanding performance of ICEMS applied as mutation screening tool was demonstrated by the successful identification of four different mutations, namely 259G>A, 286A>G, 300T>G, and 331+1G>A, in exon 5 and intron 5 of *BRCA1*, respectively. In order to reduce the length of the amplicon that contained the four mutations to a length that is more suitable for ESI-MS analysis, the 206-bp fragment was digested with the restriction enzyme *Nla*III. Because the enzyme generates four-nucleotide overhangs of the sequence CATG, the mixture to be analyzed in principle comprised four oligonucleotides of different lengths and sequences (16-, 20-, 101-, and 105-mer) as well as two reverse complementary oligonucleotides of the same length (2 × 85-mer). The reconstructed ion chromatogram of an *Nla*III restriction digest of the 206-bp amplicon generated from a heterozygous carrier of the 286A>G mutation in exon 5 of *BRCA1* is illustrated in **Fig. 13A**. The 16-mer and 20-mer oligonucleotides eluted from the column at retention times between 3.5 and 4.5 min, but they were detected as very weak signals because the mass spectrometer was tuned primarily for the sensitive detection of larger oligonucleotides. Because both fragments did not contain a mutation, no attempt was made to optimize the tuning parameters for better detection.

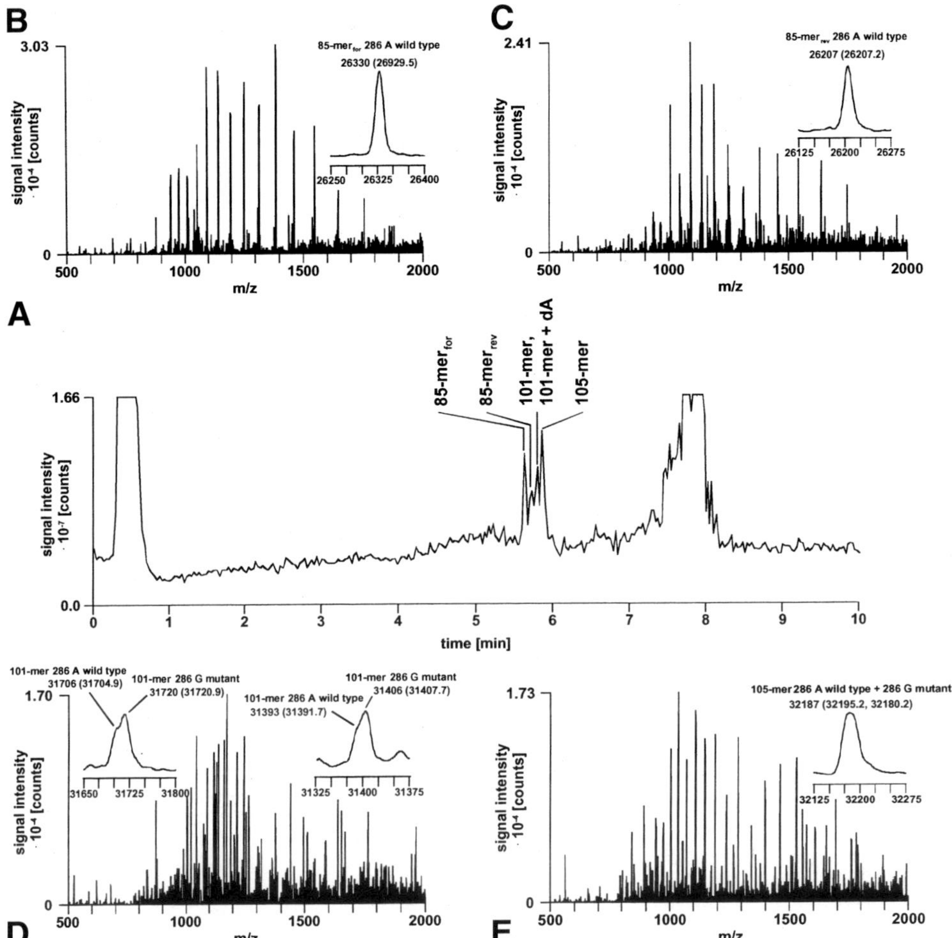

Fig. 13. Detection of the 286A>G mutation in exon 5 of *BRCA1* in a heterozygous carrier by ICEMS. (**A**) Reconstructed ion chromatogram of a *Nla*III restriction digest of a 206-bp amplicon. The 16- and 20-mer restriction fragments eluted at the beginning of the chromatogram. (**B–E**) Raw and deconvoluted (insets) mass spectra of the four peaks resolved by chromatography. Column, PS-DVB monolith, 60 × 0.20 mm i.d.; mobile phase, (A) 25 m*M* butyldimethylammonium, pH 8.40, (B) 25 m*M* butyldimethylammonium bicarbonate, pH 8.40, 40% acetonitrile; linear gradient, 5–70% B in 10 min, flow rate, 2.0 μL/min; temperature, 70°C; scan, 500–2000 amu; sheath gas, nitrogen; postcolumn addition of 3.0 μL/min acetonitrile; sample, 500 nL PCR product. (Reproduced from **ref. *28*** by permission of Wiley–VCH [copyright 2003].)

The remaining four oligonucleotides, ranging in size from 85 to 105 bp, were at least partly separated from each other. For all five single strands, raw mass spectra were extracted from the reconstructed ion chromatogram (*see* **Fig. 13B–E**) and deconvoluted to yield their intact molecular masses (*see* insets in **Fig. 13B–E**).

In comparison to the raw mass spectra of the forward (*see* **Fig. 13B**) and reverse 85-mer (*see* **Fig. 13C**), those of the 101-mer (*see* **Fig. 13D**) and 105-mer (*see* **Fig. 13E**) look quite messy. This is for good reason: The peak eluting off the column after the 85-mers did not only contain the wild-type (286A) 101-mer but also the respective mutant (286G) allele as the 286A>G mutation had been amplified from germline DNA of a heterozygous carrier. The series of multi-

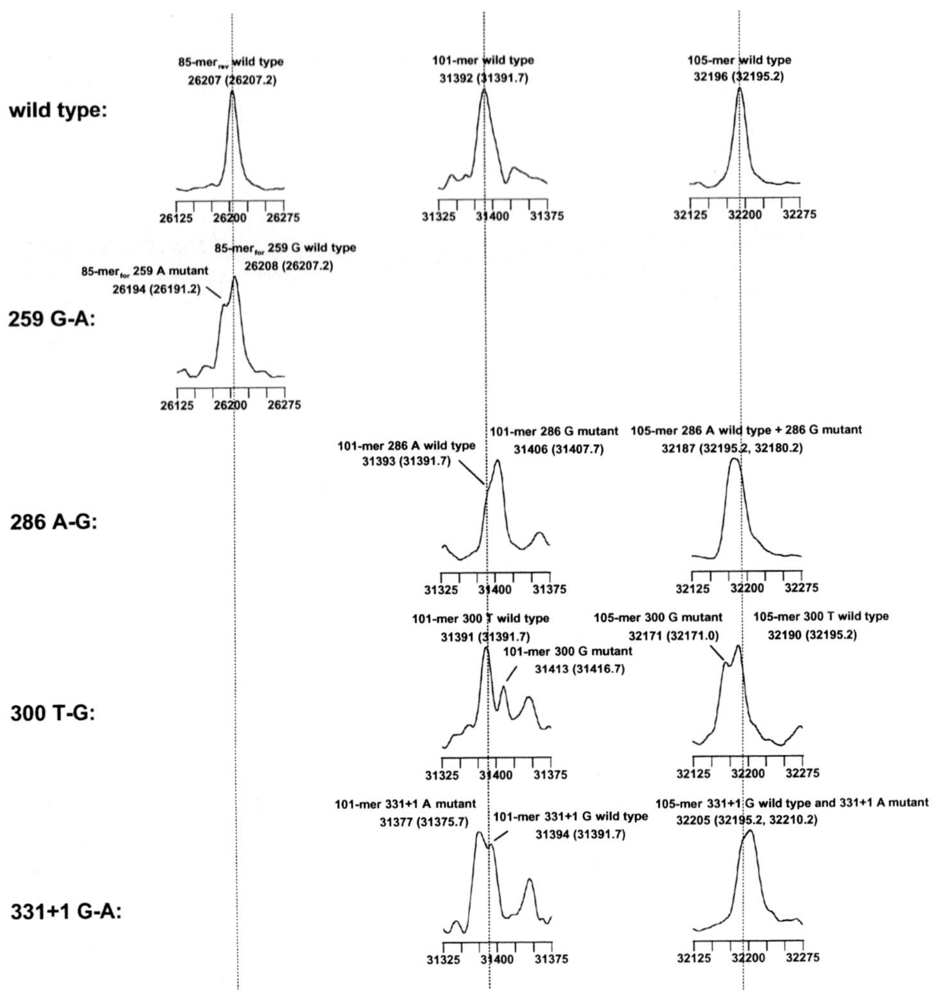

Fig. 14. Deconvoluted mass spectra of the 85-mer, 101-mer, and 105-mer restriction fragments of the wild-type and mutant alleles of the four *BRCA1* mutations 259G>A, 286A>G, 300T>G, and 331+1A>G. The top row shows the spectra of wild-type alleles only, whereas rows 2–5 show spectra obtained from heterozygous mutation carriers. Column, PS-DVB monolith, 60×0.20 mm i.d.; mobile phase, (A) 25 m*M* butyldimethylammonium bicarbonate, pH 8.40, (B) 25 m*M* butyldimethylammonium bicarbonate, pH 8.40, 40% acetonitrile; linear gradient, 5–70% B in 10 min, flow rate, 2.0 μL/min; temperature, 70°C; scan, 500–2000 amu; sheath gas, nitrogen; postcolumn addition of 3.0 μL/min acetonitrile; sample, 500 nL PCR product in each run. (Reproduced from **ref.** ***28*** by permission of Wiley-VCH [copyright 2003].)

ply charged ions of the two 85-mers were clearly visible and were deconvoluted into an almost ideal Gaussian signal on a real mass scale (*see* insets in **Fig. 13B**,**C**), whereas the deconvolutions of the 101-mers yielded signals showing more or less pronounced shoulders (*see* **Fig. 13D**). Likewise, the peak of the 105-mer fragment contained both the wild type and the mutant alleles, which manifested itself in a significant broadening of the deconvoluted mass peak (*see* **Fig. 13E**).

Figure 14 summarizes the mass spectra of all four mutations. The presence of a mutation is ideally revealed by the appearance of a second distinct peak in the mass spectrum. However, this was only the case for the 300T>G mutant in the 101-mer fragment. In all other instances, the presence of the mutant allele resulted in a shoulder (e.g., 85-mer of 259G>A) or a slight broadening of the signal (e.g., 105-mer of 286A>G). Nevertheless, in contrast to partially denaturing DHPLC (*see* **Fig. 2** of **ref. *102***), all four mutations could be readily recognized, including 286A>G and 300T>G, which clearly demonstrates the high sensitivity of ICEMS for the detection of sequence variations.

4. Conclusions

ICEMS is a powerful method for nucleic acids analysis, which is based on the hyphenation of ion-pair reversed-phase liquid chromatography to electrospray ionization–mass spectrometry. Through the use of miniaturized columns in the capillary format under optimized chromatographic and mass spectrometric conditions very small amounts of biological samples are amenable to analytical characterization. This is expected to make ICEMS applicable not only in specialized academic laboratories but also for routine analysis. The accurate molecular masses obtained from ICEMS measurements yield important information about the identity and structure of nucleic acids and will help to characterize synthetic nucleic acids as well as to detect sequence variations in genomic DNA.

References

1. Wells, R. D. (1984) High performance liquid chromatography of DNA. *J. Chromatogr.* **336,** 3–14.
2. Hecker, R. and Riesner, D. (1987) Chromatographic separation of DNA restriction fragments. *J. Chromatogr.* **418,** 97–114.
3. Thompson, J. A. (1987) A review of high performance liquid chromatography in nucleic acid research. *BioChromatography* **2,** 68–79.
4. Thayer, J. R., McCormick, R. M., and Avdalovic, N. (1996) High-resolution nucleic acid separation by high performance liquid chromaotgraphy. *Methods Enzymol.* **271,** 147–174.
5. Huber, C. G. (1998) Micropellicular stationary phases for high-performance liquid chromatography of dsDNA. *J. Chromatogr. A* **806,** 3–30.
6. Huber, C. G. (2000) Biopolymer chromatography, in *Encyclopedia of Analytical Chemistry* (Meyers, R. A., ed.), Wiley, Chichester, pp. 11,250–11,278.
7. Bartha, A. and Stahlberg, J. (1994) Electrostatic retention model of reversed-phase ion-pair chromatography. *J. Chromatogr. A* **668,** 255–284.
8. Huber, C. G., Oefner, P. J., and Bonn, G. K. (1995) Rapid and accurate sizing of DNA fragments by ion-pair reversed-phase chromatography on alkylated nonporous polystyrene/divinylbenzene particles. *Anal. Chem.* **67,** 578–585.
9. Huber, C. G., Oefner, P. J., and Bonn, G. K. (1993) High-resolution liquid chromatography of oligonucleotides on highly crosslinked poly(styrene-divinylbenzene) particles. *Anal. Biochem.* **212,** 351–358.
10. Oefner, P. J. (2000) Allelic discrimination by denaturing high-performance liquid chromatography. *J. Chromatogr. B* **739,** 345–355.
11. Andren, P. E., Emmett, M. R., and Caprioli, R. M. (1994) Micro-electrospray-zeptomole-attomole per microliter sensitivity for peptides. *J. Am. Soc. Mass Spectrom.* **5,** 867–869.
12. Fenn, J. B., Mann, M., Meng, C. K., Wong, S. F., and Whitehouse, C. M. (1989) Electrospray ionization for mass spectrometry of large biomolecules. *Science* **246,** 64–71.
13. Nordhoff, E., Kirpekar, F., and Roepstorff, P. (1996) Mass spectrometry of nucleic acids. *Mass Spectrom. Rev.* **15,** 76–138.
14. Bayer, E., et al. (1994) Analysis of double-stranded oligonucleotides by electrospray mass spectromatry. *Anal. Chem.* **66,** 3858–3863.
15. Deroussent, A., Le Caer, J.-P., and Gouyette, A. (1995) Electrospray mass spectrometry for the purity of natural and modified oligodeoxynucleotides. *Rapid Commun. Mass Spectrom.* **9,** 1–4.
16. Barry, J. P., Vouros, P., Van Schepdael, A., and Lay, S.-J. (1995) Mass and sequence verification of modified oligonucleotides using electrospray tandem mass spectrometry. *J. Mass Spectrom.* **30,** 993–1006.
17. Reddy, D. M., Rieger, R. A., Torres, M. C., and Iden, C. R. (1994) Analyses of oligodeoxynucleotides containing modified components by electrospray ionization mass spectrometry. *Anal. Biochem.* **220,** 200–207.
18. Muddiman, D. C., et al. (1996) Characterization of PCR products from Bacilli using electrospray ionization FTICR mass spectrometry. *Anal. Chem.* **68,** 3705–3712.

19. Muddiman, D. C., Anderson, G. A., Hofstadler, S. A., and Smith, R. D. (1997) Length and base composition of PCR-amplified nucleic acids using mass measurements from electrospray ionization mass spectrometry. *Anal. Chem.* **69,** 1543–1549.
20. Fuerstenau, S. D. and Benner, W. H. (1995) Molecular weight determination of megadalton DNA electrospray ions using charge detection time-of-flight mass spectrometry. *Rapid Commun. Mass Spectrom.* **9,** 1528–1538.
21. Cheng, X., et al. (1996) Molecular weight determination of plasmid DNA using electrospray ionization mass spectrometry. *Nucleic Acids Res.* **24,** 2183–2189.
22. Cole, R. B. (1997) *Electrospray Ionization Mass Spectrometry: Fundamentals, Instrumentation & Applications.* Wiley, New York.
23. Iribarne, J. V. and Thomson, A. (1976) On the evaporation of small ions from charged droplets. *J. Chem. Phys.* **64,** 2287–2294.
24. Dole, M., et al. (1968) Molecular beams of macroions. *J. Chem. Phys.* **49,** 2240–2249.
25. Mann, M., Meng, C. K., and Fenn, J. B. (1989) Interpreting mass spectra of multiply charged ions. *Anal. Chem.* **61,** 1702–1708.
26. Portier, N., Van Dorsselaer, A., Cordier, Y., Roch, O., and Bischoff, R. (1994) Negative electrospray ionization mass spectrometry of synthetic and chemically modified oligonucleotides. *Nucleic Acids Res.* **22,** 3895–3903.
27. Oberacher, H., Parson, W., Mühlmann, R., and Huber, C. G. (2001) Analysis of polymerase chain reaction products by on-line liquid chromatography-mass spectrometry for genotyping of polymorphic short tandem repeat loci. *Anal. Chem.* **73,** 5109–5115.
28. Oberacher, H., Huber, C. G., and Oefner, P. J. (2003) Mutation scanning by ion-pair reversed-phase high-performance liquid chromatography-electrospray ionization mass spectrometry (ICEMS). *Hum. Mutat.* **21,** 86–95.
29. Smith, R. D., Loo, J. A., Edmonds, C. G., Barinaga, C. J., and Udseth, H. R. (1990) New developments in biochemical mass spectrometry: electrospray ionization. *Anal. Chem.* **62,** 882–889.
30. Bleicher, K. and Bayer, E. (1994) Various factors influencing the signal intensity of oligonucleotides in electrospray mass spectrometry. *Biol. Mass Spectrom.* **23,** 320–322.
31. Stults, J. T. and Marsters, J. C. (1991) Improved electrospray ionization of synthetic oligodeoxynucleotides. *Rapid Commun. Mass Spectrom.* **5,** 359–363.
32. Brezinschek, H. P., Brezinschek, R. I., and Lipsky, P. E. (1995) Analysis of the heavy chain repertoire of human eripheral B cells using single-cell polymerase chain reaction. *J. Immunol.* **155,** 190–202.
33. Liu, C., Wu, Q., Harms, A. C., and Smith, R. D. (1996) On-line microdialysis sample cleanup for electrospray ionization mass spectrometry of nucleic acid samples. *Anal. Chem.* **68,** 3295–3299.
34. Ross, P. L., Davis, P. A., and Belgrader, P. (1998) Analysis of DNA fragments from conventional and microfabricated PCR devices using delayed extraction MALDI-TOF mass spectrometry. *Anal. Chem.* **70,** 2067–2073.
35. Huber, C. G. and Buchmeiser, M. R. (1998) On-line cation-exchange for suppression of adduct formation in negative-ion electrospray mass spectrometry of nucleic acids. *Anal. Chem.* **70,** 5288–5295.
36. Bleicher, K. and Bayer, E. (1994) Analysis of oligonucleotides using coupled high performance liquid chromatography-electrospray mass spectrometry. *Chromatographia* **39,** 405–408.
37. Apffel, A., Chakel, J. A., Fischer, S., Lichtenwalter, K., and Hancock, W. S. (1997) Analysis of oligonucleotides by HPLC–electrospray ionization mass spectrometry. *Anal. Chem.* **69,** 1320–1325.
38. Laken, S. J., et al. (1998) Genotyping by mass spectrometric analysis of short DNA fragments. *Nature Biotechnol.* **16,** 1352–1356.
39. Huber, C. G. and Oberacher, H. (2001) Analysis of nucleic acids by on-line liquid chromatography–mass spectrometry. *Mass Spectrom. Rev.* **20,** 310–343.
40. Fountain, K. J., Gilar, M., and Gebler, J. C. (2003) Analysis of native and chemically modified oligonucleotides by tandem ion-pair reversed-phase high-performance liquid chromatography/electrospray ionization mass spectrometry. *Rapid Commun. Mass Spectrom.* **17,** 646–653.
41. Hannis, J. C. and Muddiman, D. C. (1999) Accurate characterization of the tyrosine hydroxylase forensic allele 9.3 through development of electrospray ionization Fourier transform ion cyclotron resonance mass spectrometry. *Rapid Commun. Mass Spectrom.* **13,** 954–962.
42. Wunschel, D. S., Pasa Tolic, L., Feng, B., and Smith R. D. (2000) Electrospray ionization Fourier transform ion cyclotron resonance analysis of large polymerase chain reaction products. *J. Am. Soc. Mass Spectrom.* **11,** 333–337.

43. Gross, J. and Hillenkamp, F. (2000) Mass spectrometry of nucleic acids, in *Encyclopedia of Analytical Chemistry* (Meyers, R. A., ed.), Wiley, Chichester, pp. 5022–5051.
44. Ikonomou, M. G., Blades, A. T., and Kebarle, P. (1991) Electrospray–ion spray: a comparison of mechanism and performance. *Anal. Chem.* **63,** 1989–1998.
45. Banks, J. F. (1996) High-sensitivity peptide mapping using packed-capillary liquid chromatography and electrospray ionization mass spectrometry. *J. Chromatogr. A* **743,** 99–104.
46. Huber, C. G. and Krajete, A. (1999) Analysis of nucleic acids by capillary ion-pair reversed-phase HPLC coupled to negative ion-electrospray ionization mass spectrometry. *Anal. Chem.* **71,** 3730–3739.
47. Legido-Quigley, C., Smith, N. W., and Mallet, D. (2002) Quantification of the sensitivity increase of a micro-high-performance liquid chromatography–electrospray ionization mass spectrometry system with decreasing column diameter. *J. Chromatogr. A* **976,** 11–18.
48. Novotny, M. (1997) Capillary biomolecular separations. *J. Chromatogr. B* **689,** 55–70.
49. Karlsson, K.-E. and Novotny, M. (1984) A miniature gradient elution system for liquid chromatography with packed capillary columns. *HRC & CC* **7,** 411–413.
50. Kennedy, R. T. and Jorgenson, J. W. (1989) Preparation and evaluation of packed capillary liquid chromatography columns with inner diameters from 20 to 50 µm. *Anal. Chem.* **61,** 1128–1135.
51. Scott, R. P. W. and Simpson, C. F. (1982) Determination of the extracolumn dispersion occurring in the different components of a chromatographic system. *J. Chromatogr. Sci.* **20,** 62–66.
52. Novotny, M. (1988) Recent advances in microcolumn liquid chromatography. *Anal. Chem.* **60,** 500A–510A.
53. Hirata, Y. and Novotny, M. (1979) Techniques of capillary liquid chromatography. *J. Chromatogr.* **186,** 521–528.
54. Yang, F. J. (1986) Fused-silica microparticulare packed column HPLC–split flow gradient elution. *HRC & CC* **9,** 84–88.
55. Davis, M. T., Stahl, D. C., and Lee, T. D. (1995) Low flow high-performance liquid chromatography solvent delivery system designed for tandem capillary liquid chromatography–mass spectrometry. *J. Am. Soc. Mass Spectrom.* **6,** 571–577.
56. Takeuchi, T. and Ishii, D. (1982) Continuous gradient elution in micro high-performance liquid chromatography. *J. Chromatogr.* **253,** 41–47.
57. Oberacher, H., Krajete, A., Parson, W., and Huber, C. G. (2000) Preparation and evaluation of packed capillary columns for the separation of nucleic acids by ion-pair reversed-phase high-performance liquid chromatography. *J. Chromatogr. A* **893,** 23–35.
58. Premstaller, A., Oberacher, H., and Huber, C. G. (2000) High-performance liquid chromatography-electrospray ionization mass spectrometry of single- and double-stranded nucleic acids using monolithic capillary columns. *Anal. Chem.* **72,** 4386–4393.
59. Huber, C. G., Oefner, P. J., Preuss, E., and Bonn, G. K. (1993) High-resolution liquid chromatography of DNA fragments an non porous poly(styrene-divinylbenzene) particles. *Nucleic Acids Res.* **21,** 1061–1066.
60. Kebarle, P. and Tang, L. (1993) From ions in solution to ions in the gas phase. The mechanism of electrospray mass spectrometry. *Anal. Chem.* **65,** 972A–986A.
61. Tang, L. and Kebarle, P. (1993) Dependence of ion intensity in electrospray mass spectrometry on the concentration of the analytes in the electrosprayed solution. *Anal. Chem.* **65,** 3654–3668.
62. Kebarle, P. (2000) A brief overview of the present status of the mechanisms involved in electrospray mass spectrometry. *J. Mass Spectrom.* **35,** 804–817.
63. Apffel, A., Chakel, J. A., Fischer, S., Lichtenwalter, K., and Hancock, W. S. (1997) New procedure for the use of high-performance liquid chromatography–electrospray ionization mass spectrometry for the analysis of nucleotides and oligonucleotides. *J. Chromatogr. A* **777,** 3–21.
64. Huber, C. G. and Krajete, A. (2000) Sheath liquid effects in capillary high-performance liquid chromatography-electrospray mass spectrometry of oligonucleotides. *J. Chromatogr. A* **870,** 413–424.
65. Oberacher, H. and Huber, C. G. (2002) Capillary monoliths for the analysis of nucleic acids by high-performance liquid chromatography–electrospray ionization mass spectrometry. *TRAC* **21,** 166–174.
66. Oberacher, H., Oefner, P. J., Parson, W., and Huber, C. G. (2001) On-line liquid chromatography mass spectrometry: a useful tool for the detection of DNA sequence variation. *Angew. Chem. Int. Educ.* **40,** 3828–3830.
67. Harsch, A., Sayer, J. M., Jerina, D. M., and Vouros, P. (2000) HPLC-MS/MS identification of positionally isomeric benzo[c]phenanthrene diol epoxide adducts in duplex DNA. *Chem. Res. Toxicol.* **13,** 1342–1348.

68. Searcy, J. Q. and Fenn, J. B. (1976) Clustering of water on hydrated protons in a supersonic free jet expansion. *J. Chem. Phys.* **64,** 1861–1862.
69. Lin, H. Y., Gonyea, G. J., Chowdhury, S. K., and Swapan, K. (1995) Polyethylene glycol: a reagent for tuning and calibrating mass spectrometers for positive-ion atmospheric pressure chemical ionization. *J. Mass Spectrom.* **30,** 381–383.
70. Anacleto, J. F., Pleasance, S., and Boyd, R. K. (1992) Calibration of ion spray mass spectra using cluster ions. *Org. Mass Spectrom.* **27,** 660–666.
71. Moini, M. (1994) Ultramark 1621 as a calibration/reference compound for mass spectrometry. II. Positive- and negative-ion electrospray ionization. *Rapid Commun. Mass Spectrom.* **8,** 711–714.
72. Moini, M., Jones, B. L., Rogers, R. M., and Jiang, L. (1998) Sodium trifluoroacetate as a tune/calibration compound for positive- and negative-ion electrospray ionization mass spectrometry in the mass range 100–4000 Da. *J. Am. Soc. Mass Spectrom.* **9,** 977–980.
73. Premstaller, A., et al. (2001) High-performance liquid chromatography-electrospray ionzation mass spectrometry using monolithic capillary columns for proteomic studies. *Anal. Chem.* **73,** 2390–2396.
74. Oberacher, H., Walcher, W., and Huber, C. G. (2003) Effect of instrument tuning on the detectability of biopolymers in electrospray ionization mass spectrometry. *J. Mass. Spectrom.* **38,** 108–116.
75. Beaucage, S. L. and Caruthers, M. H. (1981) Deoxynucleoside phosphoramidites—a new class of key intermediates for deoxypolynucleotide synthesis. *Tetrahedron Lett.* **22,** 1859–1862.
76. Matteucci, M. D. and Caruthers, M. H. (1981) Synthesis of deoxyoligonucleotides on a polymer support. *J. Am. Chem. Soc.* **103,** 3185–3191.
77. Huber, C. G., Stimpfl, E., Oefner, P. J., and Bonn, G. K. (1996) A comparison of micropellicular ion-exchange and reversed-phase stationary phases for HPLC of oligonucleotides. *LC GC Int.* **14,** 114–127.
78. Huber, C. G. and Krajete, A. (2000) Comparison of direct infusion- and on-line liquid chromatography–electrospray ionization mass spectrometry for the analysis of oligonucleotides. *J. Mass Spectrom.* **35,** 870–877.
79. Elder, J. K., Amos, A., Southern, E. M., and Shippey, G. A. (1983) Measurement of DNA length by gel electrophoresis. *Anal. Biochem.* 1**28,** 223–226.
80. Kato, Y., et al. (1983) Separation of DNA restriction fragments by high performance ion-exchange chromatography on a non-porous ion exchanger. *J. Chromatogr.* **265,** 342–346.
81. Walcher, W., et al. (2002) Monolithic capillary columns for liquid chromatography–mass spectrometry in proteomic and genomic research. *J. Chromatogr. B* **782,** 111–125.
82. Edwards, A., Hammond, H. A., Jin, L., Caskey, C. T., and Chakraborty, R. (1992) Genetic variation at five trimeric and tetrameric tandem repeat loci in four human population groups. *Genomics* **12,** 241–253.
83. Venter, J. C., et al. (2001) The sequence of the human genome. *Science* **291,** 1304–1351.
84. International Human Genome Sequencing Consortium (2001) Initial sequencing and analysis of the human genome. *Nature* **409,** 860–921.
85. Brookes, A. J. (1999) The essence of SNPs. Gene 234, 177–186.
86. Oefner, P. J. (2002) Sequence variation and the biological function of genes: methodological and biological considerations. *J. Chromatogr. B* **782,** 3–25.
87. Dovici, N. J. and Zhang, J. (2000) Wie die Kapillarelektrophorese das menschliche Genom sequenzierte. *Angew. Chem.* **112,** 4635–4640.
88. Kristensen, V. N., Kelefiotis, D., Kristensen, T., and Borresen-Dale, A. L. (2001) High-throughput methods for detection of genetic variation. *Biotechniques* **30,** 318–322.
89. Gut, I. G. (2001) Automation in genotyping of single nucleotide polymorphisms. *Hum. Mutat.* **17,** 475–492.
90. Syvänen, A.-C. (2001) Accessing genetic variation: genotyping single nucleotide polymorphisms. *Nature Genet.* **2,** 930–942.
91. Berger, B., et al. (2002) Single Nucleotide Polymorphism genotyping by on-line liquid chromatography-mass spectrometry in forensic science with the Y-chromosomal locus M9. *J. Chromatogr. B* **782,** 89–97.
92. Oberacher, H., Oefner, P. J., Hölzl, G., Premstaller, A., and Huber, C. G. (2002) Re-sequencing of multiple single nucleotide polymorphisms by liquid chromatography–electrospray ionization mass spectrometry. *Nucleic Acids Res.* **30,** e67.
93. Oberacher, H. and Huber, C. G. (2003) Genotyping of single nucleotide polymorphisms by liquid chromatography-electrospray ionization mass spectrometry. *Anal. Bioanal. Chem.* **376,** 292–294.

94. Premstaller, A. and Oefner, P. J. (2003) Denaturing high-performance liquid chromatography. *Methods Mol. Biol.* **212,** 15–35.
95. Null, A. P., Hannis, J. C., and Muddiman, D. C. (2000) Preparation of single-stranded PCR products for electrospray ionization mass spectrometry using the DNA repair enzyme lamda exonuclease. *Analyst* **125,** 619–626.
96. Fallin, D. and Schork, N. J. (2000) Accuracy of haplotype frequency estimation for biallelic loci via the expectation-maximization algorithm for unphased diploid genotype data. *Am. J. Hum. Genet.* **67,** 947–959.
97. Underhill, P. A., et al. (1997) Detection of numerous Y chromosome biallelic polymorphisms by denaturing high-performance liquid chromatography. *Genome Res.* **7,** 996–1005.
98. Miki, Y., et al. (1994) A strong candidate for the breast and ovarian cancer susceptibility gene BRCA1. *Science* **266,** 66–71.
99. Breast Cancer Linkage Consortium (1995) Breast and ovarian cancer incidence in BRCA1-mutation carriers. Am. J. Hum. Genet. **56,** 265–271.
100. Oefner, P. J. and Underhill, P. A. (1998) DNA mutation detection using denaturing high-performance liquid chromatography, in *Current Protocols in Human Genetics* (Dracopoli, N. C. et al., eds.), Wiley, Chichester, pp. 7.10.1–7.10.12.
101. Xiao, W. and Oefner, P. J. (2001) Denaturing high-performance liquid chromatography: a review. *Hum. Mutat.* **17,** 439–474.
102. Wagner, T., et al. (1999) Denaturing high-performance liquid chromatography detects reliably BRCA1 and BRCA2 mutations. *Genomics* **62,** 369–376.

27

Comparative Genomic Hybridization in Clinical and Medical Research

Peng-Hui Wang, Yann-Jang Chen, and Chi-Hung Lin

1. Introduction

Cytogenetic research has had a major impact on the field of medicine, especially in oncology and reproductive medicine, providing an insight into the frequency of chromosomal abnormalities that occur during gametogenesis, embryonic development, and tumor development *(1)*. This is well emphasized by the continuing focus on genetic abnormalities that are associated with, as well as probably responsible for, tumor origin, tumor progression, spontaneous abortions, and congenital anomalies. However, information on recurrent chromosomal aberrations in solid tumors and in some hematological cancer is still limited. The growth of solid tumor in culture for cytogenetic analysis is poor and is compounded by low mitotic indices *(2)*. Often the specimens are contaminated with bacterial and other microbial agents and might contain large regions of necrotic tissue. In addition, the major clone that does grow might not reflect its true representation in the tumor in vivo, where multiple subclones exist with complex chromosomal alterations, making identification of primary genetic changes difficult. Furthermore, solid-tumor metaphase chromosomes often have poor morphology *(1)*. A newly described molecular-cytogenetic technique that does not rely on growth of the tumor in culture might well accelerate the rate at which perturbed chromosomal regions can be cytogenetically identified and molecular-genetically characterized in solid tumors *(2)*. In the past decade, fluorescence *in situ* hybridization (FISH) has significantly improved the cytogenetic analysis of tumors. FISH utilizes specific chromosomal probes, usually composed of cloned fragments of DNA *(3)* (see Chapter 29). These probes will anneal only to their matching complementary DNA sequences on target chromosomes and can accurately detect and target one specific gene or chromosome region at a time.

However, the application of FISH in cytogenetic analysis leaves the majority of the genome unexamined *(4)*. Now, these limitations can be circumvented through the use of molecular cytogenetic approaches referred to as new FISH-based technologies, including reverse FISH, multiplex FISH (M-FISH), spectral karyotyping (SKY), comparative genomic hybridization (CGH) analysis, and matrix or microarray–CGH (M-CGH) *(4)*. These technologies have bridged the gap between molecular genetics and conventional cytogenetics. The combination of traditional cytogenetic techniques with molecular-genetic methodologies has added a new and powerful dimension to human genetics. Although the information derived using reverse FISH is highly informative, the procedure is technically demanding and requires specialized micromanipulation equipment to microdissect the region of interest from abnormal chromosomes *(3)*. Among these technologies, CGH has provided an unparalleled insight into the nature of chromosome imbalance in disease development and progression. CGH-based technology is able to discover and map genomic regions for chromosomal gains or losses in a single experi-

From: *Medical Biomethods Handbook*
Edited by: J. M. Walker and R. Rapley © Humana Press, Inc., Totowa, NJ

ment without any prior information on the chromosomal aberration in question ***(5)***. This ability addresses many of the deficiencies of FISH and conventional cytogenetic analyses. CGH produces a map of DNA sequence copy number changes as a function of chromosomal location throughout the entire genome ***(6)***. In a typical CGH experiment, genomic DNA from tumor and normal tissue is separately labeled with different fluorochromes (green color for tumor and red for normal control); these differently labeled DNA probes are hybridized simultaneously to metaphase chromosome spreads prepared from normal individuals ***(7)***. In addition to different fluorochromes for labeling (in the direct method), haptens (in the indirect method) are also frequently used as labeling dyes because of their flexibility and cost-efficiency. CGH is also performed with differentially labeled normal DNA as a reference standard for data analysis. Detailed analysis is performed using a sensitive monochrome cooled charge-coupled device (CCD) camera and automated image analysis software. Regions of loss or gain of DNA sequences are seen as changes in the ratio of the two fluorescence intensity ratio profiles along the target chromosomes. Thus, gene amplification or chromosomal duplication in the tumor DNA produces an elevated green-to-red ratio, and deletions of chromosomal loss cause a reduced ratio ***(8–11)***. This chapter will focus on the technique of CGH and its modifications and will review the genetic perturbations revealed by CGH for a number of tumor types and its potential application in clinical practice.

2. Method

Using schematic representation, it is easy to understand that the basis of the CGH procedure involves competitive *in situ* hybridization of differentially labeled tumor DNA and normal reference DNA to normal human metaphase spreads (*see* **Fig. 1**). The CGH approach does not require mitotic tumor cells, so fresh, frozen, and paraffin-embedded tissues can be examined. In addition, only a small amount (less than 500 ng) of genomic DNA is needed for a whole-genomic analysis. However, to use CGH to determine the true incidence of chromosome abnormality in human embryos (for reproductive medicine) and to develop CGH protocols compatible with preimplantation genetic diagnosis (PGD), several modifications to existing methods are necessary, because a single cell contains only 5–10 pg of DNA. Consequently, the DNA content of the cell must be amplified prior to use for CGH. We can achieve the approx 40,000-fold amplification necessary using a polymerase chain reaction (PCR)-based strategy to enzymatically copy the single-cell genome multiple times. The amplification is so efficient that sufficient DNA can be generated from a single cell for numerous subsequent PCR analyses, as well as for CGH ***(3,12)***. Detailed methodological reviews can be found at http://www.nhgri.nih.gov/DIR/LCG/CGH/technology.html and http://www.utoronto.ca/cancyto/ or in several excellent articles ***(2–4,6,13–19)***.

2.1. Normal Metaphase Preparation

Normal metaphase slides are prepared from phytohematoglutinin-stimulated peripheral blood lymphocyte cultures from a healthy individual. Cells are arrested in mitosis by colchicines, harvested, treated with hypotonic KCl, and fixed in methanol/acetic acid. The cells are dropped onto slides in such a way that chromosomes from mitotic cells are nicely spread ***(18)***.

2.2. The Procedure of CGH

Study DNA can be obtained by any type of DNA isolation that yields high-molecular-weight DNA that is suitable for use in CGH. Normal DNA for use as a reference can be taken from blood lymphocytes from any healthy man. DNAs are extracted in phenol–chloroform using standard protocols. The extracted DNA concentrations are estimated by spectrophotometric measurement. Before beginning any labeling, the quality of the DNA must be assessed. In fact, after labeling, it is strongly suggested that the quality of the DNA be assessed again. Take a

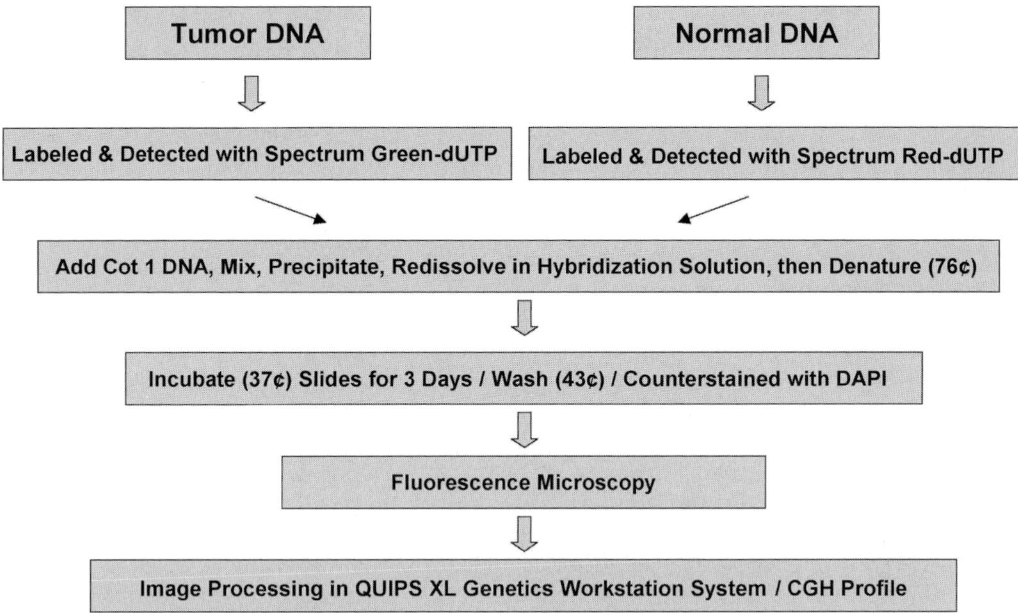

Fig. 1. Flowchart for CGH analysis. First, equal amounts (500 ng) of tumor DNA is labeled with green fluorochrome-conjugated nucleotide, such as Spectrum green–dUTP or fluorescence isothiocyanate–dUTP (FITC–dUTP) by nick translation, whereas reference DNA from peripheral lymphocytes of normal male or female donors is labeled with red fluorochrome-conjugated nucleotide, such as Spectrum red–dUTP or Texas red–dUTP. The sizes of the probes are optimized to a range from 500 bp to 2 kb. The labeled DNA (200 ng) is mixed with 10 μg of unlabeled human Cot-1 DNA for blocking ubiquitous repetitive sequences commonly detected on the centromeric and heterochromatic regions; then, it is ethanol precipitated and redissolved in 10 μL hybridization solution. The resulting probe mixture is denatured at 76°C for 7 min and reannealed at 37°C for 1 h. Metaphase chromosomes are also denatured at 76°C in 70% formamide solution for 3–4 min. The denatured probe mixture is dropped onto the metaphase chromosome slide, covered with a cover slip, sealed with rubber cement, and then allowed to hybridize by incubating the slide in a moist chamber at 37°C for 3 d. After hybridization, the slides are washed twice with 50% formamide with 2X SSC at 43°C for 10 min, followed by three changes with PN buffer at room temperature, 10 min each for removing the unbound probe. The slides are counterstained with diamidino-2-phenylindole (DAPI) and mounted in Vectashield mounting medium. Fluorescence signals from the hybridized chromosomes are captured and analyzed using CGH analysis software, such as the QUIPS XL Genetics Workstation system. Typically, 6–10 sets of metaphase chromosomes from one sample slide are observed under a fluorescence microscope equipped with a cooled CCD camera. The filter system consists of a triple-bandpass beam splitter, a triple-bandpass emission filter, and three single-bandpass excitation filters mounted on a computer-controlled filter wheel. This design allows collection of sequential, properly registered images from three fluorescence channels. The ratio profile of green (tumor) to red (control) fluorescence intensity is plotted as a function of locations along individual chromosomes using software. Data from 6–10 captured metaphases are pooled to obtain an averaged ratio profile for each sample (or patient). Genomic aberrations, such as gains (amplification) or losses (deletion), at certain chromosome regions are defined by setting the thresholds at 1.2 and 0.80, respectively.

1-µg sample of the DNA and run it on a 1% ethidium bromide stained gel. The criteria for the DNA include the following: (1) the sample must be of high molecular weight, (2) the DNA should also be quantified using an accurate spectrophotometer or a DyNA Quant 2000 fluorometer, and (3) the DNA should be dissolved in water, not TE buffer.

A successful CGH relies on the labeling of equal amounts of the high-quality DNA that is to be tested against the normal reference DNA. Test and reference DNAs are differentially labeled, and it is apparent that the size of the DNA before and after labeling is an important consideration in the overall success of the assay *(13)*. It is suggested that the labeling scheme be switched and the experiment repeated to ensure that the results are consistent.

Labeled DNAs should range in size from 500 bp to 2 kb to achieve optimal hybridization. In some cases, very high-molecular-weight DNA requires mechanical shearing before labeling, and in others, the extracted DNA can be labeled as is *(13)*. Labeling can be used with biotinylated and digoxigenin-conjugated deoxynucleotides, which require secondary detection steps before visualization posthybridization, or be used with directly fluorochrome-conjugated deoxynucleotides, which can be visualized, and require fewer posthybridization signals. **Table 1** lists the commonly used fluorochromes and their spectral characteristics. Nick translation is the method of choice for labeling the DNAs, because of its ability to regulate the nicking activity of DNAase I relative to the synthesizing activity of DNA polymerase I to achieve optimally sized fragments *(2)*. For applications on paraffin-embedded material, two publications described protocols for preparation, labeling, detection, and optimizing universal in vitro amplification of genomic DNA *(20,21)*. DOP-PCR proved to be the most effective method for amplifying and labeling genomic DNA from microdissected tissue areas *(19)*.

As described in **Fig. 1**, the procedure of CGH is briefly summarized as follows: (1) differentially label the DNAs; (2) precipitate, denature, and hybridize the DNAs; (3) posthybridization washes and detection steps; (4) fluorescence microscopy and image analysis. The result is demonstrated as **Fig. 2 A,B**.

2.3. Troubleshooting

The first ethanol–salt precipitation serves to remove the dNTPs that are not incorporated in the labeling reaction. To prevent loss of the labeled DNA, an excess of salmon sperm DNA (50 µg of salmon sperm DNA for each 1 µg of labeled DNA) is precipitated with the labeled DNA such that any loss of DNA from the precipitation procedure comes mainly from the salmon sperm DNA rather than the labeled DNA.

The best recovery of DNA results from leaving the DNA/salt/ethanol mixture at –20°C overnight before centrifugation. However, if this is not possible, 1 h at –70°C will give a decent recovery. In order to determine the amount of salmon sperm that is required, as well as the amount of water needed to yield the final concentration of 10 ng/µL, the actual amount of labeled probe must be determined.

On the one hand, chromosomes must be sufficiently denatured to permit hybridization of labeled DNAs, but, on the other hand, they must retain their morphology for unequivocal identification during karyotyping. A protocol for optimized chromosome preparations and hybridization conditions has been reported elsewhere *(13)*, which provides a guideline for troubleshooting in CGH experiments (criteria for acceptable CGH images, control experiments, quality assurance, and interpretation of ratios). Such troubleshooting is now facilitated by commercially available and standardized CGH reagents required for CGH analysis (e.g., reference metaphase spreads, labeling kits, labeled nucleotides, and labeled reference DNAs) and control tumor DNAs (labeled and unlabeled MPE 600 DNA from a breast cancer cell line) with defined chromosomal imbalances that must be visible with CGH analysis.

3. Applications

Comparative genomic hybridization is a FISH-based technique that can detect gains and losses of whole chromosomes and subchromosomal regions. Often, CGH is based on a two-

Table 1
Fluorescent Dyes Commonly Used for CGH

Fluorochrome	Color	Absorbance (nm)	Emission (nm)
DAPI	Blue	350	456
FITC	Green	490	520
Spectrum green	Green	497	524
Rhodamine	Red	550	575
Spectrum red	Red	587	612
Texas red	Red	595	615

color, competitive FISH of differentially labeled tumor and reference DNA to normal metaphase chromosomes and can scan the whole genome without prior knowledge of specific chromosomal abnormalities *(4)*. CGH accounts for all chromosomal segments in the tumor cell genome, including those present in marker chromosomes whose origin cannot be determined by conventional karyotyping. Application of the CGH procedure has permitted the identification of recurrent imbalances in a wide variety of human diseases, and so far, more than 1500 articles have been published on CGH, with approx 90% reporting the utility of CGH in delineating cytogenetic changes in cancer specimens (http://www3.ncbi.nlm.nih.gov/entrez/query.fcgi). About 6% of CGH articles have dealt with technical aspects and only a limited number have described the application of CGH in a clinical cytogenetics setting *(1)*.

3.1. Tumor Genetics

Cancer is a complex disease occurring as a result of a progressive accumulation of genetic aberrations and epigenetic changes that enable an escape from normal cellular and environmental controls *(4)*. Neoplastic cells might have numerous acquired genetic abnormalities, including aneuploidy, chromosomal rearrangements, amplifications, deletions, gene rearrangements, and loss-of-function or gain-of-function mutations *(4)*. These genetic abnormalities lead to the abnormal behavior common to all enoplastic cells, including the dysregulation of growth, lack of contact inhibition, genomic instability, and a propensity for metastases. The genes affected by aberrations in cancers are often divided into two main categories: genes that have gain-of-function (amplification) known as oncogenes, and genes that have loss-of-function (inactivation) known as tumor suppressor genes. The tumor suppressor genes often need a complete loss of function in both alleles because the chromosomes are often paired. More than 100 genes have been reported to be related to tumor development and/or tumor progression. Gene amplification is an essential mechanism of oncogene activation, in addition to structural alterations, loss of control mechanisms, insertional mutagenesis, and chromosome translocations *(22)*.

The power of CGH has been clearly proven, since the first published article by Kallioniemi et al. in 1992, as a tool to characterize chromosomal imbalances in neoplasias *(5)*. The number of studies concerning solid tumors and hematological cancers is impressive. Many excellent review articles or websites (e.g., http://www.helsinki.fi/~lgl_www/CMG.html, which contains the chromosomal locations of recurrent DNA copy number changes in 73 tumor types from 283 reports) have been found that summarize the chromosomal imbalances detected by CGH in solid tumors and in hematological diseases *(19,22–25)*. The majority of studies focus on specific types of tumor, with tumor development, progression, clinical correlation, prediction of response to therapy, and disease-free survival in a given population, with the presence of infection *(25)*. For example, Wang suggested that genes located at 6q27 might play a crucial role in the early events of ovarian tumor development and that there is a continuum in the progression model of ovarian neoplasia and the high frequency of gene amplification at 20q12-q13.2, indicating that the overpresentation of these genes might play a crucial role in the pathogenesis of

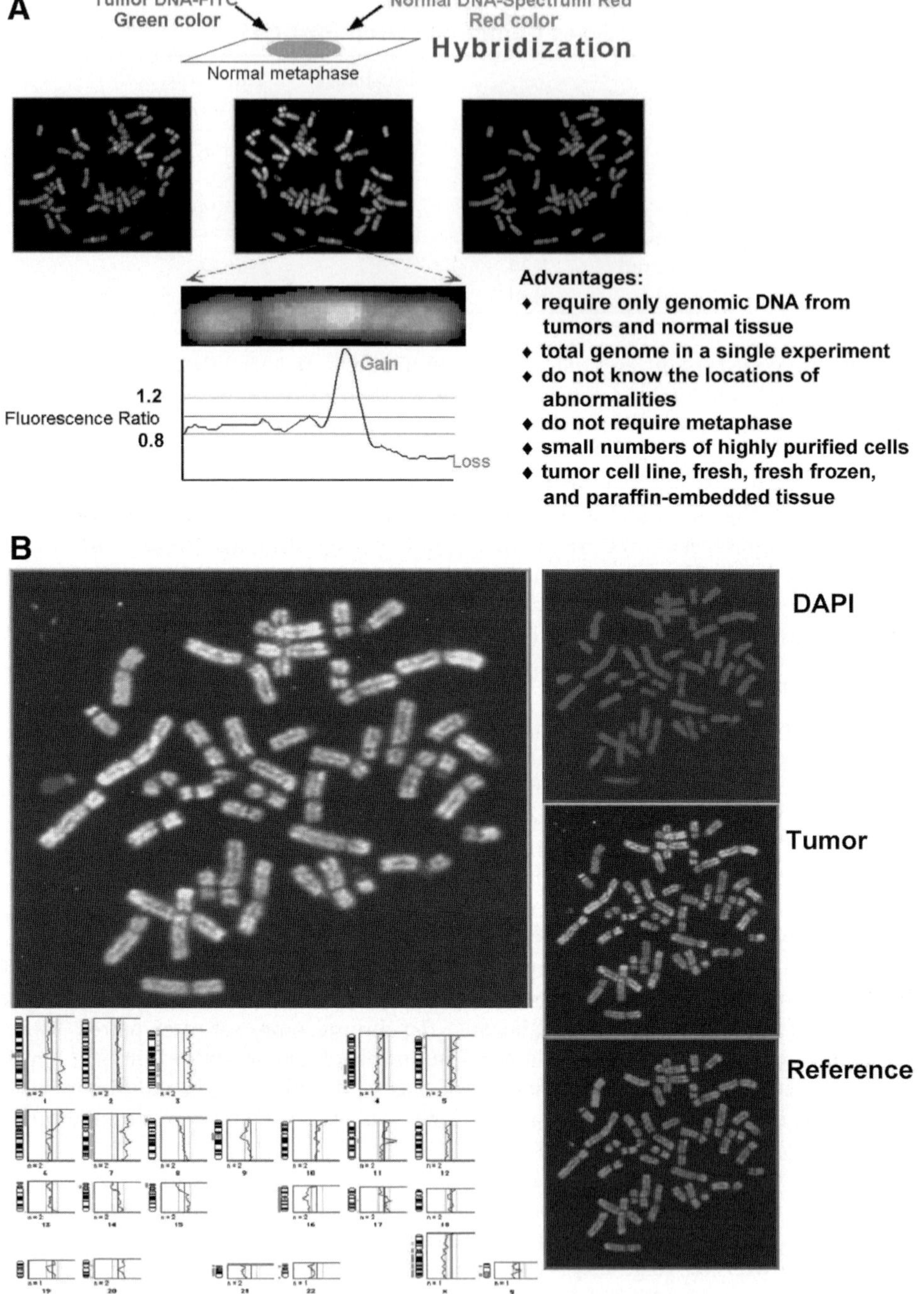

Fig. 2. (**A**) Schematic representation showing a summary of CGH at a glance. CGH offers many advantages in detecting unknown genomic imbalances from materials of interest. (**B**) Quantification of fluorescence intensity is depicted. The values of the fluorescence intensity ratio are drawn as a function of their positions along the individual chromosomes (curved lines). Ratio values less than the threshold value of 0.8 (red line to the right of the central black line) are considered gene deletions, whereas ratio values > 1.2 (green line to the right of the central black line) are indicative of gene amplifications. The potential regions of loss and gain abnormalities are also marked (red and green bars); losses are shown at the left and gains at the right of the chromosome ideograms.

ovarian cancer ***(26)***. In fact, several of the copy-number changes in ovarian cancers might be explained by known abnormalities in the genes involved in ovarian tumorigenesis (loss of 17pter-q21 and *p53*, gain on 17q and amplification of *HER2/Neu/erbB2*, amplification on 8q24 and *myc*, and gain at 3q26.3 and amplification of the p110 catalytic subunit of PI3K), which provide unique targets. The *p53* tumor suppressor has been explored as a target for gene therapy in ovarian cancer, and *HER2/neu/erB2* is being targeted by antibody-mediated therapy (herceptin) and E1A-mediated gene therapy. In the study of ovarian cancer samples from Kudoh et al., the researchers demonstrated that the presence of an increased copy number at 1q21 and 13q13 correlate with a lack of response to a chemotherapy regimen consisting of cisplatin, doxorubicin, and cyclophosphamide ***(27)***. Therefore, identification and characterization of the genes driving the copy-number abnormalities detected by CGH might provide new therapeutic targets in ovarian cancer, which might directly affect tumor cell growth or alter sensitivity to chemotherapy ***(28)***.

In addition to the application of CGH for studying solid tumors, CGH also contributes to the knowledge of chromosomal alterations in hematological diseases such as leukemia and lymphomas. With the application of CGH in lymphoma, gene amplifications are also often seen ***(19,25,29)***. Struski et al. further suggested the indications of CGH in hemopathies, including (1) a normal karyotype (a proliferation of normal cells to the detriment of tumor cells during the cell culture), (2) failure of the karyotype (low proliferative potential as multiple myeloma), and (3) uninterpretable metaphases because of poor quality (e.g., acute lymphoid leukemia) and complex karyotypes (e.g., lymphoma) ***(25)***.

Because CGH recognizes only proportional changes in copy number, the ratio profiles do not indicate the absolute copy-number change. For example, Knuutila et al. found in diploid and near-diploid cells that a ratio of 1.5 indicates at least a 100% increase in the copy number of an entire chromosome arm or of a region of a chromosome band, but the threshold is not reached when the increase is only 50% (e.g., chromosomal trisomy) ***(22)***. In addition, when a DNA copy number increase is restricted to a small chromosome area representing, for example, the amplification of a single gene, the copy number increase should be 10-fold or higher ***(22)***.

3.2. Clinical Cytogenetics

Comparative genomic hybridization has also been useful in clinical cytogenetics and has facilitated the identification and characterization of intrachromosomal duplications, deletions, unbalanced translocations, and marker chromosomes in prenatal, postnatal, and preimplantation samples ***(3,30–33)***. CGH is also useful in revising incorrectly assigned karyotypes. The ability of CGH to define more precisely the chromosomal material comprising marker chromosomes and unbalanced rearrangements has helped to further define the critical chromosomal regions that are associated with adverse phenotypic outcomes, thus providing prognostic information for genetic counseling ***(31)***. This information is also beneficial to prenatally ascertaining cases of marker chromosomes, as it might provide couples with a means to make rational and informed decisions concerning the pregnancy. CGH is also powerful in advancing molecular cytogenetics in the area of evaluating mental retardation ***(33)***. CGH has two major uses in the analysis of mental retardation, including the further characterization of unbalanced karyotypes as detected by routine banding analysis, such as additions, duplications, deletions, translocations, markers, or complex aberrations, and the screening for "hidden" chromosome aberrations in patients with apparently normal karyotypes.

Figure 3 A–D illustrates a case of congenital anomaly that was diagnosed prenatally using amniocentesis performed at the gestational age of between 14 and 20 wk. We first identified the patient as probably having an abnormal chromosome, involving chromosome 1 and another chromosome (perhaps chromosome 5 or chromosome 13), using the conventional karyotyping method. Then, we used CGH to find the possibly abnormal area—deletion of 1q terminal and amplification of 13q. Based on the findings of CGH, we suspected that the translocation chromosome might be involved with chromosome 1 and chromosome 13. Finally, we used FISH (*see* Chapter 29) to confirm the diagnosis, using whole-chromosome and specific telomere probes.

A

Patient: A female baby
Diagnosis: Congenital anomaly
G-banding: 46,XX, add(1)(q43-44?)

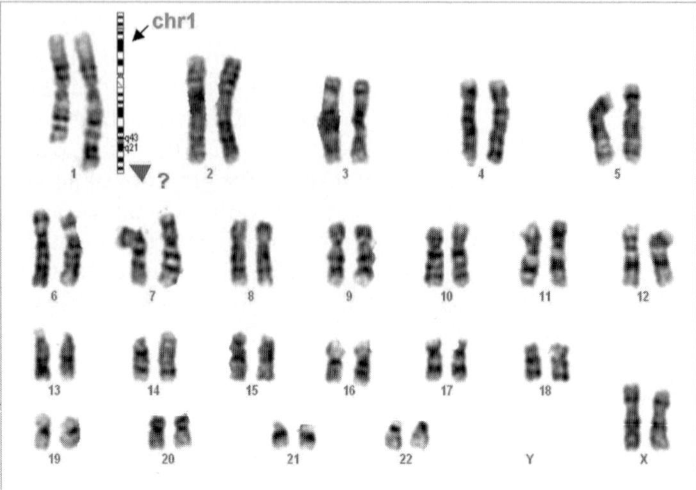

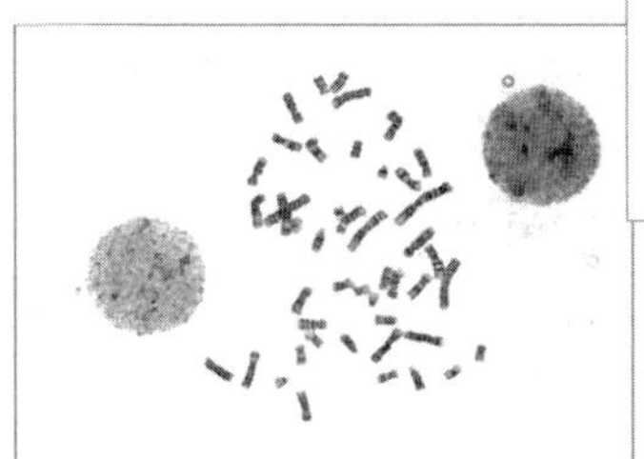

B

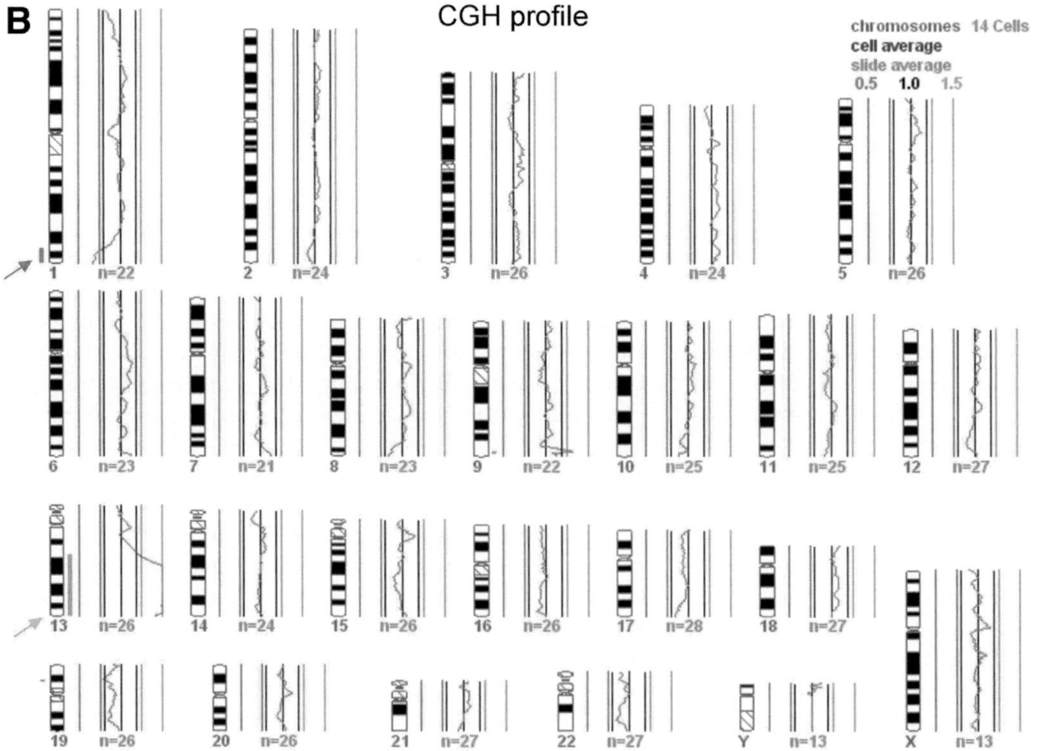

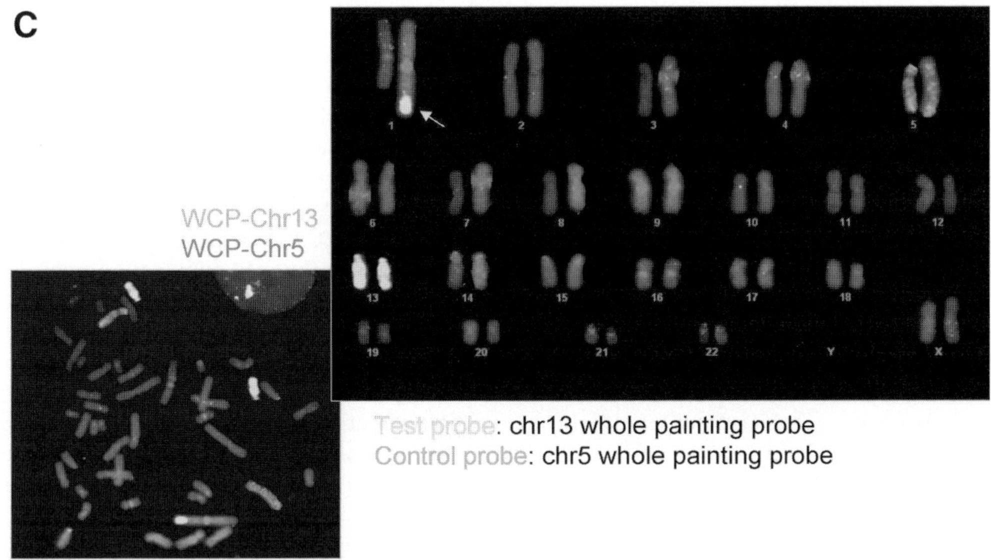

D

Test probe1: chr1 telomeric probe (green signal)
Test probe2: chr13 telomeric probe (red signal)

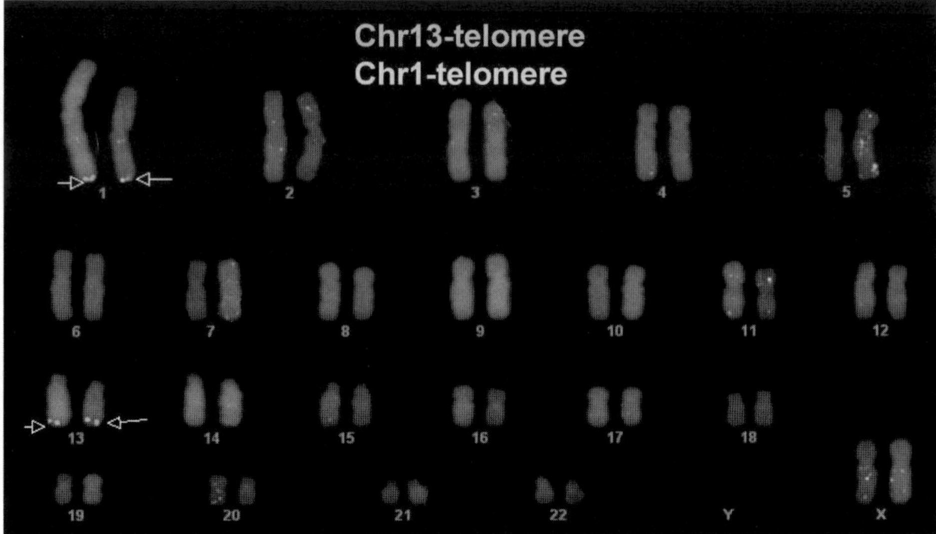

Fig. 3. The application of molecular cytogenetic tools in detecting the chromosome aberrations in a case with congenital anomaly. (**A**) Traditional G-banding analysis finds an abnormal chromosome with additional materials in the terminal end of one chromosome 1q. However, the origin of this material is unable to be determined by G-banding. (**B**) CGH is then performed and shows genomic changes with 1q terminal loss and 13q gain. (**C**) FISH with chromosome painting probes 5 and 13 (chromosome 5 painting probe is used as an internal control) is performed and shows the additional material in chromosome 1q terminal is from chromosome 13q. (**D**) FISH with 13q and 1q telomere probes is also done and detects the deletion of terminal region in one of chromosome 1. All of these FISH studies confirm the finding by CGH analysis and give us a clearer picture about this chromosome aberration.

4. Conclusion

Methods for chromosomal analysis have become increasingly powerful, benefiting enormously from the fusion of traditional cytogenetic techniques and molecular genetics. This has not only led to advances in clinical diagnosis but has also provided markers for the assessment of prognosis and disease progression. CGH might be among the most significant methodological advances and has overcome many of the technical limitations that beset earlier cytogenetic methods, allowing detailed chromosomal data to be obtained from a variety of tissues that were previously considered problematic ***(2)***. CGH has been employed for the ascertainment of chromosomal duplications, amplifications, and deletions that contribute to neoplastic transformation, revealing the chromosomal location of tumor suppressor genes and oncogenes that are central to disease development or progression and can also identify new prognostic markers. However, it should be noted that CGH has limitations. It cannot detect balanced chromosomal abnormalities such as translocations, inversions, or point mutations. In addition, pericentromeric, telomeric and heterochromatic regions cannot be evaluated, and 1p32-pter, 16p, 19, and 22 could lead to a false-positive interpretation ***(22)***. Chromosomal imbalances must be present in about 50% of cells to be detected, which means that CGH requires that tumor specimens be relatively free of surrounding normal tissue. Finally, the sensitivity and resolution of conventional chromosome-based CGH is likely limited to changes that are 5–10 Mb ***(19,20)***, although thresholds of detection for amplification might be lower (2 Mb) compared to those of detection for deletion (10–20 Mb) ***(13)***. As an example, an amplicon located at 19p, which contains *AKT2*, is frequently present in ovarian cancer cells but is too small to be detected by classical CGH ***(18)***. Similarly, the frequency of copy-number abnormalities at 3q is greater when it is analyzed by FISH with region-specific probes than when it is assessed by chromosome-based CGH ***(34)***. Because the limitations of CGH for the genetic characterization of cancer are mainly in the use of metaphase chromosomes as hybridization targets, its resolution is at the chromosomal banding level ***(22)***. Thus, a new technology—array technology adapted to CGH, designated M-CGH ***(35–38)***—will allow the resolution to increase from a cytogenetic level to a molecular level. For M-CGH, chromosome preparations are substituted by sets of well-defined genomic DNA fragments microarrayed on solid support to serve as hybridization targets ***(39)***. Thus, genomic imbalances are detected with much higher resolution, allowing copy-number changes to be associated with individual loci and genomic markers. This approach is much more demanding than the widespread application of DNA microarrays for the detection of gene expression patterns, because it requires the reliable detection of subtle differences in the fluorescence ratios of test and control genomes (e.g., the diagnosis of alterations can rely on a ratio of difference smaller than 0.1); also, the complexity of the labeled total genomic DNA probe sequence pool is several orders of magnitude higher than that of a cDNA pool ***(39)***. These requirements can be met using genomic DNA fragments cloned in bacterial, P1-derived, or yeast artificial chromosome (BAC, PAC, and YAC) vectors ***(35,36)***. Recently, M-CGH has improved significantly, because a ligation-mediated PCR BAC labeling method has improved the signal-to-noise ratios, and clone density has also been increased ***(40)***. Current human CGH arrays have a 1-Mb resolution (about 3000 features), whereas cDNA arrays are already routinely made at 10,000–15,000 features per slide ***(38)***.

Importantly, copy-number abnormalities detected by classical CGH might reflect a sum of multiple distinct small areas of copy-number abnormality. M-CGH, because of its increased ability to resolve areas of copy-number abnormalities, might exhibit an improved predictive value and also facilitate the identification of the gene or genes driving specific genomic amplifications or deletions. M-CGH is also likely to be more robust and less tedious than classical chromosome-based CGH.

In the future, the continuing improvement of CGH, such as the use of a standard reference interval instead of a fixed interval ***(41)***, or related techniques, such as M-CGH, will allow large-scale screening of abnormal changes of chromosomes in diseases and continue to provide a better understanding of cancer. Of most importance, the establishment of a pattern of genetic abnormalities for each tumor type should offer a precise histological diagnosis and assist in the prediction of which chemotherapeutic drugs will be most effective in the treatment of a given type of cancer and a single-array-based CGH experiment in a basic science laboratory could provide copy number and expression information from all human genes and a more focused DNA chip in the clinical diagnostic laboratory could assess all the specific chromosomal loci, genes, and signaling pathways involved in a given disease.

Acknowledgments

We greatly appreciate the technical support of supervisor Dr. Jeremy A. Squire at the Department of Laboratory Medicine and Pathobiology, University of Toronto, Toronto, Ontario, Canada (website: http://www.utoronto.ca/cancyto/), Miss Wen-Yuann Shyong at the Taipei Veterans General Hospital, and Dr. Wen-Ling Lee at the Department of Medicine, Cheng-Hsin Rehabilitation Center, Taipei, Taiwan. The work, which took place in the authors' laboratories, has been supported by the National Science Council (NSC 92-2314-B-075-115) and Taipei Veterans General Hospital.

References

1. Patel, A. S., Hawkins, A. L., and Griffin, C. A. (2000) Cytogenetics and cancer. *Curr. Opin. Oncol.* **12,** 62–67.
2. Houldsworth, J. and Chaganti, R. S. K. (1994) Comparative genomic hybridization: an overview. *Am. J. Pathol.* **145,** 1253–1260.
3. Wells, D. and Levy, B. (2003) Cytogenetics in reproductive medicine: the contribution of comparative genomic hybridization (CGH). *BioEssays* **25,** 289–300.
4. Boultwood, J. and Fidler, C. (eds.) (2002) *Molecular Analysis of Cancer*, Humana, Totowa, NJ.
5. Kallioniemi, A., Kallioniemi, O. P., Sudar, D., et al. (1992) Comparative genomic hybridization for molecular cytogenetic analysis of solid tumors. *Science* **258,** 818–821.
6. Baksara, B. R., Pei J., and Testa J. R. (2002) Comparative genomic hybridization analysis, in *Molecular Analysis of Cancer* (Boultwood, J. and Fidler, C., eds.), Humana, Totowa, NJ, pp. 45–65.
7. Wang, P. H., Shyong, W. Y., Lin, C. H., et al. (2002) Analysis of genetic aberrations in uterine adenomyosis using comparative genomic hybridization. *Anal. Quant. Cytol. Histol.* **24,** 1–6.
8. du Manoir, S., Schrock, E., Bentz, M., et al. (1995) Quantitative analysis of comparative genomic hybridization. *Cytometry* **19,** 27–41.
9. Piper, J., Rutovitz, D., Sudar, D., et al. (1995) Computer image analysis of comparative genomic hybridization. *Cytometry* **19,** 10–26.
10. Lundsteen, C., Maahr, J., Christensen, B., et al. (1995) Image analysis in comparative genomic hybridization. *Cytometry* **19,** 42–50.
11. du Manoir, S., Kallioniemi, O. P., Lichter, P., et al. (1995) Hardware and software requirements for quantitative analysis of comparative genomic hybridization. *Cytometry* **19,** 4–9.
12. Wells, D., Sherlock, J. K., Handyside, A. H., and Delhanty, J. D. (1999) Detailed chromosomal and molecular genetic analysis of single cells by whole genome amplification and comparative genomic hybridization. *Nucleic Acids Res.* **27,** 1214–1218.
13. Kallioniemi, O. P., Kallioniemi, A., Piper, J., et al. (1994) Optimizing comparative genomic hybridization for analysis of DNA sequences copy number changes in solid tumors. *Genes Chromosomes Cancer* **10,** 231–243.
14. Isola, J., Devries, S., Chu, L., Ghazvini, S., and Waldman, F. (1994) Analysis of changes in DNA sequence copy number by comparative genomic hybridization in archival paraffin-embedded tumor samples. *Am. J. Pathol.* **145,** 1301–1308.
15. Choo, K. H. A. (1994) *Methods in Molecular Biology: In Situ Hybridization Protocols*, Humana, Totowa, NJ.
16. Kallioniemi, A., Kallioniemi, O. P., Piper, J., et al. (1994) Detection and mapping of amplified DNA sequences in breast cancer by comparative genomic hybridization. *Proc. Natl. Acad. Sci. USA* **91,** 2156–160.
17. Kirchhoff, M., Gerdes, T., Rose, H., Maahr, J., Ottesen, A. M., and Lundsteen, C. (1998) Detection of chromosomal gains and losses in comparative genomic hybridization analysis based on standard reference intervals. *Cytometry* **31,** 163–173.

18. Tachdjian, G., Aboura, A., Lapierre, J. M., and Vigie, F. (2000) Cytogenetic analysis from DNA by comparative genomic hybridization. *Ann. Genet.* **43,** 147–154.
19. Zitzelsberger, H., Lehmann, L., Werner, M., and Bauchinger, M. (1997) Comparative genomic hybridisation for the analysis of chromosomal imbalances in solid tumours and haematological malignancies. *Histochem. Cell Biol.* **108,** 403–417.
20. James, L. and Varley, J. (1996) Preparation, labeling and detection of DNA from archival tissue sections suitable for comparative genomic hybridization. *Chromosome Res.* **4,** 163–164.
21. Kuukasjarvi, T., Tanner, M., Pennanen, S., Karhu, R., Visakorpi, T., and Isola, J. (1997) Optimizing DOP-PCR for universal amplification of small DNA samples in comparative genomic hybridization. *Genes Chromosomes Cancer* **18,** 94–101.
22. Knuutila, S., Bjorkqvist, A., Autio, K., et al. (1998) DNA copy number amplifications in human neoplasms. *Am. J. Pathol.* **152,** 1107–1123.
23. Knuutila, S., Aalto, Y., Autio, K., et al. (1999) DNA copy number losses in human neoplasms. *Am. J. Pathol.* **155,** 683–694.
24. Knuutila, S., Autio, K., and Aalto, Y. (2000) Online access to CGH data of DNA sequence copy number changes. *Am. J. Pathol.* **157,** 689–690.
25. Struski, S., Doco-Fenzy, M., and Cornillet-Lefebvre, P. (2002) Compilation of published comparative genomic hybridization studies. *Cancer Genet. Cytogenet.* **135,** 63–90.
26. Wang, N. (2002) Cytogenetics and molecular genetics of ovarian cancer. *Am. J. Med. Genet.* **115,** 157–163.
27. Kudoh, K., Takano, M., Koshikawa, T., et al. (1999) Gains of 1q21-q22 and 13q12-q14 are potential indicators for resistance to cisplatin-based chemotherapy in ovarian cancer patients. *Clin. Cancer Res.* **5,** 2526–2531.
28. Mills, G. B., Schmandt, R., Gershenson, D., and Bast, R. C. (1999) Should therapy of ovarian cancer patients be individualized based on underlying genetic defects? *Clin. Cancer Res.* **5,** 2286–2288.
29. Werner, C. A., Dohner, H., Joos, S., et al. (1997) High-level DNA amplifications are common genetic aberrations in B-cell neoplasms. *Am. J. Pathol.* **151,** 335–342.
30. Bryndorf, T., Kirchhoff, M., Rose, H., et al. (1995) Comparative genomic hybridization in clinical cytogenetics. *Am. J. Hum. Genet.* **57,** 1211–1220.
31. Levy, B., Dunn, T. M., Kaffe, S., Kardon, N., and Hirschhorn, K. (1998) Clinical applications of comparative genomic hybridization. *Genet. Med.* **1,** 4–12.
32. Levy, B., Dunn, T. M., Kern, J. H., Hirschhorn, K., and Kardon, N. B. (2002) Delineation of the dup5q phenotype by molecular cytogenetic analysis in a patient with dup5q/del 5p (cri du chat). *Am. J. Med. Genet.* **108,** 192–197.
33. Xu, J. and Chen, Z. (2002) Advances in molecular cytogenetics for the evaluation of mental retardation. *Am. J. Med. Genet.* **117,** 15–24.
34. Shayesteh L., Lu Y., Kuo W. L., et al. (1999) PIK3CA is implicated as an oncogene in ovarian cancer. *Nat. Genet.* **21,** 99–102.
35. Pinkel D., Segraves R., Sudar D., et al. (1998) High resolution analysis of DNA copy number variation using comparative genomic hybridization to microarrays. *Nature Genet.* **20,** 207–211.
36. Solinas-Toldo, S., Lampel, S., Stilgenbauer, S., et al. (1997) Matrix-based comparative genomic hybridization: biochips to screen for genomic imbalances. *Genes Chromosomes Cancer* **20,** 399–407.
37. Pollack, J. R., Perou, C. M., Alizadeh, A. A., et al. (1999) Genome-wide analysis of DNA copy-number changes using cDNA microarrays. *Nature Genet.* **23,** 41–46.
38. Armour J. A. L., Barton, D. E., Cockburn, D. J., and Taypor, G. R. (2002) The detection of large deletions or duplications in genomic DNA. *Hum. Mut.* **20,** 325–337.
39. Wessendorf, S., Fritz, B., Wrobel, G., et al. (2002) Automated screening for genomic imbalances using matrix-based comparative genomic hybridization. *Lab. Invest.* **82,** 47–60.
40. Snijders, A. M., Nowak, N., Segraves, R., et al. (2001) Assembly of microarrays for genome-wide measurement of DNA copy number. *Nature Genet.* **29,** 263–264.
41. Kirchhoff, M., Gerdes, T., Rose, H., Maahr, J., Ottesen, A. M., and Lundsteen, C. (1998) Detection of chromosomal gains and losses in comparative genomic hybridization analysis based on standard reference intervals. *Cytometry* **31,** 163–173.

28

Bioinformatic Tools for Gene and Protein Sequence Analysis

Bernd H. A. Rehm and Frank Reinecke

1. Introduction

The rapid development of efficient, automated DNA-sequencing methods has strongly advanced the genome-sequencing era, culminating in the determination of the entire human genome in 2001 *(1,2)*. An enormous amount of DNA sequence data are available and databases still grow exponentially (*see* **Fig. 1**). Analysis of this overwhelming amount of data, including hundreds of genomes from both prokaryotes and eukaryotes, has given rise to the field of bioinformatics. Development of bioinformatic tools has evolved rapidly in order to identify genes that encode functional proteins or RNA. This is an important task, considering that even in the best studied bacterium *Escherichia coli* more than 30% of the identified open reading frames (ORFs) represent hypothetical genes with no known function. Future challenges of genome-sequence analysis will include the understanding of diseases, gene regulation, and metabolic pathway reconstruction. In addition, a set of methods for protein analysis summarized under the term *proteomics* holds tremendous potential for biomedicine and biotechnology *(141)*. The large number of bioinformatic tools that have been made available to scientists during the last few years has presented the problem of which to use and how best to obtain scientifically valid answers *(3)*. In this chapter, we will provide a guide for the most efficient way to analyze a given sequence or to collect information regarding a gene, protein, structure, or interaction of interest by applying current publicly available software and databases that mainly use the World Wide Web. All links to services or download sites are given in the text or listed in **Table 1**; the succession of tools is briefly summarized in **Fig. 2**.

2. Software Tools for Bioinformatics

In the first part of this chapter, software tools will be described that mainly use algorithms and are based either on very short-sequence comparisons, physical or chemical properties of molecules, or statistics. A second group of software relies mainly on databases and will be discussed below. As so-called integrated methods are evolving and becoming more and more popular, it is difficult to divide programs into these two groups.

There are many programs routinely used to generate contiguous DNA sequences from raw data obtained from high-throughput sequencers, to assign quality scores to each base, remove contaminating sequences (such as vector DNA), and provide the means to link sequences containing applications. First, base-callers like Phred *(4,5)* extract raw sequences from raw data. There are also contig assemblers like Phrap (University of Washington, http://bozeman.mbt.washington.edu/phrap.docs/phrap.html) or CAP3 *(6)* that assemble fragments to contigs and packages like consed *(7)* or GAP4 *(8)*, which are used to finish sequencing projects. These programs are not explained in detail here.

From: *Medical Biomethods Handbook*
Edited by: J. M. Walker and R. Rapley © Humana Press, Inc., Totowa, NJ

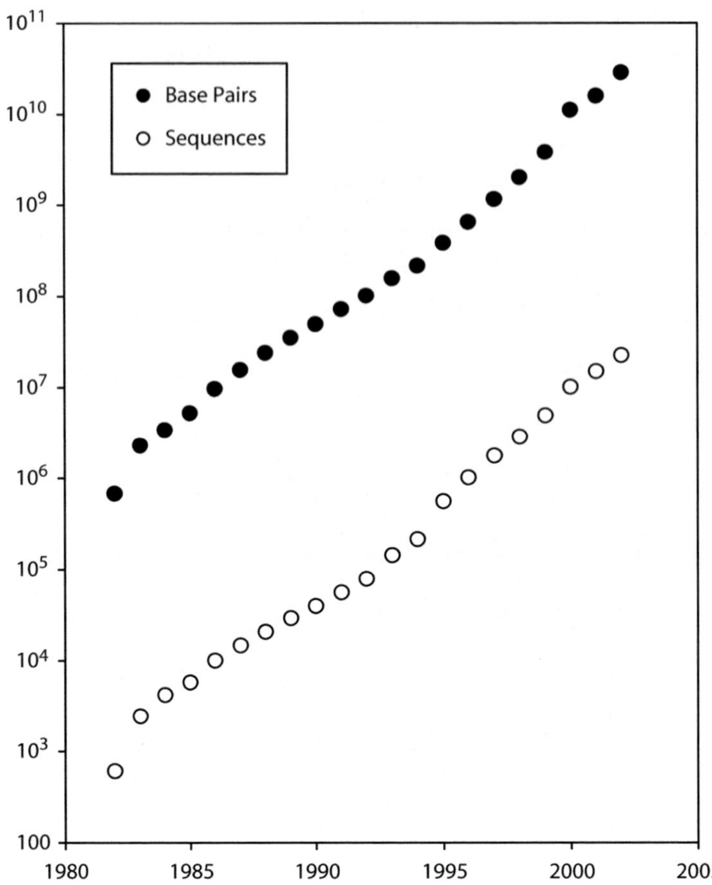

Fig. 1. Development of stored DNA-sequence information in GenBank from 1982 to 2002 (●, base pairs; ○, sequences. (From http://www.ncbi.nlm.nih.gov/Genbank/genbankstats.html.)

Table 1
Compilation of Links to Bioinformatic Services and Software

Program	Link
Gene Finding	
AMIGene	http://www.genoscope.cns.fr/agc/tools/amigene/index.html
Critica	http://www.ttaxus.com
EasyGene	http://www.cbs.dtu.dk/services/EasyGene
EUGENE'HOM	http://genopole.toulouse.inra.fr/bioinfo/eugene/EuGeneHom/cgi-bin/EuGeneHom.pl
GeneFizz	http://pbga.pasteur.fr/GeneFizz
GeneMark.hmm2	http://opal.biology.gatech.edu/GeneMark/gmhmm2_prok.cgi
GeneMarkS	http://opal.biology.gatech.edu/GeneMark/genemarks.cgi
GeneScan	http://genes.mit.edu/GENSCAN.html
Genie	http://www.soe.ucsc.edu/~dkulp/cgi-bin/genie
Glimmer	http://www.cs.jhu.edu/labs/compbio/glimmer.html
GlimmerM	http://www.tigr.org/tdb/glimmerm/glmr_form.html
Grail	http://compbio.ornl.gov/Grail-1.3
HMMgene	http://www.cbs.dtu.dk/services/HMMgene
ORF-Finder	http://www.ncbi.nlm.nih.gov/gorf/gorf.html
Procrustes	http://www-hto.usc.edu/software/procrustes
Veil	http://www.cs.jhu.edu/labs/compbio/veil.html
ZCURVE	http://tubic. tju.edu.cn/ZCURVE

Table 1 (continued)

Program	Link
Signal Finding	
MatInspector	http://transfac.gbf.de/cgi-bin/matSearch/matsearch.pl
PromotorScan	http://bimas.dcrt.nih.gov/molbio/proscan
SignalScan	http://bimas.dcrt.nih.gov/molbio/signal
TRANSFAC	http://transfac.gbf.de/TRANSFAC
Sequence Alignment	
CLUSTALW	ftp://ftp-igbmc.u-strasbg.fr/pub/ClustalW http://www.ebi.ac.uk/clustalw
CLUSTALX	ftp://ftp-igbmc.u-strasbg.fr/pub/ClustalX
DbClustal	http://igbmc.u-strasbg.fr/DbClustal/dbclustal.html
HMMER	http://hmmer.wustl.edu
T-COFFEE	http://www.ch.embnet.org/software/TCoffee.html
Phylogeny	
PAUP	http://onyx.si.edu/PAUP
GCG package	http://www.gcg.com
PHYLIP	http://evolution.genetics.washington.edu/phylip.html
Phylodendron	http://iubio.bio.indiana.edu/treeapp
TreeView	http://taxonomy.zoology.gla.ac.uk/rod/treeview.html
Protein Properties and Structure	
DAS	http://www.sbc.su.se/~miklos/DAS
ExPASy	http://www.expasy.org
META PP	http://cubic.bioc.columbia.edu/predictprotein/submit_meta.html
PredictProtein	http://cubic.bioc.columbia.edu/predictprotein
TMHMM	http://www.cbs.dtu.dk/services/TMHMM-2.0
TMpred	http://www.ch.embnet.org/software/TMPRED_form.html
Database Searching	
NCBI BLAST	http://www.ncbi.nlm.nih.gov/BLAST
WUBLAST	http://blast.wustl.edu
Protein Families and Motifs	
BLOCKS	http://www.blocks.fhcrc.org
InterPro	http://www.ebi.ac.uk/interpro/scan.html
Pfam	http://www.sanger.ac.uk/Software/Pfam
PRINTS	http://www.biochem.ucl.ac.uk/bsm/dbbrowser/PRINTS
PROSITE	http://www.expasy.org/prosite
SMART	http://smart.embl-heidelberg.de
Protein Structure	
CATH	http://www.biochem.ucl.ac.uk/bsm/cath
PDB	http://www.pdb.org
SCOP	http://scop.mrc-lmb.cam.ac.uk/scop
Structure Modeling	
CPHmodels	http://www.cbs.dtu.dk/services/CPHmodels
ESyPred3D	http://www.fundp.ac.be/urbm/bioinfo/esypred
Geno3D	http://geno3d-pbil.ibcp.fr
SWISS-MODEL	http://swissmodel.expasy.org
Protein Interaction	
BIND	http://binddb.org
CAPRI	http://capri.ebi.ac.uk
DIP	http://dip.doe-mbi.ucla.edu/dip/Main.cgi
Genome Analysis	
COG	http://www.ncbi.nlm.nih.gov/COG
NCBI Genomes	http://www.ncbi-nlm.nih.gov/genomes
SNPs and Expression Profiling	
cSNP	http://csnp.unige.ch
dbSNP	http://www.ncbi.nlm.nih.gov/Entrez
GEO	http://www.ncbi.nlm.nih.gov/geo

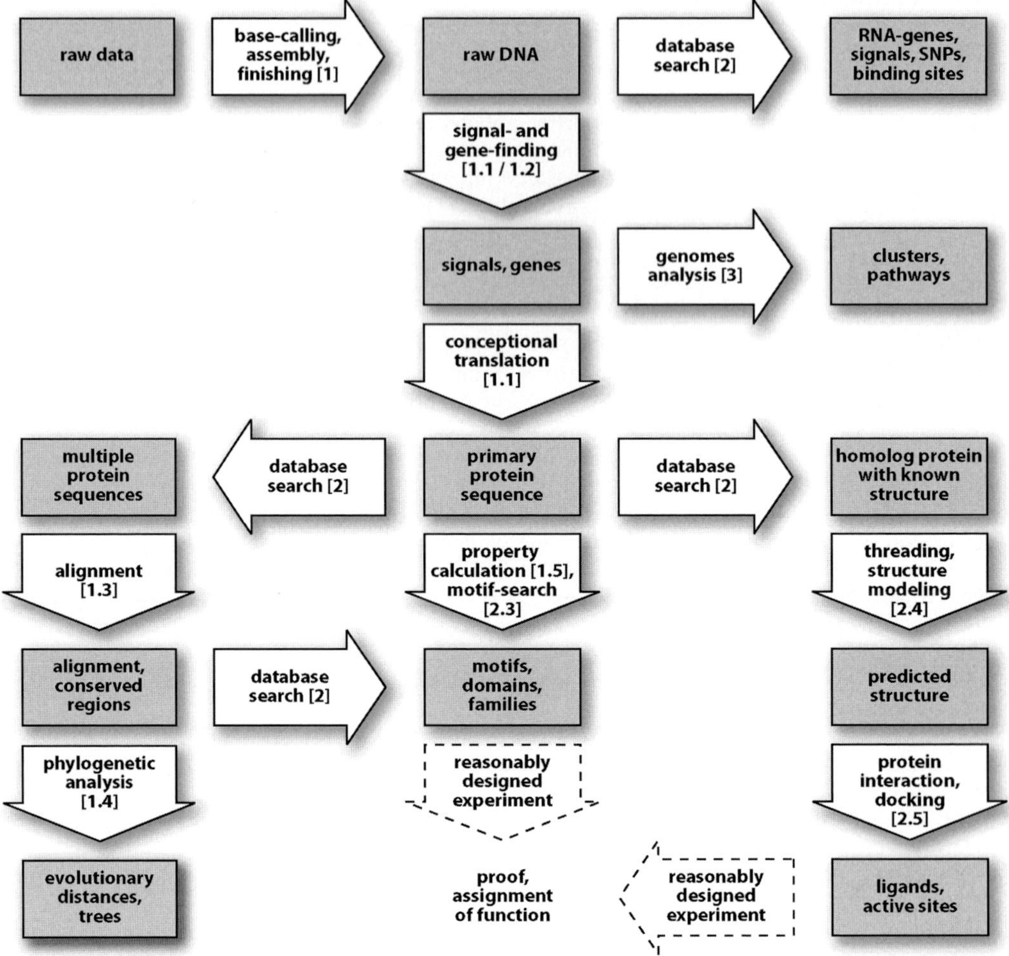

Fig. 2. Succession of programs during assignment of function to DNA or proteins. Arrows indicate action of bioinformatic tools, detailed descriptions can be found in the chapter given in brackets.

Any DNA region that can be assigned a function is of special interest. The sequence elements that can be found within them include promoters and various transcription factor-binding sites, ribosome-binding sites, start and stop codons, splice sites, and so forth. These are referred to as signals. Methods to detect them are termed *signal sensors*. In contrast, extended and variable-length regions, such as exons and introns, are termed *contents*. These are recognized by methods that can be called content sensors ***(9)***. It is a major challenge to find the functional sites responsible for gene structure, regulation and transcription. Computational methodology for finding genes and other functional sites in genomic DNA has evolved significantly over the past years ***(10–20)***.

2.1. Gene-Finding Methods

The most basic signal sensor is a simple consensus sequence or an expression that describes a consensus sequence together with allowable variations. Sophisticated types of signal sensor, such as neural nets, are extensively used ***(21,22)***. The most important content sensors are programs that predict coding regions ***(23)***. In prokaryotes, genes are identified simply by looking for long ORFs. However, these relatively simple methods cannot be transferred to higher eukaryotes. In order to discriminate coding against noncoding regions in eukaryotes, exon content sensors use statistical models of the nucleotide frequencies ***(24)*** and dependencies present in codon structures. The most commonly used statistical models are known as Markov models. Neural nets are used to combine several coding hints together with signal sensors for the flanking splice sites in exon detectors ***(22)***. Other content sensors include those for CpG sites (regions that often occur at the beginning of genes, where the dinucleotide CG appears more frequently than in the rest of the genome and sensors for repetitive DNA such as human ALU sequences ***(25)***. The latter sensors are often used as masks or filters that completely remove the repetitive DNA, leaving the remaining DNA to be analyzed.

The statistical signals that signal and content sensors try to localize are usually weak, and there are usually dependencies between signals and contents, such as the possible correlation between splice-site strength and exon size ***(26)***. During the past decade, several systems that combine signal and content sensors have been developed in an attempt to identify complete gene structures. The first program using linguistic rules and a formal grammar for the arrangement of certain signals required for gene prediction was GenLang ***(27)***. As with most integrated gene-finders to date, GenLang uses dynamic programming to combine potential exon regions and other scored regions and sites into gene prediction with a maximal total score ***(13,16,28)***. These models are called hidden Markov models (HMMs). Gene-finding HMMs can be viewed as stochastic versions of the gene structure grammars. Early gene-finding HMMs included EcoParse for *Escherichia coli* ***(29)***—also recently used in the annotation of the *Mycobacterium tuberculosis* genome ***(30)***—and Xpound ***(31)***, Veil ***(32)***, and HMMgene ***(16)*** for the human genome ***(13)***. More recent programs include GeneMark HMM ***(33)***, GLIMMER ***(34)***, and Critica ***(35)***. Different programs specializing in special detectable differences between coding and noncoding regions have been developed recently. EasyGene ***(20)*** and AMIGene ***(36)*** apply statistical methods on predicted ORFs of prokaryotes, ZCURVE ***(19)*** concentrates on nucleic acid distribution for bacterial or archaeal genomes. GeneMarkS is an improved version of GeneMark, which has been applied to identify genes in the genomes of *Bacillus subtilis* and *E. coli* with the highest accuracy described to date ***(37)***. Even physical properties of DNA are used to predict genes: GeneFizz ***(38)*** compares the physics-based structural segmentation between helix and coil domains. Measuring the spectral rotation based on a process termed discrete Fourier transform (DFT) has proven to be capable of predicting genes in *Saccharomyces cerevisiae* ***(39)***. A slightly more general class of probabilistic models, called generalized HMMs or (hidden) semi-Markov models, has roots in GeneParser ***(40)*** and is more fully developed in Genie ***(41)*** and, subsequently, GenScan ***(42)***. The above gene-finders predict gene structure based only on general features of genes, rather than using explicit comparisons to known genes and their corresponding proteins, or auxiliary information such as expressed sequence tag (EST) matches. Some gene-finding systems combine multiple statistical measures with protein database homology searches performed using the predicted gene or deduced protein ***(11,15,43)***. This homology approach was developed by Gelfand et al. ***(44)*** and is used by EUGENE'HOM ***(45)*** or Procrustes ***(15)***, which uses a "spliced alignment" algorithm, similar to a Smith–Waterman algorithm ***(46)***, to derive a putative gene structure by aligning the DNA to a partial protein homolog of the gene to be predicted. Instead of inventing and approving yet another gene-finding program, there are several approaches to combining different existing algorithms using advantages of each program to eliminate disadvantages inherent to each single method ***(47)***.

2.2. Signal-Finding Bioinformatics Methods

There is a tremendous variety of software exploiting various ways to identify structural genes in a DNA sequence, and as already outlined, there are more elements present than just structural genes. Some gene-finding approaches presented above already consider bases outside the start and stop codons, but there are specialized resources, namely MatInspector ***(48)*** and SignalScan ***(49)***, to find transcription factor-binding sites using a relatively small database called TRANSFAC ***(50)***. Another program designed to find putative eukaryotic polymerase II promoter sequences in primary sequence data is PromotorScan ***(51)***.

2.3. Sequence-Alignment Methods

An alignment refers to the procedure of comparing two or more sequences by looking for a series of individual characters or character patterns that are found in the same order in the sequences. To align sequences, identical or similar characters are grouped in the same column, whereas nonidentical characters can either be placed in the same column, resulting in a mismatch, or opposite a gap in one of the other sequences. In an optimal alignment, mismatches and gaps are distributed in such a way that matches are maximized. Two types of sequence alignment have been recognized. In a global alignment, an attempt is made to align the entire sequences with as many characters as possible. In a local alignment, stretches of sequence with the highest density of matches are given the highest priority, thus generating one or more islands of matches in the aligned sequences and producing totally mismatching regions. Alignments are the working principle of programs (e.g., BLAST) that search similar sequences by successively aligning a query with an entire sequence database; they are also useful for determining the evolutionary distance between homologous sequences of different origin.

2.3.1. Multiple-Sequence Alignment

Comparison of multiple sequences can reveal gene functions that are not evident from simple sequence homologies. As a result of genome-sequencing projects, new sequences are often found to be similar to several uncharacterized sequences, defining whole families of novel genes with no obvious function. However, such a family enables the application of efficient alternative similarity search methods. Software packages are now available that derive profiles from multiple-sequence alignments. Profiles incorporate position-specific scoring information that is derived from the abundance of a given residue in an aligned column. Because sequence families preferentially conserve certain critical residues and motifs, this information should allow more sensitive database searches. Most new profile software are based on statistical HMMs. Much more comprehensive reviews of the literature on profile HMM methods are available elsewhere ***(12,28,52–55)***. ClustalW is a well-supported and frequently used free program capable of dealing with large numbers of sequences at high processing speeds as compared to other alignment algorithms, and it is available for Macintosh, Windows, and various UNIX systems ***(56)***. There is also a graphical user interface—ClustalX ***(57)***. However, for sequences with less than 30% identity, the program T-COFFEE might be used, which is more accurate than the progressively aligning ClustalW, but slower. Once the family is defined, obtaining an acceptable multiple-sequence alignment is usually straightforward. Multiple alignments can either be generated from FASTA format files (using a ClustalW supporting website) or from DbClustal, which produces a BLAST output in which the family members can be selected and the multiple alignment subsequently produced. It is important to inspect the alignment in the graphical display of ClustalX to make sure that it appears consistent. The alignment can also be saved in multiple-sequence format (MSF), which can be read by other software for further analysis (e.g., careful editing, trimming, coloring, shading, and printing). GeneDoc available for Windows (www.psc.edu/biomed/genedoc) offers many of these editing features ***(58)***.

2.4. Phylogenetic Analysis

Phylogenetic trees can be constructed based on multiple alignments. In rooted trees, the ancestral state of the organisms, or genes, being studied is shown at the bottom of the tree, and the tree branches, or bifurcates, until it reaches the terminal branches, tips, or leaves at the top of the tree. An unrooted tree is a less intuitive, more abstract concept. Unrooted trees represent the branching order, but do not indicate the root, or location, of the last common ancestor. Ideally, rooted trees are preferable, but almost every phylogenetic reconstruction algorithm provides an unrooted tree. Molecular sequence analysis is a field in its infancy and an inexact science in which there are few analytical tools that are truly based on general mathematical and statistical principles. Consequently, many phylogenetic trees reconstructed from molecular sequences are incorrect. This is mainly caused by the following:

1. Incorrect sequence alignments.
2. The failure to properly account for site-to-site variation (all sites within sequences can evolve at different rates).
3. Unequal rate effects (the inability of most tree-building algorithms to produce good phylogenetic trees when genes from different taxa in the tree have evolved at different rates).

Of the three pitfalls, alignment artifacts are potentially the most serious. A new algorithm, paralinear (logdet) distances ***(59,60)***, provides a simple, but rigorous, mathematical solution for the third pitfall. For a discussion of many other useful algorithms currently available, including maximum parsimony likelihood and other distance methods, *see* **refs. *61*** and ***62***. Sequence alignments should be carefully checked before calculating evolutionary trees. There are several programs to help calculate trees from genome data. The best known software for reconstructing trees is the program PAUP (phylogenetic analysis using parsimony), which is part of the GCG sequence analysis package that supports logdet analysis. PAUP is user friendly and comprehensive. PHYLIP/Phylodendron are further well-known packages that contain a large variety of routines, including several that incorporate the latest theoretical developments. A stand-alone software package available for many computer platforms for viewing, editing, rearranging, and printing trees is TreeView ***(64)***.

2.5. Protein Properties and Structure Prediction

Protein sequences allow extensive calculations that help to assign function, to predict topology (subcellular localization, spanning of membranes) and structure, and to find sites that are likely to be cleaved or modified; interaction or catalytic mechanisms can be simulated. Bioinformatic resources on the WWW range from the determination of the molecular weight to complex threading and three-dimensional (3D) prediction algorithms. A huge list of tools can be found on the ExPASy proteomic tools homepage ***(65)***. Because of the great variety of programs available, several of these single tools have been integrated into one interface. Examples are PredictProtein ***(66)*** or META PP ***(67)***. These integrate resources such as SignalP ***(68)***, which predicts the presence and location of signal peptide cleavage sites in amino acid sequences from different organisms—NetOglyc [predictions of mucin type O-glycosylation sites in mammalian proteins ***(69–71)***], NetPhos [predicting potential phosphorylation sites at serine, threonine or tyrosine residues in protein sequences, ***(72)***], NetPico [predictions of cleavage sites of picornaviral proteases ***(73)***], and ChloroP [predicting chloroplast transit peptides and their cleavage sites ***(74)***]. Secondary structure prediction is performed by JPRED ***(75)***; transmembrane helices are identified using TMHMM ***(76)***, TopPred ***(77,78)***, and DAS ***(79)***. Structural databases are searched to detect similarities between remote homolog proteins too weak to be inferred from simple sequence alignment techniques by FRSVR ***(80,81)*** and SAM-T02 ***(82)*** but the detection of remote homologs represents a problem to be solved because most of the results returned are supposedly wrong. Results should be checked very carefully. Homology modeling is also covered by META PP, applying both SWISS-MODEL ***(83–85)*** and CPHmodels ***(86)*** described below.

3. Databases

A database is any collection of data or information that is specially organized for rapid search and retrieval by a computer. To cope with not only the vast amount of sequence information but also other experimental data, biological databases have been set up and are updated continuously. For biological data such as information about a protein's sequence, structure, modification, or interaction, software tools have been developed that enable searching, comparing, and retrieving these stored data.

3.1. Matching Algorithms

Basic queries like finding a key word in an article employ simple pattern-matching algorithms that do not need a statistical evaluation of the result. Bioinformatic tools started like this and most tools apply algorithms that match a query against all targets in a database, taking into account the degree and type of mismatches or gaps *(87)*. Deciding whether the observed degree of structural likeness is significant and, therefore, a hint toward functional identity is a task for statistical methods based on special biological matrices *(88,89)*. These matrices are crucial to considering the biological nature of the data. According to structural relevance, varieties in certain amino acid residues, for example, are decisive, whereas other differences can be negligible.

Sequence database matching has proven to be a remarkably useful method for assigning a function to an unknown sequence. If sequence similarity to one or more database sequences, whose function is already known, is obtained, the unknown sequence can be inferred to have the same function, biochemical activity, or structure. The strength of these inferences depends on the strength of the similarity. As a rough rule, if more than 25–30% of a protein sequence is identical in an alignment, then the sequences are homologous *(90)*. RNA genes are usually much more conserved, which is the reason why they represent suitable markers for phylogenetic analyses. Functional DNA sequences like promoters or other regulatory regions are significantly shorter than genes encoding enzyme proteins or RNA, making them hard to identify by means of sequence comparison or alignment. Identification and assignment of genes, as well as the functional and structural classification of proteins, is performed based on similarity of sequence and/or properties such as motifs or structurally conserved regions. Because DNA sequences are variable in the third base position of the codon, protein-sequence analysis is the more valuable approach. In general, sequence analysis requires the comparison of sequences from unknown genes or proteins with those of known function deposited in databases. However, the sequences of homologous proteins can diverge greatly over time, whereas the structure or function of the same proteins has diverged only slightly. Conversely, proteins with similar folds can exhibit completely different functions. However, much can be deduced about an unknown protein when significant sequence similarity is detected with a well-studied protein. Alignment provides a powerful tool to compare related sequences, and the alignment of two residues could reflect a common evolutionary origin or represent common structural and/or catalytic roles, not always reflecting an evolutionary process.

3.1.1. Substitution Matrices and Alignment Scores

In order to identify the most valuable alignments, the standard procedure is to assign scores to them. For each pair of letters that can be aligned, a substitution score is chosen. The complete set of these scores is called a substitution matrix [PAM *(91)* and BLOSUM *(92)*]. Additionally, scores are chosen for gaps, which consist of one or more adjacent nulls in one sequence aligned with letters in the other. Because a single mutational event can insert or delete more than one residue, a long gap should be penalized only slightly more than a short gap. Accordingly, affined gap costs, which charge a relatively large penalty for the existence of a gap and a smaller penalty for each residue it contains, have become the most widely used gap-scoring system. The quality of sequence comparison depends very much on the choice of appropriate substitution and gap scores. In brief, for ungapped alignments, the alignment score of a given pair of residues i and j depends on the fraction q_{ij} of true alignment positions in which these paired residues tend to appear *(88)*. Accordingly, the design of a good substitution matrix is based on estimating the target frequencies q_{ij} accurately. However, the target frequencies depend on the degree of evolutionary divergence between the related sequences of interest. Therefore, a series of matrices tailored to varying degrees of evolutionary divergence are required *(88,91,92)*. This was the intention in constructing the PAM and BLOSUM series of amino-acid-substitution matrices. These matrices are generally used unmodified for gapped local and global alignment. There is no widely accepted theory for selecting gap costs that requires adjustment for individual similarity searches *(93)*.

3.1.2. Alignment Scores and E Values

To test the biological relevance of a global or local alignment of two sequences, one needs to know how the value of an alignment score can be expected to occur by chance. Current versions of the FASTA and BLAST search programs report the raw scores of the alignments they return, as well as assessments of their statistical significance based on the extreme value distribution. Most simply, these assessments take the form of *E* values. The *E* value for a given alignment depends on its score, as well as the lengths of both the query sequence and the database sequence searched. It represents the number of distinct alignments with equivalent or superior scores that might have been expected to occur only by chance. The smaller the *E* value, the more likely that the alignment is significant and not occurring by chance ***(88,89,94)***.

3.1.3. Filtering Database Sequences

Many DNA and protein sequences contain regions of highly restricted nucleic acid and amino acid compositions and regions of short elements repeated many times. The standard alignment models and scoring systems were not designed to capture the evolutionary processes that led to these low-complexity regions. As a result, two sequences containing compositionally biased regions can receive a very high similarity score that reflects this bias alone. For many purposes, these regions are not relevant and can obscure other important similarities. Therefore, programs that filter low-complexity regions from query or database sequences will often turn a useless database search into a valuable one. For this reason, the NCBI BLAST server will remove such sections in proteins using a program termed SEG ***(95)***. Although these programs automatically remove the majority of problematic matches, some problems invariably occur. Furthermore, masking might preclude interesting hits. Therefore, it is useful to adjust the masking parameters or turn filtering off completely.

3.1.4. Database Searching

Fundamentally, performing a database search is a very simple operation: A query sequence is aligned with each of the sequences in a database and a score describing the degree of likeness is calculated using a suitable matrix. Nevertheless, sequence comparison procedures should be applied carefully. The design of a BLAST database search requires consideration of the kind of information one hopes to obtain about the query sequence of interest ***(96)***. A major constraint of database searching is that it only reveals similarity and might not indicate function. Therefore, it is better to use data that describe the natural situation as accurately as possible (e.g., comparing 3D structures of proteins with each other). This is because 3D structures rather than the primary sequence is conserved during evolution processes. However, in most cases, the information will consist of a primary sequence alone. One should, nonetheless, compare deduced protein sequences rather than DNA if the query DNA is likely to encode for a protein. This also enables the detection of remote homologs ***(97)***. In DNA comparisons, there is noise from the rapidly mutated third-base position in each codon and from comparisons of noncoding frames. In addition, amino acids have chemical characteristics that allow degrees of similarity to be assessed, rather than simple recognition of identity or nonidentity. DNA vs DNA comparison (BLASTN program) is typically used to find identical regions of sequence in a database. One should apply this search in order to find RNA-encoding or regulatory regions and/or to discover whether a protein-encoding gene has been previously sequenced or contains splice junctions. Briefly, protein-level searches are valuable for detecting evolutionarily related genes, whereas DNA searches are best for locating nearly identical regions of sequence. The following should be considered when designing a database search:

1. Search a large current database (SWISS-Prot, EMBL, Genebank).
2. Compare relevant data.
3. Filter query for low-complexity regions.
4. Interpret scores with *E* values.

5. Recognize that most homologs are not found by pairwise sequence comparison.
6. Consider slower and more powerful methods, but use iterative programs with great caution (iterative programs might indicate homology, which is not related to function).

3.2. Sequence Databases

Protein-homology searches are usually performed employing the nr (nonredundant) sequence database at the NCBI (National Center for Biotechnology Information) website. The nr database combines data from several sources, removes redundant identical sequences, and yields a collection with nearly all known proteins. A frequent update of the NCBI nr database guarantees that the most recent and complete database is used. Obviously, a search will not identify a sequence that has not been included in the database and, as databases are growing so rapidly, use of a current database is essential. Several specialized databases are also available, each of which is a subset of the nr database. One might also wish to search DNA databases at the protein level. Programs can do so automatically by first translating the DNA in all six reading frames and then making comparisons with each of these translations. The nr database, which contains the most publicly available DNA sequences, is useful to search when hunting for new genes; identified genes in this database would already be in the protein nr database. Because of the different combinations of queries and database types, there are several variants of BLAST ***(87,89,90)***. The BLAST programs can be run via the Internet or they can alternatively be downloaded from an ftp site to run locally. Another option is to use the FASTA package ***(97)***. The FASTA program is slower but can be more effective than BLAST. The package also contains SSEARCH, an implementation of the rigorous Smith–Waterman algorithm, which is slow but the most sensitive. Iterative programs such as PSI-BLAST require extreme care in their operation because they can provide misleading results; however, they have the potential to find more homologs than purely pairwise methods. The effectiveness of any alignment program depends on the scoring systems it employs ***(88,92,93)***.

3.3. Protein Family and Motif Databases

There are several collections of amino-acid-sequence motifs that indicate particular structural or functional elements. Web-based searches of these collections with a newly identified sequence allow reasonably confident functional predictions to be made. A variety of genome- and cDNA-sequencing projects is producing raw sequence data at a breathtaking speed, creating the need for a large-scale functional classification effort. On a smaller scale, the average molecular biologist can also be faced with a new sequence without any *a priori* functional knowledge. Any hint as to whether the newly identified gene encodes a transcription factor, a cytoskeletal protein, or a metabolic enzyme would certainly help to interpret the experimental results and would suggest a direction for subsequent investigations.

3.3.1. Protein vs Domain Classification

The first step is usually a database search with BLAST or a similar program. Optimally, the BLAST output would show a clear similarity to a single, well-characterized protein spanning the complete length of the query protein. However, in the worst case, the output list would fail to show any significant hit. In reality, the most frequent result is a list of partial matches to assorted proteins, most of them uncharacterized, with the remainder having dubious or even contradictory functional assignments. Much of this confusion is caused by the modular architecture of the proteins involved. An analysis of known 3D protein structures reveals that, rather than being monolithic, many of them contain multiple folding units. Each unit (domain) has its own hydrophobic core and has most of its residue–residue contacts internally. In order to fulfill these conditions, independent domains must have a minimum size of approx 50 residues unless stabilized by metal ions or disulfide bridges. Analysis of protein sequences contradicts this structural observation. Sequence pairs frequently exhibit localized regions of similarity, whereas the rest of the proteins are totally dissimilar. Folding independence for all of these so-called homology domains has not been demonstrated experimentally. For protein classification, it is important to note that the homology domains also frequently harbor independent functions. Some domains have enzymatic activity, others bind to small messenger molecules, and others specifically bind to DNA, RNA, or proteins. A multidomain protein can, therefore, have more than one function and belong to more than one protein family or class. For this reason, most of the current approaches to protein functional classification focus on domains rather than complete proteins.

3.3.2. The Protein Superfamily Assignment

Efforts have been undertaken to group protein sequences into families and superfamilies. The various approaches differ in their degree of automation, their comprehensiveness, their focus on complete proteins or protein domains, and the methodology applied. Some of these efforts are aimed exclusively at the classification of existing sequence data. Others go one step further and aim to extract the essential features from sequence families and to store them in the form of domain or motif descriptors, which can then be used for searches with user-supplied protein sequences. These searches exert high sensitivity, which has proven to be most useful for the functional assignment of unknown proteins. A parallel development, which will be discussed later, is the classification of protein 3D structures. A comprehensive discussion of this topic can be found in **refs.** ***98*** and ***99***. Some of the most popular collections, which consider the modular nature of proteins, are briefly discussed. The PROSITE pattern library was one of the pioneering efforts in collecting descriptors for important protein motifs with biological relevance ***(100,101)***. A PROSITE pattern does not describe a complete domain or even protein, but just tries to identify the functionally most important residue combinations, such as the catalytic site of an enzyme. All motifs are accompanied by extensive documentation, including references. However, the short patterns do not contain enough information to yield statistically significant matches in the large and growing protein databases. Consequently, a certain number of false-positive hits is to be expected when carrying out a database search, and any hit reported after scanning the PROSITE database with a sequence has to be treated with appropriate caution. To solve these restrictions, the PROSITE pattern library has been supplemented since 1995 by the PROSITE profile library ***(101)***. Generalized profiles are at an intermediate position between a sequence-to-sequence comparison and the matching of a regular expression to a sequence ***(102)***. The ProDom database was the first comprehensive collection of complete protein domains ***(103,104)***. It is derived from SWISS-PROT, and the domains are denoted only by cluster numbers and do not contain any biological annotation. Moreover, the automatically determined domain boundaries are unreliable and the associated search methods are not very sensitive. Pfam, which is derived from ProDom, contains HMMs, which are conceptually related to the PROSITE profiles ***(53,105)***. The Pfam models typically span complete protein domains and can be searched with the HMMER package or on a web-based server. The current release of Pfam (10.0) contains 6190 families ***(106)***. Similar to the PROSITE profiles, the Pfam models are refined iteratively, starting from clear homologs and incorporating increasingly distant family members in the process. Because of their information-rich descriptors, both collections are able to detect even very distant instances of a protein motif that are rarely found by any other method. Because Pfam models and PROSITE profiles can be interconverted ***(102)***, combination searches are available at InterPro ***(107,108)***. The current release of InterPro (7.0) contains 8547 entries describing 6416 families, 1902 domains, 163 repeats, 26 active sites, 20 binding sites, and 20 posttranslational modifications. The use of this integrated service is therefore recommended, although there are still some specialized databases that are not covered. MEROPS ***(109)*** is a catalog and classification system of enzymes with proteolytic activity (peptidases or proteases).

NIFAS is a Java applet, which retrieves domain information from the Pfam database and uses ClustalW to calculate a tree for a given domain and to enable visual analysis of domain evolution in proteins. Consideration of the evolution of certain domains might be important for functional annotation of modular proteins and for understanding the function of individual domains ***(110)***. SMART (simple modular architecture research tool) allows the identification and annotation of genetically mobile domains and the analysis of domain architectures. The SMART database is an independent collection of HMMs (domain families)—660 in version 3.5—focusing on protein domains related to signaling, extracellular, and chromatin-associated proteins ***(111–113)***. These domains are extensively annotated with respect to phyletic distributions, functional class, tertiary structure, and functionally important residues. Domains found in the nr protein database as well as search parameters and taxonomic information are stored in a relational database system. User interfaces to this database allow searches for proteins containing specific combinations of domains in defined taxa. BLOCKS ***(114)*** and PRINTS ***(63,115)*** are two motif databases that represent protein or domain families by several short, ungapped multiple alignment fragments. The current release of BLOCKS (13.0) contains 8656 blocks representing 2101 groups, which are derived from PROSITE patterns. The blocks for the BLOCKS database are made automatically by looking for the most highly conserved regions in groups of proteins documented in the PROSITE database. The Internet-based versions of the PROSITE and SWISS-PROT databases that are used are located at the ExPASy molecular biology web-server of the Geneva University Hospital and the University of Geneva. The blocks created by Block Maker are created in the same manner as the blocks in the BLOCKS database but with sequences provided by the user. Results are reported in a multiple-sequence alignment format and in the standard Block format for searching. PRINTS is a compendium of protein fingerprints. A fingerprint is a group of conserved motifs used to characterize a protein family; its diagnostic power is refined by iterative scanning of a composite of SWISS-PROT+SP-TrEMBL. Usually the motifs do not overlap, but are separated along a sequence, although they might be contiguous in 3D space. Fingerprints can encode protein folds and functionalities more flexibly and powerfully than can single motifs, their full diagnostic potency deriving from the mutual context afforded by motif neighbors. BLOCKS and PRINTS can be searched with the same programs at the website InterPro (see above). Domains important in signal transduction are likely to be found with the PROSITE profiles or SMART; Pfam emphasizes extracellular domains, and the PROSITE patterns are good at identifying enzyme classes by their active-site motif. Recently, a unified protein family resource, MetaFam, was generated to support the general classification efforts ***(116)***. MetaFam is a protein family classification built up from 10 publicly accessible protein family databases (Blocks + DOMO, Pfam, PIR-ALN, PRINTS, PROSITE, ProDom, PROTOMAP, SBASE, and SYSTERS). Meta-Fam's family "supersets" are created automatically by comparing families between the databases. However, the number of available single and combined domain descriptors should not be overestimated as a quality criterion, as the databases and the associated search methods differ in generality and sensitivity. The most promising approach to predicting the exact function of a protein is to find its characterized ortholog from a different species or a well-conserved paralog that fulfills a related but different function ***(117)***. In addition, the databases contain large amounts of incorrect annotated sequences.

3.4. Protein Structure Databases

The complexity and sophistication of biological molecular interactions are astonishing. In this context, it is essential to develop bioinformatic tools that reliably allow the prediction of protein structure, as the structure determines interaction with all kinds of small molecules (substrates, activators, repressors, drugs) and other proteins (either specific as in natural multiprotein complexes or unspecific), consequently revealing the protein's function. In the near future, representative structures for most water-soluble protein domains will be available, which will allow modeling and classification of related sequences to provide structures for all gene products. However, elucidating the function of all gene products in vivo will be a long-term challenge for biologists. The emphasis will shift to understanding of principles and control of biological function and the interactions between molecules. A 3D model of a protein can help one to understand the "docking" of ligands and proteins, which is essential to enable their rational design or modification to efficiently discover drug targets or design new drugs targeting both proteins in pathogens and disease-related human proteins.

3.4.1. Structure Classification

The Protein Data Bank, a computer-based archival file for macromolecular structures, was founded under the term "Brookhaven National Laboratory Protein Data Bank" (BNL PDB) in 1977 ***(118)***. Today, the PDB repository for the processing and distribution of 3D biological macromolecular structure data contains over 22,053 entries in a standardized file format that can be browsed and searched online ***(119)***. There are several projects taking data from PDB for further analysis. The SCOP ***(120)*** database aims to provide a detailed and comprehensive description of the structural and evolutionary relationships among all proteins whose structure is known, including all entries in PDB. It is available as a set of tightly linked hypertext documents, which make the large database comprehensible and accessible. SCOP uses three different major levels of hierarchy: family (clear evolutionarily relationship), superfamily (probable common evolutionary origin), and fold (major structural similarity). A similar approach is realized in the CATH database ***(121,122)***, which is also a hierarchical domain classification of protein structures in the PDB but only crystal structures solved to resolution better than 3.0 Å are considered, together with nuclear magnetic resonance (NMR) structures. CATH employs four major levels in this hierarchy: class, architecture, topology (fold family), and homologous superfamily.

3.4.2. Structure Modeling

Three-dimensional structure prediction (modeling) of proteins produces reliable results only using the "homology modeling" approach, which generally consists of four steps:

1. Data banks searching to identify the structural homolog.
2. Target–template alignment.
3. Model building and optimization.
4. Model evaluation.

SWISS-MODEL is a server for automated comparative homology modeling of 3D protein structures. It pioneered the field of automated modeling starting in 1993 ***(123)*** and is the most widely used free web-based automated modeling facility today. In 2002, the server computed 120,000 user requests for 3D protein models. SWISS-MODEL provides several levels of user interaction through its Internet interface ***(124)***. In the "first approach mode," only an amino acid sequence of a protein is submitted to build a 3D model. Template selection, alignment, and model building are performed completely automated by the server. In the "alignment mode," the modeling process is based on a user-defined target–template alignment. Complex modeling tasks can be handled with the "project mode" using DeepView ***(125)***, the Swiss-PdbViewer (available for PC, Macintosh, Linux and SGI, downloadable from http://www.expasy.org/spdbv), an integrated sequence-to-structure workbench. All models are sent back via e-mail with a detailed modeling report. WhatCheck analyses and ANOLEA evaluations are provided optionally. Similar homology modelers are CPHmodels ***(86)***, Geno3D ***(126)***, and ESyPred3D ***(127)***.

3.5. Protein Interaction Databases

Protein–protein interactions play important roles in nearly every event that takes place in a cell. The Biomolecular Interaction Network Database (BIND) is a database designed to store full descriptions of interactions, molecular complexes, and pathways ***(128)***. An Interaction record is based on the interaction between two objects. An object can be a protein, DNA, RNA, ligand, or molecular complex. The description of an interaction encompasses cellular location, experimental conditions used to observe the interaction, conserved sequence, molecular location of interaction, chemical action, kinetics, thermodynamics, and chemical state and can be accessed through a BLAST search against the database to gather information on the interactions of the query sequence stored in BIND *(128)*. The DIP database *(129)* catalogs experimentally determined interactions between proteins. It combines information from a variety of sources to create a single, consistent set of protein–protein interactions.

3.5.1. Protein–Protein Interaction Prediction and Docking

The protein–protein or protein–ligand docking problem started to fascinate biophysical chemists and computational biologists almost 30 yr ago ***(130,131)***. Given the 3D structures of two interacting proteins, a docking algorithm aims to determine the 3D structures of the complex by rotating and translating the proteins, generating a large number of candidate complexes in the computer, and to select favorable ones. Docking procedures are tested first on protein–protein complexes taken from the Protein Data Bank, mostly protease–inhibitor and antigen–antibody complexes. The CAPRI experiment ***(132)***, inspired by the CASP (Critical Assessment of Structure Prediction) algorithm was given atomic coordinates for protein components of several target complexes. The predicted interactions were assessed by comparison to X-ray structures and show significant success on some of the targets. However, the prediction failed with others, and progress is still needed before large-scale predictions of protein–protein interactions can be made reliably. The docking of small molecules and proteins is reviewed in **ref.** ***133***.

4. Bioinformatics Genomics and Medical Applications

4.1. Genome Analysis and Databases

Most software tools and databases presented above can be used with any kind of DNA or protein sequence. The availability of complete genome sequences of hundreds of more or less related organisms (mainly prokaryotes) allows additional approaches leading the assignment of function to thus far unknown genes and proteins that are not possible with just a subset of a genome. There are also a number of medically related databases such as OMIM (Online mendelian inheritance in man), which contains information regarding inherited and other diseases. Furthermore, attempts to correlate data regarding particular diseases are also being developed, such as the Cancer Genome Anatomy Project, which is a database of known mutations in genes arising in particular tissues.

4.2. Comparative Genomics

Methods requiring complete genome sequences include (1) mapping and alignments of entire genomes of closely related organisms to identify clusters and functional units, (2) metabolic pathway reconstruction to identify missing links and assign function (which is of special biotechnological interest), and (3) comparison of entire genomes of closely related organisms with a different phenotype. The latter technique is suitable for detecting genes that might contribute to virulence toward humans or plants. It is, therefore, of special interest for medical science, the pharmaceutic industry, and agriculture. By means of subtracting the entire genome of a harmless bacterium (e.g., a *Bacillus* strain) from the genome of a closely related virulent bacterium (e.g., *Bacillus anthracis*), genes that are not related to virulence are eliminated. This procedure yields candidates that are probably responsible for the pathogenic phenotype of *B. anthracis* ***(134,135)*** or might be suitable for use as vaccines ***(136)***. As a consequence, these genes represent promising targets for specific drugs against the virulence system of *B. anthracis* without affecting apathogenic strains. The described comparison can be performed using predicted genes only; more accurate results are obtained comparing expression profiles or applying proteomics.

Comparisons of entire genomes reveal clusters that are conserved. Clusters of orthologous groups of proteins (COGs) were delineated by comparing protein sequences encoded in 43 complete genomes, representing 30 major phylogenetic lineages. Each COG consists of individual proteins or groups of paralogs from at least three lineages and, thus, corresponds to an ancient conserved domain. The COG database ***(137,138)*** provides a phyletic pattern search web page that is available to facilitate the creation of a specific filter that, as being applied to the COGs, can filter out a COG set that will comply with the condition specified in the query.

4.3. Pharmacogenomics

Pharmacogenomics is the study of how an individual's genetic inheritance affects the body's response to drugs. The term comes from the words pharmacology and genomics and is, thus, the intersection of pharmaceuticals and genetics. Pharmacogenomics holds the promise that drugs might one day be tailor-made for individuals and adapted to each person's own genetic makeup. Environment, diet, age, lifestyle, and state of health can influence a person's response to medicines, but understanding an individual's genetic makeup is thought to be the key to creating personalized drugs with greater efficiency and safety. Pharmacogenomics combines traditional pharmaceutical sciences such as biochemistry with annotated knowledge of genes, proteins, and single-nucleotide polymorphisms.

The anticipated benefits of pharmacogenomics are as follows:

1. More powerful medicines (by a therapy more targeted to specific diseases).
2. Better, safer drugs (doctors will be able to analyze a patient's genetic profile and prescribe the best available drug therapy from the beginning).
3. More accurate methods of determining appropriate drug dosages (dosages on weight and age will be replaced with dosages based on a person's genetics).
4. Advanced screening for disease (knowledge of a particular disease susceptibility will allow careful monitoring, and treatments can be introduced at the most appropriate stage to maximize their therapy).
5. Better vaccines (vaccines made of genetic material, either DNA or RNA, promise all the benefits of existing vaccines without all the risks).
6. Improvements in the drug discovery and approval process (pharmaceutical companies will be able to discover potential therapies more easily using genome targets),
7. Decrease in the overall cost of health care.

The explosion in both single-nucleotide polymorphism (SNP) and microarray data generated from the human genome project has necessitated the development of a means of cataloging and annotating (briefly describing) these data so that scientists can more easily access and use it for their research. Database repositories for both SNP (dbSNP) and microarray (GEO) data are available at the NCBI. These databases include either descriptive information about the data within the site itself (GEO) or links to NCBI and external information resources (dbSNP). Access to these data and information resources allows scientists to more easily interpret data that will be used not only to help determine drug response but to study disease susceptibility and conduct basic research in population genetics.

4.3.1. SNP Databases

A key aspect of human genome research and pharmacogenomics is associating sequence variations with heritable phenotypes. The most common variations are SNPs, which occur approximately once every 100–300 bases (see Chapter 19). Because SNPs are expected to facilitate large-scale association genetics studies, there has recently been great interest in SNP discovery and detection. The cSNP database specializing on human chromosome 21 is a joint project between the Division of Medical Genetics of the University of Geneva Medical School and the Swiss Institute of Bioinformatics, which offers BLAST and text searches to explore their data. In collaboration with the National Human Genome Research Institute, the National Center for Biotechnology Information has established the dbSNP database ***(139)*** to serve as a central repository for both single-base nucleotide subsitutions and short deletion and insertion polymorphisms. Once discovered, these polymorphisms could be used by additional laboratories, using the sequence information around the polymorphism and the specific experimental conditions. The data in dbSNP are integrated with other NCBI genomic data and are accessible by the same tools as other NCBI databases.

4.3.2. Expression Profiling

The Gene Expression Omnibus (GEO) is a gene expression and hybridization array data repository, as well as a curated, online resource for gene expression data browsing, query, and retrieval. GEO was the first fully public high-throughput gene expression data repository and became operational in July 2000 ***(140)***. There are several ways to deposit and retrieve GEO data. The search facilities "Gene profiles," "Dataset," and "Sequence BLAST" are powerful and link to the well-known Entrez-Interface, including accession links to relevant genes and proteins.

References

1. Venter, J. C. et al (2001) The sequence of the human genome. *Science* **291,** 1304–1351.
2. Lander, E. S. et al (2001) Initial sequencing and analysis of the human genome. Nature 409, 860–921.
3. Rehm B.H. (2001) Bioinformatic tools for DNA/protein sequence analysis, functional assignment of genes and protein classification. *Appl. Microbiol. Biotechnol.* **57,** 579–592.
4. Ewing, B., Hillier, L., Wendl, M. C., and Green, P. (1998) Base-calling of automated sequencer traces using phred. I. Accuracy assessment. *Genome Res.* **8,** 175–185.
5. Ewing, B. and Green, P. (1998) Base-calling of automated sequencer traces using phred. II. Error probabilities. *Genome Res.* **8,** 186–194.
6. Huang, X. and Madan, A. (1999) CAP3: A DNA sequence assembly program. *Genome Res.* **9,** 868–877.
7. Gordon, D., Abajian, C., and Green, P. (1998) Consed: a graphical tool for sequence finishing. *Genome Res.* **8,** 195–202.
8. Staden, R. (1996) The Staden Sequence Analysis Package. *Mol. Biotech. 5,* 233–241.
9. Staden, R. (1984) Computer methods to locate signals in nucleic acid sequences. *Nucleic Acids Res.* **12,** 505–519.
10. Claverie, J.-M. (1997) Computational methods for the identification of genes in vertebrate genomic sequences. *Hum. Mol. Genet.* **6,** 1735–1744.
11. Guigo, R. (1997) Computational gene identification: an open problem. *Comput. Chem.* **21,** 215–222.
12. Krogh, A. (1998) In *Computational Methods in Molecular Biology* (Salzberg, S. L., Searls, D., and Kasif, S., eds.), Elsevier, Amsterdam.
13. Krogh, A. (1998) In *Guide to Human Genome Computing* (Bishop, M. J., ed.), 2nd ed. Academic, New York, pp. 261–274.
14. Delcher, A. L., Harmon, D., Kasif, S., White, O., and Salzberg, S. L. (1999) Improved microbial gene identification with GLIMMER. *Nucleic Acids Res.* **27,** 4636–4641.
15. Guigo, R., Agarwal, P., Abril, J. F., Burset, M., and Fickett, J. W. (2000) An assessment of gene prediction accuracy in large DNA sequences. *Genome Res.* **10,** 1631–1642.
16. Krogh, A. (2000) Using database matches with for HMMGene for automated gene detection in *Drosophila*. *Genome Res.* **10,** 523–528.
17. Shibuya, T. and Rigoutsos, I. (2002) Dictionary-driven prokaryotic gene finding. *Nucleic Acids Res.* **30,** 2710–2725.
18. Pedersen, J. S. and Hein, J. (2003) Gene finding with a hidden Markov model of genome structure and evolution. *Bioinformatics* **19,** 219–227.
19. Guo, F. B., Ou, H. Y., and Zhang, C. T. (2003) ZCURVE: a new system for recognizing protein-coding genes in bacterial and archaeal genomes. *Nucleic Acids Res.* **31,** 1780–1789.
20. Larsen, T. S., Krogh, A. (2003) EasyGene—a prokaryotic gene finder that ranks ORFs by statistical significance. *BMC Bioinformat.* **4,** 21.
21. Gelfand, M. S. (1995) Prediction of function in DNA sequence analysis. *J. Comput. Biol.* **2,** 87–115.
22. Sherriff, A. and Ott, J. (2001) Applications of neural networks for gene finding. *Adv. Genet.* **42,** 287–297.
23. Fickett, J. W. (1996) Finding genes by computer: the state of the art. *Trends Genet.* **12,** 316–320.

24. Zhang, C. T., Wang, J., and Zhang, R. (2002) Using a Euclid distance discriminant method to find protein coding genes in the yeast genome. *Comput. Chem.* **26,** 195–206.
25. Bajic, V. B. and Seah, S. H. (2003) Dragon gene start finder: an advanced system for finding approximate locations of the start of gene transcriptional units. *Genome Res.* **13,** 1923–1929.
26. Zhang, M. Q. (1998) Statistical features of human exons and their flanking regions. *Hum. Mol. Genet.* **7,** 919–932.
27. Searls, D. B. (1992) The linguistics of DNA. *Am. Sci.* **80,** 579–591.
28. Durbin, R., Eddy, S., Krogh, A., and Mitchison, G. (1998) *Biological Sequence Analysis: Probabilistic Models of Proteins and Nucleic acids.* Cambridge University Press, Cambridge.
29. Krogh, A., Mian, I. S., and Haussler, D. (1994) A hidden Markov model that finds genes in *E. coli* DNA. *Nucleic Acids Res.* **22,** 4768–4778.
30. Cole, S.T., Brosch, R., Parkhill, J., et al. (1998) Deciphering the biology of *Mycobacterium tuberculosis* from the complete genome sequence. *Nature* **393,** 537–544.
31. Thomas, A. and Skolnick, M. (1994) A probabilistic model for detecting coding regions in DNA sequences. *IMA J. Math. Appl. Med. Biol.* **11,** 149–160.
32. Henderson, J., Salzberg, S., and Fasman, K. (1997) Finding genes in DNA with a hidden Markov model. *J. Comput. Biol.* **4,** 127–141.
33. Lukashin, A. V. and Borodovsky, M. (1998) GeneMark hmm: new solutions for gene finding. *Nucleic Acids Res.* **26,** 1107–1115.
34. Salzberg, S. L., Pertea, M., Delcher, A. L., Gardner, M. J., and Tettelin, H. (1999) Interpolated Markov models for eukaryotic gene finding. *Genomics* **59,** 24–31.
35. Badger, J. H. and Olsen, G. J. (1999) CRITICA: coding region identification tool invoking comparative analysis. *Mol. Biol. Evol.* **16,** 512–524.
36. Bocs, S., Cruveiller, S., Vallenet, D., Nuel, G., and Medigue, C. (2003) AMIGene: annotation of microbial genes. *Nucleic Acids Res.* **31,** 3723–6.
37. Besemer, J., Lomsadze, A., and Borodovsky, M. (2001) GeneMarkS: a self-training method for prediction of gene starts in microbial genomes. Implications for finding sequence motifs in regulatory regions. *Nucleic Acids Res.* **29,** 2607–2618.
38. Yeramian, E. and Jones, L. (2003) GeneFizz: a web tool to compare genetic (coding/non-coding) and physical (helix/coil) segmentations of DNA sequences. Gene discovery and evolutionary perspectives. *Nucleic Acids Res.* **31,** 3843–3849.
39. Kotlar, D. and Lavner, Y. (2003) Gene prediction by spectral rotation measure: a new method for identifying protein-coding regions. *Genome Res.* **13,** 1930–1937.
40. Snyder, E. and Stormo, G. (1995) Identification of protein coding regions in genomic DNA. *J. Mol. Biol.* **248,** 1–18.
41. Reese, M. G., Eeckman, F. H., Kulp, D., and Haussler, D. (1997) Improved splice site detection in Genie. *J. Comput. Biol.* **4,** 311–323.
42. Burge, C. and Karlin, S. (1997) Prediction of complete gene structures in human genomic DNA. *J. Mol. Biol.* **268,** 78–94.
43. Xu, Y. and Überbacher, E. C. (1997) Automated gene identification in large-scale genomic sequences. *J. Comput. Biol.* **4,** 325–338.
44. Gelfand, M. S., Mironov, A. A., and Pevzner, P. A. (1996) Gene recognition via spliced sequence alignment. *Proc. Natl. Acad. Sci. USA* **93,** 9061–9066.
45. Foissac, S., Bardou, P., Moisan, A., Cros, M. J., and Schiex, T. (2003) EUGENE'HOM: a generic similarity-based gene finder using multiple homologous sequences. *Nucleic Acids Res.* **31,** 3742–3745.
46. Smith, T. E. and Waterman, M. S. (1981) Identification of common molecular subsequences. *J. Mol. Biol.* **147,** 195–197.
47. Yada, T., Takagi, T., Totoki, Y., Sakaki, Y., and Takaeda Y. (2003) DIGIT: a novel gene finding program by combining gene-finders. *Pac. Symp. Biocomput.* **2003,** 375–387.
48. Quandt, K., Frech, K., Karas, H., Wingender, E., and Werner, T. (1995) MatInd and MatInspector—new fast and versatile tools for detection of consensus matches in nucleotide sequence data. *Nucleic Acids Res.* **23,** 4878–4884.
49. Prestridge, D. S. (1991) SIGNAL SCAN: a computer program that scans DNA sequences for eukaryotic transcriptional elements. *CABIOS* **7,** 203–206.
50. Wingender, E., Chen, X., Hehl, R., et al. (2000) TRANSFAC: an integrated system for gene expression regulation. *Nucleic Acids Res.* **28,** 316–319.

51. Prestridge, D. S. (1995) Predicting Pol II Promoter Sequences Using Transcription Factor Binding Sites. *J. Mol. Biol.* **249,** 923–932.
52. Eddy, S. R. (1996) Hidden Markov models. *Curr. Opin. Struct. Biol.* **6,** 361–365.
53. Eddy, S. R. (1998) Profile hidden Markov models. *Bioinformatics* **14,** 755–763.
54. Baldi, R. and Brunak, S. (1998) *Bioinformatics: The Machine Learning Approach.* MIT Press, Boston, MA.
55. Korenberg, M. J., David, R., Hunter, I. W., and Solomon, J. E. (2000) Automatic classification of protein sequences into structure/function groups via parallel cascade identification: a feasibility study. *Ann. Biomed. Eng.* **28,** 803–811.
56. Thompson, J. D., Higgins, D. G., and Gibson, T. J. (1994) CLUSTAL W: improving the sensitivity of progressive multiple sequence alignment through sequence weighting, position-specific gap penalties and weight matrix choice. *Nucleic Acids Res.* **22,** 4673–4680.
57. Thompson, J. D., Gibson, T. J., Plewniak, F., Jeanmougin, F., and Higgins, D. G. (1997) The CLUSTAL X windows interface: flexible strategies for multiple sequence alignment aided by quality analysis tools. *Nucleic Acids Res.* **25,** 4876–4882.
58. Nicholas, K. B., Nicholas, H. B., Jr., and Deerfield, D. W., II. (1997) GeneDoc: analysis and visualization of genetic variation. *EMBNEW.NEWS* **4,** 14.
59. Lake, J. A. (1994) Reconstructing evolutionary trees from DNA and protein sequences: paralinear distances. *Proc. Natl. Acad. Sci. USA* **91,** 1451–1459.
60. Lockhart, P. J., Steel, M. A., Hendy, M. D., and Penny, D. (1994) Recovering evolutionary trees under a more realistic model of sequence. *Mol. Biol. Evol.* **11,** 605–612.
61. Brocchieri, L. (2001) Phylogenetic inferences from molecular sequences: review and critique. *Theor. Popul. Biol.* **59,** 27–40.
62. Stewart, C.-B. (1993) The powers and pitfalls of parsimony. *Nature* **361,** 603–607.
63. Attwood, T. K., Beck, M. E., Flower, D. R., Scordis, P., and Selley, J. N. (1998) The PRINTS protein fingerprint database in its fifth year. *Nucleic Acids Res.* **26,** 304–308.
64. Page, R. D. (1996) TreeView: an application to display phylogenetic trees on personal computers. *Comput. Appl. Biosci.* **12,** 357–358.
65. Gasteiger, E., Gattiker, A., Hoogland, C., Ivanyi, I., Appel, R. D., and Bairoch A. (2003) ExPASy: the proteomics server for in-depth protein knowledge and analysis. *Nucleic Acids Res.* **31,** 3784–3788.
66. Rost, B. (1996) PHD: predicting one-dimensional protein structure by profile based neural networks. *Methods Enzymol.* **266,** 525–539.
67. Eyrich, V. A. and Rost, B. (2003) META-PP: single interface to crucial prediction servers. *Nucleic Acids Res.* **31,** 3308–3310.
68. Nielsen, H., Engelbrecht, J., Brunak, S., and von Heijne, G. (1997) Identification of prokaryotic and eukaryotic signal peptides and prediction of their cleavage sites. *Protein Eng.* **10,** 1–6.
69. Hansen, J. E., Lund, O., Tolstrup, N, Gooley, A. A., Williams, K. L., and Brunak, S. (1998) NetOglyc: Prediction of mucin type O-glycosylation sites based on sequence context and surface accessibility. *Glycoconjugate J.* **15,** 115–130.
70. Hansen, J. E., Lund, O., Rapacki, K., and Brunak, S. (1997) O-glycbase version 2.0 - A revised database of O-glycosylated proteins. *Nucleic Acids Res.* **25,** 278–282.
71. Hansen, J. E., Lund, O., Rapacki, K., et al. (1995) Prediction of O-glycosylation of mammalian proteins: specificity patterns of UDP-GalNAc:-polypeptide *N*-acetylgalactosaminyltransferase. *Biochem. J.* **308,** 801–813.
72. Blom, N., Gammeltoft, S., and Brunak, S. (1999) Sequence- and structure-based prediction of eukaryotic protein phosphorylation sites. *J. Mol. Biol.* **294,** 1351–1362.
73. Blom, N., Hansen, J., Blaas, D., and Brunak, S. (1996) Cleavage site analysis in picornaviral polyproteins: Discovering cellular targets by neural networks. *Protein Sci.* **5,** 2203–2216.
74. Emanuelsson, O., Nielsen, H., and von Heijne, G. (1999) ChloroP, a neural network-based method for predicting chloroplast transit peptides and their cleavage sites. *Protein Sci.* **8,** 978–984.
75. Cuff, J. A. and Barton, G. J. (1999) Evaluation and improvement of multiple sequence methods for protein secondary structure prediction. *Proteins* **34,** 508–519.
76. Sonnhammer, E. L. L. von Heijne, G., and Krogh, A. (1998) A hidden Markov model for predicting transmembrane helices in protein sequences. in *Proceedings of the Sixth Intern Conference on Intelligent Systems for Molecular Biology (ISMB98)*, pp. 175–182.
77. von Heijne, G. (1992) Membrane protein structure prediction, hydrophobicity analysis and the positive-inside rule. *J. Mol. Biol.* **225,** 487–494.

78. Karplus, K., Barrett, C., and Hughey, R. (1998) Hidden markov models for detecting remote protein homologies. *Bioinformatics* **14,** 846–856.
79. Cserzo, M., Wallin, E., Simon, I., von Heijne, G., and Elofsson, A. (1997) Prediction of transmembrane alpha-helices in procariotic membrane proteins: the dense alignment surface method. *Protein Eng.* **10,** 673–676.
80. Fischer, D. and Eisenberg, D. A. (1996) Fold recognition using sequence-derived properties. *Protein Sci.* **5,** 947–955.
81. Elofsson, A., Fischer, D., Rice, D. W., LeGrand, S., and Eisenberg, D. A. (1996) Study of combined structure-sequence profiles. *Folding Design* **1,** 451–461.
82. Karplus, K., Karchin, R., Draper, J., et al. (2003) Combining local-structure, fold-recognition, and new-fold methods for protein structure prediction. *Proteins* **53(Suppl 6),** 491–496.
83. Peitsch, M. C. (1995) Protein modelling by E-mail. *BioTechnology* **13,** 658–660.
84. Peitsch, M. C. (1996) ProMod and Swiss-Model: internet-based tools for automated comparative protein modelling. *Biochem. Soc. Trans.* **24,** 274–279.
85. Guex, N. and Peitsch, M. C. (1997) SWISS-MODEL and the Swiss-PdbViewer: an environment for comparative protein modelling. *Electrophoresis* **18,** 2714–2723.
86. Lund, O., Frimand, K., Gorodkin, J., et al. (1997) Protein distance constraints predicted by neural networks and probability density functions. *Protein Eng.* **10,** 1241–1248.
87. Altschul, S. F., Gish, W., Miller, W., Myers, E. W., and Lipman, D. J. (1990) Basic local alignment search tool. *J. Mol. Biol.* **215,** 403–410.
88. Altschul, S. F. (1991) Amino acid substitution matrices from an information theoretic perspective. *J. Mol. Biol.* **219,** 555–565.
89. Altschul, S. F. and Gish, W. (1996) Local alignment statistics. *Methods Enzymol.* **266,** 460–480.
90. Rost, B., Schneider, R., and Sander, C. (1997) Protein fold recognition by prediction-based threading. *J. Mol. Biol.* **270,** 471–480.
91. Dayhoff, M. O., Barker, W. C., and Hunt, L. T. (1983) Establishing homologies in protein sequences. *Methods Enzymol.* **91,** 524–545.
92. Henikoff, S. and Henikoff, J. G. (1992) Amino acid substitution matrices from protein blocks. *Proc. Natl. Acad. Sci. USA* **89,** 10,915–10,919.
93. Pearson, W. R. (1995) Comparison of methods for searching protein sequence databases. *Protein Sci.* **4,** 1145–1160.
94. Karlin, S. and Altschul, S. E. (1990) Methods for assessing the statistical significance of molecular sequence features by using general scoring schemes. *Proc. Natl. Acad. Sci. USA* **87,** 2264–2268.
95. Wootton, J. C. (1994) Non-globular domains in protein sequences: automated segmentation using complexity measures. *Comput. Chem.* **18,** 269–285.
96. Altschul, S. F., Madden, T. L., Schäffer, A. A., et al. (1997) Gapped BLAST and PSI-BLAST: a new generation of protein database search programs. *Nucleic Acids Res.* **25,** 3389–3402.
97. Pearson, W. R. and Lipman, D. J. (1988) Improved tools for biological sequence comparison. *Proc. Natl. Acad. Sci. USA* **85,** 2444–2448.
98. Martin, A. C., Orengo, C. A., Hutchinson, E. G., et al. (1998) Protein folds and functions. *Structure* **6,** 875–884.
99. McGuffin, L. J., Bryson, K., and Jones, D. T. (2001) What are the baselines for protein fold recognition? *Bioinformatics* **17,** 63–72.
100. Bairoch, A. (1991) PROSITE: a dictionary of sites and patterns in proteins. *Nucleic Acids Res.* **19,** 2241–2245.
101. Bairoch, A., Bucher, P., and Hofmann, K. (1997) The PROSITE database, its status in 1997. *Nucleic Acids Res.* **25,** 217–221.
102. Bucher, P., Karplus, K., Moeri, N., and Hofmann, K. (1996) A flexible motif search technique based on generalized profiles. *Comput. Chem.* **20,** 3–23.
103. Sonnhammer, E. L. and Kahn, D. (1994) Modular arrangement of proteins as inferred from analysis of homology. *Protein Sci.* **3,** 482–492.
104. Corpet, F., Gouzy, J., and Kahn, D. (1998) The ProDom database of protein domain families. *Nucleic Acids Res.* **26,**323–326.
105. Sonnhammer, E. L., Eddy, S. R., and Durbin, R. (1997) Pfam: a comprehensive database of protein domain families based on seed alignments. *Proteins* **28,** 405–420.
106. Bateman, A., Birney, E., Cerruti, L., et al. (2002) The Pfam protein families database. *Nucleic Acids Res.* **30,** 276–280.

107. Apweiler, R., Attwood, T. K., Bairoch, A., et al. (2001) The InterPro database, an integrated documentation resource for protein families, domains and functional sites. *Nucleic Acids Res.* **29,** 37–40.
108. Mulder, N. J., Apweiler, R., Attwood, T. K., et al. (2003) The InterPro Database, 2003 brings increased coverage and new features. *Nucleic Acids Res.* **31,** 315–8.
109. Rawlings, N. D., O'Brien, E., and Barrett, A.J. (2002) MEROPS: the protease database. *Nucleic Acids Res.* **30,** 343–346.
110. Storm, C. E. and Sonnhammer, E. L. (2001) NIFAS: visual analysis of domain evolution in proteins. *Bioinformatics* **17,** 343–348.
111. Schultz, J., Milpetz, F., Bork, P., and Ponting, C. P. (1998) SMART, a simple modular architecture research tool: identification of signaling domains. *Proc. Natl. Acad. Sci. USA* **95,** 5857–5864.
112. Schultz, J., Copley, R. R., Doerks, T., Ponting, C. P., and Bork, P. (2000) SMART: a web-based tool for the study of genetically mobile domains. *Nucleic Acids Res.* **28,** 231–234.
113. Letunic, I., Goodstadt, L., Dickens, N. J., et al. (2002) Recent improvements to the SMART domain-based sequence annotation resource. *Nucleic Acids Res.* **30,**242–244.
114. Pietrokovski, S., Henikoff, J.G. and Henikoff, S, (1996) The Blocks database—a system for protein classification. *Nucleic Acids Res.* **24,** 197–200.
115. Attwood, T. K., Flower, D. R., Lewis, A. P., et al. (1999) PRINTS prepares for the new millennium. *Nucleic Acids Res.* **27,** 220–225.
116. Silverstein, K. A., Shoop, E., Johnson, J. E., and Retzel, E. F. (2001) MetaFam: a unified classification of protein families. I. Overview and statistics. *Bioinformatics* **17,** 249–261.
117. Yuan, Y. P., Eulenstein, O., Vingron, M., and Bork, P. (1998) Towards detection of orthologues in sequence databases. *Bioinformatics* **14,** 285–289.
118. Bernstein, F. C., Koetzle, T. F., Williams, G. J., et al. (1977) The Protein Data Bank. A computer-based archival file for macromolecular structures. *Eur. J. Biochem.* **80,** 319–324.
119. Berman, H. M., Westbrook, J., Feng, Z., et al. (2000) The Protein Data Bank. *Nucleic Acids Res.* **28,** 235–242.
120. Murzin, A.G., Brenner, S.E., Hubbard, T., and Chothia, C. (1995) SCOP: a structural classification of proteins database for the investigation of sequences and structures. *J. Mol. Biol.* **247,** 536–540.
121. Orengo, C. A., Michie, A. D., Jones, S., Jones, D. T., Swindells, M. B., and Thornton, J. M. (1997) CATH—a Hierarchic classification of protein domain structures. *Structure* **5,** 1093–1108.
122. Pearl, F. M. G, Lee, D., Bray, J. E, Sillitoe, I., Todd, A. E., Harrison, A. P., Thornton, J. M., and Orengo, C.A. (2000) Assigning genomic sequences to CATH. *Nucleic Acids Res.* **28,** 277–282.
123. Peitsch, M. C. and Jongeneel, V. (1993) A 3-dimensional model for the CD40 ligand predicts that it is a compact trimer similar to the tumor necrosis factors. *Int. Immunol.* **5,** 233–238.
124. Schwede, T., Kopp, J., Guex, N., and Peitsch, M. C. (2003) SWISS-MODEL: an automated protein homology-modeling server. *Nucleic Acids Res.* **31,** 3381–3385.
125. Guex, N. and Peitsch, M. C. (1997) SWISS-MODEL and the Swiss-PdbViewer: an environment for comparative protein modeling. *Electrophoresis* **18,** 2714–2723.
126. Combet, C., Jambon, M., Deleage, G., and Geourjon, C. (2002) Geno3D: automatic comparative molecular modelling of protein. *Bioinformatics* **18,** 213–214.
127. Lambert, C., Leonard, N., De Bolle, X., and Depiereux, E. (2002) ESyPred3D: prediction of proteins 3D structures. *Bioinformatics* **18,** 1250–1256.
128. Bader, G. D., Betel, D., and Hogue, C. W. (2003) BIND: the Biomolecular Interaction Network Database. *Nucleic Acids Res.* **31,** 248–250.
129. Xenarios, I., Rice, D. W., Salwinski, L., Baron, M. K., Marcotte, E. M., and Eisenberg, D. (2000) DIP: The Database of Interacting Proteins. *Nucleic Acids Res.* **28,** 289–291.
130. Levinthal, C., Wodak, S. J., Kahn, P., and Dadivanian, A. K. (1975) Hemoglobin interaction in sickle cell fibers. I. Theoretical approaches to the molecular contacts. *Proc. Natl. Acad. Sci. USA* **72,** 1330–1334.
131. Wodak, S. J. and Janin, J. (1978) Computer analysis of protein-protein interaction. *J. Mol. Biol.* **124,** 323–342.
132. Janin, J., Henrick, K., Moult, J., et al. (2003) CAPRI: a Critical Assessment of PRedicted Interactions. *Proteins* **52,** 2–9.
133. Taylor, R. D., Jewsbury, P. J., and Essex, J. W. (2002) A review of protein-small molecule docking methods. *J. Comput. Aided Mol. Des.* **16,** 151–166.
134. Read, T. D., Peterson, S. N., Tourasse, N., et al. (2003) The genome sequence of Bacillus anthracis Ames and comparison to closely related bacteria. *Nature* **423,** 81–86.

135. Ivanova, N., Sorokin, A., Anderson, I., et al. (2003) Genome sequence of Bacillus cereus and comparative analysis with Bacillus anthracis. *Nature* **423,** 87–91.
136. Smith, D. R. (1996) Microbial pathogen genomes - new strategies for identifying therapeutics and vaccine targets. *Trends Biotechnol.* **14,** 290–293.
137. Tatusov, R. L., Koonin, E. V., and Lipman, D. J. (1997) A genomic perspective on protein families. *Science* **278,** 631–637.
138. Tatusov, R. L., Natale, D. A., Garkavtsev, I. V., et al. (2001) The COG database: new developments in phylogenetic classification of proteins from complete genomes. *Nucleic Acids Res.* **29,** 22–28.
139. Wheeler, D. L., Church, D. M., Federhen, S., et al. (2003) Database resources of the National Center for Biotechnology. *Nucleic Acids Res.* **31,** 28–33.
140. Edgar, R., Domrachev, M., and Lash, A.E. (2002) Gene Expression Omnibus: NCBI gene expression and hybridization array data repository. *Nucleic Acids Res.* **30,** 207–210.
141. Rehm, B. H. A. and Reinecke, F. (2004) Evaluation of proteomic techniques: applications and potential. *Curr. Proteomics* **1**, 103–111.

29

In Situ Hybridization

Amanda D. Watters

1. Introduction

In situ hybridization is a method used to localize nucleic acid sequences in cells or tissue sections. The first *in situ* hybridization experiments were described by two groups of researchers in 1969, John et al. and Gall and Pardue (reviewed in ref. *1*). Radioactively labeled RNA probes were used to localize specific DNA sequences within the nuclei of *Xenopus laevis* *(2)*.

The *in situ* hybridization methodology exploits a fundamental property of nucleic acids: their ability to anneal (bind) to one another in a sequence-specific manner to form hybrids *(3)*. Thus, the gene sequence of interest can be detected *in situ* by labeling a known sequence of DNA or RNA to form a probe and then introducing the probe to the sample in a single-stranded form. Following annealing, the gene sequence of interest can be visualized via the probe label.

1.1. Materials Used

The most suitable type of tissue preparation for assessing chromosome or gene copy number *in situ* is frozen sections, as the material will not have been placed in fixatives that can damage DNA *(4)*. However, frozen tissue is not as easy to handle as formalin-fixed and paraffin-wax-processed tissue and long-term storage can degrade nucleotides, particularly RNA. Initial research studying genetic aberrations in interphase nuclei used cell lines or dissociated nuclei from tumors *(5–7)*. Confidence in the methodology together with improvement in probe synthesis has resulted in many studies that show the benefits of studying genetic aberrations in intact tissue sections (e.g., **refs.** *8* and *9*).

Further refinements in the methodology have led to techniques that can be applied to formalin-fixed, paraffin-wax-embedded material from the histopathology archives, which vastly increase the power of *in situ* hybridization in clinical and research environments. Several studies have adopted this approach to the study of genetic localization, chromosomal rearrangements, chromosome mapping, and numerical chromosomal and genetic aberrations in several tumor types *(10–13)*. An outline of the methodology is shown in **Fig. 1** *(14)*.

1.2. Types of In Situ *Hybridization*

There are essentially two types of *in situ* hybridization (ISH) that can be readily applied to paraffin-embedded tissue sections: fluorescence *in situ* hybridization (FISH) using DNA probes and chromogenic *in situ* hybridization (CISH) with DNA or RNA probes. One major advantage of fluorescence over nonfluorescent labels to study genetic aberrations is that multiply-labeled probes can be applied simultaneously to a tissue preparation. As a consequence, using the correct probe set, translocations or genetic aberrations relative to the chromosome copy number can be detected. Conversely, because CISH is viewed by transmitted light microscopy, tissue

From: *Medical Biomethods Handbook*
Edited by: J. M. Walker and R. Rapley © Humana Press, Inc., Totowa, NJ

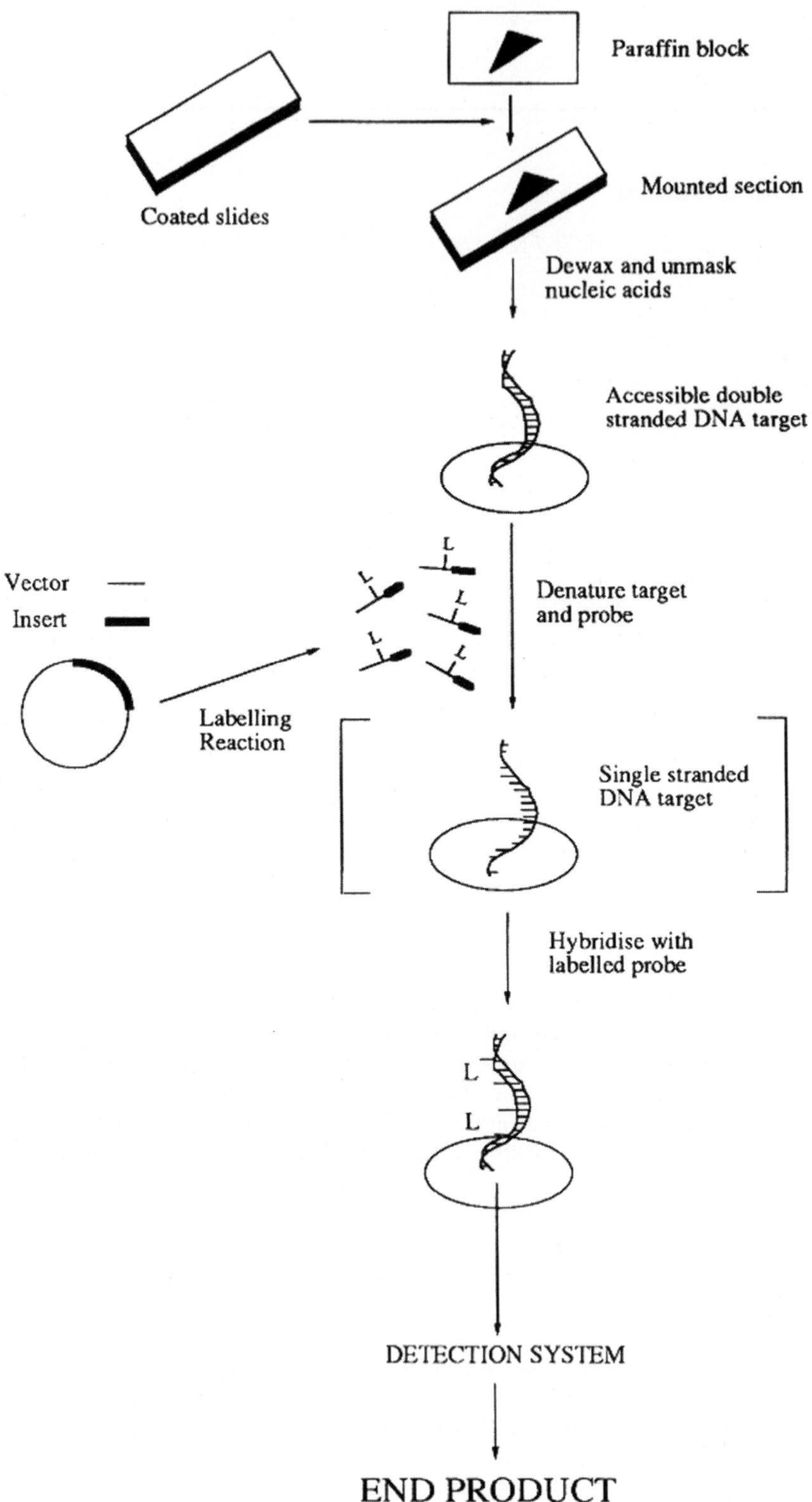

Fig. 1. Outline of *in situ* hybridisation (From **ref. *14***; reprinted by permission of Oxford University Press.)

morphology is clearly seen, and direct comparison between CISH sections and other histology techniques can be made. Indeed, this is the preferred method for studies of RNA expression where morphological details are essential *(12)*.

1.3. Dissociated Nuclei vs Intact Tissue Sections

Early articles in analyzing genetic aberrations in bladder cancer used a method involving dissociation of nuclei from 50-μm tissue sections cut from paraffin wax tissue blocks *(5–7)*. Although nuclear truncation in thin sections was avoided, the relationship between neoplastic and non-neoplastic tissue and correlation with pathology in hematoxylin and eosin sections was not possible. Subsequent research has showed, however, that it is entirely possible to glean a large amount of information using standard 5-μm tissue sections (e.g., **refs. *15*** and ***16***). Nevertheless, ISH applied to the assessment of DNA copy number using standard tissue sections requires careful evaluation to ensure that accurate assessment of ploidy is achieved because of nuclear truncation artifacts *(17)*. Control material with known ploidy must be included in the analysis.

2. Methodology

2.1. Pretreatments

As a result of fixation, unmasking of nucleic acids is a prerequisite of the *in situ* methodology with paraffin tissue sections: A compromise between optimal digestion and preservation of tissue architecture must be achieved. The recommended fixative for ISH is 10% neutral-buffered formalin, commonly used in diagnostic laboratories. The first step in any procedure that involves paraffin-wax-embedded tissue sections is to remove the wax and rehydrate the tissue. This is typically done using xylene and alcohol. Following this step, various pretreatments are used, such as weak hydrochloric acid at room temperature. This is a tissue permeabilization agent that is thought to increase hybridization signals, because acid deproteinases the tissue, thus increasing probe penetration *(18)*. This is the result of the pH-dependent fixation reaction that proceeds more rapidly at higher pH *(19)* and presumably is reversed following the introduction of an acid, such as hydrochloric acid. Generally, the use of a permeabilization step reduces the proteolytic step, thus causing less tissue damage. Pepsin is primarily used as the proteolytic enzyme in the demonstration of genomic DNA and some mRNA *(14)*, although proteinase K tends to be more popular in mRNA pretreatment protocols *(20)*. The length of this step will vary depending on the tissue used and also on the individual tissue samples *(21)*. It is possible to stain the nuclei with a nuclear stain such as 4,6-diamidino-2 phenylindole-2 hydrochloride (DAPI) and determine the extent of digestion using fluorescence before proceeding to the probe application stage *(21)*.

2.2. Prehybridization and Hybridization

Depending on the application, DNA probes can be sourced from commercial companies or synthesized in-house using enzymatic methods that incorporate haptens such as biotin or digoxigenin or fluorochromes *(22)*. Generally, the probe requires denaturation before applying to the tissue, although some commercial sources supply probes already denatured. Denaturing of both probe and target DNA renders the double-stranded DNA single stranded, thus enabling the probe and target to reanneal.

Denaturation can be performed by both chemical and physical means. Most protocols use a combination of heat, salt, and formamide, all of which have the ability to break intramolecular hydrogen bonds, thus destabilizing hybrids *(23)*. In some situations, probe and target can be denatured simultaneously by diluting the probe in the hybridization mix, which contains salt and formamide *(14)*.

Probes for the detection of RNA sequences are synthesised by in vitro transcription using one of the DNA-directed RNA polymerases *(12)*. Digoxigenin is typically added to the reaction and the resultant single-stranded probe does not generally require denaturing before applying to the tissue. However, some protocols recommend a prehybridization step that prevents background staining. The prehybridization mixture contains all of the components of the hybridization mixture minus dextran sulfate and probe *(24)*.

Hybridization is typically carried out under low-stringency conditions to enhance hybrid formation *(23)*. The hybridization mix contains reagents to allow the hybridization to proceed as efficiently as possible. The reagents in the mixture include 2X SSC (sodium saline citrate) and 50% formamide, which allow matched hybrids to form, but because this is a relatively low-stringency mix, some mismatched hybrids will also form. Other constituents of the hybridization mix are typically dextran sulfate, a macromolecule (molecular weight of greater than 8000) that increases the effective probe concentration by molecular exclusion *(14)*, and salmon sperm DNA, which can reduce nonspecific staining by blocking possible interactions with the phosphate groups of the nucleotide chain *(23)*.

2.3. Posthybridization

Posthybridization washes aim to denature and remove an unbound or weakly bound probe; for DNA probes, this is generally with a neutral high-concentration SSC combination *(14)* at a high temperature or a combination of formamide and SSC at a lower temperature. The posthybridization wash is usually of a slightly higher stringency than the hybridization conditions. Hybrids of more specific base-pairing can withstand the higher-stringency wash, and the resulting hybridization signal is specific for the DNA sequence under analysis. Posthybridization washes for RNA probes tend to be performed at lower temperatures than for DNA probes, although the stringency might need to be altered depending on the final result.

2.4. Detection

DNA sequences are detected with antibodies to the appropriate hapten and visualized with either a fluorescent or a nonfluorescent label. Directly labeled probes are usually fluorescent and can be visualized immediately following hybridization. If digoxigenin or biotin have been used to label the probe, the method proceeds to a detection system *(16)*. RNA sequences are visualized with an alkaline phosphatase-conjugated antidigoxigenin antibody, followed by a substrate of 5-bromo-4-chloro-3-indolyl-phosphate and nitro-blue tetrazolium. For visualization of nuclei using fluorescence antifade mounting medium containing 0.25–0.5 μg/mL DAPI is applied to the tissue sections; for light microscope preparations, a light nuclear stain such as hematoxylin or light green is used.

3. Applications

In situ hybridization has gained popularity over other molecular biology methods because the DNA/mRNA of interest can be viewed in the context of the tissue morphology. Over the past few years, probes and methodologies have been refined to the extent that now it is possible to perform ISH on paraffin-wax-embedded tissue sections. This is particularly relevant in a clinical setting where retrospective studies are facilitated by accessing material from the pathology archives, and results are correlated with patient outcome. Thus, ISH is rapidly becoming a technique that can be applied to a wide range of clinical situations. Studies range from the assessment of gene rearrangements in leukemia and lymphoma *(25)* to the diagnosis of B-cell lymphoma by demonstration of restriction of light-chain mRNA *(11)* to the determination of amplification of *HER2/neu* in breast cancer *(26)*.

3.1. Diagnostic FISH

Fluorescence *in situ* hybridization has been used widely in the diagnosis of various types of lymphoma. Many genetic lesions represent molecular markers of disease that have diagnostic and clinical applications *(27)*. Distinct types of non-Hodgkin's lymphomas are associated with recurrent genetic abnormalities. These include chromosomal translocations, which are involved in the pathogenesis of the disease through mechanisms including activation of a proto-oncogene via transcriptional deregulation (e.g., BCL1/CCND1, BCL2, or MYC) or the generation of a fusion product such as NPM-ALK. Using an indirect labeled protocol, FISH was successfully used to detect three chromosomal translocations, BCL1, BCL2, and MYC, in routinely fixed and processed tissue sections from mantle cell, follicular, and Burkitt's lymphomas ***(27)***. The distinct clinicopathological disease marginal zone lymphoma of mucosa-associated lymphocytic tissue type has been associated with specific numerical abnormalities. Of these, t(11:18) is the most important because it is associated with treatment selection. t(11:18) cases are resistant to *Helicobacter pylori* eradication therapy. The development of a FISH protocol to detect this abnormality is very useful in a clinical context because tissue sections generated in a diagnostic setting can be used and results evaluated on a cell-by-cell basis ***(28)***.

Until relatively recently, diagnostic FISH on formalin-fixed, paraffin-wax-embedded tissue sections was limited to the hematological malignancies. However, the emergence of a laboratory-based assay to determine levels of HER2/neu marks the beginning of a new type of patient management. One of the most common genetic abnormalities associated with breast and ovarian cancer is amplification of *HER2/neu* (*see* **Fig. 2A**), a proto-oncogene that encodes a 185-kDa transmembrane receptor protein with intrinsic tyrosine kinase activity ***(29)***. In breast cancer, detection of *HER2/neu* is now mandatory because clinical trials have shown that amplification/overexpression is a predictive marker for response to adjuvant chemotherapy ***(30)*** and a specific therapy—the humanized monoclonal antibody trastuzumab (Herceptin; Roche)—has been introduced. Trastuzumab targets the HER2/neu antigen and inhibits the growth of HER2/neu overexpressing tumor cells ***(31)***. FISH is currently being used diagnostically to detect *HER2/neu* levels in breast cancer tissue sections using a Food and Drug Administration (FDA)-approved kit marketed by Abbot Diagnostics.

3.2. FISH in Research

In research, many articles have been published applying FISH to tissue to determine tumor-associated aberrations. Human papilloma virus is widely implicated in cervical carcinogenesis, but it is thought that other abnormalities are required for biological transformation ***(32)***. By using probes for common genetic aberrations in cervical cancer, this hypothesis was confirmed and also that regional heterogeneity of genetic loss was associated with late events in tumor development and contributed to our understanding of cervical cancer ***(32)***. In the childhood brain tumor medulloblastoma, the frequency of MYCC abnormalities was studied using tissue sections. The prognostic significance was determined, albeit in a small percentage (5.2%, 4 out of 77), and an association was made between MYCC amplification and poor prognosis ***(33)***. In bladder cancer, one of the few malignancies characterized by frequent recurrence, correlations between clinical outcome and genetic aberrations have been made. Monosomy of chromosome 9 (*see* **Fig. 2B**) ***(15)*** and polysomy of chromosome 17 ***(16)*** have been associated with increased risk of recurrence. There is now a multiprobe set available from Vysis, Inc. that consists of the commonest genetic aberrations previously characterized in bladder cancer ***(34)***. The use of the probe set increased sensitivity in determining malignancy compared to conventional cytology. At present, the probe set is used on cells derived from urine specimens, but there is, undoubtedly, potential for extrapolation to tissue sections. Probe sets are now available from Vysis for prognosis and prediction of therapy response in chronic lymphocytic leukemia. A large amount of information can be gathered from one experiment, using directly labeled probes for common chromosomal abnormalities, the results of which can predict response to therapy. This approach will soon extrapolate to routine molecular diagnostics in tissue sections and will provide a powerful tool for accurate prognostication.

3.3. Diagnostic CISH

Chromogenic *in situ* hybridisation is used to diagnose melanoma and lymphoma using mRNA probes for kappa and lambda. Light-chain restriction indicates malignancy ***(11)***. CISH

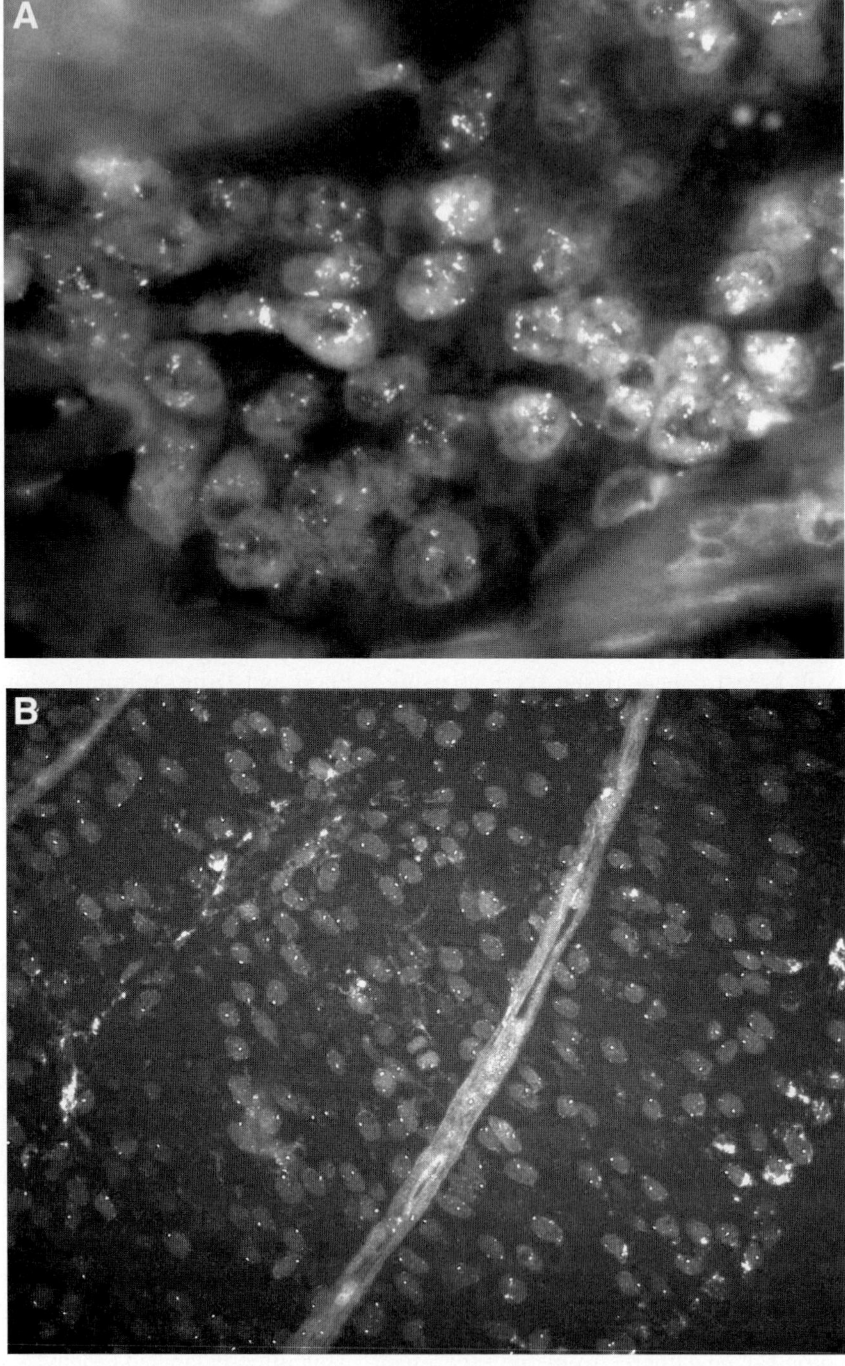

Fig. 2. (**A**) This image shows high levels of *HER2/neu* gene amplification in an invasive breast carcinoma, represented by many bright dots, some merging into large structures, within the nucleus, demonstrated by FISH; original magnification, ×1000. (**B**) This image shows a reduction to monosomy of chromosome 9 in the majority of nuclei in a low-grade transitional cell carcinoma, demonstrated by FISH; original magnification, ×400. (**C**) This image shows

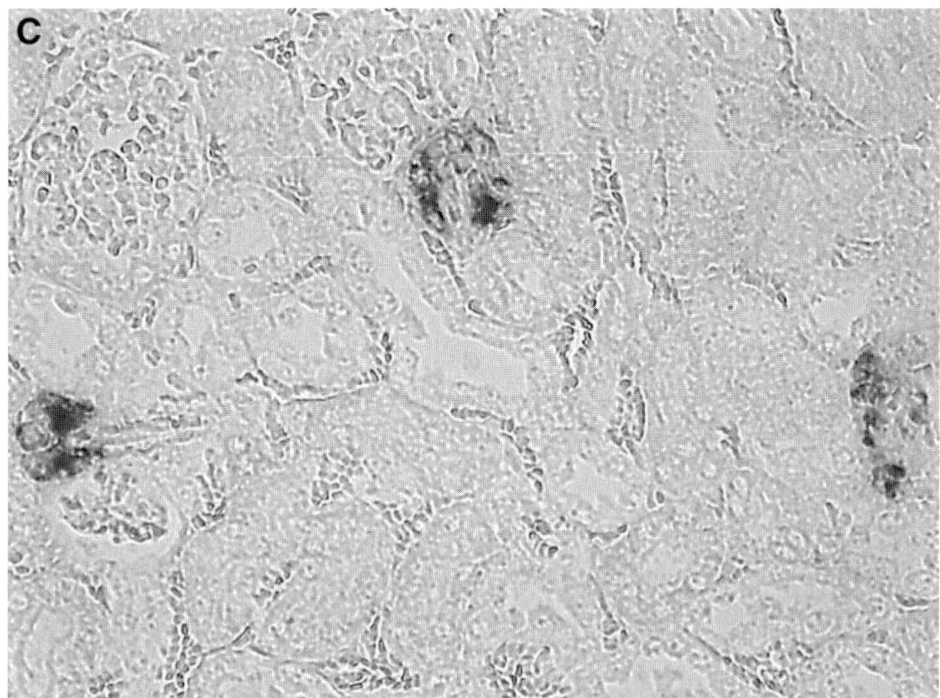

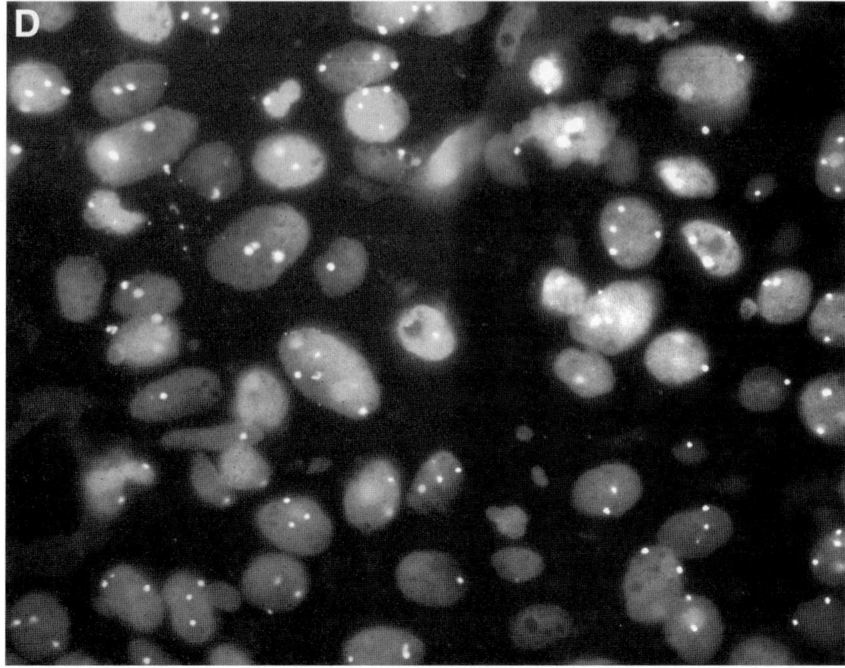

Fig. 2. *(continued)* high levels of expression of renin localized to the juxtaglomerular apparatus in a normal kidney, demonstrated by CISH; original magnification, ×400. (**D**) This image represents polysomy in a high-grade transitional cell carcinoma where the predominant cell population contains three or four dots demonstrated by FISH; original magnification, ×1000.

has also been applied to the diagnosis of hepatic tumors, using an mRNA probe for albumin. The presence of albumin differentiates between hepatocellular carcinoma and chloangiocarcinoma ***(11)***. Viral identification can be undertaken using a variety of methods, but only an *in situ* method such as CISH will provide simultaneous morphological information. Most viral *in situ* methods use probes against DNA. However, a viral mRNA transcript has been used in the demonstration of the Epstein–Barr virus ***(35)***. Human papilloma virus (HPV) is associated with the development of cervical cancer and can be used in the future as an adjunct to cervical screening. HPV DNA is readily demonstrated in tissue sections by ISH, which has the advantage over other molecular methods such as polymerase chain reaction (PCR) because the virus can be visualized with reference to tissue morphology ***(36)***.

3.4. CISH in Research

In research, CISH has been widely applied to the study of numerical chromosomal aberrations ***(6,37,38)***. For example, in adrenocortical carcinoma, an interphase cytogenetic study using probes for chromosomes commonly associated with this tumor showed that that there were less aberrations in the benign adenoma and more with tumor progression ***(39)***. The childhood tumor nephroblastoma has been associated with high levels of renin production ***(20)***, which can be demonstrated by CISH. Normal production of renin is confined to renin-secreting cells in the juxtaglomerular apparatus (*see* **Fig. 2C**). Although color separation is more difficult because chromogens are usually red and brown, dual labeling has been used successfully in conjunction with immunocytochemistry ***(40)***.

3.5. ISH to Study Preneoplastic Lesions

Many studies have demonstrated the utility of FISH/CISH to understand pathological processes of disease and also in preneoplasia. For example, data from cytogenetic studies suggest that anomalies in chromosome 17 occur early in the tumorigenesis of breast cancer and that detection of chromosomal aneusomies can be useful in the screening of high-risk individuals ***(41)***. Polysomy 17 has been reported in proliferative breast disease ***(42)***, in adjacent tissue to breast carcinoma, and in contralateral breast tissue ***(41)***. Detection of chromosomal aberrations in benign tissue can represent a novel marker of breast cancer risk ***(41)***. In bladder cancer, urothelium adjacent to malignant growth has been characterized by ISH. Numerical aberrations are hypothesized to represent an increased risk of tumor recurrence (*see* **Fig. 2D**) ***(43)***.

3.6. Future Developments

Tissue microarray technology ***(44)*** is beginning to impact on research involving tissue-based methodologies. Minute tissue cylinders (0.6 mm in diameter) are taken from hundreds of different tumor blocks and laced into an empty paraffin wax block. Slides are made that can then be rapidly assessed with DNA or mRNA probes to give rapid results of the study of a particular aberration. This leads to the acceleration of candidate tumor markers, which will rapidly become standard diagnostic practice.

In conclusion, the application of ISH to formalin-fixed, paraffin-wax-embedded tissue sections is now standardized and is part of the diagnostic procedure in several laboratories. As more prognostic markers become validated for use, there will be an increase in molecular testing with the inevitable rise in tailored therapies.

References

1. Warford, A. (1994) An overview of *in situ* hybridisation, in *A Guide to In Situ* Hybaid, UK, pp. 5–17.
2. John, H. A, Birnstiel, M. L., and Jones, K. W. (1969) RNA–DNA hybrids at the cytological level. *Nature* **223,** 582–587.
3. Herrington, C. S. (1998) Demystified...in situ hybridisation. *J. Clin. Pathol: Mol. Pathol.* **51,** 8–13.
4. Williams, C., Ponten, F., Moberg, C., et al (1999) A high frequency of sequence alterations is due to formalin fixation of archival specimens. *Am. J. Pathol.* **155,** 1467–1471.

5. Hopman, A. H. N., Ramaekers, F. C. S., Raap, A. K., et al. (1988) *In situ* hybridization as a tool to study numerical chromosome aberrations in solid bladder tumors. *Histochemistry* **89,** 307–316.
6. Hopman, A. H. N., van Hooren, E., van de Kaa, C. A., Vooijs, P. G. P., and Ramaekers, F. C. S. (1991) Detection of numerical chromosome aberrations using *in situ* hybridization in paraffin sections of routinely processed bladder cancers. *Mod. Pathol.* **4,** 503–512.
7. Sauter, G., Moch, H., Wagner, U., et al (1995) Y chromosome loss detected by FISH in bladder cancer. *Cancer Genet. Cytogenet.* **82,** 163–169.
8. Murphy, D. S., McHardy P., Coutts, J.. et al. (1995) Interphase cytogenetic analysis of *erb*B2 and topoII∝ co-amplification in invasive breast cancer and polysomy of chromosome 17 in ductal carcinoma *in situ*. *Int. J. Cancer* **64,** 18–26.
9. Wolf, N. G., Abdul-Karim, F. W., Schork, N. J., and Schwartz, S. (1996) Origins of heterogeneous ovarian carcinomas. A molecular cytogenetic analysis of histologically benign, low malignant potential, and fully malignant components. *Am. J. Pathol.* **149,** 511–520.
10. Lawrence Fox, J., Hsu, P-H., Legator, M., Morrison, L. E., and Seelig, S. A. (1995) Fluorescence in situ hybridisation: powerful molecular tool for cancer progniosis. *Clin. Chem.* **41,** 1554–1559.
11. McNicol, A. M. and Farquharson, M. (1997) In situ hybridisation and its diagnostic applications in pathology. *J. Pathol.* **182,** 250–261.
12. Poulsom, R. (2001) In situ hybridisation to localise mRNAs, in *Metastasis Methods & Protocols* (Brooks, S. A. and Schumacher, U., eds.), Humana, Totowa, NJ, pp. 177–198.
13. Poetsch, M., Dittberner, T., Woenckhaus, C., and Kleist, B. (2000) Use of interphase cytogenetics in demonstrating specific chromosomal aberrations in solid tumors—new insights in the pathogenesis of malignant melanoma and head and neck squamous cell carcinoma. *Histol. Histopathol.* **15,** 1225–1231.
14. Herrington, C. S. and McGee, J. O.'D. (1992) Principles and basic methodology of DNA/RNA detection by in situ hybridisation, in *Diagnostic Molecular Pathology, A Practical Approach, Volume I* (Herrington, C. S. and McGee, J. O.'D, eds.), Oxford University Press, Oxford.
15. Bartlett, J. M. S., Watters, A. D., Going, J. J., Ballantyne, S. A., Grigor, K. M., and Cooke, T. G. (1998) Quantitative FISH: is chromosome 9 loss a marker of disease recurrence in TCC of the bladder? *Br. J. Cancer* **77,** 2193–2198.
16. Watters, A. D., Ballantyne, S. A., Going, J. J., Grigor, K. M., Cooke, T. G., and Bartlett, J. M. S. (2000) Aneusomy of chromosomes 7 & 17 predicts recurrence of transitional cell carcinoma of the urinary bladder. *Br. J. Urol.* **85,** 42–47.
17. Watters, A. D. and Bartlett, J. M. S. (2000) Quantitation of FISH signals in archival tumours, in *Ovarian Cancer: Methods and Protocols* (Bartlett, J. M. S., ed.), Humana, Totowa, NJ, Vol. 3, pp. 253–260.
18. Syrjänen, S. (1992) Viral gene detection by *in situ* hybridisation, in *Diagnostic Molecular Pathology, A Practical Approach, Volume I* (Herrington, C. S. and McGee, J. O'D., eds.), Oxford University Press, Oxford.
19. Hopwood, D. (1996) Fixation and fixatives, in *Theory and Practice of Histological Techniques*, 4th ed., (Bancroft, J. and Stevens, A., eds.), Churchill Livingstone, London.
20. McKenzie, K. J., Ferrier, R. K., Howatson, A. G., and Lindop, G. B. M. (1996) Demonstration of renin gene expression in nephroblastoma by in situ hybridisation. *J. Pathol.* **180,** 71–73.
21. Watters, A. D. and Bartlett, J. M. S. (2002) Fluorescence in situ hybridization in paraffin tissue sections: pretreatment protocol. *Mol. Biotechnol.* **21,** 217–220.
22. Leitch, A. R., Schwarzacher, T., Jackson, D., and Leitch, I. J. (1994) In Situ *Hybridisation*. BIOS Scientific, Oxford.
23. Mitchell, B. S., Dhami, D., and Schumacher, U. (1992) *In situ* hybridisation: a review of methodologies and applications in the biomedical sciences. *Med. Lab. Sci.* **49,** 107–118.
24. Roche Diagnostics (2002) *DIG Application Manual for Non-Radioactive* In Situ *Hybridisation*, 3rd ed. Roche Diagnostics.
25. Arber, D. A. (2000) Molecular diagnostic approach to non-Hodgkin's lymphoma. *J. Mol. Diagn.* **2,** 178–190.
26. Pauletti, G., Godolphin, W., Press, M. F., and Slamon, D. J. (1996) Detection and quantitation of *HER-2/neu* gene amplification in human breast cancer archival material using fluorescence in situ hybridization. *Oncogene* **13,** 63–72.
27. Haralambieva, E., Kleiverda, K., Mason, D. Y., Schuuring, E., and Kluin, P. M. (2002) Detection of three common translocation breakpoints in non-Hodgkin's lymphomas by fluorescence *in situ* hybridisation on routine paraffin-embedded tissue sections. *J. Pathol.* **198,** 163–170.

28. Nomura K., Yoshino, T., Nakamura, S., et al (2003) Detection of t(11;18)(q21;q21) in marginal zone lymphoma of mucosa-associated lymphocytic tissue type on paraffin-embedded tissue sections by using fluorescence in situ hybridisation. *Cancer Genet. Cytogenet.* **140,** 49–54.
29. Reese, D. M. and Slamon, D. J. (1997) *HER2/neu* signal transduction in human breast and ovarian cancer. *Stem Cells* **15,** 1–8.
30. Zujewski, J. O. (2003) "Build quality in"—HER2 testing in the real world. *JNCI* **94,** 788–789.
31. Rhodes, A., Jasani, B., Couturier, J., et al (2002) A formalin-fixed, paraffin-processed cell line standard for quality control of immunohistochemical assay of HER2/neu expression in breast cancer. *Am. J. Clin. Pathol.* **117,** 81–89.
32. Herrington, C. S., Worsham, M., Southern, S. A., et al. (1996) Loss of chromosome 9 in tissue sections of transitional cell carcinomas as detected by interphase cytogenetics. A comparison with RFLP analysis. *J. Pathol.* **179,** 169–176.
33. Aldosari, N., Bigner, S. H., Burger, P. C., et al. (2002) MYCC and MYCN oncogene amplification in medulloblastoma: a fluorescence in situ hybridisation study on paraffin sections from the children's oncology group. *Arch. Lab. Med.* **126,** 540–544.
34. Halling, K. C., King, W., Sokolova, I. A., et al. (2000) A comparison of cytology and fluorescence in situ hybridisation for the detection of urothelial carcinoma. *J. Pathol.* **164,** 1768–1775.
35. Lamar Jones, M. (2002) Molecular pathology and in situ hybridisation, in *Theory and Practice of Histological Techniques*, 5th ed., (Bancroft, J. D. and Gamble, M., eds.), Churchill Livingston, London.
36. Unger, E. R. and Vernon, S. D. (2001) Detection of human papillomaviruses by polymerase chain reaction and *in situ* hybridisation, in *Methods in Molecular Medicine* (Killeen, A. A., ed.), Humana, Totowa, NJ.
37. Ramaekers, F. C. S. and Hopman, A. H. N. (1993) Detection of genetic aberrations in bladder cancer using in situ hybridization. *Ann. NY Acad. Sci.* **677,** 199–213.
38. Poddighe, P. J., Bringuier, P-P., Vallinga, M., Schalken, J. A., Ramaekers, F. C. S., and Hopman, A. H. N. (1996) Loss of chromosome 9 in tissue sections of transitional cell carcinomas as detected by interphase cytogenetics. A comparison with RFLP analysis. *J. Pathol.* **179,** 169–176.
39. Russell, A. J., Sibbald, J., Haak, H., Keith, W. N., and McNicol, A. M. (1999) Increasing genomic instability in adrenocortical carcinoma progression with involvement of chromosomes 3, 9, and X at the adenoma stage. *Br. J. Cancer* **81,** 684–89.
40. Speel, E. J. M., Jansen, M. P. H. M., Ramaekers, F. C. S. and Hopman, A. H. N. (1994) A novel triple-color detection procedure for brightfield microscopy, combining in situ hybridization with immunocytochemistry. *J. Histochem. Cytochem.* **42,** 1299–1307.
41. Botti, C., Pescatore, B., Mottolese, M., et al. (2000) Incidence of chromosomes 1 and 17 aneusomy in breast cancer and adjacent tissue: an interphase cytogenetic study. *J. Am. Coll. Surg.* **190,** 516–525.
42. Micale, M. A., Visscher, D. W., Gulino, S. E., and Wolman, S. R. (1994) Chromosomal aneuploidy in proliferative breast disease. *Hum. Pathol.* **25,** 29–35.
43. Steidl, C., Simon, R., Bürger, H., et al. (2002) Patterns of chromosomal aberrations in urinary bladder tumours and adjacent urothelium. *J. Pathol.* **198,** 115–120.
44. Bubendorf, L., Nocito, A., Moch, H., and Sauter, G. (2001) Tissue microarray (TMA) technology: minaturized pathology archives for high-throughput in situ studies. *J. Pathol.* **195,** 72–79.

30

Enzyme-Linked Immunosorbent Assay

William J. Jordan

1. Introduction

The enzyme-linked immunosorbent assay (ELISA) is typically used to detect and quantify antigen within biological fluids. Among the ELISA's attributes are a very high level of sensitivity and robustness as well as being an extremely cost-effective assay that can be performed with only basic laboratory equipment. Having been developed over 30 yr ago *(1)*, the ELISA remains an assay of choice for many routine medical and veterinary diagnostic assays, as well as being a vital assay in basic science research.

Applications of the ELISA technique are many and varied. In the clinical laboratory, the test is commonly used to detect proteins of pathogens or "markers" that are indicative of infection or disease. A common application of the ELISA in testing for infection is in the identification of pathogen-specific antibodies in serum *(2)*. Possibly the best known of the clinical diagnostic assays currently utilizing the ELISA are those for diagnosis of human immunodeficiency virus (HIV) infection. However, this technique is routinely used for diagnosis of most of the major viral, and bacterial infections including hepatitis, anthrax, and malaria.

Of the huge number of variations of the ELISA, the most commonly used are (1) the sandwich ELISA, often referred to as the "dual-antibody sandwich" or the "dual-antibody capture" ELISA, (2) the indirect ELISA, and (3) the competitive ELISA. **Subheading 2** outlines these basic techniques with brief examples of their use in the clinic.

2. Routinely Used ELISA Techniques

2.1. The Dual-Antibody Sandwich ELISA

The dual-antibody sandwich ELISA (DAS ELISA) is one of the most sensitive and specific techniques for quantifying molecules in solution. An illustration of this technique is given in **Fig. 1A**. The DAS ELISA requires two antibodies that recognize separate epitopes on the antigen to be measured such that they are able to bind to the molecule simultaneously. The "capture" antibody, which is specific for the substance to be measured, is first coated onto a high-capacity protein-binding microtiter plate (an ELISA plate). Following the coating stage, any vacant binding sites on the plate are then blocked with the use of an irrelevant protein such as bovine serum albumin (BSA). This creates a solid-phase antigen-binding surface that should not nonspecifically bind other molecules. Samples, standards, and controls are then incubated on the plate, and any antigen present subsequently binds to the capture antibody. The bound antigen is detected using a secondary antibody (recognizing a different epitope on the antigen), thus creating the "sandwich." The detection antibody is most often directly conjugated to biotin. Biotin conjugation allows an amplification process to be carried out with the use of streptavidin conjugated to an enzyme such as horse radish peroxidase (HRP). As streptavidin is a tetrameric protein, binding four biotin molecules, the threshold of detection is greatly

From: *Medical Biomethods Handbook*
Edited by: J. M. Walker and R. Rapley © Humana Press, Inc., Totowa, NJ

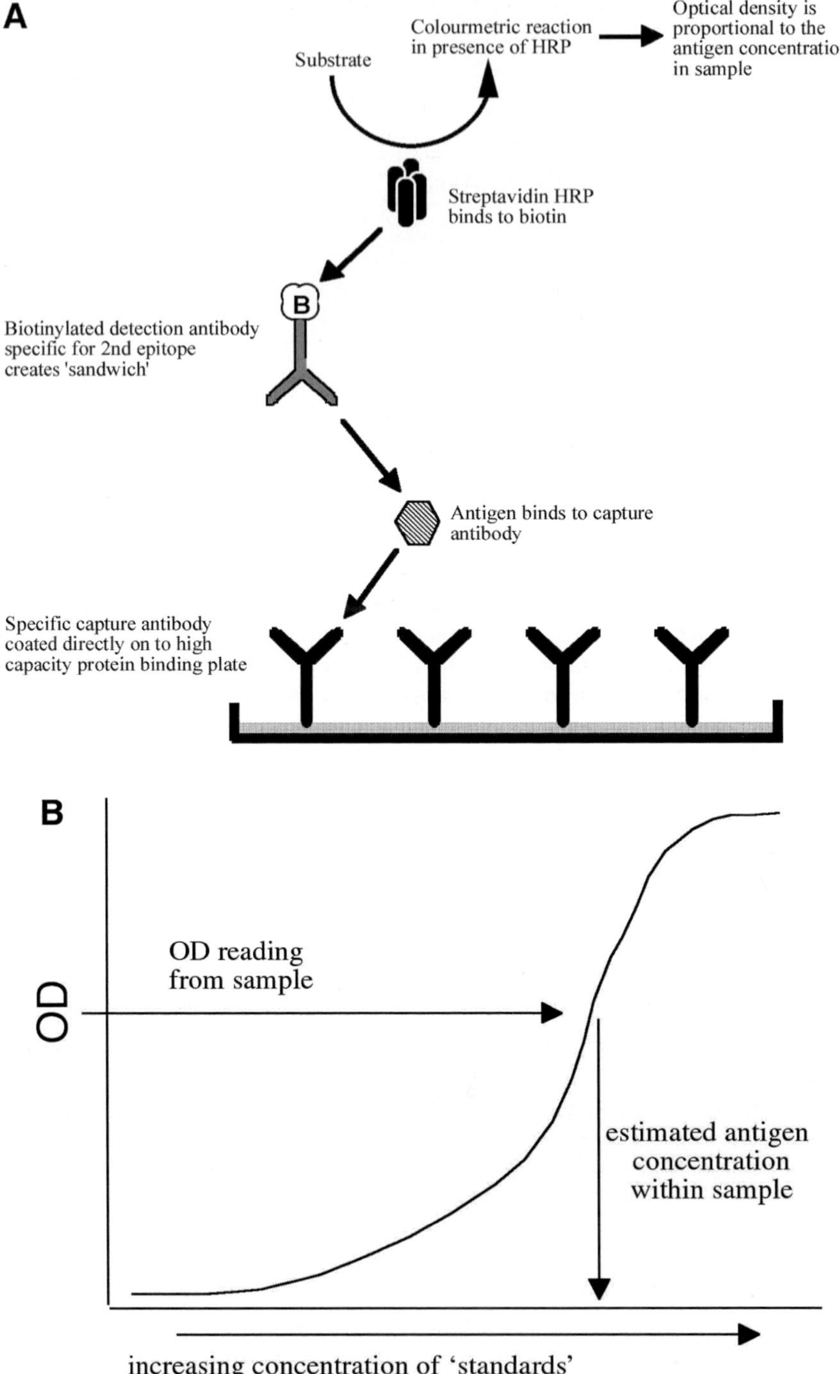

Fig. 1. (**A**) The dual-antibody sandwich ELISA. (**B**) Estimating antigen concentration using a typical DAS ELISA standard curve. In parallel with samples, a titration series of known amounts of antigen are also examined and the optical density (OD) readings are used to give a "standard curve." OD readings from samples can then be compared directly to this standard curve in order to gain an estimate of the amount of antigen in the sample. Samples often require testing at several dilution s so that the reading can be made within or as close to this part of the curve as possible.

enhanced. Developing the assay into a readable format involves the addition of a substrate such as 3,3',5,5'-tetramethylbenzidine (TMB) for the HRP enzyme. In the presence of the HRP enzyme, TMB begins a colorimetric reaction that can then be measured using a spectrophotometer. The resulting color (optical density [OD]) relates directly to the amount of antigen present within the sample. Comparison of the OD within a sample to those obtained using a standard curve of known concentrations allows an estimate of antigen concentration within that sample to be gained (*see* **Fig. 1B**)

2.1.1. Basic DAS ELISA Protocol

1. *Coating with capture antibody*: Capture antibody diluted in coating buffer is added to a high-capacity-protein binding 96-well microtiter plate. The plate is then incubated, allowing the antibody to bind to the plate; subsequently, the plate is washed to remove any excess or unbound antibody.
2. *Blocking vacant binding sites*: Vacant binding sites on the plate are then blocked with an irrelevant protein such as BSA. Following incubation, the plate is washed once more to remove excess unbound protein.
3. *Addition of samples/standards*: A titration series of known standards must be prepared. Ideally, these should be diluted in a matrix representing that of the samples to help identify false positives (e.g., if there are any substances in the matrix that bind nonspecifically to the plate that have enzyme activity). A negative control must also be included (e.g., culture medium only or serum known to be negative for the antigen to be measured). Samples and antigen standards are then incubated on the ELISA plate, allowing any antigen present to bind to the coating antibody.
4. *Addition of detection antibody*: Following the addition of samples, the biotinylated detection antibody is added, which binds to any antigen bound to the plate. Following incubation, the plate is once again washed thoroughly to remove unbound reagents.
5. *Addition of enzyme conjugate*: Streptavidin conjugated to an enzyme such as HRP is then incubated onto the plate. This binds to biotin molecules on the detection antibody. The plate is once again washed thoroughly to remove unbound reagents.
6. *Development and analysis*: The ELISA is then developed using a suitable substrate (e.g., TMB to detect the HRP enzyme). The OD can be measured using a spectrophotometer. The OD values from the standard titration of antigen are then used to determine an estimate of antigen within samples.

An example of the DAS ELISA in the clinic is the AlzheimAlert™ ELISA kit (Nymox Corporation) for measuring a neural thread protein (AD7c-NTP) in the urine of patients. The presence of high levels of AD7c-NTP has been shown to be correlated with Alzheimer's disease and is used for early diagnosis and subsequent monitoring of the progression of this disease. During the assay, microtiter plates are first coated with a monoclonal antibody reacting against the neural protein. After blocking vacant binding sites and washing the plate to remove unbound proteins, a urine sample from the patient is incubated onto the plate. Any neural protein present binds to the antibody. As this assay is quantitative, a standard titration of known amounts of recombinant neural protein is set up in parallel. The readings from these are compared to that obtained from the sample(s) to estimate the level of neural protein in the urine. The plate is then washed to remove any excess protein and a secondary (enzyme-conjugated) antibody that recognizes a different epitope of the neural peptide is added to the plate. The presence of the protein in the urine sample is thus reflected by the presence of the enzyme, and this enzyme is detected using a substrate that completes a colorimetric reaction that is read using a spectrophotometer. The presence of high levels of the neural protein AD7c-NTP is indicative of Alzheimer's disease.

2.2. The Indirect ELISA

In many instances, only one specific antibody might be available with which to create an assay to detect antigen and so a DAS ELISA is unsuitable. In such a situation, an indirect ELISA or a competitive ELISA can be established. During the indirect ELISA, the sample

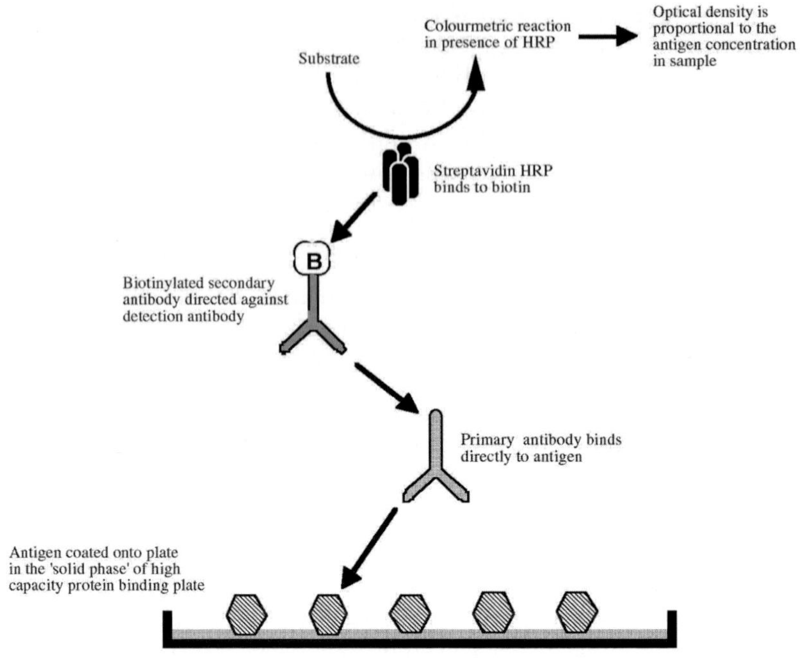

Fig. 2. The basic direct ELISA.

itself is coated directly onto the microtiter plate and is then detected using the specific antibody. An illustration of this technique is presented in **Fig. 2**.

The indirect ELISA is often used to detect antibody in samples. An example of such an indirect ELISA used in the clinic is the Murex HIV-1.2.0 ELISA kit (Abbott Laboratories). HIV-infected individuals commonly produce antibodies against the HIV protein and the presence of such antibodies in the serum is indicative of infection. The basis of this ELISA is a microtiter plate coated with a mixture of HIV proteins. This mixture includes a synthetic peptide representing an immunodominant region of HIV-1, a recombinant HIV envelope protein, and an HIV core protein. Blood is taken from patients and the serum fraction used for the test. By allowing the red blood cells from a blood sample to "settle" in a test tube, the serum layer can then be taken with a pipet. This serum sample is normally tested against control sera (both positive and negative to compare to the sample and thus give a diagnosis). During incubation of the serum on the microtiter plate, antibodies reacting against the HIV in the sample bind to the antigens. The plate is then washed to remove excess antibodies and a secondary antibody (conjugated to the enzyme HRP) that specifically recognizes human antibodies is added to the plate. This will, therefore, only bind if antibodies against the HIV proteins were originally present in the serum sample. Samples not containing specific antibody will not cause the conjugate to bind to the well. A further wash of the plate is then performed to clear any unbound secondary antibody and the substrate is then added (TMB). The wells that had sera containing specific anti-HIV antibodies develop a blue color, which is converted to yellow when the reaction is stopped with sulfuric acid. The color is read spectrophotometrically at 450 nm and is directly related to the concentration of antibody to HIV in the sample. Thus, a strong positive signal indicates that the individual has been infected with HIV.

2.3. The Competitive ELISA

There are many variations and adaptations of the competitive ELISA, although the general principle for all of these remains the same. A typical representation of this assay is shown in

Fig. 3A. As with the ELISAs described above, the initial stage of the competitive ELISA generally involves coating a high-capacity protein-binding microtiter plate with an antibody directed against the antigen to be measured. However, during a competitive ELISA, a sample that is to be analyzed is first mixed with a known amount of antigen that has been enzyme-conjugated. This mixture is then added to the coated ELISA plate. The two forms of the antigen compete for binding sites to the antibody-coated plate and this binding is proportional to their respective molar ratios in the mixture. Because only the conjugated form of the antigen allows for the colorimetric reaction to develop in the presence of the substrate, the maximal reading occurs when there is no antigen in the sample. The more antigen that is present in the sample, the lower the resulting OD reading. Thus, unlike a standard sandwich ELISA, the readout is inversely associated with the amount of antigen (*see* **Fig. 3B**).

In practice, the competitive ELISA is the least commonly used of the ELISA variations, mainly because of the increased workload and expertise required. This technique is most often used to detect antigens that are very small, such as hormones. An example of the competitive ELISA in the clinic is the BQ T4 ELISA kit (Bioquant Corporation). This ELISA is used for the quantitative measurement of total thyroxine (T4) in human serum or plasma and is used for the diagnosis of hypothyroidism and hyperthyroidism. The level of T4 is decreased in hypothyroid patients and is increased in hyperthyroid patients. The BQ T4 is a solid-phase competitive ELISA. The samples are mixed with T4 that has been enzyme-conjugated and are then added to a microtiter plate that has been precoated with an anti-T4 monoclonal antibody. T4 in the patient's serum competes with a T4 enzyme-conjugated recombinant T4 for binding sites. Plates are then washed to remove unbound T4 and T4 enzyme conjugate. Upon the addition of the substrate, the intensity of color is inversely proportional to the concentration of T4 in the samples. A standard curve is prepared relating color intensity to the concentration of the T4 and a diagnosis can be made.

A variation of this competitive ELISA is often used to measure levels of antibody in solution. In this technique, the antigen itself is coated onto the ELISA plate and an enzyme-conjugated "detection antibody" is used to generate the OD reading. Any antigen-specific antibody in the sample competes with the enzyme-conjugated antibody for binding to the plate. Again, the reading is inversely proportional to the amount of antigen present and can be cross-referenced with readings from a standard curve to gain a quantitative estimate of antibody in the sample.

2.4. The Blocking ELISA

In this variation of the competitive ELISA, the sample to be measured is not mixed with the enzyme-conjugated antigen, but is preincubated onto the coated plate prior to washing and the addition of the conjugated antigen. Thus, the sample "blocks" rather than "competes" for the sites on the plate. This can result in a greater degree of sensitivity, although it is more time-consuming because it relies on an additional step. The principle of the assay, however, remains the same as the competitive ELISA.

3. Establishing an ELISA Protocol

In order to set up a reliable and durable ELISA, it is essential to first optimize a number of the parameters mentioned above. The level of optimization will, of course, depend on exactly what is required from the assay. In some cases, a simple "yes or no" answer is desired and a simple standard procedure might be sufficient. If, however, high sensitivity is the aim with accurate quantification of the molecule in question, then carefully setting up the optimal conditions in advance saves a great deal of time in the long term.

Optimization of the following parameters is most often required: (1) pH of coating buffer, (2) concentration of capture antibody, and (3) concentration of streptavidin–HRP conjugate. Another important aspect of optimization that must be considered, especially when it is antigen rather than antibody that is to be coated onto the plate, is the type of ELISA plate used. Recent

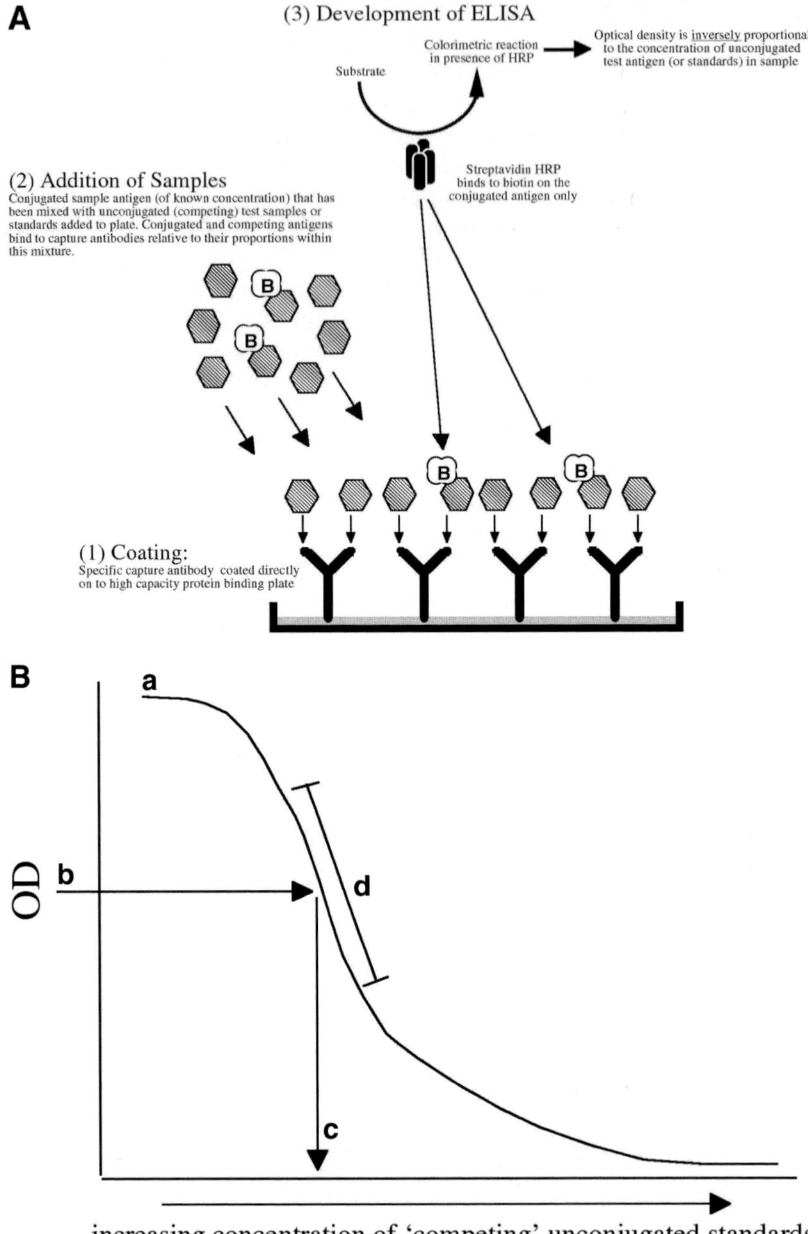

Fig. 3. (A) Principles of a typical competitive ELISA. (B) Estimating antigen concentration using a typical competitive ELISA standard curve. A constant amount of conjugated standard antigen is used to give the background OD reading (a). This reading is competed out with increasing concentrations of unconjugated antigen added, creating the standard curve. Antigen within samples competes with the constant conjugated antigen in the same manner, giving an OD reading (b) that can be read off using the standard curve to estimate the concentration within the sample (c). The most sensitive part of the curve (d) is where the smallest difference in concentration of the competing antigen has the greatest impact on the OD. Samples often require testing at several dilutions so that the reading can be made within or as close to this part of the curve as possible.

advances in ELISA-plate-binding surface technology now allow adhesion of a number of important molecules other than protein, including carbohydrates and lipids, thus expanding the array of microbial antigen that can be identified and quantified using this technique ***(3)***.

If the major aim of the ELISA is to obtain quantification of substances present in extremely low concentrations, there are a number of adaptations to the technique that can be used. Such techniques often use alkaline phosphatase (AP) enzyme systems rather than HRP, providing greater levels of sensitivity. Other technological advances that increase ELISA sensitivity can be found in the color-development stage of the technique. For example, the AP enzyme has been used to lock into a circular redox cycle producing an end product such as red formazan, which is hugely amplified in comparison to standard amplification methods ***(4)***. Chemiluminescent amplified ELISA principles have also been shown to give very high sensitivity ***(5)***. In one example, an ELISA to measure proinsulin in serum was optimized using chemiluminescence to increase the sensitivity to measure as little as 1 zmol (about 350 molecules) of alkaline phosphatase ***(6)***. Although extremely sensitive, such techniques are time-consuming to set up and optimize and more expensive than the simple colorimetric ELISAs described in this chapter, making them unsuitable for many routine diagnostic assays.

4. Clinical Applications

The ELISA is used for a huge number of clinical applications. Of these, perhaps the best known are those used for the diagnosis of HIV infection. The basic routine diagnosis of HIV infection using ELISA detects the antibodies that are produced by a patient that react against proteins of the virus. The presence of such antibodies in serum or saliva is indicative of infection ***(7–9)***. The ELISA is considered the best available screening because of its low cost, standardization, high reliability, and relatively quick turnaround. The price of each ELISA test kit can be less than $2. The ELISA's reliability and accuracy has been shown to have a sensitivity of 99.7% (i.e., 99.7% of test samples were correctly diagnosed as positive when antibodies were present). There are currently 18 Food and Drug Administration (FDA)-approved ELISAs for the detection of antibodies reacting against either HIV-1, HIV-2, or both.

The ELISA is used to identify infections from most common viral, bacterial, parasitic, and some fungal infections. Other infections that ELISA is commonly used to diagnose include hepatitis B and C viruses ***(10–13)***, parasitic infections such as *Giardia* and *Strongyloides* ***(14,15)*** and a number of bacterial infections, including anthrax ***(16)***. In most cases, these assays identify antibodies against the pathogen in the serum. However, some ELISAs are used to measure proteins from the organism themselves that might be in the blood. ELISAs are used to directly identify proteins from the circumsporozoite stage of the malaria parasite within the mosquito to identify those insects carrying the disease. The technique is also useful in early prediction, diagnosis, and tracking of the course of autoimmune disease through measurement of "rheumatoid factors" and other autoantibodies in the serum. High levels of these autoantibodies in the serum help diagnose and monitor diseases such as systemic lupus erythematosus and rheumatoid arthritis ***(17–19)***.

Other than in disease diagnosis, ELISA has a host of other important applications. The assay is used to detect hormone levels in serum—for example to examine levels of luteinizing hormone in order to determine the time of ovulation ***(20)*** or to measure human growth hormone in order to identify deficient individuals who could benefit from administration of the hormone. In food science, the ELISA is used to detect products of genes that are produced through GM technology and also to identify the presence or absence of allergens in food ***(21–23)***. In sports science, ELISAs have been developed that can detect recombinant hormones such as recombinant growth hormones or anabolic steroids that can be used illicitly by athletes or administrated to animals such as racing horses ***(24–26)***. The ELISA is also widely used in forensic drug analysis—for example to identify the presence of tetrahydrocannabinol, the active ingredient in marijuana ***(27,28)***.

References

1. Engvall, E. and Perlman, P. (1971) Enzyme-linked immunosorbent assay (ELISA). Quantitative assay of immunoglobulin G. *Immunochemistry* **8(9),** 871–874.
2. Peterson, E. M. (1981) ELISA: a tool for the clinical microbiologist. *Am. J. Med. Technol.* **47(11),** 905–908.
3. Gervay, J. and McReynolds, K. D. (1999) Utilization of ELISA technology to measure biological activities of carbohydrates relevant in disease status. *Curr. Med. Chem.* **6(2),** 129–153.
4. Johannsson, A., Stanley, C. J. and Self, C. H. (1985) A fast highly sensitive colorimetric enzyme immunoassay system demonstrating benefits of enzyme amplification in clinical chemistry. *Clin. Chim. Acta* **148(2),** 119–124.
5. Bronstein, I., et al. (1989) Chemiluminescent assay of alkaline phosphatase applied in an ultrasensitive enzyme immunoassay of thyrotropin. *Clin. Chem.* **35(7),** 1441–1446.
6. Cook, D. B. and Self, C. H. (1993) Determination of one thousandth of an attomole (1 zeptomole) of alkaline phosphatase: application in an immunoassay of proinsulin. *Clin. Chem.* **39(6),** 965–971.
7. Chassany, O., et al. (1994) Testing of anti-HIV antibodies in saliva. *Aids* **8(5),** 713–714.
8. Akanmu, A. S., et al. (2001) Evaluation of saliva-based diagnostic test kit for routine detection of antibodies to HIV. *Afr. J. Med. Med. Sci. 30(4),* 305–308.
9. Emmons, W. W., et al. (1995) A modified ELISA and western blot accurately determine anti-human immunodeficiency virus type 1 antibodies in oral fluids obtained with a special collecting device. *J. Infect. Dis.* **171(6),** 1406–1410.
10. Siddiqi, M. A. and Abdullah. S. (1988) An "antigen capture" ELISA for secretory immunoglobulin A antibodies to hepatitis B surface antigen in human saliva. *J. Immunol. Methods* **114(1–2),** 207–211.
11. Ukkonen, P., Koistinen, V., and Penttinen, K. (1977) Enzyme-immunoassay in the detection of hepatitis B surface antigen. *J. Immunol. Methods* **15(4),** 343–353.
12. Wolters, G., et al. (1976) Solid-phase enzyme-immunoassay for detection of hepatitis B surface antigen. *J. Clin. Pathol.* **29(10),** 873–879.
13. Wu, C. L., et al. (1999) Hepatitis C virus core protein fused to hepatitis B virus core antigen for serological diagnosis of both hepatitis C and hepatitis B infections by ELISA. *J. Med. Virol.* **57(2),** 104–110.
14. Conway, D. J., et al. (1993) Immunodiagnosis of *Strongyloides stercoralis* infection: a method for increasing the specificity of the indirect ELISA. *Trans. R. Soc. Trop. Med. Hyg.* **87(2),** 173–176.
15. Hopkins, R. M., et al. (1993) A field and laboratory evaluation of a commercial ELISA for the detection of *Giardia* coproantigens in humans and dogs. *Trans. R. Soc. Trop. Med. Hyg.* **87(1),** 39–41.
16. Sastry, K. S., et al. (2003) Identification of Bacillus anthracis by a simple protective antigen-specific mAb dot-ELISA. *J. Med. Microbiol.* **52(Pt 1),** 47–49.
17. Bayer, P. M., Fabian, B., and Hubl, W. (2001) Immunofluorescence assays (IFA) and enzyme-linked immunosorbent assays (ELISA) in autoimmune disease diagnostics—technique, benefits, limitations and applications. *Scand. J. Clin. Lab. Invest.* **235,** 68–76.
18. Bonagura, V. R., et al. (1989) The major rheumatoid factor cross-reactive idiotype in rheumatic disease. *Int. Rev. Immunol.* **5(2),** 139–151.
19. Griesmacher, A. and Peichl, P. (2001) Autoantibodies associated with rheumatic diseases. *Clin. Chem. Lab. Med.* **39(3),** 189–208.
20. Desai, M. P., Donde, U. M., and Khatkhatay, M. I. (2002) Improved performance of ELISAs for fertility assessment using common reagents and assay protocol as evidence from quality control studies. *J. Immunoassay Immunochem.* **23(2),** 163–180.
21. Arilla, M. C., et al. (2001) Quantification in mass units of group 1 grass allergens by a monoclonal antibody-based sandwich ELISA. *Clin. Exp. Allergy* **31(8),** 1271–1278.
22. Wei, Y., et al. (2003) A sensitive sandwich ELISA for the detection of trace amounts of cashew (*Anacardium occidentale* L.) nut in foods. *J. Agric. Food Chem.* **51(11),** 3215–3221.
23. Yamashita, H., et al. (2001) Sandwich enzyme-linked immunosorbent assay system for micro-detection of the wheat allergen, Tri a Bd 17 K. *Biosci. Biotechnol. Biochem.* **65(12),** 2730–2734.
24. Snow, D. H. (1993) Anabolic steroids. *Vet. Clin. North Am. Equine Pract.* **9(3),** 563–576.
25. Pescovitz, O. H., et al. (1986) Production of monoclonal antibodies against human growth hormone releasing hormone and their use in an enzyme-linked immunosorbent assay (ELISA). *J. Immunol. Methods* **94(1–2),** 257–262.
26. Meyer, H. H. and Hoffmann, S. (1987) Development of a sensitive microtitration plate enzyme-immunoassay for the anabolic steroid trenbolone. *Food Addit. Contam.* **4(2),** 149–160.

27. Tanaka, H. and Shoyama, Y. (1999) Monoclonal antibody against tetrahydrocannabinolic acid distinguishes Cannabis sativa samples from different plant species. *Forensic Sci. Int.* **106(3),** 135–146.
28. Kerrigan, S. and Phillips, W. H. Jr. (2001) Comparison of ELISAs for opiates, methamphetamine, cocaine metabolite, benzodiazepines, phencyclidine, and cannabinoids in whole blood and urine. *Clin. Chem.* **47(3)** 540–547.

31

Protein Therapeutics

Mouse, Humanized, and Human Antibodies

Benny K. C. Lo

1. Introduction

Antibodies play a vital role in immune defense against pathogens. They are globular glycoproteins produced by plasma cells in response to the antigenic stimulation of B-lymphocytes. Their circulation in blood and lymph contributes to the humoral component of the vertebrate immune system. Each plasma cell secretes a single clone of antibodies that bind a unique epitope of an antigen. Cooperations exist between antibodies and other immune effectors such as macrophages and complement to facilitate the removal of antigens from the body.

Because of their exquisite specificity in antigen recognition, they are used extensively as effective probes for biological markers in laboratory research. There have also been numerous attempts to exploit antibodies as specific disease-targeting agents since the early 20th century. Nevertheless, it was not until the end of the century that antibodies were finally used in patients with good levels of safety and efficacy. Breakthroughs in protein engineering and directed evolution techniques have enabled the design of "tailor-made" human antibodies to order. At the time of writing, the US Food and Drug Administration (FDA) has approved 15 therapeutic antibodies for human use and many more are in various stages of clinical and preclinical development. The recent explosion in genomic and proteomic information looks set to reveal cascades of disease targets open to antibody targeting. Undoubtedly, the continued use of antibodies in the 21st century is ensured by their unique recognition properties and, as now demonstrated, by their successful partnership with protein engineering.

In this chapter, the background to several well-established antibody engineering technologies, the major experimental steps involved, and their impact on modern medicine are described.

1.1. Antibody Structure and Function

Antibodies are Y-shaped molecules consisting of two identical heavy- and light-chains joined by disulfide and noncovalent linkages. Five major classes of heavy chains exist in humans (α, δ, ε, γ, μ) and they define the isotype of the antibody (IgA, IgD, IgE, IgG, IgM, respectively). Each heavy chain is paired with a κ or λ light chain to form a functional molecule. The different antibody subclasses possess distinct characteristics in the overall structure, tissue localization, and the ability to trigger effector functions. The following description is based on the IgG molecule that has a simple structure and is present in greatest amounts in serum.

An IgG is made up of three regions: two identical ones forming the arms (Fab, fragment antigen binding) and one other forming the base of the "Y" (Fc, fragment crystallizable) (*see* **Fig. 1A**). The Fabs and the Fc are connected through a hinge region, which confers flexibility

From: *Medical Biomethods Handbook*
Edited by: J. M. Walker and R. Rapley © Humana Press, Inc., Totowa, NJ

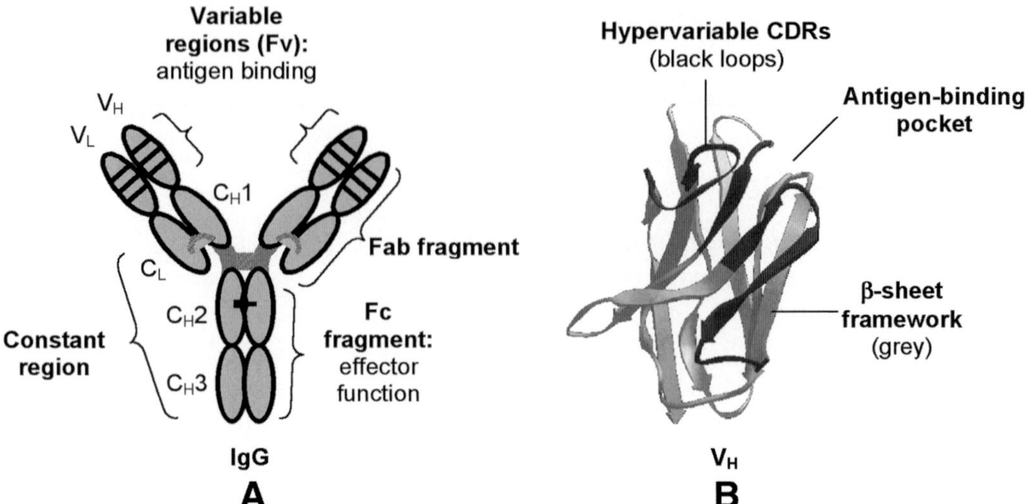

Fig. 1. Antibody structure. (**A**) Nomenclature and the domain arrangement of an IgG molecule. (**B**) The crystal structure of a V_H domain showing the β-sheet immunoglobulin fold and the relationship between the conserved framework regions and the hypervariable complementarity-determining regions (CDRs).

to the molecule. The heavy and light chains are respectively folded into four and two globular domains of around 12.5 kDa each. The variable domains from each chain (V_H and V_L, known collectively as the Fv) control the antigen-binding specificity and intrinsic affinity of the antibody. Structurally, each Fv adopts the immunoglobulin fold in which two conserved β-sheet framework scaffolds support six hypervariable loops known as the complementarity-determining regions (CDRs) (*see* **Fig. 1B**). Binding specificity and affinity are defined by the length, sequence, conformation, and relative disposition of the CDRs, and the pairing of CDRs from repertoires of V_H and V_L sequences. The presence of two Fvs in each IgG also allows simultaneous binding to two epitopes, hence increasing the functional affinity of the antibody through avidity. Genes responsible for the huge diversity of V_H and V_L sequences are the products of germline recombination of immunoglobulin variable (V), diversity (D, V_H only), and junctional (J) gene segments in B-lymphocytes, leading to billions of antibody specificities for binding diverse targets ***(1–3)***. Further refinement of the initial antibody response is enabled through the process of somatic hypermutation ***(4,5)***, yielding antibodies of even higher binding affinity.

In contrast to the Fv, the constant region of an IgG consists of eight domains (C_L, C_{H1-3}) with conserved sequences. It lacks specific structures for antigen binding but is essential for the recruitment of immune effector cells and molecules for immunological attack on the antigen. The same antigen-specific Fv can be connected to different constant regions during the course of an antibody response. This process of class switching ***(6)*** allows the functional specialization of antibodies. For instance, the Fc region of an IgG is able to bind phagocytic cells, such as macrophages and neutrophils, and complement. IgEs, on the other hand, are designed to interact with mast cells and basophils during a hypersensitivity reaction (*see* **ref. *7*** for review).

The reader is referred to **refs. *8*** and ***9*** for detailed reviews on antibody structure.

2. Antibody Engineering

2.1. Monoclonal Antibodies

Polyclonal antisera from immunized animals have traditionally been used as a convenient source of antibodies for immunological detection. They were also used in the treatment of

❶. Mice are immunized intraperitoneally and/or subcutaneously with the desired antigen. Boost injections are given at 2-4 week intervals.

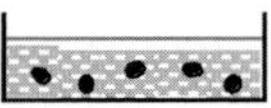

❹. Hybridomas are cultured in hypoxanthine-aminopterin-thymidine (HAT) medium to select for stable phenotypes from both fusion partners.

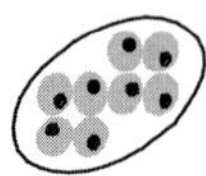

❷. Antibody-producing cells are harvested from the spleen.

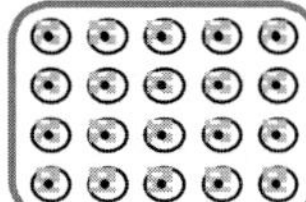

❺. Selected hybridomas are dilution-cloned in multititer plates to ensure monoclonality.

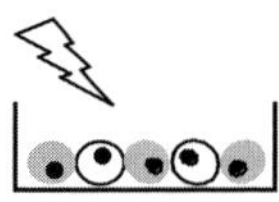

❸. Splenocytes are fused with murine myeloma cell lines by polyethylene glycol or electrofusion to form hybridomas.

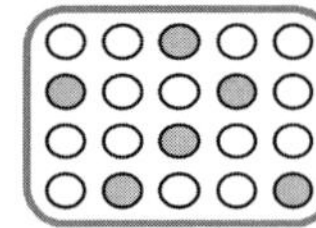

❻. Culture supernatants are screened for the secretion of specific antibodies by ELISA. Clones from positive wells are expanded, dilution-cloned and banked.

Fig. 2. Murine hybridoma technology.

infectious diseases such as pneumonia, tetanus, diphtheria, and rabies, although the occurrence of hypersensitivity reactions limited their in vivo application. Despite being easy to produce, polyclonal antibodies suffer from the disadvantage of heterogeneity with batch-to-batch variations. In addition, little control can be exerted over the fine specificity and composition of each polyclonal preparation. In 1975, Georges Köhler and César Milstein of the MRC Laboratory of Molecular Biology (Cambridge, England) developed a revolutionary method for the production of unlimited amounts of antibodies to a predefined specificity ***(10)***. To achieve this, they fused spleen cells from mice immunized with sheep red blood cells (SRBCs) with cultured murine myeloma cells. These hybrid myelomas, or "hybridomas," were propagated and screened for the production of antibodies that lysed SRBCs in the presence of complement. Each positive hybridoma clone thus secretes antibodies to a single epitope of the immunogen and can be cultured indefinitely for antibody production (*see* **Fig. 2**). This technique, for which a Nobel Prize was awarded in 1984, has led to the isolation of monoclonal antibodies to a multitude of targets. Of these, some have become indispensable research and diagnostic reagents, as symbolized by the role of monoclonal antibodies in the discovery of cluster of differentiation (CD) antigens on leukocytes. Partnered with protein engineering and molecular display technologies (*see* **Subheadings 2.2.** and **2.4.**), they are also being developed as effective therapeutic agents for a range of human diseases (*see* **Subheading 3.**).

*Experimental Steps (*See ***refs. 11** and **12** for Detailed Protocols)*

Mice (e.g., Balb/c) are immunized subcutaneously and/or intraperitoneally with the antigen of interest in Freund's adjuvant. Either soluble proteins or whole cells can be used. Following initial immunization, boost injections are given at regular intervals and sera are assayed for antibody titers for binding to the immunogen. After the final boost, mice are sacrificed and the spleen cells are isolated. The splenocyte population is then fused to myeloma cells (e.g., Sp2/0, NS0) by adding polyethylene glycol or by electrofusion, and the hybridomas are cultured in the hypoxanthine–aminopterin–thymidine selection medium. Finally, hybridoma cells are dilution-cloned and assayed individually for the secretion of antigen-specific antibodies.

2.2. Chimeric and Humanized Antibodies

Although murine hybridoma technology has been instrumental in the generation of monoclonal antibodies for clinical diagnostics and laboratory research, their application in human therapeutics is limited. This is because murine antibodies frequently elicit the human anti-murine antibody (HAMA) response ***(13,14)*** after multiple intravenous administrations. It is characterized by neutralizing serum IgG and IgM responses directed against the constant and variable regions of the murine antibody. In addition, murine constant regions are also inefficient in engaging suitable human immune effector functions for therapeutic effects. These undesirable properties posed significant challenges to the in vivo safety and efficacy of murine antibodies and, consequently, led to severe setbacks in the pursuit of developing antibodies as the "magic bullet" in the late 1970s.

Efforts to produce human antibodies by hybridoma technology ***(15)*** and Epstein–Barr virus (EBV)-mediated B-cell transformation ***(16)*** have yielded only limited successes because of the lack of robust human hybridoma fusion partners and the instability of EBV-transformed clones (*see* **refs. *17–19*** for review). To circumvent these problems, attempts were made in the early 1980s to manipulate different parts of the antibody molecule by genetic engineering. The first experiments involved the heterogeneous expression of antibodies in cultured lymphoid cell lines transfected with immunoglobulin genes ***(20,21)***. This, together with the ability to clone immunoglobulin genes by reverse transcriptase–polymerase chain reaction (RT-PCR; *see* **refs. *22–24***), paved the way for the extensive manipulations of the antibody molecule. In the mid-1980s, three groups reported methods for the production of mouse/human chimeric antibodies ***(25–27)***—murine antibodies transplanted with human antibody constant regions (*see* **Fig. 3A**). The replacement of the immunogenic murine constant regions in chimeric antibodies markedly reduced the likelihood of the HAMA response while preserving the antigen-binding specificity and affinity of the murine antibody. Chimerization also allowed the customization of effector functions by choosing constant regions from different human isotypes ***(28)***. Although the chimerization of some murine antibodies resulted in the disappearance of the HAMA response, others remained immunogenic ***(29)***. As an alternative approach, Sir Gregory Winter of the MRC Laboratory of Molecular Biology pioneered the CDR grafting technology ***(30–32)***, in which the antigen-binding specificity of a murine antibody is transferred to a human antibody by transplanting the CDR loops and selected framework region residues (*see* **Fig. 3A,B**). The reshaped, humanized antibody retains only the essential binding elements from the murine antibody (5–10% of the total sequence) and is predicted to be minimally immunogenic. Indeed, the in vivo tolerance and efficacy of many humanized antibodies have been shown to be more favorable than murine antibodies ***(33–36)*** and a number of these are now marketed as effective therapeutic drugs (*see* **Subheading 3.**). Other humanization technologies such as veneering ***(37,38)*** and guided selection ***(39)*** were also developed, but these are not discussed here.

*Experimental Steps (*See ***ref. 40*** *for Detailed Protocols)*

1. Chimerization of murine antibodies. The V_H and V_L genes of the murine antibodies are first amplified from the hybridoma cell line by RT-PCR and the amino acid sequences determined. These are subcloned into mammalian expression vectors containing the human constant regions of the desired isotypes and onto a leader sequence for export of the expressed antibodies into the extracellular medium. Finally, expression vectors for the chimeric heavy and light chains are transfected to mammalian cell lines such as COS-7 for transient expression or Chinese hamster ovary (CHO) cells for stable expression into the culture media.
2. Humanization of murine antibodies by CDR grafting. The V_H and V_L genes of the murine antibodies are first amplified from the hybridoma cell line by RT-PCR and the amino acid sequences determined. The structure of the murine variable region (crystallographic or modeled) is then examined for important interactions between framework and CDR residues. Special emphasis is placed on framework residues that regulate CDR conformations directly or indirectly. Next, human antibody frameworks are selected for CDR grafting based on their

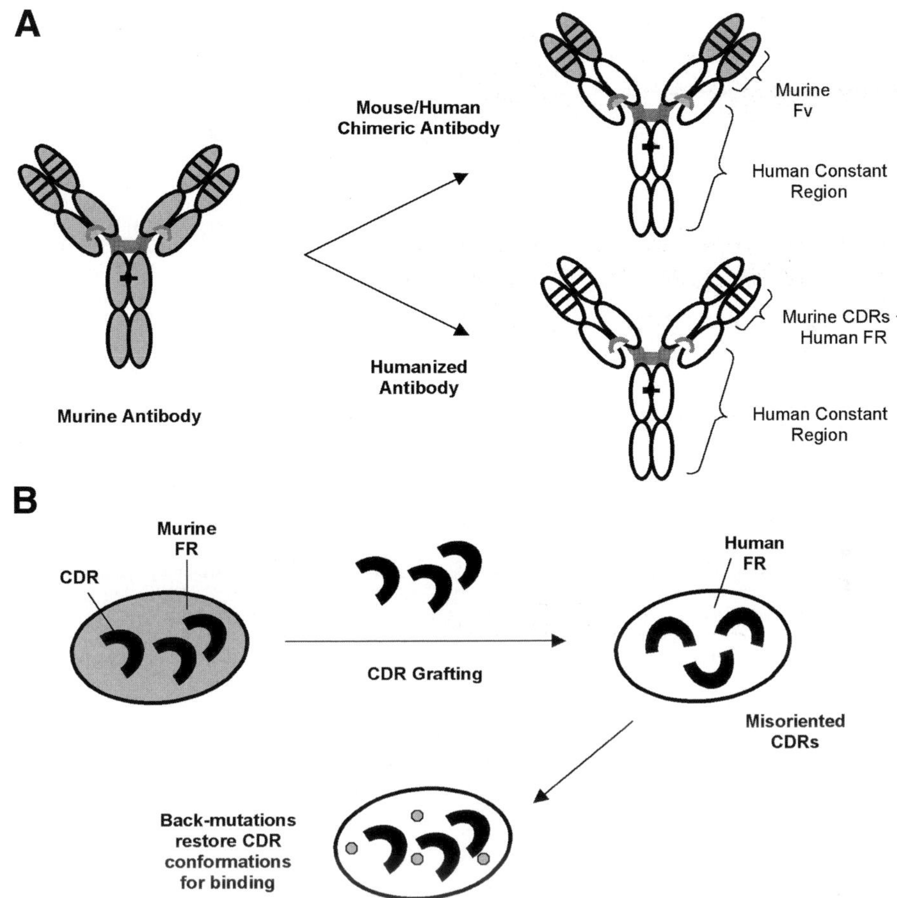

Fig. 3. Antibody humanization and CDR grafting. (**A**) Mouse/human chimeric antibodies are constructed from murine variable regions and human constant regions, making them less immunogenic than murine antibodies. Humanized antibodies contain even fewer murine sequences by retaining only the CDRs from the murine antibody. However, a small subset of murine framework (FR) residues may need to be re-introduced into the human frameworks (FR) to optimize CDR conformations for binding (see panel **B**). (Shaded ovals: murine sequences; Blank ovals: human sequences; Black lines across ovals: murine CDRs) (**B**) The transplantation of murine CDRs onto a human framework (FR) often leads to suboptimal orientations of these binding loops. Consequently, critical murine framework residues need to be re-introduced as back-mutations to restore the optimum CDR conformations for antigen binding.

ability to support the murine CDRs for antigen binding. This is achieved by a detailed sequence comparison between the candidate human frameworks and the murine antibody sequence. Molecular models can also be built to evaluate the suitability of the chosen frameworks. To produce the CDR-grafted antibody, the entire humanized V_H and V_L sequences are synthesized by PCR using overlapping primers, subcloned into expression vectors, and transfected into mammalian cell lines for expression (as described for chimeric antibodies). Alternatively, variable region genes are expressed in *Escherichia coli* as antibody binding fragments (single-chain Fv, Fab; *see also* **Subheading 2.3.**). The binding of the humanized antibody to its cognate antigen can be measured by enzyme-linked immunosorbent assay (ELISA). If optimization in binding affinity is required, one or more framework residues are usually

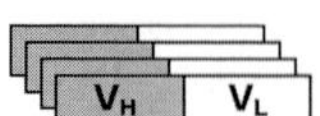

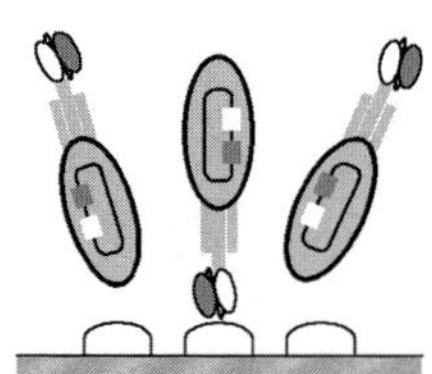

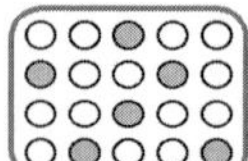

Fig. 4. Combinatorial phage antibody libraries.

mutated to the equivalent in the murine antibody, paying special attention to those positions identified as important from structural analysis. One should, however, minimize such murine mutations in order to keep the humanized antibody as "human" as possible.

2.3. Combinatorial Antibody Libraries

Chimeric and humanized antibodies, despite having desirable in vivo properties, rely on hybridoma and protein engineering procedures, which can be lengthy. In 1985, George Smith of the University of Missouri demonstrated that peptides could be displayed on the head of the filamentous bacteriophage by genetic fusion to the minor coat protein pIII ***(41)***. The bacteriophages were found to retain infectivity and display the peptides in a way accessible to antibody-binding. A few years later, it was also shown that the antibody-binding fragments Fab ***(42)*** and Fv ***(43)*** can be expressed and secreted in *E. coli* culture and even displayed as a folded single-chain Fv protein (scFv: V_H–peptide linker–V_L; *see* **refs. *44*** and ***45***) on filamentous bacteriophage ***(46)***. These seminal findings were quickly exploited for the generation of antibodies. In these systems, a combinatorial library of scFv ***(47)*** or Fab ***(48)*** antibody genes was fused to gene III and the antibody fragments were displayed as pIII fusions. Phages bearing antibodies of the desired specificity are selected by panning the library, which contains billions of different antibody-binding fragments, with the antigen. Enrichment of the positive phage population is achieved by several rounds of bacterial infection and repanning, resulting in specific antibodies of high affinity (*see* **Fig. 4**). The desired antibody phenotype is preserved by PCR-cloning the encoding gene packaged within the corresponding bacteriophage.

The antibody gene repertoire making up the phage libraries was originally amplified from B-lymphocytes of immunized mice by PCR ***(47)***. It was later extended to B-lymphocytes from humans either immunized with or exposed to the antigen during the course of disease ***(49,50)***, leading to "fully human" antibodies without the need for hybridomas. The use of naïve B-lymphocyte populations also achieved a similar level of success ***(51)***. Taking this one step further, it was found that human antibodies of high affinity can be isolated from repertoires of synthetic and randomized CDR sequences (where CDR positions are represented by a subset or all 20 amino acids) ***(52,53)***, thus completely bypassing the need for immunization and the use of animals. In addition to lead generation, phage display is also used in the optimization of antibody leads by directed evolution (reviewed in ref. ***54***). Changes to the antibody sequences are introduced as random or CDR-focused point mutations, insertions, and deletions. The pool of mutant antibody fragments is displayed on phage and those with the desired properties (affinity, on/off rates, stability, etc.) are selected using a variety of strategies ***(55)***. Using these approaches, the humoral component of the immune system is simulated in vitro; human antibodies can be isolated and affinity-matured in a matter of weeks. Antibodies to problematic antigens such as conserved and toxic molecules are readily generated. This technology has proved so versatile that a number of biotechnology companies were founded on this platform and are generating human antibody leads for clinical trials at an unprecedented rate ***(56,57)***. Other molecular display technologies such as ribosome display ***(58,59)*** and yeast display ***(60)*** are maturing and will complement phage display in this respect, as well as increasing the repertoire diversity so that high-affinity antibodies can be obtained expeditiously.

*Experimental Steps (See **Refs. 61** and **62** for Detailed Protocols)*

1. Construction of phage antibody libraries. The antibody gene repertoire is prepared by RT-PCR amplification of V_H and V_L genes from lymphocyte mRNA using specific V-gene primers for for naïve and immune libraries. For synthetic libraries and libraries for affinity maturation, the variable region template can be synthesized and the diversity introduced by randomized oligonucleotides at the desired positions using PCR. Next, the desired repertoire is subcloned into a phage or phagemid display vector and electroporated into *E. coli* expressing the F pilus (e.g., TG1). The library is then grown on large antibiotic agar plates and the library size determined. Phage antibody production is effected by inoculating the library into liquid culture (with helper phage rescue for phagemid libraries) and can be purified for panning by polyethylene glycol precipitation.
2. Panning of phage antibody libraries on antigen. To isolate antigen binders, purified phage antibodies are incubated with immobilized antigen coated on immunotube or biotinylated antigen in solution. Bound phage antibodies are isolated by thorough rinsing of the immunotube or capturing the biotinylated antigens by streptavidin-coated magnetic beads. They are then eluted from the solid support and amplified by reinfecting logarithmic-phase *E. coli* for phage production and repanning. After two to three rounds of panning, the eluted phages are used to infect *E. coli* and the infected culture is plated onto antibiotic agar plates for single colonies. These are subsequently inoculated into multititer plates for antibody expression and for antigen binding using ELISA. Antibody genes from the positive phage clones can be sequenced and cloned for large-scale expression or reformatted as IgG for in vivo applications or for other manipulations.

2.4. Transgenic Animals

In addition to protein engineering approaches, human monoclonal antibodies can also be generated by immunizing transgenic mice harboring megabase germline DNA ***(63)*** or chromosomal fragments ***(64)*** containing the human immunoglobulin loci. The endogenous murine immunoglobulin genes are inactivated so that germline rearrangement occurs predominantly with the human transgenes. Fully human monoclonal antibodies are thus obtained following conventional immunization with the desired antigen and the production of hybridomas (*see* **Subheading 2.1.**). This takes advantage of the established method of murine hybridoma production while avoiding the complex protein engineering steps associated with humanization or phage display. Currently, Xenomouse ***(65)***, HuMab-Mouse ***(66)***, TransChromo-Mouse ***(64)***, and KM-Mouse (crossbred HuMab/TransChromo mouse; *see* **ref. *67***) are the four major commercial systems available and they have generated a number of antibody candidates for clinical trials in cancer and infectious and inflammatory diseases (*see* **ref.** ***68***; http://www.abgenix.com/ and http://www.medarex.com/). In addition to mice, cloned transchromosomic cows are also being developed as a novel source for human polyclonal antibodies ***(69)***. If successful, this will clear the way for a new generation of antibody therapeutic products and reduce the dependence on human donors for intravenous immunoglobulin therapies.

Experimental Steps

Transgenic mice harboring human antibody genes are available through the following commercial suppliers: Xenomouse, Abgenix, Inc.; HuMab-Mouse and KM-Mouse, Medarex, Inc.; TransChromo-Mouse, Kirin Brewery Co., Ltd. To produce human antibodies from these animals, the protocols for producing murine hybridoma (*see* **Subheading 2.1.** and **ref.** ***70***) can be applied.

3. Clinical Applications of Therapeutic Antibodies

The selective targeting of therapeutic agents to the sites of action has been a major theme in drug design. This is particularly crucial for agents of high toxicity (e.g., cancer chemotherapeutic drugs) so that the debilitating adverse reactions are minimized. Therapeutic antibodies have been developed to meet these demands by virtue of their high target-recognition specificity, adaptability to a wide range of disease states, and the ability to recruit the host's immune effectors for a therapeutic response. Fifteen antibody-based therapeutic agents are now approved for human use in oncology, inflammation, organ transplantation, viral infection, and cardiology (*see* **Table 1**). All of these were developed using the antibody engineering techniques described in **Subheading 2**. Although their underlying mechanisms of action vary, they share the common feature of high binding specificity and affinity. In this section, a brief overview on the contribution from each antibody drug to clinical disease management,* as vivid illustrations of the impact of antibody engineering on modern medicine, is provided.

3.1. Oncology

The concept of using antibodies to seek and destroy tumor cells in vivo while sparing healthy tissues has been central to numerous research efforts since the first description of hybridoma technology. Frequently, tumors and their associated vasculature express cell surface antigens that are unique or expressed in larger quantities than normal tissues, making them attractive targets for antibody therapy. Upon binding to tumor-associated antigens, antibodies can exert a combination of cytotoxic effects through antibody-dependent cellular cytotoxicity (ADCC), complement-dependent cytotoxicity (CDC), induction of apoptosis, interference with cellular signaling pathways, or release of radioisotope or cytotoxic drug following internalization of the antibody–antigen complex ***(71,72)***.

Rituxan®, a mouse/human chimeric antibody, was the first FDA-approved anticancer antibody for relapsed or refractory CD20$^+$ B-cell non-Hodgkin's lymphoma (NHL). It binds CD20, a transmembrane protein found on the surface of normal and malignant B-cells, which regulates processes for cell cycle initiation and differentiation ***(73)*** and is expressed on over 90% of B-cell NHLs ***(74)***. The administration of Rituxan causes a rapid and sustained depletion of B-cells. It is believed to act through CDC, ADCC ***(75)***, and the induction of apoptosis ***(76)***. Results from a multicenter study ***(77)*** showed that 48% of patients with relapsed low-grade or follicular B-cell NHLs responded to four weekly doses of Rituxan (6% complete response [CR], 42% partial response) with a median duration of response (DR) of 11.2 mo. In another study in which eight weekly doses were given, the overall response rate (ORR) was 57% (14% CR) ***(78)***

*Indications and clinical data collated from prescribing information for the individual product, available through the manufacturers' websites (*see* **Table 1**).

Table 1
FDA-Approved Therapeutic Antibody Drugs

Product	Company	Construction	Target	Licensed indication(s)
1. Oncology				
Rituxan® (rituximab)	IDEC, Genentech	Chimeric IgG1	CD20	Non-Hodgkin's lymphoma
Zevalin® (ibritumomab tiuxetan)	IDEC	Murine IgG1 radio-immunoconjugate	CD20	Non-Hodgkin's lymphoma
Bexxar® (tositumomab)	Corixa, Glaxo Smithkline	Murine IgG2b-I-131 conjugate	CD20	Non-Hodgkin's lymphoma
Campath® (alemtuzumab)	ILEX	Humanized IgG1	CD52	B-cell chronic lymphocytic leukemia
Mylotarg® (gemtuzumab ozogamicin)	Wyeth	Humanized IgG4 calicheamicin conjugate	CD33	Relapsed acute myeloid leukemia
Herceptin® (trastuzumab)	Genentech	Humanized IgG1	HER2	HER2-positive metastatic breast cancer
2. Inflammation				
Remicade® (infliximab)	Centocor, Johnson and Johnson	Chimeric IgG1	Tumor necrosis factor-α	Rheumatoid arthritis and Crohn's disease
Humira® (adalimumab)	Abbott	Human IgG1	Tumor necrosis factor-α	Rheumatoid arthritis
Xolair® (omalizumab)	Genentech, Novartis	Humanized IgG1	Human IgE	Moderate to severe persistent asthma
Raptiva™ (efalizumab)	Genentech, Xoma	Humanized IgG1	Leukocyte	Psoriasis function antigen-1
3. Organ transplantation				
Orthoclone OKT® 3 (murumonab-CD3)	Ortho Biotech	Murine IgG2a	CD3	Renal, cardiac and hepatic transplant rejection
Simulect® (basiliximab)	Novartis	Chimeric IgG1	CD25	Renal transplant rejection
Zenapax® (daclizumab)	Roche	Humanized IgG1	CD25	Renal transplant rejection
4. Viral infection				
Synagis® (palivizumab)	MedImmune	Humanized IgG1	Respiratory syncytial virus F protein	Respiratory syncytial virus infection
5. Cardiology				
ReoPro® (abciximab)	Centocor, Eli Lilly	Chimeric Fab	Glycoprotein IIb/IIIa receptor	Ischemic cardiac complications and unstable angina

Note: Information compiled from product websites of the relevant manufacturer.

with a median DR of 13.4 mo. This compares favorably with results from single-agent cytotoxic chemotherapy. In addition, Rituxan was well tolerated in patients, with the majority of adverse events being low-grade toxicities. For patients refractory to Rituxan or established chemotherapeutic regimens, two recently introduced anti-CD20 antibody conjugates were found to be highly effective. **Zevalin®** (murine anti-CD20–yttrium-90 conjugate) and **Bexxar®**

(murine anti-CD20–iodine-131 conjugate) achieved impressive ORRs of 74% (15% CR) and 63% (29% CR), with median DRs of 6.4 and 25 mo, respectively ***(79,80)***. Not only do they possess cytotoxic effects similar to those from Rituxan, these radioimmunoconjugates also inflict cellular damage on target and neighboring cells by emitting ionizing β-radiation.

In addition to murine monoclonal and chimeric antibodies, two humanized antibodies are also approved for treating hematologic malignancies. **Campath®** is the first therapeutic humanized antibody constructed by CDR grafting ***(32)*** and is indicated for the treatment of B-cell chronic lymphocytic leukemia (B-CLL) in patients who have received alkylating agents and failed fludarabine therapy ***(81)***. It targets the CD52 antigen that is present on all B- and T-lymphocytes. It is thought that Campath exerts its cytotoxicity primarily through CDC, which is greatly enhanced by the large number of CD52 molecules (approx 1 million/cell) and the close proximity of the antigenic epitope to the cell membrane ***(82,83)***. Results from multicenter studies demonstrated ORRs of up to 33% (2% CR) and median DRs of up to 7 mo for B-CLL patients who had received chemotherapeutic drugs, including alkylating agents and fludarabine ***(81)***. The efficiency in which Campath depletes T-lymphocytes has also led to recent investigations into the effectiveness of the antibody as an immunosuppressant in transplant operations ***(84,85)***. Whereas Campath lyses B-CLL by CDC, **Mylotarg®** kills leukemic blasts via a conjugated cytotoxic drug. It consists of a humanized anti-CD33 antibody conjugated to the antitumor antibiotic calicheamicin and is indicated for the treatment of $CD33^+$ AML in the first relapse in patients over 60 yr of age for whom aggressive cytotoxic regiments are unsuitable. CD33 is expressed on leukemic blasts in over 80% of patients with acute myeloid leukemia (AML) ***(86)***. Upon binding to CD33, the antibody–calicheamicin conjugate is internalized into the myeloid cell and the drug is released in the lysosomes. Binding of the drug to the DNA minor groove follows, leading to DNA double-strand breaks and cell death. Studies to investigate the effectiveness of Mylotarg monotherapy in AML patients in the first relapse demonstrated an ORR upto 32% (17% CR) and a median DR upto 7.2 mo ***(86)***. A pilot study into the effect of Mylotarg administered with conventional chemotherapeutic drugs (idarubicin and cytarabine) in refractory AML has also been performed recently with encouraging results ***(87)***.

Herceptin®, a humanized antibody developed for the treatment of metastatic breast cancer, was the first therapeutic antibody approved for treating solid tumors. It targets the extracellular domain of the HER2 protein, which is overexpressed in 25–30% of primary breast cancers ***(88)*** and generally correlates with poor prognosis and decreased survival ***(89)***. HER2, also known as neu or c-erbB2, is a member of the epidermal growth factor family of tyrosine kinases. It does not have a natural ligand, but forms heterodimers with other epidermal growth factor receptors (HER1 [epidermal growth factor receptor, EGFR], HER3 and HER4). Formation of the complex initiates signaling pathways such as PI_3 kinase and Ras-MAPK (mitogen activated protein kinase), which regulate normal cell growth and development ***(90)***. HER2 overexpression leads to self-association of the receptor at the tumor cell surface, resulting in constitutive receptor activation, uncontrolled cell growth, and tumorigenesis ***(91)***. Herceptin appears to halt tumor growth by downregulating HER2 and interfering with the aberrant signaling for proliferation ***(92)***. Additional cytotoxicity can be achieved through ADCC by mononuclear cells ***(93)***. In a multicenter trial in patients with relapsed and HER2-overexpressing metastatic breast cancer and who have received one or more chemotherapeutic regimens for metastatic disease, Herceptin monotherapy achieved an ORR of 14% (2% CR) and a median DR of 9.1 mo after weekly dosing ***(94)***. In another trial in which Herceptin was given with paclitaxel in patients with metastatic breast cancer who had not received chemotherapy for metastatic disease, patients receiving Herceptin and paclitaxel achieved a significantly higher therapeutic response than those receiving paclitaxel alone (ORR 38% vs 15%, $p < 0.001$; median DR 8.3 vs 4.3 mo). The mechanism for the synergism is unclear, although Herceptin was shown to block DNA repair after cisplatin treatment on breast cancer cells in vitro ***(95)***. The role of Herceptin in conjunction with adjuvant chemotherapy is also being investigated in several trials for early-stage breast cancer ***(96,97)***. Another anti-HER2 antibody currently in phase II clinical trial, 2C4 (pertuzumab, Omnitarg™), was shown to block HER2 heterodimerization and is more effective in interfering with growth signaling than Herceptin, especially in tumors expressing low levels of HER2 ***(98)***. A number of studies have been initiated to investigate the role of 2C4 in the treatment of lung, prostate, and ovarian cancers ***(97)***. These findings underscore the importance of transmembrane signaling as a target for intervention by therapeutic antibodies ***(99)*** and could result in a wider application of anti-HER2 and other anti-EGFR therapies in cancer management.

3.2. Inflammation

Acute inflammation is a physiologic response to tissue damage. It serves to counteract the deleterious effects of the causative agent and repair the damaged area. It is characterized by rapid changes in the local circulation and the mobilization of leukocytes and inflammatory mediators. Complex interactions occur between the cellular and soluble components in the affected area, with the goal of eventual resolution of the acute episode. Nevertheless, problems arise when there is a repeated challenge from the causative agent or an imbalance between the inflammatory mediators, resulting in chronic inflammation. In addition, when a dysregulated inflammatory response is directed against a harmless or self-component, allergy, autoimmunity, and long-lasting damage could result. The large number of mediators involved in these pathological processes provides ample opportunities for antibody-based strategies in therapeutic intervention.

Tumor necrosis factor-α (TNF-α) is a cytokine produced by a wide range of leukocytes, fibroblasts, and endothelial cells in response to pro-inflammatory stimulation. It exists as both a transmembrane and a soluble form and induces the production of interleukins, expression of leukocyte adhesion factors on endothelial cells, production of tissue-degrading enzymes, and activation of polymorphonuclear leukocytes. High levels of TNF-α have been found in the joints of rheumatoid arthritis patients and the intestinal mucosa of patients suffering from Crohn's disease. **Remicade®** is a chimeric anti-TNF-α antibody developed for treating rheumatoid arthritis and Crohn's disease. It blocks the binding of the transmembrane and soluble forms of TNF-α to the TNF receptor, thereby reducing the pro-inflammatory effect of TNF-α. It is also capable of lysing cells expressing the transmembrane form of TNF-α upon binding ***(100)***.

A multicenter trial was performed where the effect of coadministering Remicade and methotrexate (MTX, a disease-modifying antirheumatoid drug [DMARD]) in rheumatoid arthritis was compared with that of MTX alone in patients who had an inadequate response to MTX ***(100)***. In all dosage schedules, the coadministration of Remicade and MTX consistently achieved better scores in terms of clinical, radiographic, and physical function responses, although the use of Remicade without coadministration of MTX is not currently indicated. In Crohn's disease, it was shown that a significantly higher proportion of patients achieved a clinical response 4 wk after receiving a single dose of Remicade compared to placebo (Remicade ORR 5 mg/kg: 81%; 10 mg/kg: 50%; 20 mg/kg: 64%; placebo: 16%; $p < 0.001$) ***(101)***. Moreover, a subsequent study in which multiple doses of Remicade were given with monitoring for up to 54 wk also demonstrated the value of eight-weekly maintenance treatment with Remicade ***(102)***. Furthermore, Remicade was shown to be effective in fistulizing Crohn's disease ***(103)*** and is indicated for reducing the number of enterocutaneous and rectovaginal fistulas and maintaining fistula closure in such patients. **Humira®**, the first fully human antibody derived from phage display technology and targets TNF-α, has recently been approved for treating rheumatoid arthritis in patients who had an inadequate response to other DMARDs. Significant benefits were demonstrated in rheumatoid arthritis patients, with an added advantage being that it is indicated for use as a monotherapy as well as in conjunction with other DMARDs ***(104)***.

Antibodies are also used to treat chronic airway inflammation in allergic asthma. **Xolair®** is a humanized anti-IgE antibody approved for moderate to persistent asthma in patients who have a confirmed sensitivity to a perennial aeroallergen and are inadequately controlled with inhaled corticosteroids ***(105)***. It blocks the antigen-independent binding of IgE antibodies to high-affinity IgE receptors (FcεRI) on mast cells and basophils, which, in turn, reduces the subsequent release of inflammatory mediators that lead to airway inflammation and bronchospasms ***(106)***. Following administration, there are marked decreases (over 10-fold) in the serum level of free IgE and the number of FcεRI expressed on basophils (twofold to threefold), with reduced responsiveness to antigen challenge ***(107)***. Results from two clinical trials indicated that patients who have received Xolair and inhaled beclometasone reported a lower number of asthmatic exacerbations and fewer symptoms than those who received beclomethasone alone ***(105)***. Recent findings that the administration of Xolair also causes the downregulation of FcεRI on dendritic cells might suggest a further role of anti-IgE antibodies in immunomodulation ***(108)***.

In addition to antibodies that target soluble inflammatory mediators, an anti-CD11a antibody has recently been approved for treating plaque psoriasis. **Raptiva®** is a humanized antibody that binds the CD11a subunit of the leukocyte-function antigen 1 (LFA-1) and is indicated for chronic moderate to severe plaque psoriasis. Psoriasis is an inflammatory disease of the skin caused by an inappropriate activation of T-lymphocytes, resulting in rapid skin shedding and plaque formation. Raptiva blocks the interactions between CD11a, which is present on all lymphocytes, and the intercellular adhesion molecule 1 (ICAM-1) on antigen-presenting cells and vascular endothelial cells. This inhibits T-lymphocyte activation, adhesion to endothelial cells, and migration to the inflamed areas and downregulates the expression of CD11a ***(109)***. Results from a multicenter study show that patients who have received 12-weekly doses of Raptiva achieved better clinical outcomes than those on placebo ***(110)***, including the 75% improvement on the psoriasis area and severity index (Raptiva: 27%; placebo: 4%; $p < 0.001$), dermatological life quality index (47% vs 14%, $p < 0.001$), itching visual analog scale (38% vs –0.2%, $p < 0.001$), and psoriasis symptom assessment and severity scales. Finally, Remicade, an anti-TNF-α antibody, has also shown promise in treating psoriasis (reviewed in **ref. *111***). This further illustrates the potential of antibodies as effective immunomodulatory agents in the treatment of diverse inflammatory conditions.

3.3. Organ Transplantation

Despite routine tissue-typing procedures to ensure a close fit between organ donors and recipients, graft rejection still occurs at a high frequency in transplant patients. Because no two individuals possess identical histocompatibility antigens (except in identical twins), T-lymphocytes that recognize non-self tissues are activated and eventually mount an immune response that result in the rejection of the transplanted tissue. To maximize the likelihood for graft survival, immunosuppressant drugs such as cyclosporin, azathioprine, and corticosteroids are given during transplantation procedures and have to be taken for life. As well as small-molecule agents, three antibody drugs are now available to manage such conditions. They have the ability to modulate specific components of the immune response and prolong graft survival. Additional benefits have been demonstrated when some of these antibodies were added to standard immunosuppressive regimens.

Orthoclone OKT®3 is a murine monoclonal anti-CD3 antibody and the first FDA-approved monoclonal antibody reagent. CD3 is a membrane protein associated with the human T-cell receptor. It is closely involved in the signal transduction process of antigen recognition by T-lymphocytes. Orthoclone reverses graft rejection by binding CD3 and subsequently interfering with T-lymphocyte function. Daily dosing of Orthoclone in patients who have undergone cadaveric renal transplants was shown to result in a higher rate of rejection reversal than high-dose corticosteroids (94% vs 75%; $p=0.009$) ***(112)***. The 1-yr graft survival rate for the Orthoclone group was also higher than the corticosteroid group (62% vs 45%; $p=0.04$), although the 1- and 2-yr patient survival rates were not significantly different. Anecdotal evidence also suggests that Orthoclone is effective in reversing cardiac and hepatic transplant rejection in patients who do not respond to corticosteroid treatment or where corticosteroids are contraindicated ***(113)***. Whereas CD3 plays a critical role in T-cell receptor function, interleukin-2 (IL-2), a soluble cytokine produced by T-helper and other cells of the immune system, is essential in the regulation of the cellular and humoral immune response. It is a key mediator in the proliferation, development, and function of T- and B-lymphocytes and natural-killer cells. **Simulect®** (a chimeric antibody) and **Zenapax®** (a humanized antibody) are immunosuppressive antibodies that bind the high-affinity interleukin-2 receptors (CD25) on activated T-lymphocytes. They block the binding of IL-2 to its receptor, thereby suppressing the cellular immune response against the newly transplanted tissue. The concurrent use of Simulect or Zenapax with standard immunosuppressants (ciclosporin, corticosteroids, and/or azathioprine) as prophylaxis in renal transplant patients was shown to reduce the incidence of acute graft rejection episodes for up to 12 mo posttransplant (see **refs. *114*** and ***115*** for results from the various trials). In particular, the use of Zenapax in conjunction with ciclosporin and corticosteroids resulted in better patient survival for up to 3 yr ***(115)***.

3.4. Viral Infection

Respiratory syncytial virus (RSV) infection of the lower respiratory tract is a serious disease that affects the newborn and young children. It is highly contagious and reported to be a leading cause for infant hospitalization in the US ***(116)***. Those who are born prematurely, immunocompromised, and with chronic lung disease and congenital heart disease are particularly at risk ***(117)***. **Synagis®** is a humanized anti-RSV F-protein antibody indicated for the prevention of RSV infections in high-risk infants. Blockade of the F-protein hampers the ability of the virus to fuse with and infect pulmonary epithelial cells and was shown to reduce RSV concentration in tracheal isolates of RSV-infected infants ***(118)***. Results from a multicenter study showed that infants with premature birth or bronchopulmonary dysplasia had a relative reduction of 55% in RSV hospitalizations when they were given five-monthly doses of Synagis compared to placebo ***(119)***. A similar study with infants who have hemodynamically significant congenital heart disease showed a relative reduction of 45% with Synagis compared to placebo ***(120)***. These exciting results illustrate the importance of passive immunization with anti-RSV antibodies in reducing morbidity and mortality in high-risk infants.

3.5. Cardiology

Anticoagulant drugs such as heparin, aspirin, and warfarin are routinely used to prevent the formation of blood clots implicated in patients at risk of myocardial infarction or stroke. **ReoPro®**, a chimeric anti-integrin Fab antibody fragment, has been developed as an adjunct to heparin and aspirin for percutaneous coronary interventions, such as balloon angioplasty, atherectomy and stent placement, and unstable angina ***(121)***. It inhibits platelet aggregation by blocking the glycoprotein IIb/IIIa receptors on blood platelets and, hence, the binding of fibrinogen and other adhesive molecules. The use of ReoPro with other anticoagulants was clinically evaluated in a series of trials and has demonstrated consistent benefits compared to placebo in a variety of end points such as death, myocardial infarction, and urgent intervention for recurrent ischemia (see **ref. *122*** for a detailed summary of clinical data). The duration of clinical benefits lasted from 30 d up to 3 yr, making ReoPro a valuable adjunct to other anticoagulant drugs to maintain vascularization and improve the quality of life in these patients.

Acknowledgments

The author thanks Sir Gregory Winter for suggestions and comments on the manuscript, Dr. James Love for assistance with Fig. 1 and the Croucher Foundation, Hong Kong for the award of a personal research fellowship.

References

1. Leder, P. (1982) The genetics of antibody diversity. *Sci. Am.* **246,** 102–115.
2. Honjo, T. (1983) Immunoglobulin genes. *Annu. Rev. Immunol.* **1,** 499–528.
3. Lansford, R., Okada, A., Chen, J., et al. (1996) Mechanism and control of immunoglobulin gene rearrangement, in *Molecular Immunology*, (Hames, B. D. and Glover, D. M., eds.), 2nd ed. IRL, Oxford.
4. Papavasiliou, F. N. and Schatz, D. G. (2002) Somatic hypermutation of immunoglobulin genes: merging mechanisms for genetic diversity. *Cell* **109,** S35–S44.
5. Diaz, M. and Casali, P. (2002) Somatic immunoglobulin hypermutation. *Curr. Opin. Immunol.* **14,** 235–240.
6. Stavnezer, J. (1996) Immunoglobulin class switching. *Curr. Opin. Immunol.* **8,** 199–205.
7. Burton, D. R. and Woof, J. M. (1992) Human antibody effector functions. *Adv. Immunol.* **51,** 1–84.
8. Alzari, P. M., Lascombe, M.-B., and Poljak, R. J. (1988) Three-dimensional structure of antibodies. *Annu. Rev. Immunol.* **6,** 555–580.
9. Padlan, E. A. (1994) Anatomy of the antibody molecule. *Mol. Immunol.* **31,** 169–217.
10. Köhler, G. and Milstein, C. (1975) Continuous cultures of fused cells secreting antibody of predefined specificity. *Nature* **256,** 495–497.
11. Harlow, E. and Lane, D. (1988) *Antibodies—a Laboratory Manual*. Cold Spring Harbor Laboratory Press, Cold Spring Harbor, NY.
12. De Prada, P. and Landry, D. W. (2004) Production and characterization of anti-cocaine catalytic antibodies. *Methods Mol. Biol.* **248,** 495–501.
13. Schroff, R. W., Foon, K. A., Beatty, S. M., Oldham, R. K., and Morgan, A. C., Jr. (1985) Human antimurine immunoglobulin responses in patients receiving monoclonal antibody therapy. *Cancer Res.* **45,** 879–885.
14. Shawler, D. L., Bartholomew, R. M., Smith, L. M., and Dillman, R. O. (1985) Human immune response to multiple injections of murine monoclonal IgG. *J. Immunol.* **135,** 1530–1535.
15. Caroll, W. L., Thielemans, K., Dilley, J., and Levy, R. (1986) Mouse × human heterohybridomas as fusion partners with human B cell tumours. *J. Immunol. Methods* **89,** 61–72.
16. Borrebaeck, C. A. K. (1989) Strategy for the production of human monoclonal antibodies using in vitro activated B cells. *J. Immunol. Methods* **123,** 157–165.
17. Carson, D. A. and Freimark, B. D. (1986) Human lymphocyte hybridomas and monoclonal antibodies. *Adv. Immunol.* **38,** 275–311.
18. Cole, S. P., Campling, B. G., Atlaw, T., Kozbor, D., and Roder, J. C. (1984) Human monoclonal antibodies. *Mol. Cell Biochem.* **62,** 109–120.
19. Niedbala, W. G. and Stott, D. I. (1998) A comparison of three methods for production of human hybridomas secreting autoantibodies. *Hybridoma* **17,** 299–304.
20. Neuberger, M. S. (1983) Expression and regulation of immunoglobulin heavy chain gene transfected into lymphoid cells. *EMBO J.* **2,** 1373–1378.
21. Oi, V. T., Morrison, S. L., Herzenberg, L. A., and Berg, P. (1983) Immunoglobulin gene expression in transformed lymphoid cells. *Proc. Natl. Acad. Sci. USA* **80,** 825–829.
22. Larrick, J. W., Danielsson, L., Brenner, C. A., Abrahamson, M., Fry, K. E., and Borrebaeck, C. A. (1989) Rapid cloning of rearranged immunoglobulin genes from human hybridoma cells using mixed primers and the polymerase chain reaction. *Biochem. Biophys. Res. Commun.* **160,** 1250–1256.
23. Orlandi, R., Gussow, D. H., Jones, P. T., and Winter, G. (1989) Cloning immunoglobulin variable domains for expression by the polymerase chain reaction. *Biotechnology* **24,** 527–531.
24. Jones, S. T. and Bendig, M. M. (1991) Rapid PCR-cloning of full-length mouse immunoglobulin variable regions. *Biotechnology* **9,** 88–89.
25. Boulianne, G. L., Hozumi, N., and Shulman, M. J. (1984) Production of functional chimeric mouse/human antibodies. *Nature* **312,** 643–646.
26. Morrison, S. L., Johnson, M. J., Herzenberg, L. A., and Oi, V. T. (1984) Chimaeric human antibody molecules: mouse antigen binding domains. *Proc. Natl. Acad. Sci. USA* **82,** 6851–6855.
27. Neuberger, M. S., Williams, G. T., Mitchell, E. B., Jouhal, S. S., Flanagan, J. G., and Rabbitts, T. H. (1985) A hapten-specific chimaeric IgE antibody with human physiological effector function. *Nature* **314,** 268–270.
28. Dyer, M. J. S., Hale, G., Hayhoe, F. G. J., and Waldmann, H. (1989) Effects of CAMPATH-1 antibodies in vivo in patients with lymphoid malignancies: influence of antibody isotype. *Blood* **73,** 1431–1439.

29. Khazaeli, M. B., Conry, R. M., and LoBuglio, A. F. (1994) Human immune response to monoclonal antibodies. *J. Immunother.* **15,** 42–52.
30. Jones, P. T., Dear, P. H., Foote, J., Neuberger, M. S., and Winter, G. (1986) Replacing the complementarity-determining regions in a human antibody with those from a mouse. *Nature* **321,** 522–525.
31. Verhoeyen, M., Milstein, C., and Winter, G. (1988) Reshaping human antibodies: grafting an antilysozyme activity. *Science* **239,** 1534–1536.
32. Riechmann, L., Clark, M., Waldmann, H., and Winter, G. (1988) Reshaping human antibodies for therapy. *Nature* **332,** 323–327.
33. Kalofonos, H. P., Kosmas, C., Hird, V., Snook, D. E., and Epenetos, A. A. (1994) Targeting of tumours with murine and reshaped human monoclonal antibodies against placental alkaline phosphatase: immunolocalisation, pharmacokinetics and immune response. *Eur. J. Cancer* **30A,** 1842–1850.
34. Sharkey, R. M., Juweid, M., Shevitz, J., et al. (1995) Evaluation of a complementarity-determining region-grafted (humanized) anti-carcinoembryonic antigen monoclonal antibody in preclinical and clinical studies. *Cancer Res.* **55,** 5935s–5945s.
35. Rebello, P. R., Hale, G., Friend, P. J., Cobbold, S. P., and Waldmann, H. (1999) Anti-globulin responses to rat and humanized CAMPATH-1 monoclonal antibody used to treat transplant rejection. *Transplantation* **68,** 1417–1420.
36. Woodle, E. S., Xu, D., Zivin, R. A., Auger, J., Charette, J., and O'Laughlin, R. (1999) Phase I trial of a humanized, Fc receptor nonbinding OKT3 antibody huOKT3gamma1(Ala-Ala) in the treatment of acute renal allograft rejection. *Transplantation* **68,** 608–616.
37. Padlan, E. A. (1991) A possible procedure for reducing the immunogenicity of antibody variable domains while preserving their ligand-binding properties. *Mol. Immunol.* **28,** 489–498.
38. Roguska, M. A., Pedersen, J. T., Keddy, C. A., et al. (1994) Humanization of murine monoclonal antibodies through variable domain resurfacing. *Proc. Natl. Acad. Sci. USA* **91,** 969–973.
39. Jespers, L. S., Roberts, A., Mahler, S. M., Winter, G., and Hoogenboom, H. R. (1994) Guiding the selection of human antibodies from phage display repertoires to a single epitope of an antigen. *Biotechnology* **12,** 899–903.
40. Lo, B. K. C. (2004) Antibody humanization by CDR grafting. *Methods Mol. Biol.* **248,** 135–159.
41. Smith, G. P. (1985) Filamentous fusion phage: novel expression vectors that display cloned antigens on the virion surface. *Science* **228,** 1315–1317.
42. Better, M., Chang, C. P., Robinson, R. R., and Horwitz, A. H. (1988) *Escherichia coli* secretion of an active chimeric antibody fragment. *Science* **240,** 1041–1043.
43. Skerra, A. and Pluckthun, A. (1988) Assembly of a functional immunoglobulin Fv fragment in *Escherichia coli*. *Science* **240,** 1038–1041.
44. Bird, R. E., Hardman, K. D., Jacobson, J. W., et al. (1988) Single chain antigen binding proteins. *Science* **242,** 423–426.
45. Huston, J. S., Levinson, D., Mudgett-Hunter, M., Tai, M., Novotny, J., and Margolies, M. N. (1988) Protein engineering of antibody binding sites: recovery of specific activity in an anti-digoxin single chain Fv analogue produced in *Escherichia coli*. *Proc. Natl. Acad. Sci. USA* **85,** 5879–5883.
46. McCafferty, J., Griffiths, A. D., Winter, G., and Chiswell, D. (1990) Phage antibodies: filamentous phage displaying antibody variable domains. *Nature* **348,** 552–554.
47. Clackson, T., Hoogenboom, H. R., Griffiths, A. D., and Winter, G. (1991) Making antibody fragments using phage display libraries. *Nature* **352,** 624–628.
48. Barbas, C. F. III, Kang, A. S., Lerner, R A., and Benkovic, S. J. (1991) Assembly of combinatorial antibody libraries on phage surfaces: the gene III site. *Proc. Natl. Acad. Sci. USA* **88,** 7978–7982.
49. Burton, D. R., Barbas, C. F., III, Persson, M. A., Koenig, S., Chanock, R. M., and Lerner, R. A. (1991) A large array of human monoclonal antibodies to type 1 human immunodeficiency virus from combinatorial libraries of asymptomatic seropositive individuals. *Proc. Natl. Acad. Sci. USA* **88,** 10,134–10,137.
50. Cai, X. and Garen, A. (1995) Anti-melanoma antibodies from melanoma patients immunized with genetically modified autologous tumor cells: selection of specific antibodies from single-chain Fv fusion phage libraries. *Proc. Natl. Acad. Sci. USA* **92,** 6537–6541.
51. Marks, J. D., Hoogenboom, H. R., Bonnert, T. P., McCafferty, J., Griffiths, A. D., and Winter, G. (1991) By-passing immunization. Human antibodies from V-gene libraries displayed on phage. *J. Mol. Biol.* **222,** 581–597.
52. Barbas, C. F., III, Bain, J. D., Hoekstra, D. M., and Lerner, R. A. (1992) Semisynthetic combinatorial antibody libraries: a chemical solution to the diversity problem. *Proc. Natl. Acad. Sci. USA* **89,** 4457–4461.

53. Hoogenboom, H. R. and Winter, G. (1992) By-passing immunisation. Human antibodies from synthetic repertoires of germline V_H gene segments rearranged in vitro. *J. Mol. Biol.* **227,** 381–388.
54. Hoogenboom, H. R. (1997) Designing and optimizing library selection strategies for generating high-affinity antibodies. *Trends Biotechnol.* **15,** 62–70.
55. Griffiths, A. D. and Duncan, A. R. (1998) Strategies for selection of antibodies by phage display. *Curr. Opin. Biotechnol.* **9,** 102–108.
56. Glennie, M. J. and Johnson, P. W. M. (2000) Clinical trials of antibody therapy. *Immunol. Today* **21,** 403–410.
57. Reichert, J. M. (2002) Therapeutic monoclonal antibodies: trends in development and approval in the US. *Curr. Opin. Mol. Ther.* **4,** 110–116.
58. Hanes, J. and Plückthun, A. (1997) In vitro selection and evolution of functional proteins by using ribosome display. *Proc. Natl. Acad. Sci. USA* **94,** 4937–4942.
59. He, M. and Taussig, M. J. (1997) Antibody-ribosome-mRNA (ARM) complexes as efficient selection particles for in vitro display and evolution of antibody combining sites. *Nucl. Acids Res.* **25,** 5132–5134.
60. Feldhaus, M. J., Siegel, R. W., Opresko, L. K., et al. (2003) Flow-cytometric isolation of human antibodies from a nonimmune Saccharomyces cerevisiae surface display library. *Nature Biotechnol.* **21,** 163–170.
61. Marks, J. D. and Bradbury, A. (2004) PCR cloning of human immunoglobulin genes. *Methods Mol. Biol.* **248,** 117–134.
62. Marks, J. D. and Bradbury, A. (2004) Selection of human antibodies from phage display libraries. *Methods Mol. Biol.* **248,** 161–176.
63. Neuberger, M. S. and Gruggermann, M. (1997) Monoclonal antibodies. Mice perform a human repertoire. *Nature* **386,** 25–26.
64. Tomizuka, K., Shinohara, T., Yoshida, H., et al. (2000) Double transchromosomic mice: maintenance of two individual human chromosome fragments containing immunoglobulin heavy and kappa loci and expression of fully human antibodies. *Proc. Natl. Acad. Sci. USA* **97,** 722–727.
65. Green, L. L. (1999) Antibody engineering via genetic engineering of the mouse: XenoMouse strains are a vehicle for the facile generation of therapeutic human monoclonal antibodies. *J. Immunol. Methods* **231,** 11–23.
66. Fishwild, D. M., O'Donnell, S. L., Bengoechea, T., et al. (1996) High-avidity human IgG kappa monoclonal antibodies from a novel strain of minilocus transgenic mice. *Nature Biotechnol.* **14,** 845–851.
67. Ishida, I., Tomizuka, K., Yoshida, H., et al. (2002) Production of human monoclonal and polyclonal antibodies in TransChromo animals. *Cloning Stem Cells* **4,** 91–102.
68. Davis, C. G., Gallo, M. L., and Corvalan, J. R. (1999) Transgenic mice as a source of fully human antibodies for the treatment of cancer. *Cancer Metastasis Rev.* **18,** 421–425.
69. Kuroiwa, Y., Kasinathan, P., Choi, Y. J., et al. (2002) Cloned transchromosomic calves producing human immunoglobulin. *Nature Biotechnol.* **20,** 889–894.
70. Davis, C. G., Jia, X.-C., Feng, X., and Haak-Frendscho, M. (2004) Production of human antibodies from transgenic mice. *Methods Mol. Biol.* **248,** 191–200.
71. Von Mehren, M., Adams, G. P., and Weiner, L. M. (2003) Monoclonal antibody therapy for cancer. *Annu. Rev. Med.* **54,** 343–369.
72. Glennie, M. J. and Van de Winkel, J. G. J. (2003) Renaissance of cancer therapeutic antibodies. *Drug Discov. Today* **8,** 503–510.
73. Tedder, T. F., Boyd, A. W., Freedman, A. S., Nadler, L. M., and Schlossman, S. F. (1985) The B cell surface molecule B1 is functionally linked with B cell activation and differentiation. *J. Immunol.* **135,** 973–979.
74. Anderson, K. C., Bates, M. P., Slaughenhoupt, B. L., Pinkus, G. S., Schlossman, S. F., and Nadler, L. M. (1984) Expression of human B cell-associated antigens on leukemias and lymphomas: a model of human B cell differentiation. *Blood* **63,** 1424–1433.
75. Reff, M. E., Carner, K., Chambers, K. S., et al. (1994) Depletion of B cells in vivo by a chimeric mouse human monoclonal antibody to CD20. *Blood* **83,** 435–445.
76. Demidem, A., Lam, T., Alas, S., Hariharan, K., Hanna, N., and Bonavida, B. (1997) Chimeric anti-CD20 (IDEC-C2B8) monoclonal antibody sensitizes a B cell lymphoma cell line to cell killing by cytotoxic drugs. *Cancer Biother. Radiopharm.* **12,** 177–186.
77. McLaughlin, P., Grillo-Lopez, A. J., Link, B. K., et al. (1998) Rituximab chimeric anti-CD20 monoclonal antibody therapy for relapsed indolent lymphoma: half of patients respond to a four-dose treatment program. *J. Clin. Oncol.* **16,** 2825–2833.

78. Piro, L. D., White, C. A., Grillo-Lopez, A. J., et al. (1999) Extended Rituximab (anti-CD20 monoclonal antibody) therapy for relapsed or refractory low-grade or follicular non-Hodgkin's lymphoma. *Ann. Oncol.* **10,** 655–661.
79. Witzig, T. E., Flinn, I. W., Gordon, L. I., Emmanouilides, C., Czuczman, M. S., and Saleh, M. N. (2002) Treatment with ibritumomab tiuxetan radioimmunotherapy in patients with rituximab-refractory follicular non-Hodgkin's lymphoma. *J. Clin. Oncol.* **20,** 3262–3269.
80. Bexxar® prescribing information, GlaxoSmithKline and Corixa Corporation, http://www.bexxar.com/.
81. Campath® prescribing information, ILEX Pharmaceuticals, http://www.campath.com/.
82. Xia, M. Q., Hale, G., Lifely, M. R., et al. (1993) Structure of the CAMPATH-1 antigen, a glycosylphosphatidylinositol-anchored glycoprotein which is an exceptionally good target for complement lysis. *Biochem. J.* **293,** 633–640.
83. Xia, M. Q., Hale, G., and Waldmann, H. (1993) Efficient complement-mediated lysis of cells containing the CAMPATH-1 (CDw52) antigen. *Mol. Immunol.* **30,** 1089–1096.
84. Tzakis, A. G., Kato, T., Nishida, S., et al. (2003) Alemtuzumab (Campath-1H) combined with tacrolimus in intestinal and multivisceral transplantation. *Transplantation* **75,** 1512–1517.
85. Chakrabarti, S., MacDonald, D., Hale, G., et al. (2003) T-cell depletion with Campath-1H "in the bag" for matched related allogeneic peripheral blood stem cell transplantation is associated with reduced graft-versus-host disease, rapid immune constitution and improved survival. *Br. J. Haematol.* **121,** 109–118.
86. Mylotarg® prescribing information, Wyeth Pharmaceuticals, http://www.mylotarg.com/.
87. Alvarado, Y., Tsimberidou, A., Kantarjian, H., et al. (2003) Pilot study of Mylotarg, idarubicin and cytarabine combination regimen in patients with primary resistant or relapsed acute myeloid leukemia. *Cancer Chemother. Pharmacol.* **51,** 87–90.
88. Slamon, D. J., Godolphin, W., Jones, L. A., et al. (1989) Studies of the HER-2/neu proto-oncogene in human breast and ovarian cancer. *Science* **244,** 707–712.
89. Sjogren, S., Inganas, M., Lindgren, A., Holmberg, L., and Bergh, J. (1998) Prognostic and predictive value of c-erbB-2 overexpression in primary breast cancer, alone and in combination with other prognostic markers. *J. Clin. Oncol.* **16,** 462–469.
90. Sundaresan, S., Penuel, E., and Sliwkowski, M. X. (1999) The biology of human epidermal growth factor receptor 2. *Curr. Oncol. Rep.* **1,** 16–22.
91. Hynes, N. E. and Stern, D. F. (1994) The biology of erbB-2/neu/HER-2 and its role in cancer. *Biochim. Biophys. Acta* **1198,** 165–184.
92. Sliwkowski, M. X., Lofgren, J. A., Lewis, G. D., Hotaling, T. E., Fendly, B. M., and Fox, J. A. (1999) Nonclinical studies addressing the mechanism of action of trastuzumab (Herceptin). *Semin. Oncol.* **26(4 Suppl 12),** 60–70.
93. Lewis, G. D., Figari, I., Fendly, B., et al. (1993) Differential responses of human tumor cell lines to anti-p185HER2 monoclonal antibodies. *Cancer Immunol. Immunother.* **37,** 255–263.
94. Herceptin® prescribing information, Genentech, Inc., http://www.herceptin.com/.
95. Pietras, R. J., Fendly, B. M., Chazin, V. R., Pegram, M. D., Howell, S. B., and Slamon, D. J. (1994) Antibody to HER-2/neu receptor blocks DNA repair after cisplatin in human breast and ovarian cancer cells. *Oncogene* **9,** 1829–1838.
96. Tan, A. R. and Swain, S. M. (2003) Ongoing adjuvant trials with trastuzumab in breast cancer. *Semin. Oncol.* **30(5 Suppl. 16),** 54–64.
97. Genentech, Inc. website, http://www.gene.com/gene/pipeline/trials/.
98. Agus, D. B., Akita, R. W., Fox, W. D., et al. (2002) Targeting ligand-activated ErbB2 signaling inhibits breast and prostate tumor growth. *Cancer Cell* **2,** 127–137.
99. Cragg, M. S., French, R. R., and Glennie, M. J. (1999) Signaling antibodies in cancer therapy. *Curr. Opin. Immunol.* **11,** 541–547.
100. Remicade® prescribing information, Centocor, Inc., http://www.remicade.com/.
101. Targan, S. R., Hanauer, S. B., van Deventer, S. J., et al. (1997) A short-term study of chimeric monoclonal antibody cA2 to tumor necrosis factor alpha for Crohn's disease. Crohn's Disease cA2 Study Group. *N. Engl. J. Med.* **337,** 1029–1035.
102. Hanauer, S. B., Feagan, B. G., Lichtenstein, G. R., et al. (2002) Maintenance infliximab for Crohn's disease: the ACCENT I randomised trial. *Lancet* **359,** 1541–1549.
103. Present, D. H., Rutgeerts, P., Targan, S., et al. (1999) Infliximab for the treatment of fistulas in patients with Crohn's disease. *N. Engl. J. Med.* **340,** 1398–1405.
104. Humira® prescribing information, Abbott Laboratories, http://www.humira.com/.

105. Xolair® prescribing information, Genentech, Inc. and Novartis Pharmaceuticals, http://www.xolair.com/.
106. Storms, W. (2002) Allergens in the pathogenesis of asthma: potential role of anti-immunoglobulin E therapy. *Am. J. Respir. Med.* **1,** 361–368.
107. MacGlashan, D. W. Jr, Bochner, B. S., Adelman, D. C., et al. (1997) Down-regulation of Fc(epsilon)RI expression on human basophils during in vivo treatment of atopic patients with anti-IgE antibody. *J. Immunol.* **158,** 1438-1445.
108. Prussin, C., Griffith, D. T., Boesel, K. M., Lin, H., Foster, B., and Casale, T. B. (2003) Omalizumab treatment downregulates dendritic cell FcepsilonRI expression. *J. Allergy Clin. Immunol.* **112,** 1147–1154.
109. Raptiva® prescribing information, Genentech, Inc. and Xoma, Inc., http://www.raptiva.com/.
110. Gordon, K. B., Papp, K. A., Hamilton, T. K., et al. (2003) Efalizumab for patients with moderate to severe plaque psoriasis: a randomized controlled trial. *JAMA* **290,** 3073–3080.
111. Weinberg, J. M. (2003) An overview of infliximab, etanercept, efalizumab, and alefacept as biologic therapy for psoriasis. *Clin. Ther.* **25,** 2487–2505.
112. Ortho Multi-center Transplant Study Group (1985) A randomized clinical trial of OKT3 monoclonal antibody for acute rejection of cadaveric renal transplants. *N. Engl. J. Med.* **313,** 337–342.
113. Orthoclone® prescribing information, Ortho Biotech Products, http://healthcareprofessionals.orthobiotech.com/products/orthoclone.jsp/.
114. Simulect® prescribing information, Novartis Pharmaceuticals, http://www.novartis-transplant.com/medpro/product_info/simulect.html/.
115. Zenapax® prescribing information, Roche Pharmaceuticals, http://www.rocheusa.com/products/zenapax/.
116. Leader, S. and Kohlhase, K. (2003) Recent trends in severe respiratory syncytial virus (RSV) among US infants, 1997 to 2000. *J. Pediatr.* **143(5 Suppl.), S127–132.**
117. Synagis® prescribing information, MedImmune, Inc., http://www.synagis.com/.
118. Malley, R., DeVincenzo, J., Ramilo, O., et al. (1998) Reduction of respiratory syncytial virus (RSV) in tracheal aspirates in intubated infants by use of humanized monoclonal antibody to RSV F protein. *J. Infect. Dis.* **178,** 1555–1561.
119. The IMpact-RSV Study Group (1998) Palivizumab, a humanized RSV monoclonal antibody, reduces hospitalization from RSV infection in high-risk infants. *Pediatrics* **102,** 531–537.
120. Feltes, T. F., Cabalka, A. K., Meissner, H. C., et al. (2003) Palivizumab prophylaxis reduces hospitalization due to respiratory syncytial virus in young children with hemodynamically significant congenital heart disease. *J. Pediatr.* **143,** 532–540.
121. ReoPro® prescribing information, Centocor, Inc., http://www.centocor.com/.
122. Ibbotson, T., McGavin, J. K., and Goa, K. L. (2003) Spotlight on abciximab in patients with ischemic heart disease undergoing percutaneous coronary revascularization. *Am. J. Cardiovasc. Drugs* **3,** 381–386.

32

Karyotyping

Avery A. Sandberg, John F. Stone, and Zhong Chen

1. Introduction

For historical background regarding karyotyping, chromosome nomenclature, and further details for some of the methodologies presented in this chapter, the reader might wish to consult a number of publications ***(1–5)***. Karyotyping is a process by which individual chromosomes are identified by a number of methodologies and then arranged accordingly into the 22 pairs of autosomes and the two sex chromosomes (X and Y) characteristic of the human genome (*see* **Fig. 1**) ***(3,4,6,7)***.

Establishment of a detailed karyotype requires metaphase cells and, hence, dividing cells in sufficient number. In some cells and tissues (e.g., bone marrow and some tumors), the mitotic rate is sufficiently high as not to require culturing for karyotypic (chromosomal, cytogenetic) preparations and analyses ***(8)***. Generally, culturing has to be performed to accomplish reliable karyotyping in materials with a low mitotic index, including blood cells (lymphocytes), most tumors, and skin (fibroblasts). The conditions of such culturing might differ from preparation to preparation and some information on this area is supplied within the sections of this chapter.

In addition to cell or tissue culture, for karyotyping to be optimal requires arrest of dividing cells in metaphase and this is usually accomplished with Colcemid (or colchicine), followed by the swelling of the cells after exposure to hypotonic solutions for spreading of the chromosomes to allow the recognition of individual chromosomes. The development of all three of the above prerequisites for optimal karyotyping (i.e., culture, metaphase arrest, and hypotonic treatment) ultimately led to the establishment of the correct number of chromosomes in the human as 46 by Tjio and Levan in 1956 ***(9)***. The recognition of individual stained chromosomes for the next 15 yr relied essentially on their size, location of the centromere, and the presence of satellites on some chromosomes. These criteria did not always reliably differentiate chromosomes, particularly those of nearly identical size and centromere location and were particularly inadequate in the identification of some translocations and marker (abnormal) chromosomes. Nevertheless, even with these shortcomings, remarkable advances in human cytogenetics were made during the nearly decade and a half following the description of the correct number of chromosomes in the human.

A major advance in karyotyping was the introduction of banding of chromosomes in 1970 ***(10,11)***. This approach afforded not only the accurate recognition of individual chromosomes based on their unique banding patterns but also of minor structural changes and the breakpoints within chromosomes, especially those involved in translocations. Banding also made possible the deciphering of the origin of some marker chromosomes, although often the origin of such markers remained unknown ***(12)***.

From: *Medical Biomethods Handbook*
Edited by: J. M. Walker and R. Rapley © Humana Press, Inc., Totowa, NJ

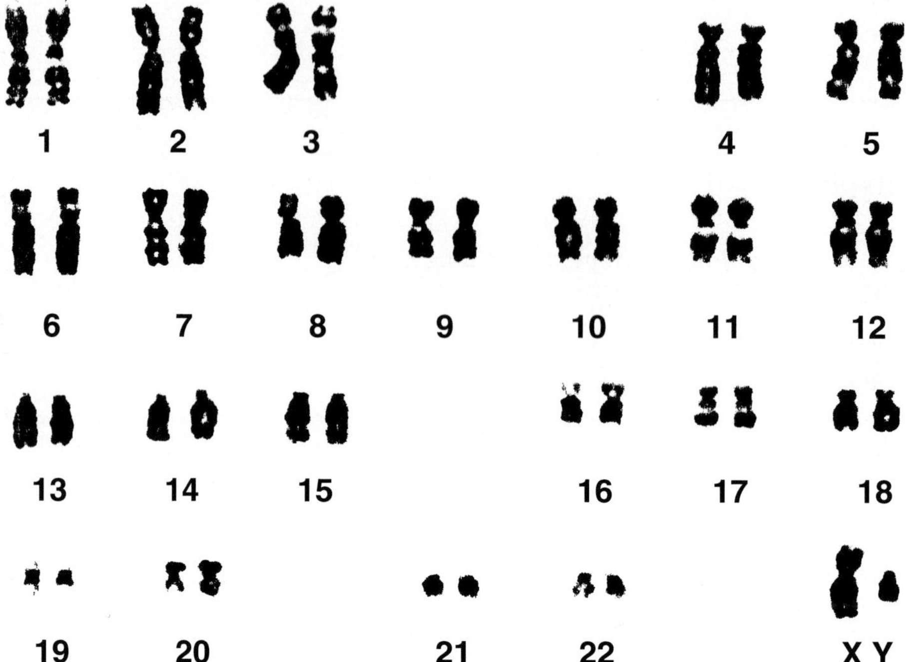

Fig. 1. G-banded normal male karyotype with 46 chromosomes (22 pairs of autosomes and X and Y sex chromosomes). Each chromosome is characterized by a unique banding pattern affording its ready recognition. Deviations from the karyotype shown in this figure, be they numerical or structural, indicate genetic abnormalities, usually of a serious nature. For more details regarding the human karyotype and its changes in various diseases and states, the reader should consult sources listed in **refs.** ***1–7***.

The development of fluorescence *in situ* hybridization (FISH) techniques was the next important advance in karyotyping ***(13)***, particularly in the recognition of subtle changes not evident by conventional cytogenetic methods, the origin of marker chromosomes, and masked structural anomalies such as translocations, inversions, and insertions. A major advantage of FISH was its usefulness in establishing some changes in interphase nuclei in archival tissues and preparations. A refinement of FISH is its application in spectral karyotyping (SKY) ***(14)*** and multifluor or multiple-color FISH (M-FISH) ***(15)*** techniques, which are especially useful in the recognition of karyotypic changes in the total chromosome set.

Other methodologies (i.e., comparative genomic hybridization [CGH] and microarrays) not related to karyotyping will be taken up in this chapter because they can be utilized for the establishment of genetic changes not recognized by the above-mentioned techniques.

2. Cytogenetic Analysis of Constitutional Disorders

Any cells or tissues that yield dividing cells in tissue culture can be utilized to establish the constitutional chromosome complement of an individual. The common sources of cells or tissues are peripheral blood (PB), skin biopsy, neonatal bone marrow, amniotic fluid, chorionic villi, and products of conception. With the exception of bone marrow (BM), each needs to be cultured for varying periods of time to obtain sufficient metaphases for analysis.

2.1. Neonatal Bone Marrow

Neonatal BM analysis is utilized in severely distressed infants with multiple congenital anomalies in order to obtain a rapid assay of aneuploidy and easily discernable structural chromosomal rearrangements; such observations can be made within 24 h. BM aspirate is placed into culture medium in the presence of Colcemid for a period of 30–90 min, after which it is exposed to a hypotonic solution and fixative as described below. The resulting metaphases are generally short and of low band resolution, hence a phytohemagglutinin (PHA)-stimulated PB sample is usually processed simultaneously (*see* **Subheading 2.2.**).

2.2. PHA-stimulated Peripheral Blood

Phytohemagglutinin-stimulated PB cultures provide the most reliable and highest-resolution cytogenetic preparations of any of the available tissues. It is utilized for the diagnosis of chromosomal abnormalities in newborns with anomalies, couples with recurrent miscarriages, infertility, and parents of fetuses with structural chromosome abnormalities as ascertained in prenatal diagnosis. In addition, high-resolution banding can distinguish subtle chromosomal rearrangements in individuals with mild malformations, certain seizure disorders, and mental retardation. These analyses have been greatly aided by FISH (*see* **Subheading 5.**) using probes specific to loci known to be associated with a disorder in question. Additionally, probes for subtelomeric sequences of individual chromosome arms have proved useful in demonstrating subtle deletions and translocations associated with some disorders.

Peripheral blood is collected in sodium heparin tubes, and 0.25–0.5 mL is inoculated into 5 mL of RPMI medium containing 15–20% fetal bovine serum (FBS) and PHA. PHA stimulation results in a parasynchronous wave of cell division at 72 h and again at 96 h, after which mitotic activity rapidly declines ***(16,17)***. Consequently, cells are harvested for analysis on the third or fourth day after culture inoculation. Cultures from newborns can be harvested at 48 h to obtain a preliminary result, but resolution is generally low, and a follow-up 72-h culture is usually needed to complete the analysis. Cells are exposed to Colcemid for 30–60 min, followed by hypotonic treatment and fixation as described below. At least 20 metaphases are analyzed from which at least four photographic or digital images are obtained, and two of these are karyotyped. In cases where mosaicism (the presence of two or more cell lines) is suspected, 50 or 100 metaphases might have to be analyzed.

Whereas the above technique generally yields metaphases of 450–550 bands per haploid complement, high-resolution techniques can yield 650–750 or more bands ***(18)***. High-resolution banding is achieved by arresting cells in late S-phase with methotrexate. Because the PHA stimulation produces parasynchronous entry into the cell cycle by T-lymphocytes, a fair number of cells can be delayed in late S-phase by treatment with methotrexate beginning on the afternoon of the second day after inoculation. Seventeen to 18 h later, on the following morning, the methotrexate block is counteracted by the addition of thymidine. Four to 5 h later, Colcemid is added, usually at reduced concentration, for 10–15 min. The shortened Colcemid time and concentration produce less chromosome contraction, resulting in more bands being visible (higher resolution). DNA intercalating agents, such as ethidium bromide or actinomycin D, can be added 1 h before harvest to further lengthen the chromosomes.

2.2.1. Procedure for Cytogenetic Analysis of Peripheral Blood Lymphocytes

For a more extensive survey of cytogenetic protocols, see **ref. *19***. Thus, detailed information on the various reagents, working solutions, specimen handling, procedures for cell culture and harvesting (including a procedure for high-resolution chromosome study), chromosome slide preparation, and staining can also be found in **ref. *19***. Here, only some details regarding key steps in these procedures used in our laboratory will be mentioned. We use RPMI 1640 medium, PHA-M (Difco) to stimulate lymphocyte growth, a working 2.5×10^{-5} *M* solution of amethopterin (methotrexate) to delay cells in late S-phase, a working ethidium bromide solution (2 mg/mL) to enhance lengthening of the chromosomes, and a 0.075 *M* solution of potassium chloride for swelling of cells.

2.3. Prenatal Diagnosis

Prenatal diagnosis of cytogenetic disorders can be achieved through the culture of cells from the amniotic fluid or the chorionic villi. For early results, amniotic fluid is generally obtained at 14–16 wk of gestation, but successful cultures can be established up to term. Optimally, 20 mL or more of amniotic fluid is aseptically obtained in two centrifuge tubes and is centrifuged and resuspended in approx 0.5 mL of culture medium. The contents of each tube is then distributed to two or three 35-mm tissue culture dishes containing a 22-mm^2 glass cover slip. After overnight incubation, the dishes are gently flooded with 1.5–2.0 mL of medium. After 5–7 d, visible colonies arising from the attachment of single cells usually yield sufficient and adequate metaphases for analysis. Colcemid is then added for 30–60 min or overnight at reduced concentration with 5-bromo-2'-deoxyuridine (BrdU), which inhibits chromosome condensation. The medium is gently removed and replaced with a hypotonic solution followed by fixation directly on the cover slip (*in situ*) ***(20,21)***. The resulting preparation has the metaphases within the colonies in which they arose, helping to distinguish in vitro artifact from true mosaicism; that is, a true fetal mosaic will have every metaphase in a given colony with an abnormality (as it was present when the sample was taken), whereas an artifact will have a single or a few cells with an abnormality (as it arose after culturing the cells).

2.3.1. Procedure for Cytogenetic Analysis of Amniotic Fluid

For detailed information regarding this procedure, see **ref. *19***. In our laboratory, we use MEMα medium with ribonucleosides and deoxyribonucleosides for culturing amniotic fluid cells. For same-day harvest, we use 1 drop of Colcemid (5 μg/mL) for cell arrest and 1 drop of 0.8 μg/mL Colcemid for next-day harvest.

2.4. Chorionic Villus Sampling

One drawback to the analysis of amniotic fluid cells for prenatal diagnosis is the requirement to wait until 14–16 wk of gestation to have a sufficient number of fetal cells for successful culturing. Because this is well into the second trimester of pregnancy, patients report considerable anxiety during the interval between amniocentesis and final reporting. An alternative approach is chorionic villus sampling (CVS), in which the chorionic villi, which are of fetal origin, are sampled at 8–10 wk gestation. At this stage, the cells of villi are actively dividing and have not yet fused with the amnion and, thus, are accessible for biopsy ***(22)***. Direct preparations (i.e., without tissue culture) can be made, but they are generally of low quality. Cultured CVS samples produce high-quality metaphases comparable to those from amniotic fluid cultures.

2.4.1. Procedure for Cytogenetic Analysis of CVS

The procedure for CVS analysis is essentially as that for amniotic fluid cells ***(19)***. If DNA studies are contemplated, check with the laboratory performing the test to determine if sodium heparin is appropriate to be included in the collection medium.

2.5. Products of Conception

Chromosome abnormalities occur in just under half of spontaneous abortions. Of these, 49% are trisomies for one of the autosomes, 19% are 45,X, 17% are triploid, 6% are tetraploid, 3% are structural abnormalities, 2% are mosaic trisomies, and the rest are of various other types ***(23)***.

Products of conception arrive in the laboratory in various conditions depending on the interval between demise of the fetus and the time of the actual miscarriage. In general, fetuses that have had abortion induced are more successfully grown than are those which have been aborted spontaneously, because of the shorter interval from demise to arrival in the laboratory. Success rates vary from approx 50% to 80% or more depending on the makeup of the specimens.

2.5.1. Procedure for Cytogenetic Analysis of Products of Conception

For details regarding this procedure, see **ref. *19***. In addition to penicillin–streptomycin (2%) used for amniotic fluid cells and CVS analysis, fungizone and gentamicin sulfate are added to the culture medium for products of conception (POCs).

3. Cytogenetic Analysis of Hematological Disorders

The establishment of karyotypes in hematological disorders, primarily the leukemias and lymphomas, must be based on the examination of the involved cells or tissues. Thus, in the case of the leukemias, BM aspirations yield optimal results in the preponderant number of patients, whereas in the lymphomas, affected tissues, usually lymph nodes, are the best source of cells carrying cytogenetic anomalies ***(24)***. In some situations, blood cells can be utilized as a source of metaphases affected by karyotypic changes, e.g., in cases with about 10% immature cells in the PB, in chronic lymphocytic leukemia, in cases where the marrow is fibrotic or extremely hypocellular, or in determining the presence of Ph+ cells in chronic myelocytic leukemia ***(1,25,26)***.

As mentioned above, a significant number of dividing cells, and hence metaphases, are usually present in the BM for cytogenetic analysis without having to resort to culture or lengthy incubation. However, in some leukemias, the number of dividing cells (especially the leukemic ones) is very low, and, hence, incubation of the marrow specimen for a number of days (2–5 d) might be necessary to generate a sufficient number of metaphases for cytogenetic study ***(27,28)***.

In cases where cytogenetic analysis reveals only abnormal metaphases, especially those with a balanced translocation, it might be necessary to rule out a constitutional anomaly. This is best established through the cytogenetic examination of PHA-stimulated lymphocytes of PB.

A number of mitogenic agents capable of stimulating lymphoid or, less frequently, myeloid cells have been introduced over the years. Outstanding among these has been PHA which is capable of stimulating the growth and division of lymphocytes of T-cell origin. However, PHA is not routinely added to BM or PB cultures in acute leukemias because PHA might interfere with the evaluation of spontaneously dividing malignant cells.

The quality of chromosome preparations has been significantly improved with some new techniques, such as the use of amethopterin for cell synchronization, the use of short exposures to mitosis-arresting agents, the use of DNA-binding agents to elongate chromosomes ***(29,30)***, improved staining procedures, and the use of conditioned culture medium containing hematopoietic growth factors—that is, GCT (giant cell tumor)-conditioned medium primarily for myeloid disorders ***(30,31)*** and PHA/IL-2 (interleukin-2) for both B- or T-cell lymphoid diseases ***(32–35)***.

The rate of successful cytogenetic analysis varies with the specific type of disease and is related to the adjustment of variables in each laboratory, such as serum concentration, medium pH, and cell concentration.

3.1. Specimens

1. Blood marrow aspirate or biopsy. One to 3 mL of BM should be aseptically aspirated into a sodium-heparinized syringe and transferred to a sterile sodium vacutainer tube. Do not freeze the specimen. Also, cell viability drops off sharply by 72 h after collection. If a marrow aspirate cannot be achieved, a BM biopsy can be used for cytogenetic analysis.
2. Peripheral blood cells. Five to 10 mL of PB should be aseptically collected and transferred, transported, and preserved in the same way as for a BM specimen.
3. Lymph node and spleen materials. Lymph node and spleen biopsies or samples should be collected aseptically and transferred to a sterile sodium vacutainer tube containing culture medium (RPMI 1640 + 5–10% FCS + 1% penicillin/streptomycin). The specimen should be transported and preserved in the same manner as for the BM specimen.

3.2. Enumeration of Cells

Establish white blood cell count (WBC) using a Coulter counter. Determine the amount of sample per culture. Optimum concentration is 10×10^6 cells/10 mL culture.

To get an estimate of hypocellularity, hypercellularity or average cellularity, an estimate can be obtained by using the thin part of the smear. A low, average, increased, or high WBC would be reflected as approx 40, 100, 200, and 400 white cells/microscopic field. Based on these estimates, the amount of sample to be added to each slide can be derived, as follows:

WBC (K)	Quick differential	Amount (mL)
<2	Low	1.0
4–10	Average	0.5
20	Increased	0.25
>40	High	0.1

3.3. Culture

Set up the specimen, according to the amount, with specimen volume adjusted so that the final concentration of white blood cells is approx 1×10^6 mL. All cultures can be set up in T-25 flasks. Cultures of BM or blood cells are set up for 24 h overnight, 48 h, and a backup culture. For lymphoid disorders 72-h PHA/IL-2 cultures can be set up. For all other details (reagents, biological, solutions, specimen handling, culture and harvest, slide preparation and chromosome staining) see **ref. *19***.

3.4. Analysis

1. In general, a minimum of 20 cells must be counted and analyzed for each case, such that a rearrangement affecting one band of any chromosome can be detected in any given cell.
2. When possible, select cells with at least 300-band resolution and few overlaps.
3. Determination must be made whether one or more clones exist. A clone is defined as two or more cells having the same rearrangements or additional chromosomes, or three cells with the same monosomy ***(36)***. However, one cell with a normal karyotype is considered adequate evidence to indicate the presence of a normal cell line.

3.5. Technical Troubleshooting

1. Clots can form in the specimen or culture (particularly of BM) because of inadequate heparin, breakdown of heparin, or abnormalities in the clotting mechanism of the patient. These clots can be broken up by aspiration through a needle or pipet and/or minced. Clots can trap cells and interfere with the harvesting procedure.
2. Cell clumps can form during harvest because of failure to resuspend pellet thoroughly prior to fixation, failure to mix cell suspension while adding fixative, or abundance of erythrocytes. The formation of these clumps can be prevented by resuspending the pellet thoroughly after hypotonic treatment. Add up to 1 mL of fixative dropwise while gently tapping the tube. For tubes containing a large proportion of erythrocytes, bring the volume of fixative to 10–15 mL immediately. It might be necessary to mix the cell suspension gently with a Pasteur pipet or to split the sample into two tubes, which can be combined after one to two changes of fixative.
3. Poor spreading of metaphase cells and the presence of cytoplasm often occur during preparation because of poor fixation and/or poor swelling of cells during hypotonic treatment. To overcome these problems, the following should be performed by repeating fixation three to four more times, increasing glacial acetic acid concentration up to 50%, dropping the cell suspension from a greater height, increasing the angle at which the slide is held, and increasing the humidity when slides are drying. An increase in humidity can result in poor fixation.
4. Poor G-bands can be a clue to the presence of cytoplasm, age of slides, or high humidity. The banding results can be improved by increasing the time in peroxide and/or trypsin, preparing fresh slides to improve the quality of metaphase spreads (slides older than 1 mo could yield

inconsistent results). The trypsin treatment might need to be lengthened, and placing slides in a drying oven or on a slide warmer (60°C) for several hours (perform banding immediately after removal).

5. Bone marrow might be a difficult specimen from which to obtain good material for analysis. In general, the more malignant the cells, the poorer the quality of chromosome preparations. It is important that the poor quality cells are not skipped over in favor of the "prettier" cells, as this could bias the results.
6. With occasional specimens, because of the patient's disease process, there might not be any or enough mitotic cells to analyze. This fact, in itself, can be informative to the physician.

4. Chromosome Abnormalities in Solid Tumors

Detailed methodologies for the examination of chromosomes in various solid tumors can be found in the literature ***(1,2,10,37–39)***.

The pathologist and tumor pathology play a key role in tumor cytogenetics. The pathologist is often responsible for the handling, selection, and disposition of the tissue samples for cytogenetic analysis. These are crucial steps, as the tissue submitted for cytogenetic analysis must be viable, representative of the tumor, and not fixed. Furthermore, the correct diagnosis and precise histologic definition of the tumor is even more crucial because they serve as a basis for correlating cytogenetic events with specific tumor types. Without the cooperation and collaboration of the pathologist, it is not possible to more fully and usefully understand and apply the cytogenetic data characterizing various tumors.

Many tumors, particularly benign ones, have a low proliferative index, necessitating cell culture that might be unsuccessful or might led to overgrowth by normal stromal cells such as fibroblasts ***(1)***. Extensive necrosis or contamination by a coexistent infection can result in a low yield of sufficient metaphases for analysis. A host of chromosomal aberrations are frequently seen in solid tumors, particularly adenocarcinomas, making it difficult to discern the primary abnormality ***(40)***. At present, it is usually impossible to obtain successful chromosome preparations from every tumor studied. The yield depends on the source of the tumor (e.g., solid tumor vs effusion), the site of the tumor, and a number of other parameters. Adequate banding might be a difficult chore in some tumors because of the condensed and fuzzy nature of the chromosomes. Despite these obstacles encountered with the karyotypic evaluation of solid tumors, refinements in cytogenetic techniques (e.g., FISH and SKY) have led to an increasing number of informative studies.

Tumor tissue can be transported in a small amount (i.e., 5–10 mL) of sterile saline solution or phosphate-buffered saline (PBS) in a sterile container suitable for transfer. For those specimens whose transport requires a greater length of time (i.e., 5–48 h), medium containing serum is preferable. RPMI-1640 medium supplemented with fetal calf serum (FCS) and antibiotics has often been used ***(41)***. L-15 medium can also serve as a transport medium ***(42)***.

Specimens are prepared for culture by disaggregation techniques, which include mechanical and/or enzymatic processes. Enzymatic disaggregation procedures with collagenase II and DNase I are preferred over the mechanical method of mincing the tumor with scissors or scalpel or digestion by trypsin. The former methods yield a larger number of viable cells and the quality of banding is superior. In addition, excellent results can frequently be obtained with long-term (i.e., 16 h to 6 d) incubation with these enzymes ***(1,38)***.

Some tumors are more difficult to culture than others, with prostatic and lung cancers being among the most difficult ***(43,44)***. An improved technique for short-term culturing of human prostate adenocarcinoma and its cytogenetic analysis has been described ***(44)***. The method is based on (1) prolonged mild collagenase treatment, (2) careful washing and repeated centrifugation and sedimentation of the disaggregated material to isolate viable prostatic epithelial cells, (3) short-term culture on collagen R-coated (Serva, Heidelberg, Germany) chamber slides with PFMR-4 medium (SBL, Stockholm, Sweden) supplemented with mitogenic factors, and (4) daily inspection of the cultured cells to determine the optimal time for harvesting.

Another type of tumor difficult to culture and analyze cytogenetically is transitional cell carcinoma of the bladder. A method that allows routine cytogenetic analysis of small cystoscopic biopsies from urothelial tumors has been described *(45)*. This method is based on prolonged mild collagenase treatment, a 12- to 16-h culture, and harvesting procedures adapted to yield maximal metaphase recovery.

The use of phorbol-12,13-dibutyrate as a mitogen appears to be useful in the cytogenetic analysis of colon tumors *(46)*. This compound, alone or in combination with other agents (e.g., calcium ionophore A23187), can be applicable to many other tumors that are difficult to karyotype because of the dearth of mitoses.

Cell culture synchronization is a method frequently used in combination with standard techniques. The general opinion holds that synchronization of cells in culture should increase the number of metaphases suitable for cytogenetic analysis. The method most widely used to date is that of methotrexate (MTX) block and thymidine release *(47)*.

The processing of the cultured cells for karyotyping follows essentially the steps described in the preceding sections of this chapter.

5. Fluorescence *In Situ* Hybridization

Cytogenetic analysis is sometimes not sufficient to detect chromosomal changes, a major reason being that it can be performed only on dividing cells. Another drawback is the limitation of cytogenetic methods in cases in which the abnormality is masked, subtle, or not visible with a conventional optical microscope. A case in point is t(12;21)(p12;q22), a not uncommon anomaly in acute lymphoblastic leukemia.

In situ hybridization was originally performed with isotopically labeled probes and subsequent autoradiological detection of specific DNA and RNA sequences in metaphase chromosomes and interphase nuclei *(48)*. A much-applied modification of this technique uses nonradioactive probes detectable with a fluorescent microscope and is, therefore, termed "fluorescence *in situ* hybridization" (FISH) *(13)*. This technique is of value in the detection of specific sequences, including small unique base sequences, not only in interphase nuclei or metaphase cells but also applied in fixed and paraffin-embedded tissues *(49)*.

The possibility of investigating numerical and structural karyotypic changes in archival materials, such as paraffin-embedded tissues, allows for retrospective studies and the simultaneous characterization of tumor cells from the surrounding stromal cells *(49)*.

The detection of structural abnormalities can be achieved with FISH in interphase or metaphase cells. As structural rearrangements could involve the short and/or long arms of chromosomes, selective staining of a specific location or region, or of the entire chromosome, is often necessary. Visualization of entire chromosomes is performed by using "chromosome-painting" probes. This approach is useful in the detection of specific translocations known to characterize a specific disorder. Unique (cosmid) sequence probes spanning a translocation breakpoint can be used to localize specific breakpoints. This is particularly useful when cytogenetic analysis is equivocal *(50–52)*. These probes can be labeled with different fluorochromes so that in normal diploid cells there are two separate signals for each chromosome pair. In cells with a translocation, two of the normal homologs show separate signals, whereas the flanking sequences are brought together, resulting in a double fluorescent spot *(53)*.

In the determination of aneuploidy, as well as structural abnormalities, an important advantage of FISH is the ability to visualize multiple targets simultaneously. Using different combinations of probes labeled with biotin, digoxigenin, or dinitrophenol, and antibodies labeled with different fluorochromes, such as fluorescein (green), rhodamine or Texas red (red), and AMCA or Cascade blue (blue), three separate chromosomal sequences can be distinguished concurrently *(54)*. The sensitivity of FISH is also useful in the detection of gene amplification, such as *ERB-B2*, *MYCI*, *INT-2*, and N-*RAS*, small chromosomal deletions, and duplications *(55)*. Notations for various FISH methodologies utilize aberrations that might be helpful to the reader (i.e., DC- or D-FISH [dual-color FISH], R×FISH [cross-species color banding

FISH], I-FISH [interphase FISH], BAC-FISH [bacterial artificial chromosomes FISH], YAC-FISH [yeast artificial chromosomes FISH], specific-sequences or cosmid FISH and *arms*FISH [combined multicolor {M-FISH} with chromosome arm-specific FISH]) ***(56)***.

For details of FISH methodologies, the reader can consult the publications of Sandberg and Chen ***(49)*** and Blancato and Haddad ***(57)***.

6. Spectral Karyotyping and M-FISH

A molecular cytogenetic technique, termed spectral karyotyping (SKY), has been developed to complement CGH, FISH, and conventional cytogenetics in karyotype analysis ***(58,59)***. SKY permits the simultaneous visualization of all human chromosomes, with each chromosome appearing in a different color. This method combines the karyotype screening ability of CGH with the ability of FISH to identify structural and numerical aberrations that might not affect copy number. The chromosome-specific probes are generated from flow-sorted analysis ***(60)*** or microdissected ***(61)*** chromosomes. Each chromosome-specific set of probes is labeled with either a single fluorochrome or a unique combination of two to three fluorochromes ***(58)***. By using a total of five fluorochromes, a pool of 24 (representing 22 autosomes and 2 sex chromosomes) uniquely labeled DNA probes can be generated, each producing a distinct color. This probe pool is subsequently hybridized to metaphase chromosome spreads.

The hybridization is visualized using spectral imaging, which combines epifluorescence microscopy, CCD (cooled charge-coupled device) imaging, and Fourier spectroscopy to measure the entire fluorescence spectrum emitted by each chromosome analysis ***(62,63)***. Because of the unique fluorochrome combinations used for labeling, each chromosome appears to be "painted" a different color, depending on the fluorochrome or multifluorochrome combination with which it was labeled. The imaging technique allows for the simultaneous detection and visualization of all of the chromosomes on the metaphase spread. The resulting spectral karyotype represents each chromosome in a different color and is, therefore, easily interpreted.

Spectral karyotyping permits the identification of abnormalities that are too subtle to detect by traditional chromosome-banding techniques. Tumors often possess numerous chromosomal abnormalities that are difficult to characterize with conventional techniques (such aberrant chromosomes are termed "marker" or "derivative" chromosomes). SKY has been shown to identify complex rearrangements and allows for a determination of the origins of virtually all aberrant chromosomes ***(64)***. SKY can also be used in conjunction with conventional banding techniques to characterize other chromosomal structures (such as double minutes and homogeneously staining regions), as well as to determine breakpoint regions more accurately ***(63,64)***.

Another technique (multifluor or multiflex-FISH) for visually discerning all human chromosomes in distinct colors is M-FISH ***(15)***. In contrast to SKY, this technique utilizes conventional filter technology and requires sequential exposures of the hybridized chromosomes with six optimal filters to obtain a "color karyotype." Whereas SKY allows for a one-step image acquisition, this technique requires multiple images to be acquired for visualization of all chromosomes.

Although SKY and M-FISH are effective in detecting chromosomal aberrations (e.g., translocations, insertions) and characterizing overall chromosomal structure, these methods do not reliably detect small chromosomal gains and deletions or pericentric and paracentric inversions. Thus, SKY and M-FISH, although powerful adjuvants to conventional cytogenetics and CGH, can be viewed as another important form of karyotype analysis that should be used in conjunction with banding-based analyses and CGH to enhance information about chromosomal structure and aberrations ***(58,64)***. Although the painting probes utilized for SKY and M-FISH are commercially available, the techniques are relatively expensive and require specialized equipment and, hence, are not used for routine cytogenetic analysis.

For further details regarding SKY and M-FISH methodologies and their application, the reader should consult the publications of Hilgenfeld et al. ***(14)***, Green et al. ***(65)***, and Bayani and Squires ***(66)***.

7. Comparative Genomic Hybridization

Comparative genomic hybridization (CGH) is a molecular cytogenetic assay that screens for DNA copy-number changes within a complete genomic complement and maps these changes onto normal chromosomes ***(65,67)***. The technique is based on the simultaneous hybridization of differentially labeled total genomic, leukemic, or tumor (including formalin-fixed, paraffin-embedded tissue) ***(68)*** DNA and normal reference DNA onto normal human metaphase chromosomes. CGH does not require culturing of leukemic or tumor cells (only DNA is needed). They are depicted in karyograms, which indicate regions of tumor-specific DNA gains and losses ***(65)***.

Comparative genomic hybridization utilizes quantitative two-color fluorescent *in situ* suppression hybridization. Total genomic DNA is isolated from a cellular or tumor specimen (test DNA) and from the cells of an individual with a normal karyotype or from the normal tissue (reference DNA) (*see* **Fig. 2**).

Quantitative measurements of fluorescence intensities are made by digital image analysis using fluorescence microscopy with a cooled CCD camera, specific optical filters, and specialized computer software ***(68–70)***. Following image acquisition, fluorescence ratio profiles are calculated along the axis of each chromosome. Average ratio profile calculations are based on the analysis of at least five metaphase spreads per tumor (*see* **Fig. 2**).

An advantage of CGH is the need for small amounts of tumor DNA (less than 1 μg). Modified methods for extracting and labeling DNA for CGH analysis of formalin-fixed, paraffin-embedded tissues are available ***(71–73)***. Using a universal DNA-amplification technique, termed "degenerate oligonucleotide primed PCR" (DOP-PCR) ***(60,74)***, as few as 2000 cells from a formalin-fixed, paraffin-embedded tumor are required, representing a tumor size of less than 1 mm^3 ***(68)***. Such a small amount of material can be easily obtained from almost any routine tissue section. Concordance between CGH analyses performed on matched fresh and formalin-fixed tissue samples is high (95%) ***(73)***. In addition, 70–90% of all archived, formalin-fixed material (including very old samples and tissues derived from autopsies) has been found to be suitable for CGH analysis ***(73)***, reiterating the great utility of this technique for retrospective genome analysis.

Chromosomal mapping by CGH can direct gene identification efforts to specific regions of the genome. The analysis of breast cancers, for instance, has revealed a number of previously undetected amplification sites on 17q and 20q ***(75)***, amplification of the short arm of the X chromosome in therapy-resistant prostate cancer, and gain of 3q in invasive cervical cancer.

Comparisons of CGH with chromosome-banding analyses ***(76)*** as well as confirmatory FISH experiments ***(77)*** have demonstrated the accuracy of this technique in identifying gains and losses in tumor DNA. The ability of CGH to detect small regions of DNA gain or loss is limited primarily by the degree of condensation of the metaphase chromosomes (the more condensed the chromosomes, the lower the resolution) and the magnitude of the copy-number changes. The resolution limit for low-level copy-number changes (loss or gain of a single copy) is estimated to be in the range of 10–20 million basepairs ***(78)***. The smallest detectable deletion is thought to be in the range of approx 3–5 million basepairs ***(78)***. In addition, CGH only detects copy-number changes relative to the average copy number in the entire tumor (e.g., diploid tumors cannot be distinguished from triploid or tetraploid tumors).

An important limitation of CGH is its inability to detect balanced or subtle genetic aberrations that fail to produce copy-number changes (such as inversions and balanced translocations). Furthermore, CGH results reflect an average of the most common aberrations found in a tumor population and can reliably detect a chromosome gain or loss only if present in greater than 50% of the cells analyzed ***(68)***. This is an important point when analyzing tumors that are intermixed with an abundant population of normal reactive and/or stromal cells or those having a great deal of clonal heterogeneity or genetic diversity.

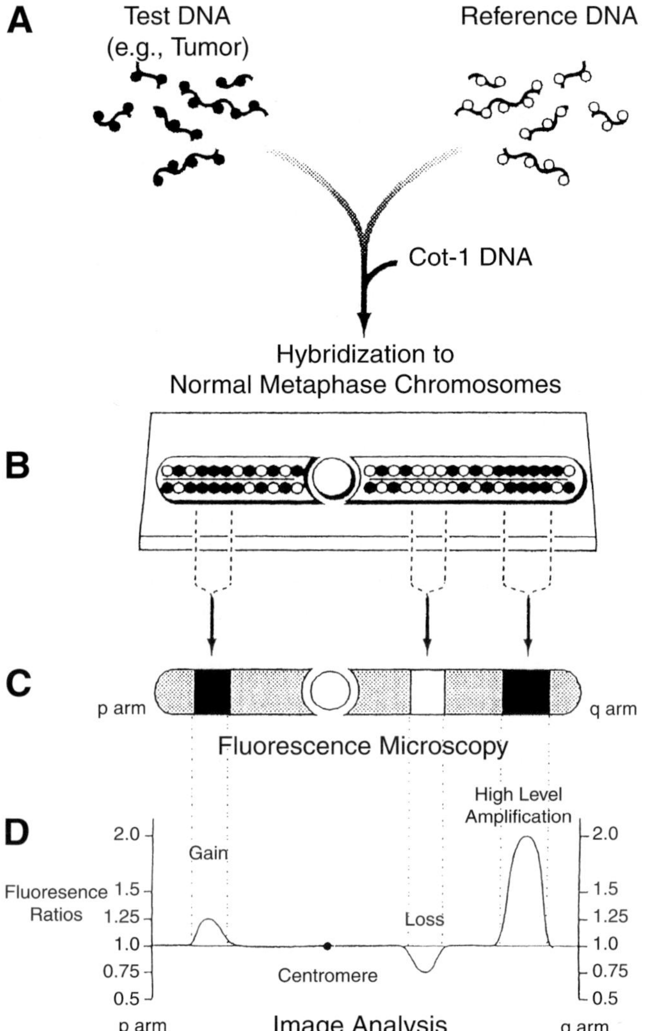

Fig. 2. Schematic diagram outlining the CGH technique *(65)*. (**A**) Test and reference DNAs are differentially labeled with fluorochromosomes (e.g., test DNA [closed circles] and reference DNA [open circles]) and (**B**) hybridized to normal metaphase chromosome spreads. (**C**) Fluorescence microscopy is used to generate a ratio image, which displays the differences in fluorescence intensities along the length of each chromosome, with green staining (depicted in black) indicative of tumor gain(s) and red staining (depicted in white) indicative of tumor loss(es). The areas depicted in gray represent regions of the chromosome that do not have copy-number changes ("normal" regions). (**D**) Fluorescence ratio profiles are calculated along the axis of each chromosome, indicative of copy-number changes in the test DNA relative to the reference DNA. A ratio of 1.0 indicates that there are no copy-number changes (e.g., a "normal" region of the chromosome), whereas ratios of 0.75 or less indicate a loss of DNA and ratios of 1.25 or greater indicate DNA gains at the corresponding location along the chromosome axis. Ratios of 2.0 or greater indicate the presence of a high-level DNA amplification.

8. Expression Profiling of Genes of Human Cells and Tumors

Expression profiling refers to the process of determining the expression of thousands of genes simultaneously in a cellular or tissue sample *(79)*. The resulting patterns of gene expression reflect the molecular basis of the sample phenotype. There are two main methods of expression profiling: hybridization based (DNA microarrays) and sequencing based (oligonucleotide microarrays). Microarray technology depends on the ability of two strands of nucleic acid, RNA or DNA, to bind to one another when they have a complementary sequence of bases; for example, one strand of DNA with the sequence ACTGG will hybridize to a strand that carries that antiparallel sequence TGACC, but not to other sequences. In DNA microarrays, RNA is extracted from the sample, converted to cDNA using reverse transcriptase, and hybridized to a DNA microarray *(79)*. DNA microarray-based gene expression profiling relies on nucleic acid hybridization and the use of nucleic acid polymers, immobilized on a solid surface (Nylon membranes, nitrocellulose filters, glass slides, or synthetic chips), as probes for complementary gene sequences *(80)*.

In oligonucleotide microarrays, probes for different genes can be deposited or synthesized directly on the surface of a silicon wafer in a patterned manner. Oligonucleotides offer greater specificity than cDNAs because they can be tailored to minimize chances of cross-hybridization, and sequences of up to 60 nucleotides in size in place of full-length cDNAs have been used effectively. Major advantages of this approach include uniformity of probe length and the ability to discern splice variants. Another advantage particular to the commonly used Affymetrix GeneChip system (Affymetrix, Santa Clara, CA) is the ability to recover samples after hybridization to a chip. This allows for a single biologic sample to be sequentially hybridized to multiple arrays, a considerable advantage when dealing with limited biologic material.

The hybridization of a test sample to an array can be detected in one or two ways. cDNA microarrays are commonly queried simultaneously with cDNAs derived from experimental and reference RNA samples that have been differentially labeled with two fluorophores to allow for the quantification of differential gene expression, and expression values are reported as ratios between two fluorescent values. Alternatively, the Affymetrix oligonucleotide system uses a single-color fluorescent label, where experimental mRNA is enzymatically amplified, biotin labeled for detection, hybridized to the wafer, and detected through the binding of a fluorescent compound (streptavidin–phycoerythrin).

9. Array-Based CGH

The clinical application of CGH has been hampered by several factors, including time-consuming procedures, resolution of the CCD camera, and sensitivity in detecting loss of fine chromosomal regions.

Array-based CGH is similar to pre-existing chromosomal CGH protocols but uses large-insert DNA clones or cDNA as chromosome-specific hybridization targets in an arrayed format instead of normal metaphase chromosomes *(81–84)*. Currently, libraries of these large-insert clones have been integrated into the draft sequence of the human genome and, thus, represent ideal targets to be used for array CGH. Copy-number alterations as detected by array CGH can be directly related to sequence information and will dramatically accelerate the identification of novel disease-causing genes. Quantitative measurements of DNA copy-number changes by high-resolution array CGH have been successfully used for oncogene delineation and detailed qualitative and quantitative analyses of chromosomal alterations of DNA copy number in cancers *(81–84)*. However, the various array techniques have the same limitations as CGH with respect to the inability to detect clonal heterogeneity and minority (<50%) cell populations.

References

1. Sandberg, A. A. (1990) *The Chromosomes in Human Cancer and Leukemia*, 2nd ed. Elsevier Science, New York.
2. Makino, S. (1975) *Human Chromosomes*, Igaku Shoin, Tokyo.

3. ISCN (1985) *An International System for Human Cytogenetic Nomenclature* (Harnden, D. G. and Klinger, H. P., eds.), March of Dimes Birth Defects Foundation, New York.
4. ISCN (1995) *An International System for Human Cytogenetic Nomenclature* (Mitelman, F., ed.), Karger, Basel.
5. Heim, S. and Mitelman, F. (1995) *Cancer Cytogenetics*, 2nd ed., Wiley–Liss, New York.
6. Mark, H. F. L. and Mark, Y. (2000) The normal human chromosome complement, in Medical Cytogenetics (Mark, H. F.L., ed.), Marcel Dekker, New York, pp. 21–43.
7. Tharapel, S. A. and Tharapel, A. T. (2000) Human cytogenetic nomenclature, in Medical Cytogenetics (Mark, H. F. L., ed.), Marcel Dekker, New York, pp. 49–87.
8. Mark, H. F. L. and Wolman, S. R. (2000) Culture and harvest of tissues for chromosome analysis, in Medical Cytogenetics (Mark, H. F. L., ed.), Marcel Dekker, New York, pp. 91–117.
9. Tijo., J.H. and Levan, A. (1956) The chromosome number of man. *Hereditas* **42,** 1–6.
10. Caspersson, T., Zech, L., Johansson, C., and Modest, E.J. (1970) Identification of human chromosomes by DNA-binding fluorescent agents. *Chromosoma* **30,** 215–227.
11. Caspersson, T., Zech, L., and Modest, E.J. (1970) Fluorescent labeling of chromosomal DNA: superiority of quinacrine mustard to quinacrine. *Science 170,* 762.
12. Mark, H.F.L. (2000) Staining and banding techniques for conventional cytogenetic studies, in *Medical Cytogenetics* (Mark, H. F. L., ed.), Marcel Dekker, New York, pp. 121–138.
13. Pinkel, D., Straume, T., and Gray, J. W. (1986) Cytogenetic analysis using quantitative, high-sensitivity, fluorescence hybridization. *Proc. Natl. Acad. Sci. USA* **83,** 2934–2938.
14. Hilgenfeld, E., Padilla-Nash, H. M., Haas, O. A., Serve, H., Schröck, E., and Ried, T. (2001) Spectral karyotyping (SKY) of hematologic malignancies, in *Hematologic Malignancies: Methods and Techniques* (Faguet, G. B., ed.), Humana, Totowa, NJ, pp. 65–79.
15. Speicher, M., Ballard, S. G., and Ward, D. C. (1996) Karyotyping human chromosomes by combinatorial multi-fluor FISH. *Nature Genet.* **12,** 368–375.
16. Nowell, P. C. (1960) Phytohemagglutinin: An initiator of mitosis in cultures of normal human leukocytes. *Cancer Res.* **20,** 462–466.
17. Moorhead, P. S., Nowell, P. C., Mellman, W. J., et al. (1960) Chromosome preparations of leukocytes cultured from human peripheral blood. *Exp. Cell Res.* **20,** 613–616.
18. Yunis, J. (1976) High resolution of human chromosomes. *Science* **191,** 1268–1270.
19. Barch, M. J., Knutsen, T., and Spurbeck J. L. (eds.) (1997) *The AGT Cytogenetics Laboratory, 3rd ed.*, Lippincott–Raven, Philadelphia.
20. Hecht, F., Peakman,D. C., Kaiser-McCaw, B., et al. (1981) Amniocyte clones for prenatal cytogenetics. *Am. J. Med. Genet.* **10,** 51–54.
21. Schmid, W. (1975) A technique for in situ karyotyping of primary amniotic fluid cell cultures. *Humangenetik* **30,** 325–330.
22. Brambati, B., Oldrini, A., Lanzani, A., and Ferrazzi, E. (1986) Chorionic villus sampling: an obstetric overview, in *Chorionic Villus Sampling (CVS)* (Pescia, G. and Nguyen The, H., eds.), Kargel, Basel.
23. Simpson, J. L. (1990) Incidence and timing of pregnancy losses: relevance to evaluating safety of early prenatal diagnosis. *Am. J. Med. Genet.* **35,** 165–173.
24. Sandberg, A. A. and Chen, Z. (2001) Cytogenetic analysis in *Hematologic Malignancies: Methods and Techniques* (Faguet, G. B., ed.), Humana, Totowa, NJ, pp. 3–18.
25. LeBeau, M. M. (1991) Cytogenetic analysis of hematological malignant diseases, in *The ACT Cytogenetics Laboratory Manual*, 2nd ed. (Barch, M. J., ed.), Raven, New York, pp. 395–449.
26. Chen, Z. and Sandberg, A. A. (2002) Molecular cytogenetic aspects of hematological malignancies: clinical implications. *Am. J. Med. Genet. (Semin. Med. Genet.)* **115,** 130–141.
27. Swansbury, J. (2003) Cytogenetic techniques for myeloid disorders, in *Cancer Cytogenetics. Methods and Protocols* (Swansbury, J., ed.), Humana, Totowa, NJ, pp. 43–57.
28. Swansbury, J. (2003) Cytogenetic studies in hematologic malignancies. An overview, in *Cancer Cytogenetics. Methods and Protocols* (Swansbury, J., ed.), Humana, Totowa, NJ, pp. 9–22.
29. Misawa, S., Horiike, S., Taniwaki, M., Abe, T., and Takino, T. (1986) Prefixation treatment with ethidium bromide for high resolution banding analysis of chromosomes from cultured human BM cells. *Cancer Genet. Cytogenet.* **22,** 319–329.
30. Yunis, J. J. (1981) New chromosome techniques in the study of human neoplasia. *Hum. Pathol.* **12,** 540–549.

31. Morgan, S., Hecht, B. K., Morgan, R., and Hecht, F. (1987) Qualitative and quantitative enhancement of BM cytogenetics by addition of giant cell tumor conditioned medium. *Karyogram* **13,** 39–40.
32. Morgan, R., Chen, Z., Morgan, S., et al. (1995) The PHA/IL2 "COCKTAIL" is an effective cytogenetic mitogen in blood and BM cells for revealing abnormal clonal karyotypes in lymphoid diseases. *Appl. Cytogenet.* **21,** 66.
33. Morgan, R., Chen, Z., Richkind, K., Roherty, S., Velasco, J., and Sandberg, A. A. (1999) PHA/IL2: an efficient mitogen cocktail for cytogenetic studies of non-Hodgkin lymphoma and chronic lymphocytic leukemia. *Cancer Genet. Cytogenet.* **109,** 134–137.
34. Sadamori, N., Matsui, S.I., Han, T., and Sandberg, A. A. (1984) Comparative results with various polyclonal B-cell activators in aneuploid chronic lymphocytic leukemia. *Cancer Genet. Cytogenet.* **11,** 25–29.
35. Wheater, R. F. and Roberts, S. H. (1987) An improved lymphocyte culture technique: deoxycytidine release of a thymidine block and use of a constant humidity chamber for slide making. *J. Med. Genet.* **24,** 113–115.
36. ISCN (1991) *Guidelines for Cancer Cytogenetics, Supplement to An International System for Human Cytogenetic Nomenclature* (Mitelman, F., ed.), Karger, Basel.
37. Sandberg, A. A. and Bridge J. A. (1994) *The Cytogenetics of Bone and Soft Tissue Tumors*, R. G. Landes, Austin, TX.
38. Limon, J., Dal Cin, P., and Sandberg, A. A. (1986) Application of long-term collagenase disaggregation for the cytogenetic analysis of human solid tumors. *Cancer Genet. Cytogenet.* **23,** 305–313.
39. Sandberg, A. A. and Bridge J. A. (1992) Techniques in cancer cytogenetics: an overview and update. *Cancer Invest.* **10,** 163–172.
40. Sandberg, A. A. and Chen, Z. (2000) Chromosome abnormalities, in *Solid Tumors: An Overview in Medical Cytogenetics* (Mark, H. F. L, ed.), Marcel Dekker, New York, pp. 413–435.
41. Gibas, L. M., Gibas, Z., and Sandberg, A. A. (1984) Technical aspects of cytogenetic analysis of human solid tumors. *Karyogram* **10,** 25–27.
42. Leibovitz, A. (1986) Development of tumor cell lines. *Cancer Genet. Cytogenet.* **19,** 11–19.
43. Gazdar, A. F. and Oie, H. K. (1986) I. Advances in cell culture. Cell culture methods for human lung cancer. *Cancer Genet. Cytogenet.* **19,** 5–10.
44. Limon, J., Lundgren, R., Elfving, P., Heim, S., Kristoffersson, U., Mandahl, N., and Mitelman, F. (1990) An improved technique for short-term culturing of human prostatic adenocarcinoma tissue for cytogenetic analysis. *Cancer Genet. Cytogenet.* **46,** 191–200.
45. Fraser, C., Sullivan, L. D., and Kalousek, D. K. (1987) A routine method for cytogenetic analysis of small urinary bladder tumor biopsies. *Cancer Genet. Cytogenet.* **29,** 103–108.
46. Eiseman, E., Luck, J. B., Mills, A. S., Brown, J. A., and Westin, E. H. (1988) Use of phorbol-12,13-dibutyrate as a mitogen in the cytogenetic analysis of tumors with low mitotic indices. *Cancer Genet. Cytogenet.* **34,** 165–175.
47. Hagemeijer, A., Smith E. M. E., and Bootsma D. (1979) Improved identification of chromosomes of leukemic cells in methotrexate-treated cultures. *Cytogenet. Cell Genet.* **23,** 208–212.
48. Pardue, M. L. and Gall, J. F. (1969) Molecular hybridization of radioactive DNA to the DNA of cytological preparation. *Proc. Natl. Acad. Sci. USA* **64,** 600–604.
49. Sandberg, A. A. and Chen, Z. (2001) FISH analysis, in *Hematologic Malignancies: Methods and Techniques* (Faguet, G. B., ed.), Humana, Totowa, NJ, pp. 19–42.
50. Chen, Z., Morgan, R., Berger, C. S., Pearce-Birge, L., Stone, J. F., and Sandberg, A. A. (1993) Identification of masked and variant Ph (complex type) translocations in CML and classic Ph in AML and ALL by fluorescence in situ hybridization with the use of bcr/abl cosmid probes. *Cancer Genet. Cytognet.* **70,** 103–107.
51. Chen, Z., Morgan, R., Stone, J.F. and Sandberg, A.A. (1994) Identification of complex t(15;17) in APL by FISH. Cancer Genet. Cytognet. 72, 73–74.
52. Mathew, S. and Raimondi, S. C. (2002) Cytogenetic techniques for myeloid disorders, in *Cancer Cytogenetics. Methods and Protocols* (Swansbury, J., ed.), Humana, Totowa, NJ, pp. 213–233.
53. Mark, H. F. L. (2000) Fluorescent in situ hybridization (FISH): Applications for clinical cytogenetics, in *Medical Cytogenetics* (Mark, H. F. L., ed.), Marcel Dekker, New York, pp. 553–571.
54. Nederlof, P. M. van der Flier, S., Raap, A.K., et al. (1989) Detection of chromosome aberrations in interphase nuclei by nonradioactive in situ hybridization. *Cancer Genet. Cytogenet.* **42,** 87–98.
55. Nederlof, P. M., Robinson, R., Abuknesha, J., et al. (1989) Three-color fluorescence in situ hybridization for the simultaneous detection of multiple nucleic acid sequences. *Cytometry* **10,** 20–27.

56. Karhu, R., Ahlstedt-Soini, M., Bittner, M., Meltzer, P., Trent, J., and Isola, J. J. (2001) Chromosome arm-specific multicolor FISH. *Genes Chromosomes Cancer* **30,** 105–109.
57. Blancato, J. and Haddad, B. R. (2000) Fluorescent in situ hybridization (FISH): principles and methodology, in *Medical Cytogenetics* (Mark, H. F. L., ed.), Marcel Dekker, New York, pp. 141–162.
58. Schröck, E., duManoir, S., Veldman, T., et al. (1996) Spectral karyotyping of human chromosomes. *Science* **273,** 494–497.
59. Garini, Y., Macville, M., duManoir, S., et al. (1996) Spectral karyotyping. *Bioimaging* **4,** 65–72.
60. Telenius, H., Pelear, A. H., Tunnacliffe, A., et al (1992) Cytogenetic analysis by chromosome painting using DOP-PCR amplified flow sorted chromosomes. *Genes Chromosomes Cancer* **4,** 257–263.
61. Guan, X.-Y., Trent, J. M., and Meltzer, P. S. (1993) Generation of band-specific painting probes from a single microdissected chromosome. *Hum. Mol. Genet.* **2,** 1117–1121.
62. Malik, Z., Cabib, D., Buckwald, R. A.,Talmi, A., Garini, Y., and Lipson S. G. Fourier transform multipixel spectroscopy for quantitative cytology. *J. Microsc.* **182,** 133–140.
63. Garini, Y., Katzir, N., Cabib, D., Buckwald, R. A., Soenksen, D., and Malik, Z. (1996) Spectral bioimaging, in *Fluorescence Imaging Spectroscopy and Microscopy* (Wang, X. F., Herman, B., eds.), Wiley, New York.
64. Veldman, T., Vignon, C., Schröck, E., Rowley, J. D., and Ried, T. (1997) Hidden chromosome abnormalities in hematological malignancies detected by multicolor spectral karyotyping. *Nature Genet.* **15,** 406–410.
65. Green, G. A., Schröck, E., Veldman, T., Heselmeyer-Haddad, K., Padilla-Nash, H. M., and Ried, T. (2000) Evolving molecular cytogenetic technologies, in *Medical Cytogenetics* (Mark, H. F. L., ed.), Marcel Dekker, New York, pp. 579–592.
66. Bayani, J. M. and Squire, J. A. (2002) Applications of SKY in cancer cytogenetics. *Cancer Invest.* **20,** 373–386.
67. DuManoir, S., Speicher, M. R., Joos, S., et al. (1993) Detection of complete and partial chromosome gains and losses by comparative genomic in situ hybridization. *Hum. Genet.* **90,** 590–610.
68. Speicher, M. R., duManoir, S., Schröck, E., et al. (1993) Molecular cytogenetic analysis of formalin fixed, paraffin embedded solid tumors by comparative genomic hybridization after universal DNA amplification. *Hum. Mol. Genet.* **2,** 1907–1914.
69. Piper, J., Rutovitz, D., Sudar, D., et al. (1995) Computer image analysis of comparative genomic hybridization. *Cytometry* **19,** 10–26.
70. DuManoir, S., Schröck, E., Bentz, M., et al. (1995) Quantification of comparative genomic hybridization. *Cytometry* **19,** 27–41.
71. Ried, T., Knutzen, R., Steinbeck, R., et al. (1996) Comparative genomic hybridization reveals a specific pattern of chromosomal gains and losses during the genesis of colorectal tumors. *Genes Chromosomes Cancer* **15,** 234–245.
72. Heselmeyer, K., Schröck, E., duManoir, S., et al. (1996) Gain of chromosome 3q defines the transition from severe dysplasia to invasive carcinoma of the uterine cervix. *Proc. Natl. Acad. Sci. USA* **93,** 479–484.
73. Isola, J., DeVries, S., Chu, L., Ghazrini, S., and Waldman, F. (1994) Analysis of changes in DNA sequence copy number by comparative genomic hybridization in archival paraffin-embedded tumor samples. *Am. J. Pathol.* **145,** 1301–1308.
74. Telenius, H., Carter, N. P., Bebb, C. E., Nordenskjold, M., Ponder, B. A. J., and Tunnacliffe, A. (1992) Degenerate oligonucleotide-primed PCR: general amplification of target DNA by a single degenerate primer. *Genomics* **13,** 718–725.
75. Kallioniemi, A., Kallioniemi, O.-P., Piper, J., et al. (1994) Detection and mapping of amplified DNA sequences in breast cancer by comparative genomic hybridization. *Proc. Natl. Acad. Sci. USA* **91,** 2156–2160.
76. Kallioniemi, A., Kallioniemi, O.-P., Sudar, D., et al. (1992) Comparative genomic hybridization for molecular cytogenetic analysis of solid tumors. *Science* **258,** 818–821.
77. Schröck, E., Thiel, G., Lozanova, T., et al. (1994) Comparative genomic hybridization of human glioma reveals consistent genetic imbalances and multiple amplification sites. *Am. J. Pathol.* **144,** 1203–1218.
78. Kallioniemi, O.-P., Kallioniemi, A., Piper, J., et al. (1994) Optimizing comparative genomic hybridization for analysis of DNA sequence copy number changes in solid tumors. *Genes Chromosomes Cancer* **10,** 231–243.

79. Ladanyi, M. and Gerald, W. C. (2003) Introduction: Present and potential impact of expression profiling studies of human tumors, in *Expression Profiling of Human Tumors: Diagnostic and Research Applications* (Ladanyi, M. and Gerald, W. L., eds.), Humana, Totowa, NJ, p. 3.
80. Ramaswamy, S. and Golub, T. R. (2001) DNA microarrays in clinical oncology. *J. Clin. Oncol.* **20,** 1932–1941.
81. Albertson, D. G., Ylstra, B., Segraves, R., et al. (2000) Quantitative mapping of amplicon structure by array CGH identifies CYP24 as a candidate oncogene. *Nature Genet.* **25,** 144–146.
82. Pinkel, D., Segraves, R., Sudar, D., et al. (1998) High resolution analysis of DNA copy number variation using comparative genomic hybridization to microarray. *Nature Genet.* **20,** 207–211.
83. Zhao, J., Roth J., Bode-Lesniewska, B., Pfaltz, M., Heitz, P. U., and Komminoth, P. (2002) Combined comparative genomic hybridization and genomic microarray for detection of gene amplifications in pulmonary artery intimal sarcomas and adrenocortical tumors. *Genes Chromosomes Cancer* **34,** 48–57.
84. Wilhelm, M., Veltman J. A., Olshen, A. B., et al. (2002) Array-based comparative genomic hybridization for the differential diagnosis of renal cell cancer. *Cancer Res.* **62,** 957–960.

33

Microsatellite Analysis

Rachel E. Ibbotson and Anton E. Parker

1. Introduction

Microsatellites, also known as STRs (short tandem repeats), are tandem repeats of a simple dinucleotide, trinucleotide, tetranucleotide, pentanucleotide, or hexanucleotide sequence (two to six bases) that occur abundantly and at random throughout most eukaryotic genomes (roughly one microsatellite every 10 kb). They are typically short (often less that 100 bp long) and embedded within unique sequence, thus being ideal for designing flanking primers for in vitro amplification by the polymerase chain reaction (PCR) (see Chapter 6). The high degree of polymorphism, as a result of variation in the number of repeat units, and the stability displayed by microsatellites make them perfect markers for use in constructing high-resolution genetic maps to identify susceptibility loci involved in common genetic diseases ***(1)***. In addition to their applications in genome mapping and positional cloning, these markers have been applied in fields as diverse as tumor biology, personal identification, population genetic analysis, and the construction of human evolutionary trees. To appreciate the impact that microsatellite analysis has had within the medical field, search PubMed for "microsatellite" on the National Center for Biotechnology Information website www.ncbi.nlm.nih.gov and thousands of entries will be found. References can be found that date back more than 20 yr. Today, microsatellites are still being used extensively to investigate the genome although rapid advances in technology have seen many modifications to the original methods.

A microsatellite is analogous to a bookmark; it highlights a particular position in the genome and it is for this reason that their use has become so widespread. A microsatellite is a marker for a particular genomic region. Each normal human cell has 22 pairs of matched (homologous) chromosomes plus a pair of sex chromosomes (either XX or XY). Each one of a pair of chromosomes is inherited either maternally or paternally and because of normal polymorphism, at any given position (locus) on the chromosome, the DNA sequence can be slightly different; that is, they are heterozygous. For example, a normal human cell might have three CA repeats on one allele and five repeats on the other; this marker would be heterozygous.

By analyzing these markers, regions that have been subjected to allelic imbalance, for instance by deletion, can be identified. This will be reflected by the loss or alteration of the markers across the region, that is, the region is tested for loss of heterozygosity (LOH) or instability. The significance of such somatically acquired genetic changes is well documented and often these manifest as neoplasia. Investigation of consistent regions of deletion has generally led to the identification of tumor suppressor genes, of note are the retinoblastoma gene (*RB-1*) on chromosome 13 ***(2)*** and the *p53* gene on chromosome 17 ***(3)***. Genetic alteration frequently results in subtle sequence modifications within microsatellites; in hereditary nonpolyposis colorectal cancers, mutated mismatch repair genes are responsible for the cell's inability to repair nucleotide mismatches and microsatellite instability (MIN) is evident ***(4)***.

From: *Medical Biomethods Handbook*
Edited by: J. M. Walker and R. Rapley © Humana Press, Inc., Totowa, NJ

The high degree of heterozygosity displayed by microsatellite markers is especially advantageous because the majority of samples will be heterozygous (otherwise said to be informative) at the locus of interest. In addition, very small amounts of tissue are required for microsatellite analysis (typically 100 ng) and many samples can be processed simultaneously with relative ease. For LOH studies, it is imperative that there is access to matched normal and tumor tissue so that direct comparison between the two can be made. In our own studies, we have analyzed tumor peripheral blood lymphocytes using granulocytes, from the same blood sample, as matched controls.

2. Experimental Design

To perform microsatellite analysis, a number of steps are involved and certain considerations must be made.

2.1. Sample Selection and Purity

Sample purity is important and so is sample availability. It should be emphasized that matched samples must be purified to greater than 85%; that is, the tumor sample should not be contaminated with normal material and vice versa, otherwise problems will be encountered during the result interpretation. Careful consideration must be given during experimental design to optimize the choice of material from which DNA will be extracted. For our own research on microsatellite markers on chromosome 7 in lymphoma patients, we have used heparinized blood samples, the target cells are in the blood and blood is relatively easy to obtain from the patient. In contrast, it might be more difficult to obtain specimens when working with, for instance, solid tumors. Histological analysis of a biopsy would be required to distinguish cancerous tissue from surrounding normal tissue, followed by microdissection to separate the selected cells for comparison. A description of tumor cell purification from blood has been included in this chapter. Methods for sample purification from other tissues can be found in the literature [e.g., recovery of tissue from slides ***(5,6)*** and microdissection ***(7)***.

Figure 1 demonstrates this point. In the unsorted experiment, two alleles are present in both the "N" constitutional lane and the "T" tumor lane; however, the smaller allele (indicated by the arrow) is less intense. After the tumor cells have been purified by FACS, two alleles are present in the "N" lane, whereas in the "T" lane, which has been amplified from DNA from pure tumor cells, only one allele is evident. DNA from contaminating normal cells in the tumor sample masked the result in the unsorted experiment.

Collecting a pure tumor sample can be achieved by a variety of methods. We have either used a flow cytometer (FACS) or magnetic microbeads, as these methods are appropriate for the blood samples that we are investigating. There are advantages and disadvantages for each of these methods.

The advantage of FACS is that you can sort on any parameter around which you can draw a gate and, in addition, more than one parameter can be selected at a time. The disadvantage is that the procedure is time-consuming and relatively expensive and the final yield of cells is low. Generally, we have sorted cells by sequential gating, first on the forward/side scatter characteristics of lymphocytes and then subsequently on CD19 positivity and either kappa or lambda, depending on the light-chain restriction.

The microbead purification method for recovery of target cells from peripheral blood is less labor-intensive and the yield is higher; however, the limitation is that it is difficult to sort on more than one parameter at a time. We have used Miltenyi (Bergisch Gladbach, Germany) CD19 microbeads for purifying tumor B-cells. Miltenyi markets a range of beads for numerous applications depending on the target cell. The beads are supplied with a detailed protocol sheet, which is simple and straightforward.

For both methods, we access the purity of our samples by labeling 1×10^6 cells, both before and after the column purification, with CD5 FITC and CD20 RPE mouse anti-human monoclonal antibodies (Dako, Denmark) and analyzing by flow cytometry (*see* **Fig. 2**).

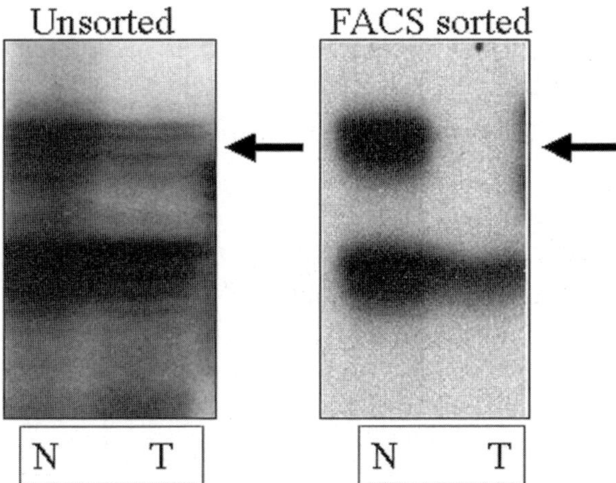

Fig. 1. Autoradiogram of ^{32}P labeled PCR products amplified from chromosome 7 and resolved by polyacrylamide gel electropheresis. Tumor (T) lymphocytes and normal (N) granulocytes from unsorted and FACS samples.

2.2. DNA Extraction

Careful consideration must be given to the method chosen to extract the DNA, depending on what sort of sample is available. For instance, a quick method in which the cells are simply boiled in buffer is really only suitable for relatively clean cell suspensions and would not be appropriate for cells that have been scraped off a fixed and stained slide. Burton et al. *(5)* discuss suitable methods for extracting DNA from slides, and commercial kits are also available (e.g., DEXPAT™ supplied by Takara Biomedicals, Japan).

2.3. Microsatellite Selection

Once the genomic region of interest has been identified, appropriate microsatellites for investigation must be selected, either from previously identified microsatellites listed in the databases or from novel microsatellites. It is possible to database search for the presence of novel microsatellites by using computer programs. However, public databases with comprehensive lists of known microsatellite loci are also available (e.g., National Center for Biotechnology Information [NCBI], http://www.ncbi.nlm.nih.gov/), which provide data on hetrozygosity rates and suggested primer sequences in addition to the locations and flanking sequences.

2.4. Microsatellite Resolution and Result Interpretation

Microsatellite polymorphism is quantified by PCR and resolved by electrophoresis to separate the alleles according to their size. There are several resolution procedures that can be used for visualizing the PCR products. Incorporation of radioactive ^{32}P followed by polyacrylamide gel electrophoresis (PAGE) has long been a standard technique and details are discussed in previous books in this series *(8)*. Advances in fluorescent chemistry and automated sequencing utilizing PAGE or, more recently, capillary electrophoresis have seen the widespread adoption of fluorescent microsatellite assays. The incorporation of fluorescence into the PCR products can be achieved by the use of fluorescent-dye-labeled primers or fluorescent deoxynucleotide triphosphates ([F] dNTPs).

The results of a microsatellite analysis will be displayed as a number of peaks on an electropherogram. A schematic representation of an electropherogram is shown in **Fig. 3**; note that the x-axis is a measure of the size of the microsatellite allele in basepairs and the y-axis is a measure of the relative amount of product detected. A microsatellite marker that shows two distinct

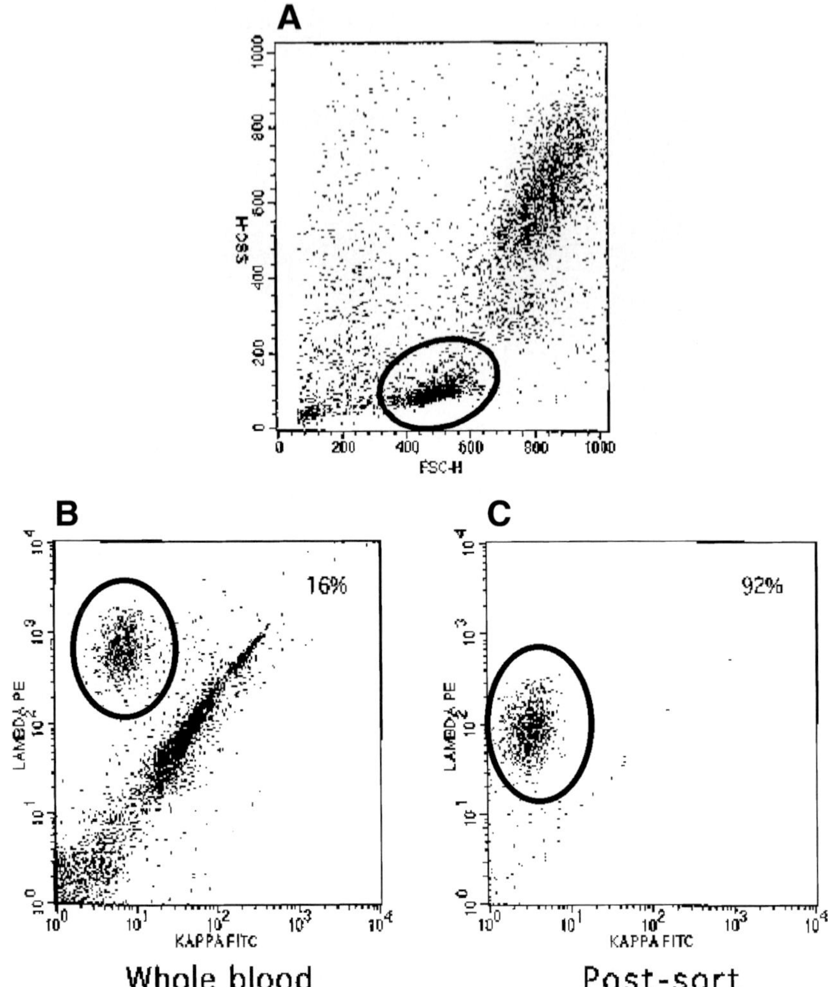

Fig. 2. FACS plots of peripheral blood from an SLVL patient: (**A**) forward scatter versus side scatter of whole blood; lymphocytes are encircled; (**B**) whole blood labeled with kappa and lambda light-chain mouse monoclonal antibodies (Dako, Denmark) before FACS sorting; (**C**) the same cells as in (**B**) but after FACS sorting for the clonal lambda population.

peaks from two alleles in the electropherogram from the constitutional DNA was considered to be informative. For LOH studies, sample pairs (i.e., control vs tumor) are scored as follows; heterozygous, not deleted (two different alleles present in both normal and tumor tissues): heterozygous, deleted (two alleles present in normal tissue but only one present in tumor tissue); homozygous, uninformative (two alleles of the same size present in both normal and tumor tissues). For MIN studies, only changes in size of the alleles is scored. The occurrence of both MIN and LOH within a tumor sample is possible.

For all informative samples, peak areas were calculated for C1 and C2, representing the short and long alleles in the constitutional (C) DNA, and for T1 and T2, representing the short and long alleles in the tumor (T) DNA (*see* **Fig. 3**). Peak area ratios for the constitutional sample (CR) and tumor sample (TR) were calculated, where CR = C1/C2 and TR = T1/T2. If the value of CR fell outside of the range 0.8–1.2, then the sample was considered unsuitable for assessment for LOH. The ratio of CR/TR was then calculated; if CR/TR was <0.25 or >4, then

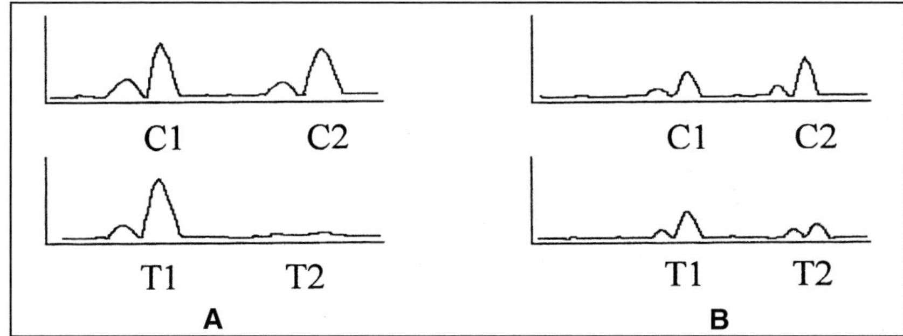

Fig. 3. Schematic representation of Genescan Eletropherogram traces showing LOH.

this indicated that the total peak area of one allele in the tumor sample was reduced by greater than 75% relative to the peak area of the remaining allele and was, thus, scored as LOH. Homozygous loss should not be determined by this method because DNA from residual normal cells in the tumor DNA fraction will interfere with interpretation of the results.

3. Applications

When a genetic investigation is implied analyzing, the defective gene might not always be possible. Initially, the defective gene must be identified, followed by gene dysfunction analysis. This might involve gene mutation tests that require knowledge not only of the nucleotide sequence of the gene but also of specific alterations within the gene. Because many different mutations can affect the function of a gene, a screening technique such as the detection of genetic instability by microsatellites can offer a more readily applicable solution. Having established that neoplastic cells can be distinguished from normal cells using molecular techniques, it would then seem feasible to look for neoplastic cells in different clinical samples. Samples such as fine-needle aspirates can be obtained from practically any site relatively noninvasively. Also, nipple aspirates can be used to screen for breast cancer, and cerebrospinal fluid to screen for cerebral neoplasms. Numerous studies analyzing different markers from a variety of different tumors have been published. Some examples are discussed.

Arzimanoglou et al. ***(9)*** attempted to correlate MIN with other biologic parameters to assess the significance of MIN in cancer by an extensive review of the medical literature. Their conclusions were that molecular diagnosis using MIN analysis had been documented in at least two types of tumor (hereditary nonpolyposis colon carcinoma [HNPCC] and sporadic bladder carcinoma), suggesting a potential role of MIN in the diagnosis and/or prognosis of other solid human tumors. Goggins et al. ***(10)*** characterized 82 carcinomas of the pancreas for DNA replication errors using a panel of microsatellite markers. Cases with MIN in at least two markers of a minimum of five tested were considered RER+. RER status was correlated with histological appearance, karyotype of the carcinomas when available, K-ras mutational status, and patient outcome. They concluded that DNA replication errors as identified by microsatellite analysis occurred in a small percentage of carcinomas of the pancreas and are associated with wild-type K-*ras* gene status and a characteristic histopathology.

The importance of genetic alterations predicting clinical outcome has also been extensively studied for human colorectal cancer. Examples include the association of metastasis to the lymph nodes and microvascular invasion with allelic deletions (LOH) of chromosome 17p ***(11)***. The work of Kato et al. ***(12)***, who indicated that the detection of LOH of *DCC* (gene deleted in colorectal cancer) in colorectal carcinoma tissue is associated with the metastatic potential of colorectal carcinoma in the liver. The incidence of allelic loss of the DCC locus was significantly greater for patients with liver metastasis than that for patients who had no liver metastasis for more than 2 yr. These results suggested that these gene alterations might be reliable biologic markers for assessing the possiblity of liver metastasis.

Moreover, recent evidence suggests that senescent tumor cells can release DNA into the circulation, which is subsequently carried by and, therefore, enriched in the serum and plasma, which provides a convenient sampling opportunity for screening. Chen et al. ***(13)*** looked for microsatellite alterations in the plasma DNA of primary small-cell lung carcinoma (SCLC) patients, knowing that 50% of patients had microsatellite alterations and that these could be detected in sputum. A microsatellite alteration was present in 76% SCLC tumors and in 71% of the plasma samples, suggesting that this might constitute a new tool for tumour staging, management, and, possibly, detection.

A novel method for the detection of circulating tumor cell DNA was presented in another microsatellite study by Nawroz et al. ***(14)***. DNA from lymphocytes (as a control) was paired with serum samples from 21 patients with primary head and neck squamous cell carcinoma (HNSCC) for the investigation. Patients were scored for alterations as defined by the presence of MIN (new alleles) or LOH in serum at each of 12 markers and then compared with primary tumor DNA. Six of 21 patients (29%) were found to have 1 or more microsatellite alterations in serum precisely matching those in the primary tumors. All six patients had advanced disease (stage III or IV). These results indicate that microsatellite markers might be useful in assessing tumor burden in cancer patients. Similarly, Spafford et al. ***(15)*** analyzed exfoliated oral mucosal cells in pretreatment oral rinse and swab samples collected from 44 HNSCC patients and from 43 healthy control subjects. A panel of 23 informative microsatellite markers were used to investigate DNA from the matched lymphocyte, tumor (from cancer cases), and oral test samples. LOH or MIN of at least one marker was detected in 38 (86%) of 44 primary tumours. Identical alterations were found in the saliva samples in 35 of these 38 cases (92% of those with markers; 79% overall). MIN was detectable in the saliva in 24 (96%) of 25 cases in which it was present in the tumor, and LOH was identified in the test sample in 19 (61%) of 31 cases. No microsatellite alterations were detected in any of the samples from the healthy control subjects, reinforcing the validity of this method for detecting and monitoring HNSCC.

In addition is the application to preclinical screening, testing asymptomatic individuals to identify those who are at risk of developing a specific disease. Microsatellite instability might prove to be a more useful screening test for some tumors; Mao et al. ***(16)*** analyzed urine samples from 25 patients with suspicious bladder lesions that had been identified cystoscopically, both by conventional cytology and molecular methods. Microsatellite changes matching those in the tumor were detected in the urine sediment of 19 of the 20 patients (95%) who were diagnosed with bladder cancer, whereas urine cytology detected cancer cells in 9 of 18 (50%) of the samples. Their results suggested that microsatellite analysis, which, they state, can be performed at about one-third the cost of cytology, might be a useful addition to current screening methods for detecting bladder cancer.

In conclusion, MIN and LOH studies will continue to extend our understanding of the underlying genetic alterations predisposing to the various cancers, where specific gene alterations are, as yet, uncharacterized. The potential of microsatellites as genetic markers for use in the screening, early diagnosis, staging, and surveillance of cancer has been demonstrated. Combining the rapid technological developments with our ever more accurate definition of the genetic alterations which characterize cancer, it is foreseen that a minimal set of molecular markers might facilitate the effective screening of asymptomatic individuals.

Acknowledgments

The Bournemouth Leukaemia Fund and the Leukaemia Research Fund funded the work in the authors' laboratory and we thank all of our colleagues in the molecular biology and cytogenetics lab for their continued support.

References

1. Koreth, J., O'Leary, J. J., and McGee, J. (1996) Microsatellites and PCR genomic analysis. *J. Pathol.* **178(3),** 239–248.

2. Lee, W.-H., Bookstein, R., Hong, F., Young, L.-J., Shew, J.-Y., and Lee, E. Y. (1987) Human retinoblastoma susceptibility gene: cloning, identification and sequencing. *Science* **235,** 1394–1399.
3. Jones, M. H. and Nakamura, Y. (1992) Detection of loss of heterozygosity at the human TP53 locus using a dinucleotide repeat polymorphism. *Genes Chromosomes Cancer* **5(1),** 89–90.
4. Kinzler, K. W. and Vogelstein, B. (1996) Lessons from hereditary colorectal cancer. *Cell* **87,** 159–170.
5. Burton, M. P., Schneider, B. G., Brown, R., Escamilla-Ponce, N., and Gulley, M. L. (1998) Comparison of Histologic stains for use in PCR analysis of microdissected, paraffin-embedded tissues. *BioTechniques* **24,** 86–92.
6. Pabst, T., Schwaller, J., Tobler, A., and Fey, M. F. (1996) Detection of microsatellite markers in leukaemia using DNA from archival bone marrow smears. *Br. J. Haematol.* **95,** 135–137.
7. Huang, Q., Choy, K. W., Cheung, K. F., Lam, D. S., Fu, W. L., and Pang, C. P. (2003) Genetic alterations on chromosome 19, 20, 21, 22, and X detected by loss of heterozygosity analysis in retinoblastoma. *Mol. Vis.* **9,** 502–507.
8. Ibbotson, R. E. and Corcoran, M. M. (2001) Detection of Chromosomal Deletions by Microsatellite Analysis, in *Molecular Analysis of Cancer* (Boultwood, J. and Fidler, C., eds.), Humana, Totowa, NJ, pp. 59–65.
9. Arzimanoglou, II., Gilbert, F., and Barber, H. R. (1998) Microsatelllite instability in human solid tumours. *Cancer* **82,** 1808–1820.
10. Goggins, M. et al. (1998) Pancreatic adenocarcinomas with DNA replication errors (RER+) are associated with wild-type K-ras and characteristic histopathology. Poor differentiation, a syncytial growth pattern, and pushing borders suggest RER+. *Am. J. Pathol.* **152,** 1501–1507.
11. Takanishi, D. M., Jr. et al. (1995) Chromosome 17p allelic loss in colorectal carcinoma. Clinical significance. *Arch. Surg.* **130,** 585–589.
12. Kato, M., Ito, Y., Kobayashi, S., and Isono, K. (1996) Detection of DCC and Ki-ras gene alterations in colorectal carcinoma tissue as prognostic markers for liver metastatic recurrence. *Cancer* **77,** 1729–1735.
13. Chen, X. Q., Stroun, M., Magnenat, J. L., et al. (1996) Microsatellite alterations in plasma DNA of small cell lung cancer patients. *Nature Med.* **2,** 1033–1035.
14. Nawroz, H., Koch, W., Anker, P., Stroun, M., and Sidransky, D. (1996) Microsatellite alterations in serum DNA of head and neck cancer patients. *Nature Med.* **2,** 1035–1037.
15. Spafford, M. F., Koch, W. M., Reed, A. L., et al. (2001) Detection of head and neck squamous cell carcinoma among exfoliated oral mucosal cells by microsatellite analysis. *Clin. Cancer Res.* **7(3),** 607–612.
16. Mao, L. et al. (1996) Molecular detection of primary bladder cancer by microsatellite analysis. *Science* **271,** 659–662.

34

Analysis of Chromosomal Translocations

Andreas Hochhaus

1. Introduction

Chromosomal translocations were the first target for the specific detection of residual tumor cells in bone marrow and peripheral blood. Some types of leukemia are regularly or generally associated with translocations. In chronic myelogenous leukemia (CML) and a proportion of patients with acute lymphoblastic leukemia (ALL), the t(9;22)(q34;q11) translocation leads to the fusion of the *ABL* gene with part of the *BCR* gene. Locating the primers such that the polymerase chain reaction (PCR) product spans the fusion point makes it possible to amplify sequences of the hybrid gene specifically (see Chapter 6). Because the breakpoints are better defined at the RNA level than at the DNA level, the primary target is the RNA, which is transcribed into cDNA by reverse transcription. (RT-PCR). In this way, a single leukemic cell can be detected from among 10^6 normal bone marrow or peripheral blood cells. Comparable approaches have been applied to other fusions such as the *PML-RAR*α fusion gene in acute promyelocytic leukemia (APL). The presence of a few cells with phenotypic or genotypic features of leukemia cells after therapeutic regimens such as stem cell transplantation is know as "minimal residual disease" (MRD).

The diagnostic and prognostic significance of MRD as detected by the amplification of translocations depends very much on the particular genetic abnormality. Most cases of APL with the t(15;17) translocation in continuous complete remission are PCR negative, and positive test results are predictive of clinical relapse. In contrast, almost all acute myeloid leukemia (AML) patients with the t(8;21) in continuous complete remission have MRD, as detected by PCR, without subsequent relapse. These extreme examples show that the diagnostic and prognostic value of residual leukemia cells detected by PCR depend on the type of leukemia, the specific treatment and the translocation analyzed.

The clinical advantage is obvious in cases in which prognosis is correlated with the presence of leukemia cells harboring a defined translocation *(1)*.

2. Leukemias with Specific Fusion Transcripts

The molecular diagnosis of leukemias began with the recognition and analysis of recurrent chromosomal translocations *(2,3)*. The genes discovered at the translocation breakpoints have drawn attention to critical regulatory pathways in hematopoietic cells that can cause cancer when they are dysregulated. In many acute leukemias, translocations fuse genes that reside on the two partner chromosomes, creating a chimeric gene with novel oncogenic properties.

Chromosomal translocations have been used to identify patients with acute leukemia with distinct clinical outcomes. In AML, for instance, the presence of a t(8;21) translocation or a inversion of the chromosome 16 identifies patients with a comparatively good prognosis,

From: *Medical Biomethods Handbook*
Edited by: J. M. Walker and R. Rapley © Humana Press, Inc., Totowa, NJ

Table 1
Selection of Translocations Associated with Hematologic Malignancies

Chromosomal translocation	Associated disease	Genes involved
t(1;6)(q11;p21)	Polycythemia vera	
t(1;9)(q10;p10)	Polycythemia vera	
t(3;21)(q26;q22)	CML-BC, AML, MDS	*EAP, AML1*
t(5;7)(q33;q11.2)	CMML	*PDGFRβ, HIP1*
t(5;10)(q33;q21.2)	CML-BC, MPD-U	*PDGFRβ, H4*
t(5;12)(q33;p13)	CEL, CMML, MPD-U, MDS	*PDGFRβ, TEL*
t(7;11)(p15;p15)	MPD, AML	
t(8;9)(q22;q34)	CEL	
t(8;13)(p11;q12)	CEL, MPD-U	*FGFR1, ZNF198*
t(9;22)(q34;q11)	CML, ALL, AML	*ABL, BCR*
t(14;18)(q32;q21),	Follicular lymphoma	*BCL2, IGH*
t(11;14)(p11;q32)	Mantle cell lymphoma	*BCL1, IGH*
t(2;5)(p23;q35)	Anaplastic large cell lymphoma	*NPM, ALK*
t(8;14)(q24;q32)	Burkitts lymphoma	*c-MYC, IGH*
t(1;19)(q23;p13)	Precursor B-ALL	*E2A, PBX1*
t(4;11)(q21;q23)	Precursor B-ALL	*MLL, AF4*
t(12;21)(p13;q22)	Precursor B-ALL	*TEL, AML1*
t(15;17)(q22;q21)	APL (AML FAB M3)	*PML, RARα*
t(8;21)(q22;q22)	AML FAB M2	*AML1, ETO*
inv(16)(p13;q22)	AML FAB M4Eo	*CBFβ, MYH11*

Abbreviations: CML, Chronic myelogenous leukemia; BC, Blast crisis; AML, Acute myeloid leukemia; MDS, Myelodysplastic syndrome; MPD-U, Myeloproliferative discorder, undefined; ALL, Acute lymphoblastic leukemia; APL, Acute promyelocytic leukemia.

whereas the t(9;22) translocation is associated with a poor outcome. Chromosomal translocations have been used to identify patients who will benefit from intensifying the dose of chemotherapy ***(4)***.

Chromosomal translocations could lead to a constitutive activation of a tyrosine kinase, which might be the target of specific therapy (e.g., *ABL* on chromosome 9q, *FGFR1* on chromosome 8p, *PDGFRα* on chromosome 4q or *PDGFRβ* on chromosome 5q ***(5)*** (**Table 1**).

Strategies to detect and quantify leukemia cells with specific translocations depend on the genomic anatomy of the fusion, i.e., the intron/exon structure and the heterogeneity of genomic breakpoints leading to variable fusion genes.

3. Chronic Myelogenous Leukemia

Chronic myelogenous leukemia (CML) constitutes a clinical model for molecular detection and therapy surveillance of malignant disease because this entity was the first leukemia shown to be associated with a specific chromosomal aberration, the Philadelphia chromosome 22q–. Rowley et al. demonstrated in 1973 that the Ph chromosome is actually the result of the recurrent translocation t(9;22)(q34;q11) ***(6,7)*** (*see* **Fig. 1**). The translocation generates two chimeric genes: *BCR-ABL* ***(8,9)*** on the derivative chromosome 22 and *ABL-BCR* on the derivative chromosome 9. *BCR-ABL* is transcribed and translated in most patients into a 210-kDa fusion protein with deregulated tyrosine kinase activity (*see* **Fig. 2**). *ABL-BCR* is expressed in about 60% of patients with CML, but it probably lacks any biological function.

The degree of tumor load reduction is an important prognostic factor for patients with CML on therapy. Response is expressed at three levels: (1) hematologic response, defined as the normalization of the peripheral blood values and of spleen size; (2) cytogenetic response,

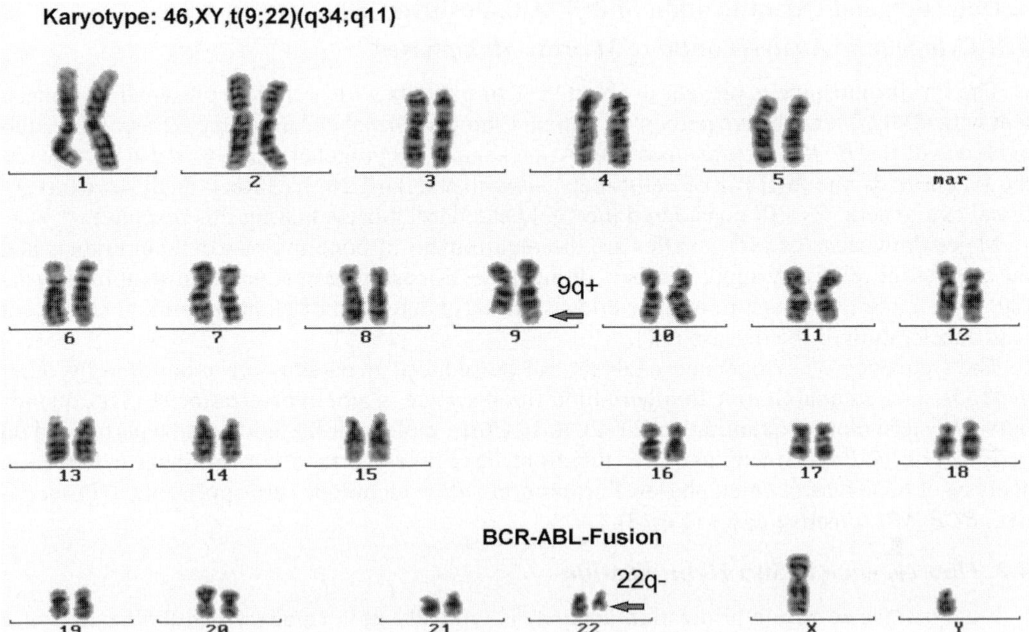

Fig. 1. Cytogenetic analysis (Giemsa banding) of a male patient with CML in chronic phase. (Courtesy of Dr. Claudia Schoch, Munich, Germany.)

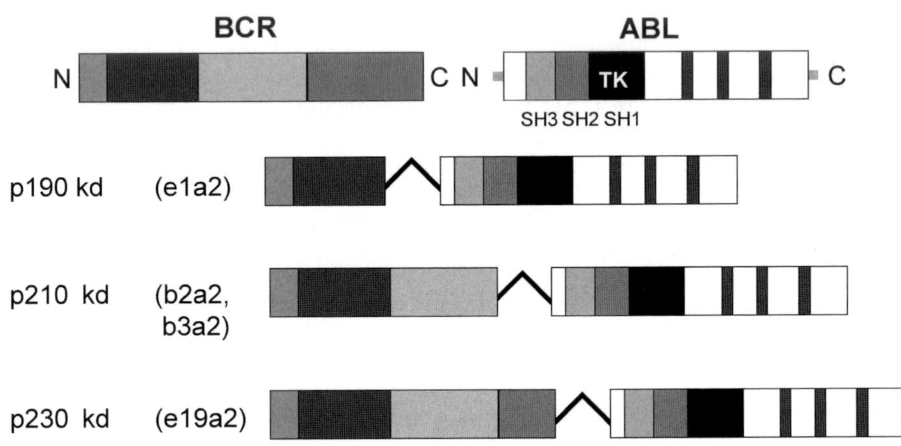

Fig. 2. Constitutive activation of the ABL tyrosine kinase domain (TK) by fusion to parts of the BCR protein.

defined as the proportion of residual Ph-positive metaphases; and (3) molecular response, defined according to the method used as the proportion of the residual *BCR-ABL* gene, transcript, or protein. The principal aim of residual disease analysis in patients with CML is to determine patient response to treatment and to enable early diagnosis of relapse.

4. Detection and Quantification of BCR-ABL-Positive Cells

4.1. Cytogenetic Analysis of Bone Marrow Metaphases

The Ph chromosome is present in about 90% of patients with a clinical presentation consistent with CML. Three to five percent of patients show a normal chromosome 22, but molecular evidence of the *BCR-ABL* translocation. At presentation cytogenetic analysis usually reveals the Ph chromosome in 100% of cells analyzed with standard 20- to 30-cell analysis. Conventional cytogenetics is still considered the "gold standard" for evaluating this response.

Major drawbacks of cytogenetics are the requirement of bone marrow cells in mitosis and the analysis of relatively small numbers of metaphases, resulting in significant sampling errors ***(10)***. The major advantage of cytogenetics is the early detection of clonal evolution consistent with acceleration of the disease ***(11)***.

The frequency of cytogenetic analysis can be reduced if patients are monitored by other methods, such as quantitative Southern blot, fluorescence *in situ* hybridization (FISH), quantitative Western blot, or quantitative RT-PCR. In CML, molecular methods can be performed on peripheral blood specimens and are, therefore, less invasive than conventional cytogenetic analysis of bone marrow metaphases. Furthermore, these techniques are applicable to Ph-negative, *BCR-ABL*-positive cases (**Fig. 3**).

4.2. *Fluorescence* In Situ *Hybridization*

Fluorescence *in situ* hybridization analysis is typically performed by cohybridization of a BCR and an ABL probe to denatured metaphase chromosomes or interphase nuclei. Probes are large genomic clones, such as cosmids or "yeast artificial chromosomes" (YACs), and are labeled with different fluorochromes. Dual-color FISH using probes for *BCR* and *ABL* genes allows the specific detection of the *BCR-ABL* gene fusion in interphase and/or metaphase nuclei. Most cells exhibit four distinct signals, two of each color corresponding to the two normal *BCR* and *ABL* alleles. CML cells are recognized by the juxtaposition of one of the *BCR* and *ABL* signals. FISH analysis does not depend on the presence of the typical Ph chromosome and will detect rare *BCR* and *ABL* variant fusions ***(13)***. The lineage of positive and negative cells can be determined in combination with conventional May–Grünwald staining or immunocytochemistry ***(14)***.

A limitation of the interphase FISH method is the background of a variant proportion of false-positive cells, depending on the probe/detection system used. In practice, the limit of detection of CML cells is typically 1–5% and depends, in part, on which probes are used, the size of the nucleus, the precise position of the breakpoint within the *ABL* gene, and the criteria used to define colocalization ***(15)***. The advantage of FISH over conventional cytogenetics is the analysis of a larger number of nuclei, resulting in smaller sampling errors. Therefore, FISH is applicable for quantification of residual disease in partial, minor, and nonresponders to interferon-α (IFN) and for determination of the BCR-ABL positivity of individual cell colonies or the proportion of *BCR-ABL*-positive cells in small samples, such as highly enriched cell fractions.

The specificity of interphase FISH can be increased by introducing an additional probe that permits identification of both the Ph chromosome and the derivative 9 chromosome in Ph+ cells, thus lowering the rate of false positivity ***(16)***. In many cases, however, large genomic deletions around the Ph translocation breakpoints preclude the use of these techniques ***(17)***.

The development of a "hypermetaphase FISH" based on improved culture techniques and computerized analysis of a large number of metaphases made it possible to distinguish different levels of Ph chromosome positivity at presentation. Where readings can be obtained on 500 cells, one can reliably estimate parameters that characterize MRD. However, hypermetaphase FISH is evaluating only cycling cells and cannot count Ph-positive cells that do not enter division ***(18)***.

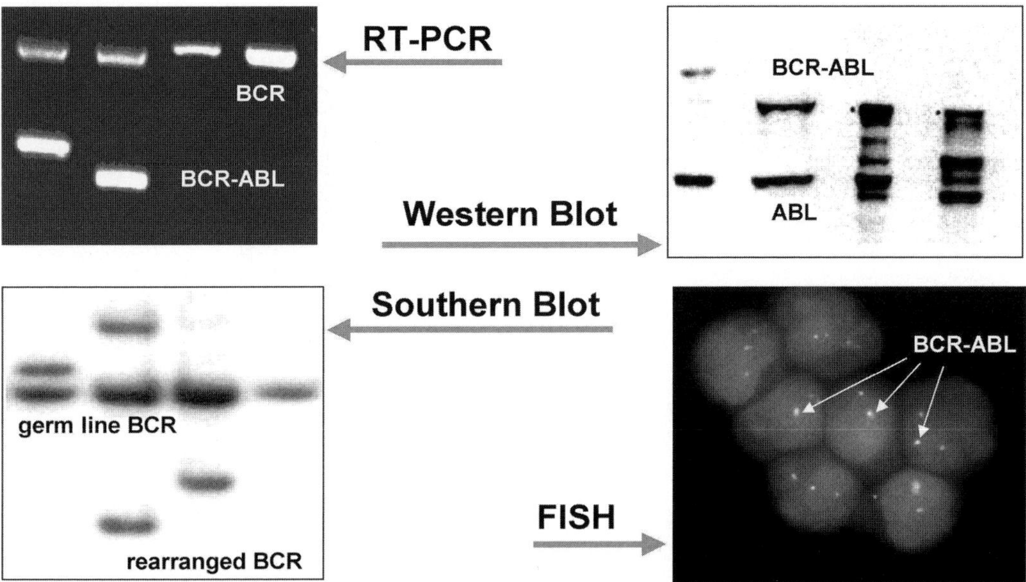

Fig. 3. Molecular methods to detect chromosomal translocations. (From **ref. *12*.**)

4.3. Southern Blot Analysis

Southern blotting exploits the fact that the breakpoint within the *BCR* gene in most cases falls in a limited area, the 5.8-kb major breakpoint cluster region (M-bcr). Genomic DNA extracted from leukocytes from a patient is digested by a set of appropriate restriction enzymes (e.g., *Bgl*II, *Xba*I, *Hin*dIII, *Eco*RI, *Bam*HI), fractionated on an agarose gel, transferred to a Nylon membrane, and hybridized to two labeled DNA probes derived from the 3' and 5' portion of M-bcr (*see* **Fig. 4**). After autoradiography, a band corresponding to the unrearranged *BCR* allele is visible; for patients with CML, one or two additional bands might reveal the rearranged *BCR* allele. Using this technique, a *BCR* rearrangement is detectable in about 98% of Ph-positive patients and in a significant proportion of Ph-negative cases. For routine use, *Bgl*II and *Xba*I are sufficient, which are informative in almost all cases. However, because there are rare restriction site polymorphisms in the M-bcr, the finding of a rearrangement with at least two restriction enzymes is generally considered necessary to exclude a false-positive result in any new patient. False-negative results can arise because the rearranged band is too large, too small, or, coincidentally, exactly the same size as the normal allele, as a result of partial deletion of *BCR* sequences on the translocated allele or of rare variants that have breakpoints outside the M-bcr. However, false-positive or false-negative results are rare.

After therapy, Southern blot analysis allows quantification of the proportion of cells with *BCR* rearrangement compared to all cells investigated. The proportion of CML cells is determined by twice the intensity of the rearranged band divided by the sum of the intensities of the rearranged plus germline bands (BCR ratio; *see* **Fig. 5**). This is because each CML cell contributes signals from one normal and one rearranged chromosome 22, whereas the normal cells contribute identical signals from two normal chromosomes. The level of disease detected in contemporaneous peripheral blood and bone marrow samples is essentially identical and, therefore, peripheral blood is usually used for analysis. Quantitative Southern blot allows the detection and quantification of as few as to 1–5% leukemic cells. To evaluate the response to treatment, the knowledge of the initial restriction pattern and intensity of the rearranged band is

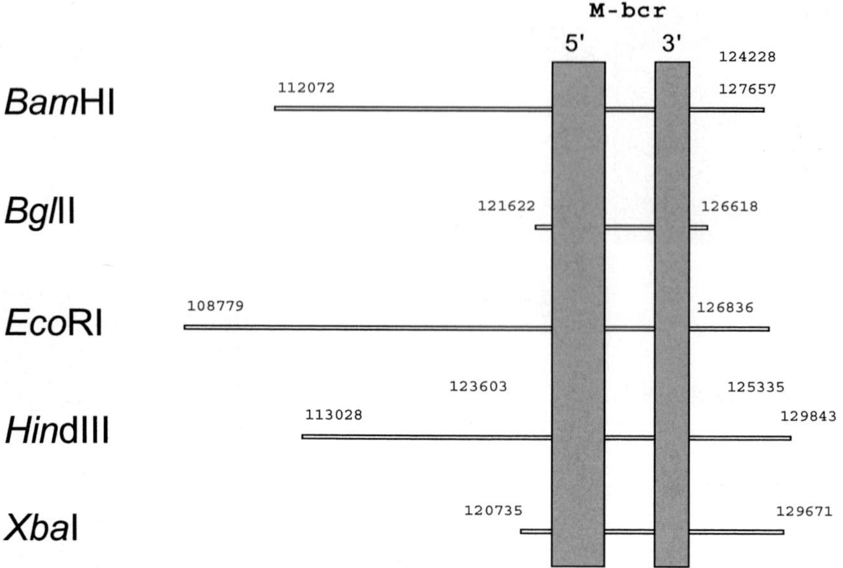

Fig. 4. Southern blot analysis. The *BCR* gene is digested by various endonucleases and labeled with two different radioactive probes. (From **ref. *19*.**)

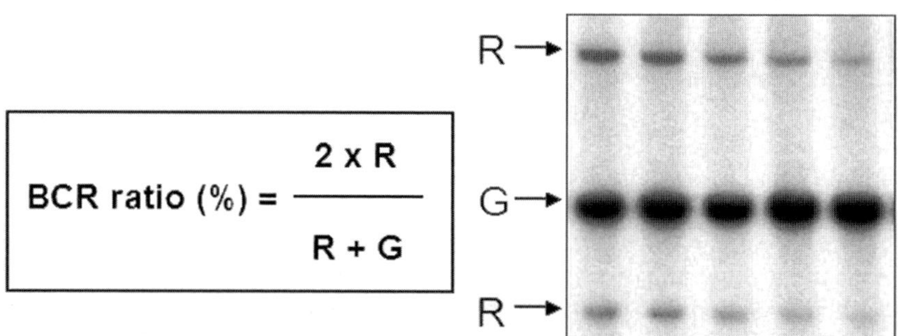

Fig. 5. Southern blot analysis after *Bgl*II digest of genomic DNA labeled with 5' and 3' M-bcr probes and autoradiographed. Arrows indicate the germline band (G) and rearranged bands (R). (From **ref. *19*.**)

mandatory. Complete cytogenetic responses are associated with disappearance of rearranged BCR bands in most cases.

The BCR ratio is usually lower than the proportion of Ph-positive metaphases. The reason for this is that Southern blotting analyzes a large number of dividing and resting cells, including *BCR-ABL*-negative lymphocytes, and cytogenetic analysis provides information only on a small number of dividing myeloid cells. Empirically derived cutoff points in the BCR ratio were introduced in order to define molecular response groups: a BCR ratio of 0% was defined as complete response, and ratios of 1–24%, 25–50%, and >50% were defined as partial, minor, and no molecular response, respectively. Using these cutoff points, a major cytogenetic response could be predicted or excluded in more than 90% of cases. The main advantage of Southern blotting over cytogenetics is the independence from dividing cells, which permits the use of peripheral blood instead of bone marrow ***(19,20)***.

4.4. Western Blot Analysis

Western blotting can be used to detect BCR-ABL proteins directly in cell extracts qualitatively and quantitatively both in bone marrow and in peripheral blood. Leukocytes are lysed in the presence of potent protease inhibitors, fractionated on a polyacrylamide gel, transferred to a Nylon membrane, and probed with an anti-ABL antibody. Different types of BCR-ABL protein (p190, p210, p230, rare variants) can be distinguished from p145 ABL by differences in migration. The limit of sensitivity is about 0.5–1%. A quantitative Western blot assay found a linear correlation between BCR-ABL/ABL protein ratios and contemporaneous conventional cytogenetics ***(21)***.

4.5. Reverse Transcriptase–Polymerase Chain Reaction

In 1989, encouraging results concerning detection of MRD by PCR in CML patients after allogeneic bone marrow transplantation were first reported ***(22)***. However, conflicting data from a comparative multicenter study revealed serious problems of the method with a high rate of false-positive results and provoked an open discussion. Over the past 15 yr, PCR has been optimized and developed. Specificity has been considerably increased by the partial standardization of methodology and the introduction of rigorous precautions to avoid contamination. Sensitivity has been improved by using nested primer pairs and performing two consecutive PCR steps. In view of the limited value of qualitative PCR for monitoring CML patients after therapy, quantitative *BCR-ABL* PCR assays were developed to monitor patients after stem cell transplantation and treatment with IFN or imatinib and are now in routine clinical use ***(12,23)***.

4.6. Screening for BCR-ABL mRNA Transcripts at Diagnosis

For diagnostic samples, the use of *multiplex PCR* has been suggested to detect simultaneously several kinds of BCR-ABL and BCR transcripts as internal controls in one reaction ***(24)*** by using three BCR primers and one ABL primer. This method allows the reliable detection of typical BCR-ABL transcripts, such as b2a2 or b3a2, and atypical types (e.g., transcripts lacking *ABL* exon a2 [b2a3 and b3a3], transcripts resulting from *BCR* breakpoints outside M-bcr, such as e1a2, e6a2 or e19a2, or transcripts with inserts between *BCR* and *ABL* exons) (*see* **Fig. 6**).

4.7. Detection of Minimal Residual Disease, "Nested" RT-PCR

Because patients with leukemia at presentation or relapse usually have a total burden of more than 10^{12} malignant cells and cytogenetics, Western blot, and conventional FISH have a sensitivity of maximum 1%, a patient with negative results might harbor as few as zero or as many as 10^{10} residual leukemic cells. At this point, the patient is judged to be in clinical and hematologic remission, although the term "remission" refers only to an arbitrary point of a continuum of residual leukemic cell numbers (*see* **Fig. 7**) ***(25)***.

Reverse transcription–PCR for *BCR-ABL* mRNA is, by far, the most sensitive assay in the context of residual disease analysis and can detect a single leukemia cell in a background of 10^5–10^6 normal cells. Therefore, PCR is up to four orders of magnitude more sensitive than conventional methods. However, patients who have no residual disease detectable by RT-PCR might still harbor up to a million malignant cells that could contribute to subsequent relapse. The sensitivity with which residual disease can be detected will be limited by the amount of peripheral blood or bone marrow that can be analyzed.

By *nested* RT-PCR with two pairs of ("nested") primers corresponding to appropriate *BCR* and *ABL* exons used in two rounds of amplification, residual CML cells after treatment can be specifically detected with a sensitivity of up to 1 in 10^6 cells (*see* **Figs. 8** and **9**) ***(26)***.

This qualitative method has been used for MRD detection after allogeneic stem cell transplantation. Most transplant centers have demonstrated that the majority of patients is PCR positive in the first 6 mo after transplantation, two-thirds of patients become PCR negative during follow-up as a result of the graft-vs-leukemia effect, persistent PCR negativity after 1 yr is a

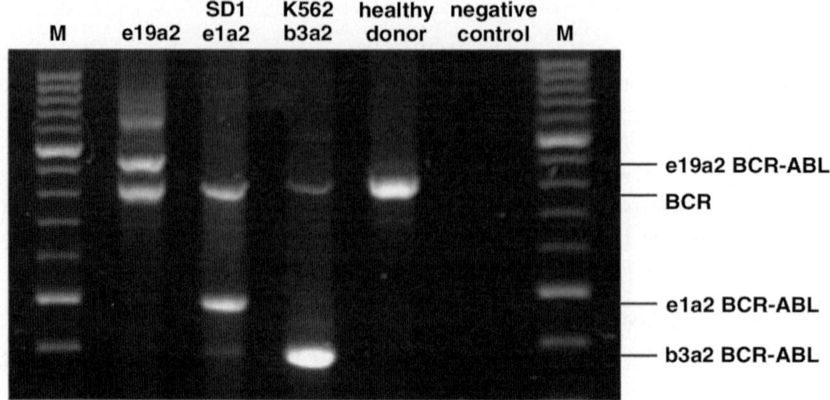

Fig. 6. Example of multiplex RT-PCR to detect *BCR-ABL* transcripts at diagnosis. BCR transcripts are amplified as internal controls. (From **ref. *24*.**)

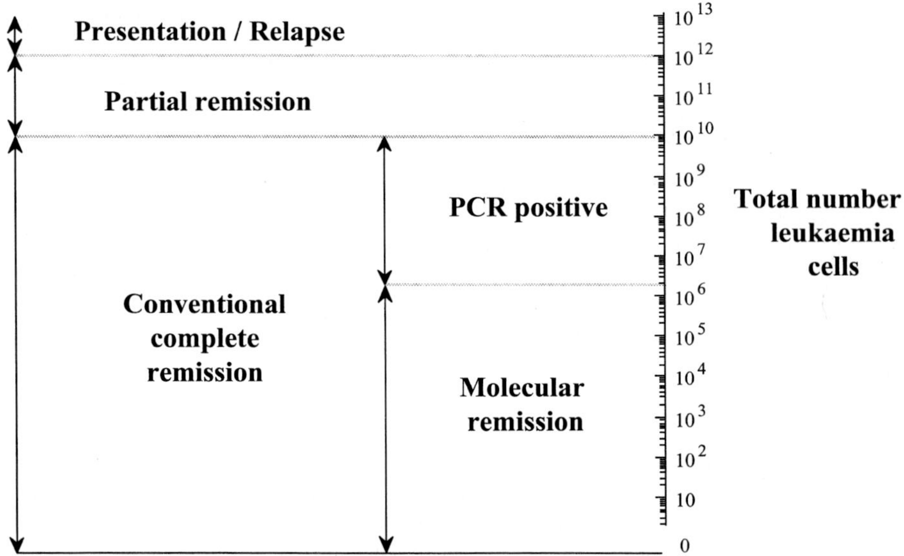

Fig. 7. Sensitivity of cytogenetic analysis vs RT-PCR in CML. Assuming that patients at diagnosis harbor about 10^{12} leukemic cells, a patient in complete cytogenetic remission might harbor up to 10^{10}, or down to no leukemic cells. RT-PCR allows the detection of minimal residual disease with a sensitivity up to 10^{-6}. (From **ref. *12*.**)

marker for good prognosis, and PCR-positive patients more than 6 mo after stem cell transplantation have a great risk of relapse. However, qualitative PCR cannot predict relapse in the individual patient. In the majority of cases after transplantation, RT-PCR and DNA-PCR (using patient-specific primers) results are concordant (i.e., that patients in remission do not generally harbor a substantial pool of CML cells that do not express *BCR-ABL* mRNA) ***(27)***.

Nested PCR is essentially useless in patients after IFN therapy even in cytogenetic remission, as almost all patients remain repeatedly positive ***(28)***.

If the RT-PCR method is pushed to the extreme, *BCR-ABL* mRNA can be detected at a very low level of 1–10 transcripts per 10^8 cells in many normal individuals, with a frequency that is

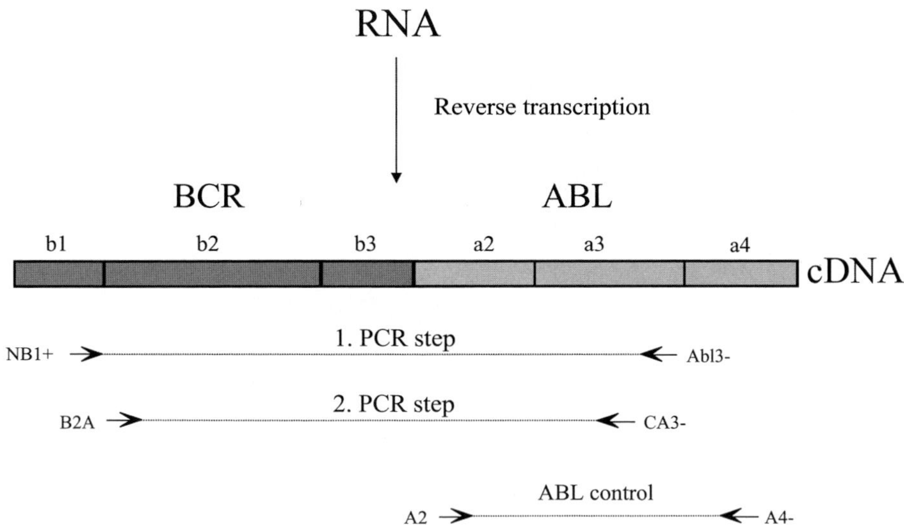

Fig. 8. Nested RT-PCR strategy for the detection of *BCR-ABL* transcripts. Primers for the *ABL* control gene span over two introns to avoid amplification of genomic DNA.

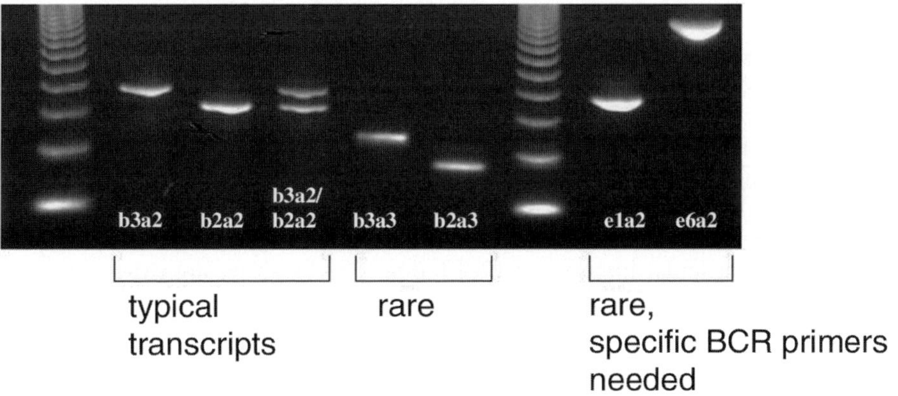

Fig. 9. Agarose gel electrophoresis of nested RT-PCR products of various types of *BCR-ABL* transcripts.

age dependent ***(29)***. It has been suggested that *BCR-ABL* and probably several other fusion genes are being continuously formed in mitotic cells in the normal bone marrow, but only the combination of an in frame *BCR-ABL* fusion in the correct primitive hematopoietic progenitor would have the selective advantage to become functional as an expanding clone. In addition, it is possible that *BCR-ABL* alone is not sufficient to result in the expansion of myeloid cell numbers and that other cooperating genetic events might be required.

4.8. Quantitative PCR

In view of the very limited value of qualitative PCR, several groups have developed quantitative PCR assays to estimate the amount of residual disease in positive specimens (see Chapter 25). Most groups have initially used *competitive PCR strategies* that can effectively control for variations in amplification efficiency and reaction kinetics ***(30)***.

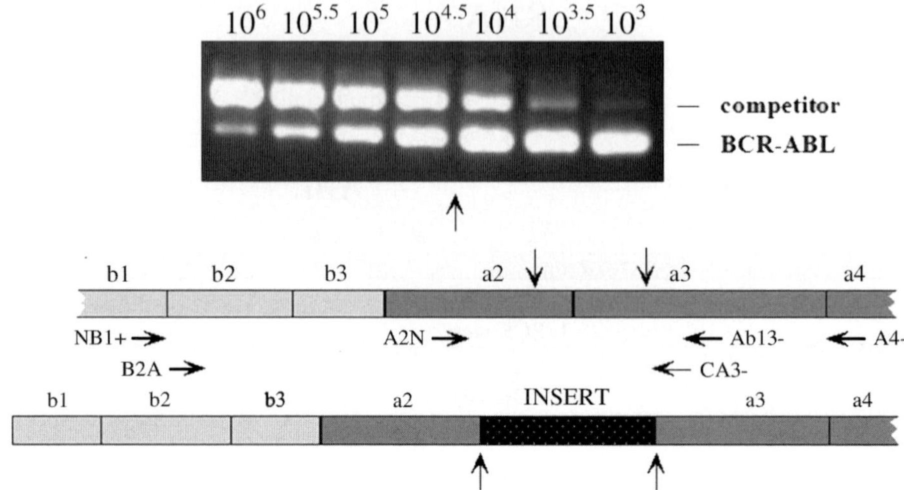

Fig. 10. Competitive PCR for the quantification of residual disease in CML. Nested PCR is performed using serial dilutions of competitor constructs added to the same volume of patients cDNA: The equivalence point (arrow) is determined by densitometry. (From **ref. *30***.)

In general, nested PCR is performed using serial dilutions of a *BCR-ABL* competitor construct added to the same volume of patients' cDNA. The equivalence point at which the competitor and sample band would be of equal intensity is determined by densitometry. In order to standardize results for both quality and quantity of blood, RNA, and cDNA quantification of transcripts of normal housekeeping genes, such as *ABL* or glucose-6-phosphate dehydrogenase (*G6PD*) has been employed (*see* **Fig. 10**). The standardized results are expressed as the ratios BCR-ABL/ABL or BCR-ABL/G6PD in percent. The quantification of the transcript level of control genes is of particular importance if different RNA qualities are expected (i.e., in particular, if samples are mailed to a specialized laboratory).

In patients after allogeneic stem cell transplantation, rising or persistently high levels of *BCR-ABL* mRNA can be detected prior to cytogenetic or hematologic relapse. Low or falling *BCR-ABL* transcript levels are associated with continuous remission, whereas high or rising *BCR-ABL* transcript levels predict relapse ***(31)***.

Quantitative PCR is the method of choice to determine the best time-point for therapeutic interventions in the case of relapse after stem cell transplantation. Quantitative PCR data have been used to determine the optimum time-point to initiate donor lymphocyte transfusions and to monitor its response ***(32)***.

Quantitative RT-PCR for *BCR-ABL* has been shown to be a reliable method for monitoring residual leukemia load in mobilized peripheral blood stem cells, particularly in Ph-negative collections. Quantitative RT-PCR allows selection of the best available collections for reinfusion into patients after myeloablative therapy (autografting) ***(33)***.

Almost all patients after IFN therapy are persistently positive for *BCR-ABL* transcripts. The median ratios of complete, partial, minor, and nonresponders differ significantly. Cytogenetic response to IFN (complete, partial, minor/none) was compared to molecular response by introducing cutoff points for the BCR-ABL/ABL ratio. Optimum cutoff points were 2% and 14%; that is, a complete response is associated with a ratio up to 2%, a partial response with a ratio of 2 and 14%, and a minor or nonresponse with a ratio of >14%. In complete cytogenetic responders, the MRD level has prognostic impact regarding stability of the response (*see* **Fig. 11**) ***(28)***.

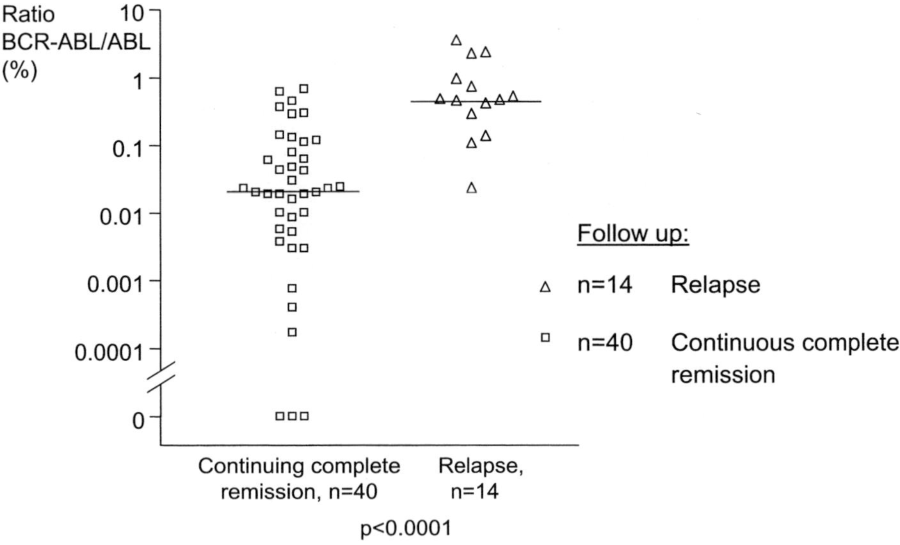

Fig. 11. Minimal residual disease in CML patients treated with IFN after achievement of complete cytogenetic remission. Residual disease spans a range over four orders of magnitude. Patients who consecutively relapsed (n=14) showed higher *BCR-ABL* values at the time of complete remission compared to patients in continuous remission. (From **ref.** ***28***.)

Recently, novel *real-time PCR procedures* have been developed that promise to simplify existing protocols (*see* **Fig. 12**) (see Chapter 25). Several procedures for quantification of *BCR-ABL* mRNA using the TaqMan system have been developed ***(34)***. The assay is based on the use of the 5' nuclease activity of *Taq* polymerase to cleave a nonextendible dual-labeled hybridization probe during the extension phase of PCR. One fluorescent dye serves as a reporter, and its emission spectra is quenched by the second fluorescent dye. The nuclease degradation of the probe releases the quenching resulting in an increase of fluorescent emission. The fluorescence is monitored by a sequence detector in real time. C_T (threshold cycle) values are calculated by determining the point at which the fluorescence exceeds a threshold limit. C_T corresponds to the amount of target transcripts in the sample.

An alternative real-time RT-PCR approach for the detection and quantification of *BCR-ABL* fusion transcripts has been established using the LightCycler technology ***(35)***, which combines rapid thermocycling with online fluorescence detection of PCR product formation as it occurs. Fluorescence monitoring of PCR amplification is based on the concept of fluorescence resonance energy transfer (FRET) between two adjacent hybridization probes carrying donor and acceptor fluorophores. Excitation of a donor fluorophore (fluorescein) with an emission spectrum that overlaps the excitation spectrum of an acceptor fluorophore results in nonradioactive energy transfer to the acceptor. Once conditions are established, the amount of fluorescence resulting from the two probes is proportional to the amount of PCR product. As a result of amplification in glass capillaries with a low volume/surface ratio PCR reaction times have been reduced to less than 30 min.

A pair of probes was designed that was complementary to *ABL* exon 3, thus enabling detection of all known *BCR-ABL* variants and also normal *ABL* as an internal control. Conditions were established to amplify less than 10 target molecules/reaction and to detect 1 CML cell in 10^5 cells from healthy donors and 1 K562 cell in 10^7 HL60 cells. To determine the utility of the assay, *BCR-ABL* and *ABL* transcripts in a series of 254 samples from 120 patients with CML after therapy were quantified. The level of residual disease was expressed as the ratio of BCR-ABL/ABL. A highly significant correlation was seen between the BCR-ABL/ABL ratios deter-

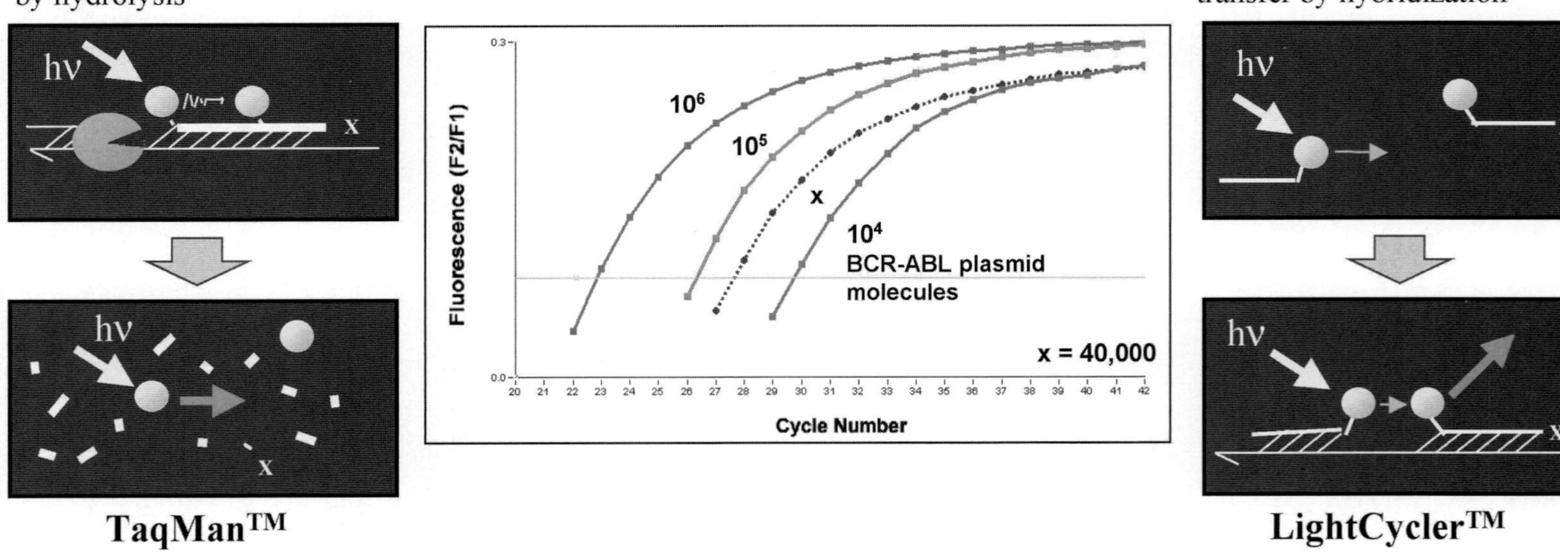

Fig. 12. Real-time PCR for the quantification of fusion gene transcripts. The TaqMan method uses a single probe labeled with two different dyes. The release of the reporter dye from the quencher during amplification by hydrolysis results in an enhanced fluorescence. The LightCycler method is based on a pair of hybridization probes that lead to an enhanced fluorescence in case of juxtaposition.

mined by the LightCycler and (1) the BCR-ABL/ABL ratios obtained by nested competitive RT-PCR performed with the same cDNA samples, (2) the proportion of Ph-chromosome-positive metaphases determined by contemporaneous cytogenetics, and (3) the BCR ratio determined by Southern blot analysis ***(36)***.

Real-time PCR approaches are reliable and sensitive methods for monitoring CML patients after therapy. The major advantages of the methodology are as follows: (1) amplification and product analysis are performed in the samc reaction vessel, avoiding the risk of contamination, and (2) the results are standardized by the quantification of housekeeping genes.

The introduction of imatinib as a selective inhibitor of the *BCR-ABL* tyrosine kinase in the treatment of CML patients dramatically demonstrates the need for a common language in the expression of MRD data. As many as 74% complete cytogenetic responders have been reported after 18 mo of imatinib therapy in newly diagnosed patients. A systematic quantitative RT-PCR analysis of the patients showed the prognostic relevance of molecular response. All patients with CCR, who showed at least a 3-log reduction (comparable with a ratio BCR-ABL/ABL <0.1%) of *BCR-ABL* transcript levels at 12 mo, remained progression-free at 24 mo. In contrast, the probability of remaining progression free was 96% at mo 24 when the patients achieved a less than 3-log reduction of *BCR-ABL* transcript levels at mo 12 ***(37,38)***.

Quantitative determination of residual disease levels after treatment for patients with CML can be achieved by various methods ***(40)***. Quantitative assessment of residual disease will help to modify and optimize treatment with imatinib for individual patients over time. However, there is a need to standardize methodologies and interpretation of results in order to reduce interlaboratory variation. Standardized parameters should be used to measure and express residual disease levels within clinical trials ***(40–42)***. The development of new real-time RT-PCR procedures offers a unique opportunity for this to be achieved and quantitative real-time PCR will shortly become a routine and robust basis for clinical decision-making, not only in CML but also in other leukemias with specific molecular markers.

The various strategies that have been developed and used in the last 20 yr to detect and quantify the *BCR-ABL* rearrangement in CML serve as a model for the detection and quantification of chromosomal translocations on the cytogenic, genomic, RNA, or protein level. The lessons learned from CML can be used to optimize strategies in other hematologic disorders characterized by specific translocations.

References

1. Wagener, C. (1997) Molecular diagnostics. *J. Mol. Med.* **75,** 728–744.
2. Rowley, J. D. (1998) The critical role of chromosome translocations in human leukemias. *Annu. Rev. Genet.* **32,** 495–519.
3. Nowell, P. C. (2002) Progress with chronic myelogenous leukemia: a personal perspective over four decades. *Annu. Rev. Med.* **53,** 1–13.
4. Bloomfield, C. D., Lawrence, D., Byrd, J. C., et al. (1998) Frequency of prolonged remission duration after high-dose cytarabine intensification in acute myeloid leukemia varies by cytogenetic subtype. *Cancer Res.* **58,** 4173–4179.
5. Cross, N. C. and Reiter, A. (2002) Tyrosine kinase fusion genes in chronic myeloproliferative diseases. *Leukemia* **16,** 1207–1212.
6. Nowell, P. C. and Hungerford, D. A. (1960) A minute chromosome in human chronic granulocytic leukemia. *Science* **132,** 1497–1501,
7. Rowley, J. D. (1973) A new consistent chromosome abnormality in chronic myelogenous leukemia detected by quinacrine fluorescence and Giemsa staining. *Nature* **243,** 290–293.
8. Groffen, J., Stephenson, J. R., Heisterkamp, N., de Klein, A., Bartram, C. R., and Grosveld, G. (1984) Philadelphia chromosomal breakpoints are clustered within a limited region, bcr, on chromosome 22. *Cell* **36,** 93–99.
9. Stam, K., Heisterkamp, N., Grosveld, G., et al. (1985) Evidence of a new chimeric bcrc-abl mRNA in patients with chronic myelocytic leukemia and the Philadelphia chromosome. *N. Engl. J. Med.* **313,** 1429–1433.
10. Hook, E. B. (1977) Exclusion of chromosomal mosaicism: tables of 90%, 95%, and 99% confidence limits and comments on use. *Am. J. Hum. Genet.* **29,** 94–97.

11. Kantarjian, H. M., Dixon, D., Keating, M. J., et al. (1988) Characteristics of accelerated disease in chronic myelogenous leukemia. *Cancer* **61,** 1441–1446.
12. Hochhaus A, Weisser A, La Rosée P, et al. (2000) Detection and quantification of residual disease in chronic myelogenous leukemia. *Leukemia* **14,** 998-1005.
13. Tkachuk, D. C., Westbrook, C. A., Andreeff, M., et al. (1990) Detection of bcr-abl fusion in chronic myelogeneous leukemia by in situ hybridization. *Science* **250,** 559–562.
14. Weber-Matthiesen, K., Winkemann, M., Müller-Hermelink, A., Schlegelberger, B., and Grote, W. (1992) Simultaneous fluorescence immunophenotyping and interphase cytogenetics: a contribution to characterization of tumor cells. *J. Histochem. Cytochem.* **40,** 171–175.
15. Chase, A., Grand, F., Zhang, J. G., Blackett, N., Goldman, J., and Gordon, M. (1997) Factors influencing the false positive and negative rates of BCR-ABL fluorescence in-situ hybridization. *Genes Chromosomes Cancer* **18,** 246–253.
16. Sinclair, P. B., Green, A. R., Grace, C., and Nacheva, E. P. (1997) Improved sensitivity of BCR-ABL detection: a triple-probe three-color fluorescence in situ hybridization system. *Blood* **90,** 1395–1402.
17. Dewald, G. W., Wyatt, W. A., Juneau, A. L., et al. (1998) Highly sensitive fluorescence in situ hybridization method to detect double BCR-ABL fusion and monitor response to therapy in chronic myeloid leukemia. *Blood* **91,** 3357–3365.
18. Seong, D. C., Kantarjian, H. M., Ro, J. Y., et al. (1995) Hypermetaphase fluorescence in situ hybridization for quantitative monitoring of Philadelphia chromosome-positive cells in patients with chronic myelogenous leukemia during treatment. *Blood* **86,** 2343–2349.
19. Reiter, A., Skladny, H., Hochhaus, A., et al. (1997) Molecular response of CML patients treated with interferon- monitored by quantitative Southern blot analysis. *Br. J. Haematol.* **97,** 86–93.
20. Verschraegen, C. F., Talpaz, M., Hirsch Ginsberg, C. F., et al. (1995) Quantification of the breakpoint cluster region rearrangement for clinical monitoring in Philadelphia chromosome-positive chronic myeloid leukemia. *Blood* **85,** 2705–2710.
21. Guo, J. Q., Lian, J. Y., Xian, Y. M., et al. (1994) BCR-ABL protein expression in peripheral blood cells of chronic myelogenous leukemia patients undergoing therapy. *Blood* 83, 3629–3637.
22. Morgan, G. J., Hughes, T., Janssen, J. W., et al. (1989) Polymerase chain reaction for detection of residual leukemia. *Lancet* **1,** 928–929.
23. Cross, N. C. P. (1997) Assessing residual leukemia. *Baillières Clin. Haematol.* **10,** 389–403.
24. Cross, N. C. P., Melo, J. V., Feng, L., and Goldman, J. M. (1994) An optimized multiplex polymerase chain reaction (PCR) for detection of BCR-ABL fusion mRNAs in haematological disorders. *Leukemia* **8,** 186–189.
25. Morley, A. (1998) Quantifying leukemia. *N. Engl. J. Med.* **339,** 627–629.
26. van Dongen, J. J. M., MacIntyre, E. A., Gabert, J. A., et al. (1999) Standardized RT-PCR analysis of fusion gene transcripts from chromosome aberrations in acute leukemia for detection of minimal residual disease. *Leukemia* **12,** 1901–1928.
27. Lin, F., van Rhee, F., Goldman, J. M., and Cross, N. C. P. (1996) Kinetics of increasing BCR-ABL transcript numbers in chronic myeloid leukemia patients who relapse after bone marrow transplantation. *Blood* **87,** 4473–4478.
28. Hochhaus, A., Reiter, A., Saußele, S., et al., for the German CML Study Group and the UK MRC CML Study Group (2000) Molecular heterogeneity in complete cytogenetic responders after interferon-_ therapy for chronic myelogenous leukemia: low levels of minimal residual disease are associated with continuing remission. *Blood* **95,** 62–66.
29. Biernaux, C., Loos, M., Sels, A., Huez, G., and Stryckmans, P. (1995) Detection of major bcr-abl gene expression at a very low level in blood cells of some healthy individuals. *Blood* **88,** 3118–3122.
30. Cross, N. C. P., Feng, L., Chase, A., Bungey, J., Hughes, T. P., and Goldman, J. M. (1993) Competitive polymerase chain reaction to estimate the number of BCR-ABL transcripts in chronic myeloid leukemia patients after bone marrow transplantation. *Blood* **82,** 1929–1936.
31. Lin, F., Kirkland, M. A., van Rhee, F., et al. (1996) Molecular analysis of transient cytogenetic relapse after allogeneic bone marrow transplantation for chronic myeloid leukemia. *Bone Marrow Transplant.* **18,** 1147–1152.
32. van Rhee, F., Lin, F., Cullis, J. O., et al. (1994) Relapse of chronic myeloid leukemia after allogeneic bone marrow transplant: the case for giving donor leukocyte transfusions before the onset of hematologic relapse. *Blood* **83,** 3377–3383.

33. Corsetti, M. T., Lerma, E., Dejana, A., et al. (1999) Quantitative competitive reverse transcriptase-polymerase chain reaction for BCR-ABL on Philadelphia-negative leukaphereses allows the selection of low-contaminated peripheral blood progenitor cells for autografting in chronic myelogenous leukemia. *Leukemia* **13,** 999–1008.
34. Preudhomme, C., Révillion, F., Merlat, A., et al. (1999) Detection of BCR-ABL transcripts in chronic myeloid leukemia (CML) using a 'real time' quantitative RT-PCR assay. *Leukemia* **13,** 957–964.
35. Wittwer, C. T., Herrmann, M. G., Moss, A. A., and Rasmussen, R. P. (1997) Continuous fluorescence monitoring of rapid cycle DNA amplification. *Biotechniques* **22,** 130–138.
36. Emig, M., Saussele, S., Wittor, H., et al. (1999) Accurate and rapid analysis of residual disease in patients with CML using specific fluorescent hybridization probes for real time quantitative RT-PCR. *Leukemia* **13,** 1825–1832.
37. Hughes, T. P., Kaeda, J., Branford, S., for the International Randomized Study of Interferon versus STI571 (IRIS) Study Group. (2003) Frequency of major molecular responses to imatinib or interferon alfa plus cytarabine in newly diagnosed patients with chronic myeloid leukemia. *N. Engl. J. Med.* **349,** 1421–1430.
38. Müller, M. C., Gattermann, N., Lahaye, T., et al. (2003) Dynamics of BCR-ABL mRNA expression in first line therapy of chronic myelogenous leukemia patients with imatinib or interferon α/ara-C. *Leukemia* **17,** 2392–2400.
39. Schoch, C., Schnittger, S., Bursch, S., et al. (2002) Comparison of chromosome banding analysis, interphase- and hypermetaphase-FISH, qualitative and quantitative PCR for diagnosis and for follow-up in chronic myeloid leukemia: A study of 350 cases. *Leukemia* **16,** 53–59.
40. van der Velden, V. H. J., Hochhaus, A., Cazzaniga, G., Szczepanski, T., Gabert, J., and van Dongen, J. J. M. (2003) Detection of minimal residual disease in hematologic malignancies by real-time quantitative PCR: principles, approaches, and laboratory aspects. *Leukemia* **17,** 1013–1034.
41. Gabert, J., Beillard, E., van der Velden, V. H. J., et al. (2003) Standardization and quality control studies of "real-time" quantitative reverse transcriptase polymerase chain reaction (RQ-PCR) of fusion gene transcripts for residual disease detection in leukemia—a Europe Against Cancer Program. *Leukemia* **17,** 2318–2357.
42. Beillard, E., Pallisgaard, N., van der Velden, V. H. J., et al. (2003) Evaluation of candidate control genes for diagnosis and residual disease detection in leukemic patients using "real-time" quantitative reverse-transcriptase polymerase chain reaction (RQ-PCR)—an Europe Against Cancer Program. *Leukemia* **17,** 2474–2486.

35

Differential Display

Theory and Applications

Irina Gromova, Pavel Gromov, and Julio E. Celis

1. Introduction

Recent data released by the Human Genome Consortium and Celera Genomics estimates that the human genome might contain about 30,000 protein coding genes ***(1***, and references therein), of which about 15% are believed to be expressed in any given cell type (see Chapter 42). The set of expressed proteins and the corresponding messenger RNAs, termed the transcriptome, defines the phenotype of a given cell, tissue, as well as the whole organism. The identification of genes that are differentially expressed under various physiological conditions is one of the major challenges in molecular biology today, as it provides an overview of the regulatory changes that take place in health and disease and highlights potential targets for drug discovery and therapeutic intervention ***(2)***. Tremendous efforts are generally required to identify those changes, as the population of altered messengers rarely comprises more than 1% of the total transcripts. Over the years, a variety of methods for the identification of differentially regulated genes in cells and tissues have been developed. Northern hybridization ***(3)***, nuclease protection ***(4)***, and subtractive ***(5)*** and differential ***(6)*** hybridization all have a number of serious drawbacks because they measure only single RNA species at a time and require a relatively large amount of starting material. The latter is particularly significant because rare transcripts, which often represent low-abundant cell cycle regulators, growth factors, and their receptors as well as signal transduction components, might be missed during the analysis.

Novel technologies that allow the screening of a whole population of mRNAs in one experiment and the detection of multiple changes in the transcriptome include differential display with its numerous modified variants ***(7–10)***, serial analysis of gene expression (SAGE) ***(11)***, and high-density hybridization-based cDNA microarrays ***(12,13)***. Differential display is perhaps the ideal method for time-course, dose response, and multiple treatment studies.

2. Differential Display: General Principles

The principle of RNA fingerprinting was introduced more then 10 yr ago with the development of two basic procedures, namely differential display (DD) ***(7)*** and RNA arbitrary primed polymerase chain reaction (PCR) (RAP-PCR) ***(14)***. Both approaches are based on the assumption that virtually every single mRNA species expressed in a living cell can be detected if a sufficiently large set of primers is utilized ***(15)***. The original concept of RNA fingerprinting is identical to the principle of random amplification of genomic DNA, in which short oligonucleotides are used as PCR primers to randomly amplify DNA fragments between targeted sites ***(16)***. The fundamental difference between DD and RAP-PCR is the way in which the initial step of reverse transcription (RT) is performed. In DD, oligo-dT-based primers initiate the

From: *Medical Biomethods Handbook*
Edited by: J. M. Walker and R. Rapley © Humana Press, Inc., Totowa, NJ

synthesis of first-strand cDNA, whereas in RAP-PCR, only arbitrary oligonucleotide primers are used. A number of basic methodologies, such as RT of total mRNA, PCR, and gel electrophoretic analysis of the amplified products followed by DNA cloning and sequencing, have been recruited from molecular biology and have been integrated in both techniques.

Experimentally, differential display, which is also known as DDRT-PCR *(17)*, comprises several steps (*see* **Fig. 1**):

- Total RNA is reverse transcribed into cDNAs using poly-A tail (anchor) primers, followed by PCR amplification in the presence of anchor as well as arbitrary oligonucleotide primers annealing to the corresponding internal sequences.
- The obtained PCR products representing the sets of mRNA populations are then resolved by gel electrophoresis generating a fingerprint, or a "snapshot" of the mRNA expression pattern. The relative intensity of a given band is proportional to the concentration of the corresponding template in the original samples *(8,18)*.
- Side-by-side comparison of fingerprints from a pair of samples (e.g., normal vs disease) detects the differentially expressed transcripts as cDNA bands of the same length but different intensity.
- The cDNA bands of interest are then excised from the gel, followed by cloning, sequencing, and further characterization.

An outline of the steps involved in the DD procedure is presented in **Fig. 1**. Results obtained by DD can be verified using a number of independent methods that include Northern hybridization *(3)*, RNA protection assay *(4)*, *in situ* hybridization *(19,20)*, nuclear run-on analysis *(21)*, and quantitative RT-PCR *(22)*.

3. Challenges of Differential Display

Since its invention in 1992, DD has quickly displaced subtractive or differential hybridization and has become the method of choice for identifying differentially expressed genes. In fact, out of the thousands of studies aimed at revealing differential mRNA expression in various biological systems, more than 50% have used DD (*see* **Fig. 2**). A number of factors summarized below illustrate the successful integration of DD in today's research:

- The method is relatively simple and inexpensive and requires only basic techniques that are routinely used in many laboratories, namely RNA purification, PCR amplification, DNA sequencing, and cloning.
- Only micrograms or even nanograms of total RNA are required as starting material. This is important if only a limited amount of sample material is available.
- The screening is independent of genome sequencing because it does not require any prior sequence information. This fact makes DD suitable for the study of highly variable bacterial or plant genomes, which are characterized by the occurrence of large families of homologous genes *(23)*.
- Many samples can be compared in parallel to monitor dynamic changes in gene expression, and multiple screening might reveal putative "molecular signatures" that characterize a particular biological state.
- The method can be effective even if only very few alterations are expected from theoretical considerations as, for example, in the case of early effects of various dose treatments or temporal stage of the process.
- The whole transcriptome of a given cell type can be screened *(17)*, but it would require the comparison of at least 600 fingerprints using 80–120 primer combinations *(24)*.

Despite the advantages enumerated above, the DD is not free from drawbacks. One of the most common problems mentioned by many authors is the relatively high frequency of false-positive bands and many investigators have reported difficulties in confirming the true differential expression of identified mRNAs. Of the bands that appear to be differentially expressed, the fraction of false positives can reach up to 50% *(25)*. These problems as well as likely solutions will be discussed in **Subheading 5.**

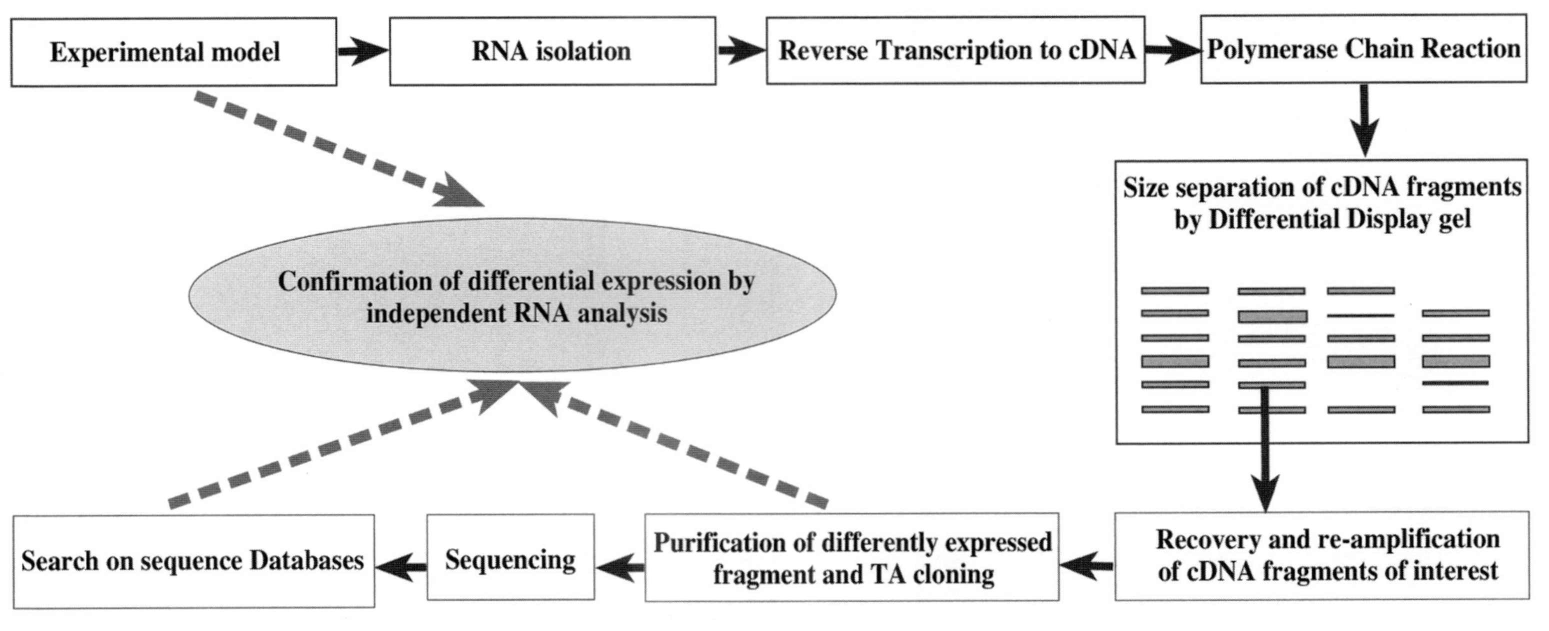

Fig. 1. Outline of the differential display procedure.

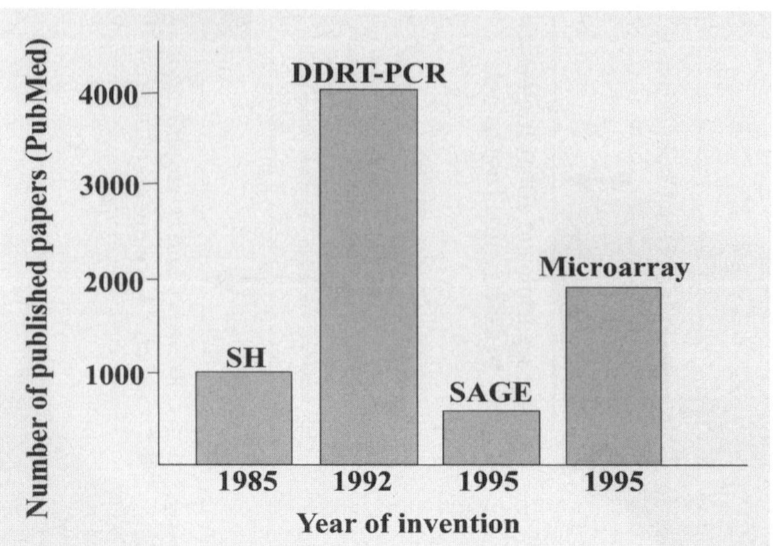

Fig. 2. Impact of a number of technologies in mRNA profiling analysis. The number of publications is based on a search in PubMed medline (September 2003, reviews are excluded). SH, subtractive hybridization; DDRT-PCR, all types of differential display; SAGE, serial analysis of gene expression; microarray, cDNA microarray.

Frequently, the stumbling blocks are not associated with the technique itself: The methodological and theoretical considerations have been efficiently improved since DD's introduction ***(21,26,27)***, but with the lack of knowledge concerning essential characteristics of the biological system ***(28)***, which will be discussed in detail in **Subheading 6.3.**

4. Alternative Methods Closely Related to Differential Display

Several modifications have been developed with the aim to improve the probability of fingerprinting rare sequences and to reduce the percent of false positives.

Nearly a dozen techniques, closely related to DD, have been developed to improve mRNA selection by applying arbitrary primers in combination with restriction enzyme digestion ***(8–10,29,30)***. The methods are essentially the same and include ordered differential display (ODD) ***(31)***, amplified differential gene expression (ADGE) ***(32)***, total gene expression analysis (TOGA) ***(33,34)***, amplification of double-stranded cDNA ends restriction fragments (ADDER) ***(35)***, gene calling ***(36)***, and restriction fragment length polymorphism (RFLP)-coupled domain directed DD (RC4D) ***(37)***. All of these methods have included a few extra steps, such as second-strand DNA synthesis, restriction digestion, and ligation of adaptor-primers before mRNA comparison. A detailed description of these methods can be found in a number of recently published reviews (***29,30***, and references therein).

Even though all of these modifications have their own particular advantages, it is difficult to evaluate their effectiveness and capability because any extra step inevitably leads to uncertainty and could increase the inter- and intra-RNA sample variability ***(9)***.

One of the most interesting formats of DD, targeted RNA fingerprinting (TRF), allows one to display the members of specific gene families using generated primers that correspond to encoding protein motifs conserved within the gene family ***(38,39)***.

5. Guide to Experimental Strategies

This section will give a step-by-step insight of experimental design and point potential steps where false positives can generate (*see* **Fig. 1**), as well as outline possible controls ensuring reproducibility and robustness of each experimental step.

5.1. Sample Preparation

Two main types of starting material are being used in DD as a source of RNA: cultured cells and tissue samples. Cultured cells are much easier to handle as compared to tissue specimens and can be readily harvested for RNA isolation. For clinically relevant samples (e.g., tissue biopsies), it is strongly advisable to shorten, as much as possible, the period of sample handling prior to freezing in liquid nitrogen. The interheterogeneity and intraheterogeneity of clinical samples is perhaps the most severe complication hampering differential expression analysis. For this reason, the sample must be carefully dissected to eliminate possible contamination by surrounding tissues. All dissecting manipulations should be performed on ice using sterile, RNAase-free instrumentation and in the presence of RNA*later* solution (Ambion) to minimize possible RNA degradation. Homogenization of nitrogen-frozen tissue sample is recommended, as it provides a higher yield of RNA compared to fresh material. mRNA isolation must be carried out in duplicate to ensure reproducibility.

5.2. Template Preparation

Several methods for isolating undegraded RNA free of chromosomal DNA contamination, which is crucial for successful RNA fingerprinting, have been developed ***(40,41)***. A number of kits suitable for RNA purification from different cell or tissue types are commercially available (Gibco-BRL, Promega, Clontech, Molecular Research Center, etc.). Isolation of poly(A)$^+$ RNA is not recommended ***(24)*** because the possible contamination of the RNA preparation by oligo(dT) might lead to high background signals. In addition, rare mRNA species could be lost during the poly(A)$^+$ RNA isolation.

Regardless of which protocol is used, careful precautions should be undertaken during all steps, including wearing DEPC-pretreated gloves and using RNA-free solutions and glassware. RNase inhibitors should be used at all steps, including the dissection procedure (Ambion, RNA*later* kit) to avoid intracellular RNA degradation as a result of the release of endogenous RNases upon cell disruption. However, one must keep in mind that inhibition complexes can be oxidized or denatured during the isolation procedure, leading to reactivation of RNases. It is also necessary to remember that the majority of the commercially available RNase inhibitors do not inactivate bacterial T1 or fungal RNases.

Even though most of the commercial kits ensure DNA-free RNA isolation, it is important to check that the RNA preparations are completely free from DNA contamination. It should be noted that even trace amounts of DNA, not detectable on an agarose gel, could work as PCR-amplification templates, contributing to the appearance of false positives on the fingerprinting patterns. To ensure the absence of genomic DNA several controls can be carried out:

- Perform PCR amplification using total RNA as a template.
- Perform PCR amplification of any constitutively expressed intron-containing gene (e.g., cyclophilin) using a set of primers that distinguish the cDNA product from the corresponding genomic fragment.

If DNase treatment is included in the RNA isolation procedure, partial hydrolysis of RNA can occur (Mg^{2+} - dependent RNA hydrolysis) despite claims that proper treatment should not affect the quality of the RNA preparation and further fingerprint analysis ***(28)***.

The quality and quantity of the isolated RNA should be carefully checked prior to subsequent steps. The ratio of 28S ribosomal RNA to 18S rRNA bands visible on agarose gel should be approx 2 : 1 and the ratio of optical densities at 260/280 nm should be in the range 1.80–1.9. It is recommended to keep the RNA at –80°C in small aliquots to avoid repeated freezing and thawing.

5.3. Reverse Transcription and PCR Amplification

According to the original DD protocol, the mRNA is converted to first-strand cDNA by using three anchored oligo-dT primers (3'-end primers), which differ from each other at the last 3' non-T base ***(7)***. This set of primers initiates reverse transcription only from a subpopulation of polyadenylated mRNAs (*see* **Fig. 3**). To display as many transcripts as possible, cDNAs are then amplified in the presence of 5'-end relatively short arbitrary primers labeled either with radioisotopes or fluorescent dyes (see below). Each arbitrary primer is randomly annealed at different distances from the 3'-end primers and can recognize 50-100 mRNAs under a given PCR condition ***(7–9)***. As a result, each pair of primers amplifies a discrete number of cDNAs of different molecular size (300–2000 bp length depending of the primer length; the average molecular size of mRNA is 1.2 kb) in a range that is optimal for gel separation (up to 100 bands per lane) ***(7)***. Based on the assumption that each arbitrary primer recognizes at least seven nucleotide stretches within the corresponding mRNA targets, several mathematic models have been worked out to evaluate the possible coverage of expressed genes in a given biological system ***(15)***. Several rules have been proposed for the optimal design of arbitrary primers:

- The primers should contain at least 50–70% G+C.
- The primers should not have inverted sequences that can create internal folding and thereby prevent the annealing process.
- The primers should be arbitrary in sequence to ensure random annealing along all mRNA species.

Some redundancy within the identified transcripts can be observed because of the fact that several primers can anneal to different parts of the same transcript, thus complicating the expression profile without extending the information.

In a classical DD that utilizes oligo-dT primers, mainly the 3'-terminal sequences of the transcripts containing untranslated regions, are amplified, a fact that can hamper their unambiguous identification. Unlike DD, RAP-PCR can ensure with high probability the amplification of coding sequences when analyzing short cDNA fragments derived from protein-coding regions or prokaryotic mRNAs, which are characterized by the lack of polyadenylated sequence ***(42,43)***. However, when the RAP-PCR is applied to eukaryotes, the arbitrary priming takes place not only on messenger RNAs but also at ribosomal and transfer RNAs templates, a fact that increases the percentage of bands deriving from not relevant amplification.

The utilization of longer primers increases the stringency of PCR, reduces background, and improves the reproducibility of the results ***(44–47)***. Detailed recommendations concerning primer design can be found in laboratory manuals ***(27,28)*** as well as several reviews ***(8,10)***.

To avoid inconsistencies of the results obtained in replicated experiments a number of precautions are recommended.

- Polymerase chain reaction amplification of cDNA populations obtained from very low amounts of total RNA could cause "statistical noise" and increase the level of false positives, although representative PCR amplification from as little as 50 ng of total RNA has been reported ***(48)***. The amount of total RNA used in the RT step should be tested in advance because very low RNA concentration could result in the display of only few bands, whereas its excess could cause smearing in the fingerprinting pattern ***(10,49)***.
- The results of PCR amplification should be verified using at least two different concentrations of the template. Only differences observed in both concentrations can be considered as reliable ***(50)***.
- It is recommended to verify the complete setup of the system by amplifying a gene that is known (if any) to be differentially regulated in the sample pair.

A large variation in the concentration of transcripts within a given mRNA sample could also hinder the detection of rare transcripts and, possibly, lead to a bias toward abundant messengers ***(51,52)***. This shortcoming could be overcome, in part, either by depleting each fingerprint and/or increasing the number of primer sets, and/or pre-fractionating the starting RNA sample (i.e., combining subtractive hybridization with DD). In the latter case, the preliminary subtraction leads to an enrichment of differentially expressed transcripts that might improve the detection of the rare transcripts by the subsequent fingerprinting step ***(53–55)***. In addition, the competition for PCR primers (and/or *Tag* polymerase) between RNA species of different abundance can also be a factor affecting the overall sensitivity of fingerprinting ***(47,51)***.

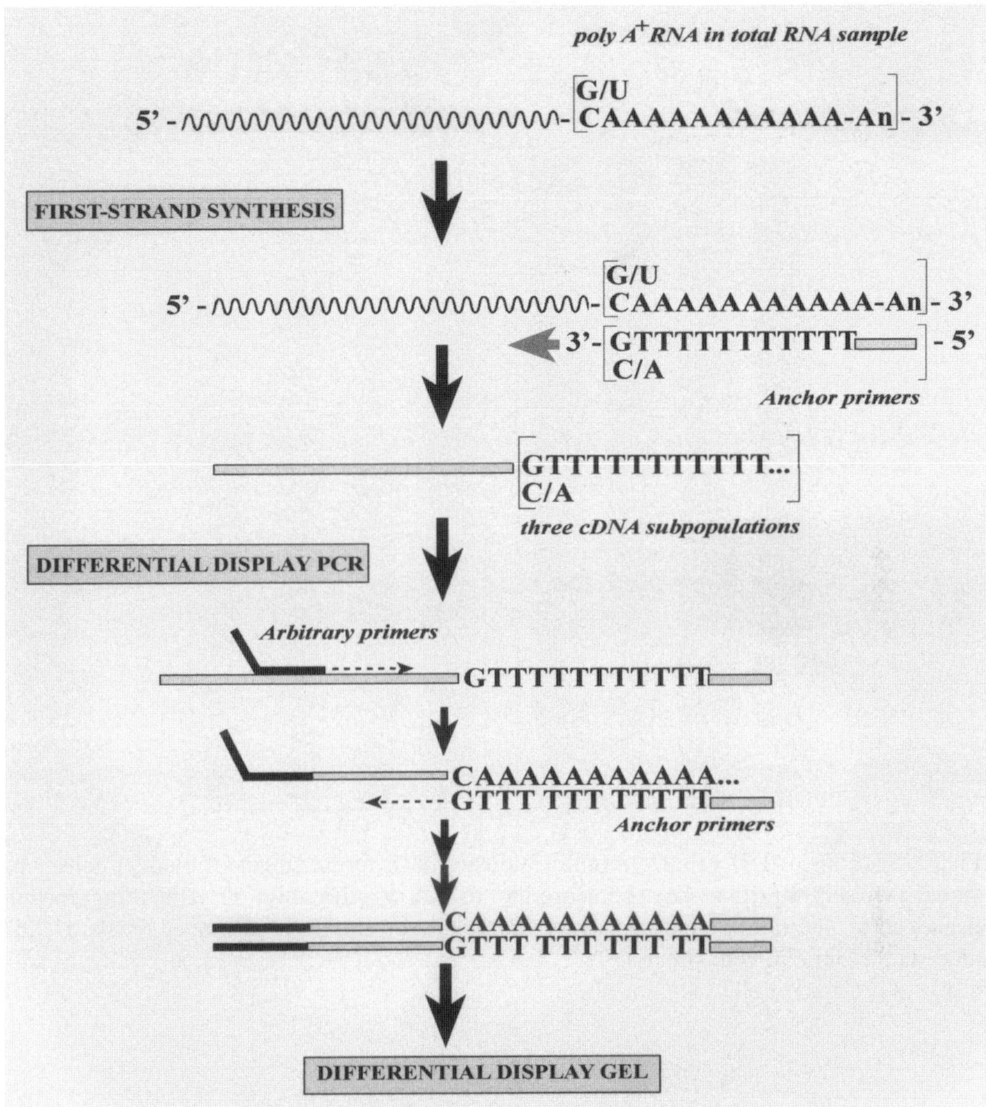

Fig. 3. Detailed flowchart of cDNA synthesis and PCR priming reactions in the original DD procedure.

5.4. Size Separation

Separation of PCR-amplified products is usually achieved by high-resolution denaturating polyacrylamide gel electrophoresis. Commercially available sequencers like GenomyxLR™ (Beckman Coulter, Inc.) or Alf Express (Pharmacia Biotech, Inc.) maintain constant temperature independently of the voltage used during the run, thus minimizing the variability in electromobility of cDNA fragments from one gel to another that is otherwise introduced by local temperature fluctuations. The RNA fingerprints can then be visualized by autoradiography (*see* **Fig. 4**) or by other methods depending on primer labeling. Detailed protocols for gel treatment after electrophoresis and gel autoradiography can be found elsewhere ***(27,28)***.

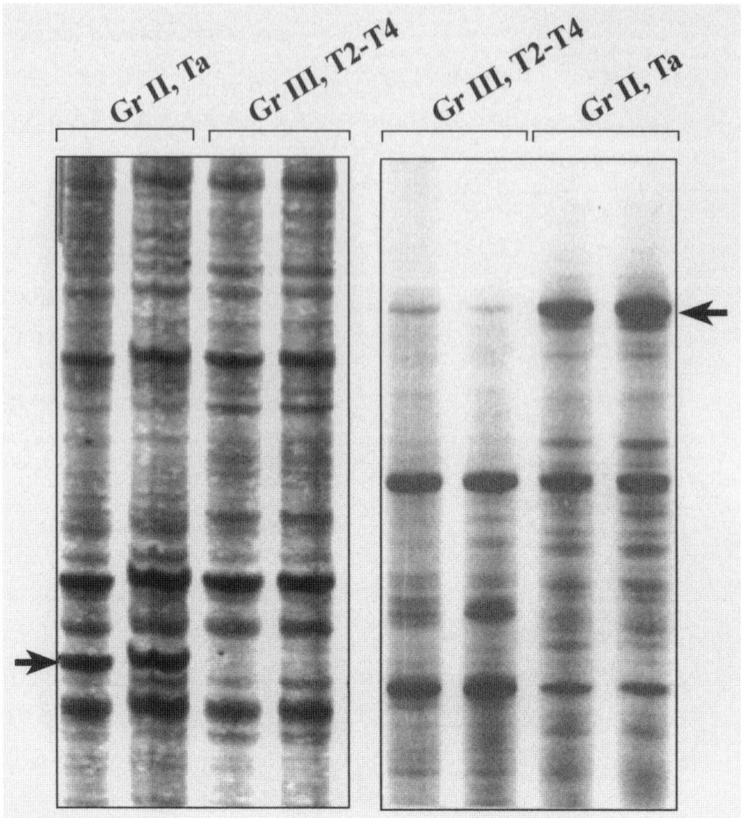

Fig. 4. Sections of DD gels from tumor biopsies at different stages of bladder cancer progression. Two sets of primers have been used for cDNA amplification. The RNA fingerprinting is visualized by autoradiography. Arrows indicate the differentially displayed bands. Detailed results can be found in **refs. *56*** and ***57***.

5.5. Fingerprint Visualization

Until recently, radioactive labeling of PCR products was the most common approach to visualizing RNA fingerprinting. This can be achieved by including either ^{32}P–, ^{35}S–, or ^{33}P-dNTP in the amplification step.

- ^{32}P provides very high sensitivity, but the resolution is not good enough because of its penetrating radiation. It is not expensive, but it requires extra care during handling.
- ^{35}S provides high sensitivity and resolution; it is not harmless because of lower energy, but it has a tendency to evaporate during the amplification step.
- ^{33}P, which is used in most cases, is a good alternative because it combines the sensitivity of ^{32}P with the ability of ^{35}S to generate high-resolution fingerprints.

Several commercial kits based on radioactive detection of PCR products can be recommended as a start point for DD analysis. These include Genomyx Hieroglyph kits (Beckman), Delta Fingerprinting kit (Clontech), RNAimage® Kits (GenHunter Corp.), displayPROFILE™ kit for prokaryotic messages (Qbiogene), as well as Differential Display Service available at GenHunter Corp. (http://www.genhunter.com) or Qbiogene's displayFIT™ website (www.gbiogene.com/services/molBio/expression.shtml).

In addition to radioactive labeling, a number of alternative methods have been developed to visualize the amplified products:

- Ethidium bromide staining in combination with agarose gel electrophoresis ***(58)*** is easy to use, but it has not received much acceptance because of the low resolution of the agarose gel electrophoresis and the low sensitivity of the staining.
- Silver staining of cDNAs separated on sequencing gels ***(59)*** is another approach that simplifies the original protocol without any significant loss in resolution and sensitivity. Compared to the radioactive method, the differentially expressed bands are easier to retrieve because of their direct visibility on the gel.
- Nonradioactive fluorescent differential display (FDD) introduced by Ito and colleagues in 1994 provides equivalent sensitivity to the original ^{33}P isotopic labeling method ***(60)***. This method gained much attention because of its cost-effectiveness, faster visualization process, and great potential in automation. There are at least two commercially available differential display kits based on fluorescent detection, namely RNAspectra Green and RNAspectra Red (GenHunter Corp.).
- More recently, an alternative nonradioactive DD method based on chemiluminescent detection of amplified products was introduced ***(61)***.

5.6. Excision and Purification of Differentially Expressed cDNA Hits

In order to minimize the retrieval of false positives, several criteria for differentially expressed cDNA bands have to be considered:

- Only bands that are reproducible in duplicate lanes should be considered for further analysis.
- In the first round, only cDNAs exhibiting a significantly differential expression level compared with the control should be considered for further analysis.
- A small central portion of the band should be excised to minimize possible cross-contaminations. As a control, for every band of interest, a "blank" should also be excised from a neighboring area containing a comparing sample. Subsequent reamplification of excised bands is done using the same pair of primers. This allows an estimation of the background of the sample and provides sufficient material for downstream analysis. In theory, the molecular size of the reamplified cDNA must correspond to the size observed on the fingerprint, but, in practice, two or more bands are often observed. Multiple bands occur as a result of the presence of more than one cDNA type in the excised band or of spurious amplification of contaminants. If the irrelevant cDNA is present at a higher concentration compared with the cDNA of interest, it would be predominantly reamplified, thus increasing the appearance of false positives. This effect can be especially pronounced when the difference in mRNA expression levels is not high, but it can be ameliorated by applying a number of extra controls/modifications (*see* **Fig. 5**).
- Northern blot affinity capturing of cDNA ***(62)*** allows the identification of correct fragments by hybridization of putative hits with original RNA samples separated by Northern blot. The corresponding differentially expressed cDNAs are captured, retrieved from the membrane, amplified once again, and, finally, cloned for further analysis. However, this procedure is not sufficiently sensitive, thus requiring substantial amount of starting RNA.
- Single-strand confirmation polymorphism (SSCP) analysis resolves cDNAs of similar size but of different sequence. The method is based on the ability of single-stranded cDNA fragments to form stable and metastable secondary structures that affect their mobility in acrylamide gels ***(63–65)***. The enriched products of interest that are separated on the basis of conformation rather than size can be then retrieved and reamplified again, followed by cloning and sequencing.
- Resolver Gold agarose gel electrophoresis separates bands with the same mobility but different sequences in the presence of a bisbenzimide derivative that preferentially binds to A+T-rich motifs ***(66)***. The procedure does not require radioactive labeling and extra RNA material.

Several other approaches aimed at reducing the occurrence of irrelevant, comigrating fragments can be found in the literature ***(67–70)***. Detailed protocols for band excision, elution from the gel, and further reamplification are described elsewhere ***(27,28)***.

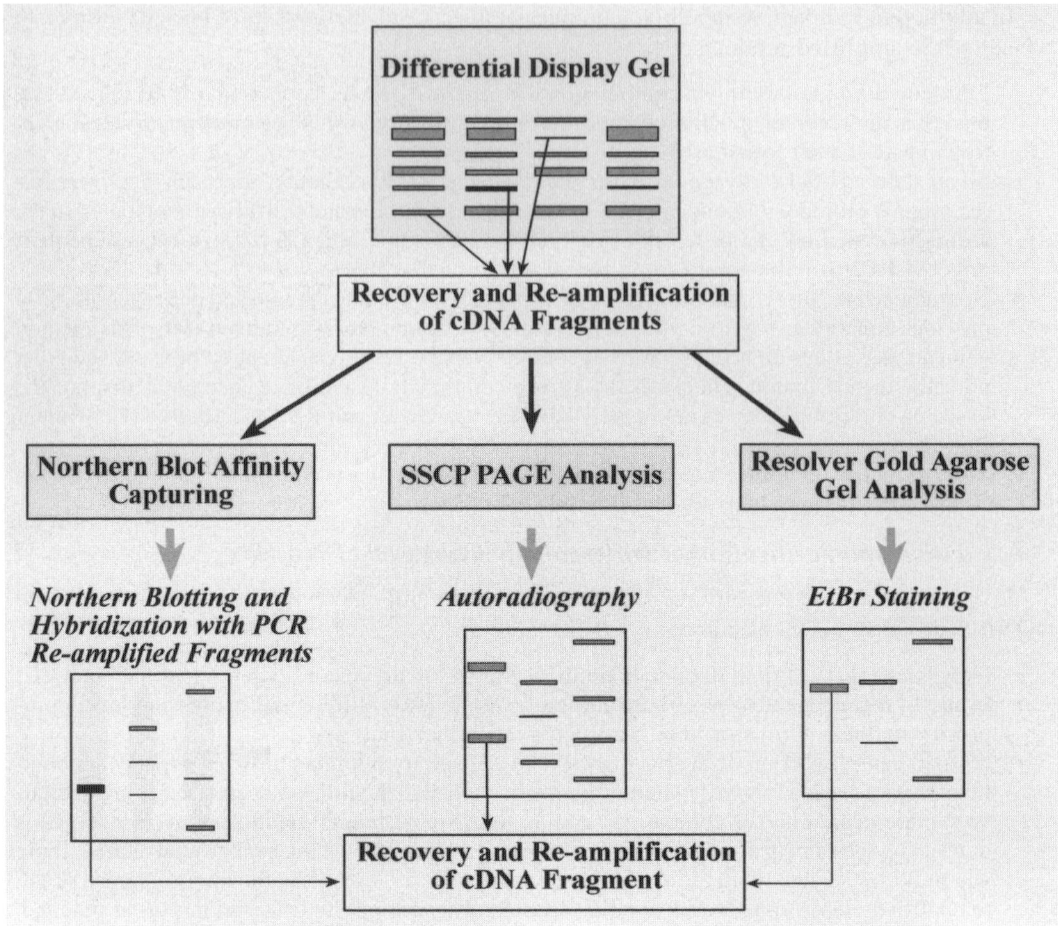

Fig. 5. Schematic of procedures that distinguish false positives from comigrating products.

5.7. Confirmation of Differentially Expressed cDNA Hits

Results obtained by the DD procedure can be verified at the RNA level using a number of methods that include Northern hybridization, *in situ* hybridization, ribonuclease protection assay, nuclear runoff assay, and RT-PCR analysis:

- Northern hybridization is not sensitive enough for rare transcripts, requires rather large amounts of RNA, and, as a result, might not always be the most appropriate.
- *In situ* hybridization can be time-consuming and tricky to perform for many hits and its sensitivity is not high enough to detect genes expressed at a low level. Only absolute variations in mRNA levels can be observed on tissue sections using this method *(71)*. However, despite the above disadvantages, this is the only method that provides the information concerning the cellular localization of the mRNA of interest.
- The ribonuclease protection assay, similar to Northern hybridization, also requires substantial amounts of RNA and needs careful optimization for every sample.
- Quantitative real-time RT-PCR analysis *(72)* is the most sensitive technique and is easy to run for multiple samples, with the lowest requirement for RNA.

5.8. Cloning and Sequencing

To facilitate the rapid cloning and analysis of PCR products, a method based on the TOPO TA cloning system, which exploits the ligation activity of topoisomerase I, is currently in use (Invitrogene). The identity of the insert can then be assessed by using a number of commercially available sequence kits or using commercial sequence services followed by searching in Gene Bank.

6. Application of Differential Display in Medicine

Profiling of clinical samples at the DNA, RNA, or protein levels is often used to detect genes that are deregulated in diseases with the aim of understanding the mechanisms underlying pathogenesis and disease progression. Because of simplicity, sensitivity, and reproducibility, DD has been used in various areas of biology that include developmental biology, neurobiology, endocrinology, immunology, cardiology, cancer, and many others. In general, the best results have been obtained when well-defined cellular systems are used for comparative analysis.

6.1. Identification of Interacting Partners

Today, the most impressive results obtained by DD are associated with the identification of targets and intracellular partners for known proteins that are involved in various cellular activities such as signal transduction, cell division, DNA repair, and apoptosis. A number of targets have been identified using DD for the *p53* tumor suppressor, the *ras* oncogenes, as well as a number of other cellular proteins. Novel p53 interactive partners identified by DD include the nuclear zinc-finger protein PAG608 ***(73)***, ei24 (alias PIG8) ***(74,75)***, the new death-domain-containing protein Pidd ***(76)***, the *p53DINP1* gene (p53-dependent damage-inducible nuclear protein 1) ***(77)***, a new member of the Bcl-2 protein family, Noxa ***(78)***, DDA3 ***(79)***, and the *Pirh2* gene encoding a RING-H2 domain-containing protein ***(80)***. DD screening has also been successfully used to identify genes controlled by the activated *ras* oncogenes, including the secreted protein IL-24 ***(81,82)***.

Many novel genes that regulate cellular growth and division, namely *DUB-1* ***(83)***, *ptk-3* ***(84)***, *P15RS* ***(85)***, the transcription factor gut-enriched Kruppel-like factor (GKLF) Id3 ***(86)***, PASG, the proliferation-associated SNF2-like gene ***(87)***, and NIMA ***(88)*** have recently been identified using various cellular systems.

6.2. Immunology

Differential display has been applied to the analysis of lymphocyte and myeloid cells derived from thymus and bone marrow ***(89)***, mononuclear cells derived from human peripheral blood ***(90–92)***, cells isolated from transgenic mice, cultured cells treated with a number of cytokines or drugs ***(93–96)***, as well as transfected cell lines ***(97)***. Genes identified include signaling molecules and molecular chaperons ***(98)***, nuclear receptors ***(92)***, transcription factors ***(99)***, secreted proteins ***(100,101)***, as well as cell surface markers ***(91,102)***. A comprehensive description of DD application in immunology has recently been published by Ali and colleagues (***103***, and references therein). A complete list of DD applications in various areas of biology can be found in the GenHanter website (http://www.genhunter.com/support/references.html). However, because the technique was originally developed to compare gene expression in cell lines ***(7)***, attempts to apply it directly to complex tissue samples have met with serious obstacles. These are described below.

6.3. Complexity of Tissue Samples

Ideally, for DD, the samples should have the same background and, if possible, only one parameter or condition should differ at any given time. This can be easily achieved with cell lines, but clinically relevant samples such as tissue biopsies are often characterized by different levels of complexity. Parameters such as age, disease status, and topological-organ localization ***(104)*** might have some impact on the expression profile of tissues and must be taken into consideration when planning an experiment. In addition, the heterogeneity of the clinical samples is very difficult to control, as tissue biopsies are usually composed of different cell types that are present in different proportions. For a example, a punched biopsy of almost any type of tumor can contain, in addition to cancer cells, blood cells, endothelial cells, infiltrating lymphocytes, adipocytes, fibroblasts, as well as other cell types. As a result, the expression profile of the sample will portray transcript populations derived from all cell types present, but only a few of the alterations might be associated with the malignant phenotype.

The internal heterogeneity of such biopsy is even more difficult to control because the cells belonging to a given cell type might exhibit various degrees of differentiation and genetic stability depending on the disease status. This could result in many phenotypical features characterized by differences in the karyotype, biochemical properties, growth potential, and so forth. Intratumor heterogeneity is well illustrated by tumor growth and progression, processes that are characterized by a multistep transformation leading to deregulation of gene expression at both the RNA and protein levels. Most of these events are the results of genomic rearrangements, mutations, and environmentally induced changes that might take place either simultaneously or in a stepwise fashion in different areas of the tumor. In the context of the DD approach, this means that different parts of the tumor (e.g., internal vs outer layer) might exhibit different mRNA profiles, a fact that has a major impact on the interpretation of the results. The problem of having an adequate "normal" control in clinical studies is also very severe because of sample heterogeneity and because of the uncertainty of the criteria by which these samples are chosen ***(104)***.

6.4. Isolation of Cell Populations Relevant for Analysis

Methods that could reduce the complexity of tissue samples to a certain extent include gross dissection of frozen tissue ***(105)***, irradiation of manually ink-stained sections to destroy genetic material ***(106)***, microdissection using manual tools ***(56,57,107,108)*** flow cytometry cell sorting applied to cell suspensions prepared from disaggregated tissues, and the recently developed technology of laser microdissection ***(109)***. The latter is frequently applied to clinically relevant samples to isolate relatively pure cells populations ***(110)***. However, it should be noted that morphological criteria alone cannot distinguish, for example, B-cells from T-cells or quiescent vs growing cells, luminal cells from myoepithelial cells. The above problem can be ameliorated by using the immunohistochemistry-guided laser dissection technique recently developed for mRNA-based profiling ***(111–114)***. This technique enhances the possibility of distinguishing subpopulation of cells that are otherwise identical.

6.5. Cancer

In this area, many efforts are being directed toward the search for genes that are deregulated during carcinogenesis and tumor progression in an attempt to reveal molecular markers for tumor classification, diagnosis, prognosis, and response to treatment. However, because of the problems associated with sample complexity, there have so far been few studies involving clinical relevant tissue samples. These are reviewed below.

6.5.1. Bladder Cancer

To our knowledge, there are two publications so far in which the purity of all tumor samples used for analysis was controlled by proteome profiling ***(57,58)***. These studies were performed using tissue biopsies from noninvasive (grade II, Ta) and invasive (grade III, T2–T4) human transitional cell carcinomas (TCCs). Two novel genes, *bc10* (*b*ladder *c*ancer, M_r *10* kDa) and a member of the glycosyltransferase family, beta *3Gn-T2*, were revealed to be almost exclusively expressed in the noninvasive lesions as judged by the analysis of a panel of 30 grade II, Ta and grade III, T2–T4 TCCs.

6.5.2. Breast Cancer

A number of novel genes that are differentially regulated during breast cancer progression have been identified by DD using clinical samples, either for comparative analysis or for confirmation of the results. Proteins identified using DD include prostate epithelium-derived Ets transcription factor (PDEF), which is over expressed in 14/20 primary tumors tested ***(115)***, the cargo selection protein (TIP47), and the signal transducer and activator of transcription (STAT3) ***(116)*** cul-4A, which is overexpressed in 14/30 cases analyzed (47%) ***(117)***, and the epithelial–stromal EPSTI1, which is upregulated in 14/14 carcinomas compared with normal breast tissue ***(118)***.

6.5.3. Gastric Cancer

The overexpression of the novel gene GCRG224 was demonstrated by comparing the expression profiles of paratumor and normal gastric mucosa tissues collected from 15 patients bearing stomach antrum adenocarcinoma ***(119)***. The expression of GCRG224 was shown to be upregulated in almost all gastric mucosal epithelium but only in a small proportion of the cancer and precancerous lesions.

6.5.4. Lung Cancer

A novel *DAL-1* gene isolated from primary lung tumors was shown to be downregulated in 39 primary non-small-cell lung carcinomas compared with patient-matched normal lung tissue (>50%) ***(120)***. By applying DD to SCLC and non-SCLC (NSCLC) cell lines, *LAMB3* and *LAMC2* genes were identified and their expression patterns were confirmed by using four primary non-SCLC and the corresponding noncancerous lung cells ***(121)***. The overexpression of RAB5A, which was identified in two human lung adenocarcinoma cell lines, was further confirmed by immunohistochemical staining of 45 patients bearing human NSCLC ***(122)***. Dihydrodiol dehydrogenase (DDH) has also been shown to be overexpressed in NSCLC but not in the normal tissue, a fact that was verified by the analysis of 381 pathological samples ***(123)*** of normal as well as metastatic lymph nodes.

Similar studies have been reported in the case of melanomas ***(124)*** and prostate cancer ***(125)***.

7. Differential Display Versus cDNA Microarray

The DD procedure has been developed in parallel with a number of genomewide expression technologies, namely SAGE ***(11)*** and hybridization-based cDNA microarrays ***(12,13)***. The latter is based on the physical hybridization of mRNA molecules with complementary cDNA or oligonucleotide probes of known sequences immobilized on a surface. cDNA microarrays provide high throughput, do not require cloning and identification, and have become one of the most powerful tools for the analysis of differentially expressed transcripts. In contrast to DD this technique depends on the availability of DNA sequencing, and, therefore, the possibility of novel gene discovery is excluded. Therefore, it follows that the number of genes subjected to analysis is restricted for genomes with a rather uncompleted sequence. A comprehensive analysis of "pro" and "contra" of differential display vs cDNA microarray has recently been presented by Liang and others ***(9,30,126)***. A brief comparison of these techniques as well as their merits and drawbacks are summarized in **Table 1**.

8. Concluding Remarks

Characterization of complex biological systems using gene-profiling technologies is the first step toward biomarker discovery. Obviously, there is no single technique currently available that can be applied to all tasks and the choice of the method should be based on the particular features of the biological system as well as the project requirements. In the end, however, only the functional characterization of the identified markers using a number of independent methods as well as animal models, irrespective of the gene-expression technologies applied, will lead to an understanding of their biological role and possible prognostic value.

Table 1
Main Features of the Major mRNA Expression-Profiling Approaches

Parameters / Technologies	SH[a]	DD	SAGE	Microarray
Detection of novel genes	Yes	Yes	No	No
Sequence dependency	No	No	Yes	Yes
Systems to be analyzed	Any	Any	Currently only eukaryotes	Currently only eukaryot
Sensitivity	High	High	Moderate	Moderate
Specificity	High	High	High	Moderate
Quantitation	Relative	Relative	High	Moderate
Sophisticated computer analysis	No	No	Yes	Yes
Cost	Low	Low-medium	Low-medium	High
Commercial Services	No	Few	No	Many

[a]SH, subtractive hybridization.

References

1. Subramanian, G., Adams, M. D., Venter, J. C., Broder, S. (2001) Implications of the human genome for understanding human biology and medicine. *JAMA* **286,** 2296–2307.
2. Gooley, A. A. and Packer, N. H. (1997) The importance of protein co- and post-translational modifications in proteome projects, in *Proteome Research: New Frontiers in Functional Genomics (Principles and Practice)* (Wilkins, M. R., Williams, K. L., Appel, R. D., and Hochstrasser, D. F., eds.), Springer-Verlag, Berlin, pp. 65–91.
3. Alwine, J. C., Kemp, D. J., and Stark, G. R. (1977) Method for detection of specific RNAs in agarose gels by transfer to diazobenzyloxymethyl-paper and hybridization with DNA probes. *Proc. Natl. Acad. Sci. USA* **74,** 5350–5354.
4. Tymms, M. J. (1995) Quantitative measurement of mRNA using the RNase protection assay, in *In vitro Transcription and Translation Protocols* (Tymms, M. J., ed.), Humana, Totowa, NJ, pp. 31–46.
5. Zimmermann, C. R., Orr, W. C., Leclerc, R. F., Barnard, E. C., and Timberlake, W. E. (1980) Molecular cloning and selection of genes regulated in Aspergillus development. *Cell* **21,** 709–715.
6. St John, T. P. and Davis, R. W. (1979) Isolation of galactose-inducible DNA sequences from *Saccharomyces cerevisiae* by differential plaque filter hybridization. *Cell* **16,** 443–452.
7. Liang, P. and Pardee, A. B. (1992) Differential display of eukaryotic messenger RNA by means of the polymerase chain reaction. *Science* **257,** 967–971.
8. Stein, J. and Liang, P. (2002) Differential display technology: a general guide. *Cell. Mol. Life. Sci.* **59,** 1235–1240.
9. Liang, P. (2002) A decade of differential display. *Biotechniques* **33,** 338–346.
10. Matz, M. V. and Lukyanov, S. A. (1998) Different strategies of differential display: areas of application. *Nucleic. Acids. Res.* **26,** 5537–5543.
11. Velculescu, V. E., Zhang, L., Vogelstein, B., and Kinzler, K. W. (1995) Serial analysis of gene expression. *Science* **270,** 484–487.
12. Schena, M., Shalon, D., Davis, R. W., and Brown, P. O. (1995) Quantitative monitoring of gene expression patterns with a complementary DNA microarray. *Science* **270,** 467–470.
13. Lipshutz, R. J., Fodor, S. P., Gingeras, T. R., and Lockhart, D. J. (1999) High density synthetic oligonucleotide arrays. *Nature Genet.* **21,** 20–24.
14. Welsh, J., Chada, K., Dalal, S. S., Cheng, R., Ralph, D., and McClelland, M. (1992) Arbitrarily primed PCR fingerprinting of RNA. *Nucleic Acids. Res.* **20,** 4965–4970.
15. Liang, P., Bauer, D., Averboukh, L., et al. (1995) Analysis of altered gene expression by differential display. *Methods Enzymol.* **254,** 304–321.
16. Welsh, J. and McClelland, M. (1990) Fingerprinting genomes using PCR with arbitrary primers. *Nucleic Acids Res.* **18,** 7213–7218.
17. Bauer, D., Muller, H., Reich, J., et al. (1993) Identification of differentially expressed mRNA species by an improved display technique (DDRT-PCR). *Nucleic Acids Res.* **21,** 4272–4280.
18. McClelland, M. and Welsh, J. (1994) DNA fingerprinting by arbitrarily primed PCR. *PCR Methods Applic.* **4,** S59–S65.
19. Gall, J. G. and Pardue, M. L. (1969) Formation and detection of RNA-DNA hybrid molecules in cytological preparations. *Proc. Natl. Acad. Sci. USA* **63,** 378–383.
20. John, H. A., Birnstiel, M. L., and Jones, K. W. (1969) RNA-DNA hybrids at the cytological level. *Nature* **223,** 582–587.
21. Rohde, M., Hummel, R., Pallisgaard, N., et al. (1997) Identification and cloning of differentially expressed genes by DDRT-PCR, In *PCR Cloning Protocols: From Molecular Cloning to Genetic Engineering* (White, B. A., ed.), Humana, Totowa, NJ, pp. 419–430.
22. Nakayama, H., Yokoi, H., and Fujita, J. (1992) Quantification of mRNA by non-radioactive RT-PCR and CCD imaging system. *Nucleic Acids Res.* **20,** 4939.
23. Breyne, P. and Zabeau, M. (2001) Genome-wide expression analysis of plant cell cycle modulated genes. *Curr. Opin. Plant Biol.* **4,** 136–142.
24. Liang, P., Averboukh, L., and Pardee, A. B. (1993) Distribution and cloning of eukaryotic mRNAs by means of differential display: refinements and optimization. *Nucleic Acids Res.* **21,** 3269–3275.
25. Wan, J. S., Sharp, S. J., Poirier, G. M., et al. (1996) Cloning differentially expressed mRNAs. *Nature Biotechnol.* **14,** 1685–1691.
26. Liang, P. and Pardee, A. B. (1995) Recent advances in differential display. *Curr. Opin. Immunol.* **7,** 274–280.
27. Colonna-Romano, S., Leone, A., and Maresca, B. (1998) Differential Display Reverse Transcription–PCR (DDRT-PCR). Lab Manual, Springer-Verlag, Berlin.

28. Medhurst, A.D., Chambers, D., Gray, J., et al. (2000) Practical aspects of the experimental design for differential display of transcripts obtained from complex tissue, in *Differential Display: A Practical Approach* (Leslie, R. A. and Robertson, H. A., eds.), Oxford University Press, Oxford, pp. 35–64.
29. Ahmed, F. E. (2002) Molecular techniques for studying gene expression in carcinogenesis. *J. Environ. Sci. Health C. Environ: Carcinog. Ecotoxicol. Rev.* **20,** 77–116.
30. Broude, N. E. (2002) Differential display in the time of microarrays. *Expert Rev. Mol. Diagn.* **2,** 209–216.
31. Matz, M., Usman, N., Shagin, D., Bogdanova, E., and Lukyanov, S. (1997) Ordered differential display: a simple method for systematic comparison of gene expression profiles. *Nucleic Acids Res.* **25,** 2541–2542.
32. Chen, Z. J., Shen, H., and Tew, K. D. (2001) Gene expression profiling using a novel method: amplified differential gene expression (ADGE). *Nucleic Acids Res.* **29,** E46.
33. Sutcliffe, J. G., Foye, P. E., Erlander, M. G., et al. (2000) TOGA: an automated parsing technology for analyzing expression of nearly all genes. *Proc. Natl. Acad. Sci. USA* **97,** 1976–1981.
34. Lo, D., Hilbush, B., and Sutcliffe, J. G. (2001) TOGA analysis of gene expression to accelerate target development. *Eur. J. Pharm. Sci.* **14,** 191–196.
35. Kornmann, B., Preitner, N., Rifat, D., Fleury-Olela, F., and Schibler, U. (2001) Analysis of circadian liver gene expression by ADDER, a highly sensitive method for the display of differentially expressed mRNAs. *Nucleic Acids Res.* **29,** E51-1.
36. Green, C. D., Simons, J. F., Taillon, B. E., and Lewin, D. A. (2001) Open systems: panoramic views of gene expression. *J. Immunol. Methods.* **250,** 67–79.
37. Fischer, A., Saedler, H., and Theissen, G. (1995) Restriction fragment length polymorphism-coupled domain-directed differential display: a highly efficient technique for expression analysis of multigene families. *Proc. Natl. Acad. Sci. USA* **92,** 5331–5335.
38. Stone, B. and Wharton, W. (1994) Targeted RNA fingerprinting: the cloning of differentially-expressed cDNA fragments enriched for members of the zinc finger gene family. *Nucleic Acids Res.* **22,** 2612–2618.
39. Tohonen, V., Osterlund, C., and Nordqvist, K. (1998) Testatin: a cystatin-related gene expressed during early testis development. *Proc. Natl. Acad. Sci. USA* **95,** 14208–14213.
40. Chomczynski, P. and Sacchi, N. (1987) Single-step method of RNA isolation by acid guanidinium thiocyanate-phenol-chloroform extraction. *Anal. Biochem.* **162,** 156–159.
41. Raha, S., Ling, M., and Merante, F. (1998) Extraction of total RNA from tissues and cultured cells, in *Molecular Biomethods Handbook* (Rapley, R. and Walker, J. M., eds.), Humana, Totowa, NJ, 1–8.
42. Bhaya, D., Vaulot, D., Amin, P., Takahashi, A. W., and Grossman, A. R. (2000) Isolation of regulated genes of the cyanobacterium Synechocystis sp. strain PCC 6803 by differential display. *J. Bacteriol.* **182,** 5692–5699.
43. Brzostowicz, P. C., Gibson, K. L., Thomas, S. M., Blasko, M. S., and Rouviere, P. E. (2000) Simultaneous identification of two cyclohexanone oxidation genes from an environmental Brevibacterium isolate using mRNA differential display. *J. Bacteriol.* **182,** 4241–4248.
44. Guimaraes, M.J., Lee, F., Zlotnik, A., McClanahan, T. (1995) Differential display by PCR: novel findings and applications. Nucleic. Acids. Res. 23, 1832-1833.
45. Zhao, S., Ooi, S. L., and Pardee, A. B. (1995) New primer strategy improves precision of differential display. *Biotechniques* **18,** 848–850.
46. Rohrwild, M., Alpan, R. S., Liang, P., and Pardee, A. B. (1995) Inosine-containing primers for mRNA differential display. *Trends. Genet.* **11,** 300.
47. Diachenko, L. B., Ledesma, J., Chenchik, A. A., and Siebert, P. D. (1996) Combining the technique of RNA fingerprinting and differential display to obtain differentially expressed mRNA. *Biochem. Biophys. Res. Commun.* **219,** 824–828.
48. Lukyanov, K., Diatchenko, L., Chenchik, A., et al. (1997) Construction of cDNA libraries from small amounts of total RNA using the suppression PCR effect. *Biochem. Biophys. Res. Commun.* **230,** 285–288.
49. Vos, P., Hogers, R., Bleeker, M., et al. (1995) AFLP: a new technique for DNA fingerprinting. *Nucleic Acids Res.* **23,** 4407–4414.
50. McClelland, M., Mathieu-Daude, F., and Welsh, J. (1995) RNA fingerprinting and differential display using arbitrarily primed PCR. *Trends Genet.* **11,** 242–246.
51. Bertioli, D. J., Schlichter, U. H., Adams, M. J., Burrows, P. R., Steinbiss, H.H., and Antoniw, J. F. (1995) An analysis of differential display shows a strong bias towards high copy number mRNAs. *Nucleic Acids Res.* **23,** 4520–4523.

52. Graf, D., Fisher, A. G., and Merkenschlager, M. (1997) Rational primer design greatly improves differential display-PCR (DD-PCR). *Nucleic Acids Res.* **25,** 2239–2240.
53. Hakvoort, T. B., Leegwater, A. C., Michiels, F. A., Chamuleau, R. A., and Lamers, W. H. (1994) Identification of enriched sequences from a cDNA subtraction–hybridization procedure. *Nucleic Acids Res.* **22,** 878–879.
54. Ariazi, E. A. and Gould, M. N. (1996) Identifying differential gene expression in monoterpene-treated mammary carcinomas using subtractive display. *J. Biol. Chem.* **271,** 29,286–29,294.
55. Fuchs, B., Zhang, K., Bolander, M. E., and Sarkar, G. (2000) Identification of differentially expressed genes by mutually subtracted RNA fingerprinting. *Anal. Biochem.* **286,** 91–98.
56. Gromova, I., Gromov, P., and Celis, J. E. (2001) A novel member of the glycosyltransferase family, beta 3 Gn-T2, highly downregulated in invasive human bladder transitional cell carcinomas. *Mol. Carcinog.* **32,** 61–72.
57. Gromova, I., Gromov, P., and Celis, J. E. (2002) bc10: a novel human bladder cancer-associated protein with a conserved genomic structure downregulated in invasive cancer. *Int. J. Cancer* **98,** 539–546.
58. Sokolov, B. P. and Prockop, D. J. (1994) A rapid and simple PCR-based method for isolation of cDNAs from differentially expressed genes. *Nucleic Acids Res.* **22,**4009–4015.
59. Lohmann, J., Schickle, H., and Bosch, T. C. (1995) REN display, a rapid and efficient method for nonradioactive differential display and mRNA isolation. *Biotechniques* **18,** 200–202.
60. Ito, T., Kito, K., Adati, N., Mitsui, Y., Hagiwara, H., and Sakaki, Y. (1994) Fluorescent differential display: arbitrarily primed RT-PCR fingerprinting on an automated DNA sequencer. *FEBS Lett.* **351,** 231–236.
61. An, G., Luo, G., Veltri, R. W., and O'Hara, S. M. (1996) Sensitive, nonradioactive differential display method using chemiluminescent detection. *Biotechniques* **20,** 342–346.
62. Li, F., Barnathan, E. S., and Kariko, K. (1994) Rapid method for screening and cloning cDNAs generated in differential mRNA display: application of northern blot for affinity capturing of cDNAs. *Nucleic Acids Res.* **22,** 1764–1765.
63. Mathieu-Daude, F., Cheng, R., Welsh, J., and McClelland, M. (1996) Screening of differentially amplified cDNA products from RNA arbitrarily primed PCR fingerprints using single strand conformation polymorphism (SSCP) gels. *Nucleic Acids Res.* **24,** 1504–1507.
64. Zhao, S., Ooi, S. L., Yang, F. C., and Pardee, A. B. (1996) Three methods for identification of true positive cloned cDNA fragment in differential display. *Biotechniques* **20,** 400–404.
65. Miele, G., MacRae, L., McBride, D., Manson, J., and Clinton, M. (1998) Elimination of false positives generated through PCR re-amplification of differential display cDNA. *Biotechniques* **25,** 138–144.
66. Gromova, I., Gromov, P., and Celis, J. E. (1999) Identification of true differentially expressed mRNAs in a pair of human bladder transitional cell carcinomas using an improved differential display procedure. *Electrophoresis* **20,** 241–248.
67. Callard, D., Lescure, B., and Mazzolini, L. (1994) A method for the elimination of false positives generated by the mRNA differential display technique. *Biotechniques* **16,** 1100–1103.
68. Shoham, N. G., Arad, T., Rosin-Abersfeld, R., Mashiah, P., Gazit, A., and Yaniv, A. (1996) Differential display assay and analysis. *Biotechniques* **20,** 182–184.
69. Liu, C. and Raghothama, K. G. (1996) Practical method for cloning cDNAs generated in an mRNA differential display. *Biotechniques* **20,** 576–580.
70. Wadhwa, R., Duncan, E., Kaul, S. C., and Reddel, R. R. (1996) An effective elimination of false positives isolated from differential display of mRNAs. *Mol. Biotechnol.* **6,** 213–217.
71. Bryant, Z., Subrahmanyan, L., Tworoger, M., et al. (1999) Characterization of differentially expressed genes in purified *Drosophila* follicle cells: toward a general strategy for cell type-specific developmental analysis. *Proc. Natl. Acad. Sci. USA* **96,** 5559–5564.
72. Wittwer, C.T., Herrmann, M.G., Moss, A.A., Rasmussen, R.P. (1997) Continuous fluorescence monitoring of rapid cycle DNA amplification. *Biotechniques* **22,** 134–138.
73. Israeli, D., Tessler, E., Haupt, Y., et al. (1997) A novel p53-inducible gene, PAG608, encodes a nuclear zinc finger protein whose overexpression promotes apoptosis. *EMBO J.* **16,** 4384–4392.
74. Lehar, S. M., Nacht, M., Jacks, T., Vater, C. A., Chittenden, T., and Guild, B. C. (1996) Identification and cloning of EI24, a gene induced by p53 in etoposide-treated cells. *Oncogene* **12,** 1181–1187.
75. Gu, Z., Flemington, C., Chittenden, T., and Zambetti, G. P. (2000) ei24, a p53 response gene involved in growth suppression and apoptosis. *Mol. Cell. Biol.* **20,** 233–241.

76. Lin, Y., Ma, W., and Benchimol, S. (2000) Pidd a new death-domain-containing protein, is induced by p53 and promotes apoptosis. *Nature Genet.* **26,** 122–127.
77. Okamura, S., Arakawa, H., Tanaka, T., et al. (2001) p53DINP1, a p53-inducible gene, regulates p53-dependent apoptosis. *Mol. Cell.* **8,** 85–94.
78. Lo, P. K., Chen, J. Y., Lo, W. C., et al. (1999) Identification of a novel mouse p53 target gene DDA3. *Oncogene* **18,** 7765–7774.
79. Oda, E., Ohki, R., Murasawa, H., et al. (2000) Noxa, a BH3-only member of the Bcl-2 family and candidate mediator of p53-induced apoptosis. *Science* **288,** 1053–1058.
80. Leng, R. P., Lin, Y., Ma, W., et al. (2003) Pirh2, a p53-induced ubiquitin-protein ligase, promotes p53 degradation. *Cell* **112,** 779–791.
81. Zhang, R., Tan, Z., and Liang, P. (2000) Identification of a novel ligand-receptor pair constitutively activated by ras oncogenes. *J. Biol. Chem.* **275,** 24,436–24,443.
82. Wang, M., Tan, Z., Zhang, R., Kotenko, S. V., and Liang, P. (2002) Interleukin 24 (MDA-7/MOB-5) signals through two heterodimeric receptors, IL-22R1/IL-20R2 and IL-20R1/IL-20R2. *J. Biol. Chem.* **277,** 7341–7347.
83. Zhu, Y., Carroll, M., Papa, F. R., Hochstrasser, M., and D'Andrea, A. D. (1996) DUB-1, a deubiquitinating enzyme with growth-suppressing activity. *Proc. Natl. Acad. Sci. USA* **93,** 3275–3279.
84. Sakuma, S., Saya, H., Ijichi, A., and Tofilon, P. J. (1995) Radiation induction of the receptor tyrosine kinase gene Ptk-3 in normal rat astrocytes. *Radiat. Res.* **143,** 1–7.
85. Liu. J., Liu, H., Zhang, X., Gao, P., Wang, J., and Hu, Z., (2002) Identification and characterization of P15RS, a novel P15(INK4b) related gene on G1/S progression. *Biochem. Biophys. Res. Commun.* **299,** 880–885.
86. Nickenig, G., Baudler, S., Muller, C., Werner, C., et al. (2002) Redox-sensitive vascular smooth muscle cell proliferation is mediated by GKLF and Id3 in vitro and in vivo. *FASEB J.* **16,** 1077–1086.
87. Lee, D. W., Zhang, K., Ning, Z. Q., et al. (2000) Proliferation-associated SNF2-like gene (PASG): a SNF2 family member altered in leukemia. *Cancer Res.* **60,** 3612–3622.
88. Wang, S., Nakashima, S., Sakai, H., Numata, O., Fujiu, K., and Nozawa, Y. (1998) Molecular cloning and cell-cycle-dependent expression of a novel NIMA (never-in-mitosis in Aspergillus nidulans)-related protein kinase (TpNrk) in Tetrahymena cells. *Biochem. J.* **334(Pt. 1),** 197–203.
89. Poirier, G. M., Anderson, G., Huvar, A., et al. (1999) Immune-associated nucleotide-1 (IAN-1) is a thymic selection marker and defines a novel gene family conserved in plants. *J. Immunol.* **163,** 4960–4969.
90. Azzoni, L., Kanakaraj, P., Zatsepina, O., and Perussia, B. (1996) IL-12-induced activation of NK and T cells occurs in the absence of immediate-early activation gene expression. *J. Immunol.* **157,** 3235–3241.
91. Ruegg, C. L., Wu, H. Y., Fagnoni, F. F., Engleman, E. G., and Laus, R. (1996) B4B, a novel growth-arrest gene, is expressed by a subset of progenitor/pre-B lymphocytes negative for cytoplasmic mu-chain. *J. Immunol.* **157,** 72–80.
92. Ishaq, M., Zhang, Y. M., and Natarajan, V. (1998) Activation-induced down-regulation of retinoid receptor RXRalpha expression in human T lymphocytes. Role of cell cycle regulation. *J. Biol. Chem.* **273,** 21,210–21,216.
93. Sun, H. B., Zhu, Y. X., Yin, T., Sledge, G., and Yang, Y. C. (1998) MRG1, the product of a melanocyte-specific gene related gene, is a cytokine-inducible transcription factor with transformation activity. *Proc. Natl. Acad. Sci. USA* **95,** 13,555–13,560.
94. Nocentini, G., Giunchi, L., Ronchetti, S., et al. (1997) A new member of the tumor necrosis factor/nerve growth factor receptor family inhibits T cell receptor-induced apoptosis. *Proc. Natl. Acad. Sci. USA* **94,** 62116–6221.
95. Babu, J. S., Sun, T., Xu, L., and Datta, S. K. (2002) B cell stimulatory effects of alpha-enolase that is differentially expressed in NZB mouse B cells. *Clin. Immunol.* **104,** 293–304.
96. Sawitzki, B., Lehmann, M., Vogt, K., et al. (2002) Bag-1 up-regulation in anti-CD4 mAb treated allo-activated T cells confers resistance to apoptosis. *Eur. J. Immunol.* **32,** 800–809.
97. Amson, R. B., Nemani, M., Roperch, J. P., et al. (1996) Isolation of 10 differentially expressed cDNAs in p53-induced apoptosis: activation of the vertebrate homologue of the drosophila seven in absentia gene. *Proc. Natl. Acad. Sci. USA* **93,** 3953–3957.
98. Semizarov, D., Glesne, D., Laouar, A., Schiebel, K., and Huberman, E. (1998) A lineage-specific protein kinase crucial for myeloid maturation. *Proc. Natl. Acad. Sci. USA* **95,** 15,412–15,417.

99. Garcia-Domingo, D., Leonardo, E., Grandien, A., et al. (1999) DIO-1 is a gene involved in onset of apoptosis in vitro, whose misexpression disrupts limb development. *Proc. Natl. Acad. Sci. USA* **96,** 7992–7997.
100. Jin, F. Y., Nathan, C., Radzioch, D., and Ding, A. (1997) Secretory leukocyte protease inhibitor: a macrophage product induced by and antagonistic to bacterial lipopolysaccharide. *Cell* **88,** 417–426.
101. Blaser, C., Kaufmann, M., Muller, C., et al. (1998) Beta-galactoside-binding protein secreted by activated T cells inhibits antigen-induced proliferation of T cells. *Eur. J. Immunol.* **28,** 2311–2319.
102. Xu, D., Chan, W. L., Leung, B. P., et al. (1998) Selective expression of a stable cell surface molecule on type 2 but not type 1 helper T cells. *J. Exp. Med.* **187,** 787–794.
103. Ali, M., Markham, A. F., and Isaacs, J. D. (2001) Application of differential display to immunological research. *J. Immunol. Methods* **250,** 29–43.
104. Cole, K. A., Krizman, D. B., and Emmert-Buck, M. R. (1999) The genetics of cancer-a 3D model. *Nature Genet.* **21,** 38–41.
105. Radford, D. M., Fair, K., Thompson, A.M., et al. (1993) Allelic loss on a chromosome 17 in ductal carcinoma in situ of the breast. *Cancer Res.* **53,** 2947–2949.
106. Shibata, D., Hawes, D., Li, Z. H., Hernandez, A. M., Spruck, C. H., and Nichols, P. W. (1992) Specific genetic analysis of microscopic tissue after selective ultraviolet radiation fractionation and the polymerase chain reaction. *Am. J. Pathol.* **141,** 539–543.
107. Emmert-Buck, M. R., Roth, M. J., Zhuang, Z., et al. (1994) Increased gelatinase A (MMP-2) and cathepsin B activity in invasive tumor regions of human colon cancer samples. *Am. J. Pathol.* **145,** 1285–1290.
108. Zhuang, Z., Roth, M. J., Emmert-Buck, M. R., Lubensky, I. A., Liotta, L. A., and Solomon, D. (1994) Detection of the von Hippel-Lindau gene deletion in cytologic specimens using microdissection and the polymerase chain reaction. *Acta Cytol.* **38,** 671–675.
109. Emmert-Buck, M. R., Bonner, R. F., Smith, P. D., et al. (1996) Laser capture microdissection. *Science* **274,** 998–1001.
110. Fend, F., Quintanilla-Martinez, L., Kumar, S., et al. (1999) Composite low grade B-cell lymphomas with two immunophenotypically distinct cell populations are true biclonal lymphomas. A molecular analysis using laser capture microdissection. *Am. J. Pathol.* **154,** 1857–1866.
111. Schmidt-Kittler, O., Ragg, T., Daskalakis, A., et al. (2003) From latent disseminated cells to overt metastasis: genetic analysis of systemic breast cancer progression. *Proc. Natl. Acad. Sci. USA* **100,** 7737–7742.
112. Murakami, H., Liotta, L., and Star, R. A. (2000) IF-LCM: laser capture microdissection of immunofluorescently defined cells for mRNA analysis rapid communication. *Kidney Int.* **58,** 1346–1353.
113. Jin, L., Thompson, C. A., Qian, X., Kuecker, S. J., Kulig, E., and Lloyd, R. V. (1999) Analysis of anterior pituitary hormone mRNA expression in immunophenotypically characterized single cells after laser capture microdissection. *Lab. Invest.* **79,** 511–512.
114. Lindeman, N., Waltregny, D., Signoretti, S., and Loda, M. (2002) Gene transcript quantitation by real-time RT-PCR in cells selected by immunohistochemistry-laser capture microdissection. *Diagn. Mol. Pathol.* **11,** 187–192.
115. Ghadersohi, A. and Sood, A. K. (2001) Prostate epithelium-derived Ets transcription factor mRNA is overexpressed in human breast tumors and is a candidate breast tumor marker and a breast tumor antigen. *Clin. Cancer Res.* **7,** 2731–2738.
116. Mellick, A. S., Day, C. J., Weinstein, S. R., Griffiths, L. R., and Morrison, N. A. (2002) Differential gene expression in breast cancer cell lines and stroma-tumor differences in microdissected breast cancer biopsies revealed by display array analysis. *Int. J. Cancer.* **100,** 172–180.
117. Chen, L.C., Manjeshwar, S., Lu, Y., et al. (1998) The human homologue for the Caenorhabditis elegans cul-4 gene is amplified and overexpressed in primary breast cancers. *Cancer Res.* **58,** 3677–3683.
118. Nielsen, H. L., Ronnov-Jessen, L., Villadsen, R., and Petersen, O. W. (2002) Identification of EPSTI1, a novel gene induced by epithelial–stromal interaction in human breast cancer. *Genomics* **79,** 703–710.
119. Wang, G. S., Wang, M. W., Wu, B. Y., You, W. D., and Yang, X. Y. (2003) A novel gene, GCRG224, is differentially expressed in human gastric mucosa. *World J. Gastroenterol.* **9,** 30–34.

120. Tran, Y. K., Bogler, O., Gorse, K. M., Wieland, I., Green, M. R., and Newsham, I. F. (1999) A novel member of the NF2/ERM/4.1 superfamily with growth suppressing properties in lung cancer. *Cancer Res* **59,** 35–43.
121. Manda, R., Kohno, T., Niki, T., et al. (2000) Differential expression of the LAMB3 and LAMC2 genes between small cell and non-small cell lung carcinomas. *Biochem. Biophys. Res. Commun.* **275,** 440–445.
122. Yu, L., Hui-chen, F., Chen, Y., et al. (1999) Differential expression of RAB5A in human lung adenocarcinoma cells with different metastasis potential. *Clin. Exp. Metastas.* **17,** 213–219.
123. Hsu, N. Y., Ho, H. C., Chow, K. C., et al. (2001) Overexpression of dihydrodiol dehydrogenase as a prognostic marker of non-small cell lung cancer. *Cancer Res.* **61,** 2727–2731.
124. Neef, R., Kuske, M. A., Prols, E., and Johnson, J. P. (2002) Identification of the human PHLDA1/TDAG51 gene: down-regulation in metastatic melanoma contributes to apoptosis resistance and growth deregulation. *Cancer Res.* **62,** 5920–5929.
125. Cole, K. A., Chuaqui, R. F., Katz, K., et al. (1998) cDNA sequencing and analysis of POV1 (PB39): a novel gene up-regulated in prostate cancer. *Genomics* **51,** 282–287.
126. King, H. C. and Sinha, A. A. (2001) Gene expression profile analysis by DNA microarrays: promise and pitfalls. *JAMA* **286,** 2280–2288.

36

Techniques for Gene Expression Profiling*

Mark P. Richards

1. Introduction

Now that the human genome and the genomes of a growing number of important model eukaryotic organisms such as yeast and mouse have been sequenced, research emphasis in the "postgenomic" era is beginning to shift to events downstream of the whole genome. Interest is now focused on the identification and characterization of individual genes and gene networks to better understand gene function at the cell, tissue and organ levels in different states of health and disease. This new approach to studying the genome has been termed "functional genomics" because efforts are directed toward understanding the connections between the expression of individual genes or groups of genes and their unique biological functions. Although every cell in the body contains the same complement of genetic material, each is distinguished by the level and the spectrum of activation or expression of a specific set of genes. Determining which genes are active in different cells and tissues under different conditions (i.e., physiological, developmental, environmental, stress, disease, etc.) aids researchers in understanding cellular and tissue function at the molecular level. Moreover, it also allows them to relate this information to a general set of characteristics observed at the cell, tissue, and organism levels, commonly referred to as the phenotype.

With a vast and growing amount of DNA sequence information now available in numerous databases, the need for genomewide expression profiling techniques has accelerated the push to understand the role of individual genes and gene networks in regulating diverse and complex biological processes. It is interesting to note that in comparison to 30–40 thousand individual genes identified or suspected in the human genome, it has been estimated that there are more than 10 times that number (estimates range from 100,000 to 10,000,000) of unique proteins produced within cells *(1)*. Genomic information (DNA sequence data) alone, although useful in predicting the existence of specific genes and providing some information about the nature of the proteins produced by them, cannot predict all of the posttranscriptional and posttranslational modifications that RNA transcripts and proteins undergo to give rise to multiple gene transcripts (mRNAs) and gene products (proteins). Thus, it is important to develop and apply a variety of techniques that can accurately measure gene structure and activity at the level of DNA, RNA, and protein *(1)*.

Figure 1 depicts what has often been referred to as the "central dogma" of molecular biology. Genetic information from the genome (the entire complement of genes present in an organism) is transferred through the transcriptome (the entire complement of mRNA transcripts)

*Mention of a trade name, proprietary product, or specific equipment does not constitute a guarantee or warranty by USDA and does not imply its approval to the exclusion of other suitable products.

From: *Medical Biomethods Handbook*
Edited by: J. M. Walker and R. Rapley © Humana Press, Inc., Totowa, NJ

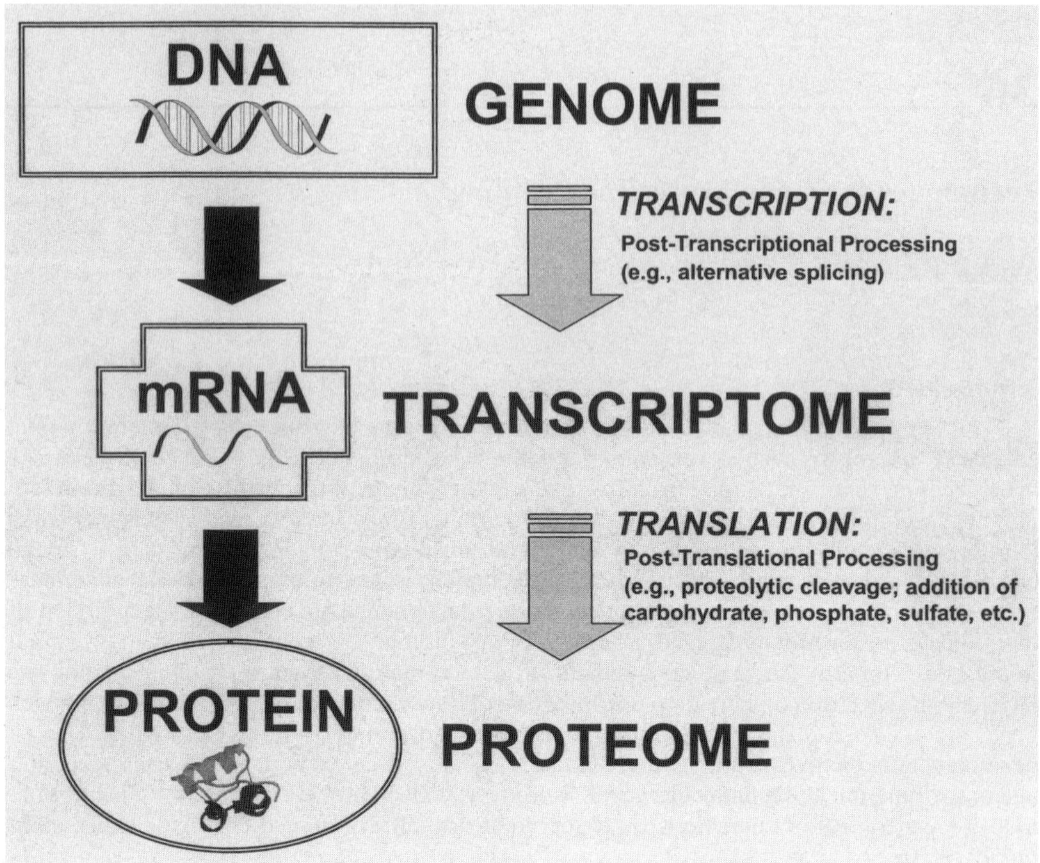

Fig. 1. Schematic representation of the "central dogma" of molecular biology describing the flow of genetic information from the DNA sequence of the genome to the protein complement of the genome.

to the proteome which contains the end products of gene expression or the complete collection of proteins produced in a particular cell. From this point of view, it is clear that gene expression profiling (i.e., the systematic identification and characterization of those genes activated or expressed in a cell) can be conducted on different levels depending on the specific research objectives. This could involve analysis of DNA, mRNA, and/or protein as a measure of gene expression. **Table 1** lists a variety of techniques that have been used to profile the genome and to assess gene activity at the mRNA or protein level. This chapter will present a brief overview of some approaches that have been taken to profile gene expression at the mRNA and protein level.

2. Methods for Profiling Gene Expression at the Transcriptome (RNA) Level

The initial step in gene expression is transfer (transcription) of the genetic information contained in genomic DNA to an RNA transcript (mRNA). The common objectives of gene expression analyses at the RNA level are to determine whether specific mRNA sequences transcribed from a particular gene(s) of interest are present in samples of cells or tissues and, if so, at what levels. Thus, transcriptional profiling, either as single-gene transcripts, small groups of related transcripts, or on a global scale using the latest high-throughput techniques to simul-

Table 1
Examples of Analytical Procedures Used for Genomic and Gene Expression Profiling

A. Genome (DNA)
- DNA sequencing
- DNA profiling/genotyping (e.g., mutations, single-nucleotide polymorphisms, repeated sequence, etc.)
- Gene mapping

B. Transcriptome (mRNA)
- Northern blot
- Nuclease protection assay
- Reverse transcription–polymerase chain reaction (RT-PCR)
- *In situ* hybridization/tissue arrays
- Differential display PCR
- cDNA/expressed sequence tag (EST) libraries
- Serial analysis of gene expression (SAGE)
- DNA microarrays

C. Proteome (Protein)
- Separation: chromatography, electrophoresis, mass spectrometry
- Immunoassays: radioimmunoassay (RIA), enzyme-linked immunosorbent assay (ELISA), Western blot, immunocytochemistry
- Function/activity assays: enzyme activity, binding affinity, etc.
- Combined techniques: high-performance liquid chromatography/mass spectrometry (LC/MS), electrophoresis/ mass spectrometry, protein tagging/mass spectrometry, tissue arrays/immunocytochemistry, protein chips/SELDI-TOF mass spectrometry
- Protein microarrays
- Direct mass imaging

taneously analyze hundreds or even thousands of unique gene transcripts collectively constitutes one avenue to explore gene activity and function. Although this is a powerful approach to understanding biological processes in tissues and cells from diverse phenotypes or physiological states, no single technique can give a complete and accurate picture of the entire transcriptome. Therefore, a combination of different techniques represents the best approach to studying gene function within cells and tissues at the level of mRNA. Gene transcripts can be studied directly as RNA species or indirectly by converting them to DNA copies (complementary DNA, cDNA) via a process known as reverse transcription (RT). The cDNA copies can then be enriched by amplification for subsequent analysis using polymerase chain reaction (PCR). Typically, researchers seek to characterize discrete portions of the transcriptome for cells or tissues from specific phenotypes related to health or disease states or from experimental treatments that cause specific physiological changes.

2.1. Profiling the Transcriptome on a Limited Scale

Detection and quantification of specific RNA sequences can be accomplished by Northern blotting *(2)*. This method involves the isolation of RNA from a cell or tissue sample and fractionation by denaturing agarose gel electrophoresis. The separated RNA species are transferred to a nitrocellulose or Nylon membrane support by blotting. The RNA transferred to the resulting membrane blot is then hybridized with a labeled (radioactive, fluorescent, chemiluminescent) cDNA or antisense RNA (cRNA or riboprobe) probe unique to the gene(s) of interest. The blot is developed and the intensity of the hybridized bands is determined by image analysis techniques. Quantification of transcripts is usually achieved by comparing or "normalizing" target transcript levels to a gene transcript that is considered to be invariant, such as a typical "housekeeping" gene (e.g., β-actin, glyceraldehyde phosphate dehydrogenase, cyclophillin, etc.) or a non-mRNA transcript such as the ribosomal RNAs (18S or 28S). Northern analysis generally requires significant quantities of RNA (either total RNA or poly-A selected), making it best suited to the analysis of moderately or highly abundant mRNA transcripts. It is relatively less sensitive than other transcript profiling techniques and fairly labor-intensive because of the multiple steps involved. The advantages of Northern analysis are that RNA species are directly analyzed and, because of that, the technique is excellent for identification of alternatively processed (e.g., alternatively spliced) gene transcripts.

Another technique used in mRNA analysis is the ribonuclease protection assay ***(3)***. Isolated RNA is hybridized to a single-stranded cDNA or an antisense RNA (riboprobe) unique to the gene(s) of interest in a buffered solution. After allowing the single-stranded probe to anneal to the target mRNA, the sample is subjected to specific enzymatic digestion to remove all single-stranded nucleic acid, leaving only double-stranded or "protected" RNA. The resulting double-stranded nucleic acid fragments are then separated on high-resolution polyacrylamide gels. Quantification is accomplished by imaging of the gel or a blot of the gel similar to the procedures for Northern analysis. This assay is much more sensitive than Northern analysis and, thus, can be applied to detect and quantify mRNAs that are expressed at low levels. One distinct advantage of ribonuclease protection assays is that they rely on solution hybridization, which permits the analysis of very low-abundance mRNAs. Also, because these assays are solution based, they are more amenable to automation to increase sample throughput.

Reverse transcription–polymerase chain reaction (RT-PCR) is one of the most sensitive techniques for mRNA analysis ***(4)***. This is largely the result of the power of the amplification step (PCR), which greatly increases the number of copies of each cDNA made against a specific gene transcript (mRNA). However, RT-PCR is an indirect analytical method for quantifying mRNA levels because RNA cannot be directly amplified. The technique consists of two parts: (1) synthesis of a cDNA from an mRNA template by RT and (2) amplification of the specific cDNA by PCR. Either during or after RT-PCR, the PCR product, a double-stranded DNA (dsDNA) referred to as an amplicon, is detected and quantified. Under carefully controlled conditions, the amount of PCR amplicon produced is directly proportional to the amount of mRNA transcript in the original sample.

A variety of methods exist for conducting RT-PCR for the quantitative analysis of gene expression. Two basic approaches are employed: (1) a kinetic approach involving the continuous monitoring of amplicon development during PCR or what is known as "real-time" PCR ***(4)*** and (2) methods that rely on postreaction separation and quantification of the PCR amplicons. A technique that combines RT-PCR and capillary electrophoresis with laser-induced fluorescence detection (CE-LIF) has been shown to be very effective in quantifying gene expression ***(5)***. As outlined in **Fig. 2**, CE/LIF directly separates and quantifies each amplicon upon completion of RT-PCR. Quantitative analysis of gene expression can be done relative to an internal coamplified gene transcript (e.g., β-actin or 18S rRNA) or it can be performed using a competitor or internal standard related to the gene of interest. Absolute quantitation can also be performed using varying quantities of a dsDNA standard amplified in parallel with the experimental samples. The PCR conditions for each gene must be optimized so that the target and standard genes are amplified at a constant rate in order for quantitative estimates of each mRNA to be accurate.

In situ hybridization (ISH) is another technique for mRNA analysis that uses radiolabeled or nonradioactive (fluorescent or enzyme-linked histochemical) nucleic acid probes to detect expression of specific genes in cell and tissue samples ***(6)***. The ISH procedure involves hybridization of a probe to specific gene sequences located in fixed specimens. This technique is especially good for localization of gene expression events to specific regions or cells within tissues. Newer approaches to ISH methodology involve the use of tissue arrays consisting of serial sections of tissue samples collected from hundreds of individual subjects mounted on glass slides to study gene expression in, for example, tumor biopsy samples for the purpose of transcriptional as well as protein profiling ***(7)***.

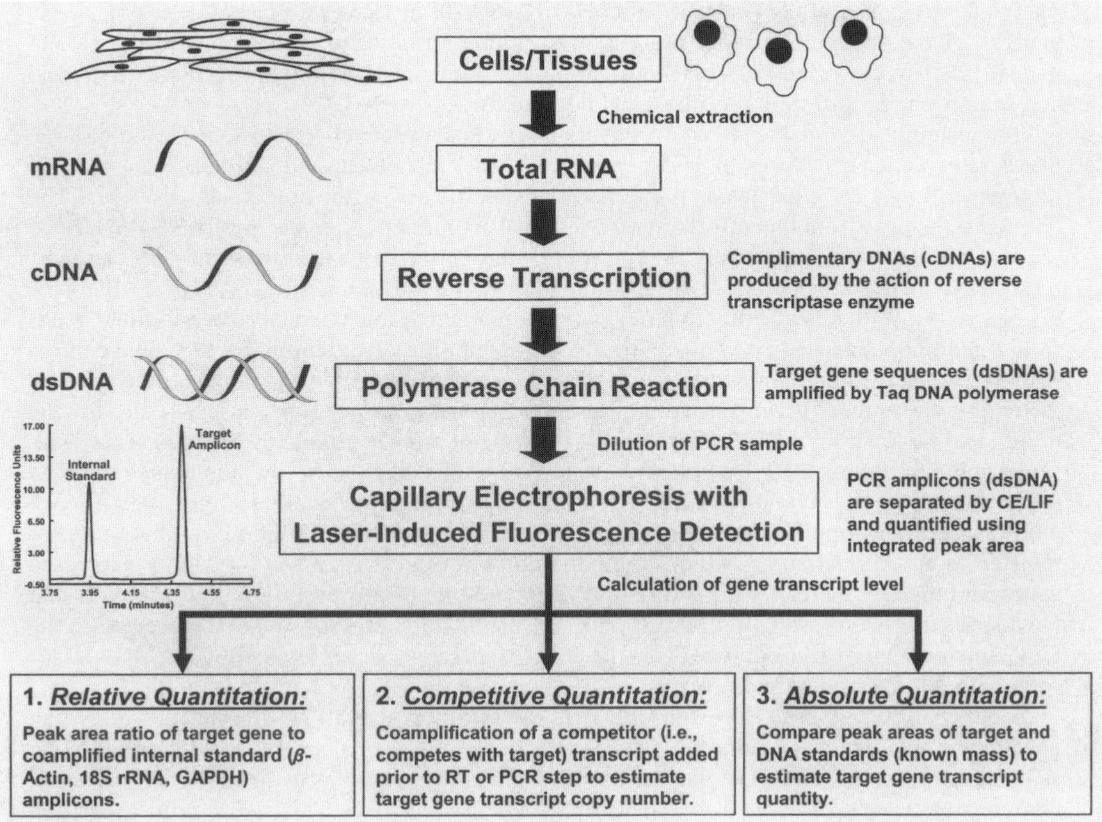

Fig. 2. Schematic representation of a technique for the quantitative analysis of gene expression (mRNA level) using RT-PCR and CE/LIF. See ref. *5*.

2.2. Global Techniques for Profiling the Transcriptome

In the past decade, a number of techniques have been developed to measure gene expression on a global scale involving the parallel assessment of the expression of hundreds to thousands of genes simultaneously in a single experiment. Included among these techniques are (1) generation and sequencing of tissue- or cell-specific cDNA or expressed sequence tag (EST) libraries, (2) differential display PCR, (3) serial analysis of gene expression (SAGE), and (4) DNA microarrays. All of these techniques rely heavily on high-throughput DNA sequencing technology, the same technology developed and employed to sequence whole genomes. Some of the most common applications are to determine differential patterns of gene expression or to compare mRNA levels in cells or tissues from two populations subjected to different experimental treatments, or between cells and tissues expressing different phenotypes, or at different developmental stages to determine changes in gene activity. The large amount of information generated by these techniques has led to the development of many different computer-based information management (software) tools (collectively referred to as bioinformatics) to analyze, interpret, and catalog the data. Although some of these global-scale techniques can produce quantitative data on gene expression (e.g., SAGE and DNA microarrays), their routine use for that purpose is still somewhat limited by technique expense, complexity, and relative insensitivity. In addition, there are difficulties associated with analyzing and interpreting the large volume of information collected in a single experiment. Instead, other techniques such as Northern analysis or RT-PCR are routinely used as follow-up techniques to confirm and more accurately quantify some of the findings obtained from SAGE or DNA microarray experiments. In general, global-scale methods, at present, are screening tools best suited for novel gene and pathway discovery, for biomarker identification, and for gene expression profiling in tissue and cell samples under different physiological conditions.

One example of global transcript profiling involves the production and sequencing of large numbers of cloned cDNAs from tissue-specific libraries in an attempt to derive a "snapshot" of the transcriptome for a particular tissue under a specific set of physiological conditions. The derived sequences from this effort are referred to as expressed sequence tags (ESTs) ***(8)***. Such sequence information has already proved useful in identifying novel genes and there are currently many databases that house large amounts of EST sequence information for a wide variety of species. With this information, it is possible to identify and characterize individual genes from a genome sequence (a process referred to as annotation). In addition, EST sequence information can be useful in conjunction with other techniques like RT-PCR in designing specific methods to accurately quantify expression of individual genes or, on a larger scale, for the design and fabrication of DNA microarrays. Because of the wide dynamic range of RNA transcript numbers among different cell and tissue types, the detection of low-abundance mRNAs using this approach is complicated. Ideally, EST libraries seek to attain a single copy of each unique gene transcript. However, in practice, such libraries need to be modified by a process referred to as "normalizing," which limits the number of cDNAs made for highly expressed genes and thereby increases the likelihood of including cDNAs corresponding to genes that have low expression levels. In doing so, the "snapshot" of expressed genes represented in the EST collection more accurately reflects the entire transcriptome for the particular cell or tissue source used to construct the library.

Serial analysis of gene expression (SAGE) is a useful tool that allows the analysis of overall gene expression patterns by digital analysis ***(9)***. Because SAGE does not require pre-existing sequence information, it can be used to identify and quantify new gene transcripts (i.e., previously uncharacterized) as well as those from known genes. Basically, what is measured by SAGE is not the actual gene transcript, but, rather, a short nucleotide sequence or "tag" of defined length located at the 3' end of each cDNA that is unique to each gene transcript. The "tag" is thus used to identify the particular gene transcript as well as to quantify it based on the assumption that the number of "tags" detected is directly proportional to the amount of mRNA in the original cell or tissue extract. Because SAGE provides descriptive information about the mRNA population present in a cell or tissue without the requirement for prior sequence information, this has particular relevance to clinical research, for example, in the discovery of new biomarkers for disease or for evaluating drug treatment effects, both cases that could involve the expression of previously uncharacterized genes.

DNA microarrays are constructed by arraying or spotting PCR-amplified cDNA (from normalized cDNA libraries, clones, or gene fragments) or synthetic oligonucleotides selected to represent as many previously identified unique gene transcripts as possible, at a high density in gridded patterns onto glass slides, silicon wafers, or Nylon membrane supports ***(10–12)***. The microarrays are utilized in a cohybridization process involving two or more fluorescently labeled probes prepared from mRNA samples obtained from cell or tissue samples representing different experimental treatments, disease states, developmental stages, or phenotypes of interest. A kinetic analysis allows gene expression levels to be determined from the locations on the array and level of hybridization (measured as fluorescence) detected for each probe ***(10–12)***. Although this approach entails a significant investment of money and effort to develop and effectively utilize the arrays, high-density array technology does have excellent potential for large-scale (genomewide) expression analyses. However, the data collected are so extensive that sophisticated bioinformatic software tools are required to analyze and draw conclusions from the array. Despite the fact that a massive number of gene transcripts can be screened simultaneously, individual arrays only provide information about events occurring within a single tissue or cell sample at a given time. Further technological developments in DNA microarray fabrication and analysis will help to identify and understand the degree of interaction in gene networks that regulate complex biological processes such as disease, stress, growth, and development.

3. Methods for Profiling Gene Expression at the Protein Level

Following transcription, genetic information contained in mRNA is transferred to proteins, the end products of gene expression, through the process of translation (*see* **Fig. 1**). Considering the wealth of information (structural and functional) that they contain, proteins have been aptly referred to as the "molecular machines of life" because it is proteins that do all of the work within cells ***(13,14)***. Expression profiling at the protein level seeks to determine whether a specific protein(s) encoded by a particular gene(s) of interest is present in samples of cells or tissues and, if so, at what level. In comparison to profiling the transcriptome, the challenges of characterizing the proteome are much greater. In part, this is the result of the fact that proteins cannot be amplified like DNA and, therefore, direct measurements must be made on native protein species. Also, because the majority of proteins are further processed or modified in many complex ways following translation (a process referred to as posttranslational processing), there is a much wider dynamic range of distinct protein species compared to the corresponding collection of mRNA transcripts. In addition to analysis of amino acid sequence, modifications of protein groups with phosphate, nitrate, sulfate, carbohydrate, lipid, and so forth, must be taken into account. In fact, it has been reported that more than 1000 variants or isoforms of a single brain protein (neurexin) can arise from the combined effects of mRNA alternative splicing (posttranscriptional processing) and differential posttranslational modification ***(15)***. Thus, from a relatively small number of genes predicted for the human genome, a potentially very large and very complex proteome can be expected. It is worth emphasizing that the full scope of this complexity has yet to be accurately defined for any organism, tissue, or cell currently under study.

3.1. Techniques for Protein Isolation, Characterization, and Quantification

Traditional protein chemistry methodologies have been employed for many years to separate and characterize individual or small groups of proteins. Proteins can be separated based on physical parameters such as mass, charge, or binding affinities. Techniques such as chromatography and electrophoresis have contributed greatly to this area. However, recent improvements in the application of advanced mass spectrometry (MS) techniques to the accurate mass analysis of proteins and peptides have proven to be invaluable to characterizing the proteome ***(16)***. Because of this, MS has become a core technology in profiling the proteomes of various species. Proteins can be fragmented by enzymatic digestion prior to or following separation by electrophoresis or chromatography and the resulting peptides can be analyzed by MS to identify the original protein by a process of "peptide mass fingerprinting." This process involves comparison of each of the peptides detected by MS against a database containing all known proteins and their component peptides ***(16)***.

Accurate detection and quantification of individual proteins can also involve use of immunological reagents such as antibodies developed to recognize discrete portions of the protein (epitopes) or compounds attached to the protein (protein side chains) such as carbohydrates ***(14)***. Immunological-based methodologies represent some of the most sensitive techniques for detecting and quantifying individual proteins, especially those present at low levels. These can involve the use of radiolabeled tracers in a radioimmunoassay (RIA) or the use of colorimetric assays involving enzymatic action on a substrate in an enzyme-linked immunosorbent assay (ELISA). However, to analyze proteins by these or other immunoassay methods, a specific antibody that recognizes each protein to be analyzed must be produced.

Other types of binding characteristics, such as metal-binding properties or the binding to other proteins (e.g., receptors), can be exploited to develop sensitive methods for the specific detection of individual proteins from a complex mixture such as a tissue or cell extract ***(16)***.

Some of these parameters, such as binding affinity, are beginning to be explored for use in characterizing larger protein subsets of the proteome ***(16)***. However, considering the projected complexity of the proteome and the fact that each of these techniques is fairly labor-intensive, these approaches remain best suited to the analysis of individual proteins or small groups of proteins.

3.2. Profiling the Proteome on a Global Scale

The newly emerging scientific discipline of proteomics ultimately seeks to isolate, characterize, and quantify all individual protein members of the proteome (the protein complement of the genome) for a particular tissue or cell type in much the same fashion as transcript profiling seeks to characterize the transcriptome (***16***). Because proteins must be analyzed in their native state, the same techniques used in analysis of individual or small numbers of proteins have been used to form the basis of global-scale techniques that analyze large numbers of proteins in parallel. What has been learned concerning the use of physical parameters such as charge, mass, enzyme activity, or binding activity for small-scale protein separations is beginning to be applied to analyses of larger portions of the proteome.

Because of the extent of complexity projected, researchers have begun to develop new technologies for profiling the entire proteome or large portions of it in single experiments. Generally, these global-scale techniques involve a variety of combinations of orthogonal analytical methodology. Current examples include the combinations of two-dimensional gel electrophoresis (2DGE) or capillary electrophoresis (CE) with MS or high-performance liquid chromatography (HPLC) with MS ***(16)***.

Like DNA microarrays, protein arrays have been developed using a variety of approaches to capture and study as many individual proteins from complex mixtures as possible ***(16–19)***. The modes of capture can involve immobilized antibodies, aptamers (i.e., a short single-stranded modified oligonucleotide that binds tightly to protein), peptides, other proteins, or via interactions (affinity based) with chemically derivatized solid surfaces. The captured proteins can then be detected by a number of methods, including fluorescence, surface-enhanced laser desorption and ionization time-of-flight (SELDI-TOF) MS, enzyme-linked colorimetric readout, or even atomic force microscopy ***(14,16–19)***. Along with such approaches comes a very large amount of raw data from the mass spectral scans of the peptides produced when individual proteins are fragmented or from other types of array detection data. This has created a need for new and more sophisticate bioinformatics to analyze, interpret, and catalog proteomic data ***(16)***.

4. Applications of Gene Expression Profiling

Biomedical researchers over much of the past several decades have endeavored to study complex biological systems one gene or one protein at a time ***(13)***. However, it is now becoming clear that physiological functions and disease states are determined by many different genes producing protein products that interact with each other in a complex functional network. In fact, less than 2% of the major human diseases are caused by changes to a single gene, whereas 98% involve multiple genes ***(1)***. Therefore, single-gene (protein) analyses are increasingly giving way to the application of global-scale, high-throughput gene expression profiling techniques. Further, the information derived from such global approaches is beginning to be integrated into "systems" models to aid our understanding of complex interactions involving gene networks and protein pathways that regulate cellular functioning ***(13)***. Transcriptional profiling alone offers an incomplete picture because it cannot account for all of the protein modifications, such as phosphorylation, or for protein–protein interactions. However, data obtained from the genome, transcriptome, and proteome are complementary to one another and thus suited to data integration strategies to form a more comprehensive functional picture of complex physiological functions. This systems approach, referred to as "molecular profiling," has been applied to the study of human cancer ***(1,14)***. Thus, the combination of genomic,

Table 2
Selected Examples of Biomedical and Clinical Research Application Areas for Gene Expression Profiling Techniques

Area	Techniques
1. Gene discovery/expression	
• Novel gene identification	Northern analysis, RT-PCR, ESTs[a], SAGE[b], DNA arrays
• Genome annotation	Comparative genomics, ESTs, DNA arrays
• Gene function	RT-PCR, DNA arrays, activity-based assays (enzyme, binding affinity, etc.), protein arrays, MS-based proteomic tools[c], RNAi[d]
2. Pathway discovery/profiling	
• Signaling	DNA arrays, MS-based proteomic tools
• Metabolic	Northern analysis, RT-PCR, DNA arrays, MS-based proteomic tools
• Regulatory	Northern analysis, RT-PCR, DNA arrays, MS-based proteomic tools
3. Biomarker discovery/profiling	
• Screening body fluids for disease biomarkers	Protein arrays, MS-based proteomic tools
• Cell/tissue biomarkers for disease	DNA arrays, tissue arrays/imaging MS, MS-based proteomic tools
• Biomarkers for tissue engineering	DNA arrays
• Monitoring gene therapy	RT-PCR, DNA arrays, protein arrays
4. Disease diagnosis/prognosis	
• Determination of clinical outcome	DNA arrays, protein arrays, MS-based proteomic tools
• Cancer (tumor) classification	DNA arrays, protein arrays, tissue arrays
• Host–pathogen interactions	DNA arrays, protein arrays
5. Therapeutic Development/Testing:	
• Therapeutic target discovery	DNA arrays, MS-based proteomic tools
• Efficacy/toxicity testing	DNA arrays, protein arrays, MS-based proteomic tools
• Vaccine development	Genomic profiling, DNA arrays, protein arrays, MS-based proteomic tools

[a]ESTs: expressed sequence tag libraries.

[b]SAGE: serial analysis of gene expression.

[c]MS-based proteomic tools: includes a variety of coupled instrument systems using mass spectrometry as the detection mode such as: 2D gels/MS, LC/MS and LC/MS/MS, CE/MS, protein chips (SELDI-TOF), and so forth.

[d]RNAi: RNA interference.

transcriptional, and proteomic profiling is a very useful strategy to understanding cellular function at the molecular level ***(1)***.

Table 2 lists selected examples of the application of a variety of expression profiling techniques to biomedical and clinical research areas. These application areas can be divided into three general categories: (1) discovery-based applications, (2) health- and disease-related applications, and (3) development and testing of new therapeutic agents such as drugs and vaccines. In the area of discovery science, applications range from the search for new genes with novel functions to the identification of pathways governing such basic functions as cellular signaling, metabolism, and gene regulation ***(16,20,21)***. Examples of disease-related applications include classifying tumors, screening for biomarkers for specific diseases such as cancer, heart disease, obesity, and so forth, characterizing host–pathogen interactions, and determining the clinical outcome for specific diseases ***(1,14,16,22,23)***. The development of new therapeutics, the evaluation of efficacy and toxicity, and the development of new vaccines increasingly involves expression profiling techniques to identify potential protein targets, determine effectiveness, and evaluate potential toxicities during development and testing *(22–26)*.

Abnormally or differentially expressed genes can serve as new drug targets or as biomarkers for genetic-based diagnosis of disease. An important challenge for validating the application of expression profiling technologies is to determine the relationship between changes in specific sets of genes and clinical features of a particular disease or physiological state *(26)*. It is also important to focus on the smallest subset of genes and gene products that accurately define the clinical manifestations of the disease or physiological state. Finally, it is important to know how well changes in gene transcription reflect distinct changes at the protein level. A lack of correlation might imply that predictions based on changes in gene activity at the RNA level might be independent of gene function as assessed at the protein level *(22,27)*.

5. A Current Perspective on Gene Expression Profiling

Although there have been great strides in profiling transcriptomes and proteomes of a variety of organisms, it must be kept in mind that a direct relationship between mRNA level and protein level does not exist for every gene *(22,27)*. Because proteins, the end products of gene expression, are ultimately responsible for all cellular functions, a comprehensive understanding of this relationship is crucial to assigning functional significance to any changes in mRNA or protein levels. Half-life, posttranscriptional modification, translational efficiency, and post-translational modification must be considered when attempting to correlate measurements of mRNA expression with protein concentration. Clearly, any comprehensive analysis of gene expression must involve a series of related analyses at the genome, transcriptome, and proteome levels. Furthermore, the proteome and transcriptome are not single-defined entities as is the case for the genome. Instead, different tissues and cells each possess a unique profile of expressed genes that sets them apart and gives rise to characteristic phenotypes. Thus, transcriptomes and proteomes are dynamic, ever-changing in relation to the environment and physiological state of the cell. In the future, much effort will be expended to determine gene expression profiles at the transcriptome and proteome levels in order to explain cellular functioning under defined conditions. Despite numerous technological advances in gene expression profiling, it must still be kept in mind that a "complete" profile of an entire transcriptome or proteome has yet to be achieved for any species, even for the relatively simple single-cell organisms such as yeast or *Escherichia coli*. This underscores the enormous challenges ahead for expression profiling.

6. Future Developments

The future will inevitably bring better and wider availability of ready-made reagents, new instrumentation, and novel approaches for gene expression profiling on a growing scale. In addition, more advances in bioinformatics to help manage and understand the large amounts of data derived from the genomes, transcriptomes, and proteomes from an expanding list of organisms under a variety of physiological and environmental conditions will be developed. Using sophisticated bioinformatics, new functional interrelationships indicating how genes interact to give rise to specific phenotypes will be discovered. The ultimate goal is to understand the flow of information within cells, tissues, or organisms by characterizing the diverse set of interconnected gene and protein networks that give rise to complex biological functions *(23)*. This will lead to a better understanding of life at the molecular level.

Finally, we are now rapidly approaching the era of the "$1000 genome;" that is, a time in which it will become feasible to sequence and screen an individual's entire genome at a reasonable cost *(28,29)*. This will require specific, rapid, and cost-effective methods for gene sequencing and expression profiling for the purpose of understanding disease processes, developing new disease biomarkers for diagnosis and early detection of disease, and for customizing drug treatments that better account for toxicity and therapeutic effectiveness unique to each individual.

As scientific advances continue to move from laboratory to bedside, what we are learning today about gene expression profiling will most certainly contribute toward achieving the ambitious goal of personalized medicine through patient-tailored diagnoses, therapies and prognoses ***(23)***.

References

1. Baak, J. P. A., Path, F. R. C., Hermsen, M. A. J. A., Meijer, G., Schmidt, J., and Janssen, E. A. M. (2003) Genomics and proteomics in cancer. *Eur. J. Cancer* **39,** 1199–1215.
2. Sabelli, P. A. (1998) Northern blot analysis, in *Molecular Biomethods Handbook* (Rapley, R. and Walker, J. M., eds.), Humana, Totowa, NJ, pp. 89–93.
3. Einspanier, R. and Plath, A. (1998) Detecting mRNA by use of the ribonuclease protection assay (RPA), in *Molecular Biomethods Handbook* (Rapley, R. and Walker, J. M., eds.), Humana, Totowa, NJ, pp. 51–57.
4. Bustin, S. A. (2002) Quantification of mRNA using real-time reverse transcription PCR (RT-PCR): trends and problems. *J. Mol. Endocrinol.* **29,** 23–39.
5. Richards, M. P. and Poch, S. M. (2002) Quantitative analysis of gene expression by reverse transcription polymerase chain reaction and capillary electrophoresis with laser-induced fluorescence detection. *Mol. Biotechnol.* **21,**19–37.
6. duSart, D. and Choo, K. H. A. (1998) The technique of in situ hybridization, in *Molecular Biomethods Handbook* (Rapley, R. and Walker, J. M., eds.), Humana, Totowa, NJ, pp. 697–720.
7. Kononen, J., Bubendorf, L., Kallioniemi, A., et al. (1998) Tissue microarrays for high-throughput molecular profiling of tumor specimens. *Nature Med.* **4,** 844–847.
8. Adams, M. D., Kelley, J. M., Gocayne, J. D., et al. (1991) Complementary DNA sequencing: expressed sequence tags and human genome project. *Science* **252,** 1651–1656.
9. Velculescu, V. E., Zhang, L., Vogelstein, B., and Kinzler, K. W. (1995) Serial analysis of gene expression. *Science* **270,** 484–487.
10. Schena, M., Shalon, D., Davis, R. W., and Brown, P. O. (1995) Quantitative monitoring of gene expression patterns with a complementary DNA microarray. *Science* **270,** 467–470.
11. Lipshutz, R. J., Fodor, S. P. A., Gingeras, T. R., and Lockhart, D. J. (1999) High density synthetic oligonucleotide arrays. *Nature Genet.* **21(Suppl. 1),** 20–24.
12. Duggan, D. J., Bittner, M., Chen, Y., Meltzer, P., and Trent, J. M. (1999) Expression profiling using cDNA microarrays. *Nature Genet.* **21(Suppl. 1),** 10–14.
13. Hood, L. (2002) A personal view of molecular technology and how it has changed biology. *J. Proteome Res.* **1,** 399–409.
14. Liotta, L. and Petricoin, E. (2000) Molecular profiling of human cancer. *Nature Rev. Genet.* **1,** 48–56.
15. Missler, M. and Sudhof, T. C. (1998) Neurexins: three genes and 1001 products. *Trends Genet.* **14,** 20–26.
16. Patterson, S. D. and Aebersold, R. H. (2003) Proteomics: the first decade and beyond. *Nature Genet.* **33(Suppl. 1),** 311–323.
17. Haab, B. B., Dunham, M. J., and Brown, P. O. (2001) Protein microarrays for highly parallel detection and quantitation of specific proteins and antibodies in complex solutions. *Genome Biol.* **2,** 0004.1–0004.13.
18. MacBeath, G. (2002) Protein microarrays and proteomics. *Nature Genet.* **32(Suppl. 2),** 526–532.
19. Sydor, J. R. and Nock, S. (2003) Protein expression profiling arrays: tools for the multiplexed high-throughput analysis of proteins. *Proteome Sci.* **1,** 1–7.
20. Brown, P. O. and Botstein, D. (1999) Exploring the world of the genome with DNA microarrays. *Nature Genet.* **21(Suppl. 1),** 33–37.
21. Patterson, S.D. (2003) Proteomics: evolution of the technology. *Biotechniques* **35,** 440–444.
22. Hanash, S. (2003) Disease proteomics. *Nature* **422,** 226–232.
23. Petricoin, E. and Liotta, L. A. (2003) Clinical applications of proteomics. *J. Nutr.* **133,** 2476S–2484S.
24. Debouck, C. and Goodfellow, P. N. (1999) DNA microarrays in drug discovery and development. *Nature Genet.* **21(Suppl. 1),** 48–50.
25. Gerhold, D. L., Jensen, R. V., and Gullans, S. R. (2002) Better therapeutics through microarrays. *Nature Genet.* **32(Suppl. 2),** 547–552.
26. Petricoin, E. F., Hackett, J. L., Lesko, L. J., Puri, R. K., Gutman, S. I., Chumakov, K., Woodcock, J., Feigal, D. W., Zoon, K. C., and Sistare, F. D. (2002) Medical applications of microarray technologies: a regulatory science perspective. Nature Genet. **32 (Suppl. 2)** 474–479.

27. Chen, G., Gharib, T. G., Huang, C.-C., et al. (2002) Proteomic analysis of lung adenocarcinoma: identification of a highly expressed set of proteins in tumors. *Clin. Cancer Res.* **8,** 2298–2305.
28. Constans, A. (2003) Beyond Sanger: toward the $1,000 genome. *The Scientist* **17,** 36–38.
29. Pennisi, E. (2003) The ultimate gene gizmo: humanity on a chip. *Science* **302,** 211.

37

Capillary Electrophoresis in Clinical Analysis

Margaret A. Jenkins and Sujiva Ratnaike

1. Introduction

Capillary electrophoresis (CE) is an exciting technique that has many clinical applications. These include serum and urine protein electrophoresis, the diagnosis of hemoglobinopathies, the quantification of drugs such as iohexol and phenobarbital, and the analysis of DNA.

Capillary electrophoresis can be defined as the separation of charged or uncharged molecules in a thin buffer-filled capillary by the application of a very high voltage. The ends of the capillary, together with the electrodes, are placed into buffer vials. The measurement of analytes takes place usually closer to the outlet of the capillary. The detector, which may be ultraviolet (UV), diode array, or laser-induced fluorescence is connected to a computer system that records separated components. A schematic view of a typical CE instrument is presented in **Fig. 1**.

The primary advantage of CE is that it has been used to automate assays that previously were labor-intensive manual procedures. In the clinical laboratory, an example of such an assay is serum protein electrophoresis. Serum protein gels were often regarded by scientists as an "art form" and were certainly attractive with their Coomassie- or naphthalene black-stained bands. However, the reproducibility of the process was often dependent on the skill of the operator. CE instruments can be programmed for each assay. This includes reproducible washing, injection, and running steps, the timing of which can be controlled to tenths of a second.

Commercial CE instrumentation is costly, the price often being between $US 32,500 and $US 65,000, depending on the complexity of the instrumentation. Dedicated protein electrophoresis commercial CE instrumentation is within a similar price range, the difference being that commercial kits of a particular manufacturer can only be run on a particular instrument.

2. Components of Capillary Electrophoresis

2.1. Capillaries

2.1.1. Forces Within the Capillary

When an electric field is imposed tangentially to the surface of the capillary, it causes the hydrated counterions in the diffuse layer to migrate toward the oppositely charged electrode and drag solvent with them. This is called electro-osmotic flow (EOF) and is represented schematically in **Fig. 2**.

The zeta potential depends on the thickness of the diffuse double layer of ions and the density of charges on the capillary surface. The EOF decreases with a number of factors *(1)*:

1. Increasing the ionic strength of the buffer used.
2. Increasing the viscosity of the buffer by the addition of an organic solvent or glucose.

From: *Medical Biomethods Handbook*
Edited by: J. M. Walker and R. Rapley © Humana Press, Inc., Totowa, NJ

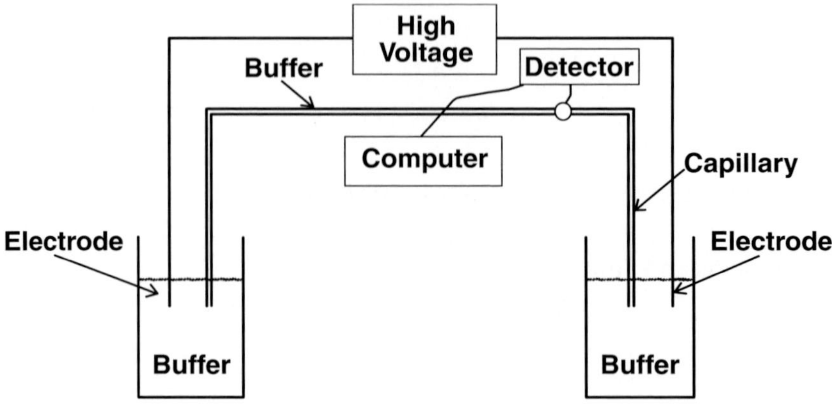

Fig. 1. Schematic diagram of a capillary electrophoresis instrument.

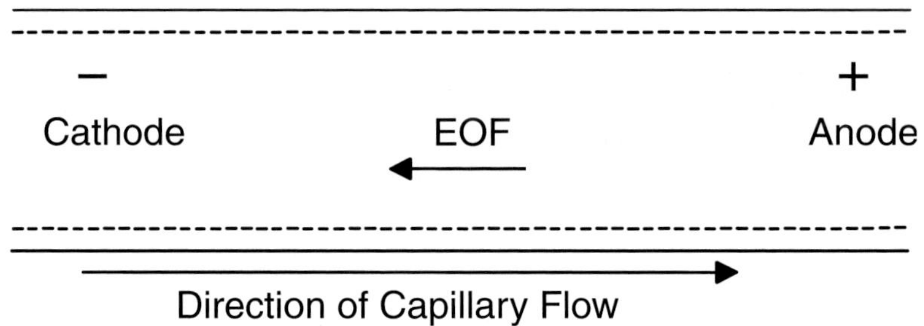

Fig. 2. Schematic diagram of electro-osmotic flow (EOF) found within the capillary when a voltage is applied.

3. Decreasing the charged density on the surface of the capillary by decreasing the pH of the buffer when using fused silica capillaries.
4. Decreasing the voltage applied to the separation.

2.1.2. Fragility of Capillaries

Originally, CE instruments were fitted with one capillary that had to be manually changed and refitted if a different type of capillary was required. Measurements of analytes in CE occurs using a detector system at the capillary window. At the capillary window, a small amount of polyimide, which normally covers the capillary, providing flexibility, has been burnt away. Because of the fragile nature of the window, changing capillaries often ended with a new capillary being required. Newer instrumentation uses cassettes into which thermostated capillaries with a fixed window are fitted. These entire cassettes can be replaced easily by lifting them out and replacing with another cassette in a matter of minutes. The life of capillaries has been substantially enhanced with the use of cassettes.

2.1.3. Internal Diameter and Length

The internal diameter of the capillary is usually between 20 and 100-µm, with many separations using a 50-µm-internal diameter capillary. The length of the capillary depends on the configuration of the instrument and is usually between 24 and 122 cm.

2.1.4. Internal Surface

Capillaries can have an internal surface that is either fused silica (uncoated) or have a coated internal surface, depending on the separation required.

2.1.5. Capillary Coatings

For coated capillaries to be effective, the coating must be chemically stable and reproducible. Modification of the surface charges should cut down EOF forces. However, EOF should not be completely inhibited when separating oppositely charged proteins.

There are a variety of capillary coatings commercially available. For those with available time, methods for coating capillaries have been extensively published by Hjerten and Kubo ***(2)*** and others. Neutral coatings such as polyacrylamide or methylcellulose are probably the commonest coatings encountered in the clinical laboratory. These neutral coatings give almost zero EOF. Methylcellulose is commonly used in the buffer for isoelectric focusing. Polyethylene glycol (PEG) has been used to give moderate EOF, whereas charged coatings include quarternary ammonium functional groups or polyethyleneimine coating for separation of basic proteins.

2.1.6. Cleanliness of the Capillary

Compared to colorimetric analysis for which often several hundred microliters of reagent are used in the reaction vessel, CE uses very small quantities of buffer, the whole capillary often containing only 2 μL of buffer. Automated analysis usually uses clean cuvets for each analysis, whereas in CE, the same reaction vessel (i.e., the capillary) is used for subsequent analysis. Hence, the capillary must be kept scrupulously clean to prevent carryover from previous analysis. This cleanliness of the capillary is obtained by the use of an automated washing cycle, which can be either at the beginning of each run or directly after each analysis.

2.2. Buffer

The most important aspect of any CE separation is the buffer used for that analysis. The buffer used for CE assays is usually quite different from the buffer used for gel separations. Most of the serum protein electrophoresis separations using gels use barbitone buffer at pH 8.6. However, the majority of the published serum protein separations by CE use borate buffer at a much higher pH (9.6 or above) ***(3–5)***. Commercial serum protein electrophoresis using dedicated CE instrumentation also use borate buffer at a similar pH ***(6)***.

The pH of the buffer used for CE separations is critical for obtaining reproducible electropherograms. By lowering the pH of a buffer by 0.5 pH unit, one or more peaks can be lost in a six-peak separation. Hence, the pH of the buffer for each CE method needs to be carefully optimized and documented and then maintained precisely at that pH when subsequent batches of buffer are prepared.

2.3. Voltage Used for CE Separations

The voltage used in CE separations is quite different from that used for gel electrophoresis of serum proteins by high-resolution agarose gel electrophoresis (HRAGE). The voltage used for CE protein separations varies according to the published methods, from 10 to 18 kV (i.e., 10,000 to 18,000 V). This compares to HRAGE separations, where the separation voltage is between 250 and 400 V.

High voltages such as these generate considerable Joule heating within the buffer of the capillary. Dissipation of this heat is essential if CE is to give reproducible results. To this end, modern CE instrumentation has included Peltier cooling, which efficiently reduces this Joule heating.

2.4. Detection Systems

These might include UV, diode array, or laser-induced fluorescence detection systems. An example of the versatility of UV detection systems is that serum protein electrophoresis can be measured at either 200 or 214 nm, whereas quantification of iohexol can be measured at 254 nm.

3. Different Approaches Using Capillary Electrophoresis

3.1. Capillary Zone Electrophoresis

Capillary zone electrophoresis (CZE) is the most widely used separation type in capillary electrophoresis, and there are many variations. The simplest of these is with fused silica capillaries and a simple one-component buffer.

An example of CZE is the use of borate buffer for serum protein electrophoresis. In this separation, after washing the capillary and rinsing with run buffer, a thin plug (i.e., a 2-s injection of sample) is introduced into the capillary. Subsequently, with the application of the high voltage, the sample separates into discrete zones within the capillary, which are measured as they pass the in-line detector. Serum and urine protein electrophoresis is discussed in detail below as examples of the clinical use of CE.

The second type of CZE uses gel- or polymer-filled capillaries. Here, solutes are separated on the basis of size, a sieving effect being created. An example of this is the rapid analysis of amplified double-stranded DNA by CE with laser-induced fluorescence detection ***(7)***. The advantages of this type of CE is the ability to distinguish between fragments that are 244 and 307 basepairs. However, the disadvantages are the need for laser-induced fluorescence detection, which would increase the cost of CE instrument by $US 32,500.

The third type of CZE is referred to as micellar electrokinetic capillary chromatography (MECC). In this type of assay, which was introduced by Terabe et al in 1985 for drug separations ***(8)*** and extended by Sepaniak et al. ***(9,10)***, the electrolyte solution contains surfactants at concentrations above their critical micelle concentrations. Examples of surfactants include sodium dodecyl sulfate (SDS), tetradecyltrimethylammonium bromide (TTAB), and cyclodextrins. The assay relies on a slow-moving micellar phase (often containing the drug of interest) to separate from the aqueous phase that moves faster than the micellar phase.

Micellar electrokinetic capillary chromatographic separations were used by Evenson and Wiktorowicz in 1992 to separate barbiturates, cardioactive drugs, and benzodiazepines using a buffer that contained 30 m*M* SDS and 30 m*M* borate (pH 9.3) with 200 mL acetonitrile per liter ***(11)***. Alternative techniques such as gas chromatography–mass spectrophotometry (GC-MS) are more precise with identifying the drug and, hence, have become the preferred method for separating barbiturate drugs.

3.2. Isotachophoresis

With this type of assay, a small water plug (often 2 s) is placed just before and immediately following the sample. This has the effect of "insulating" the sample from the buffer and augments the effect of the amount of sample introduced.

An alternative approach, within this type of CE, is to have different leading and terminating electrolytes. Although there are examples of this approach ***(12,13)***, it is not commonly used in the clinical situation.

A related example of isotachophoresis is the use of concentration of sample by the use of acetonitrile ***(14)***. An example of this type of assay is in the quantification of iohexol (*see* **Subheading 4.3.**).

3.3. Capillary Isoelectric Focusing

In this type of CE separation, there are two distinct steps: focusing and mobilization. Usually, coated capillaries are used to cut down the EOF, and the capillary is first filled with a mixture of ampholytes and protein sample. A basic solution, such as sodium hydroxide, is used as buffer at the cathode, and an acidic buffer, such as phosphoric acid, is used at the anode. On the application of the voltage, the ampholytes and proteins migrate toward a position where the pH equals the p*I* of the proteins. When focusing is complete, the current drops to a minimal level.

Bands can be mobilized past the detector in a number of ways. The most effective of these is the application of voltage plus pressure to the capillary ***(15)***. Alternatively, by changing the composition of the cathode by the addition of sodium chloride (salt mobilization) ***(16)*** or the addition of a zwitterion ***(17)***, gravity-driven flow ***(18)*** or moving the entire column past the detector will achieve mobilization ***(19)***.

The most accepted assays in CE using capillary isoelectric focusing (CIEF) are hemoglobin disorder investigations performed by Hempe and Craver ***(20)***. In 1994, these authors used CIEF to identify and quantitate major and minor hemoglobin variants. They were able to identify abnormal hemoglobin variants with similar p*I* values such as HbD and HbS. The results they published, although very impressive, used DB1-coated capillaries and a NaOH washing regime. Other users of DB1 capillaries were unable to obtain the "hundreds of runs" that Hempe and Craver reported. An alternative AAEE-coated capillary was marketed by Bio-Rad in the mid-1990s. With a different washing regime, this AAEE capillary gave vastly increased run numbers for many users ***(21)***.

4. Clinical Uses of Capillary Electrophoresis

4.1. Serum Protein Electrophoresis

The main reason for performing serum protein electrophoresis is to determine the presence of any monoclonal immunoglobulin. In multiple myeloma and its variants (Waldenstrom's macroglobulinemia, plasmacytoma, POEMS syndrome, osteosclerotic myeloma, and plasma cell leukemia), the concentration of monoclonal immunoglobulin in serum, urine, or both can be used as a diagnostic criterion ***(22,23)***. It is imperative that analysis of serum protein electrophoresis differentiates between a monoclonal and polyclonal increase in immunoglobulins. This is because a monoclonal increase is associated with a clonal process that is malignant or potentially malignant, whereas a polyclonal increase of immunoglobulins is caused by a reactive or inflammatory process.

Since 1993, many scientific papers have been published on the subject of serum protein electrophoresis by CE using both fused silica ***(3,4,24,25)*** and coated capillaries ***(26,27)***. In 1993, our laboratory developed, validated, and introduced serum protein electrophoresis by CE ***(3)***. Optimizing this method has produced a reliable, robust, and cost-effective method ***(4)***.

Capillary electrophoresis has the ability to distinguish monoclonal from polyclonal increases in immunoglobulins. The sensitivity of detection of a monoclonal band is 0.5 g/L. This quantity of monoclonal band able to be detected relies to a large extent on the amount of residual immunoglobulins in the specimen. At the other end of the scale, the optics of a CE instrument are markedly better than those of a densitometer. Hence, large monoclonal components are likely to be more accurately measured and will probably appear slightly increased compared to densitometric measurement of a monclonal band by high-resolution agarose gel electrophoresis (HRAGE).

4.1.1. Notes on Methodology

Because of the automated and open nature of most CE instrumentation, a number of methods can be written for various separations. These methods can incorporate rinsing or flushing cycles to condition the capillary, followed by injection of a specific amount of the sample, followed by a separation or voltage step that will separate the sample into its various components.

4.1.2. Principle of the Method

We have optimized the method, which calibrates the capillary on known albumin concentrations in samples, on a 1/50 dilution of serum in run buffer that is 75 m*M* boric acid containing 0.025 m*M* calcium lactate (referred to as 75 m*M* boric acid in method), pH 10.3.

Over time, we have found that the life of the capillary is increased by using 0.5 *M* NaOH to rinse the capillary instead of the previously published 0.1 *M* NaOH. Also, by using a combination of the start-up, per-sample, end-of-run, and shutdown actions, one can keep the capillary scrupulously clean between samples.

4.1.3. Electropherograms of Patient Results

The change from gels to CE is not such a drastic step. CE produces an electropherogram that is a mirror image of the tracing produced by a densitometer.

We have included three figures to illustrate the electropherograms obtained by CE. **Figure 3** shows serum from a patient who has no abnormality detected. **Figure 4** shows a serum from a patient who has a monoclonal IgG paraprotein of 29 g/L. **Figure 5** shows a serum electropherogram from a patient who has a mild increase in $\alpha 1$ and a mild polyclonal increase in gammaglobulins consistent with inflammation or infection.

4.1.4. Quantitation and Identification of Monoclonal Bands

Quantitation of monoclonal bands by CE is computer generated using software that compares the area of the monoclonal band to either an albumin standardization of the capillary ***(3)*** or to a proportion of the total protein ***(6)***. The use of CE means that a densitometer is not required for monoclonal band quantitation.

Identification of the type of immunoglobulin in a monoclonal band can either be by immunosubtraction using CE or by immunofixation using IEF (isoelectric focusing) or EP (electrophoresis) gels. A comparison of immunosubtraction and immunofixation has previously been published ***(28)***. The cost of commercial immunosubtraction is such that it can promote immunofixation as a more cost-effective method.

4.2. Urine Protein Electrophoresis

Typically, 70% of myeloma patients present with a clonal proliferation of IgG with either kappa or lambda light-chain types. IgA myeloma accounts for 15% of patients. The 15% of patients showing a monoclonal IgM usually reflect a Waldenstrom's macroglobulinemia, with only 2% being the result of a true IgM myeloma. IgD myeloma accounts for <0.5% of patients. The remainder are the result of light chain myelomas of either kappa or lambda types. These typically present with free light chains in urine (which might be present in grams/liter amount), and free light chains in serum, which are a reflection of the kidney's inability to filter the massive overproduction of light chains by the circulating plasma cells. Hence, the monitoring of urine protein electrophoresis as well as serum protein electrophoresis is extremely important in a total myeloma workup.

4.2.1. Notes on Methodology

Proteins in normal urine need to be concentrated to be detected on high-resolution gel electrophoresis. These concentrators are disposible. However, because of the manufacturing process involved, costs range from $US 4.55 to $US 5.85 per concentrator.

When we began looking at urine protein electrophoresis, we used three methods to examine concentrated urine specimens by CE. These included anion-exchange resin treatment of the urine to remove nonprotein components, the use of abnormal urine containing previously identified proteins, and the addition of known analytes to urine specimens, such as albumin, phosphate, nitrate, and oxalate. Using these techniques, albumin and Bence Jones protein were identified, and a correlation of 71 concentrated urine specimens for albumin and Bence Jones protein using CE and commercial high-resolution agarose gel electrophoresis was subsequently published ***(29)***.

In 1996, we realized that we were not utilizing the extreme sensitivity of CE in our urine protein electrophoresis method. Hence, we used the same buffer and developed a method that did not involve concentration of the urine. We found that we could see peaks for small amounts

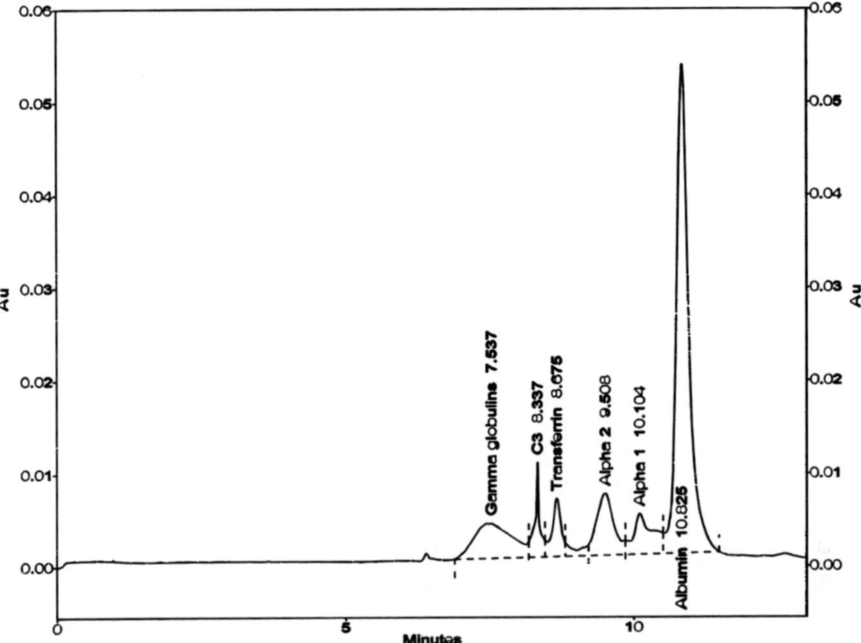

Fig. 3. Electropherogram of serum protein electrophoresis from a patient who has no abnormality detected. Albumin, α1, α2, transferrin, C3, and gammaglobulins components as indicated. Electropherogram obtained from a Beckman MDQ instrument.

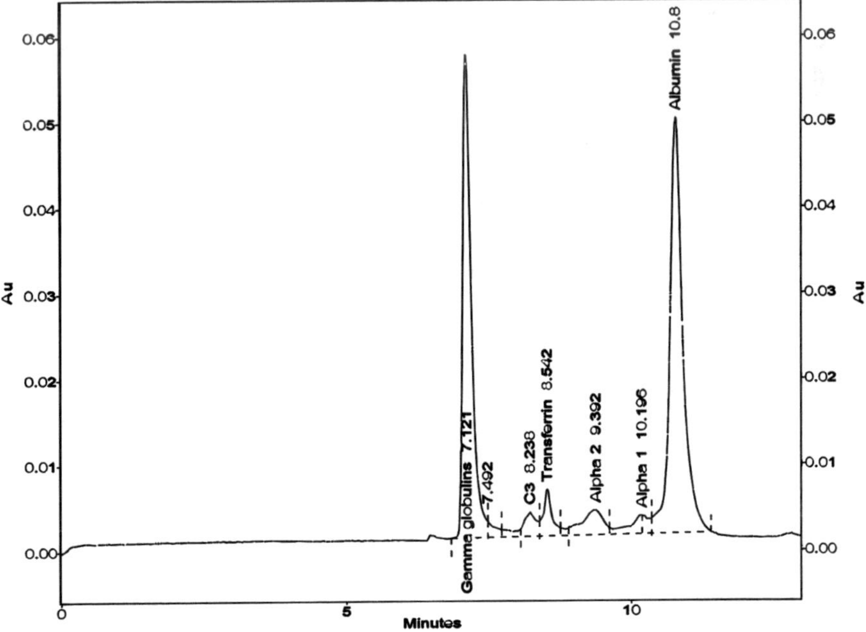

Fig. 4. Electropherogram of serum protein electrophoresis from a patient who has a 29 g/L paraprotein. Albumin, α1, α2, transferrin, C3, and gammaglobulin components as indicated. Electropherogram obtained from a Beckman MDQ instrument.

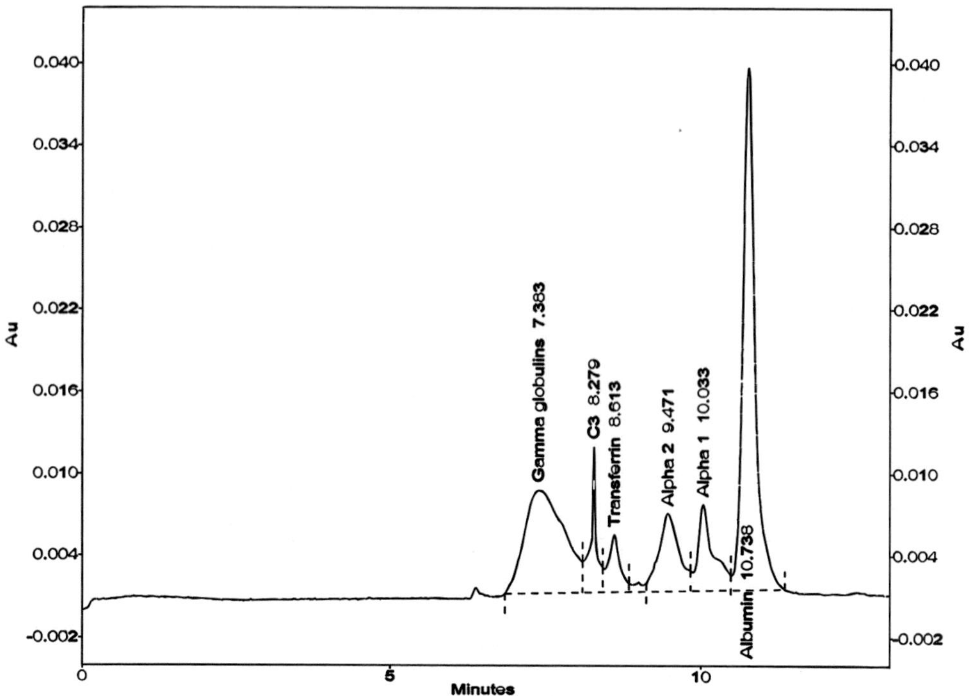

Fig. 5. Electropherogram of serum protein electrophoresis from a patient who has either an infection or inflammation. Note increase in α1, and a mild polyclonal increase in gamma globulins. Electropherogram obtained from a Beckman MDQ instrument.

of Bence Jones protein within the normal protein range. Hence, the method had the sensitivity required *(30)*.

The benefit of using unconcentrated urine specimens for analysis is the cost saved by not using a urine concentrator (>$US 4.55 per concentrator) and the time saved (approx 30 min) not concentrating the urine specimen.

4.2.2. Electropherograms of Patients Results

The first peak seen by this method has been proved by two independent research groups to be a combination of urea and creatinine *(29,31)*. Bence Jones peaks can occur from the urea/creatinine peak through to the α2 peak. However, they are usually cathodic to transferrin. Ions such as phosphate, nitrate, and oxalate are found anodic to the prealbumin peak and should be disregarded for quantitation purposes.

In setting up the CE software for urine electropherograms, it is optimal for the highest measured peak in the urine electropherogram to reach the top of the page, and the scaling on the left-hand side of the electropherogram is adjusted to the highest peak. In urine, this highest peak is often the urea/creatinine peak at about 6.6 min, shown in **Fig. 6**. Albumin is also marked in this electropherogram. Bence Jones protein is usually found between the urea/creatinine peak and transferrin. In **Fig. 7**, both albumin and Bence Jones proteins are indicated.

4.2.3. Quantitation and Reporting of Urine Proteins

If Bence Jones protein is present, the total protein should be quantitated by a manual trichloracetic acid method using a known albumin standard and a recogized quality control material, such as Bio-Rad Lyphochek (Hercules, CA).

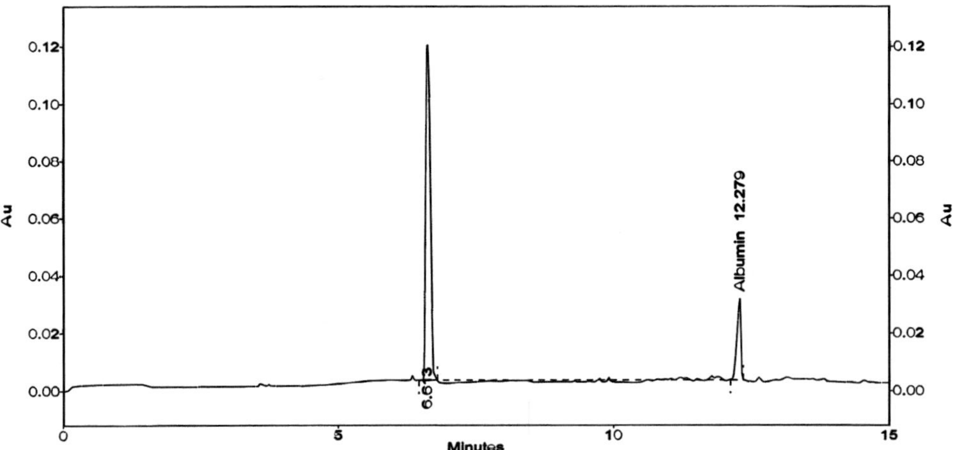

Fig. 6. Electropherogram of a urine specimen with a total urine protein of 0.16 g/L. Albumin peak at 12.3 min marked. Peak at 6.6 min is the result of urea/creatinine. Electropherogram obtained from a Beckman MDQ instrument.

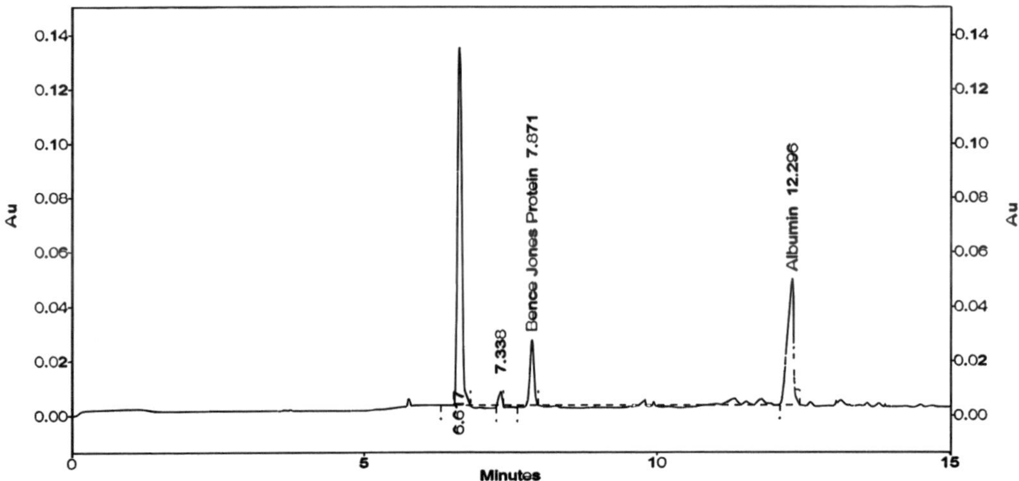

Fig. 7. Electropherogram of a urine specimen from a patient who has Bence Jones proteinuria. Total urine protein is 0.81 g/L. Albumin and Bence Jones protein peaks indicated. Electropherogram obtained from a Beckman MDQ instrument.

If there is no Bence Jones protein in the urine specimen, the total protein may be quantitated by automated methods such as Coomassie or benzethonium chloride on large analyzers such as a Hitachi 911 instrument (Hoffman-La Roche, Basel, Switzerland). However, if Bence Jones protein is present, these automated methods could underestimate the amount of total protein present. This is because some free light chains are not quantitated by these methods, and it is a specimen-related problem.

The current method for calculating the percentage of Bence Jones protein in the specimen is to manually add all of the protein peak areas and then calculate the proportion of the Bence Jones peak relative to all the protein peaks. The software of the CE instrument can do this automatically, reporting peak area percentages. This method is also used to calculate the percentage of albumin in the sample.

The total protein of the urine specimen and whether there is any Bence Jones protein present should be reported. If the total protein is greater than 0.2 g/L, the report may also indicate whether the proteinuria is glomerular or tubular in origin, or if it is a mixed glomerular/tubular proteinuria.

5.2.4. Identification of Bence Jones Protein and Intact Immunoglobulin in Electropherogram

If a small peak is found in the region between the urea/creatinine peak and the α2 region, it is important to exclude Bence Jones protein. Similarly, if the urine of a patient with a serum monoclonal band is being examined and a peak is seen between the urea/creatinine and the α2 region, it is important to follow this up and determine if any free light chains or intact immunoglobulin are present. This is done by applying 2 μL of unconcentrated urine to two tracks of either IEF or EP gel, running the IEF or EP, and then immunofixing for free kappa and free lambda light chains using Chemicon antisera (Chemicon, Melbourne, Australia).

4.3. Measurement of Serum Iohexol by CE

Iohexol is a nonionic contrast agent that has been described in recent literature as an accurate marker for the measurement of the glomerular filtration rate (GFR) ***(32)***. In the clinical situation, GFR is routinely used to measure the renal function of patients who have progressive renal disease and is especially useful for determining the dose of nephrotoxic drugs excreted by the kidneys.

Assay of iohexol using CE has several advantages over methods for iohexol employing high-performance liquid chromatography (HPLC). These include the fact that with CE there are no packed columns to equilibrate: the buffer is simple, with no pumps or mobile phases to manipulate. Also, the consumable cost of iohexol by CE is extremely low, being less than $US 0.16 per test ***(33)***.

Quality control of the assay was measured by dilution of Omnipaque 300 (647 mg iohexol/mL) (Nycomed Australia, Chatswood, Australia) and included with each run. Three controls at levels of 80, 40, and 20 mg/L diluted in iohexol-free serum were used with each run.

4.3.1. Notes on Methodology

For the measurement of GFR, patients were administered 5 mL of Omnipaque 300 by intravenous injection into the antecubital vein. Five milliliters of plain blood was removed from the patient's contralateral arm at 120, 150, 180, 210, and 240 min postinjection. Exact sample times were recorded. GFR was calculated from the plasma clearance of iohexol using the slope intercept method (one compartment model) and then approximated to a two-compartment model using the Brochner–Mortensen equation ***(34)***.

Measurement of the area under the iohexol peak is related to the area under the peak of an internal standard (IS) that is 3-isobutyl-1-methylxanthine ***(35)***. This IS is added at a concentration of 20 mg/L to acetonitrile, which is used as a deproteinizing agent for each serum and standard specimen. After the addition of the deproteinization reagent to the serum specimens, the samples were vortexed for 20 s and then spun at 12,225*g* for 5 min. Two hundred microliters of each supernatent was then placed on capped 0.5-mL polymer tubes until the supernatants were placed on the BioFocus instrument. A typical electropherogram showing the iohexol peak, X and the internal standard peak, Y is shown in **Fig. 8**.

Detection of iohexol was achieved by on-column measurement of absorbance at 254 nm, the analysis being based on a previously published method of Shihabi ***(35)***. The assay buffer was 220 m*M* boric acid, pH 8.8, separation being achieved by the application of a voltage of 6 kV. A scrupulous washing regime involving phosphoric acid, distilled water, sodium hydroxide, distilled water and assay buffer allowed 860 runs to be achieved over a 9-mo period using the one fused silica capillary.

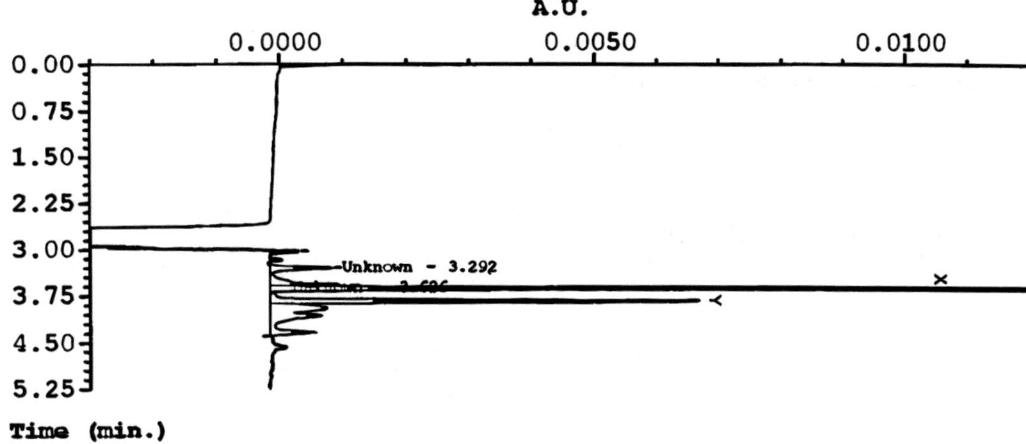

Fig. 8. Electropherogram showing iohexol peak X and internal standard peak Y. Electropherogram obtained from a BioFocus 2000 Capillary Electrophoresis System.

5. Conclusion

Clinical laboratories aim to provide accurate results with a minimal turnaround time. We have shown that capillary electrophoresis provides automated analysis giving accurate results for a variety of analytes. These include quantitation of monoclonal proteins in serum electrophoresis and diagnosis of Bence Jones protein in urine specimens. With the measurement of iohexol, the area under the iohexol peak is related to an internal standard. This is then used to calculate the GFR of renal patients.

References

1. El Rassi, Z. (1993) Capillary electrophoresis overview (theory and injection). Workshop at Conference on Capillary Electrophoresis 1993.
2. Hjerten, S. and Kubo, K. (1993) A new type of pH- and detergent-stable coating for elimination of electroendosmosis and adsorption in (capillary) electrophoresis. *Electrophoresis* **14,** 390–395.
3. Jenkins, M. A., Kulinskaya, E., Martin, H. D., and Guerin, M. D. (1995) Evaluation of serum protein separation by capillary electrophoresis: prospective analysis of 1000 specimens. *J. Chromatogr. B* **672,** 241–251.
4. Jenkins, M. A. and Guerin, M. D. (1996) Optimization of serum protein separation by capillary electrophoresis. *Clin. Chem.* **42,** 1886.
5. Dolnik, V. (1995) Capillary zone electrophoresis of serum proteins: study of separation variables. *J. Chromatogr. A* **709,** 99–110.
6. Beckman Coulter Bulletin 9246.
7. Liu, M.-S. and Chen, F.-TA. (1999) Rapid analysis of amplified double-stranded DNA by capillary electrophoresis with laser-induced fluorescence detection. In *Clinical Applications of Capillary Electrophoresis* (Palfrey, S. M., ed.), Humana, Totowa, NJ, pp. 121–126.
8. Terabe, S., Otsuka, K., Ichikawa, K., Tsuchiya, A., and Ando, T. (1985) Electrokinetic separations with micellar solutions and open-tubular capillaries. *Anal. Chem.* **56,** 834–841.
9. Balchunas, A. T. and Sepaniak, M. J. (1987) Extension of elution range in micellar electrokinetic capillary chromatography. *Anal. Chem.* **59,** 1466–1470.
10. Burton, D. E., Sepaniak, M. J., and Maskarinec, M. P. (1987) Evaluation of the use of various surfactants in micellar electrokinetic capillary electrophoresis. *J. Chromatrgr. Sci.* **25,** 514.
11. Evenson, M. A. and Wiktorowicz, J. E. (1992) Automated capillary electrophoresis applied to therapeutic drug monitoring. *Clin. Chem.* **38 (9),** 1847–1852.
12. Foret, F., Szoko, E., and Karger, B. L. (1993) Trace analysis of proteins by capillary zone electrophoresis with on-column transient isotachophoretic preconcentration. *Electrophoresis* **14,** 417–428.

13. Manabe, T., Yamamoto, H., and Okuyama, T. (1989) Fully automated capillary isotachophoresis of proteins. *Electrophoresis* **10,** 172–177.
14. Shihabi, Z. K. (1995) Sample stacking by acetonitrile-salt mixtures. *J. Capillary Electrophor.* **2(6),** 267–271.
15. Applied Biosystems Model 270A-HT Capillary Electrophoresis User Bulletin No 5. 2/92 No 127805.
16. Rodriguez-Diaz. R., Zhu, M., and Wehr, T. (1997) Strategies to improve performance of capillary isoelectric focusing. *J. Chromatogr. A* **772(1–2),** 145–160.
17. Zhu, M., Rodriguez, R., and Wehr, T. (1991) Optimizing separation parameters in capillary isoelectric focusing. *J. Chromatogr.* **559,** 479–488.
18. Rodriguez, R., Zhu, M. D., Wehr, T., and Siebert, C. (1994) Gravity mobilization of proteins in capillary isoelectric focusing, Sixth International Symposium on High Performance Capillary Electrophoresis, San Diego, CA.
19. Wu, J. and Pawliszyn, J. (1993) Fast analysis of proteins by isoelectric focusing performed in capillary array detected with concentration gradient imaging system. *Electrophoresis* **14,** 460–474.
20. Hempe, J. M. and Craver, R. D. (1994) Quantification of hemoglobin variants by capillary isoelectric focusing. *Clin. Chem.* **40(12),** 2288–2295.
21. Jenkins, M. A. and Ratnaike, S. (1999) Capillary isoelectric focusing of haemoglobin variants in the clinical laboratory. *Clin. Chim. Acta* **289,** 121–132.
22. Stringer, C. R. J. (1997) ABC of clinical haematology multiple myeloma and related conditions. *Br. Med. J.* **314,** 960–963.
23. Whicher, J. T., Calvin, J., Riches, P., and Warren, C. (1987) The laboratory investigation of paraproteinaemia. *Ann. Clin. Biochem.* **24,** 119–132.
24. Jenkins, M. A. and Guerin, M. D. (1995) Quantification of serum proteins using capillary electrophoresis. *Ann. Clin. Biochem.* **32,** 493–497.
25. Chen, F.-T.A., Liu, C.-M., and Sternberg, J. C. (1991) Capillary electrophoresis—a new clinical tool. *Clin. Chem.* **37,** 14–19.
26. Wijnen, P. A. and van Dieijen-Visser, M. P. (1996) Capillary electrophoresis of serum proteins. Reproducibility, comparison with agarose gel electrophoresis and a review of the literature. *J. Clin. Chem. Clin. Biochem.* **34,** 535–545.
27. Mazzeo, J. R. and Krull, I. S. (1991) Coated capillaries and additives for the separations of proteins by capillary zone electrophoresis and capillary isoelectric focusing. *Biotechniques* **10,** 638–645.
28. Keren, D. F. (1998) Detection and characterization of monoclonal components in serum and urine. [editorial]. *Clin. Chem.* **44(6),** 1143–1145.
29. Jenkins, M. A., O'Leary, T. D., and Guerin, M. D. (1994) Identification and quantitation of human urine proteins by capillary electrophoresis. *J. Chromatogr. B* **662,** 108–112.
30. Jenkins, M. A. (1997) Clinical application of capillary electrophoresis to unconcentrated human urine proteins. *Electrophoresis* **18,** 1842–1846.
31. Guzman, N. A., Gonzalez, C. L., Trebilcock, M. A., et al. (1993) The use of capillary electrophoresis in clinical diagnosis. In *Capillary Electrophoresis Technology* (Guzman, N. A., ed.), Marcel Dekker New York, p. 650.
32. Krutzen, E., Back, S.-E., Nilsson-Ehle, I., and Nilsson-Ehle, P. (1984) Plasma clearance of a new contrast agent, iohexol: A method for the assessment of glomerular filtration rate. *J. Lab. Clin. Med.* **104(6),** 955–961.
33. Jenkins, M. A., Houlihan, C., Ratnaike, S., Jerums, G., and Parkin, J. D. (2000) Measurement of iohexol by capillary electrophoresis: minimizing practical problems encountered. *Ann. Clin. Biochem.* **37,** 529–536.
34. Houlihan, C., Jenkins, M., Osicka, T., Scott, A,. Parkin, D., and Jerums, G. (1999) A comparison of the plasma disappearance of iohexol and ^{99m}Tc-DTPA for the measurement of glomerular filtration rate (GFR) in diabetes. *Aust. NZ J. Med.* **29,** 693–700.
35. Shihabi, Z. K. (1992) Clinical applications of capillary electrophoresis. *Ann. Clin. Lab Sci.* **22(6),** 398–405.

38

Flow Cytometry in the Biomedical Arena

James L. Weaver and Maryalice Stetler-Stevenson

1. Definition of Flow Cytometry

Flow cytometry is literally measuring cells while moving in a liquid. More specifically, a suspension of single cells is labeled with one or several fluorescent labels. In the machine, the cells are constrained into single file. These cells pass through one or more laser beams to excite the fluorescent labels. The light emitted from the fluorescent labels is collected, separated, measured, and the resulting data transmitted to the computer controlling the instrument. In addition, narrow-angle and 90° light scatter from the laser beam are measured and the data are also sent to the computer. All of these values are recorded as correlated measurements for each cell separately. The data are displayed in the computer as single-parameter histograms or two-parameter plots. The software allows populations to be identified and specific subpopulations selected for further analysis. The number and fraction of cells in specific populations can be quantitated. In addition, the amount of the fluorescent label can be calibrated, and by extension, the amount of the ligand for the label can be calculated, if needed. The patterns of expression of specific cellular proteins or changes in numbers of cells in specific populations are used to contribute to diagnosis of the patient.

2. Clinical Uses of Flow Cytometry (Overview)

Flow cytometry is used in a number of clinical situations to precisely define abnormal populations. This allows, for example, diagnosis and subclassification of malignancy or definition of the factors operative in immunodeficiency. Flow cytometry is also useful in defining physiological processes, such as enumeration of stem cells or histocompatibility testing prior to transplant. Flow cytometry is ideal in fluids, such as peripheral blood or bone marrow aspirates, where cells are naturally suspended, but it is also useful in solid tissues, from which single-cell suspensions can be obtained. The primary uses of this technology are in hematopathology, hematology, transplant medicine, and immunology.

3. A Brief History

Flow cytometry as a practical and commercially viable technology needed the convergence of four technologies: handling cells in fluidic systems, creation of specific fluorescent labels, lasers, and computers. These came together in the 1970s in the first generation of commercial instruments *(1)*. Although the history of the subsystem components can be traced back, in some cases, to the 1930s, these instruments were the first to have all of the elements of the modern flow cytometers. These instruments were large, complex, and needed special power and water cooling for the laser. Sorting was a function of some of these first-generation systems as well. In the 1980s, use of more efficient optics originally developed for use with arc lamp systems

From: *Medical Biomethods Handbook*
Edited by: J. M. Walker and R. Rapley © Humana Press, Inc., Totowa, NJ

allowed the use of lower-powered air-cooled argon ion lasers. This change converted the large, floor instrument into moderate sized and priced systems that fit on the benchtop and could be installed in any standard laboratory. The second change that occurred in the 1980s was the introduction of monoclonal anti-human cell surface protein antibodies. These antibodies were directly conjugated with fluorescent labels such as fluorescein or phycoerythrin. These were originally used in areas such as transplant patient monitoring but rapidly spread to many other areas such as classification of leukemias or lymphomas and monitoring of acquired immune deficiency syndrome (AIDS) patients. The 1990s has brought an exponential increase in the number of reagents available for clinical use. In addition to the 300+ characterized cell surface proteins, there are probes to measure many aspects of cell physiology. The instruments have evolved to incorporate improvements in electronics and computers. Clinical-grade instruments can now have two lasers and measure up to five simultaneous fluorescence parameters as well as two light scatter parameters. Specialized research instruments with three lasers measuring up to 12 fluorescence parameters have been built *(2)*.

4. Overview of the Method

Flow cytometry starts with a suspension of single cells. These cells can be from liquid specimens such as blood, cerebrospinal fluid, or bone marrow. Alternately, the cells from a solid-tissue biopsy can be released using various disaggregation methods. The next step is that the cells are labeled with fluorescent indicators. These can be monoclonal antibodies to identify specific cell populations and subpopulations or indicators of cell status or physiology. The current generation of clinical flow cytometers can measure two light scatter parameters and three to five fluorescent parameters. The labeled cells are placed on the cytometer in a tube or multiwell plate. The machine pulls the cells into a thin stream of cells, which is further hydrodynamically focused into a tight stream with cells in single file. The cells pass through a focused laser beam. The laser beam excites the fluorescent labels and the cytometer collects both the emitted fluorescent light and scattered laser light. This light is sorted using color-selective (dichroic) mirrors and filters and each selected light color is sent into a very sensitive light detector (photomultiplier tube [PMT]). The signal from each detector is amplified and converted into digital data. The values from all of the detectors are collected into a data packet for each cell and then sent to the computer for further analysis and storage (*see* **Fig. 1**). In the computer, the data for each cell are stored in a correlated list of all of the parameters for each cell. The data are displayed as one-parameter histograms (*see* **Fig. 2A**) or as two-parameter plots (*see* **Fig. 2B**). The data can be further subdivided by a process called gating. Here, the data only from a defined region in one plot are displayed in a second plot. For example, in **Fig. 3A**, the staining for all cells for CD4 are displayed, showing two populations. In **Fig. 3B**,**C**, the data for staining of the same cells for CD62L are shown. The one-parameter histogram for all of the cells in **Fig. 3A** is shown in **Fig. 3B**. In **Fig. 3C**, only the data from the cells in the High Region in **Fig. 3A** are displayed. Statistics for both number and fluorescence intensity of populations or subpopulations can be displayed. This is a very simple example of gating to evaluate specific subpopulations. Gating can be as simple as shown here or quite complex, depending on the question being evaluated.

The data from staining cells from a single patient for many markers can be combined to perform a diagnostic screening procedure. In a specific example, a panel to evaluate a suspected lymphoma could involve 5–15 tubes, with 3–5 markers being stained in each tube. The data from a panel such as this are combined with morphology and hematology data to make a specific diagnosis.

5. Preparation of Cells for Analysis

The preparation of cells for flow cytometric analysis falls into two major categories: cells from liquid tissues and cells from solid tissues. In the first class are cells from blood, bone marrow, cerebrospinal fluid, pleural effusion, or pulmonary lavage. For all liquid tissues

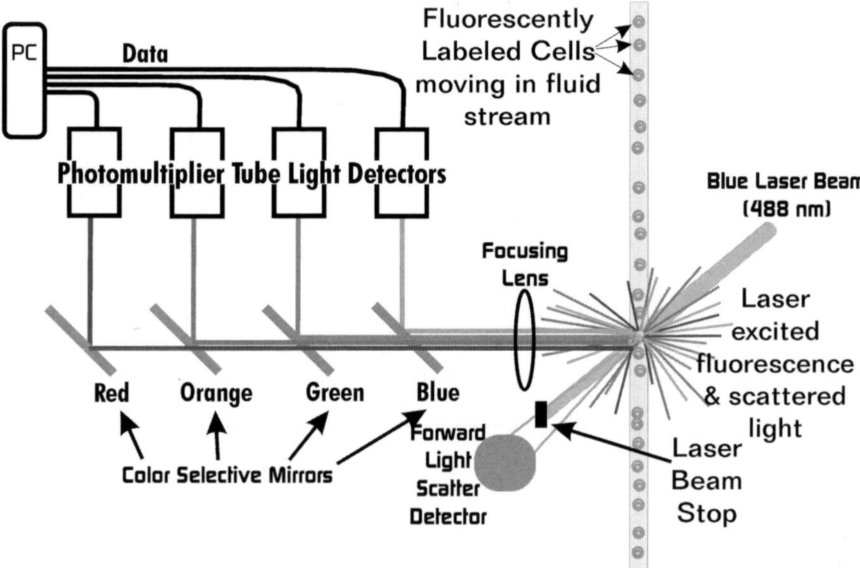

Fig. 1. General schematic of a generic clinical flow cytometer.

except blood and bone marrow, the cells are centrifuged, washed in sterile buffer, and are ready for staining. In blood and bone marrow, the optimal method is one in which there is minimal manipulation of the specimen. The large number of red cells, however, can make analysis difficult. Historically, the mononuclear cells could be purified by centrifugation on a discontinuous density gradient. However, it was discovered that certain cell populations were selectively lost and this method is no longer considered appropriate for studying leukocytes. An alternate method is to lyse the erythrocytes using one of several hypotonic solutions designed for the preparation of human samples. The leukocytes are then washed free of material from the lysed erythrocytes. Staining with monoclonal antibodies can be performed before or after lysis of red cells. Finally, the leukocytes can be stained in whole blood and data collection triggered on a fluorescence parameter rather than a light scatter parameter.

Cells from biopsy or fine-needle aspirate samples can be evaluated by flow cytometry. In fine-needle aspirates, the cells are already in suspension. Isolation of leukemia and lymphoma cells from lymph node and other hematopoietic tissues is usually achieved by mincing the tissue with a scalpel or aspirating the cells through a needle. There is even a commercially available mechanical disaggregator with an enclosed system that reduces the risk of exposure to tissue pathogens. As leukocytes are generally nonadherent cells they easily form a suspension, although filtration to remove connective tissue debris might be required. In rare cases when there is dense fibrosis, the cells are dissociated by using a mixture of proteolytic enzymes to release the individual cells from the tissue matrix. The results can be variable and quality control is difficult. These are then washed and stained as usual.

6. Stains and Markers

There are many hundreds of cellular parameters that can be measured by flow cytometry. Fortunately for the overburdened physician, the list of markers in routine clinical use is notably shorter. These fall into two major categories, markers of cell physiology and cellular proteins. The first group is evaluated using fluorescent molecules specific for the parameter being measured. The most commonly used indicator in this category is the viability markers. These are either chemicals that are impermeable to viable cells, such as propidium iodide (PI), or chemi-

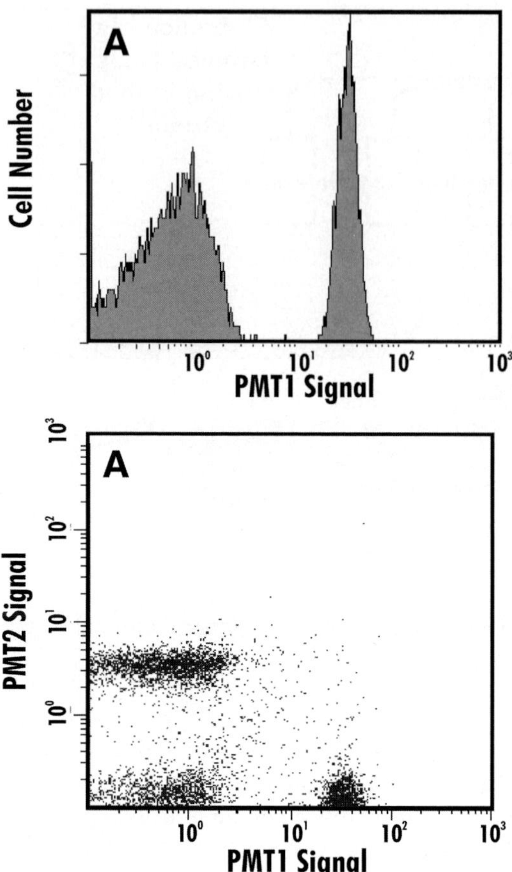

Fig. 2. Examples of basic data display types from flow cytometry. (**A**) Single-parameter histogram showing two populations of cells with low and high expressions of the fluorochrome detected in PMT1. (**B**) Dual-parameter plot (dot plot) where the position of the dot represents the fluorescence intensity of each cell for the fluorochrome detected by PMT1 and PMT2, respectively.

cals that are rapidly lost by leakage from cells that have lost cell membrane integrity, such as calcein. Cellular viability is used both as a quality control measure and to determine the status of leukemia or lymphoma cells extracted from patients on chemotherapy. There are indicators for a number of other cellular physiology parameters such as Ca^{2+}, membrane potential, oxidative burst, and many others. However, these are primarily used in research settings and will rarely be encountered in clinical practice.

The most common use of flow cytometry is to measure the presence and relative abundance of cellular proteins. This is done using fluorescently labeled monoclonal antibodies. Where the protein of interest is on the cell surface, the cells are simply mixed with the monoclonal antibodies. After a short incubation (30–60 min), the cells can be washed, fixed with paraformaldehyde to destroy possible pathogens as well as enhance stability, and then run on the flow cytometer. The fluorescence intensity is proportional to the abundance of the protein of interest. The fixation process does not affect the fluorescence intensity of the bound antibodies. In many cases, the proportion of the cells that are expressing the protein is the primary measure. In some cases, change in the level of the protein is the clinically useful marker. In the evaluation

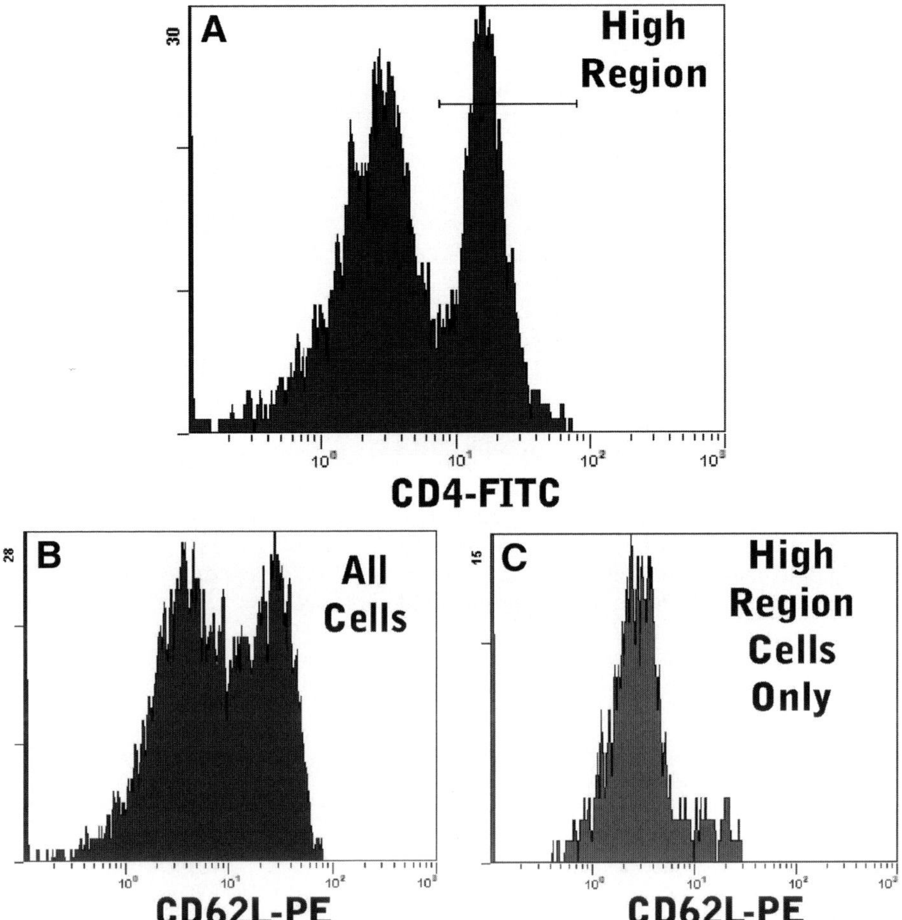

Fig. 3. Single-parameter histograms demonstrating gating. (**A**) Histogram of expression of CD4–FITC (fluorescein) with a region set on the population showing high CD4 expression. (**B**) Histogram of expression of CD62L–PE (R-phycoerythrin) for all cells shown in (**A**). (**C**) Histogram showing expression of CD62L–PE expression on only those cells in the High Region of (**A**).

of possible tumors, the aberrant expression of proteins can be used to contribute to or establish a specific diagnosis. **Table 1** lists some of the more commonly used markers in clinical flow cytometry. Flow cytometry labs will use only a subset of these markers depending on the specific clinical situation.

Monoclonal antibodies can be labeled with a variety of fluorescent molecules (fluorochromes). These fluorochromes have two important characteristics, their excitation, and emission characteristics. These are usually shown as spectra such as is **Fig. 4**. The excitation spectrum shows the proportion of light emitted as a function of varying excitation wavelength. This is usually measured at the wavelength of maximum emission. The complementary spectrum is the emission spectrum. This shows the light emitted at specific wavelengths with a fixed excitation wavelength. The fluorochromes used in clinical flow cytometry are dictated by the laser(s) installed in clinical flow cytometers. All of the clinical instruments have argon ion lasers that emit at 488 nm. Some instruments have a second HeNe or diode laser that emits at 633 nm. The more common fluorochromes for 488-nm excitation are fluorescein (FITC),

Table 1
Some Commonly Used Markers in Clinical Flow Cytometry

Marker	Normally found on	Used for
CD1a	T-Cells—immature	L/L
CD2	T-Cells	L/L ISM
CD3	T-Cells	L/L ISM
CD4	T-Cells subset	L/L ISM
CD5	T-Cells, B-CLL, MCL, and some LCL	L/L
CD7	T-Cells, AML	L/L
CD8	T-Cells subset	L/L ISM
CD43	T-Cells, NK cells, granulocytes	L/L
CD10	B-Cells, FCL, ALL, some LCL	L/L
CD11c	HCL, some B-CLL	
CD19	B-Cells	L/L ISM
CD20	B-Cells, T-cells	L/L
CD22	B-Cells	L/L
CD23	B-Cells—activated, B-CLL	L/L
Ig kappa light chain	B-Cells	L/L
Ig lambda light chain	B-Cells	L/L
CD38	Plasma cells, peripheral blood B-cells, activated T-cells	L/L
CD138	Plasma cells	L/L
CD25	Activated T- and/or B-cells	L/L
HLA-DR	Activated T- and/or B-cells	L/L
TdT	Immature lymphocytes	L/L
CD13	Myeloid	L/L
CD11b	Myeloid	L/L
CD14	Monocytic	L/L
CD33	Myeloid	L/L
CD34	Stem cell	L/L
CD36	Megakaryocytes and eyrthroid cells	
CD61	Megakaryocyte, platelets	L/L
CD64	Granulocytes, monocytes	L/L
CD16	NK Cell, T-cell, myeloid	L/L
CD56	NK Cell, T-cell, MM, AML	L/L ISM
CD57	NK Cell, T-cell	L/L
CD45	All leukocytes	ISM, PNH
CD55	All cells	PNH
CD59	All cells	PNH
CD103	B-Cells, HCL, ITL	L/L
Erythrocyte RNA	Reticulocyte counts	Reticulocyte monitoring

Abbreviations: L/L, leukemia or lymphoma screening; ISM, Immune status monitoring; PNH, Paroxymal nocturnal hemoglobinuria screening; B-CLL, B-cell chronic lymphocytic leukemia; MCL, Mantle cell lymphoma; LCL, Large-cell lymphoma; FCL, follicle center cell lymphoma; ALL, acute lymphoblastic leukemia; AML, acute myeloid leukemia; HCL, hairy cell leukemia; MM, multiple myeloma; ITL, intestinal T-cell lymphoma.

R-phycoerythrin (PE), PE–Cy5 conjugates, PerCP, and PE–Cy7 conjugates. For 633-nm excitation, allophycocyanine (APC) and APC–Cy7 are the most commonly used fluorochromes.

7. A More Detailed Description of the Technology, Including Sorting

The labeled cell suspension will be placed in a small tube or in a 96-well plate. The cells are driven into a sample pickup tubing using air pressure. The tubing delivers the cells into the flow

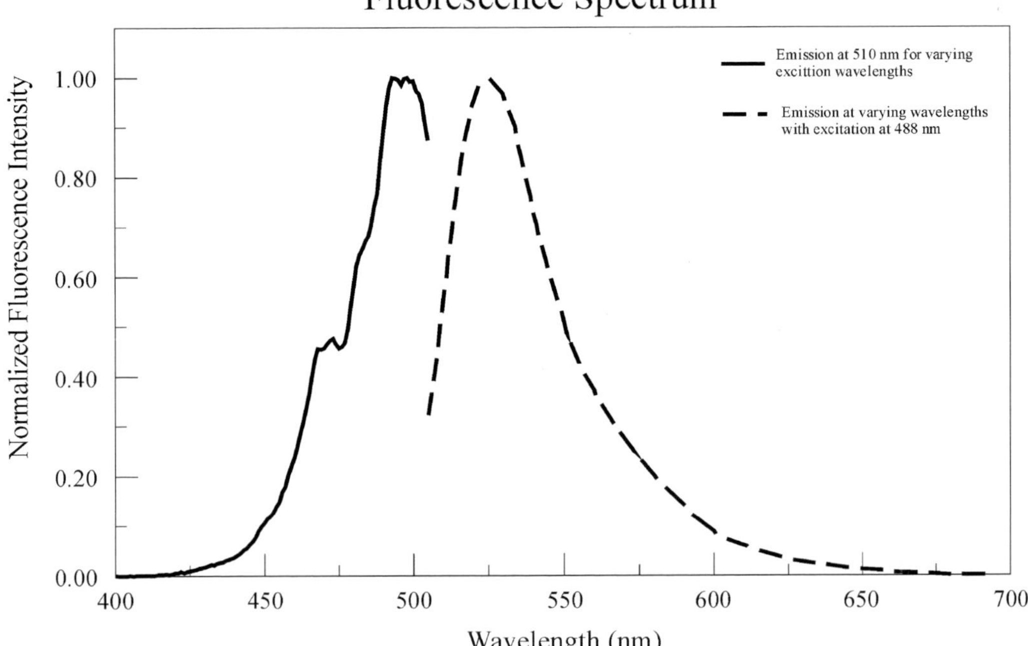

Fig. 4. Fluorescence spectrum. Solid line represents the efficiency of excitation at a fixed emission wavelength. The dashed line represents the efficiency of emission at a fixed excitation wavelength.

cell (*see* **Fig. 5**). At the same time, sheath fluid is delivered into the flow cell in a continuous flow. The conical shape at the bottom of the flow cell compresses and accelerates the sheath flow. This has the consequence of compressing and accelerating the sample stream as well. The result is that the cells in the sample stream are constrained into single file. This stream then passes through the laser beam.

As the cells pass through the laser beam, two things happen. First, the laser light is scattered at a variety of angles. Second, the fluorochromes of the various indicators or antibody conjugates are excited and emit light at their respective emission wavelengths.

The laser light that is scattered at a small angle relative to the laser beam is captured by the forward scatter detector (*see* **Fig. 1**). The laser light that is scattered at about 90° along with fluorescence light emitted in the same region is captured by the focusing lens and sent into the optical section.

The optical section uses two classes of component to split and purify the optical signal. The first class is dichroic mirrors. These are partial mirrors that have the ability to reflect light above or below a certain wavelength. A specific example is a dichroic used to separate the signal from FITC. This will reflect any wavelength below 550 nm and allow all higher wavelengths to pass (long pass). This is abbreviated as a 550 LP dichroic. There are also dichroics that have the inverse characteristics of passing all wavelengths below their cutoff and reflecting those above, these are called short pass (SP) dichroics. The other class is the bandpass filters. These are set to allow light to pass only within certain wavelengths. These are defined by a center wavelength and width; for example, a filter for fluorescein has a center of 525 nm and a width of ± 30 nm, which is to say, it admits all light between 495 and 555 nm (525/30 BP) (**Fig. 6**). There are the inverse of these that block only certain wavelengths and allow all others. These are commonly used to block scattered laser light and could have a width of only

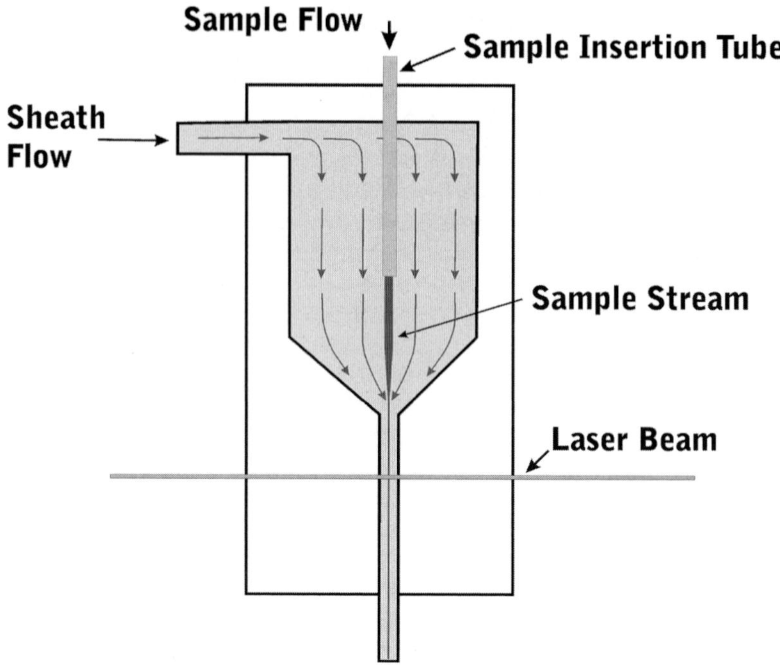

Fig. 5. Diagram of a typical flow cell. Pressurized sheath flow enters from left and is accelerated through the cone area to a high-speed laminar flow. The sample, containing the cells, enters from the sample insertion tube. The pressurization in the cone area compresses the sample stream so that cells are forced into single file. Cells in single file pass through the laser beam.

a few nanometers. **Figure 5** shows a simple optical dichroic mirror and barrier filter arrangement. There are many different possible optical arrangements depending on the space available and the relative importance of the signal strength of the various fluorochromes.

The light arrives in the photomultiplier tube (PMT) and the energy of the light photons is converted into a very small electrical signal. The light scatter signals are then amplified through either a linear amplifier or a logarithmic amplifier. In clinical cytometry, linear amps are used for light scatter signals and for DNA measurements. Measurement of cell surface proteins uses logarithmic amps. Flow cytometers provide four logs of amplification range. This allows a practical range of measurement of approx 10^3 to 10^8 molecules per cell without adjusting the amplification settings. This range is sufficient for most proteins of clinical interest. The best of the current generation of commercial systems can detect as low as a few hundred protein molecules per cell with carefully selected reagents and conditions.

For some instruments, there is additional circuitry to provide compensation, and on some instruments, all compensation is done in software. The subject of compensation is quite complex and has generated intense discussion at times among flow cytometry specialists. Compensation is needed because the light signals emitted by the fluorochromes are broad enough to bleed into adjacent filter regions. See **Fig. 7A** for an example of a simple compensation problem. Here we show the emission spectrum of FITC, a fluorochrome that will be primarily detected in the PMT1 detector. Note, however, that because of the width of the right tail of the spectrum, there is a significant part of the FITC signal that will be detected in the PMT2 detector. In order to obtain an accurate measurement of a true signal in PMT2, the unwanted signal from the FITC must be removed from the final data. This process is called compensation. In order to do proper compensation, data must be collected from samples with each fluorochrome alone. **Figure 7B** shows the appearance of a dot plot showing cells labeled only with FITC. **Figure 7B** shows the signal on the PMT2 axis from the PMT1 fluorochrome. **Figure 7C** shows

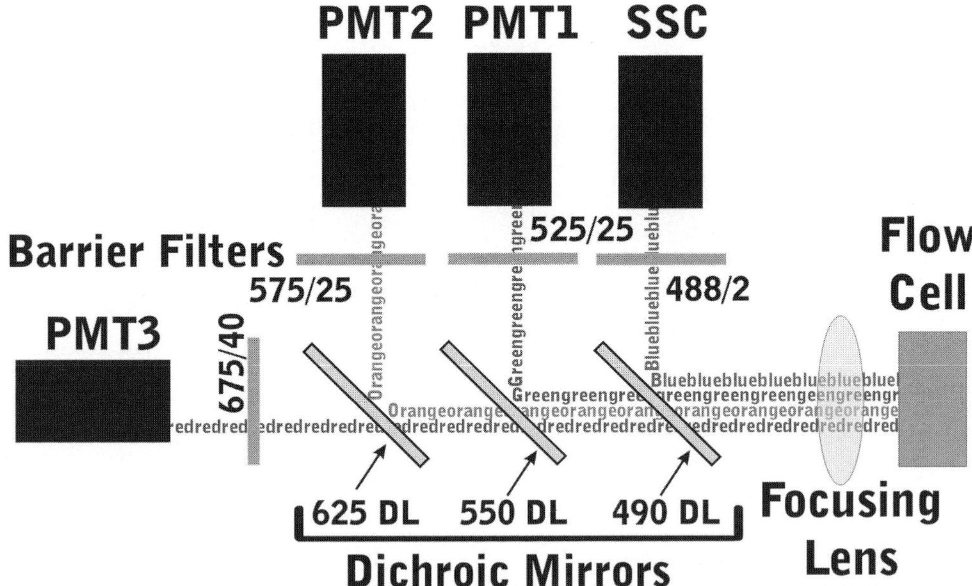

Fig. 6. Diagram of a generic optical section of a flow cytometer. Fluorescence light and scattered laser light is collected and collimated by the focusing lens. Successive colors are reflected by dichroic mirrors and purified by barrier filters before generating an electrical voltage in the photomultiplier tube.

the same data after the application of compensation. The compensated data more accurately reflects the biological levels of the markers of clinical interest.

The data from the cytometer are always sent to a computer for display, analysis, and storage. A variety of classes of computers and operating systems have been used over the years. At this time, nearly all systems will have either a Mac or PC attached. The manufacturers package acquisition and analysis software with the cytometers. In addition, there are several analysis software packages that can be used for evaluation of the data. The data are stored in a specific binary file format that cannot be read by spreadsheet programs and other simple packages. The data are usually displayed in two-dimensional (2D) plots or in one-parameter histograms. The axis in either case is the intensity of the optical signal in that channel. In the basic 2D plot, each dot represents the light intensity signal for a single cell in the two channels. For example, in **Fig. 7C**, the dots representing the cells in the left population are low for both signals, whereas the cells in the right population are bright in the PMT1 signal and low in the PMT2 signal. A one-parameter histogram is used in situations where cells are only labeled with one label or where evaluation of the dot plots has shown that a histogram display is not confounded. A specific example of this is shown in **Fig. 8**. **Figure 8A** shows the histogram for the PMT1 signal for all cells. However, examination of a dot plot (**see Fig. 8B**) shows that there are two distinct populations that overlap in their PMT1 signals. The cells only from region A are shown in **Fig. 8C** and those from region B in are shown in **Fig. 8D**. The plot is **Fig. 8E** is the two histograms from **Fig. 8C** and **Fig. 8D** displayed simultaneously, showing the two distinct populations that cannot be clearly distinguished in **Fig 8A**. **Figure 8E** is a type of plot called an overlay plot. **Figure 8B** shows rectangular regions on a dot plot. The other types of region commonly seen are polygonal and quadrants. There are other situations where a histogram plot might be quite acceptable, however, it needs to be demonstrated that populations are not being confounded. The data in **Fig. 2** are an example of data where a histogram is acceptable to differentiate populations that are low or high for the marker used on PMT1 here.

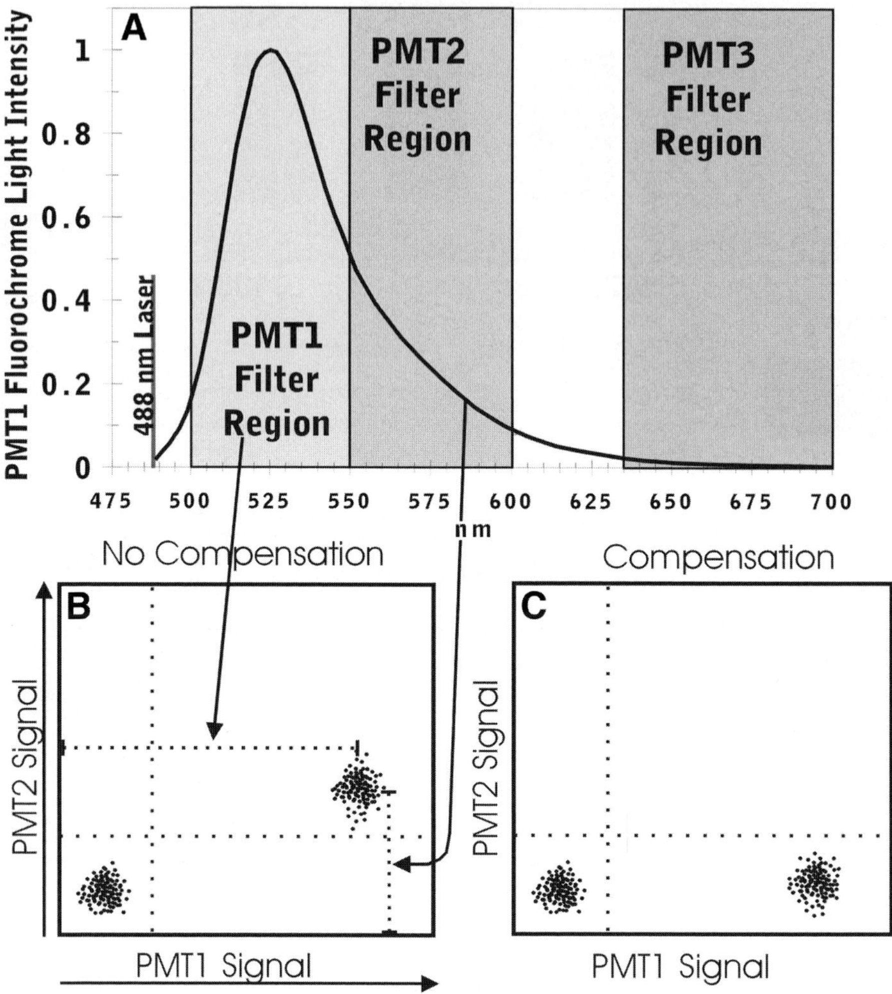

Fig. 7. Example of compensation: (**A**) emission spectrum for fluorescein with filter regions superimposed on the plot; (**B**) uncompensated fluorescence dot plot from cells labeled only with FITC, (**C**) data for the same cell population with proper compensation to remove the FITC signal from the PMT2 channel.

There are two values that are commonly used for data analysis: the percentage of cells or the fluorescence intensity of the cells, in a specific region. The percentage values might be of all cells within a viability gate or might be the percentage of cells within a specific subpopulation. As acute leukemias can have a heterogeneous pattern of antigen expression, it is important to differentiate tumor cells from normal populations so as to accurately define the immunophenotype. A blast gate can be defined in analysis of bone marrow aspirates based on the pattern of expression of CD45 vs SSC pattern (see below).

In the clinical setting, the vast majority of flow cytometry is analysis of cell populations. However, there is an extension of flow cytometry that allows cells to be physically sorted. This capability was part of the earliest systems and has advanced in speed and capacity. The current generation of high-speed sorters can separate up to 4 different populations at speeds as high as 40,000 cells per second. Sorting is conceptually simple, but the hardware and software needed to operate at these speeds are quite sophisticated. In contrast to the analyzers, the flow cell on a

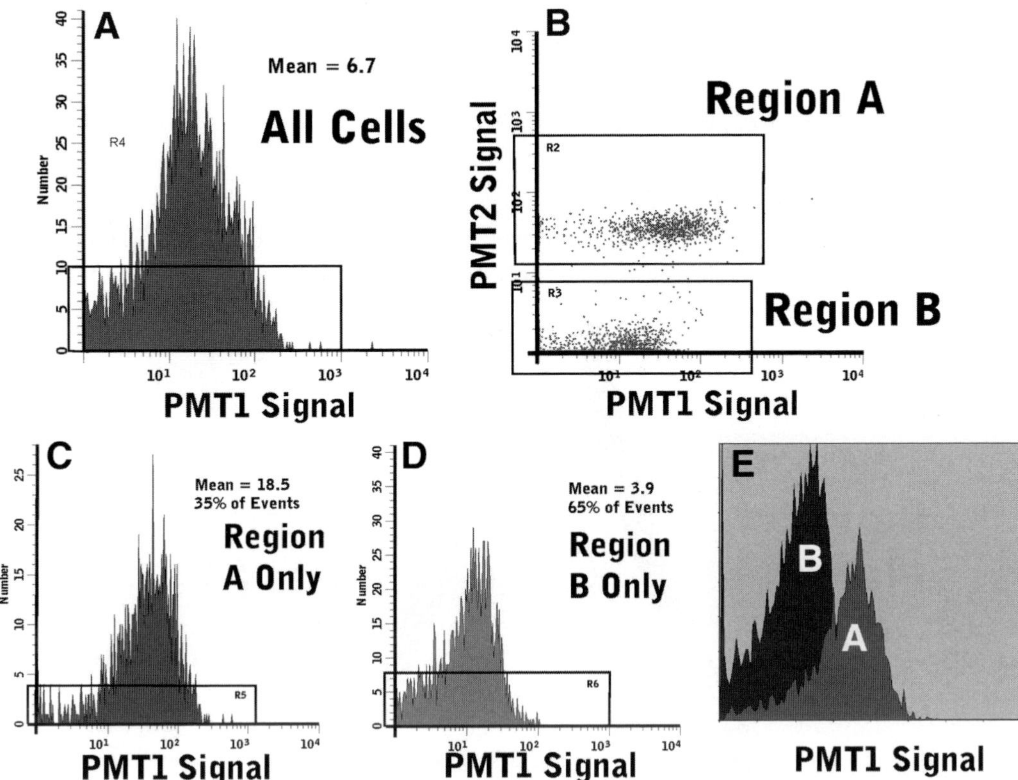

Fig. 8. Example of the use of gating to identify confounded data. (**A**) PMT1 signal for all cells; (**B**) dot plot of PMT1 and PMT2 for the same data shown in (**A**) showing two populations; (**C**) PMT1 histogram for only those cells shown in region A; (**D**) PMT1 histogram for only those cells shown in region B; (**E**) overlay plot showing histograms from (**C**) and (**D**) on the same axes.

sorter is vibrated by a piezoelectric crystal at high speeds; these are tuned to produce a stream that breaks up into droplets that can only hold one or two cells. The cells pass through the laser and the usual types of flow cytometry data are generated. The operator has previously calibrated the system so that the time between a cell being in the laser beam and being between the charge plates is exactly known. The sorter operator has specifically identified the cell populations to be sorted using one or more gates. When a cell that fits the criteria to be sorted is identified, the system initiates the sorting process. **Figure 9** is a schematic diagram of the basic sorting process. Sorting occurs because the system is able to put a specific positive or negative charge on a single droplet. These charged droplets are then attracted to the deflector plate, which has a standing charge of the opposite polarity. The cells are collected into a tube for whatever clinical use is intended.

8. Clinical Uses of Flow Cytometry

8.1. Hematopathology: Lymphoma/Leukemia Typing

The World Health Organization (WHO) classification of hematopoietic neoplasms is based on the definition of clinical entities based on morphology, immunophenotype, genetic abnormalities, and clinical features. Some hematopoietic neoplasms in the WHO classification system have a specific immunophenotype, and diagnosis is very difficult in the absence of this

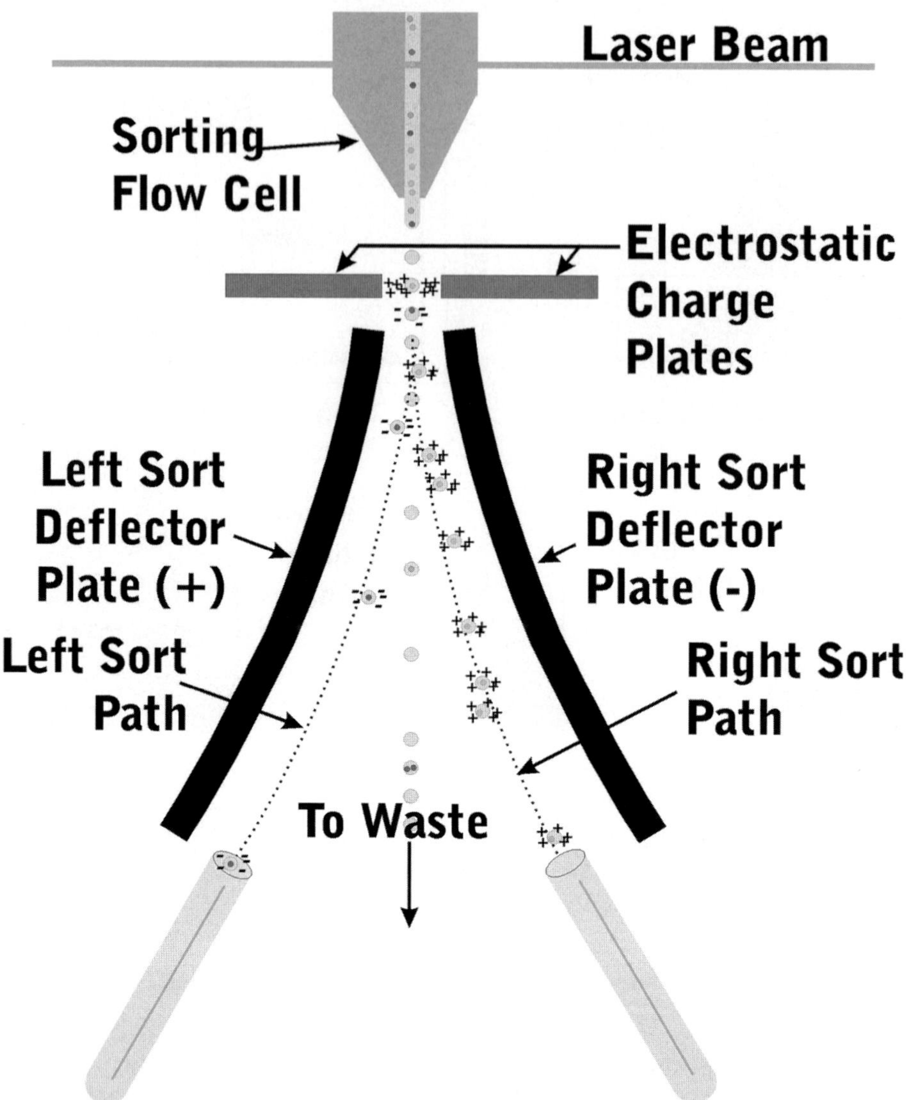

Fig. 9. Schematic demonstration of sorting. Cells pass through the flow cell and data are collected. The flow cell vibrates to break the stream into droplets containing zero or one cell. If the cell meets sort criteria, an electrostatic charge is applied as the cell passes between the charge plates. Depending on whether a positive or negative charge was applied, the deflector plates attract the charged droplets right or left, where the droplets are collected in a separate tube.

immunophenotype *(3)*. Therefore, diagnostic evaluation of a specimen suspected for a hematological malignancy should include review of the immunophenotype by flow cytometry or immunohistochemistry. Flow cytometry is clearly superior when evaluating hematopoietic neoplasms for antibody-based therapy, such as Rituximab, Campath, Mylotarg, or Zenapax, as the targeted antigen must be expressed on the cell surface for therapy to be effective. Flow cytometry has also been shown to have greater sensitivity, reduced subjectivity, and faster turnaround time compared to immunohistochemistry in the evaluation of lymphoid neoplasms

(4). In acute leukemias, flow cytometric immunophenotyping not only allows differentiation between lymphoid and myeloid lineage (crucial for prognosis and treatment) but is also useful for precise characterization of leukemias, and the pattern of antigen expression can be used to detect minimal residual disease posttherapy in most patients ***(5)***.

The majority of mature lymphoid neoplasms in North America and Europe are of B-cell origin. Diagnosis of a B-cell lymphoma requires first the identification of a neoplastic B-cell population and then subclassification into an appropriate diagnostic category. Markers of B-cell neoplasia include light-chain restriction, absence of normal antigens and presence of antigens not normally present on mature B-cells. A B-cell population with restricted light-chain expression is, with rare exceptions, considered a B-cell neoplasm. The expression of a single immunoglobulin light chain by a B-cell population results in a unimodal pattern of staining with anti-light-chain reagents; one light-chain reagent (anti-kappa or anti-lambda) is positive and the other is negative in all of the malignant cells (*see* **Fig. 10**). One of the advantages of flow cytometry is its ability to recognize monoclonal B-cells even in B-cell lymphopenia, in the presence of polyclonal B-cells or among cells with cytophillic antibody (passively adsorbed immunoglobulins from the plasma in vivo [e.g., monocytes, natural killer {NK} cells, activated T-cells, or granulocytes]) ***(6)***. In addition, in normal or benign lymph node tissue, virtually every B-cell expresses light-chain immunoglobulins and the lack of expression of surface immunoglobulins among mature B cells also suggests the presence of a monoclonal B-cell population ***(7)***. In samples containing monoclonal B-cells admixed with polyclonal B-cells, the simultaneous analysis of other markers that are differentially expressed among benign and malignant elements facilitates the detection of lymphoma cells. For example, the CD11c– B cells (region in **Fig. 11A**) are polyclonal because both positive and negative cells are seen in the kappa (*see* **Fig 11B**) and lambda (*see* **Fig 11C**) plots. In contrast, the CD11c + B-cells (region in **Fig. 11D**) in the peripheral blood specimen from a patient with hairy cell leukemia can be monoclonal (note single lambda positive population in **Fig. 11F**). Simple examination of the numbers of cells staining with kappa and lambda will not be useful in this case. Delineating an abnormal pattern of expression of antigens in B-cell neoplasia also allows subclassification into discrete diagnostic categories. Examples of antigens useful in subclassification include CD5 and CD23 in CLL and mantle cell lymphoma, CD10 in follicular lymphoma, and CD11c as well as CD103 in hairy cell leukemia ***(3)***.

Mature T-cell and NK cell neoplasms are uncommon, accounting for only 12% of all non-Hodgkin's lymphoma ***(8)***. Flow cytometric detection of T-cell neoplasia is more challenging, as one does not have as sensitive a marker for clonality. Indicators of T-cell neoplasia include subset restriction, absence of normally expressed antigens, presence of abnormal antigens, and abnormal levels of T-cell antigen expression ***(9–11)***. In normal reactive lymphoid populations, there is a mixture of CD4- and CD8-positive cells. However, mature clonal T-cell populations are restricted in CD4 or CD8 expression (majority CD4+ and CD8–, although CD4–/CD8+, CD4–/CD8–, and CD4+/CD8+ can also be observed) in a manner similar to light-chain restriction in B-cell neoplasms. Because 75% of mature T-cell neoplasms fail to express at least one T-cell antigen (*see* **Fig. 12**), analysis for the absence of a T-cell antigen is more useful than subset restriction analysis. CD7 is most frequently absent, followed by loss of CD2, CD5, and CD3. In T-cell neoplasms with antigen loss, two-thirds have greater than one pan-T-cell antigen missing ***(9,10)***. In evaluating T-cell antigen expression, it is important to remember that a small percent of peripheral blood CD3+ T-cells are CD7– and that a subset of normal gamma delta T-cells do not express CD5. Detection of abnormal or inappropriate antigen expression is useful in the diagnosis of T-cell neoplasias. Neoplastic T-cells can be detected as a homogeneous population with an abnormal level of antigen expression. For example, CD3 can be expressed at a higher or lower level than normal as measured by staining with anti-CD3 (*see* **Fig. 12**) ***(11,12)***. T-Cell large granular lymphocyte leukemias typically have abnormally dim levels of CD5 expression. CD2, CD5, CD7, and CD45 can also be expressed at abnormal levels in T-cell lymphoproliferative processes ***(13)***.

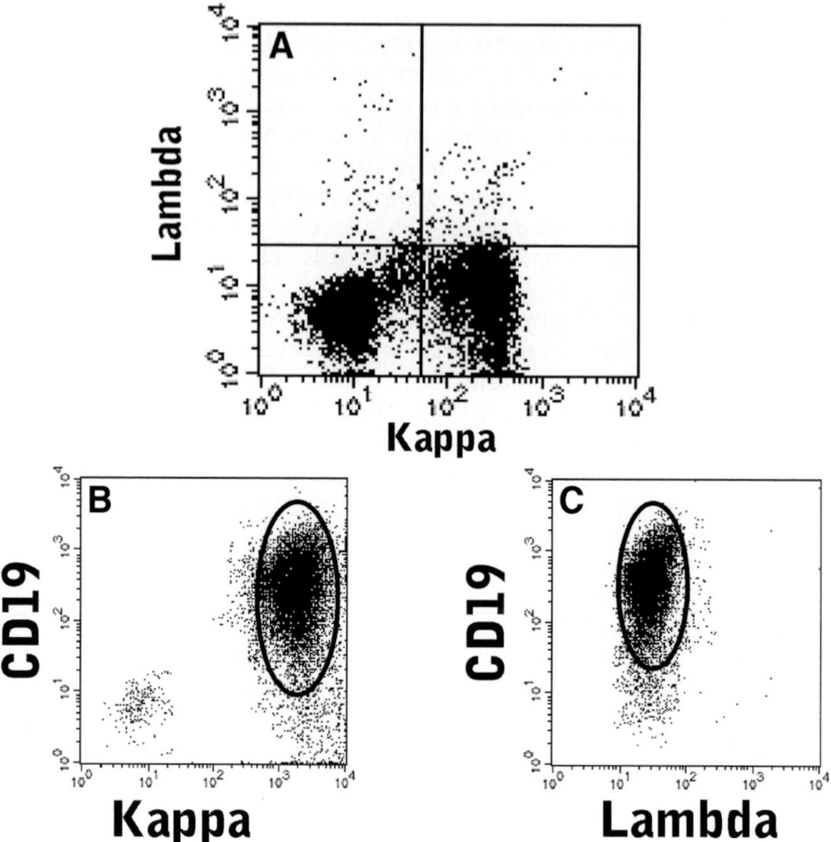

Fig. 10. Example of monoclonal light-chain population in a B-cell neoplasm: (**A**) all of the B cells are kappa positive and lambda negative; (**B**) all of the CD19-positive B-cells are kappa positive; (**C**) all of the CD19 positive B-cells are lambda negative.

A subgroup of clonal T-cell processes are not yet characterized by abnormal patterns of antigen expression but by increased numbers of T-cell subpopulations normally present in low numbers. In T-cell large granular lymphocyte (LGL) leukemia, CD8+ T-cells coexpressing CD57, CD56, or CD16 are increased. Dim CD5 expression and absence of normal T-cell antigens, such as CD7 and CD2, assist in the diagnosis. CD20, considered a B-cell antigen, is expressed by a small subgroup of normal T-cells. Detection of a significant population of CD20+ T-cells is highly abnormal. Also, a high level of gamma delta T-cells is suspicious for malignancy. Flow cytometric analysis of the TCR-V beta repertoire can be helpful in determining T-cell clonality in these cases. Recent studies indicate that clonality can be determined in the setting of increased numbers of T-cell LGLs based on expansion of a given TCR-V beta family with a sensitivity of 93% and a specificity of 80% ***(14)***.

Acute leukemias are characterized by a rapidly growing, aggressive population of neoplastic blasts. The distinction between lymphoid (ALL) and myeloid (AML) leukemia is crucial. Flow cytometric evaluation of the immunophenotypic markers of lineage and stages of differentiation allows a more precise classification of a leukemic process than morphology alone. The WHO classification of acute leukemias uses cytogenetic, immunophenotypic and molecular genetic data to define subgroups of leukemias with favorable or poor prognoses. Many of the

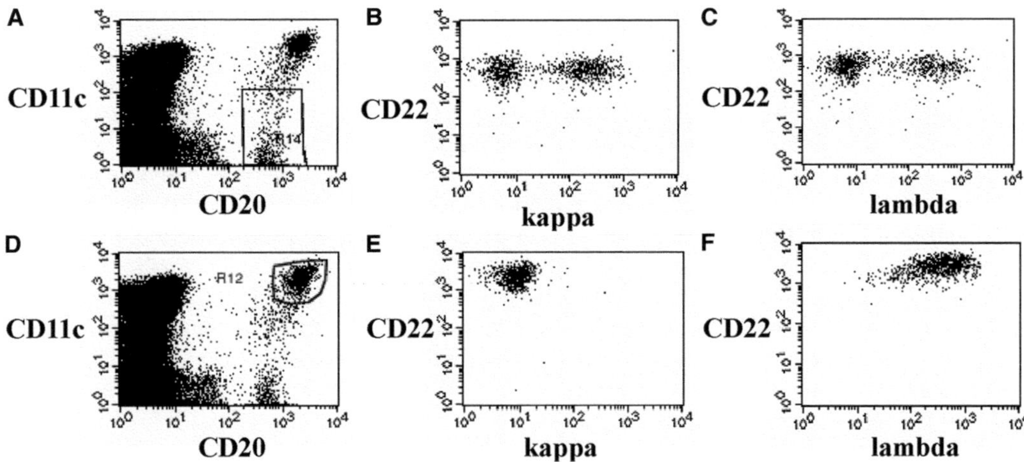

Fig. 11. Use of CD11c expression to separate malignant from normal B-cells in a patient with hairy cell leukemia: (**A**) the CD20-positive but CD11c-negative B-cell gate utilized; (**B**) the B cells in the CD20-positive CD11c-negative gate are CD22 positive and polyclonal based on kappa staining; (**C**) the B-cells in the CD20-positive, CD11c-negative gate are CD22 positive and polyclonal based on lambda staining; (**D**) the CD20-positive and CD11c brightly positive B-cell gate utilized; (**E**) the B-cells in the CD20-positive, CD11c-positive gate are CD22 positive and all kappa negative (monoclonal); (**F**) the B-cells in the CD20-positive, CD11c-positive gate are CD22 positive and all lambda positive (monoclonal).

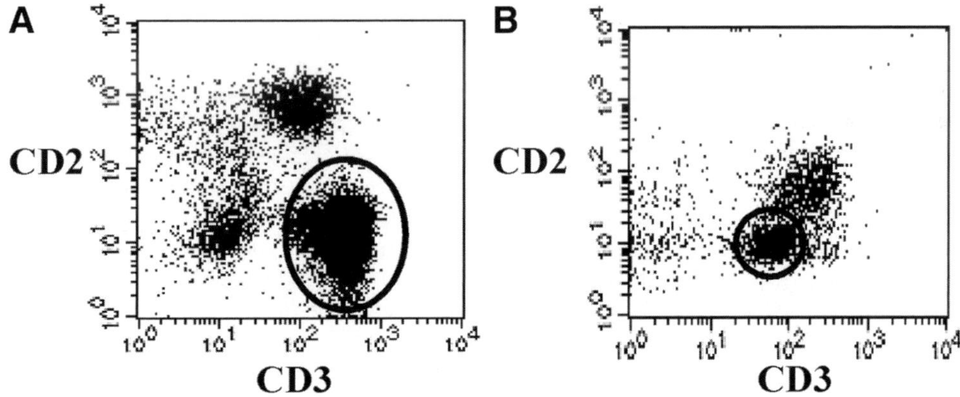

Fig. 12. Flow cytometric analysis of T-cell neoplasms: (**A**) malignant T-cells (in oval) expressing CD3 but negative for CD2; (**B**) malignant T-cells (in oval) expressing abnormally dim CD3 and negative for CD2.

genetically distinct diagnostic subgroups are closely associated with specific immunophenotypes. Therefore, flow cytometric immunophenotyping can detect antigen profiles associated with a specific molecular abnormality and prognosis (reviewed in **ref. *15***).

Because acute leukemias can have a heterogeneous pattern of antigen expression, it is important to differentiate tumor cells from normal populations so as to accurately define the immunophenotype. In the past, flow cytometric analysis of leukemia utilized FSC versus SSC to define a gate. Using this gating strategy, antigen expression was described as the percent of gated events positive for a list of antibodies. As the analysis gate also contained the normal nucleated erythroid precursors and lymphocytes, the reported percentages pertained to a mixture of leukemic blasts and normal cells. In bone marrows virtually replaced by leukemic blasts, essentially all of the cells are blasts and this strategy is useful. However, when normal elements are present in significant numbers, an analysis gate specific for blasts must be used. When the expression of CD45 vs SSC pattern for bone marrow elements is examined, cells can be segregated into distinct populations, including lymphocytes, monocytes, granulocytes, nucleated red cells, and blasts (*see* **Fig. 13**) Thus, these two parameters can be used to define a blast gate (characterized by low SSC and dim CD45) that contains few normal cells ***(16,17)***.

Flow cytometry is essential for determination of lineage in ALL. ALL cells typically occupy the blast gate on CD45 vs SSC, although in select cases, CD45 expression can be dimmer than usual and CD45-negative blasts (in the region of normal erythroid precursors on CD45 vs SSC) can be identified. ALL blasts have an immature T- or B-cell immunophenotype and can express TdT and CD34. However, expression of myeloid antigens such as CD13, CD33, and CD15 is common. For this reason, classification of ALL is based on a panel of T-cell, B-cell and myeloid antibodies.

Acute lymphoblastic leukemia blasts have an immature immunophenotype and typically express CD19 and CD10. B-ALL is subclassified into B-precursor ALL, pre-B-ALL, and mature B-cell ALL based on immunoglobulin expression. B-Precursor ALL does not express immunoglobulin, whereas pre-B-ALL is positive for cytoplasmic but not surface immunoglobulin. Mature B-cell ALL has surface immunoglobulin. Because cytoplasmic immunoglobulin is no longer regularly analyzed in the clinical flow cytometry laboratory, both B-precursor ALL and pre-B-ALL are often grouped together based on negativity for surface immunoglobulin and called B-precursor ALL. They typically express CD19, CD10, CD34, HLADR, and TdT and are usually positive for CD22 and CD24. Mature B-cell ALL, in addition to surface immunoglobulin, has brighter CD45, CD19, CD20, CD22, CD24, and CD10 but is negative for CD34 and TdT. As the immunophenotype in B-cell lineage ALL is associated with molecular abnormalities and prognosis, flow cytometric immunophenotyping can also be used as a screen to select cases for molecular analysis (for review, see **ref. *16***).

The blasts in T-cell lineage ALL often are not contained in the CD45 vs SSC blast gate and might overlap with the brighter CD45 mature lymphocyte and monocyte gates. Although the most specific marker for T-cell lineage in ALL is CD3, T-ALL is usually negative for surface CD3, necessitating intracytoplasmic staining for this antigen. As intracytoplasmic staining has a higher background, one must be careful not to overinterpret apparent dim CD3 staining as real and indicating T cell lineage. CD7 is the most sensitive marker for T-ALL but is less specific than CD3, as it is frequently expressed by myeloid leukemias. CD2, CD1a, CD5, TdT, and CD34 are also observed in T-ALL. CD4 and CD8 are typically double positive or double negative (for review, see **ref. *16***).

Flow cytometric immunophenotyping plays an important role in the WHO classification of AML. Flow cytometry is not only specific in differentiating AML from ALL but is also highly useful in identifying granulocytic, monocytic, erythroid, and megakaryocytic differentiation. Both the pattern of antigen expression as well as where the blasts fall on a CD45 vs SSC data plot help to subclassify AML. In addition, many of the specific genetically identified subgroups tend to have distinct immunophenotypic features.

Blasts in AML have an immature phenotype and can express CD13, CD33, CD117 and myeloperoxidase. CD34 and HLADR might be positive but are not lineage specific. In addition, the blasts can express the lymphoid antigens TdT, CD56, and CD7. With increasing maturation, there is decreased expression of CD33 and increased CD13, CD15, and CD11b. In AML with monocytic differentiation, the blasts can overlap with normal monocytes on the CD45 vs SSC data plot and express monocytic antigens, such as CD14 and CD36. Erythroid differentiation can be determined by bright expression of CD71 and glycophorin, whereas megakaryocytic differentiation is characterized by bright CD61 and CD41 expression (for review, see **ref. *16***).

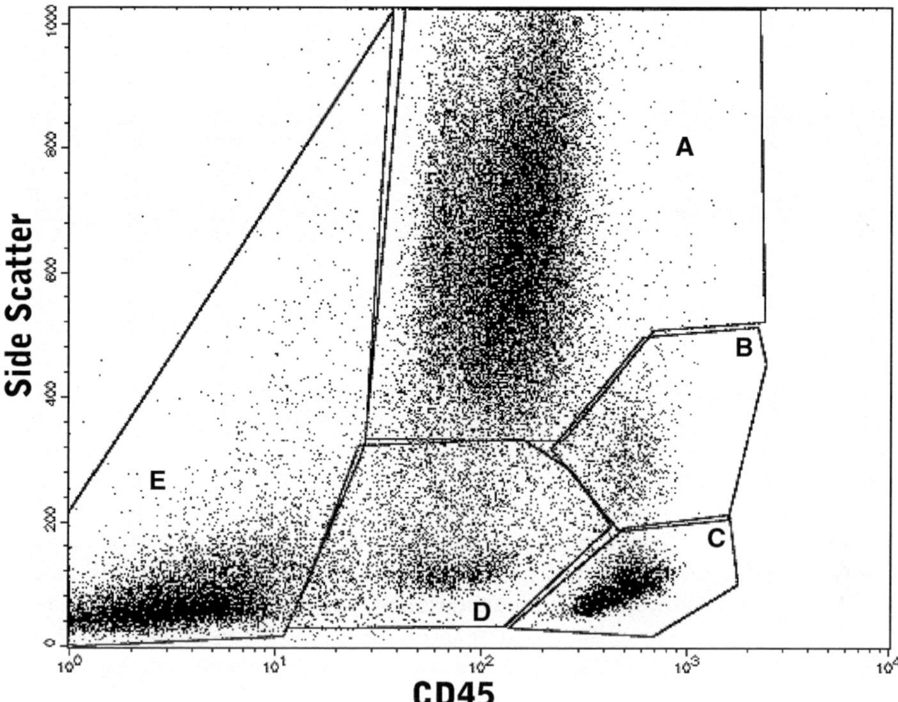

Fig. 13. Staining pattern of CD45 vs side scatter for bone marrow: (**A**) mature granulocytes; (**B**) mature monocytes; (**C**) mature lymphocytes; (**D**) blasts; (**E**) nucleated red cell precursors.

8.2. Hematology and Transfusion Medicine

Flow-cytometric-based methods are replacing older, more laborious protocols for testing of red blood cells (RBCs) (for review, see **ref.** ***18***). Flow cytometric evaluation is useful in monitoring obstetrical patients. Following pregnancy with an Rh+ fetus, Rh– mothers can become sensitized to the Rh factor and produce anti-Rh or D antigen antibodies. In future pregnancies, this can result in fetal anemia or even fetal death. Because small amounts of fetal RBCs can be detected in maternal blood in uncomplicated pregnancy and delivery, all Rh– women receive prophylactic anti-Rh, or anti-D immunoglobulin at 28 wk of pregnancy and within 72 h of delivery to prevent possible Rh alloimmunization. However, significant fetomaternal hemorrhage can occur under a number of conditions and the number of fetal RBCs must be quantitated in maternal blood to determine the appropriate dose of anti-D immunoglobulin. The Kleihauer–Betke test (KBT) for detection of fetal erythrocytes in maternal blood is based the fact that acid elutes adult hemoglobin more rapidly than fetal hemoglobin from RBCs. The intact fetal hemoglobin can be visualized by staining with erythrosine. This method, however, has been shown to be subjective, not reproducible and insensitive. Flow cytometry is a highly sensitive tool for detecting minor RBC populations. Using an antibody to fetal hemoglobin, flow cytometry has been demonstrated to have superior precision in detecting fetal RBCs. In addition, the flow cytometric method can differentiate hereditary persistence of fetal hemoglobin from fetal RBCs in the maternal blood ***(19)***.

Flow cytometry has been used to study ABH and Rh blood group antigens ***(19)***. Flow cytometry has been used to detect and quantitate RBC-bound antibody (IgG, IgM, and IgA) and complement (C3) in autoimmune hemalytic anemia. Because the standard direct antiglobulin test is sensitive and inexpensive, flow cytometry is only used in select conditions. Flow cytometry can detect small amounts of IgG bound to RBCs in patients with indications of autoimmune hemolytic anemia, but is a negative direct antiglobulin test by standard methods. In addition, flow cytometry can be used to determine the IgG subclass of RBC-bound IgG. Flow cytometry is also optimal for detecting autoimmune hemalytic anemia associated with IgM autoantibodies reacting at 37°C *(18)*.

Flow cytometry is highly useful in the clinical evaluation of platelets and allows testing with little or no isolation or manipulation (for review, *see* **ref. *20***). Because the threshold for platelet transfusions has been lowered in many institutions to 10,000 platelets/μL, it is important that platelet counts be accurate at this lower range. Nonplatelet particles can interfere with platelet counting by some methods. Enumeration of platelets by flow cytometry or immunoplatelet counting is highly precise and accurate and eliminates the problem of nonplatelet particles. In this method, platelets are labeled with an antiplatelet antibody and the ratio of platelets to RBCs is measured. The platelet count is then determined using the RBC count obtained from a hematology analyzer and the ratio of platelets to RBCs. This can be used to calibrate hematology analyzers *(21)*. In addition to quantitating platelets, flow cytometry has been used for immunophenotyping platelets. Demonstration of abnormal patterns of expression of platelet receptors has been used to define several genetic disorders, including Glanzmann's thrombasthenia and Bernard–Soulier syndrome *(20)*.

Reticulated platelets are the youngest platelets in circulation and their levels increase when thrombopoiesis is stimulated. Measuring reticulated platelets is the most sensitive and specific method to distinguish destructive or immune thrombocytopenia from marrow failure. Reticulated platelets are large and have increased quantities of RNA. Upon staining of platelets with fluorescent nuclear dyes, such as thiazole orange, all platelets take up dye in proportion to their size. Reticulated platelets take up additional dye, resulting from RNA staining, that is lost upon treatment with RNAase. This RNAase-sensitive labeling can be measured by flow cytometry for an accurate reticulated platelet count *(20)*.

Studies demonstrating immunophenotypic markers of platelet activation have markedly improved platelet function testing. P-selectin, CD63, PCA-1, and LIBS only become exposed on the platelet surface after activation. Flow cytometric immunophenotyping of platelets can reveal this activated phenotype, even when only low numbers of platelets are activated. More conventional methods of measuring platelet activation show a threshold phenomenon and are primarily suited for detecting platelet dysfunction. Flow cytometric testing, however, allows measurement of platelet hyperfunction *(20)*. Increased platelet activation in patients undergoing procedures such as coronary angioplasty can help identify patients at risk for early restenosis who require antiplatelet therapy *(22)*. This technology can even prove useful in predicting risk of cardiovascular disease, allowing early preventative intervention.

Paroxysmal nocturnal hemoglobinuriais (PNH) is a hematopoietic disorder caused by deletions, insertions, or point mutations in the phosphatidylinositolglycan complementation class A (PIG-A) gene. This results in a total or partial deficiency of surface proteins attached to the cell by a glycophosphatidylinositol (GPI) anchor. CD55 (decay accelerating factor [DAF]) and CD59 (membrane inhibitor of reactive lysis [MIRL]) are affected in PNH. Both play an important role in the regulation of complement activation and their deficiency in PNH leads to an increased susceptibility of cells to complement mediated cell lysis. Because CD55 and CD59 are normally expressed at high levels on leukocytes, RBCs, and platelets and their levels are consistently decreased in PNH hematopoietic cells, flow cytometric immunophenotypic analysis of CD55 and CD59 is a sensitive test for PNH. Testing of CD59 alone, however, leads to a high false-positive rate *(23)*. Because red cells stain with the anti-CD55 and anti-CD59 antibodies, it is difficult to achieve optimal antibody concentrations without a prelyse protocol. In a stain–lyse-and-then–wash protocol, the mean fluorescence intensity (MFI) of leukocyte staining correlates with the number of RBCs present. By optimizing the staining protocol (using a smaller blood volume and incubating longer), a nonlyse, nonwash method can be used that allows for the simultaneous analysis of CD55 and CD59 on red cells, platelets, and leukocytes *(24)*.

8.3. Histocompatibility Testing

Flow cytometry plays an important role in allogeneic transplantation *(25,26)*. It has been most useful in the detection of low-titer, or noncomplement fixing, antibodies. Originally instituted as a means for performing a "more sensitive" crossmatch, new technologies have given additional utility to this technology. The flow cytometry crossmatch detects alloantibodies that are not detectable by the standard complement-dependent cytotoxicity crossmatch. A positive flow crossmatch, even when the complement-dependent cytotoxicity crossmatch is negative, correlates with a increased risk of rejection and graft loss in solid-organ transplantation. In stem cell transplantation, it may be associated with failed engraftment.

The flow cytometric crossmatching is performed by incubating transplant recipient serum with donor lymphocytes. HLA antibodies, if present, bind to the surface of the lymphocyte and are measured using a fluorescently labeled anti-human immunoglobulin.

Multiparameter cytometry then permits the simultaneous assessment of T- and B-cell reactivities, which could help indicate the class of antibody present (human leukocyte antigen [HLA] class I or class II). It became clear, however, that the specificity of the antibodies detected in the crossmatch is important for proper interpreting a positive result. Thus, it was important to discriminate between true HLA-specific antibodies and other non-HLA antibodies.

Recently, microparticles coated with purified Class I or Class II MHC (major histocompatibility complex) proteins have been employed instead of donor cells *(27)*. This test provides sensitivity equal to the flow crossmatch, yet confirms the specificity of the antibody. In addition, this test can be used to profile patients awaiting transplantation. These data can help classify patients as "sensitized" or "nonsensitized" and aid in organ allocation or tailoring immunosuppression. In addition to being useful prior to transplantation, flow cytometry can also be used to monitor the development of donor HLA-specific antibodies posttransplant (for review, see **ref. *25***). Early detection of donor-specific reactivity can indicate the beginning of an unwanted immune response and provide an opportunity to modify immunosuppression prior to graft damage. In summary, flow cytometry crossmatching and microparticle evaluations are providing clinicians with better diagnostic information, which aids in clinical decision-making and, ultimately, will lead to better graft survival.

8.4. Immunology

Standard phenotyping studies are used as part of the diagnosis of some types of genetic immune deficiency state. For some conditions, such as severe combined immune deficiency, a very basic panel of markers will immediately identify the magnitude of the problem. Other conditions may require functional assays to identify the problem. For example, chronic granulomatous disease is routinely diagnosed using a flow cytometric assay for oxidative burst activity *(28)*.

Flow cytometry was the first method used to monitor the status of patients with human immunodeficiency virus (HIV). Initially, it was found that the level of CD4+ cells in the blood of HIV patients correlated well with clinical status *(29)*. The relationship was powerful enough that the US FDA (Food and Drug Administration) accepted the use of CD4 counts as a surrogate marker for evaluating the efficacy of treatments for HIV infection. More recently, other measurements, including direct measurement of viral RNA in blood, have been developed to complement CD4 counts *(30)*. Other markers, such as the level of the activation marker CD38 on CD8+ cells, have also been shown to be highly predictive of health status *(31)*. The expression of activation markers can also be a useful diagnostic in neonatal sepsis, where CD64 is strongly upregulated

Defects in regulation of the immune system can lead to a number of different autoimmune syndromes. As with other immune conditions, flow cytometric assays play important roles in diagnosis and in monitoring for a number of these conditions. For example, in autoimmune lymphoproliferative syndrome, one of the major diagnostic criteria is the presence of a trio of significant numbers of rare cells, specifically T-cells with TCR-beta without CD4 or CD8, CD20-postive B-cells also expressing CD5, and T-cells with both CD3 and HLA-Dr *(32)*.

A relatively new use for flow cytometry is the use of multiplexed bead assays to measure soluble proteins. This assay uses the multicolor capability of the flow cytometer to allow simultaneous evaluation of mixed populations of beads. The beads are chemically labeled with varying intensities of one or two fluorochromes and then a monoclonal antibody labeled with another fluorochrome is used to measure the test analyte. This assay can be used to simultaneously measure up to 10 proteins in a standard flow cytometer or up to 100 proteins using a dedicated system. The applications of this technology are limited only by the availability of the test reagents. Applications include evaluation of cytokines or soluble cytokine receptors to assist diagnosis of sepsis, certain autoimmune conditions, transplant rejection, some cancers, and viral infection *(33)*. The clinical usefulness of many of these new reagents is an area of active research and the number of available reagents can be expected to rapidly increase.

8.5. Stem Cells

Peripheral blood stem cells are replacing bone marrow as a source for hematopoietic progenitor cells in transplantion of cancer patients after myelosuppressive or high-dose chemotherapy. The reasons for this switch include the greater ease of collection by apheresis, elimination of the need for general anesthesia of the donor, rate of hematological reconstitution of the recipient, and decreased medical care costs. However, because the minimum number of CD34-positive hematopoietic progenitor cell required is (2–) 5×10^6/kg recipient body weight *(34,35)*, multiple aphereses must be performed to collect a sufficient number to ensure adequate engraftment. Also, because less than 0.1% of nucleated peripheral blood cells in normal controls are CD34-positive cells *(36)*, peripheral blood stem cell numbers are increased (or mobilized) by administering colony-stimulating factors. Flow cytometry is used to enumerate CD34-positive progenitor cells in the peripheral blood of donors undergoing mobilization to determine when apheresis should start and when sufficient number of hematopoietic progenitor cells are collected for engraftment (for review see **ref.** *37*). The absolute count of CD34-positive cells in the donor peripheral blood is predictive of the number of CD34-positive cells in the apheresis product and, thus, can be used to determine when apheresis should start. Decisions concerning growth factor administration and further apheresis collections are made based on the number of CD34-positive cells collected in each apheresis. As crucial treatment decisions are made based on absolute CD34 counts, it is important that methods be standardized and strict quality control be applied. Multicenter studies demonstrated variability between laboratories and served as an impetus for further standardization of methods. Propriatary kits are available that utilize optimal anti-CD34 antibodies (class III), provide for minimal manipulation, and include beads to allow direct quantification of the number of cells per unit volume. The ISHAGE multiparametric sequential gating technique showed closest agreement between labs *(38)*. This approach involves cumulative gating based on light scatter characteristics, dim expression of CD45, and expression of CD34. Only cells within the appropriate cumulative gates are counted. By including fluorescent counting beads, an absolute CD34-positive cells count can be directly determined by flow cytometry. Additional markers can be included for further characterization of the CD34-positive cells. Standardization of absolute CD34-positive cell counting using the single-platform ISHAGE method resulted in decreased differences among laboratories in a multicenter trial *(39)*.

8.6. On the Horizon

Flow cytometry has shown significant technological advances since the initial creation of the systems in the 1970s. Here are some of the advances that are on the horizon and can be found on clinical flow cytometers in the near future. The current generation of analyzers can acquire data at speeds of 3000 cells/s without discernable loss of data. The latest sorters now operate above 40,000 cells/s; it is reasonable to expect these faster electronics to appear in analyzers. A related issue is that of automation, most analyzers now require an operator to load each tube on and off of the cytometer. The latest generation of clinical cytometers automate sample loading through either tube carousels or 96-well plates. This trend is likely to continue and be extended into electronic reporting of results. An additional improvement is the development of additional fluorochromes. Some of these, such as Cy7, have emission in spectral ranges not currently used, an alternate approach is the development of nanocrystals (*see* http://www.qdots.com), which have narrower emission spectra. This should allow more markers to be analyzed in the current spectral range. Combining both approaches could lead to clinical cytometers capable of simultaneous analysis of eight or more markers in a single tube. A final flow cytometry application is the use of bead assays for soluble analytes. These are starting to appear on the market and the use of multiplexing bead strategies will allow many tests to be run with very small sample volumes. There is no doubt that other improvements and new ideas will add to the utility and versatility of this technology for many years to come.

9. Summary and Conclusions

In conclusion, flow cytometry is a powerful technology that is currently being used to provide valuable data in a variety of clinical situations. As the technology becomes more powerful, these instruments will move into new areas of clinical practice to provide data in a rapid, accurate, and cost-efficient manner.

10. Additional Resources

10.1. Books

Flow Cytometry: First Principles, by Alice Givan, Wiley, Hoboken, NJ (2001). This is a good first book to give an overview of the field with some emphasis on clinical cytometry.

Practical Flow Cytometry, 4th ed., by Howard Shapiro, Wiley, Hoboken, NJ (2003). This is the single most complete and detailed book on flow cytometry available anywhere. Not for the faint of heart or the casual reader, it is a must-have for serious flow cytometrists.

10.2. Journals

Clinical Cytometry, a journal of the International Society for Analytical Cytology (ISAC) and the Clinical Cytometry Society

Journal of Immunological Methods

10.3. Professional Societies

International Society for Analytical Cytology (http://www.isac-net.org)

The Clinical Cytometry Society (http://www.cytometry.org)

References

1. Shapiro, H. (2003) In *Practical Flow Cytometry*, 4th ed., Wiley, Hoboken, NJ, chap. 3.
2. Herzenberg, L. A., Parks, D., Sahaf, B., Perez, O., Roederer, M., and Herzenberg, L. A. (2002) The history and future of the fluorescence activated cell sorter and flow cytometry: a view from Stanford. *Clin. Chem.* **48,** 1819–1827.
3. Jaffe, E. S., Harris, N. L., Stein, H., and Vardiman, J. W. (eds.) (2001) *WHO Classification, Pathology and Genetics Tumors of Haematopoietic and Lymphoid Tissues*, IARC, Lyon.
4. Tbakhi, A., Edinger, M., Myles, J., Pohlman, B., and Tubbs, R. R. (1996) Flow cytometric immunophenotyping of non-Hodgkin's lymphoma and related disorders. *Cytometry* **25,** 113–124.
5. Weir, E. G. and Borowitz, M. J. (2001) Flow cytometry in the diagnosis of acute leukemia. *Semin. Hematol.* **38,** 124–138.
6. Fukushima, P. I., Nguyen, P. K., O'Grady, P., and Stetler-Stevenson, M. (1996) Flow cytometric analysis of kappa and lambda light chain expression. *Commun. Clin. Cytom.* **26,** 243–252.
7. Braylan, R. C., Benson, N. A., and Iturraspe, J. (1993) Analysis of lymphomas by flow cytometry. Current and emerging strategies. *Ann. NY Acad. Sci.* **677,** 364–378.
8. Anon. (1997). A clinical evaluation of the International Lymphoma Study Group classification of non-Hodgkin's lymphoma. The Non-Hodgkin's Lymphoma Classification Project. *Blood* **89,** 3909–3918.

9. Stetler-Stevenson, M., Medeiros, L. J., and Jaffe, E. S. (1995) Immunophenotypic methods and findings in the diagnosis of lymphoproliferative diseases, in *Surgical Pathology of the Lymph Nodes and Related Organs* (Jaffe, E. S., ed.), WB Saunders, Philadelphia, PA, pp. 22–57,.
10. Picker, L. J., Weiss, L. M., Medeiros, L. J., Wood, G. S., and Warnke, R. A. (1987) Immunophenotypic criteria for the diagnosis of non-Hodgkin's lymphoma. *Am. J. Pathol.* **128,** 181–201.
11. Kuchnio, M., Sausville, E. A., Jaffe, E. S., et al. (1994) Flow cytometric detection of neoplastic T cells in patients with mycosis fungoides based upon levels of T-cell receptor expression. *Am. J. Clin. Pathol.* **102,** 856–860.
12. Edelman, J. and Meyerson, H. J. (2000) Diminished CD3 expression is useful for detecting and enumerating Sezary cells. *Am. J. Clin. Pathol.* **114,** 467–477.
13. Gorczyca, W., Weisberger, J., Liu, Z., et al. (2002). An approach to diagnosis of T-cell lymphoproliferative disorders by flow cytometry. *Cytometry (Clin. Cytom.)* **50,** 177–190.
14. Lima, M., Almeida, J., Santos, A. H., et al. (2001) Immunophenotypic Analysis of the TCR-V repertoire in 98 persistent expansions of CD3+/TCR large granular lymphocytes. *Am. J. Pathol.* **159,** 1861–1868.
15. Weir, E. G. and Borowitz, M. J. (2001) Flow cytometry in the diagnosis of acute leukemia. *Semin. Hematol.* **38,** 124–138.
16. Borowitz, M. J., Guenther, K. L., Shults, K. E., and Stelzer, G. T. (1993) Immunophenotyping of acute leukemia by flow cytometric analysis: use of CD45 and right angle light scatter to gate on leukemic blasts in three color analysis. *Am. J. Clin. Pathol.* **100,** 534–540.
17. Ranier, R. Hodges, I., and Stelzer, G. (1995) CD45 gating correates with bone marrow differential. *Cytometry* **22,** 139–145.
18. Garratty, G. and Arndt, P. A. (1999) Applications of flow cytofluorometry to red blood cell immunology. *Cytometry* **38,** 259–267.
19. Davis, B. H., Olsen, S., Bigelow, N. C., and Chen, J. C. (1998) Detection of fetal red cells in fetomaternal hemorrhage using a fetal hemoglobin monoclonal antibody by flow cytometry. *Transfusion* **38,** 749–756.
20. Ault, K. A. (2001) The clinical utility of flow cytometry in the study of platelets. *Semin. Hematol.* **38,** 160–168.
21. Davis, B. and Bigelow, N. (1999) Indirect immunoplatelet counting by flow cytometry as a reference method for platelet count calibration. *Lab. Hematol.* **5,** 15–21.
22. Tschoepe, D., Schultheiss, H. P., Kolarov, P., et al. (1993) Platelet membrane activation markers are predictive for increased risk of acute ischemic events after PTCA. *Circulation* **88,** 37–42.
23. His, E. D. (2000) Paroxysmal nocturnal hemaglobinuria testing by flow cytometry—evaluation of the REDQUANT and CELLQUANT kits. *AJCP* **114,** 798–806.
24. Hernandez-Campo, P. M., Martin-Ayuso, M., Almeida, J., Lopez, A., and Orfao, A. (2002) Comparative analysis of different flow cytometry-based immunophenotypic methods for the analysis of CD59 and CD55 expression on major peripheral blood cell subsets. *Cytometry* **50,** 191–201.
25. Horsburgh, T., Martin, S., and Robson, A. J. (2000) The application of flow cytometry to histocompatibility testing. *Transplant Immunology* **8,** 3–15.
26. Kerman, R. H., Gebel, H., Bray, R., et al. (2002) HLA antibody and donor reactivity define patients at risk for rejection or graft loss. *Am. J. Transplant.* **2,** 258.
27. Bryan, C. F., McDonald, S. B., Baier, K. A., et al. (2002) Flow cytometry beads rather than the antihuman globulin method should be used to detect HLA Class I IgG antibodies (PRA) in cadaveric renal regraft candidates. *Clin. Transplant.* **16,** 15–23.
28. Lun, A., Schmitt, M., and Renz, H. (2000) Phagocytosis and oxidative burst: reference values for flow cytometric assays independent of age. *Clin. Chem.* **46,** 1836–1839.
29. Mandy, F., Nicholson, J., Autran, B., and Janossy, G. (2002) T-cell subset counting and the fight against AIDS: Reflections over a 20-year struggle. *Cytometry (Clinical Cytometry)* **50,** 39–45.
30. Fahey, J. L., Taylor, J. M. G., Manna, B., et al. (1998) Prognostic significance of plasma markers of immune activation, HIV viral load, and CD4 T-cell measurements. *AIDS* **12,** 1581–1590.
31. Giorgi, J. V., Lyles, R. H., Matud, J. L., et al. (2002) Predictive value of immunologic and virologic markers after long or short duration of HIV-1 infection. *J. Acquired Immume Deficiency Dis.* **29,** 346–355.
32. Bleesing, J. J. H. (2003) Autoimmune lymphoproliferative syndrome (ALPS). *Curr. Pharm. Design* **9,** 265–278.
33. Bienvenu, J., Monneret, G., Fabien, N., and Revillard, J. P. (2000) The clinical usefulness of the measurement of cytokines. *Clin. Chem. Lab. Med.* **38,** 267–285

34. Tricot, G., Jagannath, S., Vesole, D., et al. (1995) Peripheral Blood stem cell transplants for multiple myeloma: Identification of favorable variables for rapid engraftment in 225 patients. *Blood* **85,** 588–596.
35. Weaver, C. H., Hazelton, B., Birch, R., et al. (1995) An analysis of engraftement kinetics as a function of the CD34 content of peripheral blood progenitor cell collections in 692 patients after the administration of myeloablative chemotherapy. *Blood* **86,** 3961–3969.
36. Sutherland, D. R., Keating, A., Nayar, R., Anania, S., Stewart, A. K. (1994) Sensitive detection and enumeration of CD34+ cells in peripheral blood and cord blood by flow cytometry. *Exp. Hematol.* **22,** 1003–1010.
37. Gratama, J. W., Orfao, A., Barnett, D., et al. (1998) Flow cytometric enumeration of CD34(+) hematopoietic stem and progenitor cells. *Cytometry* **34,** 128–142.
38. Sutherland, D. R., Anderson, L., Keeney, M., Nayar, R., Chin-Yee, I. (1996) The ISHAGE guidelines for CD34 cell determination by flow cytometry. *J. Hematother.* **5,** 213–226.
39. Barnett, D., Granger, V., Kraan, J., et al. (2000) Reduction of intra and inter-laboratory variation in CD34 positive stem cell enumeration by the use of stable test material, standard protocols and targeted training. *Br. J. Haematol.* **108,** 784–792.

39

Immunocytochemistry

Patricia A. Fetsch

1. Introduction

Immunocytochemistry (aka immunohistochemistry), a technique used for the localization of specific cellular antigens, was introduced to diagnostic tumor pathology about 20 yr ago. The procedure allows for the visualization of antigens in tissue samples via the sequential application of a specific antibody to the antigen (primary antibody), a secondary antibody to the primary antibody, an enzyme complex, and a chromogenic substrate. The enzymatic activation of the chromogen results in a visible reaction product at the antigen site, which is interpreted using a light microscope. Immunocytochemistry is routinely used in hospital laboratories to diagnose cancers, identify infectious organisms, differentiate malignant from benign processes, and reveal prognostic indicators.

2. Background

A tumor (neoplasm) is an abnormal mass of tissue in which the growth is uncontrolled and progressive. Neoplasms can be benign (noninvasive) or malignant (invasive). Malignant tumors have the capability to metastasize (i.e., spread from one organ to another not directly connected with it). Metastasis of cancers can occur by seeding of body cavities whenever the neoplasm penetrates into a natural open field, such as the pleural cavity, by transport through the lymphatic system, or by dissemination through the peripheral blood.

In many cases, tumors can be identified based on their histologic characteristics using routine hematoxylin and eosin–stained tissue sections. However, malignant tumors of diverse origin might sometimes resemble each other. At times, inflammatory (reactive) lesions might be difficult to distinguish from cancerous ones. In addition, many cancer patients present with metastases. In some, the primary tumor site is obvious on the basis of clinical or radiologic features, but often, the origin of the tumor is ambiguous.

All cells in the human body, whether normal or malignant, express proteins (antigens) which, in many cases, are specific to that particular cell type. Immunocytochemical detection of tissue-specific or organ-specific antigens in the biopsy specimen of the primary tumor or metastatic deposit can lead to the identification of the tumor source. For example, prostate-specific antigen is a marker of tumors of the prostate gland. Thus, immunocytochemical stains are often employed as a means of improving diagnostic accuracy in cases where the diagnosis based on histologic appearance alone is uncertain.

3. Procedure

3.1. Sample Types

In routine hospital practice, when tissues are removed from a patient, they are either snap-frozen or placed in formaldehyde. Prior to immunocytochemical staining, formaldehyde-fixed

From: *Medical Biomethods Handbook*
Edited by: J. M. Walker and R. Rapley © Humana Press, Inc., Totowa, NJ

samples must be processed. Tissue processing includes dehydration, paraffin (wax) embedding, and sectioning (cutting) of tissues. Both paraffin-embedded and frozen tissues are routinely cut in 4- to 5-μm sections and mounted on microscopic glass slides; paraffin-embedded slides are then dried in a 60°C oven for at least 1 h. One slide is typically stained with hematoxylin and eosin for microscopic assessment of the tissue sample.

Cytologic samples can also be submitted to the laboratory for pathologic interpretation, and these include cervical smears (i.e., Pap smears), fine-needle aspirates (i.e., samples collected via the aspiration of cells from a lesion with a small bore needle), and body fluids. Processing of cytologic samples most often involves preparations such as blood smears, cytospins (a cell preparation system that uses centrifugal force to deposit cells on a microscope slide in a monolayer), or liquid-based, thin-layer preparations (ThinPreps™) ***(1–6)***. When the sample volume is sufficient, the fluid can be centrifuged to form a pellet, which can then be placed in formaldehyde and processed in a manner similar to that for a tissue sample. This is referred to as a cell block.

3.2. Fixation

Fixation prevents autolysis and preserves the antigens of excised tissues. As previously stated, the most common tissue fixative is 10% neutral-buffered formaldehyde. Other fixatives routinely used, particularly for frozen tissues and cytologic preparations, include acetone, alcohol, and paraformaldehyde ***(1,7)***. There is often a fine balance between good morphology and the need for antigen preservation. In fact, some fixatives might not be optimal for the immunocytochemical detection of certain antigens. Thus, when introducing a new antibody to the laboratory, a variety of fixatives might need to be examined to determine optimal immunoreactivity. Fixation time is also critical, as longer exposure to fixatives could decrease sensitivity in detecting certain cellular antigens. Formaldehyde fixation of tissue specimens is typically 12–24 h, whereas cytospin samples are fixed in acetone or alcohol for 5–10 min or formaldehyde/paraformaldehyde for 10–15 min ***(1)***.

3.3. Pretreatment

Many antigens are altered during formaldehyde fixation and processing, because crosslinks are formed to preserve protein structure. Antigenic determinants (epitopes) might be destroyed, denatured, or masked, which could diminish or nullify their detection. Thus, pretreatment steps are required in some instances to visualize certain antigens in formalin-fixed material. These pretreatment steps include proteolytic digestion (i.e., trypsin, pepsin, proteinase K) and heat-induced epitope retrieval, in which tissue sections are subjected to high-temperature heating in a buffer solution using microwave irradiation, autoclaving, pressure cookers, or steaming in an effort to unmask relevant antigens ***(8–13)***.

3.4. Avidin–Biotin Complex

The avidin–biotin complex (ABC) procedure is the most universally used immunocytochemical technique, with a high sensitivity and low background ***(14)***. In this method, the antigen in the specimen is recognized by a primary antibody. These primary antibodies are typically commercially manufactured and are most often anti-human antibodies produced in mice or rabbits. Next, a secondary antibody that is conjugated to biotin is applied and it attaches to the primary antibody. An avidin–biotin complex is then added. The affinity between the protein avidin and vitamin biotin is very strong. Avidin has four sites capable of binding to biotin. Three of the four binding sites are attached to biotin plus a peroxidase complex; the fourth site binds to the biotin attached to the secondary antibody. Thus, this procedure amplifies the level of staining intensity by increasing the number of potential binding sites. A chromogen is introduced that reacts with peroxide in the ABC reagent to produce a colored reaction product representing the presence of the antigen. The most commonly utilized chromogen, 3,3'-diaminobenzidene, produces a permanent brown stain and is often used with a hematoxylin (blue) counterstain. **Table 1** presents a simplified step-by-step immunocytochemical procedure.

3.5. Interpretation

When analyzing an immunostained specimen, deposits of colored chromogen indicate the presence of the antigen and represent positive staining. Depending on the cellular location of the antigen, the pattern of cell staining can be cytoplasmic, nuclear, or membrane, and focal or diffuse.

3.6. Quality Control

A positive tissue or cell preparation should be used as a control for every separate primary antibody in a run. This control is known to contain target elements of the antibody and is ideally prepared in the same manner as the patient sample from fixation through processing whenever possible.

Although every immunocytochemical run must include positive controls for each antibody, a negative control for each sample is imperative for assessing nonspecific background staining ***(15–19)***. For the negative control, a duplicate slide from the case under study is tested, and an irrelevant antibody or nonimmune serum from the sample animal (i.e., mouse or rabbit) is applied in place of the primary antibody ***(2,16)***. This slide is then used to assess nonspecific staining of the sample as a result of crossreactivity of immunogens other than those for which the primary antibody is testing. This positive staining, which is not a result of antigen–antibody binding, is termed *nonspecific background stain*. The most common cause of this is the attachment of protein to highly charged collagen in the specimen. In particular, tissues such as the kidney and liver and tumors arising from them contain a high level of endogenous biotin, which can result in background staining that is often difficult to interpret without appropriate pretreatment steps. Therefore, it is imperative to consistently include a negative specimen control on all test samples for background staining assessment. True immunoreactivity in the test samples must be of a greater intensity than that seen on the negative control sample.

4. Applications

4.1. Differential Diagnosis of Tumors

There are hundreds of antigens identifiable by the immunocytochemical technique, and a practical approach should be taken when using immunostains to help narrow the differential diagnosis.

When dealing with a neoplasm of unknown etiology, it is advisable to work up the case in a sequential manner, using discriminating antibodies first to categorize the malignancy as to the cell type of origin (*see* **Table 2**), followed by testing with confirmatory antibodies (*see* **Table 3**). There is often overlapping immunoreactivity in many tumors and expression of antigens in metastatic malignancies is often heterogenous, so it is best to use a number of antibodies to obtain the appropriate diagnosis. Monoclonal antibodies (produced in mice) provide the greatest specificity and produce the least amount of background staining compared to polyclonal antisera, which are usually made in rabbits.

The following is a brief description of the most commonly used markers in clinical laboratories.

4.1.1. Mesothelial Markers

Mesothelial cells are flattened epithelial cells that line the serous body cavities (i.e., abdomen, pericardium, thorax). Spontaneous accumulation of fluid in a serous cavity above the normally small amount is referred to as an effusion and might be caused by inflammation, fluid overload, or a malignant neoplasm. When an effusion develops in one of these cavities, mesothelial cells could exfoliate and large numbers might be found in the fluid. These cells can be benign, as in the case of a reactive/inflammatory effusion, or malignant. Malignant effusions can be the result of metastatic tumor cells or malignant mesothelial cells. Malignant mesotheliomas are tumors derived from these types of cells.

1. Epithelial membrane antigen (EMA). EMA belongs to a group of proteins known as human milk-fat globule membrane proteins. It is present in a wide variety of epithelia of both normal

Table 1
Simplified Immunocytochemical Staining Procedure

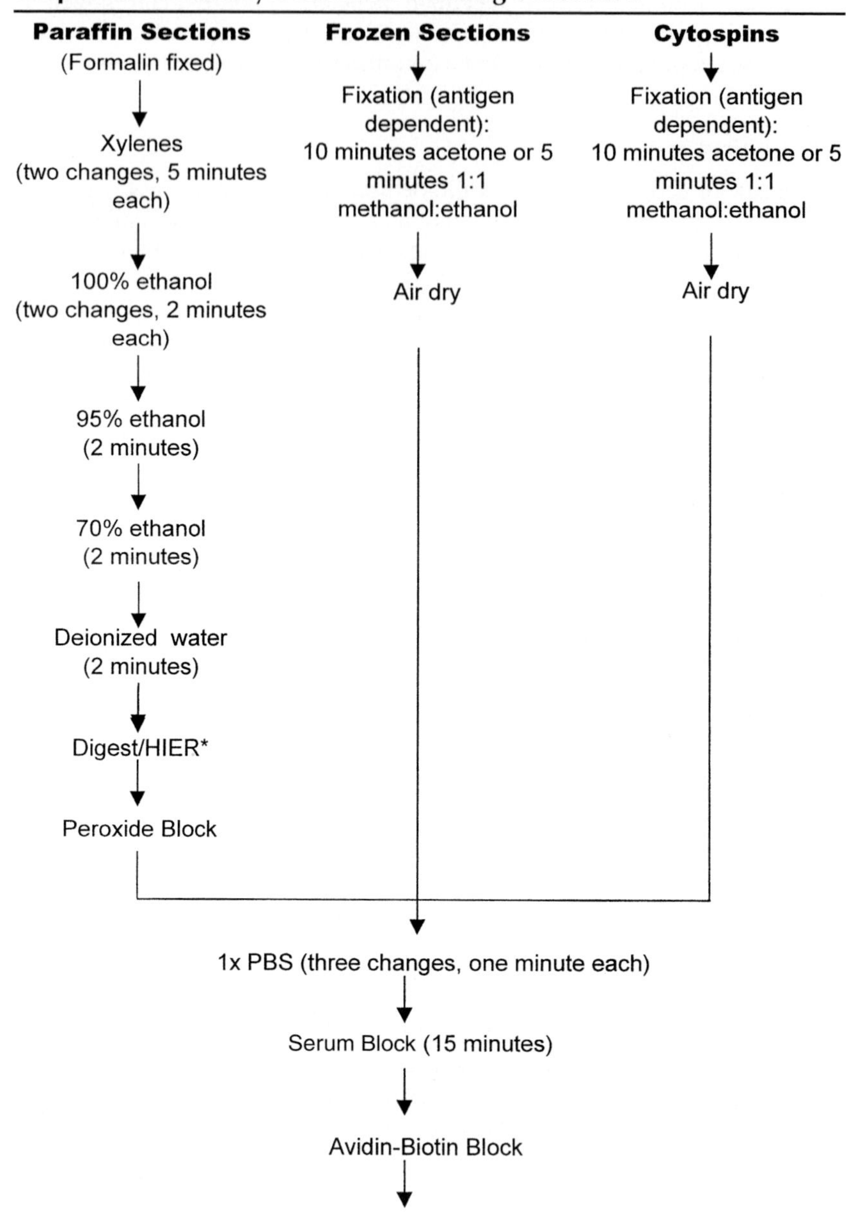

and neoplastic types *(20–25)*. A distinctive EMA pattern in malignant mesothelioma shows a characteristic "thick" cell membrane staining in the periphery of cell clusters, which highlights the long microvillus projections associated with these types of cells *(26)*.

2. Calretinin. Calretinin is a neuron-specific-calcium binding protein that is strongly expressed in neural tissues and certain non-neural cell types, including mesothelium *(27–34)*. Antibodies to this protein are strongly immunoreactive with malignant and benign mesothelial cells, as evident by both a cytoplasmic and nuclear type of staining pattern, described by one group as a "fried-egg" appearance *(35–38)*. It is of great value in differentiating effusions because of mesothelial cells origin (typically calretinin positive) from those caused by metastatic tumor cells (typically calretinin negative).

Table 1 (continued)

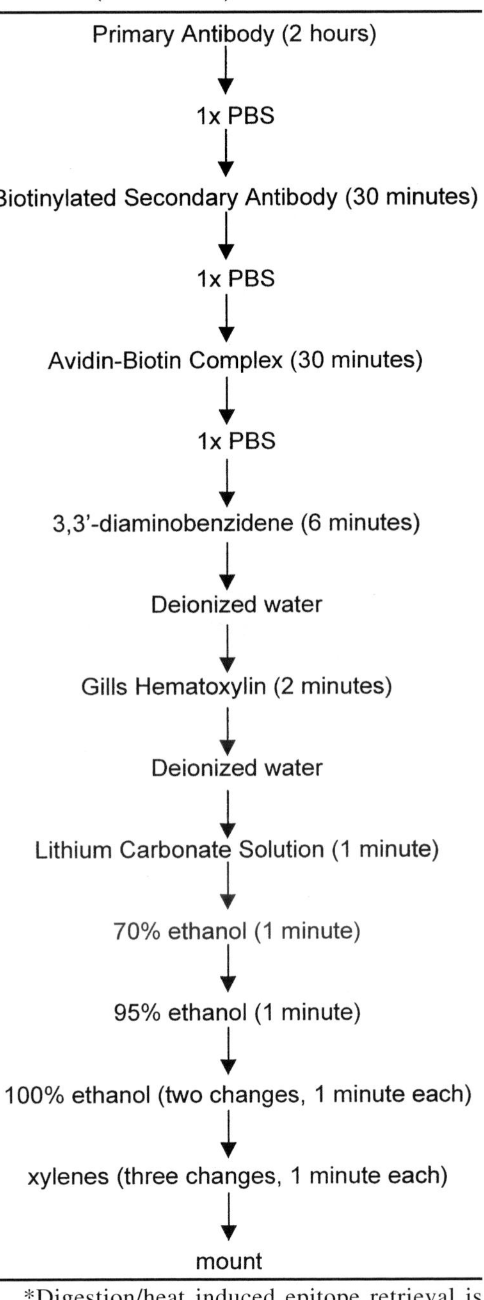

*Digestion/heat induced epitope retrieval is performed if indictated

PBS, phosphate buffered saline.

4.1.2. Carcinoma Markers

Carcinomas are malignant neoplasms derived from epithelial cells. Adenocarcinomas are a subtype of carcinoma in which the cells are in a glandular or glandlike pattern. Examples include breast, colon, ovarian, prostate and some lung cancers.

1. Cytokeratins (CKs). Cytokeratins are intermediate-sized (10-nm) monofilaments found in the cytoplasm of almost all true epithelial cell types and form part of the cytoskeletal complex in the epidermis and in most other epithelial tissues ***(20,39–41)***. The tissue-specific distribution

Table 2
Initial "Screening" Markers for Various Malignancies

Neoplasm Category	In general, immunoreactive with antibodies to
Carcinoma	Cytokeratin (CK)
Germ cell	Placental alkaline phosphatase
Lymphoid	Leukocyte common antigen (CD45)
Melanoma	S-100 protein
Mesothelioma	Calretinin
Neuroendocrine	Neuron-specific enolase
Sarcoma	Vimentin

of these intermediate filaments is generally well preserved in neoplasms and forms the basis for the use of antibodies to keratins ***(40,41)***. Antibodies to CK are used to identify normal and malignant cells of epithelial origin, with a cytoplasmic staining pattern. Most adenocarcinomas as well as mesotheliomas show intense immunoreactivity with CKs ***(33)***.

Cytokeratin-specific antibodies that are directed toward epitopes expressed in a limited number of epithelial cells can assist in identifying the site of primary or metastatic neoplasms. Antibodies to CK7 and CK20 are the most widely studied and used in this context, especially when used in combination ***(42)***.

2. Tumor-associated glycoprotein (B72.3). B72.3 is an antibody that recognizes a tumor-associated oncofetal antigen. It is present in a wide variety of adenocarcinomas, including those originating from lung, gastrointestinal tract, pancreas, breast, endometrium, and ovary ***(33,43)***. It is not expressed by leukemias, lymphomas, sarcomas, melanomas, or benign tumors ***(20,21,25,44,45)***. The antigen is present on normal secretory endometrium, but not on other normal tissues. Positive immunoreactivity is indicated by a membrane staining pattern.
3. Human epithelial antigen (BerEP4). BerEP4 is an antibody prepared by the immunization of mice with cells from the MCF7 breast cancer cell line and reacts with two glycoproteins present on the surface and in the cytoplasm of epithelial cells ***(20,21,45)***. BerEP4 stains the majority of adenocarcinomas, but in contrast to all other antiepithelial antibodies, the antibody does not label mesothelial cells ***(33,46,47)***. The antibody does not react with nerve, glial, muscle, or mesenchymal tissue, including lymphoid tissue ***(33)***. Positive reactions with BerEP4 can be evidenced by membrane staining, which can be in conjunction with cytoplasmic staining.
4. Carcinoembryonic antigen (CEA). CEA was first described as a specific marker for colon cancer ***(20,21,25,44,45,48,49)***. It is highly specific for adenocarcinoma cells and is typically negative in benign, reactive, and malignant mesothelial cells ***(21,33)***. There is little or no reactivity with nonmalignant tissues, except for granulocytes. Immunoreactivity with this antibody, indicated by a cytoplasmic staining pattern, is associated with colorectal carcinomas, as well as some lung, breast, and gastric cancers ***(33)***.
5. Thyroid transcription factor-1 (TTF-1). TTF-1 is a tissue-specific transcription factor expressed in the thyroid and lung ***(50)***. Immunoreactivity with this antibody, indicated by a nuclear staining pattern, is associated with thyroid and primary lung cancers ***(33,51)***.
6. Prostate-specific antigen (PSA)/prostatic acid phosphatase (PAP). PSA is a serine protease present in high levels in semen; PAP is an isoenzyme of acid phosphatase found in large amounts in the prostate and seminal fluid. Antibodies to PSA and PAP react with normal and neoplastic prostate tissue and are, thus, useful in the diagnosis of prostate cancer ***(33,52)***.
7. Cancer Antigen 125 (CA 125). CA 125 is a marker most strongly associated with epithelial ovarian tumors, although it is not a specific marker of ovarian cancer and can be detected immunocytochemically in breast and lung cancers as well as mesotheliomas ***(33,53–55)***. However, the morphological distinction between primary ovarian cancer and metastatic adenocarcinoma to this location can be difficult, and the addition of CA 125 to a panel of antibodies can be of benefit for that distinction.

Table 3
Confirmatory Markers for Various Malignancies

Carcinomas	
Breast	CK7+, CK20–, Ber-EP4+, B72.3+, ER+, PR+
Colon	CK7–, CK20+, CEA+, 19-9+, Ber-EP4+, B72.3+
Lung–non-small-cell	CK7+, CK20–, TTF-1+
Ovarian	CK7+, CK20+/–, B72.3+, CA125+
Prostate	CK7–, CK20–, PSA+, PAP+
Renal cell	CK7–, CK20–, EMA+, CD10+, vimentin+
Thyroid	CK 7+, CK 20–, TTF-1+
Lymphoid	
B-Cell lymphomas	CD20+, CD79a+
T-Cell lymphomas	CD3+
Hodgkin's	CD15+, CD30+
Lymphoma vs reactive lymph node	
Hyperplasia (reactive)	Bcl-2–, no clonality
Malignant lymphoma	Bcl-2+, shows either kappa or lambda clonality
Melanocytic	
Melanoma	HMB45+, MART-1+, tyrosinase+
Germ cell	
Choriocarcinoma	CK+, AFP-, HCG+
Embryonal	CK+, CD30+, CK7+, CK20–
Seminoma	CK–, AFP–, HCG–, CD30–
Yolk sac	CK+, AFP+, CD30–
Mesothelial	
Reactive	EMA+/–, BerEP4–, calretinin+
Malignant mesothelioma	EMA + (thick membrane), BerEP4–, calretinin+
Neuroendocrine	
Carcinoid	Synaptophysin+, chromogranin+, CK+, CK7–, CK20–
Small-cell-lung carcinoma	Synaptophysin+, CK+, CEA+, TTF-1+
Islet cell tumor of the pancreas	Synaptophysin+, chromogranin+, CK+, glucagon+, insulin+
Paraganglioma	Synaptophysin+, chromogranin+, S100+, CK–
Sarcoma	
Ewing's/primitive neural ectodermal tumor	S100–, CD99+
Gastrointestinal stromal tumor	CD34+, CD117+
Leiomyosarcoma	S100–, actin+, desmin+
Rhabdomyosarcoma	S100+/–, actin+, desmin+, myogenin+
Synovial	S100–, CK+, EMA+
Vascular	S100–, CD31+, CD34+, factor VIII+, CK+/–
Miscellaneous	
Infectious	CMV, SV40, HHV8, EBV LMP
Prognostic	Ki-67, p53, Her2/neu, ER/PR,CD117

4.1.3. Germ Cell Markers

Germ cell tumors are derived from the cells responsible for reproduction (i.e., ovum and spermatozoon). These tumors occur primarily in the ovary and testes and include seminomas, dysgerminomas, embryonal carcinomas, yolk sac tumors, choriocarcinomas, and teratomas.

Placental alkaline phosphatase (PLAP) is a membrane-bound enzyme normally produced by primordial germ cells and syncytiotrophoblasts of the placenta, and the detection of its expression has been useful in the diagnosis of germ cell tumors ***(33,56–58)***. α-fetoprotein (AFP), a major fetal serum protein normally produced by the fetal yolk sac, can be detected immunocytochemically in nonseminomatous germ cell tumors ***(33,59)***. Human chorionic gonadotropin (β-HCG) is synthesized by syncytiotrophoblastic cells in choriocarcinoma, and antibodies to HCG are useful in that diagnosis ***(33,60)***.

4.1.4. Malignant Melanoma Markers

Malignant melanomas are neoplasms derived from cells capable of forming the pigment melanin, and they arise most commonly in the skin or eye.

S-100 protein is a calcium-binding protein found in glial cells, melanocytes, chondrocytes, and adnexal glands of the skin ***(61–65)***. Many malignant neoplasms express S-100 protein, and practically all malignant melanomas show immunoreactivity in the nucleus and cytoplasm. Various antibodies can be used for the sensitive and specific diagnosis of malignant melanoma. These melanoma-associated antigens include the following: HMB45, an antibody directed against a premelanosome glycoprotein (gp100); MART-1 (melanoma antigen recognized by T-cells), a melanocyte-specific transmembrane protein; and tyrosinase, the rate-limiting enzyme in melanin synthesis ***(33,64–67)***. Interestingly, these melanoma-associated antigens are recognized by both CD4+ and CD8+ T-cells in a human leukocyte antigen-restricted fashion and can evoke powerful immune responses. Peptides derived from these antigens are currently being utilized as a target for T-cells in numerous melanoma immunotherapy protocols ***(68)***.

4.1.5. Lymphoid Markers

Lymphoid neoplasms are malignant proliferations of white blood cells. These tumors include leukemias (an abnormal proliferation of leukocytes in the blood and bone marrow) and lymphomas (neoplasms of lymphoid tissue, arising as discrete tissue masses). Lymphocytes are white blood cells produced by lymphoid tissue and are classified as either B-cell or T-cell. B-cell lymphomas include Burkitt, follicular, diffuse large cell and mantle cell lymphoma; T-cell lymphomas include adult T-cell leukemia/lymphoma and anaplastic large-cell lymphoma.

The cell surface antigens of leukocytes have been assigned cluster of differentiation (CD) designations. There are over 200 human leukocyte differentiation antigens. CD45 (leukocyte common antigen) is a membrane protein found on all leukocytes ***(69)***. Markers of B-cell lineage include CD19, CD20, CD22, and CD79a ***(70–73)***, whereas markers of T-cell lineage include CD3, CD4, CD5, CD7, and CD8 ***(74)***. Reed–Sternberg cells of Hodgkin's disease are marked by CD15 and CD30 ***(75–77)***.

4.1.6. Neuroendocrine Markers

Neuroendocrine tumors are neoplasms that share morphologic and biochemical features with cells of the neuroendocrine system. These tumors include carcinoids, small-cell carcinomas of the lung, and islet cell tumors of the pancreas.

Neuron-specific enolase (NSE) is found in a variety of normal and neoplastic neuroendocrine cells and predominates in the brain. Antibodies to NSE react with neuroendocrine as well as a variety of tumors, including melanomas, astrocytomas, glioblastomas, and gangliomas ***(33,78,79)***. Other neuroendocrine markers include chromogranin, a major constituent of neuroendocrine secretary granules ***(80)***, and synaptophysin, a calcium-binding glycoprotein that is the most abundant integral membrane protein constituent of synaptic vesicles of neurons ***(81,82)***. Islet cell tumors of the pancreas and carcinoids typically express both chromogranin and synaptophysin ***(33,83,84)***.

4.1.7. Sarcoma Markers

Sarcomas are connective tissue neoplasms formed by the proliferation of mesodermal cells and are usually highly malignant.

Vimentin is an intermediate filament protein that is ubiquitous in soft tissues, although not considered to be cell-type-specific ***(85)***. Desmin and actin are markers of striated and smooth muscle cells and are used in the diagnosis of tumors of muscle origin, such as rhabdomyosarcoma ***(33,86)***. Endothelial cell markers such as CD31 and factor VIII-related antigen are often expressed by vascular tumors such as angiosarcomas ***(33,87)***. Ewing's sarcoma is characterized by high MIC2 cell surface antigen expression, a glycoprotein detected by anti-CD99 ***(33,88,89)***.

4.2. Therapy-Linked Diagnostics/Prognostic Indications

The prognosis, or outcome of a disease, is dependent on many factors, including tumor type, histologic size and grade, and lymph node status. In addition, independent biologic markers have been identified that are often important prognostic determinants. All of these factors can be used to make management decisions, which direct the course of treatment for the cancer patient.

Many prognostic biologic markers can be identified immunocytochemically, and the results of these tests often have direct, immediate therapeutic implications for many types of cancer. The antigens are usually evaluated semiquantitatively, both in terms of the proportion of tumor cells showing positivity and the strength of the reaction. Hormone receptors such as estrogen and progesterone receptors on breast cancer cells can be assessed in this manner, and the presence of these receptors is a favorable prognostic indicator, because women whose tumors are estrogen and progesterone positive have a higher rate of response to endocrine therapy ***(90,91)***.

Detection of the overexpression of proto-oncogenes such as CD117 (c-Kit) and c-erb-B2 (HER-2/neu) also have prognostic value as predictors of those tumors more likely to respond to adjuvant chemotherapy. For example, antibodies to human epidermal growth factor receptor-2 (HER-2) are currently used immunocytochemically for the assessment of breast cancer patients for whom Herceptin® (trastuzumab) treatment is being considered ***(92–94)***. CD117 is an epitope on the extracellular domain of the transmembrane kinase receptor Kit, the product of the proto-oncogene c-kit, and can be detected on the cell surface of various malignant cell types. Imatinib mesylate (Gleevec®) is a tyrosinase kinase inhibitor that selectively acts on mutated Kit and is currently used as targeted molecular therapy in the treatment of chronic myelogenous leukemia and gastrointestinal stromal tumors ***(95–98)***.

4.3. Identification of Infectious Agents

Immunocytochemistry can be used for the diagnosis of infectious disease, most often in cases of viral infections such as hepatitis B, cytomegalovirus, human polyoma BK virus/simian virus 40 (SV-40), and herpesviruses (HHV8, EBV LMP). For example, Epstein–Barr virus latent membrane protein (EBV LMP) is a viral protein that serves as a marker of EBV infection, which, interestingly, is also associated with the etiology of Hodgkin's lymphoma ***(99)***. Another example of this use of immunocytochemistry is in cases of BK viral infections, which can occur in immunocompromised patients. In the kidney, this infection is associated with mononuclear interstitial inflammatory infiltrates and tubular atrophy, findings that can be difficult to distinguish from acute transplant rejection ***(100)***. Immunocytochemistry can provide rapid and sensitive results in this scenario to demonstrate BK virus infections using antibodies to SV-40 ***(101)***.

5. Conclusion

Therapeutic procedures can alter the course of disease and life-span of the cancer patient. In most instances, obtaining a correct diagnosis and appropriate treatment as early in the disease as possible improves the patient's prognosis and quality of life. In current practice, a diagnostic decision is based on tumor morphology, as well as clinical and radiologic findings. In cases where cell morphology is not definitive, the pathologist might base the diagnosis on whether a certain protein is expressed or not by the cells in question. The availability of specific monoclonal antibodies has greatly facilitated the identification of cell proteins and, thus, immunocytochemistry has become indispensable in the practice of diagnostic pathology. In fact, the combined results of morphologic and immunocytochemical examinations form the basis of most tumor classifications today.

Fig. 1. A 74-yr-old male presented with fluid (effusions) in his left and right pleural cavities. The clinical differential included reactive vs malignant etiology. The pleural fluid was tapped for diagnostic evaluation, and cytologic analysis showed an atypical mesothelial cell proliferation. Formalin-fixed, paraffin-embedded cell block sections prepared from the pleural fluid were stained with antiepithelial membrane antigen (EMA). A strong, thick membrane staining pattern is seen in the atypical cells. This distinctive pattern is often characteristic of malignant mesothelial cells and highlights their long microvillus projections. Diagnosis: malignant mesothelioma.

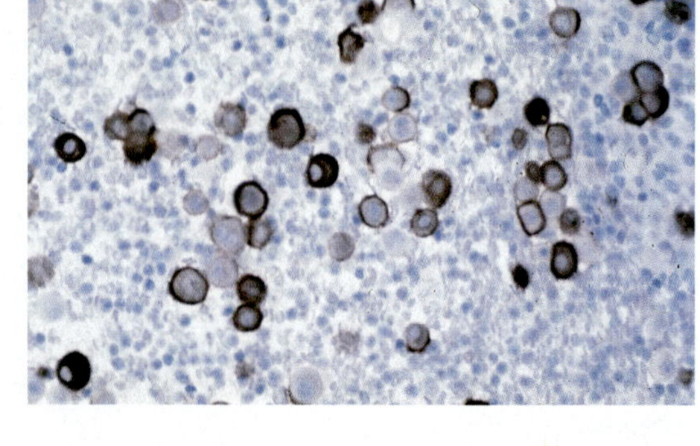

Fig. 2. A 60-yr-old female newly diagnosed with ovarian cancer and undergoing treatment developed a moderate right pleural effusion. The clinical differential included benign/reactive vs malignant etiology. A tap of the effusion showed red blood cells, reactive mesothelial cells, lymphocytes, and numerous atypical cells. Formalin-fixed, paraffin-embedded cell block sections were stained with anticalretinin. A nuclear and cytoplasmic staining pattern is shown in the reactive mesothelial cells whereas the atypical cells are negative. This case highlights the role of calretinin immunoreactivity in discriminating mesothelial cells (typically positive) from adenocarcinoma cells (typically negative) using antibodies to calretinin. Diagnosis: adenocarcinoma with ovarian primary.

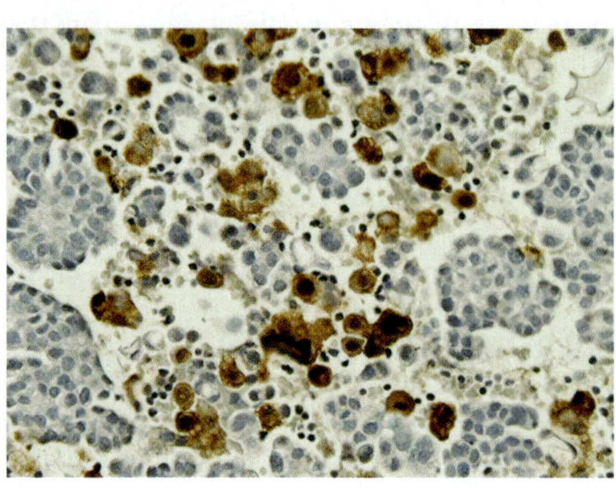

Fig. 3. A 72-yr-old female with a history of colon cancer presented with a pleural effusion. The clinical differential included infectious vs metastatic disease. The fluid was tapped and showed numerous atypical cells. Formalin-fixed, paraffin-embedded cell block sections were stained with anti-B72.3. B72.3 is an antigen expressed by a wide variety of adenocarcinomas, including 85–95% of colon cancers. A membrane staining pattern is seen in the atypical cells. Diagnosis: colorectal carcinoma with metastatic pleural effusion.

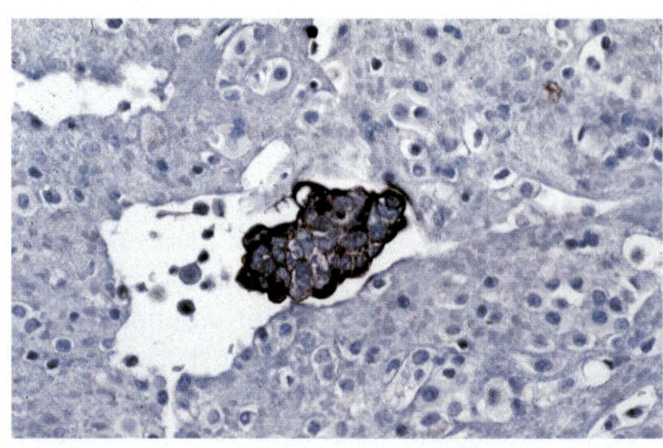

Fig. 4. A 38-yr-old female presented with a neck mass. The clinical differential included reactive vs malignant etiology. A fine-needle aspirate of the mass was performed, and cytologic examination showed red blood cells, lymphocytes, neutrophils, and atypical cells singly and in clusters. An alcohol-fixed cytospin prepared from the aspirate was stained with anti-cytokeratin. Positive staining with cytokeratin indicates an epithelial origin of the atypical cells. A cytoplasmic staining pattern is seen in the atypical cells. Diagnosis: poorly differentiated carcinoma.

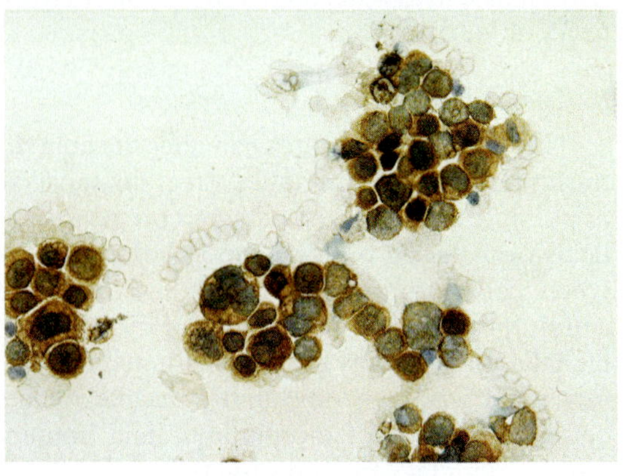

Note: All figures in this chapter are stained with the diaminobenzidene chromogen (brown stain/positive immunoreactivity) and hematoxylin (blue counterstain).

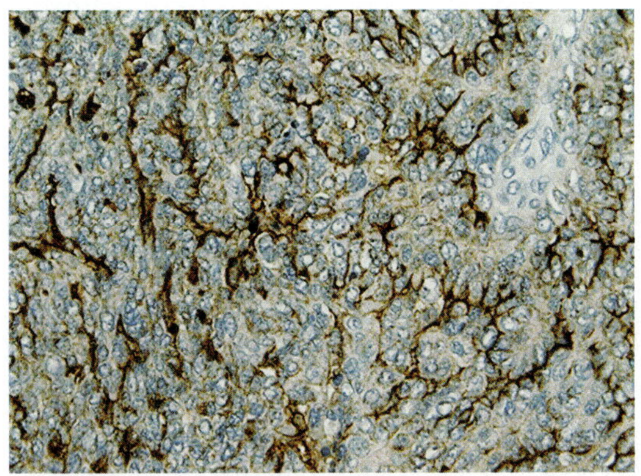

Fig. 5. A 70-yr-old female presented with a left adnexal mass. The differential diagnosis included colon vs ovarian cancer. The 14-cm mass was removed and showed histologic features characteristic of serous carcinoma of the ovary. Formalin-fixed, paraffin-embedded tissue sections were stained with anti-CA125. The malignant cells show a membrane staining pattern. This case highlights the role of CA125 immunoreactivity in discriminating between primary and metastatic carcinomas of the ovary (typically positive) and intestine (typically negative). Diagnosis: serous carcinoma of the ovary.

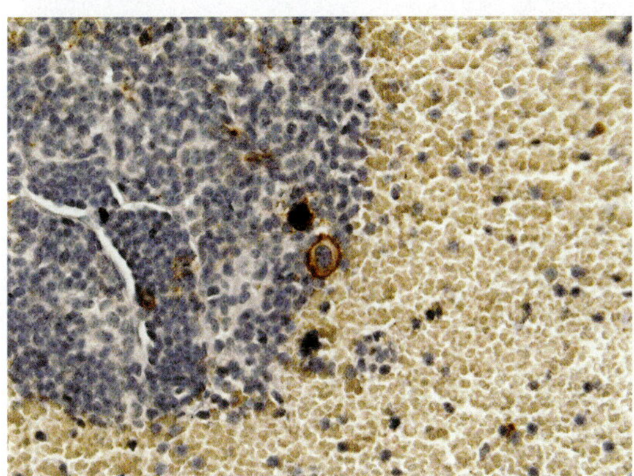

Fig. 6. A 28-yr-old male presented with multiple thyroid nodules and enlarged palpable cervical lymph nodes; he had no other previous symptoms. The clinical differential included benign goiter/reactive lymph node vs malignant thyroid/reactive lymph node vs malignant lymphoma. A fine-needle aspirate of the thyroid showed numerous lymphocytes and red blood cells, as well as large atypical cells. Formalin-fixed, paraffin-embedded cell block sections were stained with anti-CD30. CD30 is an antigen expressed on the Reed–Sternberg cells of Hodgkin's lymphoma. A strong membrane staining pattern is seen in this atypical cell. Diagnosis: Classical Hodgkin lymphoma.

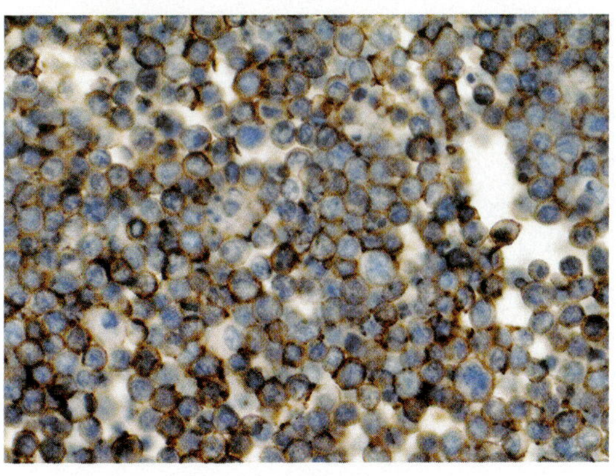

Fig. 7. A 73-yr-old male presented with enlargement of the testes. A right orchiectomy was performed and a 10-cm mass was removed. The differential diagnosis based on histologic examination included seminoma, melanoma, and carcinoma. Formalin-fixed, paraffin-embedded tissue sections were stained with anti-CD20. CD20 is an antigen acquired late in the pre-B-cell stage of maturation and remains on cells throughout most of their differentiation; it is present in almost all mature B-cell lymphomas. The atypical cells show a strong membrane staining pattern. Diagnosis: malignant lymphoma, B-cell type. *Note*: Malignant lymphoma is the most common testicular tumor in older individuals.

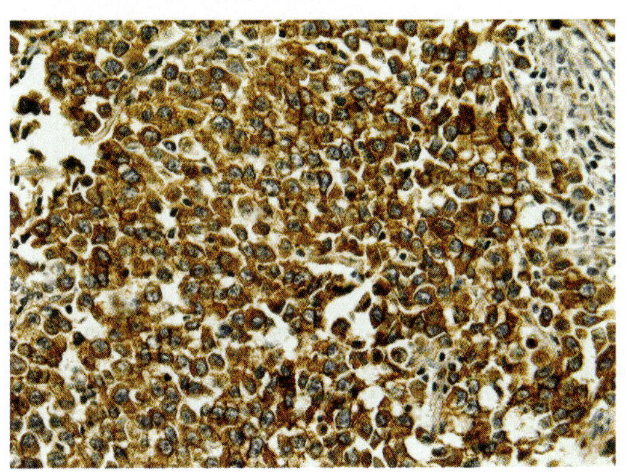

Fig. 8. A 32-yr-old male presented with acute onset of scrotal pain and swelling. An orchiectomy of a right testicular mass was performed. Based on the gross and histologic features, the differential diagnosis included germ cell tumor, lymphoma, and metastatic tumors such as melanoma and poorly differentiated carcinoma. Formalin-fixed, paraffin-embedded tissue sections were stained with antiplacental alkaline phosphatase (PLAP). PLAP is an oncofetal antigen found to be expressed by malignant tumors of germ cell origin. A strong cytoplasmic staining pattern is seen in the atypical cells. Diagnosis: germ cell tumor/seminoma. *Note*: Seminoma is the most common form of testicular germ cell tumor in younger individuals.

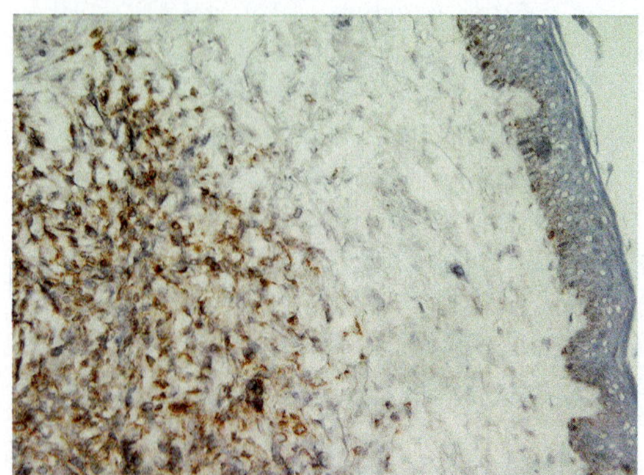

Fig. 9. A 45-yr-old male with a history of malignant melanoma presented with a suspicious nodule on the back. A skin biopsy was performed, and acetone-fixed frozen-tissue sections were stained with anti-HMB45. HMB45 is an antibody that reacts with an antigen present in premelanosomes and is a sensitive marker for melanoma. Cytoplasmic staining is seen in both the malignant cells in the dermis as well as the normal melanocytes in the epidermis of the skin. Diagnosis: metastatic malignant melanoma.

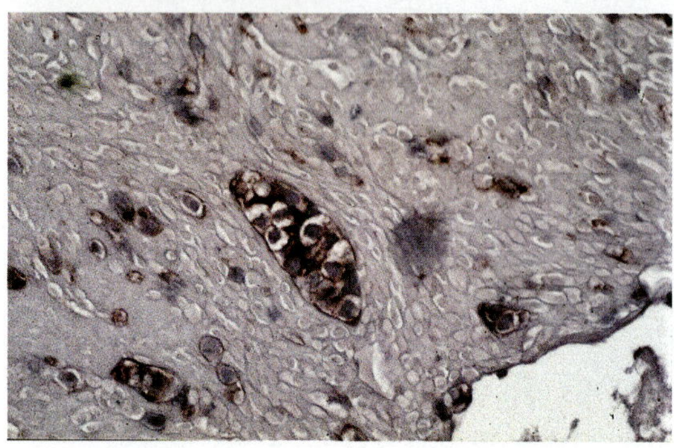

Fig. 10. A 20-yr-old female with a history of Ewing's sarcoma (primary site: left pelvis) presented with a new right inguinal nodule. The clinical differential included benign/reactive vs metastatic etiology. A fine-needle aspirate of the nodule showed blood elements and numerous atypical cells. Formalin-fixed, paraffin-embedded cell block sections prepared from the aspirate were stained with anti-CD99. CD99 is a cell surface antigen expressed in the majority of Ewing's sarcomas. A membrane and cytoplasmic staining pattern is seen in the atypical cells. Diagnosis: progressive Ewing's sarcoma/primitive neuroectodermal tumor (PNET).

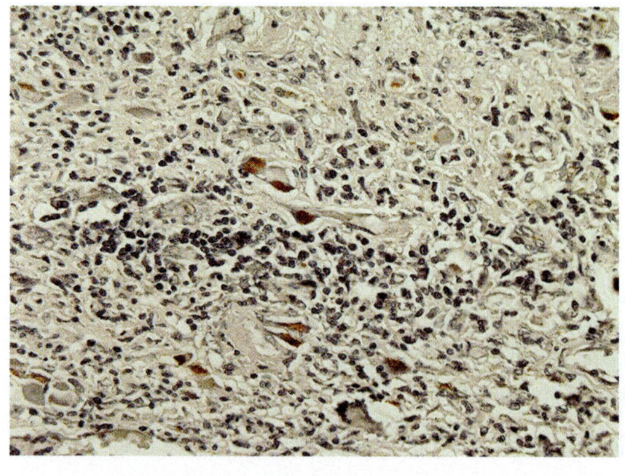

Fig. 11. A 43-yr-old Asian female presented with a 3-wk history of fever, abdominal pain, and diarrhea. A colonoscopy was done and revealed findings suggestive of carcinoma of the colon. A biopsy was then performed, and histologic examination showed scattered atypical cells as well as an inflammatory infiltrate. Formalin-fixed, paraffin-embedded tissue sections from the colon were stained with anticytomegalovirus (CMV). CMV is an important opportunistic pathogen in immunocompromised patients, and CMV colitis is not uncommon in patients with acquired immune deficiency syndrome (AIDS). The CMV inclusions in the large atypical cells scattered throughout the specimen show a nuclear staining pattern. Diagnosis: CMV colitis.

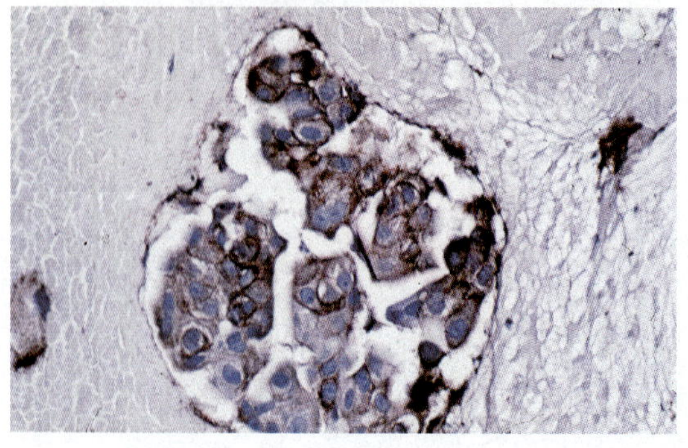

Fig. 12. A 42-yr-old female with a history of breast cancer, who 50 d after bone marrow transplant, presented with a new lesion below an existing wound (which was at the previous surgical resection site). The clinical differential included metastatic vs infectious etiology. A fine-needle aspirate of the right shoulder lesion showed blood elements and atypical cells occurring in clusters. Formalin-fixed, paraffin-embedded cell block sections prepared from the aspirate were stained with anti-HER-2/neu. The overexpression of this oncoprotein correlates with a better response to Herceptin therapy. A full membrane staining pattern is seen in the atypical cells. Diagnosis: metastatic breast cancer.

References

1. Abati, A., Fetsch, P., and Filie, A. (1998) If cells could talk. The application of new techniques to cytopathology. *Clin. Lab. Med.* **18,** 561–583.
2. Nadji, M., Ganjei, P., and Morales, A. (1994) Immunocytochemistry in contemporary cytology: the technique and its application. *Lab. Med.* **25,** 502–508.
3. Leong, A. S-Y, Suthipintawong, C., and Vinyuvat, S. (1999) Immunostaining of cytologic preparations: a review of technical problems. *Appl. Immunohistochem.* **7,** 214–220.
4. Leung, S. W. and Bedard, Y. C. (1996) Immunocytochemical staining on ThinPrep processed smears. *Mod. Pathol.* **9,** 304–306.
5. Fetsch, P.A., Simsir, A., Brosky, K., and Abati, A. (2002) Comparison of three commonly used cytologic preparations in effusion cytology. *Diagn. Cytopathol.* **26,** 61–66.
6. Han, A. C., Filstein, M. R., Hunt, J. V., Soler, A. P., Knudsen, K. A., and Salazar, H. (1999) N-cadherin distinguishes pleural mesotheliomas from lung adenocarcinomas: a ThinPrep immunocytochemical study. *Cancer* **87,** 83–86.
7. Cheepsumon, S., Leong, A. S.-Y., and Vinyuvat, S. (1996) Immunostaining of cell preparations: a comparative evaluation of common fixatives and protocols. *Diagn. Cytopathol.* **15,** 167–174.
8. Fan, Z., Clark, V., and Nagle, R. B. (1997) An evaluation of enzymatic and heat epitope retrieval methods for the immunohistochemical staining of the intermediate filaments. *Appl. Immunohistochem.* **5,** 49–58.
9. Shi, S.-R., Cote, R. J., Chaiwun, B., et al. (1998) Standardization of immunohistochemistry based on antigen retrieval technique for routine formalin-fixed tissue sections. *Appl. Immunohistochem.* **6,** 89–96.
10. Taylor, C. K., Shi, S.-R., and Cote, R. J. (1996) Antigen retrieval for immunohistochemistry: status and need for greater standardization. *Appl. Immunohistochem.* **4,** 144–166.
11. Riera, J. R., Astengo-Osuna, C., Longmate, J. A., and Battifora, H. (1997) The immunohistochemical diagnostic panel for epithelial mesothelioma: a reevaluation after heat-induced epitope retrieval. *Am. J. Surg. Pathol.* **21,** 1409–1419.
12. Shi, S. R., Key, M. E., and Kalra, K. L. (1991) Antigen retrieval in formalin-fixed paraffin-embedded tissues: an enhancement method for immunohistochemical staining based on microwave oven heating of tissue sections. *J. Histochem. Cytochem.* **39,** 741–748.
13. Miller, R. T., Swanson, P. E., and Wick, M. R. (2000) Fixation and epitope retrieval in diagnostic immunohistochemistry: a concise review with practical considerations. *Appl. Immunohistochem. Mol. Morphol.* **8,** 228–235.
14. Hsu, S. M., Raine, L., and Fanger, H. (1981) Use of avidin-biotin-peroxidase complex (ABC) in immunoperoxidase techniques: a comparison between ABC and unlabeled antibody (PAP) procedures. *J. Histochem. Cytochem.* **29,** 577–580.
15. Department of Health and Human Services, Health Care Financing Administration (1992) Clinical Laboratory Improvement Amendments of 1988; Final Rule. *Fed. Reg.* **57,** 7001–7288.
16. College of American Pathologists Laboratory Accreditation Program (2002) Standards for Laboratory Accreditation, Commission of Laboratory Accreditation Inspection Checklist, Sections 1 and 8.
17. Fetsch, P. A. and Abati, A. (1999) Overview of the clinical immunohistochemistry laboratory: regulations and troubleshooting guidelines, in *Immunocytochemical Methods and Protocols* (Javois, L. C., ed.), Humana, Totowa, NJ, pp. 405–414.
18. O'Leary, T. J. (2001) Standardization in immunohistochemistry. *Appl. Immunohistochem. Mol. Morphol.* **9,** 3–8.
19. Taylor, C. R., Shi, S.-R., Barr, N. J., and Wu, N. (2001) Techniques of immunohistochemistry: principles, pitfalls and standardization, in *Diagnostic Immunohistochemistry* (Dabbs, D. J., ed), Churchill Livingstone, Philadelphia, PA, pp. 3–43.
20. Betta, P.-G., Andrion, A., Donna, A., et al. (1997) Malignant mesothelioma of the pleura: the reproducibility of the immunohistological diagnosis. *Pathol. Res. Pract.* **193,** 759–765.
21. Shield, P. W., Callan, J. J., and Devine, P. L. (1994) Markers for metastatic adenocarcinoma in serous effusion specimens. *Diagn. Cytopathol.* **11,** 237–245.
22. Dejmek, A. and Hjerpe, A. (2000) Reactivity of six antibodies in effusions of mesothelioma, adenocarcinoma and mesotheliosis: stepwise logistic regression analysis. *Cytopathology* **11,** 8–17.
23. Dejmek, A. and Hjerpe, A. (1994) Immunohistochemical reactivity in mesothelioma and adenocarcinoma: a stepwise logistic regression analysis. *APMIS* **102,** 255–264.
24. Dejmek, A., Brockstedt, U., and Hjerpe, A. (1997) Optimization of a battery using nine immunocytochemical variables for distinguishing between epithelial mesothelioma and adenocarcinoma. *APMIS* **105,** 889–894.

25. Donna, A., Betta, P.-G., Bellingeri, D., Tallarida, F., Pavesi, M., and Pastormerlo, M. (1992) Cytologic diagnosis of malignant mesothelioma in serous effusions using an antimesothelial-cell antibody. *Diagn. Cytopathol.* **18,** 361–365.
26. Leong, A. and Vernon-Roberts, E. (1994) The immunohistochemistry of malignant mesothelioma. *Pathol. Ann.* **29(Pt. 2),** 157–179.
27. Kitazume, H., Kitamura, K., Mukai, K., et al. (2000) Cytologic differential diagnosis among reactive mesothelial cells, malignant mesothelioma, and adenocarcinoma: utility of combined E-cadherin and calretinin immunostaining. *Cancer* **90,** 55–60.
28. Ordonez, N. (1998) Value of calretinin immunostaining in differentiating epithelial mesothelioma from lung adenocarcinoma. *Mod. Pathol.* **11,** 929–933.
29. Ordonez, N. G. and Mackay, B. (1999) Glycogen-rich mesothelioma. *Ultrastruct. Pathol.* **23,** 401–406.
30. Oates, J. and Edwards, C. (2000) HBME-1, MOC-31, WT1, and calretinin: an assessment of recently described markers for mesothelioma and adenocarcinoma. *Histopathology* **36,** 341–347.
31. Cury, P. M., Butcher, D. N., Fisher, C., Corrin, B., and Nicholson, A. G. (2000) Value of the mesothelium-associated antibodies thrombomodulin, cytokeratin 5/6, calretinin, and CD44H in distinguishing epithelioid pleural mesothelioma from adenocarcinoma metastatic to the pleura. *Mod. Pathol.* **13,** 107–112.
32. Doglioni, C., Tos, A. P., Laurino, L., et al. (1996) Calretinin: a novel immunocytochemical marker for mesothelioma. *Am. J. Surg. Pathol.* **20,** 1037–1046.
33. Frisman, D. (2003) ImmunoQuery, an immunohistology query system. www.ipox.org.
34. Simsir, A., Fetsch, P. A., and Abati, A. (2001) Calretinin immunostaining in benign and malignant pleural effusions. *Diagn. Cytopathol.* **24,** 149–152.
35. Chhieng, D. C., Yee, H., Schaefer, D., et al. (2000) Calretinin staining pattern aids in the differentiation of mesothelioma from adenocarcinoma in serous effusions. *Cancer* **90,** 194–200.
36. Fetsch, P. A., Simsir, A., and Abati, A. (2001) Comparison of antibodies to HBME-1 and calretinin for the detection of mesothelial cells in effusion cytology. *Diagn. Cytopathol.* **25,** 158–161.
37. Wieczorek, T. J. and Krane, J. F. (2000) Diagnostic utility of calretinin immunohistochemistry in cytologic cell block preparations. *Cancer* **90,** 312–319.
38. Nagel, H., Hemmerlein, B., Ruschenburg, I., Huppe, K., and Droese, M. (1998) The value of anticalretinin antibody in the differential diagnosis of normal and reactive mesothelia versus metastatic tumors in effusion cytology. *Pathol. Res. Pract.* **194,** 759–764.
39. Ordonez, N. G. (1989) The immunohistochemical diagnosis of mesothelioma: differentiation of mesothelioma and lung adenocarcinoma. *Am. J. Surg. Pathol.* **13,** 276–291.
40. Cooper, D., Schermer, A., and Sun, T.-T. (1985) Biology of disease-classification of human epithelia and their neoplasms using monoclonal antibodies to keratins: strategies, applications, and limitations. *Lab. Invest.* **52,** 243–256.
41. Moll, R., Franke, W., Schiller, D., Geiger, B., and Krepler, R. (1982) The catalog of human cytokeratins: patterns of expression in normal epithelia, tumors and cultured cells. *Cell* **31,** 11–24.
42. Chu, P., Wu, E., and Weiss, L. (2000) Cytokeratin 7 and cytokeratin 20 expression in epithelial neoplasms: a survey of 435 cases. *Mod. Pathol.* **13,** 962–972.
43. Lauritzen, A. F. (1989) Diagnostic value of monoclonal antibody B72.3 in detecting adenocarcinoma cells in serous effusions. *APMIS* **97,** 761–766.
44. Kho-Duffin, J., Tao, L.-C., Cramer, H., Catellier, M. J., Irons, D., and Ng, P. (1999) Cytologic diagnosis of malignant mesothelioma, with particular emphasis on the epithelial noncohesive cell type. *Diagn. Cytopathol.* **20,** 57–62.
45. Davidson, B., Risberg, B., Kristensen, G., et al. (1999) Detection of cancer cells in effusions from patients diagnosed with gynaecological malignancies: evaluation of five epithelial markers. *Virchows Arch.* **435,** 43–39.
46. Diaz-Arias, A. A., Loy, T. S., Bickel, J. T., and Chapman, R. K. (1993) Utility of BER-EP4 in the diagnosis of adenocarcinoma in effusions: an immunocytochemical study of 232 cases. *Diagn. Cytopathol.* **9,** 516–521.
47. Chenard-Neu, M. P., Kabou, A., Mechine, A., Brolly, F., Orion, B., and Bellocq, J. P. (1998) Immunohistochemistry in the differential diagnosis of mesothelioma and adenocarcinoma. Evaluation of 5 new antibodies and 6 traditional antibodies. *Ann. Pathol.* **18,** 460–465.
48. Miedouge, M., Rouzaud, P., Salama, G., et al. (1999) Evaluation of seven tumour markers in pleural fluid for the diagnosis of malignant effusions. *Br. J. Cancer* **81,** 1059–1065.

49. Stoetzer, O. J., Munker, R., Darsow, M., and Wilmanns, W. (1999) P53-immunoreactive cells in benign and malignant effusions: diagnostic value using a panel of monoclonal antibodies and comparison with CEA-staining. *Oncol. Rep.* **6,** 433–436.
50. Holzinger, A., Dingle, S., Bejarano, P. A., et al. (1996) Monoclonal antibody to thyroid transcription factor-1: production, characterization, and usefulness in tumor diagnosis. *Hybridoma* **15,** 49–53.
51. Ordonez, N. (2000) Value of thyroid transcription factor-1, E-cadherin, BG8, WT1, and CD44S immunostaining in distinguishing epithelial pleural mesothelioma from pulmonary and nonpulmonary adenocarcinoma. *Am. J. Surg. Pathol.* **24,** 598–606.
52. Sauvageot, J. and Epstein, J. I. (1998) Immunoreactivity for prostate-specific antigen and prostatic acid phosphatase in adenocarcinoma of the prostate: relation to progression following radical prostatectomy. *Prostate* **34,** 29–33.
53. Jacobs, I. and Bast R. C. (1989) The Ca 125 tumour associated antigen: a review of the literature. *Hum. Reprod.* **4,** 1–12.
54. McCluggage, W. G. (2000) Recent advances in immunohistochemistry in the diagnosis of ovarian neoplasms. *J. Clin. Pathol.* **53,** 327–334.
55. Nap, M. (1998) Immunohistochemistry of CA 125. Unusual expression in normal tissues, distribution in the human fetus and questions around its application in diagnostic pathology. *Int. J. Biol. Markers* **13,** 210–215.
56. Fishman, W. H., Inglis, I., Stolbach, L. L., and Krant, M. J. (1968) A serum alkaline phosphatase isoenzyme of human neoplastic cell origin. *Cancer Res.* **28,** 150–154.
57. Manivel, J. C., Jessurun, J., Wick, M. R., and Dehner, L. P. (1987) Placental alkaline phosphatase immunoreactivity in testicular germ cell neoplasms. *Am. J. Surg. Pathol.* **11,** 21–29.
58. Koshida, K. and Wahren, B. (1990) Placental-like alkaline phosphatase in seminoma. *Urol. Res.* **18,** 87–92.
59. Javadpour, N. (1980) The role of biologic tumor markers in testicular cancer. *Cancer* **45(7 Suppl.),** 1755–1761.
60. Bosl, G. J. and Chaganti, R. S. (1994) The use of tumor markers in germ cell malignancies. *Hematol. Oncol. Clin. North Am.* **8,** 573–587.
61. Cochran, A. J. and Wen, D. R. (1985) S-100 protein as a marker for melanocytic and other tumours. *Pathology* **17,** 340–345.
62. Kahn, H. J., Marks, A., Thom, H., and Baumal, R. (1983) Role of antibody to S100 protein in diagnostic pathology. *Am. J. Clin. Pathol.* **79,** 341–347.
63. Takahashi, K., Isobe, T., Ohtsuki, Y., Akagi, T., Sonobe, H., and Okuyama, T. (1984) Immunohistochemical study on the distribution of alpha and beta subunits of S-100 protein in human neoplasm and normal tissues. *Virchows Arch. B.* **45,** 385–396.
64. Wick, M. R., Swanson, P. E., and Rocamora, A. (1988) Recognition of malignant melanoma by monoclonal antibody HMB-45. An immunohistochemical study of 200 paraffin-embedded cutaneous tumors. *J. Cutan. Pathol.* **15,** 201–207.
65. de Vries, T.J., Smeets, M., de Graaf, R., et al. (2001) Expression of gp100, MART-1, tyrosinase, and S100 in paraffin-embedded primary melanomas and locoregional, lymph node, and visceral metastases: implications for diagnosis and immunotherapy. A study conducted by the EORTC Melanoma Cooperative Group. *J. Pathol.* **193,** 13–20.
66. Boyle, J. L., Haupt, H. M., Stern, J. B., and Multhaupt, H. A. (2002) Tyrosinase expression in malignant melanoma, desmoplastic melanoma, and peripheral nerve tumors. *Arch. Pathol. Lab. Med.* **126,** 816–22.
67. Orchard, G. E. (2000) Comparison of immunohistochemical labelling of melanocyte differentiation antibodies melan-A, tyrosinase and HMB 45 with NKIC3 and S100 protein in the evaluation of benign naevi and malignant melanoma. *Histochem. J.* **32,** 475–481.
68. Fetsch, P. A., Marincola, F. M., Filie, A., Hijazi, Y., Kleiner, D., and Abati, A. (1999) Melanoma-associated antigen recognized by T-cells (MART-1): the advent of a preferred immunocytochemical antibody for the diagnosis of metastatic malignant melanoma in fine needle aspirations. *Cancer* **87,** 37–42.
69. Kurtin, P. J. and Pinkus, G. S. (1985) Leukocyte common antigen—a diagnostic discriminant between hematopoietic and nonhematopoietic neoplasms in paraffin sections using monoclonal antibodies: correlation with immunologic studies and ultrastructural localization. *Hum. Pathol.* **16,** 353–365.

70. Brunning, R. D., Borowitz, M., Matutes, E., et al. (2001) Precursor B lymphoblastic leukaemia/lymphoblastic lymphoma (precursor B-cell acute lymphoblastic leukaemia), in *World Health Organization Classification of Tumours. Pathology and Genetics of Tumours of Haematopoietic and Lymphoid Tissues* (Jaffe, E. S., Harris, N. L., Stein, H., and Vardiman, J. W., eds.), IARC, Lyon, pp. 111–114.
71. Chu, P. G., Chang, K. L., Arber, D. A., and Weiss, L. M. (2000) Immunophenotyping of hematopoietic neoplasms. *Semin. Diagn. Pathol.* **17,** 236–256.
72. Gatter, K. C. and Warnke, R. A. (2001) Diffuse large B-cell lymphoma, in *World Health Organization Classification of Tumours. Pathology and Genetics of Tumours of Haematopoietic and Lymphoid Tissues* (Jaffe, E. S., Harris, N. L., Stein, H., and Vardiman, J. W., eds.), IARC, Lyon, pp. 171–174.
73. Mayall, F., Dray, M., Stanley, D., Harrison, B., and Allen R. (2000) Immunoflow cytometry and cell block immunohistochemistry in the FNA diagnosis of lymphoma: a review of 73 consecutive cases. *J. Clin. Pathol.* **53,** 451–457.
74. Kikuchi, M., Jaffe, E. S., and Ralfkiaer, E. (2001) Adult T-cell leukaemia/lymphoma, in *World Health Organization Classification of Tumours. Pathology and Genetics of Tumours of Haematopoietic and Lymphoid Tissues* (Jaffe, E. S., Harris, N. L., Stein, H., and Vardiman, J. W., eds.), IARC, Lyon, pp. 200–203.
75. Stein, H., Delsol, G., Pileri, S., et al. (2001) Classical Hodgkin lymphoma, in *World Health Organization Classification of Tumours. Pathology and Genetics of Tumours of Haematopoietic and Lymphoid Tissues* (Jaffe, E. S., Harris, N. L., Stein, H., and Vardiman, J. W., eds.), IARC, Lyon, pp. 244–253.
76. Chittal, S. M., Caveriviere, P., Schwarting, R., et al. (1988) Monoclonal antibodies in the diagnosis of Hodgkin's disease. The search for a rational panel. *Am. J. Clin. Pathol.* **12,** 9–21.
77. Arber, D. A. and Weiss, L. M. (1993) CD15: a review. *Appl. Immunohistochem.* **1,** 17–30.
78. Cras, P., Federsppiel, S. S., Gheuens, J., Martin, J. J., and Lowenthal, A. (1986) Demonstration of neuron-specific enolase in nonneuronal tumors using a specific monoclonal antibody. *Ann. Neurol.* **20,** 106–107.
79. Cras, P., Martin, J. J., and Gheuens, J. (1988) Gamma-enolase and glial fibrillary acidic protein in nervous system tumors. An immunohistochemical study using specific monoclonal antibodies. *Acta Neuropathol.* **75,** 377–384.
80. Wilson, B. S. and Lloyd, R. V. (1984) Detection of chromogranin in neuroendocrine cells with a monoclonal antibody. *Am. J. Pathol.* **115,** 458–468.
81. Wiedenmann, B., Kuhn, C., Schwechheimer, K., et al. (1987) Synaptophysin identified in metastases of neuroendocrine tumors by immunocytochemistry and immunoblotting. *Am. J. Clin. Pathol.* **88,** 560–569.
82. Wiedenmann, B. and Franke, W. W. (1985) Identification and localization of synaptophysin, an integral membrane glycoprotein of Mr 38,000 characteristic of presynaptic vesicles. *Cell* **41,** 1017–1028.
83. Al-Khafaji, B., Noffsinger, A. E., Miller, M. A., DeVoe, G., Stemmermann, G. N., and Fenoglio-Preiser, C. (1998) Immunohistologic analysis of gastrointestinal and pulmonary carcinoid tumors. *Hum. Pathol.* **29,** 992–999.
84. Goldstein, N. S. and Silverman, J. F. (2001) Immunohistochemistry of the gastrointestinal tract, pancreas, bile ducts, gallbladder and liver, in *Diagnostic Immunohistochemistry* (Dabbs, D. J., ed.), Churchill Livingstone, Philadelphia, pp. 333–406.
85. Azumi, N. and Battifora, H. (1987) The distribution of vimentin and keratin in epithelial and nonepithelial neoplasms. *Am. J. Clin. Pathol.* **88,** 286–296.
86. Skalli, O., Gabbiani, G., Babai, F., Seemayer, T. A., Pizzolato, G., and Schurch, W. (1988) Intermediate filament proteins and actin isoforms as markers for soft tissue tumor differentiation and origin. II. Rhabdomyosarcomas. *Am. J. Pathol.* **130,** 515–531.
87. Ohsawa, M., Naka, N., Tomita, Y., Kawamori, D., Kanno, H., and Aozasa, K. (1995) Use of immunohistochemical procedures in diagnosing angiosarcoma. Evaluation of 98 cases. *Cancer* **75,** 2867–2874.
88. Fellinger, E. J., Garin-Chesa, P., Su, S. L., DeAngelis, P., Lane, J. M., and Rettig, W. J. (1991) Immunohistochemical analysis of Ewing's sarcoma cell surface antigen p30/32MIC2. *Am. J. Surg. Pathol.* **139,** 317–325.

89. Ozdemirli, M., Fanburg-Smith, J. C., Hartmann, D. P., Azumi, N., and Miettinen, M. (2001) Differentiating lymphoblastic lymphoma and Ewing's sarcoma: lymphocyte markers and gene rearrangement. *Mod. Pathol.* **14,** 1175–1182.
90. Pritchard, K. I. (2003) Endocrine therapy of advanced disease: analysis and implications of the existing data. *Clin. Cancer Res.* **9,** 460–467.
91. McGuire, W. L., Carbone, P. P., Sears, M. E., and Escher, G. C. (1975) Estrogen receptors in human breast cancer: an overview, in *Estrogen Receptors in Human Breast Cancer* (McGuire, W. L., Carbone, P. P., and Vollmer, E. P., eds.), Raven, New York, pp. 1–7.
92. Slamon, D. J., Leyland-Jones, B., Shak, S., et al. (2001) Use of chemotherapy plus a monoclonal antibody against HER2 for metastatic breast cancer that overexpresses HER2. *N. Engl. J. Med.* **344,** 783–792.
93. Pegram, M. D., Lipton, A., Hayes, D. F., et al. (1998) Phase II study of receptor-enhanced chemosensitivity using recombinant humanized anti-p185HER2/neu monoclonal antibody plus cisplatin in patients with HER2/neu-overexpressing metastatic breast cancer refractory to chemotherapy treatment. *J. Clin. Oncol.* **16,** 2659–2671.
94. Gancberg, D., Lespagnard, L., Rouas, G., et al. (2000) Sensitivity of HER-2/neu antibodies in archival tissue samples of invasive breast carcinomas. Correlation with oncogene amplification in 160 cases. *Am. J. Clin. Pathol.* **113,** 675–682.
95. Heinrich, M. C., Griffith, D. J., Druker, B. J., Wait, C. L., Ott, K. A., and Zigler, A. J. (2000) Inhibition of c-kit receptor tyrosine kinase activity by STI 571, a selective tyrosine kinase inhibitor. *Blood* **96,** 925–932.
96. Demetri, G. D., von Mehren, M., Blanke, C. D., et al. (2002) Efficacy and safety of imatinib mesylate in advanced gastrointestinal stromal tumors. *N. Engl. J. Med.* **347,** 472–480.
97. Mauro, M. J. and Druker, B. J. (2001) STI571: targeting BCR–ABL as therapy for CML. *Oncologist* **6,** 233–238.
98. Hernandez-Boluda, J. C. and Cervantes, F. (2002) Imatinib mesylate (Gleevec, Glivec): a new therapy for chronic myeloid leukemia and other malignancies. *Drugs Today (Barc.)* **38,** 601–613.
99. Weiss, L. M., Movahed, L. A., Warnke, R. A., and Sklar, J. (1989) Detection of Epstein–Barr viral genomes in Reed–Sternberg cells of Hodgkin's disease. *N. Engl. J. Med.* **320,** 502–506.
100. Mathur, V. S., Olson, J. L., Darragh, T. M., and Yen, T. S. (1997) Polyomavirus-induced interstitial nephritis in two renal transplant recipients: case reports and review of the literature. *Am. J. Kidney Dis.* **29,** 754–758.
101. Pappo, O., Demetris, A. J., Raikow, R. B., and Randhawa, P. S. (1996) Human polyoma virus infection of renal allografts: histopathologic diagnosis, clinical significance and literature review. *Mod. Pathol.* **9,** 105–109.

40

Ribotyping in Clinical Microbiology

Patricia Severino and Sylvain Brisse

1. Introduction

For a considerable fraction of patients who acquire a bacterial infection during their stay in the hospital (i.e., a nosocomial infection), the infection was acquired through clonal dissemination of the bacteria from another patient or from a hospital source. Alternately, the source of the infection can be the endogenous flora of the patient (e.g., through the proliferation of resistant bacteria during treatment with an antimicrobial agent). Identifying the sources and routes of infection is therefore of utmost importance in order to control infection. Bacterial strain characterization, also called typing, has become an essential tool to define which bacterial strain is, and which one is not, part of an outbreak and to identify possible sources and routes of contamination. Additionally, typing methods can help finding reservoirs of epidemic organisms as a means to prevent future contaminations and infections of patients.

Discriminatory power (i.e., the capacity to distinguish epidemiologically nonrelated strains) and reproducibility are among the most important characteristics of a typing method. Among the available typing approaches, methods that evaluate chromosomal DNA polymorphisms using genomic restriction fragment length polymorphism (RFLP) analysis stand out for their high levels of reproducibility and discriminatory power ***(1)*** (see Chapter 3). Approaches that rely on the transfer of restriction fragments onto membranes, followed by hybridization with DNA probes, have proven to be very reproducible. Their discriminatory power, however, is a function of the number of copies in the bacterial chromosome of the sequences contained in the selected probe.

Among the nucleic acid probes that are used for typing, those containing the ribosomal DNA (rDNA) sequences are used in a technique called ribotyping ***(2,3)***. Ribotyping consists on the detection of genetic variation in the genomic sequence coding for the 16S and 23S rRNA genes and in the DNA regions flanking these genes. Ribosomal RNA genes are organized in a polycistronic transcription unit, called the *rrn* operon, which is usually present in multiple copies in the bacterial genome (see, for example, rrndb.cme.msu.edu). The evolutionary conservation of genes coding for ribosomal nucleic acids makes it possible to analyze any bacterial genome with a unique probe (e.g., one derived from *Escherichia coli*). In addition, the heterogeneity of the flanking sequences is the basis for variation of the ribotype patterns, rendering ribotyping one of the most versatile strategies for pathogen typing. It has been applied to numerous species of bacteria for identification or typing purposes ***(4)***.

Recent years have witnessed a new wave of interest for ribotyping because of the full automation of the technique through the availability, since 1997, of an instrument called the RiboPrinter, manufactured by Qualicon Inc., a subsidiary of the Dupont Company.

From: *Medical Biomethods Handbook*
Edited by: J. M. Walker and R. Rapley © Humana Press, Inc., Totowa, NJ

2. Methodology

Ribotyping consists of the following steps: (1) cell lysis and DNA extraction, (2) DNA cleavage by a restriction enzyme, (3) agarose gel electrophoresis to separate the resulting fragments, (4) Southern blot transfer of DNA fragments onto a membrane, and (5) hybridization with the DNA probes and detection of the hybridization signal. The following subsections will deal with each of these steps in more detail, both for the manual and the automated procedures.

2.1. Manual Ribotyping

2.1.1. Cell Lysis and DNA Extraction

Several protocols for the extraction of high-molecular-weight DNA have been described, with slight differences when considering Gram-negative and Gram-positive bacteria ***(5,6)***. The bacterial cell membrane can be easily destroyed by detergents, such as sodium dodecyl sulfate (SDS) or sarkosyl, in combination with proteinase K. Proteins are further eliminated by phenol/chloroform extraction and the DNA is precipitated with NaCl and ethanol or isopropanol.

Commercial kits are also available for DNA extraction and purification, with results that vary according to the bacterial species analyzed.

2.1.2. DNA Cleavage with a Restriction Endonuclease

Restriction endonucleases cleave double-stranded DNA at specific recognition sites. Depending on the species analyzed and the restriction enzyme used, ribotyping can generate a variable number of distinct fingerprints, called ribotypes. This is related to the abundance of recognition sites available for the restriction enzyme and the copy number of the probe sequence in the genome. For typing purposes, high discrimination is the criterion of choice, whereas for identification purposes, an enzyme that provides less discrimination is sometimes preferred. Digestion with restriction enzymes should be performed according to instructions supplied by the manufacturer.

2.1.3. Gel Electrophoresis and Southern Blot Transfer

Appropriate agarose should be chosen for the resolution of restriction fragments in a horizontal agarose gel electrophoresis system. A low level of electroendo-osmosis is necessary in order to avoid interference with nucleic acid fragments separation.

DNA mobility is affected by the composition and the ionic strength of the electrophoresis buffer. Tris-borate buffer (TBE) and Tris-acetate (TAE) are both used for ribotyping. TAE provides faster migration and slightly better resolving power, whereas TBE has significantly higher buffering capacity. For routine work, TBE is the buffer of choice, being, additionally, less sensitive to salt contamination in samples, a common cause of band distortion. Electrophoresis should be run in constant and rather low voltage to avoid gel distortion resulting from overheating of the system. It is recommended that the same running conditions be maintained for every experiment to optimize reproducibility.

Molecular size standards must be included in at least two lanes per gel, but preferably one every four to five lanes, in order to make the comparison of band sizes possible among strains present in the same or in different gels. The presence of standards also permits the normalization of runs with the help of computer programs (*see* **Subheading 2.3.**). The choice of markers depends on the range of DNA fragments obtained in the ribotyping experiments.

Transfer of DNA fragments from agarose gels to Nylon membranes is performed by upward capillary blotting *(7)*. Nucleic acids transferred to Nylon membranes can be hybridized and dehybridized several times. However, the choice of membrane depends on the probe used.

2.1.4. Probe Labeling and Detection

Ribotyping uses labeled probes containing 23S, 16S, and 5S rRNA sequences. Originally, RNA probes either from 16S+23S *(8)* or total rRNA *(3)* were employed. Currently, DNA probes are far more common. Such probes can be obtained in a variety of ways, among which are cDNA synthesized from rRNA by reverse transcription ***(9)***, recombinant plasmid containing a cloned *rrn* DNA ***(10)***, or a synthetic oligonucleotide constructed from the 16S or 16S+23S rRNA gene sequences ***(11)***. Because of the conservation of the rRNA operon, probes derived from *E. coli* rRNA are the most common. However, rRNA sequences from other microorganisms can be used, and homologous probes, derived from the rRNA of the studied bacterium, are also employed ***(12)***. Alternatively, a probe consisting of a mixture of five oligonucleotides corresponding to conserved regions of both the small and large rRNA molecules also provides very good results ***(13)***. Because hybridization results depend on the chosen probes, comparison between studies should take into account the probe used.

A variety of labeling techniques are available for the detection of the immobilized DNA on Nylon membranes. For RNA probes and oligonucleotides, end labeling is preferred; for plasmids containing the *rrn* sequences, nick translation or random priming are the techniques of choice. Originally, probes were labeled with ^{32}P, but, currently, nonisotopic cold-labeling systems are largely employed. They are safer and easier to dispose of than the radioactive systems and provide probes that are quite stable.

Examples of nonisotopic labeling procedures include biotin-, digoxigenin-, and fluorescein-labeling systems. These chemical groups are conjugated to the probe and are detected, after hybridization, via interaction with a protein conjugated to a reporter enzyme such as alkaline phosphatase and horseradish peroxidase. The presence of an enzyme capable of hydrolyzing a chosen substrate leads to the conversion of this substrate, once the membrane is exposed to it, to a colored product or a product able to emit light. The intensity and the distribution of the product correspond to the intensity and distribution of the target DNA.

2.2. Automated Ribotyping

The RiboPrinter Microbial Characterization System (DuPont–Qualicon, Wilmington, DE, USA) performs most steps of the ribotyping procedure automatically. Bacterial colonies are picked from an agar plate, resuspended in a sample buffer, and transferred to a special sample carrier. The cell suspensions are then heated to inactivate nucleases. Lytic enzymes are added once the temperature is reduced, and the sample carrier is loaded into the instrument, together with other consumables: restriction enzyme, loading buffer, DNA molecular size markers, gel cassette, membrane, and reagents for hybridization and signal detection. The operator enters strain-related information on the system's computer, and the samples are subsequently processed automatically. The RiboPrinter allows the use of different endonucleases as long as they are active in the system's buffer conditions and at the system's digestion temperature. Restriction fragments are subjected to agarose electrophoresis in a precast gel and the fragments are captured and immobilized onto a Nylon membrane. Membrane processing includes the denaturation of DNA fragments, hybridization with a ribosomal RNA probe corresponding to the *E. coli rrnB* operon, washing, treatment with blocking agent, incubation with an antisulfonated DNA antibody/alkaline phosphatase conjugate, washing steps, and the application of a chemiluminescent substrate. The signal is detected by a charge-coupled device (CCD) camera, which detects the light intensity of the hybridized fragments. The camera converts the patterns to digital information, and the RiboPrinter software automatically processes the image. After normalization of the image and background processing, the software will generate a ribotype pattern, called RiboPrint, which is stored in the system's database. For identification, this incoming pattern will be automatically compared to those previously present in the reference identification database, which contains patterns from DuPont or from the customer. For characterization (i.e., ribotype assignment), the sample pattern is compared with all of the patterns previously stored in the database library chosen by the operator. Ribotype assignment needs to be visually checked and manually corrected by the RiboPrinter user, because high background or artifactual shifts of the band positions can alter the results. **Figure 1** exemplifies ribotypes obtained after automated ribotyping.

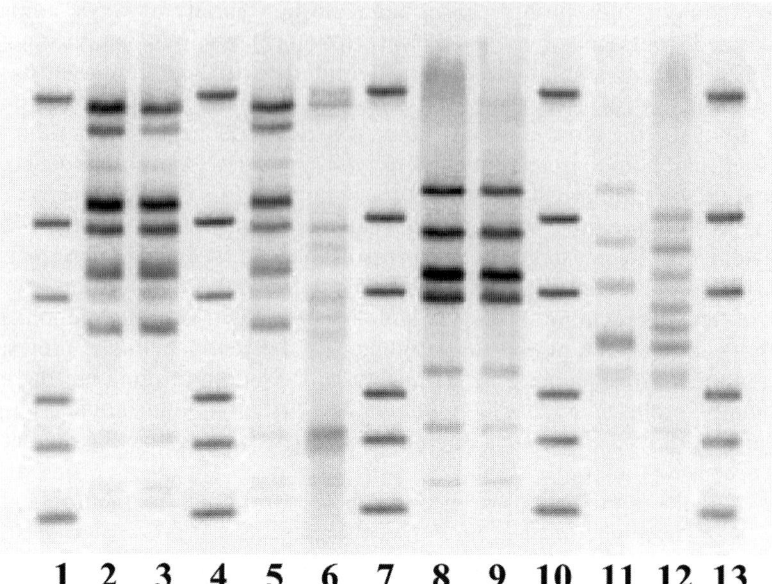

Fig. 1. Ribotype patterns obtained after automated ribotyping analysis using enzyme EcoRI in the RiboPrinter (DuPont–Qualicon, Wilmington, DE). Lanes 1, 4, 7, 10 and 13 : DNA size markers (in kb: 48, 9.6, 6.5, 3.16, 2.15, and 1.1); lanes 2, 3, 5 and 6: *Klebsiella oxytoca* strains (the patterns of the first three strains are undistinguishable); lanes 8 and 9: *Burkholderia multivorans* strains; lanes 11 and 12: *Burkholderia cepacia* genomovar III (*B. cenocepacia*) strains.

2.3. Interpretation of Ribotyping Results

The patterns generated by ribotyping must be compared with each other in order to represent valuable data. For typing purposes, the objective is to identify clonally related organisms, whereas, for identification purposes, it is to determine the species to which the strains belong. The level of similarity used needs to be adjusted to the plasticity of the organism under study and the time scale of the investigation.

In general, patterns are considered identical when they share exactly the same amount of bands and they are all of identical sizes (within given tolerance limits). As mentioned above, molecular size markers are important for the correct appreciation of these sizes. Currently, the accessibility of several software packages for data normalization tends to make such analysis more objective and reproducible. Interpretation of results always starts with the digitalization of images obtained after probe hybridization, followed by band detection and pattern comparison through an automatic system. Software such as the Windows-running BioNumerics (Applied Maths, Sint-Martens-Latem, Belgium) and the MacIntosh-running Taxotron package (Institut Pasteur, Paris, France) provide a broad range of possibilities for pattern analysis. Details on data analysis and interpretation were given recently by Grimont and Grimont ***(14)***.

3. Applicability in the Clinical Setting

Ribotyping is a robust method that exhibits excellent reproducibility and stability during the course of outbreaks. Its high stability also renders ribotyping useful for large-scale surveillance, in which the temporal and geographical distance among strains is higher. One major drawback of ribotyping is its somewhat limited discriminatory power in some bacterial groups. Indeed, because ribotyping indexes variation only at and around the ribosomal operons, but not in the rest of the genome, it is discriminatory only for those species like the *Enterobacteriaceae*, which possess a high number (around seven) of rRNA operons. Conversely, in species with few or only one rRNA operon, such as *Legionella*, ribotyping is poorly discriminatory. For a given species, the discriminatory power of ribotyping also depends on the restriction endonucleases(s) used. Literature should be consulted for the most frequently used enzymes for a given microorganism, especially if one wants to compare patterns with international databases. However, an evaluation followed by optimization should be performed by the clinical or research laboratory willing to implement the technique.

The following subsections will give practical examples of the use of ribotyping in nosocomial settings.

3.1. Ribotyping for Identification Purposes

Identifying a strain using ribotyping consists of comparing its ribotype pattern with those of reference strains. If the pattern of the query strain matches that of a strain of a known species, the query strain is attributed to that species. The value of ribotyping for identification will therefore depend on the existence of differences among the ribotype patterns of different species and on the completeness of the database used for identification. When too much ribotype pattern variation exists within a species, this approach is of limited value. For this reason, restriction enzymes used for identification purposes are selected that give little variation within species while retaining species distinction.

Identification of several pathogens to their species is often difficult using traditional methods, especially when closely related species are phenotypically very similar. However, identification is important if one is to understand the epidemiology of each species, and it is crucial in clinical microbiology when phenotypically undistinguishable species differ by their pathogenicity or transmissibility. This is the case, for example, for bacterial species belonging to the *Burkholderia cepacia* species complex ***(15,16)***. Evidence that automated ribotyping can distinguish members of the *B. cepacia* complex has been provided, but because of variation of ribotype patterns within species, a high proportion of analyzed strains could not be identified with the limited reference database available at that time ***(17)***.

Bacteria belonging to the genus *Legionella* are known to be pathogenic to immunocompromised patients, causing a pneumonia referred to as Legionnaires's disease. The genus *Legionella* currently includes 45 species. Most infections are caused by *L. pneumophila* serogroup 1, but other serogroups of this species and other *Legionella* species can cause infection. Identification of *Legionella* species is difficult because of their phenotypic resemblance, and routine identification relies on immunofluorescence tests or seroagglutination ***(18)***. Both techniques are limited because of crossreaction among species. Members of the *Legionella* genus were among the first to be investigated for the identification ability of ribotyping ***(8)***. This work was updated by analysis of 44 of the 45 currently described *Legionella* species, using enzyme *Hin*dIII and a set of five oligonucleotides (oligoMix5) as probe, and all species tested showed specific ribotype patterns ***(19)***.

Ribotyping has proven very useful for identification and taxonomic purposes in the genus *Staphylococcus* ***(20–22)***. High-quality databases of ribotype profiles that include representative strains of nearly all described species and subspecies have been constructed and are useful for the identification of unknown isolates. The capacity of ribotyping for the identification of coagulase-negative staphylococci was recently extended to automated ribotyping: using *Eco*RI, 94% of 177 non-*S. epidermidis* coagulase-negative staphylococci could be identified, and automated ribotyping was able to reliably distinguish *S. capitis* from *S. caprae*, two species that are otherwise difficult to distinguish (Carretto et al., *Clinical Microbiology and Infection*, in press).

Other bacterial groups for which ribotyping has shown ability for identification include *Acinetobacter* spp. ***(23)***, viridans group streptococci ***(24)***, aerotolerant *Campylobacter* species ***(25)***, enterococci ***(26)***, klebsiellae ***(27)***, Corynebacteria ***(28)***, pediococci ***(29)***, and Sphingomonads ***(30)***, among others. The list has expanded in the last years because of the availability of the

RiboPrinter automate, and it is likely that more applications will be developed in the near future. An interesting possibility of automated ribotyping is that the patterns obtained are standardized, which allows pattern comparison against identification databases generated by other users. In addition, availability of ribotype pattern on websites, such as the PathogenTracker website at Cornell University (www.pathogentracker.net) or the GENE database website (www.ewi.med.uu.nl/gene) render it possible to identify pathogens through the Internet ***(31)***.

3.2. Ribotyping for Typing Purposes

When ribotyping is used as a molecular typing method, high discrimination among strains is necessary, demanding a careful choice of restriction enzymes. In several instances, more than one enzyme is recommended in order to achieve proper discrimination. Outbreak investigations and epidemiological surveillance are the contexts in which ribotyping is employed as a molecular typing method.

During an outbreak investigation, the objective is to compare isolates to delineate clonally related and unrelated strains with the goal of short-term control of transmission. Important opportunistic pathogens such as *Pseudomonas aeruginosa* and *Acinetobacter* spp. have been repeatedly characterized through ribotyping ***(32–36)***. Ribotyping has often been used for corroborating epidemiological data collected during outbreaks caused by *Burkholderia* spp. ***(37)***, *Campylobacter jejuni* ***(38)***, *Corynebacterium diphtheriae* ***(39)***, *Enterobacter cloacae* ***(40)***, *Enterococcus* spp. ***(41)***, *E. coli* ***(42)***, *Klebsiella* spp. ***(43)***, *Listeria monocytogenes* ***(44)***, and *Staphylococcus aureus* ***(45)***, among others.

Automated ribotyping has been used with increasing frequency for the typing of strains involved in nosocomial outbreaks. Typing of clinically related and unrelated isolates of vancomycin-resistant *E. faecium*, for example, with automated ribotyping has been evaluated aiming to assess the utility of this tool for infection control interventions involving this micro-organism ***(46,47)***. Although ribotyping is slightly less discriminatory than pulsed field gel electrophoresis (PFGE), the high rate of interlaboratory reproducibility and the speed of the generation of results make automated ribotyping a valuable approach for characterization of vancomycin-resistant *E. faecium*. The use of automated ribotyping was also compared with PFGE for the typing of *Streptococcus pneumoniae* ***(48)***. Automated ribotyping proved to be an accurate method for genetic fingerprinting of *S. pneumoniae*, complementing PFGE. Similar approaches have shown comparable performances for the typing of *E. coli* and *Pseudomonas aeruginosa* ***(49,50)***.

In the context of epidemiologic surveillance, ribotyping can be used to monitor geographic spread of epidemic and endemic clones. The main goal here is the long-term evaluation of preventive strategies or the detection and monitoring of emerging infections. The above-mentioned standardization of the automated ribotyping system provides an important advantage over other typing systems. Automated ribotyping has been employed with epidemiological surveillance purposes for the characterization of organisms such as *Listeria monocytogenes*, methicillin-resistant *Staphylococcus aureus*, *Campylobacter* spp., *Corynebacterium diphtheriae*, and enterococci ***(39,51–54)***.

References

1. Tenover, F. C., Arbeit, R. D., Goering, R. V., and The Molecular Typing Working Group of the Society for Healthcare Epidemiology of America (1997) How to select and interpret molecular typing methods for epidemiological studies of bacterial infections: a review for healthcare epidemiologists. *Infect. Control Hosp. Epidemiol.* **18,** 426–439.
2. Grimont, F. and Grimont, P. A. D. (1986) Ribosomal ribonucleic acid gene restriction patterns as potential taxonomic tools. *Ann. Inst. Pasteur Microbiol.* **137B,** 165–175.
3. Stull, T., LiPuma, J. J., and Edlind, T. D. (1988) A broad-spectrum probe for molecular epidemiology of bacteria: ribosomal RNA. *J. Infect. Dis.* **157,** 280–286.

4. Bingen, E. H., Denamur, E., and Elion, J. (1994) Use of ribotyping in epidemiological surveillance of nosocomial outbreaks. *Clin. Microbiol. Rev.* **7,** 311–327.
5. Owen, R. J., Ahmed, A. U., and Dawson, C. A. (1987) A rapid biochemical method for purifying high molecular weight bacterial chromosomal DNA for restriction enzyme analysis. *Nucleic Acid Res.* **15,** 3631.
6. Pitcher, D. G., Saunders, N. A., and Owen, R. J. (1989) Rapid extraction of bacterial genomic DNA with guanidium thiocyanate. *Lett. Appl. Microbiol.* **8,** 151–156.
7. Southern, E. M. (1975) Detection of specific sequences among DNA fragments separated by gel electrophoresis. *J. Mol. Biol.* **98,** 503–517.
8. Grimont, F., Lefevre, M., Ageron, E., and Grimont, P. A. (1989) rRNA gene restriction patterns of *Legionella* species: a molecular identification system. *Res. Microbiol.* **140,** 615–626.
9. Garaizar, J., Kaufmann, M. E., and Pitt, T. L. (1991) Comparison of ribotyping with conventional methods for the type identification of *Enterobacter cloacae. J. Clin. Microbiol.* **29,** 1303–1307.
10. Altwegg, M. and Mayer, L. M. (1989) Bacterial molecular epidemiology based on a non-radioactive probe complementary to ribosomal RNA. *Res. Microbiol.* **140,** 325–333.
11. Romaniuk P. J. and Trust, T. J. (1987) Identification of *Campylobacter* species by Southern hybridization of genomic DNA using an oligonucleotide probe for 16S rRNA genes. *FEMS Microbiol. Lett.* **43,** 331–335.
12. Cookson, B. D., Stapleton, P., and Ludlam, H. (1992) Ribotyping of coagulase-negative staphylococci. *J. Med. Microbiol.* **36,** 414–419.
13. Regnault, B., Grimont, F., and Grimont, P. A. (1997) Universal ribotyping method using a chemically labeled oligonucleotide probe mixture. *Res. Microbiol.* **148,** 649–659. Erratum: *Res. Microbiol.* (1998) **149,** 73.
14. Grimont, P. A. D., and Grimont F. (2001) rRNA gene restriction pattern determination (ribotyping) and computer interpretation, in *New Approaches for the Generation and Analysis of Microbial Typing Data* (Dijkshoorn, L., Towner K. J., and Struelens M. J., eds.), Elsevier, Amsterdam, pp. 107–133.
15. Govan, J. R. W. and Deretic, V. (1996) Microbial pathogenesis in cystic fibrosis: mucoid *Pseudomonas aeruginosa* and *Burkholderia cepacia. Microbiol. Rev.* **60,** 539–974.
16. Pelt, C., Verduin, C.M., Goessens, W.H.F., et al. (1999) Identification of Burkholderia spp. in the clinical microbiology laboratory: comparison of conventional and molecular methods. *J. Clin. Microbiol.* **37,** 2158–2164.
17. Brisse, S., Verduin, C.M., Milatovic, D., et al. (2000) Distinguishing species of the *Burkholderia cepacia* complex and *Burkholderia gladioli* by automated ribotyping. *J. Clin. Microbiol.* **38,** 1876–1884.
18. Wilkinson, H. (1988) *Hospital-Laboratory Diagnosis of* Legionella *Infections.* US Department of Health and Human Services, Public Health Service, Centers for Diseases Control, Atlanta, GA.
19. Salloum, G., Meugnier, H., Reyrolle, M., et al. (2002) Identification of *Legionella* species by ribotyping and other molecular methods. *Res. Microbiol.* **153,** 679–686.
20. De Buyser, M. L., Morvan, A., Grimont, F., and el Solh, N. (1989) Characterization of *Staphylococcus* species by ribosomal RNA gene restriction patterns. *J. Gen. Microbiol.* **135(Pt. 4),** 989–999.
21. De Buyser, M. L., Morvan, A., Aubert, S., Dilasser, F., and el Solh, N. (1992) Evaluation of a ribosomal RNA gene probe for the identification of species and subspecies within the genus *Staphylococcus. J. Gen. Microbiol.* **138,** 889–899.
22. Chesneau, O., Morvan, A., Aubert, S., and el Solh, N. (2000) The value of rRNA gene restriction site polymorphism analysis for delineating taxa in the genus *Staphylococcus. Int. J. Syst. Evol. Microbiol.* **50,** 689–697.
23. Gerner-Smidt, P. (1992) Ribotyping of the *Acinetobacter calcoaceticus–Acinetobacter baumannii* complex. *J. Clin. Microbiol.* **30,** 2680–2685.
24. Rudney, J. D., and Larson, C. J. (1994) Use of restriction fragment polymorphism analysis of rRNA genes to assign species to unknown clinical isolates of oral viridans streptococci. *J. Clin. Microbiol.* **32,** 437–443.
25. Kiehlbauch, J. A., Plikaytis, B. D., Swaminathan, B., Cameron, D. N., and Wachsmuth, I. K. (1991) Restriction fragment length polymorphisms in the ribosomal genes for species identification and subtyping of aerotolerant *Campylobacter species. J. Clin. Microbiol.* **29,** 1679–1676.
26. Pryce, T. M., Wilson, R. D., and Kulski, J. K. (1999) Identification of enterococci by ribotyping with horseradish-peroxidase-labelled 16S rDNA probes. *J. Microbiol. Methods* **36,** 147–155.

27. Brisse, S., and Verhoef, J. (2001) Phylogenetic diversity of *Klebsiella pneumoniae* and *Klebsiella oxytoca* clinical isolates revealed by randomly amplified polymorphic DNA, gyrA and parC genes sequencing and automated ribotyping. *Int. J. Evol. Microbiol.* **51,** 915–924.
28. Bjorkroth, J., Korkeala, H., and Funke, G. (1999) rRNA gene RFLP as an identification tool for *Corynebacterium* species. *Int. J. Syst. Bacteriol.* **49,** 983–989.
29. Satokari, R., Mattila-Sandholm, T., and Suihko, M. L. (2000) Identification of pediococci by ribotyping. *J. Appl. Microbiol.* **88,** 260–265.
30. Busse, H. J., Kainz, A., Tsitko, I. V., and Salkinoja-Salonen, M. (2000) Riboprints as a tool for rapid preliminary identification of sphingomonads. *Syst. Appl. Microbiol.* **23,** 115–123.
31. Brisse, S. (2003) Automated ribotyping of bacterial strains: a standardized technology suitable for use in the construction of on-line databases and Internet exchange, in *Experimental Approaches for Assessing Genetic Diversity Among Microbial Pathogen* (van Belkum A., Duim, B., and Hays, J. P., eds.), Dutch Working Party on Epidemiological Typing, Wageningen, pp. 23–32.
32. Dijkshoorn, L., Aucken, H., Gerner-Smidt, P., Kaufmann, M.E., Ursing, J., and Pitt, T. L. (1993) Correlation of typing methods for *Acinetobacter* isolates from hospital outbreak. *J. Clin. Microbiol.* **31,** 702–705.
33. Gruner, E., Kropec, A., Huebner, J., Altwegg, M., and Daschner, F. (1993) Ribotyping of *Pseudomonas aeruginosa* strains isolated from surgical intensive care patients. *J. Infect. Dis.* **167,** 1216–1220.
34. Ferrus, M.A., Hernandez, M., and Hermandez, H.J. (1998) Ribotyping of *Pseudomonas aeruginosa* from infected patients: evidence of common strain types. *APMIS* **106,** 456–462.
35. Severino, P., Darini, A. L., and Magalhães, A. D. (1999) The discriminatory power of ribo-PCR and ribotyping for epidemiological purposes. *APMIS* **107,** 1079–1084.
36. Brisse, S., Milatovic, Fluit, A., et al. (2000) Molecular surveillance of European quinolone-resistant clinical isolates of *Pseudomonas aeruginosa* and *Acinetobacter* spp. using automated ribotyping. *J. Clin. Microbiol.* **38,** 3636–3645.
37. Chetoui, H., Melin, P., Struelens, M. J., Delhalle, E., Nigo, M. M, De Ryck, R., and De Mol, P. (1997) Comparison of biotyping, ribotyping, and pulsed-field gel electrophoresis for investigation of a common-source outbreak of *Burkholderia pickettii* bacteremia. *J. Clin. Microbiol.* **35,** 1398–1403.
38. Fayos, A., Owen, R. J., Desai, M., and Hernandez, J. (1992) Ribosomal RNA gene restriction fragment diversity amongst Lior biotypes and Penner serotypes of *Campylobacter jejuni* and *Campylobacter coli*. *FEMS Microbiol. Lett.* **74,** 87–93.
39. De Zoysa, A., Efstratiou, A., George, R. C., et al. (1995) Molecular epidemiology of *Corynebacterium diphtheriae* from northwestern Russia and surrounding countries studied by using ribotyping and pulsed-field gel electrophoresis. *J. Clin. Microbiol.* **33,** 1080–1083.
40. Bingen, E., Denamur, E., Lambert-Zechovsky, N., Brahimi, N., El Lakany, M., and Elion, J. (1992) Rapid genotyping shows the absence of cross-contamination in *Enterobacter cloacae* nosocomial infection. *J. Hosp. Infect.* **21,** 95–101.
41. Bingen, E., Denamur, E., Lambert-Zechovsky, and Elion, J. (1991) Evidence for the genetic unrelatedness of nosocomial vancomycin-resistant *Enterococcus faecium* strains in a pediatric hospital. *J. Clin. Microbiol.* **29,** 1888–1892.
42. Alos, J. I., Lambert, T., and Courvalin, P. (1993) Comparison of two molecular methods for tracing nosocomial transmission of *Escherichia coli* K1 in a Neonatal Unit. *J. Clin. Microbiol.* **31,** 1704–1709.
43. Quale, J.M., Landman, D., Bradford, P.A., et al. (2002) Molecular epidemiology of a citywide outbreak of extended-spectrum beta-lactamase-producing *Klebsiella pneumoniae* infection. *Clin. Infect. Dis.* **35,** 834–841.
44. Nadon, C. A., Woodward, D. L., Young, C., Rodgers, F. G., and Wiedmann, M. (2001) Correlation between molecular subtyping and serotyping of *Listeria monocytogenes*. *J. Clin. Microbiol.* **39,** 2704–2707.
45. Blumberg, H. M., Rimland, D., Kiehlbauch, J. A., Terry, P. M., and Wachsmuth, I. K. (1992) Epidemiologic typing of *Staphylococcus aureus* by DNA restriction fragment length polymorphism of rRNA genes: elucidation of the clonal nature of a group of bacteriophage-nontypeable, ciprofloxacin-resistant, methicillin-susceptible S. aureus isolates. *J. Clin. Microbiol.* **30,** 362–369.
46. Brisse, S., Fussing, V., Ridwan, B., Verhoef, J., and Willems, J. L. (2002) Automated ribotyping of vancomycin-resistant *Enterococcus faecium* isolates. *J. Clin. Microbiol.* **40,** 1977–1984.
47. Price, C. S., Huynh, H., Paule, S., et al. (2002) Comparison of an automated ribotyping system to restriction endonuclease analysis and pulsed-field gel electrophoresis for differentiating vancomycin-resistant *Enterococcus faecium* isolates. *J. Clin. Microbiol.* **40,** 1858–1861.

48. Quale, J. M., Landman, D., Flores, C., and Ravishankar, J. (2001) Comparison of automated ribotyping to pulsed-field gel electrophoresis for genetic fingerprinting of *Streptococcus pneumoniae*. *J. Clin. Microbiol.* **39,** 4175–4177.
49. Pfaller, M. A., Wendt, R. J., Hollis, R. J., et al. (1996) Comparative evaluation of an automated ribotyping system versus pulsed-field gel electrophoresis for epidemiological typing of clinical isolates of *Escherichia coli* and *Pseudomonas aeruginosa* from patients with recurrent gram-negative bacteremia. *Diagn. Microbiol. Infect. Dis.* **25,** 1–8.
50. Hollis, R. J. J., Bruce, J. L., Fritschel, S. J., and Pfaller, M. A. (1999) Comparative evaluation of an automated ribotyping instrument versus pulsed-field gel electrophoresis for epidemiological investigation of clinical isolates of bacteria. *Diagn. Microbiol. Infect. Dis.* **34,** 263–268.
51. Jones, R. N., Marshall, S. A., Pfaller, M. A., et al. (1997) Nosocomial enterococcal blood stream infections in the SCOPE Program: antimicrobial resistance, species occurrence, molecular testing results, and laboratory testing accuracy. *Diagn. Microbiol. Infect. Dis.* **29,** 95–102.
52. Diekema, D. J., Pfaller, M. A., Turnidge, J., et al. (2000) Genetic relatedness of multidrug-resistant, methicillin (oxacillin)-resistant *Staphylococcus aureus* bloodstream isolates from SENTRY Antimicrobial Resistance Surveillance Centers worldwide, 1998. *Microb. Drug Resist.* **6,** 213–221.
53. De Boer, P. B., Duim, B., Rigter, A., et al. (2000) Computer-assisted analysis and epidemiological value for genotyping methods for *Campylobacter jejuni* and *Campylobacter coli*. *J. Clin. Microbiol.* **38,** 1940–1946.
54. Sauders, B. D., Fortes, E. D., Morse, D. L., et al. (2003) Molecular subtyping to detect human listeriosis clusters. *Emerg. Infect. Dis.* **9,** 672–680.

41

Diagnostic Applications of Protein Microarrays

Samir Hanash

1. Introduction

The sequencing of the human genome has opened the door for proteomics by providing a sequence-based framework to mine the human proteome. Although the field of proteomics was initially dominated by two-dimensional gels and mass spectrometry, the current emphasis is on developing proteome-scale high-throughput methods, as exemplified by the development of protein microarrays. This chapter addesses the utility of protein microarrays for clinical applications. Additional information about protein microarrays can be found in several review articles that have been published in various journals *(1–4)*.

The clinical applications of protein microarrays cover a wide range from their use to detect disease or assess response to therapy based on profiling of biological fluids, to their use for profiling of disease tissue to determine disease subtypes and decide on the most appropriate therapy. Although the field of protein microarrays is still evolving, there are currently two broad classes of protein arrays. One class consists of arrays that contain protein capture agents, such as antibodies, which are utilized to assay the abundance of corresponding antigens in biological samples, and another class consists of arrays that contain proteins or peptides to be interrogated using cells, tissues, biological fluids, or single agents to uncover their interactions with specific arrayed proteins or peptides.

2. Protein Capture Arrays

Arrays containing capture agents can be used in a similar manner as sandwich assays, in which an immobilized antibody or any other type of capture agent is utilized to capture proteins of interest in a biological sample and a second labeled antibody or capture agent is used to detect and determine the abundance of the protein *(5)*. Such an approach, therefore, requires two capture agents that recognize different epitopes that need to be available for binding in order to assay any given protein(s). This dual requirement creates constraints for interrogating the abundance of large numbers of proteins simultaneously in a biological sample. Alternatively, a label-free detection method, such as mass spectrometry or surface plasmon resonance, can be utilized. Yet another alternative is to tag biological samples before they are hybridized to immobilized capture agents. The levels of the proteins in the biological sample, for which corresponding capture agents are arrayed, is determined based on quantifying the amount of tag bound to the immobilized capture agents. Haab et al. investigated this approach using a set of 115 antibodies and their corresponding protein ligands, which were labeled with Cy5 dye, and combined with a Cy3-labeled reference mixture *(6)*. They determined that some 20% of the arrayed antibodies provided specific and accurate measurements of their target

From: *Medical Biomethods Handbook*
Edited by: J. M. Walker and R. Rapley © Humana Press, Inc., Totowa, NJ

antigens at a concentration of 1.6 μg/mL or less. Some antibodies could detect ligands at concentrations of less than 1 ng/mL. In addition to determining the amount of a protein(s) in a biological sample, there is value in determining posttranslational modifications. A potentially useful approach to determine the level of phosphorylation of a protein is a dual detection system in which captured proteins are assessed with respect to their phosphorylation status. Steinberg et al. applied a novel phosphoprotein dye technology suitable for the fluorescent detection of phosphoserine-, phosphothreonine-, and phosphotyrosine-containing proteins to the determination of protein kinase and phosphatase substrate preference *(7)*. The utility of the fluorescent dye technology was demonstrated using phosphoproteins and phosphopeptides as well as with protein kinase reactions performed in miniaturized microarray assay format *(8)*. Instead of applying a phosphoamino acid-selective antibody labeled with a fluorescent or enzymatic tag for detection, a small fluorescent probe was employed as a sensor of protein phosphorylation status. The detection limit for phosphoproteins on a variety of different commercially available protein array substrates was found to be 312–625 fg, depending on the number of phosphate residues *(8)*. The development of reagents that allow assessment of protein modification such as phosphorylation, glycosylation, or other functionally relevant protein changes have substantial utility for clinical applications.

A major challenge in making biochips for global analysis of protein expression is the current lack of comprehensive sets of genome-scale capture agents, such as antibodies. Eventually, antibodies to capture proteins on a proteomewide basis, as envisioned by the Human Proteome Organization (HUPO) and others, are likely to become available. The compelling need for developing capture agents with the prerequisite specificity has led numerous biotechnology companies to devise novel strategies that include aptamers (SomaLogic, http://www.somalogic.com/), ribozymes (Archemix, http://www.archemix.com/), partial-molecule imprints (Aspira Biosystems, http://www.aspirabio.com), and modified binding proteins (Phylos, http://www.phylos.com).

Antibody microarrays with a relatively limited content as compared to DNA microarrays have become commercially available. For example, the BD Clontech Ab Microarray 500 consists of 500 distinct antibodies that detect a variety of human proteins either in circulation or in cell and tissue extract. These correspond to a broad range of functional classes involved in signal transduction, cell growth and proliferation, apoptosis, inflammation, and the immune response. A complete list of the antibodies on this array is available at http://atlasinfo.clontech.com/abinfo/ab-list-action.do. Other commercially available antibody microarrays that interrogate a more limited or a particular class of proteins are also available. The content and variety of commercially available microarrays is likely to continue to expand. A potential limiting factor for the use of commercial microarrays is their cost.

3. Arrays Containing Proteins, Peptides, or Lysates

For assays of protein interactions, arrays that contain either peptides or proteins are being produced. Peptides can be synthesized in very large numbers directly on the chip (*see* **Fig. 1**) *(9)*. Alternatively, recombinant proteins can be arrayed. There is substantial interest at the present time to assemble large sets of purified recombinant proteins for various applications, including microarray construction, as has been done with the yeast *(10,11)*. As a demonstration of the functional utility of arrays of recombinant proteins, Zhu et al. investigated 119 of the 122 predicted protein kinases in yeast to determine their substrate specificities *(12)*. They constructed miniaturized microtiter plates with 10 × 14 wells that they covalently coated with different protein or peptide substrates. A recombinant kinase, together with [γ-^{33}P]ATP, was deposited into each well. Reactivity in different wells was determined by phosphorimager analysis and they were able to determine kinase specificity. Yet another type of protein array consists of arraying lysates prepared individually from multiple clinical samples, such as tumors. Such arrays are produced in batches in which each array contains the full complement of clini-

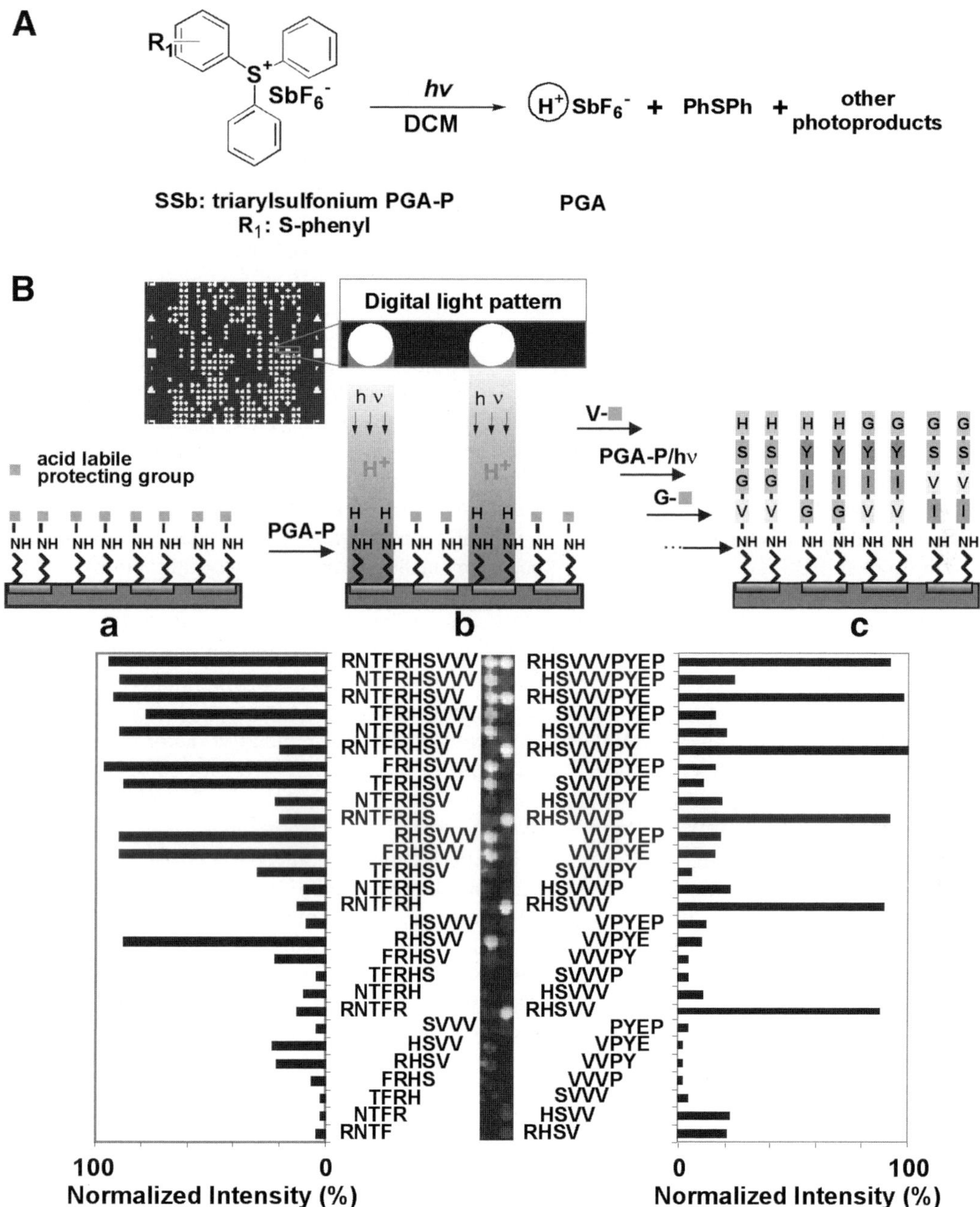

Fig. 1. Parallel on-chip peptide synthesis by light-directed photo-acid generation *(9)*. The principle is shown on top. An acid precursor molecule (PGA-P) gives an acid (PGA) at the desired location when exposed to light. The acid labile protection groups are removed at locations exposed to light enabling the growing peptide to react with the next amino acid. A digital light pattern generated by the computer is projected onto the chip to spatially control the locations at which addition of the amino acid will take place. The bottom figure shows the positional scanning of the RHSVV epitope of p53 using PAb240 antibody binding. The sequence RNTFRHSVVVPYEP is scanned with overlapping 4–10-mers. On the left side, the epitope is close to the surface, and on the right side, it is away from the surface. (Courtesy Dr. E. Gulari.)

cal samples. Each array is interrogated with a single antibody to determine the level or post-translational modification of a particular protein.

4. Array Construction, Hybridization, and Scanning

There are numerous options available for array construction, hybridization, and scanning, as illustrated in part in **Fig. 2** *(13,14)*. Various substrates, such as Nylon, polystyrene, and silica glass have been investigated, and different functional groups, such as aminosilanes, aldehydes, and epoxide groups, have been investigated with various substrates for efficiency of protein immobilization and background levels *(15–17)*. Cross-linkers have been utilized to couple proteins to a surface-modified substrate. Surface coating with various gels is another option *(18,19)*. Various approaches for oriented immobilization of proteins have also been implemented *(13)*.

In our customized procedure for array production, we have utilized readily available equipment for DNA microarray construction. Antibody solutions are loaded onto a 384-well plate at concentrations ranging from 0.1 to 0.7 mg/mL in volumes of 5 μL/well. Antibodies are arrayed using a 32-pin arrayer onto nitrocellulose membranes supported on glass slides. The spacing is set at 500 μm, and the diameter of the spots typically varies between 175 and 225 μm. The arrays are rinsed in phosphate-buffered saline (PBS), pH 7.5, containing 4% nonfat milk and 0.1% Tween-20 for 1 min to remove unbound protein. Subsequently, they are blocked with PBS supplemented with 4% nonfat milk for 40 min under agitation. The slides are then washed three times in PBS, after which they are ready for further assays. The washing and hybridization steps are performed in an automated hybridization station at 22°C. Microarrays are analyzed in a scanner using a 550-nm long-pass filter laser. Software is used to delineate the location of each spot in the array. The fluorescence intensity of each spot is calculated as the average pixel intensity of the spot. The background consisting of the median of pixel intensities from the area around the spot is subtracted. A normalization factor is utilized to eliminate differences because of experimental error or variation. The normalization factor is determined by using biotinylated–BSA (bovine serum albumin) as an internal reference sample.

5. Disease-Related Applications of Protein Arrays

Disease tissue profiling studies that have utilized protein microarrays are beginning to emerge, particularly in the cancer field. Belov et al. have used a microarray of 60 antibodies directed against cluster of differentiation (CD) antigens to immunophenotype various types of leukemia *(20,21)*. They utilized whole cells to determine expression of antigens expressed on the surface. Based on the expression pattern of CD antigens, a fingerprint was defined for chronic lymphocytic leukemia. As a model to better understand how patterns of protein expression shape the tissue microenvironment, Knezevic et al. analyzed protein expression in tissue derived from squamous cell carcinomas of the oral cavity through an antibody microarray approach for high-throughput proteomic analysis *(22)*. Utilizing laser capture microdissection to procure total protein from specific cell populations, they demonstrated that quantitative, and potentially qualitative, differences in expression patterns of multiple proteins within epithelial cells reproducibly correlated with oral cavity tumor progression. Differential expression of multiple proteins was found in stromal cells surrounding and adjacent to regions of diseased epithelium that directly correlated with tumor progression of the epithelium. Most of the proteins identified in both cell types were involved in signal transduction pathways. They hypothesized, therefore, that extensive molecular communications involving complex cellular signaling between epithelium and stroma play a key role in driving oral cavity cancer progression.

Cytokines have become a popular target for analysis using microarrays. Schweitzer et al. printed 75 cytokine antibodies on chemically derivatized glass slides and investigated cytokine secretion from human dendritic cells induced with lipopolysaccharide or tumor-necrosis factor-α *(23)*. They used isothermal rolling circle amplification *(24)* to increase the sensitivity of their assay. Some of the detection antibodies were found to crossreact with other cytokines on the

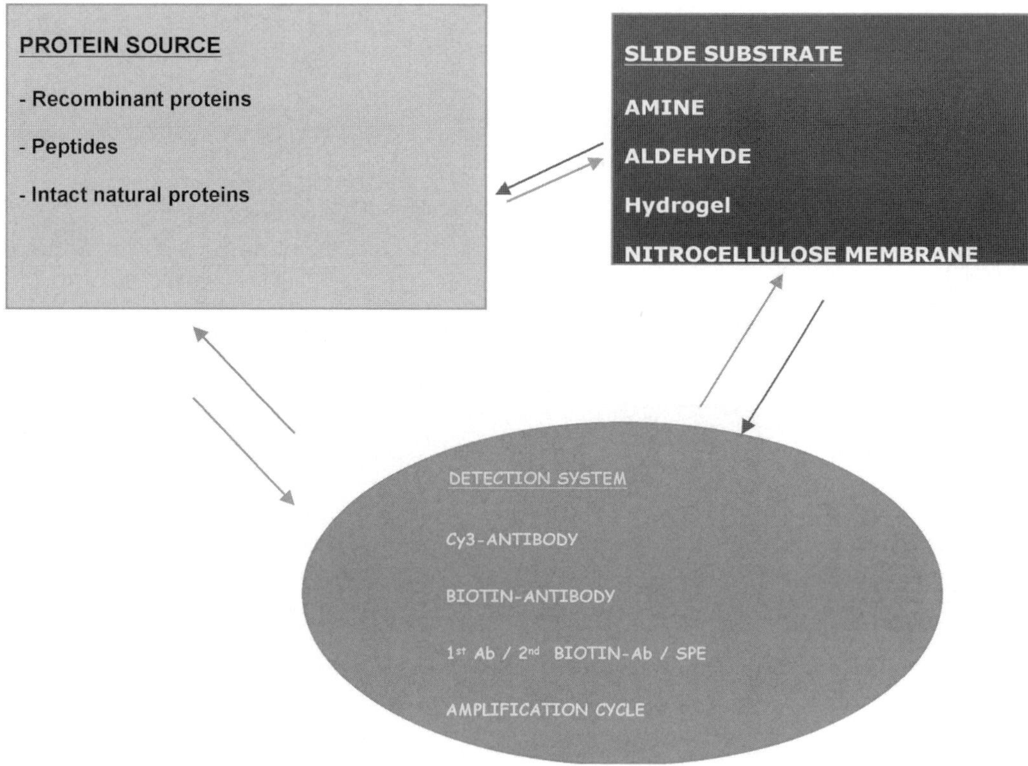

Fig. 2. Some options available for array production and signal detection

array. Therefore, they divided their collection of capture antibodies into two groups that were spotted on a different portion of the same glass slide separated by a Teflon barrier.

A reversed-phase protein array approach that immobilizes the whole repertoire of a tissue's proteins has been developed *(25)*. A high degree of sensitivity, precision, and linearity was achieved making it possible to quantify the phosphorylation status of signal proteins in human tissue cell subpopulations. Using this approach, Paweletz et al. *(25)* have longitudinally analyzed pro-survival checkpoint proteins at the transition stage from patient-matched histologically normal prostate epithelium to prostate intraepithelial neoplasia and to invasive prostate cancer. Cancer progression was associated with increased phosphorylation of Akt, suppression of apoptosis pathways, as well as decreased phosphorylation of ERK. At the transition from histologically normal epithelium to intraepithelial neoplasia, a statistically significant increase in phosphorylated Akt and a concomitant suppression of downstream apoptosis pathways preceding the transition into invasive carcinoma were observed.

A clinically relevant application of protein microarrays is the identification of proteins that induce an antibody response in autoimmune disorders *(26)*. Microarrays were produced by attaching several hundred proteins and peptides to the surface of derivatized glass slides. Arrays were incubated with patient serum and fluorescent labels were used to detect autoantibody binding to specific proteins in autoimmune diseases, including systemic lupus erythematosus and rheumatoid arthritis. In a more recent study *(27)*, a "myelin proteome" microarray was developed to profile the evolution of autoantibody responses in experimental autoimmune encephalomyelitis, a model for multiple sclerosis. Increased diversity of autoantibody responses predicted a more severe clinical course. Chronic experimental autoimmune encephalomyelitis was associated with previously undescribed extensive intramolecular and intermolecular

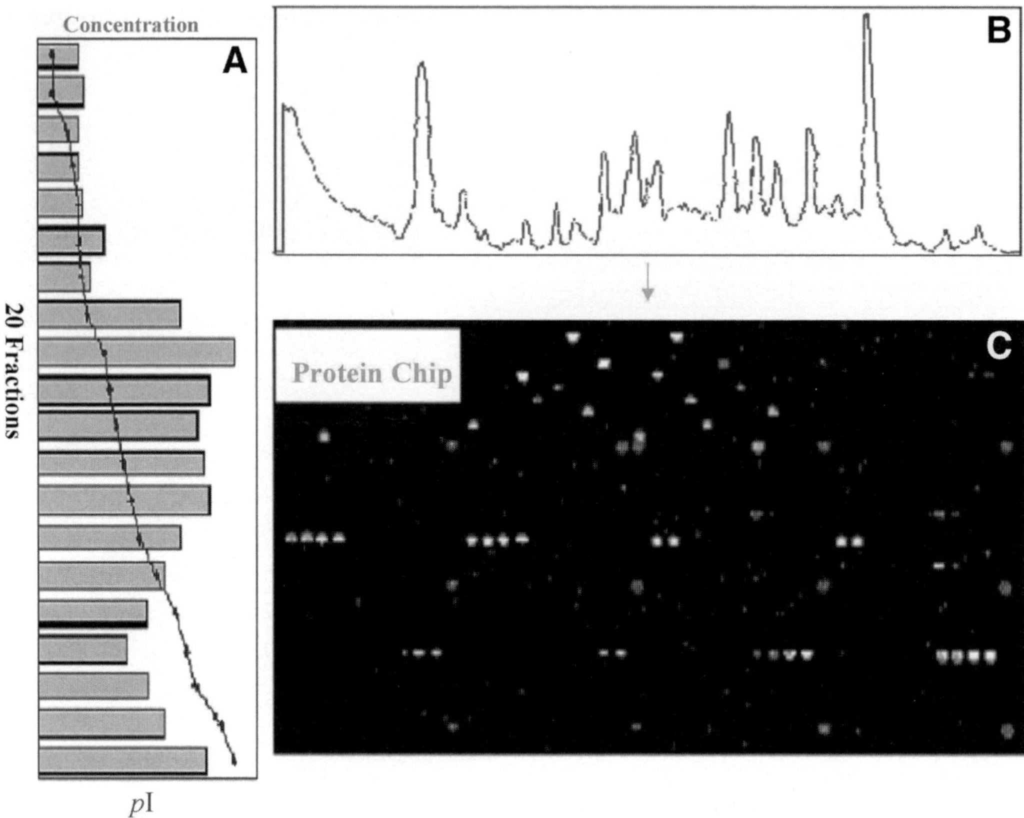

Fig. 3. Production of arrays containing natural proteins. Proteins in cell lysates are separated into 20 fractions, in a first dimension based on their charge (**A**). Each fraction is further separated in a second dimension based on hydrophobicity (**B**). Individual aliquots from each of some 2000 fractions are printed on microarray slides (**C**).

epitope spreading of autoreactive B-cell responses. Proteomic monitoring of autoantibody responses provided a useful approach to monitor autoimmune disease and to develop and tailor disease- and patient-specific tolerizing DNA vaccines.

An important consideration in protein microarrays is that proteins undergo numerous post-translational modifications that might be highly important to their functions and to disease development. However, these modifications are generally not captured using either recombinant proteins or antibodies that do not distinctly recognize specific forms of a protein. One approach for comprehensive analysis of proteins in their modified forms is to array proteins directly isolated from cells and tissues following protein fractionation schemes ***(28)*** (*see* **Fig. 3**). Fractions that react with specific probes are within the reach of chromatographic and gel-based separation techniques for resolving their individual protein constituents and of mass spectrometric techniques for identification of their constituent proteins.

A productive approach for cancer marker identification has been the analysis of serum for autoantibodies against tumor proteins. There is increasing evidence for an immune response to cancer in humans, demonstrated in part by the identification of autoantibodies against a number of intracellular and surface antigens detectable in sera from patients with different cancer types ***(29)***. The identification of panels of tumor antigens that elicit an antibody response might have utility in cancer screening, diagnosis, or establishing prognosis. Such antigens might also have utility in immunotherapy against the disease. There are several approaches for the detection of tumor antigens that induce an immune response ***(29)***. For most antigenic proteins that induce an antibody response in cancer identified using proteomics, posttranslational modifications contributed to the immune response. For example, in a study of lung cancer, sera from 60% of patients with lung adenocarcinoma and 33% of patients with squamous cell lung carcinoma, but none of the noncancer controls, exhibited IgG-based reactivity against proteins identified as glycosylated annexins I and II ***(30)***. Microarrays that contain proteins derived from tumor cells have the potential of substantially accelerating the pace of discovery of tumor antigens and yielding a molecular signature for immune responses directed against protein targets in different types of cancer ***(31,32)*** (*see* **Fig. 4**). In a study of colon cancer ***(32)***, microarrays printed with 1760 separate protein fractions, isolated from the LoVo colon adenocarcinoma cell line, were hybridized with individual sera. A fraction that exhibited IgG-based reactivity with 9/15 colon cancer sera was found to contain Ubiquitin C-terminal hydrolase L3 (UCH-L3) by tandem mass spectrometry (ESI-Q-TOF). The highest levels of UCH-L3 mRNA among the 329 tumors of different types analyzed by DNA microarrays were found in colon tumors. Independent validation by Western blotting demonstrated UCH-L3 antibodies in 19/43 sera from patients with colon cancer and in 0/54 sera from subjects with lung cancer, colon adenoma, or otherwise healthy. These data point to the utility of microarrays printed with natural proteins for the identification of cancer markers.

6. Use of SELDI Biochips for Marker Discovery

The potential of mass spectrometry to yield comprehensive profiles of peptides and proteins in biological fluids without the need to first carry out protein separations has attracted interest. In principle, such an approach would be highly suited for clinical applications because of reduced sample requirements and high throughput. This approach is currently popularized, particularly for serum analysis, by the technology referred to as surface-enhanced laser-desorption ionization (SELDI) ***(33)***. Proteins from a patient sample are captured by a variety of means, including adsorption, partition, electrostatic interaction, or affinity chromatography on various types of surface that retain proteins with distinctive properties. Although such capture surfaces are referred to as "chips," they should not be confused with microarrays, as they do not involve any type of arraying. Although SELDI has achieved some notoriety, the direct analysis of tissues or biological fluids can be simply accomplished using standard matrix-assisted laser-desorption ionization (MALDI) without the use of proprietary surfaces.

The major drawbacks of direct analysis of tissues or biological fluids by MALDI or SELDI are the preferential detection of proteins with a lower molecular mass and the difficulty in determining the identity of proteins whose masses are measured because of lack of correspondence between the masses detected and those predicted for corresponding proteins, because of posttranslational modifications. Occasionally, the masses observed match precisely the predicted masses of specific proteins. This was the case in a study of proteins secreted by stimulated CD8 T-cells, which led to the identification of the small proteins α-defensin 1, 2 and 3 as contributing to the anti-HIV-1 activity of CD8 antiviral factor ***(34)***.

Some quite noteworthy findings have been reported using SELDI. They include the ability to accurately diagnose ovarian, prostate, breast, and other types of cancer with minimal sample requirement and with high throughput. A study of ovarian cancer that has attracted considerable attention demonstrated the ability of SELDI in combination with an algorithm, to correctly identify all cancer patients, including those with limited stage I disease ***(35)***. However, there has been some concern regarding the significance of the findings because the molecules monitored in serum by SELDI-TOF proteomic patterns are likely to be present at concentrations many folds higher than traditional cancer biomarkers . It was felt that such markers would not originate from the tumor and that they are epiphenomena of cancer produced by other organs in response either to the presence of cancer or to a generalized condition of the cancer patient such as debilitation or acute-phase reaction ***(36)***. Thus, the role of MALDI and MALDI surfaces in clinical diagnosis is currently under investigation.

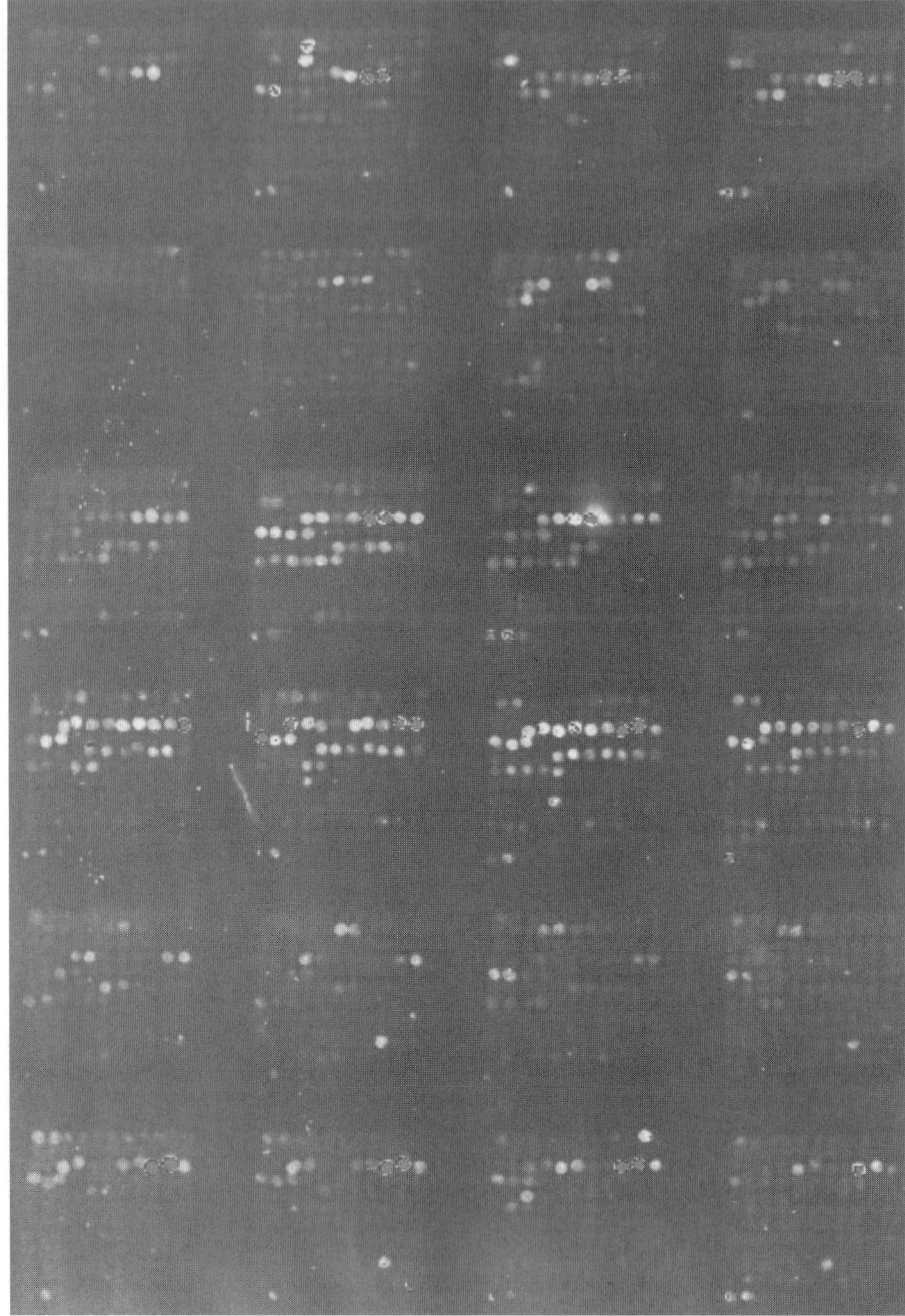

Fig. 4. A microrray containing some 2000 protein fractions obtained from a colon adenocarcinoma cell line, hybridized with serum from a patient with colon adenocarcinoma. Numerous fractions showed strong binding with IgG antibodies in the patient's serum.

7. Conclusion

Although the field of protein microarrays is in its early stages, there is potential for uncovering a wealth of valuable findings, some of which would be directly relevant to diagnostics and others might have relevance for disease classification, assessment of prognosis, and therapy. The current major bottleneck of content is likely to be gradually resolved as proteins, peptides, and their capture agents become available in increasing numbers. It might be envisioned that, in one fashion or another, protein microarrays will be established in the clinical laboratory.

References

1. Phizicky, E., et al. (2003) Protein analysis on a proteomic scale. *Nature* **422(6928),** 208–215.
2. Cahill, D. J. and Nordhoff, E. (2003) Protein arrays and their role in proteomics. *Adv. Biochem./ Eng. Biotech.* **83,** 177–187.
3. Liotta, L. A., et al. (2003) Protein microarrays: meeting analytical challenges for clinical applications. *Cancer Cell.* **3(4),** 317–325.
4. Cutler, P. (2003) Protein arrays: the current state of-the-art. *Proteomics* **3(1),** 3–18.
5. Osada, M., et al. (1998) Cloning and functional analysis of human p51, which structurally and functionally resembles p53. *Nature Med.***4,** 839–843.
6. Haab, B. B., Dunham, M. J. and Brown, P.O. (2001) Protein microarrays for highly parallel detection and quantitation of specific proteins and antibodies in complex solutions. *Genome Biology* **2(2),** RESEARCH0004.1–0004.13.
7. Steinberg, T. H., et al. (2003) Global quantitative phosphoprotein analysis using multiplexed proteomics technology. *Proteomics* **3(7),** 1128–1144.
8. Martin, K., et al. (2003) Quantitative analysis of protein phosphorylation status and protein kinase activity on microarrays using a novel fluorescent phosphorylation sensor dye. *Proteomics* **7,** 1244–1255.
9. Pellois, J. P., et al. (2002) Individually addressable parallel peptide synthesis on microchips. *Nature Biotechnol.* **20(9),** 922–926.
10. Zhu, H., et al. (2001) Global analysis of protein activities using proteome chips. *Science* **293(5537),** 2101–2105.
11. Schweitzer, B., Predki, P., and Snyder, M. (2003) Microarrays to characterize protein interactions on a whole-proteome scale. *Proteomics* **3(11),** 2190–2199.
12. Zhu, H., et al., Analysis of yeast protein kinases using protein chips. *Nature Genet.* **26,** 283–289.
13. Seong, S.-Y. and Choi, C.-Y. (2003) Current status of protein chip development in terms of fabrication and application. *Proteomics* **3(11),** 2176–2189.
14. Templin, M. F., et al. (2003) Protein microarrays: promising tools for proteomic research. *Proteomics* **3(11),** 2155–2166.
15. Plant, A. L., et al. (1991) Immobilization of binding proteins on nonporous supports. Comparison of protein loading, activity, and stability. *Appl. Biochem. Biotechnol.* **30(1),** 83–98.
16. Elia, G., et al. (2002) Affinity-capture reagents for protein arrays. *Trends Biotechnol.* **20(12 Suppl.),** S19–S22.
17. Seong, S. Y. (2002) Microimmunoassay using a protein chip: optimizing conditions for protein immobilization. *Clin. Diagn. Lab. Immunol. 9(4),* 927–930.
18. Afanassiev, V., Hanemann, V., and Wolfl, S. (2000) Preparation of DNA and protein micro arrays on glass slides coated with an agarose film. *Nucleic Acids Res.* **28(12),** E66.
19. Piletsky, S., et al. (2003) Surface functionalization of porous polypropylene membranes with polyaniline for protein immobilization. *Biotechnol. Bioeng.* **82(1),** 86–92.
20. Belov, L., et al. (2001) Immunophenotyping of leukemias using a cluster of differentiation antibody microarray. *Cancer Res.* **61,** 4483–4489.
21. Belov, L., et al. (2003) Identification of repertoires of surface antigens on leukemias using an antibody microarray. *Proteomics* **3(11),** 2147–2154.
22. Knezevic, V., et al. (2001) Proteomic profiling of the cancer microenvironment by antibody arrays. *Proteomics* **1,** 1271–1278.
23. Schweitzer, B., et al. (2002) Multiplexed protein profiling on microarrays by rolling-circle amplification. *Nature Biotechnol.* **20,** 359–365.
24. Lizardi, P. M., et al. (1998) Mutation detection and single-molecule counting usign isothermal rolling circle amplification. *Nature Genet.* **19,** 225–232.

25. Paweletz, C. P., et al. (2001) Reverse phase protein microarrays which capture disease progression show activation of pro-survival pathways at the cancer invasion front. *Oncogene* **20(16),** 1981–1989.
26. Robinson, W. H., et al. (2002) Autoantigen microarrays for multiplex characterization of autoantibody responses. *Nature Med.* **8(3),** 295–301.
27. Robinson, W. H., et al. (2003) Protein microarrays guide tolerizing DNA vaccine treatment of autoimmune encephalomyelitis. *Nature Biotechnol.* **21(9),** 1033–1039.
28. Madoz-Gurpide, J., et al. (2001) Protein based microarrays: a tool for probing the proteome of cancer cells and tissues. *Proteomics* **1(10),** 1279–1287.
29. Hanash, S. (2003) Harnessing immunity for cancer marker discovery. *Nature Biotechnol.* **21(1),** 37–38.
30. Brichory, F. M., et al. (2001) An immune response manifested by the common occurrence of annexins I and II autoantibodies and high circulating levels of IL-6 in lung cancer. *Proc. Natl. Acad. Sci. USA* **98(17),** 9824–9829.
31. Haab, B. B. (2003) Methods and applications of antibody microarrays in cancer research. *Proteomics* **3(11),** 2116–2122.
32. Nam, M. J., et al. (2003) Molecular profiling of the immune response in colon cancer using protein microarrays: occurrence of autoantibodies to ubiquitin C-terminal hydrolase L3. *Proteomics* **3(11),** 2108–2115.
33. Petricoin, E. F., et al. (2002) Clinical proteomics: translating benchside promise into bedside reality. *Nature Rev. Drug Discov.* **1(9),** 683–695.
34. Zhang, L., et al. (2002) Contribution of human alpha-dfensin 1,2 and 3 to the anti-HIV-1 activity of CD8 antiviral factor. *Science* **298(5595),** 995–1000.
35. Albert, A. S., et al. (1993) Regulation of cell cycle progression and nuclear affinity of the retinoblastoma protein by protein phosphatases. *Proc. Natl. Acad. Sci. USA* **90,** 388–392.
36. Diamandis, E. P. (2003) Point: Proteomic patterns in biological fluids: do they represent the future of cancer diagnostics? *Clin. Chem.* **49(8),** 1272–1275.

42

The Human Genome Project

Rahul Chodhari and Eddie Chung

1. Introduction

The completion of the Human Genome Project (HGP) was announced on 24 April 2003 *(1)*. It was a landmark event in the history of biology and represents one of the most remarkable achievements in the history of science and signals the beginning of a new era in biomedical research. This chapter will describe briefly the history that led to the completion of the human genome project, what we know so far, some important immediate and potential future applications, and their implications.

2. History

The completion of the HGP coincides with the 50th anniversary of the publication of the double-helical structure of DNA. It has been an amazing journey over the past 50 yr, from DNAs double helix to the 3 billion bases of the human genome. The discovery of the structure of DNA had a profound impact on biology. It planted the seed of the HGP and created a new view of biology as an information science. The digital nature of DNA structure and its complementarity account for much of its phenomenal impact on science and biology. DNA can accommodate almost any combination of sequence of the bases adenine (A), cytosine (C), guanine (G), and thymine (T). Each gene encodes a complementary RNA transcript, called messenger RNA, made up of A, C, G, and uracil (U), instead of T. The four bases of the DNA and RNA alphabets are related to the 20 amino acids of the protein alphabet by a triplet code of three letters. There are 64 different triplets or codons, only 61 of which encode an amino acid because different triplets can encode the same amino acid, and three of which signal the termination of the growing protein chain. The dictionary of DNA letters that make up the amino acids is known as the genetic code.

The ability to recreate the process of DNA replication in the laboratory led to the development of two techniques that made the HGP possible: a manual DNA-sequencing method in 1975 *(2)* and, in 1985, the discovery of polymerase chain reaction (PCR) *(3)*, whereby DNA sequences could be amplified 10^6-fold or more. The first efforts to sequence DNA, pioneered by Gilbert and Sanger in the 1970s, produced stretches of DNA sequence a few hundred bases long. The combination of technical wizardry and intensive automation in the following decade launched the "genomic era." The advances in sequencing capacity have been remarkable with the latest sequencing machines capable of sequencing over 1.5 million bases in one day.

The HGP traces its roots to an initiative in the US Department of Energy (DOE) in 1947 when the US Congress asked DOE and its predecessor agencies to develop new energy resources and technologies and to evaluate the potential health and environmental risks as a result of their production and use. In 1986, DOE announced the Human Genome Initiative,

From: *Medical Biomethods Handbook*
Edited by: J. M. Walker and R. Rapley © Humana Press, Inc., Totowa, NJ

convinced that a reference human genome sequence would help to achieve its missions. Shortly thereafter, DOE joined with the National Institutes of Health (NIH) to develop a plan for a joint HGP that officially began in 1990 ***(4)***. It soon became an international effort involving hundreds of scientists at 20 sequencing centers in China, France, Germany, Great Britain, Japan, and the United States. The five institutions that generated the most sequence were Baylor College of Medicine (Houston, TX), Washington University School of Medicine (St. Louis, MO), Whitehead Institute/MIT Center for Genome Research (Cambridge, MA), DOE's Joint Genome Institute (Walnut Creek, CA), and The Wellcome Trust Sanger Institute near Cambridge (England).

The primary goals of HGP were to discover all the human genes and make them accessible for further biological study, to determine the complete sequence of the genomes of the human and model organisms such as *Escherichia coli* and the mouse to help develop the technology and to interpret human DNA and gene function, to enhance computational resources to support future research and commercial applications, and to study human variation. In addition, a cent in every dollar was to be spent on ethical and legal issues. All of the project's goals have been completed successfully 2 yr ahead of expectation and for a cost substantially less than the original estimates.

The first genome of a living organism, *Haemophilus influenzae*, was sequenced in 1995 and those of baker's yeast (*Saccharomyces cerevisiae*), the round worm (*Caenorhabditis elegans*), fruit fly (*Drosophila melanogaster*), and mustard weed (*Arabidopsis thaliana*) followed in rapid succession between 1996 and 2000. The announcement in 1998 by Celera, a private company, of plans to sequence the human genome within 3 yr generated acrimony and competition between private and public sequencing efforts. This spurred progress and is considered a major factor in the sequencing being completed ahead of schedule. The first drafts of the human sequence were announced in June 2000 jointly by HGP and Celera at a White House ceremony and were published simultaneously in February 2001 ***(5,6)***. At an amazing speed, the "draft" sequence has since been successfully converted into a "finished" sequence that is highly accurate with less than 1 error per 10,000 letters! The finished version of the human genome now contains 99% of the gene-containing sequence, with the missing parts essentially contained in less than 400 defined gaps. These remaining gaps represent regions of DNA in the genome with unusual structures that cannot be reliably sequenced with current technology. These regions, however, appear to contain very few genes. Closing these gaps will require individual research projects and new technologies rather than the industrial-scale efforts of the Human Genome Project. The high-throughput sequencing of the human genome has, therefore, reached its natural conclusion.

3. What Do We Know About the Human Genome?

The human genome contains 3.2 billion chemical nucleotide bases (A, C, T, and G). Less than 2% of the genome encodes protein. This represents just 5% of the 28% of the sequence that is transcribed into RNA. The functions are unknown for more than 50% of discovered genes. Over half of the DNA consists of repeated sequences of various types whose function are even less well understood: 45% in four classes of parasitic DNA elements, 3% in repeats of just a few bases, and about 5% in recent duplications of large segments of DNA. Most of the "parasitic" DNA came about by reverse transcription from RNA. These stretches of repetitive sequences, sometimes wrongly labeled as "junk DNA," constitute an informative historical record of evolutionary biology ***(7)***. They provide a rich source of information for population genetics and medical genetics and are active agents for change within the genome by introducing changes into coding regions. Over time, rearrangement of the genome caused by these repeats results in entirely new genes or modification of existing genes. They might also shed light on chromosome structure and dynamics.

The coding region of a gene on average consists of 3000 bases, but genomic sizes vary greatly, with the largest known human gene being dystrophin at 2.4 million bases. Genes appear to be concentrated in random regions along the genome between vast stretches of

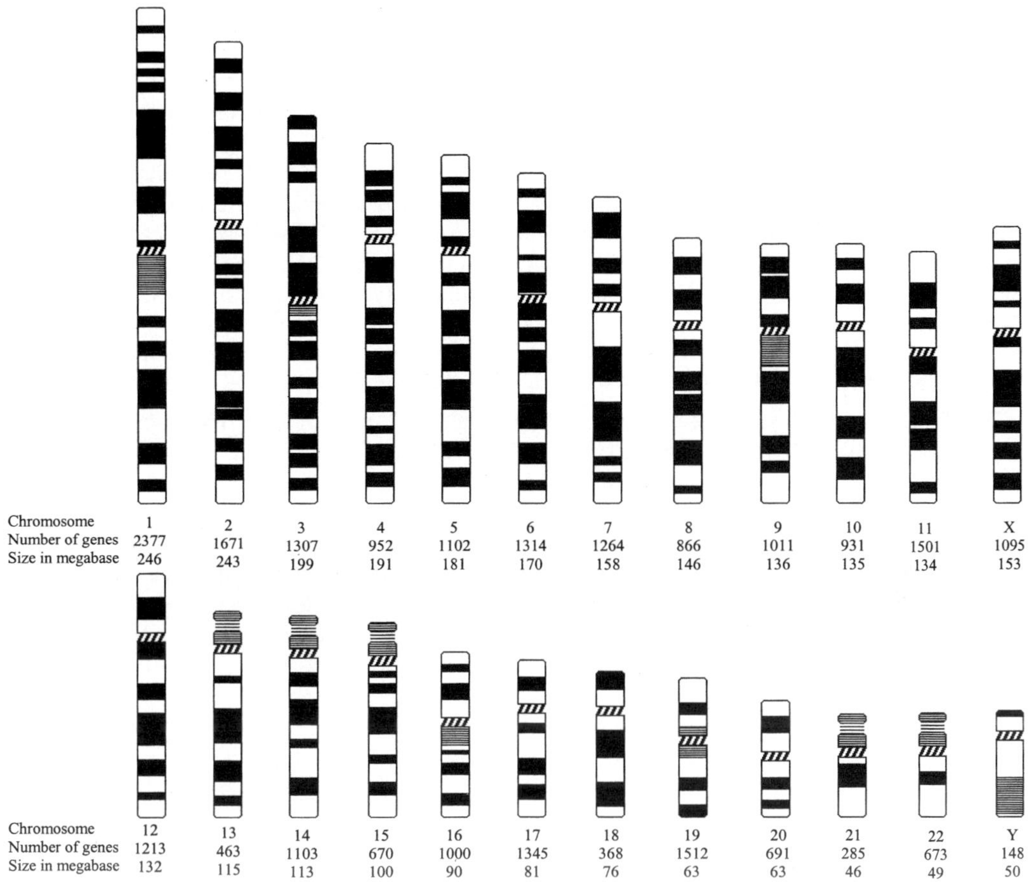

Fig. 1. Ideogram of human chromosomes with their number of genes and size in megabase (Mb).

noncoding DNA. The largest human chromosome, chromosome 1, has the most genes (almost 3000) and the Y chromosome has the fewest (*see* **Fig. 1**). There is a striking tendency for highly expressed genes to cluster in specific chromosomal regions with high gene density. Chromosome 19 is packed with genes at an average of 23 per megabase, whereas chromosome 13 is gene-poor, at just 5 genes per megabase. The three chromosomes responsible for most constitutional trisomies, 13, 18, and 21, show low gene density and low gene expression. This might account for the nonlethal effects of an extra copy of the three chromosomes. In regions of the genome rich in GC bases, the gene density is greater and the average intron size is lower. These introns, which are made up of largely repetitive sequence that breaks up the protein-coding sequences (exons) of genes, are much longer in human DNA than in many other genomes. The presence of these long stretches of repetitive intronic sequence is one factor that makes finding genes by computer difficult in human DNA.

3.1. How Many Genes Do We Have?

There are estimated to be about 25,000 protein-encoding genes in the human genome *(8)*. The exact number of genes remains uncertain. The number of coding genes in the human sequence compares with 6000 for a yeast cell, 13,000 for a fly, 18,000 for a worm, and 26,000 for a plant. None of the numbers for the multicellular organisms is highly accurate because of the limitations of gene-finding programs. It is clear that we do not gain our complexity over worms and plants by having many more genes. The major difference between humans and worms or flies lies in the complexity of our proteins with more domains (modules) per protein and novel combinations of domains. Although the number of human genes is estimated at only about 25,000, it is likely that the number of proteins or gene products encoded by the human genome is greater than this by at least twofold to threefold, as a result of alternative splicing, intergenic recombination, RNA editing, and posttranslational modifications.

Most of our genes come from the distant evolutionary past. In fact, less than 10% of the almost 1300 known protein families in our genome appear to be specific to vertebrates. The genes involved in the most elementary of cellular functions such as basic metabolism, transcription of DNA into RNA, translation of RNA into protein, DNA replication and the like appear to have evolved just once and have stayed relatively unchanged since the evolution of single-celled yeast and bacteria.

3.2. Variation in the Human Genome

The human genome sequence is approx 99.9% identical in all people. This means that there are at least 3 million locations where single-base DNA differences occur. Because a person's genotype represents the patchwork of parental genotypes, we are each heterozygous at about 3 million bases. The most common type of sequence variation is the single-nucleotide polymorphism (SNP), where, by definition, individuals differ in their DNA sequence, by a single base. The human genome has an estimated 10 million SNPs. It is believed that slight variations in our DNA sequences can have a major impact on whether or not we develop a disease and on our responses to environmental factors. Many efforts are currently under way, in both the academic and commercial sectors, to identify these SNPs and to validate them with their frequencies in the population.

4. Applications

The completion of the HGP marks the start of an exciting new era—the era of the genome in medicine and health. The human genome reference sequence provides a magnificent and unprecedented biological resource that will serve as a basis for research, discovery, and ultimately practical application *(9)*.

4.1. Bioinformatics

Continual development of experimental technologies for accumulating diverse biological data and developments in informatics technologies gave birth to the discipline of bioinformatics. The diverse range of large-scale data requires sophisticated computational methods for retrieval, handling, and analysis. The arrival of the Internet has made both user access and software development far easier than before. The availability of genomic information and the aggregated knowledge over the Internet is, indeed, the central success of the HGP.

A fundamental activity of bioinformatics is sequence analysis of DNA and proteins using various programs and databases available on the World Wide Web. The ever-increasing amount of data from the genome projects has required computer databases to develop rapid assimilation, usable formats, and algorithm software programs for efficient management of biological data. Most genome databases now feature more than just basic sequence information; they contain a wide range of annotated information on factors such as disease states, protein structure, and polymorphisms. Some of these databases can be found and accessed at the National Center for Biotechnology Information *(10)* using the Entrez browser, which is a text-based search and retrieval system. These include PubMed for biomedical literature, GenBank for genome sequence, Unigene for gene-oriented clusters of transcript sequences, and OMIM (Online Mendelian Inheritance in Man) for a comprehensive catalog of human genes and genetic disorders (*see* **Fig. 2**). Other useful sites include the human reference sequence database *(11)* and Ensembl *(12)*, which produces and maintains automatic annotation on eukaryotic genomes and sites that list disease susceptibility genetic mutations and polymorphisms and those that enable a search for disease genes given a DNA sequence. The latest 2003 annual

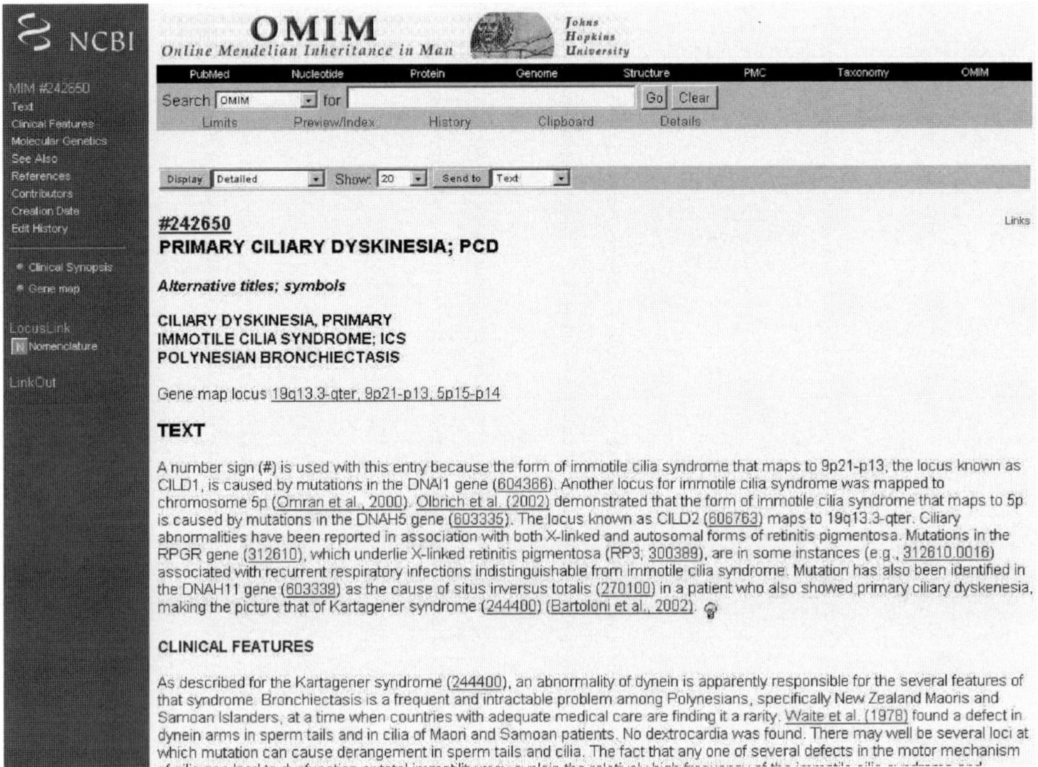

Fig. 2. A snapshot of the page on primary ciliary dyskinesia at Online Mendelian Inheritance in Man (OMIM) database.

database special issue of *Nucleic Acids Research* ***(13)*** (freely available online) contains an index of major databases with a brief summary of what each offers.

One of the simplest and best known search tools is called BLAST (Basic local alignment search tool) ***(14)***. This algorithm software is capable of searching databases for genes with similar nucleotide composition and allows comparison of an unknown DNA or amino acid sequence with hundreds or thousands of sequences from human and other organisms. Databases of known sequences can be used to identify similar sequences, which might be homologs of the query sequence. When a database is searched with the query sequence, local alignment between the query sequence and any similar sequence in the database is conducted. The result of the search is sorted in order of priority based on the degree of maximum similarity (*see* **Fig. 3**). Homology usually implies that sequences can be related by divergence from a common ancestor or share common functional aspects. If homologs or related molecules exist for a query sequence, then a newly discovered protein can be modeled and the gene product predicted.

Apart from analysis of genome sequence data, bioinformatics is now being used for a variety of other important tasks, including analysis of gene variation and expression, analysis and prediction of gene and protein structure and function, prediction and detection of gene regulation networks, simulation environments for whole-cell modeling, complex modeling of gene regulatory dynamics and networks, and presentation and analysis of molecular pathways in order to understand gene–disease interactions. Bioinformatics is an expanding field and will remain an important tool to help molecular biologists and clinical researchers to capitalize on the advantages brought by computational biology.

```
>gi|38087709|ref|XM_145621.3|   Mus musculus similar to retinoblastoma-binding protein 6 isoform 2,proliferation potential-related protein;
RB-binding
          Q-protein 1 (LOC333224), mRNA
         Length = 2598

 Score =  264 bits (133), Expect = 2e-68
 Identities = 262/305 (85%)
 Strand = Plus / Plus


Human: 1   atgtcctgtgtgcattataaatttcctctaaactcaactatgataccgtcacctttgatggctccacatctccctctgcgacttaaagaagcagattatggggagagagaagctgaaa 120
           |||||||||||||| ||||||||||||||||||||||| |||  | ||| |||||||||||||||||| ||||||| |||| | |||||||||| ||||||||||||||||||| ||||||||||
Mouse: 1   atgtcctgtgtgcactataaatttcctctaaactcagctacaacaccatcacctttgatggctccatatctctctcttctacttaaagaaacagattatggggagagaaaagctgaaa 120


Human: 121 gctgccgactgcgacctgcagatcaccaatgcgcagacgaaagaagaatatactgatgataatgctctgattcctaagaattcttctgtaattgttagaagaattcctattggaggtgtt 240
            ||||| |  | || |||||||||||||||||||||  ||||| ||||||||||||||||||| ||||| || || ||||||||||| ||||| ||||| ||||||||||||| |||| |||||| 
Mouse: 121 actgccaatagtgatctgcagatcaccaatgcagagacggaagaagaatatactgatgacaatgcactcatccctaagaattcatctgtgattgtcagaagaattcctgttgtaggtgtg 240


Human: 241 aaatctacaagcaagacatatgttataagtcgaactgaaccagcgatggcaactacaaaa gcagt 305
           || |||| | |||||||||||   ||||||  ||| |||||| ||||| |||||||||| | |||||
Mouse: 241 aagtctaaaggcaagacatatcaaataagtcacactaaaccagtgatgggaactacaaga gcagt 305
```

Fig. 3. Example of sequence comparison using BLAST. mRNA nucleotide sequence comparison of the gene retinoblastoma-binding protein 6 (*RBBP6*) between mouse and human. *RBBP6* is a gene within the critical region of the primary ciliary dyskinesia locus on chromosome 16p.

4.2. Comparative Genomics

A genome sequence contains not only a complete set of genes and their precise locations in the chromosome, but also gene similarity relationships within the genome and across species ***(15)***. Hundreds of genome sequence projects on microbes, plants, and animals have been completed since the initiation of the HGP (*see* **Fig. 4**) and many other sequencing projects are in progress. The accumulation of sequences of organisms, especially the so-called "model" organisms in which experimentation is easy, has greatly aided both our understanding and further studies of disease genes and their biological roles. The availability of the sequences of many of these organisms makes similarity searching possible such that homologous sequences in different organisms can be identified. Such comparison of genome sequences from evolutionarily diverse species has become a powerful tool for identifying functionally important genomic elements ***(16)***. Initial analyses of available vertebrate genome sequences have revealed many previously undiscovered protein-coding sequences. Mammal-to-mammal sequence comparisons have revealed large numbers of homologies in noncoding regions, few of which can be defined in functional terms. The generation of additional genome sequences from several well-chosen species is crucial to the functional characterization of the human genome. Experimental whole-genome querying has become increasingly common using tools such as RNA interference and targeted mutagenesis, particularly in organisms with smaller genomes such as baker's yeast, the nematode worm, and the fruit fly.

4.3. Advances in Genetics: Identification of Disease Genes

All diseases have a genetic component, whether inherited or resulting from the body's response to environmental stresses like viruses or toxins. The completion of the human genome sequence has had a major impact on finding genes associated with human disease. The successes of the HGP have enabled researchers to identify errors in genes that cause or contribute to a disease. The ultimate goal is to use this information to develop new ways to diagnose, treat, cure, or even prevent the thousands of human genetic diseases.

4.3.1. Single-Gene Disorders

The past two decades have witnessed an explosion in both molecular and computational technology, which has facilitated the identification of genes for a large number of inherited human disorders. These successes have been restricted largely to simple Mendelian conditions. Although important to the individuals who carry these genes, they are relatively rare and are, therefore, of limited significance in terms of public health.

There are, in general, three main approaches to mapping the genetic variants involved in a disease: functional cloning, the candidate gene strategy, and positional cloning. In functional cloning, identification of the underlying protein defect leads to localization of the responsible gene (disease→function→gene→map). In the candidate gene approach, genes with a known or proposed function with the potential to cause the disease phenotype are investigated for a direct role in disease. Positional cloning is used when the biochemical nature of disease is unknown. No knowledge of the biology of the disease or trait is required other than a secure assessment of the phenotype. Genetic markers not related to disease physiology and genomewide screens are usually the starting point for mapping the genetic factors of the disease. The aim is first to locate the genetic region within which a disease-predisposing gene lies and, once this is found, to identify the gene and determine its functional and biological role in the disease (disease→map→gene→function). Over the past decade, over 1000 genes causing human diseases or traits have been identified by positional cloning. Huntington's disease is a classic example of successful positional cloning, for which genetic linkage to chromosome 4 was reported in 1983 and the gene was cloned 10 yr later in 1993

Completion of the HGP with its associated advances in bioinformatics and comparative genomics has made both functional and positional cloning of Mendelian traits and diseases relatively straightforward. A disease gene can now be identified in a matter of months by a

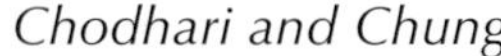

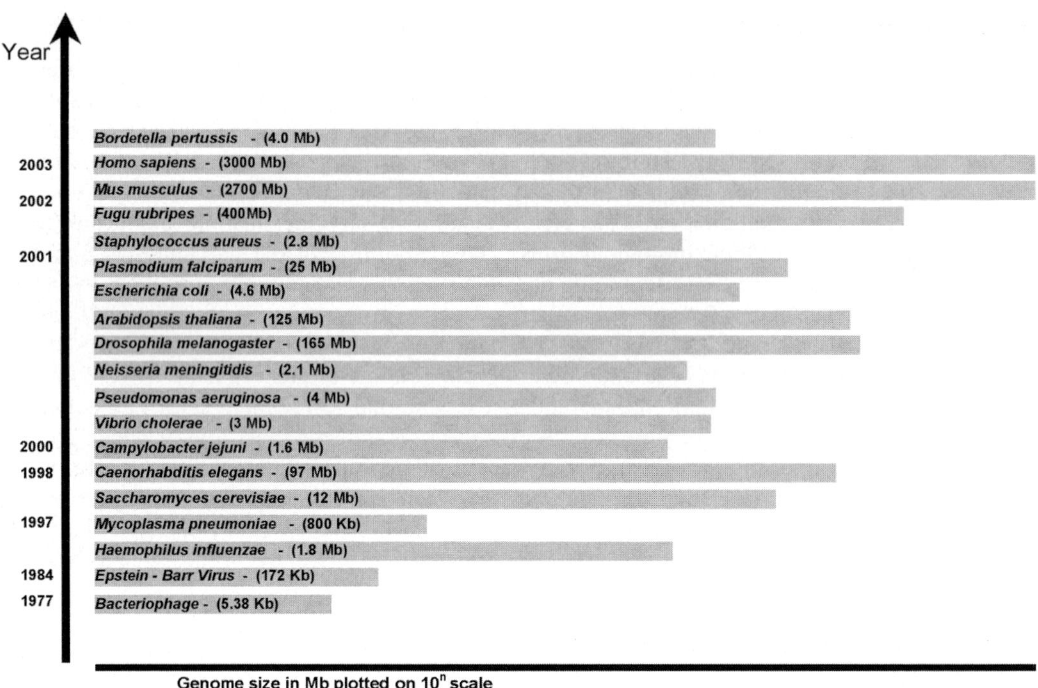

Fig. 4. A graph showing the time line (*Y*-axis) of completion of selected genomes and their respective sizes (*X*-axis) in megabases (Mb) on a logarithmic scale.

single graduate student with an appropriate family resource and access to DNA samples and associated phenotypes, an internet connection to the public genome databases, a thermal cycler, and a DNA-sequencing machine. This has been in part the result of the increased ability to identify candidate genes in a region that has already been identified by positional cloning. For example, we have recently mapped a locus for primary ciliary dyskinesia (PCD) to chromosome 16p12.1-12.2 in a group of families from the Faroe Islands. PCD is a rare autosomal recessive human condition with a clinical phenotype of recurrent respiratory tract infections, subfertility, and *situs inversus*. It is primarily a result of dysfunction of cilia that line the respiratory epithelia and sperm flagella. A critical region was defined by polymorphic DNA markers *D16S403* and *D16S3133* corresponding to a genetic distance of 2.8 cM (centimorgans) and a physical distance of 1.5 Mb. At the click of a button on a computer, the human genome sequence data of this critical region are available at the NCBI *Homo sapiens* map view (*see* **Fig. 5**) or the UCSC Genome Bioinformatics website, which contains 15 distinct known genes with strong experimental support. Further bioinformatic analysis of the genomic sequence data identified another 11 genes based on gene prediction programs with the help of expressed sequence tags data, human cDNA collections, and protein-sequence annotation. The critical region, however, contains no genes with predicted function or structure related to cilia. Flagella of *Chlamydomonas* and *Trypanosomes* share high homology in structure to human cilia, which makes these organisms ideal model systems for PCD. The genome projects of these organisms are close to completion, with different draft stages publicly available ***(17,18)***. This allows extensive comparative bioinformatics to be carried out to aid candidate gene selection within this relatively large critical region. This analysis revealed protein homology (30% identity and 40% similarities of the amino acid sequences) to both model organisms for 10 of the 26 known and predicted genes. Further prioritization will be done on the basis of predicted function and expression profile. In addition, technical advances in genetic manipulation of these model

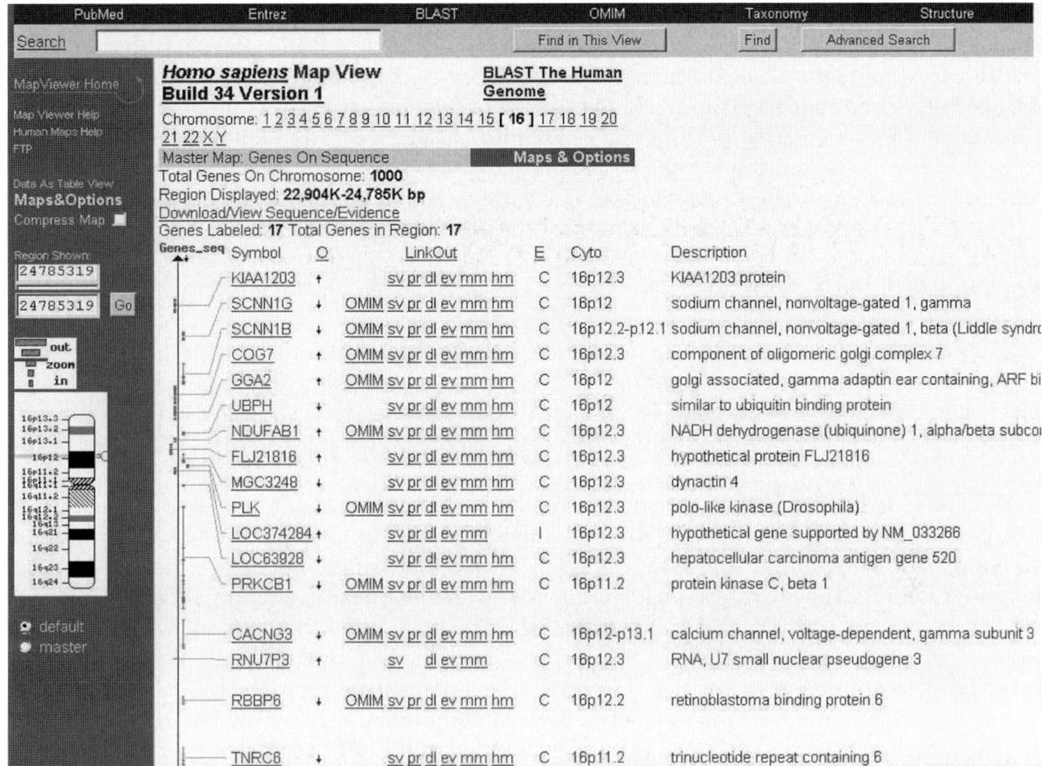

Fig. 5. A snapshot of NCBI *Homo sapiens* map view of the critical region of the primary ciliary dyskinesia locus on chromosome 16 between 22,904,000 to 24,785,000 basepairs. Genes are listed with web links to other databases such as OMIM and LocusLink, providing annotated descriptive information on the gene.

organisms such as RNA interference offers the possibility of evaluating the flagellar phenotype of mutant *Chlamydomonas* and *Trypanosoma* strains of some of the candidate genes and functional analysis.

4.3.2. Genes in Complex Disorders

Complex or multifactorial diseases result from the interaction of environmental factors and multiple genes. These include common diseases such as diabetes, epilepsy, coronary heart disease, cancers, and asthma, which are all major contributors to the morbidity and mortality of the population. The study of genomics will, therefore, most likely make its greatest contribution to health by revealing mechanisms of these common and complex diseases. However, in general, genetic variation has only subtle effects on susceptibility to common diseases, and the way in which genetic and environmental factors interact to cause disease is highly complex.

It is clear that, owing to weak linkage signals, positional cloning has limited use in the identification of lower-relative-risk, disease-associated variants. With advances in technology such as DNA microarray (DNA chip) and the presence of millions of SNPs in the human genome, it is hoped that studies using SNPs will provide new approaches to detect genetic variants which predispose to these common and complex diseases ***(19)***. Of the 1.4 million SNPs currently on the public map, only 60,000 are located in protein-coding regions, called exons, and relatively few of these change amino acids ***(20)***. The SNPs that change the amino acid sequence and the variants in gene regulatory regions that control protein expression levels are most likely to have a significant effect on the protein product of a gene. In cases where change of a single base in the genome sequence is sufficient to cause disease, it has become possible to identify this change and improve our understanding of the disease. More commonly, the SNPs are used as markers to track inheritance of genetic variants in common and complex diseases by linkage disequilibrium and epidemiological association studies (i.e., studies of affected persons and control subjects). This is because a SNP might be a marker of biological variant that happens to be associated with a disease or health because of its physical proximity to the genetic factor that is actually the cause. For this reason, recombination between the SNP and the causative genetic factor has occurred only rarely as genetic material has passed through generations. In genetic terms, the SNP and the causative genetic factor are said to be in linkage disequilibrium (LD). In the presence of LD, nearby SNPs can serve as proxies for each other in a genetic association study. Hence, a subset of SNPs spaced throughout the genome might allow a comprehensive test of common genetic variation across the entire genome. Although the exact number of SNPs needed for linkage disequilibrium studies is unknown, the 1.4 million SNPs in the public domain should offer a sufficient number to explore most regions of the genome.

Recently, it has been shown that sets of SNPs on the same chromosome are inherited together (haplotypes) separated by short segments of very low LD sometimes referred to as "recombination hot spots" *(21)*. Whereas a SNP represents a single-nucleotide variant, a haplotype represents a considerably longer sequence of nucleotides (averaging about 25,000), as well as any variants, that tend to be inherited together. In chromosomal regions where there are limited haplotype diversity, most of the chromosomes in the population can be accounted for by only a small number of distinct haplotypes known as haplotype blocks (*see* **Fig. 6**). An extension of the current efforts to catalog SNPs and correlate them to phenotype is efforts to map and use haplotypes. It is hoped that the generation of the SNP haplotype map of the human genome will dramatically decrease the number of individual SNPs to be scanned in association studies. The new map should facilitate our attempts to identify the genetic variants in complex, common diseases and how they can contribute to responses to environmental factors.

At present, SNP detection in large-scale genotyping studies is still prohibitively expensive. It will be necessary to improve our ability to detect SNPs at a lower cost. Given the large number of SNPs and the low probability that any specific one causes disease, the sample sizes in association studies need to be large enough to observe associations in a reproducible fashion and to achieve adequate statistical power. This also raises the problem of accurate definition of phenotype, because the same disease can manifest itself with different patterns in different patients. Interpretation of association studies is further complicated by the number of genes, the number of variants in each gene, and the frequency of a variant within a population. Many of these difficult technical and analytical problems need to be overcome before the promise of SNPs can be fulfilled.

4.4. Diagnosing and Predicting Disease and Disease Susceptibility

A major impact on the clinical practice of medicine as a result of the HGP will be in the area of genetic testing *(22)*. As hereditary factors in disease are identified, DNA-based tests will increasingly be used to diagnose a disease, to provide prognostic information about the course of disease, and to detect a disease in presymptomatic individuals and might, in some cases, be able to predict the risk of future disease in healthy individuals or their children.

There are several hundred genetic tests currently in clinical use. Most of these tests detect mutations associated with rare single-gene disorders such as Duchenne muscular dystrophy and cystic fibrosis. The genetic cause has been identified for more than 1000 single-gene diseases. Many of these are devastating conditions that are fatal during childhood or early adulthood. For most of these diseases, genetic testing has so far, in general, offered only the option of reproductive choice for affected families. For example, the ability to test for Duchenne muscular dystrophy either before birth or even preimplantation has enabled some individuals to avoid the devastating consequences of severe childhood disability and death for their family.

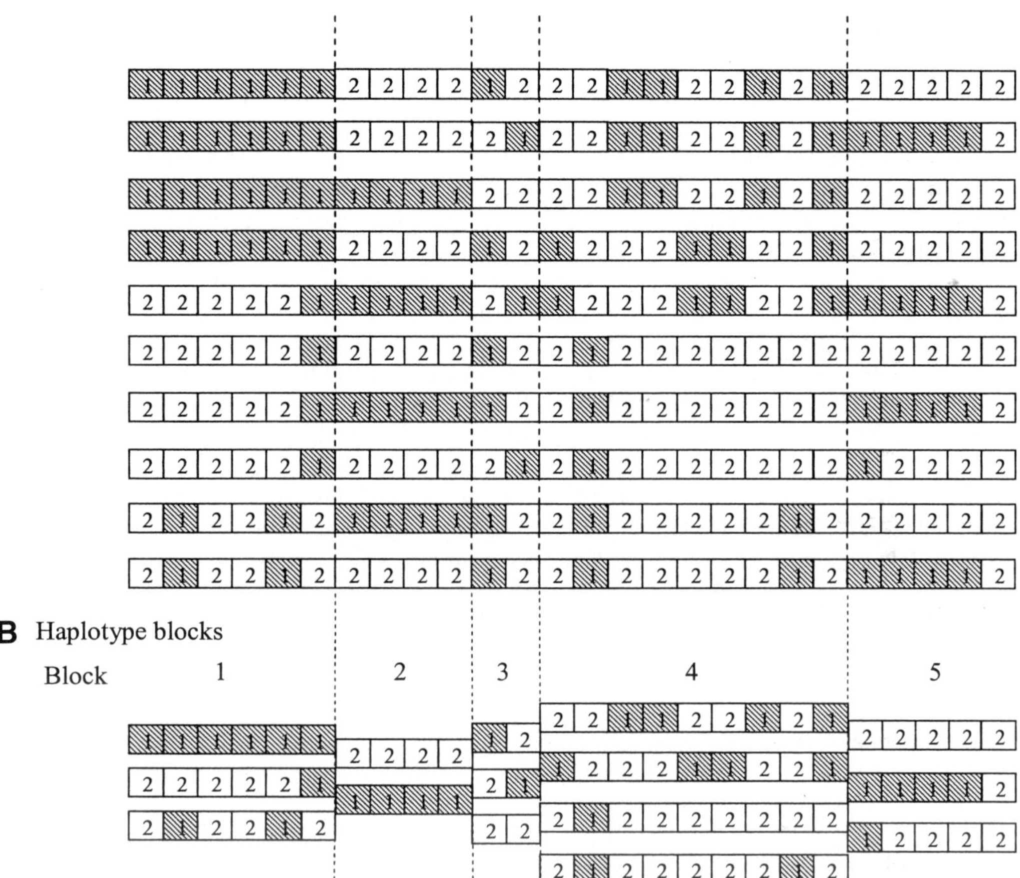

Fig. 6. Diagrammatic representation of SNP haplotypes and blocks. 1 and 2 represent SNPs with two alternative alleles 1 or 2. (**A**) Pattern of SNPs identified over a region (haplotyes) on most chromosomes in the population. (**B**) These haplotypes can be represented by a smaller number of distinct haplotype blocks separated by recombination hot spots (dashed lines).

For late-onset single-gene diseases, the emphasis falls on predictive testing for presymptomatic adults. If no treatment is available, as in the case of Huntington's disease with its severe neurological deficit and psychiatric symptoms, individuals face difficult personal dilemmas in deciding whether they really want to know their own future. This can produce very significant emotional and psychological effects on the person. For diseases where there are effective interventions to reduce their risk, it is important to identify the at-risk individuals. Strategies for identifying these people usually rely on testing individuals in families that have a strong family history of the disease or in which there is a family member known to carry a causative mutation. One example is familial hypercholesterolemia (FH), which is associated with premature development of arteriosclerosis and early onset of coronary heart disease. With the development of HMG-CoA reductase inhibitors (statins), it is now possible to reduce the risk of or even reverse the development of arteriosclerosis in patients with this condition. It is, therefore, important to identify at risk members of families with a positive history of the condition.

Recently, tests have been developed to detect mutations for a few complex disorders such as breast, ovarian, and colon cancers. Some of these tests are being used to make risk estimates in presymptomatic individuals with a family history of the disorder and, in some cases, have led to better treatment and outcome. For example, regular colonoscopies for those having mutations associated with colon cancer could potentially save thousands of deaths each year. The use of gene expression profiling in breast cancer has also shown great promise in assessing prognosis and guiding therapy.

The public and many medical communities remain unaware of the scientific and social implications of genetic testing. Although some of these tests have greatly improved and even saved lives, scientists remain unsure of how to interpret many of them. It is important not to confuse having disease genes with having the diseases that they cause. In any case, many of the tests available cannot detect every mutation associated with a particular condition. Different mutations in the same disease gene might have very different effects leading to a range of phenotypes. As a result, they can present different risks to different people and populations. In addition, results of genetic testing can affect family dynamics and pose risks for population groups if they lead to stigmatization. As genetic testing using DNA sequence becomes more common, less expensive, and more accurate, it will even be used in cases of mistaken identity, paternity dispute, and the identification of missing persons. Ascribing behavioral traits to a person's genes will remain controversial and cause many problems, not least for the courts that must resolve disputes when an individual's behavior and actions conflict with laws.

4.5. Drugs and Therapeutics Development

Pharmacogenomics is the new discipline that combines pharmacology with genomic capabilities. A virtually complete list of human gene products and how they are involved in diseases, disease pathways, and drug-response sites will lead to the discovery of thousands of new targets and drugs. Even though not all of the 25,000 or so human protein-coding genes will have products suitable as targets for drug development, there is an enormous untapped pool of human gene-based targets for therapeutic intervention.

It is well recognized that different people respond in different ways to the same drug or medical treatment. Thousands of people die each year from adverse responses to medications that might be beneficial to others. Many experience serious reactions, whereas others fail to respond at all. These differences depend on individual variation in the way in which drugs are absorbed and handled within the body as well as different reactions at target sites. Genetics contributes significantly to the variability in drug disposition and effects *(23)*. There are now numerous examples of cases in which interindividual differences in drug response are the result of sequence variants in genes encoding drug-metabolizing enzymes, drug transporters, or drug targets. One example is the cytochrome P450 multigene family, which is involved in drug metabolism and has become the focus of much current research in this area.

In the near future, new classes of medicines will be developed based on the information from gene sequence and protein structure function rather than the traditional trial-and-error method. Most drugs will be "reagent-grade pure" because they will be manufactured by recombinant DNA technology. New drugs, aimed at specific sites in the body and at particular biochemical or metabolic pathways leading to disease, will be more effective and safer. In the near future, we will be able to correlate DNA variants with individual responses to drugs, identify particular subgroups of patients, and develop drugs customized for those populations to guide treatment with the most effective drugs and to reduce adverse reactions. Ultimately, one hopes to be able to identify the right drug and dose for each patient.

4.6. Gene Therapy

Gene therapy, the potential for using genes themselves to treat disease is one of the most exciting applications of DNA science. It is the "Holy Grail" or the ultimate goal of many researchers as well as families affected by genetic diseases that information about the molecular defect involved will enable effective treatments to be developed. It has captured the imaginations of the public and the biomedical community for good reason. This rapidly developing field holds great potential for treating or even curing genetic and acquired diseases, using normal genes to replace or supplement a defective gene or to improve a normal function such as immunity. More than 600 clinical gene-therapy trials involving over 3000 patients were identified worldwide in 2002. Although most trials focus on various types of cancer, studies also involve other multigenic and monogenic, infectious, and vascular diseases. There has been one notable success of gene therapy in its use to treat a very rare genetic immunodeficiency syndrome. On the whole, however, gene-therapy cures for most genetic diseases still seem to be some way off. In the foreseeable future, we will, in general, use personalized modifications of the environment to translate genomics-based knowledge into improvements in health.

5. Social/Legal Implications

Many of the personal and social consequences as a result of a greater knowledge of the human genome remain uncertain ***(24)***. These include privacy and confidentiality of genetic information, fairness in the use of genetic information by private, commercial, and government organisations, psychological impact, stigmatization and discrimination because of an individual's genetic makeup, and uncertainties associated with genetic testing. People worry that their genetic information might be made available to others. The most commonly expressed fear is that genetic information will be used in ways that could be detrimental to people—for example, to deny them access to health insurance, employment, and education. People might react and behave very differently in the face of more precise information about their individual risk to various diseases. Some might react to the knowledge that they are susceptible to a particular disease by taking every precaution, whereas some might not want the information at all. Some might regard disease as preordained and feel that nothing can be done to alter the course of events. The arguments about these issues are very complex and have already created a large cohort of scholars in ethics, law, social science, clinical research, theology, and public policy. It is important that rational policies must be put in place to increase the public awareness, to protect against misuses of genetic information, and to avoid the possibility of discrimination and the danger of creating a "genetic underclass."

6. Conclusion

The completion of the HGP is one of the most remarkable achievements in the history of science and medicine. Its importance, however, lies in what it has begun rather than in what it has finished. It provides the complete human dictionary of DNA letters from which we will identify what the genes code for, to study what the proteins do, and what happens in the living system. The potential benefits and applications are unlimited. There are many difficult steps to be taken before the full potential of the HGP can be fulfilled. These will require collaborations among scientists, medical professionals, politicians, and many commercial and academic institutions to establish appropriate research strategy and legal and social policy, to provide adequate resource, training and education, and to develop groundbreaking technological tools. Over the next 50 yr, things will probably change faster than we expect, but usually not in the way that we anticipate. If the past 50 yr of biology is any indication of the future, the best is yet to come.

References

1. National Human Genome Research Institute. http://www.genome.gov/11006929.
2. Sanger, F. and Coulson, A. R. (1975) A rapid method for determining sequences in DNA by primed synthesis with DNA polymerase. *J. Mol. Biol.* **94,** 441–448.
3. Mullis, K., Faloona, F., Scharf, S., Saiki, R., Horn, G., and Erlich, H.. (1986) Specific enzymatic amplification of DNA in vitro: the polymerase chain reaction. *Cold Spring Harbor Symp. Quant. Biol.* **51(Pt. 1),** 263–273.

4. US Department of Health and Human Services, US DOE. (1990) *Understanding Our Genetic Inheritance. The US Human Genome Project: The First Five Years*. National Institutes of Health, Bethesda, MD.
5. International Human Genome Project Consortium (2001) Initial sequencing and analysis of the human genome. *Nature* **409,** 860–921.
6. Venter, J. C., Adams, M. D., Myers, E. W., et al. (2001) The sequence of the human genome. *Science* **291,** 1304–1351.
7. Makalowski, W. (2003) Not junk after all. *Science* **300,** 1246–1247.
8. Pennisi, E. (2003) Gene counters struggle to get the right answer. *Science* **301,** 1040–1041.
9. Guttmacher, A. E. and Collins, F. S. (2002) Genomic medicine—a primer. *N. Engl. J. Med.* **347,** 1512–1520.
10. National Center for Biotechnology Information. www.ncbi.nlm.nih.gov/Entrez.
11. Human Genome Browser Gateway. http://genome.ucsc.edu/cgi-bin/hgGateway.
12. Ensembl Genome Browser. http://www.ensembl.org/.
13. *Nucleic Acids Research* (2003) **31,** 1–516.
14. Basic Local Alignment Search Tool. www.ncbi.nlm.nih.gov/BLAST/.
15. Rubin, G. M., Yandell, M. D., Wortman, J. R., et al. (2000) Comparative genomics of the eukaryotes. *Science* **287,** 2204–2215.
16. Thomas, J. W., Touchman, J. W., Blakesley, R. W., et al. (2003) Comparative analyses of multispecies sequences form targeted genomic regions. *Nature* **424,** 788–793.
17. The JGI Genome Portal. http://genome.jgi-psf.org/.
18. The *Trypanosoma brucei* Genome Project. http://www.sanger.ac.uk/Projects/T_brucei/.
19. Botstein, D. and Risch, N. (2003) Discovering genotypes underlying human phenotypes: past successes for mendelian disease, future approaches for complex disease. *Nature Genet.* **33,** 228–237.
20. International SNP Map Working Group (2001) A map of human genome sequence variation containing 1.42 million single nucleotide polymorphisms. *Nature* **409,** 928–933.
21. Wall, J. D. and Pritchard, J. K. (2003) Haplotype blocks and linkage disequilibrium in the human genome. *Nature Rev. Genet.* **4,** 587–597.
22. Burke, W. (2002) Genetic testing. *N. Engl. J. Med.* **347,** 1865–1875.
23. Weinshilboum, R. (2003) Inheritance and drug response. *N. Engl. J. Med.* **348,** 529–537.
24. Clayton, E. W. (2003) Ethical, legal, and social implications of genomic medicine. *N. Engl. J. Med.* **349,** 562–569.

43

Prenatal Diagnosis of Inborn Errors of Metabolism

Guy T. N. Besley

1. Introduction

Inborn errors of metabolism are inherited conditions where there is a defect or block in a metabolic pathway leading to the accumulation of metabolite(s) proximal to the defect and a metabolic deficit distal to this *(1)*. The defect usually results from a specific enzyme deficiency; however, defects in enzyme cofactors or activators/stabilisers can also lead to an enzyme deficiency in vivo. There might also be defects in certain transporter systems, at the plasma membrane level or within cellular compartments, which will also lead to metabolic disorders.

Diagnosis of individual disorders is generally made initially by the identification of specific accumulating metabolites, and in most cases, urine is the most practical sample. When the affected metabolite is of relatively high molecular weight, it might accumulate within the cell and significant amounts might not escape into urine. In such storage disorders, metabolite analysis might be of limited value. The analysis and identification of abnormal levels of specific metabolites might point to a diagnosis but is not usually confirmatory, because more than one enzyme defect might be responsible for a characteristic metabolite abnormality. For example, four different enzyme deficiencies underlie Sanfilippo disease (MPS III), but all result in the accumulation of heparan sulfate. It is, therefore, important for subsequent prenatal diagnosis that the responsible enzyme be identified. However, some enzymes are cellular or tissue-specific and might, therefore, not be conveniently measured without invasive biopsy procedures. In such situations, DNA mutation analysis might help in establishing a diagnosis.

1.1. Metabolic Disorders

The range of metabolic disorders is vast. Some 5000 genetic conditions are known in man and, of these, around 400 have a confirmed underlying metabolic defect. With the recent advances in technology, particularly DNA analysis, many newly recognized disorders are described each year. There is, therefore, an expectation that prenatal diagnosis will become immediately available for all of these conditions. In practice, however, much of this pioneering work is carried out in multidisciplinary research establishments that might not be able to respond to diagnostic investigations, such as prenatal diagnosis. Sometimes, the workers themselves will have moved on by the time their important work is published. There will need to be a dialogue with these centers well in advance of any prenatal studies.

Some conditions are very rare; others are more common. However, the incidence of individual disorders will vary in different populations around the world. Maple syrup urine disease has an incidence of around 1 in 200,000 worldwide, but among the Mennonite kindred in Pennsylvania, the incidence is 1 in 176 *(1)*. Therefore, experience with different disorders, particu-

From: *Medical Biomethods Handbook*
Edited by: J. M. Walker and R. Rapley © Humana Press, Inc., Totowa, NJ

larly the rarer types, might be concentrated in different centers around the world. For many conditions, however, an incidence of 1 in 50,000–100,000 might be expected in the general population.

Many metabolic disorders are relatively mild or respond well to treatment (e.g., histidinemia, phenylketonuria, or biotinidase deficiency). Prenatal diagnosis is generally not requested or even justified in these cases. The conditions that do require prenatal diagnosis are those that are severe and life-threatening, where there is no satisfactory treatment. Some 150 of these disorders fall into this category and will require methods for prenatal diagnosis.

In some populations, carrier testing might have been carried out to identify couples at high-risk. Such is the case for Tay–Sachs disease among the Ashkenazi Jewish population *(1)*. In these families, it would be possible to offer prenatal diagnosis at the first pregnancy. However, for most "at-risk" families, prenatal diagnosis cannot be offered until they have had an infant with a confirmed diagnosis. Because each metabolic disorder results from a different underlying biochemical defect, there is no screening test to cover the whole range of disorders, as is the case for chromosome analysis. In this chapter, some general guidelines will be discussed, such as the prerequisites that must be meet, the choice of sample, and some approaches to diagnosis for different disorders. It would be totally impractical to provide methods for each of these conditions. However, examples are discussed and tables are provided, listing a number of disorders. Also, there have been a number of other reviews on this topic, to which the reader is referred *(2–4)*.

2. Prerequisites to Prenatal Diagnosis of Metabolic Disease

The first and most important requirement is to establish a firm diagnosis in the index case. This will probably take the form of a clinical diagnosis, substantiated by metabolite analysis and confirmed by enzyme diagnosis. Specific DNA mutations might also have been identified, although this information is generally not available. These findings provide the most satisfactory basis for offering prenatal testing. It is usually not sufficient to offer prenatal diagnosis unless the diagnosis is unequivocal in the index case. Even so, requests are often made to exclude disorders that have not been fully evaluated in the index case. This is dangerous and can result in the wrong condition being tested for.

In practice, most prenatal diagnoses are based on individual enzyme assays in chorionic villus samples (CV) or cultured cells. Consequently, this enzyme must have been shown to be clearly deficient in the proband. Next, the enzyme and its deficiency must be expressed in samples that can be safely taken from the fetus. Usually, this would be CV, but cultured cells (CV or amniotic) might also be used. Where the activity is not expressed in these samples, invasive procedures such as fetal liver biopsy will have been used in the past. However, with the advent of DNA-mutation analyses, alternative approaches are now potentially available.

Levels of enzyme activity and potential problems of interfering isoenzymes must be fully understood before embarking on any prenatal diagnosis. In some cases, a family might exhibit a pseudodeficiency of the enzyme under study *(5)*. This might result from a different allelic mutation, as seen for arylsulfatase A activity, the enzyme associated with the lysosomal disorder metachromatic leukodystrophy. The assay of this enzyme in CV might also be compromised by the high activity of another sulfatase, steroid sulfatase, which could mask an underlying deficiency in the fetus. It should also be remembered that for autosomal recessive disorders, 50% of tested fetuses would be expected to be heterozygous, having half the gene dosage. However, heterozygous enzyme activities can vary from 15% to 100% of normal. Low heterozygous values could lead to difficulties in interpretation, unless prior information from studies on the parents were available.

All this means is that experience in prenatal diagnosis is best retained in a limited number of centers so that each can gain sufficient experience to overcome potential difficulties. Because many metabolic disorders are quite rare, it is likely that relatively few individuals are available to undertake all of the varied assays. It is important, therefore, that early contact is made with the referral center to make sure that they are able to accept the sample and have the suitable reagents available at the time. For example, some conditions require radioisotopes with short half-lives. Ideally, the center should be the same one that made the original diagnosis and knows the family and the expected enzyme activities in affected individuals and carriers. Often, it will be necessary to send samples to other centers, perhaps abroad, for testing. It will also be important that individual laboratories unable to undertake specific tests are themselves able to recommend others.

3. Choice of Sample for Prenatal Diagnosis

The choice of sample can be dictated by which disorder is to be tested for. In most cases, direct enzyme assay of CV will be the preferred option because this sample can be obtained reasonably early into the pregnancy and the result should be available in a few days. For some disorders, there might be a choice of options; for others, a specific type of sample is required. Similarly, the amount of CV or cultured cells required will depend on the sensitivity of the assay. Discussions beforehand are, therefore, important to avoid any confusion. The advantages and disadvantages of different samples are discussed below.

3.1. Amniocentesis

3.1.1. Amniotic Fluid

Analysis of specific metabolites in amniotic fluid is useful for a number of disorders. However, the levels measured are quite low compared with that in urine from patients. For organic acid analysis, it is necessary to quantify these by gas chromatography with mass spectrometry (GC-MS) using stable isotopes as internal standards ***(6)***. Metabolite levels for some disorders are significantly increased, but for others, the increases might be only moderate. However, because concentrations of metabolites vary with gestational age, analysis is best carried out at 15–16 wk. Samples taken earlier might prove difficult to interpret. Diagnostic studies on amniotic fluid are carried out in only a few specialized laboratories. Usually, the supernatant taken from a fresh sterile sample can be sent directly, but some laboratories might require this to be dispatched frozen. Generally, about 5 mL of the supernatant will be required, but the gestational age must be known. For the analysis of sterols in the diagnosis of Smith–Lemli–Opitz syndrome, it might be necessary to analyze the total fluid-containing cells to demonstrate increased levels of 7-dehydrocholesterol. In all cases, it is advisable to take some cells to culture for confirmatory of further studies.

Recent advances in the analysis of acylcarnitines by tandem mass spectrometry might prove to be a useful approach ***(7)***. Diagnostic studies have been carried out for propionic acidemia, methylmalonic aciduria, and glutaric aciduria types I and II. This is a rapid and sensitive approach, but experience is still very limited. The diagnosis of some amino acid disorders has been achieved by direct analysis of liquor [e.g., citrullinemia and argininosuccinic aciduria ***(8)***], as well as variants of phenylketonuria resulting from defects in pterin metabolism ***(9)***.

The analysis of gycosaminoglycans in liquor has also proved to be a reliable indicator for most MPS disorders ***(10)***. Levels of specific glycosaminoglycans are clearly elevated for MPS I, II, III, and VI when analyzed by two-dimensional electrophoresis. However, the increase of keratan sulfate in MPS IV (Morquio disease) might be less evident. In all cases, it is important that confirmatory diagnostic enzyme studies always be carried out in cultured amniotic fluid cells.

3.1.2. Cultured Amniotic Fluid Cells

These have been used successfully in the diagnosis of a wide variety of metabolic disorders. If the metabolic disorder is expressed in cultured skin fibroblasts, it is likely that it will also be expressed in cultured amniotic fluid cells. Since the 1970s, there has been a wealth of experience using these cells. Most of the early studies were based on the assay of enzyme activities for the prenatal diagnosis of lysosomal storage diseases. As well as measuring specific enzyme activities, cultured amniotic fluid cells can be used to study a metabolic pathway, in order to identify a defective step. Thus, the incorporation of a radioactive precursor into cell protein has been used successfully for a variety of disorders. For example, the incorporation of ^{14}C-propionate into cell protein can be used to monitor for deficient enzyme activities in propionic acidemia or methylmalonic aciduria, or even defects in cofactor metabolism as seen in B12-dependent methylmalonic aciduria. Similarly, the release of $^{14}CO_2$ from 1-^{14}C-leucine can be used to monitor branched-chain amino acid decarboxylase activity in the diagnosis of maple syrup urine disease. Although these indirect methods are useful, there might also be limitations. For example, in early amniotic fluid cell cultures containing a significant number of epithelioid cells, the incorporation rates can be much reduced and lead to a false-positive diagnosis.

For many diagnostic tests, at least two 25-cm^2 flasks of confluent cells would be required. The number of cells required will depend on the sensitivity of the assay under test. For this reason, it is important that laboratories responsible for culture confer with those undertaking the tests exactly how many cells are required. It might also be necessary to check whether the culture medium being used is appropriate. When sending cultured cells to other laboratories, it is best to send cells that are confluent or just confluent; flasks should be filled with culture medium so as to prevent cells being sloughed off in transit. There is nothing more annoying than receiving a flask with only a few cells attached and having difficulty growing these up in time for an early result. Similarly, when overconfluent cells are sent, these might detach in transit and no longer be viable. One of the major disadvantages of using cultured amniotic fluid cells is the delay in waiting for sufficient cells to test. It will usually take 2–3 wk to grow sufficient cells from amniotic fluid taken at 16 wk gestation. Earlier amniocentesis samples generally contain fewer cells that take longer to grow and offer little advantage for biochemical analyses.

3.1.3. Chorionic Villus Tissue

Chorionic villus tissue offers great advantages over amniocentesis samples. Over the last 15 yr this has been the sample of choice for most metabolic disorders. Samples taken at around 11 wk gestation can be analyzed directly and the result obtained in a matter of days. Experience over the last decade or so has shown that CV can be used successfully for the diagnosis of some 60 different disorders ***(11–13)***.

The levels of enzyme activity in CV can differ from that in cultured amniotic fluid cells. It is important that reference values have been established for all enzymes to be used for prenatal diagnoses. The activities of the lysosomal enzymes α-neuraminidase, α-iduronidase, and sphingomyelinase are relatively low in CV, and because, theoretically, half the samples tested will be from heterozygotes with partial enzyme deficiencies, results might not always be easy to interpret. This will be especially true if experience is limited. Other enzyme activities, such as α-fucosidase and acid α-glucosidase, are relatively high in CV. The high activity of steroid sulfatase in CV might lead to complications when testing for other sulfatase deficiencies using artificial substrates. Thus, assay of arylsulfatase A activity in CV, using the standard spectrophotometric method, could result in a false-negative result because of interference from the steroid sulfatase activity. Special steps are required to overcome this when undertaking prenatal diagnosis for metachromatic leukodystrophy (MLD). Not only is this a potential pitfall, but the possible existence of a pseudodeficient allele ***(5)*** associated with low arylsulfatase A activity should always be assessed in the parents before any prenatal diagnosis for MLD.

In some situations, direct analysis of CV might not be the method of choice. For the diagnosis of mucolipidosis II (I-cell disease) or III, where, because of a processing defect, many lysosomal enzyme activities are high in plasma and extracellular fluid but low in cultured cells, direct lysosomal enzyme assay is not reliable using CV. Cultured CV cells should be used. Also, for some diagnostic tests where the incorporation of a radiolabeled precursor is studied or release of CO_2, direct analysis with CV might be less reliable than using cultured cells. There is also the practical problem of having fresh control samples available to run in parallel.

In other situations, CV might offer the only or best approach. The diagnosis of nonketotic hyperglycinemia requires the assay of glycine cleavage activity in CV. This enzyme is not expressed in cultured CV or amniotic fluid cells. It is expressed in tissue, such as the liver and placenta, and, in many cases, transformed lymphoblasts. It is, therefore, important that the enzyme deficiency is verified in the index case before any prenatal diagnosis is offered. It is also important that discussions take place with the referral laboratory before any sampling, because this test requires a relatively large amount of CV (30–50 mg).

For the diagnosis of Smith–Lemli–Opitz syndrome, direct sterol analysis of CV might be more practical than the analysis of whole amniotic fluid (with cells). Cultured cells can be used, but special steps are needed to standardize sterol levels, which can vary significantly under different culture conditions. Analysis of the ratio of 7-dehydrocholesterol to cholesterol in CV usually provides the best approach in classical cases.

When insufficient CV is available, a culture should be established if the test can be done on these cells. However, it might be more practical to request a second CV sample to avoid the delay of a further 2–3 wk for these cells to grow up sufficiently. There is also the potential problem of maternal cell contamination, especially when the sample had been small or poor at the outset. Maternal contamination is less of a problem when undertaking direct analyses, as this is unlikely to represent more than a few percent once the CV has been properly selected under the dissecting microscope, whereas in culture, maternal cells might outgrow the fetal trophoblasts.

Chorionic villus can also be used to extract DNA for those conditions where mutation analysis is to be performed. However, great care must be taken to carefully select only fetal tissue because subsequent applification studies could lead to difficulties should any maternal material be present. Molecular exclusion of maternal contamination might be necessary.

3.1.4. Fetal Blood and Fetal Tissue

Fetal blood is rarely used in the diagnosis of metabolic disease. The sample is usually taken rather late into the pregnancy and few laboratories have reference values for the various enzyme activities. Moreover, the sample size is generally insufficient for most analyses. Nevertheless, fetal blood has been used in the past, when there has been a failure in amniotic cell culture. This sample has also been used in the third trimester when fetal hydrops has been identified ***(14)***. However, it might be more practical nowadays to undertake a placental biopsy and treat this the same as CV or, if metabolite analysis of amniotic fluid is reliable, to arrange for a late amniocentesis.

In the past, when enzyme deficiencies are not expressed in CV or cultured cells, it has been necessary to study fetal liver biopsies. Such a procedure is not without risk and usually carried out late (18–24 wk gestation) into the pregnancy. Nevertheless, this approach has been used for the diagnosis of primary hyperoxaluria type I ***(15)***, ornithine carbamoyl transferase deficiency, and glycogen storage disease type I, all conditions where the enzyme deficiency is expressed in liver but not in more accessible fetal samples. Similarly, fetal skin biopsies have been used for other conditions, such as oculocutaneous albinism type I ***(16)***.

However, in all of these conditions, advances have been made that avoid invasive biopsy procedures and allow DNA to be studied either by linkage analysis or direct mutation analysis. Because DNA can be extracted from CV, this approach offers a timely and relatively safe procedure.

3.1.5. DNA

Although most metabolic disorders can in theory be diagnosed by DNA-mutation analysis, in practice this is rarely used. In most patients, the diagnosis is established from studies on metabolite concentrations in plasma and urine and the precise enzyme defect confirmed in blood or cultured cells. For most metabolic disorders, there are a wide range of different mutations responsible and few diagnostic laboratories are able to commit the time and expense to identify these. More often, mutation analysis has been a research activity that has not been able to respond to the urgent needs of a prenatal diagnosis. It is also important that any mutation that is found can be established unequivocally as being disease-causing before being used for fetal diagnosis. Also, there are potential dangers of undetected gene conversion events giving rise to false results, as reported recently for Gaucher disease *(17)*.

However, advances in DNA analysis have been significant in recent years and this is clearly the approach for those conditions where enzyme or metabolite studies cannot be carried out in a straightforward manner. Thus, for some conditions and those requiring fetal liver biopsies, DNA analysis is now the method of choice *(18–21)*. The approach can also be used where the standard method might yield an equivocal result *(22)* or where discrimination is required between an affected homozygote and heterozygote. Recent advances in microarray hybridization technology suggest that methods will become available to screen a whole range of mutations in fetal DNA collected from CV or even maternal blood. However, this is still some way off in the future.

4. Practical Aspects of Sample Collection

It is crucial that obstetricians and their staff make inquiries well in advance as to what type of sample is required and where this is to be referred for prenatal diagnosis. Unfortunately, this is not always easy even with the best forward planning. Mothers might only notify at 6–8 wk and this might not give much time to make the necessary arrangements. Good lines of communication are essential.

4.1. Amniotic Fluid

Amniotic fluid should be collected into a sterile container. The supernatant fluid can be used for metabolite analysis, but the quantity and transport requirements will need to be discussed with the referral laboratory. Usually, 5–10 mL will be sufficient and this can be sent unfrozen, without delay, to arrive the next day. Sometimes, it will be necessary to send this frozen, but not if cells are present for culture.

If cultured cells are required, it is usually best if the local cytogenetic laboratory is responsible for establishing cell culture. Again, the number of cells required for biochemical studies will vary according to the sensitivity of the assay. Usually, it is best to send two 25-cm^2 flasks of confluent cells, keeping one flask for backup. The receiving laboratory should be made aware of the culture medium used, in case this might influence their assay.

4.2. Chorionic Villus Tissue

This should be selected and checked before dispatch. Again, the quantity will vary from test to test. Ideally, approx 25 mg should be sent in sterile tissue culture medium (1–5 mL), which can include 20 units heparin/mL. Samples should reach the analytical laboratory within 24 h, although a longer period (2 d) might be acceptable when sent over a long distance. It is important that samples are not frozen at this stage, as this will result in loss of enzymes into the medium. Where possible, it is recommended that a backup culture be set up on a small amount of CV, and when an X-linked disorder is being investigated, material will be needed for fetal sexing.

Tissue received for biochemical studies must be carefully selected using a dissecting microscope and the quality and quantity carefully recorded. For most enzyme assays, the CV tissue should be washed twice in isotonic saline and, finally, pelleted by centrifugation (approx 1000*g* for 5 min) in an Eppendorf or similar plastic tube. Any excess saline should be carefully removed using laboratory tissue paper. The sample can usually be stored frozen (at –70°C) at this stage to await analysis. If insufficient CV is available for direct enzyme assay, it will be necessary to set this up in culture. However, it is likely that such small samples might carry an additional risk of maternal cell contamination.

5. Specific Metabolic Disorders

It is not practical to discuss all of the various disorders and the methods used to identify these in the fetus. Suffice to say that methods are now available for prenatal diagnosis of most conditions likely to be requested. It is probably best to contact the laboratory that diagnosed the index case first to seek advice as to what is currently available. Also, searches through the Internet for publications on diagnoses will help to locate useful contacts. There are also a number of useful reviews ***(2,3)*** that list the status and references relating to a wide variety of metabolic disorders. Some specific points are discussed next for groups of metabolic diseases.

5.1. Lysosomal Storage Disorders (See Table 1)

Most of these are serious life-threatening conditions for which there is, at present, no real effective therapy. The underlying biochemical defects have generally been known for nearly 30 yr and these conditions were some of the first to be diagnosed prenatally. They are probably the most common disorders to be diagnosed in the fetus and represent some 35 different entities. For most of the sphingolipid disorders such as Gaucher disease and Tay–Sachs disease, enzyme assays are available using simple water-soluble artificial substrates.

5.1.1. Sphigolipidoses

Laboratories will usually have a wide experience in the prenatal diagnosis of most of the more common sphingolipidoses. However, some difficulties could arise for a variety of reasons. A number of lysosomal disorders are associated with pseudodeficiencies of certain enzyme activities. The most common is that of arysulfatase A ***(23)***, the enzyme associated with metachromatic leukodystrophy (MLD). The benign pseudodeficiency allele has a frequency of around 15%, but when present with an MLD heterozygous allele, it will lead to exceptionally low enzyme activities. For Krabbe disease, a similar situation might arise; moreover, the deficient enzyme galactocerebrosidase activity is relatively low in CV and some carriers might have especially low activities. It is, therefore, recommended that the natural substrate, radioactively labeled, should be the preferred substrate for accurate discrimination of results.

The natural substrate is also recommended for Niemann–Pick diseases types A and B, where sphingomyelinase activity is deficient. In Niemann–Pick disease type C, there is a defect in cholesterol trafficking within the cell. This condition is more difficult to diagnose prenatally. Cholesterol esterification is impaired and this requires studies on cultured cells ***(24)***, leading to delays in diagnosis. Added to which, some patients do not have a sufficiently severe biochemical phenotype for accurate prenatal testing. There is an urgent need to understand the disorder in more detail, but, in the meantime, mutation testing, directly on CV, could help in some families.

5.1.2. Mucolipidoses and Glycoprotein Disorders

Many of these conditions are rare and can pose special problems. The low activity of neuraminidase (deficient in mucolipidosis I) and its instability means that studies on cultured cells are preferred. However, in mucolipidoses II and III, the secondary effect on lysosomal enzymes is normally used for diagnosis, rather than complex assay of the primary phosphorylation defect. Increased lysosomal enzyme activities in amniotic fluid have been used in the past as a diagnostic indicator ***(25)***. However, this might not always be reliable and studies on cultured cells should always be used for confirmation.

In recent years, advances in our understanding of the ceroid lipofuscinoses (Batten disease and others) have provided a more reliable approach to diagnosis. Previously, prenatal diagnosis was based on the examination of inclusion bodies by electron microscopy. For the infantile and juvenile variants, specific enzyme studies are now available ***(26)***.

Table 1
Lysosomal Disorders

Condition	Enzyme deficiency/Defect	Prenatal diagnosis
Mucopolysaccharidoses		
Type IH (Hurler, Scheie)	α-iduronidase	AF, CV
Type II (Hunter)	Iduronate sulfatase	AF, CV
Type III (Sanfilippo)		
A	Heparan *N*-sulfatase	AF, CV
B	*N*-Acetylglucosaminidase	AF, CV
C	Acetyl-CoA-glucosaminide acetyltransferase	AF, CV
D	*N*-Acetylglucosamine 6-sulfatase	AF, Poss CVC
Type IV (Morquio)		
A	Galactosamine 6-sulfatase	AFC, CV
B	β-Galactosidase	AFC, CV
Type VI (Maroteaux–Lamy)	*N*-Acetylgalactosamine 4-sulfatase	AF, CV
Type VII (Sly's disease)	β-Glucuronidase	AF, CV
Glycoproteinoses		
α-Mannosidosis	α-Mannosidase	AFC, CV
β-Mannosidosis	β-Mannosidase	AFC,CV
Fucosidosis	α-Fucosidase	AFC, CV
Aspartylglycosaminuria	Aspartylglycosaminidase	AFC, CV
Mucolipidoses		
Mucolipidosis I (Sialidosis)	Neuraminidase	AFC, CVC
Galactosialidosis	Neuraminidase and β-galactosidase protective protein	AFC, CVC
Mucolipidosis II (I-cell disease)	UDP-*N*-acetylglucosamine : lysosomal enzyme phosphotransferase	Poss AF, AFC, CVC
Mucolipidosis III (pseudo-Hurler polydystrophy)	Same as mucolipidosis II	Poss AF, AFC, CVC
Mucolipidosis type IV	Mucolipin membrane protein	DNA
Lipidoses		
GM_1 gangliosidosis	β-Galactosidase	AFC, CV
Tay–Sachs disease (variant B)	Hexosaminidase A	AFC, CV
Sandhoff disease (variant O)	Hexosaminidase A+B	AFC, CV
GM_2 activator deficiency (variant AB)	GM_2 activator protein	CVC
Metachromatic leukodystrophy	Arylsulfatase A	AFC, CV
Multiple sulfatase deficiency	Multiple lysosomal sulfatases	AFC, CV
Niemann–Pick disease		
Type A or B	Sphingomyelinase	AFC, CV
Type C	Cholesterol esterification	AFC, CVC, DNA
Farber disease	Ceramidase	AFC, CVC
Gaucher disease	Glucocerebrosidase	AFC, CV
Krabbe disease	Galactocerebrosidase	AFC, CV
Fabry disease	α-Galactosidase	AFC, CV
Schindler disease	α-*N*-Acetylgalactosaminidase	AFC, CVC
Wolman's disease/ cholesterol ester storage disease	Acid lipase	AFC, CV
Ceroid lipofuscinoses/Batten disease		
Infantile	Palmitoyl protein thioesterase	AFC. CVEdDNA
Late infantile	Tripeptidyl peptidase I	AFC. CV, DNA
Juvenile	CLN3 protein	CV, DNA
Others		
Cystinosis	Lysosomal cystine transporter	AFC, CV
Infantile free sialic acid storage disease	Lysosomal sialic acid transporter	AFC, CV
Sallas disease	Lysosomal sialic acid transporter	AF

Abbreviations: AF: cell-free amniotic fluid, usually implies metabolite analysis is useful; AFC: cultured amniotic fluid cells, usually implies cells are used for enzyme assay, or assay of pathway; CV: chorionic villus sample (direct assay), when stated, this is the preferred sample for early diagnosis; CVC: cultured chorionic villus cells, would also imply that AFC can be used; DNA: analysis on CV or other source; Poss.: indicates that test is probably available, but insufficient experience to confirm, or there are technical difficulties.

5.1.3. Mucopolysaccharidoses

Patients with these conditions excrete excessive amounts of glycosaminoglycans. This abnormality can be detected in amniotic fluid using two-dimensional electrophoresis ***(10)***. Although this provides a useful general approach, there are limitations. The test is most reliable for MPS I, II, III, and VI and when studied in liquor taken at 15–16 wk gestation. However, considerable experience is required to interpret the result, and this should always be confirmed by subsequent enzyme assays. All of the various enzyme deficiencies can be studied by direct assay on CV, although experience will be limited for the rarer conditions, such as MPS IIID and VII. The latter, a result of β-glucuronidase deficiency, has been associated with fetal hydrops ***(14,27)***. Although this is very rare and several other lysosomal disorders have also been associated with the condition ***(28)***, there are, of course, many other causes of hydrops fetalis.

In MPS II, the sex of the fetus is required to interpret the result. Some female carriers have been found to have exceptionally low activities of iduronate sulfatase. An added complication of this enzyme assay is the need to dialyze the sample prior to assay; therefore, a reasonable sized CV sample would be required for this.

5.2. Peroxisomal Disorders (See Table 2)

Some 20 different types of peroxisomal disorder are recognized; many of which are associated with severe neurological disorder and require prenatal diagnosis ***(29,30)***. For disorders of peroxisome biogenesis, such as Zellweger syndrome, direct assay of the peroxisomal enzyme, dihydroxyacetone phosphate acyltransferase (DHAP-AT), has proved useful. However, in milder forms of the disorder, the enzyme deficiency might need to be studied in cultured cells, because of variable expression in different tissues. Other studies using immunoblot analysis for specific peroxisomal β-oxidation enzymes or EM studies for peroxisomes are also useful approaches. It should also be noted that the nuchal translucency associated with Down syndrome might also be seen in Zellweger fetuses.

Increased levels of very long-chain fatty acids (VLCFAs) are associated with several types of peroxisomal disorder ***(29,30)***; however, the analysis of these directly on CV is not reliable and cultured cells are recommended ***(31)***. Even so, for X-linked adrenoleukodystrophy, VLCFA analysis is best supported by DNA-mutation analysis, as well as sexing the fetus. The immunoblot analysis of the ALD protein ***(32)*** is useful in some families. Unfortunately, VLCFA levels do not predict phenotype and the variation in expression even within a family, and the fact that female carriers might also exhibit symptoms, similar to mildly affected males, makes this a very difficult condition to monitor in the fetus.

Prenatal diagnosis of rhizomelic chondrodysplasia punctata (RCDP) can be performed in a few cases with an isolated deficiency of DHAP-AT activity by direct assay of CV. However, most cases require specific plasmalogen biosynthesis studies on cultured cells. This is quite a complex study, and when faced with ultrasound evidence of proximal limb shortening in the fetus, it will be difficult to decide how far investigations should go in the short time available. There is also a relatively common mutation associated with classical RCDP, which can be studied in some families.

Until recently, prenatal diagnosis of primary hyperoxaluria type I required fetal liver biopsy and analysis of alanine : glyoxylate aminotransferase activity. However, fetal diagnosis now is best done by DNA analysis, as long as the genotype is known ***(18)***.

5.3. Mitochondrial Disorders (See Table 3)

Prenatal diagnosis of these disorders poses major problems ***(33)***. These are heterogeneous disorders for which the underlying mechanisms are poorly understood. Many result from mutations in mitochondrial DNA (mtDNA). These mutations are not inherited in a Mendelian manner, but might be maternally inherited or sporadic. A disorder might only manifest when the number of mutations in the cell reaches a threshold, determined by the degree of heteroplasmy within cell. This can vary in different cells and, therefore, tests on CV or other fetal samples

Table 2
Peroxisomal disorders

Condition	Enzyme deficiency/defect	Prenatal diagnosis
Zellweger syndrome Neonatal adrenoleukodystrophy Infantile Refum syndrome	Peroxisome biogenesis (at least 10 complementation groups)	AFC, CV
Rhizomelic chondrodysplasia punctata	PTS 2 receptor	AFC, CVC, DNA
Pseudo–Zellweger syndrome	Peroxisomal thiolase	AFC, CVC
Pseudo-neonatal adrenoleukodystrophy	Acyl-CoA oxidase	AFC, CVC
Bifunctional enzyme defect	Bifunctional enzyme	AFC, CVC
X-Linked adrenoleucodystrophy	ALD protein	AFC, CVC, DNA
Refsum disease	Phytanic acid hydroxylase	AFC, CVC,
Hyperoxaluria type I	Alanine : glyoxylate aminotransferase	DNA, liver

Abbreviations: AF: cell-free amniotic fluid, usually implies metabolite analysis is useful; AFC: cultured amniotic fluid cells, usually implies cells are used for enzyme assay, or assay of pathway; CV: chorionic villus sample (direct assay), when stated, this is the preferred sample for early diagnosis; CVC: cultured chorionic villus cells, would also imply that AFC can be used; DNA: analysis on CV or other source; Poss.: indicates that test is probably available, but insufficient experience to confirm, or there are technical difficulties.

Table 3
Mitochondrial Disorders

Condition	Enzyme deficiency/defect	Prenatal diagnosis
Complex I	NADH-CoQ reductase	Probably no
Complex II	Succinate-CoA reductase	Probably no
Complex III	$CoQH_2$-cytochrome-*c* reductase	Probably no
Complex IV	Cytochrome oxidase	AFC, CV (difficult)
Complex V	Oligomycin-sensitive ATPase	Probably no
mtDNA/nuclearDNA mutations	DNA	Few possibly

Abbreviations: AF: cell-free amniotic fluid, usually implies metabolite analysis is useful; AFC: cultured amniotic fluid cells, usually implies cells are used for enzyme assay, or assay of pathway; CV: chorionic villus sample (direct assay), when stated, this is the preferred sample for early diagnosis; CVC: cultured chorionic villus cells, would also imply that AFC can be used; DNA: analysis on CV or other source; Poss.: indicates that test is probably available, but insufficient experience to confirm, or there are technical difficulties.

might not be representative. Furthermore, both nuclear and mitochondrial genes are involved in the expression and assembly of the mitochondrial respiratory chain enzymes. In a few cases where the enzyme deficiency is expressed in cultured cells, attempts have been made at prenatal diagnosis ***(34)***. Similarly, diagnosis by direct mutation screening has been attemped ***(35)***. However, because both approaches could result in false-positive and false-negative results, the whole procedure must be interpreted with great caution.

5.4. Amino and Organic Acid Disorders (See Tables 4 and 5)

Prenatal diagnosis is less commonly performed for some of these disorders. This is partly because many respond favorably to treatment. However, this is not true for all. Patients with these conditions generally excrete excessive amounts of specific metabolites in urine. In some cases, these metabolites can also be found in amniotic fluid ***(6–8)***. However, the levels are usually quite low and require special methods of analysis, such as stable isotope dilution methods.

Prenatal diagnosis for classical phenylketonuria (PKU) is generally not carried out because of the good prognosis on treatment. However, a few have been undertaken by DNA analysis. The more severe variants resulting from defects in pterin metabolism are quite rare but might

Table 4
Amino Acid Disorders

Condition	Enzyme deficiency	Prenatal diagnosis
'Classical' phenylketonuria	Phenylalanine hydroxylase	DNA
Tetrabiopterin homeostasis	Dihydropteridine reductase	AF, CV
Tetrabiopterin synthesis	Guanosine-triphosphate cyclohydrolase	AF
	6-Pyruvoyltetrahydropterin synthase	AF
Homocystinuria	Cystathionine synthase	AFC, CVC
Homocystinuria	Remethylation defect	AFC, CVC
Tyrosinemia I	Fumarylacetoacetate hydrolase	AF, CV, DNA
Tyrosinemia II	Tyrosine aminotransferase	No
Maple syrup urine disease	Branched-chain keto-acid dehydrogenase	AFC, CV
Nonketotic hyperglycinemia	Gylcine cleavage system	CV (not cultured cells)
Hyperlysinemia	Aminoadipic-semialdehyde synthase	AFC
Hyperprolinemia I	Proline oxidase	No.
Hyperprolinemia II	Pyrroline-5-carboxylate dehydrogenase	No.
Hyperimidodipeptiduria	Prolidase	Poss. CVC
Gyrate atrophy	Ornithine aminotransferase	Poss. CVC
Hyperornithinemia–hyperammonemia–homocitrullinaemia (HHH syndrome)	Mitochondrial ornithine transporter	CVC, AFC
Guanidinoacetate methyltransferase deficiency	Guanidinoacetate methyltransferase	AFC
N-Acetylglutamate synthetase deficiency	*N*-Acetylglutamate synthetase	No.
Carbamyl-phosphate synthetase (CPS) deficiency	Carbamyl-phosphate synthetase	DNA, liver
Ornithine carbamyltransferase (OCT) deficiency	Ornithine carbamyltransferase	DNA, liver
Citrullinemia	Argininosuccinic acid synthetase	AF, AFC, CV
Argininosuccinic aciduria (ASA)	Argininosuccinate lyase	AF, AFC, CV
Argininemia	Arginase	DNA, liver (?fetal erthrocytes)

Abbreviations: AF: cell-free amniotic fluid, usually implies metabolite analysis is useful; AFC: cultured amniotic fluid cells, usually implies cells are used for enzyme assay, or assay of pathway; CV: chorionic villus sample (direct assay), when stated, this is the preferred sample for early diagnosis; CVC: cultured chorionic villus cells, would also imply that AFC can be used; DNA: analysis on CV or other source; Poss.: indicates that test is probably available, but insufficient experience to confirm, or there are technical difficulties. No: DNA analysis may not be possible.

require prenatal diagnosis. This has been successfully done by pterin analyses in amniotic fluid ***(9)***. Similarly, the increased levels of succinylacetone associated with tyrosinaemia type I can also be studied in amniotic fluid. However, this should be backed up with either enzyme (fumarylacetoacetase) or DNA testing on cultured cells or CVS.

Prenatal diagnosis of nonketotic hyperglycinemia has always been difficult. The deficient enzyme (glycine cleavage activity) is not expressed in cultured cells, but is in liver and placenta (CV). The activity is quite low and the assay complex; therefore, a large CV biopsy (30–50 mg) is required. Added to which, false-negative results have occurred when enzyme activity is deficient in liver ***(36)***. In the future, it is hoped that DNA analysis will be the method of choice. This has been applied to families with a common mutation ***(37)***, but the complexity of the enzyme (three subunits) has restricted this approach for most families.

The urea cycle disorders are a relatively common group of amino acid disorders that are likely to require prenatal investigations. Ornithine carbamoyl transferase (OCT) deficiency is not expressed in CV or cultured cells and, in the past, has required fetal liver biopsy for enzyme

Table 5
Organic Acid Disorders

Condition	Enzyme deficiency	Prenatal diagnosis
Propionic acidemia	Propionyl-CoA carboxylase	AF, CV
Multiple carboxylase deficiency	Holocarboxylase synthetase	AF, AFC
Methylmalonic acidemia	Methylmalonyl-CoA mutase	AF, CV
Methylmalonic acidemia (B12-responsive)	Adenosylcobalamin synthesis:	AF, CVC
Isovaleric acidaemia	Isovaleryl-CoA dehydrogenase	AF, CVC
3-Methylcrotonyl-CoA carboxylase deficiency	3-Methylcrotonyl-CoA carboxylase	AFC, CV
3-Methylglutaconic aciduria type I	3-Methylglutaconic hydratase	Poss. CVC, AF
3-Methylglutaconic aciduria types II, III, IV	Not known	No
3-Hydroxy-3-methylglutaryl-CoA lyase deficiency	3-Hydroxy-3-methylglutaryl-CoA lyase	AF, CV
Mevalonic aciduria	Mevalonate kinase	AF, AFC, Poss. CV
2-Methylacetoacetyl-CoA thiolase deficiency	2-Methylacetoacetyl-CoA thiolase	Poss. CVC
Glutaric aciduria type I	Glutaryl-CoA dehydrogenase	CVC, CV, AF
Glutaric aciduria type II	Electron transfer flavoprotein (ETF)/ ETF:ubiquinone oxidoreductase	CVC, AFC
Glycerol kinase deficiency	Glycerol kinase	AF, CVC
Canavan disease	Aspartoacylase	DNA, AF
Carnitine palmitolytransferase deficiencies	CPT I and CPT II	Poss. CVC
Carnitine/acyl carnitine translocase	Carnitine/acyl carnitine translocase	Poss. CVC
SCAD deficiency	Short-chain acyl-CoA dehydrogenase	DNA, Poss. CVC,
MCADdeficiency	Medium-chain acyl-CoA dehydrogenase	DNA, CVC
VLCAD deficiency	Very long-chain acyl-CoA dehydrogenase	DNA, Poss. CVC
LCHAD deficiency	Long-chain 3-OH acyl-CoA dehydrogenase	DNA, Poss. CVC
Trifunctional protein deficiency	Long-chain 2-enoyl-CoA hydratase Long-chain 3-OH acyl-CoA dehydrogenase Long-chain 3-ketoacyl-CoA thiolase	DNA, Poss. CVC
Pyruvate dehydrogenase deficiency	Pyruvate dehydrogenase complex	DNA, Poss. CVC
Pyruvate carboxylase deficiency	Pyruvate carboxylase	CV, AFC
Fumaric aciduria	Fumarase	AFC, CVC

Abbreviations: AF: cell-free amniotic fluid, usually implies metabolite analysis is useful; AFC: cultured amniotic fluid cells, usually implies cells are used for enzyme assay, or assay of pathway; CV: chorionic villus sample (direct assay), when stated, this is the preferred sample for early diagnosis; CVC: cultured chorionic villus cells, would also imply that AFC can be used; DNA: analysis on CV or other source; Poss.: indicates that test is probably available, but insufficient experience to confirm, or there are technical difficulties.

diagnosis. Now, the diagnosis is best made by DNA analysis ***(19,38)***, although, as with most X-linked disorders, mutations are generally private and must be known beforehand. The diagnoses of citrullinemia and argininosuccinic aciduria have been made by metabolite analysis in liquor ***(8)***. However, confirmatory enzyme studies should always be conducted. This is best done by studies on the rate of ^{14}C-citrulline incorporation into cell protein using cultured CV or amniotic fluid cells. Care should be taken with the latter because epithelial cells can respond poorly and give equivocal results.

Use of radioactive precursors, such as ^{14}C-citrulline, to monitor whether a metabolic pathway is intact is a useful approach for several metabolic disorders. Both cultured cells and CV have been used; however, it might not always be possible to arrange for viable control CV to be available to assay in parallel. Few centers would be able to accept samples for direct analysis. The diagnosis of maple syrup urine disease can also be carried out using a similar radiometric assay. In this disorder, the defective decarboxylation of branched-chain amino acids can be monitored by incubating cultured cells or CV in the presence of ^{14}C-leucine and measuring the amount of radioactive CO_2 evolved. Branched-chain amino acids are not increased in amniotic fluid; therefore, metabolite analyses are not appropriate here.

Similarly, dicarboxylic acid concentrations are not raised in amniotic fluid from fetuses with fatty acid oxidation defects, such as medium-chain acyl-CoA dehydrogenase (MCAD) deficiency. Although MCAD deficiency is a relatively common metabolic disorder, with an incidence of about 1 in 10,000 newborns, few prenatal studies have been carried. This is partly because treatment to prevent life-threatening episodes is usually very effective. If prenatal diagnosis were to be carried out, this would be most likely to be by mutation analysis. Almost 90% of clinically affected patients are homozygous for the common 985A>G mutation. This percentage may be less now that many cases are identified on newborn screening. Indeed, CV and preimplantation diagnosis has now been reported ***(39,50)*** by this approach. A common mutation (1528C>G) is also found in patients with long-chain acyl-CoA dehydrogenase (LCHAD) deficiency. A similar mutation approach is applicable here also. This condition in the fetus can result in severe maternal liver disease manifesting as hemolysis, elevated liver enzymes, and low platelet levels (HELLP syndrome) in the third trimester.

For many organic acid disorders, prenatal diagnosis can be achieved by amniotic fluid metabolite analysis using stable isotope dilution mass spectrometry ***(6)***. Recent advances in the use of electrospray tandem mass spectrometry suggest that this technique could prove a powerful tool in the identification of low levels of specific metabolites. Its use in amniotic fluid acylcarnitines has already been reported ***(7)***. However, direct enzyme assay of CV in the first trimester will probably be the preferred approach. Some enzymes (propionyl-CoA carboxylase, methylmalonyl-CoA mutase) can be measured directly in CV, whereas for others, it might be more convenient to study a specific pathway using a radiometric assay.

5.5. Carbohydrate Disorders (See Table 6)

Experience in the prenatal diagnosis of these conditions is quite limited because many respond reasonably well to treatment. Galactosemia is one example. Unfortunately, the long-term outlook has proved disappointing and we might see more requests for this in the future. The deficient enzyme galactose-1-phosphate uridyl transferase is expressed in CV and cultured cells, and the metabolite galactitol can be detected in amniotic fluid ***(40)***. However, because some 75% patients are homozygous for the common Q188R mutation, it is likely that that, if applicable, mutation diagnosis might be the method choice.

Similarly, prenatal diagnosis of the glycogen storage diseases (GSDs) can also be undertaken by DNA analysis, although for GSD type II (Pompe disease), direct enzyme assay of CV is a reliable approach ***(41)***. For the hepatic forms of GSD such as types I and III, diagnosis is not straightforward. In the former, the deficient enzyme is not expressed in cultured cells or CV and, therefore, fetal liver biopsy has been used in the past. In the future, mutation testing would be the most practical approach ***(42)***. In GSD type III, the enzyme assay is complex and few prenatal diagnoses have been undertaken, as this is usually not a life-threatening condition. Again, it seems that DNA analysis might prove to be the approach used for the few cases requesting prenatal diagnosis.

The carbohydrate-deficient glycoprotein disorders (CDGs) are a relatively newly recognized group of conditions, resulting from disorders of protein glycosylation. Clinically, these are heterogeneous multisystem disorders in which a number of enzyme deficiencies have been found. The diagnosis is initially made by demonstrating abnormal patterns of glycosylation in serum transferrins. This approach is not reliable for fetal blood ***(43)***, but where the enzyme defect (e.g., phosphomannomutase) is known, this can be studied in CV or cultured cells ***(44)***.

5.6. Other Disorders (See Table 7)

It is not possible in this chapter to give a comprehensive account of all the methods used for prenatal diagnosis of all metabolic disorders. Examples from different conditions have been

Table 6
Carbohydrate Disorders

Condition	Enzyme deficiency	Prenatal diagnosis
Galactokinase deficiency	Galactokinase	AFC
"Classical" galactosaemia	Galactose-1-phosphate uridyltransferase	AFC, AF, CV
Epimerase deficiency	UDP-galactose 4-epimerase	AFC
Hereditary fructose intolerance	Aldolase B	Poss. DNA
Fructose-1,6-bisphosphatase deficiency	Fructose-1,6-bisphosphatase	Poss. DNA
Glycogen storage disease (GSD) type		
Ia Von Gierke	Glucose 6-phosphatase	DNA, liver
Ib Von Gierke	Glucose-6-phosphatase translocase (T1)	Poss. DNA
II Pompe	Lysosomal acid glucosidase	AFC, CV
III Debrancher enzyme deficiency	Amylo-1,6-glucosidase	AFC, CV
IV Brancher enzyme deficiency	1,4-Glucan 6-glycosyltransferase	AFC, CV
V McArdle	Muscle phosphorylase	No
VI Liver phosphorylase	Liver phosphorylase	No
VI Deficiency		

Abbreviations: AF: cell-free amniotic fluid, usually implies metabolite analysis is useful; AFC: cultured amniotic fluid cells, usually implies cells are used for enzyme assay, or assay of pathway; CV: chorionic villus sample (direct assay), when stated, this is the preferred sample for early diagnosis; CVC: cultured chorionic villus cells, would also imply that AFC can be used; DNA: analysis on CV or other source; Poss.: indicates that test is probably available, but insufficient experience to confirm, or there are technical difficulties.

discussed earlier. Disorders of purine and pyrimidine are relatively rare conditions but can be diagnosed in the fetus, usually by enzyme assay of CV. Similarly, some disorders of porphyrin metabolism have been successfully diagnosed prenatally. DNA analysis might be the method of choice in the future.

Prenatal diagnosis of Menkes disease, an X-linked disorder of copper transport, could pose special problems. Direct analysis of copper in CV can be used ***(45)***, but care should be taken to avoid any contamination with instruments during collection and processing of the tissue. Uptake of radioactive ^{64}Cu into cultured cells has also been used. However, the best approach is by direct mutation testing in CV, as long as the mutation is known. These tests are normally only carried out in male fetuses.

Several disorders of sterol metabolism are now known, the most significant being Smith–Lemli–Opitz sydrome. A defect in the penultimate step of cholesterol biosynthesis results in the accumulation of the precursor 7-dehydrocholesterol. Increases in the concentration of this sterol in amniotic fluid, cultured cells, or CV can be used for prenatal diagnosis ***(46,47)***. The best approach and one giving early diagnosis is direct sterol analysis in CV.

6. Future Trends

At present, most prenatal diagnoses are done by direct enzyme assay on CV. For most metabolic patients, the underlying mutations are not known. This is partly the result of cost implications and lack of sufficient specialized centers able to make a firm commitment to providing such a service for rare disorders, each with many different mutations. However, advances in this field have been considerable in recent years, and prenatal diagnosis could, in theory, be made available for most disorders by direct mutation analysis on CV or even in fetal cells from maternal blood. Although preimplantation diagnosis ***(48)*** has been undertaken for a few disorders ***(49–52)***, this procedure is still very much at an experimental stage.

Table 7
Other Disorders

Condition	Enzyme deficiency	Prenatal diagnosis
Purine and pyrimidine metabolism		
Lesch–Nyhan syndrome	Hypoxanthine phosphoribosyltransferase	AFC, CV
Adenosine deaminase deficiency	Adenosine deaminase	AFC, CV
Purine nucleoside phosphorylase deficiency	Purine nucleoside phosphorylase	AFC, CV
Dihydropyrimidine dehydrogenase deficiency	Dihydropyrimidine dehydrogenase	AF
Trace metal metabolism		
Wilson disease	Copper transporter	Poss. DNA
Menkes disease	Copper transporter	AFC, CV
Molybdenum cofactor deficiency	Molybdenum cofactor	AFC, CVC
Porphyrias		
Acute intermittent porphyria	Porphobilinogen deaminase	AFC, DNA
Congenital erythropoietic porphyria	Uroporphyrinogen III cosynthase	AF
Miscellaneous		
Carbonic anhydrase II deficiency	Carbonic anhydrase II	DNA
X-Linked ichthyosis	Steroid sulphatase	AFC, CV
Hypophosphatasia	Alkaline phosphatase	DNA
Sjögren–Larsson syndrome	Fatty alcohol NAD oxidoreductase	AFC, CVC
Smith–Lemli–Opitz syndrome	7-dehydrocholesterol reductase	CV, AF
Congenital adrenal hypoplasia	21-hydroxylase	DNA
Lowe syndrome	Phosphatidyl inositol-4,5,-bisphosphate-5-phosphatase	DNA

Abbreviations: AF: cell-free amniotic fluid, usually implies metabolite analysis is useful; AFC: cultured amniotic fluid cells, usually implies cells are used for enzyme assay, or assay of pathway; CV: chorionic villus sample (direct assay), when stated, this is the preferred sample for early diagnosis; CVC: cultured chorionic villus cells, would also imply that AFC can be used; DNA: analysis on CV or other source; Poss.: indicates that test is probably available, but insufficient experience to confirm, or there are technical difficulties.

References

1. Scriver, C. R., Beaudet, A. L., Sly, W. S., and Valle, D. (eds.) (2001) *The Metabolic and Molecular Basis of Inherited Disease*, 8th ed. McGraw-Hill, New York.
2. Besley, G. T. N. (1992) Enzyme analysis, in *Prenatal Diagnosis and Screening* (Brock, D. J. H., Rodeck, C. H., and Ferguson-Smith, M. A., eds.), Churchill Livingston, Edinburgh, pp. 127–145.
3. Kleijer, W. J. (1999) Inborn errors of metabolism, in *Fetal Medicine: Basic Science and Clinical Practice* (Rodeck, C. H. and Whittle, M. J., eds.), Churchill Livingstone, London, pp. 525–541.
4. Wraith, J. E. (2001) Prenatal diagnosis of inborn errors of metabolism, in *Fetal and Neonatal Neurology and Neurosurgery* (Levene, M. I., Chervenak, F. A., and Whittle, M., eds.), Churchill Livingstone, London, pp. 595–603.
5. Thomas, G. H. (1994) "Pseudodeficiencies" of lysosomal hydrolases. *Am. J. Hum. Genet.* **54,** 934–940.
6. Jakobs, C., Ten Brink, H. J., and Stellaard, F. (1990) Prenatal diagnosis of inherited metabolic disorders by quantitation of characteristic metabolites in amniotic fluid: facts and future. *Prenat. Diagn.* **10,** 265–271.
7. Shigematsu, Y., Hata, I., Nakai, A., et al. (1996) Prenatal diagnosis of organic acidemias based on amniotic fluid levels of acylcarnitines. *Pediatr. Res.* **39,** 680–684.
8. Mandell, R., Packman, S., Laframboise, R., et al. (1996) Use of amniotic fluid amino acids in prenatal testing for argininosuccinic aciduria and citrullinaemia. *Prenat. Diagn.* **16,** 419–424.

9. Blau, N., Kierat, L., Matasovic, A., et al. (1994) Antenatal diagnosis of tetrahydrobiopterin deficiency by quantification of pterins in amniotic fluid and enzyme activity in fetal and extrafetal tissue. *Clin. Chim. Acta* **226,** 159–169.
10. Mossman, J. and Patrick, A. D. (1982) Prenatal diagnosis of mucopolysaccharidosis by two-dimensional electrophoresis of amniotic fluid glycosaminoglycans. *Prenat. Diagn.* **2,** 169–176.
11. Poenaru, L. (1987) First trimester prenatal diagnosis of metabolic diseases: a survey in countries from the European community. *Prenat. Diagn.* **7,** 333–341.
12. Besley, G. T., Young, E. P., Fensom, A. H., and Cooper, A. (1991) First trimester diagnosis of inherited metabolic disease: experience in the UK. *J. Inherit. Metab. Dis.* **14,** 128–133.
13. Desnick, R. J., Schuette, J. L., Golbus, M. S., et al. (1992) First-trimester biochemical and molecular diagnoses using chorionic villi: high accuracy in the U.S. collaborative study. *Prenat. Diagn.* **12,** 357–732.
14. Groener, J. E., de Graaf, F. L., Poorthuis, B. J., and Kanhai, H. H. (1999) Prenatal diagnosis of lysosomal storage diseases using fetal blood. *Prenat. Diagn.* **19,** 930–933.
15. Danpure, C. J., Jennings, P. R., Penketh, R. J., Wise, P. J., Cooper, P. J., and Rodeck, C. H. (1989) Fetal liver alanine: glyoxylate aminotransferase and the prenatal diagnosis of primary hyperoxaluria type 1. *Prenat. Diagn.* **9,** 271–281.
16. Rosenmann, E., Rosenmann, A., Ne'eman, Z., Lewin, A., Bejarano-Achache, I., and Blumenfeld, A. (1999) Prenatal diagnosis of oculocutaneous albinism type I: review and personal experience. *Pediatr. Dev. Pathol.* **2,** 404–414.
17. Beutler, E., Liebman, H., Gelbart, T., and Stefanski, E. (2000) Three Gaucher-disease-producing mutations in a patient with Gaucher disease:mechanism and diagnostic implications. *Acta Haematol.* **104,** 103–105.
18. Rumsby, G. (2000) Biochemical and genetic diagnosis of the primary hyperoxalurias: a review. *Mol. Urol.* **4,** 349–354.
19. Genet, S., Cranston, T., and Middleton-Price, H. R. (2000) Mutation detection in 65 families with a possible diagnosis of ornithine carbamoyltransferase deficiency including 14 novel mutations. *J. Inherit. Metab. Dis.* **23,** 669–676.
20. Rake, J. P., ten Berge, A. M., Visser, G., et al. (2000) Glycogen storage disease type Ia: recent experience with mutation analysis, a summary of mutations reported in the literature and a newly developed diagnostic flow chart. *Eur. J. Pediatr.* **159,** 322–330.
21. Ibdah, J. A., Zhao, Y., Viola, J., Gibson, B., Bennett, M. J., and Strauss, A. W. (2001) Molecular prenatal diagnosis in families with fetal mitochondrial trifunctional protein mutations. *J. Pediatr.* **138,** 396–399.
22. Besley, G. T., Elpeleg, O. N., Shaag, A., Manning, N. J., Jakobs, C., and Walter, J. H. (1999) Prenatal diagnosis of Canavan disease—problems and dilemmas. *J. Inherit. Metab. Dis.* **22,** 263–266
23. Sanguinetti, N., Marsh, J., Jackson, M., et al. (1986) The arylsulphatases of chorionic villi: potential problems in the first-trimester diagnosis of metachromatic leucodystrophy and Maroteaux-Lamy disease. *Clin. Genet.* **30,** 302–308.
24. Vanier, M. T., Rodriguez-Lafrasse, C., Rousson, R., et al. (1992) Prenatal diagnosis of Niemann–Pick type C disease: current strategy from an experience of 37 pregnancies at risk. *Am. J. Hum. Genet.* **51,** 111–22.
25. Besley, G. T., Broadhead, D. M., Nevin, N. C., Nevin, J., and Dornan, J. C. (1990) Prenatal diagnosis of mucolipidosis II by early amniocentesis. *Lancet* **335,** 1164–1165.
26. Van Diggelen, O. P., Keulemans, J. L., Kleijer, W. J., Thobois, S., Tilikete, C., and Voznyi, Y. V. (2001) Pre- and postnatal enzyme analysis for infantile, late infantile and adult neuronal ceroid lipofuscinosis (CLN1 and CLN2). *Eur. J. Paediatr. Neurol.* **5,** 189–192.
27. Chabas, A. and Guardiola, A. (1993) Beta-Glucuronidase deficiency: identification of an affected fetus with simultaneous sampling of chorionic villus and amniotic fluid. *Prenat. Diagn.* **13,** 429–433.
28. Piraud, M., Froissart, R., Mandon, G., Bernard, A., and Maire, I. (1996) Amniotic fluid for screening of lysosomal storage diseases presenting in utero (mainly as non-immune hydrops fetalis). *Clin. Chim. Acta* **248,** 143–155.
29. Schutgens, R. B., Schrakamp, G., Wanders, R. J., Heymans, H. S., Tager, J. M., and van den Bosch, H. (1989) Prenatal and perinatal diagnosis of peroxisomal disorders. *J. Inherit. Metab. Dis.* **12,** 118–134.
30. Wanders, R. J., Schutgens, R. B., and Barth, P. G. (1995) Peroxisomal disorders: a review. *J. Neuropathol. Exp. Neurol.* **54,** 726–739.

31. Moser, A. B. and Moser, H. W. (1999) The prenatal diagnosis of X-linked adrenoleukodystrophy. *Prenat. Diagn.* **19,** 46–48.
32. Wanders, R. J., Mooyer, P. W., Dekker, C., and Vreken, P. (1998) X-Linked adrenoleukodystrophy: improved prenatal diagnosis using both biochemical and immunological methods. *J. Inherit. Metab. Dis.* **21,** 285–287.
33. Poulton, J. and Marchington, D R. (2000) Progress in genetic counselling and prenatal diagnosis of maternally inherited mtDNA diseases. *Neuromuscul. Disord.* **10,** 484–487.
34. Faivre, L., Cormier-Daire, V., Chretien, D., et al. (2000) Determination of enzyme activities for prenatal diagnosis of respiratory chain deficiency. *Prenat. Diagn.* **20,** 732–737.
35. Amiel, J., Gigarel, N., Benacki, A., et al. (2001) Prenatal diagnosis of respiratory chain deficiency by direct mutation screening. *Prenat. Diagn.* **21,** 602–604.
36. Applegarth, D. A., Toone, J. R., Rolland, M. O., Black, S. H., Yim, D. K., and Bemis, G. (2000) Non-concordance of CVS and liver glycine cleavage enzyme in three families with non-ketotic hyperglycinaemia (NKH) leading to false negative prenatal diagnoses. *Prenat. Diagn.* **20,** 367–370.
37. Kure, S., Rolland, M. O., Leisti, J., et al. (1999) Prenatal diagnosis of non-ketotic hyperglycinaemia: enzymatic diagnosis in 28 families and DNA diagnosis detecting prevalent Finnish and Israeli–Arab mutations. *Prenat. Diagn. 19,* 717–720.
38. Grompe, M., Caskey, C. T., and Fenwick, R. G. (1991) Improved molecular diagnostics for ornithine transcarbamylase deficiency. *Am. J. Hum. Genet.* **48,** 212–222.
39. Gregersen, N., Winter, V., Jensen, P. K., et al. (1995) Prenatal diagnosis of medium-chain acyl-CoA dehydrogenase (MCAD) deficiency in a family with a previous fatal case of sudden unexpected death in childhood. *Prenat. Diagn.* **15,** 82–86.
40. Holton, J. B., Allen, J. T., and Gillett, M. G. (1989) Prenatal diagnosis of disorders of galactose metabolism. *J. Inherit. Metab. Dis.* **12,** 202–206.
41. Kleijer, W. J., van der Kraan, M., Kroos, M. A., et al. (1995) Prenatal diagnosis of glycogen storage disease type II: enzyme assay or mutation analysis? *Pediatr. Res.* **38,** 103–106.
42. Wong, L. J. (1996) Prenatal diagnosis of glycogen storage disease type 1a by direct mutation detection. *Prenat. Diagn.* **16,** 105–108.
43. Clayton, P., Winchester, B., Di Tomaso, E., Young, E., Keir, G., and Rodeck, C. (1993) Carbohydrate-deficient glycoprotein syndrome: normal glycosylation in the fetus. *Lancet* **341,** 956.
44. Matthijs, G., Schollen, E., Cassiman, J. J., Cormier-Daire, V., Jaeken, J., and van Schaftingen, E. (1998) Prenatal diagnosis in CDG1 families: beware of heterogeneity. *Eur. J. Hum. Genet.* **6,** 99–104.
45. Heydorn, K., Damsgaard, E., and Horn, N. (1999) Accumulated experience with prenatal diagnosis of Menkes disease by neutron activation analysis of chorionic villi specimens. *Biol. Trace Element Res.* **71–72,** 551–561.
46. Kratz, L. E. and Kelley, R. I. (1999) Prenatal diagnosis of the RSH/Smith-Lemli-Opitz syndrome. *Am. J. Med. Genet.* **82,** 376–381.
47. Mills, K., Mandel, H., Montemagno, R., Soothill, P., Gershoni-Baruch, R., and Clayton, P. T. (1996) First trimester prenatal diagnosis of Smith–Lemli–Opitz syndrome (7-dehydrocholesterol reductase deficiency). *Pediatr. Res.* **39,** 816–819.
48. Rechitsky, S., Strom C., Verlinsky, O., et al. (1999) Accuracy of preimplantation diagnosis of single-gene disorders by polar body analysis of oocytes. *J. Assist. Reprod. Genet.* **16,** 192–198.
49. Ray, P. F., Gigarel, N., Bonnefont, J. P., et al. (2000) First specific preimplantation genetic diagnosis for ornithine transcarbamylase deficiency. *Prenat. Diagn.* **20,** 1048–1054.
50. Ioulianos, A., Wells, D., Harper, J. C., and Delhanty, J. D. (2000) A successful strategy for preimplantation diagnosis of medium-chain acyl-CoA dehydrogenase (MCAD) deficiency. *Prenat. Diagn.* **20,** 593–598.
51. Sermon, K., Henderix, P., Lissens, W., et al. (2000) Preimplantation genetic diagnosis for medium-chain acyl-CoA dehydrogenase (MCAD) deficiency. *Mol. Hum. Reprod.* **6,** 1165–1168.
52. Ray, P. F., Harper, J. C., Ao, A., et al. (1999) Successful preimplantation genetic diagnosis for sex link Lesch–Nyhan syndrome using specific diagnosis. Prenat Diagn. 19:1237–1241.

44

Gene Therapy

Methods and Application

Stella B. Somiari

1. Introduction

Gene therapy represents a set of approaches to the treatment of diseases based on the transfer of genetic material (DNA) into an individual (or animal) and is defined as the use of nucleic acid transfer, either RNA or DNA, to treat or prevent a disease ***(1–3)***. The process involves a group of technologies that enable the intentional transfer of specific exogenous genetic information into cells and the application of these technologies for pharmaceutical development ***(4)***. The gene is delivered either by direct administration of a gene-containing virus or DNA to blood or tissue or indirectly through the introduction of cells manipulated in the laboratory to harbor foreign DNA. The idea behind this technology is to treat disease by the administration of DNA (rather than a drug), which will produce an appropriate amount of gene product (usually a protein) to correct the condition. In this process, only the somatic cells and not the germ cells (eggs and sperms) are the target; therefore, such gene transfer affects only the treated individual and not the offspring. In a broad sense therefore, gene therapy can be viewed as a natural progression in the application of biomedical science to medicine. It can be employed for the correction of an underlying pathophysiological condition and can offer a one-time cure for inherited disorders for which current therapeutic approaches are ineffective or where prospective treatment appear exceedingly low.

The term "gene delivery" or "polynucleotide delivery" has been the preference of many, rather than "gene therapy," which might tend to indicate a proven therapeutic benefit and be reserved for clinical medicine ***(4)***. The aim of any gene therapy procedure is to achieve an easy nontoxic delivery of exogenous genetic material into targeted tissues. In the tissue, the genetic material should be efficiently endocytosed across the lipid bilayer and directed to the host cell nucleus. Sequences encoded by the foreign genetic material interacting with regulatory host factors should achieve long-term expression ***(5)***. Gene delivery (of a genetic material) can, therefore, require targeted integration into the host genomic DNA ***(6)***, or the existence of the material as a stable episome ***(7)***, or survival and propagation as a "minichromosome" ***(8)***.

Somatic gene therapy is a logical and natural progression in the application of fundamental medical science to medicine and offers great potential in the long term for the management and correction of inherited/acquired human disorders, cancers, and acquired immunodeficiency syndrom (AIDS). The remarkable advances in the past two decades in the area of recombinant DNA technology provided initial insights to the concept that gene therapy (gene transfer) might be used to treat disease. The major obstacles to gene therapy are (1) transfer of the exogenous genetic material, (2) use of effective vectors that would maintain high-level expression of the genes transferred to achieve tissue-specific and regulated expression of the transferred genes

From: *Medical Biomethods Handbook*
Edited by: J. M. Walker and R. Rapley © Humana Press, Inc., Totowa, NJ

to the target cell, and (3) an animal model that closely simulates disease for preclinical testing *(1–3)*. It is evident that more basic research is required to provide a better understanding of the pathophysiology of disease. This will lead to a better definition of the important target cell(s) and the design of appropriate gene delivery approaches that will have beneficial therapeutic effects *(9)*. There are two major categories of gene delivery systems or vectors, as they are commonly called: viral and nonviral vectors. Once a vector is designed, two general approaches are used for gene transfer: ex vivo transfer, where cells are removed, genetically modified and transplanted back into the same recipient, and in vivo transfer, where a foreign genetic material is directly transferred into the patient. **Figure 1** summarizes the technology of gene therapy.

2. Viral Methods of Gene Delivery

Viruses have evolved for millions of years to become highly efficient at delivering genetic material to specific cell types. They have complex structures and life cycles and many are pathogens. Several viruses are efficient gene delivery vehicles, but it is believed that they might pose some side effects, and because of previous exposure, the host might be resistant to transduction (gene transfer into the cell) and so they might not be effective for long-term use in gene therapy. The usual approach in designing viral vectors is to remove the pathogenic features while retaining the efficiency of gene delivery and expression. Therefore, the development of viral vectors requires analysis of the biology and life cycle of the parental virus and the structure of the virus and its genome. Genetic analysis to determine which viral elements are required in *cis* and which genes can be supplied in *trans* is also essential. This allows the design of complementation systems to produce viral vectors, which use generic packaging cells to supply the *trans* complementing functions. Packaging cell lines can then be used as producer cells for the generation of particular vectors. Some basic considerations for designing vectors include safety and biological activity, such as the removal of virulence genes, absence of replication competent parental virus, and host response to components of the vector. Major issues to be confronted are increasing the titer of the vector and decreasing the production of replication competent or wild-type virus.

Although vectors can be assessed by in vitro experiments, this does not usually predict their behavior in vivo. This is because many cells used for in vitro culture are transformed and even primary cells do not necessarily reflect the function of the same cells in vivo. Therefore, carrying out in vivo experiments in animals to assess organ and cell targeting, efficiency, specificity and persistence of gene expression and the likelihood of toxicity, and host immune and inflammatory responses is usually very essential. Also, animal experiments are important because the property of vectors in specific animals can vary from those exhibited by the parental vectors. Today, viral gene delivery is thought to be the most effective method *(10)* and it has been shown to transduce a variety of organ systems including the liver, skeletal muscle, heart, brain, and lung *(11)*.

2.1. Retroviral Vectors

To date, most of the retroviruses employed in clinical trials are based on mouse Moloney leukemia virus (MMLV), a well-studied and characterized retrovirus *(12,13)*. The MMLV genome encodes three polyproteins, *gag*, *pol*, *env*, which are required in *trans* for viral replication and packaging. Retroviruses are small RNA viruses that replicate through a DNA intermediate. They infect target cells through specific interaction between the viral envelope protein and a cell surface "receptor" on the target cell. The virus is then internalized and the RNA is reverse transcribed into proviral double-stranded DNA (dsDNA) by means of the viral encoded *pol* gene. Then, the dsDNA is transported to the nucleus, where it is stably integrated into the host genome *(14)*. Many types of retrovirus require mitosis and breakdown of the nuclear envelope for the viral genome to reach the nucleus. The ability of retroviruses to insert their genome into the host DNA allows for stable genetic modifications for the life of the host cell *(14)*. Retroviral vectors are packaged into viral particles, have all viral genes removed but contain

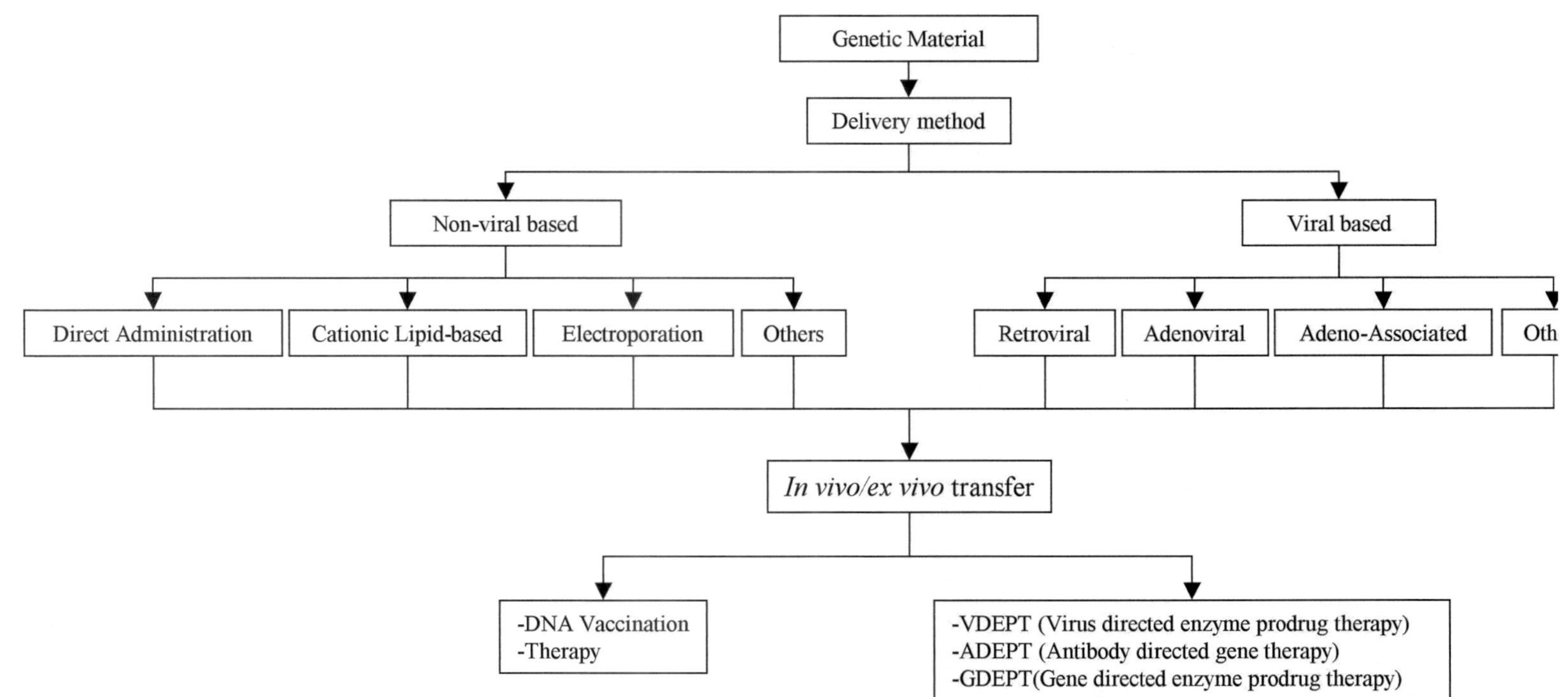

Fig. 1. The science of gene therapy. The different processes for gene delivery are based on the use of nonviral or viral vector methods to trai genetic material via in vivo or ex vivo systems for the purpose of producing a direct therapeutic effect or to boost the host immune system (vacc tion).

some of the viral regulatory sequences, and would only transduce dividing cells ***(15)*** with a high degree of efficiency and a moderate level of gene expression. To make a retrovirus vector, first a cloned DNA form of the retrovirus is isolated and coding regions for the gene products are removed. However, *cis* DNA signals required for transcription, reverse transcription, integration, and packaging of RNA are retained in the vector and genes and regulatory sequences of interest are packaged into it. This recombinant DNA form of the virus is then transferred into a mammalian cell, which is producing retrovirus gene products so that the resulting recombinant RNA can be packaged into virions for gene transfer via "infection" of new cells ***(14)***. Retroviruses can, therefore, be efficiently transferred into the genome of infected cells if the cells to be infected have the adsorbtion site, which the virion needs for attachment and entry into the cell. Mouse fibroblast cell lines transfected with DNA forms of retroviruses are usually employed as packaging cell lines. In summary, retroviral vectors are able to target dividing cells with a high degree of efficiency of gene transfer and a moderate level of gene expression, and because of their integration into the host genome, there is a more stable and long-term expression of the transgene. One of the disadvantages is their relatively small size, which limits their usefulness for gene therapy involving the transfer of genes greater than 8 kb in size. Another potential problem with retroviruses is the fact that some of the packaging cell lines might not be absolutely devoid of replicating retroviruses because of recombination events that occur with an endogenous retrovirus sequence. If the possible occurrence of such a problem is not monitored, one could be working with retrovirus vectors, that continue to be replicated and passed from cell to cell because of contaminating replicating retrovirus ***(16)***.

2.2. Adenoviral Vectors

These are the second most popular choice of viral gene delivery vectors. Adenoviruses are nonenvelope DNA viruses and members of the family Adenoviridae. They were first isolated from tonsils and adenoids, and in cell culture, they were found to be lytic for infected cells. Adenoviruses have double-stranded genomic DNA of approx 30–35 kb in length ***(17)*** embedded in a protein construct called a caspid. Adenoviruses are of approx 47 serotypes ***(18)***. They can efficiently transduce nonreplicating target cells, but viral replication occurs without integration into the host genome, leading to transient expression of the transgene. To generate a defective adenovirus for gene transfer application, the E_1 adenoviral gene important for viral gene expression and replication can be removed (*see* **Fig. 2**). E_1-deleted viruses can propagate only in a cell line that provides the E_1 gene product in *trans* ***(18)***. The current type E_1-deleted adenoviral vectors express low levels of viral antigens following infection and result in low level of DNA replication ***(12)***. A primary concern with adenoviral gene products is the stimulation of potent cellular immune and humoral responses to the injected virus cells, which result in destruction of the infected target cells and a loss of therapeutic gene expression 1–2 wk after injection ***(19,20)***. This might prevent the use of adenoviral vectors for repeated dosing. However, it has been demonstrated that the humoral response can be blocked or reduced by coadministration of immunosuppressive agents/cytokines or by the use of adenoviruses of different serotypes, which could allow repeated administration even in the presence of neutralizing antibodies ***(21)***.

Adenoviruses are being constructed by deleting certain open reading frames in their E_4 and/or E_2 gene and generating the appropriate complementing cell lines for virus production. The development of second-generation viruses with either very few or no viral genes ("gutless Ad") has produced significant reduction in host immune response. However, there still remains an unknown amount of toxicity associated with the virion structural proteins. A current report shows the development and characterization of novel empty adenovirus capsids and their impact on cellular gene expression ***(22)***.

2.3. Adeno-Associated Virus Vector

Adeno-associated virus (AAV) is a nonenvelope DNA virus that is a member of the family Parvoviridae. AAV vectors are not associated with any known human disease. The genome of

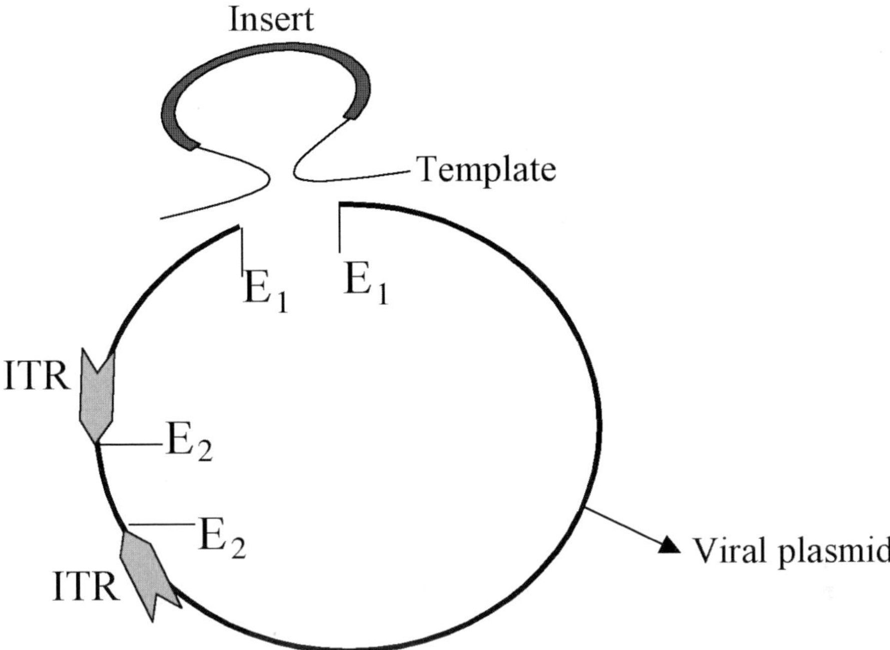

Fig. 2. Construction of a recombinant adenoviral plasmid involves the removal of the E_1 adenoviral gene and insertion of the gene of interest. Recombinant adeno-associated virus can be constructed by deleting genes that encode viral structural proteins (e.g., E_2) leaving the inverted terminal repeats (ITRs).

this virus consists of a linear DNA molecule that is 4.7 kb in length. There are both regulatory and structural genes designated as *rep* and *cap*, respectively. The *rep* genes encode four nonstructural proteins that have an essential role in DNA replication and transcription, whereas the *cap* genes encode the three structural proteins that constitute the viral capsid ***(18)***. Recombinant AAV vectors contain small single-stranded DNA genomes and are able to integrate into the host genome during replication to produce stable transduction of the target cell. These viruses can infect both dividing and nondividing cells. In creating a recombinant AAV vector, genes that encode viral structural proteins are deleted, leaving two so-called inverted terminal repeats (ITRs) plus the inserted marker gene or therapeutic gene (*see* **Fig. 2**). Several novel serotypes of AAV have been identified, each exhibiting different structural and functional characteristics ***(23,24)***. Recent studies have shown that the different serotypes have different cellular specificities; specifically, serotype 5 has been shown to transduce human pancreatic islets ex vivo with much higher efficiency than AAV serotype 2 ***(25)***. AAV vectors have not been studied to the same extent as adenoviral or retroviral systems, but they appear to be associated with fewer safety risks than the other viral systems. This is because of the elimination of all sequences coding for viral proteins, thereby reducing the risk of an immune reaction against the vector. Potential problems still facing these vectors are that of insertion mutagenesis and the generation of replication competent virus.

2.4. Other Viral Methods

Other viral vectors have been developed for use in gene therapy. These include the herpes simplex virus (HSV), the vaccinia virus (VV), and syndbis virus, the foamy virus or sumavirus. These vectors, however, have not been studied extensively and their specific advantages over the other widely used viral vectors are not well known. Some researchers are capitalizing on replicating properties of viruses to deal with some of the limitations facing gene therapy. They are finding (or creating) replicating viruses that confine their activity to the boundaries of its target tumor without infecting healthy tissues. A replicating virus has distinct advantages over nonreplicating vectors for tumor-directed applications because it leads to an increase in the percentage of cells within a tumor that expresses the therapeutic gene over time ***(26)***. Vaccina virus is an efficient replicating virus that leads to high levels of transgene expression, selectively in tumor tissue, and is capable of lysing the infected tumor cell to eradicate or reduce the tumor mass. However, it is observed that cancer patients have preformed immunological reactivity against vaccinia virus as a result of smallpox vaccination. This might limit the use of VV as a vector. Investigations are underway for alternative replicating poxvirus for cancer gene therapy. One of such investigations involves the Yaba-like disease (YLD) virus, which has been shown not to crossreact with vaccinia virus antibodies, and it replicates efficiently in human tumor cells ***(27)***. Introducing a replicating virus into a patient has its own dangers, so it is essential that the virus be purified, weakened, and modified. It must be innocuous enough to slip past a patient's immune system but infectious enough to reach all of the tumor sites.

An immune strategy that has been developed in gene therapy has been that of increasing the natural immune response of the body to a tumor. Tumors elicit an immune response from the body, but often this is not strong enough to overcome the malignant growth. Viruses can be manipulated to carry the gene for a tumor protein and then administered as tumor vaccines, which will boost the body's immune response ***(28)***. Such strategies will not only destroy the tumor but also trigger an immune-system memory that can ward off cancer recurrence. In some instances, viruses are loaded with genes for cytokines (cell signaling protein or proteins that activate elements of the immune system) and, once in the malignant cells, the genes produce a massive amount of cytokines to the point of making the tumor an agent of its own destruction.

In the area of cancer chemotherapy, toxic or suicide gene therapy has been employed. These approaches are generally called the gene-directed enzyme prodrug therapy (GDEPT) and the virus-directed enzyme prodrug therapy (VDEPT) or antibody-directed gene therapy (ADEPT). These methods involve a two-step approach. In the first step, the gene for a drug-activating enzyme is delivered into the tumor, where it is expressed. In the second step, a nontoxic prodrug, which is a substrate of the exogenous enzyme that is now expressed in the tumor, is administered systemically ***(29,30)***. The net gain is that a systematically administered prodrug can be converted to high local concentration of an active anticancer drug in tumors (*see* **Fig. 3**). VDEPT is the method where viral vectors are used to deliver a gene that encodes an enzyme, whereas ADEPT employs a tumor-associated monoclonal antibody linked to a drug-activating enzyme to create a systematically administered conjugate that only targets tumor tissues ***(29,30)***.

3. Nonviral Methods of Gene Delivery

Although virus-based systems enhance delivery efficiency, recombinant viral vector-based treatments have been associated with complications derived from highly evolved and complex viral biology. Limitations such as low titers and requirement for active cell division have led to the search for alternative viruses for gene transfer. Some of these problems have been overcome by the use of adenoviruses, but newer problems such as packaging constraint and host immune rejections have emerged. These limitations have made it necessary to develop alternate nonviral modes of DNA delivery.

3.1. Direct DNA Delivery

This is probably the simplest method of nonviral gene delivery. It involves the injection of a pure closed-circular plasmid DNA (or RNA) containing a transgene into an organ site of choice. Naked plasmid DNA (pDNA) delivered to cells in vivo results in gene expression. Direct delivery of pure plasmid DNA encoding a gene of interest can be done through different routes. The most popular methods are direct injections, such as intramuscular (i m) or intradermal (i d) and the use of a "gene gun" to bombard skin with DNA-coated gold particles (particle bombard-

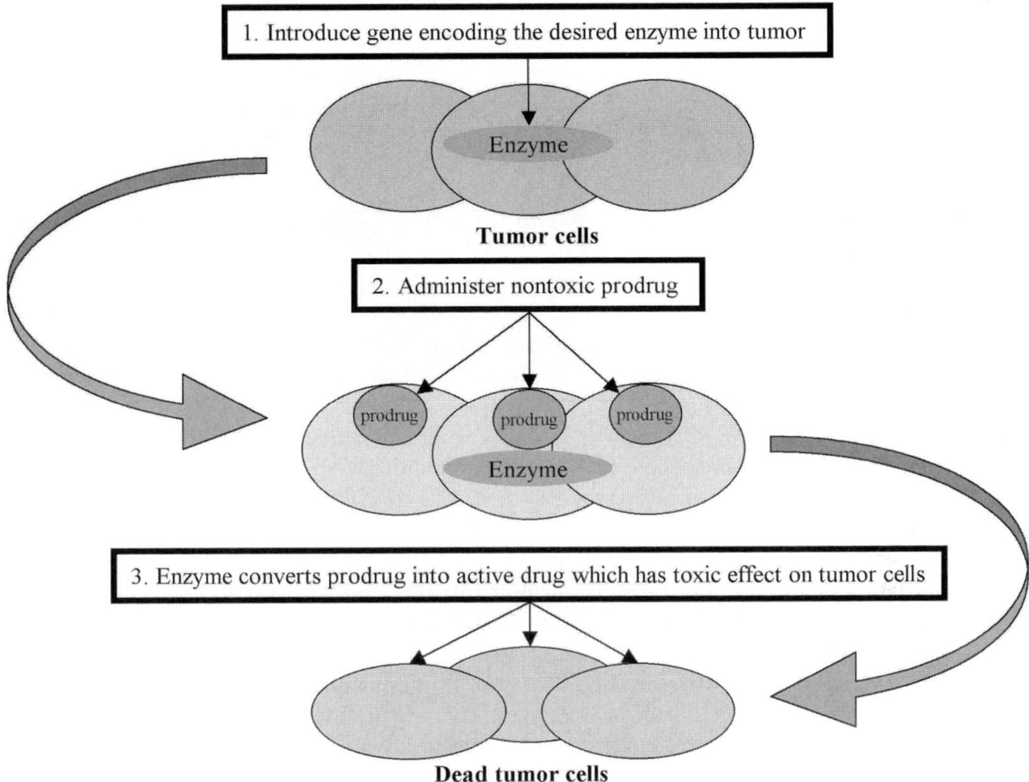

Fig. 3. Schematic representation of GDEPT, one form of suicide gene therapy.

ment). Although early investigations of direct delivery have tested different organ sites, including the liver, spleen, skin, brain, and muscle, significant expression of a reporter gene construct was reported early in mouse skeletal muscle cells *(31)*. Skeletal muscle has since been shown to be a promising target for nonviral delivery methods (of gene therapy) for both muscle and nonmuscle disorders *(32)*. The advantages muscle might have over other organs as a target for gene delivery might be because it is easily accessible, well vascularized, and able to express and process secreted proteins *(33)*. Despite the promise of im injection of plasmid vectors as gene delivery tools, several issues still remain to be addressed. One such issue is the low and highly variable level of gene expression *(34,35)*. Efficient transgene expression in hepatocytes throughout the liver has been observed following intravascular delivery of pDNA via the portal vein, the hepatic vein, or bile duct in mice *(36)*. A major advance in intravascular delivery of pDNA is the development of the tail vein delivery procedure in the Wolff and Liu laboratories *(36)*. Because the tail vein drains in the vena cava, delivery of a large bolus presumably results in a liquid volume in the vena cava that is too large for the heart to handle rapidly. The fluids back up and end up (predominantly) in the liver, resulting in gene transfer *(36)*. The tail vein method has been adopted quickly in the field of gene therapy for basic research and gene therapy evaluation. In relation to humans, it seems that similar efficiencies might be obtained by injecting pDNA into the afferent or efferent vessel of the liver by injections done via catheters *(36,37)*. Efficient transfection of skeletal muscle cells in mice, rats, dogs, and monkeys via intravascular delivery of pDNA have been reported *(38)*. It is reported that such delivery is not limited to skeletal muscle but can also target cardiac muscle (retrograde delivery via standard angioplasty catheters) and can potentially be evaluated as a method for delivery of angiogenic genes for the treatment of heart ischemia *(39)*.

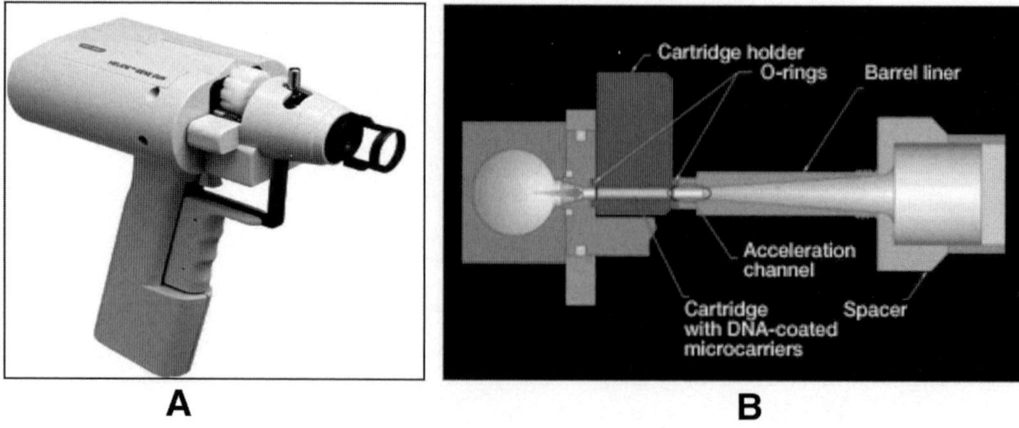

Fig. 4. (**A**) The Bio-Rad hand-held Helios gene gun and (**B**) schematic representation of the internal components of the gun. (Reprinted with permission from BioRad Laboratories, Hercules, CA.)

In the gene gun method (also referred to as the biolistic or bioballistic method), DNA coated onto the surface of 1- to 3-µm-diameter gold or tungsten beads to form a DNA–particle complex is accelerated by an electric discharge device or gas pulse and "fired" at the tissue *(5)*. A less invasive approach is the direct bombardment of the skin *(40)*. **Figure 4** shows a version of a gene gun; Bio-Rad's advanced Helios gene gun. The physical force of impact overcomes the cell membrane barrier. A wide variation in the efficiency of overall gene expression is observed for this method and this could be attributed to the variable tissue characteristics of rigidity, foreign DNA processing, and intrinsic transcriptional capacity. Analyses have shown that once inside the cell, the foreign DNA does not integrate into the genome but, rather, exists as a relatively unstable episome. Direct particle bombardment can, however, be more feasible as part of a vaccination protocol.

Herweijer and Wolff *(36)* have put forward some explanations for the loss of transgene expression following pDNA gene transfer. They observe that there is a rapid loss of expression in the first few days (of transfer to liver) followed by a slower decrease after 1 wk. The loss in the latter phase could be a result of an antigen-specific immune response in normal, immunocompetent mice, as expression is much prolonged in immunocompromised mice. Intravascular delivery results in cell damage as evidenced by pathological observation and in an increase in cell (hepatocyte) cycling. Therefore, it is likely that part of the transfected cells are destroyed or nonintegrated plasmid DNA is lost during cell cycling. However, the use of novel pDNA vectors (liver-specific albumin promoter in this case) can allow high-level sustained expression.

Direct DNA injection could have potential as a vaccination procedure because low-level gene expression is often sufficient to achieve an immunological response. Immunizations with DNA vaccines have resulted in humoral and cellular immune responses that protect against disease in preclinical models of infectious disease, cancer, and autoimmunity *(41)*. DNA vaccination differs from traditional vaccines in that just the DNA of interest is injected into the body. This DNA, encoding a desired antigen, is then delivered directly through either a hypodermal needle, a gene gun, or other delivery routes. DNA vaccines raise immune responses by expressing proteins in the vaccinated hosts and, consequently, an immune response is elicited to the expressed antigen. An added advantage of pDNA for vaccination is the presence of unmethylated CpG motifs, which have T-helper cell type 1 (Th1) immunostimulatory activity *(42)*. These DNA motifs (unmethylated form) are potent stimulators of several types of immune cell. In DNA vaccine plasmids, this immunostimulatory activity can act to mobilize the immune response against the DNA-encoded antigen. DNA vaccines possess endogenous adjuvant activity that is antigen dependent and potently induces Th1 type immune responses *(43)*.

Plasmid DNA constructs used for vaccination are similar to those used for the delivery of reporter or therapeutic genes. They have the following features: (1) a bacterial origin of replication that facilitates amplification of large quantities of plasmid DNA for purification, (2) a prokaryotic selectable marker gene, such as an antibiotic-resistance gene, (3) eukaryotic transcription regulatory elements, usually strong viral promoter/enhancer sequences that direct high levels of gene expression in a wide host cell range, (4) DNA-sequence-encoding antigenic protein/peptide of interest, and (5) polyadenylation sequence to ensure that the transcribed mRNA is appropriately terminated ***(43)***.

3.2. Cationic Lipid-Based Gene Delivery

Cationic liposome-mediated gene transfection was pioneered by Felgner and co-workers ***(44)*** and is highly efficient for transfecting DNA into a wide variety of cell lines grown in tissue culture. This technique relies upon the electrical charge properties of DNA (which is negative as a result of the phoshate backbone of the double helix), cationic lipids (positive), and cell surfaces (net negative as a result of sialic acid residues) ***(5)***. Liposomes are a heterogenous class of compounds that can be altered so that they posses a wide variety of physical and chemical properties. Liposomes can be composed of an equimolar mixture of a synthetic cationic lipid such as DOTMA (*N* [1-(2, 3,dioleyloxy)propyl]-*N*,*N*,*N*,-trimethylammonium chloride) and a helper lipid DOPE (dioleoylphospatidylethanolamine). This DOTMA/DOPE mixture (Lipofectin™) forms complexes with DNA by charge interaction upon mixing at room temperature. The resulting complex stably captures 100% of the polynucleotide, and by charge attraction and fusion properties, the lipids can adsorb to the cell membrane and deliver the nucleic acid directly into the cytoplasm, bypassing the lysosomal degradation pathway ***(5)***. There are several commercially available cationic lipids with slight differences; for example, DOTAP differs from DOSPER in that the fatty acids are connected to the 1- and 3-position of the structure's propane backbone. Other commercial products include LipofectAMINE™, Lipofectam™, and Tfx. The ratio of lipid to DNA, as well as the total concentration, must be carefully optimized to avoid toxicity.

Compared with viral vectors, cationic-lipid-based delivery systems have several advantages ***(5,15)***. DNA/lipid complexes are easy to prepare and bulk preparations can be made easily, there is no limit to the size of genes that can be delivered, and there is lower immunogenicity. More importantly, there is much less risk of inducing tumorigenic mutations by this system because genes delivered cannot replicate or recombine. However, despite the success of cationic lipid systems in transfecting cells by a local regional administration ***(45)***, they have been less efficient delivery systems compared to the viral vector-based systems. One such problem is the transport of plasmid DNA from the site of injection to the targeted cell. Lipoplexes have to circulate without being sequestered by the reticulo-endothelial system ***(46)*** in order to reach the targeted cell membrane. Physiological barriers are then encountered that prevent their intracellular trafficking. After cell membrane attachment, lipoplexes enter the cell cytoplasm either by fusion of the vector with the plasma membrane or by endocytosis ***(47)***, and then they inefficiently diffuse through the cytoplasm to the nuclear envelope. Pedroso de Lima et al. ***(48)*** have worked on different strategies aimed at improving the efficiency of gene transport to the cell nucleus. Some of such strategies include the coupling of nuclear localization peptidic sequence to plasmid DNA to increase nuclear import of the plasmid ***(49)*** and providing cationic lipid complexes that could provide transfection efficiency in the presence and absence of high serum concentration ***(50,51)***. Cationic lipid complexes are usually optimized for transfection efficiency in the absence of serum. In vivo gene transfer of lipid complexes through systemic administration have also been faced with technical difficulties such as the preparation of the DNA/lipid complexes at concentrations at which the injection volume into an animal will not be too large and large aggregates will not form. However, recent reports have shown that cationic-lipid-complex-mediated gene transfer depend not only on the formulation of the cationic liposomes and their thermodynamic phase but also significantly on the cell properties ***(52)***.

3.3. Electroporative Gene Delivery

Nonviral gene delivery methods have been associated with inefficiency, in vivo clearance, and formulation/manufacture complexities. The future of gene therapy requires the development of efficient and nontoxic polynucleotide delivery mechanism. Current evidence indicates that in vivo application of pulsed electric fields to facilitate polynucleotide uptake into cells can provide one such mechanism. Electroporation (also termed "electropermeabilization") is the application of controlled electric fields to facilitate cell permeabilization *(53)* and is a gene delivery strategy that combines naked DNA injection with pulsed electric field treatment (*see* **Fig. 5**). Electroporation allows efficient cytoplasmic uptake of large and highly charged polynucleotide molecules that normally require either specialized (and often toxic) delivery of cofactors or, in the case of naked DNA transfection, are taken up relatively inefficiently. The success of in vitro transfection with electroporation has led to the development of in vivo applications (**ref.** ***53*** provides a review of the theory and applications of in vivo electroporation). The basic concept of this method is that the application of electric pulses opens up "pores" in the cell membranes, allowing polynucleotides and other macromolecules of interest to pass down a concentration gradient into the interior of the cells. Over time after initial permeabilization, the pores close, trapping these molecules, which, in turn, can exert a biological effect. The mechanism of transportation induced by electroporation starts with an increase in membrane permeability following treatment with electric pulses *(54)* and passive diffusion through the permeabilized membrane defect.

Technical improvements in electroporation equipment have been made in recent years and better methodologies following years of experimentation have enabled increases both in gene transfer efficiency and safety *(53)*. The expression level in muscle is at least 10-fold higher compared to injection of pDNA without electroporation *(55)*. It is not clear, however, whether these increases in transgene expression (especially of secreted proteins) are the result of enhanced gene transfer into myofibers or to simultaneous transfer into different cell types (e.g., endothelial cells). These expression levels were considered sufficient to warrant investigation of this method for gene therapy for chronic anemia *(56)* or muscular dystrophies *(57)*.

4. Some Reporter Genes Used in Gene Therapy

Direct quantitation of gene expression to determine the efficiency of the gene delivery process is not always possible. The gene of interest can encode a protein that either does not have a biochemical activity or has an activity that cannot be easily assayed. Reporter genes, which are nucleic acid sequences encoding proteins with activities that can be measured in liquid assays or visualized on tissue sections or on colonies grown on agar proteins, are therefore used to determine the efficiency of the gene delivery systems. Some of the common reporter genes are Luciferase, a gene originally isolated from the fire fly. In the presence of luciferin and adenosine triphosphate (ATP), an enzyme complex is produced that reacts with molecular oxygen to produce light, carbon dioxide, and adenosine monophosphate (AMP). The light produced is easily monitored by a scintillation counter specifically designed for this purpose. The light emitted is directly proportional to the number of luciferase enzyme molecules. Another reporter gene, the green fluorescence protein (GFP), is from the jellyfish *Aequorea victoria*. This protein is activated in vivo by an energy transfer via the Ca^{2+} stimulation of the photoprotein aequorin. The blue light generated by aequorin excites GFP and results in the emission of green light. GFP is detected with long-wave ultraviolet (UV) light, absorbs light at 395 nm, and has an emission peak at 509 nm. Its nonsubstrate requirement activity makes it an attractive reporter for gene expression studies. GFP is stable, nontoxic, and resistant to photo bleaching. These properties make GFP a viable alternative to traditional reporter genes such as β-glucuronidase (GUS), chloroamphenicol acetyltransferase (CAT), firefly luciferase, or β-galactosidase (Lac Z), an enzyme that hydrolyzes lactose into galactose and glucose. Transfected cells producing high levels of β-galactosidase turn dark blue on plates containing X-gal (a substrate that can be cleaved by β-galactosidase to produce a blue color) and those that

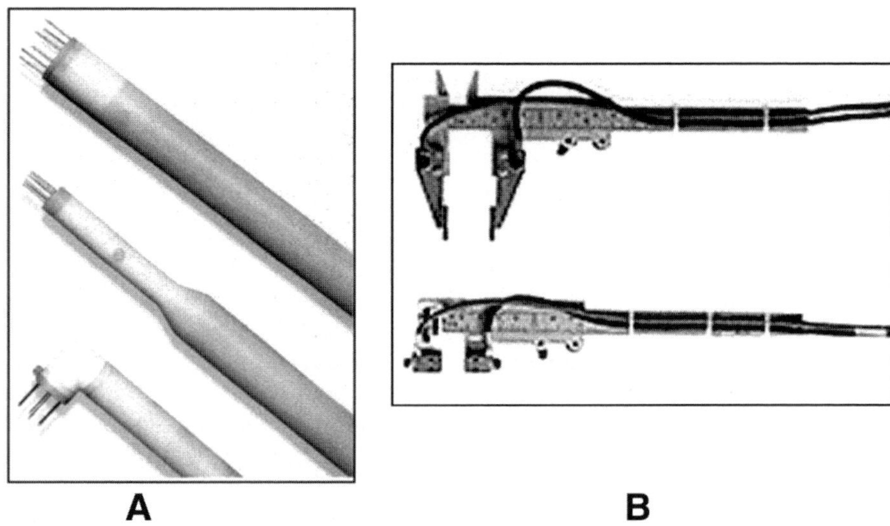

Fig. 5. (**A**) Pin electrode and (**B**) caliper electrodes. Both can be utilized for in vivo electroporative gene delivery. (From **ref. *53***).

do not produce the enzyme do not change color. Use of β-galactosidase as a reported gene is straightforward and easy to interpret. Its detection could be by a colorimetric assay.

5. Clinical Applications

More than 4000 patients have been enrolled in gene transfer experiments over the last decade. However, gene therapy research continues to produce few unambiguous results. In a phase 1 clinical study in 1999, scientists used a modified adenovirus to deliver normal versions of the *p53* gene to the lung tumors of 25 patients. Mutations in *p53* can stop malignant cells from going into the natural programmed cell death phase called apoptosis. It was observed that in 18 of these patients, the injection either reduced or arrested tumor growth ***(58)***. In that same year, a gene therapy trial aimed at treating a liver disease, which occurs when a gene on the X chromosome is missing or defective, producing too little of a liver enzyme, ornithine transcarbamylase (OCT), needed to remove ammonia from the blood, experienced a tragic outcome. An 18-yr-old volunteer died 15 d after receiving a high dose of the adenovirus carrying the gene. Follow-up studies to find an explanation for this sad outcome revealed that the volunteer might have died from a massive immune response to the adenovirus vector itself. This resulted in a systemic inflammation, which flooded the volunteer's lungs with fluid, causing acute respiratory failure and death ***(59)***. In the Necker Hospital for Children in Paris, a healthy gene was successfully introduced into the bone marrow of two children with a rare, lethal immune disorder (SCID-X1) based on the use of a complementary DNA containing a defective gammac Moloney retrovirus-derived vector and ex vivo infection of CD34+ cells. After a 10-mo follow-up period, gammac transgene- expressing T-cells and NK cells were detected in these patients, indicating the correction of the disease phenotype ***(60)***. Unfortunately, as this study progressed to include more patients, 2 out of 10 patients have recently been shown to develop leukemia within 3 yr of gene therapy ***(61)***. Cystic fibrosis (CF), a recessive genetic disease caused by mutations in the cystic fibrosis transmembrane conductance regulator (*CFTR*) gene, is yet another condition that gene therapy approaches have been developed to treat patients. Mutations in *CFTR* are the molecular cause of CF and lung disease is the most serious outcome of CF. This makes the lung the logical primary target for CF gene therapy. One of the recent CF clinical studies involved the delivery of multiple doses of a DNA/liposome formulation to the nasal epithelium of patients. Subjects received three doses, administered 4 wk apart, and results showed modest, transient gene expression with no evidence of inflammation, toxicity, or immune response toward the formulation or the expressed CFTR *(62)*. Many more clinical studies targeting different disease conditions are being carried out around the world under strict government regulations.

6. Conclusion

The goal of all gene transfer methods is to deliver genes to desired cells in tissues and organs in an efficient manner with minimal toxicity. It is apparent that although the current methods described have some disadvantages, their successful use is rapidly becoming clinically feasible for a wide range of gene correction, gene therapy, and nucleic acid vaccine applications. A limiting factor to rapid research in this area lies in the fact that direct extrapolation of animal experiments to human studies are not usually possible because animal models do not satisfactorily mimic the major manifestations of the corresponding human diseases. Because clinical studies represent both the practical implementation of basic discoveries and the experiments that will refine and define new questions, there is a clear and legitimate need for clinical studies to evaluate various aspects of gene therapy approaches. Such studies, however, must be carried out cautiously and with scientific integrity. Researchers need to focus greatly on the basic aspects of gene transfer and gene expression by improving vectors for gene delivery, enhancing and maintaining high level expression of the transferred gene, achieving tissue-specific and regulated expression of transferred genes, and directing gene transfer to specific cell types. It is unlikely that gene therapy will replace conventional treatment strategies for humans, but it might eventually become an integral part of medical science.

References

1. Crystal, R. G. (1995) Transfer of genes to humans: early lessons and obstacles to success. *Science* **270,** 404–410.
2. Miller, A. D. (1992) Human gene therapy comes of age. *Nature* **357,** 455–460.
3. Mulligan, R. C. (1993) The basic science of gene therapy. *Science* **260,** 926–932.
4. Malone, E. W. (1999) Present and future status of gene therapy, in *Advanced Gene Delivery* (Rolland, A. ed.), Harwood Academic, London.
5. Schoefield, J. P. and Caskey, C. T. (1995) Non-viral approaches to gene therapy. *Br. Med. Bull.* **51(1),** 56–71.
6. Arbones, M. L., Austin, H. A., Capon, D. J., and Greenburg, G. (1994) Gene targeting in normal somatic cells: inactivation of the interferon-γ-receptor in myoblasts. *Nature Genet.* **6,** 90–96.
7. Dorin, J. R., Dickinson, P., Alton, E. W. F. W., et al (1992) Cystic fibrosis in the mouse by targeted insertional mutagenesis. *Nature* **359,** 211–215.
8. Huxley, C. (1994) Mammalian artificial chromosomes: a new tool for gene transfer. *Gene Ther.* **1,** 7–12.
9. Orkins, S. H. and Motulsky, A. G. (1995) Report and recommendations of the panel to assess the NIH investigation in research on gene therapy. http://www.nih.gov/news/panelrep.html.
10. Wang, A. Y., Peng, P. D., Ehrhardt, A., Storm, T. A., and Kay, M. (2004) Comparison of adenoviral and adeno-associated viral vectors for pancreatic gene delivery in vivo. *Hum. Gene Ther.* **15,** 405–413.
11. Breyer, B., Jiang, W., Cheng, H., et al. (2001) Adenoviral vector-mediated gene transfer for human gene therapy. *Curr. Gene Ther.* **1,** 149–162.
12. Guild, B. C., Finer, M. H., Housman, D. E., and Mulligan, R. C. (1988) Development of retrovirus vectors useful for expressing genes in cultured murine embryonal cells and hematopoietic cells in vivo. *J. Virol.* **62,** 3795–3801.
13. Miller, A. D., Miller, D. G., Garcia, J. V., and Lynch, C. M. (1993) Use of retroviral vectors for gene transfer and expression. *Methods Enzymol.* **217,** 581–599.
14. Robbins, P. D. and Ghivizzani, S. C. (1998) Viral vectors for gene therapy. *Pharmacol. Ther.* **80(1),** 35–47.
15. Kay, A. M., Liu, D., and Hoogerbrugge, M. (1997) Gene Therapy. *Proc. Natl. Acad. Sci. USA* **94,** 12,744–12,746.

16. Duke University Institutional Biosafety Committee (1995) Retrovirus Vector Guidelines. http://www2.duke.edu/depts/gc/txtfiles/ibcretro.html.
17. Graham, F. L. and Prevec, L. (1995) Methods for construction of adenovirus vectors. *Mol. Biotechnol.* **3,** 207–220.
18. Wivel, N. A. and Wilson, J. M. (1998) Methods of gene delivery. *Gene Ther.* **12(3),** 483–499.
19. Yang, Y., Haecker, S. E., Su, Q., and Wilson, J. (1996) Immunology of gene therapy with adenoviral vectors in mouse skeletal muscle. *Hum. Mol. Genet.* **5(11),** 1703–1712.
20. Yang, Y. and Wilson, J. M. (1995) Clearance of adenovirus-infected hepatocytes by MHC class I-restricted CD4+ CTLs in vivo. *Immunology* **155(5),** 2564–2570.
21. Mack, C. A., Song, W. R., Carpenter, H., et al. (1997) Circumvention of anti-adenovirus neutralizing immunity by administration of an adenoviral vector of an alternate serotype. *Hum. Gene Ther.* **8,** 99–109.
22. Stilwel, J. L., McCarty, D. M., Negishi, A., Superfine, R., and Samulski, R. J. (2003) Development and characterization of novel empty capsids and their impact on cellular gene expression. *J. Virol.* **77(23),** 12,881.
23. Gao, G., Alvira, M. R., Wang, L., Calcedo, R., Johnston, J., and Wilson, J. M. (2002) Novel adeno-associated viruses from rhesus monkeys as vectors for human gene therapy. *Proc. Natl. Acad Sci USA* **99,** 11,854–11,859.
24. Grimm, D. and Kay, M. A. (2003) From virus evolution to vector revolution: use of naturally occurring serotypes of adeno-associated virus (AAV) as novel vectors from human gene therapy. *Curr. Gene Ther.* **3,** 281–304.
25. Flotte, T., Agarwal, A., Wang, J., et al. (2001) Efficient ex vivo transduction of pancreatic islet cells with recombinant adeno-associated virus vectors. *Diabetes* **50,** 515–552.
26. McCart, J. A., Puhlmann, M., Lee, J., et al. (2000) Complex interactions between the replicating oncolytic effect and the enzyme/prodrug effect of vaccinia-mediated tumor regression. *Gene Ther.* **7,** 1217–1223.
27. Hu, Y., Lee, J., McCart, J. A., et al. (2001) Yaba-like disease virus: an alternative replicating poxvirus vector for cancer gene therapy. *J. Virol.* **75(21),** 10,300–10,308.
28. Ribas, A., Butterfield, L. H., and Economou, J. S. (2000) Genetic immunotherapy for cancer. *Oncologist* **5,** 87–98.
29. Springer, C. J. and Niculescu-Duvaz, I. (2000) Prodrug-activating systems in suicide gene therapy. *J. Clin. Invest.* **105,** 1161–1167.
30. Xu G. and McLeod H. L. (2001) Strategies for enzyme/prodrug cancer therapy. *Clinical Cancer Research* **7,** 3314–3324.
31. Wolff, J. A., Malone, R. W., Williams P., et al. (1990) Direct gene transfer into mouse muscle in vivo. *Science* **247,** 1465–1468.
32. Liu, F., Liang, K. W., and Huang, L. (2001) Systemic administration of naked DNA: gene transfer to skeletal muscle. *Mol. Intervent.* **3,** 168–172.
33. Marshall, D. J. and Leiden, J. M. (1998) Recent advances in skeletal-muscle-based gene therapy. *Curr. Opin. Genet. Dev.* **8,** 360–365.
34. Wolff, J. A., Ludtke, J. J., Acsadi, G., Williams, P. and Jani, A. (1992) Long-term persistence of plasmid DNA and foreign gene expression in mouse muscle. *Hum. Mol. Genet.* **1(6),** 363–369.
35. Doh, S., G., Vahlsing, H. L., Hartikka, J., Liang, X., and Manthorpe, M. (1997) Spatial-temporal patterns of genes expression in mouse skeletal muscle after injection of lacZ plasmid DNA. *Gene Ther.* **4(7),** 648–663.
36. Herweijer, H. and Wolff, J. A. (2003) Progress and prospects: naked DNA gene transfer and therapy. *Gene Ther.* **10,** 453–458.
37. Eastman, S. J., et al., (2002) Development of catheter-based procedures for transducing the isolated rabbit liver with plasmid DNA. *Hum. Gene Ther.* **13,** 2065–2077.
38. Zhang, G., Budker, V., Williams, P., Subbotin, V., and Wolff J. A. (2001) Efficient expression of naked DNA delivered intraarterially to limb muscles of nonhuman primates. *Hum. Gene Ther.* **12,** 427–438.
39. McKay, M. J. and Gaballa, M. A. (2001) Gene transfer therapy in vascular disease. *Cardiovasc. Drug Rev.* **19(3),** 245–62.
40. Williams, R. S., Johnston, S. A., Riedy, M., DeVit, J. M., McElligott, S. G., Sanford, J. C. (1991) Introduction of foreign genes into tissues of living mice by DNA-coated microprojectiles. *Proc. Natl. Acad. Sci. USA* **88,** 2726–2730.

41. Donnelly, J. J., Ulmer, J. B., and Liu, M. A. (1998) DNA vaccines. *Dev. Biol. Stand.* **95,** 43–53.
42. Sato, Y., Roman, M., Tighe, H, et al. (1996) Immunostimulatory DNA sequences necessary for effective intradermal gene immunization. *Science* **273(5273),** 352–354.
43. Hanlon, L. and Argyle, D. J. (2001) The science of DNA vaccination. *Infect. Dis. Rev.* **3(1),** 2–12.
44. Felgner, P. L., Gadek, T. R., Holm M., et al. (1987) Lipofectin: a highly efficient, lipid-mediated DNA-transfection procedure. *Proc. Natl. Acad. Sci. USA* **84,** 7413–7417.
45. Nabel G. J., Nabel E. G., Yang Z. Y., et al. (1993) Direct gene transfer with DNA–liposome complexes in melanoma: expression, biologic activity and lack of toxicity in humans. *Proc. Natl. Acad. Sci. USA* **90,** 11307–11311.
46. Plank, C., Mechtler, K., Szoka, F. C., and Wagner, E. (1996) Activation of the complement system by synthetic DNA complexes: a potential barrier for intravenous gene delivery. *Hum .Gene Ther.* **7,** 1437–1446.
47. Zabner, J., Fasbender, A. J., Moninger, T., Poellinger, K. A., and Welsh, M. J. (1995) Cellular and molecular barriers to gene transfer by a cationic lipid. *J. Biol. Chem.* **270,** 18,997–19,007.
48. Pedroso de Lima, M. C., Neves, S., Filipe, A., Duzgunes, N., and Simoes, S. (2003) Cationic liposomes for gene delivery: from biophysics to biological applications. *Curr. Med. Chem.* **10(14),** 1221–1231.
49. Carrière, M., Escriou, V., Savarin, A., and Scherman, D. (2003) Coupling of importin beta binding peptide on plasmid DNA: transfection efficiency is increased by modification of lipoplex's physicochemical properties. BMC Biotechnol. http://www.biomedcentral.com/1472-6750/3/14.
50. Nchinda, G., Überla, K., and Zschörnig, O. (2002) Characterization of cationic lipid DNA transfection complexes differing in susceptibility to serum inhibition. http://biomedcentral.com/1472-6750/2/12.
51. Almofti, M. R., Harashima, H., Shinohara, Y., Almofti, A., Li, W., and Kiwada, H. (2003) Lipoplex size determines lipofection efficiency with or without serum. *Mol. Membr. Biol.* **20(1),** 35–43.
52. Caracciolo, G., Pozzi, D., Caminiti, R., and Congiu Castellano, A. (2003) Structural characterization of a new lipid/DNA complex showing selective transfection efficiency in ovarian cancer cells. *Eur. Phys. J. E: Soft Matter* **10(4),** 331–336.
53. Somiari, S., Glasspool-Malone, J., Drabick, J. J., et al. (2000) Theory and in vivo application of electroporative gene delivery. *Mol. Ther.* **2(3),** 178–187.
54. Vanbever, R., Leroy, M. A., and Preat, V. (1998) Transdermal permeation of neutral molecules by skin electroporation. *J. Control. Release* **54,** 243–250.
55. Hartika, J., Sukhu, L., Buchner, C., et al. (2001) Electroporation-facilitated delivery of plasmid DNA in skeletal muscle: plasmid dependence of muscle damage and effect of poloxamer 188. *Mol. Ther.* **4,** 407–415.
56. Terada Ytanaka, H., Okado, T., Inoshila, S., Kuwahara, M., Akiba, T., Sasaki, S., and Marumo, F. (2001) Efficient and ligand-dependent regulated erythropoietin production by naked DNA injection and in vivo electroporation. *Am. J. Kidney Dis.* **38,** S50–S53.
57. Vilquin J. T. Kennel Pf, Paturneau-Jouas M., Chapdelaine P., Boissel N., Delaere P., Tremblay J. P., Scherman D., Fiszman M. Y. (2001) Electrotransfer of naked DNA in the skeletal muscles of animal models of muscular dystrophyies. *Gene Ther.* **8,** 1097–1107.
58. Swisher, S. G., Roth, J. A., Nemunaitis J., et al (1999) Adenovirus-mediated p53 gene transfer in advanced non-cell lung cancer. *J. Natl. Cancer Inst.* **5(91),** 763–771.
59. Marshall, E. (2000) Gene therapy on trial. *Science* **288,** 951–957.
60. Cavazzana-Calvo, M., Hacein-Bey, S., de Saint Basile, G., et al (2000) Gene therapy of human severe combined immunodeficiency (SCID)-X1 disease. *Science* **28,** 288(5466), 627–629.
61. Hacein-Bey-Abina,. S., Von Kalle, C., Schmidt, M., et al., (2003) LMO2-associated clonal T cell proliferation in two patients after gene therapy for SCID-X1. *Science* **17,** 302(5644), 400–401.
62. Hyde, S. C., Southern, K. W., Gilead U., et al. (2000) Repeat administration of DNA/liposomes to the nasal epithelium of patients with cystic fibrosis. *Gene Ther.* **7,** 1156–1165.

Index